Lecture Notes in Civil Engineering 873

Lecture Notes in Civil Engineering (LNCE) publishes the latest developments in Civil Engineering—quickly, informally and in top quality. Though original research reported in proceedings and post-proceedings represents the core of LNCE, edited volumes of exceptionally high quality and interest may also be considered for publication. Volumes published in LNCE embrace all aspects and subfields of, as well as new challenges in, Civil Engineering. Topics in the series include:

- Construction and Structural Mechanics
- Building Materials
- Concrete, Steel and Timber Structures
- Geotechnical Engineering
- Earthquake Engineering
- Coastal Engineering
- Ocean and Offshore Engineering; Ships and Floating Structures
- Hydraulics, Hydrology and Water Resources Engineering
- Environmental Engineering and Sustainability
- Structural Health and Monitoring
- Surveying and Geographical Information Systems
- Indoor Environments
- Transportation and Traffic
- Risk Analysis
- Safety and Security

To submit a proposal or request further information, please contact the appropriate Springer Editor:

- Pierpaolo Riva at pierpaolo.riva@springer.com (Europe and Americas);
- Swati Meherishi at swati.meherishi@springer.com (Asia—except China, Australia, and New Zealand);
- Wayne Hu at wayne.hu@springer.com (China).

All books in the series now indexed by Scopus and EI Compendex database!

Hans Beushausen · Mahdi Kioumarsi ·
Behrouz Shafei
Editors

Proceedings of the 2nd International Conference on Net-Zero Built Environment

Innovations in Materials, Structures, and Management Practices (Net-Zero Future 2025)

Set 1

Springer

Editors
Hans Beushausen
Department of Civil Engineering
University of Cape Town
Cape Town, South Africa

Mahdi Kioumarsi
Department of Built Environment
Oslo Metropolitan University
Oslo, Norway

Behrouz Shafei
Department of Civil, Construction
and Environmental Engineering
Iowa State University
Ames, IA, USA

ISSN 2366-2557 ISSN 2366-2565 (electronic)
Lecture Notes in Civil Engineering
ISBN 978-3-032-27697-1 ISBN 978-3-032-27698-8 (eBook)
https://doi.org/10.1007/978-3-032-27698-8

This work was supported by the University of Cape Town.

This Springer imprint is published by the registered company Springer Nature Switzerland AG
The registered company address is: Gewerbestrasse 11, 6330 Cham, Switzerland

Preface

The *Proceedings of the 2nd International Conference on Net-Zero Built Environment: Innovations in Materials, Structures, and Management Practices (Net-Zero Future 2025)* took place in Cape Town, South Africa, from 5 to 7 November 2025, building on the success of the inaugural conference held in Oslo, Norway, in 2024. The conference formed part of the Net-Zero Future initiative, supported by the Research Council of Norway and the Norwegian Directorate for Higher Education and Skills. Through this initiative, a global partnership was established among Norway, the USA, Germany, South Africa, and India, enabling collaborative research and educational exchanges aimed at moving toward a net-zero built environment. The project aligns directly with several United Nation's Sustainable Development Goals, including those related to quality education, innovation, sustainable cities, climate action, and international cooperation.

Growing concerns over the deterioration of the built environment under escalating stressors, alongside the environmental burden associated with its expansion, have highlighted the urgent need to rethink conventional approaches to infrastructure development. Civil engineering, which is central to the planning, construction, and stewardship of infrastructure, plays a pivotal role in this transformation. The Net-Zero Built Environment conference series serves as a unique platform to examine the net-zero imperative through the lens of civil engineering and several related fields, underscoring the necessity of dramatically reducing the carbon footprint throughout the world.

Achieving net-zero in the built environment encompasses the implementation of new insights, practices, and policies to design, deliver, operate, and eventually retire physical infrastructure components with no net carbon emissions over their entire life cycle. Meeting this ultimate goal demands integrated strategies spanning material selection, structural innovation, and modern management practices. These three areas were therefore established as the core themes of the conference series. The Cape Town event drew substantial participation from academia, industry, and the public sector, with delegates sharing the state-of-the-art and practice across all three thematic areas.

Following a thorough peer-review process carried out by members of the conference's international scientific committee together with additional expert reviewers, 111 full-length papers were selected for inclusion in the conference proceedings. These contributions reflect a broad spectrum of authors and research teams from across the world, each offering distinct perspectives and new insights into pathways toward a net-zero built environment. Aligned with this overarching ambition, the papers showcase recent advances in: (i) low-carbon and carbon-neutral materials and manufacturing methods, (ii) robotic and automated construction techniques that minimize formwork and on-site effort, (iii) innovative structural concepts and detailing strategies that maximize performance while reducing material demand, (iv) latest condition assessment and structural health-monitoring approaches, and (v) progressive life-cycle assessment methods and infrastructure management frameworks.

We acknowledge that reaching net-zero is more than a technical or policy target; It is a ethical responsibility to safeguard a sustainable future. Lowering carbon emissions contributes to minimizing environmental stressors, mitigating the frequency and intensity of extreme weather events, and preserving fragile ecosystems. All contributors to the enclosed proceedings share a commitment to advancing the materials, structures, and management practices that benefit communities worldwide. Through the dissemination of emerging ideas and innovations, along with lessons learned from the past, the themes explored in this book provide meaningful contributions to the net-zero agenda. By equipping researchers, engineers, decision-makers, and the public with relevant tools and knowledge base, we aim to support collective progress toward a global net-zero future.

Hans Beushausen
Mahdi Kioumarsi
Behrouz Shafei

Contents

Synthesis and Characterization of Carbonatable Binders Illustrated by γ-C_2S

Nita Morina[1(✉)], Neven Ukrainczyk[1], Christiane Rößler[2], Matthias Müller[2], Thomas Sowoidnich[2], and Eduardus A. B. Koenders[1]

[1] Technical University of Darmstadt, Darmstadt, Germany
morina@wib.tu-darmstadt.de
[2] Bauhaus University Weimar, Weimar, Germany

Abstract. This study focuses on the synthesis and carbonation reactivity of γ-C_2S as a representative carbonatable binder with high potential for carbon capture and utilization (CCU). γ-C_2S was produced using different raw materials and processing routes, including conventional dry mixing, pelletization and burning, as well as a pre-synthesized C–S–H route. The resulting clinker phases were characterized with respect to mineralogy and microstructure, highlighting the influence of the synthesis pathway on phase purity and material properties. To evaluate the carbonation reactivity, various experimental approaches were applied, including direct CO_2 exposure and sodium bicarbonate-assisted reaction acceleration. Isothermal calorimetry, X-ray diffraction, and scanning electron microscopy were employed to analyze the reaction kinetics and microstructural changes. The results demonstrated that γ-C_2S exhibits significant carbonation reactivity, leading to the formation of $CaCO_3$ polymorphs while maintaining unreacted silicate phases. Furthermore, self-pulverizing behavior and color changes after firing were observed, underlining its unique processing characteristics. These findings illustrate the potential of γ-C_2S as a model system for developing carbonatable binders, providing insight into production optimization and demonstrating pathways for integrating CO_2 curing into binder technology.

Keywords: Carbonatable binders · CCU · γ-C_2S · Carbonation · Self-pulverised clinker

1 Introduction

Carbonatable binders (CB) are a recently defined class of binders that harden rapidly through chemical reaction with CO_2 under ambient conditions, typically within 24 h and in the presence of moisture [1]. They form a dense carbonated matrix with high mechanical strength while emitting significantly less CO_2 during production compared to conventional Portland cement [1–3]. Their clinker phases, such as γ-C_2S, rankinite (C_3S_2), wollastonite (CS) and gehlenite (C_2AS), are produced at lower firing temperatures and exhibit a self-pulverizing effect, eliminating the need for energy-intensive grinding [1, 4]. The self-pulverizing effect arises from a volume increase of approximately 12–14% during the transformation from β-C_2S to γ-C_2S. This phase transition

H. Beushausen et al. (Eds.): NTZR 2025, LNCE 873, pp. 1–12, 2026.
https://doi.org/10.1007/978-3-032-27698-8_1

involves a reordering of the crystal structure, which induces internal stresses that lead to spontaneous disintegration [1]. The primary reaction products of carbonatable binders are calcium carbonate ($CaCO_3$) polymorphs, such as calcite, aragonite, and vaterite, along with silica gel, which together form a dense and durable matrix [5, 6].

The key advantage of these binders lies in their high carbonation reactivity, which allows for permanent CO_2 fixation in the hardened material [5]. This property enables carbon capture and utilization (CCU) strategies, including accelerated carbonation curing or even internal CO_2 release via additives such as sodium bicarbonate ($NaHCO_3$). As a result, CB are being investigated for applications in mortars, plasters, masonry units and other unreinforced elements, where rapid carbonation curing can be efficiently applied [7, 8].

Among the CB clinker phases, γ-C_2S is of particular interest due to its high carbonation potential and self-pulverizing behavior [1]. Although it has low hydraulic reactivity, its chemical composition and microstructural transformation upon carbonation make it an ideal model system for investigating reaction mechanisms, curing strategies and microstructural development [1, 3].

This study focuses on the synthesis of γ-C_2S via different processing routes and the analysis of its carbonation reactivity under both diffusion-controlled CO_2 curing and $NaHCO_3$-assisted internal carbonation. The results provide insight into how synthesis parameters influence phase composition and reactivity and explore the potential of γ-C_2S as a carbon-storing binder for sustainable construction applications.

2 Materials and Methods

2.1 Raw Materials

The synthesis of γ-C_2S was based on a carefully selected set of raw materials that predominantly contribute CaO and SiO_2. The primary components include a high-purity limestone powder (eh_easyflow from sh minerals GmbH), which is sourced from the Calcitwerk in Heidenheim and comprises 99.2% $CaCO_3$ in the form of calcite, as confirmed by XRD analysis. Additionally, quartz sand (F36 from Quarzwerke GmbH) with a grain size of 0/0.25, and quartz powder with a similar mineralogical composition but finer particle size, were used to supply the necessary silica. To further enhance the binder's reactivity and microstructural densification, silica fume from ELKEM was incorporated, while calcium hydroxide was added to ensure the appropriate CaO content for effective carbonation.

2.2 Synthesis Procedure

To determine the target compositions for γ-C_2S, the required molar ratios of CaO and SiO_2 were first calculated. This calculation was based on the assumption that quartz sand and quartz powder provide pure SiO_2, while limestone flour contributes pure $CaCO_3$. The raw materials were weighed according to these calculations and mixed using several approaches to analyze reaction kinetics and final material properties.

The synthesis procedure was developed through iterative laboratory trials, during which two distinct preparation routes were evaluated to examine their impact on phase development and overall reactivity. In the dry-mixing approach, different combinations of raw materials were explored without predefined proportions, allowing for a flexible investigation of potential clinker phase assemblages. Initial trials employed simple manual mixing using a spoon to achieve a visually homogeneous blend. In subsequent experiments, the components were wet-milled with isopropanol (2-propanol) to facilitate rapid drying and improve the uniformity of the mixture without introducing moisture-induced agglomeration. This method proved particularly effective in the presence of quartz powder, whose finer particle size enhanced chemical interaction with the limestone component.

In addition to the dry-mixing strategy, a second route was investigated in which a precursor was formed through pozzolanic reaction prior to calcination. Following established procedures from the literature and previous laboratory experience, silica fume and calcium hydroxide were mixed with a water-to-solid ratio of 10:1. The suspension was placed in sealed plastic containers and subjected to tumbling agitation over a period of seven days to promote the formation of C–S–H. After filtration, the solid fraction was dried at 55 °C for a minimum of 24 h, subsequently milled, and finally calcined at 1350 °C for four hours in a high-temperature furnace. Phase composition of the resulting product was assessed via X-ray diffraction (XRD).

Overall, this flexible synthesis procedure, incorporating different mixing methods and thermal treatments, was essential for identifying the optimal parameters that promote the desired chemical reactions and ensure the production of a high-quality carbonatable binder with consistent performance characteristics.

2.3 Methodology

Once a satisfactory binder has been synthesized, subsequent experiments focused on assessing its reactivity. This evaluation was conducted through experimental protocols in which the binder was combined with water and subjected to controlled CO_2 exposure. Different w/b—ratios were evaluated. The primary analytical methods employed for this purpose included X-ray diffraction (XRD), thermogravimetric analysis (TGA), and scanning electron microscopy (SEM).

To investigate the influence of different CO_2 supply mechanisms, both external and internal carbonation approaches were considered. Internal CO_2 release was achieved through the direct incorporation of sodium bicarbonate ($NaHCO_3$) into the binder mixture. This method was particularly applied in isothermal calorimetry experiments, where the decomposition of $NaHCO_3$ provided an internal CO_2 source without the need for external gas supply systems. This approach allowed for the isolated observation of binder reactivity under chemically induced carbonation conditions.

In a complementary setup, externally supplied gaseous CO_2 was introduced by curing specimens in a climate chamber at 20 °C and 60% relative humidity, using a CO_2 concentration of 20%. The binder mixture in this case also contained $NaHCO_3$, enabling a combined carbonation scenario with both internal and external CO_2 sources.

For this purpose, cylindrical paste specimens (24 mm diameter, 29 mm height) were prepared using plastic molds. The binder was γ-C_2S synthesized via the C–S–H precursor route. A water-to-binder ratio of 0.55 was employed, and 1 g of $NaHCO_3$ was added per 5 g of binder. After curing in the molds for 24 h, the specimens were demolded and stored in the climate chamber for six additional days.

The findings obtained from these investigations are expected to provide insights into the functionality, reactivity, and practical applicability of the synthesized binder.

2.4 High Temperature Furnace

The synthesis experiments relied on a range of specialized equipment to achieve the controlled formation of the desired clinker phases. Central to the process was the use of a high-temperature furnace—the RHF 16/15 from Carbolite Cero—which is capable of reaching temperatures up to 1600 °C. This furnace allowed for programmable heating profiles, including adjustable ramp rates, dwell times, and cooling cycles. For this purpose, the mixtures were heated at a rate of 10 °C/min to the target temperature and held for 4 h before cooling. The raw material mixtures were placed in ceramic containers of 60 ml capacity, which facilitated uniform heating and controlled gas release during the firing process. These processing technologies were critical in replicating the precise thermal conditions required to achieve the formation of γ-C_2S.

2.5 X-Ray Diffraction (XRD)

To evaluate the quality of the produced binders, X-ray diffractometry was employed as a key analytical technique. In this work, a D2 PHASER X-ray diffractometer from Bruker was used. Featuring a copper anode and a linear Lynxeye detector, the instrument operated at 30 kV and 10 mA, scanning the samples from 5° to 70° in 0.02° increments with a 2-s measurement per step and using a wavelength of 1.5406 Å. The diffraction data were processed with DIFFRAC.EVA V4.2 software, which facilitated the identification and quantification of the mineral phases. This comprehensive equipment setup and analytical approach ensured that both the synthesis process and the resulting binder quality could be rigorously evaluated.

2.6 Isothermal Calorimetry

Isothermal calorimetry was employed to monitor the heat flow associated with hydration and carbonation reactions. For this purpose, an MC CAL isothermal reaction calorimeter from C3 Prozess- und Anlagentechnik was used. The instrument features twelve measurement channels, allowing parallel testing of multiple samples. Two of these channels serve as reference positions containing inert material (quartz sand) to provide stable baseline signals. Each sample container has a volume capacity of 20 mL, and measurements can be conducted within a temperature range of 5–90 °C. The temperature sensitivity of the system is $\pm$ 0.01 °C, while the detection limit for heat flow is 20 μW, with an overall accuracy of $\pm$ 25 μW. This setup enables precise monitoring of thermal events and facilitates the evaluation of reaction kinetics for the investigated binder systems.

2.7 Thermogravimetric Analysis (TGA)

Thermogravimetric analysis was performed using an STA 449 F5 Jupiter instrument from NETZSCH-Gerätebau GmbH. Samples were placed in aluminum oxide crucibles and analyzed under a nitrogen atmosphere with a flow rate of 20 mL/min to ensure inert conditions. The applied temperature program included an initial heating segment at 10 °C/min up to 40 °C, where the sample was held for 30 min to remove residual solvents such as isopropanol or diethyl ether commonly used during sample preparation. Subsequently, the samples were heated at a constant rate of 20 °C/min to 1000 °C. The resulting TGA curves provided mass change data as a function of both time and temperature, enabling the identification of decomposition steps and quantification of volatile components in the binder system.

2.8 Scanning Electron Microscopy (SEM)

For microstructural analysis, a Nova NanoSEM 230 (FEI, Thermo Fisher Scientific) equipped with a Through-Lens Detector (TLD) for secondary and backscattered electrons was used. The measurements were performed under high-vacuum conditions (~10^{-6} mbar) with an accelerating voltage of 2.00 kV and a working distance of 2.2 mm.

3 Results and Discussion

3.1 Binder Synthesis and Phase Composition

The synthesis of γ-C_2S was performed using the different approaches as described in Chap. 2. The resulting phase composition was analyzed by X-ray diffraction (XRD). Figure 1 compares three representative samples: a pure mixture of γ-C_2S, a dry-fired mixture using $Ca(OH)_2$ and silica fume and a sample produced via the C-S-H-based synthesis route. The diffractogram of the dry-fired mixture indicates significant quantities of unreacted SiO_2 and CaO, demonstrating that direct dry firing without prior homogenization is insufficient to achieve phase purity. In contrast, the C-S-H-based synthesis approach, although minor amounts of β-C_2S were still detected, shows a much purer result. They contain less unreacted SiO_2 and CaO. These results underline the importance of precursor preparation and repeated thermal treatment to obtain phase-pure γ-C_2S.

The effect of repeated firing of the same sample is illustrated in Fig. 2, highlighting the progressive reduction of free CaO and SiO_2 with each firing cycle. This demonstrates that, while multiple firings can improve homogeneity and conversion, careful control of the CaO content is essential, as excessive free lime would limit the usability of the product as a binder.

The macroscopic appearance of the synthesized γ-C_2S is shown in Fig. 3. Dry-fired material exhibits visible inhomogeneities (Fig. 3a) whereas the optimized synthesis route produces a homogeneous powder with a self-pulverizing behavior after cooling. This phenomenon is accompanied by a distinct color change from bright white (precursor) to a slightly greyish-blue hue (final product) which can be seen in Fig. 3b).

Two representative SEM micrographs in Fig. 4 provide insight into the particle morphology. At lower magnification, the γ-C_2S particles appear very fine and loosely

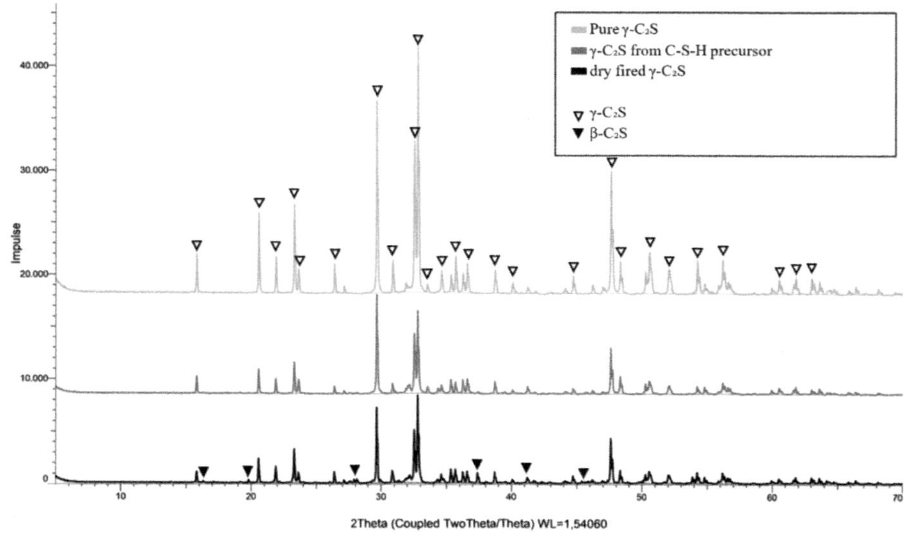

Fig. 1. XRD analysis of different methods for synthesizing γ-C_2S

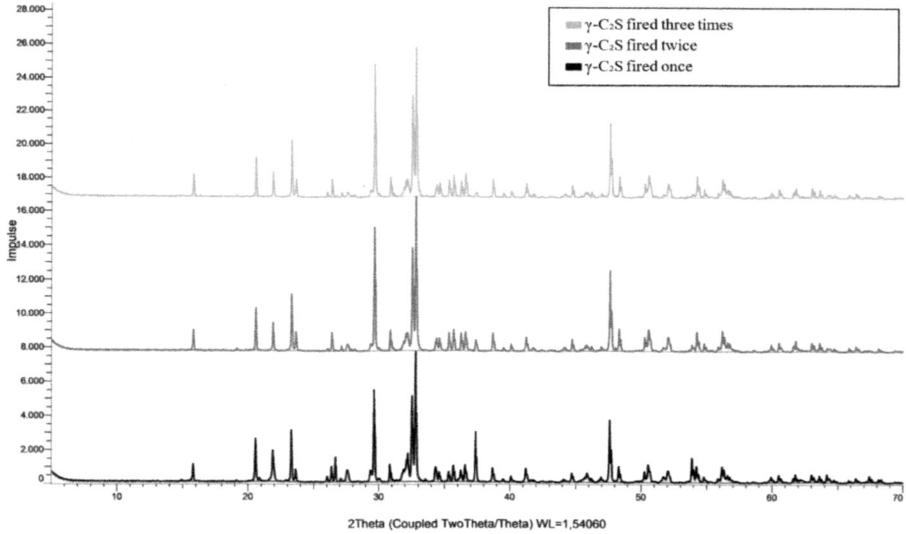

Fig. 2. XRD analysis of dry fired γ-C_2S after multiple burning cycles

agglomerated, while higher magnification shows the individual particle surfaces and characteristic morphology of the synthesized clinker. These microstructural observations, combined with the phase analysis, confirm that the synthesis strategy using C-S-H precursors yields the highest γ-C_2S purity and optimal particle characteristics for subsequent reactivity testing.

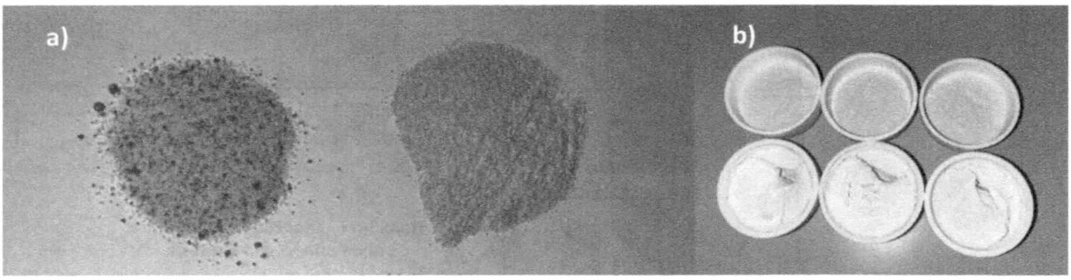

Fig. 3. **a** Comparison of dry fired γ-C_2S without (left) and with milling (right) before firing. **b** Comparison of a dry preparation of γ-C_2S before firing (lower samples) and after firing (upper samples)

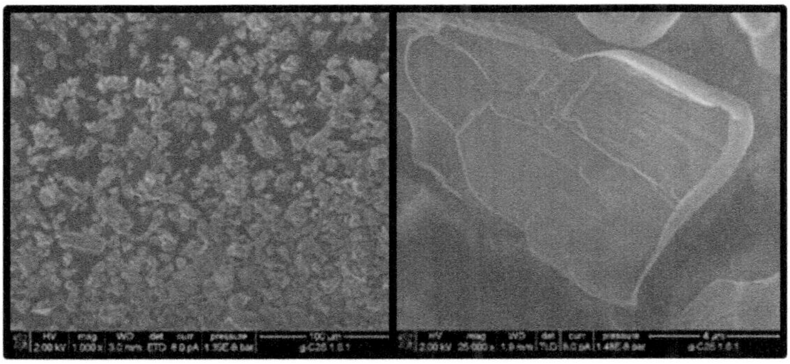

Fig. 4. SEM secondary electron (SE) images of γ-C_2S synthesized through C-S-H precursor

3.2 Reactivity and Carbonation Behavior

All reactivity and carbonation-related investigations presented in this section were carried out using γ-C_2S synthesized via the C–S–H precursor route, which had previously shown the most promising phase composition and carbonation potential.

The reactivity and carbonation behavior of the synthesized CB was initially evaluated by isothermal calorimetry. Six different paste mixtures were prepared to investigate the effect of water-to-binder ratio (w/b) and $NaHCO_3$ addition on heat evolution: (i) w/b = 1.0 with 2 g $NaHCO_3$, (ii) w/b = 1.0 with 1 g $NaHCO_3$, (iii) w/b = 0.7 with 1 g $NaHCO_3$, (iv) w/b = 1.0, (v) w/b = 0.7, and (vi) w/b = 0.5. All mixtures were based on 5 g of γ-C_2S. Prior to binder addition, $NaHCO_3$ was first mixed with water and left to stand until no further thermal activity was detected, ensuring that the initial dissolution and decomposition reaction between $NaHCO_3$ and water did not influence the recorded heat flow. The calorimetry results (Fig. 5) show distinct differences between $NaHCO_3$-containing and $NaHCO_3$-free samples. Mixtures with $NaHCO_3$ exhibit two distinct exothermic maxima, while the reference mixtures prepared with water only show a single dominant peak. Furthermore, an increase in $NaHCO_3$ dosage correlates with higher total heat release, indicating an acceleration of carbonation reactions. These findings confirm that the $NaHCO_3$ method is effective in activating and quantifying the carbonation reactivity of γ-C_2S, acknowledging that minor amounts of β-C_2S present in the synthesized binder also contribute to early hydration and carbonation.

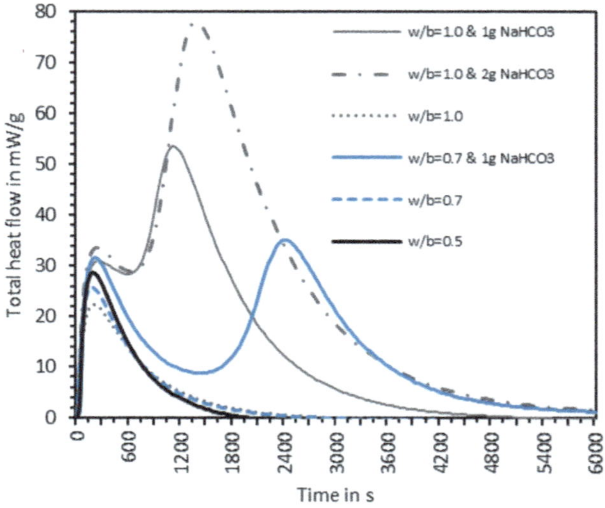

Fig. 5. Calorimetry results of six different mixtures of 5 g of γ-C_2S

To validate the phase assemblages after reaction, XRD analyses were performed on all calorimetry samples. All specimens still contained unreacted C_2S; however, the $NaHCO_3$-containing mixtures show significantly higher $CaCO_3$ peak intensities compared to the water-only references. This confirms that measurable carbonation occurred under the applied conditions.

SEM imaging (Fig. 6) provided further microstructural insights into the reacted materials. The microstructure is dominated by the presence of unreacted γ-C_2S phases and newly formed carbonate reaction products. On some places (Fig. 6a) also fibrous phases were observed, which are probably hydration products such as C-S-H. The figure shows that samples without $NaHCO_3$ contain even more hydration products (C-S-H). Samples with $NaHCO_3$ (Fig. 6b) exhibit surface coating of the γ-C_2S particles, which is presumably $CaCO_3$.

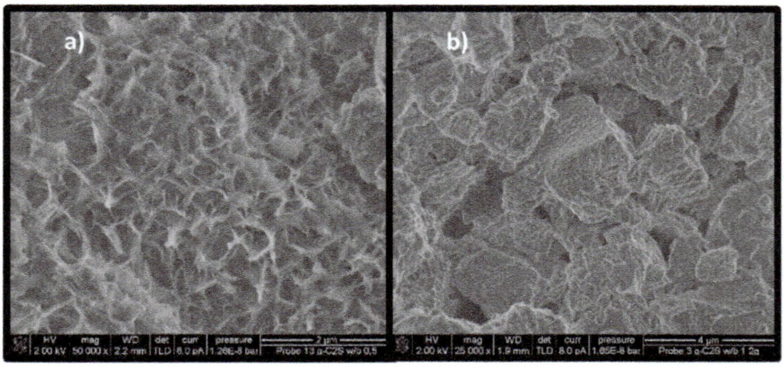

Fig. 6. SEM-SE image of γ-C_2S calorimetry sample with w/b = 0.5 (**a**) and a sample with w/b = 1,0 and 2g of $NaHCO_3$/ 5 g γ-C_2S (**b**)

To complement the paste testing, specimens were prepared as described in Sect. 2.3, using γ-C_2S synthesized via the C–S–H precursor route.

XRD analysis of this sample (Fig. 7) revealed the presence of calcite and vaterite as dominant carbonation products, along with remaining unreacted γ-C_2S. These carbonation products are typical for γ-C_2S systems, although minor amounts of aragonite can occasionally be detected depending on the curing conditions [6]. The presence of unreacted γ-C_2S suggests that diffusion limitations likely hinder complete hydration and carbonation of the bulk material. In addition, SEM-SE images (Fig. 8) reveal that the carbonation coating is not deposited directly on the C_2S surface, but that a layer of C–S–H is present in between. This suggests that the carbonation reaction might be influenced by diffusion through the C–S–H layer.

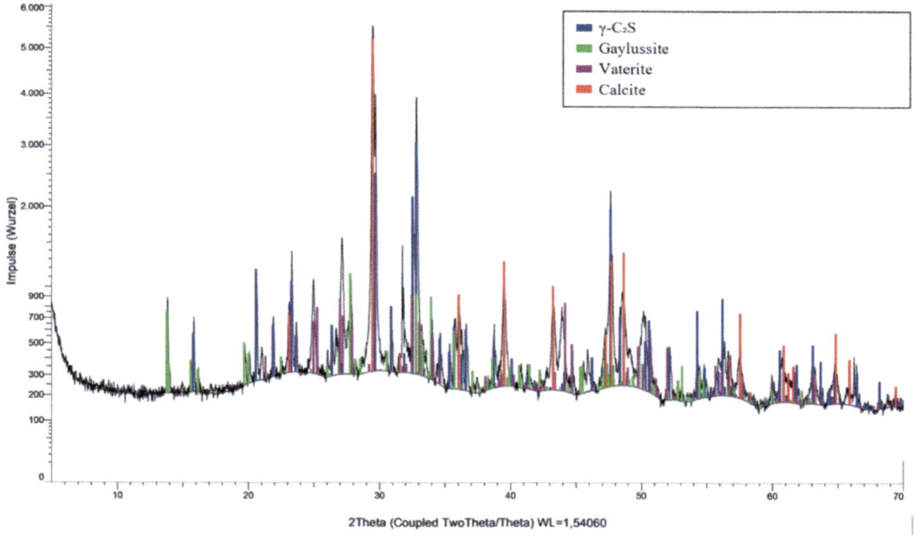

Fig. 7. XRD analysis of hardened paste sample (6 days at 20% CO_2)

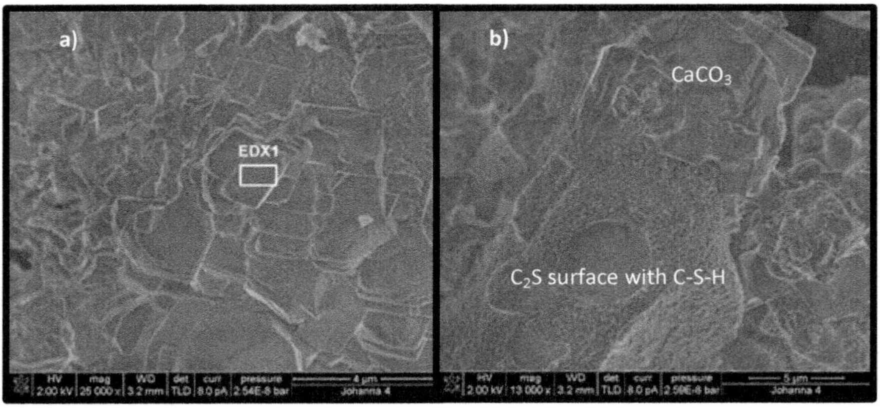

Fig. 8. SEM-SE images of hardened paste sample. **a** Shows Calcium carbonates, whereas **b** also includes an unreacted particle and C-S-H phases

Finally, thermogravimetric analysis (TGA) was performed on the reacted cylindrical specimen to estimate the amount of $CaCO_3$ and to identify potential C–S–H phases. The resulting differential thermogravimetric (DTG) curve (Fig. 9) shows distinct mass loss events, particularly a pronounced peak in the range of 600–800 °C, which is attributed to the decomposition of $CaCO_3$. Based on this decomposition step, the specimen is estimated to contain approximately 50% $CaCO_3$ by mass, indicating substantial carbonation of the γ-C_2S binder. This high degree of carbonation is consistent with the XRD and SEM/EDX findings, confirming the effective formation of carbonate phases within the reacted material.

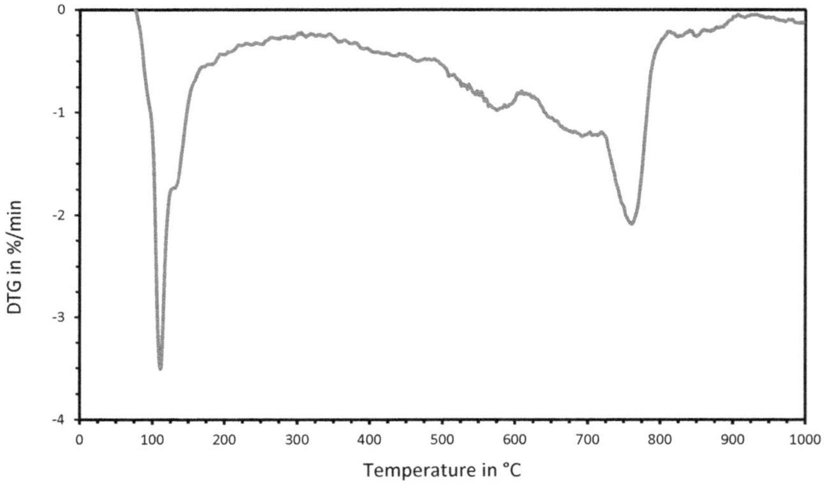

Fig. 9. Differential thermogravimetric (DTG) curve of the reacted cylindrical paste specimen showing major decomposition steps, including $CaCO_3$ decarbonation between 600 and 800 °C.

These combined results demonstrate clear carbonation reactivity of the synthesized γ-C_2S, particularly when activated with $NaHCO_3$, and provide a basis for further kinetic analysis and optimization of curing strategies.

4 Conclusions

This study investigated γ-C_2S as a model carbonatable binder, focusing on its synthesis using different routes and its carbonation behavior under different curing strategies. XRD analysis confirmed that using the C-S-H precursor route produced the highest phase purity, although minor amounts of β-C_2S remained. In contrast, dry-firing approaches resulted in significant unreacted CaO and SiO_2.

The self-pulverizing behavior of γ-C_2S was evident in all samples, especially those with high phase purity. A color change and fine particle morphology were confirmed by SEM and were consistent with previous findings [9] and are beneficial for further processing.

Isothermal calorimetry showed that $NaHCO_3$ addition markedly accelerated carbonation, producing a distinct two-peak heat flow pattern. XRD and SEM confirmed increased $CaCO_3$ formation in these systems. Under diffusion-controlled curing, calcite and vaterite formed, but unreacted γ-C_2S persisted, indicating limited carbonation depth. TGA results supported these observations.

Overall, γ-C_2S was shown to be a functional carbonatable binder, and the addition of $NaHCO_3$ proved effective for promoting and quantifying carbonation reactivity. These findings contribute to the understanding of carbonation mechanisms in calcium silicate systems and highlight the specific potential of γ-C_2S for CO_2 storage in cementitious materials. This underlines the crucial role of carbon capture and utilization (CCU) technologies in reducing the carbon footprint of the construction sector and supporting global climate change mitigation efforts [10].

Acknowledgments. This research was funded by the German Research Foundation (DFG) under the project number 542229402, titled "C-Cure: Net-Zero Concrete with Carbonatable Binders, Rapid Reactivity Testing, Internal CO2-Curing, and Upscaling Computations" as part of the Priority Programme SPP 2436: Net-Zero Concrete (project number 521082619).

References

1. Liu Z, Lv C, Wang F, Hu S (2023) Recent advances in carbonatable binders. Cement Concrete Res 173:107286. https://doi.org/10.1016/j.cemconres.2023.107286
2. Ashraf W, Olek J (2016a) Carbonation behavior of hydraulic and non-hydraulic calcium silicates: potential of utilizing low-lime calcium silicates in cement-based materials. J Mater Sci 51(13):6173–6191. https://doi.org/10.1007/s10853-016-9909-4
3. Pu Y, Li L, Shi X, Wang Q, Abomohra A (2023) Recent advances in accelerated carbonation for improving cement-based materials and CO_2 mitigation from a life cycle perspective. Constr Build Mater 388:131695. https://doi.org/10.1016/j.conbuildmat.2023.131695
4. Zhang X, Zhao M, Zhang Y (2012) Preparation and properties of self-pulverizing calcium sulfoaluminate cement. Constr Build Mater 34:107–113. https://doi.org/10.1016/j.conbuildmat.2012.02.006

5. Li G, Liu S, Hu X, Zhu J, Guan X, Shi C (2023) Tailoring self-pulverized low-calcium clinker for CO_2 sequestration. Constr Build Mater 409:134051. https://doi.org/10.1016/j.conbuildmat.2023.134051
6. Tan Y, Liu Z, Wang F (2022) Effect of temperature on the carbonation behavior of γ-C2S compacts. Cement Concrete Compos 133:104652. https://doi.org/10.1016/j.cemconcomp.2022.104652
7. Wang X, Guo M-Z, Ling T-C (2022) Review on CO2 curing of non-hydraulic calcium silicates cements: mechanism, carbonation and performance. Cement Concr Compos 133:104641. https://doi.org/10.1016/j.cemconcomp.2022.104641
8. Meyer V, de Cristofaro N, Bryant J, Sahu S (2018) Solidia cement an example of carbon capture and utilization. KEM 761:197–203. https://doi.org/10.4028/www.scientific.net/KEM.761.197
9. Lei M, Mu Y, Liu Z, Wang F, Hu S (2023) Structural evolution and characterization of γ-C2S (γ-C2S) compacts during carbonation. ACS Sustain Chem Eng 11(1):444–457. https://doi.org/10.1021/acssuschemeng.2c06623
10. Li N, Mo L, Unluer C (2022) Emerging CO2 utilization technologies for construction materials: a review. J CO2 Util 65:102237. https://doi.org/10.1016/j.jcou.2022.102237

Performance of Low-Carbon Concretes with High Amounts of Ground Limestone by Aluminate-Boosting of OPC

Alexander Gunkler[1], Thorsten Leusmann[2], Xuerun Li[3], Joachim Dengler[3], and Dirk Lowke[1](✉)

[1] Technical University of Munich, Garching, Germany
lowke@tum.de
[2] Technische Universität Braunschweig, Braunschweig, Germany
[3] BASF Construction Additives GmbH, Trostberg, Germany

Abstract. As part of the global effort to tackle climate change, Low-Carbon Concretes (LCC) are becoming increasingly important in the construction industry. They differ from standard concretes in terms of their significantly reduced carbon footprint. This can be achieved directly by (I) using CO_2-negative SCM and/or (II) using blended binders with reduced cement clinker content (e.g. substitution by SCM). Disadvantages associated with these strategies, such as reduced workability or durability, can be compensated through optimizing the particle size distribution or reducing the water-to-clinker ratio while increasing the superplasticizer dosage. Regarding limitations of the practical feasibility of these approaches and/or the availability of SCM, the substitution of cement clinker by high amounts of ground limestone in LCC seems to be particularly suitable for reducing CO_2-emissions. However, high amounts of ground limestone in LCC result in a significant reduction in mechanical properties and durability due to the lower C-S-H formation, resulting in a less dense hardened structure as well as less alkaline environment in the pore solution. A novel approach for producing LCC involves the use of Hydration Control Additives (HCA). These additives modify the hydration kinetics of OPC and enhance its strength-forming potential by converting the aluminate phases into additional solids, i.e. ettringite ('aluminate-boost'). Therefore, the use of HCA is a promising approach to realize LCC with low-clinker binders containing high amounts of limestone powder. This study investigates the application of HCA for LCC with up to 50 vol.% of ground limestone in the binder. Two reference mixes were included in the investigations: A standard concrete with 100% OPC and an identical LCC without HCA. Experimental results show that the anticipated effects of dilution due to the substantial reduction of 50 vol.% cement clinker in LCC can be fully compensated by the use of HCA: this includes mechanical properties like compressive strength as well as the durability like the freeze-thaw resistance.

Keywords: Low-carbon concrete · Eco-concrete · Aluminate-boosting of OPC

H. Beushausen et al. (Eds.): NTZR 2025, LNCE 873, pp. 13–24, 2026.
https://doi.org/10.1007/978-3-032-27698-8_2

1 Introduction

Global efforts towards a more sustainable way of life, such as those discussed at various UN conferences, emphasize the conservation of natural resources, the reduction of CO_2 emissions and the improvement of energy efficiency in the construction industry in order to protect the environment for future generations. As concrete is the most widely used building material, special attention is paid to this area. The most CO_2-emitting component of concrete is its binder, cement. Cement is a collective term that can be used to describe either Ordinary Portland Cement (OPC), which is mainly ground cement clinker, or a multi-compound or composite cement made from cement clinker and other ground supplementary cementitious materials (SCM) of varying origin and performance.

Process options for reducing CO_2 emissions in cement clinker production alone are insufficient to significantly improve the carbon footprint of cement and concrete. For this reason, the focus is on reducing the amount of cement clinker in binders or concrete itself. Such concretes that are optimized in terms of their carbon footprint are also known as Low-Carbon Concretes (LCC). However, when designing LCC, it is important to be aware that reducing the amount of cement clinker can have a negative effect on the mechanical properties and the durability of a concrete.

This paper aims to provide insight into a novel approach for the mix-design of LCC through the application of Hydration Control Additives (HCA) enabling an aluminate-boosting of OPC (one-step conversation of aluminates and anhydrite to ettringite), allowing a significant reduction of cement clinker while maintaining a good performance of the concrete in terms of mechanical properties and durability.

2 Mix-Design for Low-Carbon Concretes (LCC)

There are several approaches for the design of LCC that are partly well established in concrete technology and can also be combined with each other. In the following, the different approaches are presented and evaluated below with a special regard to current challenges for the application of SCM in LCC. Finally, a novel approach for the production of LCC via the aluminate-boosting of OPC is introduced.

2.1 State-of-the-Art

A direct approach to realize concretes with ultra-low or even zero carbon footprint that is currently the subject of intense research and media coverage is the **(I) use of CO_2 negative SCM**: Due to their manufacturing process or innate properties, certain materials, e.g. biochar or non-reactive SCM are able to bind carbon or CO_2 (Carbon Capture, Storage and Utilization, CCSU) resulting in a negative CO_2 balance. Additionally, depending on their mineral composition and the amount of bounded CO_2, non-reactive SCM can get transformed into reactive SCM able to form C-S-H phases. Consequently, CO_2-negative SCM can be used in concrete to reduce its CO_2 footprint, as well as its clinker content when being activated by recarbonisation [1].

Although this approach seems promising, the production of CO_2-negative SCMs is complex and costly. Hence, this technology is currently not transferable to the scale of the concrete industry. For this reason, common approaches for reducing the carbon footprint of concrete focus on the **(II) use of SCM**/blended binders with reduced cement clinker content [2]:

During cement hydration, Portland Cement reacts with water and forms hydrate phases that fill the pore space. In this process two main reactions can be identified: First, the silicate phases yield C-S-H and Portlandite. Secondly, the aluminate phases in the OPC (C_3A, C_4AF) react with calcium sulphate ($CaSO_4$) to form ettringite (Aft phase). Upon sulphate depletion ettringite may convert to monosulphate (sulphate AFm) or monocarbonate (carbonate AFm) depending on the total sulphate and carbonate content [3–5]. A strategy for improving the carbon footprint of OPC by the **(II) use of SCM** can be derived from each of these main reactions:

(a) **Mixing of OPC and SCM**: The SCM react with Portlandite and forms additional C-S-H phases. This results in increased strength and reduced porosity. Moreover, this additional strength-forming reaction can be used to reduce the strength-forming proportion of OPC in the binder while improving the CO_2 balance [6].
(b) **Stabilization of ettringite**: In the presence of additional $CaCO_3$ and sufficient amounts of aluminates from the SCM and OPC, carbonate AFm is formed, leaving a larger and now stable residue of ettringite, which results in a larger hydrate phase volume, lower porosity and allows a reduction of ≤ 5 wt.% clinker in the concrete [7, 8].

The desired favourable CO_2 balance of LCC generally leads to low cement contents associated with negative effects like poor workability and a decrease in durability. Such effects can be compensated by optimizing the particle size distribution [9] and/or the use of fine-tuned microfillers (e.g. silica fume) as well as reducing the water-to-clinker ratio [10]. In both cases, the use of highly effective superplasticizers is often indispensable as yield stress and viscosity of the fresh concrete are increasing due to the interlocking particles and the low water content.

2.2 Evaluation of the Different Approaches and Current Challenges Using SCM

The use of SCM for LCC or composite cements is the most widely used approach to reduce the CO_2 footprint of concretes. Main SCM used are reactive additions like ground granulated blast furnace slag (GGBS) and fly ash, as well as ground limestone as a largely inert material. In addition, the use of calcined clays in blended binders like limestone calcinated clay cements (LC3) has been continuously researched and successfully applied [11]. The various SCM differ considerably in terms of their origins and effects, which mainly affect their applicability and availability in LCC as well as their potential to contribute to reduce CO_2-emissions.

As GGBS and fly ash are by-products of steel production and coal-fired power generation, respectively, their global warming potential (GWP) is usually considered to be neutral (nom. 0 $kgCO_2/t$). GGBS is already used in high amounts of up to 95 wt.% in Blast Furnace Cements, leaving little potential for additional CO_2 reduction unless further SCM replacement is considered. The fly ash content is usually limited (e.g. $\leq$35 wt.% in Portland Pozzolana Cement) due to their low reactivity. Hence, a further potential to reduce CO_2-emissions with fly ash in LCC is questionable. Furthermore, the availability of GGBS and fly ash is expected to decline locally due to the decarbonization of steel and energy industries. As global, energy-intensive supply chains for these SCM contradict the ideas of an LCC, their future use is likely to decline locally [12].

Calcined clay and other SCMs offer promising routes for reducing cement-related emissions. However, their large-scale adoption still is limited by material availability, processing complexity, and cost [2, 13]. One of the major drawbacks for calcined clay is the reduced workability. While effective in improving strength, these methods alone may not suffice to achieve ultra-low carbon concrete targets.

In contrast, limestone is readily available in large parts of the world and not affected by this issue. Furthermore, the use of limestone is a very effective way of improving the CO_2 footprint of LCC due to its low GWP (~40 $kgCO_2/t$). The potential to substitute cement clinker by ground limestone in reinforced LCC structures is currently limited due to decreasing durability properties. Today's common Portland Limestone Cements typically contain up to $\leq$ 15–20 wt.% limestone. Regardless, using limestone in LCC is a promising widely applicable approach worthy of further research due to its assured long-term availability and high potential of reducing CO_2 emissions of concrete.

2.3 Low-Carbon Concretes Through Novel Hydration Control Additives

This paper introduces a novel approach for LCC by maximizing the binder potential of Portland cement by increasing its strength per unit cement though optimizing both the hydrate phase volume and its spatial distribution. Conversely, this approach allows the reduction of the quantity of Portland cement, accompanied by an improved CO_2 balance, while maintaining a certain strength.

Ettringite ($Ca_6Al_2(SO_4)_3(OH)_{12}\cdot 26H_2O$)), a hydrate phase that occurs in varying volume rates during hydration, possesses the properties required for the realisation of the novel approach: A large molecular volume due to its high water content and a noticeable strength. Its favourable properties, such as rapid hardening and high early-age strength are already actively used in Calcium Aluminate Cements or Calcium Sulfoaluminate Cements [14, 15]. However, the use of ettringite in such cements still widely disregards its potential for reducing CO_2 emissions.

In OPC, the formation of ettringite by the reaction of C_3A, water and added $CaSO_4$, primarily serves to regulate the setting of the cement. Therefore, the additional $CaSO_4$ slows down the C_3A reaction allowing the C_3A and C_3S reactions to occur simultaneously, while in the absence of $CaSO_4$, the cement undergoes rapid hardening and delays the C_3S reaction. The main contribution to strength and setting due to the simultaneous reactionsy of C_3A and C_3S does not come from the formation of ettringite, but from the reaction of C_3S with water to form C-S-H and $Ca(OH)_2$. Furthermore, the initial ettringite formation stops once the sulphate source is depleted ("sulfate depletion"). Moreover, the $CaSO_4$ content is optimized to prevent the formation of ettringite after the main C_3S reaction, which is accompanied by an undesired destructive expansion of the hardened cement.

As hydration progresses, the amount of ettringite decreases again due to ongoing reactions of remaining free aluminates in the absence of sulphate carriers and its thermodynamic instability upon sulphate depletion. Monosulphate (sulphate AFm) or monocarbonate (carbonate AFm) is formed with a smaller molecular volume and lower strength and pore filling capacity [3, 4, 16]. Therefore, the full strength building and volume filling potential of OPC remains untapped as only small amounts of the aluminates converts to ettringite while C_4AF mostly acts as an inert material during early hydration. Previous studies showed that certain chemical admixtures, so called Hydration Control Additives (HCA), can strongly inhibit (e.g. tartaric acid) or accelerate (e.g. alkanolamines) the reaction of the aluminates during hydration. Furthermore, studies on cement mixtures, HCA and additional quantities of $CaSO_4$ adjusted to the actual aluminate content in the cement, showed an almost complete one-step aluminate-to-ettringite conversion (C_3A, C_4AF) before the onset of the silicate reactions. Hence, the full potential of OPC through the targeted formation of ettringite can be utilized without risking destructive crystallization. The result is a binder with lower carbon emissions and enhanced strength [17–19].

This novel approach of using HCA for managing a maximizing of early ettringite formation through rapid reactions of C_3A and C_4AF with additional sulphate before the main silicate hydration reaction will henceforth be referred to as aluminate-boosting of OPC. This study examines the transferability of this approach to the scales of concrete. Therefore, investigations on aluminate-boosted LCC with high amounts of ground limestone by HCA and additional sulphate carrier were carried out with a special focus on the effects on the concrete's mechanical properties and its durability.

3 Materials and Methods

3.1 Concept of Investigation

The objective of this study is to investigate the application of the aluminate-boosting of OPC in LCC with high amounts of ground limestone. In particular, the aim is to investigate whether the aluminate-boost can compensate a decline in performance due to clinker substitution. Therefore, (1) a standard concrete with 100% OPC, (2) a non-boosted LCC with 50 vol.% of the OPC substituted by ground limestone, and (3) an identical aluminate-boosted LCC were analysed in various material tests with regard to:

- Hydration heat-flow
- Setting and hardening behaviour (development of the dynamic E-Modulus)
- Mechanical properties (development of compressive strength)
- Durability (resistance against carbonation, freeze-thaw-attack).

3.2 Materials

Table 1 shows the mix designs for the standard concrete (RF-0), the non-boosted LCC with 50 vol.% of OPC substituted by ground limestone (LS-0), and the corresponding aluminate-boosted LCC (LS-HCA).

A CEM I 52.5 N (OPC) was used as cement in all mix designs, with the addition of 10 wt.% of anhydrite ($CaSO_4 \geq 97\%$) referred to the amount of OPC. Ground limestone ($CaCO_3 \geq 97\%$) was used as SCM for the production of the non-boosted and the aluminate-boosted LCC. Accordingly, this study assumes OPC and anhydrite as part of the binder as well as ground limestone when used in the mix design.

To achieve the aluminate-boosting of the OPC in the LS-HCA mix, a commercial HCA (brand name HyCon® B) was used, which was developed and supplied by BASF Construction Additives GmbH, Trostberg, Germany [19]. It was applied in a dosage of 2.4% of its solid content, referring to the amount of OPC and anhydrite.

In addition, a PCE-superplasticizer was used in low dosages (≤ 0.15 wt.% OPC) to slightly adjust the workability. All concretes had an identical binder volume and water content (water + liquid phase HCA). As natural aggregates, sand 0/2 mm and gravel 2/8 mm were used.

Table 1. Mix designs for the various tested concretes.

	RF-0	LS-0	LS-HCA	Each mix:
Material	kg/m^3	kg/m^3	kg/m^3	
OPC	376	188	188	Binder paste vol.: 340 L/m^3 + Air void vol.: 20 L/m^3
Anhydrite	38	19	19	
Limestone	–	180	180	
Water	207	207	178	
HCA (HyCon® B)	–	–	33	
Superplasticizer	–	0.3	0.3	
Sand 0/2	929	929	929	Aggregate vol.: 640 L/m^3
Gravel 2/8	760	760	760	
w/c-ratio	0.5	1.0	1.0	

3.3 Methods

The analyses on the hydration heat-flow were carried out using a proprietary semi-adiabatic calorimeter [20]. The investigations on the setting and hardening behaviour included tests on the development of the dynamic E-Modulus, determined by quasi-continuous ultrasonic transit time measurement on fresh concrete specimens using an IP-8 measuring system by Ultratest.

The mechanical properties were examined by compressive strength test on concrete cubes (150 mm) at 2, 7, 28 and 56 d according to EN 12390–3. All specimens were demolded 1 d after casting, stored in water until the age of 7 d and then placed in a climate chamber (20 °C/65% RH) until testing.

The investigations on the durability of the various concretes covered tests on their resistance against carbonation and freeze-thaw attack.

The resistance against carbonation was tested by examining the 70 d accelerated **carbonation progress** according to EN 12390–12 at 2.0 vol.% CO_2. The carbonation depth measurements were carried out on concrete prisms ($100 \times 100 \times 300$ mm^3) at 7, 28 and 70 d. The specimens were demolded after 1 d, stored in water until the age of 28 d, placed for a further 14 d in a climate chamber (20 °C/65% RH) before being transferred to a CO_2 incubator until testing.

The resistance against freeze-thaw attack was investigated using the **capillary suction, internal damage and freeze-thaw test** method (CIF-Test) according to RILEM TC 176-IDC/BAW-Merkblatt MFB 2012 [21]. Specimens (150 × 150 × 75 mm^3) sawn-to-size from concrete cubes at the age of 28 d were used. The cubes were demolded 1 d after casting, stored in a water bath at 20 °C up to the age of 7 d and afterwards placed in a climate chamber (20 °C/65% RH). The sample preparation and test execution of 28 freeze-thaw cycles (FTC) were carried out in accordance with the standards.

4 Results and Discussion

4.1 Heat of Hydration Development

Figure 1 shows the heat flow and the cumulative hydration heat development as a function of the equivalent concrete age t_e for the various concretes. Therefore, the influence of temperature rise on concrete age was considered using the approach of Freiesleben-Hansen and Petersen [22].

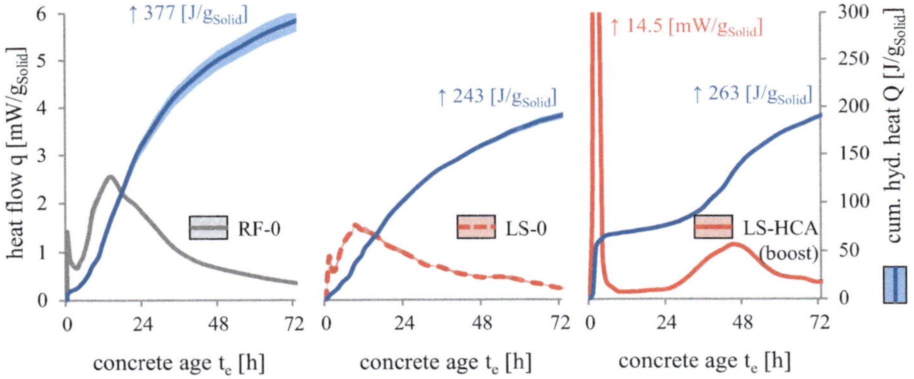

Fig. 1. Heat flow and cumulative hydration heat as a function of the equivalent concrete age t_e

The release of heat flow of the reference mix with 100% OPC in the binder (RF-0) shows an initial peak in the heat release rate, decreasing briefly with a continuous transition to a significantly higher peak and achieves a total cumulative heat of 377 J/g$_{Solid}$ after 7 d (values not displayed). In OPC binders, this behaviour is attributed to an initial aluminate reaction and the subsequent silicate reaction. The substitution of 50 vol.% OPC by ground limestone in LS-0 results in a comparable development of heat flow but at a lower degree, achieving a cumulative heat of 263 J/g$_{Solid}$ after 7 d. The aluminate-boosted LCC (LS-HCA) shows a significant peak in heat-flow during the first 2 h of hydration, followed by an almost total decline in heat-flow for the following 12 h. Afterwards, a second increase of the heat-flow was recorded between 24 h to 48 h. The cumulative heat of hydration reached a value of 263 J/g$_{Solid}$ after 7 d, which is slightly higher than the corresponding value of the LS-0 mix without hydration control agent.

4.2 Setting and Hardening Behaviour and Mechanical Properties

Figure 2 left shows the development of the dynamic E-Modulus as an indicator of the setting and hardening behaviour of the fresh concrete. The reference mix with 100% OPC binder (RF-0) shows a continuous increase of the dynamic E-Modulus reaching a value of around 34 GPa at 72 h. Likewise, the non-boosted LCC mix (LS-0) shows again a comparable but less pronounced development of the dynamic E-Modulus compared to RF-0, achieving a total value of around 26 GPa. The boosted LCC (LS-HCA) shows a sudden increase in dynamic E-Modulus within 2 h of hardening, followed by a period of hardly no further increase for up to 24 h. Thereafter, a slight but continuous increase of the dynamic E-Modulus can be noticed, corresponding to the beginning of the second peak in the heat of hydration development, see Fig. 1. Finally, the boosted LCC (LS-HCA) reaches an overall value of around 33 GPa, comparable to RF-0.

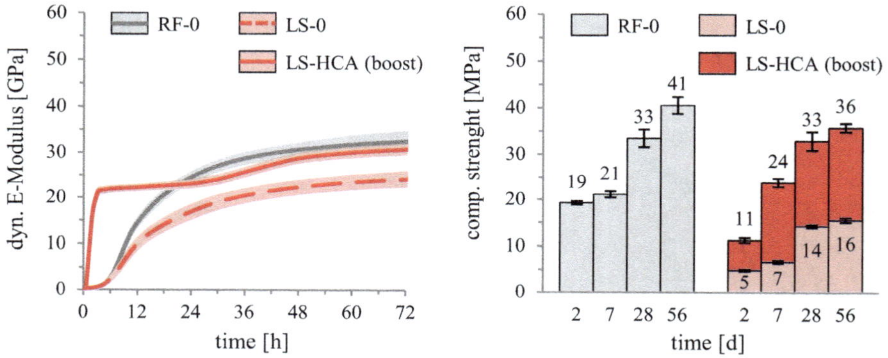

Fig. 2. Development of the dyn. E-Modulus and the compressive strength of the various mixes

The results of the compressive strength tests on the concretes, shown in Fig. 2, right, correspond to the previous findings. The reference mix with 100% OPC binder (RF-0) reaches a 28 d compressive strength of around 33 MPa. The anticipated effect of dilution due to the clinker substitution of 50 vol.% in the non-boosted LCC (LS-0) is clearly visible at all ages, resulting in a 28 d compressive strength of only 14 MPa, significantly lower (> 50%) than RF-0. The aluminate-boosted LCC (LS-HCA) demonstrates significantly higher compressive strength than the LS-0 mix at each age, performing nearly equivalently to RF-0 and achieving an identical 28 d compressive strength.

The presented results of the aluminate-boosted LCC (LS-HCA) show a significant modified setting and hardening behaviour. The heat flow is clearly separated into two reactions by an intermediate phase with almost no further release of hydration heat. The first peak is assigned to be the desired one-step aluminate-to-ettringite reaction. The increase in stiffness that must be accompanied by this sudden formation of solid can be clearly seen in the corresponding sharp rise of the dynamic E-Modulus. Furthermore, the phase of total decrease in heat-flow corresponds to the subsequent phase where the dynamic E-Modulus hardly increases. Accordingly, the second peak in heat-flow is interpreted as the silicate reaction leading to the formation of C-S-H, that occurs as a further increase in the development of the dynamic E-Modulus after the phase of hardly any increase. The compressive strength values illustrate the enormous increase in strength due to the aluminate-boost, i.e. formation of ettringite, able to fully compensate the strength losses due to clinker substitution by ground limestone.

4.3 Durability

The results of the accelerated carbonation of the concretes are shown in Fig. 3. After 70 d storage in the CO_2-incubator (2.0 vol.% CO_2), the reference mix (RF-0) shows the lowest average carbonation depth of 13.6 mm. A comparison of the results of the two LCC reveals that the boosted LCC (LS-HCA) shows a noticeable slower carbonation progress than the non-boosted LCC (LS-0) from the age of 28 d on. Finally, the measured 70 d carbonation depth of the LS-HCA mix (27.7 mm) is significantly lower than the value of the LS-0 mix (49.2 mm). It appears that the aluminate-boost causes microstructural compaction, which leads to an increase in carbonation resistance. However, the results indicate that the aluminate-boosting of OPC cannot fully compensate the reduction in carbonation resistance caused by substituting cement clinker with ground limestone. This might be due to the dilution effect of the limestone which reduced the overall portlandite content in the hardened sample.

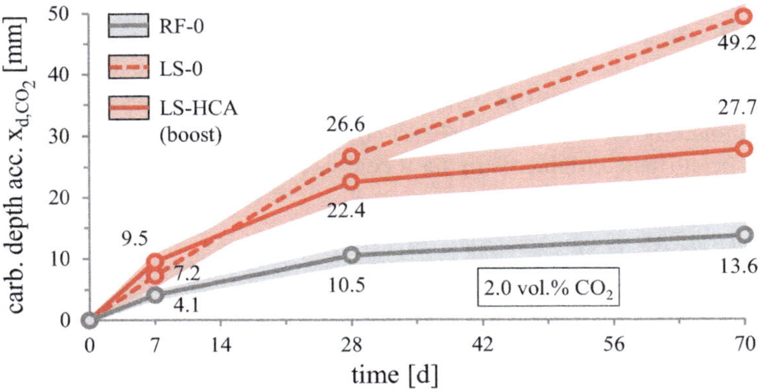

Fig. 3. Test results for accelerated carbonation in a CO_2-incubator (2.0 vol.% CO_2)

Figure 4 shows the results of the freeze-thaw tests (CIF-Test) of the various concretes. The two test criteria, the mass loss as an indicator for external structural damage and the decrease of the relative dynamic E-Modulus indicating internal structural damage, and their associated failure limits are also displayed.

The reference mix with 100% OPC binder (RF-0) shows a low mass loss of 160 g/m^2 over 28 FTC and a slight increase in relative dynamic E-Modulus $\geq$ 100% (post-hardening effect). The failure limits of 1,000 g/m^2 and 75%, respectively, are reliably met. The substitution of 50 vol.% of OPC by ground limestone (LS-0) leads to a significant increase in the mass loss up to 13,600 g/m^2. In addition, the internal structural damage is distinctive with a decrease of the relative dynamic E-Modulus down to 15% of its initial value. The failure limits are clearly not met. Despite the 50 vol.% clinker substitution, the aluminate-boosted LCC (LS-HCA) shows frost resistance almost comparable to the OPC reference (RF-0), achieving a mass loss of 500 g/m^2 with an approximately constant relative dynamic E-Modulus of $\geq$ 100% over 28 FTC. The failure limits are safely met.

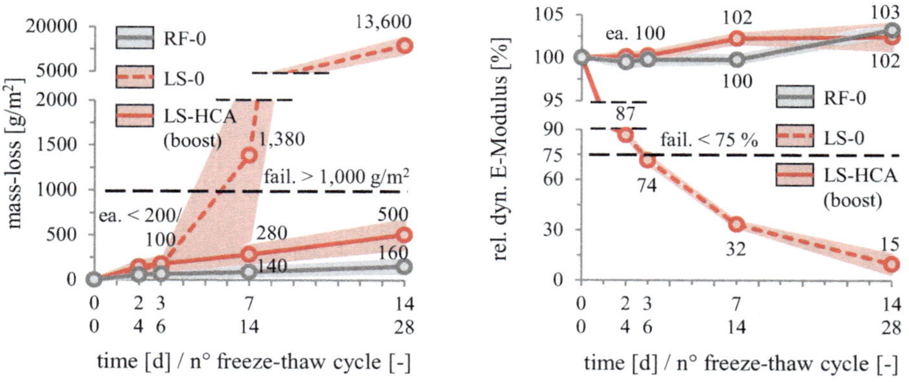

Fig. 4. Mass-loss (left) and decrease of the rel. Dyn. E-Modulus (right) due to frost exposure

All in all, the aluminate-boosting appears to provide a significant improvement in the concrete's resistance to freeze-thaw attack. Additional results of water absorption during the 28 FTC (not illustrated here) show a significantly lower water absorption of the aluminate-boosted LLC (LS-HCA) compared to the non-boosted LCC (LS-0), which can be attributed to microstructural compaction by ettringite formation.

5 Conclusion

This study aimed to investigate the application of a Hydration Control Additive (HCA) for Low-Carbon Concretes (LCC) enabling an aluminate-boosting of OPC, i.e. a one-step conversion of aluminates and anhydrite to ettringite. The HCA used in the study was a commercial product (brand name: HyCon® B), which was developed and supplied by BASF Construction Additives GmbH in Trostberg, Germany. The boosting causes microstructural compaction due to enhanced aluminate conversion forming more ettringite, allowing cement clinker substitution with high amounts of ground limestone as SCM while maintaining a good concrete performance, i.e. mechanical properties and durability.

Comparative studies on the hardening behaviour, the mechanical properties and durability of three concretes, including a standard concrete, a non-boosted LCC and an aluminate-boosted LCC, result in the following findings:

- The substitution of 50 vol.% of OPC by ground limestone in the non-boosted LCC results in a significant decrease in 28 d compressive strength (> 50%) compared to the standard concrete, as well as an increase in 70 d accelerated carbonation depth

(> 100%) and a complete loss of freeze-thaw resistance. The development of heat flow and dynamic E-Modulus are comparable in terms of the characteristic shape of the curve, but at a lower level of 7 d total heat flow (< 30%) and stiffness (< 25%).

The addition of HCA to the aluminate-boosted LCC (50 vol.% substitution of OPC by ground limestone) results in significantly improved properties:

- The aluminate-boosted LCC shows a two-step reaction in the hydration process: a fast aluminate reaction results in sudden increase in dynamic E-Modulus (corresponds to setting time) within 2 h followed by a second induction period before the silicate reaction takes places.
- The aluminate-boost leads to a significant increase in compressive strength compared to the non-boosted LCC (> 100%), enabling identical 28 d compressive strength values to the standard concrete.
- The results on durability for the aluminate-boosted LCC show a noticeable decrease of the accelerated carbonation depth (> 40%) compared to the non-boosted LCC, but still twice as high as the standard concrete with 100% OPC. Furthermore, the resistance against freeze-thaw attack improves significantly by the aluminate-boosting, comparable to the standard concrete.

Acknowledgement. The project was funded by BASF Construction Additives GmbH, Trostberg, Germany.

References

1. Li L, Wu M (2022) An overview of utilizing CO_2 for accelerated carbonation treatment in the concrete industry. J CO_2 Util 60
2. Zunino F (2023) A two-fold strategy towards low-carbon concrete. RILEM Tech Lett 8
3. Taylor HFW (1997) Cement Chemistry, 2nd edn. Thomas Telford Publishing, London
4. Locher FW (2006) Cement. Verlag Bau+Technik, Düsseldorf
5. Scrivener KL et al (2023) Advances in hydration and thermodynamics of cementitious systems. Cement Concr Res 174
6. Lothenbach B et al (2011) Supplementary cementitious materials. Cement Concr Res 41
7. De Weerdt K et al (2011) Synergy between fly ash and limestone powder in ternary cements. Cement Concr Compos 33
8. De Weerdt K et al (2011) Hydration mechanisms of ternary Portland cements containing limestone powder and fly ash. Cement Concr Res 41
9. Fennis-Huijben S (2011) Design of ecological concrete by particle packing optimization
10. Proske T et al (2013) Eco-friendly concretes with reduced water and cement contents—mix design principles and laboratory tests. Cement Concr Res 51
11. Zunino F, Scrivener KL (2023) Recent advances in understanding the hydration of limestone calcined clay cements (LC3). In: Further Reduction of CO_2-Emissions and Circularity in the Cement and Concrete Industry, 16th International Congress on the Chemistry of Cement 2023—ICCC2023, Thailand Concrete Association
12. Arrigoni A et al (2020) Life cycle greenhouse gas emissions of concrete containing supplementary cementitious materials. J Clean Prod 263

13. Dhandapani Y, Santhanam M (2017) Assessment of pore structure evolution in the limestone calcined clay cementitious system and its implications for performance. Cement Concr Compos 84
14. Burris LE, Kurtis KE (2022) Water-to-cement ratio of calcium sulfoaluminate belite cements: hydration, setting time, and strength development. Cement 8
15. Noor L et al (2023) A critical review of the role of ettringite in binders composed of CAC-PC-C$ and CSA-PC-C$. J Am Ceramic Soc 106
16. da Andrade Neto JS et al (2021) Effects of sulfates on the hydration of Portland cement—a review. Constr Build Mater 279
17. Jakob C et al (2023) Controlling ettringite precipitation and rheological behavior in ordinary Portland cement paste by hydration control agent, temperature and mixing. Cement Concr Res 166
18. Kirchberger I et al (2023) Enhancing the aluminate reaction during OPC hydration by combining increased sulfate content, triethanolamine and tartaric acid. Cement Concr Res 170
19. Li X et al (2024) Unlock the full potential of ordinary Portland cement with hydration control additive enabling low-carbon building material. Commun Mater 5
20. Hermerschmidt W et al (2012) Development of a Semi-Adiabatic calorimeter to determine heat of hydration of concrete—short report. iBMB, TU-Braunschweig (2012)
21. RILEM TC 176-IDC (2004) Test methods of frost resistance of concrete: CIF-Test: capillary suction, internal damage and freeze thaw test. Mater Struct 37
22. Freiesleben-Hansen P, Petersen EJ (1977) Maleinstrument til kontrol af betons hardening, Nordisk Betong 1. Stockholm

Development of Limestone Calcined Clay Cement (LC3) from Thermally Activated Laterite Soils: Effect of Chemical and Mineralogical Composition on Mechanical Properties

Rodrigue Cyriaque Kaze[1,2](✉) and Abdolhossein Naghizadeh[2]

[1] Department of Minerals Engineering, School of Chemical Engineering and Mineral Industries (EGCIM), University of Ngaoundéré, Ngaoundéré, Cameroon
kazerodrigue@gmail.com

[2] Department of Engineering Sciences, University of the Free State, Bloemfontein, South Africa

Abstract. Lateritic soils are widely distributed across tropical and subtropical regions globally and can exhibit pozzolanic properties when subjected to appropriate thermal activation. The present study explores the utilization of two types of thermally activated laterite soils, distinguished by their iron oxide content, clay-rich ($Fe_2O_3 \leq 10\%$) and iron-rich ($Fe_2O_3 \geq 40\%$). Both soil types were subjected to calcination at 650 °C and evaluated as potential alternative materials for the development of limestone calcined clay cement. For this purpose, ordinary Portland cement (OPC) was partially substituted with thermally activated laterite (10–30 wt.%) and limestone (5–15 wt.%) to produce mortars. Mortar mixtures were prepared at a water-to-cement and cement-to-sand ratios of 0.5 and 0.33, respectively. Mortar samples were subjected to water curing for varied durations of 7 or 28 days, after which their compressive strength was tested. The 7-day compressive strength values of mortars containing 10, 20, 25 and 30 wt.% clay-rich laterites combined with 5, 10, 12.5 and 15 wt.% limestone were 24.73, 25.14, 20.50 and 13.20 MPa, respectively. The reference samples, by comparison, showed a 7-day compressive strength of 24.4 MPa. The corresponding mixtures containing iron-rich laterites, gave strength values of 12.77, 14.58, 14.26 and 11.95 MPa. The trend was consistent in 28-day compressive strength results, whereas mortars containing 5% limestone and 10% clay-rich /iron-rich calcined laterite soils gave the maximum strength values of 36.23/27.45 MPa, respectively. Overall, mortar samples incorporating clay-rich laterite exhibited superior performance, primarily due to their higher metakaolin content. This facilitated the formation of calcium aluminosilicate hydrate (C-A-S-H) gels, which enhanced cohesion between the matrix components while reducing the presence of accessible voids and pores. However, the incorporation of clay-rich laterite soil beyond 20% resulted in an insufficient formation of portlandite (CH), which limited the extent of the pozzolanic reaction. As a consequence, the unreacted calcined clay-rich laterite remained in the mortar matrix, leading to an inhomogeneous microstructure and a subsequent reduction in mechanical performance. These findings demonstrated that thermally activated laterite can be considered as a promising material for the preparation of LC3, with

H. Beushausen et al. (Eds.): NTZR 2025, LNCE 873, pp. 25–35, 2026.
https://doi.org/10.1007/978-3-032-27698-8_3

properties that vary depending on the mineralogy of the material and replacement ratio.

Keywords: Laterite soil · Mineralogy · Thermal activation · Mechanical properties · Microstructure

1 Introduction

Portland cement is the most widely used binder the constructions worldwide. However, the production of Portland cement requires substantial energy input and the consumption of raw materials, particularly limestone, for the manufacture of clinker, which serves as its primary component. Furthermore, the production of this material accounts for approximately 5–8% of global CO_2 emissions, which significantly contributes to global warming. Additionally, it has other environmental consequences, including acid rain formation and ecological degradation [1, 2]. The International Energy Agency (IEA) estimates that by 2050, cement production will have increased by 12–23% from its current level of over 4000 million tonnes annually. As a result, immediate action is required to cut carbon emissions and slow down the ongoing climate change. To address this issue, researchers have begun developing alternative binders, known as low-carbon and sustainable cements, including limestone calcined clay cement (LC3).

LC3 is a recently developed ternary blended cement in which the alumina from calcined clay and the carbonate from limestone undergo mutual reactions, in addition to the conventional pozzolanic reaction of calcined clay and the filler effect of limestone, creating a synergistic interaction among the three principal components (clinker, calcined clay and limestone) [3]. Replacing ordinary Portland cement (OPC) with LC3 facilitates a greater clinker substitution and results in the formation of a more refined binder microstructure, leading to superior mechanical properties and durability relative to OPC [4]. Typically, LC3 is produced using a combination of limestone, marble dust, stone waste, fly ash, slag etc., and various types of clays in proportions of 1:1 or 1:2. These clays originate from diverse geological sources across different regions, including weathered rock formations and marine deposits [3, 5–8]. Since the aforementioned materials are not universally available, there is an urgent need to explore alternative resources, such as laterite soils commonly found in tropical and subtropical regions. Laterite soils primarily consist of SiO_2, Al_2O_3 and Fe_2O_3, which make them suitable ingredients for pozzolanic reaction [9].

In addition to enhancing technical performance, the incorporation of thermally activated laterite in LC3 systems presents notable environmental benefits. Partial replacement of Portland cement with calcined laterite and limestone can lead to significant reductions in CO_2 emissions associated with clinker production. Previous life cycle assessment (LCA) studies have demonstrated that LC3 formulations replacing 45–60% of clinker with calcined clay and limestone can reduce total CO_2 emissions by up to 38%, despite higher emissions from raw material transport. These systems also showed energy savings of up to 28% compared to conventional cements [10]. Such environmental advantages are particularly relevant in regions with abundant laterite resources and carbon-intensive clinker production, aligning well with the cement industry's transition toward net-zero emissions.

Several studies have investigated the use of raw and calcined laterite soils as cement replacement to produce compressed earth blocks. The characteristics of resulting hardened materials were depended on the composition of laterite used in the mixtures [11]. From the perspective of material availability, lateritic soil represents a viable clay resource for the production of LC3 in Cameroon. In the present study, the effect of chemical and mineralogical properties of different types of laterite soil on the properties of LC3 mortars have been investigated. Two types of lateritic soils from Cameroon, categorized as iron-rich and clay-rich laterites, were subjected to calcination at 600 °C and utilized in the production of LC3. The effects of mineralogy and chemical composition of calcined laterite on the compressive strength, water absorption, porosity, bulk density, and microstructure were also investigated.

2 Materials and Methods

2.1 Materials and Mortars Preparation

The laterite soils used in the study were collected from Yaoundé, Cameroon, and heated at 600 °C for 4 h with a constant heating rate of 5 °C/min. The employed calcination parameters were adopted from previous studies [12]. The X-ray fluorescence (XRF) analysis of raw clay-rich laterite revealed that the major oxides Fe_2O_3, SiO_2 and Al_2O_3 were present at 13.15, 43.20, 28.12 wt.%. The iron-rich laterite exhibited corresponding values of 49.40, 20.12, 17.40 wt.%. The CL3 binders were prepared through blending 70–100 wt.% Portland cement, with 5–15 wt.% limestone and 10–30 wt.% calcined laterite soil (clay-rich or iron-rich), whereas the proportion of calcined laterite to limestone was kept constant at 2:1 in all mixes. Table 1 presents the proportions of these materials used in the formulation of the blended LC3 binders.

Mortar mixtures were prepared through mixing LC3 cement with sand and water in a laboratory mortar mixer. In all mixtures the cement-to-sand ratio and water/cement (w/c) ratio were kept constant at 0.33 and 0.5, respectively. The LC3 fresh mortars were poured into 16 × 40 × 40 mm moulds, followed by 24 h of curing at room temperature. Subsequently, the samples were demolded and subjected to water curing until the mechanical testing was conducted at 7 and 28 days.

Table 1. Mix design and production of LC3 mortars using calcined iron-rich or clay-rich laterite, limestone, water and quartz sand.

Mixture ID	Cement (g) 42.5 R	Calcined clay-rich or iron-rich laterite (g)	Limestone (g)	Sand (g)	Water (g)
Control	1000	0	0	3000	500
85–10-5	850	100	50	3000	500
70–20-10	700	200	100	3000	500
62.5–25-12.5	625	250	125	3000	500

(continued)

Table 1. (*continued*)

Mixture ID	Cement (g) 42.5 R	Calcined clay-rich or iron-rich laterite (g)	Limestone (g)	Sand (g)	Water (g)
55–30-15	550	300	150	3000	500

2.2 Characterization Methods

In order to determine the compressive strength values, a set of five specimens of each formulation (Table 1) were subjected to compressive strength test, then the average of five readings was used as a single strength value. Also, the water absorption, porosity and bulk density measurements were done according to ASTMC1403-15 [13]. Paste mixtures were also prepared using the formulated LC3 cements and used for setting time measurements, using a Vicat needle, as described in BS standard method [14].

3 Results and Discussion

3.1 Setting Time

Table 2 gives the initial and final setting time values of the fresh LC3 pastes made using calcined clay-rich or iron-rich laterites. It can be seen that inclusion of 10% calcined iron-rich/clay-rich laterite along with 5% limestone resulted in decrease in initial setting time from 250 min to 230/210 min, respectively. Further increases in calcined iron-rich/clay-rich laterite to 30% reduced the corresponding setting times to 139 and 126 min, respectively. The observed decrease in setting time is attributed to the pozzolanic reaction between reactive phase contained in both calcined laterites with portlandite $Ca(OH)_2$. This allowed the formation of additional binder phase namely calcium aluminium silicate hydrate (C-A-S-H) gel as reported in previous studies [15, 16]. Furthermore, in the case of the mixtures containing calcined iron-rich laterite, the iron mineral present in the system might also have played a role as an accelerator in hydration process that might also have contributed to the observed setting time reduction. The flowability results of the fresh mortars indicate that the incorporation of calcined laterite and limestone enhanced the workability of the mixtures. The flowability values of mixtures increased from 161 to 189 mm or 200 mm, upon inclusion of 30% calcined iron-rich or calcined clay-rich laterite, respectively. It can be seen that the effect of calcined clay-rich laterite was more pronounced compared to that of iron-rich laterite material. The increased flow values can be attributed to the surface morphology and finer particle size of limestone compared to cement and calcined laterites. The finer limestone particles effectively fill the voids between cement grains, enhancing the packing density of the cementitious matrix and thereby improving the flowability of the resulting mortars [17]. The observed positive effect of both calcined laterite types on flowability of LC3 mortars is consistent with other studies conducted previously [16, 18].

Table 2. Initial setting time and flowability values of fresh LC3 pastes made using calcined clay/iron-clay laterite.

Mixture ID	Mixtures containing calcined iron-rich laterite		Mixtures containing calcined clay-rich laterite	
	Initial setting time (min.)	Flowability	Initial setting (min.)	Flowability
Control	250	161	250	161
85-10-5	230	169	210	172
70-20-10	183	174	154	179
62.5-25-12.5	165	185	128	189
55-30-15	139	189	126	200

3.2 Compressive Strength

Figure 1 shows the 7-day and 28-day compressive strength values of LC3 mortars made using calcined iron-rich and clay-rich laterites. Overall, LC3 mortars prepared using calcined iron-rich laterite exhibited relatively lower compressive strengths, compared to those made using calcined clay-rich laterites, regardless of the proportions of the laterite soil and limestone. It can be seen that incorporation of 10% calcined clay-rich laterite led to increase in 7-day and 28-day compressive strength, while inclusion of same proportion of calcined iron-rich laterite negatively affected the strength of mortars. The positive effect of calcined clay-rich laterite on strength development of the mortars can be attributed to a high fraction of kaolinite contained in clay-rich laterite, which was completely dehydroxylated into metakaolin upon calcination of the raw laterite at 600 °C. The presence of higher metakaolin contents in the mixtures made using calcined clay-rich laterite, enhanced the pozzolanic reaction in the system, resulting in better strength development which contributed to strength development of the mortars. This observation is consistent with other studies, who reported that the reactivity of the calcined clay was a key parameter influencing the strength development in LC3 mixtures [19]. The observed strength rise upon inclusion of 10 wt.% calcined clay-rich laterite confirms its ability to react with portlandite to form extra calcium ferri hydrate (C-F-H), calcium silicate hydrate (C-S-H), and calcium ferri silicate hydrate (C-F-S-H) phases in hardened binder system. The results show that compressive strength of the mortars decreased when proportion of calcined laterite and limestone in the mixtures increase beyond 10% and 5%, respectively. The observed negative effect on the strength development can be attributed to a dilution effect of excessive calcined clay and limestone, associated with reduction in Ca content in the mixtures that negatively affected the formation of hydrated product, which in turn resulted in strength loss.

In the case of the mortars that were made using calcined iron-rich laterite, compressive strength decreases with inclusion of the calcined clay and limestone, unlike those made using clay-rich material. The observed strength loss in the mortars upon use of iron-rich laterite can be attributed to the high content of iron mineral in the binder, which was responsible for decreasing the packing density by the loosening effect or wedging effect of iron, rather than contributing to fill the voids within the matrix [20].

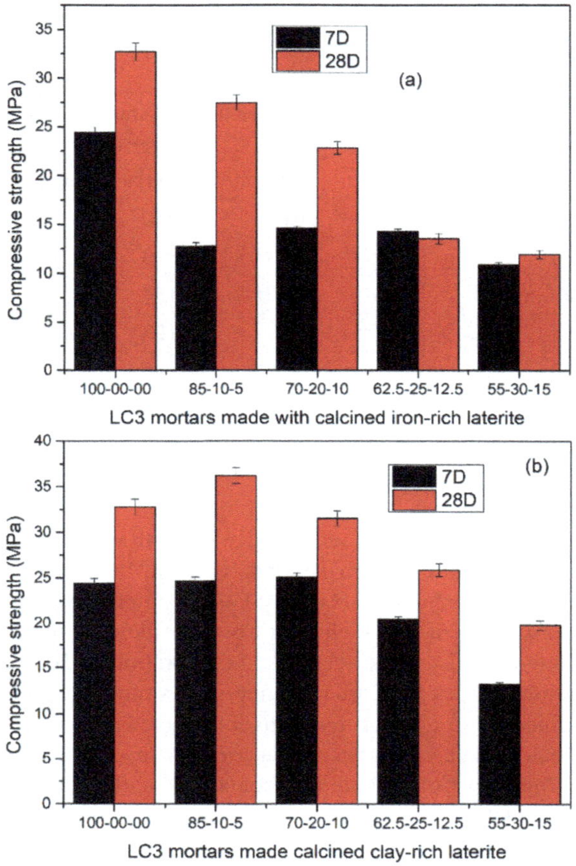

Fig. 1. Compressive strength of LC3 mortars made of calcined iron (**a**) and clay (**b**) laterites soils.

3.3 Water Absorption, Porosity and Bulk Density

Figure 2 presents the water absorption, porosity, and bulk density values recorded for samples cured for 28 days. As shown in Fig. 2a, the water absorption of the mortars increased with the rising proportion of calcined iron-rich laterite and limestone. In contrast, for mixtures incorporating calcined clay-rich laterite, the inclusion of 10% calcined clay led to a reduction in water absorption. However, further increasing the calcined laterite content from 10% to 30% adversely affected water absorption. These findings align with the compressive strength results (Fig. 1), demonstrating that the optimal incorporation of 10% calcined clay-rich laterite enhances the hydrated binder phase, leading to a denser microstructure with fewer open pores, thereby reducing water absorption.

The beneficial effect of incorporating 10% calcined clay-rich laterite in the LC3 binder is further supported by porosity (Fig. 2b) and bulk density results (Fig. 2c). Specifically, the inclusion of 10% calcined clay-rich laterite led to a reduction in porosity and an increase in bulk density, in contrast to mixtures containing 10% calcined iron-rich laterite. Regardless of the laterite soil type, increasing the proportion of calcined laterite and limestone further increased the porosity and resulted in a corresponding decrease in the bulk density of the mortars.

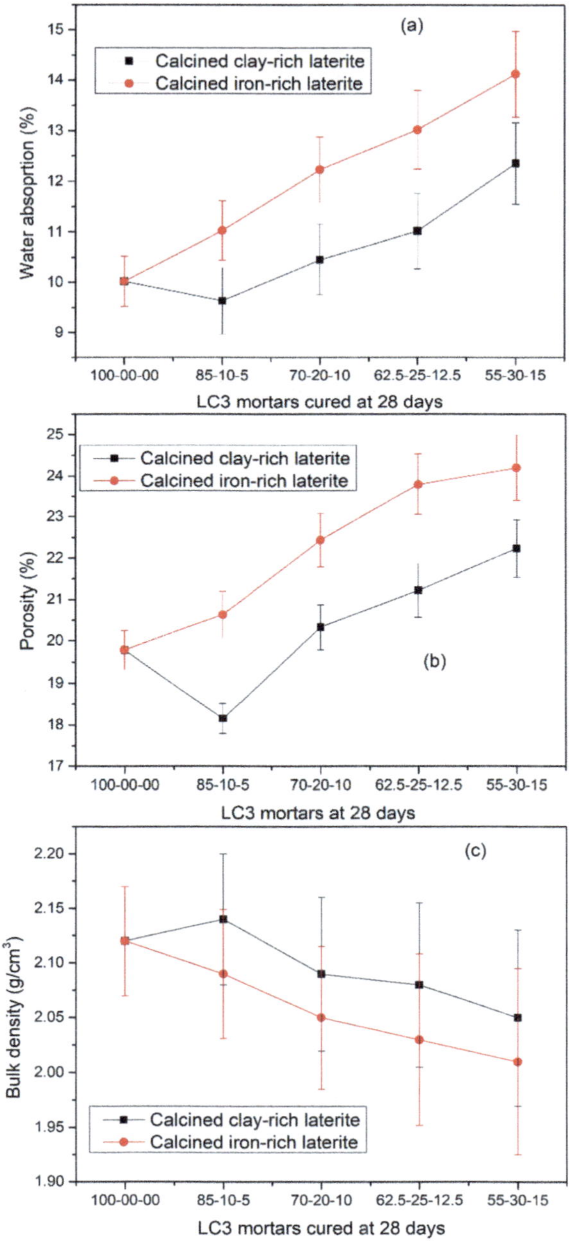

Fig. 2. Water absorption (**a**), porosity (**b**) and bulk density (**c**) of hardened LC3 mortars cured at 28 days.

3.4 Drying Shrinkage

Figure 3 shows the drying shrinkage of mortar mixtures made using calcined clay-rich or iron-rich laterite soil. It can be seen that generally LC3 exhibited relatively lower drying shrinkage values in comparison with those of the control samples made using 100% Portland cement. Indeed, LC3 mortar shrinkage decreased with increased content of both calcined iron-rich or clay-rich laterites and limestone. These observations are consistent with previous studies, who reported enhanced drying shrinkage behaviour of LC3 mortars upon increase in proportion of metakaolin content [16]. Overall, all the mixtures including control ones, complied with the maximum threshold of 0.10% established for ultimate drying shrinkage. In a comparison between two clay types, the positive effect of calcined clay-rich was more pronounced compared to that of the samples made using calcined iron-rich laterite. The high values of drying shrinkage could be attributed to filler effect and the increased water-to-cement ratio due to the replacement of cement with calcined laterite and limestone. Therefore, the combined effect of the faster hydration and finer pore structure led to higher drying shrinkage in LC3 than those observed from control. This is in line with previous studies [21–23].

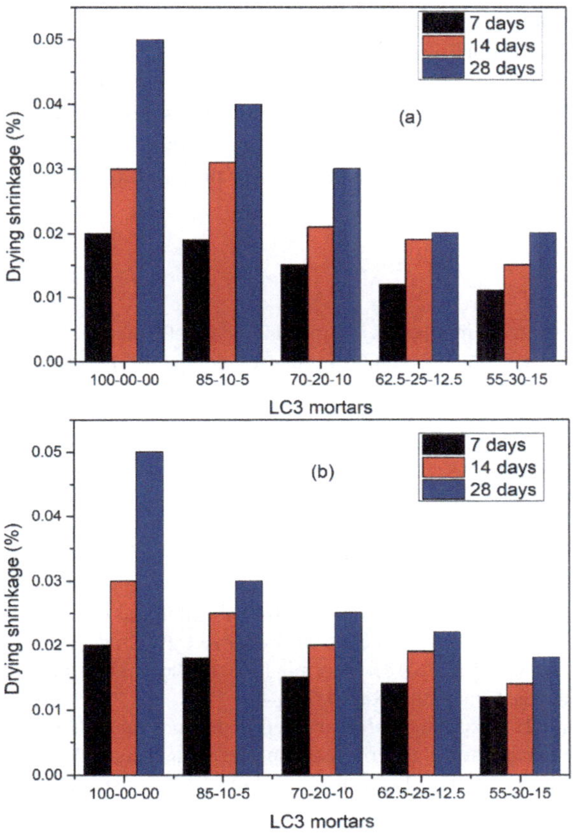

Fig. 3. Drying shrinkage LC3 mortars mixtures containing, **a** calcined iron-rich and **b** calcined clay-rich laterites.

4 Conclusions

An experimental study was conducted to evaluate the influence of the mineralogical and chemical properties of thermally activated laterites, calcined at 600 °C, as a potential calcined clay for LC3 binder mortars. Various compositions were investigated, incorporating 0–30% calcined laterite (either iron-rich or clay-rich) and 0–15% limestone into a Portland cement of CEM-I 42.5 N. The resulting blends were used to prepare mortar mixtures, and several fresh and hardened properties of the LC3 mortars were assessed. The incorporation of both calcined laterites up to 30 wt.% decreased the setting time and slightly increased the flowability.

- Mortar made with 10 wt.% of thermally activated clay rich laterite gave a higher strength value than the control samples made using 100% Portland cement, as the binder.

- Laterite is presently not being utilized as cement replacement across Africa, despite their wide availability on the continent. Suitable use of laterite soils in cement replacement, as determined in current work, would be a novel development on the continent and could contribute to industrial development while addressing socioeconomic challenges. Further research is required to thoroughly investigate laterite soils from various local sources in different countries for potential use in concretes.

References

1. Mehta KP (2001) Reducing the environmental impact of concrete. Concr Int 23(10):61–66
2. Barbhuiya S, Nepal J, Das BB (2023) Properties, compatibility, environmental benefits and future directions of limestone calcined clay cement (LC3) concrete: a review. J Build Eng 79:107794
3. Sharma M et al (2021) Limestone calcined clay cement and concrete: a state-of-the-art review. Cem Concr Res 149:106564
4. Antoni M et al (2012) Cement substitution by a combination of metakaolin and limestone. Cem Concr Res 42(12):1579–1589
5. Huang Z-Y et al (2020) Development of limestone calcined clay cement concrete in South China and its bond behavior with steel reinforcement. J Zhejiang Univ-Sci A 21(11):892–907
6. Akindahunsi AA, Avet F, Scrivener K (2020) The Influence of some calcined clays from Nigeria as clinker substitute in cementitious systems. Case Stud Constr Mater 13:e00443
7. Avet F, Li X, Scrivener K (2018) Determination of the amount of reacted metakaolin in calcined clay blends. Cem Concr Res 106:40–48
8. Lin R-S et al (2021) Experimental studies on hydration–strength–durability of limestone-cement-calcined Hwangtoh clay ternary composite. Constr Build Mater 269:121290
9. Cyriaque Kaze R et al (2022) Lateritic soils based geopolymer materials: a review. Constr Build Mater 344:128157
10. Junior ESA, de Sales Braga NT, Barata MS (2023) Life cycle assessment to produce LC3 cements with kaolinitic waste from the Amazon region, Brazil. Case Stud Constr Mater 18: e01729
11. Aziz A et al (2023) A state-of-the-art review of the structure and properties of laterite-based sustainable geopolymer cement. Environ Sci Pollut Res 30(19):54333–54350
12. Kaze CR et al (2021) Mechanical and physical properties of inorganic polymer cement made of iron-rich laterite and lateritic clay: a comparative study. Cem Concr Res 140:106320
13. ASTM, A., C1403-15 (2022) Standard test method for rate of water absorption of masonry mortars. American Society for Testing and Materials. West Conshohocken, Pennsylvania, United States
14. Institution BS (2016) Methods of testing cement: determination of setting times and soundness. BSI
15. Yu T et al (2023) Optimization of mechanical performance of limestone calcined clay cement: effects of calcination temperature of nanosized tubular halloysite, gypsum content, and water/binder ratio. Constr Build Mater 389:131709
16. Sinngu F et al (2023) Evaluation of metakaolin pozzolan for cement in South Africa. Dev Built Environ 14:100154
17. Wang D et al (2018) A review on effects of limestone powder on the properties of concrete. Constr Build Mater 192:153–166
18. Musbau KD et al (2021) Comparative performance of limestone calcined clay and limestone calcined laterite blended cement concrete. Clean Eng Technol 4:100264

19. Allalou S, Kheribet R, Benmounah A (2019) Effects of calcined halloysite nano-clay on the mechanical properties and microstructure of low-clinker cement mortar. Case Stud Constr Mater 10:e00213
20. Gu Li L et al (2023) Combined effects of micro and nano Fe_3O_4 on workability, strength, packing, microstructure and EM wave absorbing properties of mortar. Constr Build Mater 406:133407
21. Afroz S et al (2023) Shrinkage of blended cement concrete with fly ash or limestone calcined clay. Mater Struct 56(1):15
22. Dhandapani Y et al (2018) Mechanical properties and durability performance of concretes with Limestone Calcined Clay Cement (LC3). Cem Concr Res 107:136–151
23. Nguyen QD et al (2022) Autogenous and total shrinkage of limestone calcined clay cement (LC3) concretes. Constr Build Mater 314:125720

Steps Toward Sustainable Ultra-High Performance Concrete: Exploring Eco-friendly Mixture Alternatives

Saeed Bozorgmehr Nia and Behrouz Shafei(✉)

Iowa State University, Ames, USA
shafei@iastate.edu

Abstract. Ultra-high performance concrete (UHPC) has emerged as a transformative material in modern infrastructure, particularly for bridge decks, overlays, and structural repairs, due to its exceptional strength and durability. However, its widespread adoption is hindered by three critical challenges: (i) high cost driven by excessive use of ordinary portland cement (OPC), silica fume, and steel fibers; (ii) proprietary constraints, as most UHPC formulations are commercial prepackaged mixes requiring expert supervision; and (iii) significant environmental impact, with typical cement content exceeding 800 kg/m^3—contributing heavily to CO_2 emissions. Additionally, the rising cost and environmental concerns associated with silica fume, despite being a byproduct, further complicate UHPC production. This study explores the development of cost-effective, sustainable, and non-proprietary UHPC mixtures by incorporating locally available supplementary cementitious materials (SCMs) and recycled fiber alternatives. The proposed approach includes: (1) replacing OPC Type I/II with alternative cements; (2) substituting silica fume with industrial byproducts, such as fly ash or ash derived from incineration residues, agricultural combustion waste, slag, or waste glass powder; and (3) partially or fully replacing conventional steel fibers with recycled fibers sourced from polymers, end-of-life tires, or decommissioned wind turbine blades. Through a comprehensive review and performance assessment, this research identifies optimal material combinations for tailored UHPC designs, aiming to reduce the cost, enhance sustainability, and improve accessibility for global applications without reliance on proprietary systems.

Keywords: Sustainable ultra-high performance concrete · Eco-friendly mixture · Cementitious materials · Fiber alternatives

1 Introduction

The development of ultra-high performance concrete (UHPC) represents a major advancement in the evolution of cementitious composites, emerging in the 1990s to meet the growing demand for high-strength, durable materials in critical infrastructure applications. Designed to outperform conventional and high-performance concretes, UHPC is characterized by compressive strengths exceeding 120 MPa, strain-hardening tensile

H. Beushausen et al. (Eds.): NTZR 2025, LNCE 873, pp. 36–47, 2026.
https://doi.org/10.1007/978-3-032-27698-8_4

behavior, and exceptional durability against chloride ingress, freeze-thaw cycling, and abrasion. These enhanced properties have positioned UHPC as a transformative material for thin bridge overlays, long-span precast elements, blast-resistant structures, and other high-performance construction systems where structural efficiency and longevity are paramount. The initial development of UHPC was supported through intensive research and industry partnerships, resulting in several commercial proprietary products. While these products are highly optimized for performance, they rely on narrowly specified materials, strict processing protocols, and often proprietary pre-packaged formulations. This dependency on supplier-specific components and support services has limited the scalability and affordability of UHPC, particularly for public agencies. The proprietary nature of these systems also constrains innovation and adaptability, especially when exploring the use of alternative or regionally available materials.

To overcome the outlined limitations, an expanding body of research has focused on the development of non-proprietary UHPC (NP-UHPC), aiming to achieve equivalent mechanical and durability performance through open-source mix designs. These efforts emphasize cost reduction, local material utilization, and sustainability by modifying traditional UHPC formulations without sacrificing critical performance attributes. Research has demonstrated that optimizing binder composition, aggregate gradation, and fiber dosage using locally sourced or recycled constituents can result in UHPC systems that meet or exceed the required standards for structural use. Numerous studies have explored the partial or full replacement of high-energy clinker-based cements and silica fume with more sustainable binders such as portland limestone cement (PLC), calcined clays (e.g., LC^3), ground waste glass, or other industrial byproducts. These substitutions aim to reduce carbon footprint while maintaining workability and strength.

In parallel, the substitution of proprietary quartz sand with natural river sands, limestone screenings, or manufactured sands has been shown to be feasible when gradation is optimized to meet dense particle packing models. In the fiber domain, while steel fiber remains essential for enabling strain-hardening and crack-bridging behavior, recent investigations have evaluated the use of recycled steel fibers from waste tires or machining processes, synthetic fibers from plastic waste (e.g., PET and PP), and hybrid fiber systems that balance mechanical performance with environmental impact. Collectively, these studies underscore the technical feasibility of producing high-performance, non-proprietary UHPC mixtures with enhanced sustainability and regional adaptability. However, existing research often addresses one component substitution at a time, focusing either on binder, aggregate, or fiber alternatives in isolation, without holistically assessing the interaction of multiple substitutions within a unified mix design framework [1–5].

This study aims to bridge that gap by presenting a comprehensive evaluation of sustainable alternatives for all three key UHPC components: binder, aggregate, and fiber. The study synthesizes findings from recent literature to assess technical feasibility, material compatibility, and performance outcomes of integrated substitution strategies. By doing so, it seeks to provide a cohesive understanding of how eco-efficient, non-proprietary UHPC mixtures can be formulated to support broader sustainability goals in modern infrastructure development.

2 Cementitious Material Alternatives for Non-Proprietary UHPC

The binder system in conventional UHPC typically comprises a high percentage of ordinary portland cement (OPC), often exceeding 800 kg/m^3, combined with silica fume to improve packing density, reduce porosity, and promote a refined microstructure. However, this formulation contributes significantly to the embodied carbon, heat of hydration, and material costs of UHPC. As a result, the development of sustainable and non-proprietary alternatives has focused on replacing or partially substituting both OPC and silica fume with low-carbon and readily available materials.

2.1 Limestone Calcined Clay Cement (LC^3)

LC^3 is an innovative ternary blended cement composed of approximately 50% clinker, 30% calcined clay, 15% limestone, and 5% gypsum. It represents a sustainable alternative to conventional portland limestone cement (PLC), significantly reducing CO_2 emissions while maintaining comparable performance. The fineness of LC^3 typically ranges from 4500 to 6000 cm^2/g Blaine, which is slightly higher than traditional PLC, promoting enhanced early-age reactivity due to increased surface area. Its specific gravity ranges between 2.70 and 2.85 g/cm^3, slightly lower than that of pure clinker-based cement due to the incorporation of lighter aluminosilicate phases from calcined clay and limestone. Compared to PLC, LC^3 exhibits similar or slightly extended setting times depending on the clay's reactivity and kaolinite content. The presence of limestone accelerates the formation of carboaluminate phases, particularly monocarboaluminate ($C_4A\hat{S}H_{11}$), enhancing early-age strength. The chemical composition of LC^3 is tailored to leverage the synergistic reaction between metakaolin (from calcined clay) and limestone, where aluminum phases from the clay react with calcium carbonate to form stable hydration products that improve microstructural densification. Mechanically, LC^3 delivers compressive strengths equivalent to or exceeding those of PLC at 28 and 90 days, particularly when optimized for clinker substitution. The pozzolanic activity of calcined clay not only contributes to continued hydration beyond early stages but also enhances durability by refining pore structure, reducing permeability, and increasing resistance to sulfate and chloride ingress. Furthermore, the enhanced alumina content enables the stabilization of ettringite, contributing to dimensional stability and long-term performance. These properties position LC^3 as a promising low-carbon cementitious binder for structural and infrastructure applications [8–10].

2.2 Waste Glass Powder

Waste glass powder is a finely ground material produced from post-consumer or industrial soda–lime glass—primarily from bottles, windows, and screens—via crushing and milling. Its typical chemical composition consists of ~ 70–80% amorphous SiO_2, 12% Na_2O, and 8–10% CaO, enabling latent pozzolanic reactivity. The material is angular and irregular in shape, with particle sizes commonly $< 75\ \mu m$ to induce pozzolanic behavior, and a specific gravity between 2.4–2.6. Finely ground WGP ($D_{50} \leq 20\ \mu m$) can partially replace Portland cement or silica fume in UHPC at replacement levels of **10–30% by mass**, often providing optimal strength, workability, and durability. At these dosages, WGP contributes to micro-filling, densification of the matrix, and formation of secondary CSH, enhancing compressive strength and reducing permeability. Pozzolanic activity is directly related to fineness—particles < 40 μm significantly improve mechanical performance. ASR mitigation is a critical consideration for WGP. While coarse glass is reactive and deleterious, fine WGP can actually bind alkalis and reduce ASR expansion, especially in mixes with reactive aggregates or supplementary SCMs like fly ash. Rheologically, WGP pastes are more fluid than those with silica fume but slightly more viscous than fly ash systems due to crushed-angular particles. Proper admixture dosage (e.g., HRWR) ensures stable workability. Durability performance improves through pore refinement, reduced calcium hydroxide content, and improved resistance to chloride ingress and sulfate attack. Slightly extended setting times at high dosages can be advantageous in hot climates concrete casting compared to silica fume. Overall, finely ground WGP serves as an effective and sustainable supplementary binder in UHPC formulations, provided that its particle size and dosage are carefully optimized to ensure pozzolanic activity and maintain workability [6–8].

2.3 Combustion Ashes

Fly ash is a fine byproduct of coal combustion in thermal power plants, collected from flue gases. It includes two ASTM C618 classes: Class F (low calcium, < 10% CaO) and Class C (high calcium, 10–30% CaO). Class F fly ash typically contains ~ 50–70% SiO_2 and Al_2O_3 and is primarily pozzolanic, while Class C includes additional self-cementing phases due to free lime and sulfates. Fly ash particles are spherical with a particle size distribution ranging from 1–100 μm and specific gravity of 2.1–2.6. Its morphology reduces water demand and improves workability and pumpability of UHPC. Used in 10–30% cement replacement in UHPC, fly ash lowers heat of hydration, reduces autogenous shrinkage, and contributes to long-term strength, especially under heat curing. The pozzolanic reaction binds $Ca(OH)_2$, forming secondary C–S–H that densifies the matrix [4, 11–13].

Rice husk ash (RHA) is derived from controlled combustion (typically 500–800 °C) of rice husks, rich in 85–95% amorphous silica. High-quality RHA has high BET surface area (~20,000–70,000 m^2/kg) and median particle sizes of 5–25 μm, with specific gravity around 2.2–2.4. The ash is highly pozzolanic when properly processed. Mechanistically, RHA consumes $Ca(OH)_2$ and forms dense, low Ca/Si ratio C–S–H, improving strength and durability. It also acts as an internal curing agent due to its porous structure, helping reduce autogenous shrinkage. In UHPC, RHA is used at 5–15% replacement levels to enhance strength, microstructure, and resistance to carbonation and chloride penetration [4, 12–14].

Agricultural biomass ashes include sugarcane bagasse ash (SCBA), palm oil fuel ash (POFA), corn cob ash, and wood ash, primarily sourced from agro-industrial residues. SCBA typically contains 60–70% SiO_2 and 10–20% CaO/Al_2O_3, POFA has ~ 50% SiO_2 with moderate CaO and K_2O, and other ashes vary by origin. Fineness < 45 μm is essential for pozzolanic activity. Specific gravity ranges between 2.0–2.5. When used at 10–20% replacement in UHPC, these ashes can enhance late-age strength, sulfate resistance, and reduce permeability. High LOI or unburnt carbon can affect admixture compatibility and workability [4, 12–14].

Municipal solid waste incineration (MSWI) ash comprises two main types: incinerator fly ash, which requires treatment prior to use, and incinerator bottom ash (IBA), which may be rendered suitable for incorporation in concrete after appropriate processing steps such as metal extraction, washing, and aging or natural weathering in open air. Ground IBA (< 75 μm) exhibits better pozzolanic behavior and latent hydraulic activity compared to bigger particles. Used at 10–30% cement replacement, IBA improves sustainability but may reduce early strength. It has specific gravity ~ 2.4–2.6 and contains aluminosilicates and Ca-compounds. Environmental treatment should ensure heavy metal regulatory compliance. Although not widely used in UHPC yet, IBA is being researched for ultra-green concrete with reduced clinker content and embodied CO_2 [4, 12–14]. Various sustainable binder alternatives for non-proprietary UHPC formulations are illustrated in Fig. 1.

(a) (b)

(a) (d)

Fig. 1. Selected sustainable cementitious binder alternatives used in non-proprietary UHPC: (a) LC^3 cement, (b) biochar, (c) waste glass powder, and (d) incinerator bottom ash.

3 Sustainable Fine Aggregate Alternatives

Fine aggregates are critical to UHPC performance, influencing rheology, particle packing, and mechanical behavior. Conventional proprietary UHPC systems utilize high-purity quartz sand (600 μm–1 mm) to maximize packing density and matrix homogeneity. However, quartz sand is often costly, imported, and logistically constrained, limiting broader adoption. Sustainable and regionally available alternatives are increasingly explored for non-proprietary UHPC formulations. Regardless of the source, the success of these alternatives hinges on achieving optimal packing density using gradation models—such as the Modified Andreasen and Andersen (A&A) model with D_{max} less than 1.18 mm and $q = 0.22$–0.26 to minimize porosity, maximize solid volume fraction, and reduce paste demand. When properly proportioned and supported by admixture strategies, these alternatives can maintain the strength, durability, and workability required for UHPC [4–8].

3.1 Natural Fine Aggregates

Natural sands from riverbeds, dunes, or crushed rocks can replace quartz sand when they meet strict quality standards (e.g., ASTM C33), including minimal fines and organic content, particle sizes between ~ 300 μm and 1 mm, and lean washing and consistent gradation. Properly processed natural sands, optimized with the A&A model, have demonstrated compressive strengths exceeding 120 MPa and favorable rheological properties, validating their use in sustainable UHPC [4–8].

3.2 Manufactured Sands

Manufactured sands, derived from crushed granite, basalt, or limestone, offer (i) uniform gradation and angularity that enhance interparticle interlock, and (ii) fine particles that contribute to micro-filler effects and reduce paste demand. Despite potential drawbacks such as reduced flowability, blending with natural sands and adjusting HRWR dosage enables effective use in UHPC with comparable mechanical performance and improved freeze-thaw resistance [4–8].

3.3 Hybrid Aggregate Systems

Blending natural, manufactured, and masonry sand enhances particle packing and workability. Common ratios like 70:30 or 40:40:20 (manufactured: natural: masonry) improve flowability, interlock, and homogeneity. These systems reduce HRWR requirements and cost while sustaining mechanical strength, offering scalable solutions for UHPC [4–8].

3.4 Recycled Fine Aggregates

Recycled fine aggregates from construction waste align with circular economy goals but present challenges including high water absorption, variable composition, and potential contaminants. Suitable for partial replacements ($< 20\%$), particularly in hybrid blends, recycled fine aggregates contribute to sustainability without significantly compromising performance when properly treated and selected [4]. Various fine aggregate alternatives used in non-proprietary UHPC—such as silica sand, quartz sand, mason sand, and recycled concrete sand—are illustrated in Fig. 2.

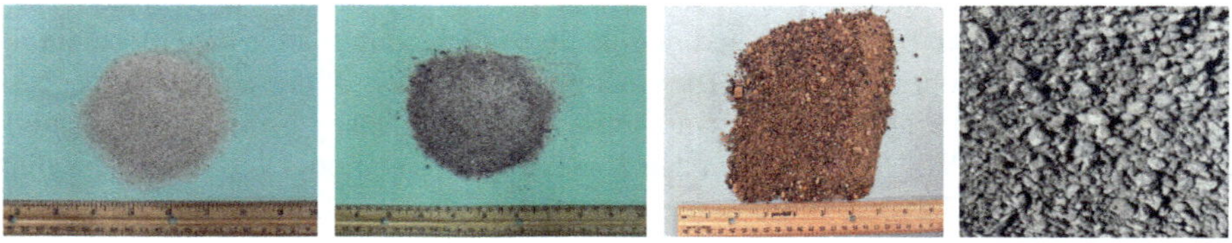

(a) (b) (c) (d)

Fig. 2. Sand alternatives for non-proprietary UHPC: (a) silica sand, (b) quartz sand, (c) mason sand, (d) recycled concrete sand.

4 Sustainable Fiber Reinforcement Alternatives

Fiber reinforcement is a fundamental component in ultra-high-performance concrete (UHPC), setting it apart from conventional high-performance concrete by enabling strain hardening, multi-crack behavior, and enhanced post-cracking tensile capacity. Traditional UHPC formulations predominantly utilize short steel fibers, typically 13 mm in length and 0.2 mm in diameter, at volumetric dosages ranging from 1.5% to 3%. These fibers contribute significantly to increasing flexural toughness, tensile strength and impact resistance. However, their use is associated with drawbacks, including high cost, energy-intensive production, and limited recyclability, which pose challenges to the sustainability of UHPC systems. To address these concerns, recent research has focused on identifying and evaluating sustainable and non-proprietary fiber alternatives that can maintain or improve mechanical performance while lowering the environmental and economic footprint of UHPC.

4.1 Recycled Steel Fibers

Recycled steel fibers, obtained from end-of-life tires or industrial metallic waste streams—such as lathe shavings, machining chips, and steel shingles—have emerged as promising alternatives to conventional virgin steel fibers in UHPC. Although these fibers typically exhibit variability in geometric and mechanical properties—such as length, aspect ratio, and tensile strength—appropriate processing techniques, including sieving, trimming, and classification, can yield fiber batches suitable for structural-grade applications. Experimental studies have demonstrated that recycled steel fibers, when used at slightly elevated volume fractions (e.g., up to 2.5%), can achieve comparable post-cracking tensile strength and toughness indices to commercial steel fibers. Moreover, their integration supports waste valorization and significantly reduces the embodied energy and carbon emissions associated with fiber production [4, 16–18].

4.2 Synthetic Fibers from Recycled Plastics

Synthetic fibers, such as polyethylene terephthalate (PET), polypropylene (PP), and polyethylene (PE), particularly those derived from recycled plastic waste, have gained attention as sustainable alternatives in fiber-reinforced UHPC. These fibers offer several environmental and functional advantages: they are corrosion-resistant and chemically inert, improving long-term durability in aggressive environments; their low density contributes to reduced unit weight of the composite; and they are sourced from recyclable feedstocks and exhibit significantly lower embodied carbon compared to steel.

Despite these advantages, synthetic fibers generally exhibit lower elastic modulus and tensile strength compared to steel fibers, limiting their effectiveness in structural load-bearing applications when used as the sole reinforcement. Furthermore, several studies have identified challenges related to their crack-bridging performance. Specifically, fibers such as polypropylene (PP) or polyester have shown a tendency to lose structural integrity over time, particularly under prolonged exposure to the highly alkaline environment of cementitious matrices. This exposure can lead to surface degradation of fibers, embrittlement, or loss of bond with the surrounding matrix, ultimately diminishing their ability to bridge and arrest microcracks. As such, the use of synthetic fibers in UHPC should be carefully evaluated based on fibers chemical stability and compatibility with the alkaline pore solution. In many cases, they are best employed as part of a hybrid fiber system, supplementing steel or more robust fibers, rather than serving as the sole reinforcing element in structural UHPC applications [4, 17–20].

4.3 Hybrid Fiber Systems (Steel + Synthetic)

To achieve an optimal balance between mechanical performance, ductility, and sustainability, hybrid fiber reinforcement systems—combining steel and synthetic fibers—have garnered significant research interest in UHPC development. These systems synergize the high tensile strength, crack-bridging efficiency, and post-cracking load resistance provided by steel fibers with the distributional uniformity, shrinkage mitigation, and cost-effectiveness of plastic-based fibers such as polypropylene (PP) or polyethylene terephthalate (PET).

Typical hybrid configurations include (i) steel fibers at 1.0–1.5% by volume, combined with 0.3–0.5% by volume of PET or PP fibers, and (ii) short synthetic fibers targeting early-age microcrack control, paired with longer steel fibers designed to resist ultimate tensile and flexural loads. Experimental investigations have demonstrated that such hybrid systems not only enhance the toughness index and energy absorption capacity of UHPC but also improve fiber dispersion and reduce the risk of fiber balling—a common issue in high-volume steel fiber mixes. The inclusion of synthetic fibers has been shown to improve workability and homogeneity of the fresh mix, thereby facilitating better orientation and distribution of steel fibers within the matrix. This improved rheology supports the development of a crack-resistant UHPC composite, especially in complex casting geometries or large-scale applications [4, 18–21].

4.4 Recycled Glass Fibers

Alkali-resistant (AR) glass fibers and basalt fibers have been investigated as partial substitutes for steel fibers in UHPC, driven by their corrosion resistance, chemical inertness, and lower density. These inorganic fibers offer improved long-term durability in aggressive environments, particularly against moisture, chlorides, and freeze-thaw damage. However, their application in structural-grade UHPC remains limited due to key performance drawbacks. Primarily, the brittle fracture behavior of these fibers results in reduced energy absorption and post-cracking toughness compared to ductile steel or certain polymeric fibers. Their relatively weak bond with the cementitious matrix, especially under high-alkalinity conditions typical of UHPC, can further compromise performance. Long-term exposure may degrade the fibers, especially non-alkali-resistant glass. To counteract these issues, strategies such as incorporating silica fume or high volumes of pozzolanic SCMs have been adopted to improve fiber-matrix adhesion and reduce alkali attack, albeit at the cost of added complexity in mix design and processing.

Recent efforts have focused on sourcing AR glass fibers from recycled materials, including post-consumer waste glass and decommissioned wind turbine blades. The latter, composed of glass fiber-reinforced polymers (GFRPs), offer an underutilized waste stream. Through mechanical shredding and thermal or chemical treatment, usable fibers can be recovered. When cleaned of residual resins, these fibers present a low-cost alternative for reinforcing UHPC in non-primary structural applications. While recycled GFRP fibers exhibit lower tensile strength and bonding efficiency compared to steel, they outperform some polymeric fibers (e.g., PET, PP) in stiffness and crack-bridging performance. However, their efficacy depends strongly on the recovery process and extent of fiber degradation. To enhance the performance of both virgin and recycled AR glass fibers in UHPC, several techniques have been explored. These include in situ chemical coating to improve alkali resistance, surface treatments to enhance moisture resistance and interfacial bonding, and autoclave curing to promote chemical adhesion via paste deposition. Additionally, active mineral admixtures such as silica fume and metakaolin mitigate alkali-induced degradation by promoting CSH formation and reducing pore solution alkalinity. These combined measures address critical degradation mechanisms—such as Si-O-Si network breakdown and interfacial weakening—while improving fiber dispersion and matrix densification. Consequently, recycled AR glass fibers remain a promising reinforcement option for sustainable, non-proprietary UHPC systems, pending further optimization. Figure 3 presents the average flexural strength of UHPC mixtures incorporating different sustainable fiber alternatives, highlighting their enhancement relative to the control mixture without fiber reinforcement [19–21].

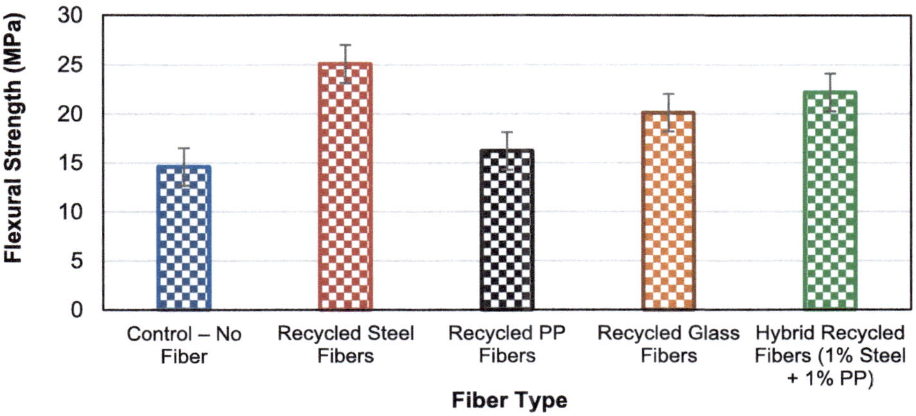

Fig. 3. Average flexural strength of UHPC mixtures incorporating different sustainable fiber alternatives.

5 Conclusions

To translate the findings of this study into practical guidance, a series of recommended ranges for each material category is outlined to support the design and implementation of sustainable non-proprietary UHPC. For cementitious materials, Ordinary Portland Cement (OPC) can be fully replaced with Portland Limestone Cement (PLC) to achieve similar performance with up to 12% reduction in CO_2 emissions. Alternatively, Limestone Calcined Clay Cement (LC^3) can be used in a ternary system, typically comprising ~ 50% OPC, 30% calcined clay, and 15% limestone, offering up to 40% emissions reduction. Supplementary cementitious materials (SCMs) such as Class F fly ash (15–30%), GGBFS (15–40%), and waste glass powder (10–25%) can be integrated to replace portions of cement or silica fume, improving long-term strength, packing density, and sustainability.

In terms of fine aggregates, natural river sand or manufactured sand can be used as 100% replacements for proprietary quartz sand, provided they are well-graded and sieved to particle sizes below 600 μm. Optimal results are often achieved using hybrid blends, typically in a 50:50 to 70:30 ratio of natural to manufactured sands, enhancing particle packing and fresh workability. Additionally, recycled fine aggregates or finely ground crushed glass sand may be used in limited amounts (5–20%) to promote circular economy goals, although their use must be carefully managed to prevent alkali-silica reactivity and performance inconsistency. Regarding fiber reinforcement, steel fibers remain essential for tensile strength and strain-hardening behavior in UHPC, with typical dosages ranging from 1.5% to 2.5% by volume. Where cost or environmental impact is a concern, recycled steel fibers (e.g., from tire processing) may serve as a viable substitute at similar or slightly higher volumes. To further enhance sustainability and reduce steel dependency, synthetic fibers derived from recycled PET or polypropylene can be incorporated at 0.3–1.0%, typically as part of hybrid systems combining 1.0–1.5% steel with 0.3–0.5% synthetic fibers, which have been shown to improve fiber dispersion, control shrinkage, and reduce mate-rial cost without compromising toughness.

The presented material substitutions and their optimized dosage ranges form the foundation for producing customizable, high-performance, and sustainable non-proprietary UHPC. Moving forward, field trials, long-term durability assessments, and full-scale structural evaluations will be critical to further test these alternatives under real-world conditions and facilitate broader adoption in public infrastructure.

References

1. Kazemian M, Shafei B (2024) Multi-scale study of the effect of superabsorbent polymers on microporosity, dimensional stability, and durability of cementitious materials. J Constr Build Mater 442:137565
2. Karim R, Shafei B (2024) Ultra-high performance concrete under direct tension: investigation of a hybrid of steel and synthetic fibers. J Struct Concr 25(1):423–439
3. Nia SB, Shafei B (2025) Carbon footprint reduction through repurposing solid wastes into sustainable construction materials: a state-of-the-art review. Clean Respons Consump 100310
4. Nia SB, Shafei B (2024) Synergistic effects of nano and micro silica on fresh and hardened properties of self-consolidating concrete. Case Stud Constr Mater 21:E03443
5. Nia SB, Nyland A, Wivast J, Kioumarsi M, Shafei B (2025) Investigations of portland limestone cement and waste glass powder for sustainable ultra-high performance concrete. Case Stud Constr Mater 22:E04425
6. Farahzadi L, Nia SB, Shafei B, Kioumarsi M (2025) Sustainability assessment of ultra-high performance concrete made with various supplementary cementitious materials. Clean Mater 15:100301
7. Wivast J, Nyland A, Nia SB, Kioumarsi M, Shafei B (2024, June) Developing eco-friendly ultra-high performance concrete by utilizing recycled alternatives. In: The international conference on net-zero built environment: innovations in materials, structures, and management practices. Springer Nature, pp 49–57
8. Kazemian M, Shafei B (2023) Effects of supplementary cementitious materials on the hydration of ultra-high-performance concrete. J Mater Civ Eng 35(11):04023388
9. Guo D, Guo M, Zhou Y, Zhu Z (2024) Use of nano-silica to improve the performance of LC3-UHPC: mechanical behavior and microstructural characteristics. Constr Build Mater 411:134280
10. ASTM International (2023) ASTM C618-23: standard specification for coal fly ash and raw or calcined natural pozzolan for use in concrete. ASTM International
11. Kang SH, Hong SG, Moon J (2019) The use of rice husk ash as reactive filler in ultra-high performance concrete. Cem Concr Res 115:389–400
12. Wu P, Dai K, Shi Y, Rong H, Zhang X (2025) Effect of biomass ash on hydration, properties and microstructure of ultra-high performance concrete. J Build Eng 104:112276
13. Shen P, Zheng H, Xuan D, Lu JX, Poon CS (2020) Feasible use of municipal solid waste incineration bottom ash in ultra-high performance concrete. Cement Concr Compos 114:103814
14. Wang X et al (2019) Optimized treatment of recycled construction and demolition waste in developing sustainable ultra-high performance concrete. J Clean Prod 221:805–816
15. Kazemian M, Shafei B (2023) Mechanical properties of hybrid fiber-reinforced concretes made with low dosages of synthetic fibers. J Struct Concr 24(1):1226–1243
16. Karim R, Shafei B (2022) Investigation of five synthetic fibers as potential replacements of steel fibers in ultra-high performance concrete. J Mater Civ Eng 34(7):04022126
17. Shi W, Shafei B (2021) Bond characteristics between conventional concrete and six high-performance patching materials. J Constr Build Mater 308:124898

18. Karim R, Shafei B (2021) Flexural response characteristics of ultra-high performance concrete made with steel microfibers and macrofibers. J Struct Concr 22(6):3476–3490
19. Karim R, Shafei B (2021) Performance of fiber-reinforced concrete link slabs with embedded steel and GFRP rebars. J Eng Struct 229:111590
20. Farahzadi L, Bozorgmehr Nia S, Shafei B, Kioumarsi M (2024, June) Preliminary environmental assessment of ultra-high-performance concrete mixtures. In: The international conference on net-zero civil infrastructures: innovations in materials, structures, and management practices (NTZR). Springer Nature Switzerland, Cham, pp 589–598
21. Zhang F, Lu Z, Wang D (2024) Working and mechanical properties of waste glass fiber reinforced self-compacting recycled concrete. Constr Build Mater 439:137172

High-Performance, Low-Carbon Concrete from Nordic Binders: A 365-Day Comparative Study of Six Supplementary Cementitious Materials

Suriyaprakash Saravanan[1](✉), Anders Selander[2], and Annika Gram[1]

[1] Division of Concrete Structures, KTH Royal Institute of Technology, Stockholm, Sweden
sursar@kth.se

[2] Heidelberg Materials Cement Sverige AB, Stockholm, Sweden

Abstract. As the construction industry seeks to reduce its carbon footprint, supplementary cementitious materials (SCMs) play a vital role in reducing the clinker content of concrete. This paper synthesizes and comparatively analyzes the 365-day compressive strength development of six locally sourced SCMs to establish their long-term performance for use in the Nordic region. The locally sourced SCMs are calcined clay (CC), fly ash (FA), natural pozzolan (NP), electric arc furnace slag (EAFS), oil shale ash (OSA), and ground granulated blast furnace slag (GGBS). A total of 31 concrete mixes were evaluated, with SCMs replacing cement at levels from 15% to 60%. While all SCM mixes exhibited lower early-age strength than the Portland cement reference, long-term strength was significantly enhanced. The most promising results include 15% FA (102 MPa), 35% NP (99.5 MPa), and 45% GGBS (96.5 MPa) at 365 days. High-volume replacement was also proven viable, with 35% oil shale ash matching the reference strength. The findings demonstrate that these regionally available SCMs offer a viable pathway towards producing high-performance concrete with significantly reduced clinker content, suitable for a wide range of structural applications.

Keywords: Supplementary Cementitious Materials (SCMs) · Long-term strength · Nordic SCMs

1 Introduction

Concrete is the most widely used construction material in the world, forming the foundation of modern infrastructure. However, the production of its primary binder, Portland cement, accounts for a significant share of anthropogenic greenhouse gas emissions, contributing approximately 5–8% of global CO_2 emissions [1]. This environmental burden has placed the concrete industry under increasing scrutiny, particularly as countries align with global climate initiatives such as the UN's 2030 Agenda and the EU's target of net-zero emissions by 2050. In this context, there is a growing demand for innovative concrete technologies that minimize environmental impact without compromising mechanical performance or durability.

H. Beushausen et al. (Eds.): NTZR 2025, LNCE 873, pp. 48–56, 2026.
https://doi.org/10.1007/978-3-032-27698-8_5

A widely adopted strategy for reducing the carbon footprint of concrete involves partially replacing cement clinker with supplementary cementitious materials (SCMs). Pozzolanic SCMs contribute to strength by reacting with the calcium hydroxide from cement hydration to form additional calcium-silicate-hydrate (C-S-H), which densifies the microstructure [2]. This process not only consumes a less desirable byproduct but also refines the pore structure, leading to enhanced long-term strength. Furthermore, this densified microstructure is critical for improving durability, as it reduces the concrete's permeability, thereby lowering chloride ingress and mitigating the risk of alkali-silica reaction (ASR). While common SCMs such as fly ash (FA) and ground granulated blast furnace slag (GGBS) have proven effective, their future availability is expected to be significantly reduced due to the global transition toward green energy and steel production [3]. As a result, attention has shifted toward alternative SCMs that are more abundant, locally available, and suitable for diverse climatic conditions, including calcined clay (CC), natural pozzolan (NP), electric arc furnace slag (EAFS), and oil shale ash (OSA).

Despite growing interest in SCMs, a significant knowledge gap remains regarding the long-term performance of SCM-rich concrete mixes. Most existing studies focus on compressive strength at 28 days, a standard benchmark that does not adequately capture the latent pozzolanic activity of many SCMs. This delayed reactivity can contribute substantially to strength development beyond 28 days, particularly at 56, 112, or even 365 days [2, 4]. Understanding long-term strength behavior is particularly relevant for infrastructure designed for service lives of 50 to 100 years, especially in cold-climate regions such as the Nordic countries, where freeze–thaw exposure, delayed hydration, and harsh environmental conditions present unique durability challenges.

This paper builds upon a comprehensive experimental program that was established in three foundational bachelor theses [5–7]. These initial studies methodically characterized the early-age and medium-term strength development up to 56 days for a suite of locally sourced SCMs. Subsequent work extended this analysis to 112 days [8]. However, the ultimate, year-long performance, a critical metric for predicting the service life and durability of modern infrastructure remained uncharacterized. The present study fills this definitive gap. By introducing and analyzing the newly obtained 365-day compressive strength data, this paper completes the long-term performance picture for the full suite of six SCMs: CC, FA, NP, EAFS, OSA, and GGBS, at replacement levels from 15% to 60%. With this complete, year-long dataset, the goal is to establish a robust framework for comparing these materials, identifying optimal replacement levels, and understanding how their chemical compositions influence ultimate strength. This work provides a conclusive basis for specifying high-performance, low-carbon concrete mixes with proven long-term strength for a wide range of structural applications in the Nordic region.

2 Materials and Methods

This study evaluates the strength development of concrete mixes where ordinary Portland cement is partially replaced by various SCMs. The methodology involved creating different concrete recipes, casting concrete cubes, and performing compressive strength tests at multiple curing ages.

2.1 Materials

A rapid hardening ordinary Portland cement (CEM I 52.5 R) was used as the reference binder in all studies. Six different SCMs, sourced from or near the Nordic region, were investigated, NP: A volcanic pozzolan sourced from open-cast mines in Iceland, CC: An illite clay from a quarry in Kunda, Estonia, which was made reactive through calcination, FA: A silica-rich, Class F fly ash compliant with SS-EN 450-1, GGBS: A by-product from the iron and steel industry, EAFS: A fine-grained Si-Mg slag, also known as silica greenstone, which is a by-product from the production of silicomanganese at Eramet in Norway. OSA: A residual material from oil shale combustion. The chemical compositions of the cement and the six SCMs, characterized by their main oxides (CaO, SiO_2, and Al_2O_3), are shown in the ternary phase diagram in Fig. 1. The aggregates used in all concrete mixes consisted of two grain sizes: 0/8 mm and 8/16 mm. A polycarboxylate-based superplasticizer was used as a flow-enhancing admixture to achieve the desired consistency for all mixes.

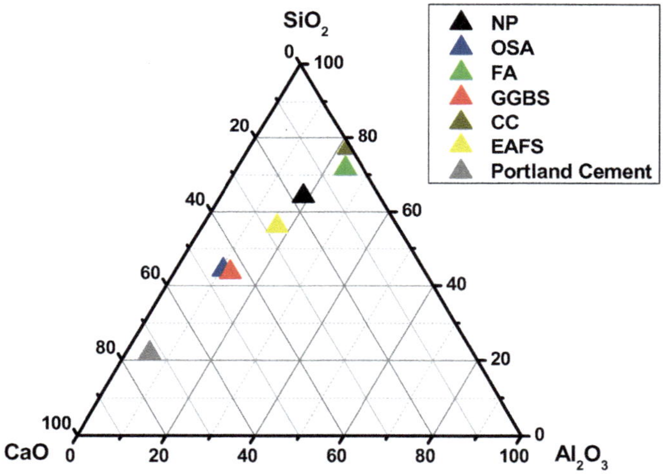

Fig. 1. Ternary Phase diagram containing the main oxides composition such as CaO, Al_2O_3 and SiO_2 for SCMs and the reference cement.

2.2 Methods

A total of 31 concrete mixes were prepared, including a 100% Portland cement reference mix and 30 mixes where cement was partially replaced by SCMs. The replacement levels investigated were 15%, 25%, 35%, 45%, and 60% by mass of the total binder. A constant water-to-binder ratio of 0.40 was maintained for all mixes, with the total water content adjusted daily to account for the moisture content of the aggregates. For each mix, standard 100 × 100 mm concrete cubes were cast for compressive strength testing. Following casting, the specimens were covered with a plastic sheet to prevent moisture loss and were kept at room temperature for 24 h. The cubes were then demolded and placed in a water bath for continuous curing until the designated testing age. The compressive strength of the concrete cubes was determined at ages of 1, 7, 28, 56, 112, and 365 days. Testing was conducted in accordance with the procedures outlined in SS-EN 12390–3. The average compressive strength value from multiple samples was recorded for each test age.

3 Results and Discussion

This section presents and analyzes the compressive strength development of concrete mixes incorporating six different SCMs over 365 days. The performance of each SCM is evaluated against a reference mix (REF) containing 100% Portland cement.

Figure 2 illustrates the compressive strength development of concrete containing NP and OSA at various replacement levels. The mixes with Natural Pozzolan (Fig. 2a) show a characteristic pozzolanic behavior, with early-age strengths (at 1 and 7 days) being lower than the reference mix across all replacement levels. This initial lag is typical for pozzolanic materials, as their reaction is slower and depends on the availability of calcium hydroxide produced during the primary hydration of cement [4, 9]. However, after 7 days, the pozzolanic activity accelerates, leading to a significant increase in the rate of strength gain. The 35% NP replacement mix emerges as the optimal dosage, consistently outperforming all other NP mixes and even the reference mix from 28 days onwards. At 112 days, this mix surpasses the reference strength by approximately 25%, demonstrating the profound contribution of the pozzolanic reaction to the densification of the cementitious matrix over time [10]. Even at a high replacement of 60%, the NP mix reaches a respectable strength of nearly 90 MPa at 365 days. Ultimately, all tested NP replacement levels outperformed the reference mix at 365 days, highlighting its excellent potential for producing high-performance, durable concrete.

The Oil Shale Ash mixes (Fig. 2b) exhibit a more moderate strength development profile. Similar to NP, all OSA mixes show lower early strength compared to the reference. The strength gain is steady across all replacement levels, with mixes containing 15% to 45% OSA converging to a similar strength band of 75–80 MPa at 365 days. Notably, 35% OSA mix achieving a compressive strength comparable to the reference mix at 365 days. While this indicates potential for using OSA to achieve satisfactory long-term performance, it is important to note the practical limitations of high replacement levels. As noted in industry practice and studies on alternative SCMs, high volumes of materials like OSA and CC can lead to significant challenges with fresh properties, particularly workability, as well as potential issues with discoloration and expansion [11]. This may explain why commercial use in blended cements, such as those conforming to the European standard CEM II/B, is often limited to lower percentages to balance strength development with practical performance requirements [9, 12].

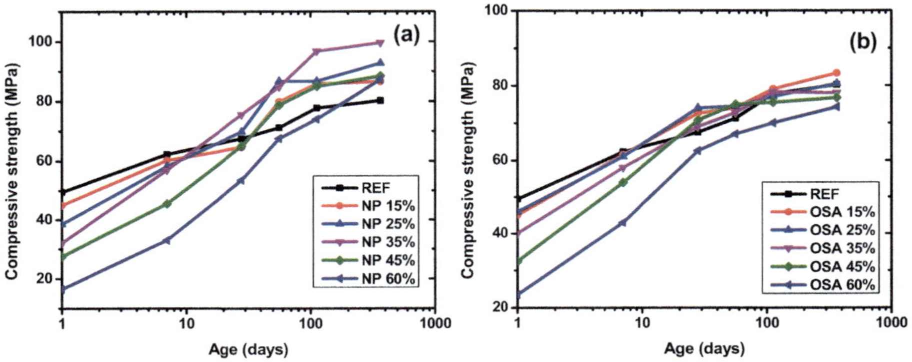

Fig. 2. Compressive strength development of **a** NP, and **b** OSA over the period of 365 days.

Figure 3 displays the compressive strength evolution for mixes containing FA and GGBS. Fly Ash (Fig. 3a) demonstrates performance consistent with its classification as a pozzolanic material. The early strength is lower than the reference, a well-documented trait attributed to the slow pozzolanic reaction of FA [4, 9]. However, the mixes with lower replacement levels show a rapid catch-up. The 15% FA mix surpasses the reference strength by 28 days and continues to gain strength, ultimately achieving a compressive strength of 102 MPa at 365 days, indicating high pozzolanic reactivity at this dosage. Mixes with higher replacement levels (25–45%) also exhibit substantial long-term gains, exceeding the reference strength at 365 days.

Ground Granulated Blast Furnace Slag (Fig. 3b) distinguishes itself with superior performance, particularly at later ages. While exhibiting a slight reduction in 1-day strength compared to the reference, the GGBS mixes show a very rapid rate of strength gain. This is because GGBS possesses both pozzolanic and inherent hydraulic (cementitious) properties, allowing it to react with water directly in addition to reacting with calcium hydroxide [4, 9]. This dual-reaction mechanism is supported by its chemical composition, which is rich in lime, silica, and alumina, similar to Portland cement [13]. As seen in the ternary diagram (Fig. 1), GGBS has significantly higher CaO content than the other pozzolanic SCMs.

The 45% GGBS replacement level appears optimal, achieving the highest strength of all mixes in this study, reaching 96.5 MPa at 365 days. This aligns with findings from other studies, which report that the greatest strengths are often achieved with GGBS blends of 40–50% [14, 15]. Even at a high replacement level of 60%, the GGBS mix slightly exceeds the strength of the reference concrete at 28 days and exceeds it by nearly 10% at 365 days.

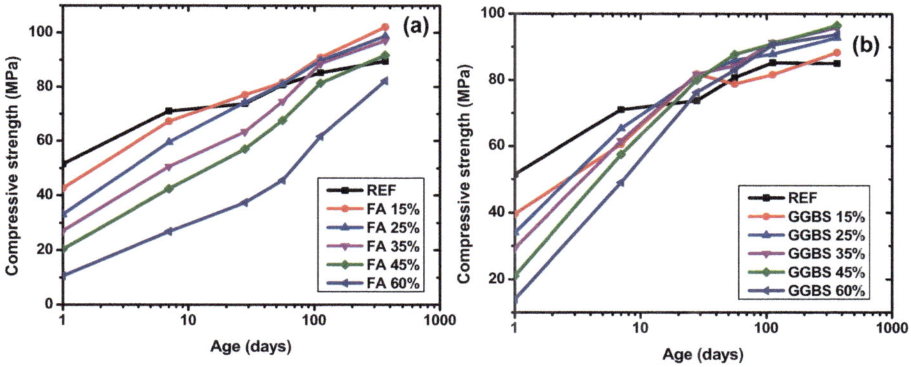

Fig. 3. Compressive strength development of **a** FA, and **b** GGBS over the period of 365 days.

The performance of concrete with CC and EAFS is shown in Fig. 4. The Calcined Clay mixes (Fig. 4a) show a significant reduction in early-age strength, which is more pronounced at higher replacement levels. The reactivity of CC stems from the amorphization of its silica and alumina content during calcination, allowing it to participate in the pozzolanic reaction [1, 2]. The 15% CC mix performs exceptionally well, recovering from its low initial strength to exceed the reference mix strength by 28 days and ultimately reaching a final strength of 94.5 MPa at 365days. However, at replacement levels of 35% and higher, the long-term strength remains considerably lower than the reference, suggesting a potential dilution effect or insufficient calcium hydroxide to activate the high volume of pozzolan [1, 4].

The Electric Arc Furnace Slag mixes (Fig. 4b) show a varied performance. EAFS is known to have slight cementitious behavior but low pozzolanic reactivity due to its crystalline nature [16]. Its effectiveness is highly dependent on its fineness and phase composition, which can include less reactive phases like monticellite and more reactive ones like merwinite [17]. As seen in the graph, all EAFS mixes show lower strength than the reference during the initial days. However, after this point, a significant acceleration in strength gain occurs. At 28 days, the 15% and 25% replacement mixes achieve strengths nearing that of the reference mix. By 365 days, the mixes with 15%, 25%, and 35% EAFS all significantly outperform the reference, with the 35% mix showing the highest strength of nearly 95 MPa. This indicates that while its early contribution is slow, EAFS becomes highly reactive at later ages, making it a very promising material for applications where long-term strength is the primary design criterion. However, it is important to note that the composition and performance of EAFS can vary significantly between producers, making direct comparisons difficult without specific characterization.

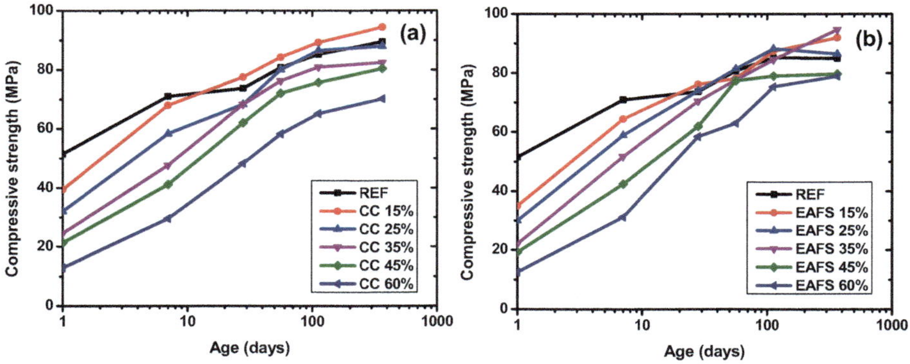

Fig. 4. Compressive strength development of **a** CC, and **b** EAFS over the period of 365 days.

A comparative analysis of all six SCMs reveals a consistent trend: while most mixes showed a reduction in early-age strength, they demonstrated significant long-term strength gains that ultimately met or exceeded the reference concrete. This performance is largely governed by the SCM's reactivity, which is linked to its chemical and physical properties as illustrated in the ternary phase diagram (Fig. 1). The SCMs can be broadly understood based on their chemical nature. While FA, NP, and CC are classic pozzolanic materials with a lower CaO content that rely on cement hydration byproducts, their highly reactive nature at optimal dosages leads to a refined microstructure and superior long-term strength [4, 18]. The pozzolanic reaction consumes the calcium hydroxide produced during cement hydration to form additional C-S-H, which densifies the matrix and enhances later-age strength [9, 10]. In contrast, the distinct performance of the two steel industry by-products is rooted in their chemistry. Based on their calcium oxide content, GGBS is classified as a hydraulic material, while EAFS is considered semi-hydraulic. The high CaO content in GGBS gives it intrinsic hydraulic properties, allowing it to contribute to strength gain independently of the cement hydration byproducts [4, 9]. The behaviors of OSA and EAFS also correlate with their compositions. Positioned near GGBS in the ternary diagram, OSA has a relatively high CaO content, suggesting it possesses both hydraulic and pozzolanic potential. This explains its solid performance at high replacement levels, where the 35% OSA mix matched the reference strength at 365 days. EAFS is chemically distinct, with high iron oxide and a moderate CaO content. However, as noted by Muhmood et al., EAFS is often highly crystalline with phases like Fe-substituted monticellite, which makes it less reactive at early ages [17]. This explains the slow initial strength gain, while the eventual hydration of its calcium- and silica-bearing phases contributes to its significant increase in strength at later ages.

4 Conclusions

This investigation evaluated the 365-day compressive strength of concrete containing six locally sourced SCMs to identify sustainable alternatives to Portland cement for the Nordic region. The study confirms that significant clinker reduction is achievable.

While this often involves a compromise in early-age strength, which can be a challenge for applications like precast production, the findings show that long-term mechanical performance can be maintained and, in several cases, enhanced compared to a 100% Portland cement reference. At optimal replacement levels, FA (15%), NP (35%), and GGBS (45%) produced concrete with superior ultimate strengths, reaching 102 MPa, 99.5 MPa, and 96.5 MPa, respectively. EAFS (35%) and CC (15%) also demonstrated excellent late-age strength gains, achieving strengths near 95 MPa. The potential for substantial clinker reduction was confirmed. Notably, concrete with 35% Oil Shale Ash matched the reference strength, while 60% GGBS surpassed it by nearly 10% at 365 days. These findings provide strong evidence that a range of regionally available SCMs can be engineered to create high-performance, low-carbon concrete. The choice of material allows for mix designs to be tailored to specific goals, from maximizing ultimate strength to maximizing sustainability. This study contributes a practical framework for adopting low-clinker binders in Nordic concrete applications, advancing the region's transition to net-zero construction.

References

1. Scrivener KL, John VM, Gartner EM (2018) Eco-efficient cements: Potential economically viable solutions for a low-CO2 cement-based materials industry. Cem Concr Res 114:2–26
2. Juenger MCG, Winnefeld F, Provis JL, Ideker JH (2011) Advances in alternative cementitious binders. Cem Concr Res 41(12):1232–1243
3. Habert G et al (2020) Environmental impacts and decarbonization strategies in the cement and concrete industries. Nat Rev Earth Environ 1(11):559–573
4. Hannesson G, Kuder K, Shogren R, Lehman D (2012) The influence of high volume of fly ash and slag on the compressive strength of self-consolidating concrete. Constr Build Mater 30:161–168
5. Moberg D, Kwiatkowski N (2024) Strength development in concrete with high incorporation of calcined clay and fly ash—a comparative study. TRITA-ABE-MBT; 24764, KTH, Stockholm
6. Brännland J, Andersson A (2023) Strength development of concrete with high proportion of mineral admixtures. In: ISRN: UTH-INGUTB-EX-B-2023/012-SE, Uppsala University, Uppsala
7. Wickman E, Jansson I (2024) Use of residual materials from the steel industry in concrete: Its impact on compressive strength development and carbon dioxide emissions. TRITA-ABE-MBT; 24627, KTH, Stockholm
8. Selander A, Bäckman A, Gram A (2025) Supplementary cementitious materials (SCMs) in Northern Europe, identifying the potential for concrete with low carbon footprint. In: Proceedings of the nordic concrete symposium 2025 (in press)
9. Oner A, Akyuz S (2007) An experimental study on optimum usage of GGBS for the compressive strength of concrete. Cement Concr Compos 29(6):505–514
10. Ganesan K, Rajagopal K, Thangavel K (2007) Evaluation of bagasse ash as supplementary cementitious material. Cement Concr Compos 29(6):515–524
11. Zuschlag P et al (2024) Hydration of composite cements containing novel SCMs. Nordic Concr Res 70(1):77–97
12. Swedish Institute for Standards: SS-EN 197-1:2011 (2011) Cement—Part 1: composition, specifications and conformity criteria for common cements. SIS-förlag, Stockholm

13. Sha W, Pereira GB (2001) Differential scanning calorimetry study of hydrated ground granulated blast-furnace slag. Cem Concr Res 31(2):327–329
14. Meusel JW, Rose JH (1983) Production of granulated blast furnace slag at sparrows point, and the workability and strength potential of concrete incorporating the slag. ACI Spec Publ 79:867–890
15. Hogan FJ, Meusel JW (1981) Evaluation for durability and strength development of a ground granulated blast furnace slag. Cement Concr Aggreg 3(1):40–52
16. Khan K, Amin MN (2017) Influence of fineness of volcanic ash and its blends with quarry dust and slag on compressive strength of mortar under different curing temperatures. Constr Build Mater 154:514–528
17. Muhmood L, Vitta S, Venkateswaran D (2009) Cementitious and pozzolanic behavior of electric arc furnace steel slags. Cem Concr Res 39(2):102–109
18. Malhotra VM, Mehta PK (2004) Pozzolanic and cementitious materials. Gordon and Breach Science Publishers, Ottawa

Fly Ash Cenosphere Incorporated Structural Lightweight Cementitious Composites: Enhancing Durability and Performance with SCMs

Irfan Hanief[1], Shayiq Rashid[1], Faisal Mukhtar[1,2], and Asad Hanif[1,2](✉)

[1] Department of Civil and Environmental Engineering, King Fahd University of Petroleum and Minerals (KFUPM), Dhahran, Saudi Arabia
asad.hanif@kfupm.edu.sa

[2] Interdisciplinary Research Center for Construction and Building Materials, King Fahd University of Petroleum and Minerals (KFUPM), Dhahran, Saudi Arabia

Abstract. This review paper explores the utilization of Fly Ash Cenosphere (FAC) as a sustainable component for structural lightweight cementitious composite (SLCC) development. Low density of FAC, spherical morphology, and pozzolanic reactivity contribute to weight reduction, improved workability, and thermal insulation. However, the porous nature of FAC results in higher water absorption, reduction in chemical resistance and causing shrinkage-induced microcracks that deteriorate their long-lasting durability against harsh environmental conditions. This paper evaluates the potential benefits of introducing supplementary cementitious materials (SCMs) which include silica fume (SF), ground granulated blast furnace slag (GGBFS) and cement kiln dust (CKD) to address limitations in lightweight cementitious composites (SLCC). SF, with its ultra fine particles and high surface area enhances pore refinement and minimizing material permeability. GGBFS improves sulfate resistance by adding hydration products that densify the microstructure while CKD acts as a stabilizing agent which improves strength through pozzolanic reactions and microstructure densification. The application of FAC together with SCMs creates an approach to produce lightweight concrete that combines durability and structural efficiency with sustainability.

Keywords: Structural light weight cementitious composite · Fly ash cenosphere · Supplementary cementitious materials

1 Introduction

Concrete remains the backbone of modern construction due to its strength, durability, and adaptability. However, its high self-weight imposes significant limitations, such as increased foundation costs, reduced seismic performance, and logistical challenges [1]. SLCC provides an effective alternative, offering reduced density (1600–2000 kg/m^3) while maintaining the mechanical performance required for structural applications [2]. Traditional SLCC relies on lightweight aggregates (LWAs) such as

H. Beushausen et al. (Eds.): NTZR 2025, LNCE 873, pp. 57–67, 2026.
https://doi.org/10.1007/978-3-032-27698-8_6

expanded clay, pumice, and sintered fly ash. Although these aggregates reduce concrete weight and improve thermal insulation, their high porosity can negatively affect strength and long-term durability.

FAC, hollow aluminosilicate microspheres formed during coal combustion, have emerged as a promising component for SLCC. With extremely low densities (~ 0.6–0.7 g/cm^3) and pozzolanic reactivity, FACs enhance workability through a ball-bearing effect and reduce concrete weight without compromising compressive strength [3]. The spherical shape of FACs enhances the water demand reduction and produces better particle arrangement and denser Interfacial Transition Zone (ITZ) which leads to improved chloride and sulfate resistance in the LWC compared to the traditional LWAs [4]. Research has shown that substituting 10–30% of fine aggregates with FAC leads to a compressive strength increase of 10–15% and thermal conductivity decrease up to 30% [5]. The fire-resistance properties of FAC based SLCC remain viable because this material maintains 80% of its compressive strength when subjected to temperatures up to 600 °C [6].

Despite these advantages, challenges remain. The higher water absorption and internal porosity of FAC may negatively affect freeze thaw durability and sulfate resistance [7]. Studies have reported that SCMs such as SF, GGBFS and CKD effectively improve the performance of FAC based SLCC. The incorporation of SF into the composite refines the pore structure alongside reduces permeability which improves strength and durability properties of the developed SLCC [8], GGBFS is reported to enhance sulfate resistance and long-term microstructural stability [9], while CKD, a cement industry byproduct, also contributes to strength development [10]. Besides effectively improving the performance of SLCC, these SCMs help in lowering carbon emissions. XRD and SEM analysis showed that FAC based SLCC and SCMs achieved matrix densification in addition to improved ITZ which resulted in enhanced mechanical properties and chemical resistance.

2 Properties and Characteristics of FAC

2.1 Physical Properties (Density, Morphology, Particle Size Distribution)

The low density of FAC is a result of the hollow spherical structure of FAC containing inert gases such as nitrogen and carbon dioxide and ranging between 0.4 to 0.8 g/cm^3 [11]. The spherical hollow microspheres of FAC have particle sizes ranging between 10 μm to 500 μm [12]. Better packing arrangement is possible because of spherical shape of FAC particles which reduce particle friction to improve concrete mix workability. FAC particle distribution patterns influence concrete behaviour because small particles boost strength while also improving packing density and larger particles lead to better insulation performance [3].

2.2 Chemical and Mineralogical Composition

The major chemical components of FAC include alumina (Al_2O_3) and silica (SiO_2) together with trace amounts of iron oxide (Fe_2O_3), calcium oxide (CaO) and magnesium oxide (MgO) [11]. The main component aluminosilicate glass provides pozzolanic activity that is beneficial in cementitious composites. The cement hydration process receives substantial support from the reactive silica and alumina components which form additional C-S-H and C-A-H phases through pozzolanic reactions [13]. FAC demonstrate both chemical stability and inertness properties that protect SLCC from detrimental reactions in harsh environments [14] (Figs. 1 and 2).

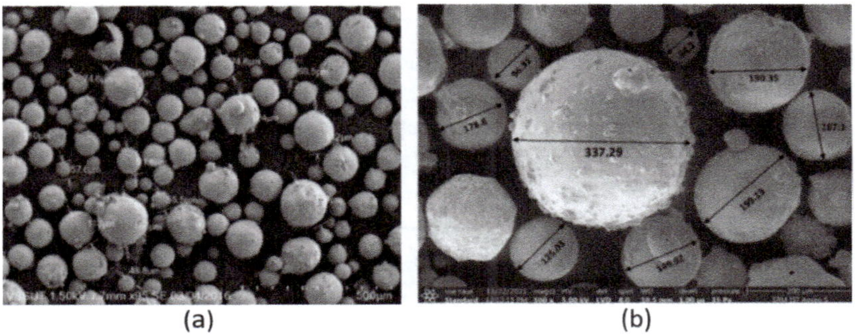

Fig. 1. SEM images (**a**) and (**b**) showing FAC [15, 16]

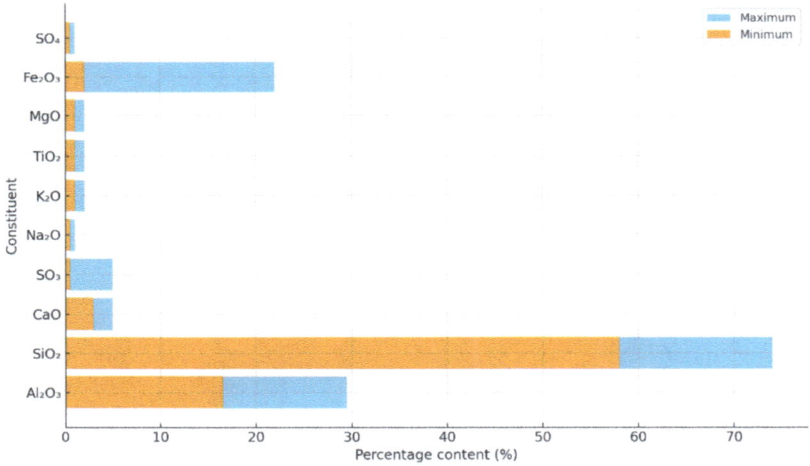

Fig. 2. Phase composition of cenosphere [17].

2.3 Thermal Insulation Properties

The thermal conductivity of FAC is low typically ranging between 0.07 to 0.15 W/m K which is the result of their lightweight and hollow structure providing excellent thermal insulation [15]. As it can be seen in Fig. 3, the thermal conductivity of lightweight cementitious composites decreases significantly with increase in the FAC content and some variation with the change in the size of FAC because of their lower inherent conductivity and increased void content.

2.4 Influence on Workability and Fresh Concrete Properties

The spherical morphology and smooth surface texture of FAC significantly enhance concrete workability by reducing internal friction between aggregates. However, the porous nature and high surface area of FAC lead to increased water demand, often necessitating adjustments in water-cement ratios or the use of superplasticizers to maintain optimal workability without compromising strength [4]. Adequate mix design optimization is critical to balancing improved workability and increased water demand, ensuring desirable fresh concrete properties. Additionally, the surface chemistry of FAC influences the dispersion of particles within the concrete matrix, affecting both setting times and rheological properties, critical for pumping and placing SLCC [18]. A decrease in the workability (measured in terms of slump) reported as presented in Fig. 4, which was attributed to the higher water absorption of the cenospheres [19].

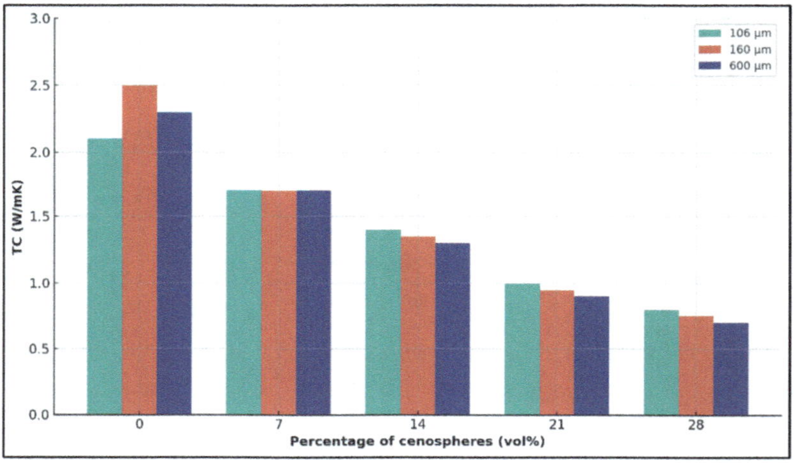

Fig. 3. Thermal conductivity of cenosphere-modified cementitious composite with different cenosphere sizes and varying content [17]

3 Challenges in FAC-Based SLCC

3.1 Increased Water Absorption

The main drawback of using FAC based SLCC stems from its high-water absorption level which results from the porous structure of fly ash cenospheres. The mixture of internal voids, large specific surface area and capillary pores found in FAC particles leads to increased water requirement during concrete preparation. The inclusion of FAC as aggregate material leads to water absorption levels which rise between 15 and 25% [20].

Hardened concrete properties like strength development and impermeability as well as durability suffer from the higher water absorption and increased porosity that FAC-based SLCC presents. The high porosity in FAC-based concrete creates penetration routes for damaging agents which speeds up deterioration mechanisms including sulfate attack and chloride ingress and carbonation and freeze-thaw damage [21].

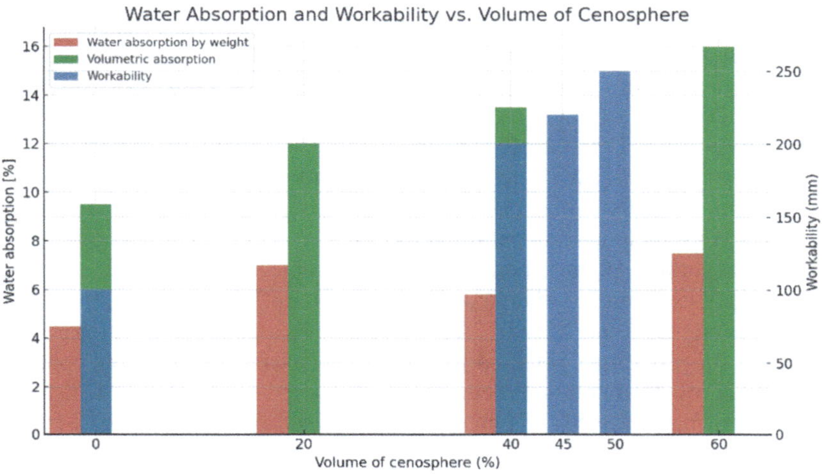

Fig. 4. Effect of cenosphere volume on water absorption and workability (measured in terms of slump, mm) [3]

3.2 Reduced Sulfate and Acid Resistance

FAC based SLCC faces a major obstacle by demonstrating an inferior resistance against sulfate and acid attack compared to traditional concrete. The strength-enhancing pozzolanic reaction provided by FAC benefits concrete microstructure and strength development but its porous nature enables rapid penetration of chemical ions including sulfates chlorides and acidic solutions [21]. The material enables fast entry for chemical degradation processes which produce ettringite and gypsum along with calcium hydroxide consumption and cementitious product dissolution that ultimately leads to severe structural damage [22]. The solution to these chemical durability problems demands specific research about how supplementary cementitious materials (SCMs) work together.

3.3 Shrinkage, Microcracking, and Durability Against Aggressive Environments

Another obstacle impacting the durability and structural performance of FAC based SLCC include Shrinkage-induced microcracking. Due to the inherently porous and lightweight characteristics of FAC, increased drying shrinkage and autogenous shrinkage frequently occur, especially in mixtures with higher FAC proportions. The excessive strains caused by shrinkage increase the formation of microcracks during hydration and drying phases that continue to expand under service conditions and environmental factors [23]. The susceptibility of FAC based SLCC to long-term degradation under cyclic wetting-drying, freeze-thaw conditions, marine exposure, and chemical attacks further amplifies durability concerns. FAC based concretes exhibited higher microcracking and more rapid degradation than conventional lightweight concretes under prolonged environmental exposure cycles [24].

3.4 Long-Term Durability

Although the field performance data is limited, laboratory tests give an idea about the long-term service life of FAC-incorporated SLCC in severe service environmental conditions. Lightweight cementitious composites incorporating FAC and reinforced with PVA fibers showed minimal microcracking and excellent fatigue resistance, retaining over 85% of flexural strength after simulated wind load cycles and exhibiting shrinkage strain below 200 microstrain at 6 days, significantly lower than conventional composites[25]. An optimized FAC-based mix (75% FAC + 25% silica sand) achieved 40.8 MPa compressive strength, 6.3 MPa flexural strength, and just 3.3% water absorption after 28 days, with shrinkage stabilized over a 301-day period, indicating minimal dimensional instability and strong long-term performance[14].

In another study, FAC-incorporated SLCC was subjected to 150 freeze–thaw cycles and 360 days of exposure to combined chloride–sulfate attack, reporting only 0.83–0.86 strength retention ratio, mass loss $< 0.71\%$, and dynamic modulus retention above 84%, all of which suggest strong potential for field applications in marine and cold-weather environments[26]. These findings demonstrate that when optimized and with additional SCMs, FAC-incorporated SLCC can maintain mechanical and durability performance throughout long service life.

4 Enhancement of FAC-Based SLCC

The FAC based SLCC demands enhancement strategies for its three main drawbacks which include increased water absorption, reduced chemical resistance and heightened susceptibility to shrinkage and microcracking. The primary methods to improve FAC based structural lightweight concrete include the addition of SCMs such as SF, GGBFS, and CKD and the latest methods include the incorporation of fibres and nanomaterials (Fig. 5).

4.1 Role of SCMs

The addition of the SF has shown positive effect towards increasing the microstructural density and reducing the permeability due to the filler effect and pozzolanic reactions. A complete consumption of some FAC particles was observed with the addition of the SF and formation of hydration products, leading to the formation of denser microstructure [11] as shown in the Fig. 6. SF additions of 5–10% resulted in a 40–50% decrease in chloride ion penetration thus demonstrating its effectiveness for decreasing permeability and increasing structural durability [25].

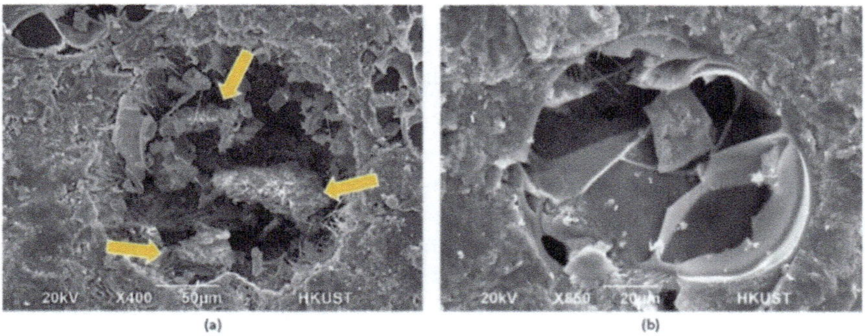

Fig. 5. FAC particle behavior in pastes **a** containing NS, and **b** without NS; (120—day age) [11].

The addition of GGBFS enhances FAC based SLCC because it possesses latent hydraulic properties which generate C-S-H and C-A-H hydration products together with other hydration products [26]. The addition of 20–30% GGBFS decreased chloride permeability by 50% and improved sulfate resistance significantly in FAC based concrete which did not contain GGBFS [27].

CKD, characterized by high calcium oxide content and alkaline properties, acts as an effective stabilizing agent when incorporated into FAC based SLCC [28]. Addition of 10–15% CKD enhanced compressive strength by approximately 15–20%, attributed to CKD pozzolanic reactions forming additional C-S-H, densifying the matrix, and significantly reducing permeability [29].

4.2 Role of Fibers

With 15% FAC and 0.5% Polypropylene (PP) fibers, compressive strength reached 23.17 MPa, tensile strength 1.88 N/mm^2, and density stayed within 1850–1920 kg/m^3, while enhancing ductility and reducing brittleness through strain-hardening [30]. At 30% FAC and 0.5% Recron fiber, the cement slurry achieved a compressive strength of 22.13 MPa and flexural strength of 1.59 MPa in 24 h for 1% Recorn fibers, with uniform fiber dispersion enhancing strength and ductility despite FAC's lightweight nature as shown in Fig. 7 [24]. In another study about green FAC based SLCC, 0.5% Polyvinyl Alcohol (PVA) or basalt fibers improved compressive strength up to 50.3 MPa and flexural strength by 11–14%, with PVA fibers maintainable up to 0.75% without loss in workability or strength, while basalt fibers beyond 0.5% caused strength reduction due to poor compactability and fiber dispersion [31, 32].

4.3 Role of Nano Materials

In FAC based SLCC, incorporating nanomaterials nano-silica (NS) and carbon nanotubes (CNTs)—along with treated FAC improved 28-day compressive and split-tensile strength by up to 17.74% and 30.18%, respectively, by enhancing hydration, C-S-H formation, and interfacial transition zone (ITZ) stiffness, despite a minor reduction in workability [33]. In another study, the binary addition of 0.15% CNTs and 0.2% NS yielded optimal results, increasing flexural strength by 89.3% (to 2.1 MPa) and compressive strength by 9.2% (to 20.1 MPa) as shown in Fig. 7, due to improved hydration and microstructure densification, while higher dosages led to strength reductions from nanoparticle agglomeration [33].

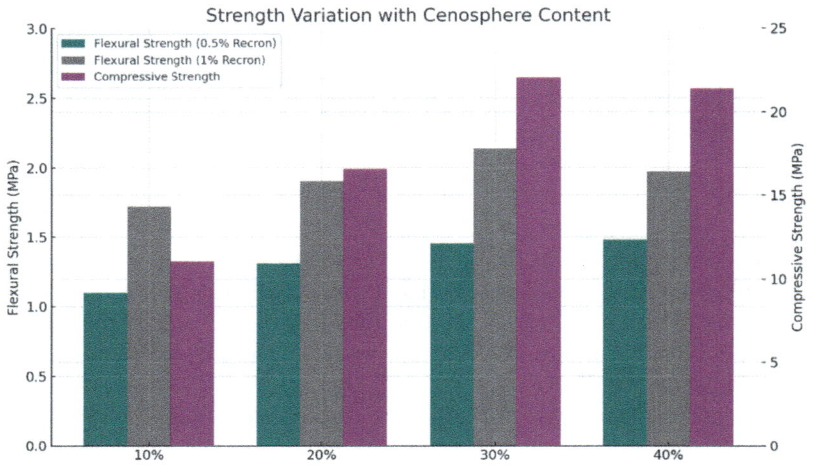

Fig. 6. Compressive strength of varying dose of cenosphere & 0.5% Recron fibers and Flexural Strength with 0.5% and 1% Recron [24].

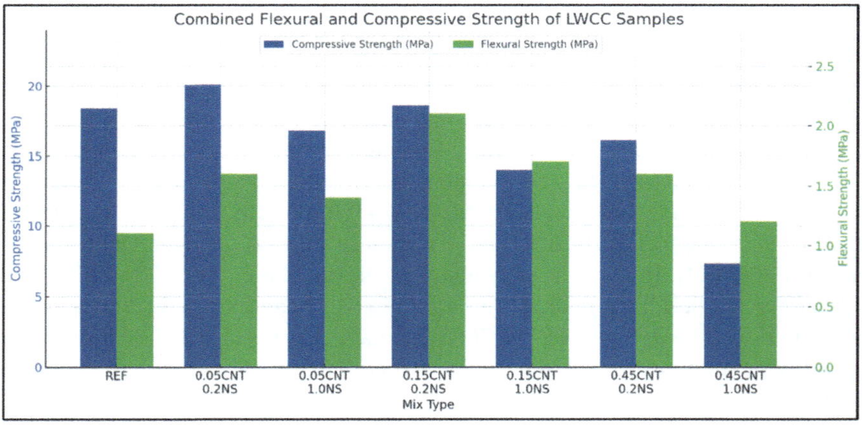

Fig. 7. Flexural and compressive strength of SLCC samples incorporating different amounts of CNT and NS.

5 Prospects and Applicability—A Way Forward

FAC based SLCC offers great potential for sustainable construction due to its lightweight nature, improved thermal insulation, and adequate strength. Its application is promising in high-rise buildings, precast elements, and fire-resistant structures. The use of SCMs like SF, GGBFS, and CKD can effectively overcome limitations such as high porosity and low sulfate resistance, improving durability and microstructural integrity. Fibers, like PP, PVA, and Recron enhance ductility and tensile strength, making the material suitable for structural use while maintaining reduced density.

Moving forward, incorporating nanomaterials such as NS and CNTs presents a powerful strategy to boost hydration, matrix densification, and ITZ quality. Optimized dosages can significantly improve strength and durability, though care must be taken to avoid agglomeration. Future research should focus on performance-based mix design, durability in aggressive environments, and life cycle sustainability. With proper enhancement strategies, FAC based SLCC can play a key role in developing low-carbon, durable, and efficient infrastructure.

The authors acknowledge the concerns regarding the long-term sustainability of FAC and SCMs such as SF. However, despite global concerns and declining amounts of generated fly ash, huge amounts are still produced in the USA, Australia, India, and China due to coal power production, which end up in landfills. Considering the prevailing trends and energy demand for the next few decades, fly ash availability may not be problematic. Also, other SCMs, such as silica fume, red mud, and GGBS, are produced as industrial byproducts from silicon, aluminium, and steel manufacturing industries, which are also being generated globally in significant amounts. Their utilization depends on regional factors, including the by-product nature, transportation distance, and associated emissions. Therefore, conducting region-specific life cycle assessments is critical for use in cementitious composites.

References

1. Anilkumar R, Prakash P, Gowda R (2019) Durability performance of structural light weight concrete. In: Lecture notes in civil engineering, Springer, pp 853–861. https://doi.org/10.1007/978-981-13-3317-0_76
2. Badogiannis E, Stratoura M, Aspiotis K, Chatzopoulos A (2021) Durability of structural lightweight concrete containing different types of natural or artificial lightweight aggregates. Corros Mater Degrad 2:554–567. https://doi.org/10.3390/cmd2040029
3. Gabriela R, Paweł O, Aleksandra P, Mariusz Ż, Filip C, Karolina K (2024) Effect of cenospheres on the proprieties of plain concrete exposed to elevated temperature. Clean Technol Environ Policy. https://doi.org/10.1007/s10098-024-03034-3
4. Danish A, Mosaberpanah MA (2021) Influence of cenospheres and fly ash on the mechanical and durability properties of high-performance cement mortar under different curing regimes, Constr Build Mater 279. https://doi.org/10.1016/j.conbuildmat.2021.122458
5. Zhou H, Brooks AL (2019) Thermal and mechanical properties of structural lightweight concrete containing lightweight aggregates and fly-ash cenospheres. Constr Build Mater 198:512–526. https://doi.org/10.1016/j.conbuildmat.2018.11.074
6. Klima KM, Schollbach K, Brouwers HJH, Yu Q (2022) Thermal and fire resistance of class F fly ash based geopolymers—a review. Constr Build Mater 323. https://doi.org/10.1016/j.conbuildmat.2022.126529
7. Hasholt MT, Christensen KU, Pade C (2019) Frost resistance of concrete with high contents of fly ash—a study on how hollow fly ash particles distort the air void analysis. Cem Concr Res 119:102–112. https://doi.org/10.1016/j.cemconres.2019.02.013

8. Mahato J, Yang J, Lee N, Kang H, Moon J (2023) Incorporation of a high volume of cenosphere particles in low water-to-cement matrix for developing high strength and lightweight cementitious composites. J Sustain Cem Based Mater 12:580–591. https://doi.org/10.1080/21650373.2022.2095678
9. Oinam Y, Ju S, Gwon S, Shin M, Pyo S (2022) Characteristics of GGBFS-based pervious concrete considering rheological properties of the binder. Int J Concr Struct Mater 16. https://doi.org/10.1186/s40069-022-00551-7
10. 26-96-r1 (n.d.)
11. Hanif A, Parthasarathy P, Ma H, Fan T, Li Z (2017) Properties improvement of fly ash cenosphere modified cement pastes using nano silica. Cem Concr Compos 81:35–48. https://doi.org/10.1016/j.cemconcomp.2017.04.008
12. Strzałkowska E, Adamczyk Z (2020) Influence of chemical composition of fly-ash cenospheres on their grains size. Int J Environ Sci Technol 17:809–818. https://doi.org/10.1007/s13762-019-02512-2
13. Golewski GL (2022) The role of pozzolanic activity of siliceous fly ash in the formation of the structure of sustainable cementitious composites. Sustain Chem 3:520–534. https://doi.org/10.3390/suschem3040032
14. Leong GW, Mo KH, Ibrahim Z, Radwan MKH, Ling TC, Sinoh SS (2023) Lightweight cementitious composites incorporating fly ash cenospheres and perlite microspheres. Constr Build Mater 404. https://doi.org/10.1016/j.conbuildmat.2023.133226
15. Satpathy HP, Patel SK, Nayak AN (2019) Development of sustainable lightweight concrete using fly ash cenosphere and sintered fly ash aggregate. Constr Build Mater 202:636–655. https://doi.org/10.1016/j.conbuildmat.2019.01.034
16. Hanif A, Lu Z, Li Z (2017) Utilization of fly ash cenosphere as lightweight filler in cement-based composites—a review. Constr Build Mater 144:373–384. https://doi.org/10.1016/j.conbuildmat.2017.03.188
17. Danish A, Mosaberpanah MA, Tuladhar R, Salim MU, Yaqub MA, Ahmad N (2022) Effect of cenospheres on the engineering properties of lightweight cementitious composites: A comprehensive review. J Build Eng 49. https://doi.org/10.1016/j.jobe.2022.104016
18. Kwan AKH, Li Y (2013) Effects of fly ash microsphere on rheology, adhesiveness and strength of mortar. Constr Build Mater 42:137–145. https://doi.org/10.1016/j.conbuildmat.2013.01.015
19. Patel SK, Satpathy HP, Nayak AN, Mohanty CR (2020) Utilization of fly ash cenosphere for production of sustainable lightweight concrete. J Inst Eng (India): Ser A 101:179–194. https://doi.org/10.1007/s40030-019-00415-6
20. Rutkowska G, Ogrodnik P, Żółtowski M, Powęzka A, Kaszewska K (2022) Analysis of the possibility of using cenospheres in the production of cement mortars for use in an elevated temperature environment. Sensors 22. https://doi.org/10.3390/s22197518
21. Jaworska-Wędzińska M, Jasińska I (2022) Durability of mortars with fly ash subject to freezing and thawing cycles and sulfate attack, Materials 15. https://doi.org/10.3390/ma15010220
22. Domagała L (2020) Durability of structural lightweight concrete with sintered fly ash aggregate. Materials 13:1–20. https://doi.org/10.3390/ma13204565
23. Chanda SS, Patel SK, Nayak AN, Mohanty CR (2023) Performance evaluation on bond, durability, micro-structure, cost effectiveness and environmental impacts of fly ash cenosphere based structural lightweight concrete. Constr Build Mater 397:132429. https://doi.org/10.1016/j.conbuildmat.2023.132429
24. Smarzewski P, Błaszczyk K (2024) Influence of cement Kiln dust on long-term mechanical behavior and microstructure of high-performance concrete. Materials 17. https://doi.org/10.3390/ma17040833

25. Hanif A, Diao S, Pei H, Li Z, Sun G (2019) Green lightweight laminated cementitious composite (LCC) for wind energy harvesting—a novel application of LCCs. Case Stud Constr Mater 10. https://doi.org/10.1016/j.cscm.2019.e00217
26. Gou H, Sofi M, Zhang Z, Zhu H, Zhu M, Mendis P (2023) Durability development of lightweight and high-strength engineered cementitious composites subject to combined sulfate–chloride attack under freeze–thaw cycles. Constr Build Mater 408. https://doi.org/10.1016/j.conbuildmat.2023.133659
27. Wang D, Zhou X, Fu B, Zhang L (2018) Chloride ion penetration resistance of concrete containing fly ash and silica fume against combined freezing-thawing and chloride attack. Constr Build Mater 169:740–747. https://doi.org/10.1016/j.conbuildmat.2018.03.038
28. Liu S, Han W, Li Q (2017) Hydration properties of ground granulated blast-furnace slag (GGBS) under different hydration environments. Medziagotyra 23:70–77. https://doi.org/10.5755/j01.ms.23.1.14934
29. Li G, Zhao X (n.d.) Properties of concrete incorporating fly ash and ground granulated blast-furnace slag. www.elsevier.com/locate/cemconcomp
30. El-Sayed AM, Faheim AA, Salman AA, Saleh HM (2022) Sustainable lightweight concrete made of cement kiln dust and liquefied polystyrene foam improved with other waste additives. Sustainability (Switzerland) 14. https://doi.org/10.3390/su142215313
31. Jeyanthi J, Revathi S (2022) Study of light weight composite concrete incorporated with polypropylene fiber and cenosphere. In: Materials today: proceedings. Elsevier Ltd, pp 4303–4309. https://doi.org/10.1016/j.matpr.2022.04.804
32. Richhariya G, Dora DT, Kumar A, Rastogi A, Pharasi S, Goyal Y, Rai M (2020) Recron fiber reinforced ultra light weight Cenosphere cement for unconsolidated formations with low fracture gradient. In: Materials today: proceedings, Elsevier Ltd, pp 6771–6775. https://doi.org/10.1016/j.matpr.2021.05.129
33. Leong GW, Pahdili EHH, Mo KH, Ibrahim Z (2022) Impacts of polyvinyl alcohol and basalt fibres on green fly ash cenosphere lightweight cementitious composite. Mater Today Proc 61:512–516. https://doi.org/10.1016/j.matpr.2021.12.519
34. Hanif A, Lu Z, Diao S, Zeng X, Li Z (2017) Properties investigation of fiber reinforced cement-based composites incorporating cenosphere fillers. Constr Build Mater 140:139–149. https://doi.org/10.1016/j.conbuildmat.2017.02.093
35. Zhang Y, Zhu X, Shah SP (2024) Reinforcing efficiency of nanomaterials on mechanical and interfacial characteristics of green concrete incorporating fly ash cenosphere. J Clean Prod 447. https://doi.org/10.1016/j.jclepro.2024.141531

Mechanical Behavior of Cementitious Composites Incorporating Silicomanganese Fume

Shayiq Rashid[1], Hammad R. Khalid[1,2], Irfan Hanief[1], and Asad Hanif[1,2](✉)

[1] Civil and Environmental Engineering Department, King Fahd University of Petroleum & Minerals, Dhahran 31261, Saudi Arabia
asad.hanif@kfupm.edu.sa

[2] Interdisciplinary Research Center for Construction and Building Materials, King Fahd University of Petroleum & Minerals, Dhahran 31261, Saudi Arabia

Abstract. This study investigates the feasibility of utilizing various grades of silicomanganese (SiMn) fume, categorized by fineness, as a sustainable supplementary cementitious material (SCM) in concrete. SiMn fume, a manganese-rich byproduct of the ferroalloy industry, was incorporated as a partial cement replacement (20%) to evaluate its effects on compressive strength at 7, 28, and 90 days under two curing regimes: water curing and steam curing. Results revealed that the finest grade of SiMn fume consistently outperformed coarser grades, demonstrating superior mechanical properties and enhanced durability. Beyond mechanical improvements, the inclusion of SiMn fume offers substantial environmental benefits. By reducing cement consumption, SiMn fume significantly lowers the carbon footprint of concrete, addressing the urgent need for sustainable and eco-friendly construction materials. This approach aligns with circular economy principles, repurposing industrial waste to minimize landfill disposal while promoting resource efficiency and waste valorisation. The findings establish SiMn fume as a viable SCM capable of delivering dual benefits: improved mechanical performance and reduced environmental impact. By leveraging SiMn fume, this study demonstrates a pathway to achieving carbon reduction targets in the construction sector while fostering the adoption of innovative, sustainable materials. This work represents a step forward in advancing SCM adoption, paving the way for greener infrastructure development and contributing to the global transition toward sustainable construction practices.

Keywords: SCMs · SiMn fume · Mechanical properties · Waste management

1 Introduction

Concrete remains the backbone of global infrastructure, with over 30 billion tonnes produced annually, making it the most consumed man-made material after water [1]. Its ubiquity stems from its versatility, durability, and relative affordability. However, this prevalence comes with severe environmental consequences, particularly due to the

H. Beushausen et al. (Eds.): NTZR 2025, LNCE 873, pp. 68–77, 2026.
https://doi.org/10.1007/978-3-032-27698-8_7

cement component in concrete mixtures. The cement industry alone is responsible for an estimated 7–8% of global CO_2 emissions [2] positioning it among the top industrial contributors to climate change.

The environmental burden of cement arises from both process emissions, primarily due to the decarbonation of limestone ($CaCO_3$) into lime (CaO), and combustion emissions, where fossil fuels are used to achieve the high temperatures (up to 1450 °C) necessary for clinker production. Beyond greenhouse gases, cement manufacturing also causes particulate emissions, depletion of natural resources, and contributes to urban heat island effects due to widespread concrete usage [3]. The sustainability of the construction sector thus hinges on innovations that reduce the environmental impact of cement without compromising structural performance.

One of the most effective and scalable approaches to reducing the environmental impact of concrete is the partial replacement of Portland cement with Supplementary Cementitious Materials (SCMs). SCMs can be broadly classified into two categories: Pozzolanic materials, which react with calcium hydroxide in the presence of water to form additional calcium silicate hydrate (C-S-H), the primary binder in concrete. Latent hydraulic materials, which exhibit cementitious behavior when activated in an alkaline environment [4]. Common SCMs include fly ash (a by-product of coal combustion) [5], ground granulated blast-furnace slag (from the iron and steel industry) [6], silica fume (a by-product of silicon metal production) [7], and natural pozzolans (like volcanic ash and calcined clay) [8–11]. When used to partially replace cement, these materials significantly reduce the carbon intensity of concrete while also enhancing properties such as durability, impermeability, sulfate resistance, and chloride ion penetration [12, 13].

With growing global interest in circular economy practices, the valorization of industrial by-products as SCMs has gained significant attention [10, 14, 15]. This approach not only mitigates the environmental burden of cement production but also diverts industrial waste from landfills and promotes resource efficiency through industrial symbiosis.

In this context, silicomanganese (SiMn) fume, a fine particulate by-product generated during the production of silicomanganese alloys in submerged arc furnaces, presents a promising but underexplored SCM option [16–18]. SiMn alloys play a critical role in steelmaking due to their deoxidizing and alloying functions [19]. The associated fume, often treated as a waste material, is rich in amorphous silica, manganese oxides, alumina, and trace metals. Its high silica content and ultrafine particle size suggest potential pozzolanic reactivity, while its industrial origin aligns well with sustainable construction goals.

Despite its potential, the integration of SiMn fume into cementitious systems has received limited research attention, particularly concerning how its chemical and physical properties interact with different curing regimes. This is a key knowledge gap, especially in construction scenarios where early strength development, long-term durability, and reduced environmental impact are all crucial performance targets.

To address this gap, the present study investigates the performance of three different grades of silicomanganese (SiMn) fume, used as 20% cement replacement by weight, on the compressive strength development of cementitious pastes [20]. The mixes were subjected to three curing regimes: 24-h steam curing at 90 °C, continuous water curing up to 90 days, and a hybrid regime combining initial steam curing followed by water curing. Compressive strength was measured at multiple ages to evaluate both early and long-term behavior. The goal is to assess the suitability of SiMn fume as a sustainable SCM under diverse curing conditions and to support its potential adoption in low-carbon concrete applications.

2 Materials and Methods

2.1 Raw Materials

Ordinary Portland Cement (OPC, Type I) conforming to ASTM C150 was sourced from a regional manufacturer, with a verified specific gravity of 3.15. Three distinct types of silicomanganese fume (SiMn), labelled SiMn-A, SiMn-B, and SiMn-C, were obtained from the Gulf Ferroalloy Company, located in Jubail, Saudi Arabia. Before experimental use, the materials were thoroughly characterized through chemical composition analysis and particle size distribution studies. Measurements performed using an Anton Paar PSA 1190 analyser revealed average particle sizes of 3.3764 μm for SiMn-A, 0.9836 μm for SiMn-B, 8.1804 μm for SiMn-C, and 30.3490 μm for OPC, as presented in Fig. 1. The elemental composition of OPC and the three fume grades is summarized in Table 1, highlighting key oxides such as SiO_2, Al_2O_3, Fe_2O_3, MnO, and CaO.

Table 1. Chemical composition of raw materials

Chemical composition /Wt %	Cement	SiMn-A	SiMn-B	SiMn-C
SiO_2	16.27	33.38	25.85	28.71
Fe_2O_3	3.96	3.58	7.49	5.71
CaO	73.97	14.79	10.77	13.43
MnO	–	43.68	50.87	48.73
K_2O	–	2.17	2.03	1.12
Al_2O_3	3.71	–	–	–
Others	2.09	2.4	2.99	2.3

A fixed replacement level of 20% SiMn fume by weight of cement was adopted in this study, based on the findings of our previous research [20]. This controlled substitution level allowed for a focused comparison across SiMn fume variants. A uniform water-to-binder (W/B) ratio of 0.30 was selected to maintain mix consistency across all batches. The individual quantities of OPC, SiMn fume, and water were measured with precision. The mixing process was conducted using a HOBART UTEST mixer, beginning with the dry blending of OPC and SiMn fume to ensure even distribution of the fine particles. Water was subsequently added, and wet mixing was carried out until a cohesive and homogeneous paste was formed. The prepared mixture was cast into pre-oiled cubic steel molds measuring 25 mm on each side, followed by brief mechanical vibration (5 s) to remove entrapped air. After casting, the specimens were sealed to limit moisture evaporation, demolded after 24 h, and subjected to two different curing regimes: conventional water curing at 23 ± 2 °C and accelerated steam curing at 90 °C for 24 h. The steam curing process was conducted in a controlled environment to replicate elevated-temperature hydration conditions. Both curing approaches are illustrated in Fig. 2.

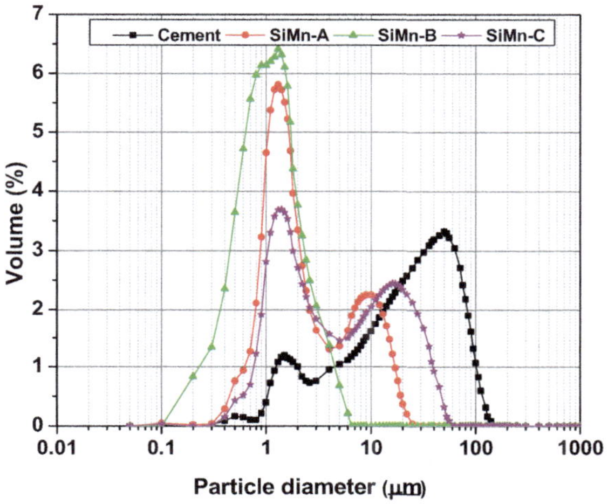

Fig. 1. Particle size distribution of raw materials mix design, specimen preparation, and curing regimes

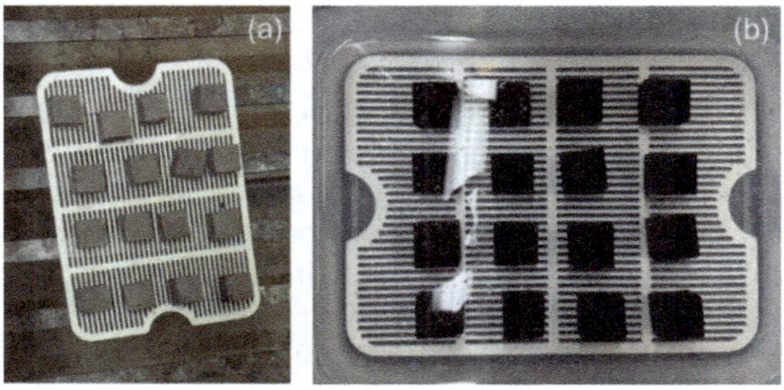

Fig. 2. Curing regimes a) steam curing and b) normal water curing

2.2 Compressive Strength

Compressive strength testing was conducted in accordance with ASTM C109 using a MATEST C55 digital compression testing machine, operating at a constant loading rate of 1 kN/s. To ensure uniform load application and minimize eccentricity, each specimen was carefully positioned at the center of the lower platen. For statistical reliability, four specimens were tested per mix, and the average compressive strength was reported as the representative value. The tests were performed after 7, 28, and 90 days of standard water curing to evaluate strength development over time.

3 Results

3.1 Compressive Strength of Water-Cured Specimens

Figure 3 illustrates the evolution of compressive strength in paste samples subjected to water curing over 7, 28, and 90 days, encompassing both the reference mix (designated as Mix C) and those incorporating a 20% substitution of cement with various grades of SiMn fume. An overall increase in compressive strength with prolonged curing duration is apparent across all mixtures, which can be primarily attributed to continued cement hydration and the progressive refinement of the internal matrix.

Among all formulations, the control mix (C), containing no SiMn fume, consistently demonstrates superior strength performance at every curing interval, achieving approximately 70 MPa at 7 days, rising to 80 MPa at 28 days, and peaking at 84 MPa by 90 days. This dominant strength behavior is aligned with its unaltered high clinker content, which promotes rapid hydration reactions known to facilitate early structural development, a well-documented outcome in cementitious systems.

On the other hand, the compressive performance of the SiMn substituted mixes varied significantly depending on the specific SiMn fume grade employed. The mix modified with SiMn-B, characterized by its extremely fine particle size, exhibited the most favorable strength progression among the SiMn fume incorporated mixes. By the 90-day mark, the compressive strength of the SiMn-B mix approached that of the C, indicating that the fineness of SiMn-B likely enhances the cement matrix. This behaviour may stem from a pronounced filler effect, improving the packing density and reducing pores within the matrix.

In contrast, mixes incorporating SiMn-A and SiMn-C showed consistently lower strength values at all ages, suggesting that coarser SiMn fume particles may limit their efficacy in contributing to mechanical performance. The early-age strength deficits observed in all SiMn fume incorporated samples, particularly at 7 and 28 days, are probably linked to the reduction in clinker content resulting from partial cement replacement. This reduction curtails the formation of hydration products that typically dominate early strength gain. Moreover, the contribution of SiMn fume at these early stages is limited, as any pozzolanic activity it may exhibit generally manifests over extended curing periods.

3.2 Compressive Strength of Steam-Cured Specimens

Figure 4 illustrates the compressive strength progression of paste specimens subjected to accelerated steam curing at 90 °C for 24 h, followed by strength testing at 24 h, 7 days, 28 days, and 90 days. The evaluated mixes included a C and three variants incorporating 20% SiMn fume, denoted as SiMn-A, SiMn-B, and SiMn-C. All specimens were exposed to steam curing for 24 h immediately after casting; thereafter, except for the

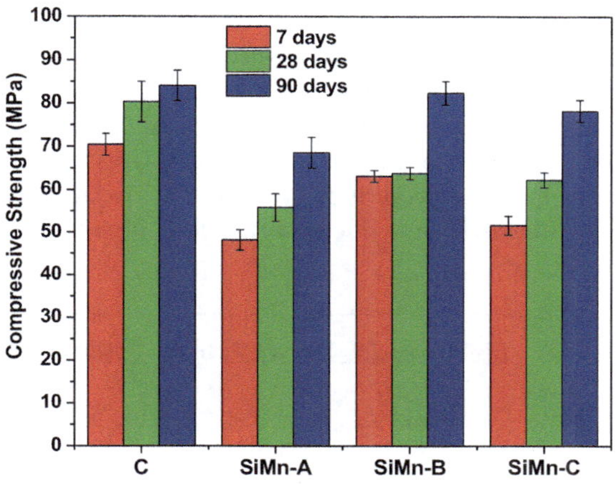

Fig. 3. Water cured specimens at 7, 28, and 90 days

24-h group, which was tested directly following steam exposure, the remaining samples were transitioned to standard water curing at 23 ± 2 °C until their designated testing ages. This hybrid curing regime was designed to simulate accelerated early strength development followed by conventional hydration conditions representative of real-world practices.

After 24 h, all mixes exhibited notably high compressive strength values, confirming the efficacy of steam curing in promoting early hydration. Among the SiMn fume mixes, SiMn-C recorded the highest early-age strength (~68 MPa), surpassing both the control (~55 MPa) and the other SiMn fume mixes. SiMn-B followed with a strength of approximately 63 MPa, while SiMn-A registered the lowest among the SiMn incorporating mixes(~50 MPa). This early performance enhancement can be attributed to the elevated curing temperature, which accelerates the dissolution of cement phases and the formation of calcium silicate hydrate (C–S–H), particularly when combined with fine and chemically active SiMn fume particles.

However, the 7-day results reveal a counterintuitive trend. A slight strength gain was observed for the control mix (~63 MPa), but most SiMn fume incorporating mixes exhibited either marginal improvement or a reduction in strength relative to their 24-h values. Notably, SiMn-C experienced a significant decline to ~50 MPa. This phenomenon may be linked to the adverse effects of high-temperature curing, including the formation of a dense outer shell around cement grains that impedes further hydration, as well as internal thermal stresses and microcracking. Such effects can suppress subsequent strength development, particularly in matrices with lower clinker content.

At 28 days, strength recovery was evident, particularly in the SiMn-B mix, which increased to ~65 MPa, likely due to delayed hydration and filler-packing effects enhancing the matrix density. SiMn-A and the control mix exhibited comparable strength (~60 MPa), while SiMn-C plateaued near ~50 MPa, suggesting limited long-term reactivity or suboptimal particle packing in that formulation.

At 90 days, the control mix continued to exhibit superior performance, reaching a peak compressive strength of ~78 MPa. SiMn-B followed closely (~72 MPa), highlighting its effective long-term contribution to strength development, potentially due to a combination of high fineness and moderate pozzolanic activity. SiMn-A and SiMn-C converged at ~68 MPa, with SiMn-A showing steady long-term improvement, while SiMn-C remained stagnant after its early peak, likely due to a lack of sustained hydration or microstructural inefficiencies introduced during early high-temperature exposure.

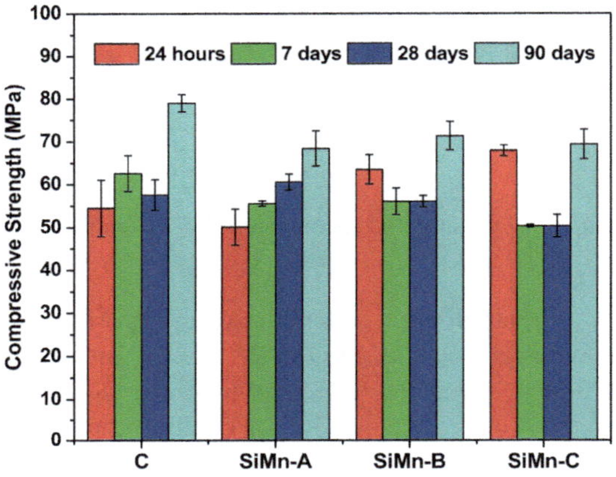

Fig. 4. Steam cured specimens 24 h, 7 days, 28 days, and 90 days

3.3 Comparison Between Water and Steam-Cured Specimens

Figures 5 and 6 illustrate the compressive strength development of all paste mixes (Control, SiMn-A, SiMn-B, and SiMn-C) under different curing conditions, specifically 24-h steam curing, extended water curing, and a combination of initial steam curing followed by water curing until the age of testing (28 and 90 days).

As expected, specimens subjected to 24-h steam curing exhibited significantly higher compressive strength at the early age across all mix types. The control mix reached approximately 55 MPa, while SiMn-C exceeded 68 MPa, highlighting the effectiveness of elevated temperature curing in accelerating hydration and early strength gain. Among the SiMn fume blends, SiMn-C outperformed all others, suggesting superior particle fineness or mineralogical compatibility conducive to rapid reaction at high temperature. In contrast, no corresponding early strength values are reported for water curing at this stage, as hydration under ambient conditions typically proceeds at a much slower rate, yielding minimal measurable strength at 24 h.

At 28 days (Fig. 5), the control mix achieved the highest strength (~80 MPa) under water curing, while its performance under steam and hybrid curing was comparatively lower. SiMn-B displayed consistent strength (~63–65 MPa) across all curing methods, indicating good compatibility with both thermal and ambient hydration environments.

SiMn-A showed moderate improvement under hybrid and steam curing, whereas SiMn-C, despite having the highest early strength, demonstrated a decline in long-term performance under steam-related curing, suggesting early thermal exposure may inhibit further hydration.

At 90 days (Fig. 6), the distinction between curing regimes becomes even more pronounced. Water-cured specimens continued to gain strength across all mixes. The control mix reached the highest compressive strength (~84 MPa), while SiMn-B achieved approximately 82 MPa, surpassing even the control. In contrast, the strength of steam-water cured specimens at 90 days remained consistently lower.

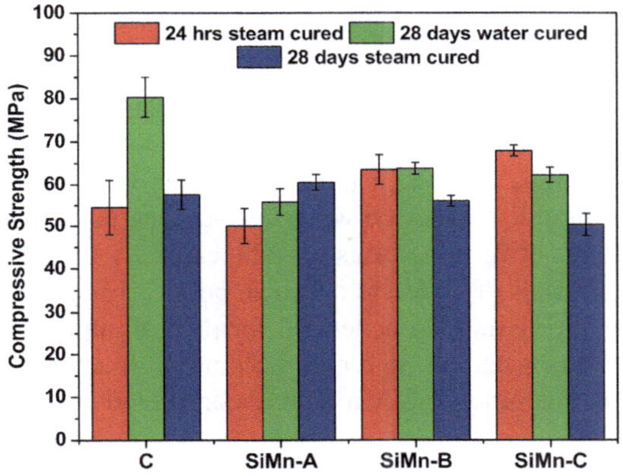

Fig. 5. Comparison of 24-h steam curing, 28-day water curing, and hybrid curing (24-h steam curing followed by water curing until 28 days)

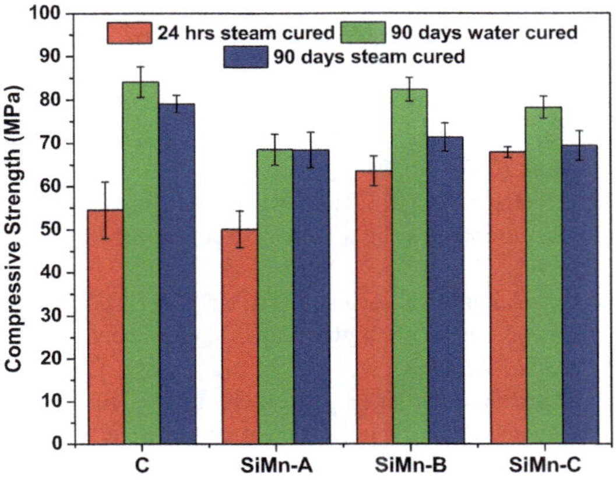

Fig. 6. Comparison of 24-h steam curing, 90-day water curing, and hybrid curing (24-h steam curing followed by water curing until 90 days)

4 Conclusions

This study examined the influence of different curing regimes, steam curing at 90 °C for 24 h, continuous water curing, and a hybrid method combining steam and water curing, on the compressive strength development of cementitious pastes incorporating three grades of silicomanganese fume (SiMn-A, SiMn-B, SiMn-C) as 20% partial cement replacements. Results demonstrated that steam curing significantly enhanced early-age strength, with SiMn-C reaching the highest 24-h value (~68 MPa). However, long-term performance was more favourable under continuous water curing, which allowed for sustained hydration and superior strength development at 28 and 90 days. Notably, the control mix and SiMn-B achieved peak strengths of ~84 MPa and ~ 82 MPa, respectively, under water curing at 90 days, outperforming their steam-cured counterparts.

The hybrid curing regime, while offering early strength benefits, often led to strength stagnation or reduction at later ages, likely due to microstructural densification or self-desiccation effects induced by early thermal exposure. Among all SiMn fume blends, SiMn-B exhibited the most consistent performance across all curing conditions and ages, indicating a favourable balance between fineness and pozzolanic reactivity. These findings emphasize that curing conditions must be carefully matched with the physical and chemical characteristics of SCMs to optimize both early-age and long-term performance. While steam curing may be beneficial for precast applications requiring rapid strength gain, continuous water curing remains the most effective method for achieving durable, high-strength outcomes in SiMn fume incorporated cementitious systems.

References

1. Scrivener KL, John VM, Gartner EM (2018) Eco-efficient cements: potential economically viable solutions for a low-CO_2 cement-based materials industry. Cem Concr Res 114:2–26. https://doi.org/10.1016/j.cemconres.2018.03.015
2. Andrew RM (2019) Global CO2 emissions from cement production, 1928–2018. Earth Syst Sci Data 11:1675–1710. https://doi.org/10.5194/essd-11-1675-2019
3. Miller SA, Horvath A, Monteiro PJM (2018) Impacts of booming concrete production on water resources worldwide. Nat Sustain 1:69–76. https://doi.org/10.1038/s41893-017-0009-5
4. Gartner E, Sui T (2018) Alternative cement clinkers. Cem Concr Res 114:27–39. https://doi.org/10.1016/j.cemconres.2017.02.002
5. Hanif A, Lu Z, Cheng Y, Diao S, Li Z (2017) Effects of different lightweight functional fillers for use in cementitious composites. Int J Concr Struct Mater 11:99–113. https://doi.org/10.1007/s40069-016-0184-1
6. Siddique S, Khan AM, Adekunle SK, Al-Osta MA, Hanif A (2025) Effect of alkaline additives on the physiochemical properties of high-volume slag cement. Struct Concr J Fib:70203. https://doi.org/10.1002/suco.70203
7. Dong C et al (2022) Fresh and hardened properties of recycled plastic fiber reinforced self-compacting concrete made with recycled concrete aggregate and fly ash, slag, silica fume. J Build Eng 62:105384. https://doi.org/10.1016/j.jobe.2022.105384
8. Eltayeb I et al (2025) Physicochemical properties of cement waste-derived portlandite-substituted limestone calcined clay cement (LC 3) binders. J Sustain Cem Mater:1–19. https://doi.org/10.1080/21650373.2025.2513664

9. Kim GM, Khalid HR, Kim HJ, Lee HK (2017) Alkali activated slag pastes with surface-modified blast furnace slag. Cem Concr Compos 76:39–47. https://doi.org/10.1016/j.cemconcomp.2016.11.009
10. Juenger MCG, Snellings R, Bernal SA (2019) Supplementary cementitious materials: new sources, characterization, and performance insights. Cem Concr Res 122:257–273. https://doi.org/10.1016/j.cemconres.2019.05.008
11. Hanif A, Parthasarathy P, Lu Z, Sun M, Li Z (2017) Fiber-reinforced cementitious composites incorporating glass cenospheres – mechanical properties and microstructure. Constr Build Mater 154:529–538. https://doi.org/10.1016/j.conbuildmat.2017.07.235
12. Thomas M (2007) Optimizing the use of Fly ash in concrete, www. CementOrg:24
13. Siddique R (2007) Wood Ash. https://doi.org/10.1007/978-3-540-74294-4_9
14. Lothenbach B, Scrivener K, Hooton RD (2011) Supplementary cementitious materials. Cem Concr Res 41:1244–1256. https://doi.org/10.1016/j.cemconres.2010.12.001
15. Snellings R, Suraneni P, Skibsted J (2023) Future and emerging supplementary cementitious materials. Cem Concr Res 171:107199. https://doi.org/10.1016/j.cemconres.2023.107199
16. Kero I, Naess MK, Tranell G (2015) Particle size distributions of particulate emissions from the ferroalloy industry evaluated by electrical low pressure impactor (ELPI). J Occup Environ Hyg 12:37–44
17. Al-Duais INA, Ahmad S, Al-Osta MA, Maslehuddin M, Saleh TA (2024) Properties of alkali-activated concrete made using the optimum combinations of precursor materials and activation parameters. J Mater Civ Eng 36. https://doi.org/10.1061/jmcee7.mteng-16944
18. Nasir M, Megat Johari MA, Maslehuddin M, Olalekan Yusuf M (2020) Magnesium sulfate resistance of alkali/slag activated silico-manganese fume-based composites. Constr Build Mater 265:120851. https://doi.org/10.1016/j.conbuildmat.2020.120851
19. Cannon WF (2014) Manganese—it turns iron into steel (and does so much more). USGS Miner Resour Progr:3087
20. Rashid S, Khalid HR, Hanif A (2025) Mechanical behavior and leaching characteristics of cementitious composites incorporating silicomanganese fume. Eur J Environ Civ Eng:1–17. https://doi.org/10.1080/19648189.2025.2494131

Development of Self-compacting Concrete Using Red Mud as a Supplementary Cementing Material

Mohamed Inayathulla[1], Shamsad Ahmad[1,2], Mohammed Al-Osta[1,2], and Asad Hanif[1,2](✉)

[1] Civil and Environmental Engineering Department, King Fahd University of Petroleum and Minerals, Dhahran 31261, Saudi Arabia
asad.hanif@kfupm.edu.sa

[2] Interdisciplinary Research Center for Construction and Building Materials, King Fahd University of Petroleum & Minerals, Dhahran 31261, Saudi Arabia

Abstract. This study investigates the use of red mud (RM), an extremely alkaline industrial byproduct of alumina, as a supplementary cementitious material (SCM) in self-compacting concrete (SCC). The tests analyze the partial cement replacement by red mud effects on SCC fresh and hardened characteristics at 0%, 10%, 20%, and 30%. Workability was tested based on slump flow, V-funnel, and U-box tests. Compressive strength was tested at various curing ages while split tensile strength, and modulus of elasticity were tested at 28 days. Results established that 10% replacement level offers the best overall performance, improving mechanical strength and stiffness due to the filler and pozzolanic effects of RM. However, lower workability and higher flow times at high RM content were seen, for which the superplasticizer dosage had to be raised. The findings validate that red mud can be effectively utilized in SCC mixes to enable green and sustainable construction and reduce cement-related environmental impacts, provided the replacement ratio is optimized.

Keywords: Self-compacting Concrete · Red Mud · Supplementary Cementitious Material · Sustainability · Industrial Waste Utilization · Fresh Properties · Hardened properties

1 Introduction

Self-compacting concrete flows by its own weight and fills the formwork completely without the application of the mechanical vibrators. This unique characteristic provides improved construction efficiency, especially in heavily reinforced complex structures, improved surface finish and reduced labor costs compared to conventional concrete [1]. However, this concrete needs more cement, fine aggregates, superplasticizers and viscosity modifying agents to achieve its excellent flowability, which can cause expenses and environmental problems due to the high carbon footprint associated with making

H. Beushausen et al. (Eds.): NTZR 2025, LNCE 873, pp. 78–88, 2026.
https://doi.org/10.1007/978-3-032-27698-8_8

cement and superplasticizer. Although this concrete offers several ways of reducing construction costs, as it does not require mechanical vibration, the labor and equipment cost is reduced. Rapid project completion leads to less overall cost. It also increases material efficiency through reduced cement consumption, especially with the use of industrial by-products. It can also improve durability and reduce long-term repair expenses. All these advantages put together make SCC an economic and sustainable construction material [2–6]. In the past decade, there has been an increasing focus on the use of industrial by-products and waste materials in concrete production, driven by the pursuit of enhanced sustainability and the reduction of environmental impacts.

To enhance sustainability, reduce costs and mitigate environmental issues, the concrete industry has targeted using industrial by-products and waste products in its production. Utilization of industrial co-products such as red mud as a partial replacement of cement in SCC. Red mud or bauxite residue is a highly alkaline by-product of alumina production from bauxite via the Bayer process [7]. On average, approximately 1.25 tons of red mud is generated per ton of alumina but can vary from 0.8 to 2.5 tons depending on the quality of bauxite. Global alumina production in 2023 produced about 177 million tons of red mud, adding to the over 4 billion tons already stored across the world. Despite research efforts, only 2–5% of the annually produced red mud is practically used, with the rest being stored [8]. The waste material poses major disposal problems, considering that it is produced in large quantities and requires huge land areas for storage. Red mud disposal can lead to environmental problems like groundwater contamination and soil contamination due to its high alkalinity and content of heavy metals. Thus, the utilization of red mud in construction material is a path towards sustainable waste management with reduced utilization of cement for concrete production [9, 10].

Red mud is primarily made up of aluminum, silica and iron oxides with minor quantities of oxide compounds of titanium and calcium [9–13]. The chemical contents imply that red mud has pozzolanic properties, indicating it can contribute to the strength increase of concrete in combination with cement. However, its high fineness and alkaline nature require careful attention to the level of replacement to attain the required workability and mechanical properties of SCC [14, 15]. Since SCC relies on its self-flowing nature to fill formwork without segregation, the incorporation of fine materials such as red mud can alter its flow and viscosity properties. Even though small quantities of red mud can enhance viscosity, excessive content can affect workability and need more superplasticizers for the desired flowability [16]. Maintaining a proper balance between sustainable cement replacement and the retention of Self-Compacting Concrete (SCC) workability remains to be a central focus of ongoing research in this domain.

Besides its engineering benefits, red mud use in SCC also contributes to environmental sustainability in that it reduces industrial waste and the carbon footprint of cement production. Cement production is among the largest sources of CO_2 emissions, contributing approximately 8% of the global greenhouse gas emissions [17, 18]. By replacing cement partially with red mud, clinker manufacturing can be avoided to a certain extent, thereby reducing the amount of energy consumed and emissions. Secondly, red mud usage in concrete-based products supports circular economic principles by transforming waste material into useful construction material [19].

A significant research gap on this topic is the use of Red Mud as a cement replacement in Self-Compacting Concrete, despite growing interest in sustainable construction. In spite of a few research studies on the mechanical properties of SCC using Red Mud, there is a noticeable lack of comprehensive research on the durability performance, rheological behavior modeling, and life cycle assessment (LCA) of SCC with Red Mud. The unique properties of Red Mud pose challenges as well as advantages. However, systematic investigation on its long-term stability in harsh conditions, quantitative rheological modeling to predict the flow behavior, and LCA to establish environmental feasibility is lacking. This gap limits the broader acceptance of Red Mud as an environmentally friendly alternative in concrete technology and highlights the need for serious studies in these important areas.

2 Materials and Methods

Cement type 1 compliant with ASTM standard was used in this study. Red mud was obtained from the Saudi Ma'aden company, and it was used in its original form without any kind of treatment, as illustrated in Fig. 1. The fine aggregate used was the locally available Dune sand with water absorption of 0.6% and specific gravity of 2.52. Coarse aggregate used was the locally available Dolomite with water absorption of 2.4% and specific gravity of 2.6. MasterGlenium ACE 507 M was used as a superplasticizer and Viscosity Modifier Agent was Sika Stabilizer-100 SA (Figs. 2, 3, and 4) (Tables 1 and 2).

Fig. 1. Ordinary Portland cement and red mud used in the study.

Table 1. Chemical composition (% by weight) of OPC versus red mud.

Chemical composition	CaO	Al_2O_3	SiO_2	SO_3	K_2O	TiO_2	Fe_2O_3	Na_2O	SrO
Cement	71.65	3.18	15.62	3.29	0.58	0.42	4.01	0.23	0.07
Red mud	7.18	24.05	16.81	1.68	0.07	7.79	21.93	7.73	0.35

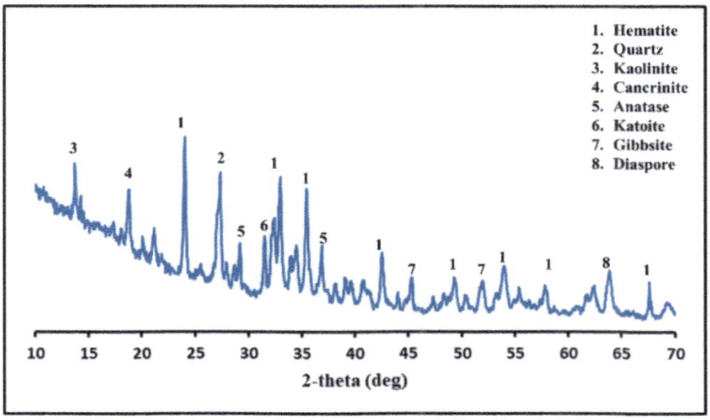

Fig. 2. X-Ray Diffractogram pattern of red mud.

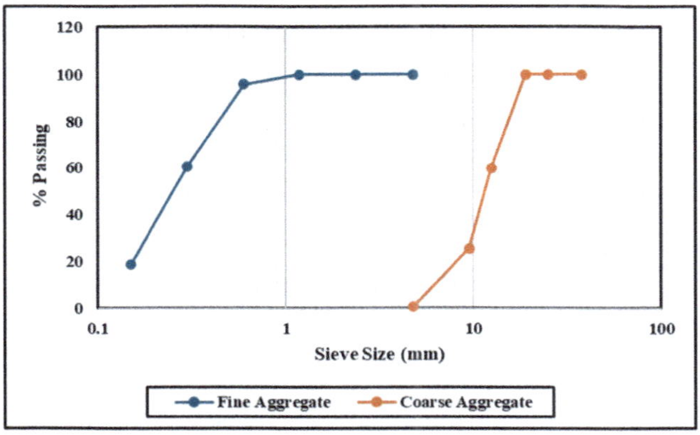

Fig. 3. Aggregate sieve analyses results.

Table 2. Mix proportion of SCC (kg/m^3).

Samples	Cement	Red mud	Dune sand (Fine aggregate)	Crushed gravel (Coarse aggregate)	Water	SP	VMA
Control	500	0	936.86	830.80	175.56	5.75	0.0
RM 10	450	50	924.19	819.57	175.21	14	6.5
RM 20	400	100	922.12	817.73	175.16	14.5	5.0
RM 30	350	150	920.04	815.89	175.10	15	3.75

Fig. 4. Cylindrical samples (75 × 150mm) prepared from a single mix batch.

3 Results and Discussions

3.1 Fresh Properties

The Slump flow table test was conducted according to ASTM C1611, to assess the flowability of the SCC mixes. The flow diameters of the four mixes ranged from 680 mm to 730 mm, as displayed in the Table 3. Fresh concrete characteristics with no indication of segregation or bleeding throughout the tests. All the mixes were classified under class SF2, which corresponds to a slump flow value between 660 mm and 750 mm. Higher RM content leads to reduced flowability that can be attributed to its fine particle size and high specific surface area, resulting in an increased requirement for water. Also, the existence of angular, porous and irregularly shaped particles in RM is responsible for high internal friction, thus decreasing the overall workability of the mix [12, 20–23]. Consequently, as the red mud content increases, the workability of SCC decreases, therefore, to meet the required flowability, we had to increase the superplasticizer dosages. The results agree with the earlier research conducted by Ghalehnovi et al. [24] and Tang et al. [25] that also documented the negative effect of RM on the fresh characteristics of concrete. Therefore, upon using RM in SCC, it is normally advisable to utilize higher dosages of super-plasticizers to satisfy the performance demands of SCC as outlined in earlier literature [23].

The V-funnel test of this study revealed the flow times as higher than the range of 6–12 s (VF1) specified by EFNARC [26] for the SCC. Referencing the Table 3 the control mix (RM0) took nearly 20 s and the RM 30 mix took 25 s. Therefore, all the mixes were classified under class VF2 (9–25 s). The mixes used for the test adhere to the criteria of acceptability based on the European standards for SCC [26]. It was apparent as a clear trend towards a radical decrease in outcomes when more quantity of RM was used, which increased the V-funnel time. Comparable findings have been reported by several researchers Ghalehnovi et al. [10] Ghalehnovi et al. [24]. According to Gesoğlu et al. [27], a finer particle size distribution results in a decrease in V-funnel test time compared to that of the control mix; however, Hunger et al. [28] demonstrated that RM incorporation tends to increase the V-funnel time due to its physical characteristics. In this study, the high fraction of fine aggregate (53%) and a low water-to-powder (w/p) ratio of 0.3 contributed to the prolonged flow times. This is in accordance with EFNARC [26], which states that the w/p ratio has a direct influence on the mix viscosity: decreasing w/p ratio increases viscosity, thereby increasing V-funnel time, also the excessive fine aggregate or too fine sand increase the paste viscosity, reducing flowability and passing ability, leading to higher V-funnel times [1, 26].

Table 3. Fresh concrete characteristics.

Samples	Flow table (mm)	V-funnel Time (sec)	U-box (mm)	Bleeding (Visual)	Segregation (Visual)	Visual stability index
Control	710	20	0	Negligible	None	1
RM 10	730	22	0	None	None	0
RM 20	705	23	2	None	None	0
RM 30	680	25	5	None	None	0

The U-box test, conducted according to EFNARC [26], gave height differences (h1–h2) ranging from 0 to 5 mm across the mixes, remaining well within the acceptable limit of 30 mm. The RM0 and RM10 mixes observed a height difference of 0 mm, while RM20 and RM30 mixes recorded values of 2 mm and 5 mm, respectively. The lower U-box values in the first three mixes reflect better workability and passing ability, which is supported by their high slump flow measurements (exceeding 700 mm). For RM30, a U-box value of 5 mm was recorded, corresponding to its comparatively lower slump flow (680 mm), yet still within acceptable performance limits.

3.2 Hardened Properties

The compressive strength of the SCC mixes, evaluated in accordance with ASTM C39/C39M-21, is presented in Fig. 5. The graph compares the performance of all the mixes, at 3, 7, and 28 days. In all the mixes, the values are increasing with time, depicting growth or improvement. With increasing RM content, compressive strength initially improves, reaching an optimum at a certain replacement level, after which a slight decline is observed. Among all mixes, RM10 exhibited the highest compressive strength at all ages, recording 42 MPa at 3 days, 44 MPa at 7 days, and a peak of 57 MPa at 28 days. The early-age strength (3 days) of the control mix (RM0) was comparable to RM20 and RM30, with RM20 and RM30 displaying nearly identical values.

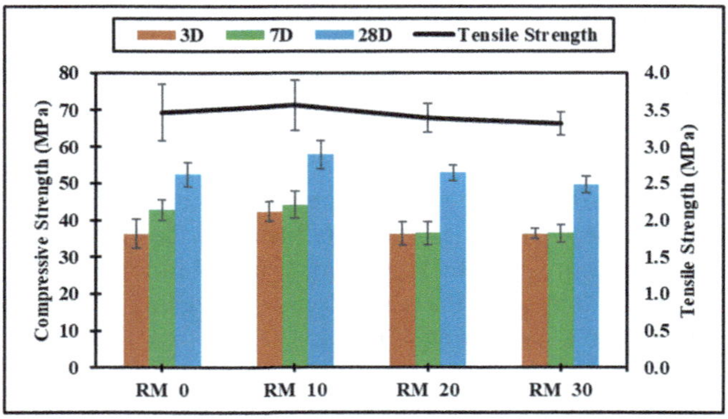

Fig. 5. Mechanical strength attributes.

At 7 days, the control mix outperformed both RM20 and RM30. At 28 days, the RM10 and RM20 mixes surpassed the control mix, showing increases of 10.42% and 0.67%, respectively. These findings align with previous studies reporting an enhancement in compressive strength with RM incorporation up to 10% replacement [13]. The development of cement mixes and cement incorporating RM was aided by the formation of C_4AF and C_3A [29, 30]. Accordingly, it can be concluded that red mud can contribute positively to the compressive strength development of SCC when used in moderate dosages [23, 31]. Additional research has also noted that limited RM content, typically up to 2–2.5%, can enhance compressive strength compared to control mixes [24, 32].

Split tensile strength testing was performed on cylindrical specimens at 28 days in accordance with ASTM C496/C496M–17. The RM10 mix achieved the highest tensile strength of 3.55 MPa, followed by the control mix at 3.5 MPa. The tensile strength of all mixes ranged from 3.3 to 3.55 MPa. The trend observed in the tensile strength results closely paralleled that of the compressive strength, as supported by Patangia et al. [33] and Putrevu et al. [34]. As the Red Mud addition increases up to 12.5% the split tensile strength of the specimens also increases, beyond which a decline in strength was observed. This improvement at moderate replacement levels is primarily attributed to the filler effect of RM, due to its finer particle size, which aids matrix densification, reduces microcrack formation, and consequently improves tensile strength [35, 36]. While the incorporation of RM did not significantly compromise the tensile strength of SCC at lower dosages [31], it is noteworthy that at higher replacement levels, a reduction in tensile performance was recorded. Specifically, Liu et al. [23] reported that the tensile strength of SCC containing RM was lower than that of the control mix at 28 days.

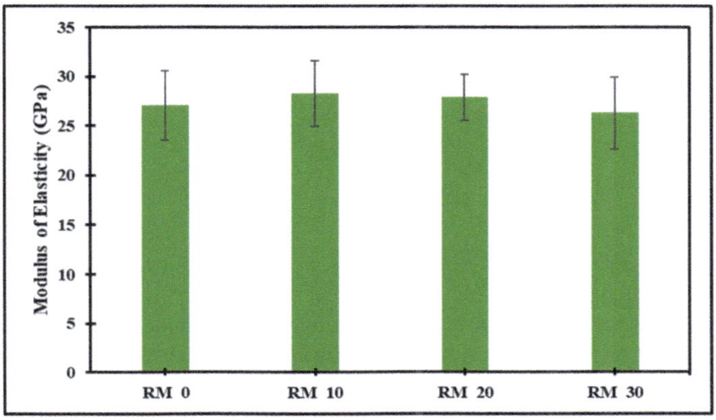

Fig. 6. Results of modulus of elasticity.

The Modulus of Elasticity (MoE) test results, conducted in line with ASTM C469/C469M–22, adhere to the same pattern as the compressive strength results, as evidenced by the strong correlation observed between the two sets of data. Figure 6 presents the MoE values for all SCC mixes at 28 days. The control mix, containing no red mud (RM), recorded an elastic modulus of approximately 26.9 GPa. Upon replacing 10% of the cement with RM, the MoE increased notably to 28.5 GPa, indicating enhanced material stiffness likely due to a denser microstructure facilitated by the filler effect of finely divided RM particles. However, if more than 10% of the cement is replaced, the MoE starts to decrease. At 20% replacement, the value drops slightly to around 27.9 GPa, and at 30% replacement, it drops even further to about 26.4 GPa. This decline happens because the amount of effective cement material is reduced or because the material becomes more porous. These findings suggest that 10% RM replacement yields the optimum improvement in elastic modulus, with 20% still being somewhat better than the control, whereas replacement beyond this affects the stiffness of the mix adversely. The observed behavior is in line with literature reports. For instance, Liu et al. [23] concluded that red mud does not have a negative impact on elastic modulus of the SCC mixtures, whereas Tang et al. [25] also did not observe any noticeable change in MoE with RM addition.

4 Conclusions

This study confirms that red mud is sustainable, effective, and performance-enhancing supplementary cementitious materials for use in self-compacting concrete. At an optimum replacement level of 10%, RM improves both compressive and tensile strength, as well as modulus of elasticity, owing to enhanced matrix densification and latent pozzolanic activity. Nevertheless, higher replacement levels are detrimental to the fresh characteristics due to elevated viscosity and water demand, therefore challenging the balance between flowability and mechanical performance. The work underscores the importance of proportion balance in the mixture as well as dosage of the admixtures to achieve desirable SCC characteristics. Furthermore, environmental benefits of RM utilization such as reducing CO_2 emissions and industrial recycling indicate its potential as a source of building infrastructure for sustainability. Future research is suggested to focus on long-term strength, environmental sustainability, and field-scale testing to enable widespread application in construction work.

References

1. Okamura H, Ouchi M (2003) Self-compacting concrete. J Adv Concr Technol 1:5–15. https://doi.org/10.3151/jact.1.5
2. Bouzoubaâ N, Lachemi M (2001) Self-compacting concrete incorporating high volumes of class F fly ash. Cem Concr Res 31:413–420. https://doi.org/10.1016/S0008-8846(00)00504-4
3. Arshad MT, Ahmad S, Khitab A, Hanif A (2021) Synergistic use of Fly ash and silica fume to produce high-strength self-compacting cementitious composites. Crystals (Basel) 11:915. https://doi.org/10.3390/cryst11080915
4. Khan MI, Usman M, Rizwan SA, Hanif A (2019) Self-consolidating lightweight concrete incorporating limestone powder and Fly ash as supplementary cementing material. Materials 12:3050. https://doi.org/10.3390/ma12183050
5. Kim Y, Hanif A, Usman M, Munir MJ, Kazmi SMS, Kim S (2018) Slag waste incorporation in high early strength concrete as cement replacement: environmental impact and influence on hydration & durability attributes. J Clean Prod 172:3056–3065. https://doi.org/10.1016/j.jclepro.2017.11.105
6. Safhi A e M, Rivard P, Yahia A, Benzerzour M, Khayat KH (2020) Valorization of dredged sediments in self-consolidating concrete: fresh, hardened, and microstructural properties. J Clean Prod 263:121472. https://doi.org/10.1016/j.jclepro.2020.121472
7. Varnavas S, Ferentinos G, Collins M (1986) Dispersion of bauxitic red mud in the Gulf of Corinth, Greece. Mar Geol 70:211–222. https://doi.org/10.1016/0025-3227(86)90003-4
8. Alumina News—General, Red mud generation trend across major countries, 2018 to 2023 (2024)
9. Khairul MA, Zanganeh J, Moghtaderi B (2019) The composition, recycling and utilisation of Bayer red mud. Resour Conserv Recycl 141:483–498. https://doi.org/10.1016/j.resconrec.2018.11.006
10. Ghalehnovi M, Asadi Shamsabadi E, Khodabakhshian A, Sourmeh F, de Brito J (2019) Self-compacting architectural concrete production using red mud. Constr Build Mater 226:418–427. https://doi.org/10.1016/j.conbuildmat.2019.07.248
11. Samal S (2021) Utilization of red mud as a source for metal ions—a review. Materials 14:2211. https://doi.org/10.3390/ma14092211
12. Sutar H, Mishra SC, Sahoo SK, Chakraverty AP, Maharana HS (2014) Progress of red mud utilization: an overview. Am Chem Soc J 4:255–279
13. Venkatesh C, Ruben N, Chand MSR (2020) Red mud as an additive in concrete: comprehensive characterization. J Korean Ceram Soc 57:281–289. https://doi.org/10.1007/s43207-020-00030-3
14. Shi C, Grattan-Bellew PE, Stegemann JA (1999) Conversion of a waste mud into a pozzolanic material. Constr Build Mater 13:279–284. https://doi.org/10.1016/S0950-0618(99)00026-4
15. Ribeiro DV, Labrincha JA, Morelli MR (2011) Potential use of natural red mud as pozzolan for Portland cement. Mater Res 14:60–66. https://doi.org/10.1590/S1516-14392011005000001
16. Wu C-R, Tang W, Zhan B-J, Kou S-C (2024) Feasibility study on using red mud as a viscosity-modifying agent for self-compacting concrete. Constr Build Mater 445:137871. https://doi.org/10.1016/j.conbuildmat.2024.137871

17. Scrivener KL, John VM, Gartner EM (2018) Eco-efficient cements: potential economically viable solutions for a low-CO2 cement-based materials industry. Cem Concr Res 114:2–26. https://doi.org/10.1016/j.cemconres.2018.03.015
18. Baloch H, Usman M, Rizwan SA, Hanif A (2019) Properties enhancement of super absorbent polymer (SAP) incorporated self-compacting cement pastes modified by nano silica (NS) addition. Constr Build Mater 203:18–26. https://doi.org/10.1016/j.conbuildmat.2019.01.096
19. The Portland cement association (PCA) (n.d.) Supplementary Cementing Materials
20. Tsakiridis PE, Agatzini-Leonardou S, Oustadakis P (2004) Red mud addition in the raw meal for the production of Portland cement clinker. J Hazard Mater 116:103–110. https://doi.org/10.1016/j.jhazmat.2004.08.002
21. Senff L, Modolo RCE, Silva AS, Ferreira VM, Hotza D, Labrincha JA (2014) Influence of red mud addition on rheological behavior and hardened properties of mortars. Constr Build Mater 65:84–91. https://doi.org/10.1016/j.conbuildmat.2014.04.104
22. Ahmadi B, Shekarchi M (2010) Use of natural zeolite as a supplementary cementitious material. Cem Concr Compos 32:134–141. https://doi.org/10.1016/J.CEMCONCOMP.2009.10.006
23. Liu RX, Poon CS (2016) Utilization of red mud derived from bauxite in self-compacting concrete. J Clean Prod 112:384–391. https://doi.org/10.1016/j.jclepro.2015.09.049
24. Ghalehnovi M, Roshan N, Hakak E, Shamsabadi EA, de Brito J (2019) Effect of red mud (bauxite residue) as cement replacement on the properties of self-compacting concrete incorporating various fillers. J Clean Prod 240:118213. https://doi.org/10.1016/j.jclepro.2019.118213
25. Tang WC, Wang Z, Liu Y, Cui HZ (2018) Influence of red mud on fresh and hardened properties of self-compacting concrete. Constr Build Mater 178:288–300. https://doi.org/10.1016/j.conbuildmat.2018.05.171
26. EFNARC (2005) The European guidelines for self-compacting concrete-specification, production and use. 63
27. Gesoğlu M, Güneyisi E, Kocabağ ME, Bayram V, Mermerdaş K (2012) Fresh and hardened characteristics of self compacting concretes made with combined use of marble powder, limestone filler, and fly ash. Constr Build Mater 37:160–170. https://doi.org/10.1016/j.conbuildmat.2012.07.092
28. Hunger M, Brouwers JH (2008) Natural Stone Waste Powders Applied to SCC Mix Design / Mehlkornfeine Abfallstoffe der Natursteinindustrie in der Projektierung von SVB. Restor Build Monum 14:131–140. https://doi.org/10.1515/rbm-2008-6206
29. Wang X, Luo Z, Zhang L, Rong H, Yang J (2016) Utilization of red mud as raw material in the production of field road cement. J Wuhan Univ Technol-Mater Sci Ed 31:877–882. https://doi.org/10.1007/s11595-016-1462-0
30. Singh M, Upadhayay SN, Prasad PM (1996) Preparation of special cements from red mud. Waste Manag 16:665–670. https://doi.org/10.1016/S0956-053X(97)00004-4
31. Tang WC, Wang Z, Donne SW, Forghani M, Liu Y (2019) Influence of red mud on mechanical and durability performance of self-compacting concrete. J Hazard Mater 379. https://doi.org/10.1016/j.jhazmat.2019.120802
32. Shetty KK, Nayak G, Shetty R (2003) Self compacting concrete using red mud and used foundry sand. Int J Res Eng Technol 3:708–711
33. Patangia J, Saravanan TJ, Kabeer KISA, Bisht K (2023) Study on the utilization of red mud (bauxite waste) as a supplementary cementitious material: pathway to attaining sustainable development goals. Constr Build Mater 375. https://doi.org/10.1016/j.conbuildmat.2023.131005
34. Putrevu M, Thiyagarajan JS, Pasla D, Kabeer KISA, Bisht K (2021) Valorization of red mud waste for cleaner production of construction materials. J Hazard Toxic Radioact Waste 25. https://doi.org/10.1061/(ASCE)HZ.2153-5515.0000629

35. Venkatesh C, Nerella R, Sri Rama Chand M (2020) Comparison of mechanical and durability properties of treated and untreated red mud concrete. Mater Today Proc 27:284–287. https://doi.org/10.1016/j.matpr.2019.11.026
36. Venkatesh C, Nerella R, Chand MSR (2020) Experimental investigation of strength, durability, and microstructure of red-mud concrete. J Korean Ceram Soc 57:167–174. https://doi.org/10.1007/s43207-019-00014-y

Mineralogical Assessment of Waste-Based SCMs in OPC Blends via XRD: Influence on Clinker Phases and Pre-hydration Phase Evolution

Kumar Shakti Srivastava[1](✉), Visalakshi Talakokula[1], Sri Kalyana Rama Jyosyula[1], and Mrittika Sengupta[2]

[1] Department of Civil Engineering, Mahindra University, Hyderabad, Telangana 500043, India
dramansrivastva2@gmail.com

[2] Department of Life Sciences, Mahindra University, Hyderabad, Telangana 500043, India

Abstract. This study investigates the mineralogical alterations induced by the partial replacement of Ordinary Portland Cement (OPC 53 grade) with a variety of waste-based supplementary cementitious materials (SCMs), using X-ray Diffraction (XRD) as the primary analytical tool. Eight SCMs, rice husk ash, silica fume, ground granulated blast furnace slag, fly ash, slime, coconut shell ash, microclay, and Concrete Enhancer (CE 10), were incorporated at 10% and 20% by weight to evaluate their impact on the unhydrated phase composition of cement. Quantitative XRD analysis using whole powder pattern fitting was employed to determine variations in the four major Clinker phases: alite (C_3S), belite (C_2S), tricalcium aluminate (C_3A), and ferrite (C_4AF). The results reveal a consistent reduction in alite content and a corresponding increase in belite across most SCM blends, indicating a potential shift in hydration behavior from early to long-term strength development. C_3A and ferrite responses varied with SCM type, with pozzolanic materials like silica fume and rice husk ash showing more favorable mineralogical transformations than inert or low-reactivity alternatives. At 10% replacement, SCMs such as silica fume, rice husk ash, and GGBFS maintained a beneficial balance between clinker retention and pozzolanic enhancement. At 20%, however, excessive dilution of clinker phases was observed in several systems, necessitating careful dosage control. Overall, the study highlights the viability of waste-derived SCMs for sustainable cement design, provided their mineralogical impact is fully understood and optimized. It is noted that the observed changes in clinker phase percentages are solely due to dilution effects, peak overlaps, and the inherent mineralogy of the SCMs, and do not indicate chemical transformation of OPC clinker minerals in the dry state.

Keywords: X-ray Diffraction (XRD) · Clinker Compounds · Supplementary Cementitious Materials (SCMs) · Cement Mineralogy · Pre-hydration Analysis

1 Introduction

The global push for sustainable construction has brought increasing attention to the environmental burdens of Ordinary Portland Cement (OPC), whose manufacture accounts for approximately 7–8% of total anthropogenic CO_2 emissions. In this context, the partial

H. Beushausen et al. (Eds.): NTZR 2025, LNCE 873, pp. 89–100, 2026.
https://doi.org/10.1007/978-3-032-27698-8_9

replacement of OPC with supplementary cementitious materials (SCMs), particularly those derived from industrial or agricultural waste, has become a central strategy to reduce the carbon footprint of concrete. However, while the mechanical and durability properties of blended systems are commonly studied in hydrated states, the mineralogical implications of SCM incorporation at the unhydrated stage are still not well understood. This gap is critical, as the mineral phases present prior to hydration govern the early chemical environment, influence the kinetics of hydration reactions, and set the stage for the development of microstructure and long-term performance.

X-ray Diffraction (XRD) offers a non-destructive and highly accurate means to quantify the crystalline phases in cementitious systems, especially the Bogue compounds: alite (C_3S), belite (C_2S), tricalcium aluminate (C_3A), and ferrite (C_4AF). Each of these phases plays a distinct role in cement hydration. Alite hydrates rapidly and is the primary contributor to early strength. Belite, although slower to react, plays a dominant role in long-term strength development [1]. C_3A influences setting time and early heat evolution, while C_4AF has a minor role in strength but contributes to sulfate resistance and heat reduction. When SCMs are introduced, they may not only dilute these primary phases by replacing cement mass but also introduce new mineral phases or promote secondary hydration products, such as calcium silicate hydrates (C–S–H) and calcium aluminate hydrates (C-A-H). Therefore, quantifying the mineralogical impact of SCMs in the unhydrated state is essential to predict both short-term and long-term behavior.

This study investigates eight diverse SCMs: rice husk ash (RHA), silica fume (SF), ground granulated blast furnace slag (GGBFS), fly ash (FA), microclay (MC), coconut shell ash (CSA), slime (fine mining waste), and CE (a commercial concrete enhancer marketed under the Spartekoncrete brand). These materials were selected for their relevance to both sustainability and performance in cementitious systems. RHA and SF are rich in amorphous silica and are known for their high pozzolanic reactivity, which allows them to react with portlandite and form additional C–S–H, thereby refining the pore structure and improving long-term durability. GGBFS is a latent hydraulic binder that activates in the alkaline environment of cement and contributes to dense microstructural development. FA, a well-established SCM, reacts slowly but provides long-term strength and enhanced workability. MC and CSA are relatively underexplored materials; MC, due to its ultrafine particle size, may influence nucleation and particle packing, whereas CSA, with variable chemical composition, may introduce phases such as Fe_2O_3, SiO_2, and CaO. Slime, while largely inert, is widely available as a mining by-product and has potential filler benefits if appropriately dosed [2].

In addition to these, CE 10 is introduced as a next-generation performance-based additive. Marketed as Spartekoncrete Enhancer, it is a mineral-based powder designed to improve concrete setting time, early strength, and workability across a range of applications, including ready-mix concrete, self-compacting concrete, and precast elements. However, while CE 10 is used in practice for its perceived functional benefits, there exists little to no data regarding its effect on cement phase composition in the unhydrated state. Given its fine particle structure and chemical complexity, CE 10 may alter the hydration environment even if it does not directly participate in crystalline phase formation.

The materials were blended with OPC at 10% and 20% by weight, and their influence on the mineralogical structure was analyzed using XRD combined with whole powder pattern fitting (WPPF) via Rietveld refinement. This approach provides precise quantification of each Bogue phase in the unhydrated blends, allowing for assessment of dilution effects, potential phase substitutions, and SCM-induced shifts in phase distribution. Importantly, this study does not aim merely to describe changes but to connect these mineralogical shifts to their likely implications on hydration kinetics, strength development pathways, and long-term durability, thereby creating a predictive framework for SCM selection based on mineral phase behavior.

Through this mineralogical lens, the research aims to provide early diagnostic insight into the suitability of different SCMs for sustainable and high-performance cement design. By establishing the extent to which each material influences the crystalline structure of unhydrated OPC, the study contributes foundational knowledge that supports more informed, scientifically optimized, and performance-driven use of alternative binders in modern concrete technology.

2 Materials and Methods

This study focuses on the mineralogical analysis of raw, unhydrated cement and its blends with various supplementary cementitious materials (SCMs), using X-ray Diffraction (XRD). Ordinary Portland Cement (OPC, 53 grade) was selected as the reference binder. Eight different SCMs were utilized for partial replacement of OPC, primarily consisting of industrial and agro-waste by-products known for their pozzolanic or filler properties. These include slime (a fine-grained mine waste), coconut shell ash, micro clay, fly ash, ground granulated blast furnace slag, silica fume, rice husk ash, and CE 10 (a commercially available concrete enhancer composed of fiber-reinforced admixtures). Each SCM was used to replace OPC at 10% and 20% by weight to form various blended cement mixes. All SCMs were sourced from regions in and around Hyderabad, Telangana, India, ensuring relevance to local material availability and promoting sustainable construction using regional waste streams.

The SCMs were dried at 105 °C for 24 h, ground if needed, and sieved through a 75 μm sieve to achieve a uniform particle size distribution. Dry blending was performed using a laboratory-scale mechanical mixer to ensure homogeneous mixing with OPC. No water or hydration activator was introduced, ensuring that the XRD analysis captured the unhydrated phase characteristics of each mix.

XRD analysis was conducted using a powder X-ray diffractometer equipped with Cu-Kα radiation ($\lambda = 1.5406$ Å), operated at 40 kV and 30 mA. Scanning was carried out in the range of 3°–100° 2θ with a step size of 0.02° and a scan rate of 1°/min. The resulting diffraction patterns were analyzed through Rietveld refinement using Whole Powder Pattern Fitting (WPPF), which enabled the quantitative estimation of Bogue's compounds, Tricalcium Silicate (C_3S), Dicalcium Silicate (C_2S), Tricalcium Aluminate (C_3A), and Tetracalcium Aluminoferrite (C_4AF).

For each blend, the percentage content of these four major cementitious phases was extracted and compared to that of OPC. This allowed a comparative understanding of how each SCM influenced the unhydrated mineralogical framework, offering early insight into potential hydration behavior, strength development, and durability trends. By eliminating the effect of water and time, the study isolates the intrinsic phase alterations caused solely by SCM addition.

3 Results and Discussion

3.1 XRD Analysis of SCM-Blended Cement (10% and 20% Replacement)

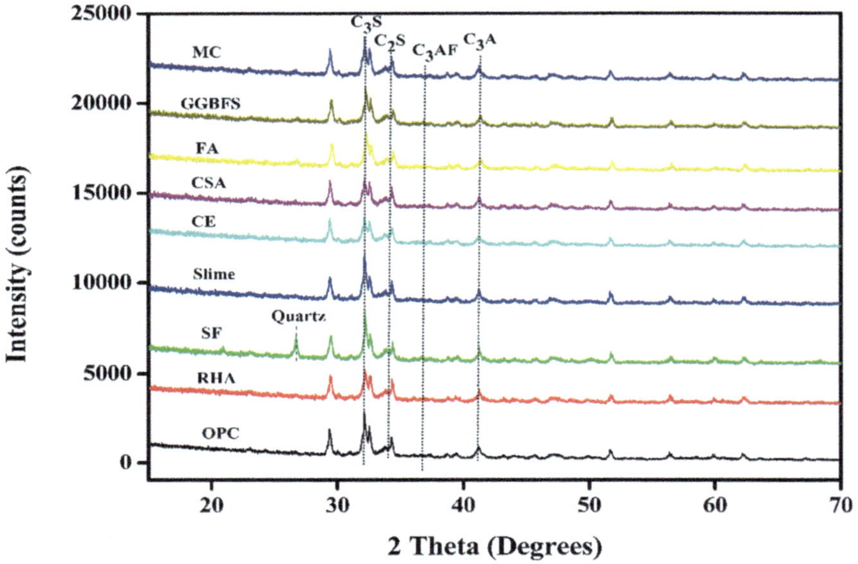

Fig. 1. X-ray diffraction (XRD) patterns of OPC blended with 10% replacement of SCMs.

Figures 1 and 2 present the X-ray diffraction (XRD) patterns for OPC blended with 10% and 20% of various SCMs, namely RHA, SF, GGBFS, slime, CE10, CSA, FA, and MC. Across all samples, a clear reduction is observed in the intensity of the major clinker peaks corresponding to C_3S, C_2S, C_3A, and C_4AF, with more pronounced suppression at 20% replacement. The degree and nature of these changes vary with the chemical composition, reactivity, and particle size of the SCMs used.

SCMs like SF and RHA lead to a significant reduction in C_3S and a visible enhancement in C_2S, especially at 20% replacement. This shift is indicative of both clinker dilution and pozzolanic activity. The sharp quartz peaks at $2\theta \approx 26.6°$ in SF and RHA suggest the presence of unreacted crystalline silica, but this is accompanied by a large quantity of amorphous silica capable of reacting with portlandite during hydration [3, 4]. The mineralogical transformation in these systems supports a hydration regime dominated by secondary C–S–H formation, leading to densification of the microstructure and improved long-term strength, albeit potentially at the cost of slower early-age reactions due to the depletion of C_3S [5, 6].

GGBFS also shows a reduction in C_3S, though less drastic than in SF or RHA. As a latent hydraulic material, GGBFS contributes to C–S–H formation upon activation by calcium hydroxide and maintains phase stability while extending long-term hydration. FA exhibits moderate dilution of C_3S and an increase in C_2S, consistent with its

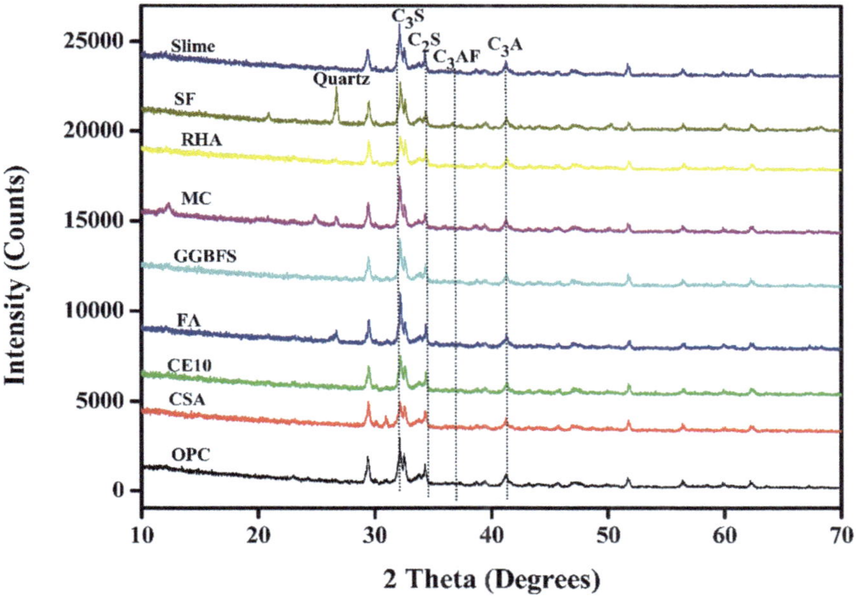

Fig. 2. X-ray diffraction (XRD) patterns of OPC blended with 20% replacement of the SCMs.

slower pozzolanic reaction. Though not highly reactive in early stages, FA contributes to long-term strength gain and improved workability [7]. In contrast, MC shows a drastic suppression of C_3S and a substantial increase in C_2S peaks. This is attributed to its ultrafine nature, which enhances particle packing and filler effects. Although MC lacks strong pozzolanic activity, its high surface area can provide nucleation sites, indirectly supporting the hydration of other cement phases. However, excessive use may reduce early strength due to the extreme dilution of alite. Slime, being largely inert, produces only marginal changes in the XRD pattern. Slight reductions in C_3S and subtle shifts in C_4AF suggest filler-type behavior, where the material contributes more to packing density than chemical reactivity. While slime does not support significant pozzolanic or hydraulic reactions, it may still assist in improving early-age compaction and reducing water demand [8, 9].

CSA introduces moderate reductions in C_3S and noticeable increases in C_4AF. The latter reflects the iron-rich nature of CSA, which promotes the formation of ferrite phases [5, 10]. This transformation may impact hydration temperature and sulfate resistance, depending on curing conditions and mix design. Meanwhile, CE10 demonstrates a broad suppression of C_3S and an elevated amorphous background in the diffractogram, indicating a largely non-crystalline contribution [11, 12]. Designed for early strength and workability, CE10 likely modifies hydration indirectly through densification or admixture interactions. However, its high dilution of clinker phases without compensating crystalline reactivity makes its use at higher dosages questionable from a mineralogical standpoint [13, 14]. Table 1 contains the details of the phase transformation in XRD analysis using SCMs.

Table 1. Expected trends in major phase transformations based on XRD analysis.

SCM	C_3S ↓	C_2S ↑	C_3A ↓	C_4AF ↓/↑
RHA	Significant ↓	Moderate ↑	Moderate ↓	Slight ↓
SF	Major ↓	High ↑	Moderate ↓	Moderate ↓
GGBFS	Moderate ↓	High ↑	Slight ↓	Slight ↓
Slime	Slight ↓	Slight ↑	Slight ↓	Minimal ↓
CE10	Moderate ↓	Moderate ↑	Moderate ↓	Moderate ↓
CSA	Moderate ↓	Moderate ↑	High ↓	Moderate ↑
FA	Slight ↓	Moderate ↑	Slight ↓	Slight ↓
MC	Major ↓	High ↑	Major ↓	Moderate ↓

These phase changes have direct implications for hydration chemistry (based on the quantified data). The reduction in C_3S reduces the early calcium silicate hydrate (C–S–H) output and may delay setting time unless compensated by admixtures or pozzolanic reaction. An increase in C_2S suggests enhanced long-term strength development, especially in SF, RHA, and MC-containing mixes [15, 16]. C_3A reductions may improve sulfate resistance but slow setting; conversely, elevated C_3A in CSA or MC can lead to rapid hydration and flash set risks if not controlled. Increases in C_4AF, as seen in CSA and CE10, alter color and may marginally influence sulfate resistance and heat evolution [17, 18].

These results collectively indicate that highly reactive pozzolanic SCMs such as SF, RHA, and GGBFS induce favorable mineralogical shifts by enhancing C_2S content while preserving or moderately reducing C_3S, supporting long-term hydration potential [19]. Inert fillers like slime and CSA offer minimal chemical benefit and are best used at low dosages to avoid excessive clinker dilution [20, 21]. CE10, though designed for performance enhancement, shows excessive suppression of clinker phases and introduces largely amorphous content, suggesting that high-level replacement with CE10 is not recommended unless supported by other activators or hybrid blends [21]. MC, despite being relatively inert chemically, significantly alters the phase distribution by promoting belite-rich compositions and heavily diluting C_3S. Its use requires careful dosage control, as it may delay early strength while offering benefits related to particle packing and long-term densification through slow secondary reactions [22].

3.2 Comparative Analysis of Bogue Compounds with OPC

Figures 3 and 4 present the percentage composition of the four major Bogue phases, C_3S, C_2S, C_3A, and C_4AF, in the SCM-blended systems relative to the reference OPC, for 10% and 20% replacement levels. This quantitative comparison, derived through whole powder pattern fitting (WPPF) from XRD data, provides a deeper understanding of how each SCM modifies the mineral framework of cement prior to hydration.

At the 10% replacement level (Fig. 3), most SCMs exhibit a moderate reduction in C_3S content, consistent with partial dilution of the clinker component. SF, RHA,

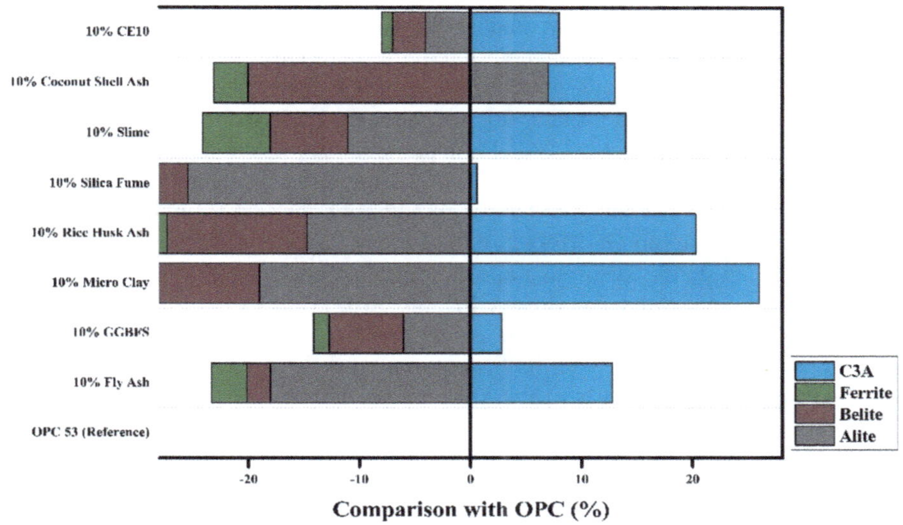

Fig. 3. Comparison of Bogue compound percentages (Alite, Belite, Ferrite, and C_3A) in various SCM-blended cements with 10% replacement relative to OPC 53.

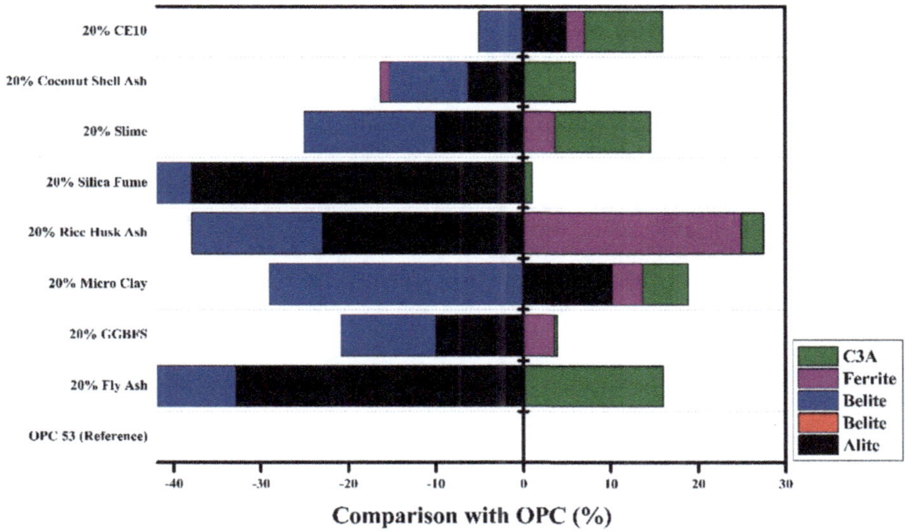

Fig. 4. Comparison of Bogue compound percentages in SCM-blended cements with 20% replacement relative to OPC 53.

and MC stand out with the largest observed C_3S decreases, exceeding 15% in some cases, indicating significant interaction with the alite-rich phase. This loss is offset by an increase in C_2S, especially in MC and SF, suggesting a mineralogical shift that supports belitic hydration pathways [23, 24]. The phase transition confirms the role of these materials in promoting delayed but stable strength development, reinforcing their suitability in performance mixes where long-term durability is prioritized over rapid early strength [25].

GGBFS and FA, though also causing reductions in C_3S, maintain a relatively balanced mineral phase distribution. GGBFS in particular shows a modest belite enhancement with minimal impact on C_3A and C_4AF, suggesting it maintains mineralogical stability even while contributing to long-term hydration through latent hydraulic activity. This behavior positions GGBFS as a reliable SCM capable of supplementing cement chemistry without sharply disrupting clinker phase integrity [26, 27].

Slime and CSA exhibit marginal reductions in C_3S and slight increases in C_2S, pointing to a predominantly inert role. Notably, CSA shows a localized increase in C_4AF content, which aligns with its higher iron oxide composition [19]. While this may contribute to slight changes in hydration temperature or color tone, it offers limited benefit in terms of reactive hydration phases. Slime, showing the least deviation across all four Bogue compounds, underscores its role as a passive filler rather than an active participant in cement chemistry. Its inclusion in the blend can aid particle packing and improve rheology, but does not appear to alter hydration chemistry significantly at this dosage [28].

CE10 shows a moderate reduction in C_3S and a corresponding increase in C_2S and amorphous background, with modest reductions in both C_3A and C_4AF. This distribution suggests that CE10 interacts with the cement matrix mainly through physical modification, altering dispersion and particle dynamics, rather than direct mineral substitution. While this may assist in improving setting time and early strength under controlled dosages, the mineralogical evidence implies that its use beyond a certain threshold may impair overall phase balance, making it less effective as a standalone replacement at high percentages [27].

At 20% replacement (Fig. 4), the mineralogical distinctions between SCMs become more pronounced. SF and MC register the highest C_3S reductions, often exceeding 30%, and the most significant C_2S increases, reflecting major shifts in hydration potential. While SF retains reactivity through its high amorphous silica content, MC's performance hinges more on physical densification and secondary gel formation at later ages. RHA maintains its belitic enhancement while minimizing C_3A and C_4AF disruption, reinforcing its potential to support sustainable, durable cement systems with optimized clinker usage.

GGBFS continues to perform favorably even at 20% replacement, exhibiting a restrained loss in C_3S and a proportionate increase in C_2S, without significant deviations in alumina- or ferrite-bearing phases. This stable behavior at higher dosage underscores its viability as a high-volume SCM, particularly for sulfate-resistant or mass concrete applications. FA, though less reactive in early stages, mirrors this trend and preserves a balanced phase structure while enhancing long-term hydration kinetics.

In contrast, CSA and CE 10 show sharper declines in C_3S with limited gains in reactive silicates, emphasizing the risk of excessive dilution without corresponding hydration benefit. CSA shows increased C_4AF content, likely from its ferric oxide-rich matrix, but minimal contribution to belitic or aluminous phases. CE 10 continues to broaden the amorphous profile with no substantial crystalline gain, reaffirming that its primary role remains performance enhancement rather than mineralogical transformation. Its use at a 20% dosage appears to overly suppress clinker reactivity and must be carefully moderated to avoid compromising structural performance.

Slime maintains minimal deviation across all clinker phases at 20%, reiterating its limited influence on cement chemistry. While its inclusion may enhance workability or early compaction through filler action, it does not support significant phase evolution, suggesting it is best suited as a minor addition in ternary blends.

Finally, C_3A behavior varies across SCMs. MC and FA exhibit elevated C_3A content at both replacement levels, which could enhance early setting but may also pose durability challenges in sulfate-rich environments. In contrast, SF and GGBFS suppress C_3A modestly, contributing to a more stable matrix in chemically aggressive exposures. These trends reinforce the importance of SCM selection not just on strength grounds but also with regard to chemical resistance and hydration heat management [20, 29].

Taken together, the quantitative XRD analysis supports several material-specific insights: SF, RHA, and GGBFS emerge as the most promising SCMs, demonstrating favorable phase shifts that preserve or enhance hydration potential while reducing reliance on high clinker content. MC and CE10 show aggressive C_3S depletion, indicating a need for dosage optimization or hybridization to avoid strength loss [30]. FA and GGBFS maintain phase integrity even at 20%, validating their role in durable blended systems. Conversely, slime and CSA, while physically beneficial, contribute minimally to hydration chemistry, underscoring the importance of pairing them with reactive SCMs in multi-blend systems [2]. These mineralogical findings enable performance prediction not through empirical strength tests alone, but from a foundational understanding of phase evolution, providing a strategic tool for mix design in sustainable cement systems.

4 Conclusions

This study evaluated the mineralogical impact of various SCMs, SF, RHA, GGBFS, FA, MC, CSA, slime, and CE 10, on OPC using XRD and WPPF analysis at 10% and 20% replacement levels. The unhydrated phase composition was examined to understand changes in C_3S, C_2S, C_3A, and C_4AF, providing insight into the early structural effects of SCM incorporation.

SF and RHA showed significant C_3S reduction and C_2S enhancement, promoting belitic profiles favorable for long-term strength. GGBFS and FA maintained balanced phase distributions while contributing to durable hydration pathways. MC caused aggressive C_3S depletion and C_2S elevation, suggesting strong packing and nucleation effects but posing risks to early strength if overdosed. CSA and slime exhibited minimal influence on phase composition, functioning mainly as fillers. CE 10 led to overall crystalline phase suppression and increased amorphous content, indicating limited mineralogical contribution at higher dosages.

These findings enable SCM selection based on phase behavior in the unhydrated state, supporting optimized mix designs with reduced clinker demand and improved durability potential. The data are particularly useful for mix designers, researchers, and RMC producers aiming for performance-based, low-carbon cement systems. The study is limited to unhydrated XRD analysis. Hydration behavior, mechanical performance, and long-term durability were not evaluated. Amorphous phase transitions and secondary products like C–S–H and AFt remain unquantified. Future work should include calorimetry, strength testing, SEM-EDS, and durability studies, alongside Life-Cycle Assessment (LCA) and CO_2 impact evaluations, to build on this phase-level foundation and link mineralogical trends to functional performance, as well as integrate these insights into mix design optimization frameworks for a low-carbon cement system.

References

1. Khan MA, Khan B, Shahzada K, Khan SW, Wahab N, Ahmad MI (2020) Conversion of waste marble powder into a binding material. Civ Eng J (Iran) 6. https://doi.org/10.28991/cej-2020-03091481
2. Mokal MP, Mandal R, Nayak S, Panda SK (2023) Efficacy of high-volume fly ash and slag on the physicomechanical, durability, and analytical characteristics of high-strength mass concrete. J Build Eng 76. https://doi.org/10.1016/j.jobe.2023.107295
3. Khan MSH, Kayali O, Troitzsch U (2016) Chloride binding capacity of hydrotalcite and the competition with carbonates in ground granulated blast furnace slag concrete. Mater Struct 49. https://doi.org/10.1617/s11527-016-0810-z
4. Poojalakshmi ES, Patel J, Sunantha B, Thomas BS, Ramaswamy KP, Ahmad Khan R (2023) Effect of mechanical activation on the properties of rice husk ash-based one part geopolymer. Mater Today Proc. https://doi.org/10.1016/j.matpr.2023.04.306
5. Karim MR, Zain MFM, Jamil M, Lai FC (2015) Development of a zero-cement binder using slag, fly ash, and rice husk ash with chemical activator. Adv Mater Sci Eng 2015. https://doi.org/10.1155/2015/247065
6. Tambara Júnior LUD, Cheriaf M, Rocha JC (2018) Development of alkaline-activated self-leveling hybrid mortar ash-based composites. Materials 11. https://doi.org/10.3390/ma11101829
7. Singh A, Mehta PK, Kumar R (2022) Performance of binary admixtures (fly ash and silica fume) on self compacting concrete. Mater Today Proc. https://doi.org/10.1016/j.matpr.2022.02.489
8. Singh RP, Vanapalli KR, Jadda K, Mohanty B (2024) Durability assessment of fly ash, GGBS, and silica fume based geopolymer concrete with recycled aggregates against acid and sulfate attack. J Build Eng 82. https://doi.org/10.1016/j.jobe.2023.108354
9. Khan MI, Sutanto MH, Napiah MB, Khan K, Rafiq W (2021) Design optimization and statistical modeling of cementitious grout containing irradiated plastic waste and silica fume using response surface methodology. Constr Build Mater 271. https://doi.org/10.1016/j.conbuildmat.2020.121504
10. Abood Habeeb G, Bin Mahmud H (2010) Study on properties of Rice husk ash and its use as cement replacement material. Mater Res 13:185–190
11. Ahmad A, Sutanto MH, Ahmad NR, Mohamad ME, Bujang M (2023) Microstructural characterization of fibric peat stabilized with Portland cement and silica fume. Materials 16. https://doi.org/10.3390/ma16010018
12. El-Gamal SMA, Amin MS, Ramadan M (2017) Hydration characteristics and compressive strength of hardened cement pastes containing nano-metakaolin. HBRC J 13. https://doi.org/10.1016/j.hbrcj.2014.11.008
13. Plando FRP, Gili MBZ, Maquiling JT (2023) Microstructural characterizations and radiation shielding quantities of rice husk ash-based self-compacting concrete and its precursors. Radiat Phys Chem 208. https://doi.org/10.1016/j.radphyschem.2023.110916

14. Saloni P, Yan Lim Y, Pham TM (2021) Influence of Portland cement on performance of fine rice husk ash geopolymer concrete: strength and permeability properties. Constr Build Mater 300. https://doi.org/10.1016/j.conbuildmat.2021.124321
15. Hossain SKS, Roy PK (2019) Fabrication of sustainable ceramic board using solid-wastes for construction purpose. Constr Build Mater 222. https://doi.org/10.1016/j.conbuildmat.2019.06.126
16. Habeeb GA, Mahmud HB (2010) Study on properties of rice husk ash and its use as cement replacement material. Mater Res 13. https://doi.org/10.1590/S1516-14392010000200011
17. Younes MM, Abdel-Rahman HA, Khattab MM (2018) Utilization of rice husk ash and waste glass in the production of ternary blended cement mortar composites. J Build Eng 20. https://doi.org/10.1016/j.jobe.2018.07.001
18. Amani A, Ramezanianpour AM, Palassi M (2021) Investigation on the sustainable use of electric arc furnace slag aggregates in eco-friendly alkali-activated low fineness slag concrete as a green construction composite. J Clean Prod 307. https://doi.org/10.1016/j.jclepro.2021.127257
19. Majhi RK, Nayak AN (2019) Bond, durability and microstructural characteristics of ground granulated blast furnace slag based recycled aggregate concrete. Constr Build Mater 212. https://doi.org/10.1016/j.conbuildmat.2019.04.017
20. Navarrete I, Vargas F, Martinez P, Paul A, Lopez M (2021) Flue gas desulfurization (FGD) fly ash as a sustainable, safe alternative for cement-based materials. J Clean Prod 283. https://doi.org/10.1016/j.jclepro.2020.124646
21. Jee AA, Pradhan B (2019) Study on development of empirical relationships between durability parameters of concrete made with different types of binder and exposed to chloride environment. Constr Build Mater 212. https://doi.org/10.1016/j.conbuildmat.2019.04.048
22. Abo-El-Enein SA, El-kady G, El-Sokkary TM, Gharieb M (2015) Physico-mechanical properties of composite cement pastes containing silica fume and fly ash. HBRC J 11. https://doi.org/10.1016/j.hbrcj.2014.02.003
23. Dittrich S, Neubauer J, Goetz-Neunhoeffer F (2014) The influence of fly ash on the hydration of OPC within the first 44 h—a quantitative in situ XRD and heat flow calorimetry study. Cem Concr Res 56. https://doi.org/10.1016/j.cemconres.2013.11.013
24. Mehta A, Siddique R (2017) Properties of low-calcium fly ash based geopolymer concrete incorporating OPC as partial replacement of fly ash. Constr Build Mater 150. https://doi.org/10.1016/j.conbuildmat.2017.06.067
25. Song H, Jeong Y, Bae S, Jun Y, Yoon S, Eun Oh J (2018) A study of thermal decomposition of phases in cementitious systems using HT-XRD and TG. Constr Build Mater 169. https://doi.org/10.1016/j.conbuildmat.2018.03.001
26. Laforest G, Duchesne J (2005) Immobilization of chromium (VI) evaluated by binding isotherms for ground granulated blast furnace slag and ordinary Portland cement. Cem Concr Res 35. https://doi.org/10.1016/j.cemconres.2004.12.011
27. Khan MSH, Kayali O, Troitzsch U (2017) Effect of NaOH activation on sulphate resistance of GGBFS and binary blend pastes. Cem Concr Compos 81. https://doi.org/10.1016/j.cemconcomp.2017.04.007
28. Srinivasa AS, Swaminathan K, Yaragal SC (2023) Effect of slag and solid activator on flowability and compressive strength of fly ash based one-part geopolymer pastes. Mater Today Proc. https://doi.org/10.1016/j.matpr.2023.03.481
29. Alexander AE, Shashikala AP (2022) Studies on the microstructure and durability characteristics of ambient cured FA-GGBS based geopolymer mortar. Constr Build Mater 347. https://doi.org/10.1016/j.conbuildmat.2022.128538
30. Sithole NT, Tsotetsi NT, Mashifana T, Sillanpää M (2022) Alternative cleaner production of sustainable concrete from waste foundry sand and slag. J Clean Prod 336. https://doi.org/10.1016/j.jclepro.2022.130399

Effect of Calcium Sulfate Adjustment on Steel Slag Composite Cements

Christopher Hoffmann(✉), Christiane Rößler, Thomas Sowoidnich, and Horst Michael Ludwig

Bauhaus-Universität Weimar, Weimar, Thuringia 99423, Germany
christopher.hoffmann@uni-weimar.de

Abstract. Annually, approximately 200 million tons of steel slag are generated globally, predominantly used in road construction or relegated to landfills. Given its composition, which includes belite—similar to those in Portland cement—steel slag might offer a potential as supplementary cementitious material (SCM). Particularly in China, the largest producer of steel slags, there is a strategic shift towards using recycled industrial by-products as SCM in concrete production, minimizing the extraction of natural resources. However, outside China the adoption of steel slags as SCM is limited due to various challenges. This study explores the hydration behavior of steel slag composite cements, with a special focus on sulfate optimization to address challenges. Previous research indicates that variations in the hydration processes of these composite cements possibly due to various calcium aluminates, known to affect early hydration stages. Therefore, the optimization of calcium sulfate levels in the cement is applied. Employing techniques such as in-situ X-ray diffraction (XRD) analysis and scanning electron microscopy (SEM), this research elucidates the hydration dynamics and evaluates the impact of calcium sulfate across several different levels. The study shows a retarding effect of Basic Oxygen Furnace (BOF) slag on Portland cement reaction, additionally the SCM affect the aluminate reaction, this can be accompanied by an additional setting regulator.

Keywords: Steel Slag · BOF-Slag · EAF-Slag · Sulfate · In-situ XRD

1 Introduction

Annually, about 200 million tonnes of steel slag are generated globally as by-products of steel production. The two predominant slag types are Basic Oxygen Furnace (BOF) slag, derived from converting pig iron into crude steel, and Electric Arc Furnace (EAF) slag, primarily produced from recycling steel scrap. These slags mainly comprise CaO and MgO, along with Al_2O_3, MnO, P_2O_5, and iron oxides in various oxidation states. Key mineral phases typically include belite (α'_H and β modifications stabilized by impurities such as phosphorus or iron [1]), iron oxides, and calcium aluminates (e.g., C_4AF). BOF slags usually contain higher lime contents, possibly with alite and free lime, whereas EAF slags have higher iron oxide levels and may contain heavy metals due to scrap variability [2].

H. Beushausen et al. (Eds.): NTZR 2025, LNCE 873, pp. 101–113, 2026.
https://doi.org/10.1007/978-3-032-27698-8_10

Currently, steel slags are predominantly used in road construction, concrete aggregates, or landfilled, though their use in concrete is restricted by expansion concerns from free lime and periclase, as well as heavy metal contamination [3]. Although research has long investigated their potential as supplementary cementitious materials (SCMs), challenges persist due to inconsistent slag compositions across sources, variable heavy metal content [4], and unpredictable influences on cement hydration. The slags can retard the cement hydration, delaying setting and strength development, with hypothesized causes including phosphorus content [5], lowered pH [6], and reactive calcium aluminates such as mayenite [7, 8]. In this respect, calcium aluminates are not unusual the Portland cement, which itself includes calcium aluminates, primarily C_3A and C_4AF. Particularly, the highly reactive C_3A has a strong influence on cement hydration, which is controlled by the addition of sulfate ions through setting regulators [9]. Similar to C_3A, mayenite or other calcium aluminates may influence cement hydration. Therefore, this study aims to investigate the hydration of two steel slag-Portland cement blends and demonstrate the effect of increased calcium sulfation across different levels.

2 Materials and Methods

A CEM I 42.5 R Portland cement (OPC) was used, previously characterized in other research [10]. This cement has a Na_2O-equivalent of 0.72, 7.6% $C_3A_{(cub)}$, 3.3% $C_3A_{(orth)}$, 4.1% calcite, and a sulfate content of 3.0%. Anhydrite (AH) as the setting regulator was used, as well as additional setting regulator. The steel (furnace) slags (SFS) originated from German steelworks, comprising one BOF slag (BOF) and one EAF slag (EAF). The chemical composition of the raw materials is displayed in Table 2. The slags were crushed and ground in a laboratory ball mill. Details concerning the particle fineness of the materials are listed in Table 4. Blended cements were produced from Portland cement, steel slag and additional anhydrite; the mixtures are shown in Table 1. The abbreviation “B” denotes a cement that contains BOF slag, while the abbreviation “E” denotes an EAF cement.

Table 1. Cement mixtures

	OPC	B0/E0	B1/E1	B2/E2	B3/E3
OPC/SFS/AH	100/0/0	70/30/0	70/29.55/0.45	70/29.1/0.9	70/28.65/1.35
$\sum SO_3$ [%]	3.0	2.1	2.4	2.6	2.9
		B4/E4	B5 / E5	E6	E7
OPC/SFS/AH		70/28.2/1.8	70/27.75/2.25	70/27.3/2.7	70/26.85/3.15
$\sum SO_3$ [%]		3.1	3.4	3.7	3.9

For the studies, each experiment was conducted with a water to binder (w/b) ratio of 0.4 by mass. To analyse the solids, the samples were stopped at hydration times of 3 and 24 h by isopropanol treatment and subsequent vacuum filtration. Following this, the samples were dried at 35 °C and stored under vacuum until measurement.

The chemical composition (Table 2) was determined using X-ray fluorescence (XRF) (S8 Tiger 1 K spectrometer, Bruker ASX). The mineralogical composition of the raw materials (Table 3), as well as the stopped samples were analyzed by X-ray diffraction (XRD) (D8 Advanced ECO, Bruker ASX), using CuKα-radiation. For In-situ XRD analysis another device (D8 Advanced, Bruker ASX) with a specialized sample holder was employed, to maintain the sample at a constant temperature of 25 °C throughout the measurement period. In addition, the sample was covered with a Kapton film to minimize atmospheric influences. Measurements spanned a range of 8–56° 2Θ with a step width of 0.015° and were conducted every 15 min over approximately 48 h. The analysis used the external standard G-factor method as per O'Connor [11, 12]. For C–S–H phase identification, the structure model from Bergold et al. [13] was used. It should be noted that only a semi-quantitative analysis of the C–S–H phase formation was conducted due to the lack of accurate reference sample. All evaluation was carried out with TOPAS software version 6.0. Stopped samples after 3 and 24 h of hydration were analysed under a scanning electron microscope (SEM) (Nova NanoSEM 230 FEG, ThermofisherScientific). Cements were also analyzed using isothermal heat flow calorimetry (TAM Air device, TA Instruments).

Table 2. Chemical composition of materials determined by XRF-spectroscopy

	SiO_2	TiO_2	Al_2O_3	Fe_2O_3	MgO	SO_3	CaO	P_2O_5	MnO	LOI
	wt. %									
OPC	21.1	0.3	5.5	2.3	1.4	3.0	61.8	0.1	–	0.8
BOF	12.6	0.5	1.8	24.9	1.35	0.2	49.3	1.6	2.3	4.2
EAF	14.1	1.1	4.9	38.9	9.6	0.1	27.7	0.2	3.2	−2.0
AH	0.2	–	0.2	–	0.6	56.4	42.6	0.1	–	n.a.

Table 3. Mineralogical composition of SFS of materials determined by XRD-Rietveld

	Belite	RO-phase	Alite	C_3A_{cub}	C_4AF/C_2F	$C_{12}A_7$	f-CaO	CH	OT	AM
	Wt. %									
BOF	29.1	4.4	3.9	–	29.4	–	6.0	4.4	5.4	17.4
EAF	27.5	37.6	–	5.2	–	3.1	–	–	15.2	11.3

RO-Phase: (Fe,Mg,Mn)O solid solution; $C_{12}A_7$: Mayenite; f-CaO: Lime; CH: Portlandite; OT: Other Phases; AM: amorphous content

Table 4. Specific surface area (Blaine and BET) and particle size (laser diffraction) of the materials

	Blaine [cm^2/g]	BET [m^2/g]	D10 [μm]	D50 [μm]	D90 [μm]
OPC	3500	1.04	1.92	14.1	41.5
BOF	5650	3.03	1.2	11.8	54.5
EAF	5750	1.59	0.83	4.8	28.8
AH	5400	1.90	1.0	7.2	35.9

3 Results

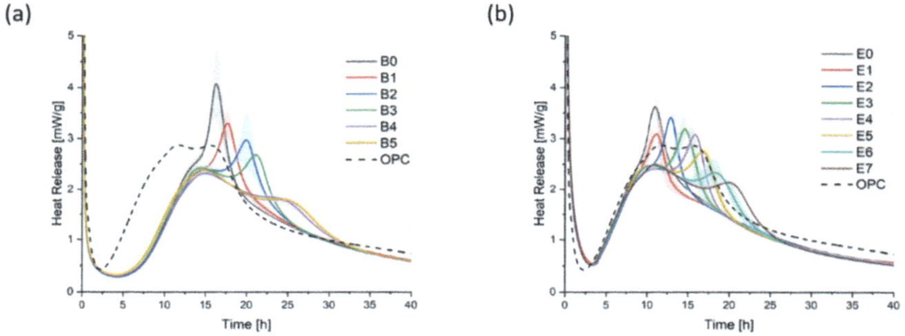

Fig. 1. Heat release of BOF-cement (a), heat release of EAF-cement (b)

Figure 1 illustrates the release of hydration heat over time. The left image (Fig. 1a) focuses on cements containing BOF slag. Initial observations indicate an extended dormant phase (0.5–6 h) compared to OPC. After this, the BOF cement develops two peaks, two peaks can also be seen in OPC. In the case of B0, the first peak is delayed compared to OPC, and the second peak is closer to the first than in OPC, appearing very tall and narrow in B0. With the addition of calcium sulfate, the second peak progressively shifts later in time, while the first peak remains visibly unchanged. The second peak also becomes flatter and broader.

For the EAF slag cement, shown in Fig. 1b, there is no comparable extension of the early dormant phase as observed with the BOF slag. Here too, the first peak is overlaid by the second peak. The addition of calcium sulfate continuously shifts the second peak later in time, while the calcium sulfate does not affect the first peak.

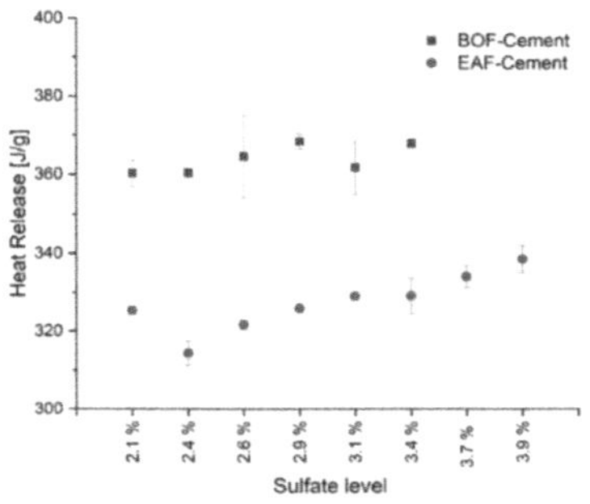

Fig. 2. Cumulative heat release after 40 h of hydration

Figure 2 displays the cumulative heat release of the cements over 40 h with different sulfation levels (2.1–3.9%, respectively B0-B5 and E0-E7). The released heat is significantly lower than those of OPC (396 J/g). The BOF cement demonstrates a relatively constant value, where the standard deviation in measurements for B3 and B4 exceeds the differences between the cements. The EAF cement initially exhibits a higher heat release with E0 compared to E1 and E2. As the addition of calcium sulfate increases, the measurements rise, with the highest heat release observed in E7.

Further investigations using in-situ XRD and SEM were conducted on samples OPC, B0, B4, E0 and E4. The results of the in-situ XRD for selected relevant phases are displayed in Fig. 3. For orientation, the phases of Portland cement are shown in the illustrations. Figure 3a and b depict the hydration process of alite.

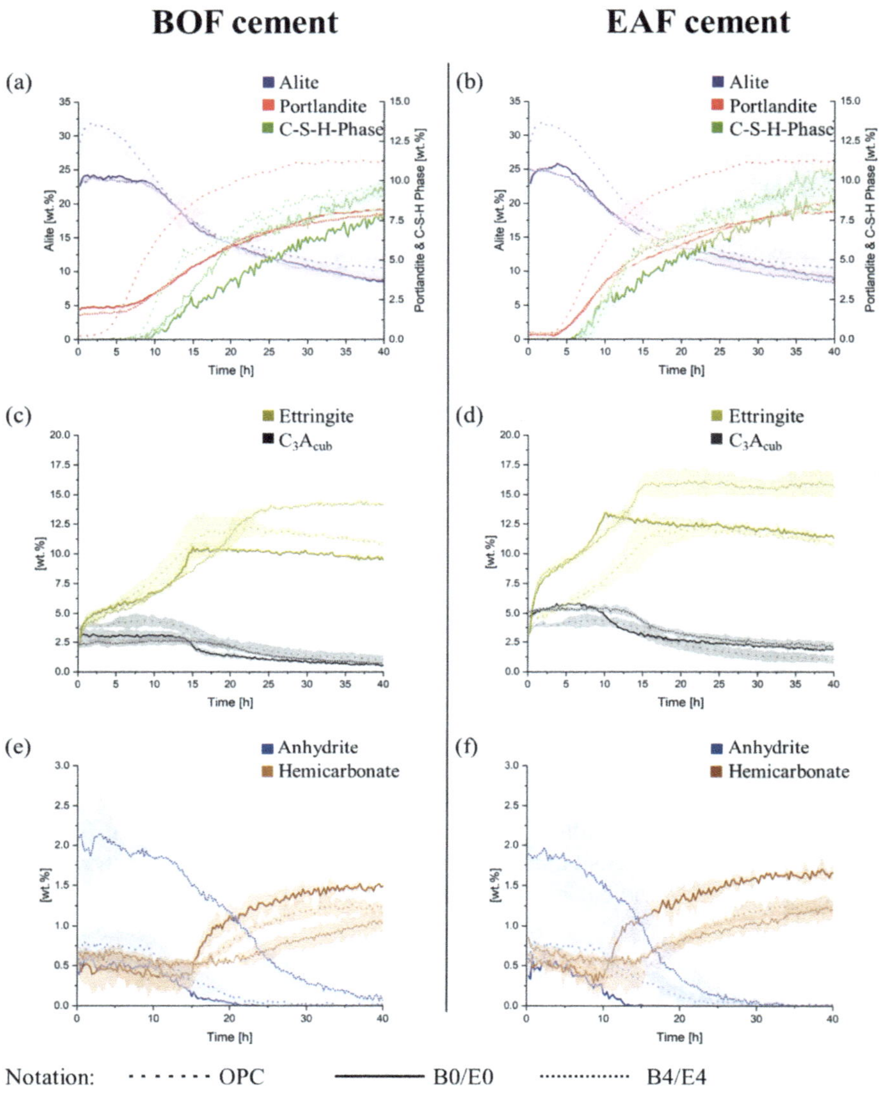

Fig. 3. Results of the in-situ XRD, BOF cement (a, c, e), EAF cement (b, d, f)

Initially, the amount of alite appears to increase, which can be attributed to settling phenomena due to measurement techniques. After a stable period, the rapid dissolution of alite begins (at 5.75 h in OPC), later flattening the curve. In sample B0, the onset of dissolution occurs later (at 9.5 h in B0) compared to OPC and EAF cement. The addition of calcium sulfate appears to have no effect on the dissolution of alite in slag cements.

Figure 3a and b also illustrates the development of the portlandite content. Initially, OPC and E0, E4 contain no portlandite, and the occurrence of portlandite then correlates with the dissolution of alite. The BOF cements contain about 2% portlandite from the start, whose content remains constant initially. The increase in portlandite in the BOF cements also correlates with the dissolution of alite, and the following curve is very similar to that of E0. Compared to OPC, less portlandite is formed in the slag cements, approximately 70% of OPC after 40 h.

The semi-quantitative detection of C–S–H using the structure described by Bergold et al. [13] occurs later than the formation of portlandite but shows a comparable increase to portlandite. The addition of calcium sulfate increases the formation of C–S–H phases in both slag cements from 74% (B0) to 91% (B4) compared to OPC, and from 81% (E0) to 110% (E4) compared to OPC.

Initially, more C_3A_{cub} is present in EAF cements than in OPC, and least in BOF cements (Fig. 3c and d). Like alite, a slight initial increase can be seen in the OPC sample, likely due to measurement techniques. Apart from that, the C_3A_{cub} content is initially constant in all samples before beginning to dissolve. In B0, the dissolution of C_3A_{cub} occurs approximately at the same time as in OPC. In E0, C_3A_{cub} dissolution occurs significantly earlier than in OPC. With the addition of calcium sulfate, the dissolution of C_3A_{cub} in both cements (B4 and E4) can be delayed.

The development of ettringite content shows that in all cements, ettringite is formed already in the first measurements, with more ettringite formed in EAF cement than in OPC and BOF cement. The addition of calcium sulfate has no effect on the initially formed amount of ettringite (first 0.5 h). After the initial formation of ettringite, there is a temporary flattening of the curve before the course becomes somewhat steeper again. This increase later abruptly stops, and the ettringite content remains constant or even slightly decreases over time. B0 and E0 show a very steep increase in ettringite formation before the last constant phase, which can be flattened by the addition of calcium sulfate, while the time during which ettringite is formed is extended, and the absolute ettringite content increases.

In all systems, hemicarbonate is formed, the formation is starting after most of the dissolution of alite and C_3A is already over. B0 and E0 form more hemicarbonate than OPC, with E0 occurring significantly earlier than in OPC. With the addition of calcium sulfate, less hemicarbonate is formed. Besides hemicarbonate, portlandite, ettringite and C–S–H, no other crystalline hydration products could be detected in any XRD measurement.

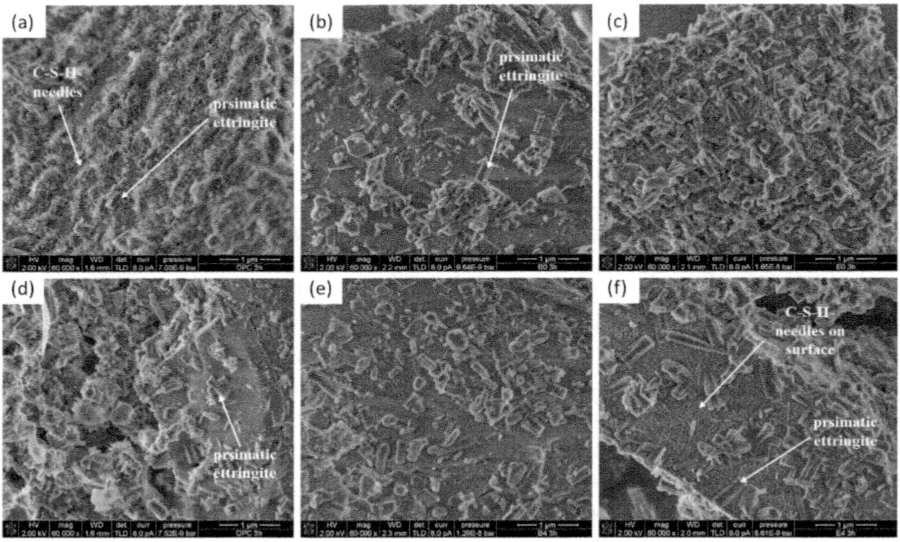

Fig. 4. SEM images of hydrated cements, stopped after 3 h: OPC (a and d), B0 (b), E0 (c), B4 (e), E4 (f)

Additional investigations using secondary electrons in SEM were carried out on cement samples stopped after 3 and 24 h of hydration. Selected images of samples stopped after 3 h at 60,000-fold magnification are presented in Fig. 4. Images of samples stopped after 24 h at 30,000-fold magnification are shown in Fig. 5. In OPC (Fig. 4a and d), surfaces exhibiting small, needle-like C–S–H crystals are observable as early as 3 h. Additionally, surfaces occupied with prismatic ettringite crystals having a length-to-width ratio of approximately 4:1 are evident (Fig. 4d). Some cement particles exhibit substantial growth of hydration products, predominantly consisting of ettringite. Conversely, certain particles display oriented C–S–H growth, which differs from the typically unordered, needle-like morphology (Fig. 4a), extending less outward from the cement surface—similar to observations made in EAF cement (Fig. 4f).

In contrast, BOF cement samples (Fig. 4b and e) exhibit no detectable C–S–H phases after 3 h. Large, uncovered surfaces are visible on the cement particles alongside regions densely occupied with ettringite. Additional sulfation in sample B4 (Fig. 4e) did not reveal significant differences compared to sample B0.

The EAF cements (Fig. 4c and f) display substantially more surfaces occupied by ettringite compared to OPC, though isolated uncovered surfaces remain visible. Occasionally, small C–S–H phases, typically oriented on the surface, are observable (e.g., Fig. 4f). No distinguishable differences between E4 and E0 were identifiable in the SEM investigations after 3 h.

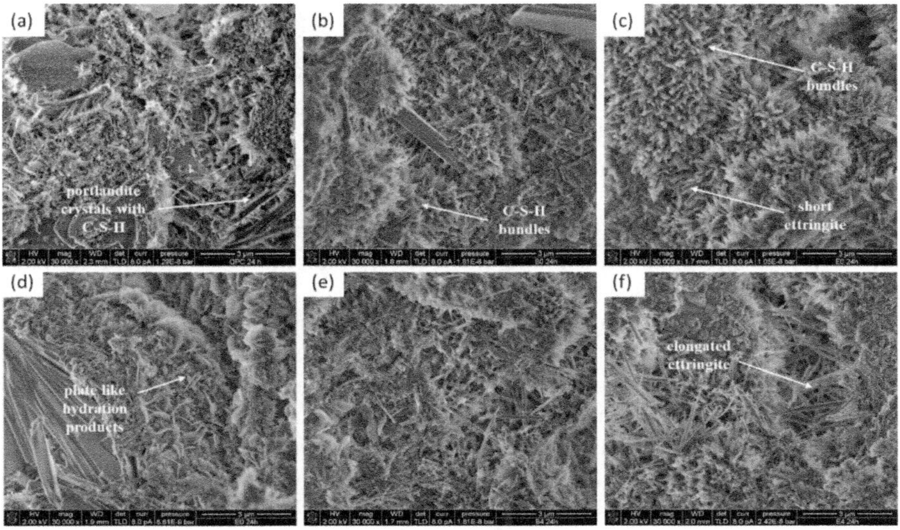

Fig. 5. SEM images of hydrated cements, stopped after 24 h: OPC (a), B0 (b), E0 (c and d), B4 (e), E4 (f)

SEM images of samples stopped after 24 h of hydration are presented in Fig. 5. In the OPC sample, hydration products comprise uniformly formed portlandite crystals, several hundred micrometers in size, partially bordered by smaller C–S–H phases. C–S–H phases extensively cover most other surfaces (Fig. 5a); however, the distinct needle-like structure observed in the 3 h samples is less pronounced, having transformed predominantly into bundles. Additionally, ettringite crystals, significantly elongated compared to the 3 h observations, are evident.

In the BOF cement sample B0 (Fig. 5b), the microstructure resembles that of OPC, with surfaces similarly coated by C–S–H phases. The presence of elongated ettringite crystals, however, is notably reduced relative to OPC. The increased sulfation in sample B4 (Fig. 5e) exhibits slightly more extensive formation of elongated ettringite crystals compared to B0.

The EAF cement sample E0 (Fig. 5c) contains abundant C–S–H phases, but notably lacks elongated ettringite crystals. Instead, prismatic ettringite crystals, clearly shorter than those in other cement samples after 24 h, can be observed interspersed among C–S–H phases (as shown in Fig. 5c, center-left-bottom). Furthermore, the EAF cement frequently displays small, plate-like hydration products, also visible in the E4 sample (Fig. 5d). With additional calcium sulfate (Fig. 5f), the formation of elongated ettringite crystals is again apparent. Apart from this, no significant differences in microstructure are observed between samples E4 and E0.

4 Discussion

Upon mixing Portland cement with water, the highly reactive C_3A phase initially dissolves rapidly, evidenced by an intense early heat release peak within the first 15 min. Sulfate ions from the setting regulator likely inhibit further dissolution of C_3A by adsorbing onto its surface, though the precise mechanisms remain under debate [14–17]. The reaction initially yields ettringite, predominantly occupying the calcium aluminate surfaces [18], as observed in SEM. In-situ XRD cannot capture this immediate dissolution, yet early ettringite detection confirms the reaction. Subsequently, ettringite formation continues at a reduced rate.

Following the initial peak, heat release diminishes significantly during the induction period. This phase ends with the dissolution of alite, leading to the formation of portlandite and C–S–H—known as the silicate reaction and corresponding to the main heat release peak. Concurrently, in-situ XRD indicates alite dissolution and portlandite formation. Initial C–S–H phases, though limited and less crystalline [13], are only identifiable by SEM after 3 h in OPC and EAF cement, but not in XRD. The oriented growth of C–S–H (Fig. 5d) indicates the influence of SCMs, particularly limestone powder, which promotes the nucleation of C–S–H [19–21]. After sulfates are depleted, either through incorporation into ettringite or adsorption onto C–S–H surfaces [22, 23], the inhibition of C_3A ends, prompting renewed dissolution of C_3A [16, 17]. This reaction, recognized as the aluminate peak following the main silicate peak, results in enhanced ettringite formation, binding adsorbed sulfate [22, 23]. The process is reflected in decreased C_3A and anhydrite levels alongside a marked increase in ettringite length, confirmed via XRD and SEM. Once sulfates are exhausted, ettringite formation halts, heat release declines, and ettringite content stabilizes. By adjusting sulfate content, the aluminate reaction can be delayed to occur post-silicate reaction, minimizing overlap [9]. Eventually, ettringite can convert into other calcium aluminates such as hemicarbonate, detectable in XRD.

4.1 BOF Cement

In BOF cement, the induction period and subsequent silicate reaction were notably delayed. Initial C_3A dissolution and ettringite formation, however, remained unaffected, as evidenced by existing ettringite in SEM and XRD. Delayed alite dissolution correlates with delayed formation of portlandite and C–S–H, aligning with known retardation phenomena in steel-slag-containing cements [5–7]. After three hours, SEM images reveal no initial C–S–H formation compared to OPC, while, apart from that, particle surfaces appear similar to other cements. Although the precise mechanism of the retardation remains unclear, delayed alite dissolution could temporally overlap with the rapid second dissolution of C_3A and swift ettringite growth, generating a pronounced, overlapping aluminate peak [9, 21]. Supplementary calcium sulfate addition delays and moderates the aluminate reaction, fostering uniform ettringite and increased C–S–H formation in B4 compared to B0, despite unchanged alite dissolution and cumulative heat release. To explain the differences in the measured C–S–H contents, the recent interpretations by Sowoidnich et al. [24] can be considered. According to these findings, insufficient sulfation leads to the incorporation of Al_2O_3 ions into C–S–H, forming of C–A–S–H phases instead of C–S–H. These C–A–S–H phases may exhibit different structural characteristics, which could result in their absence in the XRD measurements. Increased sulfation in sample B4 facilitates the incorporation of Al_2O_3 ions into ettringite rather than forming C–A–S–H, thus promoting the formation of C–S–H phases. This hypothesis could explain the differences in observed C–S–H quantities between samples B0 and B4 in the in-situ XRD measurements after approximately 15 h, coinciding with the dissolution of anhydrite.

4.2 EAF Cement

In EAF cement, increased initial ettringite formation, observed via SEM and XRD, is likely due to additional C_3A_{cub} and potentially mayenite from the EAF slag, though mayenite remained undetected in in-situ XRD due to small quantity and high solubility. Despite this elevated early ettringite formation, anhydrite levels remained comparable to BOF cement (B0). Post-induction period, alite hydration proceeds with reduced intensity compared to OPC, primarily due to the dilution effect of the slag. Unlike BOF slag, no delayed silicate reaction was observed. During alite hydration, the secondary C_3A_{cub} reaction occurs prematurely in E0, indicated by an aluminate peak accompanied by rapid ettringite formation. Limited sulfate availability likely contributes to shorter ettringite crystals. Upon sulfate exhaustion, ettringite content remains stable. Subsequent hemicarbonate formation occurs earlier than in OPC. The addition of calcium sulfate delays the secondary C_3A reaction, eliminating overlap with the silicate peak, prolonging ettringite formation, and promoting growth of longer ettringite crystals. Furthermore, additional sulfate significantly boosts C–S–H formation, surpassing OPC levels, although alite dissolution remains unchanged. Similar to BOF cement, the absence of sulfate in E0 likely induces formation of C–A–S–H phases, reducing C–S–H formation, whereas extra sulfate in E4 encourages Al_2O_3 incorporation into ettringite. Here, too, the splitting of C–S–H formation after dissolution of the anhydrite can be observed in the in-situ XRD. The observed increase in hydration products between E0 and E4 aligns with heat release data, emphasizing that strategically positioning the aluminate reaction after the silicate reaction positively impacts C–S–H formation.

5 Conclusions

In this study, two composite cements containing 30% BOF respectively EAF slag as Portland cement replacement material, were analyzed at various degrees of sulfation. Based on the results obtained, the use of additional sulfate carriers is recommended to prevent the aluminate reaction from overlaying the main silicate reaction. This has positive effects on the amount and quality of hydration products formed, although it may not necessarily be reflected in the cumulative heat release. Various phenomena can influence cement hydration when steel slags are used as SCMs, with the need for additional sulfate varying depending on the type of slag used.

BOF slag leads to a delayed silicate reaction, suppressing C–S–H formation. This is likely due to the specific properties of the slag, although the precise mechanisms of this delay remain unclear. The delay could lead to a temporal overlap of the silicate and aluminate reactions. The addition of a sulfate carrier can delay the aluminate reaction, thereby reducing the overlay and enhancing the formation of C–S–H at the expense of C–A–S–H.

In contrast, EAF slag, due to the additional introduced aluminate phases, accelerates the aluminate reaction. Their hydration products cover large parts of the surfaces at an early stage. Later, the second aluminate reaction overlaps the typical silicate reaction. Here too, the addition of additional sulfate carrier can delay the aluminate reaction to prevent this overlay, forming more and longer ettringite. While the EAF slag can potentially influence cement hydration due to the contained calcium aluminates, no typical delay effect as observed with BOF slag was detected.

References

1. Duée C, Bourgel C, Véron E et al (2015) Phosphorus speciation in dicalcium silicate phases: application to the basic oxygen furnace (BOF) slag. Cem Concr Res 73:207–214. https://doi.org/10.1016/j.cemconres.2015.03.012
2. Gautier M, Poirier J, Bodénan F et al (2013) Basic oxygen furnace (BOF) slag cooling: laboratory characteristics and prediction calculations. Int J Miner Process 123:94–101. https://doi.org/10.1016/j.minpro.2013.05.002
3. World Steel Association (2010) Steel industry by-products: project group report 2007–2009. World Steel Association, Brussels [u.a.]
4. Martins ACP, Franco de Carvalho JM, Costa LCB et al (2021) Steel slags in cement-based composites: an ultimate review on characterization, applications and performance. Constr Build Mater 291:123265. https://doi.org/10.1016/j.conbuildmat.2021.123265
5. Kollo H (1984) Untersuchungen zur Frage der zementtechnologischen Eignung einer LD-Konverterschlacke als latent-hydraulischer Zumahlstoff. Doctoral thesis, Universität der Bundeswehr München
6. Han F, Zhang Z, Wang D et al (2015) Hydration heat evolution and kinetics of blended cement containing steel slag at different temperatures. Thermochim Acta 605:43–51. https://doi.org/10.1016/j.tca.2015.02.018
7. Zhuang S, Wang Q (2021) Inhibition mechanisms of steel slag on the early-age hydration of cement. Cem Concr Res 140:106283. https://doi.org/10.1016/j.cemconres.2020.106283
8. Zhuang S, Wang Q, Luo T (2022) Effect of C12A7 in steel slag on the early-age hydration of cement. Cem Concr Res 162:107010. https://doi.org/10.1016/j.cemconres.2022.107010
9. Lerch W (1946) The influence of gypsum on the hydration and properties of Portland cement pastes. In: American society for testing materials (ed) Proceedings, Philadelphia, pp 1252–1297
10. Pott U, Crasselt C, Fobbe N et al (2023) Characterization data of reference materials used for phase II of the priority program DFG SPP 2005. Opus Fluidum Futurum—rheology of reactive, multiscale, multiphase construction materials. Data Brief 47:108902. https://doi.org/10.1016/j.dib.2023.108902
11. O'Connor BH, Raven MD (1988) Application of the Rietveld refinement procedure in assaying powdered mixtures. Powder Diffract 3:2–6. https://doi.org/10.1017/S0885715600013026
12. Jansen D, Goetz-Neunhoeffer F, Stabler C et al (2011) A remastered external standard method applied to the quantification of early OPC hydration. Cem Concr Res 41:602–608. https://doi.org/10.1016/j.cemconres.2011.03.004
13. Bergold ST, Goetz-Neunhoeffer F, Neubauer J (2013) Quantitative analysis of C–S–H in hydrating alite pastes by in-situ XRD. Cem Concr Res 53:119–126. https://doi.org/10.1016/j.cemconres.2013.06.001

14. Myers RJ, Geng G, Li J et al (2017) Role of adsorption phenomena in cubic Tricalcium aluminate dissolution. Langmuir 33:45–55. https://doi.org/10.1021/acs.langmuir.6b03474
15. Geng G, Myers RJ, Yu Y-S et al (2018) Synchrotron X-ray nanotomographic and spectromicroscopic study of the tricalcium aluminate hydration in the presence of gypsum. Cem Concr Res 111:130–137. https://doi.org/10.1016/j.cemconres.2018.06.002
16. Flatt RJ, Roussel N, Bessaies-Bey H et al (2023) From physics to chemistry of fresh blended cements. Cem Concr Res 172:107243. https://doi.org/10.1016/j.cemconres.2023.107243
17. Jansen D, Naber C, Ectors D et al (2018) The early hydration of OPC investigated by in-situ XRD, heat flow calorimetry, pore water analysis and 1H NMR: learning about adsorbed ions from a complete mass balance approach. Cem Concr Res 109:230–242. https://doi.org/10.1016/j.cemconres.2018.04.017
18. Hampson CJ, Bailey JE (1983) The microstructure of the hydration products of tri-calcium aluminate in the presence of gypsum. J Mater Sci 18:402–410. https://doi.org/10.1007/BF00560628
19. Stark J, Möser B, Bellmann F (2007) Nucleation and growth of C–S–H phases on mineral admixtures. In: Grosse CU (ed) Advances in construction materials 2007. Springer, Berlin, Heidelberg, pp 531–538
20. Berodier E, Scrivener K (2014) Understanding the filler effect on the nucleation and growth of C–S–H. J Am Ceram Soc 97:3764–3773. https://doi.org/10.1111/jace.13177
21. Zunino F, Scrivener K (2019) The influence of the filler effect on the sulfate requirement of blended cements. Cem Concr Res 126:105918. https://doi.org/10.1016/j.cemconres.2019.105918
22. Quennoz A, Scrivener KL (2013) Interactions between alite and C3A-gypsum hydrations in model cements. Cem Concr Res 44:46–54. https://doi.org/10.1016/j.cemconres.2012.10.018
23. Mota B, Matschei T, Scrivener K (2015) The influence of sodium salts and gypsum on alite hydration. Cem Concr Res 75:53–65. https://doi.org/10.1016/j.cemconres.2015.04.015
24. Sowoidnich T, Cölfen H, Rößler C et al (2023) The impact of metakaolin on the hydration of tricalcium silicate: effect of C–A–S–H precipitation. Front Mater 10. https://doi.org/10.3389/fmats.2023.1159772

Performance of Concrete Blocks Containing Recycled Sand and Blast Furnace Slag: Cost and Environmental Perspective

Eric Ababio Ohemeng[1(✉)], Molusiwa S. Ramabodu[2], and David Gyau Obour[1]

[1] Department of Building Technology, Sunyani Technical University, Sunyani, Bono Region, Ghana
ohemengababioeric@yahoo.com

[2] Department of Construction Management and Quantity Surveying, Durban University of Technology, Durban, South Africa

Abstract. This study investigates the effects of incorporating recycled sand (RS) and ground granulated blast furnace slag (GGBFS) on the strength performance, cost, and environmental impact of concrete blocks. The research was conducted in two phases. In Phase I, concrete mixtures were prepared by partially replacing ordinary Portland cement with GGBFS at 0%, 20%, 30%, 40%, and 50% while natural sand was substituted with RS at levels of 0%, 20%, 40%, and 60%. The results demonstrate a general trend in which the strength performance decreases with increasing RS content, but improves with moderate GGBFS incorporation. For instance, at 0% RS, the compressive strength increased progressively from 61.50 MPa (0% GGBFS) to a peak of 66.52 MPa (40% GGBFS), before slightly declining to 64.10 MPa at 50% GGBFS. In Phase II, cost and environmental impact analyses were performed. Results demonstrated that the inclusion of RS and GGBFS significantly reduced both production costs and ecological footprint compared to conventional concrete. These findings highlight the potential benefits of utilizing sustainable materials in concrete block production, promoting cost-effectiveness and environmental sustainability.

Keywords: flexural strength · compressive strength · sustainable concrete blocks

1 Introduction

The construction sector remains a major consumer of natural resources and a key contributor to global carbon dioxide (CO_2) emissions. The production of conventional concrete is particularly resource-intensive, relying extensively on natural sand (NS), coarse aggregate (CA), and ordinary Portland cement (OPC). The extraction of NS and CA often results in habitat degradation and ecological imbalance, while the manufacture of OPC is highly energy-demanding and responsible for approximately 7–8% of global CO_2 emissions [1, 2]. These challenges have intensified the search for more sustainable alternatives that can mitigate the environmental burden of concrete production while maintaining structural integrity and performance.

H. Beushausen et al. (Eds.): NTZR 2025, LNCE 873, pp. 114–124, 2026.
https://doi.org/10.1007/978-3-032-27698-8_11

One promising avenue for sustainable concrete development lies in the use of recycled sand (RS) sourced from construction and demolition waste. Utilizing RS not only diverts substantial volumes of waste from landfills but also reduces the depletion of natural aggregates. However, the integration of RS into concrete mixtures can negatively influence the physical and mechanical behaviour of the concretes, due to inherent differences in particle morphology, grading, and water absorption characteristics of RS [3–6]. As such, a thorough understanding of RS's impact on concrete performance is critical for facilitating its broader adoption in sustainable construction practices.

In addition, the application of supplementary cementitious materials, such as ground granulated blast furnace slag (GGBFS), has garnered significant interest as a means to enhance the environmental profile of cement-based systems [7]. GGBFS, an industrial by-product from the iron and steel manufacturing process, serves as a partial substitute for OPC in concrete formulations [8–12]. Its incorporation not only contributes to reductions in CO_2 emissions and energy use but also offers performance benefits, such as improved durability and long-term strength development, due to its latent hydraulic properties [13].

Hence, this study seeks to investigate the combined effects of RS and GGBFS on the strength, cost, and environmental performance of concrete blocks. By exploring the synergistic potential of recycled and industrial by-product materials, the research aims to contribute to the development of eco-efficient and technically sound alternatives for sustainable construction practices.

2 Experimental Procedure

2.1 Materials

The concrete mixtures were prepared using OPC (CEM I 52.5 N), GGBFS, CA, NS, RS, and water. The NS and RS employed conformed to Zone II as per IS: 383-1970 [14]. Crushed granite with maximum size of 10 mm was used for the CA. Table 1 shows the physical properties of the aggregates and the binders.

Table 1. Physical properties of the aggregates and binders.

Materials	Specific gravity	Bulk density (kg/m^3)	Fineness modulus	Moisture content (%)	Blaine fineness (cm^2/g)
NS	2.69	1695.00	2.75	1.86	–
RS	2.51	1441.00	2.61	1.83	–
CA	2.66	1600.00	1.96	1.39	–
OPC	3.15	–	–	–	3494.62
GGBFS	2.90	–	–	–	4000.00

2.1.1 Preparation of the Recycled Sand

Concrete waste sourced from a construction site was manually broken down using a hammer. RS was subsequently obtained by passing the material through a 4.5 mm sieve. To reduce the presence of fine concrete powder in the RS, the material was further sieved using a 75 μm mesh. The processed RS was then stored in sacks until it was needed for the concrete preparation.

2.2 Methods

2.2.1 Concrete Mixtures, Casting, and Curing of the Specimens

Concrete mixtures were prepared at a 1:2:1 (binder-sand-CA ratio), using a water/binder ratio of 0.5. The RS was used to replace the NS at levels of 0%, 20%, 40%, and 60%. Also, the GGBFS was used to substitute the OPC at 0%, 20%, 30%, 40%, and 50% contents. The concrete mixtures were represented as RSx-GGy, where ‘x’ is the percentage of RS and ‘y’ is the percentage of GGBFS. Table 2 shows the mixture proportions of the prepared concrete mixtures. Concrete was mixed using a concrete mixer and then used to cast 60 mm (depth) × 100 mm (width) × 200 mm (length) prism. During casting, concrete was placed in moulds in two layers, each layer compacted manually using steel rod. After casting, the specimens were carefully removed from the moulds and covered with a polyethylene sheets and stored at a room temperature for 24 h. The specimens were then cured in water for 28 days.

Table 2. Proportion of concrete mixtures.

Notation	Constituents of concrete (kg/m^3)					
	OPC	GGBFS	CA	NS	RS	water
RS0-GG0	560.70	0.00	665.3	1252.1	0.00	280.4
RS0-GG20	448.56	112.14	665.3	1252.1	0.00	280.4
RS0-GG30	392.49	168.21	665.3	1252.1	0.00	280.4
RS0-GG40	336.42	224.28	665.3	1252.1	0.00	280.4
RS0-GG50	280.35	280.35	665.3	1252.1	0.00	280.4
RS20-GG0	560.70	0.00	665.3	1001.7	212.8	280.4
RS20-GG20	448.56	112.14	665.3	1001.7	212.8	280.4
RS20-GG30	392.49	168.21	665.3	1001.7	212.8	280.4
RS20-GG40	336.42	224.28	665.3	1001.7	212.8	280.4
RS20-GG50	280.35	280.35	665.3	1001.7	212.8	280.4
RS40-GG0	560.70	0.00	665.3	751.3	425.7	280.4
RS40-GG20	448.56	112.14	665.3	751.3	425.7	280.4
RS40-GG30	392.49	168.21	665.3	751.3	425.7	280.4
RS40-GG40	336.42	224.28	665.3	751.3	425.7	280.4

(continued)

Table 2. (*continued*)

Notation	Constituents of concrete (kg/m^3)					
	OPC	GGBFS	CA	NS	RS	water
RS40-GG50	280.35	280.35	665.3	751.3	425.7	280.4
RS60-GG0	560.70	0.00	665.3	500.8	638.6	280.4
RS60-GG20	448.56	112.14	665.3	500.8	638.6	280.4
RS60-GG30	392.49	168.21	665.3	500.8	638.6	280.4
RS60-GG40	336.42	224.28	665.3	500.8	638.6	280.4
RS60-GG50	280.35	280.35	665.3	500.8	638.6	280.4

2.3 Testing of Specimens

The density and compressive strength of the concrete specimens were determined in accordance with BS 1881—Part 114 [15] and BS 6717—Part 1 [16] respectively. The flexural strength was then calculated from the formula; $\sigma = 3/2\ (LF/BD^2)$, where σ is the flexural strength (N/mm^2), L is the span length (mm), F is the maximum applied load (N), B is the average width of the specimen (mm), and D is the average thickness (mm).

3 Results and Discussion

3.1 Density of Concrete Mixtures

Figure 1 presents the density results of the concrete mixtures. The results indicate that, at each RS level, increasing the GGBFS content generally led to a marginal increase in density up to 40% replacement, followed by a slight decline at 50%. For instance, at RS0, density increased from 2583.30 kg/m^3 (GG0) to a peak of 2594.70 kg/m^3 (GG40), before slightly reducing to 2587.00 kg/m^3 (GG50). This trend may be attributed to the finer particle size and enhanced packing effect of GGBFS, which improves the matrix compactness and reduces voids. However, at higher GGBFS levels, the dilution effect of OPC may result in reduced hydration and microstructural development, thereby limiting further densification.

Conversely, an increase in RS content at constant GGBFS levels led to a consistent decline in density. For example, at 0% GGBFS, the density decreased from 2583.30 kg/m^3 (RS0) to 2325.50 kg/m^3 (RS60). This reduction may be due to the lower specific gravity of RS compared to NS (Table 1). The result is in agreement with the studies conducted by Hasbi et al. [3] and Singh et al. [4]. Overall, while the incorporation of RS tends to reduce the density of concrete, the inclusion of GGBFS up to 40% can mitigate this effect to some extent by enhancing the compactness of the matrix.

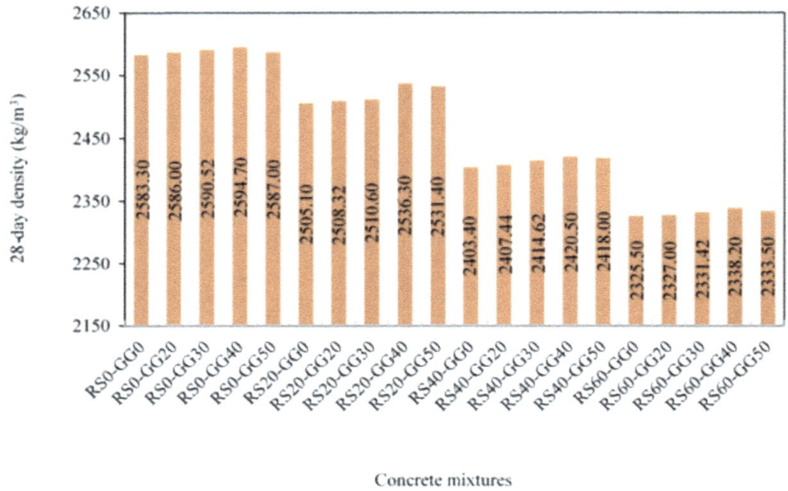

Fig. 1. Density of the concrete mixtures.

3.2 Compressive Strength of Concrete Mixtures

Figure 2 shows the compressive strength values of the concrete mixtures. As the RS content increased, the compressive strength declined significantly. For example, at GG0, strength dropped from 61.50 MPa (RS0) to 23.20 MPa (RS60). This reduction could be due to the lower mechanical quality, higher porosity, and presence of residual adhered mortar in RS, which negatively impact the interfacial transition zone (ITZ) and overall matrix integrity. Similar trend of results was obtained in [3, 4].

However, the inclusion of GGBFS partially mitigated strength loss across all RS levels. For instance, at 0% RS, the compressive strength increased progressively from 61.50 MPa (GG0) to a peak of 66.52 MPa (GG40), before slightly declining to 64.10 MPa at 50% GGBFS. This underscores the beneficial role of GGBFS in refining the pore structure and enhancing long-term strength, even in mixtures with inferior aggregate quality. The findings are in line with the results found in [8–11].

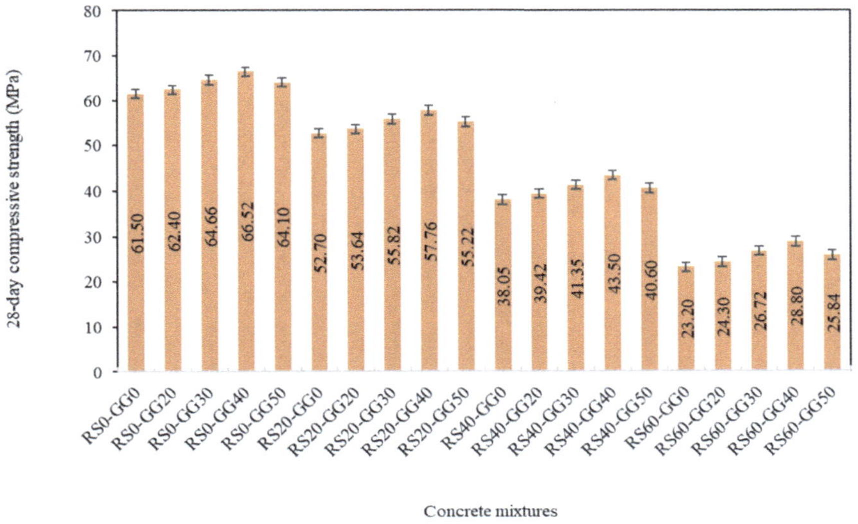

Fig. 2. Compressive strength of the concrete mixtures.

3.3 Flexural Strength of Concrete Mixtures

Figure 3 presents the flexural strength results of the concrete mixtures. The flexural strength values in the figure reveal clear trends regarding the influence RS and GGBFS on concrete performance. For mixtures containing only NS (RS0), flexural strength increased steadily with GGBFS replacement from 0% to 40%, peaking at 7.98 MPa (RS0-GG40), before slightly decreasing to 7.69 MPa at 50% substitution. This improvement may be attributed to the pozzolanic reaction of GGBFS, which refines the microstructure, reduces porosity, and improves the interfacial bond between paste and aggregate. However, excessive replacement with GGBFS may dilute the cementitious content and slow down early-age strength development, thus slightly reducing flexural performance at higher replacement levels.

However, at constant GGBFS levels, substituting NS with RS consistently reduced flexural strength. For instance, at 40% GGBFS, strength declined from 7.98 MPa (RS0-GG40) to 6.93 MPa (RS20-GG40), 5.07 MPa (RS40-GG40), and 4.34 MPa (RS60-GG40). This reduction may be attributed to the higher water absorption, residual mortar content, and weaker particle strength of RS. All of which compromise the integrity of the ITZ and lead to weaker bond strength. The result is in consistent with the studies carried out by Hasbi et al. [3] and Singh et al. [4].

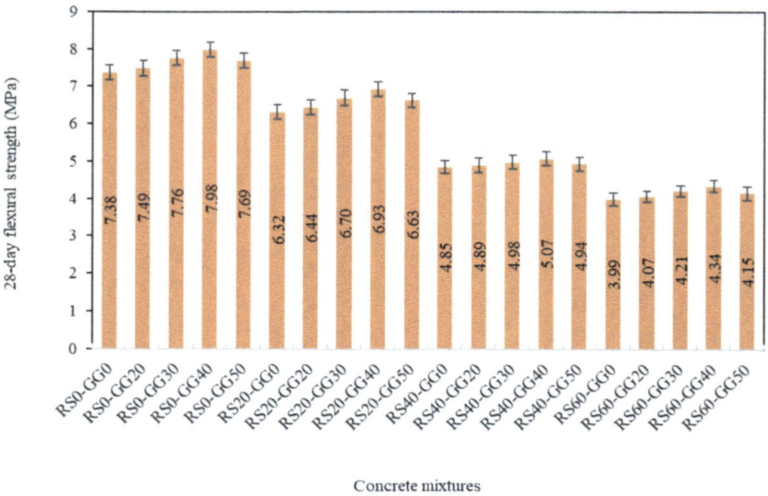

Fig. 3. Flexural strength of the concrete mixtures.

4 Cost Analysis of Concrete Mixtures

This section of the study evaluates the cost implications associated with the various concrete mixtures. The analysis was carried out using the unit costs of the constituent materials required to produce 1 m^3 of concrete. The costing approach for each concrete component followed the methodology outlined by Ohemeng [13] and Ohemeng and Ekolu [17]. Figure 4 displays the estimated cost of the concrete mixtures. The results reveal a consistent downward trend in total material cost with increasing GGBFS and RS content. For instance, at 0% RS, the cost decreased from US$132.11 (GG0) to US$100.09 (GG50), reflecting a 24% reduction. This trend can be attributed to the lower unit cost of GGBFS compared to OPC [13].

Similarly, at any given GGBFS level, increasing the proportion of RS also led to a progressive reduction in cost. For example, at GG0, the cost declined from US$132.11 (RS0) to US$127.57 (RS60). The cost savings associated with RS are primarily due to its lowers cost compared to that of NS [13]. Overall, the combined use of RS and GGBFS significantly enhances the economic efficiency of concrete production by reducing reliance on virgin materials and high-cost OPC, while simultaneously contributing to sustainable construction practices.

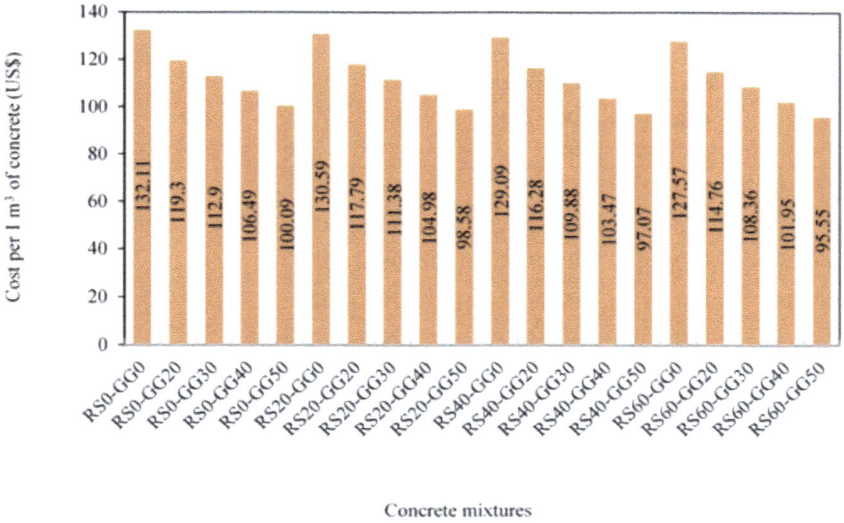

Fig. 4. Cost of the concrete mixtures.

5 Environmental Impact Analysis of Concrete Mixtures

This section presents the analysis of the environmental impact (EI) associated with the concrete mixtures. The EI data for each constituent of the concrete were sourced from existing literature [19–21]. Table 3 shows the results of the EI analysis of the concrete mixtures. The overall EI of the concrete mixtures was calculated by aggregating the environmental contributions of all components needed to produce 1 m^3 of each mixture. The mixtures were assessed across five key EI categories: abiotic depletion potential (ADP), global warming potential (GWP), ozone depletion potential (ODP), photochemical ozone creation potential (POCP), and acidification potential (AP).

A consistent and significant reduction in EI was observed with increasing GGBFS content, irrespective of the RS level. Specifically, the GWP, which is a critical indicator of climate change potential, decreased markedly from 561.86 kg CO_2 eq for the control mixture (RS0-GG0) to 300.51 kg CO_2 eq at 50% GGBFS replacement (RS0-GG50). This corresponds to an approximate 46.5% reduction in CO_2 emissions, underscoring the substantial environmental benefit of replacing OPC with GGBFS. These observations are in line with previous findings in the literature that identify GGBFS as a low-impact binder with lower energy and carbon footprints compared to OPC [18].

In contrast to the impact of GGBFS, the replacement of NS with RS at 20%, 40%, and 60% levels had a marginal influence on the environmental indicators. For instance, the GWP of RS0-GG0 (561.86 kg CO_2 eq) and RS60-GG0 (559.19 kg CO_2 eq) showed a negligible difference, despite a complete 60% substitution of NS. Similar patterns were observed for other categories, including ADP and ODP, all of which remained relatively unchanged across different RS levels. The mixture RS60-GG50 recorded the lowest values across all five EI categories.

Table 3. Environmental impact for producing the various concrete mixtures.

Notation	Environmental impact category				
	ADP (kgSbeq)	GWP ($kgCO_2$ eq)	ODP (kgCFC-11 eq)	POCP (kgC_2H_4eq)	AP ($kgSO_2eq$)
RS0-GG0	2.24	561.86	6.13×10^{-5}	0.06	1.71
RS0-GG20	1.83	457.32	4.96×10^{-5}	0.05	1.44
RS0-GG30	1.62	405.05	4.38×10^{-5}	0.04	1.30
RS0-GG40	1.41	352.79	3.79×10^{-5}	0.04	1.18
RS0-GG50	1.21	300.51	3.21×10^{-5}	0.03	1.04
RS20-GG0	2.24	560.97	6.13×10^{-5}	0.06	1.71
RS20-GG20	1.83	456.43	4.96×10^{-5}	0.05	1.44
RS20-GG30	1.62	404.16	4.38×10^{-5}	0.04	1.30
RS20-GG40	1.41	351.90	3.79×10^{-5}	0.04	1.18
RS20-GG50	1.21	299.62	3.21×10^{-5}	0.03	1.04
RS40-GG0	2.24	560.08	6.13×10^{-5}	0.06	1.70
RS40-GG20	1.83	455.54	4.97×10^{-5}	0.05	1.43
RS40-GG30	1.62	403.27	4.38×10^{-5}	0.04	1.29
RS40-GG40	1.41	351.01	3.80×10^{-5}	0.04	1.17
RS40-GG50	1.21	298.73	3.21×10^{-5}	0.03	1.03
RS60-GG0	2.24	559.19	6.14×10^{-5}	0.06	1.70
RS60-GG20	1.83	454.65	4.97×10^{-5}	0.05	1.43
RS60-GG30	1.62	402.38	4.38×10^{-5}	0.04	1.29
RS60-GG40	1.41	350.12	3.80×10^{-5}	0.04	1.17
RS60-GG50	1.21	297.84	3.22×10^{-5}	0.03	1.03

6 Conclusions

The findings of this study highlight the viability of incorporating RS and GGBFS as sustainable alternatives in the production of concrete blocks. While higher proportions of RS may compromise mechanical properties, the inclusion of GGBFS was found to effectively mitigate these drawbacks. The key conclusions drawn from the study are as follows:

i. Increasing the replacement of NS with RS results in a reduction in density, compressive strength, and flexural strength of the concrete mixtures.
ii. The incorporation of GGBFS up to 40% improves the strength performance of the concrete mixtures, demonstrating its potential as a partial substitute for OPC.
iii. The combined use of RS and GGBFS contributes to a notable reduction in the overall cost of concrete production.

iv. Simultaneous utilization of RS and GGBFS leads to a significant decrease in carbon emissions, thereby supporting low-carbon construction practices and sustainable development goals.

Overall, the study affirms that the strategic combination of RS and GGBFS can play a pivotal role in producing environmentally friendly and cost-effective concrete, without substantially compromising performance when used within optimal replacement levels.

References

1. Boden TA, Andres RJ (2017) Global CO_2 emissions from fossil-fuel burning, cement manufacturing and gas flaring. In: Carbon dioxide information analysis center (CDIAC), Oak ridge National laboratory (ORNL), Oak Ridge, TN, USA, pp 1751–2014. https://doi.org/10.3334/CDIAC/00001_V2017
2. Ohemeng EA, Naghizadeh A (2023) Alternative cleaner production of masonry mortar from fly ash and waste concrete powder. Constr Build Mater 374:130859. https://doi.org/10.1016/j.conbuildmat.2023.130859
3. Hasbi Y, Huseyin YA, Ilhami D, Osman S, Gokhan D (2011) Effects of the fine recycled concrete aggregates on the concrete properties. Int J Phys Sci 6(10):2455–2461. http://www.academicjournals.org/IJPS
4. Singh R, Nayak D, Pandey A, Kumar R, Kumar V (2022) Effects of recycled fine aggregates on properties of concrete containing natural or recycled coarse aggregates: a comparative study. J Build Eng 45:103442. https://doi.org/10.1016/j.jobe.2021.103442
5. Delsaute B, Staquet S (2020) Impact of recycled sand and gravels in concrete on volume change. Constr Build Mater 232:117279. https://doi.org/10.1016/j.conbuildmat.2019.117279
6. Yu L, Huang L, Ding H (2019) Rheological and mechanical properties of ultra-high-performance concrete containing fine recycled concrete aggregates. Materials 12
7. Ohemenga EA, Ramabodu MS, Nena TD, Kancheva Y (2024) Performance of sustainable mortars containing blast furnace slag and fine concrete waste: an environmental perspective. Cog Eng 11(1):2313053. https://doi.org/10.1080/23311916.2024.2313053
8. Vittalaiah A, Ravinder R, Kumar CV (2020) Study on effect of strength and durability parameters and performance of self compacting concrete replacement with GGBS at different dos ages. E3S Web Conf 184:01106
9. Lodhi U, Choudhary KK, Ghritkaushik N (2021) Determination of strength of concrete by using ground-granulated blast-furnace slag (GGBS). Int J Adv Eng Manag 3(10):1325–1329
10. Nagendra V, Sashidhar C, Kumar SMP, Venkata Ramana V (2020) Effect of particle size on compressive strength of concrete with ground granulated blast furnace slag as admixture along with microstructural analysis. Int J Emerg Technol 11(3):1179–1184
11. Prakash S, Kumar S, Biswas R, Rai B (2022) Influence of silica fume and ground granulated blast furnace slag on the engineering properties of ultra-high-performance concrete. Innov Infrastructure Solut 7:117. https://doi.org/10.1007/s41062-021-00714-7
12. Topcu IB, Boga AR (2010) Effect of ground granulate blast-furnace slag on corrosion performance of steel embedded in concrete. Mater Des 31:3358–3365. https://doi.org/10.1016/j.matdes.2010.01.057
13. Ohemeng EA (2023) Potentials of waste concrete elements for production of construction materials. PhD thesis, University of Johannesburg, Johannesburg, South Africa. Available from: http://hdl.handle.net/102000/0002
14. IS: 383-1970. Specifications for coarse and fine aggregates from natural sources for concrete. Bureau of Indian Standards, New Delhi, India

15. British standard institution (1983) Method for determination of density of hardened concrete. BS 1881, Part 114, BSI London
16. British standard institution (1986) Method for testing compressive strength of concrete. BS 6717, Part 1, BSI London
17. Ohemeng EA, Ekolu SO (2020) Comparative analysis on costs and benefits of producing natural and recycled concrete aggregates: a south African case study. Case Stud Constr Mater 13:e00450. https://doi.org/10.1016/j.cscm.2020.e00450
18. Majhi RK, Nayak AN (2020) Production of sustainable concrete utilising high-volume blast furnace slag and recycled aggregate with lime activator. J Clean Prod 255:120188
19. Blengini GA (2006) Life cycle assessment tools for sustainable development: case studies for mining and construction industries in Italy and Portugal. PhD thesis, Universidade Tecnica de Lisboa, Instituto Superior Tecnico, Portugal
20. Chen C, Habert G, Bouzidi Y, Jullien A, Ventura A (2010) LCA allocation procedure used as an incitative method for waste recycling: an application to mineral additions in concrete. Resour Conserv Recycl 54:1231–1240. https://doi.org/10.1016/j.resconrec.2010.04.001
21. Braga A (2015) Comparative analysis of the life cycle assessment of conventional and recycled aggregate concrete (in Portuguese). Master thesis in civil engineering, Instituto Superior Tecnico—University of Lisbon, Portugal

Effect of Recycled Sand and Blast Furnace Slag on the Performance of Mortars: Environmental and Cost Evaluation

Eric Ababio Ohemeng[1](✉) and Molusiwa S. Ramabodu[2]

[1] Department of Building Technology, Sunyani Technical University, Sunyani, Bono Region, Ghana
ohemengababioeric@yahoo.com

[2] Department of Construction Management and Quantity Surveying, Durban University of Technology, Durban, South Africa

Abstract. The use of sustainable materials in construction is essential for reducing environmental impact and optimizing costs. This study examines the effect of incorporating recycled sand and ground granulated blast furnace slag (GGBFS) on the performance of mortars. Experimental tests were conducted to evaluate the density, compressive strength, and flexural strength of the mortars. Also, comparative cost and environmental impact analyses were performed to determine the feasibility of these alternative materials. The results indicate that while higher proportions of recycled sand reduce mortar strength, the inclusion of GGBFS helps enhance the overall performance. For instance, replacing 40% of the cement with GGBFS increased the 28-day compressive strength of the mortar mixture comprising 100% cement, 80% natural sand, and 20% recycled sand from 33.2 to 35.7 MPa. Additionally, the combined use of recycled sand and GGBFS significantly lowers production costs and carbon emissions compared to conventional mortars. This research highlights the potential of these materials in developing sustainable and cost-effective mortars, promoting resource efficiency and reducing the environmental footprint of construction practices.

Keywords: Flexural strength · Compressive strength · Sustainable mortar · Cost-efficient mortar

1 Introduction

The production of conventional mortar relies heavily on natural sand (NS) and ordinary Portland cement (OPC), both of which have substantial environmental footprints. NS extraction leads to habitat destruction and ecological imbalance, while cement production is energy-intensive and accounts for approximately 7–8% of global CO_2 emissions [1, 2]. Consequently, there is a growing demand for sustainable alternatives that can reduce the environmental impact of mortar production without compromising its performance.

H. Beushausen et al. (Eds.): NTZR 2025, LNCE 873, pp. 125–135, 2026.
https://doi.org/10.1007/978-3-032-27698-8_12

One promising approach involves the partial replacement of NS with recycled sand (RS) derived from construction and demolition waste. RS not only diverts waste from landfills but also conserves natural aggregates. However, its incorporation may affect the physical and mechanical properties of mortars [3–5]. Therefore, understanding its influence on mortar performance is essential for promoting its broader application in construction.

Also, supplementary cementitious materials such as ground granulated blast furnace slag (GGBFS) have gained attention for their potential to enhance the sustainability of cement-based materials. GGBFS, a by-product of the iron and steel industry, can partially replace cement in mortar mixtures [6–9]. Its use contributes to a reduction in greenhouse gas emissions and energy consumption, while potentially improving durability and long-term performance of mortars due to its latent hydraulic properties [10].

The preceding discussion clearly indicates that RS and GGBFS are effective alternatives to NS and cement, respectively, in mortar production. Their combined application offers a promising pathway toward reducing the environmental impact of construction activities. Hence, the current study seeks to investigate the combined effects of incorporating RS and GGBFS on the performance of mortar. The RS is used as a partial replacement for NS, and GGBFS is used to partially substitute cement in varying proportions. By exploring the synergy between recycled aggregates and industrial by-products, this research aims to contribute to the development of environmentally responsible construction materials.

2 Experimental Procedure

2.1 Materials

The mortar mixtures were prepared using OPC (CEM I 52.5 N), GGBFS, NS (pit type), RS, and water. The density of the cement and GGBFS were 3.14 g/cm^3 and 2.92 g/cm^3, respectively. Their Blaine fineness values were 3494.60 cm^2/g and 4000.00 cm^2/g, respectively. All sand types employed complied with the grading specifications outlined in [11]. The physical characteristics of the sands are detailed in Table 1. Potable water was utilized for both mixing and curing of the mortar samples.

Table 1. Physical properties of the different sands.

Material	Specific gravity	Bulk density (kg/m^3)	Fineness modulus	Moisture content (%)
Natural sand	2.68	1595	2.76	1.85
Recycled sand	2.50	1357	2.62	1.82

2.1.1 Preparation of the Recycled Sand

Concrete waste obtained from a construction site was fragmented using a hammer. The RS was obtained by using a 4.5 mm sieve. The waste concrete powder in the RS was minimized by sifting it through a 75 μm sieve. The prepared RS was then put in sacks until the day of the mortar preparation. Currently, coarse recycled concrete aggregates (CRCAs) are produced on a commercial scale. To ensure consistency and facilitate standardization, the RS processing method could be incorporated into the established commercial production workflow of CRCAs.

2.2 Methods

2.2.1 Mortar Mixtures, Casting, and Curing of the Specimens

Mortar mixtures were prepared at a 1:3 (binder to sand ratio), using a water/ binder ratio of 0.70. The RS was used as a NS replacement in proportions of 0%, 20%, 40%, and 60%. Also, the GGBFS was employed as a substitute for OPC at 0%, 20%, 30%, 40%, and 50% levels. The mortar mixtures were denoted as NS_X-RS_Y-GG_Z, where x is the percentage of NS, y is the percentage of RS, and z is the percentage of GGBFS. Table 2 shows the mixture proportions of the prepared mortars. Mortars were mixed using a mortar mixer and then used to cast 50 mm cubes and 40 mm × 40 mm × 160 mm prisms. During casting, mortar was placed in moulds in two layers, each layer compacted on a vibrating table. After casting, specimens were covered with a polyethylene sheet and stored at room temperature for 24 h. The specimens were then demoulded and cured in water for 28 days.

Table 2. Proportions of mortar mixtures.

Notation	Constituents of mortar (g)				
	OPC	GGBFS	Natural sand	Recycled sand	Water
NS100-RS0-GG0	879.0	0.0	2637.0	0.0	615.3
NS100-RS0-GG20	703.2	175.8	2637.0	0.0	615.3
NS100-RS0-GG30	615.3	263.7	2637.0	0.0	615.3
NS100-RS0-GG40	527.4	351.6	2637.0	0.0	615.3
NS100-RS0-GG50	439.5	439.5	2637.0	0.0	615.3
NS80-RS20-GG0	879.0	0.0	2109.6	527.4	615.3
NS80-RS20-GG20	703.2	175.8	2109.6	527.4	615.3
NS80-RS20-GG30	615.3	263.7	2109.6	527.4	615.3
NS80-RS20-GG40	527.4	351.6	2109.6	527.4	615.3
NS80-RS20-GG50	439.5	439.5	2109.6	527.4	615.3
NS60-RS40-GG0	879.0	0.0	1582.2	1054.8	615.3
NS60-RS40-GG20	703.2	175.8	1582.2	1054.8	615.3
NS60-RS40-GG30	615.3	263.7	1582.2	1054.8	615.3
NS60-RS40-GG40	527.4	351.6	1582.2	1054.8	615.3
NS60-RS40-GG50	439.5	439.5	1582.2	1054.8	615.3

(continued)

Table 2. (*continued*)

Notation	Constituents of mortar (g)				
	OPC	GGBFS	Natural sand	Recycled sand	Water
NS40-RS60-GG0	879.0	0.0	1054.8	1582.2	615.3
NS40-RS60-GG20	703.2	175.8	1054.8	1582.2	615.3
NS40-RS60-GG30	615.3	263.7	1054.8	1582.2	615.3
NS40-RS60-GG40	527.4	351.6	1054.8	1582.2	615.3
NS40-RS60-GG50	439.5	439.5	1054.8	1582.2	615.3

2.3 Testing

The Blaine air fineness levels and density of the binders were determined in accordance with ASTM C204 [12] and ASTM C188 [13], respectively. The density, compressive strength, and flexural strength were evaluated in accordance with BS 1881—Part 114 [14], ASTM C109 [15], and BS EN 1015 [16], respectively. Three specimens were used to obtained the average value of each hardened mortar property test.

3 Results and Discussion

3.1 Density of the Mortars

Figure 1 presents the 28-day density results of the mortars. Obviously, RS and GGBFS have influence on the density of the mortars. At a constant GGBFS level, the mortars' density generally decreases as the RS content increases. At 0% GGBFS, the density of the control mortar (i.e., NS100-RS0-GG0) decreased from 2164.30 to 2155.94 kg/m^3 when 60% of the NS was replaced with RS. The reduction in density could be attributed to the differences in specific gravity of the two materials (Table 1). Also, the table shows that at a fixed RS content, the density of the mortars experienced a slight increase up to 40% GGBFS, following by a minor drop at 50%. For example, the density of the control mortar was increased from 2164.30 kg/m^3 to 2164.88 kg/m^3 and 2165.75 kg/m^3 for 30% and 40% GGBFS replacement, respectively. The control mortar's density then dropped to 2164.12 kg/m^3 when 50% of the cement was substituted with GGBFS. GGBFS is finer than cement (Sect. 2.1) and can help fill voids between particles, leading to increasing packing density, especially up to 40%. Beyond 40%, excess GGBFS could lower the cementitious reactivity, reducing the formation of dense hydration product [7, 17].

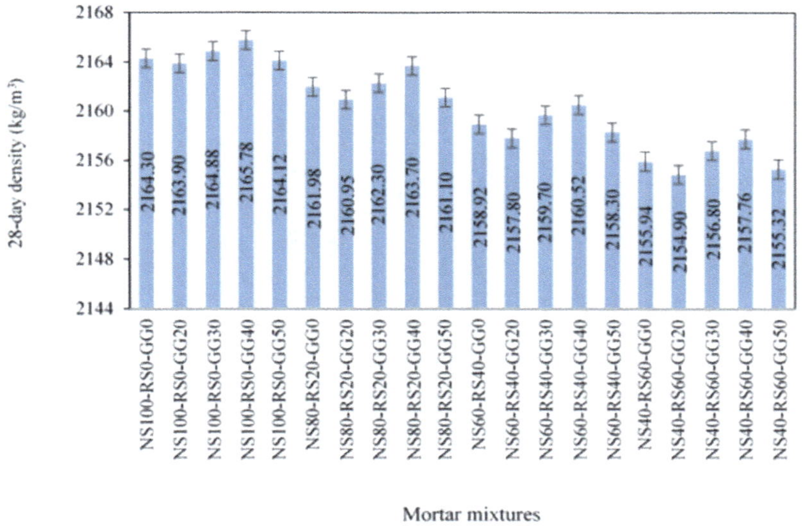

Fig. 1. Density of the mortars.

3.2 Compressive Strength of the Mortars

Figure 2 presents the 28-day compressive strength of the mortars. Evidently, RS and GGBFS have effect on the 28-day compressive strength of the mortars. The highest compressive strength values were obtained at 0% RS, ranging from 37.10 to 41.50 MPa depending on GGBFS content. As RS content increases to 20%, 40%, and 60%, compressive strength consistently decreases regardless of GGBFS level. At 40% RS, the 28-day compressive strength of the control mortar (i.e., NS100-RS0-GG0) drops from 37.10–41.50 MPa to 25.68–28.64 MPa depending on the GGBFS content. This reduction in compressive strength could be attributed to the lower quality of RS. RS typically contains impurity, weaker particles, and higher porosity. These qualities of RS can negatively affect the bond with cement paste [3–5]. Again, the higher water demand of RS could lead to decline in compressive strength. RS absorbs more water, which can reduce effective water-cement ratio and leads to weaker hydration. The result is consistent with the studies conducted by Ghorbel et al. [3] and Kepniak and Lukowski [5].

Also, it can be observed from Fig. 2 that at each RS level, increasing GGBFS content from 0 to 40% improves compressive strength up to a certain point. However, beyond 40% GGBFS, compressive strength tends to decrease. For instance, at NS100-RS0-GG0, compressive strength increased from 38.60 MPa (0% GGBFS) to 41.50 MPa (40% GGBFS), then drops to 38.12 MPa (50% GGBFS). The improvement in compressive strength with GGBFS (up to 30–40%) could be due to the latent hydraulic properties of GGBFS. GGBFS reacts with calcium hydroxide to form additional C-S-H gel, enhancing long-term strength [7]. However, addition of high-volume of GGBFS (beyond 40%) could lead to a significant reduction in hydration products and strengths [17].

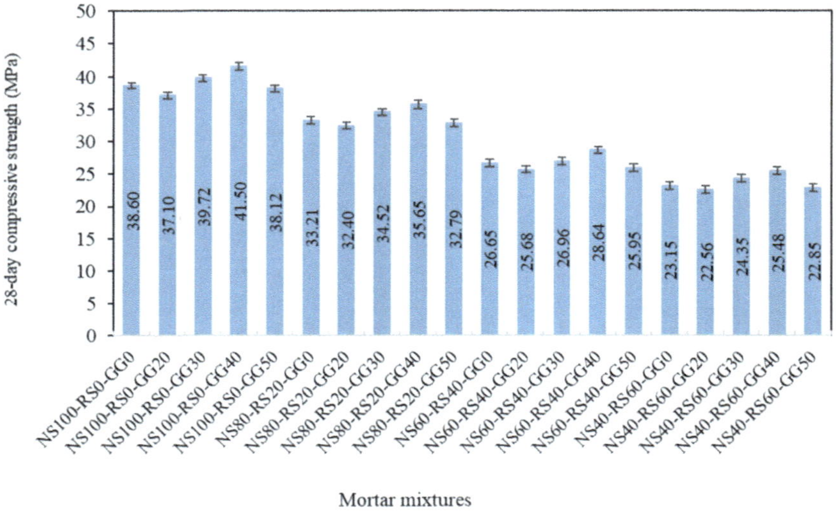

Fig. 2. Compressive strength of the mortars.

3.3 Flexural Strength of the Mortars

Figure 3 shows the 28-day flexural strength of the mortars. With 0% RS, the highest flexural strengths were recorded from 7.95 to 8.90 MPa, depending on GGBFS content. The flexural strength decreased with increasing RS content. The flexural strength of the control mortar fell from 8.30–8.90 MPa to around 7.04–7.70 MPa at 20% RS depending on the content of GGBFS. The lowest flexural strengths values from 5.15–5.78 MPa were observed at 60% RS. The result is in agreement with the experimental studies carried out by Ghorbel et al. [3] and Kepniak and Lukowski [5].

Also, Fig. 3 demonstrates the effect of GGBFS on the performance of the mortars. At each RS level, the flexural strength of the mortars tends to increased gradually up to 30–40% GGBFS, then decline slightly at 50%. For example, the flexural strength of the mortar mixture NS100-RS0-GG0 increased from 8.30 to 8.90 MPa when 40% GGBFS was employed. However, the 8.90 MPa was dropped to 8.20 MPa for 50% GGBFS. The explanation for the trend of the results is likely to be the same as the ones already given for the compressive strength in Sect. 3.2.

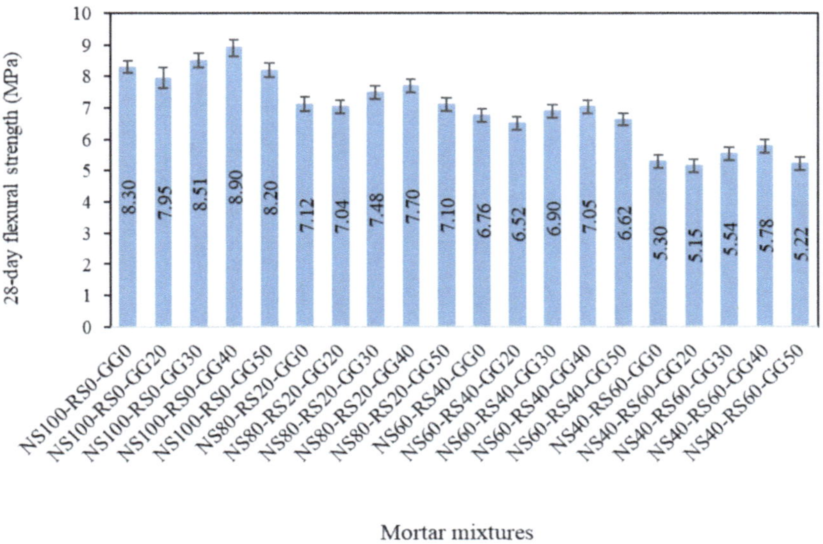

Fig. 3. Flexural strength of the mortars.

4 Cost Analysis

This part of the study discusses the cost implications of the mortar mixtures. The analysis was conducted based on the cost of the individual materials used to produced 1 m^3 of the mortars. The cost of each element of the mortar was obtained in a similar manner described by Ohemeng [10] and Ohemeng and Ekolu [18]. Figure 4 presents the cost of the various mortars. Clearly, GGBFS and RS have effect on the cost of the mortars. Within each fixed NS-RS composition, increasing the GGBFS content consistently reduces the cost of the mortars. For instance, the cost of NS100-RS0-GG0 mixture was reduced from US$ 97.30 to US$ 73.16 when 50% of the cement was replaced by GGBFS, representing a decline of 24.8%. Cement is typically the most expensive component in mortar. GGBFS, a by-product of the steel industry, is generally cheaper [10]. Replacing cement with GGBFS thus lowers the overall cost of the mortars. The higher the replacement level, the more significant the cost savings.

It can also be seen from Fig. 4 that at each fixed GGBFS content level, increasing the amount of RS levels leads to a progressive cost reduction. For instance, the cost of the mortar mixture NS100-RS0-GG0 was decreased from US$ 97.30 to US$ 93.53 for mortar mixture NS40-RS60-GG0. RS is generally less expensive than NS due to its waste-based origin and lower processing cost [18]. Using more RS reduces the demand for the more-costly natural resource, thereby lowering the total cost of the mortars. The combined use of GGBFS and RS significantly lowers raw material expenses, which reflected in the overall cost of the mortars.

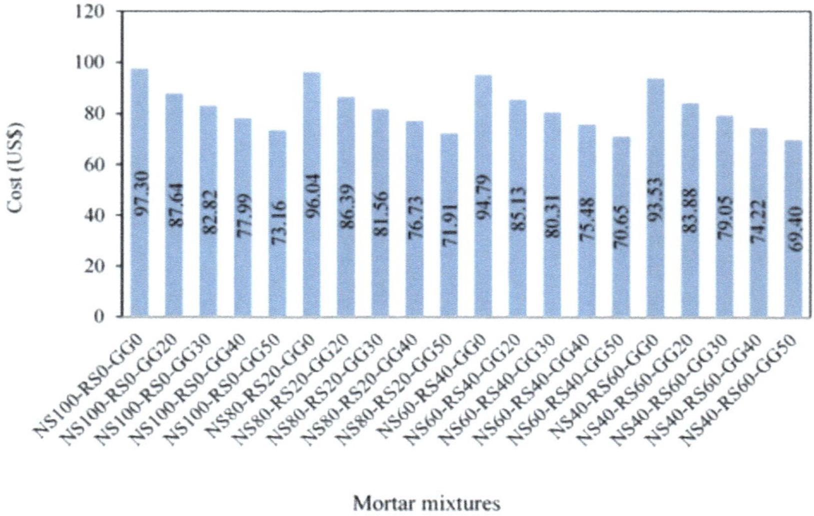

Fig. 4. Cost of the mortars.

5 Environmental Impact Analysis

The environmental impact (EI) analysis of the produced mortar mixtures is discussed in this section. Data on the EI of each mortar's component was gathered from the literature [19–21]. Table 3 shows the results of the EI analysis of the mortars. The EI of the mortar mixtures was determined by summing the environmental contributions of all the components required to produce 1 m^3 of each mortar mixture. Each mixture was evaluated for five environmental categories: abiotic depletion potential (ADP), global warming potential (GWP), ozone depletion potential (ODP), photochemical ozone creation potential (POCP), and acidification potential (AP).

Across all NS-RS ratios, increasing GGBFS content consistently reduces all five EI indicators. For the case of NS100-RS0-GG0 mortar mixture, the GWP was reduced from 458.15 kgCO_2 eq to 284.76 kgCO_2 eq when 40% of the cement was substituted with GGBFS. Cement production is highly energy-intensive and emits a large amount of CO_2 and other pollutants [1, 22, 23]. GGBFS being an industrial by-product has lower embodied energy and emissions. Thus, replacing cement with GGBFS reduces environmental burdens.

Also, the analysis shows that within each fixed GGBFS level, changing the NS-RS ratio has minimal or insignificant impact on the environmental indicators. For instance, the ADP, GWP, and ODP of the mortar mixture NS100-RS0-GG0 were 1.86 kgSbeq, 458.15 kgCO_2 eq, and 5.07×10^{-5} kgCFC-11eq, respectively. For the case of mortar mixture NS60-RS40-GG0, environmental indicator values of 1.86 kgSbeq, 456.59 kgCO_2 eq, and 5.08×10^{-5} kgCFC-11eq were obtained for ADP, GWP, and ODP, respectively.

Table 3. Environmental impact for producing 1 m^3 of the various mortars.

Notation	Environmental impact category				
	ADP (kgSbeq)	GWP ($kgCO_2$ eq)	ODP (kgCFC-11eq)	POCP (kgC_2H_4eq)	AP ($kgSO_2eq$
NS100-RS0-GG0	1.86	458.15	5.07×10^{-5}	0.04	1.35
NS100-RS0-GG20	1.51	371.45	4.10×10^{-5}	0.04	1.14
NS100-RS0-GG30	1.34	328.10	3.62×10^{-5}	0.04	1.02
NS100-RS0-GG40	1.17	284.76	3.13×10^{-5}	0.03	0.91
NS100-RS0-GG50	1.00	241.41	2.64×10^{-5}	0.03	0.80
NS80-RS20-GG0	1.86	457.36	5.08×10^{-5}	0.04	1.35
NS80-RS20-GG20	1.51	370.66	4.10×10^{-5}	0.04	1.14
NS80-RS20-GG30	1.34	327.31	3.62×10^{-5}	0.04	1.02
NS80-RS20-GG40	1.17	283.97	3.13×10^{-5}	0.03	0.91
NS80-RS20-GG50	1.00	240.62	2.64×10^{-5}	0.03	0.80
NS60-RS40-GG0	1.86	456.59	5.08×10^{-5}	0.04	1.35
NS60-RS40-GG20	1.51	369.89	4.10×10^{-5}	0.04	1.14
NS60-RS40-GG30	1.34	326.54	3.63×10^{-5}	0.04	1.02
NS60-RS40-GG40	1.17	283.20	3.14×10^{-5}	0.03	0.91
NS60-RS40-GG50	1.00	239.85	2.65×10^{-5}	0.03	0.80
NS40-RS60-GG0	1.86	455.80	5.09×10^{-5}	0.04	1.35
NS40-RS60-GG20	1.51	369.10	4.11×10^{-5}	0.04	1.14
NS40-RS60-GG30	1.34	325.75	3.63×10^{-5}	0.04	1.02
NS40-RS60-GG40	1.17	282.41	3.14×10^{-5}	0.03	0.91
NS40-RS60-GG50	1.00	239.06	2.65×10^{-5}	0.03	0.80

6 Conclusions

The findings of this study affirm the potential of incorporating RS and GGBFS as sustainable alternatives in mortar production. Although higher levels of RS reduce the strength properties of the mortars, the inclusion of GGBFS mitigates these effects by enhancing the overall performance. The main findings of this study are:

i. Increased replacement of NS with RS leads to a reduction in density, compressive strength and flexural strength of the mortars.
ii. Utilisation of GGBFS up to 40% enhances density and strength properties of mortars.
iii. The employment of RS and GGBFS leads to reduction in mortar production cost.
iv. The combined use of RS and GGBFS contributes to substantial reduction in carbon emissions.

v. While the incorporation of RS yields marginal reductions in both environmental impact and production cost relative to NS, its adverse effect on mechanical performance warrants careful consideration. To preserve the strength characteristics of the mortars, the RS replacement ratio should not exceed 30%.

References

1. Boden TA, Andres RJ (2017) Global CO_2 emissions from fossil-fuel burning, cement manufacturing and gas flaring. Carbon Dioxide Information Analysis Center (CDIAC), Oak National Laboratory (ORNL), Oak Ridge, TN, USA, 1751–2014. https://doi.org/10.3334/CDIAC/00001_V2017
2. Ohemeng EA, Naghizadeh A (2023) Alternative cleaner production of masonry mortar from fly ash and waste concrete powder. Constr Build Mater 374:130859. https://doi.org/10.1016/j.conbuildmat.2023.130859
3. Ghorbel E, Wardeh G, González-Fonteboa B (2023) Effect of 10% recycled sands from various storage sites on the mortar's properties. Eur J Environ Civ Eng 27(14):4067–4085. https://doi.org/10.1080/19648189.2023.2169766
4. Jan A, Ferrari L, Bortolotti V, Mikanovic N, Ben-Haha M, Franzoni E (2024) Enhancement of mortar's properties by combining recycled sand and limestone calcined clay cement. Constr Build Mater 442:137591. https://doi.org/10.1016/j.conbuildmat.2024.137591
5. Kepniak M, Lukowski P (2024) Multicriteria analysis of cement mortar with recycled sand. Sustainability 16:1773. https://doi.org/10.3390/su16051773
6. Awang H, Aljoumaily ZS (2017) Influence of granulated blast furnace slag on mechanical properties of foam concrete. Cogent Eng 4:1409853. https://doi.org/10.1080/23311916.2017.1409853
7. Ohemenga EA, Ramabodu MS, Nena TD, Kancheva Y (2024) Performance of sustainable mortars containing blast furnace slag and fine concrete waste: an environmental perspective. Cogent Eng 11(1):2313053. https://doi.org/10.1080/23311916.2024.2313053
8. Oleiwi SM (2021) Compressive strength of mortar with partial replacement of cement by fly ash and GGBFS. Diyala J Eng Sci 14(4):147–156. https://doi.org/10.24237/djes.2021.14412
9. Thakur LC, Kumar S, Singh JP (2016) Assessment of the properties of cement and mortar using GGBS. Int J Innov Res Sci Eng Technol 5(8):15224–15231. https://doi.org/10.15680/IJIRSET.2016.0508098
10. Ohemeng EA (2023) Potentials of waste concrete elements for production of construction materials, PhD Thesis, University of Johannesburg, Johannesburg, South Africa. Available from: http://hdl.handle.net/102000/0002
11. ASTM C144 (2003) Standard specification for aggregate for masonry mortar. ASTM, International, West Conshohocken, PA 19428-2959, USA
12. ASTM C204 (2007) Standard test methods for fineness of hydraulic cement by air—permeability. ASTM International
13. ASTM C188 (1995) Standard test method for density of hydraulic cement. ASTM International
14. BS 1881–114 (1983) Method for determination of density of hardened concrete. British Stand ards Institution
15. ASTM C109 (2013) Standard test method for compressive strength of hydraulic cement mortar. ASTM International
16. BS EN 1015-11 (1999) Methods of test for mortar for masonry: determination of flexural and compressive strength of hardened mortar. British Standards Institution

17. Prosek Z, Nezerka V, Hluzek R, Trejbal J, Tesarek P, Karraa G (2019) Role of lime, fly ash, and slag in cement pastes containing recycled concrete fines. J Constr Build Mater 201:702–714. https://doi.org/10.1016/j.conbuildmat.2018.12.227
18. Ohemeng EA, Ekolu SO (2020) Comparative analysis on costs and benefits of producing natural and recycled concrete aggregates: a South African case study. Case Stud Constr Mater 13:e00450. https://doi.org/10.1016/j.cscm.2020.e00450
19. Blengini GA (2006) Life cycle assessment tools for sustainable development: case studies for mining and construction industries in Italy and Portugal. PhD Thesis, Universidade Tecnica de Lisboa, Instituto Superior Tecnico, Portugal. Available from: https://fenix.tecnico.ulisboa.pt
20. Chen C, Habert G, Bouzidi Y, Jullien A, Ventura A (2010) LCA allocation procedure used as an incitative method for waste recycling: an application to mineral additions in concrete. Resour Conserv Recycl 54:1231–1240. https://doi.org/10.1016/j.resconrec.2010.04.001
21. Braga A (2015) Comparative analysis of the life cycle assessment of conventional and recycled aggregate concrete (In Portuguese) (Master thesis in Civil Engineering). Instituto Superior Tecnico—University of Lisbon, Portugal
22. Majhi RK, Nayak AN (2020) Production of sustainable concrete utilising high-volume blast furnace slag and recycled aggregate with lime activator. J Clean Prod 255:120188
23. Mehta PK, Monteiro PJM (2014) Concrete: microstructure, properties, and materials, 4th edn. McGraw-Hill Education, New York

Enforced Carbonation of Basic Oxygen Furnace (BOF) Slag: Experimental Study and Thermodynamic Modeling

Natalia Pires Martins(✉), Hadi Kazemi Kamyab, Yury Villagran Zaccardi, Peter Nielsen, and Mieke Quaghebeur

Materials and Chemistry, VITO, Boeretang 200, 2400 Mol, Belgium
natalia.piresmartins@vito.be

Abstract. Many recent studies show that the enforced carbonation of basic oxygen furnace (BOF) slags is a promising method to sequester CO_2 in the material while improving their volume stability, which may enhance their suitability for construction applications. Some potential applications include the use of carbonated BOF slags as a replacement for natural aggregates in concrete, in the production of shaped monoliths, or, after particle size reduction, as an alternative binder and a supplementary cementitious material (SCM). In this study, experiments and thermodynamic modelling are used to investigate phase changes and the CO_2 uptake of two different size fractions of a BOF slag (0–3 mm and 0.5–3 mm) upon enforced carbonation, in addition to the effect of pre-conditioning duration (2 or 24 h) at 10 wt.% moisture. The enforced carbonation process was carried out on the moist BOF slag in an autoclave at 0.1 barG overpressure and 40 °C, using a gas mixture containing 30% CO_2 and 70% N_2 for 24 h. The experiments show that the uncarbonated size fractions have considerable differences in phase composition, which ultimately define their final CO_2 binding capacity. Pre-conditioning at 10 wt.% moisture had little impact on CO_2-uptake and phase assemblage. This was attributed to the pre-hydrated initial state of the materials, as revealed by XRD and TGA. The reaction sequence and the water release during carbonation in the autoclave were not accurately predicted by thermodynamic modelling. However, this could potentially be addressed by the implementation of kinetic models and adequate consideration of the reactivity of the amorphous fraction of the BOF slag.

Keywords: Mineral carbonation · Artificial aggregates · BOF slag

1 Introduction

Basic oxygen furnace (BOF) slag is a by-product of steel production, which shows potential as a precursor for CO_2 sequestration through mineral carbonation (with CO_2 uptake capacities in the range of 53–182 kg CO_2/ton BOF being reported [1–3]. Such an approach offers the dual environmental benefit of valorizing industrial waste and mitigating greenhouse gas emissions. The carbonation process converts reactive calcium

H. Beushausen et al. (Eds.): NTZR 2025, LNCE 873, pp. 136–147, 2026.
https://doi.org/10.1007/978-3-032-27698-8_13

and magnesium phases into stable carbonates. The maximum CO_2 uptake will depend on the mineralogical composition of the BOF slag and the efficiency of the carbonation curing process. Curing can be performed under various conditions of temperature, pressure, and moisture content, including wet and semi-dry methods [3, 4]. Significant progress has been made in understanding the carbonation products formed during the mineral carbonation of BOF slags [3]. Pre-conditioning BOF slag particles can play a key role in carbonation kinetics by controlling the formation of a water film and initiating the dissolution of reactive phases [5], potentially promoting the precipitation of hydration products. Moreover, particle size effects [6] play an important role in achieving effective carbonation degrees. To this aim, critical knowledge gaps persist regarding thermodynamic modelling of the carbonation potential of BOF slags, as well as comparisons between the resulting theoretical and effective CO_2 uptakes resulting from specific particle engineering, pre-conditioning protocols, and carbonation conditions.

The present study couples thermodynamic modelling with experimental data to evaluate the reaction degree of BOF slag phases. A two-step process with moisture pre-conditioning followed by semi-dry carbonation in a reactor is examined. The results indicated that the effects of pre-conditioning duration were more limited than anticipated, potentially linked to the pre-hydration degree of the BOF slag (as received). Challenges in the modelling of carbonated phase assemblage and assessment of the reactivity of the amorphous fraction of the BOF slag with CO_2 are identified.

2 Materials and Methods

The BOF slag used in this study contains particles in the 0–3 mm range and was supplied by Arcelor Mittal. Its characterization data is shown in Table 1. The BOF slag was used in experiments: (a) as received (whole), and (b) after sieving above 500 μm (BOF-C, the coarse fraction). The phase compositions of both fractions are also presented in Table 1, showing that they differ, most remarkably, in the content of portlandite. This phase is concentrated in the fine fraction; hence, sieving has resulted in a lower portlandite content in BOF-C. Descriptions of the used analytical methods are made in Sect. 2.1.

Table 1. Chemical composition and phase composition of the BOF slag (whole; 0–3 mm) and BOF-C (> 500 μm). ACn: amorphous and non-quantified crystalline phases.

Chemical composition (wt.%)

	CaO	Fe_2O_3	SiO_2	MgO	Al_2O_3	LOI (%)
BOF	46.3	23.7	10.3	4.2	2.2	13.6

Phase composition (wt.%)

	Wustite	Magnetite	Quartz	Iron	Lime	Srebrodolskite	Portlandite	γ-C_2S	β-C_2S	Calcite	ACn
BOF	3.6	2.7	0.5	0.9	1.1	10.4	18.5	1.0	10.2	5.9	45.2
BOF-C (> 500 μm)	4.9	3.9	0.4	1.0	2.5	12.4	10.5	2.3	9.4	8.6	43.9

The presence of an amorphous phase in BOF slags is a matter of debate in the literature. While BOF slags are expected to be completely crystalline after solidification [7], the existence of an amorphous fraction is reported in multiple studies. As seen in Table 1, the amorphous content quantified in both BOF and BOF-C is above 40 wt.%; considerably higher than the 14–17 wt.% reported in other studies [8, 9], but within the wider 10–55 wt.% range reported elsewhere [10]. A very high initial amorphous content could point to the weathered state of the initial materials due to natural hydration of the slag in the slag yard, which is supported by the detection of pyroaurite ($Mg_6Fe_3(CO_3)(OH)_{16}\cdot 4(H_2O)$) and hydrogarnet in the XRD patterns. Both hydrated phases were detected as broad reflections, and their accurate quantification in the samples was not possible using the Rietveld method. For that reason, they are not included individually in Table 1 and are instead quantified as part of the amorphous fraction.

2.1 Experimental Study

The BOF (whole) and BOF-C were pre-conditioned, carbonated, and characterized after each processing step to determine the phase composition and the CO_2 uptake. Pre-conditioning consisted of adding 10 wt.% di-water and storing the moist material at RH > 90% for 2 h and 24 h. After the designated pre-conditioning duration, the materials were placed in a tray and static carbonation was carried out in a lab-scale in-house reactor at 0.1 barG overpressure, 40 °C, using a gas mixture containing 30% CO_2 and 70% N_2 for 24 h. The CO_2-uptake after carbonation was determined based on the variation in the CO_2 content quantified by total carbon (TC) analysis (multi EA 4000 analyzer, Analytik Jena).

The phase composition of the samples was determined by X-ray diffraction (XRD) using the Rietveld method on HighscorePlus and the PDF-5 database. Powders below 63 μm were measured at 40 kV and 45 mA on a PANalytical Empyrean with a Co source using a step size of 0.0131° 2θ and a counting time of 0.02 s/step. The samples were vacuum-packed and stored for a few days before analysis. The amorphous fraction was estimated using the external standard method (pure, crystalline rutile). Finally, the absolute phase abundances were rescaled to a common basis, i.e., g/100g of initial BOF slag, taking into account the mass changes resulting from pre-conditioning and subsequent carbonation.

Thermogravimetric analysis (TGA) was carried out on a NETSCH—STA 449 F3 Jupiter. 50mg powder samples (< 63 μm) were analyzed from room temperature to 1000 °C at 10 °C/min using a nitrogen flow of 250 ml/min.

The chemical composition of the amorphous fraction of the BOF slag was estimated using Eq. (1), based on the total chemical composition of the sample, determined by X-ray Fluorescence (XRF) on a fused bead, and the chemical composition of the crystalline fraction. The latter was determined based on XRD assuming stoichiometric compositions of BOF slag phases. This simplification implies that the presence of low concentrations of V_2O_5, MnO, CrO_3, and TiO_2, which have been reported in some of the BOF phases [8, 11], is disregarded here. For ferrite and wustite, Fe:Al = 0.80:0.20 and Fe:Mg = 0.85:0.15 were considered, respectively, because they resulted in lower error, i.e., negative oxide concentrations (Y_i), after the application of Eq. (1).

$$Y_i = \frac{W_i - (1 - Acn) * X_i}{Acn} \tag{1}$$

where W_i is the total content of oxide i in wt.%, determined by XRF; X_i is the content of oxide i relative to the crystalline fraction, extracted from XRD/Rietveld refinement; and ACn is the amorphous fraction.

2.2 Thermodynamic Modeling

The open-source software GEM-Selektor was used for the thermodynamic simulations. Since in semi-dry systems the availability of moisture is limited, focus was given to gas-solid equilibria, and the aqueous electrolyte model was disabled. The simulations consisted of the step-wise addition of (0.5 g of) CO_2 to systems equilibrated at 40 °C, and 1.1 barG. Preliminary simulations were used to identify phases that, despite being predicted as thermodynamically stable reaction products, are not expected to form in the system under the experimental conditions used or in a relevant timescale, e.g., zeolites, SiO_2 minerals, and anhydrous calcium silicates (wollastonite, alite). Those phases were manually prohibited from forming.

3 Results and Discussion

3.1 CO_2-Uptake

To investigate the effect of pre-conditioning experimentally, the BOF size fractions were pre-conditioned by adding 10% di-water for 2 and 24 h, after which they were carbonated. The CO_2-uptake of BOF and BOF-C after carbonation, as determined by TC analysis, was 147.2 ± 0.0 and 30.8 ± 0.3 kg/ton for 2 h of pre-conditioning, and 147.2 ± 0.5 and 31.0 ± 1.6 kg/ton of dry initial solids for 24 h, respectively. The BOF-C fraction reaches considerably lower CO_2-uptake than the BOF (whole), which is an expected result attributed to the coarser particle size of the former. The prolonged duration of the pre-conditioning step at 10 wt.% moisture (from 2 h to 24 h) resulted in no improvement in the CO_2-uptake.

The maximum theoretical CO_2 uptake of the BOF slag, assuming full carbonation of the initial CaO content, and correcting for the initial content of calcite determined by XRD, is 340 kg CO_2 per ton of initial BOF. The results of TC analysis indicate that less than 50% of the maximum theoretical CO_2 uptake was achieved.

3.2 XRD

The phase composition of the initial, pre-conditioned, and carbonated samples, determined by XRD/Rietveld, is shown in Fig. 1. Most of the crystalline phases present in the original samples (BOF, BOF-C) do not present a substantial change in content ($\geq$ 3 wt.%) throughout pre-conditioning and carbonation, with the exception of portlandite, β-C_2S, and calcium carbonates. Although the calcium ferrite phase, i.e., srebrodolskite is considered reactive it did not show a clear decrease in content, in agreement with another study [12]. The content of the amorphous fraction also varies throughout the processes, especially in the BOF-C. Generally, no precipitation of crystalline hydration products is detected after pre-hydration of the BOF and BOF-C, regardless of the pre-hydration duration. Moreover, the formation of carbonated phases was comparable for samples pre-conditioned for 2 h and 24 h. These results suggest that the anticipated carbonation-enhancing chemical effect of prolonged pre-conditioning was not verified in the samples, which is in agreement with the constant CO_2 uptake determined by TC analysis (Sect. 3.1). However, the variation in the amorphous content is not clearly explained based on the analysis of the crystalline phases and demands further investigation/verification by alternative methods. The amorphous content determinations in this study were made using the external standard method, which is considered generally accurate compared to other methods [13].

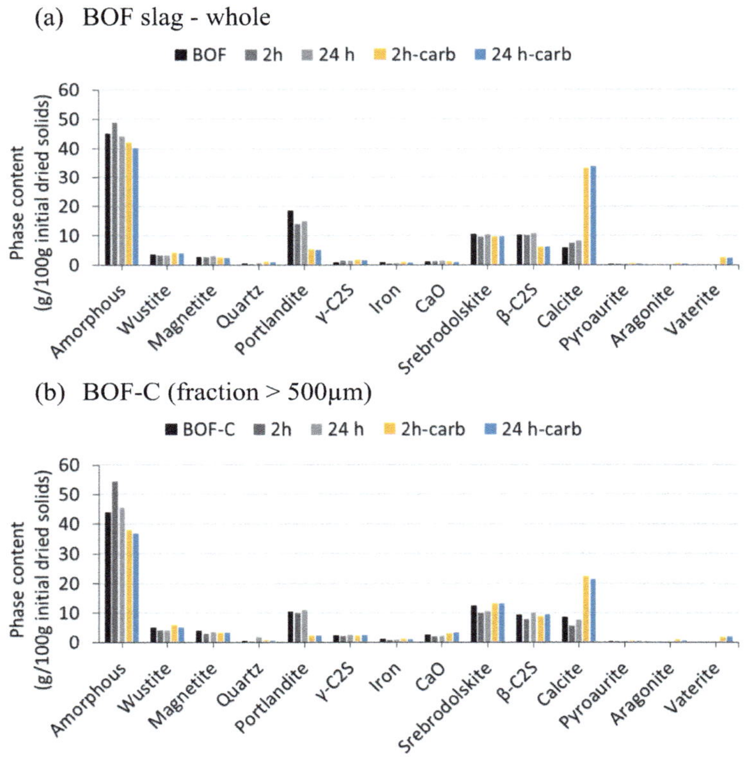

Fig. 1. Rescaled phase contents (to g/100g of initial dried solids) of samples of (**a**) BOF and (**b**) BOF-C: initial (black), pre-conditioned at 10% moisture for 2 (dark gray) and 24 h (light gray), and the corresponding carbonated samples that had been pre-conditioned for 2 (yellow) and 24 h (blue).

When the pre-conditioned BOF samples are carbonated in the autoclave, a significant decrease in the portlandite content is detected, accompanied by the reaction of β-C_2S and the extensive formation of calcite. Minor amounts of vaterite and aragonite were also detected after carbonation. Interestingly, those changes take place while there is a simultaneous reduction in the amorphous content, suggesting that partial carbonation of the amorphous fraction might occur.

Despite being a carbonatable phase, the reaction of γ-C_2S was not detected based on XRD/Rietveld analysis. This phase is generally considered more reactive to carbonation than β-C_2S [14]. In light of that, it is reasonable to assume that the little/no reaction of this phase in the materials investigated is limited by physical barriers. Such barriers could be associated to: (1) insufficient surface exposure to CO_2 when γ-C_2S is present as inclusions in large BOF particles, and/or (2) the formation of a calcite reaction layer around the γ-C_2S surface due to the eventual pre-carbonation of this phase during storage, prior to the use of the materials in this study. The formation of product layers around the undissolved particle core is known to work as a shield, preventing further reaction [15]. This effect can be even more critical to the extent of carbonation in materials with a larger particle size and is likely the main factor behind the limited reaction of β-C_2S in the BOF-C samples.

3.3 TGA

Selected samples were measured by TGA to validate XRD results and evaluate the evolution of the bound water content throughout the processes. The TG and DTG curves are shown in Fig. 2, where the main mass loss events are indicated: the decarbonation of $CaCO_3$ (calcite, vaterite, aragonite); the dehydroxilation of portlandite ($Ca(OH)_2$); and the dehydration of the amorphous phase(s). The gradual mass loss in the temperature range 30–375 °C was assigned to amorphous phases because the only abundant crystalline phase containing bound water identified in the BOF and BOF-C is portlandite (see Fig. 1 and Table 1).

The TG and DTG curves of the initial and pre-conditioned materials (black and gray), which largely overlap, especially in BOF-C, show that the pre-conditioning resulted in little changes in the contents of portlandite, $CaCO_3$, and bound water. This suggests that the extent of hydration and carbonation during pre-conditioning is negligible for the BOF-C, while only minor carbonation of $Ca(OH)_2$ occurred in the BOF slag during pre-conditioning. After enforced carbonation (yellow curves), intense conversion of portlandite to $CaCO_3$ is observed. The quantifications based on TGA are shown in Table 2. The carbonation process resulted in a substantial reduction in the content of bound water (BW) in the samples, predominantly due to the carbonation of portlandite. Most of the BW that is bound in the amorphous phases, however, remains bound in the amorphous phases after carbonation. In the samples investigated here, this corresponds to 58–68 wt.% of the total BW. It should be noted that the BW determinations may be somewhat overestimated due to the possible, though unconfirmed, presence of amorphous carbonates in the samples, as these decompose at temperatures below 525 °C. The TGA results generally confirm what was suggested based on the initial characterization of the BOF and BOF-C. Both fractions already presented an elevated level of pre-hydration prior to the imposition of controlled pre-conditioning (10 wt.% moisture, for 2 or 24 h). Hence, the pre-conditioning resulted in little additional changes to the material.

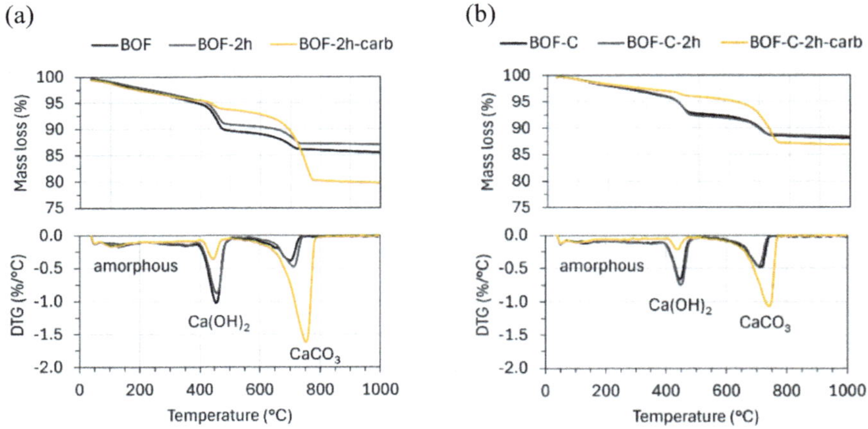

Fig. 2. TG and DTG curves of samples of (**a**) BOF and (**b**) BOF-C: initial (black), pre-conditioned at 10% moisture for 2 h (dark gray), and the corresponding carbonated samples that had been pre-conditioned for 2 h (yellow).

Table 2. Values of total bound water, including CH (mass loss below 525 °C); CH (calculated based on mass loss between 375 and 525 °C, using the tangential method), bound water, excluding CH (the difference between total bound water and the water bound in CH), and total $CaCO_3$ (calculated based on mass loss between 525 and 800 °C), determined by TGA. CH refers to portlandite ($Ca(OH)_2$). Values were rescaled to a common basis (g/100g initial dried solids) based on the mass changes verified during pre-conditioning and carbonation.

g/100g initial dried solids		Total bound water, incl. CH	Bound water, excl. CH	CH	$CaCO_3$
BOF	Initial	10.1	6.0	17.0	8.0
	2 h	8.9	5.2	15.2	8.0
	2 h carb	6.4	5.2	4.7	32.6
BOF-C	Initial	7.1	4.8	9.6	9.3
	2 h	7.5	4.7	11.4	8.0
	2 h carb	4.0	3.5	2.0	20.0

3.4 The Amorphous Fraction of the BOF Slag

The composition of the amorphous fraction, estimated using Eq. (1), is given in Fig. 3 for the BOF slag before and after carbonation. The concentration of CO_2 in the amorphous fraction for both samples fluctuated around zero ± 2 wt.%, which is within the error margin of the methodology. In general, Fig. 3 shows that the individual oxide concentrations of the amorphous fraction of the BOF slag do not change substantially after carbonation. Nevertheless, the estimated simultaneous slight reduction in CaO and increase in SiO_2 leads to a decrease in the Ca/Si ratio of the amorphous fraction from 2.5 to 2.0. This could be explained by one or a combination of the following: (1) the formation of silicious gel due to the reaction of C_2S, leading to a Si-enrichment of the amorphous fraction; (2) the formation of $CaCO_3$ due to the reaction of a fraction of the amorphous calcium, for instance from C-S-H, leading to a Ca-enrichment of the crystalline fraction; or (3) the conversion of amorphous $CaCO_3$ (if present at all) to crystalline $CaCO_3$.

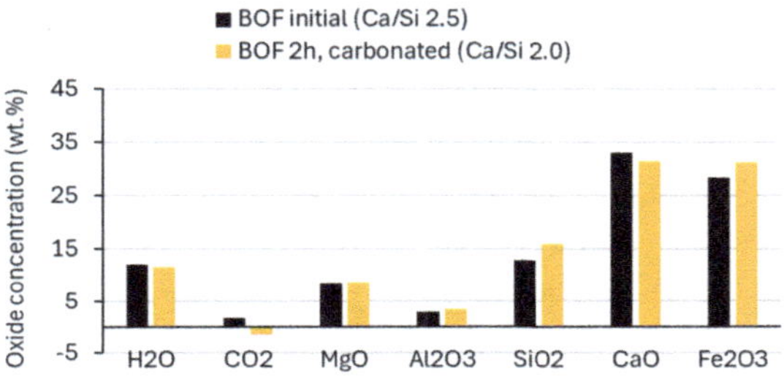

Fig. 3. Estimated oxide composition of the amorphous fraction of initial BOF slag before (black) and after carbonation (yellow).

Based on previous studies, the XRD-amorphous fraction of the BOF slag corresponds mainly to C_2S [9] or C-S-H [12]. While the hydration of crystalline C_2S is typically slow [9], its amorphous counterpart is expected to be more reactive towards hydration [8] and might generate C-S-H during prolonged storage in the presence of moisture. The high initial Ca/Si ratio of the amorphous fraction of BOF slag (2.5), i.e., higher than the Ca/Si ratios measured in C-S-H gels, and the estimated concentrations of Fe_2O_3, MgO, and Al_2O_3 indicate the presence of more than one amorphous phase. It is suggested that the amorphous fraction comprises several amorphous hydration products formed during storage and pre-conditioning, such as C-S-H and AFm phases (hydrogarnet, pyroaurite). The latter could have formed due to pre-hydration of srebrodolskite [9], while the former could be a reaction product of the pre-hydration of C_2S.

The reactivity of at least part of the amorphous fraction of the BOF slag was suggested based on XRD, given the decrease in amorphous content after carbonation. This aspect is further investigated by mass balance, taking into account the degree of reaction (DoR) of crystalline reactive phases determined by TGA (CH) and XRD (CH and β-C_2S) after carbonation in the autoclave. For the BOF slag, the DoR of CH and β-C_2S is 70% and 40%, respectively. For the BOF-C, the DoR of CH was estimated at 80%, while no substantial reaction was observed for β-C_2S based on XRD quantification. For both starting materials, similar determinations ($\pm$ 2%) were obtained for the DoR of CH by TGA. Those values were worked out together with the total $CaCO_3$ content in the carbonated samples, assuming direct gas-solid carbonation reactions for CH (Eq. 2) and β-C_2S (Eq. 3) [15], resulting in the breakdown of the $CaCO_3$ "precursors" shown in Fig. 4.

$$Ca(OH)_2 + CO_2 \rightarrow CaCO_3 + H_2O \tag{2}$$

$$Ca_2SiO_4 + 2\ CO_2 \rightarrow 2\ CaCO_3 + SiO_2 \tag{3}$$

The results in Fig. 4 show that, both in BOF and BOF-C, the direct carbonation reactions of CH and β-C_2S are not sufficient to make up for the total detected $CaCO_3$ content. The determinations made by TGA and XRD are in close agreement and suggest that a substantial fraction of the total $CaCO_3$ formed in the carbonated BOF and BOF-C must be a product of the carbonation of other phase(s). Given that no other crystalline phase has shown an unambiguous decrease in content after the enforced carbonation experiments, the formation of this fraction of $CaCO_3$ is attributed to the partial reaction of the amorphous fraction. Since Eq. 3 accounts for the full carbonation of the calcium present in the reacted fraction of β-C_2S in each sample, skipping the intermediate formation of amorphous C-S-H from β-C_2S hydration, the estimation is rather conservative.

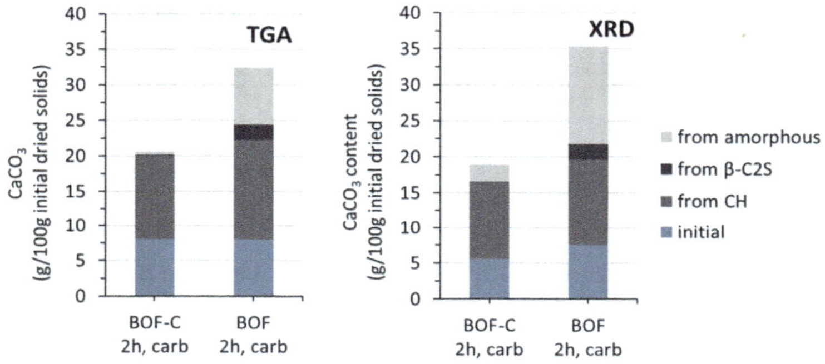

Fig. 4. Breakdown of the total $CaCO_3$ content of the carbonated samples, determined by TGA and by XRD. The portion assigned as "from CH" was calculated based on the DoR of CH (determined by XRD and TGA). The portion assigned as "from β-C_2S" was calculated based on the DoR of β-C_2S (determined by XRD only). The portion assigned as "from amorphous" was determined based on the subtraction of the former and the initial $CaCO_3$ in the uncarbonated sample from the total $CaCO_3$ content in the sample (determined by XRD and TGA).

3.5 Thermodynamic Modelling

Crystalline carbonatable phases that displayed reactive behavior (based on XRD analysis: CH and β-C_2S) were used as input for the equilibrium calculations of phase assemblage as a function of the addition of CO_2; other phases were considered inert/unreacted. The results are shown in Fig. 5. The 10 wt.% moisture was not added to the system definition because preliminary simulations showed that this would result in the full consumption of the moisture by the hydration of β-C_2S, which is not realistic within the duration of the carbonation experiment due to the slow kinetic of β-C_2S hydration. The simulations of both materials were performed without imposing kinetic restrictions for CH and β-C_2S. Obviously, the CO_2-uptake reaches a plateau after full carbonation of CH and β-C_2S. Given the lower initial CH content in BOF-C, this maximum theoretical CO_2-uptake plateau is lower for the BOF-C. The results indicate the occurrence of two simultaneous reactions: the reaction of CH with CO_2, forming $CaCO_3$ and releasing free water, as indicated by Eq. (2); and (b) the reaction of β-C_2S with the free water released by Eq. (2), leading to the formation of C-S-H and CH. While the fresh CH instantly reacts with the added CO_2 and forms $CaCO_3$, the start of the carbonation of C-S-H is only predicted after CH depletion and is characterized by a progressive decrease in the Ca/Si ratio until full decalcification occurs. At that late stage, the only carbonation products predicted are Si-gel and $CaCO_3$.

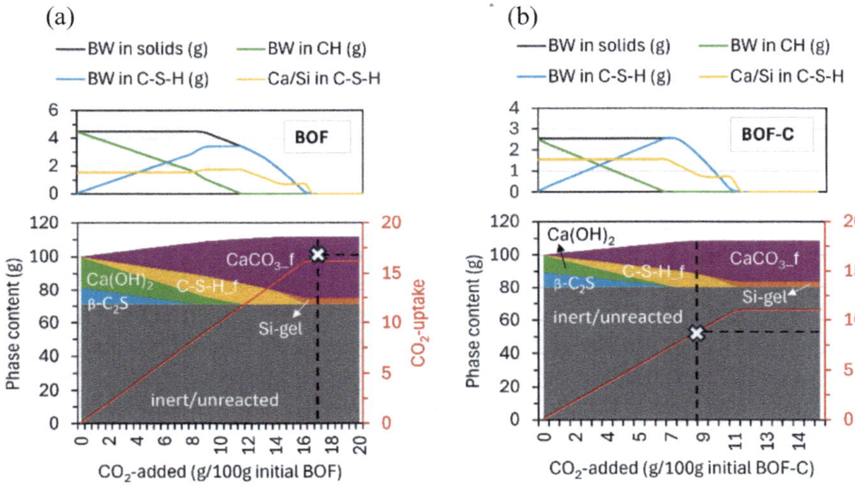

Fig. 5. Thermodynamic modelling of the carbonation of (**a**) BOF slag (whole); and (**b**) BOF-C. The phase contents are given in g per 100g of uncarbonated, pre-hydrated material. The contents of total bound water (BW) in the solids, in CH, and in C-S-H are shown above, as well as the Ca/Si ratio in C-S-H. The continuous red line (secondary y-axis) indicates the CO_2 uptake of the simulated phase assemblage (g/100g initial BOF or BOF-C). The dashed black lines and white cross indicate the rescaled CO_2 uptake (by TGA) of BOF and BOF-C: 16.6 and 8.80 g/100g of initial material. "_f" was added to C-S-H and $CaCO_3$ to highlight that both were formed during the carbonation process; not to be mistaken with the initial C-S-H and $CaCO_3$ contained in the starting material, which in the model are part of the fraction designated as "inert/unreacted".

Based on the simulations, the total BW in the solids remains constant despite CH carbonation, due to the binding of the "released" water by the hydration of β-C_2S. The experiments show that, in fact, the total BW in the system is substantially reduced after carbonation for 24 h, and the actual DoR of β-C_2S is considerably lower than what is predicted in the simulations at the value of CO_2-uptake determined by TGA (indicated with a white cross in Fig. 5). The implementation of a kinetic model of β-C_2S reaction would likely reduce the observed discrepancies between simulation outcomes and experimental data, since it could impose an appropriate DoR for β-C_2S instead of allowing its hydration to be controlled by the generation of free water in the system. The simultaneous carbonation of CH and C-S-H is also not captured by the thermodynamic simulations alone, thus requiring coupling with the kinetic modelling of C-S-H carbonation. Contrary to what is suggested by the model, in semi-dry carbonation, carbonatable phases can react simultaneously with minimal interference or mutual impact on their carbonation rates, as the process is primarily governed by CO_2 diffusion [16]. However, their reaction rates are not uniform. Experimental evidence shows that CH carbonates more rapidly than β-C_2S and C–S–H, largely due to its higher solubility [12], as reflected in its significantly higher DoR after 24 h.

Finally, the model supports what was already suggested by mass balance calculations and XRD results: the initial amorphous phase(s) in the BOF slag, particularly the C–S–H fraction, undergo partial carbonation. This conclusion is more clearly illustrated in Fig. 5a, which shows that the experimental CO_2 uptake matches a highly advanced reaction stage in the simulation, one that can only be reached if the carbonation of non-crystalline phase(s) is taken into account.

4 Conclusions and Perspectives

This study investigated the pre-conditioning and carbonation of two BOF slag particle size fractions using experiments and thermodynamic modeling. It was found that pre-conditioning did not have a significant impact on the phase assemblage of the samples nor on their CO_2-uptake. This was attributed to the pre-hydrated state of the initial materials, which already presented 4.8–6.0 wt.% of bound water (excluding portlandite). The results suggested that multiple amorphous phases were present in the initial materials, including hydrated phases (likely hydrogarnet, pyroaurite, and C-S-H). Based on mass balance calculations, at least part of the amorphous fraction reacts with CO_2 and forms additional $CaCO_3$, leading to a decrease in the Ca/Si ratio of the amorphous fraction ($2.5 \rightarrow 2.0$).

Thermodynamic modeling of phase assemblages must be integrated with complementary approaches to accurately reflect experimental observations. Notably, it has become evident that the hydration of C_2S requires regulation through a kinetic model. Simulations indicate that the water produced from the carbonation of $Ca(OH)_2$ remains largely unreactive during the carbonation process, rather than contributing to C_2S hydration as predicted thermodynamically. Secondly, the simultaneous carbonation of C-S-H and CH, observed experimentally, also requires the implementation of kinetic considerations in the model. Without them, the model assumes an unrealistic scenario in which C-S-H carbonation only begins after complete carbonation of CH.

Beyond the scope of this study, reactive transport models are also welcome to fully account for the impact of particle size, layer formation, and mass transport on the carbonation of BOF slag for use as artificial aggregates. Moreover, although elements such as Mg, Al, and Fe were excluded from the simulations performed in this study, they are present in the systems and might potentially affect their ultimate CO_2-uptake. Those elements are often hosted in phases for which thermodynamic properties have still not been determined. Database expansion is therefore warranted to encompass the complexity of the systems.

Acknowledgments. The present research was funded by VLAIO (Flanders Innovation & Entrepreneurship): ICON Project Eco2Chape (HBC.2023.0680). Special thanks to Senne Smets for carrying out the carbonation experiments, Zsófia Kovácska and Myrjam Mertens for XRD data acquisition, and Roger Hameleers for XRF analyses.

References

1. Quaghebeur M et al (2015) Accelerated carbonation of steel slag compacts: development of high-strength construction materials. Front Energy Res 3(52):12
2. Reddy KR et al (2018) CO2 sequestration using BOF slag: application in landfill cover. In: PREXIV, Thessaloniki, Greece
3. Wang S et al (2024) A review on the carbonation of steel slag: properties, mechanism, and application. Materials (Basel) 17(9)
4. Wang S, Kim J, Qin T (2024) Mineral carbonation of iron and steel by-products: state-of-the-art techniques and economic, environmental, and health implications. J CO2 Util 81
5. Zhang X et al (2024) Use of steel slag as carbonation material: a review of carbonation methods and evaluation, environmental factors and carbon conversion process. J CO2 Util 88
6. Polettini A, Pomi R, Stramazzo A (2016) Carbon sequestration through accelerated carbonation of BOF slag: influence of particle size characteristics. Chem Eng J 298:26–35
7. Thomas C et al (2019) 7—Steel slags, in new trends in eco-efficient and recycled concrete. In: J de Brito, F Agrela (eds). Woodhead Publishing, pp 169–190
8. Santos WF et al (2023) Quantitative analysis and phase assemblage of basic oxygen furnace slag hydration. J Hazard Mater 450:131029
9. Kaja AM et al (2021) On the optimization of BOF slag hydration kinetics. Cem Concr Compos 124
10. Lyza JE et al (2023) Challenges of quantitative phase analysis of iron and steel slags: a look at sample complexity. Powder Diffr 38(2):119–138
11. Wang D, Chang J, Ansari WS (2019) The effects of carbonation and hydration on the mineralogy and microstructure of basic oxygen furnace slag products. J CO2 Util 34, pp 87–98
12. Chang J, Wang D, Fang Y (2018) Effects of mineralogical changes in BOFS during carbonation on pH and Ca and Si leaching. Constr Build Mater 192:584–592
13. Scarlett NVY, Madsen IC (2018) Effect of microabsorption on the determination of amorphous content via powder X-ray diffraction. Powder Diffr 33(1):26–37
14. Jiang T, Cui K, Chang J (2023) Development of low-carbon cement: carbonation of compounded C2S by β-C2S and γ-C2S. Cement Concr Compos 139:105071
15. Biava G, Depero LE, Bontempi E (2024) Accelerated carbonation of steel slag and their valorisation in cement products: a review. Span J Soil Sci 14
16. Song Q, Chai SYW, Ling T-C (2025) Role of calcium compounds in BOFS on phase evolution and microstructural changes upon high-gravity carbonation. Cem Concr Res 195:107913

Balancing Performance and Sustainability: A Study on Rice husk in Innovative Plaster Formulations

Chiara Chiatti[1], Giorgio Baldinelli[2], and Francesco Asdrubali[1](✉)

[1] Department of International Human and Social Sciences, Perugia Foreigners' University, Piazza Fortebraccio 4, 06122 Perugia, Italy
francesco.asdrubali@unistrapg.it

[2] Department of Engineering, University of Perugia, Via G. Duranti 93, 06125 Perugia, Italy

Abstract. Pursuing higher energy efficiency and reduced environmental impact is a key strategy in building sector renovation. Sustainable building design increasingly promotes the use of materials incorporating natural or recycled components. Within this framework, the present study investigates the hygrothermal performance of three innovative plaster formulations to assess the effect of partially re-placing expanded amorphous silica - known for its energy-intensive production process - with a more sustainable agricultural by-product, i.e., rice husk. The three formulations differ in the percentage of rice husk content: 5.0%, 7.5% and 10.0%, respectively, substituting equivalent amounts of expanded silica. Experimental measurements of thermal conductivity, heat capacity and thermal diffusivity are carried out to assess the absolute behavior of each mix under controlled laboratory conditions. Results show how the introduction of rice husk slightly reduces the overall thermal performance of the plaster, particularly in terms of thermal conductivity, but brings advantages in terms of breathability and moisture regulation, contributing to improved indoor comfort and environmental sustainability. This research contributes to the growing body of knowledge on sustainable building technologies by proposing practical solutions to reduce the environmental footprint of construction materials, while maintaining a balanced compromise between technical performance and ecological criteria.

Keywords: Insulation · Sustainable materials · Thermo-physical properties · Building envelope

1 Introduction

The construction sector has a significant environmental impact, primarily due to the intensive use of non-renewable resources. This contributes to land degradation, solid waste generation, air and noise pollution, and elevated greenhouse gas emissions [1]. It is estimated that in Europe buildings are responsible for around 40% of total energy consumption and 36% of energy-related CO_2 emissions, while in Italy buildings cause almost 19% of total GHG emissions, primarily from heating and cooling [2]. The outdated building stock (81% constructed before 1990) and a low annual renovation rate of 0.85% have further driven up buildings energy consumption in recent years [3].

H. Beushausen et al. (Eds.): NTZR 2025, LNCE 873, pp. 148–157, 2026.
https://doi.org/10.1007/978-3-032-27698-8_14

Improving energy efficiency and reducing environmental impacts are central strategies in the ongoing transition toward more sustainable construction practices. To address these challenges, researchers are focusing on developing innovative thermal insulation systems that reduce heat exchange between buildings and their external environments while simultaneously lowering ecological footprints [4]. Within this context, there is growing interest in sustainable materials that incorporate natural and recycled components, particularly low-cost agricultural residues [5]. Natural fibers – derived from industrial and agricultural by-products – offer not only effective thermal insulation but also contribute to a significant reduction in environmental impact [6]. By utilizing such materials, the construction industry can advance material reuse, conserve energy, and support the broader objectives of sustainable development and resource efficiency.

Rice is one of the most widely cultivated crops globally, feeding nearly half of the world's population. Rice husk, a by-product of rice milling, accounts for approximately 20% of rice production, generating an estimated 200 million tons of agricultural waste annually [7]. Due to its slow biodegradation, rice husk presents environmental disposal challenges, reinforcing the importance of its valorization as a functional construction material. Moreover, rice husk is characterized by a high silica content, which distinguishes it from other agricultural residues and enhances its technical performance in building applications [8]. Recent research has focused on the development of insulation materials from rice husk, including panels produced through pulping methods, achieving thermal conductivity values of 0.037 W/mK and densities around 137 kg/m^3 [9]. Composite materials incorporating rice husk with different binders have demonstrated favorable thermal and mechanical properties, with thermal conductivity ranging from 0.0045 to 0.108 W/mK and densities between 107 and 378 kg/m^3 [10–12]. Additionally, rice husk ash has been used as partial replacement for Portland cement to enhance concrete performance, and hybrid insulation blocks made with rice husk and aluminous cement have also been studied [13]. When applied using the blowing technique, rice husk fiber is evenly distributed within building surfaces, preserving internal porosity and improving the thermal performance of the final product. This method enables insulation to be installed without removing interior or exterior finishes, creating air pockets whose thermal behavior varies according to the fiber type, ultimately enhancing the building's overall energy efficiency [14]. The same application technique allows for rapid installation, reducing on-site labor requirements, and it is well suited for modular construction contexts [15]. Moreover, the fibers are often more cost-effective than conventional insulation materials, offering a financially viable option for sustainable housing developments. From the durability point of view, it has been shown that rice husk helps in reducing capillary water absorption and improve resistance to water erosion, thereby enhancing moisture resistance [16]. Rice husk fibers also contribute to the reduction of shrinkage and cracking during drying, with reported values typically below 3–4% [17]. This effect is attributed both to the reinforcement action of the rough fibrous particles and to chemical bonding phenomena occurring within the matrix. Although specific data on biological stability of rice husk plasters are limited, available research indicates that rice husk additions can mitigate crack propagation and surface erosion, while fermentation processes may release organic compounds that further stabilize the mixture [18].

In this framework, the present study investigates the hygrothermal behavior of plaster mixtures incorporating rice husk as a partial replacement for expanded amorphous silica. The substitution aims to reduce the environmental burden of traditional plasters without significantly compromising thermal performance. By combining laboratory testing and sustainability evaluation, this research contributes to the growing topic of knowledge on construction innovative technologies by:

- Demonstrating the viability of rice husk as a functional component in plaster formulations;
- Offering a low-impact alternative to energy-intensive materials like expanded silica;
- Supporting the use of agricultural by-products in industrialized construction systems and techniques.

Ultimately, this work promotes a balanced approach to sustainability and material development, in which environmental criteria and technical performance are optimized in tandem, advancing the growth of more resilient and eco-friendly building materials [19].

2 Materials and methods

2.1 Materials and samples preparation

This study investigates the hygrothermal performance of three innovative plaster formulations designed to evaluate the effects of partially replacing expanded amorphous silica, i.e., an industrial material with a highly energy-intensive production process, with a more sustainable alternative: rice husk. The plasters are based on a lightweight mineral-cork matrix composed of cork granules (0–3 mm), expanded amorphous silica, perlite, and pumice. For the experimental campaign, three formulations were produced by replacing 5%, 7.5% and 10% by weight of expanded silica with equivalent amounts of rice husk. The mixing water content ranged between 0.8 and 1.0 l/kg, resulting in an average fresh density of 250 ± 15% kg/m^3. Two samples for each plaster formulation were realized (20 cm × 20 cm, Fig. 1), accounting for a thickness that could be representative of a realistic layer, i.e., ~ 3 cm.

The proposed material has a lower carbon footprint than the conventional one, since expanded amorphous silica is produced through an energy intensive process, while rice husk, a by-product of rice milling, is a natural material which incorporates biogenic carbon and that, instead of being disposed of, is recycled as a new material.

Fig. 1. Samples (20 cm × 20 cm) of the three investigated plasters with different contents of rice husk (5%, 7.5%, 10%).

2.2 Thermal investigation

Hot Disk apparatus. Thermal conductivity (λ, W/mK), thermal diffusivity (α, mm^2/s), and specific heat (c, MJ/m^{3}K) of the three plaster formulations were determined using the Transient Plane Source (TPS) method, an iterative approach based on transient heat conduction analysis. Measurements were performed with a Hot Disk 2500S (Fig. 2a), in accordance with ISO 22007-2:2022 [20]. The experimental setup consisted of placing the Hot Disk sensor n. 5501 between the flat surfaces of two identical samples. The sensor, which acted simultaneously as a heat source and temperature probe, was made of a nickel spiral resistor through which an electric current was applied, causing a controlled temperature increase. The generated heat dissipated symmetrically through both samples at a rate dependent on the thermal transport properties of the investigated material. The transient temperature rise over time, $\Delta T(t)$, could be expressed as:

$$\Delta T(t) = \Delta T_i(t) + \Delta T_s(t)$$

where $\Delta T_i(t)$ represents the initial temperature rise across the insulating layer of the sensor, and $\Delta T_s(t)$ corresponds to the temperature increase at the sample surface. By analyzing the transient temperature response, λ and α were determined. The thermal diffusivity was estimated iteratively using the dimensionless specific time function $D(\tau)$, defined by $\tau = \left(t \cdot \alpha \cdot r^{-2}\right)^{1/2}$, where r is the sensor radius. The relation between $\Delta T_s(t)$ and $D(\tau)$ was fitted using a least-squares method, adjusting α and the time correction parameter t_c until a linear correlation was achieved. The thermal conductivity λ was then calculated from the slope of the linear relation:

$$\Delta T_s(\tau) = \frac{P_0}{\pi^{\frac{3}{2}} r \lambda} D(\tau)$$

where P_0 is the power emitted by the sensor. Finally, the volumetric heat capacity C was calculated as:

$$C = \lambda/\alpha$$

For each plaster formulation, five measurements were carried out by repositioning the sensor at different locations along the sample edges. The average values and their standard deviation were finally considered. All tests were conducted under controlled laboratory conditions, at 26 °C and 36% relative humidity.

Heat Flow Meter apparatus. Thermal conductivity was also measured using a Heat Flow Meter (HFM 446 Lambda Large Eco-Line [21], Fig. 2b) under steady-state conditions, in accordance with ASTM C518-21 [22]. The sample was placed between the instrument's upper and lower plates, which were subsequently maintained at two different temperatures to establish a controlled thermal gradient. The latter induced a vertical heat flux through the sample, from the hot plate to the cold plate, governed by Fourier equation for one-dimensional heat conduction:

$$q = \frac{Q}{A} = \lambda \frac{\Delta T}{\Delta x}$$

where A is the sample area, ΔT is the temperature difference across the sample, and Δx is the thickness of the sample. During the test, embedded sensors in both plates continuously recorded temperature and heat flux. The voltage signals (E, in mV) generated by the sensors were converted into heat flux values using pre-established calibration constants. The heat flux q was directly proportional to the measured voltage, based on separate calibration factors for the upper (S_u) and lower (S_l) plates:

$$q = S_{u/l} \cdot E_{u/l}$$

Calibration of each plate was conducted over a defined temperature range using reference materials with known thermal conductivities traceable to a National Metrology Institute. Once thermal equilibrium was reached and all readings stabilized, the measured values are used to calculate λ as follows:

$$\lambda = q \cdot \frac{\Delta T}{\Delta x} = \frac{1}{2}(S_u E_u + S_l E_l) \frac{\Delta x}{\Delta T}$$

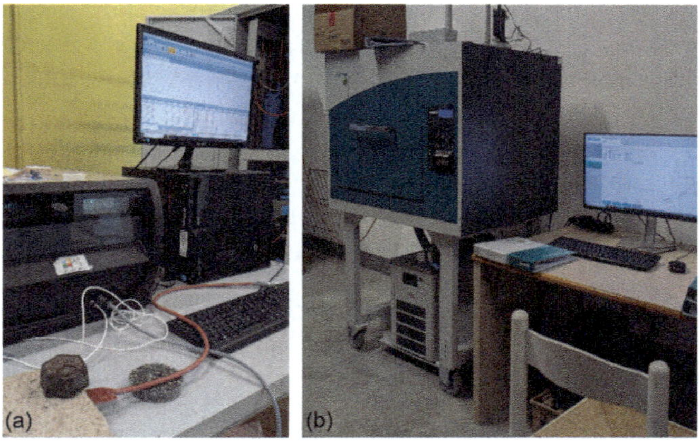

Fig. 2. Laboratory equipment used for the thermal investigation of the plasters: (**a**) hot disk 2500S, and (**b**) heat flow meter HFM 446 lambda large eco-line.

3 Results from thermal investigation

In this section, the results of the thermal tests are presented and compared to determine the thermophysical properties of the three plaster formulations incorporating different amounts of rice husk. Measurements from both the hot disk (Fig. 3a) and heat flow meter (Fig. 3b) methods demonstrated significantly lower conductivity values than those of conventional cement-based plasters (typically 0.7-1.0 W/mK). Table 1 summarizes the results obtained from the two in-lab equipment.

3.1 Results from the Hot Disk apparatus

Clear differences can be detected in the thermal performance of the samples as the rice husk content increased. Specifically, thermal conductivity (λ) increases with the percentage of rice husk, rising from 0.102 W/mK for the formulation containing 5% rice husk to 0.125 W/mK for the 10% formulation (Fig. 3a). This trend suggests that the addition of rice husk, while potentially increasing porosity, also introduces materials with higher intrinsic conductivity compared to the original matrix. Moreover, the plaster with an intermediate content of rice husk (7.5%) exhibits a thermal conductivity value between the other two, though more closely aligned with the 5% formulation. This indicates a non-linear change in the rate of increase of thermal conductivity with higher rice husk content, which can be interpreted as the result of competing microstructural effects. At low substitution levels (5 and 7.5 %) rice husk particles fit into the existing packing, resulting in roughly constant λ. At 10% content, however, a threshold is crossed: rice husk particles may begin to form more continuous contact networks or bridge gaps between high-conductivity solid components, raising the effective conduction. Alternatively, rice husk may simply have higher intrinsic λ: replacing a critical fraction of silica removes enough low-λ material to increase bulk thermal conductivity.

As per Table 1, a decreasing trend in thermal diffusivity (α) can be noted for increasing rice husk content (from 0.200 mm^2/s for the 5% formulation up to 0.176 mm^2/s for the 10% formulation). This indicates a slower heat propagation in the samples as more rice husk is added, which is desirable for thermal lag and buffering in building envelopes. A lower α value means that the material takes longer to respond to an external temperature change, enhancing thermal comfort and energy efficiency under summer dynamic conditions of exposure. Finally, specific heat (c) increases considerably with rice husk content, moving from 0.511 MJ/m^{3}K (5% formulation) to 0.708 MJ/m^{3}K (10% formulation). This implies that the material can store more heat per unit volume with higher rice husk content. Combined with the decreasing diffusivity, it suggests that rice husk addition enhances the thermal inertia of the plaster.

3.2 Results from the Heat Flow Meter apparatus

Figure 3b represents the thermal conductivity values of the three plaster formulations as measured using the HFM method. Consistent with the results obtained from the Hot Disk, thermal conductivity increases with higher rice husk content. However, the values recorded by the HFM are systematically lower: λ is measured at 0.071 W/mK for the 5% formulation, 0.073 W/mK for the 7.5% formulation, and 0.081 W/mK for the 10%

formulation. This discrepancy between the two methods can be attributed to differences in the measurement configuration. In the Hot Disk setup, the sensor is sandwiched between the flat surfaces of two identical samples, while in the HFM method the sample itself is positioned between the upper and lower plates of the device, leading to a reduced contact between the sample and the plate because of the roughness and irregularities of the sample surface. HFM results in Table 1 are reported together with their maximum deviation, as specified by the instrument technical sheet (0.5% of the measured λ value).

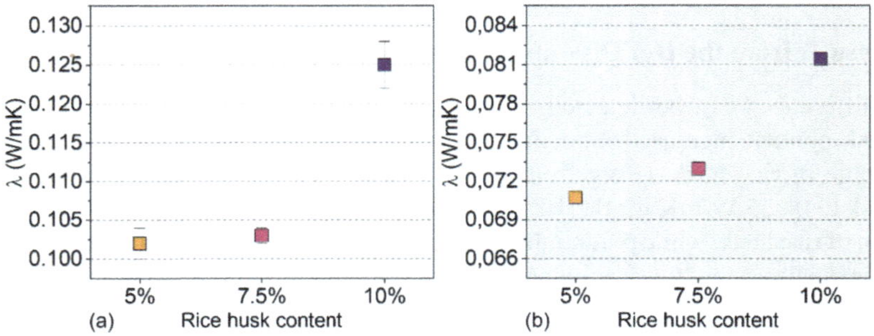

Fig. 3. Thermal conductivity values measured using (**a**) the hot disk and (**b**) the heat flow meter approach.

Table 1. Summary of thermal conductivity values obtained from the whole thermal investigation on the three plasters.

Husk rice content (%)	Hot disk			Heat flow meter
	λ [W/mK]	α [mm^2/s]	c [MJ/m^{3}K]	λ [W/mK]
5	0.102±0.002	0.200±0.010	0.511±0.035	$0.071 \pm 3.55 \times 10^{-4}$
7.5	0.103±0.001	0.185±0.003	0.556±0.009	$0.073 \pm 3.65 \times 10^{-4}$
10	0.125±0.003	0.176±0.008	0.708±0.042	$0.081 \pm 4.05 \times 10^{-4}$

4 Conclusions

In this work, the use of rice husk as a viable alternative to expanded amorphous silica in innovative plasters is investigated in terms of thermal insulation. Thermal conductivity of three plaster formulations is measured according to a Hot Disk and a Heat Flow Meter approach. Both measurement techniques consistently indicate an increase in thermal conductivity with higher rice husk content, although the rate of change is not strictly linear. Specifically, the Hot Disk results show a rise from 0.102 W/mK at 5% rice husk to 0.125 W/mK at 10%, with the intermediate 7.5% formulation exhibiting values closer to the 5% mixture. Similarly, HFM measurements confirm the same trend, but yield systematically lower values, ranging from 0.071 to 0.081 W/mK. The discrepancy between the two methods can be attributed to differences in the measurement setup and sample-device contact conditions.

Conventional gypsum- and cement-based plasters generally exhibit thermal conductivity values above 0.7 W/mK, while lightweight insulating plasters based on expanded perlite, vermiculite or aerogel typically range between 0.07 and 0.12 W/mK [23, 24]. Therefore, the rice husk-enhanced plasters tested in this study fall within or just above the range of commercial insulating plasters. Although they do not compete with high-performance insulators, they represent a notable improvement in thermal insulation compared to traditional plasters, especially for eco-sustainable applications. Moreover, rice husk plasters behavior suggests potential for use in hybrid systems where moderate insulation and enhanced thermal inertia are desirable, such as interior plasters in warm climates or buffer layers in multi-layer wall assemblies. In this framework, future research will focus on the application of one selected plaster formulation in a real-scale prototype building, enabling in-situ monitoring of its thermal behavior under realistic environmental conditions, as well as on the development of a Life Cycle Assessment study [25], in order to quantify the environmental benefits of the proposed material. Major benefits can be preliminary outlined: amorphous silica is known to have a high embodied energy and embodied carbon due to the energy-intensive processes required for its production (~ 3–5 kg CO_2eq per kg of material) [26]. By contrast, rice husk is an abundant agricultural by-product with negligible productions burdens when considered as a waste stream, and its use contributes to waste valorization and circular economy strategies.

Acknowledgements. This study was possible thanks to the Project "Choisis- Characterization of Innovative and Sustainable Insulating Solutions", funded within the Italian National Recovery and Resilience Plan (PNRR), Mission 4 – Education and Research- Measure M4C2 – Investment 1.1 PRIN – Research programs of national interest – project code 2022372TM9, CUP F53D23001540006.

References

1. Dai S, Ma Y, Zhang K (2022) Land degradation caused by construction activity: investigation, cause and control measures. Int J Environ Res Public Health 19(23):16046
2. D'Agostino D, Congedo PM, Albanese PM, Rubino A, Baglivo C (2024) Impact of climate change on the energy performance of building envelopes and implications on energy regulations across Europe. Energy 288:129886
3. Schmidt M, Jia Z, L. Weber Z (2024) Assessment of the alignment of CDP's international energy-related portfolio and policies with the Paris Agreement. Perspectives Climate Research, Freiburg
4. Hafez FS, Sa'di B, Safa-Gamal M, Taufiq-Yap Y, Alrifaey M, Seyedmahmoudian M, Stojcevski A, Horan B, Mekhilef S (2023) Energy efficiency in sustainable buildings: a systematic review with taxonomy, challenges, motivations, methodological aspects, recommendations, and pathways for future research. Energy Strat Rev 45:101013
5. Aldeek Z, Ayyoub S (2025) Exploring the potential of agricultural waste products as innovative building materials. Int J Des Nat Ecodynamics 20(4):795–801
6. Zhang R, Shen Z, Park B, Feng T, Aldykiewicz A Jr, Desjarlais A, Hun D, Shrestha S (2024) Natural fibers as promising core materials of vacuum insulation panels. Constr Build Mater 453:138890

7. FAO (2025) Crop prospects and food situation. FAO, Rome
8. Soltani N, Bahrami A, Pech-Canul M, Gonzalez L (2015) Review on the physicochemical treatments of rice husk for production of advanced materials. Chem Eng J 264:899–935
9. Rodriguez Neira K, Cardenas-Ramirez JP, Rojas-Herrera CJ, Haurie L, Lacasta AM, Torres Ramo J, Sanchez-Ostiz A (2024) Assessment of elaboration and performance of rice husk-based thermal insulation material for building applications. Buildings 14(6):1720
10. Buratti C, Belloni E, Lascaro E, Merli F, Ricciardi P (2018) Rice husk panels for building applications: thermal, acoustic and environmental characterization and comparison with other innovative recycled waste materials. Constr Build Mater 171:338–349
11. Marques B, Tadeu A, Antonio J, Joao A, de Brito J (2020) Mechanical, thermal and acoustic behaviour of polymer-based composite materials produced with rice husk and expanded cork by-products. Constr Build Mater 239:117851
12. Pascual A, Romero M, Serra E, Guerrero J, Perez R (2024) Sustainable insulation panel for buildings made of rice husks and posidonia. Constr Build Mater 445:137983
13. Ma W, Kolb T, Ruther N, Meinlschmidt P, Chen H, Yan L (2024) Physical, mechanical, thermal and fire behaviour of recycled aggregate concrete block wall system with rice husk insulation. Energy Build 320:114560
14. Soto M, Rojas C, Cardenas-Ramirez JP (2023) Characterization of a thermal insulating material based on a wheat straw and recycled paper cellulose to be applied in buildings by blowing method. Sustainability 15(1):58
15. Romero Quidel G, Soto Acuna MJ, Rojas Herrera CJ, Rodriguez Neira K, Cardenas-Ramirez JP (2023) Assessment of modular construction system made with low environmental impact construction materials for achieving sustainable housing projects. Sustainability 15(10):8386
16. Angue VMM, Dela Cruz CEM, Kahano FAMP, Magsino JA, Abaya ESDP (2020) Property characterization of plaster mortar utilizing sawdust and rice hull ash. Int J Civ Struct Eng Res 8(1):30–39
17. Dighe B, Singh MR (2025) From ancient practices to contemporary green building: analyzing organic additives in Pitalkhora earthen plasters for sustainable construction. Green Technol Sustain 3(3):100195
18. Savadogo N, Binta Traore Y, Lankoande N, Nshimiyimana P, Messan A (2024) Investigation of the fermentation mode of rice husk for the stabilization of earth plaster. J Miner Mater Charact Eng 12(1)
19. Asdrubali F, Grazieschi G, Gandola DM (2025) The Role of environmental product declarations in the decarbonization of building materials and components. Energies 18(5):1308
20. International Organization for Standardization (2022) ISO 22007-2:2022 Plastics—determination of thermal conductivity and thermal diffusivity—part 2: transient plane heat source (hot disc) method
21. NETZSCH (2025) HFM 446 Lambda large eco-line. NETZSCH, [Online]. Available: https://analyzing-testing.netzsch.com/en-US/products/heat-flow-meter-hfm-and-guarded-heat-flow-meter-ghfm/hfm-446-lambda-eco-line-large. (Consultato il giorno 31 July 2025)
22. ASTM C518-21 (2021) Standard test method for steady-state thermal transmission properties by means of the heat flow meter apparatus, p 16
23. EN 1745 (2020) Masonry and masonry products. Methods for determining thermal properties. European Committee for Standardization, Brussels
24. Marusiak S, Pavlikova M, Szafraniec M, Pavlik Z (2022) Thermal properties of natural hydraulic lime-based plastering repair mortars with expanded glass granulate. In: AIP Conference Proceedings, vol 2488, no 1
25. Asdrubali F, Fronzetti Colladon A, Segneri L, Gandola DM (2024) LCA and energy efficiency in buildings: mapping more than twenty years of research. Energy Build 321:114684
26. Hammond G, Jones C (2008) Embodied energy and carbon in construction materials. Proc Inst Civ Engineers—Energy 161(2):87–98

Investigation of the Optimal Use and Dosage of Biochar in Concrete Applications

Amir Ramezani, Saeed Bozorgmehr Nia, and Behrouz Shafei(✉)

Department of Civil, Construction, and Environmental Engineering, Iowa State University, Ames, IA 50011, USA
shafei@iastate.edu

Abstract. Concrete, an essential construction material for modern infrastructure, significantly contributes to global carbon emissions, primarily through cement production. As sustainability becomes a critical objective in construction, biochar has emerged as a viable solution to enhance concrete properties while addressing environmental challenges. Biochar is a carbon-rich material obtained from the pyrolysis of organic waste, capable of sequestering atmospheric carbon dioxide when incorporated into concrete. This study explores the potential of biochar as a partial substitute for cement or aggregates, emphasizing its impact on workability, mechanical performance, and durability. Sourced globally from agricultural, forestry, and municipal wastes, biochar's performance in concrete depends on properties such as particle size, porosity, and water absorption capacity. Studies consistently show that low dosages (1–5% by weight of cement) can improve mechanical performance and microstructure through internal curing and filler effects, while higher dosages may compromise strength due to dilution and porosity. Additionally, biochar incorporation enhances durability by refining pore structure, reducing permeability, and improving resistance to chloride and carbonation. This study provides an analysis of biochar-enhanced concrete, identifies optimal usage strategies, and highlights opportunities for future research in developing carbon-negative construction materials.

Keywords: Biochar · sustainable concrete · cement replacement · internal curing · carbon sequestration · durability · pyrolysis.

1 Introduction

Concrete, as one of the most used construction materials for civil infrastructure, is also a major contributor to global carbon emissions. Cement production alone is responsible for about 5–7% of worldwide CO_2 emissions [1]. This environmental burden, coupled with the depletion of natural resources, has motivated researchers to find sustainable alternatives for concrete ingredients [2, 3]. Biochar, a carbon-rich material obtained from the pyrolysis of biomass, has emerged as a promising sustainable additive or partial replacement in cementitious composites [4–6]. Biochar is produced by thermochemically converting organic waste (such as agricultural residues, wood chips, and so on)

H. Beushausen et al. (Eds.): NTZR 2025, LNCE 873, pp. 158–166, 2026.
https://doi.org/10.1007/978-3-032-27698-8_15

under limited oxygen, providing a stable form of carbon [7]. Importantly, this process not only creates a usable material but also sequesters carbon that the biomass absorbed from the atmosphere, potentially making biochar-enhanced concrete carbon-negative. In other words, using biochar in concrete can store atmospheric carbon in buildings, mitigating climate change.

Beyond carbon storage, biochar offers practical benefits in concrete. Its porous structure and high surface area can improve cement hydration by serving as internal curing reservoirs, observed in materials such as natural zeolite [8]. Studies have shown that adding a small amount of biochar (often a few percent by weight of cement) can increase compressive and flexural strength of concrete, refine pore structure, and enhance durability [9, 10]. At the same time, being lightweight and carbon-rich, biochar particles can reduce the density of concrete and partially replace conventional aggregates or cement, thereby reducing the use of non-renewable materials [6]. All these attributes align with the principles of green building and circular economy, converting waste into a value-added construction material and reducing landfill load [11, 12].

However, the effectiveness of biochar in concrete depends on many factors, including the type of biomass source, production conditions, and its incorporation method within the concrete mixture. Although, multiple studies have explored biochars from different sources, each with different physical and chemical properties, the optimal biochar properties (such as particle size, porosity, and surface chemistry) for concrete applications, as well as the proper dosage and preparation methods (e.g. pre-soaking or surface treatment) to maximize benefits without compromising workability or mechnical properties, remain unexamined [13]. Moreover, understanding the long-term performance and environmental impact of biochar-concrete is crucial for its acceptance in industry.

This study provides a comprehensive investigation of the role of biochar in concrete applications, with an emphasis on both material performance and environmental impact. Firstly, the availability of biochar feedstocks worldwide, including common sources and their characteristics, as well as methods for biochar characterization, are explored. Then the effects of biochar on concrete properties are evaluated, covering fresh mixture workability, hardened mechanical properties, and durability aspects in biochar-modified concrete. Experimental results and trends from existing studies are summarized, and practical applications are highlighted. Finally, recommendations and future directions are discussed, including guidelines on optimal biochar usage (dosage, particle size, and pre-treatment), selection of suitable biochar sources, expected performance under different exposure conditions, and emerging trends for sustainable construction with biochar.

2 Availability of Biochar Sources Worldwide

Biochar production has been expanding globally in recent years, driven by interests in waste management and carbon sequestration. Large quantities of agricultural and forestry residues that were once burned or left to decompose can now be converted into biochar around the world [1, 4, 5, 14]. Countries with significant agricultural sectors or forestry industries, such as China, India, the United States, Brazil, and many EU nations, have seen a rise in biochar production. For instance, China is reported as one of the leading countries in biochar research and production, reflecting its focus on carbon mitigation and waste reuse [15]. Similarly, research programs in North America, Europe, and Australia are utilizing local biomass feedstocks, including wood chips and crop husks [1, 6, 14]. Industrial organic wastes and even municipal solid wastes can also be pyrolyzed to provide biochar, as some studies have used biochar made from sewage sludge or food waste digestate [9, 10, 16]. This global distribution of biochar production means that a variety of biochar types are available, each derived from the local biomass prevalent in the region. Table 1 provides information about biochar sources that have been investigated for concrete applications, highlighting the diversity of materials.

Table 1. Common biomass sources for biochar used in concrete applications.

Biochar feedstock	Region	Notable characteristics
Wood waste (sawdust, timber off-cuts)	Europe, North America, East Asia	High carbon content, moderate ash; can achieve a high surface area when pyrolyzed optimally
Agricultural residues (rice husk, sugarcane bagasse, corn stover)	Asia, America, Africa	Silica-rich ash (especially rice husk), porous structure, can have significant water absorption
Poultry litter or animal manure	Oceania (e.g., New Zealand), USA	Typically, lower fixed carbon and very high mineral ash content can influence early-age concrete properties differently
Municipal solid waste or digestate (mixed food and green waste)	East Asia (e.g., China)	Variable composition, moderate ash and nutrient content; carbon-negative when derived from waste treatment
Coconut shells, palm shells	South Asia, Southeast Asia	Hard biomass provides biochar with high carbon and high porosity, which can be used as lightweight aggregate particles

3 Physical Properties and Concrete Applications

The physical and chemical properties of biochar play a critical role in how it interacts with cement paste. Important properties include particle size and fineness, specific surface area, porosity and water absorption capacity, bulk density, and ash content or composition. utilize a range of analytical techniques to characterize biochar prior to its application in concrete, ensuring a proper understanding of these parameters.

Biochar can be produced as a coarse material (granules several millimeters in size) or a fine powder (if ground after pyrolysis). In the research field of concrete technology, biochar has been used in forms ranging from fine powder (passing a 75 μm sieve) up to sand-sized granules (2–4 mm) [10, 14]. Sieve analysis is often performed to determine the particle size distribution of a biochar sample, especially if it is to be used as a fine aggregate replacement. Finer biochar generally has more surface area and can fill voids in the cement matrix better (filler effect), but it also tends to demand more water and can increase the viscosity of the mixture. On the other hand, coarser biochar can be used as lightweight aggregate replacement, as in the case of Chen et al. [10] utilized 2–4 mm biochar particles as part of lightweight aggregate concrete. Another approach is to use biochar particle sizes similar to cement (10–50 μm). This method can partially act like a cementitious additive. Laser diffraction particle size analysis is another technique occasionally used for fine biochar to get a detailed distribution of particle sizes.

Biochar's surface area (usually measured by Brunauer-Emmett-Teller (BET) nitrogen adsorption), reflects its porosity. A higher BET surface area (m^2/g) often correlates with an enhanced capacity to absorb water and potentially accommodate cement hydration products on its surface. For example, a wood-derived biochar showed a surface area of 196 m^2/g, whereas a food-waste biochar from the same study had only 9.7 m^2/g, indicating that the wood char is far more porous [9]. Biochars produced at lower temperatures (300–500 °C) often have less surface area but more remaining volatile matter, whereas those produced at higher temperatures (600–800 °C) are more carbonized and can develop extensive microporosity [17]. Measurement of surface area enables researchers to predict the water retention capacity of the biochar (by adsorption in pores) and its influence on setting, as a highly porous biochar can expedite hydration by offering internal curing.

The most crucial characteristic for practical application is the water absorption capacity of the biochar. Biochar's porous structure can absorb, and this is advantageous when utilized appropriately, as pre-wetted biochar can gradually release water, functioning as an internal curing reservoir for concrete. However, this characteristic can have negative effects if the biochar is dry, as it can absorb water. The phenomenon results in a diminished effective water-to-cement (w/c) ratio and reduced workability. A food waste digestate biochar showed 45–50% water absorption by mass, while some wood biochars can absorb over 2 g of water per g of biochar [10, 18]. In addition, Wang et al. [19] reported a Wood-derived biochar with an absorption capacity of 1.3 g water per g biochar. Such data is crucial for mix design, high absorption biochars are often pre-soaked to avert the depletion of mixing water.

Biochar possesses a lower density compared to sand or cement. This difference is utilized to produce lighter concrete. Zou et al. [20] used corncob biochar with a bulk density of only 224 kg/m^3 as a substitute for normal-weight aggregate, significantly reducing concrete density. Knowing the density is important for volumetric replacement calculations and assessing how substituting heavier materials with lightweight biochar impacts the concrete volume produced per batch. Additionally, biochar contains a fraction of inorganic ash, minerals that do not burn off during pyrolysis. The ash can include silica, alumina, iron oxide, calcium, potassium, etc., depending on the feedstock. For example, biochar made from food waste digestate had an ash content exceeding 50% [10].

Scanning electron microscopy (SEM) images reveal the surface texture of biochar, which typically exhibits a porous cell-wall structure (as shown in Fig. 1 for biochar sourced from oat stover and corn stover). SEM analysis confirms the presence of macropores and surface roughness, features that contribute to improved mechanical interlocking with the cement paste. Several studies have presented SEM micrographs illustrating the formation of cement hydration products, such as calcium silicate hydrate (C–S–H) gel and calcium hydroxide crystals, on the surface of biochar particles, indicating strong integration with the cement matrix [21–24].

(a) (b)

Fig. 1. SEM images from biochar based on (a) oat stover and (b) corn stover.

4 Effects of Biochar on Concrete Properties

In this section, findings from studies about the impacts of biochar on the fresh characteristics, hardened mechanical performance, durability of concrete, and its capacity to sequester carbon are presented. A primary aspect of including biochar into a concrete mixture is ensuring sufficient workability. Biochar, being highly porous and often fine, tends to decrease the slump or flow of concrete [25]. The porous morphology causes biochar particles to absorb mix water and also increases internal friction in the concrete mixture [15]. Almost all studies report a decrease in slump with the addition of biochar, particularly as the dosage increases. In this regard, Sirico et al. [1] observed slump dropping from about 200 mm for the control concrete mixture to 30 mm with 10% biochar addition. In another case, Wang et al. [17] reported that where sand was fully replaced by pre-wetted biochar, the slump was still about 200 mm (240 mm control), a relatively small 16% reduction. It is important to note that many researchers either soak biochar in water for 24 h prior to mixing or adjust the mix water upwards to account for char absorption [10, 18].

Another fresh property influenced by biochar is the setting time. The effect has been variable in the literature, but several studies indicated that biochar accelerates the setting (shortens set times). Javed et al. [14] found that increasing biochar dosage from 0% to 5% led to reduced initial and final setting times, regardless of biochar type.

This was attributed to the filler effect and reduced effective w/c due to the biochar's physical properties (internal pores) [14]. Fresh density is another factor, since biochar is lightweight, concrete mix density in the fresh state decreases as biochar content increases. This feature can be beneficial for producing lightweight concrete. Sirico et al. [1] reported 5% decrease in the density of the concrete mixture with 10% biochar compared to the normal mixture. The reduced mass can slightly increase air content. However, it predominantly indicates that each cubic meter of concrete has a lower weight, which can reduce formwork pressure and transportation loads.

Compressive strength is often the main metric examined when incorporating biochar into concrete. A consistent trend has been observed based on the previous studies. There is an optimum biochar content (usually low) at which compressive strength is maximized, and beyond which strength decreases due to the "dilution" effect (biochar replacing some of the cementitious material) [1, 14]. Many studies identify this optimum around 1–5% biochar by weight of cementitious material [1, 14, 18, 25]. At these dosages, biochar can act as filler, packing into the pores and refining the microstructure (fewer large voids) [16]. It can also enhance hydration by providing internal curing from water released by biochar leads to more complete cement hydration, contributing to increased compressive strength [9]. Additionally, the rough and porous surface of biochar provides a good interfacial transition zone (ITZ) for cement paste to bond [10]. Lorenzoni et al. [26] showed via X-ray CT that at 25 vol% replacement of cement with biochar, the biochar particles can act as flaws initiating microcracks under stress. Chen et al. [6] reported that a 30% fine aggregate replacement by biochar reached 13.1 MPa at 28 days, which was approximately 35% lower than a reference mixture without biochar. Notably, these higher replacements typically prioritize the maximization of carbon storage rather than enhancing mechanical performance. As a result, the biochar component should be restricted to a small percentage of the binder mass when mechanical performance is an important factor. It is worth noting that curing conditions and particle size of biochar influence these results. Sirico et al. [1] found that under dry curing, certain biochar dosages performed better, whereas under continuously wet curing, the differences were smaller. This result indicates that biochar's internal curing capacity is more pronounced when external curing is restricted, as it facilitates low-humidity curing by supplying internal moisture.

The permeability and water absorption of concrete are frequently dictated by its pore structure, which influences its resistance to the penetration of water and aggressive substances. As observed, biochar at optimal levels typically enhances the pore structure by reducing the number of huge pores. In this regard, Gupta et al. [9] found that adding biochar decreased the water absorption coefficient, making it less permeable than the control mixture, even comparable with mixtures containing silica fume alone. The internal curing also contributes to more complete hydration, which can reduce the connectivity of pores. Chloride ion and carbonation penetration were shown to drop in concretes with biochar up to 3% due to the internal curing and promoted hydration degree [5]. Rashid et al. [4] observed that a concrete mixture with 10% biochar resulted in a 19.5% improvement in the carbonation resistance. This further highlighted the potential of biochar, when used in optimal proportions, to enhance concrete's durability, mechanical properties, and environmental performance without compromising its workability or structural integrity.

5 Conclusions

The increasing global emphasis on sustainable construction practices has driven the search for materials that can reduce concrete's environmental footprint without compromising its performance. Among these alternatives, biochar stands out due to its dual benefit, improving concrete properties and sequestering atmospheric carbon. As a byproduct of biomass pyrolysis, biochar offers a practical and scalable way to convert waste into a high-value additive for cementitious composites. This study illustrated that biochar can significantly influence fresh and hardened concrete properties. Its effects, however, depend on various parameters such as feedstock source, pyrolysis temperature, particle size, porosity, and incorporation method. Biochars derived from wood, agricultural residues, and municipal waste vary widely in chemical and physical characteristics. For example, wood-based biochars tend to offer high surface area and low ash content, while food-waste biochars may contain substantial inorganic residue. These factors affect water absorption, particle interaction, and reactivity within the cement matrix.

Fresh concrete properties are notably affected by biochar's porous morphology and water absorption capacity. Unsoaked biochar may absorb mixing water, leading to decreased workability and reduced effective w/c ratio. However, this challenge can be mitigated through pre-wetting or adjusting the mixing water. Research has shown that while slump generally decreases with increasing biochar content, proper handling can maintain acceptable workability. Additionally, biochar tends to reduce setting time due to its internal porosity and filler effect, accelerating early hydration reactions.

When used at optimal dosages (typically 1–5% by weight of cement), biochar enhances mechanical performance. These improvements stem from multiple mechanisms, including densification of the matrix (filler effect), enhanced hydration due to internal curing, and improved bonding at the ITZ between cement paste and biochar particles. SEM analyses confirm that biochar surfaces support the formation of CSH gels, strengthening the composite structure. However, excessive biochar (above 10%) leads to some extent compressive strength reductions, attributed to dilution of cementitious materials and increased internal porosity that may act as stress concentrators under load. As a result, careful optimization is essential to balance mechanical performance with sustainability goals.

Durability is another crucial benefit associated with biochar-modified concrete. Low levels of biochar contribute to reduced water permeability and improved resistance to chloride ingress and carbonation, thereby extending service life. These durability improvements suggest that biochar incorporation could be particularly valuable in aggressive exposure environments or applications requiring enhanced long-term performance.

Beyond material performance, biochar's environmental significance cannot be overstated. Its ability to permanently store carbon absorbed by biomass offers a unique avenue for climate change mitigation. Some advanced applications, such as biochar encapsulation and CO_2 curing, enable concrete mixtures to achieve carbon-negative footprints.

These findings align with circular economy principles by transforming agricultural or urban waste into a high-performance construction input, reducing landfill burden, and lowering embodied carbon. Future research should focus on developing standardized mix design guidelines incorporating biochar properties, exploring novel pre-treatment methods (e.g., surface activation or nano-coating), and evaluating structural-scale applications. Life cycle assessments and cost-benefit analyses will also be essential to facilitate industry adoption.

References

1. Sirico A et al (2021) Biochar from wood waste as additive for structural concrete. Constr Build Mater 303:124500
2. Ramezani A, Moodi F, Mirabrishami M, Zolfagharnasab A, Bahman-Zadeh F (2024) Experimental and thermodynamic investigating of carbonation behavior in alkali-activated slag-calcined clay materials with different binder constituent. Constr Build Mater 446:137987
3. Farahzadi L, Nia SB, Shafei B, Kioumarsi M (2025) Sustainability assessment of ultra-high performance concrete made with various supplementary cementitious materials. Cleaner Mater 15:100301
4. Rashid S, Raghav A, Goyal A, Danie Roy AB, Singh M (2024) Biochar as a sustainable additive in cementitious composites: a comprehensive analysis of properties and environmental impact. Ind Crop Prod 209:118044
5. Ling Y, Wu X, Tan K, Zou Z (2023) Effect of biochar dosage and fineness on the mechanical properties and durability of concrete. Materials 16:2809
6. Chen L et al (2022) Biochar-augmented carbon-negative concrete. Chem Eng J 431:133946
7. Sun LM et al (2025) Biochar production, activation, and applications: a comprehensive technical review, carbon capture. Sci Technol:100421
8. Kazemian M, Shafei B (2022) Internal curing capabilities of natural zeolite to improve the hydration of ultra-high performance concrete. Constr Build Mater 340:127452
9. Gupta S, Kua HW (2020) Combination of biochar and silica fume as partial cement replacement in mortar: performance evaluation under normal and elevated temperature. Waste Biomass Valoriz 11:2807–2824
10. Chen L, Zhu X, Zheng Y, Wang L, Poon CS, Tsang DCW (2024) Development of high-strength lightweight concrete by utilizing food waste digestate based biochar aggregate. Constr Build Mater 411:134142
11. Dashti P, Ranjbar S, Ghafari S, Ramezani A, Nejad FM (2023) RSM-based and environmental assessment of eco-friendly geopolymer mortars containing recycled waste tire constituents. J Clean Prod 428:139365
12. Nia SB, Shafei B (2024) Synergistic effects of nano and micro silica on fresh and hardened properties of self-consolidating concrete. Case Stud Constr Mater 21:e03443
13. Room S, Bahadori-Jahromi A (2025) Hydration kinetics of biochar-enhanced cement composites: a mini-review. Buildings 15:2520
14. Javed MH, Sikandar MA, Ahmad W, Bashir MT, Alrowais R, Wadud MB (2022) Effect of various biochars on physical, mechanical, and microstructural characteristics of cement pastes and mortars. J Build Eng 57:104850
15. Room S, Bahadori-Jahromi A (2024) Biochar-enhanced carbon-negative and sustainable cement composites: a scientometric review. Sustainability 16:10162
16. Akhtar A, Sarmah AK (2018) Novel biochar-concrete composites: manufacturing, characterization and evaluation of the mechanical properties. Sci Total Environ 616:408–416

17. Zhao Z, El-Naggar A, Kau J, Olson C, Tomlinson D, Chang SX (2024) Biochar affects compressive strength of Portland cement composites: a meta-analysis. Biochar 6:21
18. Dixit A, Gupta S, Dai Pang S, Kua HW (2019) Waste valorisation using biochar for cement replacement and internal curing in ultra-high performance concrete. J Clean Prod 238:117876
19. Wang D, Jantwal A, Kaynak E, Sas G, Das O (2025) Promoting internal curing in concrete by replacing sand with sustainable biochar. Case Stud Constr Mater 22:e04542
20. Zou S, Chen X, Sham ML, Lu J-X, Poon CS (2025) Carbon sequestration in aggregate and concrete by encapsulated biochar and carbonation: experiment and simulation. Cem Concr Compos 159:105990
21. Senadheera SS et al (2023) Application of biochar in concrete–a review. Cem Concr Compos 143:105204
22. Gupta S, Kashani A, Mahmood AH (2022) Carbon sequestration in engineered lightweight foamed mortar–effect on rheology, mechanical and durability properties. Constr Build Mater 322:126383
23. Gupta S (2021) Carbon sequestration in cementitious matrix containing pyrogenic carbon from waste biomass: a comparison of external and internal carbonation approach. J Build Eng 43:102910
24. Gupta S, Kashani A, Mahmood AH, Han T (2021) Carbon sequestration in cementitious composites using biochar and fly ash–effect on mechanical and durability properties. Constr Build Mater 291:123363
25. Du J, Wang Y, Bao Y, Sarkar D, Meng W (2023) Valorization of wasted-derived biochar in ultra-high-performance concrete (UHPC): pretreatment, characterization, and environmental benefits. Constr Build Mater 409:133839
26. Lorenzoni R, Cunningham P, Fritsch T, Schmidt W, Kruschwitz S, Bruno G (2024) Microstructure analysis of cement-biochar composites. Mater Struct 57:175

Coal Bottom Ash as a Precursor for Alkali-Activated Bricks

Dylan Maluleke(✉), Areej Gamieldien, and Hans Beushausen

Concrete Materials and Structural Integrity Research Unit (CoMSIRU), Department of Civil Engineering, University of Cape Town, Cape Town, South Africa
dylanmaluleke@gmail.com

Abstract. This study investigates the feasibility of utilizing coal bottom ash (BA) with particle sizes smaller than 325 μm as a binder in alkali-activated concrete (AAC) to produce non-structural bricks (i.e., 7 MPa bricks). BA was sourced from a Sasol petrochemical plant in Secunda, South Africa (SA). In addition to BA, the AAC also incorporated gasification ash (GA)—another waste produced as part of the Sasol petrochemical process flow—as an aggregate in place of natural aggregates (NA). The activator used was sodium hydroxide (NaOH) at concentrations of 10 M and 15 M, and curing temperatures of 50, 80, and 105 °C were evaluated. The findings of the study indicated that AAC bricks of compressive strengths up to 38 MPa can be manufactured using BA cured at 105 °C. This result suggests that, depending on mix composition, AAC bricks can be manufactured with these materials not only for non-load-bearing applications but also for load-bearing applications.

Keywords: Alkali-activated concrete · Bottom ash · Industrial waste · Reuse · Non-structural · Concrete.

1 Introduction

Coal ash, a by-product of electricity generation, is often discarded in landfill facilities as waste. These facilities are costly to build/maintain and often pose a danger of polluting the environment. A small amount of this ash, particularly fly ash (FA), is recycled and used as a cement extender. However, a large proportion of ash, such as bottom ash (BA), is often discarded because it is less reactive and has bigger particle sizes when compared to FA.

This study investigated an alternative use of BA as a binder in non-structural alkali-activated bricks. These bricks were required to meet the minimum compressive strength requirements for non-load bearing purposes as defined by ASTM–129, 2017, and the general standard of 7 MPa set out for masonry bricks in South Africa. The use of BA in non-structural bricks (non-load-bearing wall construction) can help reduce reliance on conventional construction materials, such as clay and concrete bricks, which contribute to resource depletion and carbon emissions. This would also help reduce the amount of 'waste' sent to landfill sites.

H. Beushausen et al. (Eds.): NTZR 2025, LNCE 873, pp. 167–176, 2026.
https://doi.org/10.1007/978-3-032-27698-8_16

1.1 Background

South Africa currently generates 78% of its electricity from coal through its utility company, Eskom (CSIR, 2022). As a result, 34.4 million tons of coal ash are produced annually (Reynolds-Clausen and Singh 2019). Other industrial role players, such as Sasol, a South African energy and chemicals company, add to this number by producing 9 million tons of coal ash per annum (Matjie et al., 2017). It is estimated that approximately 50 million tons of ash are produced in South Africa annually, of which less than 7% is recycled (Reynolds-Clausen and Singh 2019). This results in significant capital and environmental costs to build and maintain ash disposal facilities (ADFs). There are no public records of the amount of ash currently stockpiled in ADFs, but it can be estimated that in the last 30 years, 1, 5 billion tons of ash have been disposed in South Africa.

The problem of an oversupply of coal ash unfolds against the backdrop of a country with an infrastructure supply deficiency, particularly in housing. In SA, 2.5 million low-income households lack formal housing. As such, this puts the country in a unique position to pursue low-cost and environmentally friendly construction materials such as AAC bricks and blocks. To put this into context, a two-bedroom 'RDP' style house will require approximately 3300 ($220 \times 110 \times 75$ mm) bricks (RDP = Reconstruction and Development Programme). Thus, to build 2.5 million RDP houses would require 8.25 billion bricks. Considering a brick made up of 100% ash, a total of 26 million tons of ash would be required to meet the housing demand (assuming a brick weight of around 3 kg), equivalent to half of the country's annual ash production.

Alkali-activation involves the addition of a strong base to an aluminosilicate precursor, resulting in the formation of a cementitious material (Kearsley 2021). Various precursors—including ashes (e.g., FA, BA, rice husk ash, volcanic ash), clays (e.g., illite, halloysite, kaolinite), slags, and mine tailings—are suitable for this process (Ahmad and Rashid 2022).

Zhang et al. (2016) described the activation of aluminosilicate as characterized by four (4) main phases.

(i) Dissolution of vitreous/amorphous aluminosilicates under high pH conditions.
(ii) Precipitation of various aluminate and silicate species into various oligomers shortly after the commencement of the reaction.
(iii) The formation of alkaline aluminosilicate hydrate gel ($M_2O.Al_2O_3.nSiO_2.mH_2O$, M is Na and/or K).
(iv) Development of Si-richer (C-(N)-A-S-H) gels by incorporating more Si into the reaction products during curing.

Historically, FA has been the preferred aluminosilicate precursor for AAC over BA due to its finer particle size and higher reactivity (Provis 2015). Using BA typically requires additional grinding to enhance its reactivity and achieve good compressive strength results. For example, Sathonsaowaphak et al. (2008) demonstrated that ground BA exhibits high reactivity, achieving compressive strengths between 24 to 58 MPa when sodium hydroxide (NaOH) is used as the activator.

This study explores the feasibility of using unprocessed BA for lower-strength (non-structural) concrete such as brick and block manufacturing, potentially eliminating the need for energy-intensive grinding. It also proceeds to compare AAC brick manufacturing with traditional brick manufacturing methods (i.e., Portland Cement Concrete and clay-fired bricks), emphasizing environmental, economic, and technical (strength) considerations. This study is intended to contribute to the body of knowledge that will lead to the development of standardized practices for AAC production, advancing its potential as a sustainable alternative in the construction industry.

2 Experimental Plan

The experimental process of this investigation involved the following main steps:

1. The constituent ash materials were characterized in terms of particle size and mineralogical composition.
2. The precursor material (BA), aggregates (GA + BA), and the alkali activator were mixed.
3. The mix was cast into 50 mm cube moulds.
4. Curing was carried out in two phases. Phase 1 involved allowing samples to cure in the laboratory at ambient temperatures in the moulds for 24 h. Phase 2 of curing was done at elevated temperatures (50, 80 or 105°C) for 24 h in an oven.
5. Compressive strength testing was conducted 3 days after oven curing in accordance with SANS 5863:2006.

2.1 Materials

The coal ash used in this investigation was sourced from the Sasol Secunda factory in Mpumalanga, South Africa. This factory utilizes low-rank bituminous coal in its production value chain, where the coal is divided into two processing streams: electricity generation and gasification, based on the coal's size and quality. The gasification process employs coarser coal particles (greater than 5 mm), while higher-quality coal is directed to boilers for electricity generation, producing a by-product known as bottom ash (BA) collected from the base of boilers. BA consists of a fine to coarse particle size distribution, as shown in Fig. 1.

Fig. 1. Coal bottom ash (BA) <325 µm–used as the binder (left) and gasification ash (GA) used as the aggregate (right) sourced from the Sasol Secunda factory.

The coarse coal undergoes high temperatures and pressure (gasification) to produce a mixture of carbon monoxide and hydrogen known as synthesis gas (Du Plessis et al. 2008). The process produces gasification ash (GA) as a by-product of the combustion process. GA takes the form of granular particles between 2 mm and 9 mm, as illustrated in Fig. 1, and was utilized as the coarse aggregate.

The BA used in this research was sourced from the Sasol Secunda factory on the crest of the fine coal ash dam 5, which remains in operation. The coarser aggregate material (CA) was sourced from the crest of the coarse ash stack, which was also operational at the time of collection, implying that this is fresh and unweathered ash.

2.2 Material Classification

Material Particle Size Distribution A sieve analysis was conducted in accordance with SANS 201:2008 to determine the particle grading of the two precursor materials, BA and GA. Prior to the sieve analysis, the samples were oven-dried at 50°C for 24 h to ensure the complete removal of moisture, as stipulated by SANS 201:2008. After drying, the samples were allowed to rest for 3 h before proceeding with the sieve analysis. The results of the material grading are shown in Fig. 2.

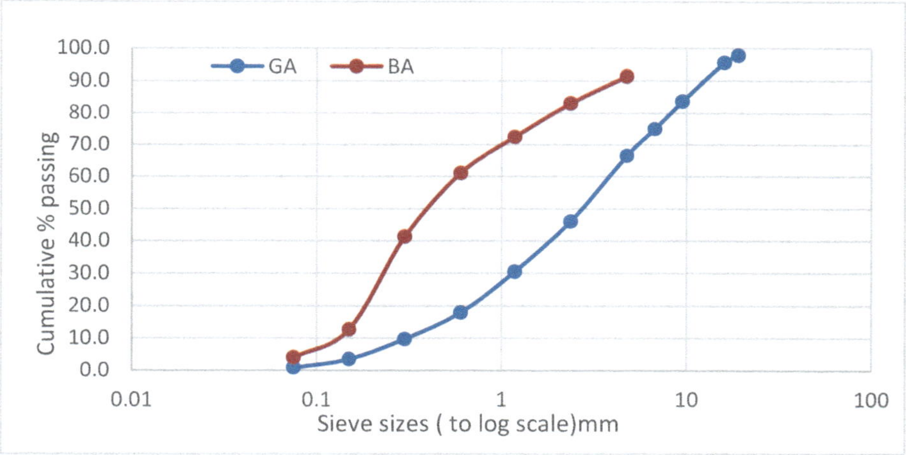

Fig. 2. Particle size distribution curve for Sasol bottom ash (BA) and gasification ash (GA).

In this study, BA of particle sizes smaller than 325 μm (as shown in Fig. 1) were used as the binder, as applied in a reference study using low-reactivity precursor material by Madani et al. (2022). BA particles larger than 325 μm were categorized and used as fine aggregates. These aggregate particles were excluded from the binder content calculations of the two mix designs. However, it is acknowledged that while particles above 325 μm were considered solely as aggregates in the mix design, they may contribute to the strength development of the AAC bricks. This potential contribution was not quantified in this research.

Material Mineralogical Composition Matjie et al. (2005) conducted a detailed compositional analysis of both BA and GA produced by Sasol at both the Secunda and the Sasolburg sites. This analysis assessed the physical, mineralogical, and chemical composition of the said ash streams. The study employed computer-controlled scanning electron microscopy (CCSEM) and X-ray diffraction (XRD) to identify the mineral phases of the ash, and X-ray fluorescence spectrophotometry (XRF) to determine its elemental composition. As shown in Fig. 3, the results indicate that Sasol BA has a high aluminosilicate composition, similar to that of South African FA, as per SANS 1491–2:2005 (Du Plessis et al. 2008). This theoretically suggests that Sasol BA should be suitable for alkali activation. The ash also contains less than 10% CaO, classifying it as a low-calcium ash. The lack of sufficient CaO indicates that additional energy from external activators, such as NaOH, is required to initiate hydration.

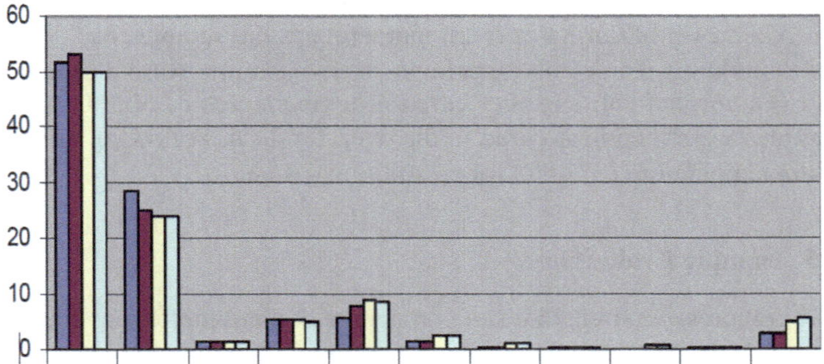

	SiO2	Al2O3	TiO2	Fe2O3	CaO	MgO	K2O	Na2O	P2O5	S	LOI
Sasolburg 1 GA	51.7	28.6	1.5	5.4	5.8	1.3	0.5	0.3	0.2	0.4	2.7
Sasolburg Bottom Ash	53.1	24.9	1.4	5.4	8.0	1.6	0.5	0.3	0.1	0.2	2.8
Secunda GA Phase 1	50.1	24.0	1.4	5.3	8.9	2.5	1.0	0.4	0.7	0.5	5.0
Secunda GA Phase 2	50.0	24.0	1.4	5.1	8.4	2.5	1.0	0.4	0.7	0.5	5.8

Fig. 3. Elemental analysis of Sasol GA and BA sampled from Secunda and Sasolburg (Matjie et al. 2005).

2.3 Alkali Activator Selection

In this study, sodium hydroxide (NaOH) was used as the alkali activator, as it is the most commonly used activator in alkali activation. As a result, it provides a substantial body of literature as a reference point for the mix designs and evaluation. Additionally, it offers practical benefits, including high solubility in water, widespread availability, and low cost, making it one of the most economical alkali activators available (SEDIRA, 2017). Activator molarities of 10 and 15 M were selected based on literature findings that suggest the appropriate value of NaOH to range from 5–15 M (Madani et al. 2020).

This activator was prepared to be used as a two-part mix (i.e., activator solution prepared separately and added to precursor material).

2.4 Mixing Procedure

All mixing was conducted at room temperature (~23°C) to ensure consistency in results by eliminating variations due to differences in mixing conditions. The BA and GA were mixed in a pan mixer until a uniform mix was achieved, following a standardized mixing time of 2 min. Subsequently, activator solutions were added to the pan mixer and mixed for an additional 2 min until a workable mortar was achieved. The alkali-activator to binder ratio used in this study was 1:2 in accordance with most literature.

Following the mixing process, the AAC mortar was poured into 50 × 50 × 50 mm cubes and compacted on a vibrating table for 10–20 s. The cubes were allowed to set at room temperature for 24 h, after which they were demolded.

After 24 h of curing at room temperature, the samples had set and were prepared for demolding. Three cube specimens from each mix were placed on oven trays, open to oven air, and subsequently cured at temperatures of 50, 80, and 105°C. The cube samples were then heat-cured in the oven for 24 h. Following this, they were cured at room temperature (~23 °C) for an additional 48 h.

2.5 Sample Evaluation

The compressive strength testing procedure of the samples was done in accordance with SANS 5863:2006, using the mean strength of 3 identical samples as the test result. The achieved compressive strength test results of the cube samples were evaluated in accordance with ASTM C127 (Table 1) to determine whether the bricks meet the minimum requirements for non-load-bearing masonry units. Samples that outperform the non-load-bearing requirement were also assessed against the ASTM C90 strength requirements for load-bearing bricks as presented in Table 2.

Table 1. Strength requirements for non-load-bearing masonry units (ASTM—129, 2017)

Criteria	Compressive strength, min, psi (MPa)
Average of 3 units	600 (4.14)
Individual unit	500 (3.45)

Table 2. Strength, absorption, and density classification requirements for load-bearing bricks (ASTM—C90, 2017)

Density classification	Oven-dry density of concrete, lb./ft^3 (kg/m^3)	Maximum water absorption, lb./ft^3 (kg/m^3)	Minimum net area compressive strength, lb./in^2 (MPa)
	Average of 3 units	Average of 3 units	Individual units
Light weight	Less than 105 (1680)	18 (288)	20 (320)
Medium weight	105 to less than 125 (1680–2000)	15 (240)	17 (272)
Normal weight	125 (2000) or more	13 (208)	15 (240)

3 Results and Discussion

3.1 Compressive Strength Results

The compressive strength results of the two mix designs at curing temperatures of 50, 80, and 105°C were observed and recorded, as shown in Table 3. The results presented in Table 3 enable the study of the effect of activator molarity and the effect of curing temperatures on the compressive strength development for BA-based AAC.

The results show a general trend where the compressive strength of BA based AAC increases with increasing curing temperature. This aligns with the findings from literature, which indicated that BA had low reactivity, thus making curing at elevated temperatures necessary (Zhang et al. 2016). The maximum compressive strength observed in this study was 36.8 MPa (for MX1 cured at 105°C).

Table 3. Mix design and compr. strength for AAC samples cured at 50, 80, and 105°C.

Mix	Curing temp. (°C)	Activator	Binder (BA) kg/m3	Density (kg/m3)	Avg compr. strength fcu, 3 (MPa)	StdD of fcu (σ, fcu)
MX1	50	15 M NaOH		1961	4.1	0.04
	80			1967	19.1	1.72
	105		600	1861	36.8	2.80
MX2	50	10 M NaOH		1744	2.1	0.01
	80			1709	8.8	1.07
	105			1696	11.1	1.61

The results also show an increase in compressive strength of the AAC with an increase in the molarity of NaOH when used as the alkali-activator. This finding is supported by most literature and is due to the improved rate of reaction that results from introducing surplus Na + ions necessary to form C-(N)-A-S-H gel required for hydration because of the higher NaOH molarity (Sathonsaowaphak et al. 2008). Fig. 4 shows a head-to-head comparison of the compressive strength development of MX1 (15 M NaOH) versus MX2 (10 M NaOH).

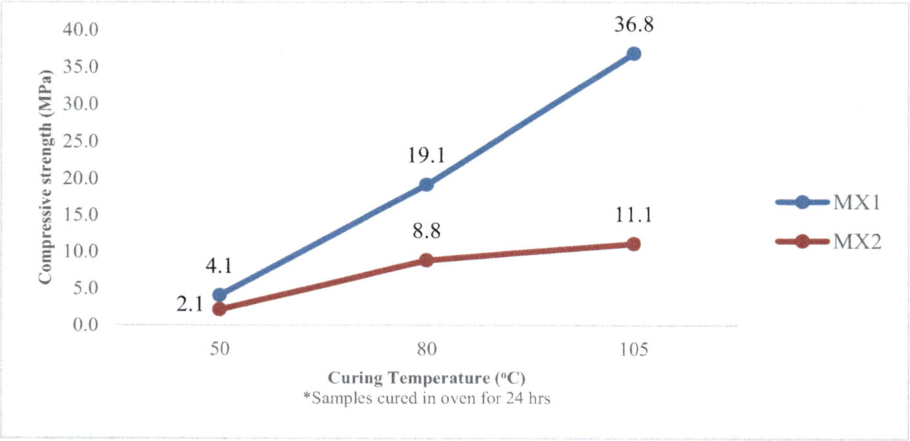

Fig. 4. Compressive strength development of AAC samples versus curing temperature

The compressive strengths of the MX1 samples increased progressively from 4.1 MPa at 50°C to 19.1 MPa at 80°C and 36.8 MPa at 105°C. These findings are consistent with the findings from Madani et al. (2022) on low-reactivity precursors where the increase in NaOH molarity from 5 M to 15 M resulted in increasing compressive strength.

The ASTM–C90 (2017) specifies a minimum compressive strength of 13.8 MPa for load-bearing masonry brick units. MX1 samples cured at 80 °C, achieved an average compressive strength of 19.1 MPa, surpassing this requirement, suggests that they would be suitable for use as load-bearing masonry brick units.

Furthermore, MX1 samples cured at 105°C achieved a compressive strength of 36.8 MPa, which exceeds the ASTM—C90 requirement for load-bearing bricks whilst being on par with industry standards for structural concrete (~ 30 MPa). This level of performance indicates that there is potential for the application of AAC in precast concrete structural elements such as slabs, bridge culverts, and concrete sewer pipes.

MX2 was designed as an optimized version of MX1 which had demonstrated good results. The optimization involved reducing the NaOH molarity from 15 M to 10 M, representing a 33% reduction in the activator molarity. This adjustment resulted in compressive strengths of 2.1 MPa, 8.8 MPa, and 11.1 MPa for curing temperatures 50°C, 80°C, and 105°C respectively. Keeping all other variables constant; this represented an average compressive strength reduction of 58% compared to MX1. However, MX2 samples cured at 80°C meet the minimum strength requirement for load-bearing bricks ASTM–129 (2017) and the 7 MPa target—industry standard for low-to-medium load-bearing wall construction bricks in South Africa.

4 Conclusion

This study aimed to develop an alkali-activated brick mix using BA as the binder precursor material and GA as the aggregate material to produce bricks with a compressive strength of 7 MPa. Based on the research findings, the following conclusions can be drawn:

- BA can be effectively activated to produce alkali-activated concrete suitable for various strength classifications, including structural and non-structural applications.
- The optimal mix design to achieve the target compressive strength of 7 MPa was identified as MX2, utilising 10 M NaOH as the activator and cured at 80°C.

Overall, the findings of this investigation highlight significant untapped potential in the alkali-activation of BA. Future research could explore the following avenues:

- Exploring the utilisation of less aggressive chemical activators at lower molarities.
- Investigating the feasibility of alkali activation of BA at ambient temperature.
- Exploring using BA for 100% cement replacement for structural applications such as precast concrete slabs, sewer pipes, etc.

In conclusion, this study demonstrated that alkali activation of low-reactivity precursors, such as coal bottom ash, can be achieved. While the study provided valuable insights, it only began to explore the extensive potential of alkali-activated concrete derived from bottom ash. The findings highlighted both the challenges and advantages, reinforcing the potential of alkali-activated concrete as a viable alternative to traditional Portland cement concrete.

References

Ahmad M, Rashid K (2022) Novel approach to synthesize clay-based geopolymer brick: optimizing molding pressure and precursors' proportioning. Constr Build Mater 322:126472

Alexander M (2017) Durability of concrete design and construction, 1st edn. Taylor & Francis Group, Boca Raton

Busch P, Kendall A, Murphy CW, Miller SA (2022) Literature review on policies to mitigate GHG emissions for cement and concrete. Resour Conserv Recycl 182:106278

Builders (2024) Bricks. Available at: https://www.builders.co.za/Building-Materials/Bricks/c/11204. Accessed 25 Nov 2024

Du Plessis H, Kearsley EP, Matjie RH (2008) Effect of grinding time on the particle size distribution of gasification ash and Portland cement clinker. J South Afr Inst Civ Eng 50(2):18

Frasson A Jr, Casali JM, Oliveira AL, Prudêncio LR Jr (2012) A mix design methodology for concrete block units, proceedings of the 15th international brick and block masonry conference. Florianópolis, Brazil, pp 24–28

Hassan A, Arif M, Shariq M (2019) Influence of microstructure of geopolymer concrete on its mechanical properties—a review. Adv Sustain Constr Mater Geotech Eng Sel Proc TRACE 2018:119–129

Kearsley E (2021) Innovations in concrete. In: Alexander M (ed) Fulton's concrete technology. Midrand, Cement and Concrete SA, pp 1041–1057

Madani H, Malakooti S, Ramezanianpour AA, Joshaghani A, Shah SP (2020) Geopolymer bricks made from less active waste materials. Constr Build Mater 247:118441

Matjie RH, Bunt JR, van Heerden JHP (2005) Extraction of alumina from coal fly ash generated from a selected low-rank bituminous south African coal. Miner Eng 18(3):299–310

Provis JL, Palomo A, Shi C (2015) Advances in understanding alkali-activated materials. Cem Concr Res 78:110–125

Reynolds-Clausen K, Singh N (2019) South Africa's power Producer's revised coal ash strategy and implementation Progress. Coal Combust Gasif Prod 11(1):10–17. https://doi.org/10.4177/CCGP-D-18-00015.1

Sathonsaowaphak A, Chindaprasirt P, Pimraksa K (2009) Workability and strength of lignite bottom ash geopolymer mortar. J Hazard Mater 168(1):44–50

Sedira N, Castro Gomes J, Kastiukas G, Zhou X, Vargas A (2017) A review on mineral waste for chemical-activated binders: mineralogical and chemical characteristics

Thannimalay L, Yusoff S, Zawawi NZ (2013) Life cycle assessment of sodium hydroxide

The Brick Centre (2025) Cavity Walls. Available at: https://thebrickcentre.co.za/cavity-walls/. Accessed 5 Aug 2025

Zhang Z et al (2016) Compositional, microstructural, and mechanical properties of ambient condition cured alkali-activated cement. Constr Build Mater 113:237–245

Potential Use of Heavy Oil Fuel Ash for Smart and Sustainable Infrastructure

Redwan-Ul-Islam(✉), Saheed Adekunle, Mesfer M. Al-Zahrani, and Kalimur Rahman

King Fahd University of Petroleum and Minerals, Academic Belt Road, Dhahran 31261, Kingdom of Saudi Arabia

{g202313630,saheedka,mesferma,mkrahman}@kfupm.edu.sa

Abstract. Amid the urgent global demand for sustainable, low-carbon construction materials, innovative strategies are essential for utilizing industrial by-products. Self-sensing materials enable structures to monitor their own conditions in real time, a capability driven by their electrical conductivity, where stress or damage induces measurable changes in resistivity. Heavy Oil Fuel Ash (HOFA), a largely underutilized residue of oil-fired power plants, poses significant environmental disposal challenges but also offers promising potential for reuse. This study investigates the feasibility of partially substituting Ordinary Portland Cement (OPC) with both untreated and water-treated HOFA. Mortar cubes (50 mm) and cylinders (75 × 150 mm) were prepared with 2, 4, and 6% replacement of OPC by HOFA. Compressive strength and electrical resistivity of HOFA-infused mortars were studied. Results showed that untreated HOFA incorporation provided moderate strength enhancement with decreasing resistivity (increasing conductivity). To further improve reactivity, HOFA was soaked in water for 4 h, dried, and reused. This treatment, by and large, improved compressive strength and increased conductivity. The findings demonstrate that HOFA, in both untreated and treated forms, has promising potential as a conductive filler for self-sensing without compromising strength. Its use not only diverts hazardous waste from landfills but also provides functionalization of cementitious composites. The study underscores the importance of optimizing treatment and dosage to balance strength and conductivity, paving the way for greener construction materials.

Keywords: Heavy Oil Fuel Ash · conductive filler · conductivity · self-sensing

1 Introduction

Today's world is moving towards sustainable, smart, and greener materials to reduce the impact of harmful elements like CO_2. Among the other byproducts, like red mud and silica fume, heavy oil fly ash is one of the underutilized byproducts that has promising potential for reuse. Fuel oil, a broad category that encompasses several refined products with densities ranging from kerosene to residual fuel oil, is categorized into grades (No. 1–No. 6) based on viscosity [1]. Typically, "heavy oil fuel ash" denotes residual fuel oil or fuel oil No. 6. It is the densest of all fuel oil grades and ranks among the least valuable

H. Beushausen et al. (Eds.): NTZR 2025, LNCE 873, pp. 177–186, 2026.
https://doi.org/10.1007/978-3-032-27698-8_17

petroleum products produced as a byproduct. Heavy oil fuel ash is also referred to in the literature as heavy fuel fly ash [2], heavy oil fuel ash [3], oil ash [4], fuel ash [5], heavy fuel oil fly ash [6], oil fly ash [7], oil-fired fly ash [8]. In this paper, all terms will be denoted as heavy oil fuel ash (HOFA).

Saudi Arabia, where the present study was conducted, consumes over 320 million barrels of heavy fuel oil and crude oil for power generation each year, resulting in the daily production of roughly 400 tons of HOFA, and the disposal of it in landfills increases the danger of contamination in the atmosphere, soil, and groundwater [9]. Therefore, to utilize the full potential of HOFA, researchers are investigating various scenarios. The metals in HOFA, which are inherently detrimental, possess a significant impact and may be repurposed for many uses. It has been used in the synthesis of lucrative carbon-based, value-added products, including carbon nanotubes and activated carbon [10].

Massarweh et al. [11] used HOFA as a replacement for OPC and concluded that HOFA acts as a set retarder but can increase strength if it is replaced by OPC up to 3%. They reported that for 3% replacement of OPC by HOFA, the 28-day compressive strength rises from 48.4 MPa (control case) to 52.1 MPa. A similar investigation was conducted by Camilleri et al. [3], revealing that the 28-day compressive strength was around 51.3 MPa for specimens with 5% HOFA, which was lower than the value of 54.3 MPa obtained for the control specimens. Similarly, Mofarrah and Husain [12] used HOFA as a pigment element in mortar with a 2-5% substitution and reported that the compressive strength was maintained. To produce activated carbon, Uddin et al. [13] used HOFA in the process of pyrolysis of a combination of fly ash and road paving asphalt that was dissolved in kerosene at a temperature of 650 °C in an environment of nitrogen dioxide. HOFA was also investigated as a stabilized fill material [14], as a natural adsorbent for the removal of contaminants from wastewater [15].

HOFA contains a significant amount of unburned carbon [16], along with varying levels of vanadium (V), nickel (Ni), and other trace elements [17]. As the amount of carbon is significant in HOFA, increasing its concentration in cement is expected to increase the composite's electrical conductivity (or decrease its resistivity). This increasing conductivity can impart self-sensing behavior to HOFA-based cementitious composites, transforming them into smart and sustainable materials, provided that care is taken to preserve their mechanical properties. Such an interesting application of HOFA has not been explored. Unlike earlier research on HOFA, which has largely emphasized its utilization as a supplementary cementitious material, as a feedstock for producing activated carbon and carbon-based nanomaterials, etc., the novelty of this study lies in advancing HOFA as a conductive filler for self-sensing cementitious composites. This paper presents the initial results of an ongoing experimental study aimed at producing mortars with good electrical conductivity without compromising strength.

2 Experimental Program

2.1 Materials and Methods

For the experiment investigations, ASTM Type I cement (OPC), with a specific gravity of 3.16, was used to prepare mortar cubes and cylinders. HOFA, collected from a local Saudi electric power plant, was characterized by a specific gravity of 1.4 and an average particle size of about 22 µm. Local sand with a specific gravity of 2.65 was used as the fine aggregate. For the untreated case, HOFA was used as received, having no clumps. For the treated case, the material was immersed in water for 4 h, filtered, and then dried at 40 °C to a constant weight. It was subsequently mixed with cement, sand, and water. To maintain a good mortar flow, polycarboxylate superplasticizer (PCE) was used in the mortar mixtures. Figure 1 illustrates the step-by-step procedure for the pre-treatment and incorporation of HOFA to produce mortar mixtures.

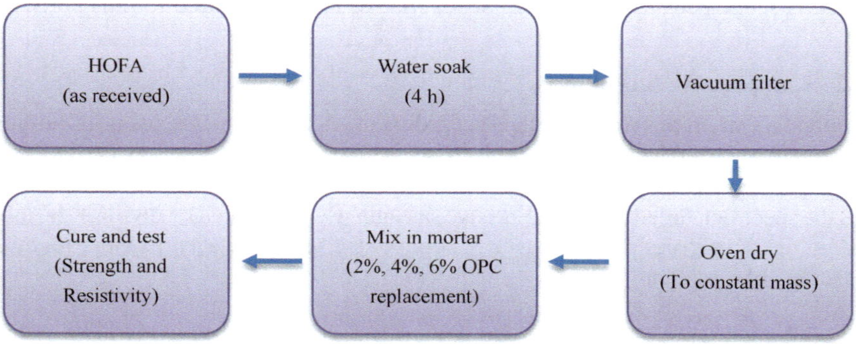

Fig. 1. HOFA pre-treatment and experimental workflow

2.2 Preparation of Test Specimens

To prepare the test specimens, OPC was replaced by HOFA at 2, 4and 6% by mass. The water-to-binder ratio was 0.6, while the sand-to-binder ratio was 2.75. To study the development of compressive strength and electrical resistivity, 50 mm mortar cubes and 75 × 150 mm mortar cylinders were prepared, respectively. For both untreated and treated cases, cement, sand, and HOFA were mixed using a laboratory mortar mixer. After the dry constituents were properly blended, a mixture of water and PCE was added gradually. The wet mixture was then mixed thoroughly again. The mixes were cast into molds and then vibrated to ensure better compaction. The untreated HOFA mixes were denoted as HOFA 0, 2, 4, and 6%, while the treated cases were denoted as HOFA 4 h 2, 4, and 6%. The mix design of mortars for the treated HOFA cases is shown in Table 1.

Table 1. Mix design of mortar cubes for HOFA 4 h

Cases	Cement (g)	HOFA (g)	Sand (g)	Water (g)	SP (% of binder)
HOFA 4 h 0%	446.121	0.000	1226.832	267.672	0.00%
HOFA 4 h 2%	437.198	8.922	1226.832	267.672	0.10%
HOFA 4 h 4%	428.276	17.845	1226.832	267.672	0.22%
HOFA 4 h 6%	419.353	26.767	1226.832	267.672	0.44%

2.3 Material Characterization

2.3.1 XRD and SEM-EDS

To study the properties of untreated and treated HOFA, X-ray diffractometry (XRD) was conducted on HOFA samples. The XRD scanning of both untreated and treated powdered HOFA was performed with Ultima IV RIGAKU X-ray diffractometer over 2θ range of 5°–80°. Scanning electron microscopy coupled with energy dispersive spectroscopy (SEM-EDS) was employed to examine the morphological and chemical properties of HOFA. A Jeol electron microscope, model JSM 6610 LV, was used for this purpose.

2.3.2 Compressive Strength

Compressive strength was determined according to ASTM C109 [18] using an automatic hydraulic MATEST® compression machine. After curing, the test was carried out by applying compressive loading on a 50 mm cube specimen at a constant rate of 1.0 kN/s until the specimen failed. The compressive strength was calculated by dividing the failure load by the cube cross-sectional area. The recorded value for each mixture represents an average of 3 replicas.

2.3.3 Electrical Resistivity

For electrical resistivity, saturated 75 × 150 mm cylindrical specimens were tested according to ASTM C1876 [19] using an RCON® bulk electrical resistivity meter. The meter applies a known AC current across the ends of the specimen via water-saturated contact sponges at a frequency of 1 kHz to determine the electrical resistivity of the specimen.

3 Results and Discussion

3.1 SEM-EDS Analysis

Figure 2 (left) shows the surface morphology of untreated HOFA. The micrograph indicates that HOFA particles are nearly spherical with significant porosity, which may indicate the release of volatiles during combustion. Further, the EDS results presented in Fig. 2 (right) indicate that the analyzed ash particles are primarily unburnt carbon spheres with a notable content of sulfur-containing compounds.

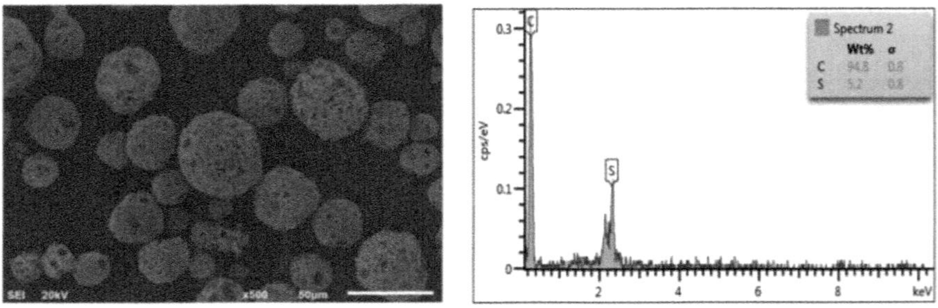

Fig. 2. (Left) SEM micrograph and (right) EDS analysis of untreated HOFA

3.2 XRD Analysis

Figure 3 presents the XRD spectra of treated and untreated HOFA samples. The spectra are composed of characteristic peaks of amorphous carbon at 2θ of about 25.5°–43.5°, with similar patterns reported for carbon black [20], activated carbon [21], and carbon nanotube [22]. Following a 4-h soaking period (HS4), the carbon peak intensities increased significantly compared to the untreated case (HS0), indicating that some soluble constituents of HOFA have been removed by water treatment, and the predominance of carbon content relative to other components. There is also a possibility that the constituents removed by the water treatment are the strength-reducing elements of HOFA.

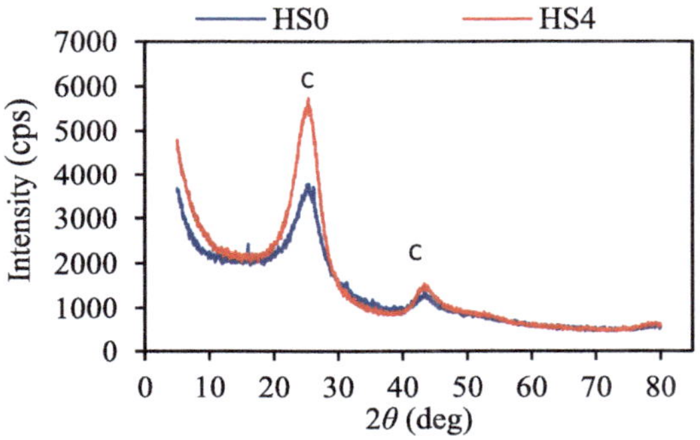

Fig. 3. XRD of untreated (HS0) and treated (HS4) HOFA.

3.3 Compressive Strength

The results of compressive strength are shown in Fig. 4.

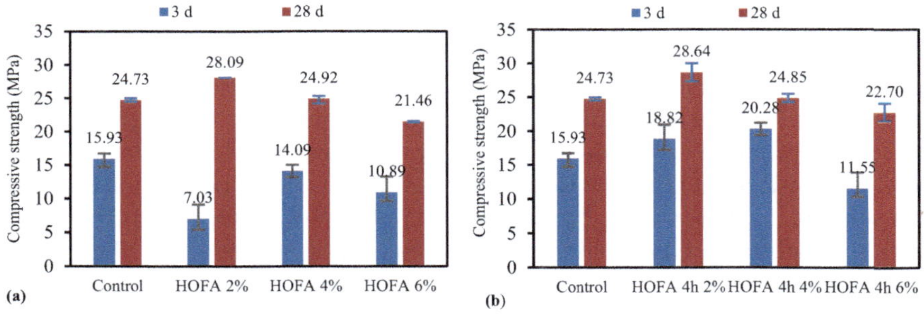

Fig. 4. 3-and 28-day compressive strengths of (a) untreated HOFA and (b) HOFA 4 h mortars

For untreated HOFA, 3-day strength values exhibited mixed behavior; however, for 28-day strength, a gradual reduction was observed. Compared to the control case, 4% replacement seems better as its strength is higher than that of the control case. After soaking in water for 4 h, the HOFA 4 h case showed good improvement for 3 days but a similar effect for 28 days. For the 2% soaked case, the compressive strength is slightly higher than that of the untreated one, increasing from 28.09 MPa to 28.64 MPa. These findings highlight the beneficial role of pre-treatment in enhancing reactivity, possibly through the removal of soluble salts (as confirmed by the XRD results) and improved particle surface activity, which facilitates cement hydration.

3.4 Electrical Resistivity

Electrical resistivity results are illustrated in Fig. 5, showing a very strong negative linear correlation with HOFA contents.

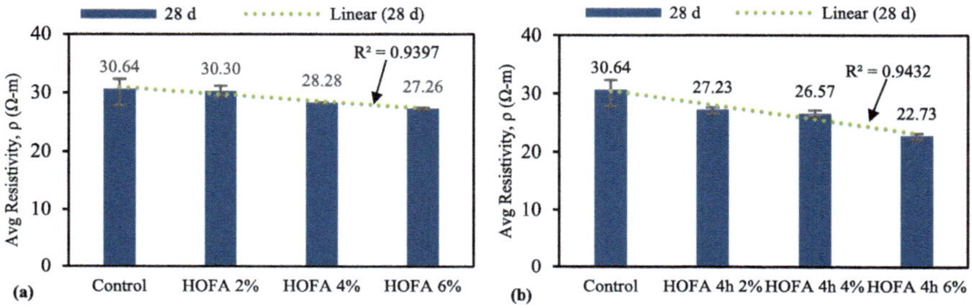

Fig. 5. 28-day electrical resistivity of (a) untreated HOFA and (b) HOFA 4 h mortars

At 28 days, a clear contrast emerged between the electrical resistivity of mortars containing untreated and treated HOFA. Untreated HOFA mixes showed only a moderate decline from the control (30.64 Ω·m) to 30.30 Ω·m (2%), 28.28 Ω·m (4%), and 27.26 Ω·m (6%). In contrast, HOFA 4 h mortar cylinders exhibited a significant reduction in resistivity, dropping from the control value of 30.64 Ω·m to 27.23 Ω·m (2%), 26.57 Ω·m (4%), and 22.73 Ω·m (6%). This suggests that while pre-treatment enhanced compressive strength by improving the reactivity of HOFA, it also increased ionic conductivity, likely due to modifications in pore connectivity.

The gradual reduction in electrical resistivity observed in both untreated and water-treated HOFA mortars directly indicates the establishment of a more continuous and interconnected conductive network within the cement matrix, which is essential for self-sensing behavior. In traditional mortars, elevated resistivity constrains the piezoresistive response, as current flow is primarily governed by ionic conduction through the pore solution. However, with the incorporation of HOFA, especially following a 4-h water pretreatment, the carbon-rich particles function as a conductive filler, reducing bulk resistivity and approaching the electrical percolation threshold. Table 2 presents a benchmarking of electrical resistivities of cementitious composites incorporating various conductive fillers under saturated measurement conditions.

Table 2. Comparative benchmarking of different conductive fillers

Reference	Filler type	Dosages (% of binder)	Composite studied	Method	Resistivity (Ω·m)
[23]	Graphite	4–25 wt%	40 × 40 × 160 mm prismatic paste	2-probe, saturated	5.00–1.00
[24]	Graphene	0.25 wt%	38 × 12.7 × 160 mm prismatic mortar	2-probe, saturated	6750
[25]	GNP + CNT	1–5 vol%	20 × 20 × 40 mm prismatic paste	2-probe, saturated	53–3.75
[26]	Carbon black	1–10 wt%	40 × 40 × 160 mm cement composite	4-probe, oven dry	1620–4.54
This study	HOFA untreated	2–6 wt%	75 × 150 mm cylindrical mortar	2-point, saturated	30.30–27.26
This study	Water-treated HOFA	2–6 wt%	75 × 150 mm cylindrical mortar	2-point, saturated	27.23–22.73

4 Durability and Environmental Implications

The disposal issue of HOFA is becoming a burning problem for oil-rich countries. On the other hand, the high carbon and metals present in HOFA negatively affect the environment through water and ground contamination [14]. Though it is possible to produce carbon nanoparticles from HOFA, it requires significant energy input, accompanied by high processing cost and CO_2 footprint. Additionally, the usage levels of carbon nanoparticles are very low, which is unfavorable to industrial waste utilization. Therefore, utilizing HOFA as a cement replacement not only solves disposal issues but also lowers the carbon footprint and increases sensitivity towards structural health monitoring.

From a durability perspective, while the increased conductivity imparted by HOFA, as with all conductive fillers, could accelerate corrosion propagation when combined with oxygen, chloride ingress, and elevated temperature, the self-sensing capability offers a counterbalancing advantage by enabling early detection of structural flaws before catastrophic failure. The early indication of structural distress offers opportunities for early retrofitting and strengthening the existing structure before catastrophic consequences. Nevertheless, the long-term performance of HOFA-based composites must be evaluated under key environmental stressors such as moisture fluctuations, temperature variation, and cyclic mechanical loading.

5 Conclusion

In this study, HOFA was explored as a conductive filler for formulating self-sensing cementitious composites. The main conclusions of the study are:

- At 28 days, 2% HOFA replacement achieved the highest compressive strength, rising from 24.73 MPa (control) to 28.09 MPa for untreated and 28.64 MPa for water-treated HOFA. Strength decreased progressively at higher dosages (4–6%), though water-treated mixes consistently outperformed untreated ones at the same HOFA dosages.
- For 6% replacement in water-treated case, the electrical resistivity dropped from 30.64 Ω·m to 22.73 Ω·m, a reduction of nearly 26% compared to the control. However, at 6% HOFA, the compressive strength was lower than the control.
- Water treatment of HOFA can be extended for 4% replacement, as the compressive strength is higher (24.85 MPa) and electrical resistivity is lower (26.57 Ω·m) than the control case. Therefore, it can be concluded that around 4% replacement is an optimal condition for both compressive strength and self-sensing.

6 Recommendations

While the laboratory-scale results are promising, several challenges must be acknowledged before HOFA-based mortars can be widely adopted. The investigation was done in a laboratory with mortars only. However, concrete is widely used in practical fields, which is a scalability issue. Moreover, the variability of HOFA composition and the quality control of the pre-treatment process must be considered in terms of scalability. Integration into existing practices also requires an assessment of workability, admixture compatibility, and curing regimes, as these factors can impact both conductivity and strength under field conditions. Future work should investigate various pre-treatment processes, such as heating, mixing with other substances, and other conditions. Mixing HOFA with other conductive fillers and eventually making hybrid materials for both normal and reinforced concrete can also be addressed. Additionally, life-cycle assessment (LCA) and techno-economic studies are crucial for benchmarking HOFA composites against conventional conductive fillers, ensuring both sustainability and practical viability.

References

1. ASTM D396-19a, Standard specification for fuel oils. [Online]. Available: https://store.astm.org/d0396-18a.html. Accessed: 27 Aug 2025
2. Georgiou A, Chousidis N, Ioannou I (2022) Self-compacting Cementitious composites with heavy fuel fly ash replacement. Constr Mater 2(4):276–296. https://doi.org/10.3390/constrmater2040018

3. Camilleri J, Anastasi M, Torpiano A (2013) The microstructure and physical properties of heavy oil fuel ash replaced Portland cement for use in flowable fill concrete and the production of concrete masonry units. Constr Build Mater 38:970–979. https://doi.org/10.1016/j.conbuildmat.2012.09.054
4. Zummo SA King Fahd university of petroleum & minerals Dhahran 31261, Saudi Arabia deanship of graduate studies master of science in civil engineering
5. Rabah MA, Anadolly MBE, Shereif RAE, Atrees MS, Agamy HME (2017) Preparation of valuable products from cleaned carbon of fuel ash. AIMS Mater Sci 4(5):1186–1201. https://doi.org/10.3934/matersci.2017.5.1186
6. Al-Osta MA, Baig MG, Al-Malack MH, Al-Amoudi OSB (2016) Study of heavy fuel oil fly ash for use in concrete blocks and asphalt concrete mixes. Adv Concr Constr 4(2):123–143. https://doi.org/10.12989/acc.2016.4.2.123
7. Shawabkeh R, Khan MJ, Al-Juhani AA, Al-Abdul Wahhab HI, Hussein IA (2011) Enhancement of surface properties of oil fly ash by chemical treatment. Appl Surf Sci 258(5):1643–1650. https://doi.org/10.1016/j.apsusc.2011.07.136
8. Hsieh YM, Tsai MS (2003) Physical and chemical analyses of unburned carbon from oil-fired fly ash. Carbon N Y 41(12):2317–2324. https://doi.org/10.1016/S0008-6223(03)00283-5
9. Husain T, Ahmad M Environmental engineering and computer application: proceedings of the 2014. [Online]. Available: https://books.google.com.sa/books?hl=en&lr=&id=FY7YCQAAQBAJ&oi=fnd&pg=PP1&dq=T.+Husain,+M.+Ahmad,+Environmental+Engineering+and+Computer+Application-+Chan+(Ed.).+2015&ots=zbx3BU2jMj&sig=hr8WI1xG7Sy3mkg5ILHDEXokDhA&redir_esc=y#v=onepage&q&f=false. Accessed: 27 Aug 2025
10. Basha SI, Aziz A, Maslehuddin M, Ahmad S, Hakeem AS, Rahman MM (2020) Characterization, processing, and application of heavy fuel oil ash, an industrial waste material – a review. Chem Rec 20(12):1568–1595. https://doi.org/10.1002/tcr.202000100
11. Massarweh O, Maslehuddin M, Al-Dulaijan SU, Shameem M (2021) Performance evaluation of heavy oil fly ash as a retarder of Portland cement hydration. J Build Eng 34:101881. https://doi.org/10.1016/j.jobe.2020.101881
12. Mofarrah A, Husain T (2013) Use of heavy oil Fly ash as a color ingredient in cement mortar. Int J Concr Struct Mater 7(2):111–117. https://doi.org/10.1007/s40069-013-0042-3
13. Azhar Uddin M, Shinozaki Y, Furusawa N, Yamada T, Yamaji Y, Sasaoka E (2007) Preparation of activated carbon from asphalt and heavy oil fly ash and coal fly ash by pyrolysis. J Anal Appl Pyrolysis 78(2):337–342. https://doi.org/10.1016/j.jaap.2006.09.004
14. Mofarrah A, Husain T, Danish EY (2012) Investigation of the potential use of heavy oil fly ash as stabilized fill material for construction. J Mater Civ Eng 24(6):684–690. https://doi.org/10.1061/(ASCE)MT.1943-5533.0000442
15. Mofarrah A, Husain T (2013) Evaluation of environmental pollution and possible management options of heavy oil fly ash. J Mater Cycles Waste Manage 15(1):73–81. https://doi.org/10.1007/s10163-012-0090-9
16. Vitolo S, Seggiani M, Filippi S, Brocchini C (2000) Recovery of vanadium from heavy oil and orimulsion fly ashes. Hydrometallurgy 57(2):141–149. https://doi.org/10.1016/S0304-386X(00)00099-2
17. Ali S et al (2020) Evaluation of alumina reinforced oil fly ash composites prepared by spark plasma sintering. Int J Appl Ceram Technol 17(4):1948–1958. https://doi.org/10.1111/ijac.13491
18. Standard test method for compressive strength of hydraulic cement mortars (using 2-in. or [50-mm] cube specimens) 1. [Online]. Available: www.astm.org
19. ASTM C1876 Standard test method for bulk electrical resistivity or bulk conductivity of concrete (2019). https://doi.org/10.1520/C1876-19

20. Lee S-M, Lee S-H, Roh J-S (2021) Analysis of activation process of carbon black based on structural parameters obtained by XRD analysis. Crystals (Basel) 11(2):153. https://doi.org/10.3390/cryst11020153
21. Liu X-Y et al (2010) Preparation of a carbon-based solid acid catalyst by Sulfonating activated carbon in a chemical reduction process. Molecules 15(10):7188–7196. https://doi.org/10.3390/molecules15107188
22. Abdulrazzak F, Saleh HM, El-Sheikh SMM (eds) (2019) Behavior of X-ray analysis of carbon nanotubes. IntechOpen, Rijeka. https://doi.org/10.5772/intechopen.85156
23. Uher V, Černý V, Drochytka R, Baránek Š (2022) The effect of exposure conditions on the properties of cementitious composites with reduced electrical resistivity. Buildings 12(12):2124. https://doi.org/10.3390/buildings12122124
24. Liu H, Deshmukh A, Salowitz N, Zhao J, Sobolev K (2022) Resistivity signature of graphene-based Fiber-reinforced composite subjected to mechanical loading. Front Mater 9. https://doi.org/10.3389/fmats.2022.818176
25. Wang Y et al (2024) Advances in self-sensing cement-based composites containing nano materials for smart civil infrastructures. Measurement 230:114514. https://doi.org/10.1016/j.measurement.2024.114514
26. Monteiro AO, Cachim PB, Costa PMFJ (2017) Self-sensing piezoresistive cement composite loaded with carbon black particles. Cem Concr Compos 81:59–65. https://doi.org/10.1016/j.cemconcomp.2017.04.009

Durability Performance of Concrete Made with Gold Tailings from Barberton Mine, South Africa

Rhoda A. Adeyeye[1](✉), Jacob O. Ikotun[1], Solomon Oyebisi[2], and Zesizwe Ngubane[1]

[1] Civil Engineering Department, Durban University of Technology, Midland, South Africa
22172827@dut4life.ac.za, {jacobi,zesizwen}@dut.ac.za
[2] Civil Engineering and Geomatics Department, Durban University of Technology, Durban, South Africa
solomon0@dut.ac.za

Abstract. Recently, reprocessing of primary tailings to recover leftover metals and minerals has gained popularity in the mining industry, promoting sustainable resource supply and environmental management of accumulated waste. However, a huge volume of reprocessed residual tailings is left indiscriminately discarded after reprocessing, resulting in environmental degradation. Thus, this study investigates the durability performance of concrete modified with gold tailings (GT) from Barberton Mine, South Africa, as a sand substitute. Concrete of C35 MPa was prepared with GT substituting crusher sand at varying levels from 0 to 100% by weight, using a water/cement ratio of 0.5, and cured for 7 to 90 days. The durability results demonstrated that GT-based concrete exhibited comparable resistance to oxygen permeability, water sorptivity, and chloride conductivity as a reference concrete, indicating high-quality performance. These findings highlight the advantage of gold tailings as a viable alternative to crusher sand, contributing to reduced dependency and enhanced resource sustainability.

Keywords: Chloride Conductivity · Compressive Strength · Concrete · Durability · Gold Tailings · Oxygen Permeability and Water Sorptivity

1 Introduction

Tailings are fine-grained solid residue generated during the extraction of valuable metals and minerals [1]. With advancements in mineral extraction technologies coupled with the decreasing ore grades in primary deposits, tailings are increasingly being reprocessed as a valuable source of minerals. This approach has recently gained traction in the mining industry, supporting sustainable resource use and enhancing the environmental management of accumulated waste [1–5]. Gold tailings are one of the tailings undergoing reprocessing, a practice implemented in South Africa for over a decade [6]. The residual tailings remaining after reprocessing are typically disposed of on land via dry stacking. Although the toxicity of these residual gold tailings is reduced, the stacked deposits continue to occupy a vast expanse of land that could otherwise support

H. Beushausen et al. (Eds.): NTZR 2025, LNCE 873, pp. 187–198, 2026.
https://doi.org/10.1007/978-3-032-27698-8_18

socio-environmental and economic development [1, 7]. In addition, dry-stacked tailings contribute to air pollution, negatively impacting air quality in mining regions. Therefore, repurposing residual gold tailings is essential for promoting the circular economy, environmental sustainability, and more efficient land use [1, 8, 9].

The growing interest in repurposing mine tailings has led researchers to explore their potential use as construction materials across various industries, notably in concrete production [10, 11]. Concrete is fundamental to modern construction and ranks as the second most consumed material worldwide, following water [12]. It is typically composed of binders, aggregates, water, and, when necessary, admixtures. Concrete, whether cement-based or geopolymer-based, requires a substantial volume of aggregates. Incorporating tailings as a partial or complete replacement offers a dual benefit, such as supplementing the aggregate supply for concrete production and reducing the volume of land-disposed tailings [12–14].

Given their fine-grained characteristics, gold tailings have been explored as a viable substitute for fine aggregates in concrete [10]. For example, Song et al. [15] investigated the strength and durability performance of concrete in which river sand was replaced with gold tailings. They reported that the gold tailings-based concrete showed enhanced mechanical properties, reduced water absorption, and improved resistance to freeze-thaw cycles and chloride ion penetration, attributed to the tailings' pozzolanic activity and filler effect. Luo and Luo [16] replaced quartz sand with gold tailings in ultra-high-performance concrete (UHPC) and examined the influence of different curing periods on mechanical and microstructural properties. Their findings indicated increased strength, also due to the pozzolanic activity of the tailings. Lokesha et al. [13] assessed the durability of geopolymer concrete incorporating gold tailings as a partial replacement for river sand, reporting reduced chloride ion penetration compared to conventional concrete. However, most existing studies have focused on primary gold tailings rather than the residual tailings from reprocessing, evidenced by the consistently high pozzolanic activity observed. The chemical composition of reprocessed residual gold tailings likely differs due to the prior extraction of valuable minerals, which may influence their behavior in concrete. Therefore, further research is required to evaluate the durability performance of concrete incorporating reprocessed residual gold tailings.

Therefore, this study examines the durability performance of cement concrete incorporating reprocessed residual gold tailings as a replacement for crusher sand. The assessed durability properties include oxygen permeability, water sorptivity, porosity, and chloride conductivity.

2 Materials and Methods

2.1 Materials

In this study, the cementitious material used was CEM I 52.5 R grade Portland cement (PC) produced by Afrisam Plant in Johannesburg, following the manufacturer's specifications. Type C fly ash (FA), conforming to the specifications outlined in SANS 50450-1:2014, was used as a cement extender [17]. Andesite crusher sand and uniformly graded andesite stone with a nominal size of 19 mm were used as fine and coarse aggregates, respectively. The aggregates were obtained from the Afrisam Plant in Johannesburg, South Africa, and met the manufacturer's specifications, complying with SANS 1083:2014 [18]. Local tap water was used for mixing. To improve workability, Sika ViscoCrete 3088, a modified polycarboxylate-based superplasticizer (SP) compliant with BS EN 934-2:2009, was used [19].

Gold tailings

In this study, the gold tailings used as a substitute for sand were sourced from the Fairview Mine, located in Barberton, Mpumalanga, South Africa. This mine is part of the Barberton Mines complex, owned and operated by Pan African Resources (PAR) Group. Since the commissioning of its tailings reprocessing plant in 2013, the operation has processed approximately 1.2 million tonnes of gold tailings annually, with an average gold grade of 0.7 g/t. As a result, millions of tonnes of residual gold tailings have accumulated over the years [6]. These reprocessed residual gold tailings were collected for the utilization in this study as a crusher sand replacement.

2.2 Fine Aggregates Characterization

Gold tailings and andesite crusher sand presented in Fig. 1 were the fine aggregates used in this study. Characterization tests were conducted to assess the differences between the two materials. The physical properties, summarized in Table 1, show a degree of similarity between the gold tailings and the andesite crusher sand. However, particle size distribution analysis presented in Fig. 2 shows that gold tailings are finer than crusher sand, indicating a higher specific surface area [12].

Reprocessed residue gold tailings   Andesite crusher sand

Fig. 1. Image of fine aggregates used in this study

Elemental composition, determined using energy-dispersive X-ray (EDX) analysis, is presented in Table 2. The primary constituents of the gold tailings are silica and some metal oxides, although their concentrations are lower than those reported in previous studies [7, 10, 15, 20]. This suggests that certain chemical components have been

Table 1. Physical properties and chemical composition of fine aggregates

Properties	SG[1]	BD[2]	FM[3]	WA[4]	ST[5]	Shape
GT[6]	2.76	1522	0.66	5.87	Rough	Angular
ACS[7]	2.94	2011	3.19	3.90	Rough	Angular

[1] Specific gravity
[2] Bulk density
[3] Fineness modulus
[4] Water absorption
[5] Surface texture
[6] Gold tailings
[7] Andesite crusher sand

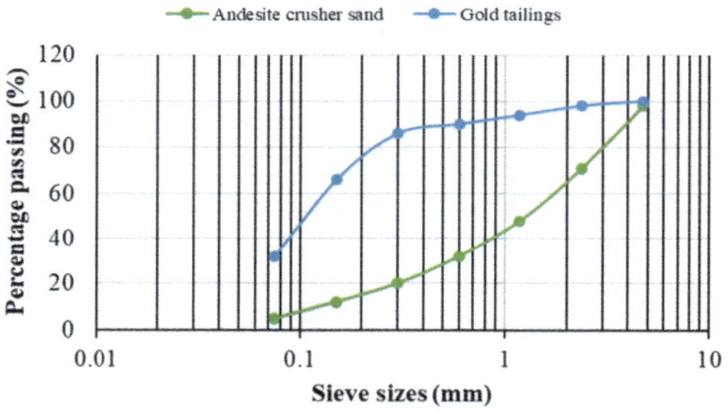

Fig. 2. Particle size distribution of fine aggregates

extracted during reprocessing, resulting in reduced pozzolanic activity and supporting the suitability of residual gold tailings primarily as a sand replacement rather than a reactive material.

Table 2. Elemental composition of fine aggregates

Element	Si	Al	Fe	Mg	K	Na	Ca	O	Cl	C	S
GT[6]	20.96	8.11	0.54	0.35	8.06	0.00	0.00	59.33	0.00	2.65	0.00
ACS[7]	8.78	1.58	2.56	0.8	0.56	0.4	26.58	42.34	0.1	15.76	0.58

Residual gold tailings from reprocessing are generally alkaline and contain heavy metal concentrations below regulatory limits [21, 22], indicating minimal environmental risk. The gold tailings used in this study exhibit similar characteristics, with no risk of leachable contaminants, as confirmed by the absence of acidic sulfide compounds. Figure 3 presents the X-ray diffraction (XRD) pattern of the gold tailings, showing the mineralogical composition. The analysis identified soft minerals such as clinochlore, muscovite, dolomite, and siderite. A high content of quartz and muscovite was observed, which can be attributed to their prevalence in the Barberton host rock and their strong resistance to both physical and chemical weathering [21, 23]. The negligible presence of gypsum and absence of pyrite indicated by the XRD pattern correlate with the negligible CaO and sulfide content, respectively. This shows the material's inert nature and low leaching potential. Additionally, the lack of detectable uranium suggests that the residual gold tailings are non-radioactive. These findings collectively indicate that several metals and minerals have already been extracted during reprocessing, further supporting the inertness and environmental safety of the gold tailings used in this study.

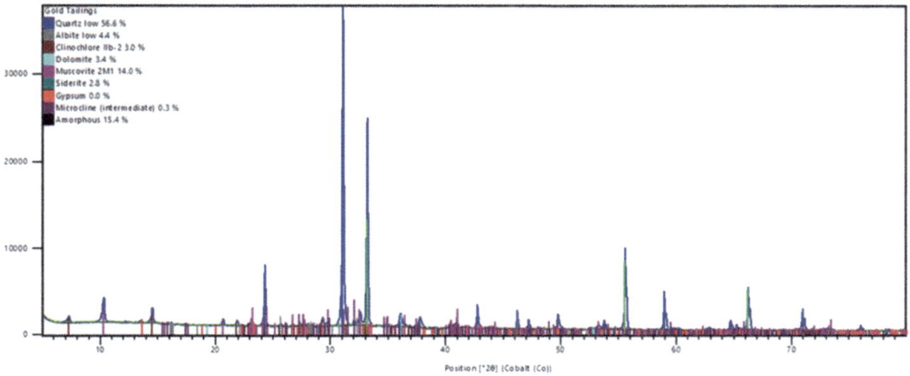

Fig. 3. XRD pattern of gold tailings

2.3 Mix Design and Curing Conditions

The concrete mix for a target strength of 35 MPa (reference concrete) was designed using the volumetric mix method founded by the South African Cement and Concrete Institute (C&CI). Six mixtures were prepared using 0.5 w/c, a constant quantity of crushed stone, and a fixed 70–30% blend of PC and FA, selected based on South African practice and proven performance in past studies [24]. The target slump specified in the design ranged from 75 mm to 150 mm. To achieve this, the superplasticizer (SP) dosage was adjusted in proportion to the increasing gold tailings content. The SP dosage remained within the standard limit of 5% by mass of cement [19]. As presented in Table 3, concrete mixtures designated as T0 (reference concrete), T10, T25, T50, T75, and T100 were formulated by replacing crusher sand with gold tailings at 0 to 100% levels. Concrete mixtures were prepared by first mixing the dry materials for 1 min, then adding water and superplasticizer, and continuing mixing for another 1 min. Subsequently, a slump test was conducted, after which 100 × 100 × 100 mm^3 samples were cast for durability testing. After 24 h, the samples were demolded and cured in water, with testing carried out after 7, 28, 56, and 90 days of curing. Both concrete mixing and water curing were performed at a controlled temperature of 22–25 °C, complying with SANS 5861-1:2006 [25].

Table 3. Mix proportion of concrete samples (kg/m^3)

Mix	PC	FA	Aggregates			Water	SP (% of total binder content)
			Andesite crushed stone	Gold tailings	Andesite crusher sand		
T0	294	126	1084	–	842	210	0.06
T10	294	126	1084	84.2	757.8	210	0.078
T25	294	126	1084	210.5	631.5	210	0.85
T50	294	126	1084	421	421	210	1.23
T75	294	126	1084	631.5	421	210	1.58
T100	294	126	1084	842	–	210	1.96

2.4 Test Methods of Concrete

The slump test for each concrete mix was performed, complying with SANS 5862-1:2006 [26], using a slump cone measuring 300 mm in height, with a top diameter of 100 mm and a base diameter of 200 mm. Furthermore, durability tests, based on the South African durability index (DI), including oxygen permeability, water sorptivity, and chloride conductivity, were conducted on concrete specimens after 28, 56, and 90 days of curing, following South African standards [27]. For testing, four concrete discs (70 ± 2 mm diameter, 30 ± 2 mm thickness) were cut from $100 \times 100 \times 100$ mm^3 specimens for each mix and curing age. The discs were oven-dried at 50 ± 2 °C for 7 days, cooled to 23 ± 2 °C, and then tested, with average values recorded as results. The oxygen permeability index (OPI), water sorptivity index (WSI), and chloride conductivity index (CCI) tests was performed at each curing age, complying with South African standards [27].

3 Results and Discussion

3.1 Workability

The slump result presented in Fig. 4 shows that the workability of concrete decreased as andesite crusher sand was progressively replaced with SGTs. This is attributed to the fine size of gold tailings particles and the high muscovite content in the gold tailings [22]. The presence of muscovite could significantly affect the workability because it is a non-swelling clay with a high-water absorption capacity [28]. The result showed that the concrete mix lost its workability completely after 10% replacement of the crusher sand with gold tailings. This loss in workability is attributed to the increased muscovite content introduced by the gold tailings. As shown in Table 3, the superplasticizer dosage increases with a higher proportion of gold tailings in the concrete mix. This is due to muscovite reducing the dispersion efficiency of the superplasticizer in the cement paste, indicating higher superplasticizer adsorption in mixes with higher gold tailings content

[28]. The mixes became increasingly rubbery and spongy with higher gold tailings content, increasing slump values. This behaviour is likely due to the higher muscovite content, which increases the viscosity of fresh concrete and contributes to the observed rise in slump [29–32].

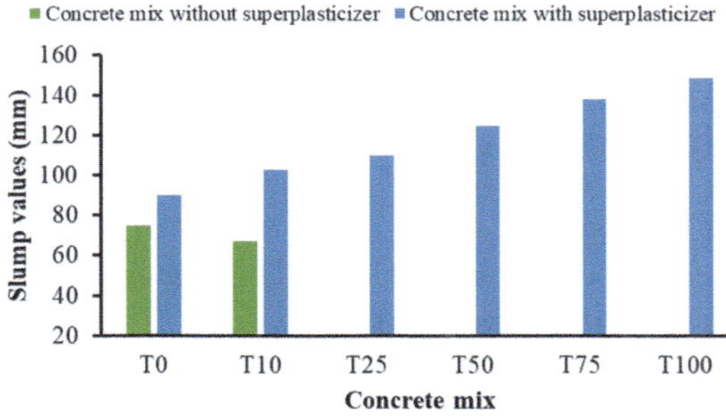

Fig. 4. Slump value of the concrete mix

3.2 Durability Properties

The durability indices, namely oxygen permeability, water sorptivity, and chloride conductivity, were used to assess the transport of gases, moisture, and chloride ions in concrete and concrete porosity. These indices, influenced by curing type and duration, were calculated from the average of four specimens tested at 28, 56, and 90 days. Concrete quality was evaluated using grading ranges specified in the South African Research Monograph [33].

Oxygen permeability

According to the South African Research Monograph [33], the higher the OPI value, the lower the permeability. The OPI results in Fig. 5 show that T100, T0, and T50 have the highest OPI at 28, 56, and 90 days, respectively. After 90 days, T25 and T50 showed higher OPI than the reference concrete, attributed to the pore-refining filler effect of gold tailings. Additionally, OPI values increased over time, reflecting improved resistance to oxygen permeability as curing progressed.

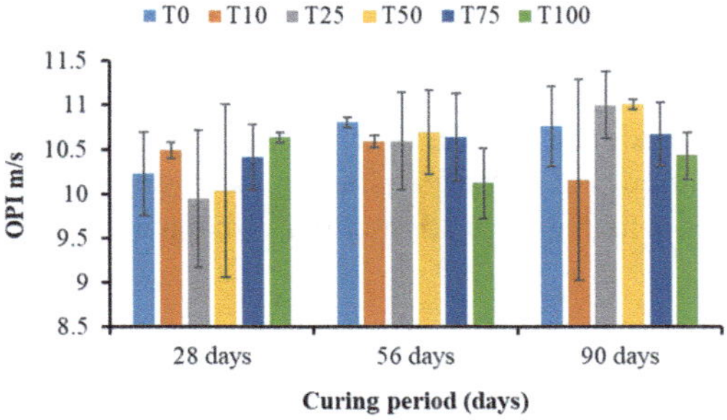

Fig. 5. Oxygen permeability indices of concrete samples.

OPI differences between concrete samples were relatively small, suggesting limited test sensitivity to gold tailings variation. OPI values ranging from 9.95 to 11.01 signify concrete quality classified as good to excellent, based on the South African Research Monograph [33]. Therefore, it can be concluded that residual gold tailings from reprocessing do not adversely affect the permeability resistance of concrete. Moreover, the high muscovite content in the gold tailings does not impair the concrete's resistance to permeability.

Water sorptivity and porosity

The WSI values of concrete represent the water absorption rate through capillary suction, adjusted for porosity; lower WSI values indicate higher resistance to moisture penetration [33]. WSI results in Fig. 6 show that T25, T50, T75, and T100 have lower WSI values compared to the reference concrete. This might be attributed to the pore-refining filler effect of gold tailings. It was observed that T10 has the highest WSI value, possibly from uneven gold tailings distribution, having minimal impact on disrupting capillary formation, resulting in a higher capillary absorption rate [34]. According to the South African Standard [27] WSI values below 6 and between 6 and 10 are classified as "excellent" and "good" water resistance, respectively. Therefore, after 90 days, T75 and T100 demonstrated excellent concrete quality, while T0, T10, T25, and T50 achieved good concrete quality. This suggests that gold tailings-based concrete exhibits satisfactory resistance to water penetration.

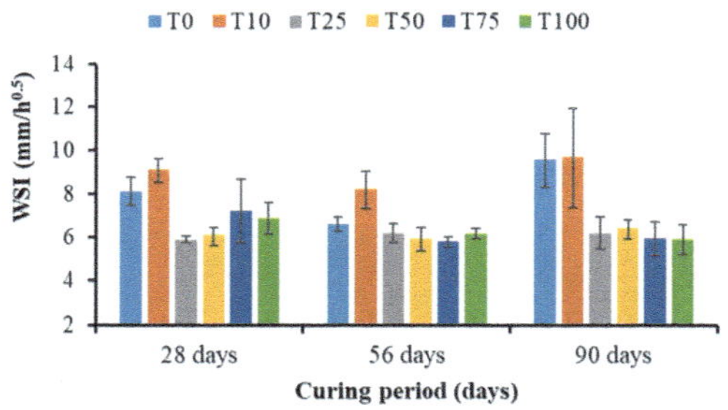

Fig. 6. Water sorptivity indices of concrete samples

Moreover, Fig. 7 illustrates the porosity results of concrete samples saturated with calcium hydroxide, showing a consistent decrease in porosity across all mixes with longer curing durations. The reduction confirms that the chemical composition and high muscovite content of gold tailings do not hinder cement hydration but rather contribute to refining the pore structure of the concrete. It further indicates that the presence of high muscovite does not impede the concrete's resistance to water. At 90 days, T25 exhibited the lowest porosity among all samples, possibly because of the uniformly distributed gold tailings particles, which enhanced the filler effect and minimized pore spaces.

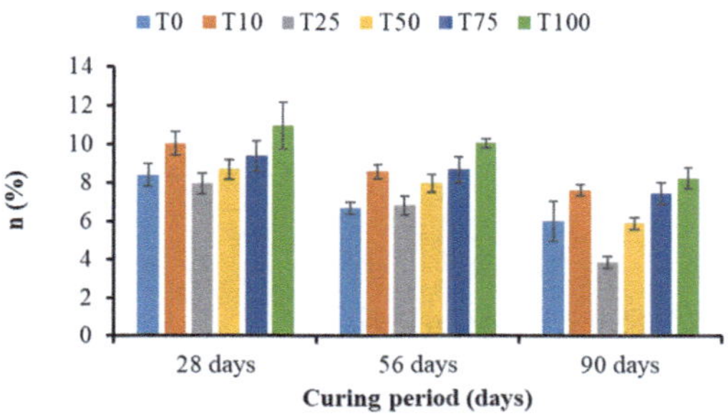

Fig. 7. Porosity of concrete specimens saturated with calcium hydroxide

Chloride conductivit

According to the South African Standard [27] lower CCI values signify higher chloride conductivity resistances. The CCI results in Fig. 8 indicate that all gold tailings-based concrete samples, except T25 and T50 at 90 days, exhibited higher CCI values compared to T0. This increase may be attributed to the higher specific surface area of the gold tailings, which reduced cement paste coverage on individual particles. As a result, aggregate adhesion was weakened, leading to lower chloride penetration resistance [35].

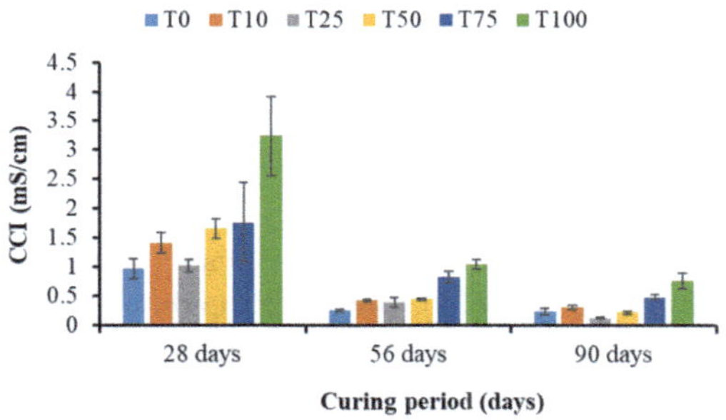

Fig. 8. Chloride conductivity indices of concrete specimens.

In contrast, the lower CCI values observed in T25 and T50 may be due to the uniform distribution of gold tailings particles, which contributed to pore refinement and improved resistance to chloride penetration [36]. However, a decrease in CCI values was observed with longer curing durations. Moreover, all concrete samples at 56 and 90 days exhibit satisfactory resistance to chloride penetration, ranging from good to excellent concrete quality based on the South African Standard [27].

4 Application

This study demonstrates that concrete made with residual gold tailings from reprocessing exhibits satisfactory durability performance, supporting its use as a sustainable alternative to crusher sand. Gold tailings-based concrete can be used in various structural components, including foundations, walls, beams, and slabs. Its use helps reduce stacked tailings, conserves natural resources by lowering demand for crusher sand, and cuts production costs, particularly in mining areas. A partnership between the mining and concrete industries could mutually benefit both sectors by addressing waste management and material sourcing, thereby supporting sustainable development and environmental conservation.

5 Conclusion and Recommendations

This study assessed the durability performance of reprocessed residual gold tailings as a sustainable alternative to crusher sand in concrete. Based on the findings, the following conclusions and recommendations are presented:

1. **Material Properties and workability**: Gold tailings are environmentally friendly and relatively inert, with finer particles, lower density, and higher water absorption than andesite crusher sand. They contain soft minerals and a high muscovite content, which may affect the concrete's performance characteristics. However, their similar characteristics make them a viable substitute.
2. **Workability**: Replacing crusher sand with gold tailings reduced workability due to finer particles, higher water demand, and high muscovite content. However, the addition of superplasticizers improves workability.
3. **Durability**: Gold tailings-based concrete samples achieved good to excellent quality in oxygen permeability index (OPI), water sorptivity index (WSI), and chloride conductivity index (CCI), indicating good resistance to gas, moisture, and chloride penetration. Moreover, the use of gold tailings does not adversely affect the pore structure, demonstrating that gold tailings-based concrete maintains satisfactory durability.
4. **Practical Benefits**: Utilizing reprocessed residual gold tailings as fine aggregate in concrete decreases dependence on conventional sand, reduces concrete production costs (especially in mining regions), minimizes carbon emissions associated with sand processing, aids in tailings waste management, and mitigates environmental impact. As gold tailings require no additional processing, they present a practical and environmentally sustainable alternative.
5. **Future Research**: Further studies are needed to evaluate other durability aspects such as sulfate resistance, carbonation depth, alkali-silica reaction, shrinkage, corrosion, and thermal behavior.
6. **Application Potential**: Given their durability and environmental benefits, reprocessing residual gold tailings can be applied in construction, supporting sustainable concrete production and resource conservation.

References

1. Sarker S, Haque N, Bhuiyan M, Bruckard W, Pramanik B (2022) Recovery of strategically important critical minerals from mine tailings. J Environ Chem Eng 10(3):107622
2. Mesa D, Barriga D, Berríos P, Rodríguez E, Amelunxen R (2020) Reprocessing historic tailings—three Chilean case studies. In: Proceedings of the procemin geomet, pp 1–11

3. Wivast J, Nyland A, Nia S, Kioumarsi M, Shafei B (2024) Developing eco-friendly ultra-high-performance concrete by utilizing recycled alternatives. In: The international conference on net-zero civil infrastructures: innovations in materials, structures, and management practices (NTZR). Springer Nature Switzerland, Cham, pp 49–57
4. Farahzadi L, Nia S, Shafei B, Kioumarsi M (2025) Sustainability assessment of ultra-high performance concrete made with various supplementary cementitious materials. Cleaner Mater 15:100301
5. Nia S, Nyland A, Wivast J, Kioumarsi M, Shafei B (2025) Investigations of Portland limestone cement and waste glass powder for sustainable ultra-high performance concrete. Case Stud Constr Mater 22:e04425
6. Mutimutema P (2021) Value recovery from mine calcine tailings doctoral dissertation. Stellenbosch University, Stellenbosch
7. Guan Y, Li Y, Zhang H, Sun R, Tian J, Yang Y (2023) Preparation of low-cost green engineered cementitious composites (ECC) using gold tailings and unoiled PVA fiber. J Build Eng 78:107455
8. Quan X, Wang S, Liu K, Xu J, J., Zhao, N., Liu, B. (2022) Influence of molybdenum tailings by-products as fine aggregates on mechanical properties and microstructure of concrete. J Build Eng 54:104677
9. Balegamire C, Dable P, Kone K, Nkuba B (2022) Use of a design of experiments (DOE) for studying the substitution of natural river sand (NRS) by gold mine tailings (GMT) in concrete manufacturing. Recycling 7(4):50
10. Li S, Chen J, Gao W, Lyu X, Liang Z, Zhou W (2024) Current situation and prospects for the clean utilization of gold tailings. Waste Manag 180:149–161
11. Ikotun J, Adeyeye R, Otieno M (2022) Application of mine tailings sand as construction material – a review. MATEC Web Conf 364:05008
12. Adiguzel D, Tuylu S, Eker H (2022) Utilization of tailings in concrete products : a review. Constr Build Mater 360:129574
13. Lokesha E, Aruna M, Reddy S (2025) Durability characteristics of geopolymer concrete produced using gold ore tailings along with recycled coarse aggregates. Sci Rep 15(1):12051
14. Garcia-Troncoso N, Baykara H, Cornejo M, Riofrio A, Tinoco-Hidalgo M, Flores-Rada J (2022) Comparative mechanical properties of conventional concrete mixture and concrete incorporating mining tailings sands. Case Stud Constr Mater 16:e01031
15. Song Q, Zou Y, Bao J, Zhang P (2024) Disposal of solid waste as building materials: a study on the mechanical and durability performance of concrete composed of gold tailings. J Mater Res Technol 30:2111–2124
16. Luo B, Luo Z (2022) Influence of curing regimes on the mechanical properties and microstructure of ultra-high-performance concrete blended with gold mine tailings. Mater Lett 311:131527
17. SANS 50450-1:2014: South African National Standard (2014) Portland cement extenders. Part 1 fly ash
18. SANS 1083:2014: South African National Standard (2014) Aggregates from natural sources-aggregate for concrete
19. BS EN 934-2:2009 (2009) Admixture for concrete, mortar and grout. Concrete admixtures. Definitions, requirement, conformity, making and labelling
20. Wang J et al (2021) A novel design of low carbon footprint ultra-high performance concrete (UHPC) based on full scale recycling of gold tailings. Constr Build Mater 304:124664
21. Mashifana T, Sithole T (2021) Clean production of sustainable backfill material from waste gold tailings and slag. J Clean Prod 308:127357
22. Ahmed T, Elchalakani M, Basarir H, Karrech A, Sadrossadat E, Yang B (2021) Development of ECO-UHPC utilizing gold mine tailings as quartz sand alternative. Cleaner Eng Technol 4:100176

23. Yuan J, Ma H, Guo R, Ma X, Komarneni S (2020) Sustainable materials by mimicking natural weathering. ACS Sustain Chem Eng 8(29):10920–10927
24. Alexander M, Santhanam M, Ballim Y (2010) Durability design and specification for concrete structures—the way forward. Int J Adv Eng Sci Appl Math 2:95–105
25. SANS 5861-1:2006: South African National Standard (2006) Concrete tests part 1-mixing fresh concrete in the laboratory
26. SANS 5862-1:2006: South African National Standard (2006) Concrete tests part 1-CONSISTENCE of freshly mixed concrete—slump test
27. Alexander M (2018) Durability index testing procedure manual Ver 4.5.1, April 2018
28. Chi H et al (2024) Unraveling polycarboxylate superplasticizer (PCE) compatibility in muscovite-blended cement paste through aggregation mechanisms. J Build Eng 95:110133
29. Mueller O (1971) Some aspects of the effect of micaceous sand on concrete. Civ Eng Siviele IngSwese 1971(9):313–315
30. Nosrati A, Addai-Mensah J, Skinner W (2012) Muscovite clay mineral particle interactions in aqueous media. Powder Technol 219:228–238
31. Sposito R (2022) Investigations on the impact of metaphyllosilicates and calcined common clays on the rheology and early hydration of cements admixed with different superplasticizers doctoral dissertation. Dissertation, Neubiberg, Universität der Bundeswehr München
32. Zografou A, Heath A, Walker P (2014) China clay waste as aggregate in alkali-activated cement mortars. Proc Inst Civ Eng Constr Mater 167(6):312–322
33. Alexander M, Mackechnie J, Ballim Y (1999) Guide to the use of durability indexes for achieving durability in concrete structures, research monograph no. 2. Department of Civil Engineering, University of Cape Town, Cape Town
34. Siddique S, Jang J (2020) Assessment of molybdenum mine tailings as filler in cement mortar. J Build Eng 31:101322
35. Ghazi A, Jamshidi-Zanjani A, Nejati H (2022) Utilization of copper mine tailings as a partial substitute for cement in concrete construction. Constr Build Mater 317:125921
36. Ince C (2019) Reusing gold-mine tailings in cement mortars: mechanical properties and socio-economic developments for the Lefke-Xeros area of Cyprus. J Clean Prod 238:117871

Development and Characterization of Eco-efficient Magnesium Oxysulfate Composites: Influence of Carbonation on Phase Formation and Material Properties

Adriano Azevedo(✉) and Holmer Savastano

Department of Biosystems Engineering, Research Center on Materials for Biosystems (BioSMat), University of São Paulo (USP), São Paulo, Pirassununga, Brazil
adrianogalvao@usp.br

Abstract. This study investigates the development of eco-efficient composites based on magnesium oxysulfate (MOS) cement reinforced with lignocellulosic fibers and subjected to accelerated carbonation. The aim is to evaluate how carbonation time influences phase formation and the physical–mechanical properties of the material. Composite samples were exposed to a CO_2-rich atmosphere (10% CO_2, 90% RH, 20 °C) for 24 h and 7 days. Mechanical tests, X-ray diffraction (XRD), infrared (FTIR), and Raman spectroscopy were employed to monitor phase evolution and microstructural changes. The results indicate that short-term carbonation promotes microstructural densification and increases toughness, despite a slight reduction in mechanical strength due to partial transformation of hydration products. Prolonged carbonation, however, led to a significant decline in strength parameters as the mechanically active 5-1-7 phase was consumed, replaced by hydrated magnesium carbonates such as nesquehonite and hydromagnesite. These findings underscore the critical role of carbonation in tailoring composite performance. Optimization of CO_2 curing parameters is essential to enhance durability while preserving structural integrity. The study contributes to the advancement of sustainable construction materials with lower carbon footprints and improved long-term behavior.

Keywords: Magnesium Oxysulfate cement · lignocellulosic fibers · accelerated carbonation · hydrated magnesium carbonates · Eco-efficient composites

1 Introduction

The growing demand for construction materials that are lighter, more durable, and possess superior mechanical performance has driven the development of fiber-reinforced cementitious composites using natural fibers. These materials have gained prominence by combining low weight, reduced environmental impact, and good mechanical properties, representing a promising alternative to widely used synthetic reinforcements such as glass and polypropylene fibers [1–3]. Among natural fibers, lignocellulosic ones offer

H. Beushausen et al. (Eds.): NTZR 2025, LNCE 873, pp. 199–211, 2026.
https://doi.org/10.1007/978-3-032-27698-8_19

significant advantages due to their biodegradability, abundance, economic viability, and contribution to the sustainability of the construction sector [4]. However, the durability of these fibers in Portland cement matrices remains a challenge due to the high alkalinity of the medium, which promotes degradation through mineralization and alkaline hydrolysis. These processes reduce the adhesion between the fiber and the matrix, compromising the tensile strength and integrity of the composite over time [5, 6]. To overcome this limitation, cementitious systems based on magnesium oxide and magnesium sulfate, known as magnesium oxysulfate cements (MOS), have proven effective in protecting plant fibers against alkaline degradation, due to their lower pH values and chemical stability, which preserve fiber integrity [7, 8].

MOS cement is produced from the reaction between magnesium oxide (MgO) and a magnesium sulfate solution. This cementitious matrix exhibits advantageous characteristics compared to traditional cements, such as lower density, good adhesion, thermal insulation capacity, and fire resistance [9–12]. Moreover, it stands out for its low energy consumption during production, since MgO can be calcined at temperatures around 750 °C, significantly lower than those required for Portland clinker production, which reach up to 1450 °C [13, 14]. The essential raw material for MOS cement, MgO (magnesium oxide), can be obtained through different technological routes, giving the material a versatile and sustainable character. These routes include: (1) calcination of magnesium carbonate ($MgCO_3$), known as the dry route; and (2) calcination of precipitated magnesium hydroxide [$Mg(OH)_2$], derived from seawater or brines rich in magnesium, using quicklime or dolomitic lime, a process known as the wet route. In addition to these conventional sources, MgO can also be extracted from magnesium silicate-rich rocks, such as serpentinite and olivine, which contain minerals capable of capturing CO_2 and forming stable magnesium carbonates. Several studies have investigated the synthesis of high-purity $Mg(OH)_2$ from these sources, highlighting the potential of silicate rocks as a promising raw material [15, 16]

The environmentally favorable character of MOS cement is reinforced by the incorporation of natural fibers, which act as carbon-sequestering agents and facilitate accelerated carbonation processes, promoting the formation of carbonated compounds that enhance the durability and mechanical strength of the composites [17–20]. Accelerated carbonation techniques are particularly effective in MOS matrices, promoting transformations in the hydrated phases and improving the composite microstructure [13, 21]. MgO carbonation can generally be described as the formation of magnesite from MgO by absorbing CO_2 ($MgO + CO_2 \rightarrow MgCO_3$), or by incorporating H_2O to form needle-like nesquehonite ($MgCO_3 \cdot 3H_2O$), acicular artinite ($MgCO_3 \cdot Mg(OH)_2 \cdot 3H_2O$), and disk-like hydromagnesite ($4MgCO_3 \cdot Mg(OH)_2 \cdot 4H_2O$) [12, 18–23]. The hydrates formed, depending on the curing conditions, lead to a gain in mechanical strength due to their inherent resistance and their capacity to bind aggregate particles [24–26]. The use of accelerated carbonation in MOS cement-based materials can lead to an enhancement in the physical and mechanical performance of the materials. These materials can be produced with the addition of vegetal fibers and supplementary minerals, resulting in a reduced overall carbon footprint of the composites when compared to those produced with OPC [9, 10, 27]. Nevertheless, technical challenges persist regarding the improvement of the fiber–matrix interface and the durability of lignocellulosic reinforcements in aggressive environments [22, 23].

Unlike previous studies that investigated carbonation in MOS systems under a single curing regime or focused on specific reinforcement types, this work uniquely compares short-and long-term carbonation durations in MOS composites reinforced with bleached eucalyptus pulp. It provides a comprehensive correlation between phase transformation and mechanical behavior, thereby offering insights for prefabricated eco-efficient applications.

2 Materials and Methods

2.1 Production of MgO-Based Composites

The production of the magnesium oxide-based cementitious composites was carried out through the formulation of an inorganic matrix composed of highly reactive magnesium oxide (MgO) and a concentrated aqueous solution of magnesium sulfate ($MgSO_4$), with a concentration of 25% by weight/volume (Fig. 1(a)). The proportions of the components were adjusted to achieve an approximate molar ratio of 10:1:20 for MgO, $MgSO_4$, and water, respectively, following the methodology proposed by Azevedo et al. (2025) [15].

Fig. 1. MgO-based composites production.

To enhance the mechanical performance of the composite, cellulosic pulp (bleached eucalyptus pulp) was incorporated as a reinforcement at 8% by mass, relative to the amount of MgO in the formulation. This addition promoted the formation of a magnesium oxysulfate (MOS) cement matrix, contributing to a more cohesive internal structure. The mixing stage followed a standardized protocol to ensure the homogeneity of the composite, resulting in a uniform distribution of the cellulosic reinforcement within the matrix. For molding, the composites were prepared by pouring the fresh cementitious paste into molds. After this initial step, the molds were vibrated for 35 s to achieve adequate matrix compaction. Subsequently, the molds were completely filled with the remaining paste and sealed with plastic film, ensuring controlled curing conditions and reproducibility of the manufacturing process.

2.2 Curing and Carbonation Process

Following an initial curing period at room temperature, the cementitious composites (Fig. 1(b)) were subjected to accelerated carbonation to investigate the effects of exposure to a carbon dioxide (CO_2)-enriched atmosphere on their final properties. Two different exposure durations were adopted—24 h and 7 days—to establish the relationship between carbonation time and the resulting changes in material behavior. The durations of 24 h and 7 days were selected to simulate accelerated curing protocols commonly employed in industrial precast production and to capture both early-stage and more advanced carbonation effects. This range also aligns with previous optimization studies for CO_2 curing in MgO-based composites [9, 21, 24]. The carbonation process (referred to as CARB) was carried out under controlled environmental conditions: CO_2 concentration of 10%, relative humidity (RH) of approximately 90%, and a constant temperature of 20 °C. As a reference group (REF), non-carbonated specimens were stored under identical temperature conditions (20 °C) but at a relative humidity of 60%, in the absence of CO_2. After the curing phase, the composite plates were cut into prismatic specimens with standardized dimensions of 160 × 40 × 12 mm (length × width × thickness), which were subsequently used for physical and mechanical characterizations. For each experimental condition (REF, CARB-24 h, and CARB-7d), ten specimens were prepared and tested, ensuring statistical reliability of the results. Statistical analysis was carried out to compare values for physical and mechanical properties, curing regimens, and accelerated carbonation using the Tukey test at a significance level of 5%. S.A.S 9.3 (Statistical Analysis System) software was used during this comparison.

2.3 Physical-Mechanical and Microstructural Characterizations

Mechanical performance was assessed by means of four-point bending tests using an INSTRON universal testing machine, equipped with a 5 kN load cell and a 135 mm support span, in accordance with RILEM (1984) guidelines [27]. The parameters evaluated included: limit of proportionality (LOP), modulus of rupture (MOR), modulus of elasticity (MOE), and specific energy (SE). These values were obtained after 28 days of curing and following the carbonation exposure cycles. In addition to mechanical testing, physical properties such as bulk density, water absorption, and total porosity were determined for both carbonated and non-carbonated specimens, following ASTM C948–81 (2016) [28]. X-ray diffraction (XRD) analysis was performed using a Rigaku miniFlex 600 diffractometer, with a scanning range of 5–80° 2θ, operating at 50 kV voltage and 100 mA current. The scan speed was set to 1°/min with a step size of 0.02°. Fourier Transform Infrared Spectroscopy (FTIR) analyses were carried out using a PerkinElmer Spectrum One spectrometer with an ATR accessory. Scans were performed in the range of 4000–500 cm^{-1}, with a resolution of 4 cm^{-1}. A total of 32 scans were applied for each analyzed sample. Raman spectroscopy was conducted using a Renishaw Raman spectrometer equipped with a 738 nm laser and a charge-coupled device (CCD) detector. The laser, with a power of ~ 6 mW, was focused through a Newport M-5X microscope objective lens (N.A. = 0.10) to a spot size of ~ 4 μm in diameter. Raman spectra of MOS cement samples were obtained to identify characteristic peaks of different hydration phases and the formed carbonation products. Each spectrum was recorded with an exposure time of 50 s and two accumulations.

3 Results

3.1 Mechanical and Physical Properties

The mechanical performance of the MgO-based cementitious composites, as summarized in Table 1, highlights the significant influence of the carbonation process on material behavior. The reference sample (Ref), which was not subjected to carbonation, exhibited the highest values for modulus of rupture (MOR = 7.23 MPa), limit of proportionality (LOP = 6.80 MPa), and modulus of elasticity (MOE = 4.64 GPa). These results are consistent with the presence of hydration products, particularly the 5–1-7 phase, known for its substantial contribution to mechanical strength in magnesium oxysulfate (MOS) cement systems. Carbonation for 24 h (Carb1) resulted in a moderate decrease in mechanical performance (MOR = 6.05 MPa; LOP = 5.06 MPa; MOE = 2.87 GPa). This behavior can be attributed to the partial transformation of alkaline phases, mainly brucite ($Mg(OH)_2$), into hydrated magnesium carbonates.

Table 1. Mechanical properties of the carbonated and non-carbonated MgO-based composites produced.

Sample	MOR (MPa)	SD	LOP (MPa)	SD	MOE (GPa)	SD	SE (kJ/m^2)	SD
Ref	7.23**a**	1.02	6.80**a**	1.02	4.64**a**	0.7	0.33**b**	0.05
Carb1	6.05**b**	0.67	5.06**b**	0.67	2.87**b**	0.68	0.45**a**	0.12
Carb7	3.84**c**	0.56	3.26**c**	0.56	2.59**b**	0.59	0.44**ab**	0.14

MOR: Modulus of Rupture, SD: Standard Deviation, LOP: Limit of Proportionality, MOE: Modulus of Elasticity, SE: Specific Energy. Identical letters in the same column indicate no significant differences according to Tukey's test ($p < 0.05$, $n = 10$).

Although these carbonation products do not directly enhance mechanical strength, they promote pore refinement and microstructural densification. This is corroborated by the slight reduction in porosity and water absorption, and increase in density observed in Table 2, suggesting that short-term carbonation provides a balance between matrix densification and preservation of mechanically active hydrated phases. In contrast, prolonged carbonation for 7 days (Carb7) led to a more pronounced decrease in mechanical properties (MOR = 3.84 MPa; LOP = 3.26 MPa; MOE = 2.59 GPa). This decline is associated with the extensive carbonation not only of alkaline phases but also of the 5-1-7 hydrated phases, as presented in Eq. 1 [29].

$$(n+1)Mg(OH)_2.MgSO_4.(m+1)H_2O + CO_2 \rightarrow MgCO_3 + nMg(OH)_2.MgSO_4.mH_2O \quad (1)$$

The consumption of these critical strength-contributing phases impairs the composite's mechanical performance, despite the benefits of pore filling and reduced porosity (from 38.8% in Ref to 29.35% in Carb7), and water absorption, as presented in Table 2. The formation of magnesium carbonate hydrates results in a denser, yet mechanically weaker, matrix.

Interestingly, despite the drop in strength-related parameters, the specific energy (SE), which is linked to the material's toughness, increased from 0.33 kJ/m^2 (Ref) to 0.44 kJ/m^2 (Carb7). The increase in specific energy (SE) may be attributed to improved energy dissipation mechanisms within the densified microstructure. The partial formation of hydrated magnesium carbonates likely enhanced the fiber–matrix interfacial bonding, reducing stress concentrations and promoting more stable crack propagation, as observed in other carbonated MOS systems [17, 25]. This indicates that the carbonated matrix, while less strong, may be more capable of energy dissipation, possibly due to a more uniform and compact microstructure.

Table 2. Physical properties of the carbonated and non-carbonated MgO-based composites produced.

Sample	Density (g/cm^3)	SD	Porosity (%)	SD	Water Absorption (%)	SD
Ref	1.15**c**	0.01	38.8**a**	0.97	33.57**a**	1.2
Carb1	1.19**b**	0.01	37.67**ab**	0.67	31.46**b**	0.74
Carb7	1.35**a**	0.02	29.35**c**	1.08	21.67**c**	1.09

Identical letters in the same column indicate no significant differences according to Tukey's test ($p < 0.05$, $n = 10$).

3.2 Microstructural Characterization

X-ray diffraction (XRD). The evolution of crystalline and amorphous phases in MgO-based cementitious materials subjected to different curing environments can be effectively examined through X-ray diffraction (XRD) analysis. As illustrated in Fig. 2, the diffraction patterns of both carbonated and non-carbonated samples reveal characteristic peaks corresponding to brucite ($Mg(OH)_2$, COD 96-100-0055), periclase (MgO, COD 96-900-6786), and notably, the 5-1-7 phase—an essential hydrated phase in magnesium oxysulfate (MOS) cement systems that plays a critical role in imparting mechanical strength to the composites.

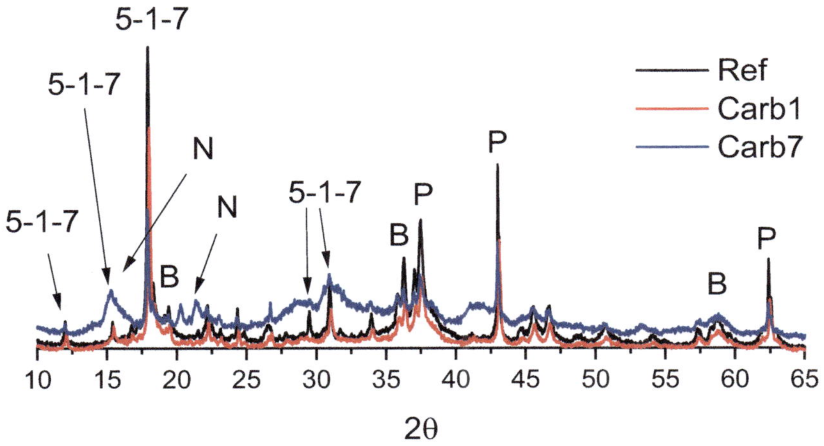

Fig. 2. Diffractograms of carbonated and non-carbonated samples. 5-1-7 = Phase 5-1-7, N = nesquehonite ($MgCO_3{\cdot}3H_2O$), B = Brucite ($Mg(OH)_2$ and P = Periclase (MgO).

A noticeable attenuation in the diffraction peak intensity of the 5-1-7 phase, particularly near 26.7° 2θ, was observed in carbonated samples, suggesting the transformation of this phase due to carbonation reactions. This phenomenon is closely linked to the formation of amorphous or poorly crystalline carbonation products on the surface, which can mask the diffraction signals of original crystalline phases. These findings are consistent with the study by Azevedo et al. (2024), which documented the development of low-crystallinity hydrated magnesium carbonates (HMCs) following CO_2 exposure [30]. Additionally, a broad amorphous halo extending from 10 to 55° 2θ further supports the formation of these disordered carbonate phases. Notably, in the composites carbonated for 7 days, specific diffraction peaks were identified near 21.4° and 21.8° 2θ, which align with the reference pattern for nesquehonite ($MgCO_3{\cdot}3H_2O$, PDF 00-020-0669).

The emergence of these peaks indicates the crystallization of nesquehonite as a product of prolonged carbonation. Moreover, an increase in peak intensity near 16.5° 2θ suggests the presence of $MgCO_3{\cdot}3H_2O$ in a less ordered state, further reinforcing the hypothesis of heterogeneous carbonate formation depending on local physicochemical conditions. The transformation of brucite and the concurrent decline in the intensity of 5-1-7 phase peaks indicate that carbonation affects not only superficial layers but also the internal microstructure of the material, progressively altering the phase composition throughout the matrix.

Infrared Spectroscopy (FTIR). The infrared spectroscopy (FTIR) profiles of the MgO-based cementitious composites are depicted in Fig. 3.

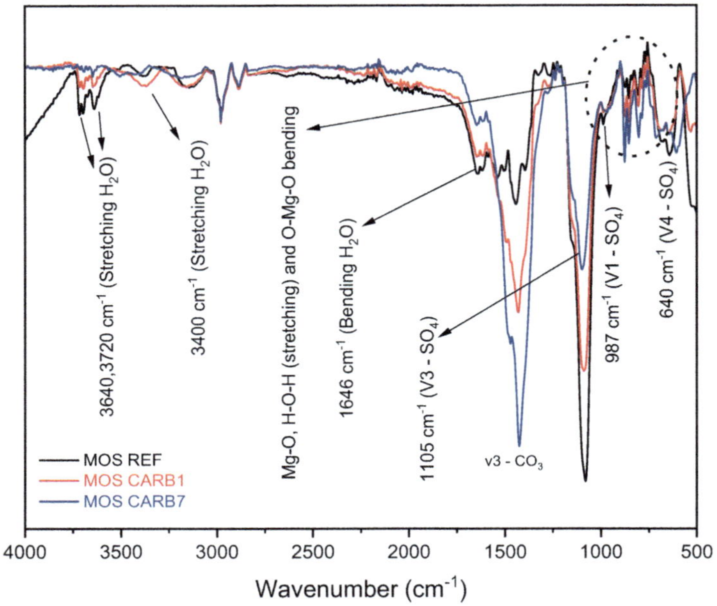

Fig. 3. FTIR transmittance bands of carbonated and non-carbonated samples.

The high-frequency region (4000–2500 cm^{-1}) features bands attributed to stretching vibrations of hydroxyl anions and molecular water. A broad and asymmetric absorption band spanning from 3750 to 2750 cm^{-1}, with a transmittance minimum near 3400 cm^{-1}, is indicative of water molecule stretching vibrations. In contrast, the sharper bands centered at 3720 and 3640 cm^{-1} correspond to vibrations of hydroxyl groups (–OH). Shifting focus to the lower frequency domain (2500–500 cm^{-1}), the spectrum reveals a distinct band at 1646 cm^{-1}, which is associated with H–O–H bending vibrations of water molecules. Moreover, three prominent bands relate to sulfate ion (SO_4^{2-}) vibrations: a strong band at 1105 cm^{-1} representing the asymmetric stretching mode ν3(SO_4), another at 987 cm − 1 corresponding to the symmetric stretching ν1(SO_4), followed by bending vibrations at 640 cm^{-1} (ν4). These sulfate vibration patterns are characteristic of free, uncoordinated SO_4 tetrahedra present in 5-1-7 crystalline structure. The decrease in these bands during accelerated carbonation can be associated with a reaction between CO_2 and crystalline 5-1-7 phases, as described in Eq. 1 and previously discussed. A notable feature appears near 1460 cm^{-1}, where an intensified band consistent with the asymmetric stretching mode ν3 of carbonate groups (CO_3^{2-}) is observed. This intensification correlates positively with increased CO_2 exposure during curing, suggesting the progressive formation of hydrated magnesium carbonates (HMCs). Such spectral shifts support previously discussed findings on the chemical transformation induced by carbonation. The diminishing intensity of water-related bands in the 3400–3800 cm^{-1} range, especially after prolonged carbonation, further indicates dehydration within the matrix [31]. Carbonation reactions are initiated by the consumption of residual $Mg(OH)_2$, a hydration product of MgO, leading to early-stage HMC formation in the matrix's voids and pores. These newly formed products improve the interface between the matrix and reinforcing materials, enhancing energy transfer efficiency and deformation capacity, as evidenced by the stress–strain profiles, as seen in the SE values presented in Table 1.

Raman Spectroscopy. Figure 4 presents the Raman spectroscopy analysis of the studied materials. In the spectral range of 1200–500 cm^{-1}, characteristic Raman-active modes related to various structural units were observed. Notably, three active vibrational modes of the $SO_4{}^{2-}$ tetrahedron were identified, confirming their unbound state within the 5-1-7 crystal lattice. Specifically, the asymmetric stretching $\nu_3(SO_4)$ was detected as a low-intensity band at 1017 cm^{-1}, while the symmetric stretching $\nu_1(SO_4)$ exhibited the highest intensity at 987 cm^{-1}. Additionally, the bending mode $\nu_4(SO_4)$ was observed at 625 cm^{-1}, validating the presence of free $SO_4{}^{2-}$ groups within interstitial sites of the crystal structure. Other bands were attributed to brucite ($Mg(OH)_2$), particularly a prominent peak at 1040 cm^{-1}, as confirmed through the RRUFF database (Brucite R040077). The presence of this peak in reference samples and its subsequent reduction following CO_2 curing supports the notion that carbonation consumes $Mg(OH)_2$, contributing to HMCs formation [32]. Moreover, two peaks exclusive to carbonated samples were observed. The band near 1100 cm^{-1} is assigned to the symmetric stretching of $CO_3{}^{2-}$ ions, indicative of nesquehonite formation, according to Davies and Bubela (1973) and corroborating with XRD results. The carbonation reaction forming nesquehonite is described by Eq. 2 [33]:

$$Mg(OH)_2 + CO_2 + 2H_2O \rightarrow MgCO_3 \cdot 3H_2O \text{ (Nesquehonite)} \tag{2}$$

With continued carbonation, nesquehonite can transform into hydromagnesite, as illustrated in Eq. 3:

$$4MgCO_3 \cdot 3H_2O + Mg(OH)_2 + H_2O \rightarrow 4MgCO_3 \cdot Mg(OH)_2 \cdot 4H_2O \text{ (Hydromagnesite)} \tag{3}$$

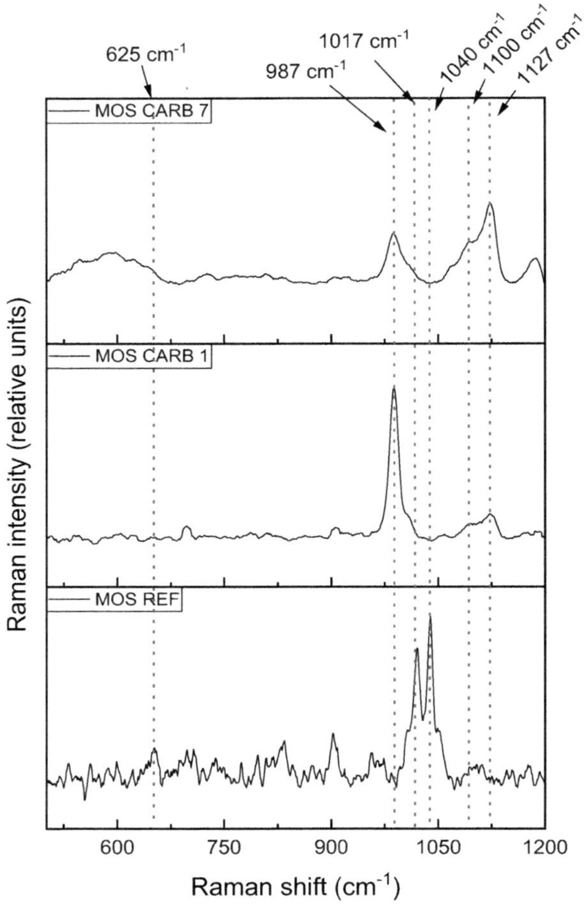

Fig. 4. Raman spectra of carbonated and non-carbonated samples.

These findings are consistent with the study by Mi et al. (2024), which showed that magnesium-based cements initially form nesquehonite in the early carbonation stages, eventually giving way to hydromagnesite as the predominant product. The disappearance of the 625 and 1017 cm^{-1} peaks after carbonation indicates the possible consumption of the 5-1-7 phase during HMC formation, further supporting the results discussed in previous sections [32].

4 Conclusions

The results of this study demonstrate that accelerated carbonation significantly alters the chemical and structural characteristics of MgO-based cementitious composites. The FTIR and Raman spectroscopy analyses confirmed the progressive transformation of the matrix during CO_2 exposure, particularly the formation of hydrated magnesium carbonates (HMCs) such as nesquehonite and, subsequently, hydromagnesite. These carbonation products were associated with increased matrix density and reduced total porosity. The disappearance of characteristic SO_4^{2-} vibrational bands suggests that carbonation consumes both hydration products and the crystalline 5-1-7 phase. This phase, crucial for mechanical strength, was progressively depleted with longer carbonation durations, as confirmed by both spectroscopic data and XRD analysis. These findings highlight the importance of optimizing carbonation parameters—such as CO_2 concentration, relative humidity, curing duration, and temperature—to promote HMCs formation while preserving the structural integrity of the 5-1-7 phase. These findings suggest that short-duration carbonation can be strategically applied in the manufacturing of prefabricated components to enhance toughness and dimensional stability without compromising structural integrity. Tailoring carbonation exposure during early-stage curing may offer an effective pathway for producing lightweight, eco-efficient panels suitable for non-load-bearing applications.

Acknowledgements. The authors would like to express appreciation for the support of the sponsors. FAPESP processes nº: 24/02445-4, 2025/11339-6, and 2025/27072-9. CNPQ – INCT process nº 408791/2024-1. The University of Manchester.

References

1. Correia VC, Santos SF, Tonoli GHD, Savastano H (2020) Characterization of vegetable fibers and their application in cementitious composites. In: Nonconventional and Vernacular Construction Materials, pp 141–167
2. Santos RF et al (2021) Nanofibrillated cellulose and its applications in cement-based composites: a review. Constr Build Mater 288:123122
3. da Costa Correia V, Santos SF, Savastano H (2018) Vegetable fiber as reinforcing elements for cement-based composite in housing applications—a Brazilian experience. MATEC Web Conf 149:01007
4. Li Q, Ibrahim L, Zhou W, Zhang M, Yuan Z (2021) Treatment methods for plant fibers for use as reinforcement in cement-based materials. Cellulose
5. Filho JDAM, Silva FDA, Toledo Filho RD (2013) Degradation kinetics and aging mechanisms on sisal fiber cement composite systems. Cem Concr Compos 40:30–39
6. Wei J, Meyer C (2014) Improving degradation resistance of sisal fiber in concrete through fiber surface treatment. Appl Surf Sci 289:511–523
7. Mármol G, Savastano H (2017) Study of the degradation of non-conventional MgO-SiO_2 cement reinforced with lignocellulosic fibers. Cem Concr Compos 80:258–267
8. Rodier L, da Costa Correia V, Savastano Junior H (2020) Elaboration of eco-efficient vegetable fibers reinforced cement-based composites using glass powder residue. Cem Concr Compos 110:103599
9. Ba M, Xue T, He Z, Wang H, Liu J (2019) Carbonation of magnesium oxysulfate cement and its influence on mechanical performance. Constr Build Mater 223:1030–1037
10. Marmorato GCE, Gladis C (2014) Magnesium oxysulfate fiber cement. Key Eng Mater 600:308–318
11. Shand MA (2020) Magnesium oxysulfate cement. In: Magnesia Cements, pp 75–83

12. Walling SA, Provis JL (2016) Magnesia-based cements: a journey of 150 years, and cements for the future? Chem Rev 116:4170–4204
13. Chau CK, Chan J, Li Z (2009) Influences of fly ash on magnesium oxychloride mortar. Cem Concr Compos 31:250–254
14. Qin L, Gao X, Chen T (2018) Recycling of raw rice husk to manufacture magnesium oxysulfate cement based lightweight building materials. J Clean Prod 191:220–232
15. Azevedo AGS, Baltazar L, Faria P, Savastano H (2025) Durability assessment of MgO-based fiber cement reinforced with lignocellulosic fibers and cured in a CO_2-rich atmosphere. J Build Eng 104:112327
16. Barbieri V, Lassinantti Gualtieri M, Manfredini T, Siligardi C (2021) Lightweight concretes based on wheat husk and hemp hurd as bio-aggregates and modified magnesium oxysulfate binder: microstructure and technological performances. Constr Build Mater 284:122751
17. Hu Z, Guan Y, Chang J, Bi W, Zhang T (2020) Effect of carbonation on the water resistance of steel slag–magnesium oxysulfate (MOS) cement blends. Materials 13:5006
18. Wu C, Zhang H, Yu H (2016) Preparation and properties of modified magnesium oxysulfate cement derived from waste sulfuric acid. Adv Cem Res 28:178–188
19. Papanicolaou CG (2016) Applications of textile-reinforced concrete in the precast industry. In: Textile fibre composites in civil engineering. Elsevier, pp 227–244
20. Boulos L, Foruzanmehr MR, Robert M (2018) Evolution of the interfacial transition zone and the degradation mechanism of zirconia treated flax fabric reinforced cementitious composites. Constr Build Mater 190:120–130
21. Tonoli GHD et al (2019) Influence of the initial moisture content on the carbonation of fibers in cementitious composites
22. Mejia-Ballesteros JE, Rodier L, Filomeno R, Savastano H Jr, Fiorelli J, Rojas MF (2021) Influence of the fiber treatment and matrix modification on the durability of eucalyptus fiber reinforced composites. Cem Concr Compos 124:104280
23. Azevedo AGS, Strecker K (2017) Brazilian fly ash based inorganic polymers production using different alkali activator solutions. Ceram Int 43:9012–9018
24. Azevedo AGS et al (2024) Effect of the rehydration method on the physical–mechanical properties of CO_2-cured magnesium-based fiber cement boards. Civil Eng 5:247–264
25. Filomeno RH, Fioroni CA, Azevedo AGS, Savastano H (2023) Influence of different curing regimes and accelerated carbonation on the physical, mechanical, and aging performances of fiber cement. J Build Eng 78:107791
26. Azevedo AGS et al (2023) Assessment of the physical-mechanical performance of magnesium oxide-based fiber cement submitted to accelerated carbonation. Cerâmica 69:312–317
27. RILEM Technical Committee 49TFR (1984) Testing methods for fibre reinforced cement-based composites. Mater Constr 17:441–456
28. ASTM C948-81 (2016) Standard test method for dry and wet bulk density, water absorption, and apparent porosity of thin sections of glass-fiber reinforced concrete. ASTM Int 81:1–2
29. Azevedo AGS, Savastano H (2024) Assessment of carbonation as a complementary strategy to increase the durability of magnesium oxysulfate (MOS)-based fiber cement boards. Constr Build Mater 438:137086
30. Azevedo AGS, Filomeno R, Gonçalves MM, Faria P, Savastano H (2024) Innovative MOS-based fiber cement boards: effect of kraft pulp mills waste and curing by accelerated carbonation. Constr Build Mater 431:136525
31. Hay R, Celik K (2020) Accelerated carbonation of reactive magnesium oxide cement (RMC)-based composite with supercritical carbon dioxide ($scCO_2$). J Clean Prod 248:119282
32. Mi T, Chen X, Yang EH, Unluer C (2024) Quantification of carbonated Mg-based cement pastes by Raman spectroscopy. Cem Concr Res 178:107454
33. Grünhauser Soares E, Castro-Gomes J (2021) Carbonation curing influencing factors of carbonated reactive magnesia cements (CRMC) – A review. J Clean Prod 305:127210

Use of Macro Fibers to Enhance the Structural Integrity of Environmental Engineering Concrete Structures

Raymond Pepera and Behrouz Shafei(✉)

Department of Civil, Construction, and Environmental Engineering, Iowa State University, Ames, IA 50011, USA
{rpepera,shafei}@iastate.edu

Abstract. Environmental engineering concrete structures are critical in managing liquids and solid wastes in facilities like wastewater plants and storage tanks. These structures must be durable to withstand harsh conditions that can degrade their functionality, affect public health, and increase maintenance costs. This study develops high-performance fiber-reinforced concrete (FRC) mixes designed to enhance the durability and longevity of such infrastructures. Achieving a compressive strength of 55 MPa, the mix features a low water-to-binder ratio of 0.37 to improve resistance to chloride ingress. It incorporates various macro fibers, including polyvinyl alcohol (PVA), polypropylene (PP), and alkaline-resistant (AR) glass, at different dosages, along with consistent dosages of PP microfibers, to assess each fiber's impact on concrete properties. The experimental framework cast the FRC into a set of beams, differentiated by fiber type and dosage, and supplemented by control mixes for comparison. Tests of uncracked beams included analyses for flexural capacity and water absorption. Cracked beams underwent sustained loading in a saline solution using modified old creep frames, simulating long-term environmental exposure. This setup applied continuous loads, submerging the beams to emulate severe conditions. Post-loading analysis of cracked beams for residual flexural strength, chloride penetration depth, and chloride content provided invaluable insights into the FRC's performance under simulated conditions. The results underscore the potential of fibers in improving critical structural integrity and durability of environmental engineering concrete structures.

Keywords: Fiber-reinforced concrete · Environmental engineering structures · Durability · Sustainable construction · Synthetic fibers

1 Introduction

Concrete remains the most widely used construction material globally, essential to civil infrastructure development due to its compressive strength, versatility, and economic efficiency. However, structures operating in aggressive environments such as wastewater treatment facilities, chemical containment tanks, and storage systems experience substantial durability challenges. These challenges arise primarily from chloride ingress

H. Beushausen et al. (Eds.): NTZR 2025, LNCE 873, pp. 212–221, 2026.
https://doi.org/10.1007/978-3-032-27698-8_20

and sustained mechanical loading, which degrade concrete and reinforcement over time [1]. For environmental engineering structures, where structural integrity directly affects public health and operational safety, durability is not merely a design consideration but a necessity.

The American Concrete Institute (ACI), through ACI 350, has highlighted performance-based criteria to guide the design of environmental structures. These criteria emphasize reducing permeability, minimizing crack widths, and selecting materials that resist long-term chemical and mechanical degradation [2]. Yet, traditional steel-reinforced concrete systems often fail prematurely in these conditions due to corrosion-induced cracking and loss of structural capacity. Maintenance and repair costs continue to rise worldwide; a 2021 American Society of Civil Engineers (ASCE) report estimates that failing water and wastewater infrastructure in the United States alone will require over $434 billion in investments by 2029 [3].

Recent research has increasingly focused on FRC as a viable enhancement to conventional mixes. FRC introduces synthetic, natural, or metallic fibers into the cementitious matrix to improve post-cracking behavior, impact resistance, and overall durability [4, 5]. Synthetic macrofibers, including PVA, PP, and AR fibers, offer distinct advantages such as corrosion resistance, high tensile strength, and low density while reducing the carbon footprint associated with steel production. Studies by Lee et al. [6] and, more recently, by Kazemian and Shafei [7] have shown that PVA fibers form strong bonds with the cementitious matrix and significantly reduce crack width under load, contributing to self-healing and prolonged durability.

This study experimentally investigates high-performance concrete's (HPC) mechanical and durability performance, incorporating three types of macro fibers: PVA, PP, and AR. Positioned at the intersection of durability, sustainability, and structural resilience, the research quantifies how fiber type and dosage affect performance under service-like conditions. The findings provide critical guidance for future design codes and material selection in environmental engineering concrete structures by evaluating behavior under sustained loading and chloride exposure in pre-cracked specimens.

2 Materials and Methods

2.1 Concrete Mix Design and Specimen Preparation

All concrete mixtures in this study were developed as HPC targeting a compressive strength of 55 MPa, with a water-to-binder (w/b) ratio of 0.37 to reduce permeability and enhance long-term durability. The binder consisted of 325 kg/m^3 of Type I portland limestone cement (PLC) with 15% limestone replacement and 81 kg/m^3 of Class F fly ash (corresponding to a 20% replacement by mass of the total binder content), following [8, 9], respectively. Natural sand (921 kg/m^3) and crushed gravel (922 kg/m^3) were used as fine and coarse aggregates, conforming to [10]. A polycarboxylate-based water reducer (superplasticizer) compliant with [11] was incorporated to ensure consistent workability.

The fiber selection for this study was guided by the findings of a recent investigation by Kazemian and Shafei [7], which evaluated the mechanical benefits of various synthetic fibers in cementitious composites. Based on their performance in improving tensile behavior and crack control, three macrofibers were selected: PVA, PP, and AR fibers (Fig. 1). Each was added at two dosage levels of 0.25% and 0.50% by volume of concrete. In addition, a constant 0.0625% volume of PP microfibers was included in all FRC mixes to improve microcrack resistance. Specimen IDs were structured to reflect the macro fiber type and dosage. For example, PV050U refers to a mix with 0.50% PVA fibers, whiles PV025U indicates 0.25% PVA fibers. This naming convention was applied consistently across all fiber types for clarity and comparison. A control mix without fibers was prepared to enable a comparative evaluation of fiber effects. This served as a baseline for isolating fiber type and dosage effects. The base mix proportions were held constant across all variations to ensure that observed differences in behavior could be directly attributed to the added fibers.

HPC was mixed in a 1.78 kg/m^3 capacity pan mixer. Dry materials, cement, fly ash, and aggregates, were mixed for 5 min, followed by staged water addition over three intervals with 2 min of mixing between each. Fibers were then gradually introduced and dispersed for an additional 5 min. Workability was adjusted with the superplasticizer to achieve a slump range of 12–15 cm. Beams were cast into 102 × 102 × 356 mm wooden molds. The HPC mixes were filled in two layers and compacted using a vibrating table. Specimens were demolded after 24 h and cured at 23 °C and 95% RH until testing.

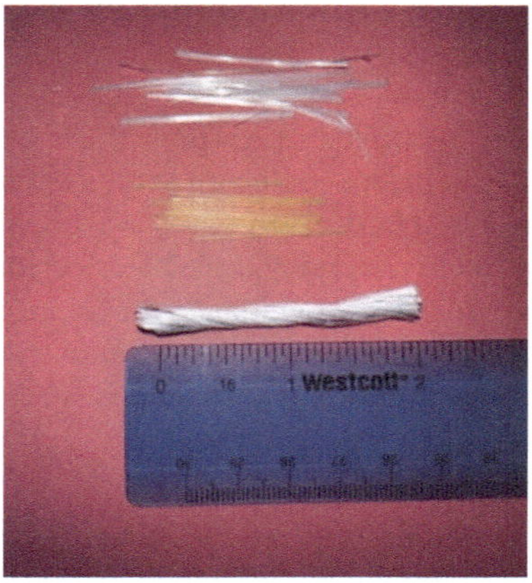

Fig. 1. AR, PVA, and PP fibers macro fibers (placed from top to bottom).

2.2 Crack Simulation and Exposure Protocol

The experimental program was divided into two primary specimen groups: uncracked beams and cracked beams. Uncracked beams were tested in their original, intact state to assess their baseline mechanical and durability performance, including flexural strength and water absorption tests. In contrast, cracked beams were intentionally pre-cracked under controlled conditions to replicate service-level cracking before sustained loading and environmental exposure. This was achieved using a three-point bending configuration [12]. The loading was applied using a SATEC testing machine at a constant rate of 0.01 mm/min. The test was halted when the crack width at midspan reached a controlled range of 0.2–0.3 mm, measured using a precision crack width gauge. This crack width range reflects service-level damage reported in the literature to accelerate chloride ingress while preserving structural continuity significantly. Cracks were introduced in the exact location and orientation for all beams to ensure consistency across specimens. This dual approach enabled a direct evaluation of the role of pre-existing cracks on chloride ingress, flexural performance degradation, and the long-term durability of FRC structures.

After pre-cracking, these specimens were subjected to sustained mechanical loading for 21 days. A custom test setup was developed using repurposed steel creep frames to apply continuous load while immersed in a corrosive environment (Fig. 2). Each frame was constructed using four 381 × 381 mm steel plates connected by 914 mm threaded rods, providing rigidity and adjustability. High-strength steel springs (recycled from decommissioned train assemblies) were the primary load application mechanism.

Fig. 2. Sustained loading setup.

Each sustained loading setup was coupled with a saline exposure system to simulate aggressive environmental conditions representative of environmental engineering concrete structures. Cracked beams were placed in 127 × 127 × 381 mm plexiglass containers with two stainless-steel supports placed 300 mm apart and submerged halfway into a 5% NaCl solution, chosen to reflect severe chloride exposure typical of environmental engineering concrete structures. The circulation system ensured constant flow around each specimen, preventing chloride stratification or depletion. Tight-fitting lids were used to cover the reservoirs, and openings for hose passage were sealed with waterproof tape to reduce evaporation.

A stainless-steel roller was placed at the midspan of each beam to replicate a standard three-point bending configuration for sustained load application. The applied load levels were selected based on the initial cracking thresholds determined from prior flexural testing. The beams were subjected to a sustained load of 6.7 kN. This load magnitude was carefully chosen to maintain the pre-existing crack openings without advancing further fracture or compromising structural stability. To ensure consistent loading throughout the 21-day exposure period, load cells were installed directly above the roller at the load application point. The sustained load on each specimen was monitored daily, and adjustments were made as needed to maintain force stability across all setups.

2.3 Mechanical and Durability Testing

2.3.1 Flexural Strength Testing

Flexural strength testing was performed to assess the load-carrying capacity of the concrete beams before and after environmental exposure. Both uncracked and cracked beams were tested using a three-point bending configuration [12]. A clear span of 300 mm was used, with the load applied centrally through a steel roller aligned at midspan. For uncracked beams, testing was conducted after 28 days of curing to determine the baseline flexural behavior of each FRC mix. For cracked specimens, residual flexural strength tests were performed immediately following the 21-day sustained loading and saline exposure period to quantify the extent of strength degradation due to long-term environmental stress. During testing, a linear variable differential transformer (LVDT) was mounted at the midspan to measure vertical deflection, with readings captured every 0.1 s to generate accurate load-deflection curves. The modulus of rupture (MOR) was calculated using the maximum applied load and specimen geometry.

2.3.2 Water Absorption Test

Water absorption testing was conducted to evaluate the permeability and pore structure of the uncracked FRC beams, serving as an indirect measure of long-term durability. The procedure followed [13] involves oven-drying and immersion stages to determine the mass of absorbed water relative to the dry weight of the specimen. Before testing, three beam specimens from each FRC mix were selected and air-dried. Each specimen was oven-dried at 110 °C for 24 h until mass stabilization was achieved. After cooling to room temperature, the dry mass was recorded, and the specimens were submerged in water maintained at 21 °C. Saturated mass readings were taken at 24 and 48 h, and water absorption percentage was calculated accordingly.

2.3.3 Chloride Penetration Test

Qualitative chloride penetration tests were conducted after the 21-day sustained loading and saline exposure period to assess the chloride resistance of cracked FRC specimens. A 0.1 M silver nitrate ($AgNO_3$) solution was used to visually determine chloride penetration depth, following the [14] procedure. Beams were sectioned along pre-existing, vertical midspan or diagonal shear cracks. After brushing away surface debris, the solution was sprayed onto the exposed concrete surfaces. Within minutes, the Ag^+ ions reacted with free Cl^- ions in the concrete to form a white AgCl precipitate, delineating areas of chloride ingress. Regions without chloride turned brown due to the formation of Ag_2O. Penetration depth was measured using a digital caliper at 10 points per surface, and average values were used to compare chloride ingress across fiber types and crack geometries.

3 Results

3.1 Flexural Strength Testing

The flexural strength evaluation of FRC specimens revealed apparent differences based on fiber type and dosage. As shown in Fig. 3, tests on uncracked beams showed that all fiber-reinforced mixes outperformed the control, with PVA delivering the highest improvements in MOR. At a 0.25% dosage, PVA increased MOR by 43%, followed by PP (28%) and AR (17%). Increasing the dosage to 0.50% further enhanced strength, especially for PP, which saw a 10.5% gain. PVA showed strong performance even at lower dosages, while AR's improvement was modest due to weaker matrix bonding.

After 21 days of sustained loading in saline conditions, residual strength testing confirmed that PVA again provided the highest post-cracking capacity. The PVA mix recorded the highest residual MOR, with PP and AR trailing behind (Fig. 4). The increase in performance with higher fiber dosage emphasized the role of fibers in maintaining load capacity under environmental stress.

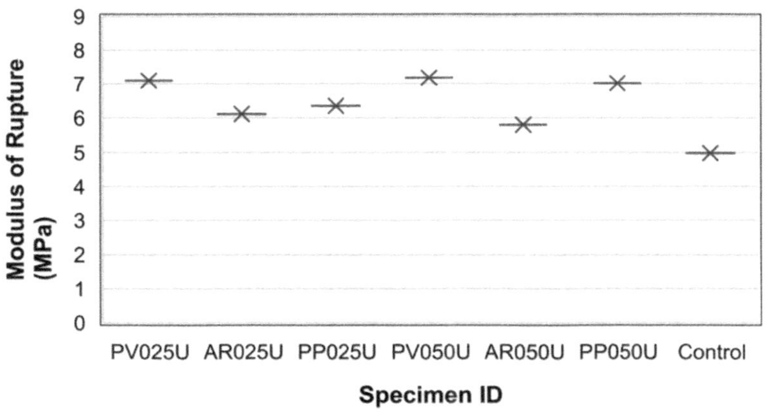

Fig. 3. MOR calculated for the developed uncracked FRC beams.

3.2 Water Absorption Test

The water absorption results, presented in Fig. 5, highlight apparent differences in permeability among various FRC mixes, consistent with the past studies [15–20]. The control mix recorded a water absorption rate of 1.273%, serving as the baseline. PP fibers were the most effective in reducing absorption, with the 0.50% dosage achieving the lowest rate at 0.867%, a 31.9% reduction, attributed to their hydrophobic and fibrillated structure. PVA fibers also showed good performance at 0.25% dosage, reducing absorption by

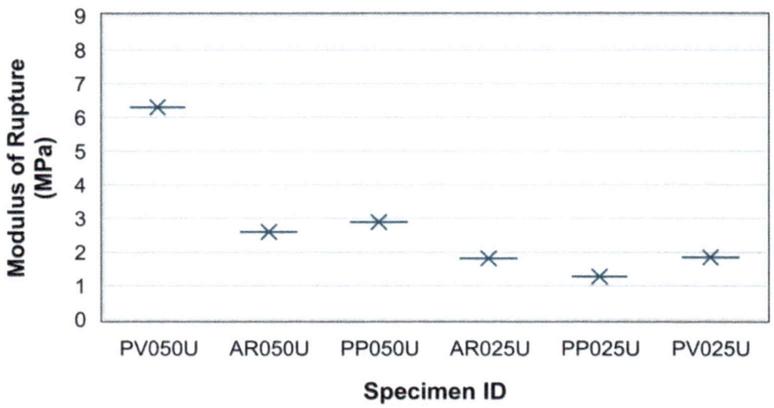

Fig. 4. MOR of FRC beams after the residual flexural capacity test.

22.8%, while the 0.50% dosage (PV050U) showed a modest 7.15% reduction, possibly due to fiber overdosing affecting matrix uniformity. In contrast, AR fibers performed poorly, with the 0.50% dosage (AR050U) resulting in the highest absorption at 1.562%, a 22.7% increase over the control, likely due to weak bonding and fiber brittleness. Overall, PP fibers demonstrated the most consistent and effective reduction in permeability, contributing to enhanced durability.

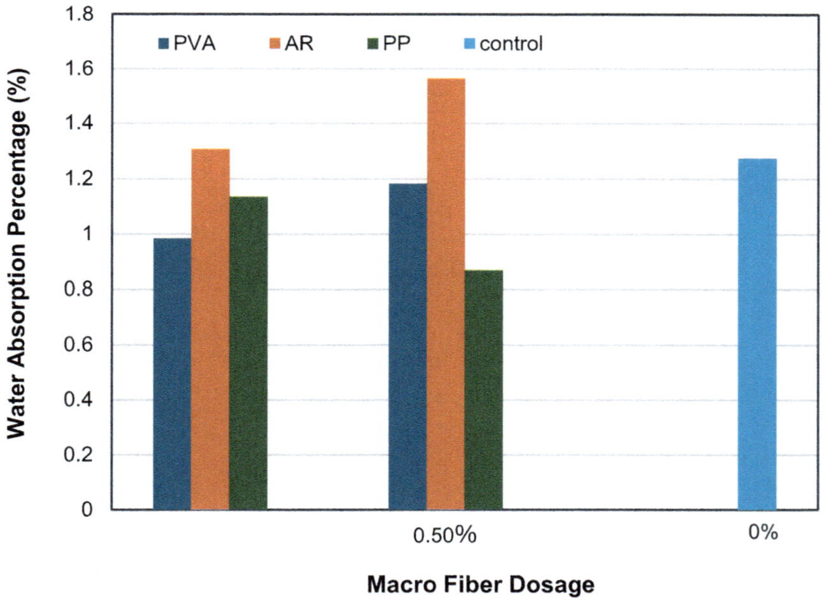

Fig. 5. Water absorption percentage for FRC mixes at different macro fiber dosages.

3.3 Chloride Penetration Depth Test

Chloride penetration test results showed that fiber type and dosage significantly influenced resistance to chloride ingress in specimens under sustained load and saline exposure. PVA fibers exhibited the best performance. PV050U recorded the lowest penetration, 8.13 mm (middle) and 7.98 mm (slant), representing 43.8% and 33.6% reductions compared to PV025U. This confirms the dosage-dependent effectiveness of PVA in limiting ion transport through tight crack control. PP fibers were less effective. PP025U had the highest middle penetration (15.92 mm), while PP050U showed moderate improvements, 11.66 mm (middle) and 9.08 mm (slant), but still lagged behind PVA, reflecting weaker fiber-matrix bonding. AR fibers showed a balanced but limited effect. AR050U reduced ingress by 13.3% (middle) and 12.4% (slant) compared to AR025U. Though AR fibers promoted uniform crack distribution, their brittle behavior and lower pull-out resistance limited their performance.

4 Conclusions

This study demonstrated that incorporating synthetic macro fibers into HPC significantly enhances mechanical strength and durability in aggressive environmental conditions. Under sustained loading and saline exposure, all FRC mixes outperformed the control, varying performance by fiber type and dosage.

PVA fibers delivered the most consistent and superior performance across all evaluations. PVA-reinforced mixes exhibited the highest flexural strength and retained the most excellent residual capacity after sustained loading. At a 0.50% dosage, PVA reduced chloride ingress by up to 43.8%, outperforming other fiber types due to its strong bond with the cement matrix and ability to maintain tight crack widths. While PVA also contributed to reduced water absorption, the effect was more pronounced at lower dosages, suggesting that very high content may affect matrix uniformity. Overall, PVA fibers are highly effective in enhancing load capacity, limiting permeability, and mitigating chloride penetration.

PP fibers were most effective in reducing water absorption, with the 0.50% dosage decreasing permeability by 31.9% which was the best among all mixes. This is attributed to PP's hydrophobic nature and fibrillated structure. PP fibers also improved flexural strength, though not as significantly as PVA. Their performance in chloride resistance was moderate, reflecting their limited ability to control fine cracking due to lower bond strength with the cement paste. Nevertheless, PP fibers offer clear benefits in reducing permeability and boosting toughness.

AR glass fibers showed moderate, balanced improvements across all tests. They contributed to modest gains in flexural and residual strength and offered uniform crack distribution. However, they were the least effective in reducing water absorption and chloride ingress. AR fibers' brittle behavior and low pull-out resistance limited their effectiveness under sustained loading and environmental stress.

Acknowledgements. The research study, the results of which are reported in this paper, was sponsored by the American Concrete Institute NEx Program. The authors would like to acknowledge and thank the sponsor for this support. Opinions, findings, and conclusions expressed in this paper are of the authors and do not necessarily reflect those of the sponsor.

References

1. Weiss J, Couch J, Pease B, Laugesen P, Geiker M (2017) Influence of mechanically induced cracking on chloride ingress in concrete. J Mater Civ Eng 29(9)
2. American Concrete Institute (ACI) (2021) Code requirements for environmental engineering concrete structures and commentary, Farmington Hills, MI
3. American Society of Civil Engineers (ASCE) (2021) Report card for America's infrastructure, Reston, Virginia
4. Zheng Y, Lv X, Hu S, Zhuo J, Wan C, Liu J (2024) Mechanical properties and durability of steel fiber reinforced concrete: a review. J Build Eng 82:108025
5. Paul S, van Zijl G, Šavija B (2020) Effect of fibers on durability of concrete: a practical review. Materials 13(20):4562
6. Lee SK, Jeon MJ, Cha SS, Park CG (2017) Mechanical and permeability characteristics of latex-modified fiber-reinforced roller-compacted rapid-hardening-cement concrete for pavement repair. Appl Sci 7(7)
7. Kazemian M, Shafei B (2023) Mechanical properties of hybrid fiber-reinforced concretes made with low dosages of synthetic fibers. Struct Concr 24(1):1226–1243
8. ASTM International (2021) Specification for blended hydraulic cements. West Conshohocken, PA
9. ASTM International (2022) Specification for coal Fly ash and raw or calcined natural Pozzolan for use in concrete. West Conshohocken, PA
10. ASTM International (2018) Specification for concrete aggregates, West Conshohocken, PA
11. ASTM International (2017) Specification for chemical admixtures for concrete, West Conshohocken, PA
12. ASTM International (2016) Test method for flexural strength of concrete (using simple beam with Center-point loading), West Conshohocken, PA
13. ASTM International (2021) Test method for density, absorption, and voids in hardened concrete, West Conshohocken, PA
14. Nordtest (1999) NT build 492: concrete, mortar and cement-based repair materials—chloride migration coefficient from non-steady-state migration experiments, Espoo, Finland
15. Karim R, Shafei B (2024) Ultra-high performance concrete under direct tension: investigation of a hybrid of steel and synthetic fibers. Struct Concr 25(1):423–439
16. Karim R, Shafei B (2022) Investigation of five synthetic fibers as potential replacements of steel fibers in ultra-high performance concrete. J Mater Civ Eng 34(7):04022126
17. Veigas MG, Najimi M, Shafei B (2022) Cementitious composites made with natural fibers: investigation of uncoated and coated sisal fibers. Case Stud Constr Mater 16:E00788
18. Karim R, Shafei B (2021) Flexural response characteristics of ultra-high performance concrete made with steel microfibers and macrofibers. Struct Concr 22(6):3476–3490
19. Shafei B, Kazemian M, Dopko M, Najimi M (2021) State-of-the-art review of capabilities and limitations of polymer and glass fibers used in fiber-reinforced concrete. Materials 14(2):409
20. Dopko M, Najimi M, Shafei B, Wang X, Taylor P, Phares B (2020) Strength and crack resistance of carbon microfiber reinforced concrete. ACI Mater J 117(2):11–23

Towards Sustainable Infrastructure Using Self-healing Ultra-high-performance Concrete (UHPC)—A Review

Tuba Iqbal[1](✉), Shamsad Ahmad[1,2], Subhan Ahmad[2], Saheed Kolawole Adekunle[1,2], and Mohammed A. Al-Osta[1,2]

[1] Civil and Environmental Engineering Department, King Fahd University of Petroleum and Minerals, Dhahran 31261, Saudi Arabia
g202391490@kfupm.edu.sa

[2] Interdisciplinary Research Center for Construction and Building Materials, King Fahd University of Petroleum & Minerals, Dhahran 31261, Saudi Arabia

Abstract. To achieve net-zero targets in construction, it is essential to utilize low-carbon materials and highly durable systems that minimize lifecycle emissions and depletion of resources. Ultra-high performance concrete (UHPC), developed for achieving the superior strength and durability, offers significant sustainability benefits when enhanced with self-healing capabilities. This review critically examines major self-healing mechanisms in UHPC, including autogenous healing, chemical admixtures, encapsulated healing agents, and microbial calcite precipitation. Each approach is evaluated based on healing kinetics, microstructural improvements, durability enhancement, environmental performance, and overall sustainability contributions. Key factors influencing healing efficacy: crack geometry, healing duration, curing conditions, exposure environment, binder composition, moisture availability, internal curing, and material optimization, are thoroughly discussed. Recent studies demonstrate that self-healing UHPC effectively restores both mechanical and transport properties, achieving crack closure efficiencies exceeding 90% and significantly reducing chloride ion penetration and permeability. These advancements substantially extend infrastructure service life, lower maintenance requirements, and reduce embodied carbon, directly enabling resilient, climate-adapted, low-impact, and sustainable infrastructure solutions aligned with long-term climate goals.

Keywords: Ultra-High-Performance Concrete · Self-Healing Concrete · Durability Enhancement · Lifecycle Emissions · Sustainable Infrastructure

1 Introduction

The rapid development of global infrastructure has intensified the demand for high-performance, durable, and sustainable construction materials. Conventional infrastructure often deteriorates under aggressive environmental exposure and mechanical loads, resulting in frequent maintenance, elevated lifecycle costs, and high carbon emissions [1, 2]. Improving durability and sustainability is therefore essential to meet global net-zero targets [3].

H. Beushausen et al. (Eds.): NTZR 2025, LNCE 873, pp. 222–233, 2026.
https://doi.org/10.1007/978-3-032-27698-8_21

UHPC has emerged as a promising solution due to its exceptional compressive strength (>150 MPa), extremely low permeability, dense microstructure, and resistance to chloride ingress, carbonation, and freeze-thaw cycles [4]. Its durability makes it ideal for critical infrastructure such as bridges, tunnels, nuclear containment, and marine structures. However, UHPC remains susceptible to shrinkage, thermal, and load-induced cracking, which threatens long-term performance [5, 6].

To address this, self-healing technologies have been introduced to enable UHPC to autonomously repair microcracks, restoring structural integrity and extending service life [7, 8]. Self-healing in UHPC typically falls into four categories: autogenous (intrinsic) healing, chemical admixtures, encapsulated healing agents, and microbial methods [9–11], see Fig. 1 for a visual representation of these mechanisms. Autogenous healing relies on continued hydration of unhydrated cement particles and calcium carbonate formation via carbonation [12, 13]. Chemical admixtures, such as crystalline admixtures (CA), improve this process by accelerating the formation of healing products [14]. Encapsulation involves embedding agents like epoxy or silica gels in polymeric or mineral capsules that rupture upon cracking, effectively sealing wider cracks [15, 16]. Microbially Induced Calcite Precipitation (MICP) uses bacteria to biologically deposit calcium carbonate, enhancing crack closure particularly in chloride-rich environments [17, 18].

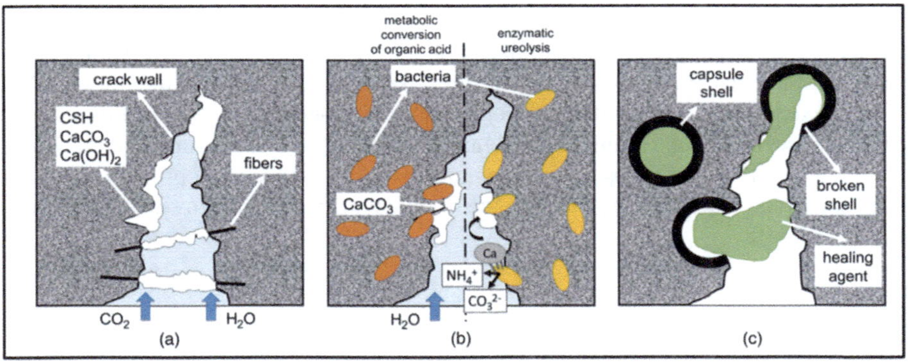

Fig. 1. Self-healing mechanisms: (a) autogenous; (b) autonomous bacteria-based; and (c) autonomous capsule-based [19].

Healing effectiveness depends on crack geometry (ideally < 100 μm), healing time, exposure conditions, and binder composition [5, 10, 15, 19]. Crack geometry, particularly crack width, critically affects healing potential, with optimal self-healing generally occurring in cracks smaller than 100 μm [5, 10]. Healing duration also significantly impacts self-healing performance, with extended curing periods substantially improving mechanical and durability recovery [15, 19]. Moisture availability and ion concentrations strongly influence healing kinetics [6]. Supplementary cementitious materials (SCMs) and nanoparticles can further enhance healing potential [15].

Overall, integrating self-healing strategies into UHPC improves durability, reduces maintenance, and lowers lifecycle carbon emissions [4, 10]. Studies have demonstrated remarkable recovery rates of mechanical properties (>90% crack closure efficiency) and substantial reductions in chloride permeability, directly translating into enhanced structural durability and resilience [7, 8]. This review synthesizes key mechanisms, performance parameters, and sustainability implications to support the broader adoption of self-healing UHPC in infrastructure.

2 Overview of Ultra-High-Performance Concrete

UHPC is defined as a cementitious composite characterized by superior mechanical properties, including compressive strengths generally exceeding 150 MPa, splitting tensile strengths above 5 MPa, and exceptional durability [1, 2]. Its composition includes cement, SCMs such as silica fume, fly ash, and slag, fine aggregates, superplasticizers, and steel fibers for enhanced ductility and toughness [4]. A low water-to-binder (w/b) ratio (<0.25) is essential for achieving a dense microstructure and superior strength development [15]. Compared to conventional concrete, UHPC offers significantly improved durability, mechanical performance, and long-term sustainability. Its dense matrix minimizes chloride ion diffusion and carbonation, increasing resistance to chemical attack and freeze-thaw cycles [20]. The inclusion of steel fibers enhances tensile strength, crack resistance, and ductility, enabling thinner, lighter, and longer-lasting structural components. These features contribute to sustainability by reducing repair frequency, lowering lifecycle emissions, and improving material efficiency [5, 7].

UHPC has been successfully applied in bridge decks, highway infrastructure, marine structures, nuclear facilities, and architectural facades [2]. Its high strength and durability make it particularly suitable for aggressive environments and structures requiring extended service life and minimal maintenance interventions. However, broader adoption is limited by high initial costs, complex curing requirements, susceptibility to autogenous shrinkage cracking, and the absence of widely accepted design standards in many regions [10].

3 Self-Healing Mechanisms in UHPC

3.1 Autogenous (Intrinsic) Healing

Autogenous healing in UHPC is its intrinsic ability to repair microcracks without external agents, driven by continued hydration and carbonation [20]. Water ingress activates unhydrated cement particles, forming C–S–H gels that fill voids [15]. While dissolved atmospheric CO_2 reacts with $Ca(OH)_2$ to precipitate $CaCO_3$, sealing cracks [5, 10]. Moisture availability, internal curing (via Superabsorbent polymers [SAP] or lightweight aggregates), and binder composition significantly influence healing [6]. SCMs like silica fume, fly ash, and slag refine pore structure and enhance pozzolanic activity, improving autogenous healing performance [7].

3.2 Chemical Admixture-Based Healing

Chemical admixture-based healing involves integrating specialized chemical admixtures, primarily Crystalline admixtures (CA), to augment UHPC's intrinsic healing ability. CA mainly contains calcium oxide (CaO) and tricalcium silicate (C_3S), which react with water to form calcium silicate hydrate (C–S–H) crystals [8, 9]. These crystals fill cracks and voids, considerably reducing concrete permeability and enhancing durability [10, 11]. Studies indicate that CA significantly improves self-healing in UHPC by promoting rapid sealing of microcracks, particularly in aggressive environments like chloride-rich or freeze-thaw conditions [12, 13]. The reaction between CA and water produces additional hydration products, enhancing mechanical recovery and durability characteristics. This approach has demonstrated improved tensile and flexural properties post-healing, along with substantial reductions in chloride ion penetration and permeability [19].

3.3 Encapsulated Agents-Based Healing

Encapsulated healing agents in UHPC involve embedding micro or macro-capsules filled with healing agents such as epoxy, polyurethane, or mineral-based materials that rupture upon cracking to release sealants and promote autonomous crack repair [14]. Polymeric capsules, especially those using epoxy or polyurethane, effectively seal wider cracks (up to 300 μm), restoring mechanical properties and reducing permeability [21]. Silica-based mineral capsules offer greater chemical stability and durability under harsh environmental conditions [15, 16]. Healing efficiencies exceeding 90% have been reported in aggressive exposure scenarios [17]. However, challenges remain in achieving uniform capsule dispersion, long-term stability, and minimizing negative effects on fresh UHPC properties [15, 18]. A detailed comparison of healing agents, their efficiency ranges, healing durations, and sustainability benefits across all mechanisms, including encapsulated systems, is summarized in Table 1.

3.4 Microbially Induced Calcite Precipitation (MICP)-Based Healing

MICP in UHPC employs ureolytic bacteria such as *Bacillus subtilis* and *Bacillus pasteurii* to biologically precipitate calcium carbonate, sealing cracks and enhancing long-term durability [20]. When water and oxygen enter through cracks, bacteria metabolize nutrients (urea and calcium), producing $CaCO_3$ that fills cracks up to 500 μm [6, 19]. This method improves chloride resistance, restores compressive and tensile strengths, and supports sustainability by reducing chemical use and maintenance [5]. However, challenges include maintaining bacterial viability in UHPC's high-alkaline environment, ensuring even distribution of bacteria and nutrients, and achieving consistent long-term performance. Encapsulation strategies using hydrogel, silica gel, or diatomaceous earth improve bacterial protection, enhance survival, and optimize healing efficiency for aggressive exposure conditions [7, 8].

Table 1. Overview of healing mechanisms, agents, efficiency, and performance metrics in UHPC systems.

Healing mechanism type	Healing agent	Crack closure efficiency (%)	Healing time (days)	Sustainability metrics	Source references
Autogenous & Stimulated CA	CA	~ 100% efficiency ($\leq$100 μm)	30–365	Reduced chloride penetration, enhanced mechanical and durability properties, sustained long-term structural performance	[7, 22, 23]
Autonomous (SAP-based)	SAPs	Up to 100% (pure water), moderate-low (chlorides/sulfates)	28–280	Improved corrosion resistance, sustained hydration, and effective durability enhancement under moisture-rich environments	[8, 9, 24, 25]
Autonomous (Bacteria-based)	Bacteria-based Healing Agents	50–460 μm cracks effectively healed	7–180	Reduced permeability, improved strength (up to 42%), lower lifecycle maintenance, improved sustainability	[5, 6, 26–28]
Autonomous (Encapsulation)	Encapsulated Chemical Agents	Effective up to 300 μm	28–90	Improved crack sealing, durability, reduced maintenance costs	[5, 26]
Autogenous & Stimulated (Fibers)	Fibers (Steel, Glass, Polypropylene, Basalt)	Effective healing $\leq$100 μm	28–180	Enhanced mechanical performance, durability under cyclic/aggressive environments, improved ductility and strain recovery	[14, 29]
Autogenous (Nanoparticles-enhanced hydration)	Nanoparticles (nano-silica, nano-$CaCO_3$, etc.)	High efficiency typically <200 μm	28–90	Improved mechanical properties, durability, reduced permeability, refined microstructure	[30]

(continued)

Table 1. *(continued)*

Healing mechanism type	Healing agent	Crack closure efficiency (%)	Healing time (days)	Sustainability metrics	Source references
Carbonation-induced	Carbonation-based (Pressurized CO_2 curing)	High carbon uptake (~ 80 kg CO_2/m^3), effective carbonation healing	1–28	Significant CO_2 sequestration, dense microstructure, reduced permeability despite slight strength reduction (~10–15%)	[12]
Autogenous (Pozzolanic Reaction)	Pozzolanic Materials (Silica fume, Fly ash, etc.)	Excellent (> 90% at ~ 100 μm)	28–180	Enhanced impermeability, refined pore structure, mechanical recovery, improved sustainability	[11]
Autogenous (Fiber-based healing)	Polypropylene (PP) Fibers	> 65% strength recovery post-high temperature	Varied post-fire curing	Improved durability under fire damage, significantly reduced permeability	[15, 31]
Autogenous (Waste-based hydration)	Waste Glass Powder (WGP)	Effective healing via wollastonite formation	Varied curing periods	Significant reuse of waste materials, reduced porosity, improved structural integrity	[18]
Autonomous (Expansive Agent-induced)	Silicate-modified Calcium Oxide	Significant visible crack healing	7–28	Major improvements in durability, reduced permeability	[17]
Autonomous (Nano additives)	Alumina & Cellulose Nanofibers	High efficiency ($\geq$90%, cracks <100 μm)	28–365	Complete mechanical restoration, enhanced structural durability, improved predictive modeling	[16, 23]
Autogenous (hydration, carbonation, etc.)	Autogenous healing (hydration, carbonation)	High for cracks $\leq$ 100 μm, limited for larger cracks	Varied 7–365	Continued hydration and carbonation improve durability, permeability reduction, chloride resistance, enhanced lifecycle	[22, 28]

4 Factors Affecting Self-healing Performance of UHPC

Several factors govern the self-healing efficiency of UHPC. Crack geometry, especially width, critically affects healing efficiency in UHPC. Microcracks under 100 μm generally heal effectively through continued hydration and carbonation, while cracks above 150 μm exhibit reduced healing due to limitations in autogenous mechanisms [20]. Wider cracks require encapsulated or microbial agents for efficient sealing [5, 10]. See Fig. 2 for crack width ranges across methods. Studies confirm optimal healing performance occurs in cracks $\leq$ 100 μm [7]. Healing duration also impacts performance. While significant sealing often occurs within 28 days, continued healing can progress up to 180 days under sustained moisture exposure [8, 9]. Extended curing times enhance hydration, crack sealing, and mechanical recovery, which is essential for durability in long-service structures [11]. Environmental conditions, including moisture, temperature, and chemical exposure, directly influence healing kinetics. Moist conditions promote hydration and carbonation, while moderate temperatures (20–30 °C) improve reaction rates [13]. Extreme or chemically aggressive environments (e.g., chloride or sulfate exposure) may impair healing, though CA and microbial systems have shown some resilience [14, 21].

Internal curing using SAPs, lightweight aggregates, or cellulose-based materials improves moisture retention, promoting sustained hydration and better healing outcomes. These materials have shown particular promise when combined with healing agents [15, 16]. Binder composition and mix design also play a critical role. Incorporating SCMs like silica fume, slag, or fly ash refines pore structure and accelerates hydration, improving healing potential [17]. Nanomaterials such as nanosilica and nanoclay enhance hydration kinetics and microstructure densification, leading to better mechanical performance and durability [18]. Sustainable binder design contributes directly to extended service life and reduced maintenance demands [15, 20].

5 Sustainability Analysis of Self-Healing UHPC

Self-healing UHPC enhances sustainability by improving durability, reducing permeability, chloride ingress, and carbonation, which extends service life [5]. Figure 2b compares healing efficiency across autogenous, microbial, encapsulated, and nano-based systems. Structures with self-healing UHPC require fewer repairs, cutting lifecycle maintenance costs and embodied emissions [9]. Quantitative assessments show significant environmental and economic advantages. Replacing OPC with Portland limestone cement and substituting silica fume with waste glass powder reduces embodied CO_2 by 25–30% [32]. Mixes using ground granulated blast furnace slag (GGBS) or recycled concrete aggregates (RCA) achieve 20–35% CO_2 reductions while maintaining durability [33, 34]. These environmental benefits come with economic gains. Lifecycle analyses report 20–35% lower maintenance costs over 50 years due to fewer repairs and greater resistance to aggressive environments [32, 35]. Microbial healing and encapsulated agents, especially in nano-modified UHPC, further reduce material use [7–9, 11]. Self-healing also improves resource efficiency and reduces reliance on high-range water-reducing admixtures [32]. Although standardizing evaluation methods remains a challenge, current data support integrating self-healing UHPC into sustainable infrastructure. Its ability to lower emissions, extend service life, and reduce costs makes it a strong candidate for climate-resilient, low-maintenance construction.

6 Key Research Insights and Existing Gaps

A synthesis of existing literature on self-healing UHPC highlights distinct mechanisms and performance outcomes. Autogenous healing, via continued hydration and carbonation, effectively seals microcracks (<100 μm), reducing permeability and improving durability [20]. Crystalline admixtures promote additional C–S–H formation, especially in chloride environments [7]. Encapsulated agents, polymeric and mineral-based can seal larger cracks (100–300 μm), restoring mechanical integrity [5, 10]. Microbial methods such as MICP heal cracks up to 500 μm through bacterial $CaCO_3$ precipitation, enhancing both strength and durability [6]. Nano-engineering and internal curing further boost performance: nanomaterials (e.g., nanosilica, nanoclay) refine microstructure and accelerate hydration [16], while SAPs and lightweight aggregates sustain moisture for prolonged healing [8, 9]. However, challenges persist particularly the stable integration of healing agents in UHPC's alkaline matrix [11–13], along with issues of cost, scalability, and sparse field data [14, 21]. The lack of standardized testing protocols continues to hinder practical adoption [15]. Despite encouraging lab-scale results, microbial systems often underperform in highly alkaline UHPC environments, where bacterial survival and $CaCO_3$ precipitation are compromised [6, 28]. Similarly, encapsulated agents suffer from uneven distribution, unpredictable rupture, and potential weakening of the UHPC matrix if used in high dosages [7]. CA-based systems also show inconsistent healing under freeze-thaw cycles and high-chloride exposure [7]. Crucially, gaps between lab findings and real-world structural applications remain. Few studies validate long-term self-healing under actual loading and environmental conditions. Overcoming these challenges requires standardization, field trials, and potentially hybrid approaches combining multiple healing strategies to improve reliability and adoption.

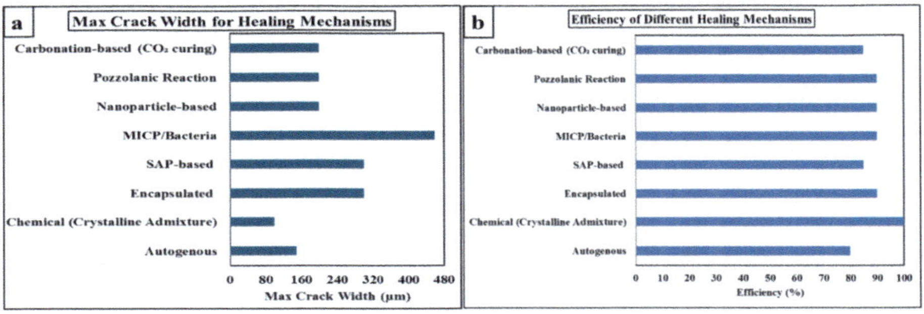

Fig. 2. a. Maximum crack width across healing methods, **b.** Healing efficiency across healing methods [4–9, 11, 12, 14–18, 21–31].

7 Practical Considerations and Recommendations

Implementing self-healing UHPC in infrastructure involves several practical challenges. Ensuring uniform distribution and long-term stability of healing agents, encapsulated or microbial, remains difficult due to dispersion issues and environmental sensitivity [20]. Bacterial viability in high-alkalinity UHPC requires advanced encapsulation strategies, limiting real-world adoption [6]. High initial costs for specialized materials, production, and quality control also hinder broader industry use [4]. Optimizing UHPC for self-healing requires integrating SCMs such as silica fume, fly ash, and slag, which refine pore structure and enhance hydration, improving healing efficiency [7]. Internal curing using superabsorbent polymers (SAP) or lightweight aggregates supports sustained moisture availability and prolonged healing [8]. Nanoscale additives like nano-silica and nano-clay further improve hydration rates and microstructure, increasing durability and strength [9, 10]. For encapsulated systems and MICP, robust encapsulation via polymeric or mineral-based capsules is essential for controlled release and durability under aggressive conditions [11, 12].

Policy frameworks must evolve, introducing standardized testing protocols and incentivizing sustainable materials through regulatory credits [13].Despite higher upfront costs, lifecycle analyses show significant long-term savings due to reduced maintenance and extended service life, reinforcing the economic viability of self-healing UHPC [14, 21].

8 Conclusions and Future Directions

Integrating self-healing technologies into UHPC significantly enhances mechanical properties, durability, reduces maintenance requirements, and aligns with global net-zero sustainability goals. Autogenous and chemically-induced healing are highly effective for microcrack repair, while encapsulated agents and microbial approaches address wider cracks with promising results. Innovations in nano-engineering and internal curing techniques have improved healing kinetics, microstructural refinement, strength recovery, and long-term durability. These enhancements contribute to extended service life, lower lifecycle emissions, and improved resource efficiency in concrete infrastructure. Despite initial higher costs, self-healing UHPC offers long-term economic and environmental benefits. Future research should emphasize standardized performance metrics, long-term field validation, exposure-driven durability assessments, and scalable, cost-effective production strategies to accelerate practical adoption in diverse and aggressive service environments.

References

1. (ACI 239R-18), American Concrete Institute (2018) Ultra-high-performance concrete: an emerging technology report
2. Russell HG (n.d.) Ultra-high performance concrete: a state-of-the-art report for the bridge community
3. The IPCC (n.d.) Special Report on Global Warming of 1.5°C. www.quno.org.

4. Akeed MH et al (2022) Ultra-high-performance fiber-reinforced concrete. Part IV: durability properties, cost assessment, applications, and challenges, case studies. Constr Mater 17. https://doi.org/10.1016/j.cscm.2022.e01271
5. Hou S, Li K, Wu Z, Li F, Shi C (2022) Quantitative evaluation on self-healing capacity of cracked concrete by water permeability test—A review. Cem Concr Compos 127:104404. https://doi.org/10.1016/j.cemconcomp.2021.104404
6. Nodehi M, Ozbakkaloglu T, Gholampour A (2022) A systematic review of bacteria-based self-healing concrete: biomineralization, mechanical, and durability properties. J Build Eng 49:104038. https://doi.org/10.1016/j.jobe.2022.104038
7. Cappellesso V, Ferrara L, Gruyaert E, Van Tittelboom K, De Belie N (2024) Resilient crystalline admixture in ultra-high performance self-healing concrete under cyclic freeze-thaw with de-icing salts. Cem Concr Res 181. https://doi.org/10.1016/j.cemconres.2024.107524
8. Hassi S, Javanmardi A, Lai Z, Rashidi M (2024) Corrosion resistance of healed UHPC with superabsorbent polymers subjected to chloride seawater attack. Constr Build Mater 451:138683. https://doi.org/10.1016/j.conbuildmat.2024.138683
9. Alameri M, Mohamed Ali MS, Elchalakani M, Sheikh A, Fan R (2024) Self-healing and mechanical behaviour of fibre-reinforced ultra-high-performance concrete incorporating superabsorbent polymer under repeated and sustained loadings. Fibers 12. https://doi.org/10.3390/fib12110095
10. Xi B, Al-Obaidi S, Ferrara L (2023) Effect of different environments on the self-healing performance of ultra high-performance concrete – a systematic literature review. Constr Build Mater 374. https://doi.org/10.1016/j.conbuildmat.2023.130946
11. Zheng Q et al (2025) Influence of ultra-fine pozzolanic materials on the self-healing capabilities of ultra-high performance concrete under carbonation conditioning. Cem Concr Res 187. https://doi.org/10.1016/j.cemconres.2024.107712
12. Dixit A, Du H, Pang SD (2021) Carbon capture in ultra-high performance concrete using pressurized CO_2 curing. Constr Build Mater 288. https://doi.org/10.1016/j.conbuildmat.2021.123076
13. Kannikachalam NP et al (2023) Assessment of sustainability and self-healing performances of recycled ultra-high-performance concrete. ACI Struct J 120:117–132. https://doi.org/10.14359/51737336
14. Cui K, Liang K, Chang J, Lau D (2022) Investigation of the macro performance, mechanism, and durability of multiscale steel fiber reinforced low-carbon ecological UHPC. Constr Build Mater 327. https://doi.org/10.1016/j.conbuildmat.2022.126921
15. Qian Y, Yang D, Xia Y, Gao H, Ma Z (2021) Transport properties and resistance improvement of ultra-high performance concrete (UHPC) after exposure to elevated temperatures. Buildings 11:416. https://doi.org/10.3390/buildings11090416
16. Huang Z, Cuenca E, Ferrara L (2025) Optimized data-driven method to study the self-healing and durability of ultra-high performance concrete. Eng Appl Artif Intell 143. https://doi.org/10.1016/j.engappai.2025.110043
17. Wang YS, Lee HS, Lin RS, Wang XY (2022) Effect of silicate-modified calcium oxide-based expansive agent on engineering properties and self-healing of ultra-high-strength concrete. J Build Eng 50. https://doi.org/10.1016/j.jobe.2022.104230
18. Zhao X, Lu JX, Tian W, Cyr M, Tagnit-Hamou A, Poon CS (2025) Self-healing performance of thermally damaged ultra-high performance concrete: rehydration and recovery mechanism. Cem Concr Res 191. https://doi.org/10.1016/j.cemconres.2025.107825
19. Rajczakowska M, Habermehl-Cwirzen K, Hedlund H, Cwirzen A (2019) Autogenous self-healing: a better solution for concrete. J Mater Civ Eng 31. https://doi.org/10.1061/(asce)mt.1943-5533.0002764

20. Kim S, Yoo D-Y, Kim M-J, Banthia N (2019) Self-healing capability of ultra-high-performance fiber-reinforced concrete after exposure to cryogenic temperature. Cem Concr Compos 104:103335. https://doi.org/10.1016/j.cemconcomp.2019.103335
21. Yoo D-Y, Shin W, Chun B, Banthia N (2020) Assessment of steel fiber corrosion in self-healed ultra-high-performance fiber-reinforced concrete and its effect on tensile performance. Cem Concr Res 133:106091. https://doi.org/10.1016/j.cemconres.2020.106091
22. Al-Obaidi S, He S, Schlangen E, Ferrara L (2023) Effect of matrix self-healing on the bond-slip behavior of micro steel fibers in ultra-high-performance concrete. Mater Struct Mater Constr 56. https://doi.org/10.1617/s11527-023-02250-5
23. Davolio M, Al-Obaidi S, Altomare MY, Lo Monte F, Ferrara L (2023) A methodology to assess the evolution of mechanical performance of UHPC as affected by autogenous healing under sustained loadings and aggressive exposure conditions, Cem Concr compos 139. https://doi.org/10.1016/j.cemconcomp.2023.105058
24. Yang Z, Miao C, Deng W, Liu D, Zhang J, Gu J (2024) Multi-scale characterization of SAP impact on self-healing behavior in UHPC under varied crack widths and environments. Constr Build Mater 421. https://doi.org/10.1016/j.conbuildmat.2024.135649
25. Hassi S, Javanmardi A, Menu B, Lai Z, Huang F (2025) Effect of super absorbent polymers on the self-healing capability of macrocracked ultra-high performance concrete under highly aggressive environments. Colloids Surf A Physicochem Eng Asp 705. https://doi.org/10.1016/j.colsurfa.2024.135540
26. Amran M et al (2022) Self-healing concrete as a prospective construction material: a review. Materials 15. https://doi.org/10.3390/ma15093214
27. Doostkami H, de Jesús Estacio J, Cumberbatch S, Formagini P, Serna MR-F (2023) Self-healing capability of conventional, high-performance, and ultra high-performance concrete with commercial bacteria characterized by means of water and chloride penetration. Constr Build Mater 401. https://doi.org/10.1016/j.conbuildmat.2023.132903
28. Mitra A (2022) Bacteria in ultra-high-performance concrete- review of factors affecting durability. Interantional Journal Of Scientific Research In Engineering And Management 06. https://doi.org/10.55041/ijsrem11845
29. Kravanja G, Mumtaz AR, Kravanja S (2024) A comprehensive review of the advances, manufacturing, properties, innovations, environmental impact and applications of ultra-high-performance concrete (UHPC). Buildings 14. https://doi.org/10.3390/buildings14020382
30. Mahmood RA, Kockal NU (2021) Nanoparticles used as an ingredient in different types of concrete. SN Appl Sci 3. https://doi.org/10.1007/s42452-021-04461-3
31. Qian Y, Yang D, Xia Y, Gao H, Ma Z (2023) Properties and improvement of ultra-high performance concrete with coarse aggregates and polypropylene fibers after high-temperature damage. Constr Build Mater 364. https://doi.org/10.1016/j.conbuildmat.2022.129925
32. Nia SB, Nyland A, Wivast J, Kioumarsi M, Shafei B (2025) Investigations of Portland limestone cement and waste glass powder for sustainable ultra-high performance concrete. Case Stud Constr Mater 22:e04425. https://doi.org/10.1016/j.cscm.2025.e04425
33. Farahzadi L, Bozorgmehr Nia S, Shafei B, Kioumarsi M (2025) Sustainability assessment of ultra-high performance concrete made with various supplementary cementitious materials. Cleaner Mater 15:100301. https://doi.org/10.1016/j.clema.2025.100301
34. Wivast J, Nyland A, Nia SB, Kioumarsi M, Shafei B (2025) Developing eco-friendly ultra-high-performance concrete by utilizing recycled alternatives, in, pp 49–57. https://doi.org/10.1007/978-3-031-69626-8_5

35. Farahzadi L, Bozorgmehr Nia S, Shafei B, Kioumarsi M (2025) Preliminary environmental assessment of ultra-high-performance concrete mixtures, in, pp 589–598. https://doi.org/10.1007/978-3-031-69626-8_49

Self-healing and Mechanical Property of UHPC Incorporating Sporosarcina Pasteurii

Yuxiang Tan[1], Alessandro Mattedi[2], Enrico Marsili[2], and Weizhuo Shi[1](✉)

[1] Department of Civil Engineering and New Materials Institute, University of Nottingham Ningbo China, Taikang Road 199, Ningbo 315100, People's Republic of China
{Yuxiang.tan,weizhuo.shi}@nottingham.edu.cn

[2] Life Science and Healthcare Centre, CBI, University of Nottingham Ningbo China, Taikang Road 199, Ningbo 315100, People's Republic of China
{Alessandro.MATTEDI,Enrico.Marsili}@nottingham.edu.cn

Abstract. The low water-to-binder ratio of ultra-high performance concrete (UHPC) can result in early-age shrinkage and cracking due to self-desiccation. Bacterial self-healing of cracks represents an environmentally friendly and sustainable approach compatible with concrete. Sporosarcina Pasteurii can form endospores which are resistant to the alkaline environment of the concrete matrix. As a porous cementitious supplementary material, natural zeolite is used to immobilize bacteria without disrupting the matrix. This study investigates the effect of incorporating natural zeolite-immobilized S. Pasteurii into non-proprietary UHPC on mechanical properties and self-healing performance via compressive strength tests and crack healing assessments. Four mixtures containing different BZP content levels were prepared and assessed up to 56 days of age. Results show that cracks up to 0.38 mm can be completely healed by 28 days of age. Moreover, S. Pasteurii in UHPC remained viable after 56 days.

Keywords: ultra-high performance concrete · self-healing · bacteria

1 Introduction

Ultra-high performance concrete (UHPC), known for its exceptional strength, ductility, and durability, was developed in response to the demand for advanced construction materials (Richard and Cheyrezy 1995). The typical compressive strength of UHPC ranges from 120 to 200 MPa, with tensile and flexural strength reaching 20 MPa and 30 MPa, respectively (Zhang et al. 2018; Mishra and Singh 2019). To achieve the desirable performances, UHPC typically utilizes a particular combination of cement, fine-grained silica fume, selected gradation of sands, water, superplasticizer, steel fibers, and additional supplementary cementitious materials (Bajaber and Hakeem 2021). Despite variations in UHPC definitions, a low water-to-binder (w/b) ratio in the range of 0.14–0.25 is generally accepted (Azmee and Shafiq 2018; Farzad et al. 2019; Mishra and Singh 2019).

However, the low w/b ratio can lead to autogenous shrinkage, which can reach up to 2000 to 3000 microstrains due to self-desiccation (Liu et al. 2021), potentially leading to

H. Beushausen et al. (Eds.): NTZR 2025, LNCE 873, pp. 234–245, 2026.
https://doi.org/10.1007/978-3-032-27698-8_22

micro-cracking and durability concerns (Kheir et al. 2021). To prevent micro-cracking from expanding, enhancing the autogenous healing capability of concrete is one of the effective approaches. Among the enhanced autogenous healing techniques (chemically based healing agents, microcapsule-based agents, mineral admixture-based agents and bio-based agents), bio-based self-healing is deemed more environmentally friendly and compatible with concrete matrix (Souradeep et al. 2017; Vijay et al. 2017; Jogi and Lakshmi 2021; Nodehi et al. 2022). It has been reported that bacteria induced calcium carbonate precipitation can reduce permeability and penetration of water and chloride into concrete matrix by closing cracks, thus improving concrete durability (De Muynck et al. 2008; Qian et al. 2009; Wang et al. 2014; Muhammad et al. 2016; Wang et al. 2016).

There are various bacteria strains that can survive the alkaline environment of concrete matrix and precipitate calcium carbonate by urease activity. Sporosarcina Pasteurii is one of the most commonly used gram-positive bacteria isolated from soil. S. Pasteurii can form dormant endospores when subjected to unfavorable conditions (concrete mixing and hardening) and germinate when living requirements are met (Omoregie et al. 2019; Tayebani and Mostofinejad 2019). Good compatibility has been found between S. Pasteurii and concrete, studies have reported that the incorporation of S. Pasteurii into concrete can improve compressive strength up to 35% (than the control group) and reduce water absorption and chloride penetrability of concrete (Chahal et al. 2012; Tayebani and Mostofinejad 2019).

Direct addition of bacteria to cementitious material has been reported to improve compressive strength and reduce pore volume, however, substantial portion of unprotected bacteria experience premature death due to high pH environment of cementitious material (Ersan et al. 2015; Li et al. 2019). It has been recommended to use capsules or immobilizing medium to facilitate viability of bacteria when added to cementitious materials (Wang et al. 2012). However, capsule materials, shape and size can affect the rheology of concrete. Spherical microcapsules may have lubricating effects. Notably, capsule material should not absorb water to avoid competing with cementitious materials for water (Lucas et al. 2010).

Natural zeolite is an abundant mineral resource that is known for its porous microstructure and high pozzolanic activity (Elaiopoulos et al. 2008; Kazemian and Shafei 2022). In past studies, Bhaskar et al. (2017) reported that the percentage of urea decomposed by bacteria in high pH slurry has increased from 5 to 70% with the protection of zeolite particles. Zeolite immobilization of bacteria has shown potential in enhancing bacteria viability and mechanical properties of mortar in a sustainable way. However, there are few studies in the current literature that focus on the self-healing capacity of ultra-high performance concrete (UHPC) incorporating zeolite immobilized S. Pasteurii. To explore the possibility of maintaining the durability of UHPC subject to early-age cracking, it is important to investigate the self-healing effect of UHPC incorporated with bacteria immobilized by natural zeolite.

This study aims to investigate the self-healing effect of UHPC incorporating zeolite-immobilized bacteria. The self-healing capability will be evaluated using compressive strength tests and the ratio of healed crack area. The outcome of this study will shed light on effective approaches for mitigating cracking caused by early-age shrinkage and inform the design of more sustainable UHPC.

2 Materials and Methods

2.1 Raw Materials

Portland cement used in this project is the P52.5 provided by Hailuo Cements Co., Ltd. Silica fume is simi-compacted particles with a bulk density of 350 kg/m^3 provided by the Elkem Silicones Co., Ltd. Fine sand is river sand sieved to have size between 80 um and 400 um. Coarse sand is quartz sand sized from 400 um to 600 um. Natural zeolite is excavated from Dabie mountain and is used as the Sporosarcina Pasteurii carrier. The chemical compositions of the binders, i.e., cement, silica fume and natural zeolite, are shown in Table 1. The bulk density of raw materials is shown in Table 2. The particle size distribution of the raw materials is shown in Fig. 1.

Table 1. Chemical compositions of binders.

Compositions (%)	SiO_2	Al_2O_3	Fe_2O_3	CaO	SO_3	K_2O
Cement	18.006	4.566	5.69	63.044	3.807	1.492
Silica fume	94.599	0.496	0.175	1.228	0.332	1.166
Natural zeolite	70.785	14.016	1.95	4.394	0.022	5.906

Table 2. Bulk density of raw materials.

Raw materials	Cement	Silica fume	River sand	Quartz sand	Natural zeolite
Bulk density (g/cm^3)	1.111	0.330	1.426	1.299	0.815

2.2 Bacteria Cultivation

Bacterial zeolite particles (BZP) were prepared using natural zeolite and Sporosarcina Pasteurii. The S. Pasteurii was LMG22257 (BCCM code), acquired from China center for type culture collection (CCTCC) Adapting to Bhaskar et al. (2017)'s method, S. Pasteurii was cultivated using a nutrient media: nutrient broth powder (18.0 g/L) and urea (20.0 g/L) filtered through 0.22 μm sterile filter. The S. Pasteurii was first inoculated from a − 80 °C cryovial to a petri plate prepared with nutrient media with agar. The petri

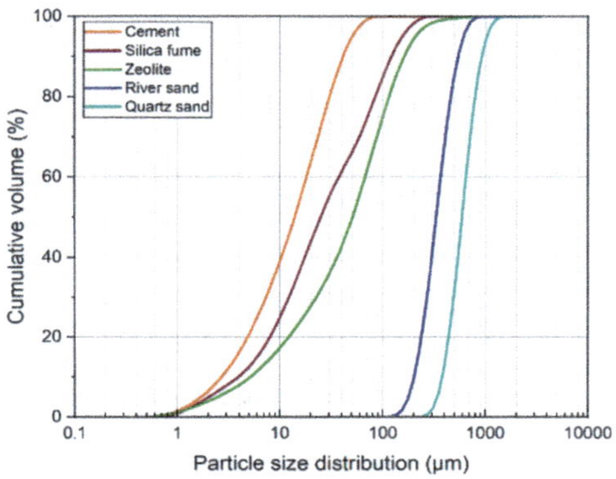

Fig. 1. Particle size distribution of raw materials.

plate was incubated at 30 °C for 24 h and checked for culture morphology before the S. Pasteurii was further inoculated to a 50 mL falcon tube with 20 mL of nutrient media. After 18 h in the shaker incubator (30 °C and 180 rpm), the fully activated S. Pasteurii was transferred from the tube to two flasks each containing 150 mL of nutrient media. After another 18 h in the shaker incubator (30 °C and 180 rpm), the OD_{600} of the flasks was tested and the S. Pasteurii solution was diluted with sterile water to acquire 750 mL of S. Pasteurii solution at OD_{600} of 0.6.

2.3 Bacterial Zeolite Particles Preparation

BZP was prepared by adding sterile natural zeolite powders to SP solution at a proportion of 300 g of zeolite and 750 mL of SP solution with an OD_{600} of 0.6. Excessive moisture was removed by vacuum drying the BZP suspension in an oven at 40 °C to constant weight. Dissolved and filtered urea (90 g), calcium nitrate (90 g) and nutrient powder (15 g) were added to the BZP powder and mixed before the suspension was vacuum dried to constant weight. The BZP was then ground to uniform sized powders for concrete mixing.

2.4 Mix Proportions

Four groups of specimens were cast and tested in this study to investigate the effect of the addition of BZP to UHPC at various levels on mechanical and durability properties. The control group contains plain natural zeolite without bacteria. The 'BZP4' group contains bio-zeolite particles at 4% of total mass, replacing quartz sand. The 'BZP6' and 'BZP8' groups contain BZP at 6% and 8% of total mass, respectively. The mix proportions are shown in Table 3.

Table 3. Mix proportions of UHPC mixtures by mass percentages (%).

Groups	Cement	SF	RS	QS	HRWR	Water	Steel fiber	Plain zeolite	BZP
Control	33.6	5.1	13.2	27.0	0.5	7.4	5.9	6.0	/
BZP4	33.6	5.1	13.2	29.0	0.5	7.4	5.9	/	4.0
BZP6	33.6	5.1	13.2	27.0	0.5	7.4	5.9	/	6.0
BZP8	33.6	5.1	13.2	25.0	0.5	7.4	5.9	/	8.0

* SF stands for silica fume, RS is river sand, QS is quartz sand, HRWR is high rank water reducer, BZP is bio zeolite particles.

2.5 Compressive Strength Test

The compressive strength was conducted following ASTM C109/C109M-24 (2024). For each group, cubic samples (50 × 50 × 50 mm) were made and tested at the age of 3d, 7d and 28d, respectively. Three duplicates were used in every test.

2.6 Ratio of Crack Area Healed by Optical Microscopy

The cracks on the cylindrical specimens (50 mm height and 100 mm diameter) were induced by split compression. The compression was removed when appropriate cracks appeared. A steel bracelet was used to restrain the specimen from being completely split under compression. The artificial cracking setup is shown in Fig. 2. The samples were cured under 12 h wet-dry cycles using Calcium Nitrate (0.2 M) and urea (0.2 M) solution. Self-healing of cracks on the surface of mixture BZP6 was observed and captured using a micro-lens at fixed angle, lighting conditions, focus distance and height. This self-healing performance was evaluated by the crack area healed using image processing techniques. The software Image Pro Plus was employed to quantify the crack area of specimens through time. Images were taken at specimen age of 1d, 3d, 7d, 14d, 28d and 56d.

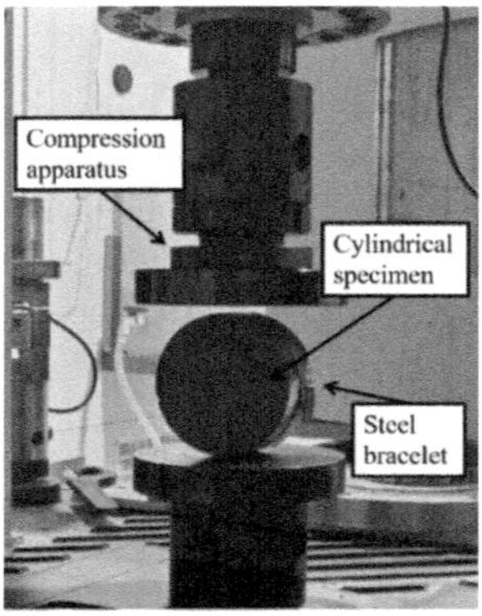

Fig. 2. Split compression setup.

2.7 Scanning Electron Microscopy (SEM) and Energy Dispersive Spectroscopy (EDS)

SEM was utilized to confirm the immobilization of SP by zeolite powder and verify the morphology of the MICP product calcite. EDS was employed to verify the chemical composition of the microbial induced carbonate precipitation. Samples were collected from the substance precipitated on the surface of UHPC specimens at the age of 28d.

3 Results and Discussion

3.1 Compressive Strength

As shown in Fig. 3, higher BZP content generally led to lower compressive strengths across all tested concrete ages. This indicates that increased biological mass within the UHPC matrix might have a negative impact on strength development. Specifically, the mixture with BZP content at 4% had the highest strength among all test groups. Compared to the control group, BZP4 showed an averaged compressive strength of 103.2%, 97.9% and 97.8% at the age of 3d, 7d and 28d, respectively. The higher strength at 3d indicated that BZP4 had faster hydration process at early age. On the other hand, BZP 8 exhibited only 84.1%, 74.0% and 88.9% of strength compared to the control group at 3d, 7d and 28d, respectively. BZP8 had the slowest strength development and the lowest strength among the test groups. This could be due to that the viable S. Pasteurii cells competed with the cementitious materials for water in the matrix that inhibited the hydration process and thus the strength development. Although the microbial induced carbonate precipitation can contribute to strength development (Bhaskar et al. 2017; Ye et al. 2025), the ultimate strength of BZP8 was still 11.1% lower than the control group. Ye et al. (2025) reported that the normal concrete incorporated with zeolite immobilized S. Pasteurii first showed an increase in compressive strength followed by a decrease, as the zeolite content increased from 10 to 50% in replacement of fine aggregate by mass. In a separate effort that used fly ash as the bacteria carrier, Jagannathan et al. (2018) also reported positive results of compressive strength of normal concrete incorporating S. Pasteurii compared to the control group. The increases could be due to that the matrix of normal concrete can provide sufficient mixing water for both bacterial activity and cementitious materials, thus it benefited from the microbial induced carbonate precipitation that strengthened the pore structure without reducing hydration intensity. Besides the biological factor, other properties of BZP that differ from natural zeolite including larger particle size, mineral coating and smaller specific surface area of BZP could also result in the decrease in compressive strength of bio-UHPC. Notably, having the same zeolite content at 6%, the compressive strength of BZP6 was 9.9% lower than that of the control group (28d), indicating that the introduction of S. Pasteurii to UHPC by zeolite immobilization can result in a minor reduction in ultimate strength.

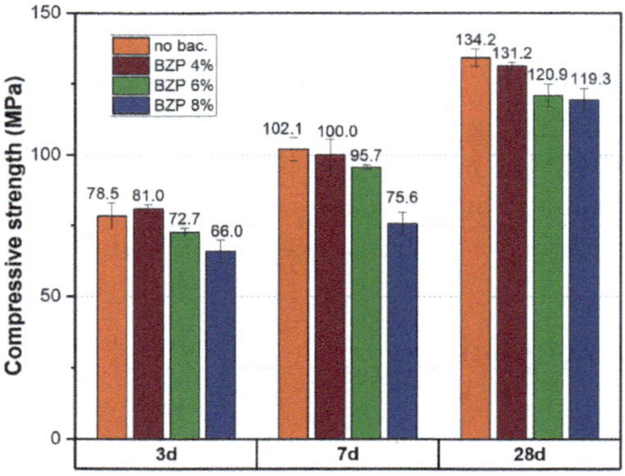

Fig. 3. Compressive strength of UHPC mixtures with different levels of S. Pasteurii content through time.

3.2 Ratio of Crack Area Healed

The crack-healing behavior has been achieved by microbial induced carbonate precipitation. To obtain a convincing and representative interpretation of the healing performance, this investigation evaluated the change in crack area instead of crack width by utilizing the image processing technique. For mixture BZP6, the evolution of crack area healing ratio within the testing period is shown in Fig. 4. By the age of 56d, both 0.30 mm and 0.38 mm cracks were completed healed where 0.58 mm and 0.94 mm cracks were only largely healed (around 80%). In a separate effort, cracks up to 0.417 mm were found to be nearly 100% closed by the age of 28d using S. Pasteurii encapsulated by calcium sulphoaluminate cement (Xu and Wang 2018). Bhaskar et al. (2017) reported that 0.10 mm cracks in cementitious mortar had been completed repaired by the S. Pasteurii immobilized in zeolite after 4 months of healing under water. Wang et al. (2022) also reported the complete healing of cracks under 0.30 mm within a week and subsequent sealing of cracks up to 0.60 mm after 4 weeks of healing. It is obvious that thinner cracks were healed faster than wider cracks in terms of area percentage, however, it is not sure that wider cracks heal slower in terms of absolute healing rate given by area healed per day (mm^2/day). Fig. 5 shows the change of absolute healing rate of cracks with different widths through time (BZP6 mixture). It is obvious that longer cracks can achieve higher absolute healing rate due to the extra length. To facilitate a reliable comparison, it was assumed that all cracks had a fixed length of 1.0 mm.. The crack area repaired every day can be calculated. It can be observed that the healing rate of all crack widths generally reduced overtime. The initial healing rate (within the first week) of larger cracks (0.58 mm and 0.94 mm) was higher than that of smaller cracks. But this phenomenon was not observed after the first 7 days. The decrease in healing rate of 0.30 mm and 0.38 mm cracks after 28d was due to the completion of healing of the cracks. The decreasing healing rate of 0.58 mm and 0.94 mm cracks indicated the reduction of bacterial activity and viability through time. Similarly, Basaran (2013) reported a rapid decrease in number of viable S. Pasteurii cells in cement paste since d0. 20% (1%),

1% (28%) and 0.4% (46%) of total viable cells (sporulation rate) survived d1, d7 and d28, respectively. Notably, in the mentioned study the S. Pasteurii was added directly to cement paste without immobilization or encapsulation (Basaran 2013). Bundur et al. (2017) and Xu and Wang (2018) also reported similar results. Considering the remaining healing rate of 0.58 mm and 0.94 mm cracks by the age of 56 days, S. Pasteurii cells immobilized by zeolite remained viable in the alkaline concrete matrix for an extended period, validating the effective protection from bacteria immobilization as confirmed by Fronczyk et al. (2023).

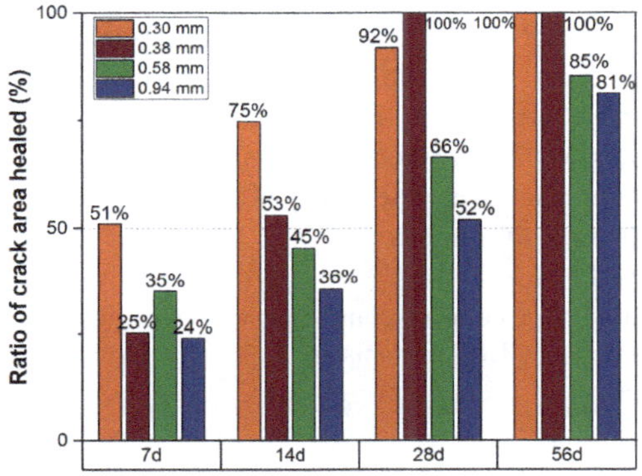

Fig. 4. Crack healing ratio of cracks with different initial widths through time.

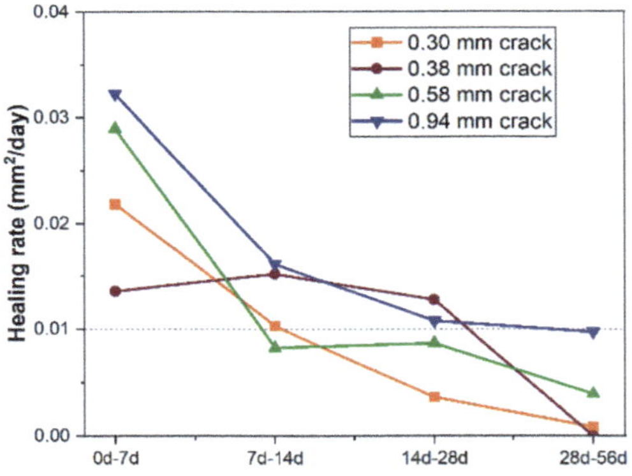

Fig. 5. Absolute healing rate of cracks with different widths through time.

3.3 Semi-Quantitative Analysis of Precipitation Using EDS

SEM and EDS were employed to evaluate the self-healing products formed at the crack area of UHPC specimens. There was a thin layer of white precipitation on the surface of the control group and a much heavier precipitation along the cracks on BZP6 samples. Both precipitations on 28d samples were analyzed using EDS mapping technique to provide an initial understanding of their chemical compositions. Table 4 shows the EDS results. It can be seen that the BZP6 precipitation had much higher Ca content (183.3%) lower O content (62.8%) than the control group. The O/Ca ratios were 3.58 and 1.23 for the control group and BZP6, respectively. These values are close to the O/Ca ratios of $Ca(NO_3)_2$ (2.4) and $CaCO_3$ (1.2) in terms of molar mass, despite the impurity and interference from urea (one of the curing media components). Also considering the higher content of C and lower content of N in BZP6 than the control group, it can be inferred that the BZP6 precipitation was $CaCO_3$-based and control group precipitation was $Ca(NO_3)_2$-based. Additionally, the BZP6 precipitation also exhibited 2.52 Wt% K content, this could be due to the zeolite presence in the collected sample, indicating the good bonding between S. Pasteurii and zeolite particles. The natural zeolite used in this study contained 5.906% Wt% of K_2O, as shown in Table 1. The well-developed calcium carbonate crystals at the crack area acted as an effective crack filler that can help prevent the ingression of corrosive substances.

Table 4. Chemical composition of precipitations on samples by EDS analysis.

Elements	Control group (Wt%)	BZP6 (Wt%)
C	12.33	18.18
N	20.18	17.61
O	52.58	33.03
Ca	14.70	26.95
Si	0.21	0.47
K	0.00	2.52
Na	0.00	1.25
Total	100.00	100.00

4 Conclusions

This study investigated the effect of incorporation of natural zeolite immobilized S. Pasteurii into a non-proprietary UHPC on mechanical property and self-healing performance by compressive strength test and crack healing ability. Four mixtures containing different levels of BZP content were prepared and assessed up to the age of 56 days. Based on the discussion of the results, the following conclusions can be drawn:

1. The introduction of S. Pasteurii to UHPC by zeolite immobilization can result in a minor reduction in the ultimate compressive strength. This can be attributed to that the viable S. Pasteurii cells competed with the cementitious materials for water in the matrix that inhibited the hydration process and thus the strength development.

Compared to natural zeolite, the larger particle size, mineral coating and smaller specific surface area of bacterial zeolite particles could also result in the decrease in compressive strength.

2. Immobilized S. Pasteurii in UHPC showed good self-healing efficiency. Cracks up to 0.38 mm can be completely healed by the age of 28d under mineral medium curing. 0.94 mm crack can be largely (81% by area) healed by the age of 56d. And a complete healing can be expected, since S. Pasteurii remained viable for an extended period under the protection of zeolite immobilization.
3. The microbial induced precipitation at the crack area of UHPC was $CaCO_3$-based substance by the EDS reports. Good bonding between S. Pasteurii and zeolite particles was achieved by vacuum impregnation. Chemical element K exclusively found in zeolite particles was observed in S. Pasteurii precipitation sample.

References

ASTM C109/C109M-24 (2024) Standard test method for compressive strength of hydraulic cement mortars (using 2-in. Or [50 mm] cube specimens). American Society for Testing and Materials, pp 1–12

Azmee NM, Shafiq N (2018) Ultra-high performance concrete: from fundamental to applications. Case Stud Constr Mater 9

Bajaber MA, Hakeem IY (2021) UHPC evolution, development, and utilization in construction: a review. J Mater Res Technol 10:1058–1074

Basaran, Z. (2013). Biomineralization in cement based materials: inoculation of vegetative cells.

Bhaskar S, Hossain KMA, Lachemi M, Wolfaardt G, Kroukamp MO (2017) Effect of self-healing on strength and durability of zeolite-immobilized bacterial cementitious mortar composites. Cem Concr Compos 82:23–33

Bundur ZB, Bae S, Kirisits MJ, Ferron RD (2017) Biomineralization in self-healing cement-based materials: investigating the temporal evolution of microbial metabolic state and material porosity. J Mater Civ Eng 29(8)

Chahal N, Siddique R, Rajor A (2012) Influence of bacteria on the compressive strength, water absorption and rapid chloride permeability of concrete incorporating silica fume. Constr Build Mater 37:645–651

De Muynck W, Debrouwer D, De Belie N, Verstraete W (2008) Bacterial carbonate precipitation improves the durability of cementitious materials. Cem Concr Res 38(7):1005–1014

Elaiopoulos K, Perraki T, Grigoropoulou E (2008) Mineralogical study and porosimetry measurements of zeolites from Scaloma area, Thrace, Greece. Microporous Mesoporous Mater 112(1–3):441–449

Ersan YC, Da Silva FB, Boon N, Verstraete W, De Belie N (2015) Screening of bacteria and concrete compatible protection materials. Constr Build Mater 88:196–203

Farzad M, Shafieifar M, Azizinamini A (2019) Experimental and numerical study on bond strength between conventional concrete and ultra high-performance concrete (UHPC). Eng Struct 186:297–305

Fronczyk J, Janek M, Szel M, Pyzik A, Franus W (2023) Immobilization of (bio-)healing agents for self-healing concrete technology: does it really ensure long-term performance? Compos Part B-Eng 266

Jagannathan P, Narayanan KSS, Arunachalam KD, Annamalai SK (2018) Studies on the mechanical properties of bacterial concrete with two bacterial species. Mater Today: Proc 5(2):8875–8879

Jogi PK, Lakshmi TVSV (2021) Self healing concrete based on different bacteria: a review. Mater Today: Proc 43:1246–1252

Kazemian M, Shafei B (2022) Internal curing capabilities of natural zeolite to improve the hydration of ultra-high performance concrete. Constr Build Mater 340

Kheir J, Klausen A, Hammer TA, De Meyst L, …De Belie, N. (2021) Early age autogenous shrinkage cracking risk of an ultra-high performance concrete (UHPC) wall: modelling and experimental results. Eng Fract Mech 257

Li LW, Zheng QF, Li Z, Ashour ARA, Han BG (2019) Bacterial technology-enabled cementitious composites: a review. Compos Struct 225

Liu JH, Farzadnia N, Khayat KH, Shi CJ (2021) Effects of SAP characteristics on internal curing of UHPC matrix. Constr Build Mater 280

Lucas S, Senff L, Ferreira VM, Aguiar JB, Labrincha JA (2010) Fresh state characterization of lime mortars with Pcm additions. Appl Rheol 20(6)

Mishra O, Singh SP (2019) An overview of microstructural and material properties of ultra-high-performance concrete. J Sustain Cem-Based Mater 8(2):97–143

Muhammad NZ, Shafaghat A, Keyvanfar A, Majid MZA et al (2016) Tests and methods of evaluating the self-healing efficiency of concrete: a review. Constr Build Mater 112:1123–1132

Nodehi M, Ozbakkaloglu T, Gholampour A (2022) A systematic review of bacteria-based self-healing concrete: biomineralization, mechanical, and durability properties. J Build Eng 49

Omoregie AI, Ngu LH, Ong DELK, Nissom PM (2019) Low-cost cultivation of Sporosarcina Pasteurii strain in food-grade yeast extract medium for microbially induced carbonate precipitation (MICP) application. Biocatal Agric Biotechnol 17:247–255

Qian CX, Wang JY, Wang RX, Cheng L (2009) Corrosion protection of cement-based building materials by surface deposition of CaCO3 by bacillus pasteurii. Mater Sci Eng C-Biomim Supramol Syst 29(4):1273–1280

Richard P, Cheyrezy M (1995) Composition of reactive powder concretes. Cem Concr Res 25(7):1501–1511

Souradeep G, Dai PS, Wei KH (2017) Autonomous healing in concrete by bio-based healing agents—A review. Constr Build Mater 146:419–428

Tayebani B, Mostofinejad D (2019) Penetrability, corrosion potential, and electrical resistivity of bacterial concrete. J Mater Civ Eng 31(3)

Vijay K, Murmu M, Deo SV (2017) Bacteria based self healing concrete—A review. Constr Build Mater 152:1008–1014

Wang J-Y, De Belie N, Verstraete W (2012) Diatomaceous earth as a protective vehicle for bacteria applied for self-healing concrete. J Ind Microbiol Biotechnol 39(4):567–577

Wang JY, Ersan YC, Boon N, De Belie N (2016) Application of microorganisms in concrete: a promising sustainable strategy to improve concrete durability. Appl Microbiol Biotechnol 100(7):2993–3007

Wang JY, Snoeck D, Van Vlierberghe S, Verstraete W, De Belie N (2014) Application of hydrogel encapsulated carbonate precipitating bacteria for approaching a realistic self-healing in concrete. Constr Build Mater 68:110–119

Wang XZ, Xu J, Wang ZP, Yao W (2022) Use of recycled concrete aggregates as carriers for self-healing of concrete cracks by bacteria with high urease activity. Constr Build Mater 337

Xu J, Wang XZ (2018) Self-healing of concrete cracks by use of bacteria-containing low alkali cementitious material. Constr Build Mater 167:1–14

Ye NS, Liu Z, Wang P, Sun YS, He XL (2025) Self-healing of concrete crack based on modified zeolite immobilizing microorganisms. Biochem Eng J 213

Zhang HR, Ji T, Zeng XP, Yang ZX, Liang YN (2018) Mechanical behavior of ultra-high performance concrete (UHPC) using recycled fine aggregate cured under different conditions and the mechanism based on integrated microstructural parameters. Constr Build Mater 192:489–507

Effects of Curing Temperature and Silicate Modulus on Compressive Strength of Volcanic Ash Geopolymer Mortars

L. Tchadjie[1], S. O. Ekolu[2], and A. Naghizadeh[3](✉)

[1] Department of Civil, Environmental and Geomatic Engineering, University College London, London, UK
l.tchadjie@ucl.ac.uk

[2] Department of Civil Engineering, Nelson Mandela University, Port Elizabeth, South Africa

[3] Department of Engineering Sciences, University of the Free State, Bloemfontein, South Africa
naghizadeha@ufs.ac.za

Abstract. This study investigated how curing temperature and silicate modulus (SiO_2/Na_2O) affect the compressive strength of geopolymer mortars. Volcanic ash was used as the aluminosilicate material, while water glass solutions with silicate moduli of 1.50, 1.63, and 1.78 were used as the activator. Geopolymer mortars were cast, then thermally treated at 23, 45, 60 and 80 °C for 24 h and kept at room temperature until the testing age. The 7-day compressive strengths of the mortars ranged from 8.7 to 31.9 MPa, while the 28-day strengths varied between 14.1 and 32.7 MPa. The maximum mortar strength was obtained at 80 °C curing temperature using an activator with a silicate modulus of 1.50. Nearly all the ultimate strength for these mortars was attained within 7 days of curing. Results show that elevated curing temperatures rapidly increase geopolymer binders or products' early-age strength development.

Keywords: Volcanic ash · Silicate modulus · Curing temperature · Geopolymer · Compressive strength

1 Introduction

Geopolymer binders result from the chemical reaction involving a fine aluminosilicate powder with an activator solution, usually potassium or sodium silicate [1]. These binders develop attractive engineering properties, including fast hardening, high early compressive strength, low shrinkage, thermal stability, long-term durability, and low carbon dioxide (CO_2) footprint. When synthesised under optimal conditions, geopolymer binders may reduce about 80% of the CO_2 released in the production of conventional Portland cement [2]. It is worth noting that the Portland cement manufacturing process accounts for nearly 8% of the global CO_2 emissions [3]. Due to the significant concerns over CO_2 emission worldwide, intensive investigations by the engineering fraternity are being carried out to develop alternative binders to Portland cement, of which geopolymer cements are foremost.

H. Beushausen et al. (Eds.): NTZR 2025, LNCE 873, pp. 246–258, 2026.
https://doi.org/10.1007/978-3-032-27698-8_23

Aluminosilicate materials of various origins, such as fly ash, metakaolin, slag, granite waste, and pyroclastic rocks, such as volcanic ash, etc., can be utilised to prepare geopolymer cements [2, 4–8]. Various investigations have been reported on the use of volcanic aluminosilicate materials for the manufacture of geopolymer cements and concretes [9–11]. Volcanic ash deposits are present in enormous quantities in many countries and can be mined at low cost [11, 12].

Generally, geopolymer binder characteristics are determined by several synthesis parameters including the quantity of reactive phases in the raw material, fineness and composition, silicate modulus of the activator, curing conditions, mixing parameters and activator-to-powder ratio [1]. Liew et al. [13] investigated the effect of powder-to-liquid ratio and of alkaline activator ratios, during the synthesis of metakaolin-based geopolymers. The powder-to-liquid ratio represents the mass ratio of the aluminosilicate raw material to alkaline solution. Water glass solution was used as the liquid activator and was prepared by combining sodium hydroxide (NaOH) with commercial water glass in different proportions to achieve a silicate modulus (SiO_2/Na_2O) varying from 0.17 to 0.33. They reported that both the powder-to-liquid ratio and SiO_2/Na_2O molar ratio significantly affected the mechanical properties of the resulting material. Maximum strength was attained at a powder-to-liquid ratio and a SiO_2/Na_2O ratio of 0.80 and 0.20, respectively. A study by Naghizadeh et al. [4] on fly ash geopolymer mortars showed that optimum compressive strength was achieved using the mixtures made with 12 M NaOH, sodium silicate to NaOH ratio of 1.5 and powder-to-liquid ratio of 0.5. Recently, Firdous et al. [14] investigated the effect of silicate modulus on the geopolymeric reactivity of six different volcanic materials. They reported that the extent of geopolymerization increased with reduction of the silicate modulus. However, the optimum silicate modulus giving the best compressive strength was related to the mineralogical composition of the volcanic aluminosilicate material used.

Bewa et al. [15] investigated geopolymer cements made using calcined lateritic soil. The activator solution used was a 10 M solution of phosphoric acid. The fresh geopolymer paste samples were cured for 24 h at different elevated temperatures comprising 23, 40, 50, 60, 70, 80, and 90 °C. They reported that increasing the curing temperature from 40 to 90 °C significantly reduced the compressive strength of geopolymer pastes. The optimum 28-day compressive strength of 83 MPa was attained with 40 mm cube paste samples stored at room temperature. In the contrary, Shoaei et al. [16] reported that increasing the curing temperature from 60 to 90 °C improved the compressive strength of waste ceramic geopolymer mortars. However, curing temperatures beyond 90 °C had a negative effect on the strength development. Falah et al. [17] observed that curing temperature and the Na_2O content of the activator solution, majorly altered the microstructural characteristics of a low-alumina mine tailings-based geopolymer binder. In their study, sodium silicate solution of varied concentrations 10, 15, 20, 25, and 30 wt%, was used as the activator. Paste samples consisting of 20 × 20 × 80 mm prisms were prepared and kept in an oven at 40 or 60 °C for different durations of up to 28 days. Results showed that increasing the concentration of liquid activator from 10 to 30 wt%, improved compressive strength and decreased the apparent porosity. A higher sodium silicate content and/or curing temperature, promotes silica dissolution from the aluminosilicate material, leading to greater geopolymerization and to a more compact microstructure. Balczàr et al. [18]

prepared geopolymer mortars made from metakaolin using 6–18 M NaOH solutions as the activator. They found that the hardening process was controlled by the concentration of the alkali. They also reported that increasing the concentration of NaOH significantly improved compressive strength, partly due to decrease in water content with increase in the alkali concentration.

The aim of this work was to evaluate the influence of curing temperature and SiO_2/Na_2O molar ratio of the liquid activator, on compressive strength development in volcanic ash geopolymer mortars. A sodium silicate activator of different silicate moduli comprising 1.50, 1.63 and 1.78 was used. Geopolymer mortars were prepared and thermally treated at 23, 45, 60 and 80 °C for 24 h, then stored at room temperature until the age of testing. Compressive strength tests were done at the ages of 7 and 28 days.

2 Methodology

2.1 Materials

Volcanic material was collected from Cameroon and used as the aluminosilicate source material for geopolymer synthesis. The volcanic material consisted of blackish particles of irregular shape with an average size of 6.5 mm. These particles were highly porous and had low density (Fig. 1). After collection, the volcanic material was oven-dried at 105 °C for 24 h, pulverised in a ball mill and sieved through a 75 µm (No. 200) sieve. The particle size distribution (PSD) of the volcanic ash material is shown in Fig. 2. The particle sizes can be seen to range between 0.1 and 1000 µm, with a mean diameter (d50) of 26.3 µm. The volume distribution (Fig. 2) of the milled volcanic ash was unimodal, indicating its homogeneity. The Blaine specific surface area of the milled material was 300 m^2/kg, and its specific gravity was 2.89. Composition of the volcanic ash material is given in Table 1. The most abundant oxides in the volcanic ash were SiO_2 (45.94%), Al_2O_3 (15.58%) and Fe_2O_3 (13.68%). Its SiO_2/Al_2O_3 molar ratio was 5.0. The negative value of the LOI of volcanic ash might be caused by oxidation of iron-bearing phases during the thermal treatment, resulting in the formation of new phases and in a gain of weight [19].

Fig. 1. Volcanic material

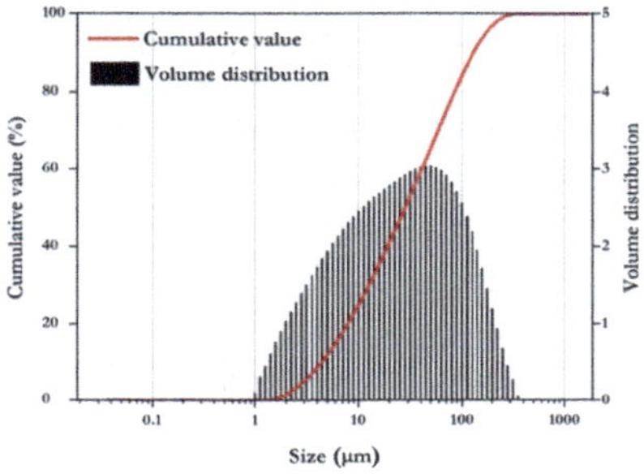

Fig. 2. Particle size distribution of the milled volcanic ash

Table 1. Chemical composition of the raw material used [9].

Oxides (wt%)	SiO_2	Al_2O_3	Fe_2O_3	CaO	TiO_2	MgO	K_2O	Na_2O	P_2O_5	LOI
Volcanic ash	45.94	15.58	13.68	9.00	2.86	7.10	1.43	3.75	0.60	−0.52

Mineralogical composition of the raw material was analyzed using an Empyrean PANALYTICAL diffractometer set at 40 kV and 40 mA. The step size used was 0.026 deg. with monochromated CuKα radiation ($\lambda = 1.54056$ Å). A scanning rate of 0.1 deg. per sec was used over the 2θ range of 5–90°. Fig. 3 gives the X-ray diffraction pattern of the volcanic ash. It shows Anorthite sodian ($Na_{1.92}Ca_{2.08}Si_{10}Al_6O_{32}$), Forsterite ($Si_4Mg_{5.68}Fe_{2.32}O_{16}$), Diopside ($Ca_{3.52}K_{0.48}Mg_{3.32}Al_{0.76}Si_{7.92}O_{24}$) and Hedenbergite ($Na_{0.40}Ca_{3.60}Fe_4Si_8O_{24}$) as the major crystalline phases of the volcanic ash. The hump spread between 20° and 40° is associated with the presence of an amorphous phase.

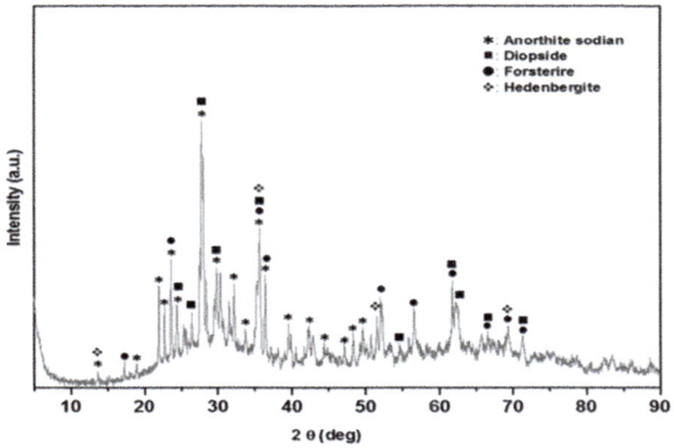

Fig. 3. XRD pattern of volcanic ash

Local silica sand was prepared by combining three different particle sizes in suitable proportions, in order to meet the requirements for standard sand as specified in EN 196-1 [20]. The resulting sand with a specific gravity of 2.65 and fineness modulus of 2.5, was used to prepare the geopolymer mortar mixtures.

The alkaline activator used was prepared by mixing the NaOH solution (SH) of various concentrations with water glass (SS) at a SS/SH mass ratio of 2.4, to achieve the silicate moduli of 1.50, 1.63 and 1.78. The activator solution prepared was kept at room temperature for at least 24 h before it was used to prepare fresh mortar mixtures. The storage duration of 24 h is necessary to stabilize the silicate species in the activator solution. Details of the activator solutions prepared, are presented in Table 2. The ratios given in Table 2 are molar ratios of SiO_2/Na_2O and H_2O/Na_2O in the alkaline solution. These ratios were calculated according to the characteristics of the SS and SH. For example, if SS contains 8.81% Na_2O and 29.05% SiO_2 by mass, 1 L of 12 M NaOH with a density of 1.39 g/mL is obtained by dissolving 480 g of NaOH pellets in distilled water. NaOH solutions were prepared by dissolving the pellets in deionized water. The SS used was of industrial grade supplied by PQ Silicas South Africa (Pty) Ltd. Its composition and physical properties are summarized in Table 3. The alkaline activator is obtained by mixing SH with SS at SS/SH mass ratio of 2.4. The SiO_2/Na_2O and H_2O/Na_2O molar ratio in the resulting mixture is obtained from Eqs. (1–7):

Mass of SH solution:

$$m_{SH} = \rho_{SH} \times V = 1.39 \times 1 \times 10^3 = 1390\, g \tag{1}$$

Mass of SS solution:

$$m_{SS} = 2.4 \times m_{SH} = 3336\, g \tag{2}$$

Total moles of Na_2O:

$$n_{Na_2O} = \frac{W_1 \times m_{ss} + \left(\frac{m_{NaOH}}{2 \times M_{NaOH}}\right)}{M_{Na_2O}} = \frac{0.0881 \times 3336 + \left(\frac{480}{2 \times 40}\right)}{62} = 10.74\ \text{mol} \tag{3}$$

Total moles of SiO_2:

$$n_{SiO_2} = \frac{W_2 \times \text{m}_{\text{SS}}}{M_{SiO2}} = \frac{0.2905 \times 3336}{60} = 16.15\ \text{mol} \tag{4}$$

Total moles of H_2O (water comes from SH and SS solutions):

$$\begin{aligned} n_{H_2O} &= \frac{(1-(W_1+W_2)) \times m_{SS} + \ (m_{SH} - m_{NaOH})}{M_{H_2O}} \\ &= \frac{(1-(0.2905+0.0881)) \times 3336 + \ (1390-480)}{18} = 165.72\ \text{mol} \end{aligned} \tag{5}$$

SiO_2/Na_2O molar ratio:

$$\frac{n_{SiO_2}}{n_{Na_2O}} = \frac{16.15}{10.74} = 1.50 \tag{6}$$

H_2O/Na_2O molar ratio:

$$\frac{n_{H_2O}}{n_{Na_2O}} = \frac{165.72}{10.74} = 15.43 \tag{7}$$

Where:

m_{SH}: Mass of SH solution (g)	W_1, W_2: Mass fractions of Na_2O and SiO_2 in SS (Table 3)
ρ_{SH} : Density of 12 M NaOH solution (g/mL)	M_{Na_2O}= 62 g/mol, M_{SiO2} = 60 g/mol, M_{H_2O} = 18 g/mol
m_{SS}: Mass of SS solution (g)	n_{Na_2O}, n_{SiO_2}, n_{H_2O}: Moles of Na_2O, SiO_2, H_2O

Table 2. Mix proportions of the activator solutions

Mix ID	Concentration of SH (M)*	SS/SH*	Oxide breakdown (%)			H_2O/Na_2O	SiO_2/Na_2O
			SiO_2	Na_2O	H_2O		
S_1	8	2.4	20.51	11.87	67.40	19.14	1.78
S_2	10	2.4	20.51	13.02	66.47	17.06	1.63
S_3	12	2.4	20.51	14.09	65.40	15.43	1.50

*SS—sodium silicate, SH—sodium hydroxide

Table 3. Composition and physical characteristics of the water glass used

Product name	SiO_2/Na_2O (Wt. ratio)	Oxide breakdown (%)			Density	Viscosity (cP)	pH
		Na_2O	SiO_2	H_2O			
Silchem 3379	3.3	8.81	29.05	62.14	1.5	378	11.73

2.2 Methods

Geopolymer mortar mixtures were prepared by mixing volcanic ash with the alkaline activator for 3 min, using a laboratory mortar mixer. Silica sand fine aggregate was then added and mixing continued for 2 min. The sand-to-volcanic ash mass ratio of 2.25 and activator-to-volcanic ash mass ratio of 0.6 were kept constant for all mixtures. The aggregate-to-volcanic ash ratio was chosen to make highly workable mixes for ease of preparation and handling. Also, Irshidat et al. [21] recommended an aggregate to binder mass ratio in the range of 2.25–2.75 for the synthesis of geopolymer mortars [21]. The mixture design details of the geopolymer mortars are shown in Table 4. The flow test was performed according to ASTM C 1437. The fresh mortars prepared were immediately poured into 50 mm cube moulds and compacted according to ASTM C 109. After casting, the cubes were wrapped in plastic bags to prevent moisture loss, then cured at 23, 45, 60 and 80 °C for 24 h. The maximum curing temperature was limited to 80 °C according to previous findings from the literature that reported the negative effects of high curing temperatures on strength development of geopolymer products [15, 16]. Thereafter, the cubes were demoulded and stored at room temperature under a relative humidity of 80–85% while sealed in plastic bags until the age of testing. It was observed that the geopolymer samples that were kept at room temperature and at 45 °C did not set within 24 h and were thus demoulded after 14 days and after 2 days, respectively.

Table 4. Mix designs of the geopolymer mortars

Activator solution	Mix-ID	Curing temperature (°C)	*VA (g)	Liquid /Powder	H_2O /Binder	Silica sand (g)	*Agg/Powder	Molar ratios in alkaline solution	
								SiO_2/Na_2O	H_2O/Na_2O
S_1	GP11	23	310	0.6	0.40	698.0	2.25	1.78	19.14
	GP12	45	310	0.6	0.40	698.0	2.25	1.78	19.14
	GP13	60	310	0.6	0.40	698.0	2.25	1.78	19.14
	GP14	80	310	0.6	0.40	698.0	2.25	1.78	19.14
S_2	GP21	23	310	0.6	0.39	698.0	2.25	1.63	17.06
	GP22	45	310	0.6	0.39	698.0	2.25	1.63	17.06
	GP23	60	310	0.6	0.39	698.0	2.25	1.63	17.06
	GP24	80	310	0.6	0.39	698.0	2.25	1.63	17.06
S_3	GP31	23	310	0.6	0.38	698.0	2.25	1.50	15.43
	GP32	45	310	0.6	0.38	698.0	2.25	1.50	15.43
	GP33	60	310	0.6	0.38	698.0	2.25	1.50	15.43
	GP34	80	310	0.6	0.38	698.0	2.25	1.50	15.43

*VA-volcanic ash, Agg-aggregates

The compressive strengths of the geopolymer specimens stored at 45, 60 and 80 °C were determined at 7 and 28 days. For mortar samples kept at room temperature, compressive strength was determined only at 28 days since the samples could not harden by the age of 7 days. In Table 4, the liquid/powder ratio is calculated as the mass ratio of the alkaline solution to volcanic ash. The H_2O/binder ratio is calculated as the mass ratio of the total H_2O in the mixture to volcanic ash. The amount of total H_2O includes H_2O from the alkaline solution and extra water, if any. In our study, no extra water was added. So, the total amount of water is obtained from Eq. (5). This explains why the H_2O/binder mass ratio reduces with decrease in the silicate modulus.

3 Results and Discussion

Fig. 4 shows the 7-day and 28-day compressive strengths of 50 mm cube mortars as influenced by the varied silicate modulus and curing temperature. The numerical values for compressive strength are presented in Table 5. It can be seen that for specimens cured at 45, 60 and 80 °C, the 7-day compressive strengths varied between 8.7 and 31.9 MPa, while the corresponding 28-day strengths ranged from 14.1 to 32.7 MPa. The maximum compressive strength obtained was 32.7 MPa, achieved at the age of 28 days with the mortar that was synthesised using an activator with a silicate modulus of 1.50, and then cured at 80 °C. In general, increasing the curing temperature improved strength development at all curing ages, irrespective of the silicate modulus. Workability, measured via flow values (Table 5), increased from 139 mm to 164 mm as the SiO_2/Na_2O molar ratio rose from 1.50 to 1.78. This trend corresponds to the higher H_2O/binder mass ratio observed at increased SiO_2/Na_2O ratios (Table 2), as the presence of water enhances the flow workability of geopolymer mortars [22]. Although all mortar mixtures were highly workable, it is important to note that, since geopolymerization is not a hydration process, excess water may adversely affect strength development in geopolymer binders [23]. Practically, higher flow aids casting and in-situ use; lower flow suits precast via vibration or admixtures.

Table 5. Compressive strengths of the geopolymer mortar mixtures made at varied silicate moduli and curing temperatures

Molar ratio in alkaline solution	Curing temperature (°C)	Workability (mm)	Compressive strength (MPa)		Gain in compressive strength (%)
SiO_2/Na_2O			7 days	28 days	
1.78	23	164	–	1.8	–
	45		6.0	14.1	61.5
	60		17.1	22.8	75.0
	80		24.0	24.2	99.3
1.63	23	148	–	1.6	–
	45		7.08	14.7	62.4
	60		18.7	21.7	86.2
	80		28.8	29.1	98.7
1.50	23	139	–	1.8	–
	45		8.3	15.5	65.8
	60		22.9	24.6	92.9
	80		30.3	31.8	97.7

It is evident in Fig. 4 that elevated temperature curing leads to an exponential increase in compressive strength as a power function, expressed in Eq. (8). This relationship aligns with studies reporting non-linear strength gain in geopolymer binders cured at elevated temperatures [15, 17]. For example, Bachtiar et al. [24] reported that the compressive strength of fly ash-based geopolymer mortars cured at 25, 65, 85 and 105 °C followed a non-linear regression model with a correlation coefficient (R^2) close to 1.

$$f_{cu} = mT^2 + bT + c \tag{8}$$

Where, $\boldsymbol{f_{cu}}$ is the compressive strength of mortar, T is the curing temperature, **m** and **b** are coefficients of T, while c is a constant. The equation is valid within the tested temperature range of 23–80 °C for volcanic ash geopolymers, where strength increases without adverse effects. Extrapolation beyond 80 °C may not hold due to potential strength reduction from microcracking [15, 17]. The coefficients m, b and c should be empirically derived for specific geopolymer mix designs and aluminosilicate sources.

Furthermore, decreasing the silicate modulus from 1.78 to 1.50 for mortar samples cured at 45, 60 and 80 °C led to improvements in compressive strength. Moreover, the 28-day strength values obtained for the samples cured at room temperature were very low, ranging from 1.6 to 1.8 MPa.

The increase in strength with higher curing temperatures may be attributed to improvement in the rate of dissolution of the reactive aluminosilicate phases during geopolymerization. This process leads to the formation of a greater amount of polymeric gel, which, in turn, forms a more compact microstructure. A similar trend was reported by Sindhunata et al. [25], who noted the beneficial influence of heat curing on strength development in fly ash geopolymer binders. Similarly, Rovnanik [26] observed that elevated curing temperatures of 60 and 80 °C promoted the formation of a hardened microstructure in metakaolin-based geopolymers, especially at the early age.

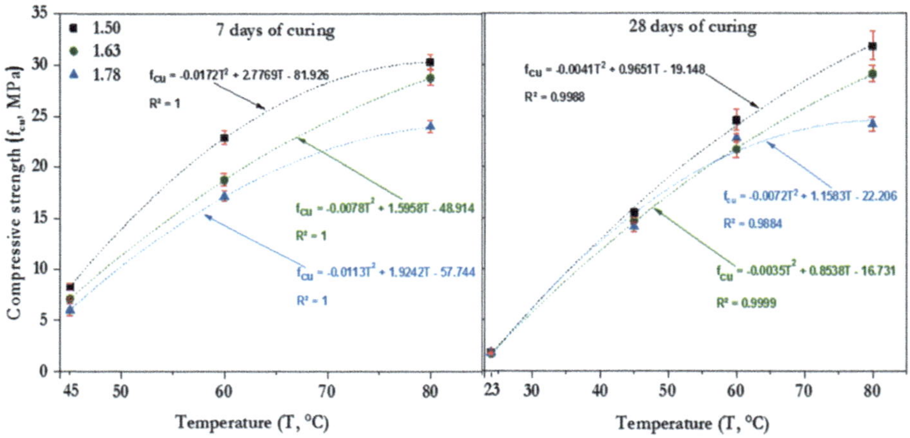

Fig. 4. Compressive strengths of mortar specimens prepared at varied silicate moduli and cured at varied temperatures

The improvement in compressive strength with decrease in silicate modulus of the activator solutions may be explained by the higher Na_2O content, which consequently increases the alkalinity of the reactional system, thus promoting the dissolution of volcanic ash particles. Moreover, the SiO_2/Na_2O molar ratio in water glass is a key parameter that has a major influence on geopolymerization kinetics and the resulting characteristics of the binder product. It has been reported that greater amounts of monomeric silicate species are formed when the SiO_2/Na_2O molar ratio decreases [27]. The silicate species formed react easily with aluminate species to generate the polymeric network [28, 29]. Recently, Firdous et al. [14] also mentioned an enhancement in the degree of geopolymerization as the silicate modulus decreased.

The low strength development for mortar samples cured at 23 °C relative to those cured at 45, 60 and 80 °C, shows the slow geopolymeric reactivity of volcanic ash used in the study, regardless of the silicate modulus. These observations also indicate the predominant effect of curing temperature over that of the silicate modulus, during geopolymerization of the materials of low reactivity.

Figure 5 illustrates the gain in compressive strength of geopolymer mortars over time, expressed as a percentage of the 28-day strength. The numerical values for the gain in compressive strength are presented in Table 5. For samples cured at 45 and 60 °C, the 7-day strength was between 61.5 and 92.9% of the 28-day strength, compared with 97.7–99.3% for mortars cured at 80 °C.

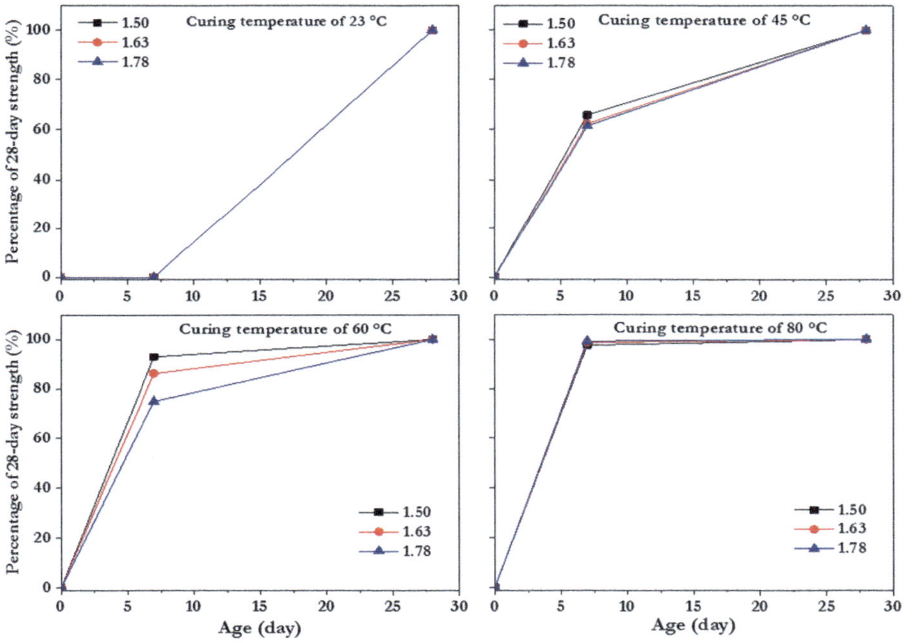

Fig. 5. Gain in compressive strength of geopolymer mortars at varied silicate moduli and curing temperatures

These results show that geopolymer mortars cured at 80 °C almost attained their ultimate 28-day compressive strength in 7 days. Under these curing conditions, it appears that the polymeric structure formation is generally completed within 7 days. These observations also agree with the findings of Rovnanik [26]. High early strength development, as observed in the present study, favours using geopolymer binders for rapid construction works and concrete precasting.

4 Conclusion

Compressive strength development in volcanic ash geopolymer mortars synthesised at a varied silicate modulus of the alkali activator and cured at various elevated temperatures was investigated in this study. Results showed that volcanic ash exhibits very slow geopolymeric reactivity at room temperature, regardless of the activator's silicate modulus. However, elevated curing temperatures rapidly increased compressive strength. Moreover, the effect of curing temperature in promoting geopolymerization was greater than that of the silicate modulus. Maximum mortar strength was attained at 80 °C curing temperature using an activator with a silicate modulus 1.50. The 7-day compressive strength was 97.7–99.3% of the 28-day compressive strength, implying that almost all the ultimate strength was attained within 7 days. The observed rapid early strength development is beneficial for applying the volcanic ash geopolymer binders in precasting and fast-track construction works or projects. Future research should investigate long-term durability properties, such as shrinkage, water resistance and acid resistance, of these optimised volcanic ash geopolymer mortars to further validate their potential for construction applications.

Acknowledgements. The National Research Foundation (NRF) of South Africa in collaboration with The World Academy of Sciences (TWAS), sponsored this research through the Doctoral Scholarship grant No. 99993. Leonel N. Tchadjie wishes to thank the NRF-TWAS program for this study grant. The authors would like to thank PQ Silicas South Africa (pty) Ltd. for supporting this research by donating the sodium silicate activator used in the study.

References

1. Tchadjie LN, Ekolu SO (2018) Enhancing the reactivity of aluminosilicate materials toward geopolymer synthesis. J Mater Sci 53:4709–4733
2. Davidovits J (2011) Geopolymer chemistry and applications, 3rd edn. Institut Géopolymère, France
3. Andrew RM (2018) Global CO_2 emissions from cement production. Earth Syst Sci Data 10:1–52
4. Naghizadeh A, Ekolu SO (2018) Effect of mix parameters on strength of Geopolymer mortars-experimental study. In: Sixth international conference on durability of concrete structures. Leeds. United Kingdom, pp 315–320
5. Naghizadeh A, Welman-Purchase M, Lagrange L, Ekolu SO, Tchadjie LN (2025a) Re–activation of recycled fly ash–based geopolymer binder to enhance mechanical properties. Case Stud Constr Mater:e04967
6. Naghizadeh A, Ekolu SO, Welman-Purchase M, Lagrange L, Tchadjie LN (2025b) Effective recycled binder produced upon thermal activation to enhance mechanical properties of fly ash–based geopolymer mortars. Dev Built Environ:100688
7. Naghizadeh A, Ekolu SO (2019) Method for comprehensive mix design of fly ash geopolymer mortars. Constr Build Mater 202:704–717
8. Tchadjié LN, Djobo JNY, Ranjbar N, Tchakouté HK, Kenne BBD, Elimbi A (2016) Potential of using granite waste as raw material for geopolymer synthesis. Ceram Int 42:3046–3055
9. Tchadjie LN, Ekolu SO (2018) Effect of bauxite content and fineness on compressive strength of volcanic ash-based geopolymer mortars. In: International conference on interdisciplinary approaches for cement-based materials and structural concrete. RILEM publications S.A.R.L., Funchal, Portugal, pp 419–424
10. Djobo JNY, Elimbi A, Tchakoute HK, Kumar S (2016) Reactivity of volcanic ash in alkaline medium, microstructural and strength characteristics of resulting geopolymers under different synthesis conditions. J Mater Sci 51:10301–10317

11. Tchakoute HK, Elimbi A, Yanne E, Djangang CN (2013) Utilisation of volcanic ashes for the production of geopolymers cured at ambient temperature. Cem Concr Compos 38:75–81
12. Ekolu SO, Thomas MDA, Hooton RD (2006) Studies on Ugandan volcanic ash and tuff. In: Proceedings of the first international conference on advances in engineering and technology. Elsevier Ltd, Entebbe (Uganda), pp 75–83
13. Liew YM et al (2012) Optimisation of solids-to-liquid and alkali activator ratios of calcined kaolin geopolymeric powder. Constr Build Mater 37:440–451
14. Firdous R, Stephan D (2019) Effect of silica modulus on the geopolymerization activity of natural pozzolans. Constr Build Mater 219:31–43
15. Bewa CN, Tchakouté HK, Rüscher CH, Kamseu E, Leonelli C (2019) Influence of the curing temperature on the properties of poly (phospho—ferro—siloxo) networks from laterite. SN Appl Sci 1
16. Shoaei P, Musaeei HR, Mirlohi F, Narimani Zamanabadi S, Ameri F, Bahrami N (2019) Waste ceramic powder-based geopolymer mortars: effect of curing temperature and alkaline solution-to-binder ratio. Constr Build Mater 227:116686
17. Falah M, Obenaus-Emler R, Kinnunen P, Illikainen M (2019) Effects of activator properties and curing conditions on alkali-activation of low-alumina mine tailings. Waste Biomass Valorization:1–13
18. Balczár I, Korim T, Dobrádi A (2015) Correlation of strength to apparent porosity of geopolymers—understanding through variations of setting time. Constr Build Mater 93:983–988
19. Vandenberghe RE, De Resende VG, Da Costa GM, De Grave E (2010) Study of loss-on-ignition anomalies found in ashes from combustion of iron-rich coal. Fuel 89:2405–2410
20. Ekolu SO (2014) Potential south African standard sand for cement mortar testing and research. In: International conference on advances in construction materials and structures, Johannesburg, pp 253–260
21. Irshidat MR, Abdel-Jawad YA, Al-Sughayer R (2018) Feasibility of producing sustainable geopolymer composites made of locally available natural pozzolan. J Mater Cycles Waste Manag 20:1751–1760
22. Sathonsaowaphak A, Chindaprasirt P, Pimraksa K (2009) Workability and strength of lignite bottom ash geopolymer mortar. J Hazard Mater 168:44–50
23. Nuruddin MF, Demie S, Shafiq N (2012) Effect of mix composition on workability and compressive strength of self-compacting geopolymer concrete. Can J Civ Eng 1203:1196–1203
24. Bachtiar E, Marzuki I, Erdawaty MF, Junaedy F, Ramadhan AM, Setiawan FR (2020) The relationship of temperature and compressive strength on geopolymer mortar using fly ash-based. IOP Conf Ser Mater Sci Eng 885:0–9
25. Sindhunata JSJ, VanDeventer GC, Lukey H, Xu (2006) Effect of curing temperature and silicate concentration on Fly-ash-based Geopolymerization. Ind Eng Chem Res 45:3559–3568
26. Rovnaník P (2010) Effect of curing temperature on the development of hard structure of metakaolin-based geopolymer. Constr Build Mater 24:1176–1183
27. Svensson IL, Sjoberg S, Ohman L-O (1986) Polysilicate equilibria in concentrated sodium silicate solutions. J Chem Soc Faraday Trans 1(82):3635
28. Vidal L et al (2016) Controlling the reactivity of silicate solutions: a FTIR, Raman and NMR study. Colloids Surfaces A Physicochem Eng Asp 503:101–109
29. Weng L, Sagoe-Crentsil K (2007) Dissolution processes, hydrolysis and condensation reactions during geopolymer synthesis: part I-low Si/Al ratio systems. J Mater Sci 42:2997–3006

Effect of Slag Incorporation on Durability Index Performance of Ambient-Cured Fly Ash Geopolymer Concrete

A. Naghizadeh[1](✉), C. Wu[2], and L. Tchadjie[3]

[1] Department of Engineering Sciences, University of the Free State, Bloemfontein, South Africa
naghizadeha@ufs.ac.za

[2] Institute of Port, Coastal, and Offshore Engineering, Ocean College, Zhejiang University, Zhoushan, China

[3] Department of Civil, Environmental and Geomatic Engineering, University College London, London, UK

Abstract. This study investigates the influence of ground granulated blast furnace slag (GGBS) incorporation on the mechanical and durability performance of ambient–cured fly ash–based geopolymer concrete. Two types of sodium silicate–based alkali activators, distinguished by their silicate modulus (SiO_2/Na_2O = 1.31 and 1.43), were used to prepare concrete and paste mixtures with varying GGBS contents (0–40%). Setting time, compressive strength, and durability indices, including Oxygen Permeability Index (OPI), Water Sorptivity Index (WSI), Chloride Conductivity Index (CCI), and total porosity, were systematically evaluated. The results demonstrated that mixtures activated with the lower silicate modulus (MS = 1.31) exhibited superior long–term strength, achieving compressive strengths up to 90.8 MPa at 90 days. Incorporation of GGBS significantly reduced setting times and enhanced strength development, attributed to the synergistic co–formation of N–A–S–H and C–A–S–H gels. Durability performance improved notably with increasing GGBS content, with all mixtures exceeding OPI, WSI, and CCI thresholds for good to excellent durability. However, diminishing returns were observed beyond 30% GGBS content, indicating an optimal balance between strength gain and matrix densification. These findings underscore the critical role of both activator composition and GGBS dosage in optimising the performance of geopolymer concretes under ambient curing conditions.

Keywords: Geopolymer concrete · Sorptivity · Oxygen permeability · Chloride conductivity · Durability index · GGBS · Fly ash

1 Introduction

Replacing ordinary Portland cement (OPC) with low–carbon binders has emerged as a key objective within the construction industry due to growing concerns over the environmental footprint of OPC. Alternative eco–friendly binders such as geopolymers have gained significant attention as viable, low–carbon substitutes [1, 2]. Geopolymer binders

H. Beushausen et al. (Eds.): NTZR 2025, LNCE 873, pp. 259–273, 2026.
https://doi.org/10.1007/978-3-032-27698-8_24

are synthesised inorganic polymers that can be derived through alkali activation of aluminosilicate materials [3]. While the manufacture of alkali activators entails energy input, the overall carbon emissions associated with geopolymer production are substantially lower than those of OPC [2]. This environmental benefit is primarily attributed to the elimination of lime calcination from the production process of the binder material, the use of industrial by-products rather than virgin raw materials, as well as higher recyclability capacity compared to normal cement [4–6]. Beyond sustainability, geopolymer concretes often exhibit superior mechanical and durability performance, contingent upon the type of precursor material and curing protocol applied [7–9].

In fact, numerous studies have reported superior durability characteristics of geopolymer concretes relative to their OPC counterparts, particularly in terms of resistance to chemical attack, thermal degradation, and permeability [10]. Hamidi et al. [11] demonstrated that increasing the concentration of sodium hydroxide in fly ash–based geopolymers enhanced flexural strength up to 12 M, beyond which excessive OH^- ions led to strength deterioration. The effect of silicate modulus (SiO_2/Na_2O molar ratio) was explored by Gao et al. [12] for metakaolin–based geopolymers. Results indicated that higher silicate modulus prolonged setting time and increased compressive strength up to a ratio of 1.5. However, beyond this value, increased viscosity impaired workability and reduced strength.

Bellum et al. [13] found that incorporating up to 40% GGBS improved compressive strength and reduced porosity, though workability declined with higher slag content. In another study, Nagajothi et al. [14] reported enhanced resistance of fly ash/GGBS-based geopolymer to sulphate and acid exposure, along with reduced water absorption relative to OPC concrete. However, higher chloride ion penetration was observed in the geopolymer system. Despite the promising attributes of geopolymer binders, research on the long–term durability performance of fly ash–GGBS systems remains limited. Notably, durability index (DI) testing, a robust methodology routinely applied to OPC concretes, has not yet been extensively used to evaluate geopolymer concretes [15–18].

The objective of this study was to evaluate the influence of GGBS incorporation on the performance characteristics of fly ash–based geopolymer concrete, with a particular focus on DI performance, compressive strength, and setting time. To this end, geopolymer concrete mixtures were formulated using varying proportions of GGBS and different types of alkali activators. Durability–related performance was assessed using standardised DI tests, including water sorptivity, oxygen permeability, and chloride conductivity. Microstructural analysis was conducted using scanning electron microscopy (SEM) to provide insight into the morphological characteristics of the binder matrix.

2 Experiments

2.1 Materials

A low–calcium (Class F) fly ash was utilised as the primary aluminosilicate precursor in the formulation of geopolymer concrete, either individually or blended with GGBS. The fly ash had a Blaine fineness of 207.3 m^2/kg and a specific gravity of 2.24. The chemical compositions of both fly ash and GGBS are provided in Table 1.

To activate the precursor materials, a binary alkali activator comprising sodium silicate and sodium hydroxide (NaOH) solutions was employed. The sodium silicate solution had a silicate modulus (SiO_2/Na_2O) of 3.3 and contained 36% solids by weight. SA polycarboxylate-based superplasticiser (density: 1.07 kg/L; pH: 5.5) was added to improve the workability of the fresh mixtures. Natural aggregates were used in all mixtures.

Table 1. Chemical composition of the fly ash and GGBS materials used in the mixtures.

Oxides (%)	Al_2O_3	CaO	Fe_2O_3	MgO	P_2O_5	K_2O	SiO_2	Na_2O	TiO_2	LOI*
Fly ash	30.27	4.59	3.58	1.06	0.38	0.77	56.45	0.14	1.57	0.42
GGBS	15.77	34.54	0.84	8.21	0.01	1.02	41.36	–	0.58	−0.57

*LOI—loss on ignition

2.2 Preparation of Concrete Mixtures

Two distinct alkali activator solutions, designated as *LS* and *HS*, were formulated with silicate moduli (MS) of 1.31 and 1.43, respectively. These were prepared by blending sodium hydroxide solutions of varying concentrations with sodium silicate in specific proportions. The resulting activators had total solids contents of 34.6% and 31.4%, respectively. The formulation approach was adapted from prior research [15]. Geopolymer concrete mixtures were prepared using a combination of fine and coarse aggregates, along with binder systems comprising fly ash alone or blended with GGBS. The proportion of GGBS was 0, 20, 30 or 40% of the total raw material. Detailed mixture proportions are provided in Table 2. The fresh concrete mixtures were cast in 100 mm size cube moulds and cured at room temperature for an additional 27 days.

At 28 days, compressive strength tests were performed on concrete cube specimens in accordance with ASTM C109 [19]. For each mix, three cubes were tested to determine the average strength. Additionally, four cubes per mixture were cored and cut into disc–shaped specimens (70 mm diameter × 30 mm thickness) for DI testing according to the South African National Standard SANS 3001-CO3-1 [16]. Oxygen Permeability Index (OPI) and Water Sorptivity Index (WSI) tests were conducted in accordance with SANS 3001-CO3-2 [17], while Chloride Conductivity was evaluated based on SANS 3001-CO3-3 [18]. Total porosity was measured separately following the ASTM C642 [19] test method.

Table 2. Properties of geopolymer concrete mixtures.

Mix ID	Alkali activator		Raw material		Aggregate		Additive
	LS (g)*	*HS* (g)*	Fly ash (g)	GGBS (g)	Stone (g)	Sand (g)	SP** (g)
0%LS	1600	0	3200	0	7623	6937	64
20%LS	1600	0	2560	640	7623	6937	64
30%LS	1600	0	2240	960	7623	6937	64
40%LS	1600	0	1920	1280	7623	6937	64
0%HS	0	1600	3200	0	7623	6937	64
20%HS	0	1600	2560	640	7623	6937	64
30%HS	0	1600	2240	960	7623	6937	64
40%HS	0	1600	1920	1280	7623	6937	64

**LS*—sodium silicate alkali activator with $SiO_2/Na_2O = 1.31$ and solids content = 34.6%, *HS*—sodium silicate alkali activator with $SiO_2/Na_2O = 1.43$ and solids content = 31.4%

In parallel, geopolymer paste mixtures were prepared to evaluate the effects of alkali activator type and GGBS content on setting time. To maintain consistency with the binder phase of the concrete mixtures, the activator–to–binder (A/B) ratio was fixed at 0.5, while the proportion of FA: GGBS was varied (i.e., 100:0, 80:20, 70:30, and 60:40). Also, paste mixtures were prepared manually, and setting time was measured using an automatic Vicat apparatus in accordance with ASTM C191 [21]. Table 3 presents the detailed proportions of the paste mixtures.

Table 3. Properties of geopolymer paste mixtures.

Mix ID	Alkali activator		Raw material	
	LS (g)*	*HS* (g)*	Fly ash (g)	GGBS** (g)
P-0%LS	150	0	300	0
P-20%LS	150	0	240	60
P-30%LS	150	0	210	90
P-40%LS	150	0	180	120
P-0%HS	0	150	300	0
P-20%HS	0	150	240	60
P-30%HS	0	150	210	90
P-40%HS	0	150	180	120

*Properties of alkali activators have been given in Table 2
*GGBS – ground granulated blast-furnace slag

3 Results and Discussion

3.1 Setting Times

Fig. 1 presents the initial and final setting times of fly ash–based geopolymer paste mixtures incorporating varying amounts of GGBS, specifically 0%, 20, 30, and 40%, activated using two different alkali activators. Across all GGBS levels, mixtures activated with *HS activator* consistently exhibited shorter setting times compared to those prepared with *LS activator*.

This reduction in setting time with *HS activator* can be attributed to its higher silicate modulus ($SiO_2/Na_2O = 1.43$) in comparison to *LS activator* (1.31). These findings are consistent with results reported by Ling et al. [22], who observed that increasing the SiO_2/Na_2O ratio from 1.0 to 1.5 significantly decreased the setting time of FA/GGBS–based systems. In the current study, the lower silicate modulus and solids content of the *LS activator* are associated with increased solution viscosity [23]. This higher viscosity impedes the dissolution of aluminosilicate species from both fly ash and GGBS during the early stages of geopolymerization, thereby delaying the setting process [24]. It is important to note, however, that although *HS activator* resulted in faster initial reactivity, *LS activator* produced mixtures that demonstrated superior long–term strength development, suggesting more sustained geopolymerization beyond the initial setting phase.

The incorporation of GGBS into the paste mixtures had a significant influence on setting time, regardless of activator type. Adding 20% GGBS to fly ash paste mixtures reduced the initial setting time from 1900 min to 388 min with *LS activator*, and from 1390 min to 255 min with *HS activator*. Further increases in GGBS content to 30% and 40% led to additional, albeit smaller, reductions in setting time. For mixtures using *LS activator*, the initial setting times dropped to 371 and 229 min for 30% and 40% GGBS, respectively. Corresponding values for the *HS activator* were 163 and 134 min, respectively.

The pronounced decrease in setting time with increasing GGBS content is largely due to the high calcium (Ca) content in GGBS, which accelerates early–stage solidification. Generally, geopolymer binders predominantly form sodium aluminosilicate hydrate (N–A–S–H) gel. However, when GGBS is included, calcium silicate hydrate (C–S–H) gel can also form alongside N–A–S–H in the initial reaction stages, contributing to faster setting [25, 26]. Consequently, higher GGBS contents lead to shorter initial and final setting times due to this dual gelation mechanism [27].

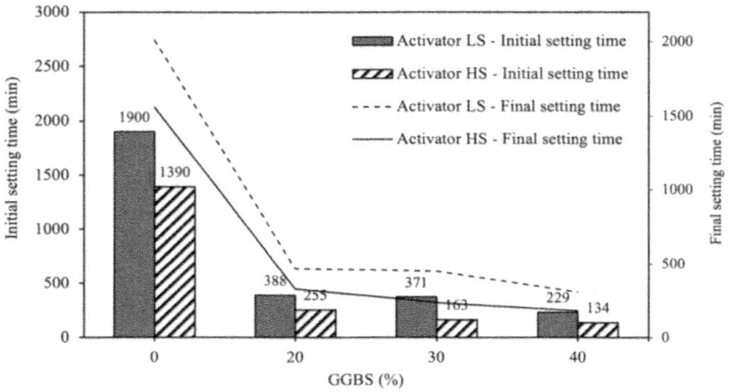

Fig. 1. Effect of GGBS content on initial and final setting time of fly ash geopolymer concrete mixtures activated using: *LS*—$SiO_2/Na_2O = 1.31$ and solids content = 34.6%, *HS*—$SiO_2/Na_2O = 1.43$ and solids content = 31.4%

3.2 Compressive Strength

Figure 2 illustrates the 28-day and 90-day compressive strength results of ambient–cured fly ash–based geopolymer concrete mixtures incorporating varying contents of GGBS and activated using two distinct alkali activators with differing silicate moduli (MS). Overall, the results underscore the significant influence of both the alkali activator composition and GGBS incorporation on the mechanical performance of the geopolymer concretes.

It was observed that mixtures activated using the low-silicate modulus activator (*LS*, with MS = 1.31) consistently outperformed those activated with the high–silicate modulus solution (*HS*, with MS = 1.43), irrespective of GGBS content or curing duration. For instance, control mixtures without GGBS (0%LS and 0%HS), which were solely based on fly ash, recorded 28-day compressive strengths of 58.7 MPa and 53.9 MPa, respectively. Upon extending the curing period to 90 days, these strengths increased to 83.1 MPa and 70.1 MPa. These findings are in alignment with previous research [15, 26], which reported enhanced strength development at lower silicate modulus values due to increased alkalinity. This increase in alkalinity, stemming from a higher Na_2O content, promotes the more extensive dissolution of the aluminosilicate source material, thereby accelerating geopolymerization and resulting in more extensive gel formation and denser microstructure.

In addition to the activator effect, the incorporation of GGBS had a pronounced positive impact on compressive strength across all mixtures. A clear trend was observed where increasing GGBS content led to notable strength gains at both 28 and 90 days, regardless of the type of activator used. At the highest GGBS content of 40%, the 90-day compressive strengths reached 90.77 MPa for the 40%LS mixture and 97.57 MPa for the 40%HS mixture. The corresponding 28-day values were 83.3 and 72.5 MPa, respectively. The strength development with GGBS addition is attributed to the synergistic effect of calcium, which not only participates in the early formation of calcium-aluminosilicate-hydrate (C–A–S–H) gel but also complements the typical sodium-aluminosilicate-hydrate (N–A–S–H) gel matrix formed from fly ash activation.

These findings reinforce the well-documented role of calcium-rich precursors such as GGBS in improving the reaction kinetics and microstructure of alkali-activated systems [12]. The co-formation of C–A–S–H and N–A–S–H gels leads to a more compact and interconnected binder matrix, which translates into enhanced mechanical performance [27–30]. Furthermore, the results reveal that while both activators benefit from GGBS addition, the *LS* activator offers better long–term performance due to its higher alkalinity and increased driving force for continued geopolymerization.

These trends collectively underscore the importance of optimising both the activator composition and GGBS dosage to balance early–age and long-term strength development in ambient-cured geopolymer concretes.

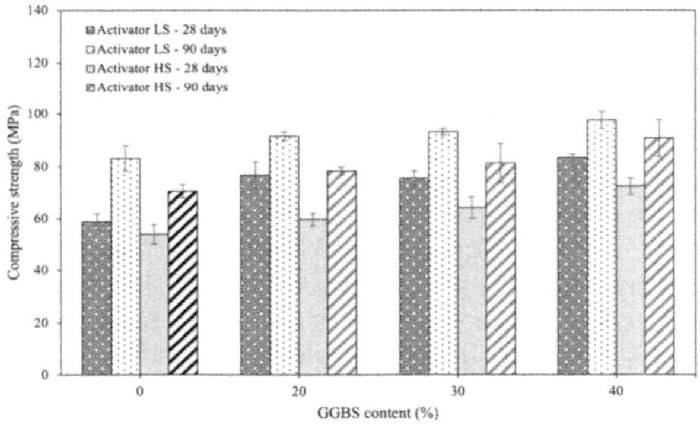

Fig. 2. 28—day and 90—day compressive strength of geopolymer concrete samples containing different GGBS contents and activated using different alkali activators of: (a) *LS* - SiO_2/Na_2O = 1.31 and solids content = 34.6%, (b) *HS* - SiO_2/Na_2O = 1.43 and solids content = 31.4%.

3.3 Durability Indexes

Figures 3, 4, 5 and 6 present the DI results of the control geopolymer concrete specimens, including the OPI, WSI, and CCI. These indices are widely used to characterise concrete in terms of permeability, capillary absorption, and ionic diffusion resistance, respectively [31].

Across all mixtures, OPI values exceeded the recommended durability threshold of 9.5, irrespective of SiO_2/Na_2O ratio or GGBS content. A marginal increase in OPI was observed as GGBS content rose from 0 to 30%, suggesting improved microstructural densification and reduced oxygen permeability. However, a slight decline in OPI was recorded at 40% GGBS, indicating that excessive GGBS content may introduce heterogeneities that compromise uniform pore structure. Despite this reduction, all OPI values still fell within the range classified as "good to excellent durability" [31]. This trend mirrors the compressive strength results, where gains between 30% and 40% GGBS were also relatively minor (less than 8%).

WSI results exhibited a nonlinear response to GGBS content. For mixtures activated with *LS* activator ($SiO_2/Na_2O = 1.31$), WSI decreased from 3.9 mm/h$^{0.5}$ at 0% GGBS to 3.1 mm/h$^{0.5}$ at 20%, reflecting improved resistance to capillary suction. However, further GGBS additions increased WSI to 3.6 mm/h$^{0.5}$ at 30% and 4.4 mm/h$^{0.5}$ at 40%, indicating a possible reduction in matrix compactness. A similar trend was observed in samples that were made using *HS* activator ($SiO_2/Na_2O = 1.43$), where WSI declined from 3.6 mm/h$^{0.5}$ at 0% to 3.2 mm/h$^{0.5}$ at 20%, remained stable at 3.7 mm/h$^{0.5}$ at 30%, and rose to 4.9 mm/h$^{0.5}$ at 40%. Although all mixtures remained below the 6 mm/h$^{0.5}$ limit for "excellent" durability [31]. However, as highlighted by Moore et al. [32], WSI cannot be interpreted in isolation from porosity due to their interdependence. Total porosity (Fig. 6, Table 4) decreased consistently with increasing GGBS content, from 13.4 to 10.2% for LS mixtures and 13.3–11.6% for HS mixtures. The most significant reductions occurred up to 30% GGBS, after which only marginal gains were observed, suggesting a plateau in matrix densification. The rise in WSI beyond 30% GGBS despite lower porosity suggests enhanced pore connectivity [32]. This observation is supported by SEM analysis (Fig. 7), which shows a denser matrix at 40% GGBS, likely contributing to refined but more interconnected pores. These findings are in line with previous research highlighting the role of GGBS in enhancing gel homogeneity and reducing permeability through densified C–A–S–H and N–A–S–H phases [33, 34].

CCI values decreased substantially with increasing GGBS content for both activator types. As shown in Fig. 5, CCI dropped from 3.0 mS/cm (*LS*) and 4.0 mS/cm (*HS*) at 0% GGBS to 1.0 mS/cm for both activators at 30 and 40% GGBS. According to Alexander et al. [31], CCI values below 1.5 mS/cm indicate good resistance to chloride ingress. This benchmark was achieved by all mixtures incorporating $\geq$ 30% GGBS. These reductions are attributed to the refined pore structure and matrix densification facilitated by GGBS incorporation [35, 36]. Reduced porosity and connectivity, as well as the formation of dense reaction gels, effectively limit ion mobility. The comparable CCI values at 30 and 40% suggest a saturation point beyond which additional GGBS yields diminishing returns in chloride resistance. While these results align with compressive strength and durability trends, it is important to consider the influence of free alkalis and distinct pore chemistry in geopolymer systems, which differ from Portland cement–based concretes [15].

The key performance indicators of all geopolymer concrete mixtures, including setting time, compressive strength, durability indices, and total porosity, are summarised in Table 4, providing a direct comparison of trends and the identification of the optimal GGBS content.

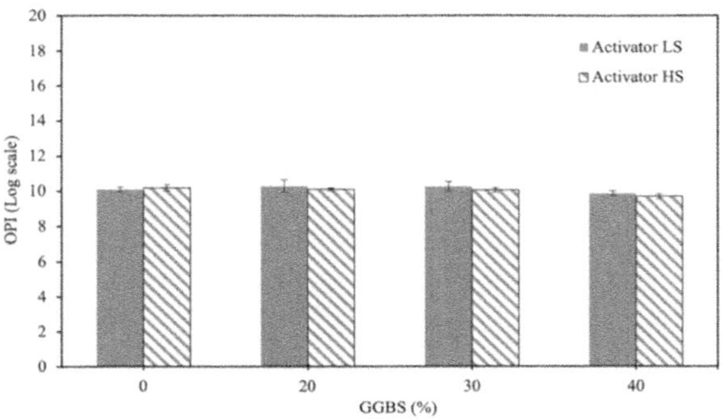

Fig. 3. Effect of GGBS content on oxygen permeability index (OPI) of fly ash geopolymer concrete mixtures activated using: *LS* – SiO_2/Na_2O = 1.31 and solids content = 34.6%, *HS* – SiO_2/Na_2O = 1.43 and solids content = 31.4%.

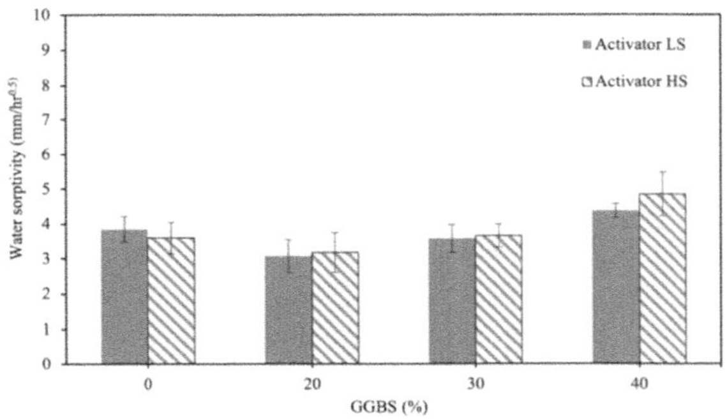

Fig. 4. Effect of GGBS content on water sorptivity index (WSI) of fly ash geopolymer concrete mixtures activated using: *LS* – SiO_2/Na_2O = 1.31 and solids content = 34.6%, *HS* – SiO_2/Na_2O = 1.43 and solids content = 31.4%.

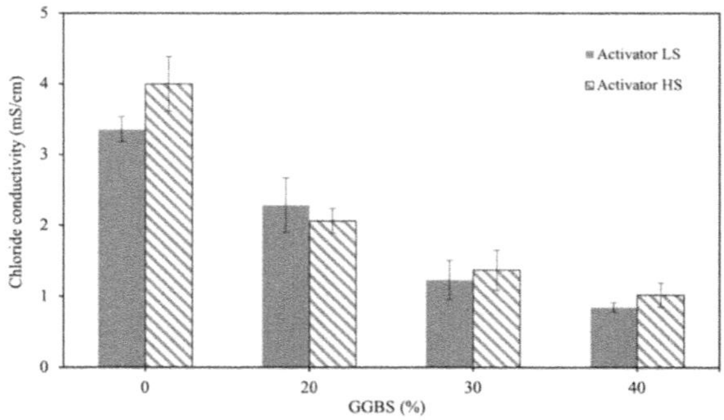

Fig. 5. Effect of GGBS content on chloride conductivity index (CCI) of fly ash geopolymer concrete mixtures activated using: *LS* – $SiO_2/Na_2O = 1.31$ and solids content = 34.6%, *HS* – $SiO_2/Na_2O = 1.43$ and solids content = 31.4%.

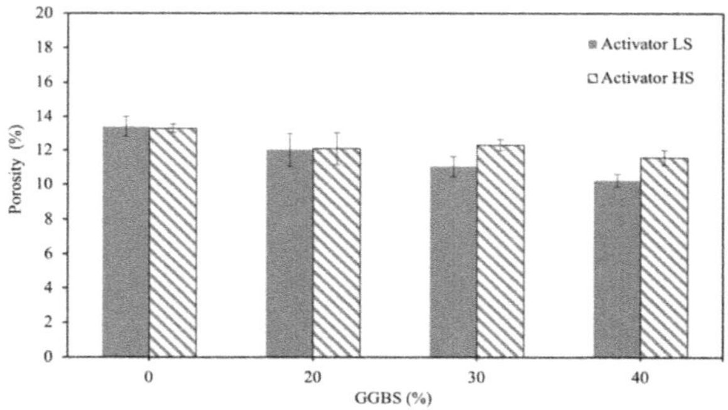

Fig. 6. Effect of GGBS content on porosity of fly ash geopolymer concrete mixtures activated using: *LS* – $SiO_2/Na_2O = 1.31$ and solids content = 34.6%, *HS* – $SiO_2/Na_2O = 1.43$ and solids content = 31.4%

Table 4. Summary of the geopolymer concrete mixtures' performance

Mix ID	Initial setting time (min)	28-day compressive strength (MPa)	90-day compressive strength (MPa)	OPI	WSI (mm/h0.5)	CCI (mS/cm)	Total porosity (%)
0%LS	1900	58.7	83.1	9.6	3.9	3.0	13.4
20%LS	388	70	85	9.7	3.1	2.0	11.5
30%LS	371	77	88	9.8	3.6	1.0	10.5
40%LS	229	83.3	90.8	9.6	4.4	1.0	10.2
0%HS	1390	53.9	70.1	9.5	3.6	4.0	13.3

(continued)

Table 4. (*continued*)

Mix ID	Initial setting time (min)	28-day compressive strength (MPa)	90-day compressive strength (MPa)	OPI	WSI (mm/h0.5)	CCI (mS/cm)	Total porosity (%)
20%HS	255	63	80	9.6	3.2	2.0	12.5
30%HS	163	68	85	9.7	3.7	1.0	11.8
40%HS	134	72.5	97.6	9.6	4.9	1.0	11.6

**LS*—sodium silicate alkali activator with $SiO_2/Na_2O = 1.31$ and solids content = 34.6%, *HS*—sodium silicate alkali activator with $SiO_2/Na_2O = 1.43$ and solids content = 31.4%

4 Scanning Electron Microscopy (SEM) Analysis

Figure 7 presents the SEM micrographs of ambient–cured fly ash geopolymer concrete specimens prepared without GGBS (Fig. 7a—0%LS) and with 40% GGBS replacement (Fig. 7b—40%LS), both activated using the LS solution ($SiO_2/Na_2O = 1.31$). The microstructure of the 0%LS sample reveals a relatively coarse binder matrix interspersed with unreacted fly ash particles, which is a typical feature of fly ash–based geopolymer systems due to incomplete dissolution under ambient conditions.

In contrast, the 40%LS mixture exhibits a significantly denser and more homogeneous matrix, with a noticeably lower fraction of unreacted fly ash. This refined microstructure can be attributed to the incorporation of GGBS, which promotes the formation of calcium-rich reaction products such as C–S–H and C–A–S–H gels, in addition to the conventional geopolymeric N–A–S–H gel [34, 35]. These additional gel phases contribute to enhanced binder packing, reduced pore connectivity, and the development of a denser network.

These microstructural observations corroborate the mechanical and durability performance trends previously discussed. As shown in Figs. 1 and 6, the inclusion of 40% GGBS resulted in notable increases in compressive strength and a marked reduction in total porosity. Similar trends have been reported in previous studies [26, 27], further supporting the beneficial role of GGBS in enhancing the microstructural integrity of ambient–cured geopolymer binders.

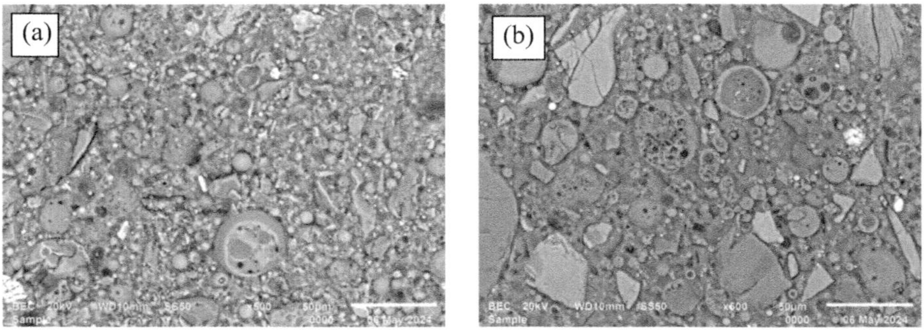

Fig. 7. Scanning electron microscopy (SEM) features of geopolymer concrete samples of (a) 0%LS, (b) 40%LS.

5 Conclusions

This study comprehensively examined the effects of GGBS content and alkali activator type on the setting time, mechanical strength, and durability of ambient–cured fly ash geopolymer concrete. The findings can be summarised as follows:

The inclusion of GGBS significantly accelerated setting, especially when used with high–silicate modulus activators, due to the formation of calcium–rich phases that facilitate early–stage solidification.

Mixtures with low–silicate modulus activator (MS = 1.31) consistently achieved higher long–term compressive strength than those with higher modulus (MS = 1.43). Strength gains were proportional to GGBS content up to 40%, with peak 90-day values exceeding 97 MPa.

Incorporation of GGBS notably enhanced durability by reducing permeability, sorptivity, and chloride conductivity. The most substantial improvements were observed up to 30% GGBS content, beyond which further enhancements were marginal. Total porosity decreased with increasing GGBS content, supporting observed improvements in durability and strength. A densification plateau was reached beyond 30% GGBS, suggesting limited benefits from higher replacement levels.

Overall, findings show that partial substitution of fly ash with GGBS in geopolymer concretes can significantly enhance mechanical and durability performance under ambient conditions. An optimal GGBS content of approximately 30% is recommended for achieving balanced performance without compromising matrix integrity or cost efficiency.

These findings also have practical implications for large-scale and structural applications. The enhanced strength and durability at the optimum GGBS content suggest that the developed geopolymer concrete can meet structural-grade requirements. Ambient curing and the use of industrial by-products enable feasible on-site or precast production without energy-intensive thermal curing.

References

1. Maddalena R, Roberts JJ, Hamilton A (2018) Can Portland cement be replaced by low–carbon alternative materials? A study on the thermal properties and carbon emissions of innovative cements. J Clean Prod 186:933–942
2. Turner LK, Collins FG (2013) Carbon dioxide equivalent (CO_2) emissions: a comparison between geopolymer and OPC cement concrete. Constr Build Mater 43:125–130
3. Kaze RC, Naghizadeh A, Tchadjie L, Cengiz Ö, Kamseu E, Chinje FU (2024) Formulation of geopolymer binder based on volcanic-scoria and clay brick wastes using rice husk ash-NaOH activator: fresh and hardened properties. Sustain Chem Pharm 40:101627
4. Naghizadeh A, Welman-Purchase M, Lagrange L, Ekolu SO, Tchadjie LN (2025a) Re–activation of recycled fly ash–based geopolymer binder to enhance mechanical properties. Case Stud Constr Mater e04967
5. Naghizadeh A, Ekolu SO, Welman-Purchase M, Lagrange L, Tchadjie LN (2025b) Effective recycled binder produced upon thermal activation to enhance mechanical properties of fly ash–based geopolymer mortars. Dev Built Environ 100688
6. Naghizadeh A, Tchadjie LN, Ekolu SO, Welman-Purchase M (2024) Circular production of recycled binder from fly ash-based geopolymer concrete. Constr Build Mater 415:135098
7. Hardjito D, Wallah SE, Sumajouw DMJ, Rangan BV (2004) Brief review of development of geopolymer concrete, George Hoff symposium. American Concrete Institute, Los Vegas, USA, pp 1–10
8. Naghizadeh A, Ekolu SO (2017) Pozzolanic materials and waste products for formulation of geopolymer cements in developing countries: a review. J Concr Soc South Afr 151:22–31
9. Das SK et al (2020) Fresh, strength and microstructure properties of geopolymer concrete incorporating lime and silica fume as replacement of fly ash. J Build Eng 32:101780
10. Duxson P, Fernandez-Jimenez A, Provis JL, Lukey GC, Palomo A, van Deventer JSJ (2007) Geopolymer technology: the current state of the art. J Mater Sci 42(9):2917–2933
11. Hamidi RM, Man Z, Azizli KA (2016) Concentration of NaOH and the effect on the properties of fly ash based geopolymer. Procedia Eng 148:189–193
12. Gao K et al (2014) Effects SiO_2/Na_2O molar ratio on mechanical properties and the microstructure on nano—SiO_2 metakaolin—based geopolymers. Constr Build Mater 53:503–510
13. Bellum RR, Al Khazaleh M, Pilla RK, Choudhary S, Venkatesh C (2022) Effect of slag on strength, durability and microstructural characteristics of fly ash-based geopolymer concrete. J Build Pathol Rehabil 7(1):25
14. Nagajothi S, Elavenil S, Angalaeswari S, Natrayan L, Mammo WD (2022) Durability studies on fly ash based geopolymer concrete incorporated with slag and alkali solutions. Adv Civ Eng 7196446
15. Naghizadeh A, Ekolu SO, Tchadjie LN, Solomon F (2023) Long-term strength development and durability index quality of ambient-cured fly ash geopolymer concretes. Constr Build Mater 374:130899
16. SANS (South African National Standard). SANS 3001-CO3-1 2015. Civil engineering test methods part CO3–1: concrete durability index testing. Preparation of test specimens. Pretoria: SABS Standards Division
17. SANS (South African National Standard) 3001-CO3-2 2015. Civil engineering test methods part CO3-2: concrete durability index testing: oxygen permeability test. Pretoria: SABS Standards Division
18. SANS (South African National Standard) 3001-CO3-3 2015. Civil engineering test methods part CO3–3: concrete durability index testing: chloride conductivity test. Pretoria: SABS Standards Division

19. ASTM C 642 (2013) Standard test method for density, absorption, and voids in hardened concrete. American Society for Testing and Materials, 1–3
20. ASTM C 109 (2020) Standard Test Method for Compressive Strength of Hydraulic Cement Mortars (Using 2-in. or [50-mm] Cube Specimens). American Society for Testing and Materials, 1–11. https://doi.org/10.1520/C0109_C0109M-20
21. ASTM C 191 (2021). Standard test methods for time of setting of hydraulic cement by Vicat needle. American Society for Testing and Materials, 1–8
22. Ling Y, Wang K, Wang X, Hua S (2019) Effects of mix design parameters on heat of geopolymerization, set time, and compressive strength of high calcium fly ash geopolymer. Constr Build Mater 228:116763
23. Deb PS, Nath P, Sarker PK (2014) The effects of ground granulated blast-furnace slag blending with fly ash and activator content on the workability and strength properties of geopolymer concrete cured at ambient temperature. Mater Des 1980–2015(62):32–39
24. Malkawi AB, Nuruddin MF, Fauzi A, Almattarneh H, Mohammed BS (2016) Effects of alkaline solution on properties of the HCFA geopolymer mortars. Procedia Eng 148:710–717
25. Saha S, Rajasekaran C (2017) Enhancement of the properties of fly ash based geopolymer paste by incorporating ground granulated blast furnace slag. Constr Build Mater 146:615–620
26. de Klerk D, Naghizadeh A, Ekolu SO, Welman-Purchase M (2025) Recycled cement use to produce fly ash–based geopolymer binders suitable for ambient curing: comparison with slag effects. Constr Build Mater 468:140394
27. Magugu K, Ekolu S, Naghizadeh A, Quainoo H (2025) Mixture-related factors influencing setting for ambient curing of Fly ash–based Geopolymer binders. Comput Eng Phys Model 8(1):50–70
28. Bernal SA, Rodríguez ED, Mejia de Gutiérrez R, Provis JL, Delvasto S (2012) Activation of metakaolin/slag blends using alkaline solutions based on chemically modified silica fume and rice husk ash. Waste Biomass Valorization 3:99–108
29. Hasnaoui A, Ghorbel E, Wardeh G (2019) Optimization approach of granulated blast furnace slag and metakaolin based geopolymer mortars. Constr Build Mater 198:10–26
30. Lemougna PN et al (2020) Effect of slag on the improvement of setting time and compressive strength of low reactive volcanic ash geopolymers synthetized at room temperature. Mater Chem Phys 239:122077
31. Alexander MG, Mackechnie JR, Ballim Y (1999) Guide to the use of durability indexes for achieving durability in concrete structures. Research monograph 2:102
32. Moore AJ, Bakera AT, Alexander MG (2021) A critical review of the water Sorptivity index (WSI) parameter for potential durability assessment: can WSI be considered in isolation of porosity? J South Afr Inst Civ Eng 63(2):27–34
33. Zhu Z, Huo W, Sun H, Ma B, Yang L (2023) Correlations between unconfined compressive strength, sorptivity and pore structures for geopolymer based on SEM and MIP measurements. J Build Eng 67:106011
34. Yilmazoglu A, Yildirim ST, Yıldız S, Behçet ÖF (2024) Effect of GGBS content and water/geopolymer solid ratio on the mechanical, elevated temperature resistance, and sorptivity properties of FA/GGBS-based geopolymer concrete. J Mater Civ Eng 36(4):04024032
35. Hasan MJ, Hossain KMA (2021) Assessing suitability of geopolymer composites under chloride exposure. In: Canadian society of civil engineering annual conference. Springer Nature Singapore, Singapore, pp 375–387
36. Duży P, Hager I, Choińska-Colombel M, Amiri O (2023) Chloride diffusion and mechanical performances of Geopolymer concrete with blended precursor. In: International congress on polymers in concrete. Springer Nature Switzerland, Cham, pp 220–229

Effect of Calcination Temperature on Mechanical and Microstructural Properties of Gibbsite-Rich Laterite-Based Geopolymer Binder

Rodrigue Cyriaque Kaze[1,2](✉) and Abdolhossein Naghizadeh[2]

[1] Department of Minerals Engineering, School of Chemical Engineering and Mineral Industries (EGCIM), University of Ngaoundéré, Ngaoundéré, Cameroon
kazerodrigue@gmail.com

[2] Department of Engineering Sciences, University of the Free State, Bloemfontein, South Africa

Abstract. Gibbsite-rich laterite soil was thermally activated at 300, 400, 500, and 700 °C with a constant heating rate of 5 °C/min and then held at the peak temperature for 2 h to enhance its reactivity for geopolymer binder synthesis. Each thermally activated sample was mixed with an alkaline solution (sodium silicate with silica modulus of 1.4) at calcined powder-to-alkaline solution ratio of 0.6. The resulting geopolymer pastes were cured for 28 days at room temperature (23 ± 5 °C), followed by testing of flexural strength, water absorption, porosity and bulk density. Results showed that flexural strength increased from 5.23 to 11.45 MPa, when calcination temperature increased from 300 to 500 °C, while further increase in calcination temperature to 700 °C led to reduction in strength to 6.12 MPa. The samples made using gibbsite-rich laterite soil calcined at 500 °C exhibited significant formation of sodium aluminosilicate hydrate (N-A-S-H) gel, leading to a densified microstructure with relatively low water absorption (7.60%) and porosity (12.02%). The observed decrease in mechanical performance due to increase in calcination temperature beyond 500 °C is attributed to the onset of particle sintering, which coarsens the material and hinders dissolution. These results suggest that geopolymers derived from optimally calcined laterite soil hold promise for use in structural and engineering applications.

Keywords: Laterite · Gibbsite · Heat treatment · Mechanical properties · Microstructure

1 Introduction

Laterite soils found in tropical and subtropical zones around the world are mainly composed of sesquioxides such as hydrated iron and aluminium oxides. These soils are formed from alteration of kaolinite in which a high proportion of Al atoms in the kaolinite mineral in octahedra site are replaced by iron Fe^{2+} or Fe^{3+} providing by iron minerals like ilmenite and hematite [1, 2]. This phenomenon, known as laterization, makes the laterite soil vulnerable to chemical reagents due low the crystallinity of kaolinite as reported

H. Beushausen et al. (Eds.): NTZR 2025, LNCE 873, pp. 274–284, 2026.
https://doi.org/10.1007/978-3-032-27698-8_25

in previous study [3–5]. Some studies reported that alkali activation of raw laterite (non-calcined), has shown superior mechanical strength and stability compared to raw kaolin clay under similar conditions [4–6]. Several studies were conducted on thermally activated laterite soils for the geopolymer synthesis. The obtained characteristics depended on the synthesis condition. Kaze et al. [7] varied the effect of calcination temperature from 500 to 700 °C on hardened properties of iron-rich laterite based geopolymer. Their results showed that calcination of iron-rich laterite at optimum temperature of 500 °C resulted in the highest flexural strength of 6.0 MPa. Recently, Venyite et al. [3] reported that the maximum compressive strength of pure raw laterite was 18.96 MPa, while it was increased to 29 MPa with incorporation of 20 wt% metakaolin as a co-binder.

The current study investigates the feasibility of using thermally activated gibbsite-rich laterite soil at 300, 400, 500, and 700 °C as solid precursor materials for the synthesis of geopolymer binders using sodium silicate alkaline activator, as well as to understand the effect of calcination temperature on the physicochemical and mechanical properties of the resulting binders. The geopolymerization products were characterized using setting time, and environmental scanning electron microscopy (ESEM). Mechanical properties were measured using flexural strength, water absorption, porosity, and bulk density. Laterite-based geopolymer characteristics are interpreted in terms of microstructure and flexural strength. In addition, the structure of the produced laterite-based geopolymer binders were characterized using X-ray diffraction (XRD) and Fourier transform infrared spectroscopy (FTIR) techniques.

2 Experimental Program

2.1 Materials and Method Characterization

Gibbsite-rich laterite soil used as a precursor in geopolymer binders was characterized in previous work [3]. It was mainly composed of kaolinite, quartz, gibbsite, goethite and hematite as the main minerals. The collected laterite was oven dried at 105 °C for 24 h and then heated at 300, 400, 500 and 700 °C for 2 h at peak temperature in an electrical furnace, while the heating rate was kept constant at 5 °C/min. Each calcined powder was mixed with a sodium silicate solution at a liquid-to-solid ratio 0.6 with an electrical mixer for 3 min. The resulting fresh paste was poured into silicon mould having diameter 100 × 20 × 20 mm. After 24 h samples were demoulded and then stored at room temperature until mechanical testing was performed at 28 days. All mixtures were prepared with a calcined laterite-to-alkaline solution mass ratio of 0.6 (500 g of laterite and 300 g of alkaline solution) for each calcination temperature (300, 400, 500, and 700 °C).

The chemical bonds within the geopolymer binders were determined using a Thermo Scientific Nicolet 6700 FTIR equipment with a wavenumber range of 40–4000 cm^{-1}. While mineral phases were identified using a Bruker D8 ADVANCE XRD operating, with a 2θ range of 5° to 70° and a scanning speed of 5°/min. The microstructural observation was carried out using JEOL IT300LV SEM coupled with energy-dispersive spectroscopy (EDS), with an acceleration voltage of 40.0 kV. The samples were coated with a thin Pt layer to improve their surface conductivity.

3 Results and Discussion

3.1 Phase Evolution

Figure 1 shows the infrared spectra of geopolymer made with thermal laterite soil at 300, 400, 500 and 700 °C. The lower absorption bands appearing at 3694 and 3617 cm^{-1} in all infrared spectra are attributed to vibrational modes of O–H bonds belonging to kaolinite suggesting their incomplete transformation into metakaolinite [8–10]. The absorption bands located at interval of 3388–3430 cm^{-1} and 1630–1644 cm^{-1} are assigned to stretching modes of O-H belonging to water molecule linked to geopolymer network [8–10]. The absorption bands appearing between 1410 and 1430 cm^{-1} are attributed to the stretching modes of C–O bonds. These absorption bands are formed from the reaction of CO_2 with free Na^+ in the geopolymer network [11, 12]. The pronounced absorption bands located at 983, 978, and 961 cm^{-1} are assigned to stretching modes of Si-O-Al and Fe-O-Si-O-Al bonds belonging to geopolymer network [13, 14]. These bands also confirms that geopolymerization occurred during the synthesis [15]. The bands appearing at 798, 739, 652 and 532 cm^{-1} are assigned to vibration modes of Al-O bending, stretching of Si-O bonds, Si-O-Fe bonds [14, 16, 17]. The mineral crystalline phases identified on the diffractograms recorded on hardened geopolymers are reported in Fig. 2. All the X-ray patterns are dominated with the reflection peaks of crystalline phases. The identified mineral phases are kaolinite (JCPD, $Al_2O_3{\cdot}2SiO_2\ 2H_2O$), quartz (JCPD, SiO_2), Goethite (JCPD, FeOOH), Hematite (JCPD, Fe_2O_3) and illite (JCPD, $K,Mg,Fe)Al_2Si_3AlO(OH)_2$). The reflection peaks of kaolinite mineral are well pronounced on samples made with thermally activated laterite at 300, 400 and 500 °C. This indicates their non total conversion upon heating. Therefore, this remained kaolinite would have reacted with alkaline solution to form hydrated hydrosodalite through the Eq. (1) as reported in previous studies using untreated clay as main solid precursor for the geopolymer synthesis [8, 10].

$$3(Al_2O_3 \cdot 2SiO_2 2H_2O) + 4Na_2SiO_3 \rightarrow Na8[AlSiO_4]_6(OH)_2 \cdot nH_2O) \quad (1)$$

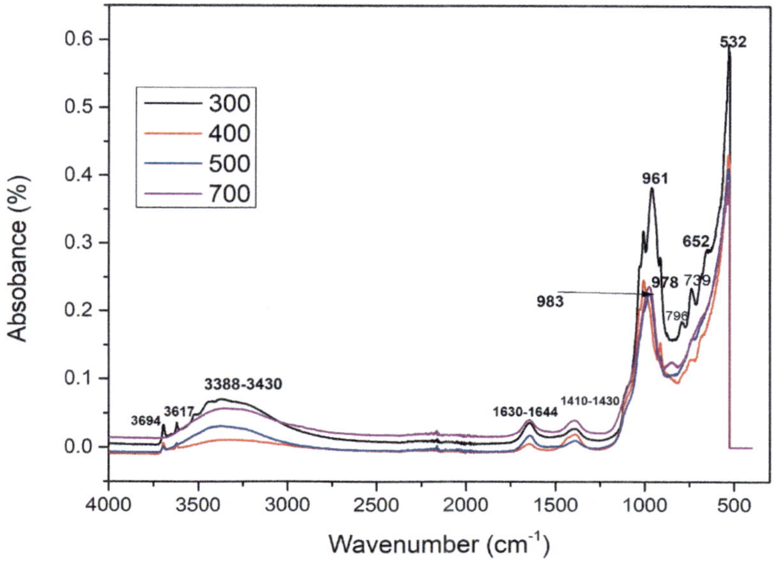

Fig. 1. Infrared spectra of geopolymer binders made using thermally activated laterite calcined at 300, 400, 500 and 700 °C.

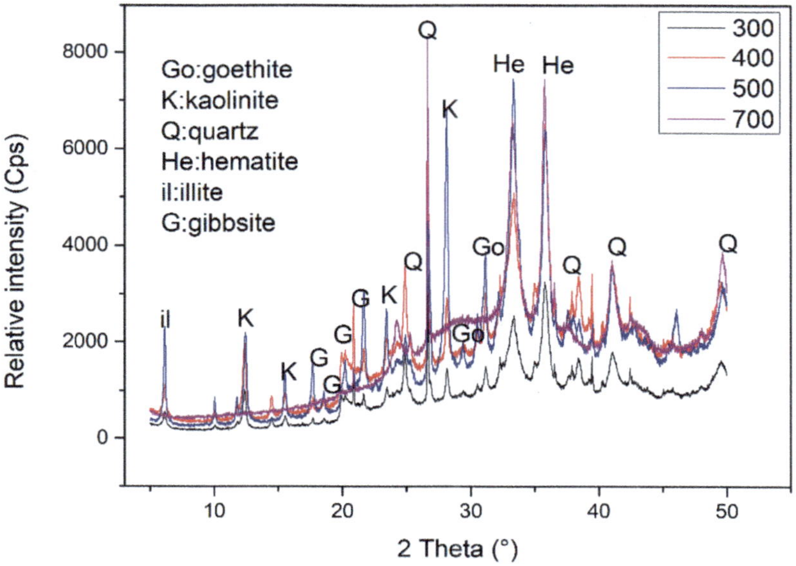

Fig. 2. X-ray patterns of consolidated geopolymer made using thermally activated laterite calcined at 300, 400, 500 and 700 °C.

3.2 Mechanical Properties

Figure 3 displays the flexural strength of hardened geopolymer cured for 28 days. The samples made using laterite calcined at 300, 400, 500 and 700 °C gave strength values of 5.23, 7.13, 11.6 and 8.45 MPa, respectively. It can be seen that increase in the calcination temperature up to 500 °C improved the strength development indicating the progressive formation of reactive phase within the calcined powders favourable for the geopolymerization reaction [18]. The optimum strength of 11.6 MPa was achieved on sample heated at 500 °C suggesting the most released of Al- and Si-oligomers favourable for high polycondensation ensuring high compacity with few voids justifying the high strength. The observed strength loss due to increase in calcination temperature beyond 500 °C suggest the onset of sintering in laterite particles which had not dissolve in alkaline medium as reported in [7]. Further studies such as wet dry cycling is required to evaluate the stability of such cementitious systems. The decreased strength of the samples containing laterite that was calcined at 700 °C could be linked to excessive released Al-species as a result of conversion of gibbsite into Al_2O_3. Therefore, the greater release of Al species in alkaline medium would affect the Si/Al molar ratio known as one of the key parameters governing the geopolymerization reaction as reported in [19, 20]. Hence, the high Si/Al ratio inhibits the precursor's dissolution and release in an alkaline environment, resulting in a large number of unreacted laterite particles in geopolymer products. The agglomeration phenomenon interferes with the geopolymer reaction, reducing the final strength of the geopolymer products. This can be confirmed using an electron microscope, as seen in Fig. 4, which reveals a higher quantity of unreacted laterite particles. The low strength recorded samples heated at 300 and 400 °C are in agreement with other studies suggesting the lower transformation of kaolinite into metakaolinite required for the development of high strength [16, 21].

Overall, the laterite-based binders achieved flexural strengths ranging between 5 and 12 MPa are lower than the 15–25 MPa typically reported for metakaolin-based geopolymers [19]. This lower performance is linked to the heterogeneous and partially crystalline nature of laterite, which limits amorphous aluminosilicate availability. Nonetheless, its low cost and abundance make laterite a promising precursor for applications requiring moderate strength, and performance could be enhanced by blending with reactive materials such as metakaolin or volcanic ash.

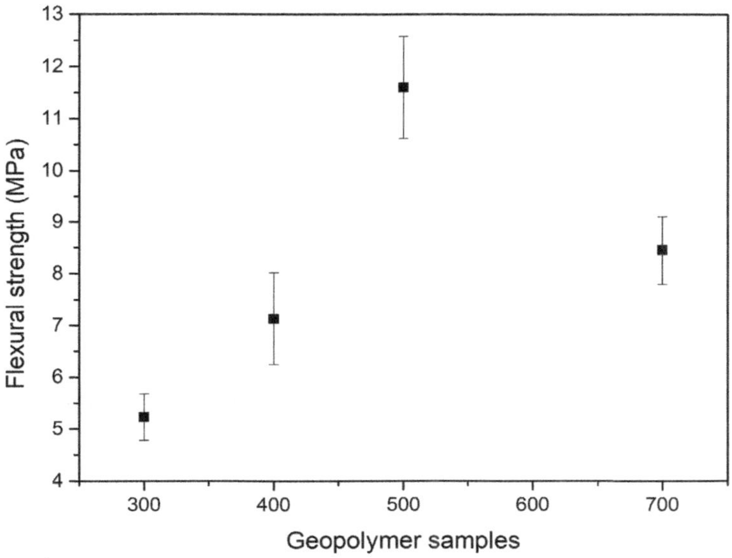

Fig. 3. Flexural strength of hardened laterite-based geopolymer cured at 28 days.

3.3 Microstructure

Figure 4 displays the SEM images of consolidated geopolymer binders cured for 28 days. It can be seen that all the micrographs appear heterogeneous, with coarse particles dispersed throughout within the geopolymer matrix. This indicates the complexity of mineral phases present in this type of clay in comparison with standard kaolin clay. These coarse grains appearing within the different geopolymer matrices likely related to the formation of mineral phases such as quartz and hematite, as discussed in Sect. 3.1. However, the micrographs of geopolymer samples made with thermally activated laterite at 500 (Fig. 4c) and 700 °C (Fig. 4d) seem exhibiting relatively fewer voids and pores compared to those made at 300 (Fig. 4a) and 400 °C (Fig. 4b). This is in line with the flexural strength development, indicating an enhancement in sodium aluminosilicate hydrate (N-A-S-H) formation upon increase in calcination temperature to 500 and 700 °C. Alkali activation of laterite soils calcined at 500 or 700 °C developed densified binder structures allowing better connectivity between different components within the geopolymer matrix. Furthermore, the samples containing laterite calcined at 500 °C or 700 °C exhibited undissolved mineral phases such as quartz and hematite. This observation, also reported in previous studies, indicates their resistance or partial reactivity in alkaline environments. In other research, the presence of such unaltered minerals was found to enhance the strength and densify the geopolymer matrix, thereby contributing to improved performance [22]. The progressive densification observed from micrographs with increased calcination temperature suggests the thermal decomposition of gibbsite mineral [23] releasing the Al-species in alkaline which reacted with Si-species favourable for polycondensation establishing the Si-O-Al bonds. However, an excessive amount of aluminum involved in the geopolymerization process can adversely affect the reaction, leading to a reduction in the strength of the hardened binder. This was likely the case for samples containing laterite calcined at 700 °C.

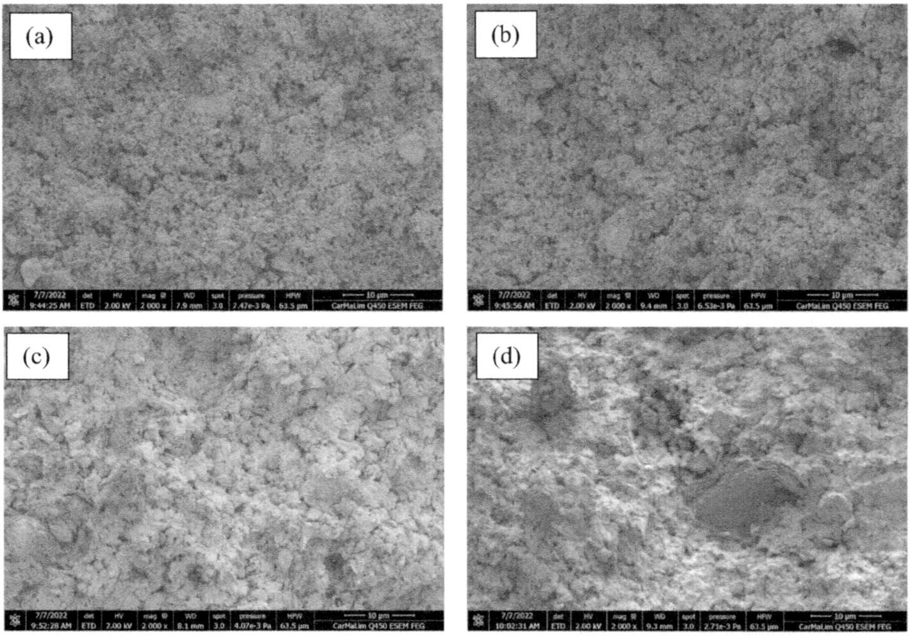

Fig. 4. SEM images of geopolymer binder made of thermally activated laterite at 300 (**a**), 400 (**b**), 500 (**c**) and 700 (**d**) cured at 28 days

3.4 Water Absorption and Porosity

Figure 5 shows the water absorption and porosity values of the calcined laterite-based geopolymers cured at 28 days. It is observed that both water absorption and porosity values decreased with increased calcination temperature. This indicates the formation of densified and compact geopolymer matrix with few accessible voids capable of retaining water molecules when samples are immersed in water for 24 h before the measurements. Water absorption and porosity decreased from 11.05 to 7.60% and from 20.12 to 12.02%, respectively, as the calcination temperature increased from 300 °C to 500 °C. This observation is consistent with the trend of mechanical properties confirming the progressive densified structure reliable on higher produced geopolymer binder phase allowing the better connection among different components. The lower values of water absorption and porosity recorded on sample made with laterite heated at 500 °C indicating the formation of a compact structure with fewer open voids, which is in line with the highest mechanical property achieved. However, it was noticed a slightly increase beyond 500 °C suggesting increased porosity that weakened strength development as explained earlier. Finally, the high values recorded on samples made with thermal activated laterite at 300 and 400 °C could be attributed to presence of none dehydroxylated kaolinite which allowed the retention of water molecules when samples are immersed in water for the measurement [5, 7].

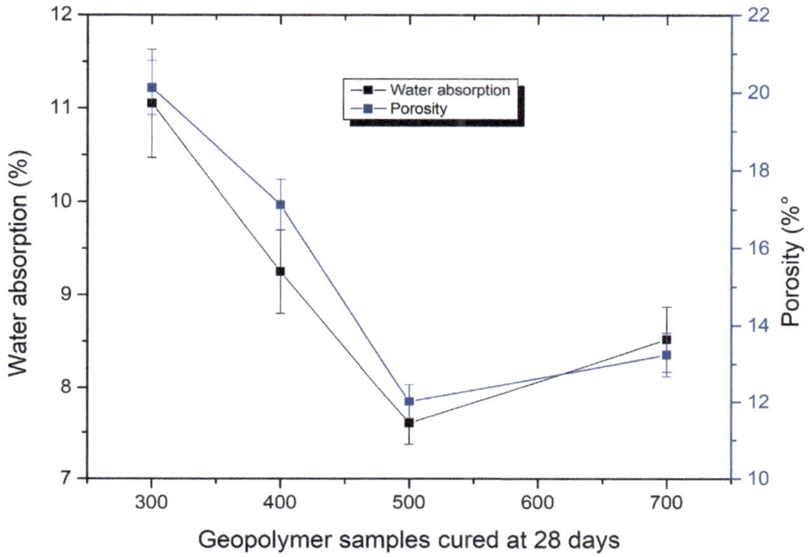

Fig. 5. Water absorption and porosity of hardened laterite based geopolymer cured at 28 days.

3.5 Setting Time

Figure 6 displays the initial and final setting time recorded on fresh geopolymer pastes made with thermally activated gibbsite-rich laterite powder heated at varied temperatures (300, 400, 500 and 700 °C). It appears that the setting time varies between 125 and 100 min, 75 and 62 min, respectively for initial and final setting time. This decrease correlates with the degree of amorphisation of metakaolinite. In fact, the dihydroxylation of kaolinite occurred progressively when the heating temperature was increased from 500 to 800 °C as reported in previous studies [16, 21]. In addition, during the thermal activation, the dihydroxylation of gibbsite favoured the additional release of Al-species in alkaline medium, able to react with Si-species, resulting in shorter setting. The slight increase recorded on the sample heated at 700 °C may be attributed to the beginning of sintering particles upon thermal heating as reported in [7]. Although the precise Si/Al molar ratio of each thermally activated laterite sample could not be determined in this study, the measured setting times provide an indirect assessment of powder reactivity, which correlates with the observed strength development trends.

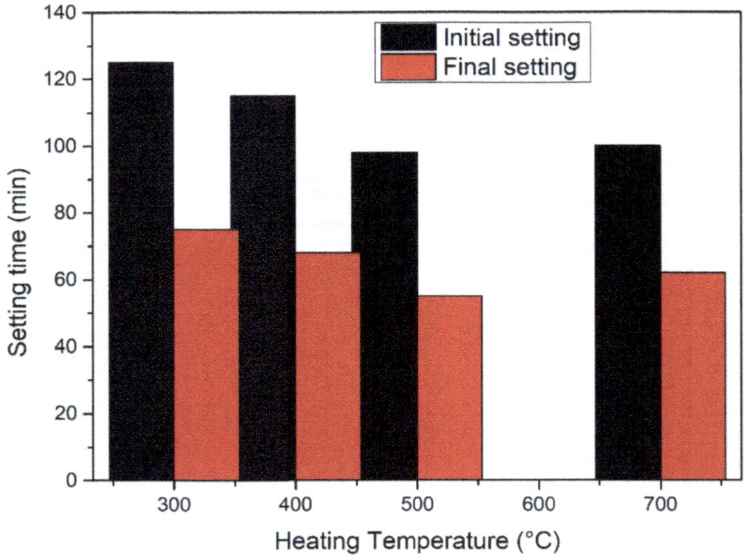

Fig. 6. Setting times of fresh geopolymer pastes made with thermally activated gibbsite laterite powders at different temperature.

4 Conclusion

The mechanical and microstructural properties of geopolymer using thermally activated gibbsite-rich laterite soil calcined at varied temperatures of 300, 400, 500 and 700 °C was studied. The following conclusions can be drawn based on the results.

The increase of calcination temperature from 300 to 500 °C improved the flexural strength of produced geopolymer products. This suggests an enhancement in formation of geopolymerization product ensuring better connectivity between different components within the matrix. The slightly decrease recorded beyond 500 °C is likely attributed to the onset of sintering in laterite particles of laterite which did not dissolve easily and affected the development of geopolymer binder resulting in low strength. The obtained properties from the produced geopolymer products could be used in engineering applications where high strength is required. However, further studies in terms of durability such as wet-dry cycling and carbonation should be conducted to well assess the stability of these specimens.

References

1. Gidigasu M (2012) Laterite soil engineering: pedogenesis and engineering principles. Elsevier, vol 9
2. Schwarz T, Germann K (1995) Weathering surfaces, laterite-derived sediments and associated mineral deposits in north-east Africa. In: Palaeoweathering, palaeosurfaces and related continental deposits, pp 367–390

3. Venyite P et al (2022) Effect of combined metakaolin and basalt powder additions to laterite-based geopolymers activated by rice husk ash (RHA)/NaOH solution. SILICON 14(4):1643–1662
4. Kamseu E et al (2020) Ferrisilicates formation during the geopolymerization of natural Fe-rich aluminosilicate precursors. Mater Chem Phys 240:122062
5. Duna LL et al (2023) Performances of laterite-based geopolymer matrice filters on the bacteriological filtration of contaminated groundwater and microstructure. J Water Process Eng 53:103699
6. Rodrigue Kaze C et al (2021) Synergetic effect of rice husk ash and quartz sand on microstructural and physical properties of laterite clay based geopolymer. J Build Eng 43:103229
7. Kaze RC et al (2017) The corrosion of kaolinite by iron minerals and the effects on geopolymerization. Appl Clay Sci 138:48–62
8. Esaifan M et al (2016) Hydrated lime/potassium carbonate as alkaline activating mixture to produce kaolinitic clay based inorganic polymer. Appl Clay Sci 126:278–286
9. Aldabsheh I et al (2015) Dissolution behavior of Jordanian clay-rich materials in alkaline solutions for alkali activation purpose. Part I. Appl Clay Sci 115:238–247
10. Esaifan M et al (2015) Development of inorganic polymer by alkali-activation of untreated kaolinitic clay: reaction stoichiometry, strength and dimensional stability. Constr Build Mater 91:251–259
11. Djinsi Vaïmata G et al (2024) The effect of combining rice husk ash with laterite-based inorganic polymers activated by potassium-rich shea pellet ash. Sustain Chem Pharm 39:101607
12. Sontia Metekong JV et al (2024) Feasibility of production fired bricks based lateritic soil at very low temperature. Int J Ceram Eng Sci 6(4):e10225
13. Onisei S et al (2018) Inorganic polymers made of fayalite slag: On the microstructure and behavior of Fe. J Am Ceram Soc 101(6):2245–2257
14. Kaze CR et al (2020) Meta-halloysite to improve compactness in iron-rich laterite-based alkali activated materials. Mater Chem Phys 239:122268
15. Kaze RC et al (2024) Formulation of geopolymer binder based on volcanic-scoria and clay brick wastes using rice husk ash-NaOH activator: fresh and hardened properties. Sustain Chem Pharm 40:101627
16. Kaze CR et al (2020) Reaction kinetics and rheological behaviour of meta-halloysite based geopolymer cured at room temperature: effect of thermal activation on physicochemical and microstructural properties. Appl Clay Sci 196:105773
17. Naghizadeh A et al (2024) Circular production of recycled binder from fly ash-based geopolymer concrete. Constr Build Mater 415:135098
18. Onyelowe KC et al (2023) Characterization of net-zero pozzolanic potential of thermally-derived metakaolin samples for sustainable carbon neutrality construction. Sci Rep 13(1):18901
19. He P et al (2016) Effects of Si/Al ratio on the structure and properties of metakaolin based geopolymer. Ceram Int 42(13):14416–14422
20. Wang Y et al (2020) Effects of Si/Al ratio on the efflorescence and properties of fly ash based geopolymer. J Clean Prod 244:118852
21. Elimbi A, Tchakoute HK, Njopwouo D (2011) Effects of calcination temperature of kaolinite clays on the properties of geopolymer cements. Constr Build Mater 25(6):2805–2812
22. Tchakoute HK et al (2015) Influence of gibbsite and quartz in kaolin on the properties of metakaolin-based geopolymer cements. Appl Clay Sci 107:188–194
23. Viktor Zsolt B, Kristály F, Szücs I (2013) Influence of the short time grinding on the thermal decomposition processes of gibbsite produced by the Bayer process. Mater Sci Eng 38(1):15–27

A Review on the Use of Shell Waste in Geopolymer Concrete: Toward Sustainable and Eco-friendly Construction Materials

L. Tavakoli[1], M. R. Shiravand[1], S. Mahboubi[1](✉), and M. Kioumarsi[2]

[1] Faculty of Civil, Water and Environmental Engineering, Shahid Beheshti University, Tehran, Iran
Sh_mahboubi@sbu.ac.ir
[2] Oslo Metropolitan University (OsloMet), Oslo, Norway

Abstract. The construction industry generates substantial environmental harm because Portland cement production releases more than 4 billion tons of CO_2 every year. The worldwide issues of climate change and industrial waste buildup require immediate development of sustainable alternatives for conventional construction materials. Geopolymer concrete made from alkali-activated fly ash and slag industrial by-products presents a sustainable concrete alternative that produces fewer carbon emissions than conventional cement-based materials. The construction industry has started to consider marine and terrestrial shell waste containing calcium carbonate ($CaCO_3$) as a beneficial additive for producing geopolymer and lightweight concretes. Scientific studies demonstrate that shell waste improves geopolymer concrete performance by delivering better mechanical properties along with enhanced durability and environmental advantages from waste valorization and CO_2 emission reduction. The research paper evaluates recent studies about utilizing shell waste in geopolymer concrete for material performance improvement as well as sustainability benefits. The review investigates the potential of shell-derived materials to function as geopolymer precursors which supports both circular economy principles and construction sector net-zero emission targets.

Keywords: Seashell waste · Geopolymer concrete · Waste utilization

1 Introduction

The building industry, the largest user of natural resources in the world, most importantly in Portland cement production and aggregate extraction [1], is one of them. The most voluminous building material, concrete, is a vital element in infrastructure building. This notwithstanding, growing demand for building has hugely escalated its impact on the environment [2]. Production of Portland cement consists of the calcination of clay and limestone above 1450 °C, releasing very high amounts of CO_2. The sector alone accounts for more than 8% of the world's CO_2 emissions at over 4 billion tonnes annually [3].

The original version of the chapter has been revised. The affiliation Tajrish has been mistakenly written as have been corrected to Tehran. A correction to this chapter can be found at https://doi.org/10.1007/978-3-032-27698-8_110

H. Beushausen et al. (Eds.): NTZR 2025, LNCE 873, pp. 285–297, 2026.
https://doi.org/10.1007/978-3-032-27698-8_26

Global warming has therefore become one of the most pressing issues of the past decades, attracting interest from scientific as well as engineering communities [4]. According to the World Meteorological Organization 2021 report, the level of the greenhouse gases has increased substantially. Emissions in the period between 2000 and 2010 grew by 24%, which was three times the increase in the previous decade [5]. Extraction of natural aggregates (sand and gravel) causes severe environmental consequences, including soil erosion, riverbank degradation, and destruction of natural scenery. Unchecked extraction of fly ash has resulted in land subsidence, loss of biodiversity, and social conflicts in many countries [6]. Scientists have thus given increased attention to recycled and other materials for the use in concrete to reduce these issues. These include fly ash, blast furnace slag, red mud, rice husk ash, pumice, calcined clay, glass powder, and industrial by-products. Fly ash. Geopolymer concrete, capable of reducing CO_2 emissions by 40–80% compared to conventional concrete, has emerged as a prospective and sustainable alternative. The binders can potentially replace 60% of virgin raw materials in concrete mixtures and are more durable against heat, acid, sulfate, freezing, and chloride penetration, along with improving durability and reducing water absorption [7]. One of the new trends that have grown over the past few years is to utilize seashell waste as a replacement of part of the aggregate or the cement in concrete [8]. Studies have revealed that use of ground seashell powder at 5–15% replacement levels can provide up to 55% savings of CO_2 compared to traditional mortars. This waste stream is particularly relevant in marine regions and is consistent with the principles of the blue circular economy. Yet, seashell addition studies remain underdeveloped relative to other alternative binders, with nearly 89% being concentrated in Asia (i.e., China, South Korea, and Taiwan) and significantly fewer from Oceania, Africa, Europe, and the Americas [8]. The aim of the present paper is to conduct a systematic review on the utilization of seashell waste in geopolymer concrete focusing attention towards both environmental benefits and mechanical properties. In the process, the review will search for the gaps in available research such as:

- The lack of standard pretreatment and processing protocols for shell waste;
- The limited experience with long-term durability in actual service conditions;
- The absence of high-scale deployments and code conformity studies.

By discussing these issues, this paper contributes to the development of a better understanding of how waste seashells can be a sustainable material in advancing towards net-zero construction.

2 Seashells as Geopolymeric Material

Seashell waste is an industry by-product of the aquaculture and seafood processing industries. It is primarily made up of calcium carbonate ($CaCO_3$) and is usually dumped along shorelines as waste. Based on the BeCoCoH Report (2022), from each 1 kg edible shellfish meat, there is about 700 g of shell waste produced; in some areas, like Arcachon Bay, France, that is over 5,000 tonnes a year [9]. Uncontrolled disposal not only causes environmental issues like bad odors and soil and water pollution but also is the loss of a great carbonate resource. Shells with their high hardness and dense internal structure are of good use in concrete [10]. While their pozzolanic activity is limited by the presence of $CaCO_3$, shells are nevertheless useful in heat of hydration regulation and chemical stability improvement in cement matrices [11]. They also enhance paste-aggregate bonding due to their angular particle size and rough surface texture [11]. However, because of their low density, high porosity, and large water absorption, direct use of shells as aggregates requires careful mix redesign and granular skeleton optimization [12]. From a classification perspective, the shells are usually divided into two broad categories. The first comprises marine shells, which are harvested from crops such as oysters, snails, scallops, and pearl shells mainly as seafood industry waste or field collection off coastal areas. The second is what is locally referred to as "mountain shells" in commercial or regional usage. The term commonly refers to carbonate-based natural materials like fossiliferous limestone or shell powders collected from sedimentary deposits. Both practice groups consist of the same chemical composition ($CaCO_3$), but marine shells usually include salts and organic impurities that must be pretreated carefully, while mountain shells are more stable and higher purity, harder in the long term for use in concrete [13]. Shells may perform a variety of functions in concrete technology depending on their processing and particle size. They can be used as a partial replacement for cement (after grinding and, in certain cases, calcination), as a filler to increase packing density and reduce porosity, or as fine or coarse aggregate in mortar and concrete mixes. To highlight these differences, Table 1 presents a comparative summary of the most significant features of marine and mountain shells.

Table 1. Comparison between Marine Seashells and Mountain Shells for Use in Concrete.

Property	Marine Seashells	Mountain Shells
Origin	Marine organisms	Sedimentary marine fossils (limestone)
Main Composition	$CaCO_3$ with impurities (salts, organic matter)	Purified $CaCO_3$ with higher hardness
Processing Required	Thorough washing to remove salts and organics	Typically requires crushing and grinding
Specific Gravity	Lower (lighter)	Higher (heavier and more resistant)
Application in Concrete	Suitable for lightweight and non-structural concrete	Suitable for structural concrete elements such as beams and columns
Corrosion Risk	Higher (due to chloride presence)	Lower (due to greater purity and density)
Durability	Moderate (more susceptible to aggressive environments)	High (long-term chemical and physical stability)

2.1 Studies on Geopolymer Concrete Incorporating Marine Seashells

One of the earliest systematic investigations on the use of oyster shells (OS) as an alternative to natural sand in cement mortar was conducted by Yun et al. [15]. The intention of this study was to assess the feasibility of recycling shell waste from the sea in construction for alleviating pressure on natural resources. The oyster shells were crushed, pulverized, and sieved and employed as a partial replacement of fine aggregates in mortar. The key parameters which were compared and assessed were compressive strength at 7 and 28 days age, rheological properties, workability, drying shrinkage, and volume stability. The results indicated that sand could be partially replaced by finely ground oyster shells to a specific proportion. Though there was some reduction in compressive strength, workability and lightness of the mix were also improved. The study concluded that oyster shells as a cost-effective and sustainable substitute for sand in non-structural mortars and light architectural uses can be utilized particularly in regions where natural aggregate sources are scarce (Fig. 1).

Fig. 1 Is an oyster shell (Crassostrea gigas) collected from the Arcachon Bay region in France [14].

2.2 Environmental Assessment of Oyster Shell Recycling Using Life Cycle Analysis

Alvarenga et al. [15] conducted a Life Cycle Assessment (LCA) of recycling oyster shell (OS) and concluded that the use of OS as agricultural lime is far less environmentally impactful than landfilling or incineration. They went on to point out that minimal processing such as washing, drying, and grindingnis needed to get the shells ready for reuse. Khyau et al. [16] investigated the use of waste oyster shells (WOS) in controlled low-strength materials (CLSM). They found that partial replacement to 20% improved workability, reduced specific gravity, and maintained desired strength suitable for trench backfilling and sub-base courses. Elviya et al. [17] demonstrated that partial replacement of fine aggregates by seashells (10–20%) produced acceptable compressive strength lightweight concrete, though higher substitution rates led to loss of strength. Eyo and Yee [18] analyzed concrete mixtures with 0–100% crushed oyster shells (COS). Based on their findings, replacement up to 30% provided acceptable performance, including lower density and improved workability. But shrinkage growth also increased and modulus of elasticity reduced. Moneron-Giorietz et al. [19] employed finely ground seashells as additive materials in metakaolin geopolymers and observed the addition of the shells improved the formation of C–A–S–H and N–A–S–H gels, which enhanced compressive strength and reduced porosity. Silva et al. [20] documented various applications of oyster shell waste (OSW) in industries such as concrete, agriculture, ceramics, water treatment, and biomaterials, and also saw its potential to be employed in geopolymer systems due to the large concentration of $CaCO_3$. Tayeh et al. [21] assessed more than 50 studies and proved that incorporation of 5–15% marine shell waste improved strength, durability, as well as resistance to sulfate and chloride attack, in addition to reducing emissions and construction costs. Chen et al. [22] mixed COS with supplementary cementitious materials (SCMs) such as fly ash and ground granulated blast furnace slag (GGBFS), which improved microstructure, increased compressive strength, and reduced the overall carbon footprint. Arizzi et al. [23] demonstrated that calcined marine shells (scallops, clams, and oysters) produce reactive lime with good properties for heritage conservation and green mortars. Rana et al. [24] reported that the substitution of coarse aggregate by up to 15% COS in M30-grade concrete yielded increased compressive, tensile, and flexural strength. Hasnaoui et al. [25] prepared alkali-activated seashell powder-based mortars and reported dense microstructures and appropriate compressive strength. Komuola et al. [26] explored COS-based concrete with polypropylene fibers (PPF) and observed that 15% COS with PPF enhanced mechanical strength and resistance to cracks. Ravirdan et al. [27] observed that 15% WOS coarse aggregate gave maximum mechanical performance and reduced density, suitable for lightweight concrete. Liao et al. [28] studied white oyster shell powder (WOSP) in cement mortars and demonstrated that 10–15% replacement improved compressive strength, reduced porosity, and diminished environmental resistance to degradation. Syed et al. [29] reported that 10% COS replacement offered the best mechanical performance in concrete. Ramly et al. [30] demonstrated that substituting 5–10% seashell powder (SSP) improved strength, reduced permeability, and improved overall durability. Bellei et al. [31] also touched upon the use of WOS in cementitious composites and highlighted the importance of pretreatment procedures that are standardized and the use of it in carbon savings. Liao et al. [22] demonstrated in another piece of research that replacing 10–30% of the fine aggregates by waste oyster shell powder favorably enhanced compressive and flexural strength, chloride penetration resistance, and microstructure and durability of mortar composites. Bellei et al. [32] showed that incorporation of powdered oyster shells into lime provides better thermal insulation, albeit lower strength with higher dosages. These papers collectively highlight the promise of oyster shell waste (OSW) as a green component of cement and geopolymer systems. Either partially used as cement, fine/coarse aggregate, or mineral filler replacement, OSW has demonstrated its ability to enhance mechanical strength, improve durability, reduce density, and lower permeability when applied at the optimal dosage level (typically 10–15%). Moreover, OSW addition promotes circular economy, exhibiting significant environmental benefits such as the reduction of CO_2 emissions, energy savings, and effective waste valorization. Notwithstanding the above specifications, the performance of OSW-filled composites is influenced by conditions such as shell shape, treatment method, replacement content, and matrix type. Future research should therefore seek to standardize processing procedures, establish long-term sustainability, and increase scale of application with a view to enhancing utilization of OSW in green construction activities. Figure 2 indicates some examples of seashells as calcium carbonate sources of geopolymer concrete: (a) Oyster shell (Crassostrea gigas) with rough outer surface and smooth inner lining; (b) Mussel shell (Mytilus edulis) dark, elongated structure, best for minimizing porosity and weight; (c) Cockle shell with steep radial ridges and concave surface, enhancing interfacial bonding in geopolymer mixtures; and (d) Clam shell with smooth, layered surface, natural filler in sustainable concrete composites.

(a) (b) (c) (d)

Fig. 2. Seashells **a** Oyster shell, **b** Mussel shell, **c** Cockle shell, **d** Clam shell

3 Mechanical Properties in Cementitious and Geopolymer Composites Incorporating Seashell Powder

Experimental studies have always confirmed that the incorporation of waste oyster shell powder (WOSP) as a partial cement or aggregate replacement can significantly affect the mechanical properties of geopolymer and cementitious composites. The observed effects are primarily subject to the level of replacement, processing and pretreatment of the shells, and binder matrix chemical composition. For instance, Liao et al. [28] explained that substitution of 10% river sand by 10% WOSP enhanced compressive strength from the control mix value of approximately 25.5 MPa to the modified mix value of approximately 27.5 MPa, which was an enhancement by nearly 2 MPa. The improvement was also attributed to superior durability characteristics, particularly greater resistance against chloride ion penetration. Similarly, Kim and Lee [33] investigated the microstructural effect of WOSP and found that the replacement up to 15% either slightly enhanced or maintained compressive strength. Importantly, they observed a significantly enhanced internal microstructure with reduced pore size and improved interfacial bonding between shell particles and the cement matrix, leading to lower permeability and greater service life of the composites. On the other hand, in slag-based geopolymer system, Okoro and Oyebisi [34] reported that employing 10% WOSP did not show an effective improvement of the compressive strength, which was maintained constant at around 21.8–22.2 MPa. This finding demonstrates that the mechanical contribution of OSW to alkali-activated systems may be limited and highly controlled by the chemical character of the raw material. Moreover, Liao [28] and Kim and Lee [33] further presented microstructural evidence from their SEM and XRD investigations. The observations showed enhanced pore filling, formation of additional C–S–H gels, and overall matrix densification. These alterations not only worked towards strength development but also permeability reduction, hence improved long-term durability. While these benefits are present, several studies caution that replacement levels greater than 15% can prove to be harmful in poor interfacial bonding, porosity, and decreased long-term compressive strength. Specifically, Kim and Lee [33] revealed that replacement ratios of greater than 25% led to decreased compressive strength and internal degradation caused by unsuccessful bonding or poor interfacial transition zones. Based on the evidence that is currently present, certain key recommendations can be highlighted:

- The optimum replacement level of WOSP is approximately 5–15%, which maximizes strength gain and durability.
- Proper pretreatment in terms of washing, drying, and grinding is essential for effective use of shells.
- The chemical composition of the binder (cement or geopolymer) is important because the mechanical outcomes vary considerably as a function of the matrix [28–33].

Table 2 gives compressive strength values at various replacement levels of seashell powder, and Fig. 3 shows these graphically.

Table 2. Comparison of Compressive Strength with Shell Powder

Author	Control Mix (MPa)	5% Shell Powder	10% Shell Powder	Remarks
Liao et al. [28]	25.5	≈26.5–27	≈27.5	Strength by 2 MPa; better chloride resistance
Kim and Lee [33]	28–30	≈28–30	≈28–30	Stable or strength; improved microstructure
Okoro and Oyebisi [34]	21.8–22.0	≈21.8–22.0	≈21.8–22.2	No significant strength change in geopolymer

4 Durability and Environmental Performance

High amounts of research have been carried out to assess the durability and environmental performance of cementitious and geopolymer composites containing waste oyster shell powder (WOSP). Durability was directly through chemical and physical testing in some studies carried out, while in others microstructural assessments (e.g., SEM, XRD) or site-based environmental performance testing were performed. The depth and level of such research were dictated by research objectives and methodologies. Liao et al. [28] in their study examined samples of mortar containing 0–30% WOSP substitution for river sand. The authors noted considerable improvements, such as a tremendous reduction in water absorption, sorptivity reduction, and lower chloride diffusion coefficient value. SEM and XRD examination confirmed a denser pore structure, improved pore filling, and increased content of hydration products like C–S–H gels, all contributing to enhanced long-term durability [28]. Similarly, Liu et al. [35] studied ternary supplementary cementitious material (SCM) mixes with the addition of oyster shell powder. Effects included as much as a 30% reduction in water absorption and an impressive 81% decrease in RCPT charge that indicates noteworthy improvement in chloride penetration resistance and transport properties [35]. In applications of recycled aggregate porous concrete (RAPC), Yoon et al. [26] demonstrated that partial cement replacement by WOS levels of up to 40% lowered the chloride migration coefficient by 17–34% and electric flux by 17–26%. Moreover, they reported that CO_2 emissions decreased by approximately 52%, highlighting the environmental benefit of WOS use [26]. However, they also showed that durability gains reduced at higher levels of replacement, which was an issue for long-term performance. A comprehensive review by Bamigboye et al. [37] also noted that lower replacement levels (< 15%) improve water resistance, freeze–thaw durability, and carbonation resistance, whereas higher replacement levels (> 25%) can impact performance negatively via increased porosity and reduced interfacial bonding strength [37] (Table 3).

Table 3. Summary of Durability Performance in Seashell-Based Systems

Reference	Durability Indicator	Key Findings
Liao et al. [28]	Water absorption, chloride diffusion, sorptivity	10–30% WOSP led to reduced water absorption and chloride diffusion; denser microstructure observed via SEM/XRD
Liu et al. [35]	Absorption and RCPT in ternary SCM system	Water absorption decreased by up to 30%; chloride charge passed (RCPT) reduced by 81% in SCM system
Yoon et al. [26]	Chloride migration, electric flux, CO_2 emission	Chloride penetration reduced by 17–34%; CO_2 emissions reduced by up to 52%; long-term durability may decrease at high substitution levels
Bamigboye et al. [37]	General durability review (water, freeze–thaw, carbonation)	Improved durability observed with < 15% replacement; performance may decline at replacement levels > 25%

5 Environmental Impacts and Sustainability Benefits

The incorporation of seashells (marine and "mountain" shells) as additives in construction materials particularly geopolymer concrete has been an effective recycling method for the disposal of biogenic wastes. Not only does it enhance the mechanical behavior of composites, but it is also crucial for reducing the environmental footprint of conventional concrete, enabling the transition to low-carbon and sustainable construction.

Several studies have shown the environmental benefits of oyster shell waste (OSW). For example, Lopes da Silva et al. (2023) and Capuana et al. [33] reported that the replacement of 10–15% cement with seashell powder caused a reduction in CO_2 emission by 15–25%, largely due to the avoidance of limestone decarbonation during clinker production. Seashells are also one of the major wastestreams of the fishing industry, typically disposed of along the coastal areas. Their recycling in concrete, especially in high-production countries such as China, the Philippines, Iran, and Brazil, reduces environmental pollution, disposal cost, and saves dependence on virgin limestone quarrying [19].

In the energy aspect, Rana et al. [24] demonstrated that pretreatment of the shells (grinding, drying, and washing) takes far less energy than the calcination of limestone in cement kilns. Ramelee et al. [30] also reported that the addition of seashell powder reduces permeability and enhances moisture resistance, which benefits not only mechanical durability, but also reduced soil and water degradation from leaching. A simplified LCA conducted by Capuana et al. [33] also showed that seashell powder concretes exhibited a Global Warming Potential (GWP) saving of as much as 18% and Non-Renewable Energy Use (NREU) saving of 22%.

Figure 3 illustrates the relative contributions of the environmental indicators, i.e., natural resource consumption reduction (25%), non-renewable energy consumption (22%), global warming potential (18%), CO_2 emissions (20%), and soil/water pollution (15%). Taken together, these findings confirm the eco-efficiency of OSW in geopolymer concrete.

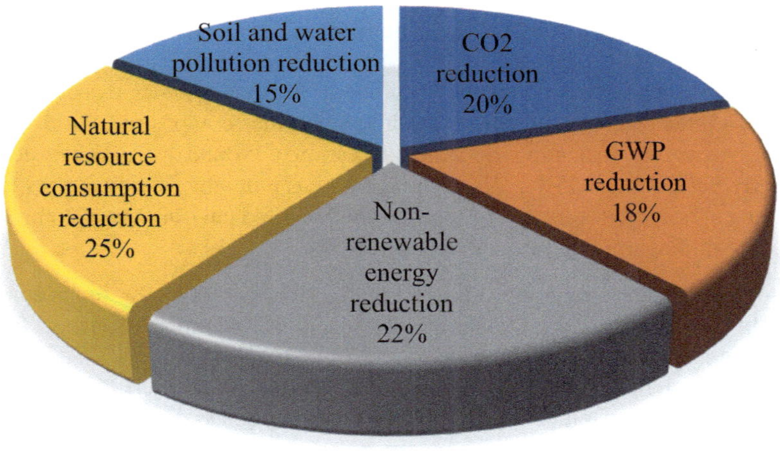

Fig. 3. Relative contribution of environmental indicators to the overall benefits of seashell powder

Table 4 summarizes the key findings reported across the literature, highlighting mechanical, durability, and environmental benefits of OSW at different replacement levels.

Table 4. Summary of Mechanical, Durability, and Environmental Benefits of Seashell Waste (OSW) in Concrete

Notes	Observed Effect at Optimal Replacement (5–15%)	Property/Benefit
Optimal at ~ 10% replacement	3–5 MPa compared to control mixes	Compressive strength
Dependent on shell fineness	Moderate improvement	Flexural/Tensile strength
Denser microstructure observed (SEM/XRD)	Significant reduction in absorption and permeability	Durability (water absorption, chloride resistance)
Less degradation than control	Improved stability at $\geq$ 250 °C	Thermal resistance
Due to avoided clinker decarbonation	15–25%	CO_2 emissions
Lower processing energy vs. cement calcination	Up to 22% (NREU)	Energy consumption
$>$ 15% leads to dilution and reduced cohesion	5–15%	Resource consumption

6 Conclusion

This review extensively explored the viability of seashell powder as a green partial cement or regular aggregate replacement in geopolymer composites. On the basis of systematic reviewing of existing literature, utilization of ground or crushed oyster shells, particularly within the 5–15% replacement level, has a number of advantages in terms of mechanical properties, durability, thermal stability, and environmental compatibility.

The key findings can be summarized as follows:

- A 3–5 MPa increase in compressive strength at the level of 10% replacement was noted by the majority of the studies, thereby specifying the range as the most favorable for strength increase.
- Thermal resistance was increased with the addition of seashell powder, with the samples having high temperatures ($\geq$ 250 °C) showing less degradation than control mixes.
- SEM and XRD tests confirmed that shell $CaCO_3$ content is responsible for the development of dense and stable gel phases (e.g., C-(A)-S-H) to improve microstructure.
- Statistically uniform water absorption and permeability decrease was achieved, demonstrating added protection against environmental conditions and improved long-term durability.

- Environmental benefits included up to 20% reduction in CO_2 emissions, 25% saving in natural resources, and substantial GWP and NREU improvement.
- The positive effects are greatly reliant on the fineness of particles, purity of $CaCO_3$, calcination procedure, and intraparticle porosity structure of the shells.
- Substitution of over 15% commonly led to binder dilution, decreased paste cohesion, and poorer interfacial zones.
- Future research areas should include standardization of pretreatment processes, estimation of long-term durability at real conditions of service, and investigation on large-scale applications that still ensure compliance with building codes and regulations. This will be central to the expansion of seashell waste practical application towards net-zero and sustainable construction.

References

1. Durastanti C, Morett L (2020) Environmental impacts of cement production: a statistical analysis. Appl Sci 10:8212
2. Benhelal E, Zahedi G, Shamsaei E, Bahadori A (2013) Global strategies and potentials to curb CO_2 emissions in cement industry. J Clean Prod 51:142–161
3. Andrew RM (2019) Global CO_2 emissions from cement production. Earth Syst Sci Data 10:195–217
4. Mccaffrey R, Magazine L (2002) Climate change and the cement industry 15:19
5. I. IEA (2021) Greenhouse gas emissions from energy data explorer, in IEA
6. Ramesh V, Muthramu B, Rebekhal D (2025) A review of sustainability assessment of geopolymer concrete through AI-based LCA. AI Civil Eng 4(3):1
7. Sing Wong L. Durability performance of geopolymer concrete: a review
8. Liao Y, Fan J, Li R, Da B, Chen D, Zhang Y (2022) Influence of the usage of waste oyster shell powder on mechanical properties and durability of mortar. J Clean Prod 357:131872
9. BeCoCoH, Beach-clean coast healthy: Report on shellfish co-product valorization and environmental impacts
10. Zhu Y, Chen D, Yu X, Liu R, Liao Y (2024) Properties of seashells. Encyclopedia
11. Silva AL, d Kohlman Rabbani ER, Shakouri M (2025) Seashell powder as a sustainable alternative in cement-based materials: a systematic literature review. Sustainability 17(2):592
12. Martínez-García C, González-Fonteboa B, Martínez-Abella F, Carro-López D (2017) Performance of mussel shell as aggregate in plain concrete. Constr Build Mater 139
13. Eziefula UG, Ezeh JC, Eziefula BI (2018) Properties of seashell aggregate concrete: a review. Constr Build Mater 192:287–300
14. Yoon H, Park S, Lee K, Park J (2004) Oyster shell as substitute for aggregate in mortar. Constr Build Mater 149:403–414
15. Alvarenga RAF, Giesen G, Dewulf J, Huijbregts MAJ (2012) Life-cycle assessment of sea shell waste management options. Int J Life Cycle Assessment 17:1328–1339
16. Kuo W-T, Wang H-Y, Shu C-Y, Su D-S (2013) Engineering properties of controlled low-strength materials containing. Constr Build Mater 46:128–133
17. Elviya A, Sulaiman SA, Prawoto Y (2015) Effect of partial fine aggregate replacement with seashells on concrete properties. Constr Build Mater 95:448–458
18. Eyo EU, Yee SS (2015) Utilization of crushed oyster shell as partial replacement for fine aggregate in concrete. Int J Eng Res Appl 5(3):48–52
19. Moneron-Giurizz M, Paya J, Borrachero MV, Soriano L, Monzo J (2018) Use of seashell waste as an alternative raw material in metakaolin-based geopolymer binders. Constr Build Mater 187:371–383

20. Silva RV, de Brito J, Dhir RK (2019) Performance of cementitious materials with recycled aggregates containing seashells: A review. J Clean Prod 214:616–628
21. Tayeh BA, Jaafar MS, Voo YL, Ali AAA (2019) Utilization of seashell powder in sustainable construction materials: a review. Constr Build Mater 216:364–376
22. Chen HJ, Chen YY, Lee WC (2019) Performance evaluation of concrete incorporating recycled sea shells and supplementary cementitious materials. J Clean Prod 229:1252–1263
23. Arizzi A, Cultrone G, Brümmer M, Sebastián-Pardo E (2020) Recycling seashell waste into eco-friendly lime mortars for architectural heritage conservation. Constr Build Mater 257:119533
24. Rana S, Singh A, Bhardwaj A, Gupta N (2021) Mechanical and durability performance of concrete incorporating waste seashells as coarse aggregates. J Clean Prod 284:124759
25. Hasnaoui A, Mellas M, El Gherabi N, Houari H (2021) Valorization of seashell waste in alkali-activated materials: physico-mechanical and microstructural properties. Mater Today: Proc 45:7719–7724
26. Komuola TA, Okon EE, Olanipekun EA, Akinwumi II (2021) Mechanical properties of concrete incorporating crushed oyster shell and polypropylene fibre. Case Stud Constr Mater 15:e00694
27. Ravirdan R, Balamurugan P, Venkatasubramanian C (2021) Mechanical and durability characteristics of concrete incorporating waste oyster shell as coarse aggregate. Mater Today: Proc 45:7185–7190
28. Liao KY, Huang R, Lin WT, Chang TP (2022) Mechanical properties and microstructure of cement mortar containing white oyster shell powder. Constr Build Mater 323:126586
29. Syed R, Farhan SA, Khan K (2022) Utilization of seashells as partial replacement of coarse aggregate in concrete. J Mater Civ Eng 34(5):04022032
30. Ramly ZTA, Safiee NA, Zainudin NF, Razi P (2022) Strength, permeability and durability performance of concrete containing seashell powder as cement replacement. Mater Today: Proc 66:2422–2428
31. Bellei P, Torres I, Solstad R, Flores-Colen I (2023) Potential use of oyster shell waste in the composition of construction composites: a review. Buildings 13(6):Article 1546
32. Bellei P, Magalhães F, Pereira M, Torres I, Solstad R, Flores-Colen I (2023) Innovative thermal renders incorporating oyster shells for sustainable insulation. Sustainability 15(22):15952
33. Kim MO, Lee MK (2023) Strength and microstructural changes in cementitious composites containing waste oyster shell powder. Buildings 13(12):3078
34. Okoro W, Oyebisi S (2023) Mechanical and durability assessment of steel slag-seashell powder-based geopolymer concrete. Heliyon e13188
35. Liu S, Wang Y, Liu B, Zou Z, Teng Y, Ji Y, Zhou Y, Zhang L, Zhang Y (2022) Sustainable utilization of waste oyster shell powders with different fineness levels in a ternary supplementary cementitious material system. Sustainability 14(10):5981
36. Yoon J, Park S-M, Jang J-G, Kim J (2021) Effects of waste oyster shell on durability and biological performance of recycled aggregate porous concrete. Materials 15(17):6117
37. Anthony, Bassey, Denise (2021) Mechanical and durability assessment of concrete containing seashells: a review. Cogent Eng 8(1):Article 1883830

Model for Predicting Flexural Strength of Sustainable Mortars Containing Waste Concrete Powder

Eric Ababio Ohemeng[1(✉)] and Molusiwa S. Ramabodu[2]

[1] Department of Building Technology, Sunyani Technical University, Sunyani, Bono Region, Ghana
ohemengababioeric@yahoo.com

[2] Department of Construction Management and Quantity Surveying, Durban University of Technology, Durban, South Africa

Abstract. This study focuses on developing predictive model for the flexural strength of mortars incorporating waste concrete powder (WCP) as a sustainable material. The proposed model estimates the flexural strength based on the 28-day compressive strength of control mortar and WCP content. The model was validated using independent data sets, demonstrating its accuracy and reliability in predicting the flexural strength of WCP mortars. The results indicate that the model provides realistic and consistent predictions, making it a valuable tool for assessing the strength performance of sustainable mortars containing WCP. This innovative approach supports the broader adoption of WCP as a sustainable alternative in mortar production, promoting resource efficiency and environmental sustainability.

Keywords: Flexural strength · Waste concrete powder · Sustainable mortar

1 Introduction

The construction industry has witnessed significant advancements in sustainable materials and technologies aimed at minimizing environmental impacts. One of the primary concerns in this sector is the excessive consumption of natural resources and the generation of construction and demolition waste. Among the various waste materials, waste concrete powder (WCP) has gained attention as a potential supplementary material in mortar and concrete production. WCP is a fine by-product obtained from the crushing and grinding of concrete waste, which can be utilized as a partial cement replacement, contributing to waste management and sustainability [1–6].

The incorporation of WCP in mortar production presents a promising approach to achieving sustainability by reducing the demand for natural resources, lowering carbon emissions associated with cement production, and improving waste reutilization [7]. Understanding the influence of WCP on mortar properties requires robust predictive models that can accurately estimate flexural strength based on mix proportions and material characteristics. Several machine learning and statistical modelling techniques

H. Beushausen et al. (Eds.): NTZR 2025, LNCE 873, pp. 298–305, 2026.
https://doi.org/10.1007/978-3-032-27698-8_27

have been applied in civil engineering to predict mortar properties with high accuracy [8–12]. These models help optimize mix designs and assess performance without extensive experimental trials, thereby saving time and resources. Developing an accurate predictive model for the flexural strength of sustainable mortars incorporating WCP can facilitate its practical application and promote confidence in its use among engineers and researchers.

Hence, the present study aims to develop a reliable model for predicting the flexural strength of sustainable mortars containing WCP. The model will consider various influencing factors, including the WCP replacement ratio, water-to-binder ratio, and other mix parameters. By leveraging advanced predictive techniques, this research seeks to enhance the understanding of WCP's effects on mortar properties and provide a valuable tool for optimizing sustainable mortar formulations.

2 Methodology

2.1 Retrieval of Data Set for the Development of the Model

The data for the development of the model was obtained from Scopus, Google Scholar, and the Web of Science, as they are among the most widely used scientific databases [13]. On Scopus, the search query '*waste concrete powder or recycled concrete powder or recycled cement and mortar*' was used. A similar search was conducted on Google Scholar and the Web of Science.

2.2 Development of the Model for Predicting Flexural Strength of WCP-Based Mortars

First, the relationships between the 28-day compressive and flexural strengths of WCP mortars reported by various authors were identified, as illustrated in Fig. 1. It is clear that these two properties share a direct linear relationship. Equation (1) expresses the average equation that links the two strength properties. With an average R^2 value of 0.898, the equation indicates that an increase in the compressive strength of mortars is associated with a corresponding rise in their flexural strength.

$$c_{s28} = pf_{t28} - q \tag{1}$$

where c_{s28} denotes compressive strength at day 28, f_{t28} represents flexural strength at day 28, average $p = 11.902$, and average $q = 67.329$. From Eq. (1), f_{t28} was expressed in terms of c_{s28} as found in Eq. (2).

$$f_{t28} = 0.084c_{s28} + 5.66 \tag{2}$$

Ohemeng et al. [14] developed the model presented in Eq. (3) for estimating the 28-day compressive strength of WCP mortars.

$$c_{s28} = c_{tr28} - 0.595wcp + 0.335 \tag{3}$$

where, c_{s28} is the 28-day compressive strength of WCP mortar, c_{tr28} is the 28-day compressive strength of the control mortar at 28 days, and wcp is the WCP content. By

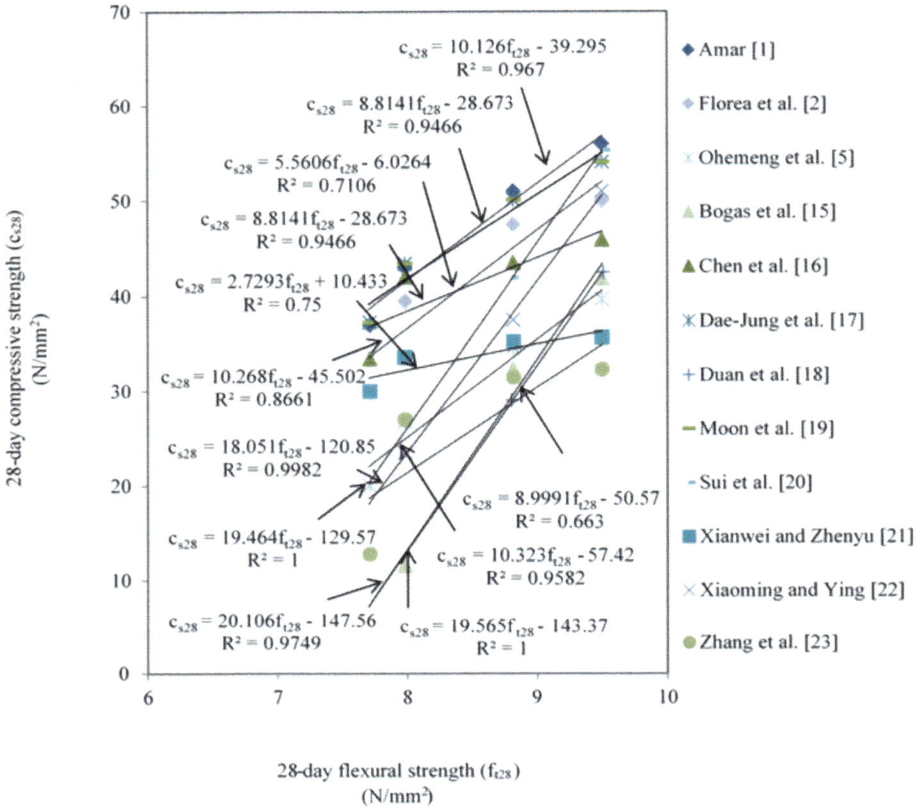

Fig. 1. Relationship between compressive and flexural strengths of WCP mortars.

substituting Eq. (3) into Eq. (2), the model for predicting the 28-day flexural strength was derived, as shown in Eq. (4).

$$f_{t28} = 0.084c_{tr28} - 0.05wcp + 5.69 \tag{4}$$

Equation (4) was validated using data from multiple literature sources. It was found that the model produced more accurate results when the constant was decreased by 1.19. Consequently, Eq. (4) was modified accordingly, resulting in the development of Eq. (5).

$$f_{t28} = 0.084c_{tr28} - 0.05wcp + 4.50 \tag{5}$$

3 Verification of the Predictive Model

Data taken from the various literature [1, 2, 15–23] was employed for the verification of the model. Figure 2 is a plot of the estimated flexural strength values for WCP mortars against real measured results. There is a strong correlation between the two values, as the data points closely follow the line of equality. Table 1 shows the individual outcomes of the 28-day flexural strengths that were predicted and attained. In addition, the ratio of real value (RV) to estimated value (EV) is given in Table 1. With an average of 0.82, the RV/EV values fall between 0.59 and 1.04. Obviously, these RV/EV ratios are near 1.0, reflecting a strong alignment between the real and estimated flexural results. Table 1 also presents the residuals (RV-EV), indicating that the errors in predicting flexural strength generally fall within $\pm$ 1.5 N/mm^2. The coefficient of variation, which illustrates the degree of data variability relative to the population mean, was found to be 9.80%, which is low and suggests that the model's prediction accuracy is good [24–26].

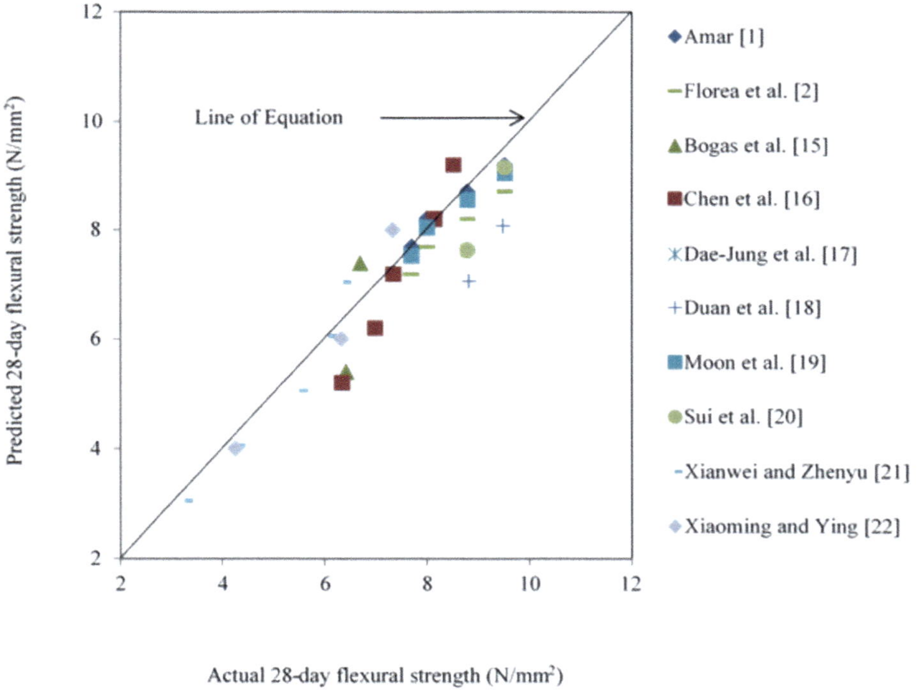

Fig. 2. Predicted values against actual flexural strength results of WCP mortars obtained by different authors.

3.1 Robustness of the Predictive Model

The data employed for the model validation cover a wide spectrum of variables, such as different WCP sources, a range of WCP particle sizes, various cement-to-sand ratios, and varying water-to-binder ratios, as reported in the literature. The formulated equations demonstrated excellent accuracy and reliability by effectively capturing the impact of these factors.

Table 1. Source of literature employed for the flexural strength model's evaluation.

Data source	WCP level (%)	28-day compressive strength (N/mm^2)	28-day flexural strength (N/mm^2)			
			Real value (RV)	Estimated value (PV)	RV/EV (ratio)	Residual (RV-EV)
Amar [1]	0	56.00	9.50	9.20	1.03	0.30
	10	51.00	8.81	8.70	1.01	0.11
	20	43.00	7.98	8.20	0.97	– 0.22
	30	37.00	7.70	7.70	1.00	0.00
Florea et al. [2]	0	50.00	8.20	8.70	0.94	– 0.50
	10	47.50	7.80	8.20	0.95	– 0.40
	20	39.50	6.30	7.70	0.82	– 1.40
	30	30.00	5.40	7.20	0.75	– 1.80
Bogas et al. [15]	0	41.90	6.70	8.02	0.84	– 1.32
	20	32.40	6.40	7.02	0.91	– 0.62
Chen et al. [16]	0	45.90	8.50	8.36	1.02	0.14
	10	43.50	8.15	7.86	1.04	0.29
	20	42.00	7.35	7.36	1.00	– 0.01
	30	33.50	7.00	6.86	1.02	0.14
Dae-Jung et al. [17]	0	54.10	8.90	9.04	0.98	– 0.14
	10	50.10	8.50	8.54	1.00	– 0.04
	20	43.40	7.90	8.04	0.98	– 0.14
	30	37.30	7.30	7.54	0.97	– 0.24
Duan et al. [18]	0	42.50	7.70	8.07	0.95	– 0.37
	20	29.00	6.40	7.07	0.91	– 0.67
Moon et al. [19]	0	54.10	8.92	9.04	0.99	– 0.12
	10	50.10	8.54	8.54	1.00	0.00
	20	43.45	7.90	8.04	0.98	– 0.14
	30	37.35	7.39	7.54	0.98	– 0.15
Sui et al. [20]	0	55.34	8.79	9.15	0.96	– 0.36
	30	41.91	4.63	7.64	0.61	– 3.01

(*continued*)

4 Limitations

This study utilized secondary data to develop a predictive model based on a linear regression approach. Consequently, future research should explore the development of predictive models for the strength properties of WCP mortars using experimental primary

Table 1. (*continued*)

Data source	WCP level (%)	28-day compressive strength (N/mm^2)	28-day flexural strength (N/mm^2)			
			Real value (RV)	Estimated value (PV)	RV/EV (ratio)	Residual (RV-EV)
Xianwei and Zhenyu [21]	0	35.70	6.35	7.50	0.85	– 1.15
	10	34.85	6.10	7.00	0.87	– 0.90
	20	34.00	5.52	6.50	0.85	– 0.98
	30	30.10	4.30	6.00	0.72	– 1.70
	40	26.00	3.25	5.50	0.59	– 2.25
Xiaoming and Ying [22]	0	51.00	7.30	8.78	0.83	– 1.48
	20	37.50	6.30	7.78	0.81	– 1.48
	40	23.50	4.24	6.78	0.63	– 2.54

data. These models should incorporate a wider range of variables, including different sources of WCP, varying particle fineness, water-to-binder ratios, cement-to-sand ratios, and chemical compositions. Additionally, advanced modelling techniques such as multiple regression or machine learning should be employed. Such studies would provide a valuable complement to the predictive model proposed in this study.

5 Conclusions

The present study developed a predictive model for estimating the flexural strength of sustainable mortars incorporating WCP as a supplementary material. The predictive model exhibited high accuracy in estimating flexural strength, as confirmed by validation results. Key insights from this study include:

i. The incorporation of WCP influences the flexural strength of mortars, with the optimal replacement level ensuring a balance between strength development and sustainability.
ii. The predictive model offers a reliable tool for engineers and researchers to assess the flexural strength of mortars made with WCP, minimizing the need for extensive experimental testing.
iii. The application of the developed model would facilitate the design of sustainable mortars containing WCP.
iv. This study underscores the potential of WCP as a viable material for sustainable mortar production.

References

1. Benyahia A (2020) Bond behavior of self-compacting mortar containing construction and demolition waste under elevated temperatures. Asian J Civil Eng 22(3):405–415. https://doi.org/10.1007/s42107-020-00321-4

2. Florea MVA, Ning Z, Brouwers HJH (2014) Activation of liberated concrete fines and their application in mortars. J Constr Build Mater 50:1–12. https://doi.org/10.1016/j.conbuildmat.2013.09.012
3. Gastaldi D, Canonico F, Capelli L, Buzzi L, Boccaleri E, Irico S (2015) An investigation on the recycling of hydrated cement from concrete demolition waste. Cement Concr Compos 61:29–35. https://doi.org/10.1016/j.cemconcomp.2015.04.010
4. Kim YJ, Choi YW (2012) Utilization of waste concrete powder as a substitution material for cement. J Constr Build Mater 30:500–504. https://doi.org/10.1016/j.conbuildmat.2011.11.042
5. Ohemeng EA, Ekolu SO, Quainoo H, Naghizadeh A (2022) Economical and eco-friendly masonry mortar containing waste concrete powder as a supplementary cementitious material. Case Stud Constr Mater 17:e01527. https://doi.org/10.1016/j.cscm.2022.e01527
6. Ohemeng EA, Ramabodu MS, Nena TD, Kancheva T (2024) Performance of sustainable mortars containing blast furnace slag and fine concrete waste: an environmental perspective. Cogent Eng 11(1):2313053. https://doi.org/10.1080/23311916.2024.2313053
7. Ohemeng EA, Naghizadeh A (2023) Alternative cleaner production of masonry mortar from fly ash and waste concrete powder. Constr Build Mater 374:130859. https://doi.org/10.1016/j.conbuildmat.2023.130859
8. Ahmad SA, Ahmed HU, Mohammed BK, Rafiq SK, Gul-Mohammed JF (2024) Sustainable construction analytics: smart modeling for compressive strength prediction in Arabic gummodified mortar. Mater Circular Econ 6(1):16. https://doi.org/10.1007/s42824-024-00108-x
9. Armaghani DJ, Panagiotis GA (2021) A comparative study of ANN and ANFIS models for the prediction of cement-based mortar materials compressive strength. Neural Comput Appl 33:4501–4532. https://doi.org/10.1007/s00521-020-05244-4
10. Gayathri R, Rani SU, Cepová L, Rajesh M, Kalita KA (2022) Comparative analysis of machine learning models in prediction of mortar compressive strength. Processes 10:1387. https://doi.org/10.3390/pr10071387
11. Nguyen NH, Kien TT, Seunghye L, Armagan K, Thuc PV (2022) Prediction compressive strength of cement-based mortar containing metakaolin using explainable categorical gradient boosting model. Eng Struct 269:114768. https://doi.org/10.1016/j.engstruct.2022.114768
12. Sevim UK, Bilgic HH, Cansiz OF, Ozturk M, Atis CD (2020) Compressive strength prediction models for cementitious composites with fly ash using machine learning techniques. Constr Build Mater 271:121584. https://doi.org/10.1016/j.conbuildmat.2020.121584
13. Li L, Sun W, Hu W, Sun Y (2021) Impact of natural and social environmental factors on building energy consumption: based on bibliometrics. J Build Eng 37(2021):102136. https://doi.org/10.1016/j.jobe.2020.102136
14. Ohemeng EA, Ramabodu MS, Nana-Addy E (2025) Predictive models for strength properties of sustainable mortars containing waste concrete powder: an empirical approach. J Discov Sustain 6(815):1–14. https://doi.org/10.1007/s43621-025-01575-1
15. Bogas JA, Carrico A, Pereira MFC (2019) Mechanical characterization of thermal activated low-carbon recycled cement mortars. J Clean Prod 218:377–389. https://doi.org/10.1016/j.jclepro.2019.01.325
16. Chen X, Li Y, Bai H, Ma L (2021) Utilization of recycled concrete powder in cement composite: strength, microstructure and hydration characteristics. J Renew Mater 9(12):2190–2208. https://doi.org/10.32604/jrm.2021.015394
17. Dae-Jung M, Yang-Bae K, Jae-Suk R (2018) An approach for the recycling of waste concrete powder as cementitious materials. J Ceram Process Res 9(3):278–281
18. Duan Z, Hou S, Xiao J, Li B (2020) Study on the essential properties of recycled powders from construction and demolition waste. J Clean Prod 253:119865. https://doi.org/10.1016/j.jclepro.2019.119865

19. Moon DJ, Moon HY, Kim YB (2005) Fundamental properties of mortar containing waste concrete powder. Geosyst Eng 8(4):95–100. https://doi.org/10.1080/12269328.2005.10541243
20. Sui Y, Chuping O, Shu L, Zhang J, Qingbo T (2020) Study on properties of waste concrete powder by thermal treatment and application in mortar. Appl Sci 10(998):1–10. https://doi.org/10.3390/app10030998
21. Xianwei MW, Zhenyu W (2013) Effect of ground waste concrete powder on cement properties. Adv Mater Sci Eng 918294. https://doi.org/10.1155/2013/918294
22. Xiaoming K, Ying L (2019) Effect of recycled fine powder and mineral admixture on the properties of the cement mortar. IOP Conf Ser Earth Environ Sci 219:012026. https://doi.org/10.1088/1755-1315/219/1/012026
23. Zhang Y, Tongheng Z, Zhiwu L, Chaoqun M (2021) Effect of recycled powder on properties of cement mortar. E3S Web Conf 299:02009. https://doi.org/10.1051/e3sconf/202129902009
24. Lifecon DD (2003) Service life models: life cycle management of concrete infrastructures for improved sustainability. In: Final report by Dipl-Ing, Sascha Lay, Technical Research Centre (VTT) p 169
25. Ohemeng EA (2023) Potentials of waste concrete elements for production of construction materials. PhD Thesis. University of Johannesburg, Johannesburg, South Africa. Available from: http://hdl.handle.net/102000/0002
26. Ohemeng EA, Naghizadeh A, Ramabodu MS (2024) Empirical model for predicting elastic modulus of CRCA concrete: an approach towards sustainable concrete design. Adv Sci Technol 145:87–96. https://doi.org/10.4028/p-yOR735

Optimizing In-Situ Strengthening of Low-Strength Porous Mortar Using a Definitive Screening Design Study

Ren Wei[1], Edgar Rodriguez Alfaro[1], Tian Wang[2], and Min Wu[1](✉)

[1] Department of Civil and Architectural Engineering, Aarhus University, Aarhus C 8000, Denmark
mnwu@cae.au.dk
[2] Department of Mechanical and Production Engineering, Aarhus University, Katrinebjergvej 89G-F, Aarhus N 8200, Denmark

Abstract. Carbon Capture and Storage (CCS) is an essential technology to achieve the decarbonisation goals of our society. The use of non-producing oil and gas wells for CCS is appealing, while the secure sealing of the wells for long-term integrity remains a critical challenge. Scale deposition in oil and gas production pipes poses significant challenges in the plugging/sealing phase. The current practice requires milling and extraction of scale from the underground pipes, which is both costly and time-consuming. This study explores an innovative approach aiming to keep the mineral scale in-situ and use it as a (part of) barrier material for sealing the wells. However, this approach is hindered by the inherent low mechanical strength and high porosity of the scale. We investigate a method to chemically consolidate a low-strength limestone-based mortar, designed to simulate the scale from the oil and gas industry. The treatment employs diammonium hydrogen phosphate (DAP) to promote the formation of strengthening mineral phases within the porous matrix. A Definitive Screening Design (DSD) was used to evaluate the effects of six key treatment factors. The DSD model successfully identified the significant factors and their interactions influencing the compressive strength and water absorption of the mortar. Optimization based on the model predicted notable strength improvement. Validation experiments confirmed these predictions, achieving a 46.65% increase in compressive strength and a reduction of over 22% in water absorption. These findings provide strong proof-of-concept for in-situ scale strengthening, offering a potentially viable and cost-effective pathway for sustainable well plugging/sealing and abandonment.

Keywords: Carbon Capture and Storage (CCS) · Definitive screening design · Phosphate treatment · Material consolidation

1 Introduction

As the global community intensifies its efforts towards decarbonisation, Carbon Capture and Storage (CCS) has become a vital strategy to achieve the "Net-Zero" emissions goal [1, 2]. In this context, geological storage of CO_2 in non-producing oil and gas

H. Beushausen et al. (Eds.): NTZR 2025, LNCE 873, pp. 306–316, 2026.
https://doi.org/10.1007/978-3-032-27698-8_28

wells/reservoirs is an increasingly attractive option, repurposing existing infrastructure and utilising geological formations with proven capacities for long-term carbon sequestration. The success of these projects, however, fundamentally depends on maintaining the long-term integrity of all legacy wellbores, where effective plugging and sealing to prevent leaking of the stored CO_2 is critical [2, 3]. Mineral scale, which is the accumulation of various materials on the inner surface of the wellbore in critical sections of the oil and gas production pipes, poses significant challenges not only during the production phase but also in the plugging (as well as the abandonment) phase of the wells [4]. If left in the pipes, this scale can form a weak, porous interface with the plugging and barrier material (typically cement sheath), leading to compromised seals that endanger the safe storage of CO_2. To ensure long-term integrity of the CCS wells, section milling is normally required to establish a cross-sectional barrier directly towards formations. Milling of the deposited scale and/or the steel casing underground is time-consuming and costly, especially in regulatory regimes with substantial required plug lengths [5]. Retrieving the milled materials to surface is another challenge, and disposal of the materials can also cause environmental concerns [6].

This study investigates an innovative alternative aiming at transforming the scale in-situ into a competent barrier component. Such an approach presents a promising opportunity to avoid or reduce the milling operations, delivering significant cost and time savings while lessening environmental impacts through decreased cement use and waste generation. While promising, this concept faces a primary challenge: the inherent mechanical weakness of the scale itself. Usually characterised by a porous, layered structure rich in calcium carbonate, scale exhibits low strengths and high permeability, rendering it unsuitable as a reliable barrier without enhancement. Therefore, in-situ consolidation is crucial. Drawing inspiration from conservation science for cultural heritages [7], where diammonium hydrogen phosphate (DAP) is utilised to consolidate weathered carbonate stones, this study explores a chemical treatment based on the similar mechanisms. The principle involves the reaction of DAP with calcium carbonate in the scale to precipitate hydroxyapatite (HAP, $Ca_{10}(PO_4)_6(OH)_2$)—a biomimetic and highly durable mineral found in bone and tooth enamel—within the pore structure and defects, thereby densifying and strengthening the scale. First proposed in 2011 for stone preservation, phosphate (e.g. DAP) treatments have since been successfully applied to a variety of substrates, including sandstone, concrete, and murals [7]. Despite the potential of this phosphate treatment, its application to materials like wellbore scale remains largely unexplored, and critical knowledge gaps persist. Most existing research centres on single-variable effects, neglecting the complex interactions between multiple treatment parameters. Furthermore, the effectiveness of this method for significantly improving the strength of low-strength, high-porosity materials has not been systematically optimised. The second innovation of this study is, therefore, to address this gap by conducting a systematic, multi-factorial investigation to understand and optimise treatment for a scale-analogous material. Additionally, obtaining and preparing real scale from the oil and gas industry into standard test samples for characterising strengthening effects is a challenge.

Building on these insights, this study aims to address this complexity efficiently. We employed a Definitive Screening Design, a powerful statistical approach capable of modelling main effects, two-factor interactions, and critical quadratic effects with a minimal number of experimental runs [8, 9]. A key advantage of definitive screening designs compared to traditional screening designs is their ability to keep main effects orthogonal, avoiding aliasing with quadratic effects and two-way interactions. This allowed for a rigorous analysis of six key process parameters in this study: DAP concentration, $CaCl_2$ concentration, treatment duration (reaction time), pH, temperature, and ethanol content. Due to the difficulty in obtaining homogeneous real-world scale (stringent authority approval needed), a low-strength limestone-based mortar was developed to simulate its mechanical and chemical properties for this systematic study. This study, for the first time, proposes a comprehensive optimization of the DAP treatment for strengthening a simulated wellbore scale. The main objectives include to: (1) investigate the potential of this inorganic chemical route to significantly improve the properties of a low-strength, porous scale-analogue material; (2) quantify the improvements in key mechanical and physical properties, namely compressive strength and water absorption; (3) develop a robust, optimised treatment process by identifying the critical parameter interactions that govern the strengthening effect, which lays the foundation for further optimisation and provides valuable insights for future improvement.

2 Experimental

2.1 Preparation of the Low-Strength Scale-Analogue Mortar

Due to the difficulty in sourcing homogeneous real-world scale for standardized testing, a low-strength limestone-based mortar was developed to serve as a consistent analogue in this explorative study. The sample was prepared using a Type II/A-LL 52.5 N Portland cement and a fine limestone aggregate. Table 1. shows the chemical compositions of the cement. Following a series of mix design trials, a final composition with a water-to-cement (W/C) ratio of 1.0 and a limestone-to-cement (L/C) ratio of 6.0 was selected. After casting, the samples were sealed with plastic sheeting to minimize moisture evaporation and cured in a temperature-controlled environment maintained at 25 ± 1 °C for 24 h. Afterwards, the samples were removed from their moulds and immersed completely in tap water, where they underwent further curing at 75 ± 1 °C for a period of 14 days (the purpose was to mature the samples rather quickly, and our previous tests showed that the strength development afterwards was very slow). After the curing process, the samples were taken out of the water, and any surface moisture was promptly removed using a dry cloth. The saturated mass ($W_{\_sat,\ initial}$) of each sample was measured with an accuracy of 0.01 g. Subsequently, the samples were oven-dried at 105 ± 2 °C for 24 h, cooled in a desiccator, and weighed to obtain their dry mass ($W_{\_dry,\ initial}$). The above formulation consistently produced $5 \times 5 \times 5\ cm^3$ cubic specimens with an initial average compressive strength of 8.12 MPa and a water absorption of 11.32% (the properties were close to the real scale extracted from an oilfield in the North Sea [10]), establishing a reliable baseline for the strengthening experiments. The dried samples were stored in sealed plastic bags at room temperature until the commencement of the DAP treatment.

Table 1. Chemical compositions of the cement (wt.%).

Oxide	CaO	SiO_2	Al_2O_3	Fe_2O_3	SO_3	K_2O	MgO	P_2O_5
Cement	67.80	17.0	5.16	3.21	4.52	0.49	0.87	0.37

2.2 Definitive Screening Design (DSD) and the Treatment Procedure

A Definitive Screening Design (DSD) was employed to systematically and efficiently investigate the effects and interactions of six key process parameters on the strengthening effects on the properties of the samples. The chosen factors were DAP concentration, $CaCl_2$ concentration, treatment duration, pH, temperature, and ethanol content. The rationale for these factors is to promote the formation of calcium phosphate (CaP) phases within the pores of the samples; for instance, elevated temperature can accelerate ion transfer, while ethanol can influence the microstructure of the precipitates. The experimental parameters and their respective levels are summarised in Table 2.. A total of 17 runs were conducted, which included one centre point and four additional replicate centre points to enhance the estimation of experimental error, see Table 3..

Table 2. Experimental parameters and levels for the definitive screening design.

Parameters		Unit	Minimal level	Central level	Maximal level
X_1	DAP concentration	M	0.10	1.55	3.00
X_2	$CaCl_2$ concentration	M	0.00010	0.00155	0.00300
X_3	pH	–	8	10	12
X_4	Temperature	°C	25	53	80
X_5	Treatment duration	Days	1	4	7
X_6	Ethanol content	%	0	5	10

For each of the 17 experimental runs, the treatment solutions were prepared by dissolving DAP and $CaCl_2$ powders in deionized water. The pH was then adjusted potentiometrically using 0.01 M NaOH, measured with a low-alkali-error glass pH electrode (LAQUA, HORIBO, Japan). These solutions were maintained at their target temperatures in a calibrated oven, with sealed containers to prevent evaporation. Fully submerged cubic specimens in sealed plastic containers were immersed in the solutions, maintaining a 5:1 solution-to-mortar weight ratio. This immersion method ensured uniform DAP distribution and penetration. To differentiate the effects of chemical treatment from continued further cement hydration, a control set of samples was immersed in deionized water under the identical temperature and duration conditions for each run. After the treatment period, the samples were removed, rinsed with deionized water to remove surface residues, and dried in an oven at 105 °C until reaching constant weight. For each run, three samples were tested to compare results and calculate an average.

Table 3. Definitive-screening design for the phosphate treatment.

Runs	X_1	X_2	X_3	X_4	X_5	X_6
1	−1	−1	−1	−1	1	0
2	1	−1	1	−1	0	1
3	0	0	0	0	0	0
4	1	1	0	−1	1	1
5	0	0	0	0	0	0
6	0	1	−1	−1	−1	−1
7	−1	1	1	0	1	−1
8	−1	1	−1	1	0	1
9	0	0	0	0	0	0
10	1	0	−1	1	1	−1
11	−1	0	1	−1	−1	1
12	0	0	0	0	0	0
13	−1	−1	0	1	−1	−1
14	0	0	0	0	0	0
15	0	−1	1	1	1	1
16	1	1	1	1	−1	0
17	1	−1	−1	0	−1	1

Note "−1", "0" and "1" indicate the "minimal", "central" and "maximal" levels, respectively

2.3 Sample Characterisation

To quantify the efficacy of the DSD treatment protocols, both the treated and the control samples were characterized for their primary mechanical and physical properties. For each experimental condition, the average result from three replicate specimens was reported.

(1) Water Absorption: water absorption was calculated to evaluate changes in the accessible pore volume potentially induced by the treatment. The dry mass (Wdry) and saturated mass (Wsat) of the samples were determined based on the procedure described in Sect. 2.1. The water absorption percentage was calculated using the formula:

$$Water\,Absorption(\%) = [(Wsat - Wdry)/Wdry] \times 100 \quad (1)$$

Measurements were performed on three samples per condition, and the mean value was recorded. Both initial (before treatment) and final (after treatment) water absorption values were determined.

To isolate the effect of the chemical treatment from changes caused by further hydration, the final water absorption values were adjusted. The "Water absorption_Excluding Hydration" was calculated by taking the initial water absorption of the untreated mortar and adding the difference between the water absorption of the treated samples and the water absorption of its corresponding control samples with further hydration, as shown in the following Equation:

$$\begin{aligned} &\textit{Water absorption\_Excluding Hydration} \\ &= \textit{Water absorption\_Initial} \\ &\quad + (\textit{Water absorption\_Treated} - \textit{Water absorption\_Hydrated}) \end{aligned} \tag{2}$$

This adjusted metric ensures that the analysis focuses solely on the changes induced by the DAP treatment. The final performance was evaluated using the Increment Ratio (as shown below), which measures the relative change from the original state:

$$\begin{aligned} \textit{Increment Ratio}\ (\%) = &\ (\textit{Water absorption\_Excluding Hydration} \\ &/\textit{Water absorption\_Initial}) \times 100 \end{aligned} \tag{3}$$

(2) The uniaxial compressive strength was measured to assess the impact of the DAP treatment on the load-bearing capacity of the treated samples. In accordance with ASTM C642, specimens were conditioned to a saturated-surface-dry state before testing. The tests were conducted on a 250 kN electro-hydraulic servo universal testing machine (FORM+TEST Seidner & Co., Germany) at a constant loading rate of 250 N/s until failure. The final compressive strength of the treated samples was compared to both their initial condition and the corresponding control samples with further hydration. A similar correction for hydration effects, comparable to that used for water absorption, was applied to the compressive strength data before modelling.

3 Results and Discussions

The experimental data for the tested samples were analysed using the Fit Definitive Screening platform in JMP software. This platform employs an "Effective Model Selection" methodology specifically designed for DSDs, which automatically identifies and models the significant main effects, two-factor interactions, and quadratic effects to build predictive models for the responses of the materials. All results presented have been adjusted to eliminate the effects of further hydration of cement during the phosphate treatment periods.

3.1 Analysis of the Compressive Strength

The DAP-based treatment demonstrated a highly variable impact on the compressive strength of the samples, with outcomes ranging from an increase of 53.13% to a notable decrease of −14.26%. The DSD generated a statistically significant and robust predictive model for the compressive strength ($R^2 = 0.86$, $p = 0.0028$), the quality of which is visualized in the Actual by Predicted plot (Fig. 1).

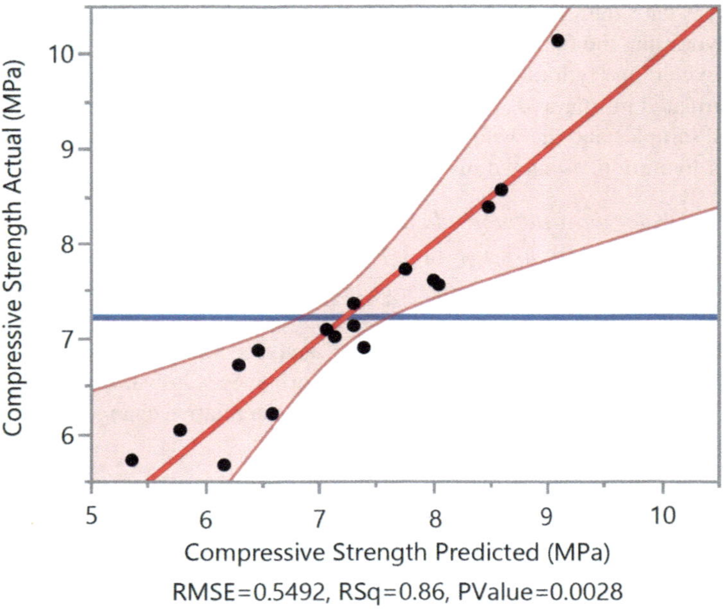

Fig. 1. Actual by predicted plot for the compressive strength model, demonstrating high prediction accuracy.

The summary of the effects by the model (Fig. 2) revealed that the strengthening mechanism is governed primarily by complex interactions. A significant negative interaction between DAP and $CaCl_2$ concentration (coefficient = −0.809, LogWorth = 2.975) and a strong positive interaction between temperature and $CaCl_2$ concentration (coefficient = +0.604, LogWorth = 2.202) were the most influential terms. A significant quadratic effect for pH was also identified, indicating a non-linear response where an optimal alkaline environment is required to maximise the strength. This aligns with the chemical principle that pH governs the deprotonation of phosphate ions, a key step in the precipitation of strengthening mineral phases. Using the Prediction Profiler (Fig. 3), the optimal parameter settings were identified as 3 M DAP, 0.0001 M $CaCl_2$, 25 °C, and a pH of 12, which was predicted to increase the compressive strength by 40.84%.

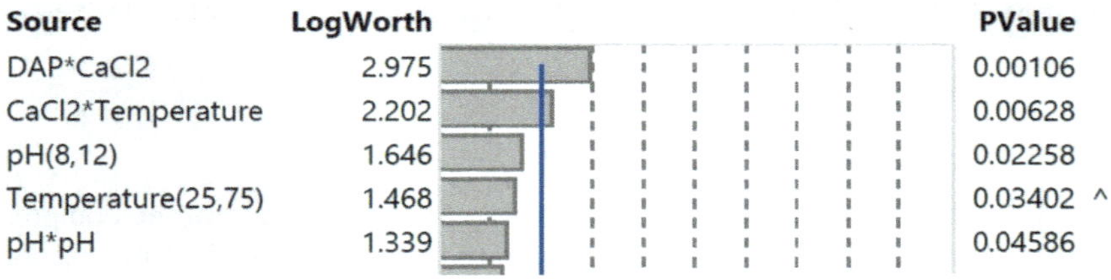

Source	LogWorth	PValue
DAP*CaCl2	2.975	0.00106
CaCl2*Temperature	2.202	0.00628
pH(8,12)	1.646	0.02258
Temperature(25,75)	1.468	0.03402 ^
pH*pH	1.339	0.04586

Fig. 2. Effect summary of the studied parameters on the compressive strength.

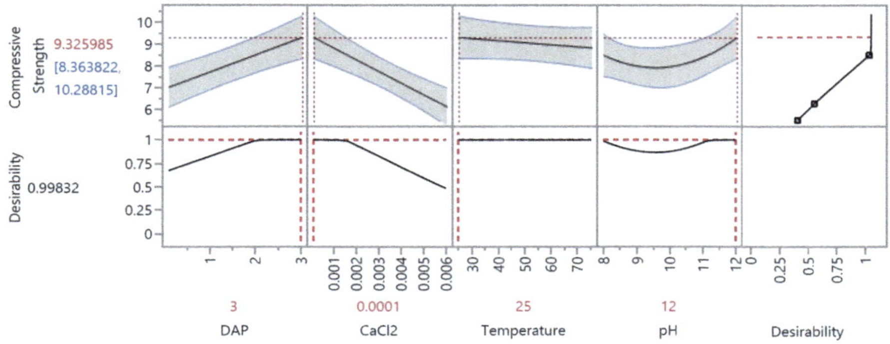

Fig. 3. Prediction profiler showing the maximized desirability for the compressive strength.

3.2 Analysis of the Water Absorption

Water absorption, a key indicator of permeability and porosity, was analysed alongside the compressive strength. The treatment generally lowered the water absorption, with the most effective run achieving a 22.29% ratio reduction. However, consistent with the compressive strength results, some factor combinations proved harmful, increasing water absorption ratio by up to 12.64%. The DSD model for water absorption (Fig. 4) identified pH as the most influential factor (LogWorth = 1.215), followed by treatment duration/time (LogWorth = 0.883). Unlike the compressive strength model, the water absorption model indicated that simpler, main effects were more significant than complex interactions. Longer treatment times and higher DAP concentrations primarily contributed to decreasing water absorption, supporting the idea that increased reaction duration and reactant levels promote more complete mineral precipitation and pore filling. The optimal conditions for minimising water absorption, according to the model (Fig. 5), were 3 M DAP, pH 8, and a 7 day treatment, predicted to reduce the water absorption ratio by 19.19%.

Source	LogWorth	PValue
pH(8,12)	1.215	0.06100
Time(1,7)	0.883	0.13104
DAP(0.1,3)	0.485	0.32710

Fig. 4. Effect summary for the water absorption, showing the influence of significant factors.

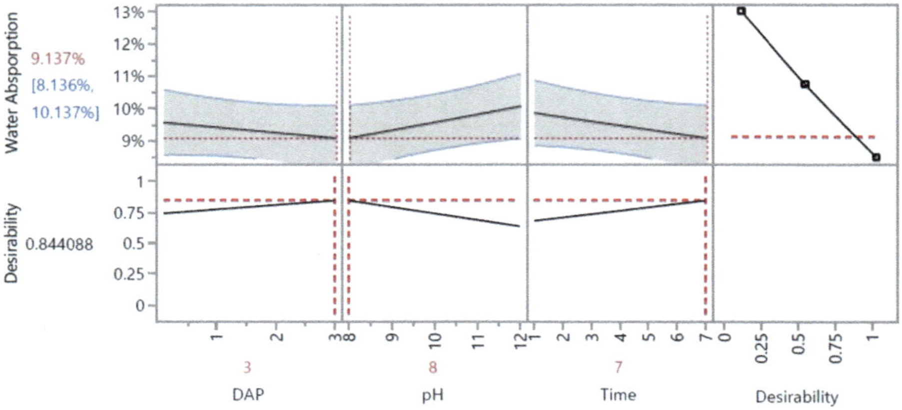

Fig. 5. Prediction profiler showing the maximized desirability for the water absorption.

3.3 Discussions and Validation

A direct comparison of the compressive strength and the water absorption analyses reveals a strong inverse correlation, which is fundamental to the success of this consolidation treatment. The experimental runs that produced the highest gains in compressive strength consistently resulted in significant reductions in water absorption (e.g., Run 2: + 53.13% compressive strength, −17.63% water absorption ratio). This powerful link confirms the underlying mechanism of the DAP treatment: the precipitation of calcium phosphate phases within the pore network of the samples. This pore-filling action simultaneously reduces porosity (lowering the water absorption) and creates a denser, more consolidated microstructure (increasing the compressive strength). Interestingly, the models show that the optimal conditions for maximizing strength (high pH, short treatment duration) differ from those for minimizing water absorption (neutral pH, long treatment duration). This suggests that different precipitation kinetics may favour either structural reinforcement or simple pore-sealing. To resolve this, a final corroboration test was performed using the conditions optimized for compressive strength, as this is a primary requirement for a barrier material. The validation experiment was highly successful. The treatment not only achieved a 46.65% increase in compressive strength, surpassing the prediction of the model, but also delivered a substantial and highly beneficial 22.64% ratio reduction in water absorption. This confirms that optimizing for the primary mechanical property also yields excellent results for the secondary physical property. These findings validate the DSD-optimized DAP treatment as a highly effective method for transforming a weak, porous, and permeable material into a more robust, and lower-permeable barrier, paving the way for potentially using the scale as (part of) a plugging and sealing material.

However, it should be mentioned that despite the promising results obtained in this proof-of-concept study, there is more work to be conducted: (1) the simulated scale should be further strengthened to align with the performance requirements of barrier materials intended for sealing, plugging, and abandonment applications in oil and gas wells; (2) the current results include mechanical strength and water absorption; to further clarify the strengthening mechanisms of the approach, microstructural and mineralogical analyses (e.g. SEM and XRD) are to be performed to confirm the precipitation of hydroxyapatite within the mortar pores; (3) the strengthening effects on real scale will need to be verified; (4) durability testing under simulated downhole conditions (e.g. CO_2 exposure or thermal cycling) are necessary to assess the long-term performance and applicability of the proposed treatment; (5) scale-up plans need to be developed for applying the approach in actual well conditions, considering factors including temperature, pressure, and chemical compatibility.

4 Conclusions and Perspectives

This study successfully demonstrated that a DAP-based chemical treatment, optimized through a Definitive Screening Design (DSD), can significantly enhance the mechanical and physical properties of a low-strength mortar simulating oil and gas wellbore scale. The key conclusions are as follows:

(1) High Efficacy of Optimized Treatment: A systematic, multi-factorial investigation was essential for navigating the complex parameter interactions. The DSD-optimized treatment was experimentally validated, achieving a remarkable 46.65% increase in compressive strength and a concurrent 22.64% reduction ratio in water absorption.
(2) Mechanism Confirmation: The strong inverse correlation between increased compressive strength and decreased water absorption confirms that the primary mechanism of the treatment is the precipitation of mineral phases within the pore network, creating a denser, stronger, and less permeable material.
(3) Validation of a Novel Plugging and Abandonment Strategy: The findings provide strong proof-of-concept for the in-situ consolidation of wellbore scale. This approach indicates the possibility of a novel strategy to transform a waste material into a competent barrier component, offering a potentially more cost-effective and sustainable pathway for plugging and abandonment of oil and gas wells.

Future work should focus on further improving the strengthening effects, e.g. possibly combining the phosphate treatment with other methods, and validating the effects on scale materials. The long-term durability of the treatment under simulated downhole conditions should also be verified to establish the basis for potential field applications.

Although targeting at strengthening the mineral scale from the oil and gas industry to develop innovative solutions for sealing CCS wells (as well as general well abandonment), the current work remains highly relevant in a broader sense for enhancing almost all calcium-containing systems, e.g. cementitious materials. Strengthening low-strength cementitious systems, such as degraded cement and concrete materials, would help to extend the service lives of existing concrete structures and promote the reuse of concrete elements, among others [10]. Such strategies are important to facilitate the decarbonisation of the carbon-intensive built environment.

Acknowledgements. The authors kindly acknowledge the support received from the Danish Offshore Technology Centre (DTU Offshore) and COWI Foundation for this work.

References

1. Lin Q, Zhang X, Wang T, Zheng C, Gao X (2022) Technical perspective of carbon capture, utilization, and storage. Engineering 14:27–32
2. Shi X, Zhang C, Gupta KK, Ambat R, Wu M (2025) Risks of cement and rock-cement-metal interface degradation in geological carbon sequestration reservoirs: Mechanisms, influencing factors and mitigation measures. Carbon Capture Sci Technol 100419
3. Shi X, Wu M (2023) Cement degradation in geological carbon sequestration conditions. In: 16th International Conference on Durability of Building Materials and Components (XVI DBMC), p 12. Beijing, China
4. Li J, Tang M, Ye Z, Chen L, Zhou Y (2017) Scale formation and control in oil and gas fields: a review. J Dispers Sci Technol 38(5):661–670
5. Vrålstad T, Saasen A, Fjær E, Øia T, Ytrehus JD, Khalifeh M (2019) Plug & abandonment of offshore wells: Ensuring long-term well integrity and cost-efficiency. J Pet Sci Eng 173:478–491
6. Gao J, Liu Q, Song L, Shi B (2019) Risk assessment of heavy metals in pipe scales and loose deposits formed in drinking water distribution systems. Sci Total Environ 652:1387–1395
7. Sassoni E (2018) Hydroxyapatite and other calcium phosphates for the conservation of cultural heritage: a review. Materials 11(4):557
8. Jones B, Nachtsheim CJ (2011) A class of three-level designs for definitive screening in the presence of second-order effects. J Qual Technol 43(1):1–15
9. Erler A, de Mas N, Ramsey P, Henderson G (2013) Efficient biological process characterization by definitive-screening designs: the formaldehyde treatment of a therapeutic protein as a case study. Biotechnol Lett 35(3):323–329
10. Miah MJ, Thygesen J, Simonsen ME, Nielsen RP, Wu M (2025) Strengthening the bulk properties of cement mortar through promoted in-situ formation of hydroxyapatite: Feasibility and perspectives. Constr Build Mater 459:139773

GWP Assessment Approach for UHPC Mixtures Using Ingredient-Level EPDs

Leila Farahzadi(✉) and Mahdi Kioumarsi

Department of Built Environment, OsloMet—Oslo Metropolitan University, Pilestredet 35, 0166 Oslo, Norway
leilafar@oslomet.no

Abstract. Ultra-High-Performance Concrete (UHPC) is increasingly used in modern infrastructure due to its exceptional mechanical and durability properties. However, the environmental performance of UHPC remains difficult to assess accurately because standardized Environmental Product Declarations (EPDs) are lacking. This is primarily due to the diversity and novelty of UHPC mix designs, which differ significantly from conventional concrete. To address this gap, this study developed a transparent and flexible Excel-based platform to calculate the Global Warming Potential (GWP) of UHPC mixtures based on the life cycle data of individual ingredients. Using unit GWP values derived from Ecoinvent v3.9.1 and manufacturer-specific EPDs, the tool estimated the cradle-to-gate GWP for a functional unit of 1 m^3 of UHPC. Several previously studied UHPC mixes were reassessed using the new platform, and the results were compared with earlier GWP values obtained from SimaPro tool. The findings revealed that the developed tool enables consistent and reproducible assessment of UHPC environmental performance, supporting more informed decisions in sustainable construction.

Keywords: Ultra-High-Performance Concrete (UHPC) · Global Warming Potential (GWP) · Environmental Product Declaration (EPD) · Life Cycle Assessment (LCA) · Sustainability · Excel Tool

1 Introduction

Ultra-High-Performance Concrete (UHPC) is an advanced class of cement-based composites that has gained growing attention in structural engineering due to its exceptional mechanical and durability properties [1, 2]. With compressive strengths exceeding 120 MPa [3], high tensile ductility, and low permeability [4], UHPC exhibits outstanding resistance to abrasion, freeze-thaw cycles, and chemical deteriorations [5]. These characteristics make it highly suitable for infrastructure applications such as bridges, marine structures, high-rise buildings, and precast façade elements, where long-term performance and material efficiency are critical. The use of UHPC allows for thinner structural elements, longer spans, and reduced maintenance over the service life of structures, thus offering both functional and economic advantages.

Despite these benefits, the environmental footprint of UHPC remains a concern. UHPC typically incorporates three times more cement than conventional concrete [6]

H. Beushausen et al. (Eds.): NTZR 2025, LNCE 873, pp. 317–326, 2026.
https://doi.org/10.1007/978-3-032-27698-8_29

(Fig. 1), and often includes additional high-impact components such as steel fibers. As cement and steel are among the highest contributors to global CO_2 emissions, approximately 8% and 7–8%, respectively [7–9], the carbon intensity of concrete materials [10, 11] and UHPC formulations can be significant. Thus, while UHPC holds potential for long-term sustainability through durability and material reduction, this must be balanced against its high embodied emissions at the production stage [12–14].

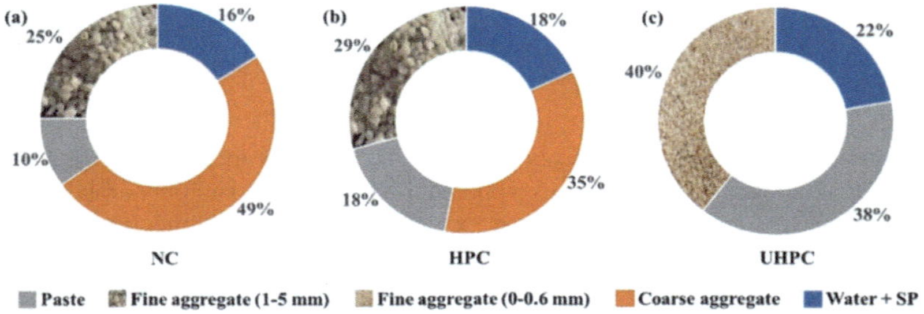

Fig. 1. Components of different types of concrete: **a)** Normal Concrete (NC), **b)** High-Performance Concrete (HPC), **C)** Ultra-High-Performance Concrete (UHPC) [15].

Life Cycle Assessment (LCA) provides a robust framework for evaluating environmental impacts such as Global Warming Potential (GWP) [16–18], but applying it to UHPC is particularly challenging. Due to the material's relatively recent emergence and wide variation in proprietary mix designs. Furthermore, data and standardized Environmental Product Declarations (EPDs) for UHPC are lacking. This absence limits the ability of engineers, researchers, and decision-makers to evaluate and compare the sustainability of different UHPC formulations. Furthermore, most LCA databases and software tools do not include detailed datasets for unique UHPC components like steel fibers or different and novel Supplementary Cementitious Materials (SCMs), requiring either proxy data or time-intensive manual modeling [19, 20]. As a result, existing studies often rely on simplified assumptions or fragmented data, limiting their reliability and applicability.

To address this gap, the study presents a transparent, adaptable, and user-friendly Excel-based platform for calculating the cradle-to-gate GWP of UHPC mixtures. The tool enables users to input mix proportions and link them with verified GWP values from recognized sources such as Ecoinvent or product-specific EPDs. By calculating the environmental impact at the ingredient level, the platform facilitates consistent and reproducible comparisons between UHPC formulations. To validate its effectiveness, several previously studied UHPC mixes are reassessed using this new method, revealing discrepancies with earlier estimates and identifying the key drivers of GWP in each case. The mixes were selected from UHPC formulations designed and texted at Iowa State University Concrete Lab, each with compressive strength exceeding 120 MPa and documented SimaPro results available for comparison.

This approach provides a practical and scalable methodology for evaluating UHPC's environmental performance, especially in early-stage design and research settings. While not a substitute for full LCA modeling or standardized EPDs, the proposed Excel tool offers an interim solution that promotes transparency, comparability, and informed material selection in low-carbon construction.

2 Materials and Methods

This study applied an LCA framework to quantify the GWP of various UHPC mixtures. The methodology was structured into three main components: data collection and inventory development, the design and development of an Excel-based calculation platform, and the comparative assessment of previously analyzed UHPC mixtures using the new tool. The assessment adheres to ISO 14040 and ISO 14044 standards [21, 22], with system boundaries defined according to the EN 15804 + A2 standard [23], focusing on cradle-to-gate stages (A1–A3) (Fig. 2). This includes raw material extraction, transportation, and manufacturing, providing an appropriate scope for evaluating UHPC's production-phase emissions before construction or end-of-life stages.

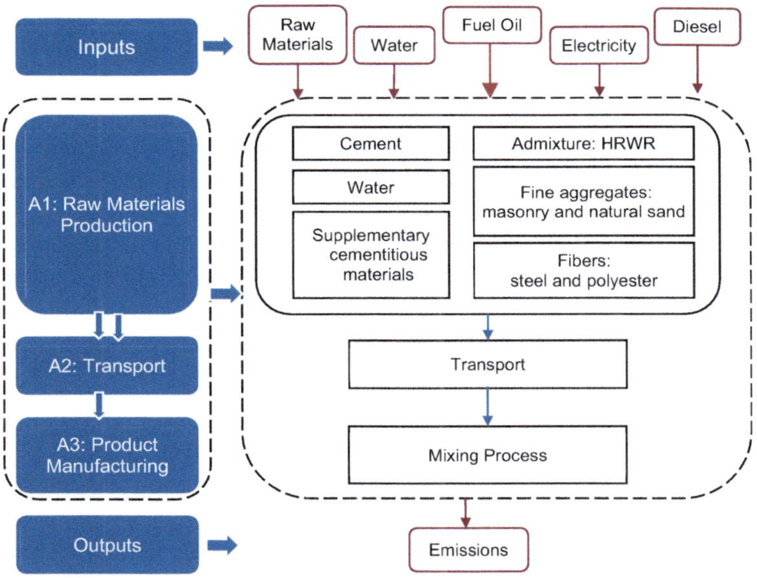

Fig. 2. System boundary considered for this study, adapted from [24].

The functional unit for this study was defined as 1 cubic meter (1 m^3) of UHPC. This volume standardizes the environmental comparison across different UHPC mixtures, regardless of their specific compositions, provided that they deliver comparable performance. The selected mixes achieved characteristic compressive strengths above 120 MPa, aligning with ASTM C1856 [3], thereby ensuring a consistent performance baseline. All material inputs and emissions were normalized to this unit, allowing for an equitable evaluation of environmental burdens across formulations with varying binder content, additives, or fibers.

2.1 Data Collection and Inventory Development

The UHPC mixtures analyzed in this study were selected based on laboratory-tested mixtures published in previous studies [13, 25]. These included both conventional high-cement UHPC formulations and alternative versions incorporating SCMs such as recycled glass powder and Portland Limestone Cement (PLC). Fine aggregates primarily consisted of masonry and natural sand. Masonry sand refers to clean, natural fine aggregate typically used in masonry and concrete aggregate. The binder system was typically based on ordinary Portland cement (OPC), partially replaced by SCMs to reduce clinker content. Steel and alternative (polyester) fibers were incorporated to enhance tensile strength and ductility, while high-range water-reducing admixtures (HRWR) are used to achieve low water-to-binder ratios without compromising workability.

Cradle-to-gate environmental data for each UHPC constituent were compiled as GWP values expressed in kilograms of CO_2 equivalent per kilogram of material. These values were primarily sourced from the Ecoinvent database (v3.9.1) [26], with data sets aligned to the EN 15804 system model to ensure consistency with European standards for EPDs. Where available, manufacturer-specific EPDs were also integrated. In cases where direct EPDs were unavailable, proxy data were used based on comparable materials or validated literature values.

In addition to material-related emissions, this study included energy consumption during factory mixing as part of the GWP assessment. UHPC typically requires an extended mixing duration due to its low water content and dense particle packing; in industrial-scale production, the mixing process lasts between 25 to 30 minutes. The energy demand for this process depends on mixer size, batch volume, and motor efficiency. For the purposes of this study, the electricity consumption per cubic meter of UHPC was estimated based on typical factory mixer specifications and was added as a separate input in the Excel platform. The corresponding GWP was calculated by multiplying the estimated energy consumption (in kWh) by the country-specific electricity emission factor, based on Norway's electricity mix. The GWP assessment accounted for both direct and indirect energy consumption. Indirect energy use, including raw material extraction, processing, and upstream transport, is embedded in the A1–A3 stages via the EPDs for each ingredient. In addition, energy consumed during the mixing process at the production site was explicitly calculated based on estimated electricity usage (kWh). This value was then multiplied by Norway's average electricity grid emission factor (17 g CO_2-eq/kWh), reflecting the country-specific low-carbon electricity mix. Although this contribution is minor due to Norway's clean energy profile, its inclusion enhances the completeness of the cradle-to-gate boundary and ensures alignment with EN 15804 guidelines. Transportation emissions were also considered. An average transport distance of 100 kilometers was assumed. Transport emissions were estimated using an emission factor of 0.1 kg CO_2-eq per ton-kilometer, based on average standard truck transport values in Europe [26].

2.2 Excel Platform Development and Calculation Framework

To support transparent and flexible assessment, an Excel-based platform was developed. The platform was designed to calculate the total GWP associated with 1 m^3 of UHPC based on the specific mix proportions and the environmental impact of each constituent. The core structure of the tool includes three interactive components: a material database, a mix design input sheet, and a results dashboard. The material database lists all UHPC ingredients along with their respective GWP values and data sources. It is designed to be updatable, allowing users to revise values when new EPDs become available. The GWP values for each material were obtained from verified Environmental Product Declarations (EPDs) when available, and from the Ecoinvent v3.9.1 database for materials lacking specific EPDs, all reflecting cradle-to-gate (A1–A3) emissions. The input sheet allows users to define the mass of each component in the mix (in kg/m^3), using dropdown menus to select from the material list. Input fields are also provided for transport distances and energy use. The tool employs embedded functions to link input values with the environmental database. It also includes conditional formatting and automated charts (e.g., pie and bar graphs) that help visualize the contribution of each ingredient to the total GWP, facilitating interpretation and comparison.

The results dashboard calculated the total GWP for each UHPC mixture using the formula:

$$\begin{aligned} GWP_{total} = & \sum(m_i \times GWP_i) + \text{Transport Impact} \\ & + \text{Mixing Energy Impact} \end{aligned}$$

where:

- m_i is the mass of component i in the UHPC mix (kg/m^3)
- GWP_i is the Global Warming Potential of component i (kg CO_2-eq/kg)
- Transport Impact is the optional transport-related emission, calculated as $m_i \times Distance_i \times EF_{transport}$, where Distance is in km and $EF_{transport}$ is the emission factor
- Mixing Energy Impact is the GWP from electricity used during UHPC production mixing, calculated as:

$$\text{Mixing Energy Impact} = E_{mix} \times EF_{electricity}$$

where:

E_{mix} is the estimated electricity consumption for mixing 1 m^3 of UHPC (in kWh)
$EF_{electricity}$ is the emission factor for the local electricity grid (in kg CO_2-eq/kWh)

In this study, the mixing process was assumed to last 25–30 minutes based on standard factory-scale UHPC production protocols. The electricity demand per batch was estimated using industrial mixer specifications and scaled to 1 m^3 of UHPC output. For Norway, where most of the production scenarios are assumed, a low-carbon electricity mix was applied; however, the Excel tool allows for this value to be customized to reflect regional variations.

2.3 Comparative Assessment of UHPC Mixtures and Validation

To demonstrate the utility and reliability of the platform, several previously assessed UHPC mixtures were re-evaluated using the Excel tool. These included conventional UHPC, mixes incorporating SCMs as partial cement replacements, and fiber-reinforced

UHPC variants using different types and quantities of reinforcement. The recalculated GWP values were then compared to those originally reported using legacy tools or simplified assumptions. Table 1 illustrates the UHPC alternatives with different components' type and dosage.

Table 1. Mixture designs of UHPC, with data collected for 1 m^3 of UHPC.

	OPC	PLC	Silica fume	Glass powder	Natural sand	Masonry sand	Water/binder	HRWR	Steel fibers	Polyester fibers
Conventional UHPC	930		70		800	400	186.6	45	156	0
UHPC 1	850	–	–	150	800	400	186.6	45	156	0
UHPC 2	850	–	–	150	800	400	186.6	45	117	6.7
UHPC 3	850	–	–	150	800	400	186.6	45	78	13.4
UHPC 4	–	850	–	150	800	400	186.6	45	156	0
UHPC 5	–	850	–	150	800	400	186.6	45	117	6.7
UHPC 6	–	850	–	150	800	400	186.6	45	78	13.4

To ensure the robustness of the results, several validation steps were undertaken. Manual recalculations of selected mixes were performed to verify the automated outputs of the Excel platform. Additionally, the results were compared with available LCA studies on similar UHPC systems in the literature.

2.4 Limitations

While the proposed platform offers a practical and transparent approach to assessing UHPC's environmental impact, certain limitations remain. Data availability continues to be a challenge, particularly for emerging materials and novel additives. In such cases, the use of proxy data introduces uncertainty that should be acknowledged in interpretation. The cradle-to-gate scope does not account for energy consumption during curing, onsite construction activities, or end-of-life treatment, all of which can influence UHPC's full life cycle impact. Transport distances were assumed as average values; however, in real-world applications, these can vary based on material sourcing and production location, potentially influencing the overall GWP results.

3 Results and Discussions

The developed Excel-based assessment tool (Fig. 3) achieved consistent, yet systematically lower GWP values compared to the SimaPro evaluations for all examined UHPC mixtures, as presented in Table 2 The conventional UHPC mix demonstrated the highest absolute GWP at 1239.1 kg CO_2-eq/m^3 in our assessment, which was 1.82% lower than the SimaPro result. This pattern of slightly reduced GWP values persisted across all optimized mixtures, with differences ranging from 14.7 to 32.3 kg CO_2-eq/m^3

(1.64% to 2.68% variation), suggesting a small but consistent methodological divergence between the two approaches. The differences between the Excel-based and SimaPro GWP results primarily stem from discrepancies in data sources and modeling detail. The Excel tool relies on updated, material-specific EPDs, while SimaPro may use more generic or region-based LCI entries. Additionally, the Excel model applies cradle-to-gate boundaries (A1–A3) and a simplified aggregation method, whereas SimaPro includes broader background processes and indirect emissions. Subtle differences in transportation assumptions and energy modeling may also contribute. Despite these variations, the GWP differences, the close agreement between methods ranging from 1.64% to 2.68%, remain small, confirming the Excel platform's reliability for early-stage sustainability assessments and comparative analysis. The Excel tool's ability to rapidly evaluate "what-if" scenarios proved valuable for iterative design by modifying mix designs, materials' dosage, and EPD data,

Table 2. Comparative analysis of GWP assessments in excel platform and SimaPro software

Mixtures	Total GWP (kg CO_2eq/m^3) assessed in the excel file	Total GWP (kg CO_2eq/m^3) assessed in SimaPro	Difference (kg CO_2eq/m^3)	Percentage change (%)
Conventional UHPC	1239.1	1262	−22.9	−1.82
UHPC 1	1174.7	1207	−32.3	−2.68
UHPC 2	1104.5	1132	−27.5	−2.43
UHPC 3	1034.3	1057	−22.7	−2.15
UHPC 4	1021.7	1047	−25.3	−2.42
UHPC 5	951.5	971	−19.5	−2.01
UHPC 6	881.3	896	−14.7	−1.64

Analysis of impact contributions revealed cement production as the dominant emissions source, accounting for 68–72% of total emissions in conventional mixes, with optimized formulations showing reduced contributions of 58–65% through effective cement replacement strategies. The progressive GWP reduction from conventional UHPC to UHPC 6 (28.9% decrease) demonstrates the significant potential for emissions reduction through mix design optimization, where each 100 kg/m^3 cement reduction typically yielded a 75–85 kg CO_2-eq/m^3 GWP decrease. However, this relationship became non-linear at lower cement contents as other constituents like steel fibers (12–15% of total GWP) and superplasticizers (5–8%) gained relative importance in the overall impact profile.

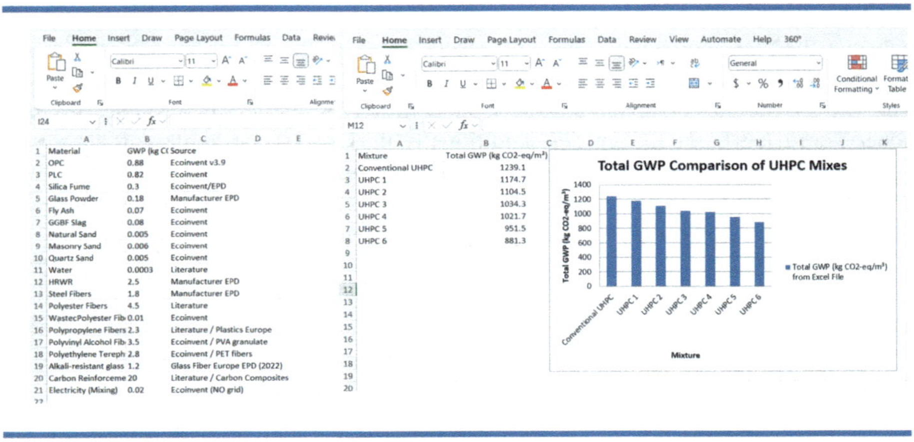

Fig. 3. The interface of the excel platform for GWP calculation

4 Conclusion

This study developed and validated an Excel-based assessment tool for evaluating the GWP of UHPC mixtures, addressing a significant gap in sustainability assessment where standardized EPDs are lacking. The tool allows for flexible, ingredient-level calculations using up-to-date EPDs from databases like ecoinvent or manufacturer-specific declarations, particularly beneficial for innovative materials such as recycled glass powder and waste-derived fibers, which are often absent or poorly represented in conventional LCA tools like SimaPro.

The comparative analysis showed strong agreement between the Excel platform and SimaPro, with GWP differences ranging from 1.64% to 2.68%. These small discrepancies are attributable to methodological distinctions such as definitions, data granularity, and the use of proxy values in commercial databases. Notably, the Excel tool provided more accurate results for formulations containing non-traditional supplementary cementitious materials, underscoring its value in assessing novel UHPC designs.

Results confirmed that UHPC's environmental footprint is primarily driven by cement content, and that optimized mixtures, employing SCMs and lower-impact fibers, can reduce GWP by up to almost 29%. The platform also proved effective for sensitivity analyses, scenario testing, and iterative mix design during early project phases. While the tool may slightly underestimate absolute GWP values, applying a 2–3% correction factor can help mitigate this for decision-critical applications.

For future work, the continued expansion of the tool's material database, particularly with verified EPDs for emerging binders, alternative aggregates, and regional energy mixes, will enhance its precision and applicability. This work contributes a transparent and adaptable methodology to the broader field of sustainable construction materials and supports the advancement of low-carbon UHPC for resilient and climate-conscious infrastructure.

References

1. Ghafari, E, et al (2012) Enhanced durability of Ultra high performance concrete by incorporating supplementary cementitious materials
2. Imam A et al (2022) A review study on sustainable development of Ultra high-performance concrete. AIMS Mater Sci 9(1):9–35
3. ASTM C1856/C1856M-17 (2017) Standard Practice for fabricating and testing specimens of Ultra-high performance concrete, vol 04.02. ASTM International, West Conshohocken, PA, USA
4. Chen Y et al (2019) Optimization of ultra-high performance concrete, quantification of characteristic features. Cogent Eng 6:1–12
5. Azmee NM, Shafiq N (2018) Ultra-high performance concrete: from fundamental to applications. Case Stud Constr Mater 9:e00197
6. Azmee N, Nuruddin M (2017) Impact performance of low cement Ultra-high-performance concrete, vol 223, pp 481–488
7. Khaiyum M, Sarker S, Kabir G (2023) Evaluation of carbon emission factors in the cement industry: an emerging economy context. Sustainability 15:15407
8. Peplow M (2021) The race for green steel. C&EN Glob Enterp 99(22):22–29
9. Kim J et al (2022) Decarbonizing the iron and steel industry: a systematic review of sociotechnical systems, technological innovations, and policy options. Energy Res Soc Sci 89:102565
10. Farahzadi L, Kioumarsi M (2023) Application of machine learning initiatives and intelligent perspectives for CO2 emissions reduction in construction. J Clean Prod 384:135504
11. Farahzadi L, Kioumarsi M (2022) Intelligent initiative to reduce CO2 emissions in construction. In: World congress in computational mechanics and ECCOMAS congress. Scipedia: Oslo, Norway
12. Farahzadi L, Kioumarsi M, Shafei B (2023) Life cycle assessment of concrete mixes for global warming mitigation and carbon footprint reduction. In: Global cleaner production conference (JCPC). Shanghai, China
13. Farahzadi L et al (2024) Life-cycle environmental assessment of Ultra-high-performance concrete with sustainable materials and fiber substitutions. Clean Eng Technol 23:100846
14. Farahzadi L et al (2025) Sustainability assessment of ultra-high performance concrete made with various supplementary cementitious materials. Cleaner Mater 15:100301
15. Wang X et al (2021) Design of sustainable ultra-high performance concrete: a review. Constr Build Mater 307:124643
16. Hosseini SA, Farahzadi L, Pons O (2021) Assessing the sustainability index of different post-disaster temporary housing unit configuration types. Build Eng:42
17. Farahzadi L, Kioumarsi M (2024) Preliminary life cycle assessment and environmental impact evaluation of RC bridge deck: a case study in Norway. In: ICCEA 2023, Lecture notes in civil engineering. Springer, pp 233–244
18. Farahzadi L et al (2024) Preliminary environmental assessment of ultra-high performance concrete mixtures. In: Lecture notes in civil engineering. In: Kioumarsi M, Shafei B (eds) Net-Zero future 2024- 1st international conference on net-zero built environment, vol 237. Springer: Oslo, Norway
19. Olanrewaju OI, Enegbuma WI, Donn M (2024) Challenges in life cycle assessment implementation for construction environmental product declaration development: a mixed approach and global perspective. Sustain Prod Consum 49:502–528
20. Zhang X et al (2024) Trends toward lower-carbon ultra-high performance concrete (UHPC)—a review. Constr Build Mater 420:135602

21. ISO 14044 (2006) ISO 14044—environmental management—life cycle assessment—requirements and guidelines. International Organization for Standardization, London, UK
22. ISO 14040 (2006) ISO 14040—environmental management—life cycle assessment—principles and framework. International Organization for Standardization, London, UK
23. EN 15804+A2 (2019) Sustainability of construction works: environmental product declarations; core rules for the product category of construction products. British Standards Institution, London, UK
24. Norwegian EPD Foundation (2002) EPD-Norway: program operator for Environmental Declaration (EPD). https://www.epd-norge.no/.
25. Bozorgmehr Nia S et al (2025) Investigations of portland limestone cement and waste glass powder for sustainable ultra-high performance concrete. Case Stud Constr Mater 22:e04425
26. Ecoinvent 3.9.1. Ecoinvent Database 2022

R^3 Framework for Sustainable Concrete: Resilience, Refinement, and Reduction with Ultrafine Silica Glass Powder

Ahsanullah Khan(✉) and Kizhakkumodom Venkatanarayanan Harish

Department of Civil Engineering, Indian Institute of Technology, Kanpur, Kanpur, India
kahsan@iitk.ac.in

Abstract. This study presents the *R^3 framework* (Resilience, Refinement, and Reduction) to evaluate the potential of Ultrafine Silica Glass Powder (USGP) as a sustainable supplementary cementitious material (SCM) in high-performance concrete (HPC). Produced by grinding waste glass into ultrafine particles, USGP enhances pozzolanic reactivity and mitigates the risk of alkali-silica reaction (ASR), addressing key limitations of conventional glass powder. HPC mixes with 10 and 20% USGP replacement were investigated for mechanical, durability, and environmental performance. The 10% USGP mix significantly improved early-age strength, while both USGP blends achieved ~ 5% higher 28-day compressive strength compared to the control. Although the 20% mix showed a slight strength reduction at 90 days, both mixes met high-strength concrete criteria. Durability performance improved substantially, with chloride ion penetrability reduced by 70–90% and bulk resistivity increasing over five times, indicating refined pore structure and enhanced matrix densification as observed in microstructural analysis. Carbon assessment revealed CO_2 emissions reductions of 9.07% and 17.56% for the 10% and 20% mixes, respectively. Overall, this study highlights USGP as a promising alternative SCM for producing high-strength, durable concrete while advancing sustainability goals.

Keywords: Waste glass powder · High-performance concrete (HPC) · Ultra-fine supplementary cementitious · Durability · Sustainability · Low-carbon concrete

1 Introduction

According to the World Green Building Council, construction materials can directly support several Sustainable Development Goals (SDGs) outlined in the UN's 2030 Agenda for a sustainable built environment [1]. Concrete, with an annual global consumption of 20–25 Gt, remains the most widely used construction material due to its central role in meeting housing and infrastructure demands [2]. Given its essential role in societal development, reducing the environmental impact of concrete, particularly CO_2 emissions, is crucial. This necessitates the development of eco-efficient alternatives such as "green concrete" [3].

H. Beushausen et al. (Eds.): NTZR 2025, LNCE 873, pp. 327–339, 2026.
https://doi.org/10.1007/978-3-032-27698-8_30

High-performance concrete (HPC) represents a new generation of advanced cementitious materials designed to meet stringent mechanical and durability requirements while offering environmental advantages [3–6]. HPC typically features a low water-to-cement (w/c) ratio and incorporates chemical admixtures and supplementary cementitious materials (SCMs). Although HPC contains a higher cement content per unit volume, its optimized material use allows for the creation of slender sections, reduced weight, and enhanced durability factors that contribute to lower lifecycle emissions and maintenance requirements [3, 5, 7, 8]. Cement is the dominant source of CO_2 emissions in concrete, making its partial replacement with SCMs a key mitigation strategy. Among emerging SCMs, waste glass powder (WGP) shows promise due to its pozzolanic reactivity, filler effect, and local availability. Studies have indicated that WGP can enhance workability, strength, and durability through microstructural densification. These attributes underline the WGP's potential as a sustainable and widely applicable SCM [9–11]. The performance of a cement system containing waste glass powder is largely determined by particle size and its chemical composition [9, 12].

The primary limitation in using waste glass (WG) in concrete lies in the potential for alkali-silica reaction (ASR), where the reactive silica in WG interacts with alkalis (Na^+, K^+, Ca^{2+}), leading to the formation of expansive gels that can deteriorate structural integrity. Moreover, WGP typically exhibits lower pozzolanic reactivity and may retard cement hydration when compared to conventional SCMs such as ground granulated blast furnace slag (GGBS) and fly ash. These characteristics often result in delayed setting times and reduced early-age strength [13–15]. However, challenges such as ASR, delayed hydration, and reduced early strength limit its broader application. The particle size of WGP is a critical factor influencing both ASR mitigation and pozzolanic activity. Grinding WG in a ball mill improves reactivity and reduces ASR risk by increasing surface area and disrupting expansive gel formation, making it a viable SCM for durable and sustainable concrete [4].

This study introduces ultrafine silica glass powder (USGP), a highly refined form of WGP, as a sustainable SCM. USGP addresses the limitations of conventional WGP and enables a tri-functional approach termed the **R^3 framework**.

- **Resilience** through enhanced durability and resistance to aggressive ions,
- **Refinement** through pore structure densification and improved packing, and
- **Reduction** in environmental impact via waste utilization and cement replacement.

This paper evaluates the influence of USGP on mechanical strength, durability, and microstructural development along with environmental impact, supporting its potential as a viable material for net-zero and circular construction practices.

2 Material and Methods

2.1 Material Properties

The characteristics of cement and USGP used in this study were evaluated through chemical and physical analysis. Chemical composition was determined using X-ray fluorescence spectroscopy (XRF), particle size distribution (PSD) was assessed using a laser particle analyzer, surface area was measured by the Brunauer–Emmett–Teller

(BET) method, and phase characterization was conducted using X-ray diffraction (XRD). Table 1 presented the chemical and physical properties of both the materials. The Blaine fineness of USGP was 890 m^2/kg, significantly higher than that of OPC (350 m^2/kg), indicating its much finer texture. The mean particle size of USGP was 4.47 μm (< 8 μm), compared to 18.04 μm for OPC. Figures 1 and 2 illustrate the particle size distribution and XRD patterns of OPC and USGP, respectively. The XRD spectrum of USGP exhibited a broad hump around 22°, indicating the absence of distinct crystalline peaks and confirming its predominantly amorphous nature [16, 17]. The morphology of the cementitious materials was examined using Scanning Electron Microscopy (SEM) in secondary electron (SE) mode. As shown in Figs. 3a and b, USGP particles exhibit a wide particle size range, consistent with the PSD results, and display angular and irregular edges. In contrast, OPC particles exhibit a more defined polyhedral morphology. Natural fine aggregates (150 μm to 4.75 mm) and coarse aggregates (4.75–10 mm), conforming to the specifications of IS:383, were used in the mix [18]. A polycarboxylate ether (PCE)-based high-range water-reducing admixture (HRWRA) was used to get the desired fresh properties for high-performance concrete (HPC). Table 2 presented the mix proportions adopted in the study.

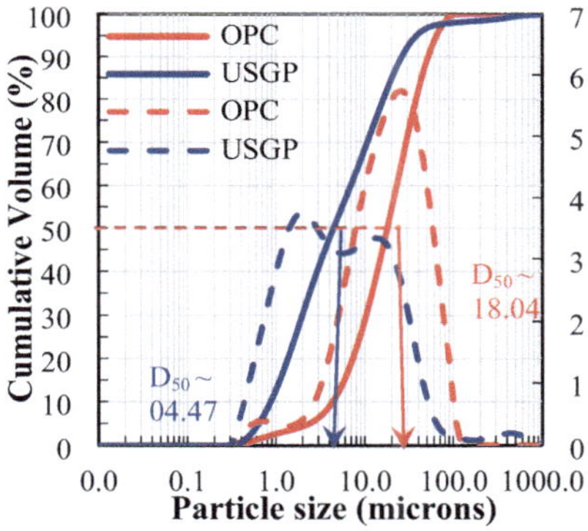

Fig. 1. Particle size distribution of OPC and USGP

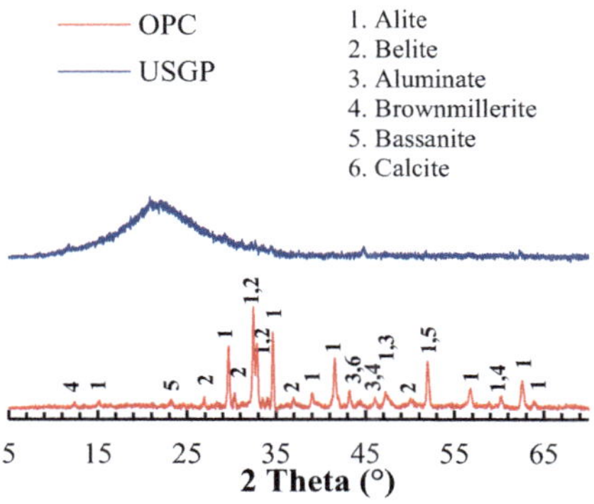

Fig. 2. X-ray diffraction spectrum of OPC and USGP

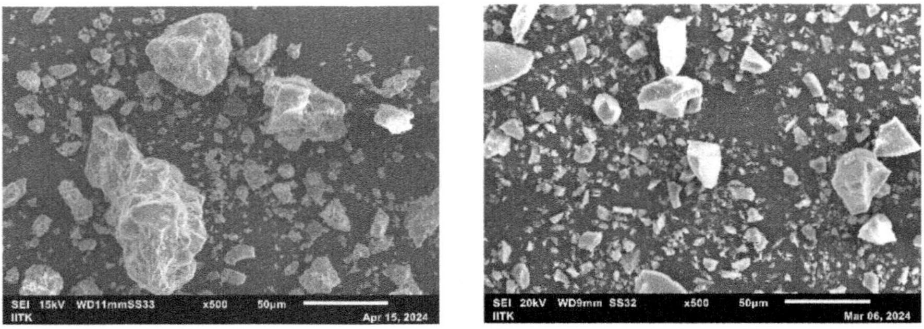

(a) OPC (b) USGP

Fig. 3. Scanning electron microscopic images of OPC and USGP

Table 1. Chemical and physical properties of cement and USGP

Material	CaO (%)	SiO_2 (%)	Al_2O_3 (%)	Fe_2O_3 (%)	MgO (%)	SO_3 (%)	K_2O (%)	Na_2O (%)	Cl (%)	LOI (%)	SG	D_{50} (μm)	SSA (m^2/kg)
OPC	58.47	22.23	3.18	3.85	3.42	2.38	0.97	0.62	0.01	3.96	3.15	18.04	350
USGP	0.03	91.84	2.96	0.54	0	0	0.21	4.28	0	0	2.17	4.47	890

Note OPC: Ordinary Portland Cement (53 grade), USGP: Ultrafine Silica Glass Powder, LOI: Loss on Ignition, SG: Specific gravity, SSA: Specific surface area

Table 2. Mix proportions for high-performance concrete

Mix IDs	Quantity of materials (per kg/m^3)						
	Binder		Aggregates		Water	Chemical admixture	
	OPC	USGP	Fine	Coarse		S_p	VMA
M0	640	0	868	742	192	4.48	0
M1	576	64	868	742	192	4.48	0
M2	512	128	868	742	192	3.84	1.28

2.2 Test Methods

To evaluate the mechanical and durability properties of concrete, standardized tests were conducted on specimens of varying shapes and sizes. The mechanical performance of concrete was assessed through compressive strength, split tensile strength, and flexural strength tests. Compressive strength tests were performed on 100 mm cube specimens at 1, 3, 7, 28, and 90 days of curing. Split tensile strength tests were conducted on cylindrical specimens (100 mm × 200 mm) at 28 and 90 days, while flexural strength tests were carried out on prism specimens (100 mm × 100 mm × 500 mm) at the same curing age. Mechanical properties of concrete were performed as per IS 516 (part 1) [19]. The durability properties of concrete were evaluated using the rapid chloride permeability test (RCPT) and bulk electrical resistivity tests, test set up as shown in Fig. 4. Test performed as per ASTM C1202 and ASTM C1876, respectively [20, 21]. RCPT was conducted on cylindrical disc specimens (100 mm diameter × 50 mm height) at 28 and 90 days of curing. Bulk electrical resistivity measurements were taken on cylindrical specimens (100 mm × 200 mm) at 7, 28, 56, and 90 days. For each mix and testing age, four replicate specimens were cast to get statistical reliability.

2.3 Environmental Impact Assessment

In the concrete production, greenhouse gas (GHG) emissions predominantly arise from the combustion of fossil fuels during the manufacturing and transportation of materials. Accordingly, this study focuses on carbon dioxide (CO_2), methane (CH_4), and nitrous oxide (N_2O) as the primary GHGs associated with the construction industry, consistent with earlier research. To ensure uniformity and comparability, all GHG emissions are converted into carbon dioxide equivalents (CO_2-eq) using their respective Global Warming Potential (GWP) values [22–24].

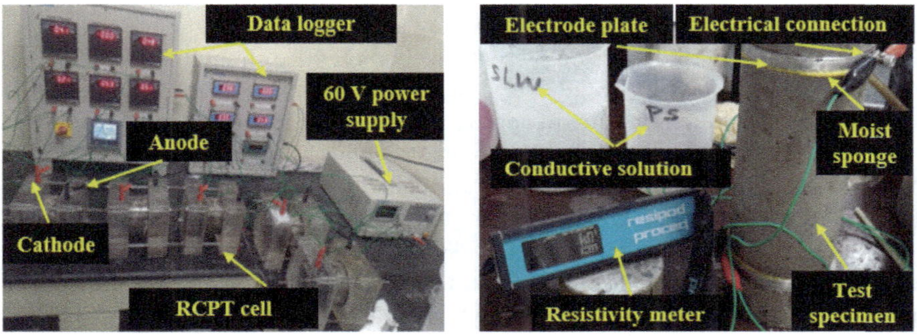

(a) Rapid chloride permeability test (b) Bulk resistivity

Fig. 4. Photographs of rapid chloride ion permeability test and bulk resistivity set-ups used

This enables a standardized evaluation of the environmental impact across different concrete mixtures. The emission factors adopted for each material used in HPC are presented in Table 3. The CO_2 emissions associated with each HPC mix, expressed in kg per cubic meter ($kgCO_2/m^3$), are calculated using the Eq. 1.

$$CO_2\,\text{emission} = \sum_{i}^{n} W_i(p_i) \tag{1}$$

In this equation, W_i quantity of material '*i*' required to produce one cubic meter of concrete, 'n' indicates the total raw material in mix, and p_i indicates the CO_2 emissions factor (kg).

Table 3. Equivalent emission factor of raw material used in HPC

Material	Emission factor kgCO_2/kg	References
Ordinary Portland Cement (OPC)	0.83	[25]
Ultrafine Silica Glass Powder (USGP)	0.016	[26, 27]
Fine aggregate	0.0155	[28]
Coarse aggregate	0.0408	[28]
Water	0.0003	[28, 29]
Superplasticizer (SP)	2.388	[29]
Viscosity modifying admixture (VMA)	2.388	[29]

3 Results and Discussion

3.1 Mechanical Properties

Compressive strength. Figure 5a presents the compressive strength of concrete mixes containing 0, 10, and 20% USGP over time. At early ages (1, 3, and 7 days), the 10% USGP mix showed strength gains of approximately 38% and 4% at 1 and 3 days, respectively, compared to the control. By 7 days, its strength was nearly equivalent to control. This early-age enhancement is primarily attributed to the nucleation effect of ultrafine USGP particles, which accelerate cement hydration [30]. In contrast, the 20% USGP mix showed strength reductions of 16%, 8%, and 15% at 1, 3, and 7 days, respectively, due to the dilution effect, where higher cement replacement lowers clinker content and increases the effective water-to-cement (w/c) ratio, both of which hinder strength development [16]. At 28 days, both 10% and 20% USGP mixes reached compressive strengths of 66 MPa and 67 MPa, respectively, approximately 5% higher than the control. This improvement is attributed to continued hydration, the pozzolanic reactivity of USGP, and the filler effect of ultrafine particles, which together enhance matrix densification [17]. At 90 days, the 10% mix retained a strength comparable to the control (80 MPa), whereas the 20% mix showed a ~ 10% reduction (72 MPa), likely due to limited $Ca(OH)_2$ availability for continued pozzolanic activity and lower overall binder content [5].

Split tensile strength. Figure 5b presents the split tensile strength of the same mixtures. At 28 days, the 10% and 20% USGP mixes achieved 4.53 MPa and 4.59 MPa, respectively, 7–8% higher than the control. This enhancement is due to improved matrix integrity resulting from pozzolanic activity, continued cement hydration, and enhanced particle packing [30]. At 90 days, the tensile strength declined to 4.74 and 4.64 MPa for the 10 and 20% mixes, reflecting 20–21% reductions compared to the control. These losses are attributed to the lower clinker content and potential interference in bond development due to excess glass powder [31].

Flexure strength. Figure 5c shows the flexural strength results. At 28 days, the 10% and 20% USGP mixes recorded values of 5.97 MPa and 6.38 MPa, respectively. The 10% mix was comparable to the control, while the 20% mix exhibited ~ 5% higher strength. This gain is attributed to the combined pozzolanic and micro-filler effects of USGP, leading to a denser and more crack-resistant matrix.

At 90 days, the flexural strengths were 8.14 and 8.34 MPa for the 10 and 20% mixes, showing slight reductions of ~ 4 and ~ 2% compared to the control. These minor losses are associated with dilution effects and reduced long-term hydration efficiency at higher USGP levels [16, 32].

Overall, both 10 and 20% USGP mixes enhanced early-age and 28-day mechanical properties due to the synergistic effects of pozzolanic reactivity and micro-filler action. The 10% USGP mix consistently demonstrated balanced strength development across all ages, while the 20% mix showed higher mid-term strengths but exhibited slight reductions at 90 days due to dilution and reduced long-term hydration. Thus, both replacement levels are effective, with 10% USGP offering optimal mechanical performance over time.

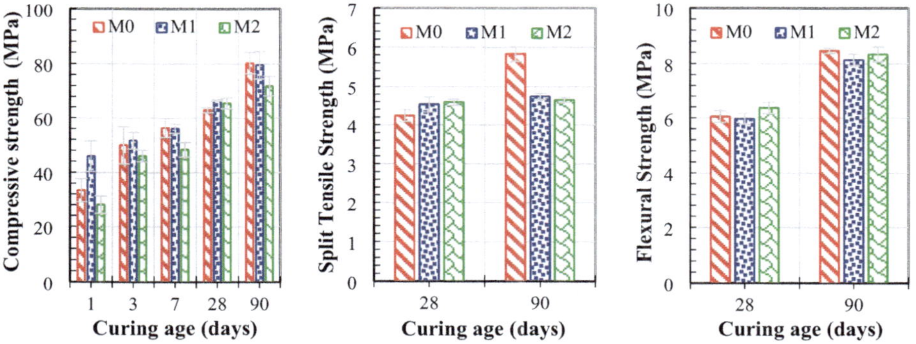

(a) Compressive strength with time (b) Split tensile strength with time (c) Flexure strength with time

Fig. 5. Mechanical properties of HPC containing USGP

3.2 Durability Properties

Rapid chloride penetration test (RCPT). Figure 6a presents the total charge passed (in coulombs) through concrete specimens, representing chloride ion penetrability as per ASTM C1202 [20]. Higher charge values indicate lower resistance to chloride ingress. A notable decrease in charge passed was observed with increasing curing age: 43.5%, 21.3%, and 51.1% reductions for the control, 10% USGP, and 20% USGP mixes, respectively. These reductions resulted in an upgrade in chloride ion penetrability classification, shifting from "Low" to "Very Low" and from "Very Low" to "Negligible."

At 28 days, the 10% and 20% USGP mixes showed significant reductions in charge passing, 80.5% and 88.9%, respectively, compared to the 28-day control. Relative to the 90-day control, the 10% and 20% mixes demonstrated even greater reductions of 73% and 91%, respectively. These enhancements are attributed to the pozzolanic reactivity and micro-filler effect of ultrafine USGP, which contribute to refined pore structure, reduced ionic conductivity, and improved matrix densification. Continued hydration, the consumption of $Ca(OH)_2$, and the formation of secondary C–S–H further explain the progressive improvement in chloride resistance over time [30, 31].

Overall, the incorporation of 10% and 20% USGP significantly improves the chloride penetration resistance of HPC. The observed reductions in total charge passed and upgrades in CIP classification highlight USGP's effectiveness in refining the pore structure, enhancing matrix densification, and sustaining long-term durability. These findings affirm USGP's potential as a sustainable SCM for durable concrete in chloride-rich environments.

Bulk resistivity (BR). Figure 6b presents the bulk electrical resistivity of concrete specimens, indicating classifications for chloride ion penetrability according to ASTM C1202 [20]. Higher resistivity values reflect greater resistance to chloride ingress. At 7 and 28 days, mixes containing 10% and 20% USGP exhibited increases of 11% and 92%, and 194% and 444%, respectively, over the control. This upward trend continued at 56 and 90 days, with both USGP mixes showing more than fivefold increases compared to the control. From 28 days onward, all USGP mixes fell under the "Very Low" class for chloride ion penetrability, indicating substantial improvement in durability. These enhancements are attributed to the pozzolanic activity and micro-filler effect of ultrafine USGP, which refine the pore structure, reduce ionic conductivity, and densify the microstructure. The consistent increase in resistivity with curing age further reflects ongoing hydration, secondary C–S–H formation, and progressive consumption of $Ca(OH)_2$ [30, 31, 33].

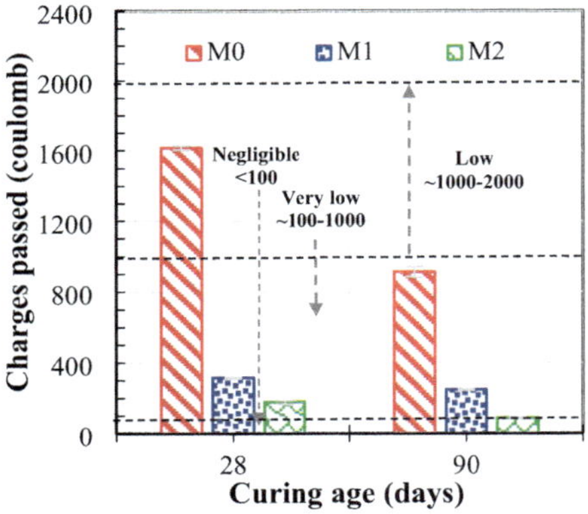

(a) Charges passed with time

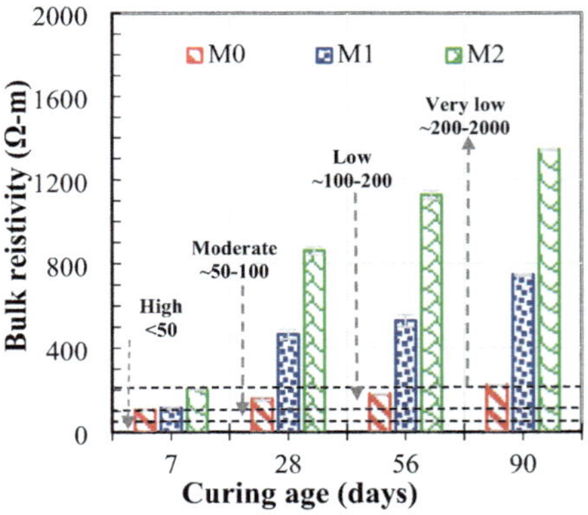

(b) Resistivity with time

Fig. 6. Durability properties of HPC containing USGP

3.3 BSE Image Analyses

Figure 7 compares the backscattered electron (BSE) images of mixtures M0 and M2. The microstructural features observed include fine aggregates (pale grey regions with irregular shapes and distinct interfacial transition zone boundaries), entrapped air voids (dark to light grey spheres), unhydrated cement grains (white spots dispersed across the matrix), voids (spherical black regions), and the cementitious matrix (light grey areas surrounding unhydrated grains). A higher proportion of unhydrated cement grains is evident in the M0 specimen compared with the M2 specimen, which contains 20% USGP. This can be attributed to the higher cement content in M0 and the enhanced pozzolanic reactivity of USGP in M2. Furthermore, the matrix in the USGP-blended mixture appears denser, owing to both the filler effect of ultrafine glass powder and its pozzolanic contribution. Overall, microstructural investigation confirms that USGP incorporation improves the pore structure of concrete by refining pores and generating additional binding phases, resulting in a denser, more homogeneous matrix with reduced porosity and enhanced durability.

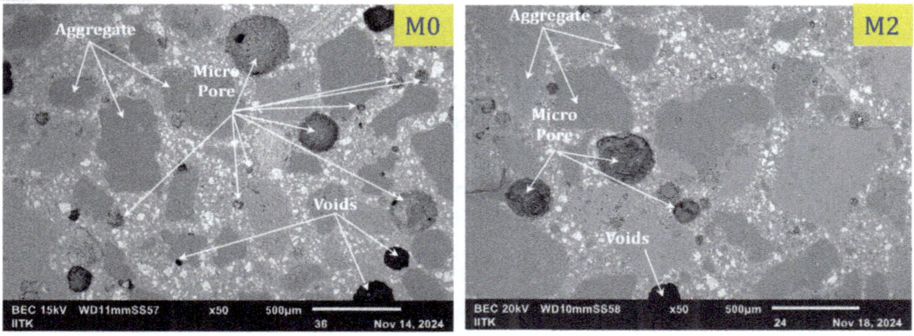

Fig. 7. Back-scattered electron images of M0 and M2 mixes showing compactness in matrix porosity

3.4 Environmental Impact Assessment (EIA)

The CO_2 emission factors for the constituent materials used in producing HPC were adopted from relevant literature, as summarized in Table 3. A detailed assessment of total CO_2 emissions per cubic meter of concrete, following the methodology outlined in Sect. 2.4, is presented in Table 4. Component-wise analysis reveals that cement alone contributes approximately 90% of the total emissions in the control mix (M0), reaffirming its dominant environmental impact. This underscores the substantial mitigation potential of partial replacement of cement with USGP.

Table 4. CO_2 emissions from HPC mix

Constituent	CO_2 emission per meter cubic of concrete ($kgCO_2/m^3$)		
	M0	M1	M2
Cement (OPC)	531.20	477.08	424.96
USGP	–	1.02	2.05
Fine aggregate (FA)	13.45	13.45	13.45
Coarse aggregate (CA)	30.27	30.27	30.27
Superplasticizer (Sp)	10.69	10.69	9.17
Water	0.06	0.06	0.06
Viscosity modifier (VMA)	–	–	3.05
Total CO_2	585.68	532.58	483.02
% Reduction	0.00	– 9.07	– 17.56

Mixes incorporating 10% and 20% USGP achieved CO_2 emission reductions of 9.07% and 17.56%, respectively, relative to the control mix. These reductions are primarily attributed to the partial replacement of OPC, the most carbon-intensive component in concrete. The trend further demonstrates a consistent decrease in embodied carbon with increasing USGP content. To evaluate performance-based CO_2 intensity, emissions were normalized by corresponding performance metrics (e.g., compressive strength, chloride resistance, and bulk resistivity). When assessed in conjunction with these parameters, the reduced emissions per unit performance further validate the sustainable potential of USGP-blended concretes [34].

4 Conclusions

1. The incorporation of USGP mitigates the early-age strength limitations typically observed with waste glass powder. HPC mixes that contain 10% and 20% USGP replacement either meet or exceed the high-strength requirements specified in IS 456:2000.
2. The incorporation of USGP significantly enhances durability performance. Resistance to chloride ion penetration improved by 70–90% in the RCPT test, while bulk resistivity increased by a factor of 3–6 compared to the control mix. These improvements are attributed to the synergistic effects of USGP's pozzolanic reactivity and micro-filler action, which refine the pore structure and reduce permeability.
3. Concrete reduced its CO_2 emissions per cubic meter by approximately 9.07% and 17.56%, respectively, for M1 and M2 mixes compared to the control mix. These reductions are attributed to the partial replacement of cement.
4. Overall, this study confirms the USGP as a promising alternative SCM to produce high-strength and durable concrete while advancing sustainability goals. Additionally, the use of USGP mitigates CO_2 emissions, valorizes non-biodegradable waste, supports the principles of a circular economy, and contributes to low-carbon concrete development.

References

1. World Green Building Council, Sustainable built environments & the UN's Sustainable Development Goals (2019). https://worldgbc.org/sustainable-development-goals/
2. Randl N, Steiner T, Ofner S, Baumgartner E, Mészöly T (2014) Development of UHPC mixtures from an ecological point of view. Constr Build Mater 67:373–378. https://doi.org/10.1016/j.conbuildmat.2013.12.102
3. Matos AM, Milheiro-Oliveira P, Pimentel M (2024) Eco-efficient high performance white concrete incorporating waste glass powder. Constr Build Mater 411. https://doi.org/10.1016/j.conbuildmat.2023.134556
4. Mehta A, Ashish DK (2020) Silica fume and waste glass in cement concrete production: a review. J Build Eng 29:100888. https://doi.org/10.1016/j.jobe.2019.100888
5. Teng S, Lim TYD, Sabet Divsholi B (2013) Durability and mechanical properties of high strength concrete incorporating ultra fine ground granulated blast-furnace slag. Constr Build Mater 40:875–881. https://doi.org/10.1016/j.conbuildmat.2012.11.052
6. Soliman NA, Tagnit-Hamou A (2017) Partial substitution of silica fume with fine glass powder in UHPC: filling the micro gap. Constr Build Mater 139:374–383. https://doi.org/10.1016/j.conbuildmat.2017.02.084
7. Habert G, Denarié E, Šajna A, Rossi P (2013) Lowering the global warming impact of bridge rehabilitations by using ultra high performance fibre reinforced concretes. Cem Concr Compos 38:1–11. https://doi.org/10.1016/j.cemconcomp.2012.11.008
8. Abbas S, Nehdi ML, Saleem MA (2016) Ultra-high performance concrete: mechanical performance, durability, sustainability and implementation challenges. Int J Concr Struct Mater 10:271–295. https://doi.org/10.1007/s40069-016-0157-4
9. Jiang Y, Ling TC, Mo KH, Shi C (2019) A critical review of waste glass powder—multiple roles of utilization in cement-based materials and construction products. J Environ Manage 242:440–449. https://doi.org/10.1016/j.jenvman.2019.04.098
10. Khan MNN, Saha AK, Sarker PK (2020) Reuse of waste glass as a supplementary binder and aggregate for sustainable cement-based construction materials: a review. J Build Eng 28:101052. https://doi.org/10.1016/j.jobe.2019.101052
11. Lu JX, Shen P, Zhang Y, Zheng H, Sun Y, Poon CS (2021) Early-age and microstructural properties of glass powder blended cement paste: improvement by seawater. Cem Concr Compos 122:104165. https://doi.org/10.1016/j.cemconcomp.2021.104165
12. Bignozzi MC, Saccani A, Barbieri L, Lancellotti I (2015) Glass waste as supplementary cementing materials: the effects of glass chemical composition. Cem Concr Compos 55:45–52. https://doi.org/10.1016/j.cemconcomp.2014.07.020
13. Yang S, Lu JX, Poon CS (2020) Recycling of waste glass in dry-mixed concrete blocks: evaluation of alkali-silica reaction (ASR) by accelerated laboratory tests and long-term field monitoring. Constr Build Mater 262:120865. https://doi.org/10.1016/j.conbuildmat.2020.120865
14. Elaqra H, Rustom R (2018) Effect of using glass powder as cement replacement on rheological and mechanical properties of cement paste. Constr Build Mater 179:326–335. https://doi.org/10.1016/j.conbuildmat.2018.05.263
15. Du H, Tan KH (2017) Properties of high volume glass powder concrete. Cem Concr Compos 75:22–29. https://doi.org/10.1016/j.cemconcomp.2016.10.010
16. Kameche ZA, Djelil M, Dahmani B (2023) Effects of incorporating silica glass powder as a supplementary cementitious material on selected properties of portland cement mortar. J Build Eng 78:107550. https://doi.org/10.1016/j.jobe.2023.107550
17. Maddalena R, Hall C, Hamilton A (2019) Effect of silica particle size on the formation of calcium silicate hydrate [C-S-H] using thermal analysis. Thermochim Acta 672:142–149. https://doi.org/10.1016/j.tca.2018.09.003

18. Bureau of Indian Standards (1970) Specification for coarse and fine aggregates from natural sources for concrete (IS: 383-1970). Indian Stand, pp 1–24
19. IS 516 (Part 1/Sec 1): 2020 (2020) Hardened concrete—methods of test. Is 54:1–20. www.standardsbis.in
20. ASTM C1202 (2012) Standard test method for electrical indication of concrete's ability to resist chloride ion penetration. Am Soc Test Mater 1–8. https://doi.org/10.1520/C1202-25.2
21. ASTM IC1876 (2024) Standard test method for bulk electrical resistivity or bulk conductivity of concrete. ASTM Int I:98–100. https://doi.org/10.1520/C1876-24
22. Bhagath Singh GVP, Subramaniam KVL (2019) Production and characterization of low-energy Portland composite cement from post-industrial waste. J Clean Prod 239:118024. https://doi.org/10.1016/j.jclepro.2019.118024
23. Cao Z et al (2016) Toward a better practice for estimating the CO_2 emission factors of cement production: an experience from China. J Clean Prod 139:527–539. https://doi.org/10.1016/j.jclepro.2016.08.070
24. Ali MB, Saidur R, Hossain MS (2011) A review on emission analysis in cement industries. Renew Sustain Energy Rev 15:2252–2261. https://doi.org/10.1016/j.rser.2011.02.014
25. Turner LK, Collins FG (2013) Carbon dioxide equivalent (CO_2-e) emissions: a comparison between geopolymer and OPC cement concrete. Constr Build Mater 43:125–130. https://doi.org/10.1016/j.conbuildmat.2013.01.023
26. Sathiparan N, Subramaniam DN (2024) Potential use of crushed waste glass and glass powder in sustainable pervious concrete: a review. Clean Waste Syst 9:100191. https://doi.org/10.1016/j.clwas.2024.100191
27. Letelier V, Henríquez-Jara BI, Manosalva M, Parodi C, Ortega JM (2019) Use of waste glass as a replacement for raw materials in mortars with a lower environmental impact. Energies 12. https://doi.org/10.3390/en12101974
28. Chiaia B, Fantilli AP, Guerini A, Volpatti G, Zampini D (2014) Eco-mechanical index for structural concrete. Constr Build Mater 67:386–392. https://doi.org/10.1016/j.conbuildmat.2013.12.090
29. Thilakarathna PSM, Seo S, Baduge KSK, Lee H, Mendis P, Foliente G (2020) Embodied carbon analysis and benchmarking emissions of high and ultra-high strength concrete using machine learning algorithms. J Clean Prod 262:121281. https://doi.org/10.1016/j.jclepro.2020.121281
30. Lam WL, Cai Y, Sun K, Shen P, Poon CS (2024) Roles of ultra-fine waste glass powder in early hydration of Portland cement: hydration kinetics, mechanical performance, and microstructure. Constr Build Mater 415:135042. https://doi.org/10.1016/j.conbuildmat.2024.135042
31. Zong H et al (2023) The role of ultra-fine supplementary cementitious materials in the durability and microstructure of airport pavement concrete. Constr Build Mater 392:131954. https://doi.org/10.1016/j.conbuildmat.2023.131954
32. Lashari AR, Kumar A, Kumar R, Rizvi SH (2023) Combined effect of silica fume and fly ash as cementitious material on strength characteristics, embodied carbon, and cost of autoclave aerated concrete. Environ Sci Pollut Res 30:27875–27883. https://doi.org/10.1007/s11356-022-24217-9
33. Lam WL, Cai Y, Sun K, Shen P, Poon CS (2024) Roles of ultra-fine waste glass powder in early hydration of Portland cement: hydration kinetics, mechanical performance, and microstructure. Constr Build Mater 415. https://doi.org/10.1016/j.conbuildmat.2024.135042
34. Reis R, Camões A, Ribeiro M, Malheiro R (2024) Eco-Efficient concrete for sustainable construction. https://doi.org/10.20944/preprints202404.0957.v1

Eco-efficiency Performance Analysis of Recycled Concrete Aggregates as Sustainable Alternative Material in Road Pavement: An Eco-value/Cost Ratio Modeling Approach

Ademilade Olubambi[1](✉), Oluwabukunmi Ogunsanya[2], Clinton Aigbavboa[3], and Bolanle Ikotun[4]

[1] Department of Civil Engineering Science, University of Johannesburg, Auckland Park Campus, 2006, Johannesburg 2092, South Africa
ajoeolubambi@gmail.com

[2] Department of Civil Engineering, Surveying and Construction, Kingston University London, London, UK

[3] cidb Centre of Excellence & Sustainable Human Settlement and Construction Research Centre, Faculty of Engineering and the Built Environment, University of Johannesburg, Johannesburg 2092, South Africa

[4] Department of Civil, Environmental Engineering and Building Science, University of South Africa, Florida Campus, Johannesburg, Gauteng, South Africa

Abstract. In the construction industry, eco-efficiency refers to increasing the value generated through modifications to processes while minimizing resource or material consumption and lowering the environmental impact over the material or service's life. Furthermore, eco-efficiency applies to all facets of construction, from material procurement and construction to marketing and distribution. Hence, this study compares the efficiency of applying natural coarse aggregate and recycled concrete aggregate in road pavement systems. Evaluations of the literature first demonstrate the effects of both natural and recycled concrete aggregate in the various road surface projects that are now underway. Also, measures were selected to evaluate eco-efficiency. This comprises the ratio of environmental value added to environmental impact, economic value to raw material costs, and environmental impact to economic benefits. Results show that both natural and recycled aggregates are utilized and consumed in ways that are both sustainable and unsustainable. The utilization of natural aggregates in concrete is decreasing, and they have a bigger environmental impact than recovered aggregates.Therefore, in so as to achieve sustainable construction and maximize profitability, greater emphasis needs to be given to maximizing the potential of recycled aggregates in road pavement construction. This can be achieved by implementing government incentives for projects that utilize recycled materials. Additionally, raising awareness through educational campaigns about the environmental and economic benefits of using recycled aggregates can encourage more widespread adoption. Furthermore, investing in research to improve the quality and performance of recycled materials can make them a more attractive option for construction projects.

H. Beushausen et al. (Eds.): NTZR 2025, LNCE 873, pp. 340–351, 2026.
https://doi.org/10.1007/978-3-032-27698-8_31

Keywords: Circular material · Eco-efficiency performance · Eco-value/cost modeling · Recycled concrete aggregates · Road pavement · Sustainable construction

1 Introduction

At least 25% of all solid waste in every country is made up of construction and demolition (C&D) waste, which is included in the solid waste stream. In the USA, more than 375 million tons of waste were produced, of which 21.7 million tons came from construction and 353.6 million tons from demolition. Additionally, the demolition of roads and bridges accounted for 157.4 million of these trash. However, concrete—the most crucial of all C&D materials—makes up more than 70% of the material produced after construction [1]. Materials from C&D operations can be utilized as recycled and reusable materials due to the deficiency of natural mineral resources, the percentage increase in material demand, and the rising costs of waste management. However, the use of natural aggregates (NA) in construction has a significant impact that results in resource depletion, energy use, and environmental harm. The increasing speed of industrialization and civilization necessitates a vast number of primary aggregates [2–4].

The use of recycled aggregates (RA) is crucial for sustainable development since it preserves natural aggregates and reduces the amount of landfill space needed for the disposal of demolition trash. Moreover, it reduces the consumption of energy. Given the rising demand on landfills, governments continue to be interested in the diversion of C&D materials in general and concrete in particular [5]. Recycling concrete reduces expenses by 34–41% because of its market value. Additionally, cement reduces greenhouse gas (GHG) emissions by 23–28% because it may be reclaimed from the recycling process and utilized again. As a result, there are fewer greenhouse gas emissions during the manufacturing of fresh concrete. RA is therefore a green building material. In the present scenario, crushed hardened concrete is being recycled as aggregate [6–8].

According to studies, RA concrete's compressive, tensile, and flexural strengths are between 85 and 95% of those of NA concrete. It can be utilized in a variety of ways and is sufficiently durable. The results of the experiment, RA may replace 30% of natural coarse aggregates without compromising any important properties. Therefore, there are numerous economic and environmental benefits to recycling recycled materials [9]. As a result, using recycled concrete aggregates (RCA) in building is a relatively recent technique, particularly in developing nations. Although RCA processing and application have significantly improved in concrete road construction, the latter is mostly utilized in foundations and sub-foundations. But as the likely potential of these RCA is not fully harnessed, this is considered 'downcycling' [10, 11]. Road pavement is one of the most energy-intensive infrastructure elements that is dependent on non-renewable natural resources.

Consequently, the use of RA made from C&D waste in the building of The use of unbound materials in road foundations, sub-bases, and embankments is a successful management technique for increasing recycling rates and creating a secondary market opportunity for RA [12]. Currently, the majority of pavement construction uses RA generated from C&D waste. From minor percentages to complete material substitution, RA made from C&D waste is often employed in pavement layers. When Leite et al. [13] investigated the viability of employing RA for pavement applications, they compared the behavior of the recycled materials to that of limestone aggregates. The results showed that RA from C&D waste, which has higher densities and has been used as a coarse base and sub-base layer for low-volume roads, performs outstandingly. Furthermore, RA can assist in enhancing layer performance and is abundant in cementitious materials [14–16].

However, the results indicate that a mixture of 75% NA and 25% RCA produced the same permanent deformation characteristics and resilient reaction as a densely graded coarse aggregate base, which is used in the road pavement's base and sub-base layers [17]. The RA uses in roads are presently being standardized, despite the fact that their use of materials for road construction has already been assessed. The majority of nations are enforcing uniform restrictions on the use of NA on public roads. One limiting element, for instance, is the presence of sulfur compounds in both concrete and road materials [18, 19]. There is a need for proactive actions to ensure the dimensional stability of the section and avoid potential pathologies in adjacent concrete structures or cement-treated pavement layers [20]. The first step in the RCA process is the creation of urbanite, or concrete debris items, which can be recycled by choosing, moving, and utilizing the pieces as slabs or bricks, as shown in Fig. 1. These are often made, perhaps with a lot of effort, as with a chisel.

In addition to metals like rebar, magnets, and other equipment, crushing also involves getting rid of garbage, wood, and paper. The mixture must be sorted by size, crushed by a crushing machine, and alternate particles must be eliminated by methods such water flotation and hand-picking in order to be recycled separately [21]. RAC durability is influenced by a number of characteristics, including moisture content, W/C ratio, concrete age, and aggregate replacement ratio. Concrete is less resistant to corrosion, carbonation, and chloride penetration. Because of the additional inter-facial transition zone in RAC, the carbonation depth is greater [22–24]. As a result, RAC develops excessive porosity, decreasing durability. According to the results of the freeze-thaw test, RA losses for coarse aggregates can be 1–12% higher than NA, while RA losses for fine aggregates can be 10–33% higher than NA. The cured cement paste's and the residual material, such as mortar, air entrainment determines this resistance. Concrete's soundness test indicates that RAC has a higher volume loss [25–27]. This loss is caused by mortar fragments that are present in RA breaking apart and de-bonding from the aggregate. There may be less volume loss if the mortar is very strong. Since recycled aggregate is composed of natural aggregate and mortar, the qualities of the mortar have a significant impact on the properties of RAC [28–30].

The significance of eco-efficiency in construction materials lies in its ability to reduce environmental impact while maintaining or improving performance. By utilizing recycled concrete aggregates (RCA), the construction industry can decrease the depletion of natural resources and lower carbon emissions. This approach not only supports sustainable development but also promotes cost-effectiveness in large-scale infrastructure projects [31–34]. Use of fly ash, slag cement, and silica fume can increase durability. This study aims to provide insight into the eco-efficiency performance of RCA through a denotative investigation conducted as a case scenario, even though other studies highlight experimental comparisons between RCA and NA. Given the effects of employing RCA in road pavement construction as opposed to virgin aggregate, this study attempts to address this issue [35–38].

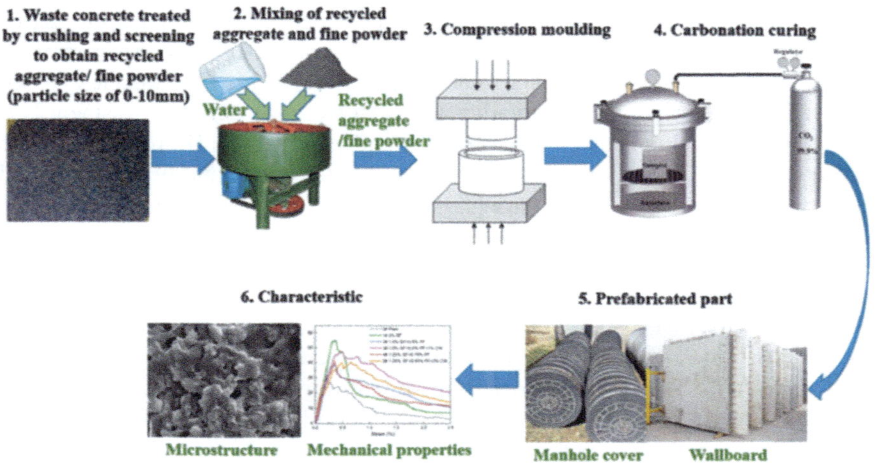

Fig. 1. Process of recycled concrete aggregate production from a 500-ton-per-day plant [34]

2 Literature Review

As countries and regions all over the world began growing economically, it gradually became clear that degradation of the environment is an inevitable consequence of both industrialization and economic development. In an effort to deal with this issue, the primary means of encouraging a shift from unsustainable to sustainable development is the suggested use of eco-efficient materials [39]. Eco-efficiency is a management strategy that seeks to achieve more with less, asserted by Glavič et al. [40]. It is based on the notion that more goods and services may be produced with less resources and with less waste and pollution. Furthermore, eco-efficiency is a sustainability metric that combines economic and environmental performance. However, as of right now, no widely accepted definition exists [41, 42]. Consequently, eco-efficiency can be viewed as a marketing strategy for sustainable development or as a metric for environmental performance.

Koskela and Vehmas [43] described the most significant eco-efficiency in 2012 as follows: the desire to produce more goods and services with less resources, which reduces waste and pollution, is the foundation of the eco-efficiency concept. According to Vogtländer et al. [44], eco-efficiency of the product is defined as the ratio of the (added) environmental impacts of the product or service (e.g., SO_2 emissions) to the (added) production values (e.g., GDP). The ideal way to achieve eco-efficiency, however, is to offer competitively priced goods and services that satisfy requirements and add value to life while gradually lowering the environmental effects of products and guaranteeing resource efficiency over the course of their whole life cycle [45].

2.1 Evaluating the Eco-efficiency of Circular Materials

There are four (4) primary ratio categories are used to evaluate a material's eco-efficiency. Considering the field of production, the first two are environmental productivity and its inverse, environmental intensity of production [46]. While environmental cost-effectiveness is determined from the perspective of environmental improvement measures, the second pair, environmental improvement cost, is inverse. Furthermore, since sustainability assessments are a crucial component of the industry for optimizing current products and for supplying evidence-based arguments for strategic and policy discussions, a validated and quantitative approach to sustainability assessment is necessary for the development of new products and processes [47]. To be able to find more sustainable solutions, it is essential to evaluate the sustainability of materials. The effects and advantages that arise throughout the end-of-life usage of materials, particularly when taking cradle-to-grave into account, are crucial to analyze, even if a number of methodologies have been created for evaluating environmental sustainability. Eco-efficiency analysis is one of the modern methods for evaluating sustainability that takes into account both environmental and economic factors [48, 49].

The eco-efficiency analysis tool facilitates strategic decision-making along the entire value chain. It also enables industries to drive innovative product development, which focuses on bringing more sustainable products to the marketplace. This methodology identifies the factors whose optimization will directly translate into improvements in its sustainability profile, even during the early stages of product development. It facilitates a clear understanding of trade-offs and helps prevent the inadvertent shifting of environmental impacts from one area to another or between the economic and environmental pillars [50, 51]. The eco-efficiency analysis compares the life cycles of products or manufacturing processes in a holistic approach from the assessment including raw materials sourcing, product manufacture, and use, to disposal or recycling. Furthermore, it is used to identify ways to make improvements in terms of environmental impact and cost and takes the consumption behavior of end-users into account, as well as various recycling and disposal options [52].

However, when analyzing, it is very important to look at the environmental impact in proportion to a product's cost-effectiveness. Recently, eco-efficiency analysis was validated which follows ISO 14944:2006 for environmental life cycle assessments and life cycle costs which are based on ISO 14045:2012 [53]. Since eco-efficiency analysis aims to evaluate the environmental performance of a product or process in relation to its economic value. By comparing the environmental impacts and costs, it will assist the industry to identify opportunities in reducing resource use and emissions while maintaining or improving profitability. This approach encourages sustainable development by promoting practices that balance ecological and economic considerations [54]. Consequently, eco-efficiency aims to improve economic performance, for example, by lowering waste and pollutants as well as energy, water, and resource inputs [55]. Despite their essential nature, these kinds of operations are nevertheless predicated on the idea that industrial systems are made up of one-way linear material flows. Eco-efficient methods can lower the linear material flow's volume, velocity, and toxicity, but they still have an effect on the environment. Since the goal of eco-efficiency techniques is to minimize the negative consequences of industrial production on the environment, zero impact is the optimum achievable goal [57] (Fig. 2).

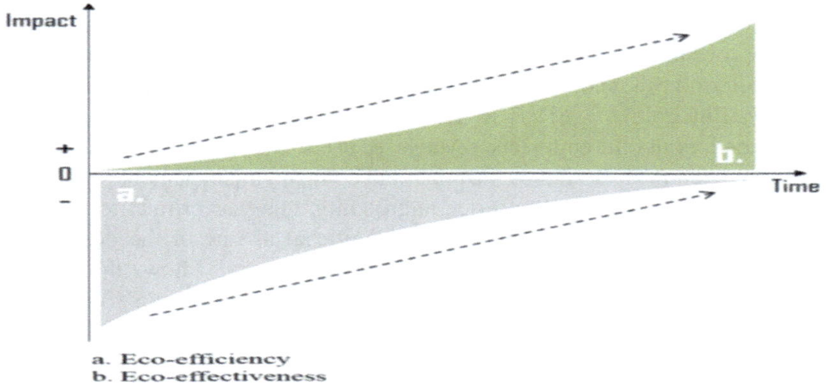

Fig. 2. An eco-efficiency versus eco-effectiveness diagram [58]

2.2 Eco-cost/Value Ratio Modeling and Application

To determine if production will be successful or whether the material can be used efficiently, the eco-cost/value ratio (EVR) combines eco-cost and value. As a result, the material should have a low eco-cost (low environmental impact) and an appealing value for customers. Here, the market value—also known as the fair price or perceived consumer value—is the value. The value, cost, and eco-cost are the three aspects of a product itself. Residents of a country are increasingly refusing to tolerate high levels of industrial pollution [58]. Consequently, countries impose stricter regulations (such as tradable emission rights, eco-taxes, and enforcement of the best technology). The internal manufacturing costs will subsequently include eco-costs [45]. The industry may face opportunities as well as risks as a result of the internalization of eco-cost. my product can survive more stringent government requirements when it has a lower environmental impact than my rival's.

Therefore, this characteristic of the low eco-cost of my product is a competitive edge. A product or materials' short- and long-term market potential can be analyzed through putting each material in the portfolio matrix. The fundamental tenet of the product portfolio matrix is that the material service is distinguished by its long-term market need (low eco-cost) and short-term market potential (high value/cost ratio). Regarding product strategy, the matrix yields strategic directions that include increasing the green design's value/cost ratio to expand the market, reducing the eco-cost of successful products to make them suitable for future markets, and discontinuing products with low eco-cost and low value/cost ratios (small market, little profit). The typical issue with many "green designs" is that their present value/cost ratio is low. Production costs are typically greater than those of the classic solution; in certain instances, the perceived quality is even lower [59, 60].

3 Material and Methods

This study's eco-efficiency variables were obtained from the literature reviewed, while data were collected quantitatively. In an effort to determine the options utilized between NA and RCA in road pavement construction, surveys from three construction businesses' most recent projects were contacted. The investigation was conducted using the following formula to determine the eco-cost/value ratio: Eco-efficiency of a products/or materials = $\sum EVA/EIR$, But, $EVA = \sum MV/CR$, $EIR = \sum EI/EC$. Three small groups of people—clients, contractors and site engineers—were used in the experiment, and their purchasing patterns were examined. Both the consumer and company management groups seemed to choose their products based on price and quality, this was critical to the analysis. Environmental considerations only become significant as one of the deciding factor when there is no choice based on price and quality. Figure 3 shows the relationship between production costs, the selling price, and a successful offering's Willingness to Pay (WTP) by the clients. It is believed that only clients who believe the offering has excess value are going to consider procuring it. The Long and short market prospects of the NA and RCA were considered.

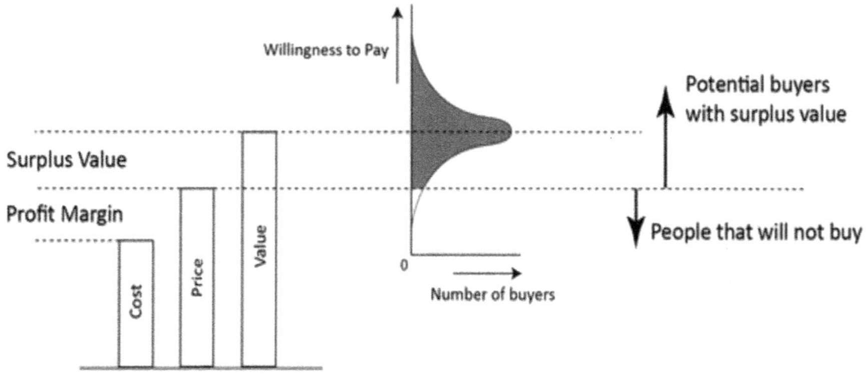

Fig. 3. The relation between costs of production, the price at point of sale, and the Willingness to Pay (WTP)

4 Results and Discussion

The tabulated results of the eco-efficiency performance analysis are shown in Table 1. These results indicate that the production of NA has a greater environmental impact and a significantly higher long-term cost than that of RCA. The application of RCA in road pavements has a good sustainability performance, as the total environmental value added surpasses the environmental impact. As a result, RCA is a sustainable and eco-friendly material that may be used to build road pavement. Using RCA in construction not only reduces the need for NA extraction but also decreases landfill waste by recycling existing materials. Furthermore, RCA tends to have lower transportation costs since it can be sourced from nearby demolition sites. This makes it a cost-effective and environmentally responsible choice for sustainable construction practices. Reducing landfill waste is crucial for minimizing environmental degradation and conserving natural resources. By utilizing RCA in construction, we effectively repurpose materials that would otherwise contribute to landfill overflow, promoting a circular economy.

Table 1. Eco-efficiency evaluation of RCA and NA applied in road pavement construction

	Scenarios	MV ($per/tonnes)	CR ($per/tonnes)	EVA = $\sum$MV/CR	EI (μ)	EC	$\sum$EI/EC	$\sum EVA/EIR$
NA	1	22.3	18.0	1.238	0.738	1.000	0.738	1.721
	2	23.5	22.5	1.044	0.645	1.000	0.645	
	3	21.8	20.3	1.074	0.567	1.000	0.567	
RCA	1	34.7	18.0	1.927	0.747	1.000	0.747	2.417
	2	45.0	20.5	2.195	0.879	1.000	0.879	
	3	38.7	18.5	2.092	0.945	1.000	0.945	

5 Conclusion and Recommendations

The utilization of sustainable materials in the construction of road pavement is made possible in large part by the consistency of government policy. Proactive companies with good product strategy suffer when government restrictions are delayed or even dropped. Players may become reluctant to take proactive measures in the future as a result, and the transition process may suffer significant harm. The ideal design strategy is to increase value where value is high and decrease eco-cost where eco-cost is high. This study indicate the most successful design possibilities. Eco-efficiency metrics were developed from a comparison of the durability and different applicability of RCA and NA in road pavement construction in order to assess RCA's performance. The scenario of the eco-efficiency of both NA and RCA is shown through the application of this formula to calculate and evaluate its eco-efficiency. While a scenario with lower eco-efficiency would demonstrate that it is not sustainable, one with higher eco-efficiency would be more sustainable, meaning it would provide more economic value per unit environmental effect.

Therefore, the eco-efficiency performance analysis can assists in determining the genuine comparison between NA and RCA uses in road pavement construction. For road construction project to achieve a more sustainable development, eco-efficiency is a crucial concept that takes into account both the environmental effects and the added value of their operations. Despite recent research and publications, the study of eco-efficiency remains a challenging undertaking because there is currently no standard, widely recognized definition or methodology for measuring it, and it is primarily concerned with eco-efficiency at the corporate level. However, the concept can be used to its fullest potential in order to assess the economic and environmental performance of goods and services as well as the use of circular materials in various projects, particularly in road construction, in order to lower costs and make material recycling easier. In conclusion, RCA must be taken into account as an environmentally friendly material when building road surfaces. One of the challenges associated with implementing RCA is ensuring the consistent quality and performance of recycled materials. Therefore, there should be regulatory and technical barriers that need to be addressed to facilitate widespread adoption.

References

1. Aboginije AJ (2020) An assessment of construction and demolition waste management systems in the Nigerian construction industry. Master Dissertation, University of Johannesburg (South Africa), pp 1–10
2. Aboginije A, Aigbavboa C, Thwala WA (2021) Holistic assessment of construction and demolition waste management in the Nigerian construction projects. Sustainability 13(11):6241
3. Purchase CK, Dhafer MA, Matthew JK et al (2021) Circular economy of construction and demolition waste: a literature review on lessons, challenges, and benefits. Materials 15(1):76
4. Shiran J, Chaminda G, Arun K (2014) Assessment of recycled concrete aggregates as pavement material. Geomech Eng 6(3):235–248
5. Kou S, Poon C, Agrela F (2011) Comparisons of natural and recycled aggregate concretes prepared with the addition of different mineral admixtures. Cem Concr Compos 33(8):788–795
6. Jayakody S, Gallage C, Ramanujam J (2017) Effects of reclaimed asphalt materials on geotechnical characteristics of recycled concrete aggregates as a pavement material. Road Mater Pavem Des 20(1):754–772
7. Ebenezer OF, John TK, Adewumi JB, Jian L (2023) A comprehensive review on the use of recycled concrete aggregate for pavement construction: properties, performance, and sustainability. Clean Mater 9:100199
8. Jayakody S, Gallagher C, Kumar A (2014) Assessment of recycled concrete aggregates as a pavement material. Geomech Eng 6(3):235–248
9. Jiménez JR, Ayuso J, Agrela F, López M, Galvín AP (2012) Utilisation of unbound recycled aggregates from selected CDW in unpaved rural roads. Resour Conserv Recycl 58:88–97
10. Verian KP, Whiting N, Olek J, Jain J, Snyder M (2013) Using recycled concrete as aggregate in concrete pavements to reduce materials cost. Publication FHWA/IN/JTRP-2013/18. Joint Transportation Research Program, Indiana Department of Transportation and Purdue University, West Lafayette, Indiana
11. Pusks A, Corbu O, Szilgyi H, Moga LM (2014) Construction waste disposal practices: the recycling and recovery of waste. WIT Trans Ecol Environ 191:1313–1321
12. Benjamin K (2013) Performance of concrete made with commercially produced recycled coarse and fine aggregate in Cape Peninsula. Master Thesis, Department of Civil Engineering, University of Cape Town, Cape Town, South Africa
13. Leite MB (2001) Evaluation of mechanical properties of concrete produced with recycled aggregates from construction and demolition waste. PhD. Thesis, University of Rio GrandedoSul (in Portuguese)
14. Bester J, Kruger D, Miller B (2017) South African construction and demolition waste procedure and its sourced material effects on concrete. MATEC Web Conf 02008:1–9
15. Martinho FC (2018) Feasibility assessment of the use of recycled aggregates for asphalt mixtures. Sustainability 10:1737
16. Hossain MU, Poon CS, Lo IM, Cheng JC (2016) Comparative environmental evaluation of aggregate production from recycled waste materials and virgin sources by LCA. Resour Conserv Recycl 109:67–77

17. US EPA (2016) Advancing sustainable materials management: 2014 fact sheet, United States Environmental Protection Agency, Office of Land and Emergency Management, Washington, DC 20460
18. Geis GS, Quattrone M, Ambrós WM, Grigore CB, Hoffmann SC (2021) Current applications of recycled aggregates from construction and demolition: a review. Materials 14:1700
19. Contreras Llanes M, Romero Pérez M, Gázquez González MJ et al (2022) Construction and demolition waste as recycled aggregate for environmentally friendly concrete paving. Environ Sci Pollut Res 29:9826–9840
20. Musa SS, Al-Mukaram N, Mahdi MB (2021) Assessment of asphalt mixture behaviour containing recycled concrete aggregates. In: Key engineering materials, vol 895. Trans Tech Publications, Ltd., pp 139–146
21. El-Tahan D, Gabr A, El-Badawy S et al (2018) Evaluation of recycled concrete aggregate in asphalt mixes. Innov Infrastruct Solut 3:20
22. Al-Bayati HKA, Tighe SL, Achebe J (2018) Influence of recycled concrete aggregate on volumetric properties of hot mix asphalt. Resour Conserv Recycl 130:200–214
23. Adebanji O, Ayobami B, Emmanuel A (2018) Pavement interlayer material improvement using industrial waste: review of literatures. Int J Civ Eng Technol 9(6):1114–1122
24. Tam VWY, Soomro M, Evangelista ACJ (2018) A review of recycled aggregate in concrete applications (2000–2017). Constr Build Mater 172:272–292
25. Aytekin B, Mardani-Aghabaglou A (2021) Sustainable materials: a review of recycled concrete aggregate utilization as pavement material. Transp Res Rec 2676(3):468–491
26. Barbudo A, Agrela F, Ayuso J, Jiménez JR, Poon CS (2012) Statistical analysis of recycled aggregates derived from different sources for sub-base applications. Constr Build Mater 28(1):129–138
27. Zhang Z, Shen B, Ren H, Wang J, Li S, Liu H (2017) The variability and evaluation method of recycled concrete aggregate properties
28. Abass BJ, Albayati AH (2020) Influence of recycled concrete aggregate treatment methods on performance of sustainable warm mix asphalt. Cogent Eng 7(1):1–20
29. Panda B, Imran NT, Samal K (2021) A study on replacement of coarse aggregate with recycled concrete aggregate (RCA) in road construction. In: Das B, Barbhuiya S, Gupta R, Saha P (eds) Recent developments in sustainable infrastructure. Lecture notes in civil engineering, vol 75. Springer, Singapore. https://doi.org/10.1007/978-981-15-4577-1_91
30. Gabr AR, Cameron DA, Andrew R, Mitchell PW (2011) Comparison of specifications for recycled concrete aggregate for pavement construction. J ASTM Int 1 8(10):1–15
31. Bennert T, Papp WJ, Maher A, Gucunski N (2007) Utilization of construction and demolition debris under traffic-type loading in base and subbase applications. Transp Res Rec J Transp Res Board 1714(1):33–39
32. Arulrajah A, Disfani MM, Horpibulsuk S, Suksiripattanapong C, Prongmanee N (2014) Physical properties and shear strength responses of recycled construction and demolition materials in unbound pavement base/subbase applications. Constr Build Mater 58:245–257
33. Surya M, VVL KR, Lakshmy P (2013) Recycled aggregate concrete for transportation infrastructure. Proc Soc Behav Sci 104:1158–1167
34. Vieira CS, Pereira PM (2015) Use of recycled construction and demolition materials in geotechnical applications: a review. Resour Conserv Recycl 103:192–204
35. Ohemeng EA, Ekolu OS (2020) Comparative analysis on costs and benefits of producing natural and recycled concrete aggregates: a South African case study. Case Stud Constr Mater 13:1–13
36. Makul N, Fediuk R, Amran M, Zeyad AM. Klyuev S, Chulkova I, Ozbakkaloglu T, Vatin N, Karelina M, Azevedo A (2021) Design strategy for recycled aggregate concrete: a review of status and future perspectives. Crystals 11(6):695 (1–21)

37. Chakravarthi S, Shankar S (2025) Feasibility analysis of recycled asphalt pavement and recycled concrete aggregate in emulsified asphalt-treated bases. Int J Pavement Eng 26(1):1–18
38. Kilani A, Adelani AA, Oladejo OS, Ikotun BD, Olubambi A (2024) Analyzing an efficient mix design for the production of quality asphalt concrete: a means of reducing roads' maintenance cost. J Sustain Constr Mater Technol 9(2):177–198
39. Sugihara K (2019) Multiple paths to industrialization: a global context of the rise of emerging states. In: Otsuka K, Sugihara K (eds) Paths to the emerging state in Asia and Africa. Emerging-economy state and international policy studies. Springer, Singapore. https://doi.org/10.1007/978-981-13-3131-2_1
40. Glavič P, Lesjak M, Hirsbak S. European training course on eco-efficiency. Paper presented at 15th European roundtable on sustainable consumption and production (Bregenz, Austria. Available at: https://vbn.aau.dk/en/publications/european-training-course-on-eco-efficiency. Accessed 8 Dec 2024
41. Bianchi M, Valle ID, Tapia C (2020) Measuring eco-efficiency in European regions: evidence from a territorial perspective. J Clean Prod 276:123246 (1–14)
42. Aboginije A, Aigbavboa C, Thwala W (2023) Modeling and usage of a sustainametric technique for measuring the life-cycle performance of a waste management system: a case study of South Africa. Front Sustain 3:943635
43. Vogtlander JG, Scheepens AE, Bocken MP, Peck D (2017) Combined analyses of costs, market value and eco-costs in circular business models: eco-efficient value creation in remanufacturing. J Remanuf 7:1–17
44. Vogtländer JG, Hendriks ChF, Brezet JC (2001) Allocation in recycle systems: an integrated model for the analyses of environmental impact and economic value. Int J Life Cycle Assess 6(6):344–355
45. Vogtländer J (2002) Communicating the eco-efficiency of products and services by means of the eco-costs/value model. J Clean Prod 10:57–67
46. Koskela M, Vehmas J (2012) Defining eco-efficiency: a case study on the Finnish forest industry. Bus Strateg Environ 21(8):546–566
47. Vogtländer J (2010) LCA-based assessment of sustainability: the eco-costs/value ratio (EVR), VSSD
48. Wever R, Vogtlander JG (2013) Eco-efficient value creation: an alternative perspective on packaging and sustainability. J Packag Technol Sci 26(4):229–248 (46)
49. John I, Kwofie EM, Ngadi M (2020) Two decades of eco-efficiency research: a bibliometric analysis. Environ Sustain 3:155–168. https://doi.org/10.1007/s42398-020-00105-1
50. Nikolaou IE, Matrakoukas SI (2016) A framework to measure eco-efficiency performance of firms through EMAS reports. Sustain Prod Consump 8:32–44
51. Caiado RG, De Freitas Dias R, Mattos LV, Quelhas OLG, Leal Filho W (2017) Towards sustainable development through the perspective of eco-efficiency—a systematic literature review. J Clean Prod 165:890–904
52. Xu T, Umair M, Cheng W, Hakimova Y, Mang G (2024) Evaluating eco-efficiency as a metric for sustainable urban growth: a comparative study of provincial capital cities in China. Ecol Ind 169:112959
53. Dyllick T, Hockerts K (2002) Beyond the business case for corporate sustainability. Bus Strat Environ 11:130–141
54. Burnett RD, Skousen CJ, Wright CJ (2011) Eco-effective management: an empirical link between firm value and corporate sustainability. Account Public Interest 11:1–15
55. Braungart M, McDonough W, Bollinger A (2007) Cradle-to-cradle design: creating healthy emissions—a strategy for eco-effective product and system design. J Clean Prod 15:1337–1348

56. Levidow L, Lindgaard-Jørgensen P, Nilsson A, Skenhall SA, Assimacopoulos D (2016) Process eco-innovation: assessing meso-level eco-efficiency in industrial water-service systems. J Clean Prod 110:54–65
57. Yang L, Zhang X (2018) Assessing regional eco-efficiency from the perspective of resource, environmental and economic performance in China: a bootstrapping approach in global data envelopment analysis. J Clean Prod 173:100–111
58. Borrello M, Pascucci S, Cembalo L (2020) Three propositions to unify circular economy research: a review. Sustainability 12(10):4069 (1–22)
59. Al-Shami S, Rashid N (2022) A holistic model of dynamic capabilities and environment management system towards eco-product innovation and sustainability in automobile firms. J Bus Indus Mark 37(2):402–416
60. Hendriks C, Janssen GM, Vogtländer J (2002) The eco-costs/value-ratio. Advances in building. Technology 2:1391–1398

Eco-Efficient Tannin-Based Insulation Foams

Dino Clarke[1](✉), Paulina Faria[2], Leon van Kralingen[1], and Gülşah Balamut Arslan[3]

[1] NTE Company (Pty) Ltd, Pietermaritzburg, South Africa
{dino,leon}@nte.co.za

[2] CERIS, Department Civil Engineering, NOVA University Lisbon, Caparica, Portugal
paulina.faria@fct.unl.pt

[3] Kastamonu Entegre Ağaç San. Tic. A.Ş., R&D Center, İstanbul, Türkiye
gulsah.balamut@keas.com.tr

Abstract. Insulation foams are used for many applications in the building and other industries. Many are synthetic-based and perform well in terms of thermal conductivity, but not so well for acoustic insulation, in case of fire and in terms of environmental impact. Tannins are a natural product extracted from the bark or wood of certain tree species. Some tannin-based foams have shown to have good fire-resistant properties. With the aim to achieve competitive bio-based insulation foam and sandwich panel that can be economic, ecological and efficient, with good performance both for thermal and acoustic insulation and safe in case of fire, tannin-based foams were formulation and samples production was optimized. The work was developed within international project MoreWood, focused on introducing more bio-based resources for greener building products. Four formulations working towards replacement of the formaldehyde crosslinker with eco-friendlier glutaraldehyde were defined, containing 100% formalin, 50%/50% formalin/glutaraldehyde, 25%/75% formalin/glutaraldehyde and 100% glutaraldehyde as the crosslinking agent. Lab scale foam samples of 200 x 200 x 50 mm^3 were produced and tested. Results show these foams have competitive density and thermal conductivity when compared to literature studies and commercial synthetic-based foams in the market, justifying further characterization for future scale-up production.

Keywords: Bio-based building product · Samples production method · Thermal insulation

1 Introduction

Due to growing worldwide concerns around climate change, sustainable development, and reducing reliance on products derived from petrochemical resources, there is an increased focus on the use of renewable, bio-based resources, as well as the recycling of various agricultural wastes into sustainable alternatives as building materials to produce building products [1]. The aim is to achieve building products with lower embodied energy and environmental impact [2], but both economically and technically efficient

H. Beushausen et al. (Eds.): NTZR 2025, LNCE 873, pp. 352–363, 2026.
https://doi.org/10.1007/978-3-032-27698-8_32

for diverse applications in building elements. However, for some types of applications, reducing the risk in case of fire and improving fire resistance of these bio-based products is also a major concern, both in terms of flaming and smoldering combustion [3]. For others, contribution for thermal insulation and acoustic insulation are fundamental. Under this framework, project Morewood, as part of the Eurogia 2030 initiative, seeks to support green building efforts by developing viable alternatives that introduce more wood and bio-based materials into building products. This would allow for decreased consumption of oil-derived raw materials, increasing the recyclability of the products, and reducing CO_2 and greenhouse gas emissions.

Tannins are polyphenolic compounds found in variable amounts in most plants. They are especially abundant in the leaves, wood and bark of various tree species [4]. Historically, they have been extracted industrially for the manufacture of leather, as well as the production of wood adhesives. More recently they have also found applications in other areas such animal feed additives, mining, pharmaceuticals and water treatment [5]. Tannins are grouped into two categories: hydrolysable tannins (polyesters of gallic- and ellagic-acids) and condensed tannins (proanthocyanins, polymers of flavan-3-ol units) [6].

There has also been interest in using tannins for the development of tannin-furanic based bio-foams for thermal and acoustic insulation as well as other uses [7, 8]. The foams are generally produced by combining tanning with furfuryl alcohol, a crosslinking agent, a blowing agent (low boiling point solvent) and an acid catalyst. Addition of the acid catalyst leads to an auto-polymerization of the furfuryl alcohol, as well as reaction with the tannin. The heat thus generated causes the blowing agent to boil, resulting in the foam forming. Addition of a surfactant can help with the foaming process. The heat generated also facilitates reaction of the tannin with the crosslinking agent, causing hardening of the foam. The foaming process therefore requires the correct balance between the foaming process and the hardening process. If hardening occurs too soon the foam will not reach adequate height, but if it occurs too slowly the foam can reach its maximum height whilst still unstable and collapse in on itself. Foams have been made using condensed tannins from numerous sources including quebracho, mimosa, and pine. Some of these foams have been found to have high porosity (95–98%), good thermal performance and a large range of compressive strength between 0.015 to 0.175 N/mm^2 [9–11]. Tannins extracted from different trees have been tested [10]. Some studies have been researching on how to improve fire safety [12] and test it for tannin-based foams [13], showing they do not burn and so, do not emit toxic gases when burning, contrary to what happen with totally synthetic foams, such as polystyrenes and polyurethanes. Others compared rigid polymeric foams with advantages for tannin-furanic foams [14].

Within international project MoreWood, focused on introducing more bio-based resources for green building products, work is ongoing towards the development of tannin-based insulation foam panels which are expected to be used for thermal and acoustic insulation.

2 Materials and Methods

2.1 Aims and Preliminary Experiments

With the aim of developing ecological, economic and technically efficient tannin-based insulation foam panels, the focus has been on reducing the amount of formaldehyde in the formulation by replacing it with a more environmentally friendly crosslinker. The tested crosslinkers were glutaraldehyde, glyoxal and hexamine.

2.2 Materials

The following materials were used to produce the tannin-based foams. The tannin used was a condensed tannin from the Black Wattle (Mimosa, *Acacia Mearnsii*) tree. This tannin is supplied as a commercial extract called Bondtite 945, produced by NTE Company (South Africa), as a powder. Furfuryl alcohol, glyoxal, hexamine, Tween-80®, polyethylene glycol (PEG-400), glutaraldehyde and pentane were purchased from Merck/Sigma Aldrich (South Africa) and used as supplied. The formalin (37% formaldehyde) was purchased from Protea Chemicals (South Africa). Para-toluene sulfonic acid (p-TSA) was purchased as a solid and made up to a 65% solution with distilled water.

2.3 Foam Formulation, Preparation and Samples

Surfactant Effects. The work started by developing a Standard (Std) foam formulation using formalin (37%) as the crosslinking agent and pentane as the low boiling point blowing agent. Water (6.0 g), furfuryl alcohol (10.5 g), formalin (7.4 g) and PEG-400 (4.0 g) were mixed together in a 250 ml plastic beaker. The tannin powder (30.0 g) was then added, and the mixture stirred with a plastic stirring rod until the tannin was dissolved. The pentane (3.0 g) was then added, and the mixture stirred vigorously for 1 min until homogenous. The p-TSA catalyst (11.0 g) was finally added, the mixture stirred vigorously for 30 s, and then quickly poured into a cylindrical plastic mold (8 cm x 8 cm). Foaming generally started within 1 min and was allowed to continue until complete and the foam had hardened. It was then removed from the mold and left at room temperature (RT) overnight. The foams thus formed had a fairly non-uniform open cell structure with large open cells on the bottom (Fig. 1, left).

A surfactant was then incorporated, at 2.5% (m/m based on tannin powder added). Initial surfactants evaluated were natural, plant derived, and highly biodegradable surfactants commonly available and used in the cosmetics and laundry industry (decyl and lauryl glucoside). However, these had no effect on the foaming characteristics. The non-ionic surfactant, Tween-80®, was then added to the formulation (just before the pentane addition) and this led to notable improvements (Fig. 1, right). Final foam height doubled, and the foam cells were significantly smaller and more uniform.

Initial Crosslinker Comparison. Five foam formulations were produced in order to test the effects of different crosslinkers on foam formation (Table 1).

The water, furfuryl alcohol, crosslinker and PEG-400 were mixed together in a 250 ml plastic beaker. The tannin powder was then added, and the mixture stirred with a plastic

Fig. 1. Tannin foams with (right) and without (left) surfactant

Table 1. Tannin foam formulations with glyoxal and hexamine.

Mass (g)	100% formalin	50/50 formalin/glutaraldehyde	100% Glutaraldehyde	100% Glyoxal	100% Hexamine
Tannin	30.0	30.0	30.0	30.0	30.0
Water	6.0	4.2	2.5	6.0	6.0
Furfuryl alcohol	10.5	10.5	10.5	10.5	10.5
Formalin 37%	7.4	3.7	–	–	–
Glutaraldehyde 25%	–	5.5	10.9	–	–
Glyoxal 40%	–	–	–	6.8	–
Hexamine	–	–	–	–	2.7
PEG-400	4.0	4.0	4.0	4.0	4.0
Tween-80®	0.75	0.75	0.75	0.75	0.75
Pentane	3.0	3.0	3.0	3.0	3.0
p-TSA 65%	11.0	11.0	11.0	11.0	11.0

stirring rod until the tannin was dissolved. The Tween-80® and pentane were then added, and the mixture stirred vigorously for 1–2 min until homogenous. The p-TSA catalyst was finally added, the mixture stirred vigorously for 30 s, and then quickly poured into a similar cylindrical plastic mold.

Foaming started after 30–180 s and was allowed to continue until complete and the foam hardened into a dark brown solid. The foam was then allowed to cure at RT overnight. Table 2 and Fig. 2 show the foaming characteristics of the various crosslinker systems.

Table 2. Foaming characteristics of crosslinking systems.

	Foam start (min)	Foam finish (min)	Foam height (cm)	Final height post curing (cm)	Shrinkage %
Standard (100% Formalin)	1:10	2:45	15.0	14.5	–
50/50 Formalin/Glutaraldehyde	1:00	3:40	15.5	14.0	3.4
100% Glutaraldehyde	1:50	6:08	14.5	11.0	24.1
100% Glyoxal	3:00	8:05	13.2	4.5	65.5
100% Hexamine	None	n/a	n/a	n/a	n/a

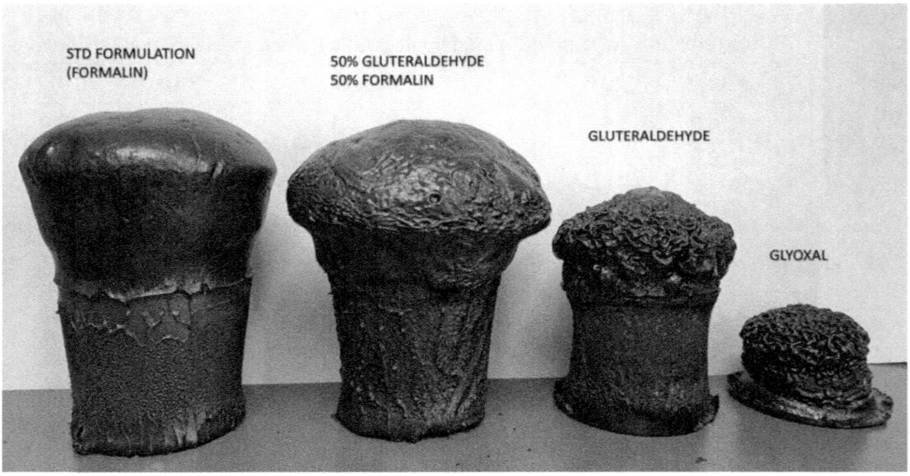

Fig. 2. Tannin foams showing final form after overnight curing

As can be seen from Fig. 2 and Table 2, the 50/50 Formalin/Glutaraldehyde formulation was very similar in final height to the standard formulation, with only slightly more shrink back during curing to give a more wrinkled appearance on the foam surface.

The 100% Glutaraldehyde formulation had a slightly lower initial foam height compared to the first two but had a significant shrink back during the curing process. However, it was thought this would still be a viable option to use if additional heat was added, which would allow for faster curing before the foam can shrink back. The longer time to initiate foaming could also be beneficial when scaling up.

The formulation using glyoxal produced significantly less foam than the others and was also less stable and collapsed quite significantly before curing. The use of hexamine did not even produce any foam with the original formulation. Addition of more acid catalyst (up to 200% extra) produced some foaming but this immediately collapsed on itself before any significant height could be achieved. Glutaraldehyde was therefore chosen as the crosslinker to continue working on as a substitution for formalin.

Scale Up into Foam Panels. Four formulations, containing increasing substitution of formalin with glutaraldehyde, were produced and are displayed in Table 3.

Table 3. Tannin foam formulations, in g, with increasing glutaraldehyde content.

Mass [g]	Formulation I	Formulation II	Formulation III	Formulation IV
Tannin	80.0	80.0	80.0	80.0
Water	16.0	11.2	10.0	7.0
Furfuryl alcohol	28.0	28.0	28.0	28.0
Formalin 37%	19.0	9.8	6.0	–
Glutaraldehyde 25%	–	14.6	19.2	28.1
PEG-400	11.0	11.0	11.0	11.0
Tween-80	2.0	2.0	2.0	2.0
Pentane	8.0	8.0	8.0	8.0
p-TSA 65%	29.0	29.0	29.0	29.0

The water, furfuryl alcohol, crosslinker and PEG-400 were first mixed together in a 500 ml plastic beaker which was then fitted with an overhead stirrer. The tannin powder was then added portion-wise over 5 min, and the resulting mixture was then stirred for 15 min to ensure complete dissolution of the tannin. The Tween-80® was then added and stirred for 2–5 min, followed by the pentane, continuing stirring for 1 min. The beaker was then removed from the overhead stirrer, the p-TSA solution added in one portion, and the resulting mixture stirred vigorously with a thick (10 mm) plastic stirring rod for 30 s, before being poured into a stainless-steel mold (21 cm × 25 cm). Once foaming had started, the mold was then placed into an oven at 105 °C for 10 min, before being allowed to cool for 20 min. The hardened foam was then demolded and placed back in the oven for 10 min to finish curing, before being allowed to cool to RT overnight (Fig. 3).

Three triplets of samples of the four tannin-based insulation foams were prepared to be tested. The foams were named:

- Formulation 1—100% Formalin → 100F
- Formulation II—50% Formalin and 50% Glutaraldehyde → 50F + 50G
- Formulation III—25% Formalin and 75% Glutaraldehyde → 25F + 75G
- Formulation IV—100% Glutaraldehyde → 100G

Fig. 3. Stainless-steel mold used and examples of final foam panel samples produced.

2.4 Test Methods

Visual observations and Apparent Density—Visual observation and handling was performed on all the foams. After mass stabilization at 24°C and 60% relative humidity (RH), density was assessed by geometrical method using a digital caliper and balance with precision of 0.001 g. Each dimension of each specimen was determined twice and the average dimension considered.

Thermal conductivity and specific heat—The thermal properties were assessed from using a portable ISOMET2114 with an IPS1105 surface probe with 60 mm diameter and measurement range 0.04–0.3 W/(m.K). As a surface probe was used, both sides of the specimens were tested. All the measurements were performed at a temperature of 25.1 ± 1 °C and 58.4 ± 2.4% relative humidity, in controlled laboratory conditions.

3 Results and Discussion

3.1 Visual Observation and Density

The visual observation and handling showed that the foams are very dark (almost black), they leave a slight odor and some fine particles when handled. Apparent density is presented in Table 4.

Comparing the density of the developed tannin-based foams, it seems percentages of glutaraldehyde of 75% or more reduces the density by 11%.

Table 4. Volume, mass and density (average and standard deviation) of the tannin-based foams.

Foam	Sample	Volume [dm^3]	Mass [g]	Density [kg/m^3]		
100F	1	1.81	114.899	63.57	61.29	3.59
	2	1.63	104.613	64.08		
	3	1.66	93.136	56.23		
50F + 50G	1	1.70	109.715	64.65	61.48	4.59
	2	1.69	109.520	64.79		
	3	1.81	99.280	54.99		
25F + 75G	1	1.88	97.124	51.76	54.95	2.91
	2	1.88	110.313	58.80		
	3	1.64	88.895	54.30		
100G	1	1.91	103.761	54.24	54.31	0.94
	2	1.97	109.301	55.50		
	3	1.88	99.955	53.20		

Comparing the apparent density results with commercial common foams, such as polyurethane and polystyrene, values vary significantly for low, medium and high densities. Comparing with densities of tannin-based foams from literature, the developed foams are within or below the limits presented by Delgado-Sánchez et al. [7] using tannin from Quebracho and furanic (22–69 kg/m^3), Marie et al. [15] not specifying the tannins used (55–75 kg/m^3), Tondi et al. [9] using tannin from Mimosa (56–115 kg/m^3), Celzard et al. [16] also using tannins from Mimosa (40–140 kg/m^3), Lacoste et al. [17] using tannin from Mimosa also with formaldehyde (40–190 kg/m^3), from pine with formaldehyde (40–190 kg/m^3) and from pine without formaldehyde (35–70 kg/m^3), Cop et al. [18] using tannins from Pinus pinaster and Picea abies L. (100–102 kg/m^3) and Rodrigues et al. [13] using tannin from Acacia mearnsii (128 $\pm$ 3 kg/m^3).

Therefore, the values of the developed tannin-based foams seem to be acceptable.

3.2 Thermal Conductivity and Specific Heat

The thermal conductivity and specific heat results on face 1, face 2 and the average of both faces, considering the triplets of each foam, are presented in Tables 5, 6 and 7, together with the temperature and relative humidity of the laboratory conditioning.

All results are consistent on both surfaces of the samples. Considering specific heat, it is higher for the glutaraldehyde-based foams. In terms of thermal conductivity, all foams present very similar results and with very low standard deviation, for temperatures of 25 ºC and 58% RH.

Comparing with thermal conductivities of tannin-based foams from literature, the developed foams are: higher than the results presented by Tondi et al. [9] using tannin from Mimosa (0.023–0.030 W/(m.K)) and Lacoste et al. [17] using tannin

Table 5. Thermal conductivity and specific heat—average and standard deviation in grey (side 1).

Foam	Sample	T [°C]	HR [%]	Thermal conductivity			cp		
				[W/(m.K)]			$\times 10^{-6}$ [J/(m^3.K)]		
100F	1	23.8	59	0.041	0.041	0.000	0.139	0.119	0.018
	2	24.0	60	0.042			0.123		
	3	24.1	59	0.041			0.095		
50F + 50G	1	24.2	57	0.041	0.041	0.000	0.120	0.117	0.005
	2	24.3	56	0.041			0.109		
	3	25.0	59	0.041			0.121		
25F + 75G	1	24.8	54	0.041	0.041	0.001	0.104	0.138	0.029
	2	24.9	55	0.041			0.134		
	3	25.0	55	0.042			0.175		
100G	1	25.1	56	0.041	0.042	0.001	0.103	0.142	0.036
	2	25.1	56	0.041			0.136		
	3	25.2	57	0.043			0.189		

from Mimosa with formaldehyde (0.037 W/(m.K)), from pine with formaldehyde (0.034 W/(m.K)) and from pine without formaldehyde (0.033 W/(m.K)); within the results presented by Delgado-Sánchez et al. [7] using tannin from Quebracho and furanic (0.036–0.045 W/(m.K)) and Marie et al. [15] not specifying the tannins used (0.032–0.050 W/(m.K)); lower than the results presented by Kain et al. [19] using tannins from Mimosa and from Quebracho (0.077–0.087 W/(m.K)).

Comparing thermal conductivity results with commercial common foams, such as polyurethane and polystyrene ones, with values circa 0.035 W/(m.K), the developed tannin-based foams seem to be competitive.

Table 6. Thermal conductivity and specific heat—average and standard deviation in grey (side 2).

Foam	Sample	T [°C]	HR [%]	Thermal conductivity			cp		
				[W/(m.K)]			$\times 10^{-6}$ [J/(m^3.K)]		
100F	1	25.5	56	0.041	0.041	0.000	0.117	0.119	0.018
	2	25.3	62	0.042			0.142		
	3	25.3	62	0.041			0.097		
50F + 50G	1	25.3	62	0.041	0.041	0.000	0.110	0.118	0.008
	2	25.3	62	0.041			0.129		
	3	25.6	61	0.041			0.115		

(continued)

Table 6. *(continued)*

Foam	Sample	T [°C]	HR [%]	Thermal conductivity [W/(m.K)]			cp × 10^{-6} [J/(m^3.K)]		
25F + 75G	1	25.6	60	0.041	0.042	0.001	0.115	0.131	0.021
	2	25.6	60	0.041			0.117		
	3	25.6	60	0.043			0.161		
100G	1	25.9	59	0.042	0.042	0.000	0.150	0.158	0.038
	2	26.0	58	0.042			0.208		
	3	26.1	57	0.042			0.117		

Table 7. Average and standard deviation (StDv) of thermal conductivity and specific heat, considering both sides of the samples.

Foam	Thermal conductivity [W/(m.K)]		cp ×10^{-6} [J/(m^3.K)]	
	Average	StDv	Average	StDv
100G	0.041	0.000	0.119	0.018
50F + 50G	0.041	0.000	0.117	0.007
25F + 75G	0.041	0.001	0.134	0.026
100G	0.042	0.000	0.150	0.038

4 Conclusions

The tannin-based foams present interesting results. Considering production methods, they seem viable for industrial scaling up. They are dark, with slight odor and fine particles release and lightweight. The thermal conductivity is circa 0.04 W/(m.K) for all foams, being all classifiable as thermal insulation. The glutaraldehyde-based foams have a lower density, without significative change on the thermal conductivity but an increase on the specific heat.

Compared to commercial foams, such as synthetic-based polyurethane and polystyrene ones, the tannin-based foams certainly have environmental advantages, to be quantified, and other technical advantages, namely in terms of safety in case of fire, to be confirmed.

Therefore, further work is justified to achieve complementary characterization of the foams, based on European and other international standards, and validation of all the results obtained by comparison with other bio-based foams and synthetic foams from literature and on the market. Complementary tests are being performed, namely to assess mechanical performance, water vapor permeability, water absorption and drying capacity, adsorption-desorption, flammability and release of toxic gases.

Within a South African context, the opportunities of extra tannin demand would provide socio-economic and environmental benefits to the Eastern Cape, here there are large areas where black wattle is mostly growing wild and uncontrolled ("jungle wattle"), impacting water resources. The clearing out of this invasive jungle wattle and conversion of suitable areas into well managed plantations should have beneficial impacts on communities, the environment and water resources. There are many small grower communities and subsistence farmers who may benefit, both from the direct financial gain, as well as the technical skills and knowledge imparted in the process. The wood from the trees can also be utilized, either exported for paper pulp or sold locally for charcoal production.

Acknowledgements. This work is supported within the scope of the EUROGIA 2030 project Morewood, ID number EU2030-19-2. The authors would like to thank the Department of Science and Innovation (DSI) of South Africa for funding (DSI/CON C3213/2023), as well as the Portuguese Agency for Innovation (ANI) for funding (COMPETE2030-FEDER-00365700, operation 12706), and Tânia Santos for helping with the experimental campaign. The Portuguese Foundation for Science and Technology is also acknowledged for funding CERIS research unit (UIDB/04625/2025).

References

1. Le DL, Salomone R, Nguyen QT (2023) Circular bio-based building materials: a literature review of case studies and sustainability assessment methods. Build Environ 244:110774
2. Mouton L, Allacker K, Röck M (2023) Bio-based building material solutions for environmental benefits over conventional construction products—life cycle assessment of regenerative design strategies (1/2). Energy Build 282:112767
3. Yang Y, Haurie L, Wang DY (2022) Bio-based materials for fire-retardant application in construction products: a review. J Therm Anal Calorim 147:6563–6582
4. Das AK, Islam MN, Faruk MO, Ashaduzzaman M, Dungani R (2020) Review on tannins: extraction processes, applications and possibilities. S Afr J Bot 135:58–70
5. Pizzi A (2019) Tannins: prospectives and actual industrial applications. Biomolecules 9:344
6. Khanbabaee K, van Ree T (2001) Tannins: classification and definition. Nat Prod Rep 18:641–649
7. Delgado-Sánchez C, Santiago-Medina F, Fierro V, Pizzi A, Celzard A (2018) Optimisation of "green" tannin-furanic foams for thermal insulation by experimental design. Mater Des 139:7–15
8. Pizzi A (2019) Tannin-based biofoams—a review. J Renew Mater 7(5):474–489
9. Tondi G, Pizzi A, Olives R (2008) Natural tannin-based rigid foams as insulation for doors and wall panels. Maderas Ciencia y Tecnologia 10(3):219–227
10. Martinez de Yuso A, Lagel M, Pizzi A, Fierro V, Celzard A (2014) Structure and properties of rigid foams derived from quebracho tannin. Mater Design 63:208–212
11. Li X, Pizzi A, Lacoste C, Fierro V, Celzard A (2013) Physical properties of tannin/furanic resin foamed with different blowing agents. BioResources 8(1):743–752
12. Chen X et al (2022) Flame-retardant and thermally-insulating tannin and soybean protein isolate (SPI) based foams for potential applications in building materials. Constr Build Mater 315:125711
13. Rodrigues MBB et al (2024) Bio-based tannin foams: comparing their physical and thermal response to polyurethane foams in lightweight sandwich panels. Compounds 4(1):1–16

14. Borrero-López AM, Nicolas V, Marie Z, Celzard A, Fierro V (2022) A review of rigid polymeric cellular foams and their greener tannin-based alternatives. Polymers 14(19):3974
15. Marie Z, Celzard A, Fierro V, Dam S (2022) A review of rigid polymeric cellular foams and their greener tannin-based alternatives. Polymers 14(19):3974
16. Celzard A, Zhao W, Pizzi A, Fierro V (2010) Mechanical properties of tannin-based rigid foams undergoing compression. Mater Sci Eng, A 527(15):4438–4446
17. Lacoste C, Basso MC, Pizzi A, Laborie M-P, Celzard A, Fierro V (2013) Pine tannin-based rigid foams: mechanical and thermal properties. Ind Crops Prod 43:245–250
18. Čop M, Lacoste C, Conradi M, Laborie M-P, Pizzi A, Sernek M (2015) The effect of the composition of spruce and pine tannin-based foams on their physical, morphological and compression properties. Ind Crops Prod 74:604–612
19. Kain G et al (2015) Effects of different flavonoid extracts in optimizing tannin-glued bark insulation boards. Wood Fiber Sci 47(3):258–269

Mechanical Properties of Lunar Concrete: Experimental and Simulation Perspectives

Shima Pilehvar[1(✉)], Aya Kassem[2], Erivan Hussein Osman[2], Maroush Imtiaz Malik[2], Sumayo Osman Ashkir[2], and Mahdi Kioumarsi[1,2]

[1] Department of Engineering, Østfold University College, Fredrikstad, Norway
Shima.Pilehvar@hiof.no

[2] Department of Built Environment, OsloMet—Oslo Metropolitan University, Oslo, Norway

Abstract. Lunar construction requires durable and in situ materials capable of withstanding extreme environmental conditions such as vacuum, radiation, and thermal cycling. Lunar concrete, commonly as *mooncrete*, has been investigated using regolith simulants combined with binders including geopolymers, sulfur, polymers, and aggregates. This paper presents a systematic literature review of the mechanical properties of lunar concretes, along with findings from both experimental studies and simulation-based analyses. The main discussion in this work is on compressive and flexural strengths, durability under freeze–thaw cycles, and performance in vacuum conditions. Geopolymer and sulfur concretes emerge as the most promising alternatives, demonstrating compressive strengths up to 70 MPa, while polymer-modified and regolith-aggregate concretes offer complementary benefits but face challenges related to brittleness and durability. A critical outcome of this review is the identification of gaps in current research, particularly the reliance on regolith simulants and the lack of standardized testing protocols. By synthesizing advancements and challenges, this review provides a roadmap for developing effective lunar construction materials that support sustainable and cost-efficient extraterrestrial habitats.

Keywords: Mooncrete · Lunar concrete · Mechanical properties

1 Introduction

In the past few decades, the idea of constructing extraterrestrial habitats on planets beyond earth has captured the interest of scientists and engineers. The Moon, as Earth's closest neighbor, is considered the most promising extraterrestrial location for initial human settlement. However, transporting construction materials from Earth is highly energy-intensive, making the development of in-situ resources essential [1].

One major challenge is the severe vacuum on the Moon, which accelerates water evaporation during the hydration process [2]. This water loss consequently leads to increased porosity in the concrete, causing a loss in strength and reduced radiation shielding [3]. Despite this limitation, lunar soil remains highly attractive due to its abundance and favorable mechanical properties, making it a key candidate for in-situ construction

H. Beushausen et al. (Eds.): NTZR 2025, LNCE 873, pp. 364–371, 2026.
https://doi.org/10.1007/978-3-032-27698-8_33

materials. Slyuta [4] highlighted variations in these properties across different lunar landscapes. Lunar soil can be employed either in its raw state or after treatments to create construction materials such as lunar cement, lunar concrete, and lunar bricks [3]. As mentioned in Kobaka et al. [5] only 400 kg of lunar regolith has been brought back from the Apollo and Luna lunar missions to earth. Therefore, researchers have relied on producing lunar regolith simulants as a substitute to experiment with the feasibility of creating mooncrete from lunar regolith. These stimulants are designed to replicate the chemical, physical, and morphological characteristics of authentic lunar regolith [6]. Furthermore, Wang and Snoeck [7] explored sulfur concrete, tracing its historical evolution and emphasizing its viability as a construction material for extraterrestrial bodies. While not entirely focused on lunar construction, Xiao et al. [8] explored the concept of rebar-free concrete construction (RFCC), which could be valuable in extreme conditions where conventional reinforcement is unsuitable.

The objective of this review is to conduct a systematic analysis of the mechanical properties of mooncrete. Although previous reviews have explored various aspects of extraterrestrial construction, a comprehensive review of challenges and advancements related to lunar concrete is lacking. As lunar construction is a relatively new field with unique environmental constraints such as severe vacuum conditions, extreme temperature fluctuations, and limited water availability, it is crucial to consolidate existing knowledge and identify gaps that require further research. By examining historical developments alongside recent advancements, this review aims to provide researchers and engineers with a more precise roadmap for developing effective lunar construction materials, ultimately paving the way for sustainable and cost-effective lunar habitats.

2 Method

The methodology used in this paper is a systematic literature review to collect and summarize the key mechanical properties of various types of Lunar concrete. The Preferred Reporting Items for Systematic Reviews and Meta-Analyses (PRISMA) method proposed by [9] was adopted to conduct this review. A qualitative investigation was conducted using the Scopus database. The steps in the methodology are as follows: (1) selection of database and relevant keywords; (2) systematically searching, screening, and filtering relevant articles; (3) investigating trends in repeated keywords/topics through bibliographic analysis; (4) analyzing and categorizing the articles into key findings. Figure 1 presents a Map of co-occurrence of keywords by VOSviewer software.

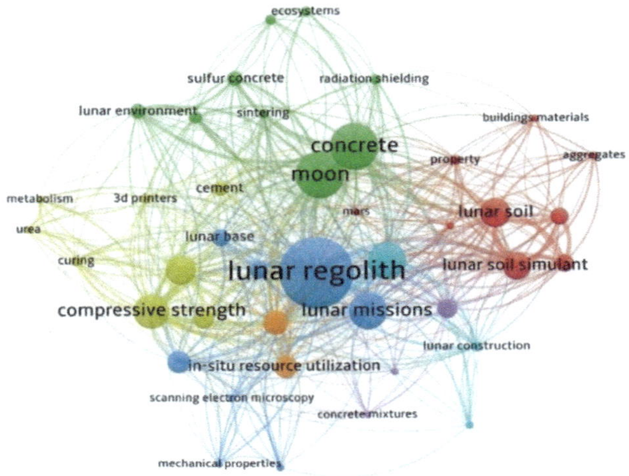

Fig. 1. Map of co-occurrence of keywords by VOSviewer software.

3 Mechanical Properties of Lunar Concrete

This chapter presents findings and discussions from the systematic studies on the main mechanical properties of lunar concrete.

3.1 Geopolymers

Geopolymer concrete has gotten significant attention due to its potential use for lunar construction. Understanding the impact that porosity, curing conditions, and alkaline-solution concentration have on compressive strength is essential for optimizing geopolymer concrete production for lunar construction applications.

It is important to find an optimal concentration of alkaline activator to be used when manufacturing the geopolymer. Zhang et al. [10] prepared specimens using BH-1 lunar regolith simulant combined with NaOH solutions in three different NaOH concentrations of 6, 8, and 10 mol/L. The utilization of 6 M NaOH was insufficient to complete the geopolymerization due to inadequate alkalinity, whereas the 10 M NaOH samples contained an excessive concentration of Na + ions, which led to side reactions that damaged the geopolymer structure. The optimal performance was observed at 8 M NaOH, which provided a balance between sufficient alkalinity and structural stability.

Wang et al. [11] also used NaOH as an alkaline solution. It was found that specimens containing an excess amount of NaOH alkaline activator led to the formation of cracks, consequently reducing compressive strength. Thus, the optimal molar ratio of Na_2O/Al_2O_3 was determined to be 1.25, which resulted in a slight reduction in compressive strength under the influence of 24h vacuum conditions and exposure to 30 freeze-thaw cycles in liquid nitrogen. Despite the decrease in compressive strength, the value remains suitable for construction on the moon.

Pilehvar et al. [12] investigated the extreme temperature variations and vacuum of the lunar environment, as well as different amounts of urea as a superplasticizer were examined with respect to the effect on the mechanical properties of lunar regolith geopolymers for 3D printing. They utilized NaOH as an alkaline solution DNA-1 as lunar regolith simulant. As shown in Fig. 2, they found that the compressive strength increased for all samples after the lunar day–and–night cycle, compared to the samples precured for 3 h at 80 °C. The increase in compressive strength after the lunar cycle is much higher at ambient conditions, which demonstrates that the vacuum causes the geopolymer deterioration. Adding urea decreased the compressive strength at both vacuum and ambient pressure.

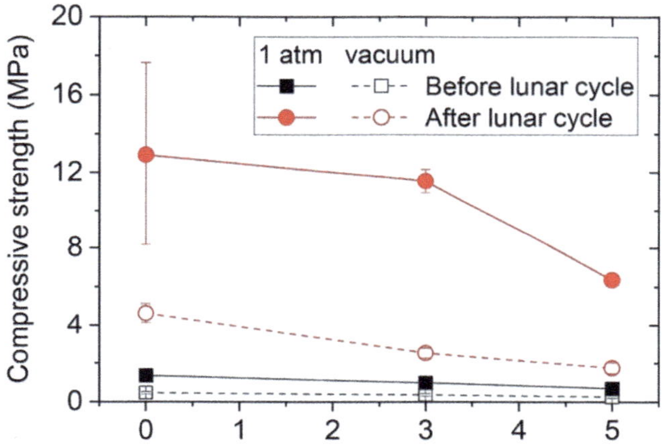

Fig. 2. Compressive strength of lunar geopolymer before and after a lunar cycle at vacuum and at ambient pressure. Adopted from Pilehvar et al. [12]

Wang et al. [13] investigated the use of tektite powder as a lunar regolith simulant utilizing two alkali-activators: sodium hydroxide (NaOH) and sodium silicate (Na_2SiO_3). It was found that the compressive strength of the samples with NaOH remained almost unchanged after 30 freeze-thaw cycles, whereas the samples containing Na_2SiO_3 experienced a 3.76 MPa reduction in compressive strength. SEM analysis and water measurements in vacuum suggested that sodium hydroxide could be a more effective alkali-activator compared to sodium silicate. The sodium silicate samples exhibited high residual water content, which worsened the durability of the freeze-thaw. Cai et al. [14] presented an autoclaved concrete made from lunar regolith simulant and calcium hydroxide as an alkali-activator. They found that increasing lunar regolith simulant (LRS) fineness caused an increase in pore volume and reduced compressive strength. Furthermore, the authors observed a 33% increase in compressive strength when the briquetting pressure was increased from 25 to 50 kN, suggesting a positive correlation between briquetting pressure and compressive strength. The observed increase in strength was explained by the consequential reduction in total pore volume induced by increased briquetting pressure. The study also revealed that an increase in calcium hydroxide content improved the compressive strength. This is because calcium hydroxide promoted the generation of target hydration products.

Research by Zhang et al. [10] on geopolymer concrete, produced by using BH-1 lunar regolith simulant and various NaOH concentrations, demonstrated a drastic reduction in flexural strength after exposure to high and cryogenic lunar temperatures. The flexural strength was initially 5.7 MPa. Subsequent exposure to cryogenic temperatures resulted in a notable decrease of 49%, reducing the flexural strength to 2.9 MPa. Further exposure to high temperatures above 90 °C caused the formation of microcracks. This led to an additional 21% reduction in flexural strength, resulting in a final strength of 1.7 MPa.

Lunar and Martian geopolymer concrete incorporated with NaOH and K_2SiO_3 into the mixture were studied by Alexiadis et al. [15]. The lunar and Martian regolith simulants utilized were JSC-LUNAR-1A and JSC MARS-1A, respectively. The results revealed that lunar geopolymer concrete exhibited higher flexural strength than both Martian geopolymer concrete and traditional cement concrete. Upon fracture, lunar geopolymer concrete displayed a clean breakage into a few fractured pieces, unlike Martian and cement samples, which disintegrated into a powdery state.

3.2 Sulfur Concrete

Sulfur, identified as one of the most abundant volatiles on the lunar surface, offers a promising alternative binder for lunar concrete, particularly advantageous since it can be extracted from regolith at moderate temperatures and does not require water, a scarce resource on the Moon [16]. Toutanji et al. [17] explored the use of sulfur concrete for the mentioned purpose. They discovered that a concrete mixture consisting of 65% JSC-1 lunar regolith simulant and 35% sulfur resulted in a compressive strength of 31 MPa, which was higher than that of normal Portland cement concrete with a water-to-cement ratio of 0.43, which achieved only 24.2 MPa.

3.3 Polymer-Modified Concrete

Studies focusing on polymer-based lunar concrete have shown promising results regarding compressive strength [18–21].

Roberts and Scrutton [18] proposed using ordinary starch as a binder with extraterrestrial regolith simulants for creating lunar and Martian concrete named "StarCrete". This technology yielded promising strength results in which the lunar StarCrete reached a compressive strength of 91.7 MPa, which can be characterized as a high-strength concrete. Based on finding in [18], the utilization of bio-polymer-based concrete resulted in a satisfactory flexural strength of 2.1 MPa for lunar StarCrete, which is comparable to ordinary concrete at 2.5–4.5 MPa.

Lee et al. [19] also employed polyethylene as a polymer with a simulant named Korea-Hanyang Lunar Simulant-1 (KOHLS-1). This study investigated the effectiveness of bottom-up heating in achieving reduced heating time and energy during the production of polyethylene lunar concrete. The experimental results revealed that the compressive strength achieved by bottom-up heating was 5.7 MPa, roughly half of the strength achieved using top-down heating, which was 12.9 MPa. Nevertheless, the bottom-up heating time was twice as fast as the top-down heating method, while also operating at a lower temperature of 200 °C instead of 230 °C reducing thermal energy consumption and achieving a sufficient compressive strength for lunar construction.

Studies [20, 21] highlighted the superior compressive strength of lunar-simulant polymer concrete compared to conventional concrete, making it a viable option for lunar construction projects. Caluk et al. [20] investigated the behavior and material properties of LHS-1 lunar soil simulant with seven different epoxy percentages. The results showed that the optimal epoxy content was 18%, yielding the highest compressive strength of about 124 MPa.

Sik Lee et al. [21] developed a lunar concrete consisting of lunar regolith simulant and polyethylene as a polymer. This polymer-based concrete exhibited a compressive strength ranging between 12.6–12.9 MPa. After considering the disparities in gravity between the moon and the earth, the equivalent compressive strength in the earth was determined to be approximately 70 MPa. This strength is notably higher than that of typical structural members, where the resistance to external loading caused by gravitational forces in the earth lies between 30 and 40 MPa.

Oh et al. [22] investigated the ultralow-binder content (UBC) composites produced by mixing thermoplastic binders and JSC-1 lunar soil simulant filler. The binders studied were polyetherketoneketone (PEKK), polyimide (PI), and a polyamide nanoclay composite (PAnc). The PEKK-based UBC composite exhibited the highest flexural strength out of the rest of the binders. It was even stronger than typical steel-reinforced concrete despite containing only 4 wt.% of PEKK binder. Furthermore, the application of higher pressure during the compaction process was found to facilitate the structural integrity of the composite which is especially important when using ultralow binder amounts. Thus, compaction effectively enhances the flexural strength of the concrete. However, curing at higher temperatures led to a decline in flexural strength, but it remained satisfactory for lunar construction.

3.4 Regolith-Based Concrete

A new concept of incorporating lunar aggregate simulant into ordinary cement concrete rather than utilizing lunar soil was introduced by Zarzycki et al.[23]. They suggested that lunar concrete could be produced with lunar aggregates rather than lunar soil, which is the material predominantly employed in existing research on lunar concrete. Two cement composites were made, one using standardized sand and the other with Ilmenite as a lunar aggregate simulant (LAS). Both cement composites exhibited a consistent pattern of increasing compressive strength throughout the curing period. The sample with sand reached a higher compressive compared to the sample with LAS, with roughly 54 MPa and 49 MPa, respectively.

Lauermannova et al. [24] employed magnesium oxychloride cement (MOC) with graphene as an additive to enhance its mechanical properties. A high amount of lunar regolith simulant OPRL2N was utilized, to minimize the need to import raw materials from the earth. The flexural strength of samples containing graphene, magnesium oxychloride and lunar regolith simulant was higher than the reference sample with pure magnesium oxychloride, even for the samples containing a high amount of regolith after periodic heating and freezing. This suggests that graphene efficiently enhances the flexural strength despite the amount of regolith simulant used. However, an essential aspect regarding experiments with high amounts of regolith was highlighted by the researchers of this article: utilization of a higher amount of regolith simulant leads to the greater discrepancy in mechanical properties between the actual lunar regolith and the lunar regolith simulant.

4 Challenges and Future Work

Current research primarily focuses on developing mooncrete based on lunar regolith simulants. Below, the research gaps identified in the literature review are outlined, along with potential directions for future research.

- Future research should prioritize the advancement of mooncrete through experimentation with authentic lunar soil samples, emphasizing the utilization of in-situ resources for lunar construction. However, the scarcity of lunar regolith samples retrieved from missions highlights a significant challenge.
- Varying testing methods for mooncrete properties have been observed throughout the articles, indicating a lack of standardized research. Therefore, the next critical phase of research should entail the development of standardized methods.

5 Conclusion

Mooncrete is a promising solution for lunar construction. Lunar concretes can achieve compressive strengths up to 70 MPa and flexural strengths up to 10 MPa, achieving strengths comparable to high performance terrestrial concretes. Sulfur and geopolymer binders emerge as the most promising candidates, though each has limitations. Polymer-modified and aggregate regolith concretes provide unique advantages but also face durability and brittleness challenges. Future work should prioritize durability assessment under realistic lunar conditions and the development of hybrid binders.

References

1. Khitab A et al (2016) Lunar concrete: prospects and challenges. Astron Rep 60(2):306–312
2. Wilhelm S, Curbach M (2014) Review of possible mineral materials and production techniques for a building material on the moon. Struct Concr 15(3):419–428
3. Toklu YC, Akpinar P (2022) Lunar soils, simulants and lunar construction materials: an overview. Adv Space Res 70(3):762–779
4. Slyuta EN (2014) Physical and mechanical properties of the lunar soil (a review). Sol Syst Res 48(5):330–353
5. Kobaka J et al (2023) A study of lunar soil simulants from construction and building materials perspective. Case Stud Construct Mater 18:e02082
6. Isachenkov M et al (2022) Characterization of novel lunar highland and mare simulants for ISRU research applications. Icarus (New York, N.Y. 1962), vol 376, pp 114873
7. Wang Q, Snoeck D (2024) To boldly go where no one has gone before: sulfur concrete, a promising construction material fulfilling the demands for a sustainable future on celestial objects: a review. Mater Today (Kidlington, England) 72:301–317
8. Xiao J et al (2024) Rebar-free concrete construction: concept, opportunities and challenges. J Build Eng 86:108933

9. Liberati A et al (2009) The PRISMA statement for reporting systematic reviews and meta-analyses of studies that evaluate health care interventions: explanation and elaboration. PLoS Med 6(7):e1000100–e1000100
10. Zhang R, Zhou S, Li F (2022) Preparation of geopolymer based on lunar regolith simulant at in-situ lunar temperature and its durability under lunar high and cryogenic temperature. Constr Build Mater 318:126033
11. Wang K-T et al (2017) Lunar regolith can allow the synthesis of cement materials with near-zero water consumption. Gondwana Res 44:1–6
12. Pilehvar S et al (2021) Investigation of severe lunar environmental conditions on the physical and mechanical properties of lunar regolith geopolymers. J Mater Res Technol 11:1506–1516
13. Wang KT et al (2016) Development of near-zero water consumption cement materials via the geopolymerization of tektites and its implication for lunar construction. Sci Rep 6(1):29659
14. Cai L et al (2019) Preparation of autoclave concrete from basaltic lunar regolith simulant: effect of mixture and manufacture process. Constr Build Mater 207:373–386
15. Alexiadis A, Alberini F, Meyer ME (2017) Geopolymers from lunar and Martian soil simulants. Adv Space Res 59(1):490–495
16. Gracia V, Casanova I (1998) Sulfur concrete: a viable alternative for lunar construction. In: Space 98. pp 585–591
17. Toutanji HA, Evans S, Grugel RN (2012) Performance of lunar sulfur concrete in lunar environments. Constr Build Mater 29(1):444–448
18. Roberts AD, Scrutton NS (2023) StarCrete: A starch-based biocomposite for off-world construction. Open Eng 13(1)
19. Lee J et al (2018) Bottom-up heating method for producing polyethylene lunar concrete in lunar environment. Adv Space Res 62(1):164–173
20. Caluk N, Seisdedos G, Azizinamini A (2023) Destructive and non-destructive testing of potential lunar polymer concrete for future lunar habitable infrastructure. Constr Build Mater 405:133395
21. Sik Lee T, Lee J, Yong Ann K (2015) Manufacture of polymeric concrete on the Moo*n*. Acta Astronautica 114:60–64
22. Oh K et al (2020) Impact formation of ultralow-binder-content composite "lunar cement." CEAS Space J 13(2):183–187
23. Zarzycki PK, Katzer J (2020) A proposition for a lunar aggregate and its simulant. Adv Space Res 65(12):2894–2901
24. Lauermannova A-M et al (2021) Regolith-based magnesium oxychloride composites doped by graphene: novel high-performance building materials for lunar constructions. FlatChem 26:100234

Unlocking Subsurface Resilience: Geochemical Evidence of Shale Reactivity for Permanent CO_2 Storage Under Supercritical Conditions

Loic Bethel Dje and Mileva Radonjic(✉)

Barrier Materials and Geomimicry Lab, School of Chemical Engineering, Oklahoma State University, Stillwater, OK 74078, USA
mileva.radonji@okstate.edu

Abstract. Long-term carbon storage requires more than just pore space; it fundamentally depends on the geochemical stability of the host rock. This study investigates the mineralogical transformations in Caney Shale after exposure to supercritical CO_2 ($scCO_2$), offering critical insights into its potential for permanent carbon sequestration. We tested two distinct shale zones: B2 (reservoir) and C2 (caprock); by exposing them to $scCO_2$ at 17.2 MPa and 60 °C for 30 days in a closed-loop, flow-through system. Analysis using SEM-EDS and Raman spectroscopy revealed contrasting reactivity between the zones. The B2 reservoir facies showed stable carbonate and illite content, moderate elemental redistribution, and increased organic disorder, indicating active CO_2-mineral interactions and effective mineral trapping. In contrast, the C2 caprock facies exhibited only minor carbonate and feldspar loss, the formation of secondary kaolinite and anorthite, and limited elemental mobility, which is consistent with its function as a seal. These results underscore the importance of geochemical processes in determining CO_2 retention. We conclude that reservoir facies are well-suited for mineral trapping, while caprock facies can serve as robust reactive barriers. Integrating these geochemical evaluations into CCS planning is essential for enhancing site reliability and achieving the goals of a net-zero infrastructure.

Keywords: Supercritical CO_2 · Mineralogical Sequestration · Shale

1 Introduction

Achieving net-zero emissions demands a multi-pronged approach that integrates both surface and subsurface carbon management strategies. Renewables, electrification, and efficiency help reduce emissions, but scalable carbon removal is essential to offset CO_2 from hard-to-abate sectors [1, 2]. Carbon capture and storage (CCS) is a central strategy for reducing atmospheric CO_2, and relies on geological storage in saline formations, depleted reservoirs, and shales, yet organic-rich shales remain understudied despite their abundance and dual role as seals and storage media. The mineralogical heterogeneity of shales significantly influences their reactivity with CO_2, determining whether chemical alterations enhance or diminish porosity and permeability [3]. These reactions can either

H. Beushausen et al. (Eds.): NTZR 2025, LNCE 873, pp. 372–383, 2026.
https://doi.org/10.1007/978-3-032-27698-8_34

bolster or undermine the physical integrity and sealing capacity of the storage formation over time. Previous studies have often treated shales as homogeneous units, reporting generalized findings such as feldspar dissolution and clay transformations under $scCO_2$ or CO_2–brine conditions. Such formation-scale averages, however, can obscure critical facies-specific responses, particularly in formations with high depositional and diagenetic variability like the Caney Shale of Oklahoma [4, 5]. This study evaluates $scCO_2$–shale interactions at a facies-specific resolution, comparing clay-rich caprock (C2) and carbonate-rich reservoir (B2) facies. By combining SEM-EDS mineralogical mapping with micro-CT porosity and fracture analysis, we aim to: Quantify mineral-specific reactivity under dry $scCO_2$ conditions, identify which facies are most susceptible to chemo-mechanical alteration and provide design-relevant insights for CCS in shale formations.

Ultimately, this work demonstrates how facies-level mineralogical and elemental changes govern seal integrity, storage efficiency, and mineral trapping. In doing so, it reinforces the importance of subsurface chemical stability as a core design parameter in the transition to a net-zero carbon future.

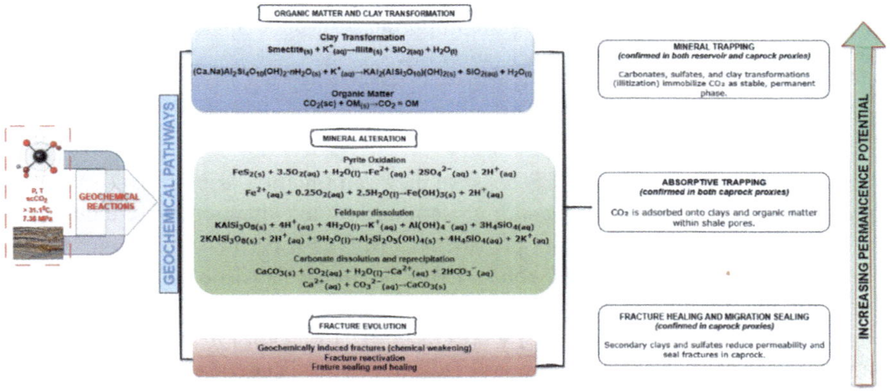

Fig. 1. Schematic illustration of mineralogical CO_2 sequestration pathways in shale formations, emphasizing geochemical interactions. The diagram highlights the dual influence of scCO on organic$_2$ matter reactivity and fracture propagation, with a focus on their implications for long-term carbon immobilization. This study centers on the geochemical signatures of $scCO_2$–shale interactions in support of Net-Zero initiatives

2 Geochemical Sequestration

2.1 Geochemical Pathways in CO_2 Sequestration

In shale formations, the efficiency of CO_2 trapping depends heavily on mineral assemblages and pore structure. Mineral trapping is the most permanent sequestration mechanism, involving the dissolution of primary silicates and carbonates, which releases cations (Ca^{2+}, Mg^{2+}, Fe^{2+}, K^+, Na^+) that then precipitate as stable secondary minerals [6, 7].

For instance, feldspar dissolution can drive secondary carbonate precipitation, while primary carbonate dissolution is often balanced by re-precipitation. The extent and rate of these reactions are highly sensitive to in-situ conditions, including temperature, pressure, fluid composition, brine salinity, and the surface area of reactive minerals [8]. Supercritical CO_2 ($scCO_2$) is particularly effective at driving these processes due to its high diffusivity and low viscosity, which allow it to penetrate micro- and nano-scale pore spaces [9]. The transformation of clays is time-dependent in CO_2-rich environments, especially the transition from smectite to illite [10–12]. In shale systems, pyrite oxidation and subsequent secondary mineral formation, along with alteration of clay minerals, have been noted, underscoring the importance of understanding mineral-specific interactions during long-term sequestration [5].

2.2 Shale Formations as Dual-Function CCS Media

Shales are uniquely suited for CCS because they can function as both a reservoir and a caprock. Their low permeability and high capillarity make them effective natural seals, while their chemical heterogeneity provides pathways for permanent storage; with key mineral components and specific roles [13]. Clay minerals (illite, smectite, kaolinite) which dominate the fine-grained matrix and govern the rock's ductile behavior. Carbonates (calcite, dolomite, ankerite) serves as buffer for acidity from carbonic acid and contribute directly to mineral trapping. Feldspars (K-feldspar, albite) act as significant long-term sources of cations (K^+, Na^+) for various alteration-centered reactions. Sulfides (pyrite) which can oxidize to release iron and sulfate, leading to the formation of secondary minerals like jarosite [14, 15]. Organic matter is another critical component, as shales with high total organic carbon (TOC) can adsorb large quantities of CO_2, providing an additional trapping mechanism. Therefore, facies characterization is essential, as ductile, illite-rich facies tend to act as stable seals while heterogeneous reservoir facies drive both adsorption and mineral trapping (Fig. 2). However, physico-chemical changes to this organic matter, such as aromatization or wettability alteration, could compromise the rock's mechanical [16, 17].

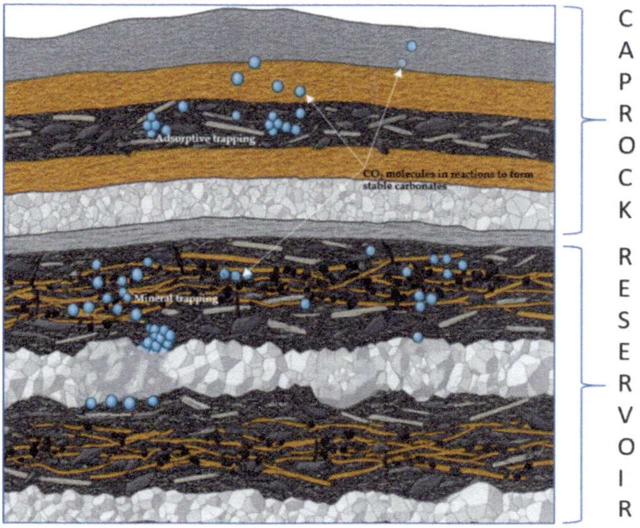

Fig. 2. Schematic representation of shale's dual role in CO_2 sequestration. An illustration of both adsorptive trapping in organic-rich layers and mineral trapping through geochemical reactions with reactive minerals such as feldspar and carbonates. Layered heterogeneity enables simultaneous storage functions, highlighting shale's capacity to act as both a reservoir and a seal in carbon capture and storage (CCS) strategies.

3 Case Study: Geochemical Evolution of the Caney Shale Under $scCO_2$ Conditions

The Caney Shale in southern Oklahoma provides a representative example of a self-sealing shale unit containing both reservoir and caprock analogs within one formation. Its mineralogical heterogeneity makes it an ideal system to assess facies-specific responses to CO_2 [16]. Two facies were examined: B2, a carbonate-rich brittle shale representing the reservoir interval, and C2, a clay-rich ductile shale serving as a caprock analog. This comparison evaluates their reactivity under supercritical CO_2 ($scCO_2$), simulating subsurface carbon storage conditions and assessing facies-specific trapping mechanisms.

3.1 Experimental Methodology

Polished blocks (1 cm × 1 cm × 0.5 cm) of both B2 and C2 facies were exposed to $scCO_2$ at 17.2 MPa and 60 °C for 30 days in a closed-loop, flow-through reactor (Fig. 3). This setup maintained continuous mineral–fluid interaction at reservoir-relevant temperature and pressure conditions.

Pre- and post-exposure characterization was performed to track mineralogical, elemental, and structural changes. Imaging and elemental analysis were conducted using SEM-EDS (low vacuum, 10 kV) with elemental maps acquired at approximately 0.5 μm pixel resolution. Complementary phase identification was performed using Raman spectroscopy (532 nm laser) to analyze both inorganic and organic phases.

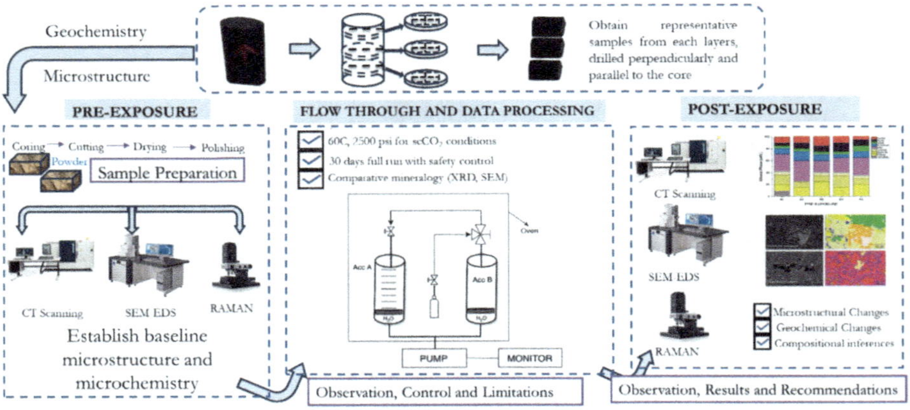

Fig. 3. Experimental setup for shale samples exposed to $scCO_2$ (60 °C, 17.2 MPa, 30 days). Pre- and post-exposure analyses (SEM-EDS, Raman) tracked mineral, elemental, and organic transformations.

3.2 Results and Interpretation

The reservoir proxy (B2) exhibited distinct geochemical changes after $scCO_2$ exposure. Calcite remained stable at ~ 25%, indicating effective buffering of acidity from CO_2 dissolution. Illite content also held at ~ 20%, while framework silicates (K-feldspar, albite) showed minimal alteration. A new phase, jarosite (~5%), appeared in post-exposure samples, most likely from partial pyrite oxidation that released ferric iron and sulfate [14, 18]. Elemental mapping revealed the mobilization and redistribution of calcium, iron, and potassium, which supported localized mineral precipitation and potential secondary trapping (Fig. 4). Additionally, Raman spectroscopy indicated a marked increase in the disorder of the organic phase, with the I_D/I_G ratio increasing from 0.82 to 0.91 (Fig. 6). This shift implies enhanced reactivity and potential for sorptive trapping due to increased surface area and molecular complexity within the carbon matrix [19].

In contrast, the C2 caprock proxy displayed a more passive geochemical response. Calcite content declined slightly from 15% to 13%, and feldspar decreased from 5% to 4%, indicating limited dissolution [20]. However, new phases such as kaolinite and anorthite emerged, each contributing around 4%, likely derived from feldspar weathering. Illite experienced a modest reduction from 25% to 22%, suggesting mild clay transformation rather than full breakdown, as shown in Fig. 5. Elemental mobility in C2 was comparatively subdued, with little evidence of significant cation transport or re-precipitation.

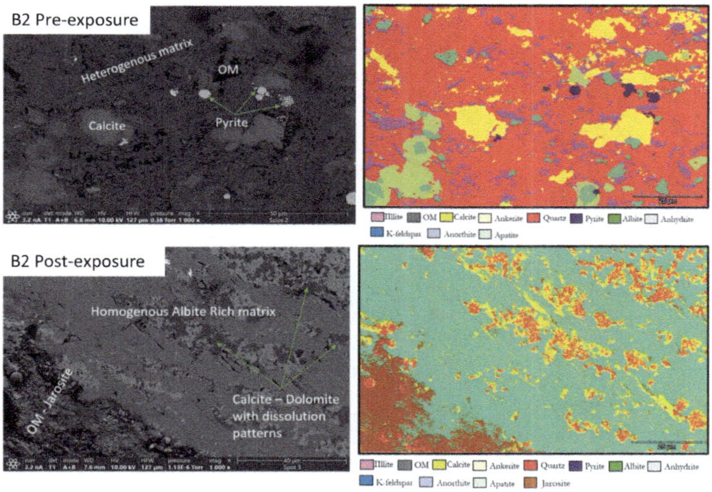

Fig. 4. Mineralogical evolution of reservoir proxy B2 before and after $scCO_2$ exposure. SEM-EDS micrographs (left) and elemental maps (right) show a pre-exposure matrix of calcite, pyrite, organic matter, and minor feldspars. Post-exposure, the matrix became albite-dominated, with carbonate dissolution features and new jarosite from pyrite oxidation. These changes confirm active geochemical reactions, secondary mineral formation, and elemental redistribution, supporting both dissolution and trapping mechanisms during CO_2 sequestration.

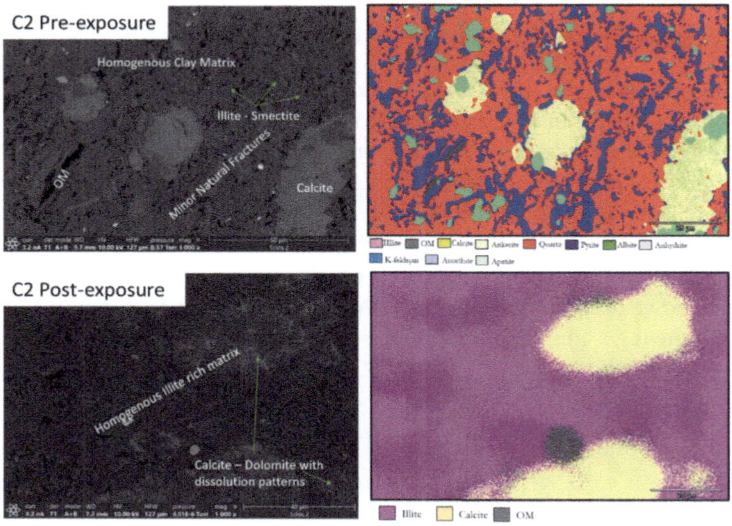

Fig. 5. Mineralogical evolution of caprock proxy C2 before and after $scCO_2$ exposure. SEM-EDS micrographs (left) and elemental maps (right) show a pre-exposure matrix of illite, smectite, dispersed calcite, and organic matter. Post-exposure, the matrix became illite-rich with clear carbonate dissolution. This illitization reduces swelling, lowers permeability, and enhances sealing integrity, reinforcing C2's role as an effective barrier for CO_2 storage.

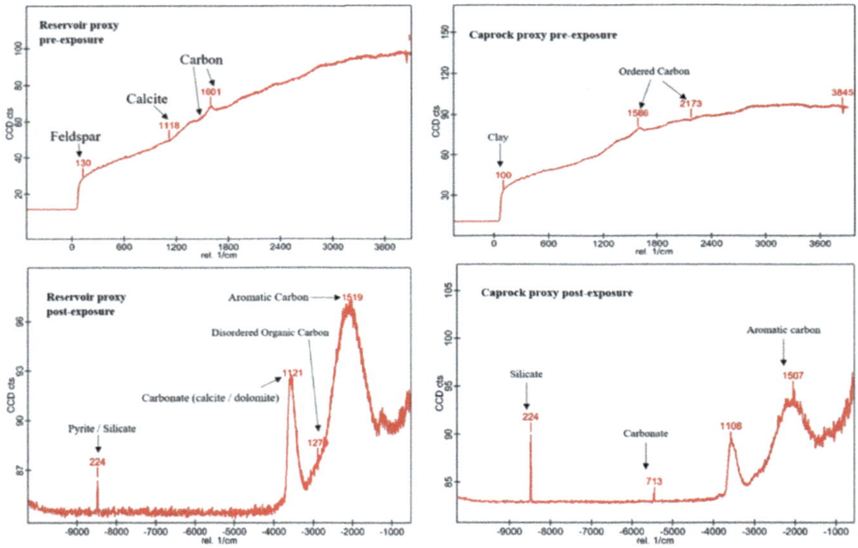

Fig. 6. Raman spectra of B2 (reservoir proxy) and C2 (caprock proxy) before and after $scCO_2$ exposure. Post-exposure spectra show increased aromaticity and disorder in B2, indicated by a rise in the I_D/I_G ratio (0.82 → 0.91), reflecting enhanced surface reactivity of the organic matrix. In contrast, C2 shows minimal structural alteration (I_D/I_G: 0.64 → 0.72), with persistent silicate and carbonate peaks, confirming clay stability and sealing integrity under $scCO_2$ conditions.

3.3 Inferences from the Experiments

The geochemical behavior of B2 and C2 underscores the need for facies-specific evaluation in CCS site selection. B2 exhibited favorable conditions for CO_2 immobilization through buffering, mineral precipitation, and increased organic reactivity. This reactivity supports its role as an effective reactive storage zone. Conversely, the limited interaction observed in C2 confirms its function as a chemically stable seal; ideal for maintaining long-term containment integrity [21]. Recognizing such distinctions within shale formations enhances the reliability of CO_2 storage predictions and informs targeted engineering design. Table 1 shows the comprehensive outcome of our analysis, with Fig. 7 showing a compositional chart of the elemental changes pre and post $scCO_2$ exposure.

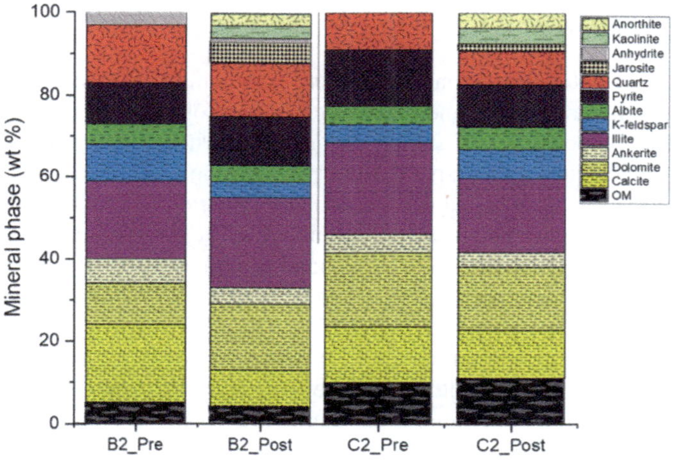

Fig. 7. Mineralogical phase distribution in B2 (reservoir proxy) and C2 (caprock proxy) pre- and post-scCO_2 exposure. Stacked columns illustrate the relative abundances of mineral phases before and after exposure to supercritical CO_2. B2 exhibits notable shifts in carbonate and feldspar phases, while C2 shows increased illite dominance post-exposure. These mineralogical changes reflect geochemical reactivity in B2 and the enhanced sealing potential of C2 through clay stabilization.

Table 1. Summary of mineralogical changes across both facies pre and postexposure to scCO_2

Mineral phase	Caprock proxy (C2)	Reservoir proxy (B2)
K-feldspar	Illite-Smectite (~40% clay conversion)	Fe-rich Illite, Illite-Smectite (~30 transformation)
Albite	Illite-Smectite	Illite-Smectite
Calcite/Dolomite	Calcite dissolved or undetected	Dolomite/Ankerite → Ankerite (~50% preserved/reprecipitated)
Ankerite	Ankerite (Stable or minor change)	Ankerite (Recrystallized/reprecipitated)
Apatite	Apatite (Stable)	Apatite (~10–20% density increase)
Quartz	Low	No significant change
Pyrite	Not observed post-exposure	Jarosite (Oxidation and Fe–S phase transformation)
Organic matter (OM)	Mildly altered (I_D/I_G: 0.64 → 0.72; ~13% increase)	Increased disorder (I_D/I_G: 0.82 → 0.91; ~11% increase)

4 Geochemistry as a Pillar in CCS Design

While petrophysical properties are important, geochemical screening is essential for verifying the mineral stability and reactivity required for long-term CCS security. The mineralogical composition of a storage formation is critical to its performance. Reactive phases like calcite and feldspar dissolve in carbonic acid, releasing cations ($Ca2^+$, $Fe2^+/Fe3^+$, $Mg2^+$, Na^+, K^+) that drive secondary mineral precipitation: the key to permanent mineral trapping. The formation of minerals like siderite, magnesite, or jarosite is a direct indicator of successful geochemical immobilization of CO_2 [22, 23]. For example, the formation of siderite, magnesite, or jarosite signals successful geochemical immobilization of CO_2.

Clay mineral transformations, such as the smectite-to-illite transition or interlayer collapse, also require close monitoring, as they can alter fluid pathways and compromise sealing efficiency [24, 25]. Simultaneously, structural changes in organic matter, tracked by Raman I_D/I_G ratios, influence CO_2 sorption and retention capacity [26]. Our facies-resolved analysis moves beyond bulk-rock averages to reveal distinct functional roles within a single formation. Brittle, carbonate-rich reservoir facies (B2) promote mineral trapping, while ductile, clay-rich caprock facies (C2) evolve toward improved sealing integrity. This reframes ductile shales as chemically dynamic, rather than static, barriers.

Our facies-resolved analysis demonstrates distinct functional roles: brittle, carbonate-rich reservoir facies promote mineral trapping through dissolution-enhanced porosity, while ductile, clay-rich caprock facies evolve toward improved sealing integrity. We further identify a two-stage reactivity model under $scCO_2$: initial feldspar dissolution enhances percolation pathways, followed by secondary clay transformation and precipitation that potentially progressively reduce permeability. Supported by quadrant-based SEM-EDS and micro-CT mapping, this approach moves beyond formation-averaged observations, providing a mechanistic explanation for heterogeneity and reframing ductile shales as chemically dynamic rather than static barriers.

These geochemical insights can be directly applied to CCS operations. Reactive transport models can forecast mineral-fluid interactions and their geomechanical impacts, informing well placement and risk mitigation strategies. Engineering choices, such as cement and casing materials, can be tailored to the local mineralogy to prevent degradation [17, 27].

Embedding geochemistry into CCS planning is essential for building resilient net-zero infrastructure. Stable facies such as C2 act as natural seals, while reactive zones like B2 can be harnessed for storage when coupled with predictive monitoring [28, 29]. Geochemical stability limits reactivity and long-term alteration, while mineral insights inform durable materials, zoning, and injection policies, positioning reservoirs as assets in carbon-neutral strategies [30]. By embedding geochemistry into the design process, we can transform CCS reservoirs from passive containers into active, reliable agents for achieving net-zero goals.

5 Conclusion

Our results demonstrate that geochemistry is central to central to designing secure CO_2 storage in shale formations. In the Caney Shale, carbonate-rich reservoir facies (B2) effectively buffered acidity and redistributed cations, confirming their suitability as reactive storage zones. In contrast, the clay-rich caprock facies (C2) remained largely stable, evolving to enhance its sealing integrity. These contrasts highlight the necessity of facies-specific evaluation to ensure both efficient trapping and long-term containment in CCS projects.

Some uncertainties remain, as short-term laboratory experiments cannot fully replicate long-term reservoir kinetics, brine-mediated reactions, or stress-coupled processes. Addressing these gaps is possible through extended flow-through experiments, the use of reactive-brine systems, and integrated geomechanical testing. Ultimately, will improve the predictive capability of our models and further establish geochemical stability as a reliable asset in the global transition to a net-zero future.

Acknowledgments. This work wouldn't be possible without funding from Mileva Radonjic's Samson Petroleum Professorship Chair Funds. Parts of this work were carried out in the Microscopy Laboratory, Oklahoma State University, which received funds for purchasing the equipment from the NSF MRI and DOE NETL programs. The authors would like to acknowledge Jim Puckette for training and granting access to his laboratory space and equipment for sample preparation; Brent Johnson and Lisa Whitworth, for training and assistance during materials characterization at the OSU Microscopy Lab; and all the members of the Barrier Materials and Geomimicry lab, specifically Ali Ettehadi and Bruce Mutume; and my beloved family for being a daily motivation and inspiration.

References

1. Raihan A, Tuspekova A (2023) Towards net zero emissions by 2050: the role of renewable energy, technological innovations, and forests in New Zealand. J Environ Sci Econ 2:1–16. https://doi.org/10.56556/jescae.v2i1.422
2. Oladapo BI, Olawumi MA, Olugbade TO, Ismail SO (2025) Data analytics driving net zero tracker for renewable energy. Renew Sustain Energy Rev 208:115061. https://doi.org/10.1016/j.rser.2024.115061
3. Oyewole KA, Okedere OB, Rabiu KO, Alawode KO, Oyelami S (2023) Carbon dioxide emission, mitigation and storage technologies pathways. Sustain Environ 9:2188760. https://doi.org/10.1080/27658511.2023.2188760
4. Leng J, Bump A, Hosseini SA, Meckel TA, Wang Z, Wang H (2024) A Comprehensive review of efficient capacity estimation for large-scale CO_2 geological storage. Gas Sci. Eng. 126:205339. https://doi.org/10.1016/j.jgsce.2024.205339
5. Dje LB, Radonjic M (2025) Reactivity of shale to supercritical CO_2: insights from microstructural characterization and mineral phase evolution in caney shales for CCUS applications. Materials 18:3382. https://doi.org/10.3390/ma18143382
6. Xiong F, Wang Y, Puckette J, Grammer M, Awejori G, Radonjic M (2024) Geochemical and microstructural characterization of shale rock: a workflow for site evaluation for energy-fluids storage. Energy Fuels 38:20875–20891. https://doi.org/10.1021/acs.energyfuels.4c03197

7. Li S, Zhang S, Xing H, Zou Y (2022) CO2–brine–rock interactions altering the mineralogical, physical, and mechanical properties of carbonate-rich shale oil reservoirs. Energy 256:124608. https://doi.org/10.1016/j.energy.2022.124608
8. Sanguinito S et al (2018) Quantifying dry supercritical CO_2-induced changes of the utica shale. Fuel 226:54–64. https://doi.org/10.1016/j.fuel.2018.03.156
9. Xu J, Yuan G, Cao Y, Jin Z, Liu K (2025) Feldspar dissolution rates and controlling factors under near-equilibrium conditions in subsurface settings. Sci China Earth Sci. https://doi.org/10.1007/s11430-024-1608-7
10. Chang X, Lin S, Yang C, Wang K, Liu S, Guo Y (2024) A critical review of ScCO2-enhanced gas recovery and geologic storage in shale reservoirs. Gas Sci Eng 125:205317. https://doi.org/10.1016/j.jgsce.2024.205317
11. Awejori GA, Dong W, Doughty C, Spycher N, Radonjic M (2024) Mineral and fluid transformation of hydraulically fractured shale: case study of caney shale in Southern Oklahoma. Geomech Geophys Geo-Energy Geo-Resour 10:128. https://doi.org/10.1007/s40948-024-00835-0
12. Olabode A, Radonjic M (2014) Shale caprock/acidic brine interaction in underground CO_2 storage. J Energy Resour Technol 136. https://doi.org/10.1115/1.4027567
13. Goodman A, Sanguinito S, Kutchko B, Natesakhawat S, Cvetic P, Allen AJ (2020) Shale pore alteration: potential implications for hydrocarbon extraction and CO_2 storage. Fuel 265:116930. https://doi.org/10.1016/j.fuel.2019.116930
14. Song Y, Yang L, Wang H, Sun X, Bai S, Wang N, Liang J, Zhou L (2021) The coupling reaction of Fe2+ bio-oxidation and resulting Fe3+ hydrolysis drastically improve the formation of iron hydroxysulfate minerals in AMD. Environ Technol 42:2325–2334. https://doi.org/10.1080/09593330.2019.1701564
15. Dje LB, Awejori GA, Radonjic M (2024) Comparison of geochemical reactivity of marcellus and caney shale based on effluent analysis. In: Proceedings of the 58th U.S. rock mechanics/geomechanics symposium, ARMA, Golden, Colorado, USA, June 23 2024, pp D041S053R004
16. Radonjic M, Luo G, Wang Y, Achang M, Cains J, Katende A, Puckette J, Grammer M, King GE (2020) Integrated microstructural characterisation of Caney Shale, OK. In: Proceedings of the proceedings of the 8th unconventional resources technology conference. American Association of Petroleum Geologists: Online
17. Fitts JP, Peters CA (2013) Caprock fracture dissolution and CO2 leakage. Rev Mineral Geochem 77:459–479. https://doi.org/10.2138/rmg.2013.77.13
18. Zhou Q et al (2024) Pyrite characteristics in lacustrine shale and implications for organic matter enrichment and shale oil: a case study from the triassic yanchang formation in the Ordos Basin NW, China. ACS Omega 9:16519–16535. https://doi.org/10.1021/acsomega.4c00259
19. Carolina Sparavigna A (2024) Graphene, graphene oxide and carbon nanotubes in raman spectroscopy. Int J Sci 13:1–26. https://doi.org/10.18483/ijSci.2773
20. Duan G, Ram R, Xing Y, Etschmann B, Brugger J (2021) Kinetically driven successive sodic and potassic alteration of feldspar. Nat Commun 12:4435. https://doi.org/10.1038/s41467-021-24628-1
21. Characterization of Shale Cap-Rock Nano-Pores in Geologic CO_2 ContainmentShale Cap-Rock Nano-Pores in Geologic CO_2 Containment | Environmental & Engineering Geoscience | GeoScienceWorld Available online: https://pubs.geoscienceworld.org/aeg/eeg/article-abstract/20/4/361/60529/Characterization-of-Shale-Cap-Rock-Nano-Pores-in (Accessed on 24 Sept 2025)
22. Fentaw JW, Emadi H, Hussain A, Fernandez DM, Thiyagarajan SR (2024) Geochemistry in geological CO_2 sequestration: a comprehensive review. Energies 17:5000. https://doi.org/10.3390/en17195000

23. Olabode A, Radonjic M (2017) Geochemical markers in shale-CO_2 experiment at core scale. Energy Proc 114:3840–3854. https://doi.org/10.1016/j.egypro.2017.03.1516
24. Du H, Radonjic M, Chen Y (2020) Microstructure and micro-geomechanics evaluation of pottsville and marcellus shales. J Pet Sci Eng 195:107876. https://doi.org/10.1016/j.petrol.2020.107876
25. Olabode A, Radonjic M (2014) Diagenetic influence on fracture conductivity in tight shale and CO_2 sequestration. Energy Procedia 63:5021–5031. https://doi.org/10.1016/j.egypro.2014.11.532
26. Dong M, Gong H, Sang Q, Zhao X, Zhu C (2022) Review of CO_2-kerogen interaction and its effects on enhanced oil recovery and carbon sequestration in shale oil reservoirs. Resour. Chem. Mater. 1:93–113. https://doi.org/10.1016/j.recm.2022.01.006
27. Loring JS et al (2021) Synergistic coupling of CO_2 and H_2O during expansion of clays in supercritical CO_2–CH_4 fluid mixtures. Environ Sci Technol 55:11192–11203. https://doi.org/10.1021/acs.est.1c00275
28. Zhou X et al (2021) Changes of multiscale surface morphology and pore structure of mudstone associated with supercritical CO_2—water exposure at different times. Energy Fuels 35:4212–4223. https://doi.org/10.1021/acs.energyfuels.0c03920
29. Shi Z, Driba DL, Lopez Rivera N, Kariminasab M, Beckingham LE (2024) A review of coupled geochemical–geomechanical impacts in subsurface CO_2, H_2, and air storage systems. Energies 17:2928. https://doi.org/10.3390/en17122928
30. Boudreau K, Robinson M, Farooqi Z (2022) IPCC sixth assessment report: climate change 2021: the physical science basis summary for policymakers. Can J Emerg Manag 2.https://doi.org/10.25071/6sw6za31

CO_2-Reducing and Energy Storing Textile Carbon Concrete Heating Plate Using Phase Change Material for Smart Home Systems

Ralf Gliniorz(✉), Marco Lindner, Sandra Gelbrich, and Henrik Funke

Chemnitz University of Technology, Professorship of Lightweight Structures and Polymer Technology, Research Group Lightweight Constructions in Civil Engineering, Chemnitz, Germany

ralf.gliniorz@mb.tu-chemnitz.de

Abstract. The construction industry is obligated to make significant efforts to ensure a sustainable future, particularly with regard to reducing CO_2 emissions. A predominant element of this is the conservation, assurance and storage of energy, with a concurrent reduction in CO_2 emissions. It is evident that this will not occur spontaneously; a multitude of technological advancements are imperative. The most prevalent form of energy is heat. It is very important to reduce and save this power, but also to store it. The present study investigates a carbon fiber textile concrete heating plate that incorporates phase change material. The plate exhibits the capacity to convert electrical energy, whether self-generated by photovoltaic (PV) panels or obtained from the electrical grid, directly and efficiently into thermal energy. This thermal energy can be released to the environment in a targeted manner, controllable via integration into smart home systems. The effect of phase change material is also discussed. As these store heat latently, a special adjustable temperature plateau can be set in the heating process. The requisite temperature increase, in relation to the energy storage capacity, was determined through a combination of calculations and empirical testing. This analysis considered the thickness of the plates and the proportion of phase change material. The concrete used consists of CO_2-reduced cement and local resources. Combined with presence recognition and artificial intelligence for lifestyle habits, these smart homes will be able to efficiently distribute rare and precious energy to where it is needed.

Keywords: carbon concrete heating · storing energy · smart home integration · phase change material · reducing CO_2 concrete · CO_2 reduction storage

1 Introduction

The construction industry contributes significantly to the release of greenhouse gases worldwide. According to the UN Environment Program's Global Status Report for Buildings and Construction 2022, buildings and the construction sector are responsible for around 37% of global energy-related CO_2 emissions [1]. In addition to the high energy requirements during operation, the production of energy-intensive building materials

H. Beushausen et al. (Eds.): NTZR 2025, LNCE 873, pp. 384–394, 2026.
https://doi.org/10.1007/978-3-032-27698-8_35

such as cement is a major source of emissions. In view of the global climate targets—in particular the commitments under the Paris Climate Agreement—it is imperative to drastically reduce the CO_2 footprint in the construction industry.

One promising strategy is the development of multifunctional building materials which, in addition to their primary function as a load-bearing structure or envelope material, also take on energy-related tasks. In this context, the integration of energy storage systems directly into building components is becoming increasingly important. One innovative solution is the combination of textile-reinforced, carbon-heated concrete plates with phase change materials (PCM). Textile-reinforced concrete, an innovative composite material made of high-performance fine-grained concrete and corrosion-free reinforcement, enables lightweight and resource-efficient construction methods with a lower carbon footprint compared to reinforced concrete [2]. Through the additional integration of electrical heating systems in conjunction with PCM, thermal energy can be efficiently stored and released again as required, which offers considerable advantages, particularly in the context of load management and the use of renewable energies (e.g. photovoltaics) [3].

Phase change materials enable latent heat storage by changing between solid and liquid aggregate states, whereby large amounts of energy are absorbed or released at an almost constant temperature. Their combination with carbon fiber-heated textile concrete elements allows targeted load shifting and smoothing of consumption peaks, which can both increase the self-consumption of regeneratively generated energy and reduce CO_2-intensive energy imports [4]. The development and application of such hybrid systems in the construction industry therefore represents an important contribution to the decarbonization of the building sector.

In addition to the publications [5, 6], this paper reports on the influence of the PCM on the strength properties of the textile concrete slabs and contains a simplified description of the energy storage possibilities as a function of the plate thickness and the necessary temperature increase. The functional principles shown can also be transferred to other matrix systems, such as polymers or clay [7].

2 Materials and Methods

2.1 Components for Heating Plates

Concrete. The base material of the plates is a fine-grained concrete, more precisely mortar, but called concrete in the following. The concrete gives the structure the necessary stability, keeps the heating roving insulated in position and conducts the heat energy consistently to the outside.

A natural stone heater made of marble could be used as a comparison, but its lower efficiency due to milled and glued copper heating on the back, limited surface design, and hard-to-reach curves are disadvantages.

The used gray cement gives the plates a gray appearance without any color. A wide range of additional powdered color is available to meet almost any customer requirement (see Fig. 1).

Fig. 1. Concrete coloring powder.

The concrete is developed as self-compacting concrete. Here "X0" was specified as the exposure class. Table 1 contains the qualitative and quantitative compositions of the concrete. The mixture contains a water/cement ratio of 0.60.

Table 1. Composition of gray and white concrete

Component	Proportion
CEM II/A-LL 42.5 N	500 kg/m^3
Limestone powder	150 kg/m^3
Quartz sand 0–2 mm	1300 kg/m^3
Superplasticizer	2.2 kg/m^3
w/c	0.60

Reinforcement. There are various reasons for reinforcement. Positioning the carbon heating during casting and relieving the concrete during thermal stress are the most important. Ultimately, practicality and price are the deciding factors. A large selection of fibers is available. As described in [5], "Vertex G 120" is utilized in the application. It is a glass fiber thermoset with a grid spacing of approx. 40 mm and a basis weight of 145 g/m^2.

Carbon Roving. As in previous publications with carbon heating [5–7], the roving can be formed without any problems and does not lose any property during tailored fiber placement process. In this study, the roving used is TENAX-E STS40, which has 48K filaments and a linear density of 3200 tex.

Phase Change Material (PCM). The advantages of PCM have been clearly demonstrated and discussed in detail in [5]. For PCMs, it is essential to understand that they function as latent heat accumulators. When they change phase from solid to liquid, heat continues to be stored without an increase in temperature. Once the material's capacity is reached, its temperature continues to rise. The process reverses in the same way when the stored heat energy is released. The PCM material used here is "Advanced PCM-Granulat" with a plateau temperature of 28 °C. It contains a rubber carrier material and the PCM itself. Supplied as loose granules, it can be applied to concrete structures.

2.2 Preparation of Samples

Specimens Concrete Strength and Compressive Strength. Two different materials were investigated. The first is pure concrete; the second is concrete with 4 wt.% of PCM added. All components were prepared in an "Eirich R05T" mixer. To assess the flow properties and workability, the mixtures were tested according to DIN EN 13395-1 slump test [8].

For mechanical characterization of strength, the two mixes are placed in special prismatic forms, each is $40 \times 40 \times 160$ mm^3.

For mechanical characterization of compressive strength, the broken prismatic elements are used from strength characterization.

All specimens were stored underwater for 28 days.

Carbon Heating Manufacturing. To produce the meander automatically, Tailored Fiber Placement (TFP) is used to place the carbon roving on a thin thermoplastic fleece. Next, the carbon roving can be contacted semi-automatically (robot-assisted screwing with a vacuum gripper), and the sensor is applied and wired with all necessary fixings. Finally, it is placed on the reinforcement grid (see Fig. 2).

The carbon fleece and fiber reinforcement are now melted together by hot air. Finally, the mold is prepared with a release agent, spacer, and carbon heating elements.

The PCMs are distributed during the concrete casting process.

Fig. 2. Carbon roving onto thermoplastic fleece and reinforcement grid.

2.3 Test Set-Up

Flexural Strength. The tests were based on DIN EN 12390-5:2019 for 4-point flexural strength [9].

This method determines the material parameters more realistically, especially for plate-shaped samples. These parameters are less influenced by possible inhomogeneities in the sample [10–13].

The test is executed using the Toni Technik TONINORM SERIE 2000 with a 20 kN load frame. The support with (L) is 135 mm and the support distance (L/3) 45 mm (see Fig. 3).

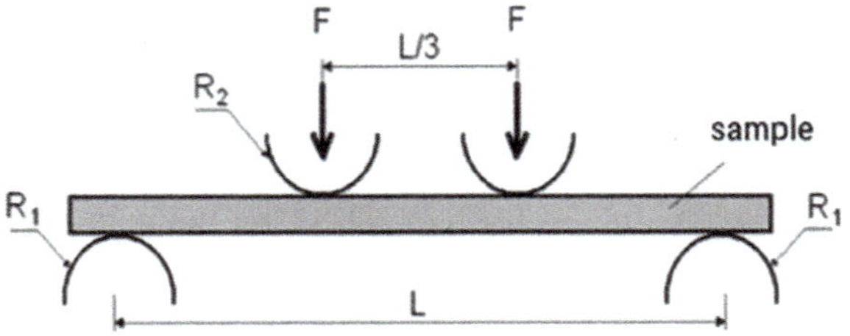

Fig. 3. Schematic overview of the 4-point flexural test (force F, support width L, support radius R1, impactor radius R2) [7].

Compressive Strength. Six samples were tested in each case (concrete pure and with PCM). The Toni Technik ToniPACT II was used to determine the compressive strength (see Fig. 4), which was done in accordance with DIN EN 196-1 [14].

Fig. 4. Sample during compressive test.

Calculating the PCM Effect on Heat Energy Storage. The theoretical supposition that an electrical storage system attains a storage capacity Cs of 0.9 kWh/m^2 constitutes the foundation of the calculation for the thermal energy storage system composed of concrete and PCM, as it was started in the previous publication [5]. The new feature is that the PCM content can be adjusted to vary the capacity according to plate thickness (from 20 to 60 mm), using the volume correction factor f. Table 2 shows all the constants used for the calculation.

Here are the improved formulas currently in use:

Volumetric heat capacity of concrete:

$$v = c^* \rho \left[Wh / \left(m^3 \times K \right) \right] \tag{1}$$

Specific energy storage concrete plate (thickness d [m], PCM volume correction factor f):

$$Es = v^* d^* f \left[Wh / \left(m^2 \times K \right) \right] \tag{2}$$

Weight per unit area PCM: ms [g/m^2].
Energy per square meter PCM:

$$Em = e^* ms \left[Wh / m^2 \right] \tag{3}$$

Temperature range required for energy storage:

$$\Delta T = (Cs - Em)/Es[K] \tag{4}$$

Table 2. Constants for the calculation

Name	Value
Specific heat capacity of concrete **c**	1 kJ/(kg × K)
Density of concrete $\boldsymbol{\rho}$	2100 kg/m^3
Storage capacity (Specific energy) PCM **e**	240 kJ/kg

Effect of PCM on Cooling Down Process. To understand how these PCM materials work, an experiment was conducted as shown in [5]. Two concrete plates were compared: one with 4 wt.% PCM and one without. Both plates were heated to 40 °C and placed next to each other to cool down. This process was documented by an infrared camera.

Components for Smart Home. To make a system smart home capable, there are some intelligent devices needed [5]. The core is a pair of Shelly controllers ("1PM Pro" and "Plus 1PM"), which communicate via WLAN and these distribute and monitor the energy flow, that means they decide where energy is needed. To run the system self-sufficient, an energy source (two solar panels, called balcony power station) is used.

3 Results and Discussion

3.1 Mechanical Characterization via Flexural and Compressive Strength

The 4-point flexural tests and compression tests were carried out on the concrete material with and without PCM. As shown in Fig. 5, adding 4 wt.% PCM to the concrete decreases the strength (flexural tensile strength by ~ 10% to 5 MPa and compressive strength by ~ 16% to 25 MPa).

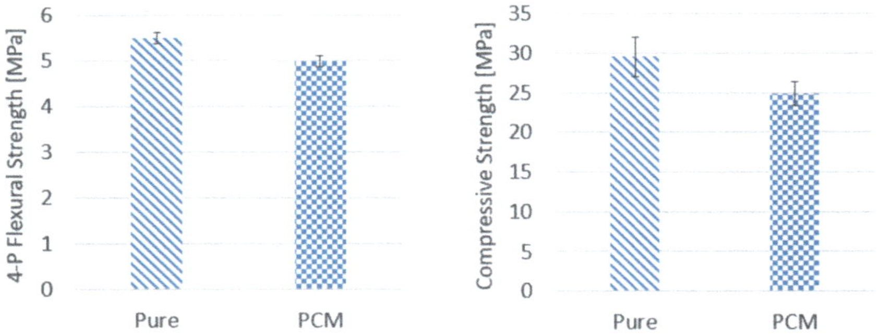

Fig. 5. 4-point flexural tensile strength (left) and compressive strength (right) results for concrete with and without PCM.

Although this does not result in a significant loss of strength, proof by calculation is required for its use as a self-supporting wall or ceiling plate in private or public buildings.

At present, no results from a long-term study on the durability of the panels are available due to the novelty of the technology. However, all components have been carefully tested for alkali resistance. This results in a long service life as long as the concrete remains stable. Thermal cycling effects and associated thermal stresses have been simulated and do not pose a risk if the heating power does not exceed 300 W/m^2. The carbon meanders generate an even heat distribution across the entire surface, and the fiber grids help to stabilize against cracking.

3.2 Calculating the PCM Effect on Heat Energy Storage

There are two scenarios for calculating the required energy storage of 0.9 kWh/m2: The first scenario is the process without the addition of PCM. The second scenario is the process with increasing PCM content with plate thickness. For each calculation, there are two graphs: one showing the resulting weight per unit area of the plate, and the other showing the temperature increase required to achieve the energy storage (see Fig. 6).

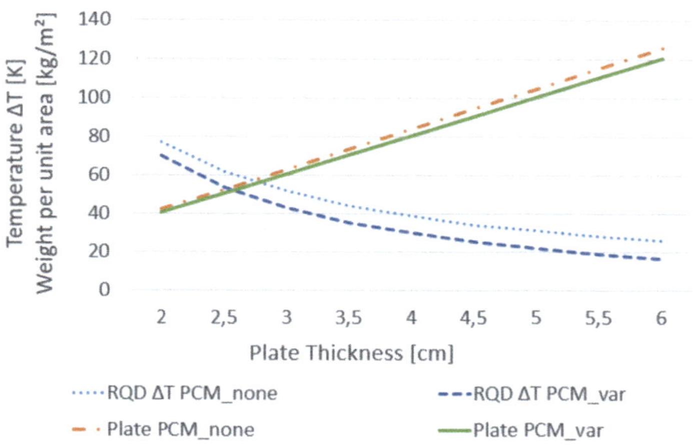

Fig. 6. Required temperature increase for storage capacity and weight per unit area of the plates.

The weight per unit area of the plate increases linearly with its thickness. As more PCM is added, the weight of the plate decreases because concrete is replaced with PCM, which has a lower density.

To reach the default energy storage, the required temperature difference decreases as a nonlinear function of plate thickness and approaches zero when the plate is infinitely thick. The added phase-change material (PCM) shifts the required temperature downwards compared to the concrete itself. The greater the amount of PCM added, the greater the shift and the better the heat energy storage characteristics.

3.3 CO_2-Reducing Effects

Carbon Fiber. The goal is to use CO_2-neutral, "green" carbon fibers produced using renewable energy and raw materials. These "green" carbon fibers are currently being developed in the USA and Australia, and will soon be available at a research and development factory in Germany [15]. The heating elements consist of carbon fixed inside the concrete. As long as the concrete is stable, the carbon is safe and cannot burn into CO_2.

Concrete. The use of CEM II/A-LL reduces the carbon footprint up to 17.4% compared to CEM I [16]. CEM III and CEM V have an even higher potential (35.9% and 39.4%, respectively), but these cements were not considered in this field of investigation. However, they may be considered in the future. Another way to reduce the carbon footprint is to use local resources, such as quartz sand and limestone powder. Using these resources reduces emissions significantly because transportation routes are shorter.

PCM Energy Storing. Saving energy instead of losing it has a high potential for reducing the carbon footprint because stored energy does not need to be produced. This development saves only a small amount of heat energy for a limited time (several hours), which is favored and enhanced by the use of PCM. Despite this limited capacity, it is progress that can be built upon.

Smart Home Application. The components mentioned in Sect. 2.3 generate energy and intelligently distribute it to consumers in the house. One example of these is the carbon heating plate, which serves as a reference object at the project partner's facility (surface 1 m^2 and 300 W power). It uses colored backlighting to indicate the heating status (blue = cold, green = heating, red = temperature reached), see Fig. 7.

Fig. 7. Reference object with an energy-monitoring backlight.

This comes with an intelligent kind of CO_2 reduction through energy distribution where and when it's needed. This is influenced by human presence, position, movement, and long-term behavior intelligently evaluated by AI algorithms in a smart home.

The choice of database format depends a lot on the specific needs of the smart home system. Time series databases like InfluxDB and TimescaleDB are great for storing sensor data efficiently. NoSQL databases like MongoDB and Cassandra offer flexibility and scalability, while relational databases like PostgreSQL can be useful for more complex data models. Edge databases and cloud-based solutions are also relevant options, especially when processing and storing data close to the source or in distributed systems. Each of these systems offers APIs (Application Programming Interfaces) and integration options for various smart home frameworks, enabling seamless connection to existing systems (e.g., Amazon Alexa, Google Assistant, Apple HomeKit).

This means a house is not heated everywhere and at all times; rather, it is heated only at particular times and to a particular intensity via the infrared heating from the carbon heated plates. This saves much energy and significantly lowers the carbon footprint of buildings, though this is not quantified here.

4 Conclusion

This paper describes the potential for effective CO_2 reduction through the use of carbon fiber-heated concrete plates in combination with PCM and their integration into smart homes. The use of CO_2-reduced concrete, "green" carbon fibers, energy-storing PCM, and intelligent energy distribution paves the way for a future with improved resource efficiency for private and public buildings in the modern construction industry.

The goal of combining the high-performance lightweight material concrete with new integrated functions (damage prediction, etc.) and a wide range of design options within an automatable value chain has been achieved and confirmed by a multitude of results. This marks an important milestone for the establishment of lightweight construction in concrete building. In addition to cement, other important resources such as sand and additives can be saved during production. Furthermore, the lighter construction method enables lower transport costs and thus a better CO_2 balance. This project serves to establish lightweight construction and thus offers, in addition to the aspect of environmental protection, considerable market potential for numerous companies in the field of ecological lightweight construction solutions.

Acknowledgements. The authors would like to thank the following sponsors for their financial support «BMFTR—Bundesministerium für Forschung, Technologie und Raumfahrt» (Federal Ministry of Research, Technology and Space) within the initiative «WIR! Wandel durch Innovation in der Region—SmartERZ» for the project "CoCoHeat", funding indicator 03WIR1414A. Many thanks also to our industry partners predEVOLUTION technologies GmbH and Betonstein Flöha GmbH.

References

1. UNEP (2022) 2022 Global status report for buildings and construction. United Nations Environment Programme. https://globalabc.org
2. Curbach M, Jesse F, Ortlepp R, Scheerer S (2016) Carbon concrete—a vision for the future. Beton-und Stahlbetonbau 111(10):658–667. https://doi.org/10.1002/best.201600051
3. Zhou D, Zhao CY, Tian Y (2012) Review on thermal energy storage with phase change materials (PCMs) in building applications. Appl Energy 92:593–605. https://doi.org/10.1016/j.apenergy.2011.08.025
4. Cabeza LF, Castell A, Barreneche C, de Gracia A, Fernández AI (2015) Materials used as PCM in thermal energy storage in buildings: a review. Renew Sustain Energy Rev 15(3):1675–1695. https://doi.org/10.1016/j.rser.2010.11.018
5. Gliniorz R, Lindner M, Gelbrich S, Funke H (2025) Energy storing carbon concrete heating element with phase change material for integration into smart home systems. In: MATEC Web of Conferences 2025, vol. 409, pp. 13004. EDP Sciences
6. Gliniorz R (2023) Energy-efficient reinforced heating system implemented as a carbon concrete form-work. Springer Nature Switzerland AG 2023 Zembaty Z et al (eds), ECCE 2022, LNCE 322, pp 258–272
7. Gliniorz R (2024) Multi-functional integration in clay-cardboard sandwich structure with semi-automated manufacturing. Springer Nature, Switzerland AG 2025 ECCE 2024, pp 130–143, ISBN 978-3-031-73775-6

8. DIN EN 13395-1 (2002–09) Prüfverfahren; Bestimmung der Verarbeitbarkeit—Teil 1: Prüfung des Fließverhaltens von thixotropem Mörtel (2002)
9. DIN EN 12390-5 (2009-07) Prüfung von Festbeton—Teil 5: Biegezugfestigkeit von Probekörpern
10. Lindner M, Rudolph E, Gliniorz R, Funke H, Gelbrich S (2023) Anisotropic properties of additively manufactured concrete elements. Appl Res Article https://doi.org/10.1002/appl.202300035; https://onlinelibrary.wiley.com/doi/10.1002/appl.202300035 [Google Scholar]
11. Lindner M, Vanselow K, Gliniorz R, Gelbrich S (2020) Additiv gefertigte Leichtbaustrukturen mit Beton und Faserbewehrung. In: Fachtagung Werkstoffe und Additive Fertigung, https://additive-fertigung-2020.dgm.de/home [Google Scholar]
12. Lindner M, Rudolph E, Gliniorz R, Funke H, Gelbrich S (2022) Combined additive manufacturing technology for continuous production of precision concrete elements for modular construction. In: Proceedings of the 14th fib PhD symposium in civil engineering. ISBN 978-2-940643-17-2; Seiten 787–794 [Google Scholar]
13. Lindner M, Rudolph E, Gliniorz R, Funke H, Gelbrich S (2022) Tool development for robot assisted flow production processes for the fabrication of thin walled fibre reinforced concrete elements. In: Proceedings of the 14th fib PhD symposium in civil engineering. ISBN 978–2–940643–17–2; Seiten 771–778
14. DIN EN 196-1:2016-11: Prüfverfahren für Zement, insbesondere die Bestimmung der Festigkeit (2016)
15. Naumann M (2024) Carbon LabFactory Sachsen soll ein Leuchtturm für nachhaltige Carbonfaser-Herstellung in Europa werden. https://www.tu-chemnitz.de/tu/pressestelle/aktuell/12375
16. Rudnicki T, Stałowski P (2024) Performance research of cement concrete pavements with a lower carbon footprint. Materials 17(13):3162 11. https://doi.org/10.3390/ma17133162

Neuroarchitecture and Net Zero Carbon Building Design: A Comparative Analysis to Illustrate the Potential of Neuroarchitecture as a Promoter for the Wider Adoption of Net Zero Carbon Design

François Nortjé(✉)

Cape Town, South Africa
FrancoisN@0-1.co.za

Abstract. Architecture faces the dual challenge of advancing human well-being in buildings, while mitigating environmental degradation caused by construction and operation of buildings. Two recent design paradigms—Neuroarchitecture and Net Zero Carbon (NZC) design—offer complementary strategies toward this goal. Neuroarchitecture draws on cognitive neuroscience to optimize spatial design for human perception, emotion, and behaviour, while NZC design emphasises the reduction of operational and embodied carbon to achieve climate neutrality. Despite their distinct origins, both fields converge on the principle of applying evidence-based design, and that can integrate human and ecological performance. This paper proposes that Neuroarchitecture and NZC design should be pursued as synergistic frameworks, each reinforcing the effectiveness and motivation for the application of the other. A review of contemporary literature identifies overlapping interventions, such as daylight optimisation, biophilic design, and material selection, which simultaneously enhance cognitive health and reduce environmental impact. Case studies, including *The Edge* (Amsterdam) and *CHUM* (Montreal), offer insights into design strategies informed by neuroscience and carbon reduction, and how closely it aligns. The paper highlights opportunities for research, education, and policy to advance this integration. Ultimately, it argues that future-ready architecture requires a dual commitment: buildings must be both climate-resilient and neurobiologically attuned to their occupants.

Keywords: Neuroarchitecture · Net Zero Carbon design · Environmental psychology · Cognitive psychology of place

1 Introduction

Architecture today operates under unprecedented pressures. On one hand, climate change demands a rapid transition toward buildings with radically reduced carbon footprints. On the other hand, an expanding body of research demonstrates that the built environment has profound effects on cognitive function, mental health, and social behaviour. These

H. Beushausen et al. (Eds.): NTZR 2025, LNCE 873, pp. 395–410, 2026.
https://doi.org/10.1007/978-3-032-27698-8_36

two imperatives—environmental sustainability and human well-being—are often treated separately in design discourse. Yet both address fundamental aspects of architecture's purpose: shaping the conditions of life.

Net Zero Carbon (NZC) design has emerged as a central framework in the fight against climate change. By targeting reduction in both operational and embodied carbon, NZC strategies aim to balance energy use with on- or off-site renewable energy generation, achieving carbon neutrality across the building life cycle (RIBA: 2021; Goldstein et al.: 2020). The focus is systemic and quantitative, emphasising material selection, energy efficiency, and lifecycle assessment.

In parallel, Neuroarchitecture applies insights from neuroscience and psychology to the design of space, emphasising how environmental stimuli influence human perception, stress, attention, and social interaction (Eberhard: 2009). Through tools such as fMRI, EEG, and behavioural studies, it demonstrates that spatial qualities—light, acoustics, geometry, and biophilic elements, amongst others—directly affect human well-being and performance in the short and long term.

While the drivers differ—ecological crisis in the case of NZC, cognitive science in the case of Neuroarchitecture—the design implications frequently converge. This is not generally realised in either of these design fields, for example:

Daylighting reduces energy demand while enhancing circadian health and cognitive performance.

Biophilic design integrates vegetation and natural materials, supporting carbon sequestration and building climate control while reducing stress and improving attention.

Material choices such as timber not only lower embodied carbon but also provide sensory warmth and psychological comfort.

Despite these overlaps, the two paradigms have rarely been studied together. Sustainability discourse often prioritises measurable energy metrics at the expense of the lived experience, while Neuroarchitecture risks neglecting the ecological contributions of its interventions. As a result, opportunities for cross-fertilisation remain underdeveloped.

This paper addresses this gap by proposing synergy that integrates Neuroarchitecture and NZC design. It argues that the future of architecture does lie in choosing between ecological and human-centered priorities, but in aligning them. To develop this claim, the paper pursues three objectives:

Conceptual Comparison – outlining the origins, aims, and methods of NZC and Neuroarchitecture, identifying both tensions and synergies.

Case Study Analysis – examining projects such as The Edge (Amsterdam) and Bosco Verticale (Milan) to demonstrate practical intersections between neuroscientific and carbon-neutral design.

Conclusion: Implications for Practice – exploring how this integration can inform professional design methodologies.

The discussion proposes a discourse in architectural research that seeks to reconcile human-centered and ecological design. By situating Neuroarchitecture and NZC within a shared evidence-based paradigm, the paper suggests that a path toward future-ready architecture, one that reduces environmental impact while actively supporting cognitive health and well-being, can be found.

2 Conceptual Foundations: The Current State of Architecture

2.1 The Problem with the Modernist Movement in Architecture

Modernist architecture, with its characteristic curtain walled glazed facades, flat roofs, standardised plans, and reliance on mechanical climate control, continues to exert a profound influence on the built environment. This architectural legacy is however increasingly scrutinised in relation to both Net Zero carbon imperatives and occupant well-being. Architectural historians and building physicists have noted that many Modernist typologies perform poorly in most climates zones, often experiencing overheating and high operational energy demands (Coburn et al. 2017). This design ethos normalised energy-intensive solutions such as ubiquitous air conditioning and thus embedded unsustainable practices into the architectural mainstream. At the same time, advances in neuroarchitecture scientifically proves that spatial attributes—such as material richness, scaled complexity, daylight modulation, and biophilic integration—support cognitive performance and emotional health. The reductive tendencies of Modernism, which prioritise functional efficiency over sensory engagement, neglect these features and thereby compromise the psychological and physiological needs of building occupants (Heatherwick: 2023).

2.2 Modernism as Style and Ideology

Emerging in the aftermath of World War II, Modernism advanced the idea of buildings as machines: stripped of ornament, symbolism, and cultural narrative. Architect Louis Sullivan's late-19th-century dictum "form ever follows function" became a guiding slogan, later adopted and amplified as a rejection of decorative or expressive architecture (Coburn: 2017). This ethos spread globally, shaping a "way of life" as much as a design language. Critics argue that, over time, Modernism devolved into intellectual rigidity and mechanistic problem-solving, distancing architecture from the emotional and sensory dimensions of human experience (Condia: 2020). Heatherwick (2023) emphasises that such reductive design has left generations of building users without conscious architectural appreciation, reducing their engagement with space to by and large subconscious level and not positive at that. Public criticism of Modernist architecture has intensified in recent decades, highlighting its neglect of emotionally supportive atmospheres (Heatherwick 2023; Biederman and Wessel 2006). These critiques underscore a persistent undervaluing of human-centered design considerations, particularly in commercial development, where financial imperatives often override architectural meaning (Condia: 2020). Developers frequently prioritise mass-produced, cost-efficient materials, minimal natural light planning, and deep-plan layouts, thus maximizing return on investment but diminishing experiential quality. Consequently, the century-old axiom "form ever follows function" arguably now reflects an economic reality— "form follows money."

2.3 Human Experience and the Critique of Modernist Aesthetics

Heatherwick (2023) identifies seven recurring characteristics of Modernist buildings—flatness, plainness, linearity, shininess, monotony, anonymity, and excessive seriousness—that collectively result in environments perceived as boring, intimidating, or emotionally sterile. While some architects and developers attempt to counteract this sterility through "iconic" high-tech " structurally gymnastic ", architectural expressions, these interventions prioritise visual impact and symbolic status over the lived human experience of onlookers and occupants. Alexander (2023) observes that such monumentality produces striking yet emotionally hollow spaces, privileging innovation and image-making over a sense of life or belonging.

2.4 Positive Contributions of Modernism

Despite these criticisms, Modernism's influence cannot solely be reduced to shortcomings. Historically, the movement represented a radical departure from pre-industrial building traditions and introduced several significant contributions that remain relevant. Modernism democratised architecture by advocating mass housing, prefabrication, and new construction techniques, making functional dwellings accessible to wider populations (Birch: 2024). The movement also championed structural rationalism and technological experimentation. Advances in reinforced concrete, steel framing, and curtain wall construction expanded the possibilities of architectural form, facilitating the rise of skyscrapers and complex public buildings (Boeri: 2018). Finally, Modernism's emphasis on openness, light, and minimalism reflected the period's cultural aspirations for transparency and progress (Condia: 2020). Two examples highlight this dual legacy. Le Corbusier's *Unité d'Habitation* (1952, Marseille) pioneered mixed-use vertical living, integrating housing, shops, and communal facilities to foster social interaction. Yet its austere concrete form and rigid modularity limited adaptability and sensory richness. Conversely, the Pruitt-Igoe housing project in St. Louis (1954–1972) illustrates the consequences of Modernism combined with economic underfunding: its demolition symbolised the failure to create humane and sustainable urban environments (Mehaffy and Salingaros: 2013).

2.5 Modernism and Sustainability Challenges

From a sustainability perspective, Modernist design principles prove problematic. Mehaffy and Salingaros (2013) argue that Modernism is "inherently hard to green" due to its reliance on expansive curtain walls, deep plan geometries, dependence on artificial lighting and ventilation, and use of high-embodied-carbon materials such as aluminum and steel. These attributes limit the potential for passive environmental control and contribute to substantial operational and embodied carbon footprints. They contend that no "bolt-on" green technologies—such as photovoltaic systems or more efficient HVAC equipment—can fully resolve these systemic flaws. Furthermore, technological fixes may exacerbate energy consumption through rebound effects, such as the Jevons paradox, whereby increased efficiency lowers costs but encourages higher overall consumption.

Thus, Modernist architecture faces a dual crisis: its spatial and aesthetic reductiveness undermines human well-being, while its material and operational logics conflict with urgent Net Zero carbon requirements. Addressing these shortcomings requires a fundamental reconceptualisation of architectural priorities, integrating ecological responsibility with neuroscientific insights into human experience (Abbas et al. 2021; Eberhard 2009) (Fig. 1).

Fig. 1. Example of a the generic typology of modernist "Iconic" building and neurologically sensitive building: "Falling Water" house by architect Frank Lloyd Wright. *Source* AI-generated and Google images.

3 Neuroarchitecture

3.1 Definition and Scope

Neuroarchitecture is an evidence-based approach to design that integrates insights from neuroscience, environmental psychology, and architecture to understand how the built environment affects human brain function, cognitive performance, and stress levels (Abbas et al. 2021). It incorporates natural lighting (Mostafavi et al. 2024), biophilic design (Szewrański et al. 2024), spatial configuration (Xylakis et al. 2021), and multisensory stimuli to create environments that are supportive, restorative, and capable of eliciting awe and wonder. As such, neuroarchitecture represents a scientific methodology for linking design decisions to measurable human outcomes.

The following table serves as a brief introduction on what Neuroarchitecture studies and measures (Table 1).

Table 1. Summary of recent studies highlights the influence of specific architectural elements on human responses:

Reference and Year	Design elements studied	Measured outcomes
Finnigan (2024)	Visual, auditory, thermal, olfactory stimuli	Psychophysiological restoration, improved focus
Thampanichwat et al. (2024)	Natural light, plants, wood textures	Reduced neuroinflammation, lower stress (qEEG)

(continued)

Table 1. *(continued)*

Reference and Year	Design elements studied	Measured outcomes
Valentine et al. (2025)	Physiological step responses, allostasis	Enhanced attention, reduced mental fatigue
Xylakis et al. (2021)	Curved vs. basic spatial forms, lighting transitions	Increased emotional arousal, inducing awe
Elghor et al. (2024)	Curvilinear geometries, organic forms	Reduced brain stress, sustained cognitive fluency
Mostafavi et al. (2024)	Views, shading, interior design elements	Time-sensitive cognitive performance shifts
Zhang and Anderson (2020)	Green views, VR nature exposure, wooden interiors	Reduced cortisol, improved attentional recovery

3.2 Historical Context

Although neuroarchitecture concepts are increasingly prominent in the 21st century, their intellectual roots trace back to early 20th-century research by Wilhelm Wundt, Kurt Lewin, and Abraham Maslow. At that time, the insights on human perception, cognition, and environmental influence did not integrate into mainstream architectural practice, as it coincided with the dominance of the rise of Modernism. Initial studies relied on empirical observation and interviews with occupants to understand spatial effects, forming the foundation of evidence-based design (EBD). Eagleman (2011) notes that conscious processing accounts for less than 1% of the brain's total processing capacity, implying that users are largely unaware of the subtle physiological effects of their surroundings. Coburn et al. (2017) further assert that verbal accounts of spatial experience often reflect rationalisation rather than accurate reflection of subconscious responses. Modern neuroarchitecture builds upon these foundations using scientific tools that measure both neurological and physiological responses to the built environment. Common methods include:

Brain imaging: EEG, fMRI, fNIRS, MEG.

Physiological arousal sensors: EDA/GSR, heart rate (HR), heart rate variability (HRV), cortisol levels.

Behavioural and visual tracking: Eye-tracking, EMG/facial coding (Lee et al. 2022).

Since 2010, neuroscientific research has expanded to examine architectural styles (Choo et al. 2017), spatial contours, enclosure, interior vs. natural environments, lighting, colour, and their impacts on memory and perception (Banaei 2017).

3.3 Principles of Human-Centered Design

Human-centered architectural designs share several characteristics that promote emotional engagement and well-being:

Beyond function and aesthetics: While functionality and visual appeal remain foundational, effective design also evokes "delight," that is an additional subconscious layer of emotional engagement.

Positive emotions: Environments that induce joy, awe, or wonder can elevate mood and subjective well-being.

Proportion and harmony: Well-proportioned and interest driven façades , balanced compositions, and harmonious colour palettes including principles such as fractals as found in nature , contribute to visual delight . Psychological and Physiological responses are measurable using virtual reality and MRI-based studies.

Sense of discovery: Spatial sequences, unexpected views, hidden features and elements of surprise , stimulate exploration and reinforce subconscious value. In Neuroarchitecture, referred to as "prospect and refuge" features and "affordances".

Connection to nature: Integrating physical natural elements—gardens, water features, natural materials, or patterns derived from nature—enhances cognitive restoration and emotional health.

Many principles of human-centered design inherently align with carbon-conscious strategies, as designs that optimise daylight, passive ventilation, and natural materials simultaneously reduce operational and embodied carbon loads, as will be explored further in this paper (Fig. 2).

Fig. 2. Current example of neuro sensitive and carbon friendly architecture: maggie centre, leeds university, *Source* Architects Thomas Heatherwick.

4 Net Zero Carbon (NZC) Building Design

4.1 Definition

NZC building design focuses on reducing or neutralising carbon emissions through passive design strategies, energy-efficient technologies, and sustainable construction materials. It integrates renewable energy sources, such as solar and wind power, while minimising reliance on fossil fuels in manufacture and operation. The basic concept underpinning Net Zero Carbon buildings is a zero-energy balance. More recent definitions expand the term to include embodied carbon and life-cycle performance, reflecting a more holistic sustainability agenda (AIA 2021). Techniques for achieving Net Zero include passive solar design (Building orientation relative to insolation), high-performance envelopes, on-site renewable energy generation, and optimised HVAC systems. Importantly, many of these interventions intersect with comfort and usability—bridging technical and experiential domains (IEA 2020). The push for Net Zero Carbon is not merely technical but also ethical. Buildings contribute nearly 40% of global greenhouse gas emissions; thus, they present a critical frontier for mitigation (UNEP 2020).

5 Synergy Between Neuroarchitecture and Net Zero Carbon Design

The comparative framework presented in the following table highlights the shared attributes of neuroarchitecture and net-zero carbon design. While neuroarchitecture originates from cognitive neuroscience and focuses on the modulation of human well-being through spatial stimuli, net zero design is driven by ecological imperatives, lifecycle carbon accounting, and regulatory frameworks. These diverse origins, are shown in Columns A and C below. Several overlaps exist that provides simultaneous benefits to human health and environmental performance as reflected in column B below (Table 2):

Table 2. Synergies between Neuroarchitecture and Net Zero Carbon Design.

A. Neuroarchitecture (Human-Centered)	B. Synergies (Integration Zone)	C. Net Zero Carbon Design (Ecology-Centered)

(continued)

Table 2. *(continued)*

A. Neuroarchitecture (Human-Centered)	B. Synergies (Integration Zone)	C. Net Zero Carbon Design (Ecology-Centered)
Drivers: • Cognitive neuroscience • Well-being goals • Human environmental perception research	Shared interventions → Dual benefits: • Daylighting → Supports circadian rhythms; reduces artificial lighting loads (Edwards: 2002) • Biophilic design → Enhances psychological restoration; sequesters carbon via vegetation (ARK: 2019, Kaplan: 1989) • Natural materials (e.g., timber) → Provide sensory warmth; lower embodied carbon (Doodoo: 2010) • Acoustic and thermal control → Improves focus; boosts envelope energy performance (Cole: 2005)	Drivers: • Climate change mitigation • Regulatory frameworks (Cole: 2005) • Lifecycle carbon emissions accounting
Methods: • Neuro-physiological measurement (EEG, HRV, fMRI, etc.) • Behavioural assessments • VR simulation of spatial stimuli	Shared Outcomes: • Occupant health and productivity gains • Dual-performance architecture (carbon + well-being) • Educational tools for integrated design	Methods: • Energy simulation and modeling • Passive design (orientation, insulation, daylighting) • Low-carbon materials and assembly techniques
General Neuroarch Outcomes: • Reduced stress and anxiety • Enhanced cognition, mood, and focus • Stronger emotional connection to place (Joy: 2014)	Shared general outcome. Overall improved human habitat and reduced carbon footprint	General NZC Outcomes: • Lower operational and embodied carbon • Energy-efficient and resilient buildings • Alignment with NZC targets
Intervention: Daylighting and Lighting Psychology: – Natural daylight optimises mood and circadian function; reduces artificial lighting – Adaptive lighting supports cognition and reduces energy (Boyce: 2014)	→ Synergy: Natural light serves both physiological health and energy savings	Intervention: Passive glazing, daylight harvesting systems, high-performance windows (Reinhart et al.: 2006; Li 2001)

(continued)

Table 2. (*continued*)

A. Neuroarchitecture (Human-Centered)	B. Synergies (Integration Zone)	C. Net Zero Carbon Design (Ecology-Centered)
Intervention: Biophilic Integration – Vegetation, textures, views promote stress recovery (Kaplan: 1989)	→ Synergy: Biophilic spaces enhance emotional well-being and reduce HVAC demand	Intervention:Green walls, rooftop gardens, planting schemes to offset energy demand and capture carbon (Beatley: 2011)
Intervention: Material Selection and sensory experience – Wood, fractal patterns support comfort (Taylor: 2006)	→ Synergy: Natural materials promote occupant comfort and sequester embodied carbon	Intervention:Low-carbon materials like mass timber; bio-based insulation (Cole: 2005)
Intervention: Spatial Legibility and Acoustic Comfort – Clear wayfinding, proportion, and sound control support cognitive clarity	→ Synergy: Optimised layout reduces circulation energy and enhances cognitive load management	Intervention:Passive zoning, optimised facades, high-performance envelopes (Cole: 2005)
Intervention: Restorativeness and Stress Recovery – Nature views accelerate recovery (Berto: 2014; Ulrich et al.: 1991)	→ Synergy: Recovery spaces reduce stress and lower HVAC and lighting demands	Intervention: Courtyard gardens, atria as natural thermal buffers

6 Examples of Built Projects Featuring a Synergy Between Neuroarchitecture and Net Zero Carbon Design Principles

6.1 The Edge, Amsterdam

The following table lists the specific intervention implemented at the EDGE building in Amsterdam under column A and indicates the Human well-being improvement under column B and environmental improvement under C (Table 3):

Table 3. The Edge Amsterdam: Zones of Synergy: Both Neuroarchitecture and NZC goals achieved with design interventions listed in column A

A. Design lever	B. Neuroarchitecture benefit	C. Carbon/energy benefit
Daylight-first atrium and façade design	Circadian support, visual comfort, legible wayfinding via natural light gradients	Reduced electric-lighting demand; lower plug/lighting load aligns with PV generation profile. (EDGE)

(*continued*)

Table 3. (*continued*)

A. Design lever	B. Neuroarchitecture benefit	C. Carbon/energy benefit
User-level control via app + PoE LEDs	Greater perceived control reduces stress and boosts satisfaction and task engagement	Setpoint flexibility and task-tuned illuminance cut hours-at-full-load; granular data enables continuous commissioning. (Philips lighting, New Atlas)
Sensor-rich adaptive operation	Stable thermal/acoustic/Indoor Air Quality conditions support attention and memory	Demand-controlled ventilation/conditioning and smart zoning reduce kWh/m^2 without comfort complaints. (PLP Architecture, Schneider Electric)
ATES + envelope tuning	Thermal stability reduces thermal discomfort and associated cognitive distraction	Major reduction in HVAC energy intensity; facilitates electrification and high BREEAM scoring. (Schneider Electric Blog, Schneider Electric)

CHUM Hospital, Montreal: The following table lists the specific intervention implemented at the Centre hospitalier de l'Université de Montréal under column A and indicates the Human well-being improvement under column B and environmental improvement under C (Table 4).

Table 4. The CHUM hospital Montreal: Zones of Synergy: Both Neuroarchitecture and NZC goals achieved with design interventions listed in column A

A. Design lever	B. Neuroarchitecture effect	C. Carbon/energy benefit
Daylight-rich atriums and corridors	Enhances mood, supports orientation, reduces stress	Lowers reliance on electric lighting; thermal load managed via high-performance glazing
Green views and biophilic elements	Facilitates healing, reduces anxiety, improves patient outcomes	Provides passive cooling via shading and evaporation; supports healing gardens as passive buffers
Acoustic and spatial comfort	Reduces cognitive and emotional fatigue for occupants	Sound attenuators can double as thermal buffers; better envelope performance reduces energy loss

(*continued*)

Table 4. (*continued*)

A. Design lever	B. Neuroarchitecture effect	C. Carbon/energy benefit
Wayfinding clarity	Reduces cognitive overload, confusion, and stress	Efficient layout reduces corridor lengths and HVAC zoning complexity
Advanced HVAC control	Stable thermal comfort improves cognitive focus and reduces agitation	Demand-controlled ventilation reduces energy use in unused zones
High-efficiency envelope	Enhanced visual comfort and thermal consistency	Reduced heating/cooling loads; lower overall carbon use

The Bosco Verticale , Milan: The following table lists the specific intervention implemented at the Bosco Verticale under column A and indicates the Human well-being improvement under column B and environmental improvement under C (Table 5).

Table 5. The Bosco Verticale, Milan: Zones of Synergy: Both Neuroarchitecture and NZC goals achieved with design interventions listed in column A.

A. Design lever	B. Neurological effect	C. Carbon/energy benefits
Green façades simultaneously provide restorative nature exposure for residents while sequestering CO_2 and improving insulation	Biophilic immersion: Extensive façade greenery connects occupants with nature, reducing stress and enhancing psychological well-being (Ulrich 1984; Kellert and Calabrese (2015)	Carbon sequestration: Vegetation captures ~ 30 tons of CO_2/year and produces ~ 19 tons of oxygen/year
Solar modulation improves both circadian health and reduces cooling loads in summer	Daylight optimization: Balconies and greenery filter natural light, reducing glare and creating dynamic light environments beneficial to circadian rhythms (Li et al. 2022)	Energy efficiency: green façades reduce solar gain, lowering air conditioning demand by ~ 30%
Dynamic façades promote human sensory engagement while reducing mechanical climate control reliance	Multi-sensory engagement: seasonal variation of plants stimulates visual, olfactory, and tactile senses, enhancing place attachment and reducing cognitive fatigue	Thermal regulation: vegetation maintains stable microclimates, reducing peak heating/cooling needs
Shared gardens function as both community-building amenities and urban biodiversity habitats	Community well-being: shared green spaces foster social cohesion, improving mental health outcomes (Hartig et al. 2014)	Biodiversity support: Bosco Verticale hosts over 1600 birds and insects, increasing ecological resilience

(*continued*)

Table 5. (*continued*)

A. Design lever	B. Neurological effect	C. Carbon/energy benefits
Human-centered sustainability aligns ecological resilience with occupant comfort and desirability	Restorative design ethos: architecture supports human recovery, productivity, and mental health	Urban carbon reduction: dense vertical greenery offsets embodied carbon through ecological services

7 Proposed Future Studies on Synergy Between Neuroarchitecture NZC

A study is proposed to compare the design features of two buildings from a Neuroarchitecture and Net Zero Carbon point of view, to determine how many aspects not intentionally designed for, are coincidentally present. Also whether adding minor additional effort on the NZC and / or Neuroarchitecture side, these designs could be made more compliant on both aspects.

Suggested buildings:

Neuroarchitecture examples that were not explicitly designed to be carbon Neutral:

The Well Living Lab (Minnesota, USA): A research facility dedicated to studying how indoor environments affect human health, using Neuroarchitecture principles like optimised lighting and spatial configurations.

Maggie's Centres is at various locations in the UK. These cancer care centers are designed with biophilic elements, natural lighting, and stress-reducing layouts to support emotional well-being.

Net-zero carbon buildings not explicitly designed to incorporate Neuroarchitecture elements:

The Edge (Amsterdam, Netherlands), as described above.

Bullitt Center (Seattle, USA): A net-zero energy building using rainwater harvesting, on-site solar power, and ultra-efficient insulation to maintain carbon neutrality.

8 Conclusion

This paper has examined the relationship between Neuroarchitectural science and Net Zero Carbon (NZC) design, two fields that emerge from distinctly different intellectual considerations but are illustrated to have a significant and co-beneficial overlap. While primary objectives differ—cognitive and emotional well-being on the one hand, and carbon reduction on the other—many design interventions, such as daylighting, biophilic integration, and spatial clarity, serve both. Realisation of such overlaps, as argued, would have architects and developers find greater incentive to adopt NZC principles. To enhance human health, productivity, and long-term building desirability generates both social and economic value in commercial property development. This adds user value and thus better financial returns that is the goal of commercial property development. To learn that carbon-reducing design elements are simultaneously achievable should serve as a significant motivation to also add the brief of designing for Net Zero Carbon goals. Nonetheless, both fields face critical challenges. In Neuroarchitecture, concerns of technocratic determinism warn against overreliance on neuroscientific data at the expense of cultural and poetic dimensions of design (Robinson and Pallasmaa 2015). Similarly, NZC privileges energy metrics without sufficient attention to human comfort and psychological needs. The future of sustainable design, therefore, lies in harmonizing quantitative carbon-reduction methods with the qualitative demands of architectural empathy. The synergy between Neuroarchitecture and NZC represents more than methodological alignment: it constitutes a shared vision for future-ready design. Empirical evidence of future research should confirm that through the psychological benefits of sustainable interventions, carbon-neutral design strategies can be argued to be human-centered. The deliberate cooperation between these design paradigms could generate an architectural movement that surpasses Modernism and restores the central purpose of architecture: Promoting human well-being through occupation and at least expense to the nature that ultimately enables all life forms.

References

1. Abbas S et al (2024) Neuroarchitecture: how perception of our surroundings impacts the brain. Biology 13:202
2. AIA (2021) In: Framework for design excellence. American Institute of Architects
3. Ark Architects (2019) Neuroscience, a tool for modern architecture, http://ark-architects.com/neuroscience-a-tool-for-modern-architecture-arl-architects/
4. Banaei M et al (2017) Walking through architectural spaces: the impact of interior forms on human brain dynamics. Front Hum Neurosci 11(477):1–3
5. Beatley T (2011) Bioliphic cities: integrating nature into urban design and planning. Island Press, Washington, DC
6. Berto R (2014) The role of nature in coping with psycho-physiological stress: a literature review on restorativeness. Behavioural Sci 4(4):394–409
7. Birch D (2024) Neuroarchitecture as healing design in a changing climate. Design Intelligence Media
8. Boyce PR (2014) Human factors in lighting. CRC Press, Boca Raton
9. Condia B (ed) (2020) Meaning in architecture: affordances, atmosphere and mood. New Prairie Press, Kansas State University Libraries, Manhattan
10. Coburn A, Vartanian O, Chatterjee A (2017) Buildings, beauty and the brain: a neuroscience of architectural experience. J Cogn Neurosci 29(9):1521–1531
11. Choo H, Nasar J, Nikrahei B et al (2023) Neural codes of seeing architectural styles. Sci Rep 7:40201. https://doi.org/10.1038/srep40201
12. Cole RJ (2005) Building environmental assessment methods: redefining intentions and roles. Build Res Inf 33(5):455–467
13. Dodoo A, Gustavsson L, Sathre R (2010) Lifecycle primary energy implication of retrofitting a wood-framed apartment building to passive house standard. Resource Conserve Recycle 54(12):1152–1160

14. Eagleman D (2011) In: Incognito: the secret lives of the brain. Pantheon Books, New York. ISBN 978-0307377333
15. Eberhard JP (2009) Applying neuroscience to architecture. Newton 62(6):753–756
16. EDGE. Edge.tech/portfolio/the-edge?utm-source=chatgpt.com
17. Edwards L, Torcellini P (2002) A literature review of the effects of natural light on building occupants. NREL Report
18. Elghor AA, Shafik Z, Ali AA, El-Husseiny M (2024) Building for the mind: a neuro-architectural approach of spatial experience for sustainable mental health and wellbeing. J Renew Energy Sustain Dev 10(2):876–888
19. Finnigan KA (2024) Sensory responsive environments: a qualitative study on perceived relationships between outdoor built environments and sensory sensitivities. Land 13, Article 636
20. Goldenstein B, Journaridis D, Newell JP (2020) The carbon footprint of household energy use in the United States. Proc Natl Acad Sci 117(32):19122–19130
21. Hartig et al (2014) Nature and health. Annual Rev Public Health 35:207–28
22. Heatherwick T (2023) Humanise. Penguin Books, London
23. IEA (2020) Energy efficiency. International Energy Agency, Paris
24. Joye Y, van den Berg A (2014) Restorative environments. In: Gifford R (ed) Environmental psychology. Wiley, Hoboken, pp 57–68
25. Kaplan R, Kaplan S (1989) The experience of nature: a psychological perspective. Cambridge University Press, New York
26. Kellert SR, Calabrese EF (2015) The practice of biophilic design. Terrapin Bright Green, New York
27. Dawes MJ, Lee JH, Ostwald MJ (2023) Intelligibility and cellularity in the villas of Palladio and Le Corbusier: examining Rowe's observations. Arch Net-IJAR 17(2)
28. Li DHW, Lam TNT (2001) An investigation of daylighting performance and energy saving in a classroom. Build Environ 36(2):137–143
29. Mehaffy M, Salingaros NA (2013) Why modernist design may be inherently hard to green. Fifth Estate 1(3):7–15
30. Mostafavi A, Xu TB, Kalantari S (2024) Effects of illuminance and correlated color temperature on emotional responses and lighting adjustment behaviours. J Build Eng 108, Article 108833
31. Philips Lighting, New Atlas. Slc.philips.com/support/cases/office/edge
32. PLP Architecture (2025) Smart building adaptive operations report, PLP Architecture, Schneider Electric
33. Reinhart CF, Mardaljevic J, Rogers Z (2006) Dynamic daylight performance metrics for sustainable building design. LEUKOS 3(1):7–31
34. RIBA (2021) Royal institute of british architects. RIBA 2030 Climate Challenge, Version 2. RIBA London
35. Robinson S, Pallasmaa J (2015) Mind in architecture: neuroscience, embodiment, and the future of design. MIT Press, Cambridge
36. Szewrański S, Mrówczyńska MM, van Hoof J (2024) Biophilia in contemporary design: navigating future opportunities and challenges. Indoor and Built Environ
37. Taylor RP (2006) Reduction of physiological stress using fractal art and architecture. Leonardo 39(3):245–251
38. Thampanichwat C, Wongvorachan T, Bunyarittikit S, Chunhajinda P, Phaibulputhipong P, Wongmahasiri R (2024) Development of a building evaluation framework for biophilic design: enhancing psychological and physiological comfort. Buildings 14(10), Article 3254
39. Ulrich RS et al (1991) Stress recovery during exposure to natural and urban environments. J Environ Psychol 11(3):201–230

40. UNEP (2020) Global status report for buildings and construction. United Nations Environment Programme, Nairobi
41. Valentine C, Mitcheltree H, Sjövall IAK, Khalil MH (2025) Architecturally mediated allostasis and neurosustainability: a proposed theoretical framework for the impact of the built environment on neurocognitive health. Brain Sci 15(2), Article 201
42. Xylakis E (2021) Integrating spatial psychology in Juvenile rehabilitation centers: a design framework. In: Proceedings of the 2021 international conference on architecture and urban planning, IEEE, New York, pp 45–58
43. Zhang Z, Andersen M (2020) A review of the effectiveness of metrics for assessing human responses to biophilic environments involving views, shading, and interior design elements. Build Environ 175:106798

Net Zero Carbon Potential of Bio-based Building Materials

Coralie van Reenen(✉) and Yerdashin Padayachi

Council for Scientific and Industrial Research, Meiring Naude Rd, Pretoria, South Africa
cvreenen@csir.co.za

Abstract. This research paper explores the net zero carbon potential offered by bio-based building materials in South Africa, specifically, mass timber and two bio-based concretes, hempcrete and nonCrete. As the construction industry grapples with the urgent need to reduce its carbon footprint, bio-based materials offer a promising path toward sustainability. Hempcrete, known for its excellent insulation properties and carbon sequestration capabilities; nonCrete, making use of alien invasive species, thereby restoring natural systems; and mass timber, lauded for its strength and renewability, are analysed for their environmental benefits and practicality in modern construction. The study examines the cradle to gate carbon emissions of these materials, as well as fitness for purpose. Findings suggest that bio-based materials not only contribute to significant carbon savings during their production and usage phases but also enhance building performance. The study methodology compares the embodied carbon of a comparable section of wall of each material. Calculations are based on data available in embodied carbon databases, case studies and environmental product declarations. The paper finds that despite limited standardised data, the selected materials have net negative embodied carbon and concludes with recommendations for integrating these materials into mainstream construction practices to drive the industry towards a more sustainable future.

Keywords: Bio-based concrete · Embodied carbon · Mass timber

1 Introduction

The construction industry is facing the challenge of reducing its carbon footprint. Traditional building materials, such as concrete and steel, contribute significantly to global carbon emissions due to their energy-intensive production processes [1]. As sustainability becomes more critical, bio-based building materials offer a promising alternative, advancing climate resilience, low emissions, and better lives in line with South Africa's just transition goals. This article examines the net zero carbon potential of bio-based materials, particularly focusing on mass timber and bio-based concretes.

Bio-based materials can potentially lower the embodied carbon and energy demand of buildings relative to conventional materials [2]. Bio-based materials are partially or fully derived from renewable biological products or by-products [3]. In this study, materials that are produced using vegetative biomass in part or in whole are considered. The carbon

H. Beushausen et al. (Eds.): NTZR 2025, LNCE 873, pp. 411–421, 2026.
https://doi.org/10.1007/978-3-032-27698-8_37

that is sequestered by vegetation through photosynthesis is permanently stored in the building material, only being released if the material is burned or decomposed—even when this happens, the biomass is considered to be emissions neutral. There are various types of bio-based construction materials that are used for structural elements, envelope materials, insulation and other components. This study focusses on selected building envelope materials that are available in South Africa, specifically, mass timber and two types of bio-concrete—hempcrete and nonCrete (see Fig. 1).

Fig. 1. Images of mass timber construction, Hempblock infill walls, and nonCrete test wall (from left to right).

2 Literature Review

To provide a global backdrop for this study within the South African context, the literature was reviewed to understand the status of the use of biobased building materials globally. Ten key papers were identified that examined bio-based building materials, of which two were studies relating to residential buildings, two were on office buildings, and two were on multi-storey buildings, reflecting the potential broad application of bio-based materials in buildings.

The selected papers included a diverse range of bio-based products, such as CLT, bamboo, bio-concrete, lime hempcrete composites, hemp shiv, straw, bio-mass based insulation and general wood products. The studies covered a range of locations and climates, indicating that bio-materials are diverse in type and application and are finding some uptake all over. Eight of the ten studies reported reductions in net embodied carbon impact compared to conventional materials and two reported net negative embodied carbon.

None of the literature in this exercise referenced the use of bio-materials in Africa or South Africa. A further search for literature referencing the use of bio-materials specifically in South Africa identified papers about wood, hemp, bamboo, straw, cork, nonCrete and salvaged timber. The findings reveal that timber is widely used in South Africa [4, 5], however, this refers to the common practice of using timber for roof trusses, while the interest of the current study is the use of mass timber for the building envelope. The references to hemp [6], bamboo [7], and nonCrete [8] indicate that these materials are emerging technologies, while straw, cork, and salvaged timber lack documented use. Furthermore, straw and cork are only mentioned as insulation materials and not structural or envelope materials; hemp is mentioned as masonry infill and insulation. According to da Silva [9], the production of bio-based building materials in South Africa is not mainstreamed, despite an abundance of renewable raw materials from timber plantations, agricultural crops and Invasive Alien Plants (IAPs).

The literature cites several advantages of bio-based materials, including low embodied carbon [10], superior strength of some [11, 12], and superior thermal insulation [13]. However, no single product is suitable for all contexts and uses [8].

The review of literature produced two key findings, specifically in the South African context. Firstly, there is a low uptake of bio-based construction materials in South Africa, as evidenced by the low number of research papers available. Secondly, there is a lack of knowledge regarding the true environmental implications of bio-based materials and the carbon risk impacts of sustainable construction materials in general [14].

3 Research Objectives and Methodology

The goal of this study is to establish whether the lower embodied carbon of emerging bio-based construction materials in South Africa justifies their use instead of conventional methods. The objective was to compare the embodied carbon of three bio-based building envelope materials that are available in South Africa—hempcrete, nonCrete, and mass timber—to establish which has the greatest potential to reduce embodied carbon in buildings.

The exercise only considers the cradle to gate life cycle assessment (LCA) data (stages A1 to A3, as defined by EN 15978). Although this represents a limitation, this scope was set to eliminate variables associated with transport and construction emissions, which could not be established. In each scenario, 1 m^2 of the installed wall was considered, including any joining and structural elements required to constitute the wall.

As far as possible, emissions data from the South African manufacturers were used. However, only one of the manufacturers (non-Crete) of the three products analysed was able to provide embodied carbon data. The data for mass timber and hempcrete was sourced from literature and international manufacturers. Data sourced was in most cases not verified by a third party and the LCA scope is not always clear—for example, in some cases, Stage A1 to A3 data is given and in other cases Stage A1 to A5 figures are used. This research attempts to account for these factors to present a fair comparison between the three local products. The lack of actual and reliable embodied carbon data for local products was a limitation, however, the study examines the theoretical case for each bio-material, formulating a methodology for application in future studies when actual data becomes available, and discussing other considerations for each material.

4 Understanding Bio-based Building Materials

The three selected materials are considered here in terms of their embodied carbon, as well as other qualities. A square meter of walling of mass timber, hempcrete blocks and nonCrete blocks is considered, using standard available elements to establish the respective wall thicknesses of 165, 300 and 230 mm.

4.1 Mass Timber

Mass timber (MT) products are engineered for strength and durability. These products include cross-laminated timber (CLT), laminated veneer lumber (LVL), and glue-laminated timber (glulam), creating massive timber planes or framing elements for constructing walls, floors, roofs, partitions and other elements [12], as shown in Fig. 1.

MT displays similar structural properties as concrete and steel, while requiring less energy to be produced [15] and can be manufactured in larger and longer sections than conventional construction timber [11]. Elements are typically manufactured off site in a factory where precise measurement and cutting reduces material waste and enables rapid erection on site [12]. Prefabrication results in less noise, dust and construction waste on site [11]. This also means that much of the emissions normally associated with construction occur off site and fall into Stage A3 of a life cycle assessment. This could potentially skew a comparative LCA with other products, depending on the scope of the assessment.

Embodied Carbon of Mass Timber. MT is considered renewable, since trees can be re-grown; carbon sequestered during their growth offsets the emissions generated during production, with studies showing a reduction in embodied emissions of up to 90.8% relative to conventional steel structure construction, considering cradle to site emissions [15]. Such claims, however, are inconsistent across different studies and are highly dependent on the accuracy and scope of emissions assessments, and the context and location [16]. Data varies across different studies and little to no data is available for South African products.

An online search for mass timber manufacturers in South Africa yielded three main manufacturers, one of which makes no claims regarding emissions, another claiming 0.9 tonnes of carbon dioxide (CO_2) sequestered per cubic metre, and the third similarly claiming a 1:1 ratio of stored carbon per volume of timber. However, these figures appear to only refer to sequestered carbon and do not account for production emissions. According to the ICE database (version 4.0)[1], CLT has embodied carbon of 0.25 $kgCO_2e/kg$ (for scoped boundaries A1 to A3), or -1.39 $kgCO_2e/kg$ if sequestered carbon is accounted for. However, the ICE database notes that this could vary depending on the timber species, density, moisture content and several other variables; these variables make definitive claims based on the ICE data impossible. It is not clear in the literature whether the off-site cutting and preparation of prefabricated elements is included in this, however, it is assumed that it is not.

In the context of this study, solid elements used for exterior wall panels are considered, assuming load-bearing CLT wall elements with a thickness of 165 mm. The characteristic density of South African (SA) Pine is stated to be 360 kg/m^3 (for S5 timber), according to the local manufacturer's data [17], which means that 1 m^2 of wall weighs 59.4 kg, resulting in embodied carbon of -82.566 $kgCO_2e$, if using the ICE database value of -1.39 $kgCO_2e/kg$.

[1] https://circularecology.com/embodied-carbon-footprint-database.html.

4.2 Hempcrete

Hempcrete is a bio-composite material made from the inner woody core (hemp hurds or shiv) of the hemp plant mixed with a lime-based binder to make a concrete-like compound. It has good insulation properties and possesses significant carbon sequestration capabilities. Hempcrete can be used in various ways in construction, such as in-situ casting, precast blocks, plaster or spray [18, 19]. In this study, hempcrete blocks (600 mm × 300 mm × 308 mm) are considered, based on products available in South Africa.

Hempcrete has low compressive strength and is often used as infill material within timber columns [19]. The strength and density of hempcrete varies depending on the exact mix and method of application, which is also true of the thermal conductivity properties [18, 19].

Embodied Carbon of Hempcrete. Hemp grows quickly and has a high yield, making it a rapidly renewable resource with high carbon sequestration potential [20]. The hemp plant absorbs CO_2 during its growth, and the lime binder continues to absorb CO_2 from the atmosphere after production in a process known as carbonation [18, 20]. This dual carbon sequestration process makes hempcrete a highly attractive option for sustainable, low-carbon buildings.

The energy embodied in the field production of hemp is low relative to other crops [21], however the process of manufacturing lime requires heating of calcium carbonate to high temperatures and about 41% of the volume is emitted as CO_2, resulting in an initial high embodied carbon as the product it leaves the gate (LCA stage A3); this is off-set by the CO_2 that is absorbed through carbonation over a 100-year lifetime [21] but technically falls out of the scope of cradle to gate.

Only two local suppliers of hempcrete blocks were found, neither of which were able to provide embodied carbon data for their products. Data obtained directly from a manufacturer in the United States of America[2] suggest that 1 m^2 of a Hempblock wall holds −56 $kgCO_2e$ of embodied carbon, including lime render and perlite cement; it is unclear whether this includes the structural timber framing, and the size of the blocks and thickness of the wall is not given.

Ip and Miller [21] calculate that 1 m^2 of hempcrete wall installed (cast in-situ with timber framing) has a carbon footprint of −36.08 $kgCO_2e$ for a 300 mm thick non-load bearing wall, including the timber frame. This considers the case for the installed product over a 100-year lifetime and thus accounts for carbon sequestered by the lime beyond the construction phase. If the sequestration through carbonation of lime (−28.55 $kgCO_2e$) and construction and haulage (3.76 $kgCO_2e$) are excluded, the resultant carbon footprint of the product at LCA stage A3 calculates to −11.29 $kgCO_2e$. A similar case for a load-bearing wall is presented by Ip and Miller [21], showing emissions of −35.53, or −22.8 $kgCO_2e$ if lime carbonation (−13.6 $kgCO_2e$) and construction and haulage (0.87 $kgCO_2e$) are excluded.

Emanuele et al. [22] demonstrate that an in-situ cast hempcrete wall has lower carbon emissions associated with its production (as per IPCC method) than hempcrete blocks (excluding the impact of carbonation) due to the energy required for the manufacture of

[2] https://hempblockinternational.com/.

the blocks. However, the CO_2 emissions cited for the manufacture of the blocks is very similar to that of the in-situ wall cited by Ip and Miller.

These examples demonstrate the complexities of calculating the embodied carbon, which is dependent on the interpretation of the scope and factors such as varying agricultural practices, transport distances, lime processing methods and mix design [23].

For the sake of this exercise, the data provided by Ip and Miller [21] for a non-load-bearing wall will be assumed, which is in the range cited by various studies [23].

4.3 NonCrete

NonCrete is an innovative biomass concrete developed in South Africa. It uses chipped biomass from IAPs, such as Eucalyptus, as a supplementary ingredient in the concrete mix. The volume of ordinary Portland cement (OPC) in the mix is reduced by the inclusion of a proprietary binder, and the mix is specifically designed, considering the size and moisture content of the wood chips.

The concrete mix can be used for various precast elements; in this paper the load-bearing bricks are considered, which are the same size as conventional clay bricks and have properties at least equivalent to conventional masonry construction.

Embodied Carbon of NonCrete. The carbon sequestered during the growth of the trees is retained in the wood chips within the concrete mix for bricks, resulting in a net negative embodied carbon emission intensity. Furthermore, the control of IAPs contributes to the restoration of indigenous ecosystems [24], supporting a circular economy.

According to the manufacturer, the load-bearing bricks considered for this study have an embodied carbon value of -96 $kgCO_2e/m^3$ or -21 $kgCO_2/m^2$ (for Stages A1 to A3) for a double skin load-bearing wall. If conventional cement mortar with a 1:4 cement to sand ratio is used for building the wall (using a carbon emission intensity factor of 0.146 $kgCO_2e/kg$, as per ICE v4), the resulting overall embodied carbon of 1 m^2 wall is -15.7 $kgCO_2e$.

4.4 Comparative Analysis Discussion

The above theoretical calculations of the embodied carbon associated with construction of 1 m^2 of wall using each of the materials is summarised in Table 1, considering only LCA stages A1 to A3. The results show that MT has the lowest embodied carbon and that all products have net negative emissions, with potential to reduce the overall carbon footprint of a building.

Table 1. Summary of embodied carbon associated with each construction material.

Material	Embodied carbon per m^2 constructed
Mass timber	-82.566 $kgCO_2e$
Hemp blocks	-11.29 $kgCO_2e$ (including lime carbonation) -36.08 $kgCO_2e$ (excluding lime carbonation)
NonCrete bricks with mortar	-15.7 $kgCO_2e$

However, there are several limitations with these values, in particular the fact that in all cases, the cited emissions data are not clear and transparent, and rigorously calculated local data is lacking.

Apart from embodied carbon, other characteristics of the materials are also relevant to consider when selecting materials. These are discussed below and summarised in Table 2. Some key metrics (based on Agrément SA fitness for purpose requirements for alternative building systems) are used to evaluate further qualities of each of the three materials, including thermal performance (insulation properties), acoustic performance (sound transmission), compressive strength, and fire resistance.

Table 2. Characteristics of materials.

Metric	Mass timber	Hempcrete block	NonCrete brick[3]
Thermal performance (U-value)	0.12 W/mK [17]	0.23 [25]	TBC
Acoustic performance (sound transmission)	35 dB [26]	51 dB [27]	TBC
Compressive strength	16–18 MPa [17]	1 MPa [27]	10 MPa
Fire rating (minutes)	60 min [17]	120 min [27]	180 min

Since these are emerging technologies, much of the data is not yet published, limiting the ability to rigorously compare these products. NonCrete demonstrates the longest fire resistance (tested in accordance with SABS 10177-2), although the other two products may have stopped testing and only reported on the minimum requirements for fire resistance. All three products demonstrate at least a 60-min fire resistance.

Hempcrete is porous, giving it unique attributes of thermal and acoustic insulation; it has a high noise reduction coefficient [19] and hygrometric properties, enabling it to control humidity [28]. Of the products with known acoustic performance, hempcrete shows superior performance, outperforming standard brick wall performance, which is typically has sound transmission index of about 40 dB, and meeting most requirements set by Agrément SA, which varies between 33 and 52 dB, depending on the use case. The acoustic performance of CLT walls alone (i.e., without additional insulation layers) is not well-documented. Literature seems to indicate a weighted sound reduction index (R_w) of 33–35 dB for a 100 mm thick panel [26, 29]. For the current study, we accept the R_w of 35 dB for the 165 mm panel, and highlight that further research is needed regarding this aspect. The sound reduction index of nonCrete was not available at the time of writing.

The nonCrete manufacturer claims that their bricks have good thermal insulation properties, although specific data could not be sourced; both CLT and hempcrete show good thermal insulation, meeting the minimum requirements for thermal insulation specified by the National Building Regulations through SANS 10400-XA:2021.

Although no data on the durability of all three materials could be sourced, it was found that hempcrete is susceptible to freeze-thaw degradation, resulting in a weakening

[3] https://noncrete.com/.

of the material [18]. Hempcrete has a low compressive strength and is suitable for infill walls only, while the nonCrete load-bearing bricks have a compressive strength of 10 MPa. MT is noted for its strength and durability, making it suitable for a variety of construction applications, including high-rise buildings, and is relatively light weight, reducing transportation and installation costs, further enhancing its practicality. The load-bearing capacity of each material is based on data from the manufacturers or literature and could not be verified against specific local test requirements.

An important consideration in terms of the environmental impact associated with each product is the renewability and recyclability of each product. The growth period for the bio-based components of each product is a factor of how quickly the resource can be replenished. The growth period for South African (SA) Pine used in locally produced CLT is 25 years [30], compared to the four-month growth period of hemp. According to Ip and Miller [21], 30 kg of hemp is used in 1 m^2 of 300 mm thick wall and 45.82 $kgCO_2e$ is effectively stored over the lifetime of the plant—this means that hemp sequesters more carbon per weight than SA Pine used in CLT (which is a 1:1 ratio). However, further factors that are beyond the scope of this study need to be considered, such as the carbon that is released from the soil during the frequent ploughing for re-planting of hemp.

Studies show that Eucalyptus stores more carbon over a typical plantation tree life span than Pine trees [31], although this depends on the region and specific sub-species. A review of literature did not yield data specific to South African plantation pines (*Pinus radiata*) and non-plantation alien invasive species used by nonCrete (*Eucalyptus camaldulensis*). A direct comparison between these two species, used in mass timber and nonCrete respectively, is not truly possible because the growth period of the wild invasive Eucalyptus tree is not fixed. A further consideration is the renewability of the bio-mass, which is not directly applicable in the case or IAPs; although the Eucalyptus is preferably not re-grown because it is considered invasive, the production of nonCrete has the advantage of addressing circularity in the restoration of natural ecosystems rather than renewability.

All three products claim to be recyclable at end of life. Hempcrete can be broken down and worked into the soil and is biodegradable and recyclable [18] leaving behind a nutrient rich compos or waste-to-energy incineration; nonCrete can be crushed and reused as aggregate in new nonCrete products; mass timber can be reused or chipped and recycled in composite wood products.

5 Findings and Discussion

The findings suggest that bio-based materials contribute to significant carbon savings, demonstrating net negative embodied carbon. However, embodied carbon data is limited, and several other critical factors need to be considered when evaluating the suitability of each material for a particular project.

In terms of environmental impact and renewability, hemp grows rapidly enabling rapid replenishment, supporting frequent manufacturing cycles, and sequestering more carbon per weight than SA Pine used in CLT, which has a much longer growth cycle and requires heavy harvesting and processing equipment. To ensure environmental sustainability, plantations must be managed responsibly and support biodiversity. Eucalyptus is an invasive species and its use in nonCrete results in environmental benefits associated with the restoration of natural ecosystems.

All three bio-based materials considered claim recyclability: hempcrete is biodegradable, nonCrete can be crushed and reused as aggregate, and mass timber can be reused or recycled in composite products.

Selecting the optimal bio-based material for a construction project involves weighing a series of trade-offs based on project-specific needs and the unique characteristics of hempcrete, nonCrete, and MT. Local accessibility of the products may ultimately be a key factor in selecting a material, which impacts delivery time, costs, and transport emissions. Project teams should conduct a holistic assessment, considering performance, environmental impact, durability, cost, and regulatory context to make an informed selection. The development of a decision-matrix that considers the various factors is recommended.

An important observation in this study is the lack of standardised and verified data on each product, making a rigorous comparison impossible. Even though standards exist for establishing the embodied carbon of materials (such as ISO 14067, EN 15978, and PAS2050), the data is not presented in a uniform way in literature and product data sheets. Further research and standardisation will enhance the comparability of these promising bio-based materials, supporting their broader adoption in sustainable construction.

5.1 Recommendations for Integrating Bio-based Materials

While the environmental benefits of the bio-based materials discussed here are evident, the products and market are at a low readiness level. To integrate bio-based materials into mainstream construction practices, the following recommendations are proposed.

Evidence-based policies and regulations should be developed and implemented in South Africa to enforce transparent environmental product declarations, disclosing the embodied carbon of products in a standardised metric.

Continued research and development is essential to improve the performance and cost-effectiveness of bio-based materials. This is possible through collaborative efforts between manufacturers, research bodies and regulators.

6 Conclusion

The goal of this study is to establish whether the lower embodied carbon of bio-based construction materials justifies their use instead of conventional methods. Despite inconsistent and non-standardised data, it can be concluded that bio-based building materials, such as mass timber, hempcrete and nonCrete, offer a promising path toward achieving net zero carbon targets in the construction industry. The integration of bio-based materials into mainstream construction practices can drive the industry towards a more sustainable future, reducing its carbon footprint and contributing to global efforts to combat climate change.

References

1. Anderson J, Moncaster A (2020) Embodied carbon of concrete in buildings, Part 1: analysis of published EPD. Build Cit 1(1):198–217. https://doi.org/10.5334/BC.59
2. Chen L et al (2024) Biomaterials technology and policies in the building sector: a review. Environ Chem Lett 22(2):715–750. https://doi.org/10.1007/S10311-023-01689-W
3. Le DL, Salomone R, Nguyen QT (2023) Circular bio-based building materials: a literature review of case studies and sustainability assessment methods. Build Environ 244:110774. https://doi.org/10.1016/J.BUILDENV.2023.110774
4. Crafford PL, Blumentritt M, Wessels CB (2017) The potential of South African timber products to reduce the environmental impact of buildings. S Afr J Sci 113:9–10
5. Windapo A, Olugboyega O, Pomponi F, Moghayedi A, Emuze F, Inggs M (2022) Causality between challenges, availability, and extent of use of local building materials. S Afr J Sci 118:8
6. Goliath Y (2020) Hemp building materials in the South African market. SSRN Electr J 2:29. https://doi.org/10.2139/SSRN.3689029
7. Ross S, Alexander M, Viability AM, Inggs M (2022) Viability of whole-culm bamboo construction in South Africa: a preliminary assessment. S Afr J Sci 118(7/8):12468
8. Goswein V et al (2022) Wood in buildings: the right answer to the wrong question. IOP Conf Ser Earth Environ Sci 17:12067. https://doi.org/10.1088/1755-1315/1078/1/012067
9. da Silva EJ (2022) Analysis of raw material availability for the production of bio-based composites in South Africa. University of Hamburg. https://ediss.sub.uni-hamburg.de/handle/ediss/10087
10. Weiss M et al (2012) A review of the environmental impacts of biobased materials. J Ind Ecol 16:S169–S181. https://doi.org/10.1111/J.1530-9290.2012.00468.X
11. Abed J, Rayburg S, Rodwell J, Neave M (2022) A review of the performance and benefits of mass timber as an alternative to concrete and steel for improving the sustainability of structures. Sustainability 14(9):5570
12. Smith RE, Griffin G, Rice T, Hagehofer-Daniell B (2018) Mass timber: evaluating construction performance. Arch Eng Des Manag 14(1–2):127–138. https://doi.org/10.1080/17452007.2016.1273089
13. Bourbia S, Kazeoui H, Belarbi R (2023) A review on recent research on bio-based building materials and their applications. Mater Renew Sustain Energy 12(2):117–139. https://doi.org/10.1007/S40243-023-00234-7
14. Moghayedi A, Massyn M, Le Jeune K, Michell K (2022) Evaluating the impact of building material selection on the life cycle carbon emissions of South African affordable housing. In: IOP conference series: earth and environmental science, institute of physics. https://doi.org/10.1088/1755-1315/1101/2/022021
15. Greene JM, Hosanna HR, Willson B, Quinn JC (2023) Whole life embodied emissions and net-zero emissions potential for a mid-rise office building constructed with mass timber. Sustain Mater Technol 35:e00528. https://doi.org/10.1016/J.SUSMAT.2022.E00528
16. Crawford RH, Cadorel X (2017) A framework for assessing the environmental benefits of mass timber construction. Proced Eng 196:838–846. https://doi.org/10.1016/J.PROENG.2017.08.015
17. Mass Timber Technologies (2023) Cross-laminated timber and glue-laminated timber design guide. https://www.masstimbertech.co.za/designguide. Accessed 30 May 2025
18. Muhit IB, Omairey EL, Pashakolaie VG (2024) A holistic sustainability overview of hemp as building and highway construction materials. Build Environ 256:111470. https://doi.org/10.1016/J.BUILDENV.2024.111470

19. Yadav M, Saini A (2022) Opportunities and challenges of hempcrete as a building material for construction: an overview. Mater Today Proc 65:2021–2028. https://doi.org/10.1016/J.MATPR.2022.05.576
20. Ansari H, Tabish M, Zaheer MM (2025) A comprehensive review on the properties of hemp incorporated concrete: an approach to low carbon footprint construction. Next Sustain 16:100075. https://doi.org/10.1016/j.nxsust.2024.100075
21. Ip K, Miller A (2012) Life cycle greenhouse gas emissions of hemp–lime wall constructions in the UK. Resour Conserv Recycl 69:1–9. https://doi.org/10.1016/J.RESCONREC.2012.09.001
22. Emanuele S, Capua DI, Paolotti L, Moretti E, Rocchi L, Boggia A (2021) Evaluation of the environmental sustainability of hemp as a building material, through life cycle assessment. Environ Clim Technol 25(1):1215–1228. https://doi.org/10.2478/rtuect-2021-0092
23. Shanbhag SS, Dixit MK, Sideris P (2024) Examining the global warming potential of hempcrete in the United States: a cradle-to-gate life cycle assessment. Develop Built Environ 20:100572. https://doi.org/10.1016/J.DIBE.2024.100572
24. Göswein V et al (2021) Invasive alien plants as an alternative resource for concrete production: multi-scale optimization including carbon compensation, cleared land and saved water runoff in South Africa. Resour Conserv Recycl 167:105361
25. Hempblock (2025) Hempcrete block internal solid wall insulation. https://hempblock.co.uk/wp-content/uploads/2022/03/Hemp-Blocks-Solid-Wall-Insulation-lo-res.pdf
26. Loriggiola F, Dindustria LD, Granzotto N, Di Bella A (2019) Acoustics behaviour of CLT structure: transmission loss, impact noise insulation and flanking transmission evaluations. Proceed Acoust 12:785
27. Afrimat (2022) AfrimatHemp technical data sheet. https://www.afrimathemp.co.za/wp-content/uploads/2022/02/AFRIMATHEMP_Technical-Data_Block-System.pdf. Accessed 12 May 2025
28. Lupu ML, Isopescu DN, Baciu I-R, Maxineasa SG, Pruna L, Gheorghiu R (2022) Hempcrete-modern solutions for green buildings. IOP Confer Ser Mater Sci Eng 8:1242. https://doi.org/10.1088/1757-899X/1242/1/012021
29. Pagnoncelli L, Morales F (2016) Cross-laminated timber system (CLT): laboratory and in situ measurements of airborne and impact sound insulation. EuroRegio 16:785
30. Sawmilling South Africa (2025) Why wood—sawmilling South Africa. https://sawmillingsa.co.za/why-wood/#benefits
31. Olmedo GF et al (2020) Baseline of carbon stocks in *Pinus radiata* and *Eucalyptus* spp. plantations of chile. Forests 11(10):1063

Earth-Based Materials for Urine-Diverting Dry Toilets

T. D'Andrea[1], M. S. Smit[2](✉), A. Van der Merwe[2], and E. Tilley[1]

[1] ETHZurich, Raemistrasse 101, 8006 Zurich, Switzerland
[2] University of Pretoria, Lynnwood Rd, Pretoria, South Africa
phia.smit@up.ac.za

Abstract. Sanitation in informal settlements in South Africa has been a persistent issue for decades. A promising alternative to conventional pit latrines is urine-diverting dry toilets (UDDTs). Though waterless and appropriate for hot, dry conditions, UDDTs require significantly more materials than pit latrines, especially cement. However, earth-based materials made from local soil can be used to construct different parts of UDDT systems, thus lowering costs and resource requirements. In this research project, the feasibility of using earth-based materials for UDDT was investigated. Mechanical and water erodibility properties of earth-based materials were tested. The efficacy of cement-stabilisation and various surface treatments was also tested. All results were compared to low-strength cement concrete. The earth-based materials tested were less expensive than low-strength cement concrete, but they exhibited substantially lower mechanical performance. Regarding water erosion, earth-based materials performed well when coated with industrial brick sealer or environmentally friendly linseed oil. Notably, the cement-stabilised variant, when coated, showed erosion resistance comparable to low-strength cement concrete.

Keywords: Earth concrete · Cement stabilisation · Water erodibility · Dry sanitation · South Africa

1 Introduction

Unlined pit latrines constructed by households are the most common sanitation system used in low-income or informal communities. Pit toilets pose severe risks, including flooding, groundwater contamination, health hazards, and safety concerns for young children. They also require labour-intensive maintenance, and space becomes a challenge as new pits must be dug once old ones fill up.

A promising alternative to conventional pit latrines is urine-diverting dry toilets (UDDTs), which reduce health risks and allow for the re-use of human waste as fertilisers. UDDTs are waterless sanitation systems designed to separate urine and faeces at the source. Sealed containers are typically used to collect diverted urine, while the faeces fall into a chamber where it undergoes dehydration. UDDTs operate independently of sewer infrastructure. One of the drawbacks of UDDTs is the requirement of a more robust

H. Beushausen et al. (Eds.): NTZR 2025, LNCE 873, pp. 422–435, 2026.
https://doi.org/10.1007/978-3-032-27698-8_38

and properly constructed structure, which increases initial costs compared to simple pit latrines [1].

Earth-based construction is not only known to offer environmental advantages but also has the potential to decrease the initial costs of structures. Earth-based materials are made from local earth using a variety of manufacturing methods, such as adobe and rammed earth. Earth-based materials are often stabilised using additives like cement, lime or cow dung to improve mechanical strength [2, 3].

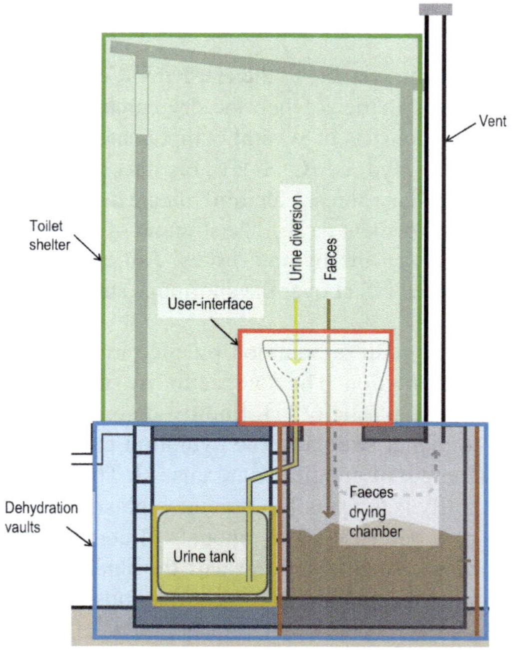

Fig. 1. Schematic of UDDT (adapted from Tilley [4])

Generally, a UDDT structure consists of three parts; the toilet shelter, the dehydration vaults and the user interface (Fig. 1). The urine is stored in plastic containers inside or outside of the dehydration vaults. The structural and durability requirements of the parts are different. And thus, it is reasonable to use different materials to minimize construction costs.

The aim of this research was to investigate the suitability of earth-based concrete for application in the different parts of the UDDT structure. The objectives were:

- To evaluate the mechanical strength and durability of earth concrete, cement-stabilised earth concrete and cement concrete.
- To evaluate the effect of surface treatments on the durability of earth concrete and cement-stabilised earth concrete.
- To do a cost analysis of the earth concrete, cement-stabilised earth concrete and cement concrete.

2 Background

A large portion of construction materials consists of granular materials that have been compacted and/or bound. Granular material consists of silt, sand, and larger aggregates. Cement concrete consists of coarse and fine aggregate bound together by hardened cement paste, while cement stabilisation of pavement substructure layers, typically consists of granular material, Portland cement and water mixed and compacted. Earth concrete consists of granular materials that are bound together by clay [5].

2.1 Strength of Earth-Based Materials

The strength of cement concrete is due to the Hardened Cement Paste (HCP) that binds the aggregate. Cement paste hardens when the dry cement is hydrated and hydration products are formed. HCP consists of several components, which include the hydrates (primarily Calcium Silicate Hydrate (C–S–H)), crystals of portlandite, the residue of unhydrated cement and capillary pores. Calcium Silicate hydrate (C–S–H) is responsible for the strength of hardened cement paste. The strength of C–S–H derives from the Van der Waals-type bonds between the hydrate fibres. Although each individual bond is relatively weak, the integrated effect over the enormous surface area (200,000 m^2/kg) is considerable [6].

Cement stabilisation of earth is common practice in road construction. It is also used in earth building construction. The compactibility of clayey soils is improved by adding cement to reduce the plasticity [7]. Improved compactability results in higher density which results in higher strength. The hydration products of cement and water also contribute to the strength of granular materials [8]. The cement dosage is between 3 and 10% and this quantity is too small to provide high strength cementing action [9]. Compressive strength may decrease when amounts of cement stabilisation lower than 5% are used. This is because cement interferes with the binding force of clay. The greater the clay content, the higher the amount of cement that must be added [5].

Earth concrete consists of granular materials bound by clay. Clay is a by-product of the erosion of feldspar and other minerals [5]. Minke [5] relates plastic strength and dry strength to the interlamellar binding force of clay minerals in earth concrete.

Various bonding mechanisms contribute to soil strength [10]. These mechanisms are Van der Waals electromagnetic bonding, capillary adhesion, cementation with compaction and cementation by deposition. The contribution of each mechanism to tensile strength is dependent on particle diameter. The mechanisms are all influenced by the presence of water and water content [10].

When earth concrete is continuously exposed to water, it does not perform well because water dissolves the Van Der Waals electromagnetic bonding, and the capillary adhesion stress reduces as the water content increases. This results in reduced strength, which could have immediate structural implications (failure) or make building elements (walls) more susceptible to erosion.

The range of compressive strength of adobe earth concrete is 1.0–2.5 MPa, and for cement-stabilised earth concrete is 0.75–3.5 MPa [5]. The typical compressive strength of fired clay bricks and cement bricks are 7 MPa and 3.5 MPa respectively [11], while low-strength cement concrete typically has a strength lower than 15 MPa [12]. Note that the upper limit of the strength of earth concrete is 1 MPa weaker than cement bricks, while fired clay bricks are more than twice as strong as earth concrete bricks.

2.2 Water Erosion

The mechanism of water erosion is explained using Terry [13]. Water drop impact possesses kinetic energy, which is a function of mass and velocity. The domain of soil sustaining this energy is concentrated around the drop periphery. This results in the stress applied to the soil surface being much greater on the periphery than elsewhere on the projected area of a drop. Values in the range of 2–6 MPa after 50 μs have been measured.

During drop impact, the water drop penetrates the soil surface and spreads out at the same time. The vertical stresses form a compression pit and a compensatory bulge around the pit perimeter. The vertical stresses transform to lateral shear stresses as the drop collapses and flows radially outward. The collapsed drop is emitted outward into numerous ejection droplets.

Splash detachment mechanisms consist of aggregate breakdown and cratering. Craters result from the combination of vertical compressive and lateral shearing forces. Cratering is a micro-scale soil deformation process, and the resistance of soil to cratering is determined by soil shear strength, soil deformation behaviour and surface microtopography. Soil material with low shear strength tends to have wide and shallow craters, while soil material with high shear strength tends to have narrow and deep craters. However, craters are also highly dependent on soil water repellence or the presence of water film. The size of craters is highly dependent on the drop diameter and impact velocity.

Splash transport can be of soil particles loosened by splash detachment or soil particles loosened by other processes. There are three splash transport mechanisms: splashing, splash saltation and splash creep. Splashing is the entrainment and dispersal of soil particles in ejection droplets. Surface water films influence splashing. It decreases the proportion of splashes which are loaded with soil particles and increases the splash emission angle toward the vertical. Splash saltation is the casting out of soil particles due to the velocity of impact, instead of particles being entrained in the water that is splashing out. Splash creep is when a water drop impact pushes and rolls materials sideways without vertical displacement. Soil particles that are too large to be thrown into parabolic trajectories are prone to this effect.

The resistance to water erosion can be determined using accelerated tests or simulated tests. These tests relate the performance of the material directly to the degradation mechanism. The water resistivity of a material can also be related indirectly to properties such as strength or surface hardness [14]. Accelerated and simulation tests consist of spray tests and drip tests. Accelerated tests increase the degradation factors to compensate for the reduced time frame. Simulation tests attempt to model in-service conditions exactly. Water erosion is influenced by volume of impacting water, time of exposure, impact velocity, angle of impact drops, drop diameter and antecedent moisture content. Numerous configurations of the spray test have been reported [15–20]. Table 1 shows spray test setup details and results from two papers [17, 19]. The results are of the poorest performing material in each paper.

The resistance to water erosion of the spray test is reported in different forms, such as erosion rate (mm/h), pitting depth and mass loss percentage. Typically, the erosion

Table 1. Spray test setup details and selected results from literature

Paper	Test Details	Material	Result type	Result value
Ola and Mbata [17]	Distance: 140 mm Test time: 3 h Pressure: 150 kPa	Reddish-brown clay sand, 20.3% clay Stabilisation: 5% cement and compressed to 1 MPa	Mass loss (%)	0.665%
Saranya et al. [19]	Distance: 180 mm Test time: <2 h Pressure: 147 kPa	6.2% fines (clay and silt) Stabilisation: 30% coal ash, 3% cement and rammed	Pitting depth	5 mm in 2 h (2.5 mm/h)

rate is taken as the pitting depth divided by the test duration. When samples completely disintegrate, the sample thickness divided by the spray time to disintegration is used as the erosion rate.

Earth concrete can be altered to reduce water erosion by adding binders such as cement and lime. Alternatively, the surface of earth concrete can be coated/treated with various materials, usually in liquid form of natural or synthetic origin. Examples of natural coatings are double-boiled linseed oil and cooked starch. Examples of synthetic coatings are acrylic resins and silicone resins. These coatings have water-repelling properties and inhibit the absorption of water into earth concrete surfaces [5].

3 Materials and Methods

3.1 Testing Methodology

Earth concrete, cement-stabilised earth concrete and low-strength cement concrete were tested for mechanical strength and durability. The effect of two surface treatment types, linseed oil and DIY brick sealer, on the durability of earth concrete and cement-stabilised earth concrete was also tested.

Mortar prisms (40 × 40 × 160 mm) were used to test flexural and compressive strength. Between nine and twelve prisms were prepared for each material. The strength was tested after 28 days. Durability in terms of water erodibility was tested. A pressurised spray test was used. Samples consisted of 40 mm thick slabs each with an area of 150 by 150 mm. Three samples of each combination of material type and surface treatment type were tested. The effect of surface treatment on low-strength cement concrete was not tested.

Table 2 shows the mix design used for the concrete types in the format of kilograms per cubic meters. The mix designs of the earth concrete and cement stabilized earth concrete were based on volume parts mix design and loose bulk density. The volume parts mix design of all the concrete types is included in Table 2. A water/cement ratio of 0.8 was used for the low-strength cement concrete. For the cement-stabilised earth concrete, ±4.5% cement by mass of the entire mix was used, which resulted in a very high water/cement ratio of 3.63.

Table 2. Mix design of concrete in conventional format

	Mix design (kg/m^3)		
	Earth concrete*	Cem. Stab. Earth concrete#	Cement concrete†
Water	288	344	300
Cement (CEM I 52.5 N)	NA	95	375
Earth	624	548	NA
Coarse silica sand	378	332	513
Fine silica sand	846	744	1026
Water/cement ratio	NA	3.63	0.8
% Cement of total mass	NA	4.59	16.9

Based on volume parts Water:Cement:Earth:Sand
*1.5:0:2:3 #1.5:0.5:2:3 †1.5:1.9:0:7.7

3.2 Materials

Earth from outside the "Engineering 4.0 Laboratory" at the University of Pretoria was used. It was a reddish brown and had a 23% clay content. According to the Unified Soil Classification System, the material has a CL, or lean clay, classification. The relative density is 2.66 and the particle size distribution of the earth is shown in Fig. 2.

Portland cement (CEM I 52.5 N), produced by Afrisam Ltd, with a relative density of 3.14 was used. Two silica sands with different particle size distributions were used. The relative density of silica sand is 2.65. The particle size distributions of the cement and the two silica sands are also shown in Fig. 2.

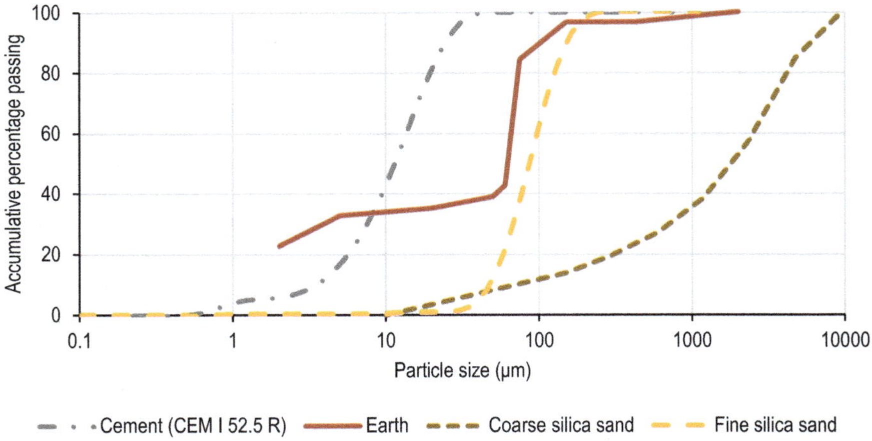

Fig. 2. Particle size distribution of materials

3.3 Mixing, Casting and Curing

All concrete types were mixed in a rotatable pan mixer. The dry constituents were mixed before water was introduced. After introducing the water constituents were mixed until all the materials were dispersed thoroughly.

To cast the earth concrete and cement stabilised earth concrete mortars, plywood moulds with no bottoms were used. The moulds were placed on a plywood board. The fresh mortar mixture was placed and compacted in three layers in the moulds. The moulds were then lifted upwards, leaving the formed samples (prisms and slabs) on the board. The mortar prisms of the cement concrete were cast in conventional steel moulds, while the water erodibility slabs were cast in plywood moulds with the required dimensions.

For earth concrete, and cement stabilised earth concrete, the samples must dry to gain strength. The samples were placed in a room with a relative humidity of 55% and a temperature of 25 °C. The samples were not covered. The cement concrete samples were demoulded after 24 h and then exposed to the same conditions as the other concrete types.

3.4 Testing of Strength and Water Erodibility

The flexural strength and compressive strength of the prisms were determined according to the procedure in SANS 50196-1:2006 Part 1: Determination of strength [21]. The 40 × 40 × 160 mm prisms were loaded at three points, 50 mm apart, until failure. A compression load was applied to each half of the 40 × 40 × 160 mm prism until failure. Between nine and twelve samples were tested after 28 days.

The spray test was used to determine the water erodibility. A custom test setup was designed and built. Figure 3(a) shows a side view of the test setup. Although the sample size was 150 × 150 × 40 mm, a screen was used to expose only a 100 mm diameter sample surface. Figure 3(b) shows the exposed sample surface and screen. The projection distance was 230 mm. A constant pressure pump was used to deliver a flow rate of 90 l/min at a pressure of 150 kPa. Figure 3(c) shows the showerhead with a diameter of 45 mm with 1.5 mm diameter holes. Three samples were tested concurrently.

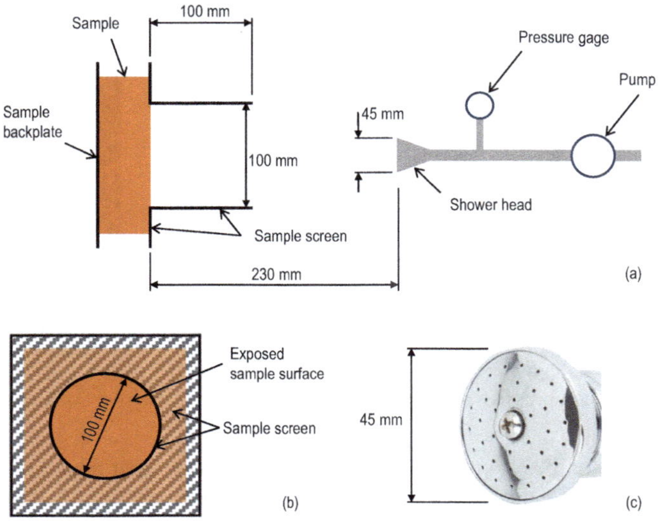

Fig. 3. Spray test setup (a) side view, (b) screen and exposed surface and (c) showerhead

To quantify water erodibility, the volume change of the samples subjected to the spray test was determined. A high-definition, multi-functional handheld 3D scanner was used [22]. The accuracy of the scanner is up to 0.04 mm, depending on the scanning mode used. Samples were scanned before and after being subjected to the spray test.

4 Results and Discussion

4.1 Mechanical Strength and Density

Figure 4 shows the compressive strength and density of the respective materials. The coefficient of variance is also shown. The compressive strength of the two respective earth concrete types was almost an order of magnitude smaller than that of the cement concrete. The cement stabilisation at 4.5% did not improve the strength of the earth concrete notably. In fact, the compressive strength of the cement-stabilised earth concrete was lower than that of the earth concrete. The coefficient of variance of the strength properties of both the earth concrete and cement stabilised earth concrete was high in comparison to that of the cement concrete. The flexural strength followed the same trend but is not shown.

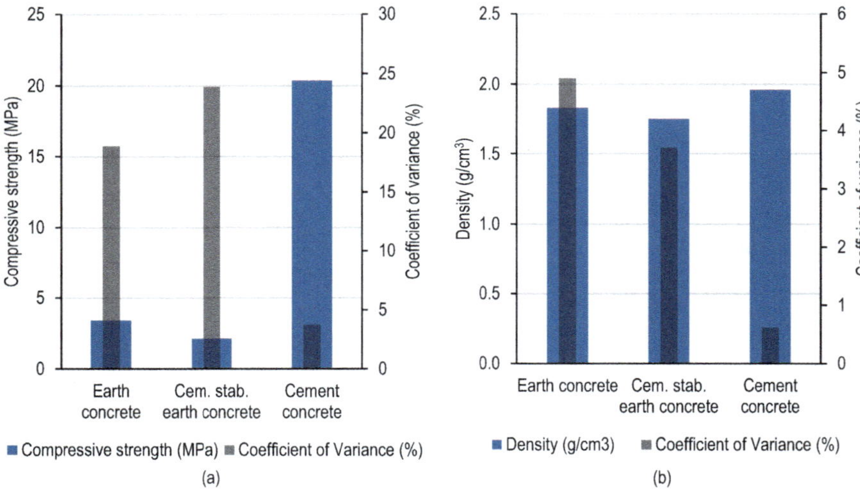

Fig. 4. Material (a) compressive strength and (b) density with respective coefficients of variance

The variability of the compressive strength results was greater than that of the density results. The average density of the earth concrete, and cement stabilised earth concrete were 1.83 and 1.75 g/cm^3, respectively, and lower than the density of the cement concrete of 1.96 g/cm^3. Note that the theoretical density of the earth concrete and cement stabilized earth concrete based on Table 4 were 2.17 and 2.09 g/cm^3.

4.2 Water Erodibility

The rate of percentage volume loss was used to quantify water erodibility. Table 3 shows results of all samples tested. Three samples were tested for each combination of concrete type and surface treatment. The average is not used because only three samples were tested and because of the variability of the results.

The raw earth concrete test was stopped after 1 min of water exposure. All other samples were exposed for 4 h. Cement stabilisation of earth concrete improved the water erosion properties. Surface treating earth concrete with linseed oil made its water erodibility performance comparable to that of cement-stabilised earth concrete. Surface treating earth concrete with brick sealer made the water erodibility performance comparable to cement concrete. The brick sealer on cement stabilised earth concrete seemed to have a less beneficial effect on performance.

4.3 Cost Analysis

Table 4 shows the cost per kilogram of material and the cost per coat of surface treatment. The cost of the cement and building sand is based on commercial prices. It is assumed that the earth material is free and does not account for the time cost of acquiring and processing it. Table 5 shows the cost of material surface treatment of a load-bearing wall that is 5 m long, 2.4 m high and 0.212 m thick. It is assumed that only the outer wall is surface-treated using three coats of liquid.

Table 3. Water erodibility as rate of percentage volume loss (%/h)

		Earth concrete			Cem. stab. earth concrete			Cement concrete
		Surface treatment						
		Raw*	Linseed oil	DIY brick sealer	Raw	Linseed oil	DIY brick sealer	Raw
Rate of volume loss (%/hr)	Samples 1	50.41	0.038	0.005	0.029	0.018	0.007	0.011
	2	38.29	0.036	0.002	0.264	0.013	0.015	0.002
	3	42.23	0.136	0.011	0.030	0.010	0.014	0.012
Legend		0.0	8.3	16.7	25.0	33.3	41.7	50.0

*Test stopped after 1 min

Table 4. Material and surface treatment cost

Material	Cost (R/kg)
Cement	0.92
Earth	NA
Building sand	0.55
Surface treatment	**Cost (R/m^2)**
Linseed oil	12.0
Brick sealer	15.4

Table 5. Material cost of $5 \times 2.4 \times 0.212$ m ($2.54\ m^3$) wall

			Wall cost	Legend
Earth concrete	Surface treatment	Raw	R1 713	R1 600
		Linseed oil	R2 145	R1 900
		DIY brick sealer	R2 267	R2 100
Cem. stab. earth concrete		Raw	R1 728	R2 400
		Linseed oil	R2 159	R2 700
		DIY brick sealer	R2 281	R3 000
Cement concrete		Raw	R3 031	R3 200

The cost of raw earth concrete and cement stabilised earth concrete is approximately half that of cement concrete. The additional cost of surface treating the two earth concrete types results in the cost of the earth concrete types being approximately 75% of that of the cement concrete.

4.4 Discussion

The strength of the earth concrete and cement stabilised earth concrete was lower and more variable than that of the cement concrete. Earth concrete is a more variable material than cement concrete, but sample preparation and curing could have contributed to the variability by influencing the sample geometry. Additionally, the strength test method used in this study was developed to test cement mortar prisms. Other strength test methods should be explored, such as the Brazilian disc test [23].

Cement stabilisation did not improve the strength of the earth concrete. This is likely because the cement dosage was lower than 5% by mass. It did, however, improve the water erosion resistance of earth concrete approximately 400 times in terms of the rate of percentage volume loss.

For the spray test, water was ejected at a pressure of 150 kPa over a distance of 230 mm before it hit the sample surface. The earth concrete samples disintegrated within one minute of exposure. Two papers from Table 1 also used pressures of 150 kPa [17, 19]. The poorest performing samples from these papers did not disintegrate completely over a period of more than 2 h, but the sample materials were modified using cement and/or coal ash and underwent compaction (applied pressure of 1 MPa or rammed). It can be argued that the spray test is more of an impact test, especially at a pressure of 150 kPa. It would be more appropriate to test the earth concrete using the drip test.

Applying a DIY brick sealer improved the resistance to water erosion of earth concrete to such an extent that it was comparable to that of cement concrete. The cost of raw earth concrete is approximately half of the cost of cement concrete, but surface treatment is essential for its resistance to water erosion. The additional cost of surface treatment results in relatively little savings in comparison to cement concrete.

In terms of which parts of UDDTs can be constructed using earth-based materials the following is observed regarding earth concrete with no surface treatment. The toilet shelter has the most potential to be constructed using earth concrete. For a wall height of 2.4 m, the compressive strength of 3.5 MPa can accommodate the self-weight of the wall. A more thorough structural analysis is, however, required and the mix design needs to be optimised for higher strength. Large roof overhangs should be used for protection against rain.

The vault that houses the urine tanks can be constructed from earth concrete. The appropriate precautions have to be put in place to protect the earth concrete walls from ground moisture, as would be done for most buildings. Whether the walls of the faeces drying chamber can be built using earth concrete is dependent on how fast the faeces dry. There will be a range of humidity that the earth concrete will be able to accommodate; however, if it can be avoided, the integrity of the UDDT structure should not depend on managing the humidity levels in the faeces drying chamber. The effect of surface treatments on water absorption should be explored. The user interface cannot be made from raw earth concrete and surface treatment has to be used.

5 Conclusion and Recommendations

The aim of this research was to investigate the suitability of earth-based concrete for application in the different parts of the UDDT structure. It was found that the strength of earth concrete and cement-stabilised earth concrete was almost an order of magnitude smaller than that of cement concrete. The resistance of earth concrete to water erosion was very poor, but if the earth concrete is stabilised with cement or surface treated it improves to being comparable to cement concrete. The cost analyses showed that the cost of earth concrete and cement stabilised earth concrete is almost half of the cost of cement concrete. When surface treatments are applied, the cost becomes comparable to that of cement concrete.

The use of earth concrete and traditional building methods could therefore be considered to reduce the cost of UDDT construction, which remains one of the most significant barriers to implementing UDDTs. However, care must be taken to ensure the durability of earth concrete by coating or stabilising the material to reduce the risk of structural degradation due to water erosion.

References

1. Mkhize N, Taylor M, Udert KM et al (2017) Urine diversion dry toilets in eThekwini municipality, South Africa: acceptance, use and maintenance through users' eyes. J Water Sanit Hyg Develop 7:111–120. https://doi.org/10.2166/washdev.2017.079

2. Millogo Y, Aubert JE, Séré AD et al (2016) Earth blocks stabilized by cow-dung. Mater Struct Materiaux et Constructions 49:4583–4594. https://doi.org/10.1617/s11527-016-0808-6
3. Gamrani N, Chaham KR, Ibnoussina M et al (2012) The particular "rammed earth" of the Saadian sugar refinery of Chichaoua (XVIth century, Morocco): mineralogical, chemical and mechanical characteristics. Environ Earth Sci 66:129–140. https://doi.org/10.1007/s12665-011-1214-6
4. Tilley E, Ulrich L, Luthi C, et al (2014) Compendium of sanitation systems and technologies, 2nd revised edition. Swiss Federal Institute of Aquatic Science and Technology (Eawag)
5. Minke G (2006) Building with earth: design and technology of sustainable architecture. Birkhauser, Basel
6. Soutsos M, Domone P (2018) Construction materials: their nature and behaviour. CRC Press, Boca Raton
7. Van Damme H, Houben H (2018) Earth concrete. Stabilization revisited. Cem Concr Res 114:90–102. https://doi.org/10.1016/j.cemconres.2017.02.035
8. O'Flaherty CA (1967) Highway engineering volume II, 2nd ed. Edward Arnold Ltd, London
9. Firoozi AA, Guney Olgun C, Firoozi AA, Baghini MS (2017) Fundamentals of soil stabilization. Int J Geo-Eng 8:9. https://doi.org/10.1186/s40703-017-0064-9
10. Ingles OG (1962) Bonding forces of soils part III: a theory of tensile strength for stabilized and naturally coherent soils. In: Proceedings of the 1st conference of the Australian road research board, pp 1025–1047
11. AfriSam Bricks and blocks. www.afrisam.com, last accessed 2025/09/09
12. Owen G (2009) Fulton's concrete technology, 9th ed. Cement & Concrete Institute, Midrand
13. Terry JP (1998) A rainsplash component analysis to define mechanisms of soil detachment and transportation. Aust J Soil Res 36:525–542. https://doi.org/10.1071/S97078
14. Heathcote KA (1995) Durability of earthwall buildings. Constr Build Mater 9:185–189. https://doi.org/10.1016/0950-0618(95)00035-E
15. Guettala A, Abibsi A, Houari H (2006) Durability study of stabilized earth concrete under both laboratory and climatic conditions exposure. Constr Build Mater 20:119–127. https://doi.org/10.1016/j.conbuildmat.2005.02.001
16. Kariyawasam KKGKD, Jayasinghe C (2016) Cement stabilized rammed earth as a sustainable construction material. Constr Build Mater 105:519–527. https://doi.org/10.1016/j.conbuildmat.2015.12.189
17. Ola SA, Mbata A (1990) Durability of soil-cement for building purposes: rain erosion resistance test. Constr Build Mater 4:182–187. https://doi.org/10.1016/0950-0618(90)90038-3
18. Venkatarama Reddy BV, Jagadish KS (1987) Spray erosion studies on pressed soil blocks. Build Environ 22:135–140. https://doi.org/10.1016/0360-1323(87)90033-3
19. Saranya RS, Sharma AK, Anand KB (2018) Performance appraisal of coal ash stabilized rammed earth. J Build Eng 18:51–57. https://doi.org/10.1016/j.jobe.2018.03.001
20. Stazi F, Nacci A, Tittarelli F et al (2016) An experimental study on earth plasters for earthen building protection: the effects of different admixtures and surface treatments. J Cult Herit 17:27–41. https://doi.org/10.1016/j.culher.2015.07.009
21. SANS 50196-1:2006 (2006) SANS 50196-1:2006 methods of testing cement part 1: determination of strength
22. SHINING 3D Technology (2025) EinScan Pro HD high definition, multi-functional handheld 3D scanner
23. Gaspar TAV (2017) The tensile behaviour of three unsaturated soils subjected to the Brazilian Disc Test. Master's dissertation, University of Pretoria

Dredged Sediment Reimagined: A Nature-Based Approach for Coastal Infrastructure

Marc Antoun[1(✉)], Jip Koster[1], Hugo Ekkelenkamp[1], Amine el Mahdi Safhi[2], Shima Pilehvar[2], and Mahdi Kioumarsi[3]

[1] NETICS B.V., Alblasserdam, Netherlands
marc@netics.nl
[2] Faculty of Engineering, Østfold University College, Halden, Norway
[3] Department of Built Environment, Oslo Metropolitan University, Oslo, Norway

Abstract. The coastal and riverine areas worldwide produce enormous volumes of dredged sediment (DS) every year. Meanwhile, traditional dike and shoreline reinforcement methods continue to rely on high-carbon, resource-intensive materials like basalt or conventional concrete. This paper presents a nature-based solution (NBS) that reimagines DS as a circular material for coastal infrastructure. The approach DS-based pavement blocks convert locally sourced DS into robust, ecologically beneficial revetment blocks, thus reducing primary raw-material consumption, transport emissions, and overall environmental impact. Through a multiphase research and demonstration project in Hansweert, Zeeland (the Netherlands), it shows how high-pressure compaction of sediment (via *GEOWALL®* technology) coupled with optimized binders delivers compressive strengths of 5–15 MPa. Key factors include locally appropriate sediment selection, sand or granulate additions, and reduced cement clinker contents (CEM/GGBS binder). Laboratory assessments verified freeze-thaw durability, abrasion resistance, and flexural strength. A 100 m^2 dike pilot confirmed performance under tidal and weather exposures. Early ecological monitoring indicates rapid colonization by algae, barnacles, and other organisms, underscoring the potential for synergy with existing ecosystems. A life cycle analysis shows a 30–50% drop in the Milieu Kosten Indicator compared to concrete blocks. Cost calculations, factoring in reduced disposal fees, suggest that local sediment-based blocks can be cost competitive with standard solutions if produced near the dredging site. This study positions DS-based pavement blocks as a practical, NBS approach to coastal protection, merging circular economy principles with robust engineering criteria for resilient, habitat-friendly infrastructure.

Keywords: Dredged sediments · Nature-based solutions · Circular economy · Dike reinforcement · Coastal infrastructure

H. Beushausen et al. (Eds.): NTZR 2025, LNCE 873, pp. 436–447, 2026.
https://doi.org/10.1007/978-3-032-27698-8_39

1 Introduction

1.1 Background and Motivation

Large portions of coastal and riverine regions rely on dredging to maintain navigation channels and harbors. This activity generates millions of cubic meters of dredged sediment (DS) annually [1, 2]. In many countries, much of this material is considered waste and is either stored in landfills or confined disposal facilities [3]. At the same time, infrastructure projects—particularly dike reinforcements—require substantial amounts of primary resources, such as quarried basalt, rockfill, or cement-based concrete, resulting in a markedly linear supply chain [4, 5].

Sustainability imperatives have sparked interest in nature-based solutions (NBS) that not only reduce carbon footprints but also integrate more seamlessly with local ecosystems. While some NBS rely on living elements (marshes, mangroves), semi-hard approaches can still be considered nature-based when they minimize negative ecological impacts and create conditions conducive to biodiversity [6–8]. The project introduced in this paper seeks to transform DS into a robust yet ecologically attuned pavement block that can protect coastal zones and riverbanks while reducing reliance on primary raw materials. Previous studies proved the feasibility of recycling DS in coastal infrastructure at the laboratory scale [9–12].

1.2 Objectives

The main objective of this research is to demonstrate that local DS can be reimagined as a nature-based building material for coastal infrastructure. Specifically, the paper aims to: develop a revetment block using high-pressure compaction of DS, requiring minimal cementitious binders; Verify mechanical properties through rigorous laboratory tests; Conduct a pilot installation on a dike slope in Hansweert, the Netherlands, to evaluate real-world performance under tidal, wave, and weather conditions; Assess the ecological benefits, including habitat creation, biodiversity, and minimal chemical leaching, aligning with NBS principles; Compare the life cycle impact (ECI score: environmental cost indicator) and financial cost of DS-based blocks with conventional revetment options. By engaging a circular economy perspective—linking dredging activities to civil construction needs—we aim to reduce transport emissions, curb cement usage, and ultimately limit the environmental toll of coastal defense works.

1.3 Project Site: Hansweert Dike in Zeeland

The Westerschelde estuary in the southwestern Netherlands experiences semidiurnal tides with ranges of 3–5 m. Hansweert is one of the sites selected for water safety enhancements due to insufficient dike crest height and aging revetments. Meanwhile, local water authorities regularly dredge channels to maintain navigability, storing the sediment in depots nearby (Fig. 1). The pilot location offers a synergy between sediment availability, short transport distances (~5 km), and the need for dike reinforcement.

Fig. 1. Aerial view of Hansweert, Zeeland.

1.4 Nature-Based Elements and Novelty

While some might equate NBS purely with soft or bio-based systems, this study argues that DS-based pavement blocks align with a more inclusive definition, incorporating the use of local, area-specific sediments that mirror the native mineral and organic composition; Habitat creation potential, as the resulting blocks' surfaces and interstices foster macroalgae, barnacles, and other marine life, similar to natural rock surfaces; Reduced resource footprint (less clinker, shorter transport, and no new quarrying), consistent with ecological design and circular economy goals.

The synergy of mechanical, chemical, and physical stabilization of DS at a pilot scale sets this research apart from earlier small-scale trials and conventional block-making. By arriving at TRL 6 (technology readiness level), we build confidence among coastal authorities, designers, and policymakers considering transitions toward nature-based or nature-inclusive solutions.

2 Materials and Methods

2.1 Research Framework

The overall project followed three phases:

- Phase 1—Laboratory research: which covers the physical, chemical, and mechanical characterization of DS. It also includes the production of test blocks for strength, durability, and limited ecological tests.
- Phase 2—Scale-up and feasibility: including large-scale press design and prototyping (*GEOWALL*®), preliminary revetment design for the pilot, and life cycle analysis (LCA) and cost modeling.
- Phase 3—Pilot installation and monitoring: including the production of full-sized blocks (~0.125 m^3 each), the installation on a 100 m^2 dike slope, and monitoring of physical stability and ecological colonization for +12 months.

2.2 Dredged Sediment Selection and Characterization

A disposal depot managed by the local water authority (*Waterschap Scheldestromen*) supplied the DS. The material consisted of brackish silt with moderate organic content. Five composite samples of ~50 kg each were taken from different sections/depths to capture potential variability.

Basic laboratory analyses included: grain size distribution (laser diffraction + sieve tests), organic matter (loss-on-ignition at 550 °C), salinity and chlorides (leachate analysis), plasticity (Atterberg limits), and contaminant screening (heavy metals, PAHs), as per Dutch AP04 guidelines for DS reuse.

The key results included a grain size predominantly silty sand (silt fraction ~40–50%), 3–5% of organic matter, a salinity of ~0.5–0.8 *wt.%* (mildly brackish), and chemical contamination below reuse thresholds, clearing the DS for beneficial use [3].

2.3 Stabilization Methodology

Converting wet, cohesive DS into load-bearing revetment blocks relied on three complementary strategies (Fig. 2).

- Physical stabilization–Minor sand or granulate addition (~10–15%) to optimize the silt–sand ratio for better compaction and reduced shrinkage.
- Chemical stabilization–Cement (CEM III/B) and ground granulated blast furnace slag (GGBS) to develop strength. The aim was to reduce clinker content as much as possible.
- Mechanical stabilization–High-pressure compaction (*GEOWALL®*) to minimize voids and trigger a dense, stable matrix.

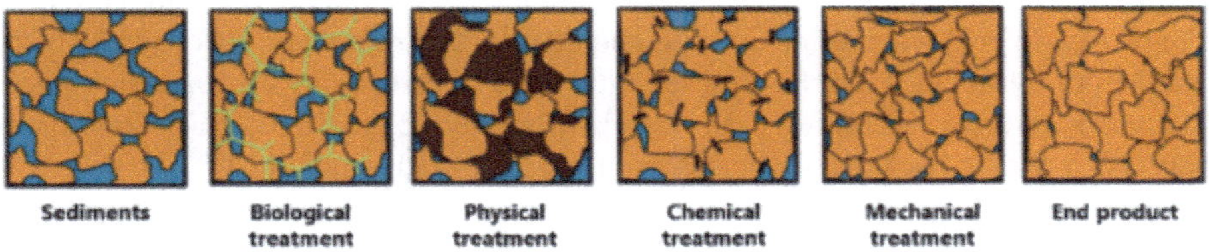

Fig. 2. Simplified stabilization concept integrating physical, chemical, and mechanical components.

Physical Stabilization. Preliminary laboratory trials established that a small fraction of coarse sand improved compaction outcomes. Else, recycled masonry or crushed concrete fines were tested in some recipes to enhance the circular economy aspect, but the final field blocks used locally sourced sand to ensure consistent quality.

Chemical Stabilization. Two binders were tested: a binder with high slag content for enhanced sulfate/chloride resistance (CEM III/B), and up to 50% GGBS replacement of cement to further reduce clinker usage. For the mixing, the DS, binder, and sand were blended into a horizontal pugmill for 2–3 min. Optimal water content ensured a workable, homogenous mix for compaction.

Mechanical Stabilization (*GEOWALL®*). Following pilot experience, a new 200-ton hydraulic press was used that shapes ~50 × 50 × 50 cm blocks under pressures of 5–10 MPa (Fig. 3). The synergy of high compaction and carefully chosen binders allowed for a significant drop in cement dosage compared to typical concrete. Pressed blocks remain dimensionally stable upon ejection, requiring 7–28 days of curing before final placement.

Fig. 3. Pilot-scale *GEOWALL®* press in operation.

2.4 Laboratory Testing

Strength Testing. Mechanical properties were tested. for compressive strength, cubic samples of 15 cm^3 were tested at 7, 14, and 28 days following a protocol adapted from NEN-EN 12390-3 [13]. For flexural strength, three-point bending on prismatic samples (40 × 10 × 10 cm) according to the standard NF EN 12390-5 [14]. Each formulation was tested with at least five replicates to ensure statistical reliability.

Durability Testing. For freeze-thaw, specimens underwent 28 cycles from +20 to −5 °C with partial water immersion, simulating natural winter exposure according to EN 12390-9 [15]. Strength was measured in pre- and post-cycling. For abrasion testing, a rotating cylinder test with a standard silica-based abrasive measured mass loss, reflecting potential wear from tidal-driven sediment transport. The testing was conducted according to EN 12390-9 [16].

Water Absorption. The testing was conducted in accordance with NF EN 772-21 [17]. The prepared blocks were submerged until water uptake stabilized, yielding a percentage by mass. This informs both potential leaching and the resilience of blocks over repeated wetting-drying cycles.

2.5 Field Pilot Setup

Dike Section and Loading Conditions. The test section in *Hansweert* occupies ~100 m^2 on a 1:4 dike slope partially below means high tide. With typical wave heights around 0.5–1.0 m (up to ~1.5 m in storms), the pilot experiences daily tidal flux and moderate wave action. The existing structure allowed the pilot to be placed atop an older basalt revetment layer, ensuring no compromise to dike safely even if any test blocks failed.

Block Production and Installation. A dedicated dredging factory near the sediment depot housed the *GEOWALL®* press. Once the mix design was finalized, ~250 large blocks (~0.125 m^3 each) were manufactured over three weeks, followed by 28 days of air-curing on-site. Afterwards, blocks were trucked ~5 km to the dike (Fig. 4).

The installation largely mirrored conventional revetment block placement: starting with the preparation of a gravel or sand bedding layer, followed by the interlocking the blocks in a running-bond pattern, then leaving small gaps unfilled to encourage natural infiltration of water and sediment.

Fig. 4. Placement of the blocks on the *Hansweert* dike pilot site

Monitoring Strategy. The project included monthly visual inspections to check for cracks, block displacement, or spalling. Strength sampling involved extracting cores or small cubes from select blocks at 3, 6, and 12 months for laboratory testing. Topographic scans were conducted before and after storm events to identify micro-level changes, while ecological observations involved taking periodic photographs—every three months—of algae, barnacles, and benthic organisms colonizing the blocks.

3 Results and Discussion

3.1 Laboratory Results

Mix Design Outcomes. Table 1 illustrates four representative mix designs (R_1–R_4). Cement + GGBS usage varied from 60 to 120 kg/m^3. Sand additions ranged 10–15% to optimize compaction. A fixed w/b ratio of 0.50 was used across all formulations in the initial recipe development phase.

Table 1. Selected mix compositions and 28-day compressive strength

Mix ID	Sediments		Sand		Cement	GGBS	28-d *fc*
Unit	(wt.%)	(kg/m^3)	(wt.%)	(kg/m^3)	(kg/m^3)	(kg/m^3)	(MPa)
R_1	85	748	10	88	120	0	7.5 ± 0.9
R_2	80	656	15	123	90	90	10.2 ± 1.1
R_3	83	706	12	102	80	70	8.8 ± 0.7
R_4	85	748	10	88	60	60	5.9 ± 0.9

The mixture R_2 offered the highest 28-day compressive strength (~10 MPa), while the mixture R_4 achieved ~6 MPa with the lowest cement usage. All used significantly less cement than typical concrete (~180–220 kg/m^3).

Early Versus Long-Term Strength. At 7 days, many samples reached over half their final 28-day strength. Mixes with more GGBS (R_2, R_3) had slower early strength gain but benefited from improved longer-term hydration and sulfate resistance, especially in brackish environments. None of the mixes showed detrimental effects from the salinity present in the DS.

Durability Testing. For freeze-thaw, strength reduction was modest (0–10%) over 28 cycles. R_2 displayed the least scaling. For the abrasion test, mass losses stayed under 2% for all mixes, signifying suitable performance for a tidal environment with potential suspended sediments.

3.2 Field Pilot Observations

Placement and Initial Performance. Blocks were installed using standard revetment-laying equipment. The interlocking pattern functioned similarly to basalt or concrete blocks. After 6 months in situ, no block displacements or corner spallings were detected. Visual inspections post-moderate storms (wave heights ~1 m) revealed only superficial weathering—an expected result of normal in-service conditions (Fig. 5).

In Situ Strength Verification. Core samples from pilot blocks at 3, 6, and 12-month affirmed lab results: compressive strength increased slightly over time, topping out near 11 MPa (for R_2). The slight disparity with lab data was attributed to real-site curing conditions—possibly slower but ultimately yielding similar or slightly higher final strengths.

Ecological Colonization. A prominent benefit of a nature-based approach is the creation of habitat niches. At 6 and 12 months, the submerged surfaces of the blocks showed macroalgae (particularly *Fucus vesiculosus* and *Ulva* species) covering as much as 70–80% of the area. Barnacles (*Balanus improvisus*) and small gastropods took advantage of the gaps between blocks, indicating favorable conditions for invertebrate colonization. Additionally, chemical leaching appeared negligible, as suggested by a stable local pH and the absence of discoloration or surface film.

These findings suggest the blocks mesh well with existing brackish ecology, paralleling natural rock surfaces and thus reflecting the NBS component. If future designs incorporate additional texturing or small crevices, the potential for biodiversity might be enhanced [18].

Fig. 5. Blocks on the slope, partially submerged at mid tide.

3.3 Life Cycle and Cost Analyses

Environmental Impact (ECI Score). An LCA was carried out to compare (1) standard basalt blocks or conventional concrete revetment elements with (2) DS-based pavement blocks, which incorporate local DS, partial sand, and a reduced cement content. The system boundaries included raw material extraction (or dredging), transport, production, installation, and potential end-of-life scenarios (landfill or reuse via crushing). The resulting ECI score was 30–50% lower for the DS-based blocks, largely due to reduced cement clinker consumption, elimination of basalt or gravel quarrying, short transport distances (~5 km from depot to site), and avoided disposal fees for DS.

Cost Comparison. Major cost drivers included binder usage (cement/GGBS), compaction energy, labor, and transport. When factoring in disposal fees saved (~€15–30 per ton of DS), the net cost of producing DS-based pavement blocks at scale (≥20–30 m^3/day) approached or slightly undercut typical revetment block prices. Sensitivity analyses indicated that transport distances above ~30 km could erode the economic advantage, emphasizing the local nature-based approach.

Based on the detailed cost breakdown (Fig. 6), the production cost of one DS-based pavement block measuring 50 cm^3 amounts to ~€153. In contrast, a conventional block element of the same size ranges between €50 and €80. This significant price gap clearly indicates that, under current conditions, blocks made from DS are not yet economically competitive with traditional concrete alternatives.

The high cost is primarily driven by production-related expenses, including equipment investment (€43), location rental (€60), and personnel costs (€15), which together account for over 85% of the total cost. These fixed and operational costs are difficult to offset unless economies of scale are achieved, or production processes are optimized.

Despite the higher cost, DS-based blocks may offer added environmental and circular economy benefits. However, from a purely economic standpoint, further innovation and scaling are required before such materials can viably compete with established concrete products in the market.

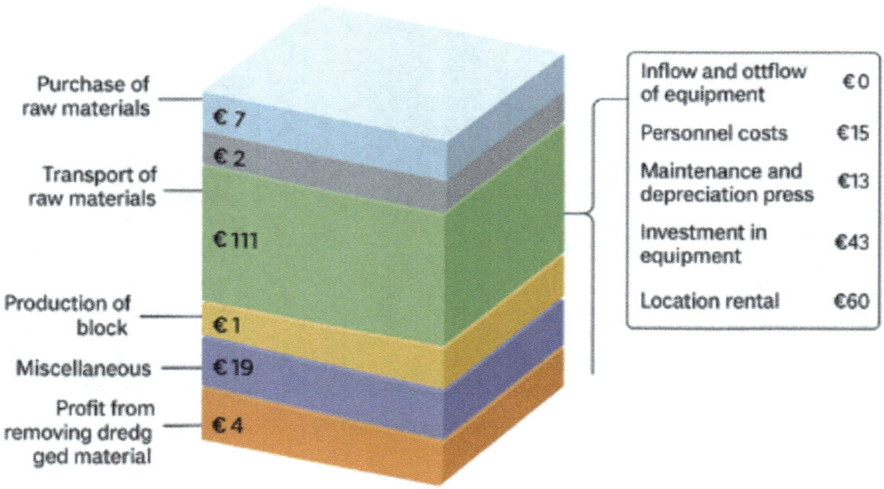

Fig. 6. Cost breakdown of dredged material block.

3.4 Technical Discussion and Challenges

The first consideration is organic content: if future sites produce DS with higher organic levels or very fine-grained material, additional pre-treatment (such as dewatering) or an increased binder dosage might be necessary. Another factor is salt/chloride effects; although mild brackish conditions posed minimal issues here, extremely high salinity may require specialized cement types or a fresh water rinsing step.

In terms of scalability, the production rate of about 30 blocks per day is sufficient for demonstration purposes but may be too low for extensive dike reinforcements; industrial-scale deployment would likely demand automated lines or partnerships with precast factories. Regarding regulatory frameworks, the *Dutch Besluit Bodemkwaliteit* (*Soil Quality Decree*) applies to the beneficial use of DSs, and any design must accommodate consistent sampling and QA/QC (Quality assurance/Quality control) procedures, given the variable contamination profiles of each dredge site. Lastly, to enhance the nature inclusiveness of such solutions, additional block shaping or surface texturing can boost biodiversity, by merging strong engineering performance with ecological benefits in line with NBS goals.

Overall, the successful pilot indicates that a synergy of chemical, physical, and mechanical stabilization can yield a durable, lower-impact revetment. With thorough site-specific characterization and engineering design, DS can seamlessly integrate into coastal infrastructure.

4 Conclusions

This paper outlines how DS-based pavement blocks represent an NBS by transforming DS into a practical revetment block for coastal infrastructure. This multi-phase research concludes:

- Technical feasibility: By combining partial sand addition, blended binders, and high-pressure compaction (GEOWALL®), this study achieved compressive strengths of 5–15 MPa. Durability tests (freeze-thaw, abrasion) and in situ pilot monitoring confirmed robust performance under tidal conditions.
- Nature-inclusive benefits: Rapid colonization by local algae and invertebrates underscores the ecological compatibility of the blocks. DS-based construction arguably maintains region-specific mineral profiles that promote biodiversity.
- Environmental advantages: A LCA yielded a 30–50% lower ECI score versus conventional concrete revetment, driven by reduced cement use, shorter transport, and avoided sediment disposal.
- Cost competitiveness: When large volumes of DS are available near the site, the approach can be cost-competitive or cheaper than standard blocks, especially if disposal fees are factored in.
- Outlook: Expanding beyond the 100 m^2 pilot toward full-scale dike reinforcements requires further automation, standardized QA/QC, and proactive ecological design features. Nonetheless, this pilot demonstrates the potential for bridging dredging and civil works in a circular, nature-based paradigm.

Acknowledgments. The gratitude is extended to *Waterschap Scheldestromen*, the *High-Water Protection Programme* (HWBP), *NETICS* B.V., and *Deltares* for funding and facilitating this research. The authors also appreciate the lab and field teams at *Hansweert* for their unwavering commitment, as well as the constructive feedback from local stakeholders and scientific peers on integrating ecological components into this approach. Safhi, Pilehvar, and Kioumarsi acknowledge receiving funding via the EU-H2023 MSCA (101155413, UP-RISE). The authors' views and opinions are their own and do not represent those of the European Union or its affiliated institutions.

References

1. Safhi AM, Mejjad N, El-FadilI H, Bortali M (2024) Dredged materials in Morocco: current practices, policies, and roadmap for sustainable management. Case Stud Construct Mater 20:e03045. https://doi.org/10.1016/j.cscm.2024.e03045
2. Safhi AEM, Rivard P, Benzerzour M, Abriak N-E (2021) Artificial rocks made from dredged sands of the Magdalen Islands (Canada): preliminary study. In: Escalante-Garcia JI, Castro-Borges P, Duran-Herrera A (eds) Proceedings of the 75th RILEM annual week 2021. Springer, Cham, pp 994–1003
3. Ekkelenkamp HHM, Pieterse JJ, Koster J, Van Doornik T (2023) Eindrapportage Zetsteen van Baggerspecie: Onderzoek En Ontwikkeling van Een pro-Efvak Met Zetsteen Gemaakt van Baggerspecie. NETICS, p 98
4. Alvarez-Guerra M et al (2008) Sustainable management options and beneficial uses for contaminated sediments and dredged material. Fresen Environ Bull 17:1539–1553
5. Saathoff F, Cantré S, Sikora Z (2015) South Baltic guideline for the application of dredged materials, coal combustion products and geosynthetics in dike construction: [project] DredgDikes. Universität Rostock, Univ., Agrar- und Umweltwiss. Fakultät [u.a.]: Rostock
6. European Commission (2015) Directorate general for research and innovation. Towards an EU research and innovation policy agenda for nature-based solutions and re-naturing cities: final report of the horizon 2020 expert group on "nature based solutions and re naturing cities". Publications Office: LU
7. Robotham J et al (2023) Nature-based solutions enhance sediment and nutrient storage in an agricultural lowland catchment. Earth Surf Process Landforms 48:243–258. https://doi.org/10.1002/esp.5483
8. Almström B, Danielsson P, Göransson G, Hallin C, Larson M (2022) Experiences of nature-based solutions for mitigating ship-induced erosion in confined coastal waters. Ecol Eng 180:106662. https://doi.org/10.1016/j.ecoleng.2022.106662
9. Dubois V, Abriak NE, Zentar R, Ballivy G (2009) The use of marine sediments as a pavement base material. Waste Manag 29:774–782. https://doi.org/10.1016/j.wasman.2008.05.004
10. Limeira J, Agullo L, Etxeberria M (2010) Dredged marine sand in concrete: an experimental section of a harbor pavement. Constr Build Mater 24:863–870. https://doi.org/10.1016/j.conbuildmat.2009.12.011
11. El-Shinawi A, Kramarenko V (2015) Assessment of stabilized dredged sediments using portland cement for geotechnical engineering applications along Hurghada Coast, Red Sea, Egypt. Asian J Appl Sci 3:562
12. Parson LE, Swafford R (2012) Beneficial use of sediments from dredging activities in the Gulf of Mexico. J Coast Res 60:45–50. https://doi.org/10.2112/SI_60_5
13. NEN-EN 12390-3 (2019) Testing hardened concrete—part 3: compressive strength of test specimens
14. AFNOR NF EN 12390-5 (2019) Testing hardened concrete—part 5 : flexural strength of test specimens
15. AFNOR EN 12390-9 (2020) Testing hardened concrete—part 9: freeze-thaw resistance with deicing salts—scaling
16. AFNOR NF EN 13892-3 (2015) methods of test for screed materials—part 3: determination of wear resistance—Böhme
17. AFNOR NF EN 772-21 (2011) Methods of test for masonry units—part 21: determination of water absorption of clay and calcium silicate masonry units by cold water absorption
18. Meijer M, van Grunsven R, Meininger P, Persijn A, Meijer AJM, van Grunsven R, Meininger PL, Persijn A (2011) Planten en wiergemeenschappen op de westerscheldedijken. Projectbureau Zeeweringen, Bureau Waardenburg Rapport Nr. 11-037. Middelburg/Culemborg

Sustainable Sediment Management Strategies: Comparative Lessons from Norwegian and Baltic Ports

Amine el Mahdi Safhi[1(✉)], Solveig Kristiansen[2], Shima Pilehvar[1], and Mahdi Kioumarsi[2]

[1] Faculty of Engineering, Østfold University College, Halden, Norway
aesafhi@hiof.no
[2] Department of Built Environment, OsloMet—Oslo Metropolitan University, Oslo, Norway

Abstract. Contaminated sediments present significant environmental and operational challenges for ports, particularly in regions with long industrial legacies like Norway and the Baltic Sea. This research explores sustainable approaches to managing these sediments, specifically focusing on three major Norwegian ports: Oslo, Borg, and Trondheim. The findings are compared to similar practices employed in neighboring countries, including Sweden, Finland, and Denmark. Norway has adopted a proactive stance by combining stabilization and solidification (STSO), sediment washing, and environmentally controlled dredging to reduce contamination and promote reuse. Technical reports provide quantitative data to support the feasibility of reusing treated sediments for port infrastructure. In contrast, neighboring Baltic countries prioritize containment and risk mitigation through capping or confined disposal, often limited by stricter regulatory interpretations. This comprehensive analysis examines the technological prowess, ecological security, and regulatory landscape of these approaches. It is based on more than a decade of field trials and environmental studies. It proposes solutions for aligning regulatory frameworks, fostering innovation, and expanding the scale of sediment valorization.

Keywords: Contaminated sediments · Stabilization/solidification · Sediment washing · Environmental dredging · Circular economy · Sustainability

1 Introduction

Ports globally serve crucial roles in trade and transportation but frequently accumulate various contaminants from industrial operations, shipping activities, urban drainage, and accidental spills [1, 2]. Such contaminants often include heavy metals, hydrocarbons, pesticides, and persistent organic pollutants (POPs), including polycyclic aromatic hydrocarbons (PAHs), polychlorinated biphenyls (PCBs), and tributyltin (TBT), known for their persistence and toxicity [3], posing serious environmental threats to aquatic ecosystems and public health [4]. Traditional sediment management methods, including ocean disposal and landfill storage, face increasing scrutiny due to their adverse environmental impacts, such as marine ecosystem degradation, groundwater pollution, and reduced land availability [5–7].

H. Beushausen et al. (Eds.): NTZR 2025, LNCE 873, pp. 448–455, 2026.
https://doi.org/10.1007/978-3-032-27698-8_40

For a long time, dredged sediments were considered waste and were either sent to landfills or confined to aquatic disposal. However, due to the implementation of stricter environmental regulations, such as the EU Water Framework Directive, the need for circular economy solutions, and the growing volume of sediments, there is now a growing interest in sustainable alternatives. In this context, the Baltic countries have pioneered approaches that combine remediation with resource recovery, particularly through stabilization/solidification (STSO) techniques (Fig. 1).

Norway, renowned for its rigorous environmental standards and proactive sustainability approaches, offers valuable insights into advanced sediment management techniques. Ports such as Oslo, Borg, and Trondheim demonstrate various approaches to addressing contamination challenges using innovative technologies, providing viable solutions for sediment remediation and valorization.

This report presents three case studies from Norway, comparing them to practices in Nordic countries. This research synthesizes these experiences, providing valuable knowledge for the worldwide improvement of sediment management practices.

2 Methodology

The study includes a comprehensive review and analysis of various sources, such as government regulations, detailed reports by Norwegian port authorities, in-depth studies by renowned research institutions (NGI, Multiconsult), and academic initiatives like earthresQue. The methodology combines both qualitative and quantitative assessments. It examines detailed sediment composition, contaminant levels, technological efficiencies and environmental monitoring data. By employing this comprehensive analytical system, it offers valuable perspectives on the appropriateness and impact of distinct approaches to managing sediments.

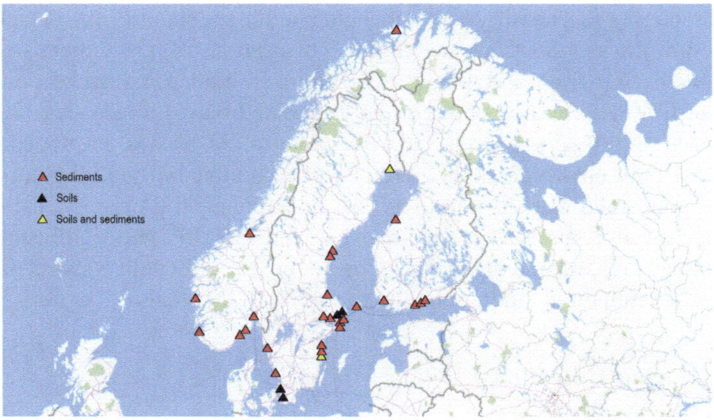

Fig. 1. Map of the Nordic region showing locations of sampling test points for STSO projects in the Nordic countries: Norway, Sweden and Finland [8].

3 Case Studies

In this section, three case studies are presented as summarized in Table 1.

Table 1. Summary of case study sites and contaminants.

Port	Volume dredged (m^3)	Contaminant types	Risk class	Dominant grain size	Historical cause
Oslo	180,000	Heavy metals, PAHs, PCBs	IV–V	Fine-grained	Industrial runoff, shipping
Borg	3,300,000	TBT, PAHs, PCBs	III–IV	Sand & silt (20–60%)	Port activities
Trondheim	Not specified	Chromium	IV	Heterogeneous	Historical industrial waste

3.1 Oslo Harbor Case Study

The Kongshavn Bay area of Oslo was subjected to significant rehabilitation efforts, paving the way for upcoming harbor expansion projects. About 180,000 m^3 of sediment was dredged, of which 105,000 m3 were classified as contaminated (Classes IV—V). For decades, the Oslo Harbor suffered from severe sediment contamination due to industrial activity and urban runoff. Oslo Harbor historically faced significant sediment contamination due to extensive industrial activities, leading to an estimated 0.50 to 0.65 mm^3 of contaminated sediments [8]. This issue was addressed by choosing an STSO technique that has been proven effective at immobilizing contaminants and stabilizing sediment. This approach aligns with findings from field studies that confirm the effectiveness of STSO for reducing leachability and enhancing sediment stability [9–11]. A thorough examination of the sediment revealed high concentrations of heavy metals and organic contaminants in fine-grained sediment. An optimized binder mixture of ordinary Portland cement (30%) combined with ground granulated blast furnace slag (70%) with a binder content of 120 kg/m^3 and a water-to-binder ratio of 5 was developed. This mixture achieved both environmental safety and structural stability. Laboratory and field evaluations confirmed marked improvements in mechanical properties, including substantial increases in compressive strength exceeding 1500 kPa and significant decreases in permeability. This method made it possible to reuse sediment in construction projects in a safe and compliant manner, in line with sustainability goals.

3.2 Borg Harbor Case Study

Borg Harbor was significantly contaminated, primarily by persistent organic pollutants (PAHs, PCBs, TBT), requiring a targeted management approach. Approximately 3.3 million cubic meters of sediment required remediation, with a significant portion classified as heavily contaminated. Sediment washing, which targets sediments containing 20%

to 60% sand and silt, emerged as the chosen remediation technology due to their heterogeneous composition. Sediment washing is particularly effective when applied to sand-rich sediments, where selective fractionation enables partial reuse [12–14]. This process involved sequential mechanical separation techniques that efficiently isolated cleaner sand fractions suitable for reuse. This significantly reduced the disposal volume of heavily contaminated sediments. Continuous environmental surveillance during the cleaning procedure revealed negligible pollution emissions, validating their security and effectiveness. The successful use of the cleaned sediment fractions in construction projects has further demonstrated the practicality and sustainability of the approach.

3.3 Trondheim Harbor Case Study

The port of Trondheim was faced with uncommon pollution issues, mainly due to chromium from historical industrial activities. To resolve this issue, a combination of mechanical dredging with tailored chemical and physical remediation methods was used. Detailed sediment characterization identified significantly elevated levels of chromium, which required a precise treatment strategy to reduce its bioavailability and environmental mobility. Following implementation of this technique, follow-up testing revealed significant decreases in chromium content, enabling safe disposal or possible reuse.

3.4 Baltic Region Comparison

The approach to managing sediment in Sweden, Finland, and Denmark is more risk-averse and focused on containment, in contrast to Norway's strategy, which prioritizes reuse and valorization (Table 2). The ports of Gothenburg and Oskarshamn in Sweden have given top priority to dredging and capping to remediate sediments contaminated with polychlorinated biphenyls (PCBs) and mercury. The capping process typically involves the use of geotextile membranes and clean sand or gravel layers to physically isolate the contaminants and prevent their dispersal. The Swedish environmental authorities prioritize long-term storage and environmental monitoring, with limited involvement in sediment reuse. Due to public concerns about the ecological integrity of treated sediments, the use of land-based disposal or reuse strategies has been hampered. Although pilot studies on sediment dewatering and stabilization have been carried out, they have not yet advanced to a commercial level.

In Finland, major ports, such as Turku and Kotka, primarily use confined aquatic disposal to manage contaminated sediments. In lower-risk areas, natural recovery processes supported by routine environmental monitoring can sometimes be used. However, reuse remains rare because of the combination of strict regulations and public opposition, especially concerning perceived health and environmental risks. Consequently, the Finnish model strongly prioritizes risk mitigation over the appreciation of sediment, regardless of its composition.

Denmark imposes one of the most stringent sediment reuse regulations among Baltic Sea countries. Reuse is generally allowed only for sediments classified as Class I or II, which are considered to have minimal or no contamination. In any case, the sediments must be used exclusively in non-sensitive areas, such as offshore backfill or road embankments. To allow for reuse in Denmark, there is a thorough requirement to test not only the overall concentration of contaminants, but also their ability to seep out, or "leach", as part of an exhaustive pre-approval process. The national policy is very similar to that of the European Union Water Framework Directive, which requires the protection of water bodies against chemical degradation. This strict regulatory interpretation has created high barriers for circular approaches to sediment management and leaves minimal room for experimentation or reuse of moderately contaminated materials.

Table 2. Comparative overview of regulatory frameworks.

Country	Regulation stringency	Reuse allowance	Public perception barrier	Monitoring emphasis	Leaching tests required
Norway	Moderate	Yes (Class IV–V)	Moderate	Yes	Yes
Sweden	High	Rare (pilot-scale only)	High	Strong	Yes
Finland	High	No	Very high	Strong	Yes
Denmark	Very high	Only class I–II	High	Very strong	Extensive

4 Discussions

The study underscores the adaptability and resilience of the sediment management strategies implemented by Norwegian ports. Each instance highlights the crucial need for customized approaches that consider detailed sediment analysis and contaminant profiles. Oslo Harbor's STSO methodology exemplifies the effectiveness of contaminant immobilization, enabling the reuse of sediment in construction applications and demonstrating significant potential for a circular economy. The sediment washing technique used by Borg Harbor demonstrates efficient and safe methods for resource recovery. It effectively reduces the volume of contaminated sediment and promotes reuse.

Trondheim Harbor's focus on chromium remediation highlights the need for targeted techniques to manage site-specific contaminants effectively. In all cases, rigorous environmental monitoring and adaptive management were essential for validating the methods' environmental compliance and sustainability. These results highlight the critical importance of integrated sediment management frameworks, which combine technical expertise, regulatory compliance, and sustainability considerations.

These cases also demonstrate significant potential for further research and innovation in sediment treatment technologies. Future studies could focus on optimizing resource recovery, minimizing environmental impact further, and developing cost-effective, scalable, and globally applicable sediment remediation solutions.

Norway's recent efforts at sediment remediation illustrate a pragmatic shift toward remediation and reuse. The STSO approach in Oslo and sediment washing in Borg both demonstrate the technical and environmental viability of valorization when conditions are favorable, particularly with moderately contaminated sediments containing recoverable granular material.

Numerous obstacles hinder widespread adoption, such as sediment variability complicating standardized treatment design, high initial costs for treatment facilities (such as washing facilities and binder materials), regulatory uncertainty regarding reuse thresholds and long-term liability, and public perception as a barrier to deploying treated sediments in urban areas. Conversely, Sweden, Finland, and Denmark tend to favor precautionary management, which often involves avoiding reuse altogether. This ensures low risk, but misses out on opportunities for a circular economy, such as reducing the need for virgin aggregates or minimizing landfill volumes. Reusing sediments aligns with circular economy principles and offers measurable ecological benefits in terms of resource savings and reduced environmental burdens [1, 15].

5 Conclusions and Recommendations

The analysis highlights the success of Norwegian ports in implementing a variety of sustainable sediment management methods. These experiences have the potential to be widely adopted on a global scale, promoting the implementation of circular economy principles in sediment management practices. STSO, sediment washing, and targeted mechanical dredging all have their own unique advantages, which depend on the specific site conditions and contamination profile. This highlights the need for adaptable, customized management approaches. The Norwegian examples provide important reference points for global sediment management, promoting sustainable environmental protection and resource use.

To improve sediment management practices in the Baltic Sea, we recommend the following:

- Create unified guidelines for recycling at the HELCOM/EU level, considering both the contaminant's ability to leach out and its mechanical properties.
- Expand regional pilot programs that experiment with comprehensive sediment washing and STSO under diverse sediment types.
- Encourage innovation in sediment reuse by providing funding for green infrastructure and circular economy initiatives.
- Establish regional treatment centers that have the capability to handle dredged materials from multiple ports.
- Involve communities in reuse initiatives to foster trust through openness and environmental surveillance.

Acknowledgement. This study was funded via the EU-H2023 MSCA (101155413, UP-RISE). The authors' views and opinions are their own and do not represent those of the European Union or its affiliated institutions.

References

1. Safhi AM, Mejjad N, El Fadill H, Bortali M (2024) Dredged materials in morocco: current practices, policies, and roadmap for sustainable management. Case Studies Construct Mat 20:e03045. https://doi.org/10.1016/j.cscm.2024.e03045
2. Birafane M, Liu W, Khalikov S (2020) The strategic positioning of Moroccan seaports: an application of the Boston consulting group growth-share matrix. Open Transportat J 14. https://doi.org/10.2174/1874447802014010133
3. Eggleton J, Thomas KV (2004) A review of factors affecting the release and bioavailability of contaminants during sediment disturbance events. Environ Int 30:973–980. https://doi.org/10.1016/j.envint.2004.03.001
4. Mejjad N, Safhi AEM, Laissaoui A (2025) Insightful analytical review of potential impacts of microplastic pollution on coastal and marine ecosystem services. J Hazardous Mater Adv 17:100578. https://doi.org/10.1016/j.hazadv.2024.100578
5. Safhi AEM (ed.) (2022) CRC press: Boca Raton, 2022 valorization of dredged sediments as sustainable construction resources. ISBN 978-1-003-31555-1
6. Bouzar B, Mamindy-Pajany Y, Hurel C (2023) Innovative reuse of fly ashes for treatment of a contaminated river sediment: synthesis of layered double hydroxides (LDH) and chemical performance assessments. Waste Biomass Valor. https://doi.org/10.1007/s12649-023-02056-0
7. Dorleon G, Rigaud S, Techer I (2024) Management of dredged marine sediments in southern France: main keys to large-scale beneficial re-use. Environ Sci Pollut Res. https://doi.org/10.1007/s11356-024-33129-9
8. Lindh P, Lemenkova P (2023) Effects of water—Binder ratio on strength and seismic behavior of stabilized soil from Kongshavn. Port Oslo Sustain 15:12016. https://doi.org/10.3390/su151512016
9. Moghrabi I, Ranaivomanana H, Bendahmane F, Amiri O, Deneele D (2023) Solidification study of an organic fine sediment using geopolymer binders properties. Int J Geomech 23:04023186. https://doi.org/10.1061/IJGNAI.GMENG-8187
10. Lang L, Chen B (2022) Stabilization of dredged sediment using activated binary cement incorporating nanoparticles. J Mater Civ Eng 34:04021381. https://doi.org/10.1061/(ASCE)MT.1943-5533.0004017
11. Todaro F et al (2023) Recycling of contaminated marine sediment and industrial by-products through combined stabilization/solidification and granulation treatment. Materials 16:2399. https://doi.org/10.3390/ma16062399
12. Mulligan CN, Yong RN, Gibbs BF (2001) An evaluation of technologies for the heavy metal remediation of dredged sediments. J Hazard Mater 85:145–163. https://doi.org/10.1016/S0304-3894(01)00226-6
13. Doni S et al (2018) Combination of sediment washing and bioactivators as a potential strategy for dredged marine sediment recovery. Ecol Eng 125:26–37. https://doi.org/10.1016/j.ecoleng.2018.10.009
14. Lumia L, Giustra MG, Viviani G, Di Bella G (2020) Washing batch test of contaminated sediment: the case of Augusta Bay (SR, Italy). Appl Sci 10:473. https://doi.org/10.3390/app10020473
15. Crocetti P, González-Camejo J, Li K, Foglia A, Eusebi AL, Fatone F (2022) An overview of operations and processes for circular management of dredged sediments. Waste Manage 146:20–35. https://doi.org/10.1016/j.wasman.2022.04.040

Novel Geotechnical Seismic Isolation System Using Recycled Rubber-Sand Mixture

Subhadeep Banerjee(✉), J. S. Dhanya, and A. Boominathan

Indian Institute of Technology Madras, Chennai 600036, India
subhadeep@civil.iitm.com

Abstract. Usages of natural and improvised geo-materials are getting popular every day. For example, cement-admixed with natural soil is well-known for its ability to gain in strength with a relatively low setting time along with its workability characteristics. Compared to other ground improvement methods, cement treatment has many advantages such as rapid strength increase, higher shear strength, reduction in settlement, suitable for inland, onshore and offshore environment. However, the usage of cement always increases the carbon footprint. More recently recycling of scrap tires is necessitated to ease out the stockpiling. Studies have been carried out to use shredded scrap tire mixed with sand for the applications such as retaining wall backfill, embankment material, buried pipeline backfill, and base isolation to ensure the bulk usage of scrap tires. Geotechnical seismic isolation (GSI) systems, utilizing geomaterials, have recently emerged as an innovative and low-cost solution for protecting structures from earthquake damage. The GSI systems typically incorporate a specialized energy-dissipating layer consisting of sand-rubber mixture (SRM), which provides high damping and reduced stiffness, effectively dissipating energy beneath building foundations. A series of laboratory-model tests were conducted in a test tank setup to investigate the response of a scaled model footing resting on GSI system subjected to cyclic loading. The results were analyzed in terms of settlement and bearing pressure to evaluate the effectiveness of the footing on the GSI system. Further to substantiate the experimental observations, a three-dimensional finite-element analysis of a model footing resting on a GSI bed is presented.

Keywords: Rubber-sand mixture · Seismic isolation · Recycle material

1 Introduction

Over the past few decades, much emphasis is given on the construction of earthquake-resistant buildings based on structural control concepts. Aseismic design of structures was achieved through increased strength and stiffness of structural elements to withstand earthquake-generated inertial forces on the building. Though the strengthening method leads to safeguarding the structure from collapse at, the functionality of the structure after an earthquake cannot be guaranteed due to excessive damages. An alternative strategy is to decouple the structure from the footing to minimize the intensity of the strong earthquake vibrations transmitted into the structure through base isolation techniques.

H. Beushausen et al. (Eds.): NTZR 2025, LNCE 873, pp. 456–463, 2026.
https://doi.org/10.1007/978-3-032-27698-8_41

Seismic isolation systems reduce the magnitude of energy transferred to the structure by shifting the natural period of the structure such that the peak earthquake spectral acceleration value zone is crossed.

In contrast, the Geotechnical Seismic Isolation (GSI) system utilizes geomaterials like rubber, geosynthetics, sand, and gravel to create a flexible and durable barrier between the building foundation and the ground. This approach provides a cost-efficient and environmentally sustainable option compared to conventional isolation systems [1–6]. A widespread approach for geotechnical seismic isolation involves the utilization of Sand-Rubber Mixture (SRM), which consists of sand mixed with rubber particles from crushed recycled tyres. SRM offers a blend of low stiffness, high flexibility, and damping properties that effectively dissipate the energy of seismic waves below the foundation level [7]. One of the major steps of proposing alternative materials is comprehensive life cycle assessment [8, 9]. There is currently no study reported on life cycle assessment of SRM materials.

In general, SRM exhibits a higher degree of compressibility compared to pure sand due to the presence of elastic and deformable rubber particles. This results in immediate settlement like that of sand during static loading. However, the settlement response under repeated loading needs to be thoroughly examined when SRM has been utilized as a GSI layer below foundations when considering building safety under seismic and other ground-borne vibrations. Several researchers have conducted extensive laboratory studies to investigate the cyclic behavior of SRM and the influence of rubber content and particle size on the shear strength and deformation of the mixture [10, 11]. However, model studies on the response of footings resting on GSI layers under repeated loading are limited. Therefore, the current study aims to experimentally investigate the behavior of a scale model footing on a sand-rubber mixture GSI bed under cyclic loading conditions, followed by numerical simulations.

2 Scaled Model Study

2.1 Materials

In this research, river sand procured from Chennai, India, was utilized and classified as poorly graded sand (SP) according to the Indian Standard, IS 1498 (BIS 1970). The scrap tire was initially subjected to a shredding process, which resulted in the separation of rubber chips and steel fiber reinforcements through a series of cutting and granulation stages. The steel fibers were removed, and the rubber pieces were further granulated into angular-shaped particles of size less than 4.75mm. These granulated rubbers were then classified as an equivalent of poorly graded sand (SP).

Recent studies [3, 5, 11], found that SRM with a 30% rubber content exhibited a high damping ratio and adequate shear strength and can play a crucial role in reducing dynamic vibrations. Therefore, SRM with a rubber content of 30% was used in this study. Undrained triaxial tests were conducted on the sand of the SRM sample, where peak friction angles of 37.5° and 28.5°, respectively, were obtained (Fig. 1).

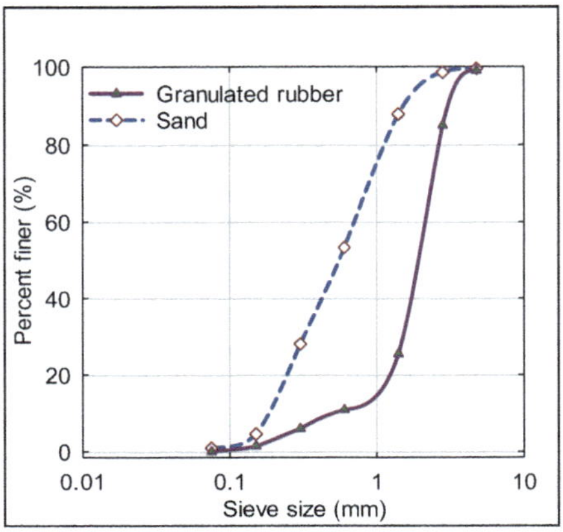

(a)

(b)

Fig. 1. **a** Particle size distribution curve for sand and granulated rubber **b** typical granulated rubber, sand and sand-rubber mixture.

2.2 Test Setup and Procedure

Load tests were carried out on a model footing resting on geo-base isolation layer composed of sand rubber mixtures in a sand bed tank setup (Fig. 2). Scaling laws of similitude by Wood (2004) were adopted for geometry and materials for the present study [12]. The test tank used for the study was made of steel and has an inside dimension of 1 m × 1 m × 1 m. The model square footing is made of square steel plates having a size (B) of 100 mm and thickness of 10 mm. A scale factor (λ) of 10 was used to model a prototype footing of size 1 m × 1 m. The thickness of footing of 12mm arrived by equating the scaling factors of stiffness (K) of the soil with the flexural rigidity of the footing (EI) to maintain similar values of stresses and strains in modelling [12]. For the study, the prototype footing is made of concrete with Young's modulus (E) of 20 GPa while the model footing is made of steel with E of 200 GPa. If the average particle size of the soil/SRM used in the model and prototype structure is identical, the scale factor for density of sand and SRM is chosen as 1 since the same material is used for the model and prototype. To avoid overestimation of test results of model geogrid reinforcement, the scale factors proposed for 1 g model test by Viswanadham and Konig (2004) were used which accounts for the scaling effects of tensile load-strain behaviour and identical frictional bond behaviour [13]. The tensile strength-strain characteristics of the model geogrid represent the commercially available prototype geogrid Fortrac R (tensile strength 800 kN/m).

The test tank was filled with sand up to a depth of 0.9 m for all the tests. As figure suggests, the load tests were performed on a model footing that was placed on a GBI layer made of SRM, in a laboratory setup that simulates a sand-bed tank. The loading was applied on the load plate in a displacement-control mode using a computer-controlled pneumatic actuator. The load was measured using an electronic load cell with a capacity of 20 kN and the displacement was measured by a displacement transducer (LVDT). For more detailed information on the arrangements of the testing apparatus and procedure, please refer to the work of Dhanya et al. (2019) [3].

Initially, the load tests were conducted on a footing resting on a sand bed compacted at a relative density of 85%. The sand bed was prepared in layers of 0.1m each, using the sand pluviation/rainfall technique [14] where the sand was poured from a fixed height to achieve the desired density. The next series of tests were conducted on a footing placed on a GSI layer, which was prepared by excavating the required depth and size of the GSI layer in the sand bed and backfilling it with SRM. Considering the static capacity requirement, the depth of the GSI layer was varied as 0.1 B and 0.2 B, where B is the width of the footing while the width of the GSI layer was kept constant as 3B. A vertical compressive load was applied to the top of the footing using the actuator at a loading rate of 0.5 mm/min. The load was applied until a distinct ultimate bearing pressure was observed, or a considerable settlement of the footing occurs for a small increment in vertical load. Readings from the load cell, LVDT and strain gauges were collected simultaneously at a sampling rate of 5s for the entire experiment.

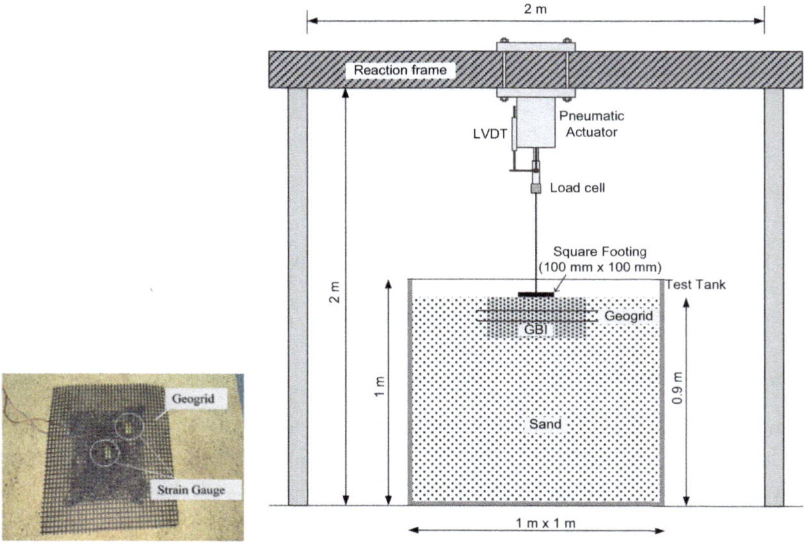

Fig. 2. General arrangements of model footing placed on geocell-GSI bed in the test tank

3 Results and Discussion

The results obtained from different series of tests are plotted as bearing pressure against the footing settlement or settlement ratios. To quantify the bearing capacity improvement, due to geogrid reinforcement, the results were also discussed in terms of the bearing capacity improvement factor. Further, the strain measured along the geogrids were also discussed in the study.

3.1 Test Setup and Procedure

The load-settlement curves obtained for the footing resting on GSI layer reinforced with geogrid reinforcement is illustrated in Fig. 3. It is evident from the figure that footing on geogrid reinforced GBI system exhibits a sharp increase in bearing capacity and reduction in the settlement. Furthermore, it is found from Fig. 3 that the geogrid reinforced GSI system fails at higher values of settlement ratio indicating a significant increase in ultimate capacity of footings due to reinforcement. The figure also shows that the ultimate bearing capacity of footing (corresponding to s/B ratio of 20%) on GSI system increases by three times for double geogrids and two times for single geogrid reinforcement.

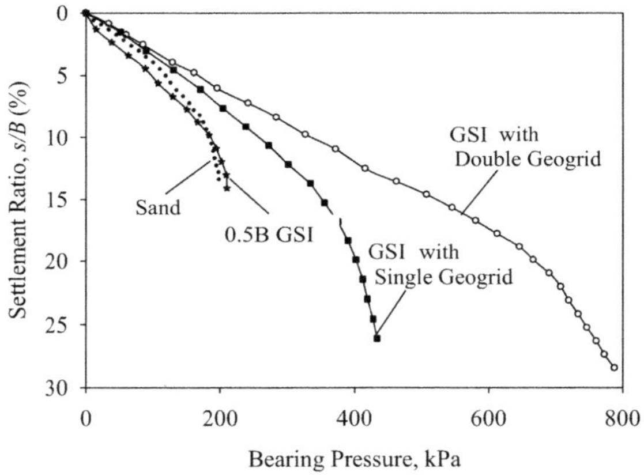

Fig. 3. Pressure-settlement plot for the footing load tests

3.2 Numerical Simulations

The numerical modelling of the footing on the GSI system was carried out using the Finite Element (FE) code ABAQUS. The numerical modelling of the footing on the GSI system was carried out using the Finite Element (FE) code ABAQUS 6.14 (2014). Though two-dimensional (2D) plane strain models are ideal for strip footings, they can be adeptly extended to simulate the response square footings with the bearing capacity being approximately 5% lower in such cases as well as for rectangular footings. In principle, the 2D plane strain model is adequate to analyze the footing-GSI response within the sand bed tank setup with suitable boundary conditions. Figure 4 shows that the numerical studies using ABAQUS can simulate the response of geogrid reinforced GSI system reasonably well.

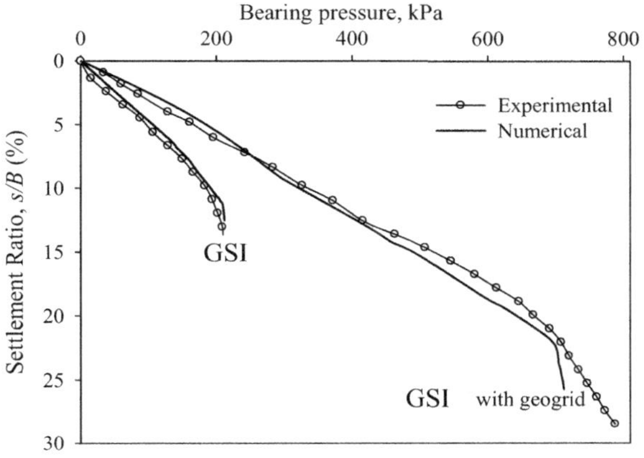

Fig. 4. Comparison of experimental and numerical results

3.3 Seismic Settlement

In the present study, the settlement aspects of the low-rise building placed on GSI layer under seismic loading conditions was numerically analysed considering the effect of geogrid inclusion on GSI layers. The maximum depth of replacement of soil with GSI layer was limited to 2 m (0.2B) considering practicality and cost efficiency under field conditions. The width of the GSI layer was maintained as 1.2B constantly throughout. For the study, the thickness of GSI layer was considered as 0.2B below the footing of the low-rise building. Upto three layers of geogrid, placed at a spacing of 0.05B were included between the GSI layers. The first layer of geogrid was placed at a depth of 0.5 m below the ground surface. Input motions considered for the dynamic analysis involves 2015 Nepal Earthquake (Mw-7.8; PGA – 0.14g; Freq – 0.2Hz).

From this point onwards, the notations GSIur and GSIr will be used to represent the reinforced and unreinforced GSI cases respectively. Similarly, the unreinforced natural soil will be represented as Sur. Figure 5 shows the vertical settlement of the building at the centre of the footing level throughout the 2015 Nepal earthquake input motion. The settlement analysis points out that building with GSIur system experiences a higher vertical settlement by around 10% compared to Sur case. On the other hand, the presence of geogrid has arrested settlement arising due to seismic load to a considerable extent for GSIr and Sr cases. The compressive nature of SRM further ensures higher interlocking of the SRM with geogrid. Compared to the GSIur system, the percentage reduction in the maximum settlement for the GSIr system is about 35% for single geogrid, 55% for double geogrid and 60% for three geogrid layer system. Though three-layered geogrid reinforcement shows a better reduction in the settlement, the performance of two layers and three-layered system are not much different indicating that double layer geogrid system itself is sufficient for effective seismic settlement reduction for GBI system as observed from Fig. 5.

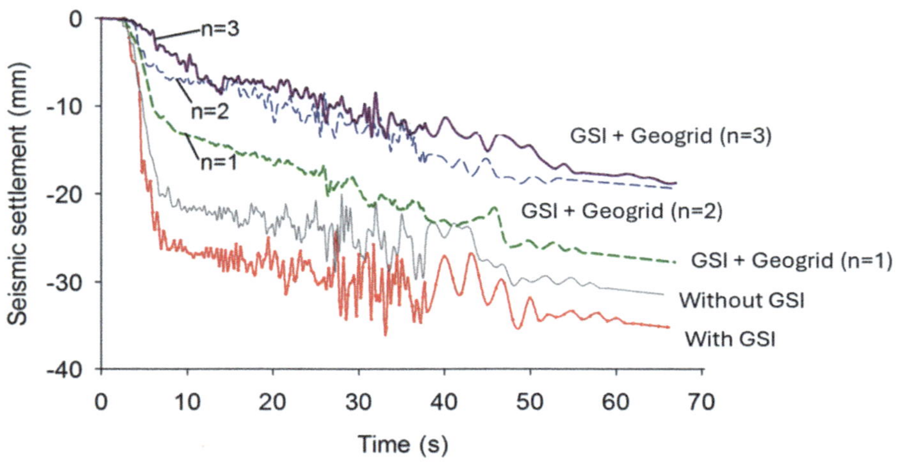

Fig. 5. Seismic settlement of the footing subjected to the 2015 Nepal earthquake input motion

4 Conclusions

The research presented here demonstrates the performance of the sand-rubber mixture (SRM) GSI system under both static and dynamic loadings. The results show the introduction of the proposed GSI layer between the footing and the underlying sandy strata reduces bearing capacity and increases settlement. Provision of geogrids to the geo-base isolation system was found to increase the bearing capacity by two times for single geogrid layer and three times for double geogrid layer. The finite element analysis carried out on footing on geogrid reinforced GSI system indicate that the number, position and length of geogrids along with the width of GSI layer are the primary parameters influencing the increase in bearing capacity and reduction in settlement of the footing. In any case, the length of reinforcement shall be higher than the width of the GSI layer. The most prominent finding to emerge from this study is that double layer reinforcement within the GSI layer is suitable for satisfactory performance in terms of bearing capacity and settlement of the footing. Furthermore, it can also be observed that the geogrid inclusion arrests the seismic settlement arising in GSI layer significantly. Double layered geogrid systems are recommended for effective settlement reduction of GSI layers under static and seismic loading. The next step of this ongoing research is currently focusing on the field implementations followed by a comprehensive life cycle assessment in line with the recent studies carried out for cementitious materials [8, 9]. It should be noted that the approach is more suited for mid to low-rise buildings as the static capacity requirement for the larger structures may not be adequate with the proposed GSI system.

References

1. Tsang H (2008) Seismic isolation by rubber–soil mixtures for developing countries. Earthquake Eng Struct Dyn 37(2):283–303

2. Pitilakis K, Karapetrou S, Tsagdi K (2015) Numerical investigation of the seismic response of RC buildings on soil replaced with rubber—Sand mixtures. Soil Dyn Earthquake Eng 79:237–252
3. Dhanya JS, Boominathan A, Banerjee S (2019) Performance of geo-base isolation system with geogrid reinforcement. Int J Geomech 19(7):04019073
4. Dhanya JS, Boominathan A, Banerjee S (2020) Response of low-rise building with geotechnical seismic isolation system. Soil Dyn Earthquake Eng 136:106187
5. Dhanya JS, Boominathan A, Banerjee S (2022) Investigation of geotechnical seismic isolation bed in horizontal vibration mitigation. J Geotech Geoenviron Eng 148(12):04022108
6. Forcellini D (2020) Assessment of geotechnical seismic isolation (GSI) as a mitigation technique for seismic hazard events. Geosciences 10(6):1–14
7. Tsang HH, Tran DP, Hung WY, Pitilakis K, Gad E (2021) F: Performance of geotechnical seismic isolation system using rubber-soil mixtures in centrifuge testing. Earthquake Eng Struct Dyn 50(5):1271–1289
8. Farahzadi L, Nia SB, Shafei B, Kioumarsi M (2025) Sustainability assessment of ultra-high performance concrete made with various supplementary cementitious materials. Clean Mat 15:100301
9. Farahzadi L, Nia SB, Shafei B, Kioumarsi M (2025) Preliminary environmental assessment of ultra-high-performance concrete mixtures. Lecture Notes Civ. Eng. 237:589–508
10. Ehsani M, Shariatmadari N, Mirhosseini SM (2015) Shear modulus and damping ratio of sand-granulated rubber mixtures. J Cent South Univ 22:3159–3167
11. Senetakis K, Anastasiadis A, Pitilakis K, Souli A (2012) Dynamic behavior of sand/ rubber mixtures. Part II J ASTM Int 9(2):1–13
12. Wood DM (2004) Geotechnical modelling. Spon Press, Abingdon, UK
13. Cresswell A, Barton ME, Brown R (1999) Determining the maximum density of sand by pluviation. Geotech Test J 22(4):324–328
14. Yetimoglu T, Wu JTH, Saglamer A (1994) Bearing capacity of rectangular footings on geogrid-reinforced sand. J. Geotechn. Eng. 120:2083–2099

Predicting Compressive Strength of 3D Printed Earthen Building Materials Using Soft Computing Methods

Maryam Bypour[1(✉)], Aryan Baibordy[2], Mohammad Yekrangnia[2], and Mahdi Kioumarsi[3]

[1] Department of Civil Engineering, Semnan University, Semnan, Iran
mbypour@gmail.com
[2] Department of Civil Engineering, Shahid Rajaee Teacher Training University, Tehran, Iran
{aryanbaibordy,yekrangnia}@sru.ac.ir
[3] Department of Built Environment, OsloMet—Oslo Metropolitan University, Oslo, Norway
mahdi.kioumarsi@oslomet.no

Abstract. The use of 3D printing technology has introduced earth-based materials as a sustainable alternative for eco-friendly construction. However, predicting their mechanical properties, such as compressive strength, remains a challenge due to the complex interactions between material composition, printing parameters, and curing conditions. This study investigates the application of machine learning (ML) techniques to predict the compressive strength of 3D-printed earthen materials using a diverse set of features derived from existing experimental research. Six ML models were implemented: KNN, SVM, Decision Tree (DT), Random Forest (RF), XGBoost, and a novel ensemble model, referred to as the Proposed Regressor (PR). Initially, the performance of five baseline models was assessed. Subsequently, an ensemble model was developed by integrating the best-performing models from both the training and test datasets. The final ensemble consisted of KNN, RF, and XGBoost. The results indicate that the PR model achieved the highest predictive accuracy across both training and test datasets which demonstrates its superior generalization capability. This approach enhances the accuracy and efficiency of compressive strength estimation, and contributes to the advancement of sustainable construction practices.

Keywords: 3D printing · Earthen materials · Machine learning · Sustainable construction

1 Introduction

The construction industry is currently facing significant challenges, including pollution from carbon emissions, depletion of natural resources, and the lack of proper disposal sites for demolished concrete structures. A promising solution to these issues is the utilization of earth-based materials [1–3]. With the growing emphasis on sustainable and cost-effective construction due to climate change and rapid urbanization, earthen additive manufacturing presents an innovative approach for developing safe, high-performance, and scalable housing solutions [4].

H. Beushausen et al. (Eds.): NTZR 2025, LNCE 873, pp. 464–472, 2026.
https://doi.org/10.1007/978-3-032-27698-8_42

Recently, 3D printing (3DP) has emerged as a groundbreaking and sustainable technology in the construction industry. Advanced manufacturing techniques, such as 3DP, are essential for improving the quality of earth-based construction by enhancing detailing, accuracy, and reproducibility. Additionally, this technology minimizes labor requirements while accelerating the construction process [5]. Research findings have confirmed the feasibility of using materials such as clay for 3D printing in construction, which highlights their potential to achieve high print quality and promising option for sustainable building practices [6].

Optimizing the 3DP process involves adjusting key parameters such as layer height, printing speed, and extrusion rate. Research has demonstrated that the optimal parameters for achieving high dimensional accuracy include a printing speed of 20 mm/s, an extrusion multiplier of 1.5, and a print layer height of 2 mm. Furthermore, thermal conductivity analysis indicates that increasing the percentage of date palm fibers (DPF) from 0 wt% to 5 wt% leads to a 72% reduction in thermal conductivity, significantly enhancing the insulation properties of printed structures [7].

Studies on the mechanical properties of 3D-printed soil-based materials reveal that an increase in soil content results in higher water demand, reduced densification, and increased porosity, ultimately leading to lower compressive strength [8]. However, incorporating 1% sisal fibers by weight of soil and stabilizing the mix with aqueous starch gels (up to 5% w/w) improves workability, minimizes cracking, and enhances the feasibility of using these materials for 3D printing applications [9].

Biopolymer stabilizers have also shown promise in improving the rheological and structural properties of earthen materials for 3D printing. Research on the influence of sodium alginate on kaolinite-based mixtures indicates that the addition of alginate enhances strength and reduces shrinkage in printed structures [10].

Establishing reliable processing methods for 3D printing locally sourced earthen materials across different environmental conditions is crucial to validating large-scale earthen additive manufacturing as a promising solution for global housing demands and climate-resilient architecture. Despite the rapid advancement of digital manufacturing technologies, challenges remain in adapting 3D printing systems for earth-based materials such as cob [11].

Extensive research has been conducted on 3D printing in construction [12]. A study on the use of local clay for fabricating terracotta components using 3D printing demonstrated the feasibility of achieving high-quality print, thereby establishing clay as a viable material for construction-based additive manufacturing [6].

Although earth-based materials exhibit distinct structural and thermal properties compared to concrete, most existing 3D-printed mortar structures function as temporary formwork for traditional reinforced concrete systems. However, 3D-printed earth offers a sustainable alternative as recyclable, fire-resistant, and optimized formwork for reinforced concrete and insulation materials, positioning it within an innovative low-carbon design framework [13].

Due to the complex nature of earth-based materials, there is currently no standardized design code for their application. Consequently, understanding their mechanical properties requires extensive and continuous experimental research. To address this challenge, researchers have increasingly turned to machine learning (ML) techniques to develop predictive models using datasets derived from existing literature. Although ML methods have been widely applied in civil engineering research [14–18], their use in predicting the properties of earth-based materials remains limited.

To bridge this research gap, this study employs various ML models, including K-Nearest Neighbors (KNN), Support Vector Machine (SVM), Decision Tree (DT), Random Forest (RF), XGBoost, and a newly proposed ensemble model named the Proposed Regressor (PR). Initially, the performance of five baseline ML models was evaluated. Based on these findings, the ensemble model was developed by integrating the best-performing models from both the training and testing phases. The final proposed model combines KNN, RF, and XGBoost to enhance predictive accuracy. The novelty of this research lies in utilizing ML models to predict the compressive strength of 3D-printed earthen building materials which contributes to the advancement of sustainable construction practices.

2 Data Collection

To evaluate the compressive strength of the 3D-printed earthen materials, dataset containing a total of 51 specimens was gathered from five experimental studies. The collected dataset consisted of five features namely: soil content (% by weight of earthen mixture), plasticity index of the soil (PI, %), fiber content (% by weight of earthen mixture), stabilizer content (% by weight of earthen mixture), and curing duration (days). The target was the compressive strength (CS) of the specimens, measured in MPa. The collected dataset is elaborated in Table 1. Specifically, in this table, the range of each variable is indicated. Accordingly, as shown in Table 1, the soil percentage in the selected specimens ranges from 0% to 100% and were cured for 3, 7, and 28 days. The liquid limit (LL) of the materials varies between 22 and 55.

Table 2 illustrates the basic statistics of the gathered dataset including the mean, standard deviation (SD), minimum, Q1 (25%), Q2 (50%), Q3 (75%), and maximum of each feature. For example, for the feature PI (with 51 data points), the mean is 12.33% with a standard deviation of 6.70%. The minimum, quartiles, and maximum values are 6.00%, 6.00%, 10.80%, 16.00%, and 30.50%, respectively. These statistics allow for an analysis of the variation and range of each variable in the dataset.

A heatmap of the correlation matrix for the variables is depicted in Fig. 1, where different colors indicate the strength and direction of correlations. This visual representation provides precise correlation coefficients which makes it easier to distinguish variables with strong positive or negative relationships from those with minimal or no correlation. Based on the figure, fiber content exhibits the most significant influence on compressive strength, suggesting that a higher fiber percentage enhances the target value. In contrast, curing time appears to have little to no effect on the compressive strength.

Table 1. Dataset collected from literature.

Study	Features						Target
	Soil* (%)	LL (%)	PI (%)	Fiber* (%)	Stabilizer* (%)	Curing (day)	CS (MPa)
(Faleschini et al. 2023) [19]	45.72, 46.22	22.00	6.00	0, 0.50	8.25–11.00	28.00	0.39–1.40
(Bhusal, Sedghi, and Hojati 2023) [20]	92.50–100	33.60	13.70	0, 1.50	0–6.00	7.00, 28.00	0.50–6.15
(Bhusal, Uviña, and Hojati 2022) [21]	94.00–97.00	52.10	30.50	1.00, 1.50	2.00, 4.00	7.00	0.56–7.00
(Silva et al. 2022) [9]	66.00	27.00	10.80	0, 1.00	0–5.00	3.00, 7.00, 28.00	0.30–2.15
(Sahana et al. 2024) [22]	0–33.33	35.00	16.00	9.00–31.00	33.33	28.00	9.00–31.00

* By weight of earthen mixture

Table 2. Descriptive statistics of the gathered dataset.

	Count	Mean	SD	Min	25%	50%	75%	Max
Soil (%)	51	54.53	29.33	0.00	45.72	46.22	66.00	100.00
PI (%)	51	12.33	6.70	6.00	6.00	10.80	16.00	30.50
Fiber (%)	51	4.53	8.14	0.00	0.00	1.00	1.50	31.00
Stabilizer (%)	51	12.14	12.30	0.00	2.50	8.25	11.00	33.33
Curing (day)	51	22.00	9.89	3.00	7.00	28.00	28.00	28.00
CS (MPa)	51	5.15	7.93	0.30	0.59	1.16	6.58	31.00

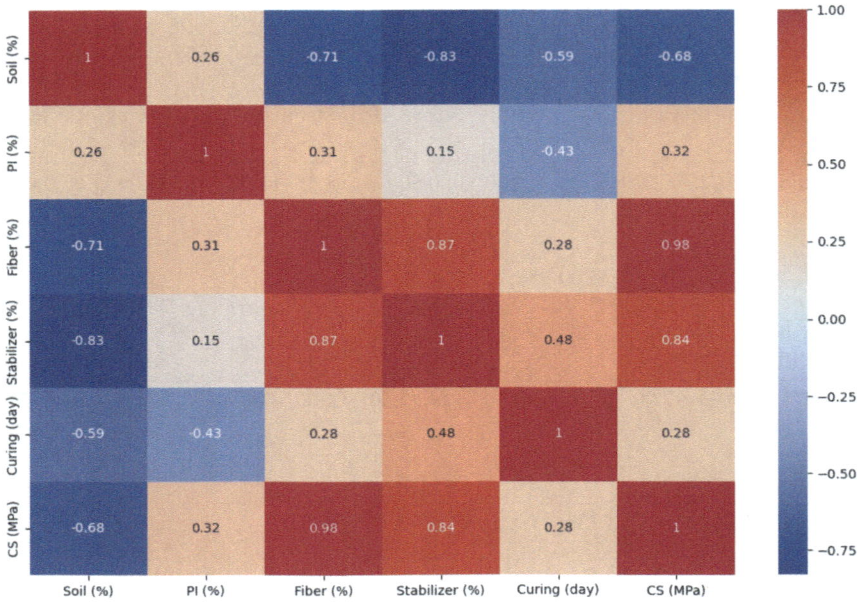

Fig. 1. Correlation matrix of the dataset.

3 Machine Learning Methods

This study implemented six machine learning (ML) models, including KNN, SVM, decision tree (DT), random forest (RF), XGBoost, and a novel ensemble model proposed by the authors named as a proposed regressor (PR) model. These models were implemented so as to predict the compressive strength of 3DP mixtures based on various target variables, namely soil content, PI, fiber content, stabilizer content, and curing duration. First, the performance of the five baseline ML models was evaluated. Then, the authors developed the ensemble model by selecting the best-performing models from both the training (80% of gathered dataset) and test (20% of gathered dataset) datasets. Based on this approach, the final proposed method combined KNN, RF, and XGBoost into an ensemble.

To further enhance the performance of model, weighted contributions were assigned to each ML model within the proposed ensemble model using trial and error method. Specifically, XGBoost and KNN were given a weight of 2, while RF was assigned a weight of 1.

4 Results and Discussion

Figure 2 illustrates the performance of various ML models analyzed in this study, including KNN, SVM, DT, RF, XGBoost, and PR. The models were evaluated using test data resulted in R2 scores of 0.96, 0.58, 0.76, 0.94, 0.94, and 0.99, with corresponding RMSE values of 1.63, 5.47, 4.10, 2.03, 1.90, and 0.93, respectively.

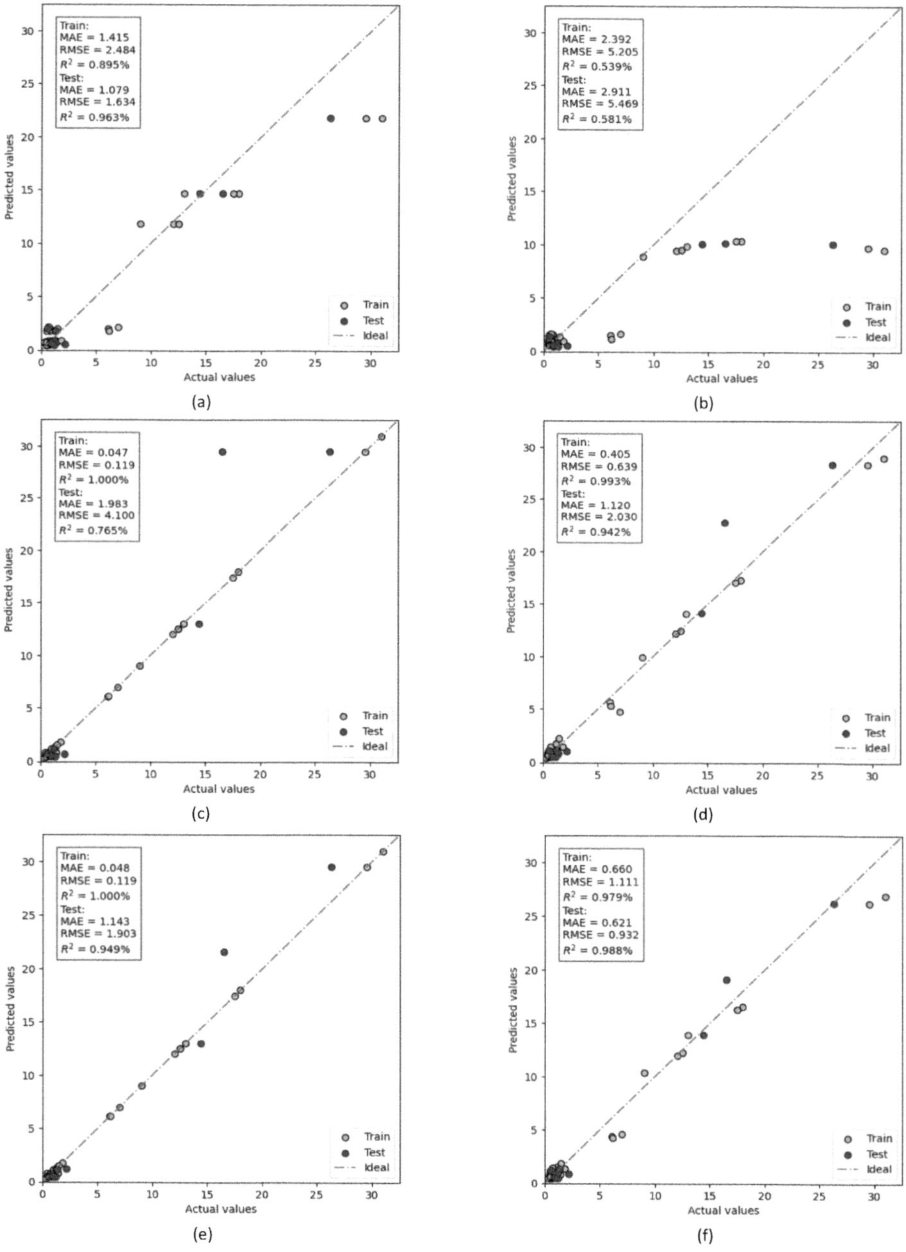

Fig. 2. The results of ML models: **a** KNN, **b** SVM, **c** DT, **d** RF, **e** XGBoost, and **f** PR.

As shown in Fig. 2, the PR model demonstrates the best performance in predicting both the training and test datasets, particularly the test dataset, which plays a crucial role in assessing the generalizability of the model.

5 Future Work

The proposed model can be embedded in a decision-support tool for selecting optimal printing parameters or material compositions during the design phase of 3D-printed earthen structures. Additionally, it could serve as a rapid evaluation system for onsite quality control, reducing the need for destructive testing. These enhancements are now clearly discussed to better align the study with practical construction workflows.

6 Conclusion

This study utilized machine learning models to predict the compressive strength of 3D-printed earthen materials. Initially, the performance of five baseline ML models was assessed. Subsequently, an ensemble model was developed by selecting the top-performing models based on both training and test dataset evaluations. The key findings are as follows:

- Performance analysis on the test dataset revealed that KNN, RF, XGBoost, and PR achieved superior accuracy, with R^2 scores of 0.96, 0.94, 0.94, and 0.99, respectively.
- Among all models, the PR model exhibited the highest predictive accuracy across both the training and test datasets which demonstrates its strong generalization capability.

References

1. Rückrich S, Agranati G, Grobman YJ (2023) Earth-based additive manufacturing: a field-oriented methodology for evaluating material printability. Archit Sci Rev 66(2):133–143
2. Baibordy A, Yekrangnia M, Jahromi SG (2025) A comprehensive study on the mechanical properties of natural fiber reinforced stabilized rammed earth using experimental and data-driven fuzzy logic-based analysis. Clean Mat 2025:100300
3. Baibordy A, Yekrangnia M, Khodabakhshian F (2024) Analysis of the mechanical behavior of rammed earth mixes using soft computing methods with an emphasis on sustainability. The international conference on net-zero civil infrastructures: innovations in materials, structures, and management practices (NTZR). Springer
4. Curth A et al (n. d.) 3d printing earth: local, circular material processing and life cycle assessment. Circular Material Processing and Life Cycle Assessment
5. Pajonk A et al (2022) Multi-material additive manufacturing in architecture and construction: a review. J Build Eng 45:103603
6. El Aabbas M et al (2024) Advancing sustainable construction: terracotta component development through extrusion-based 3d printing with local clay. Constr Build Mater 442:137549
7. Yousaf A, Khan SA, Koç M (2024) 3DP for sustainable built environment–Synthesis and thermomechanical characterization of composite materials based on local soil and date palm fiber waste. J Clean Prod 481:144050

8. Soda PRK, Dwivedi A, Gupta S (2024) Development of 3D printable stabilized earth-based construction materials using excavated soil: evaluation of fresh and hardened properties. Sci Total Environ 924:171654
9. Silva G et al (2022) Eco-friendly additive construction: analysis of the printability of earthen-based matrices stabilized with potato starch gel and sisal fibers. Constr Build Mater 347:128556
10. Maierdan Y et al (2024) Rheology and 3D printing of alginate bio-stabilized earth concrete. Cem Concr Res 175:107380
11. Gomaa M et al (2021) 3D printing system for earth-based construction: case study of cob. Autom Constr 124:103577
12. Pemas S et al (2024) Enhancing clay-based 3D-Printed mortars with polymeric mesh reinforcement techniques. Polymers 16(15):2182
13. Curth A et al (2024) 3D printing earth: Local, circular material processing, fabrication methods, and Life cycle assessment. Constr Build Mater 421:135714
14. Bypour M, Yekrangnia M, Kioumarsi M (2025) Machine learning-driven optimization for predicting compressive strength in fly ash geopolymer concrete. Clean Eng Technol 2025:100899
15. Bypour M et al (2024) Explainable tuned machine learning models for assessing the impact of corrosion on bond strength in concrete. Clean Eng Technol 23:100834
16. Mahmoudian A, Bypour M, Kontoni D-PN (2024) Tree-based machine learning models for predicting the bond strength in reinforced recycled aggregate concrete. Asian J Civil Eng 25(8):5899–5924
17. Mahmoudian A, Bypour M, Kioumarsi M (2024) Explainable boosting machine learning for predicting bond strength of FRP rebars in ultra high-performance concrete. Computation 12(10):202
18. Bypour M, Yekrangnia M, Kioumarsi M (2024) Evaluation of the compressive strength of fly ash-based geopolymer concrete using machine learning. The international conference on net-zero civil infrastructures: innovations in materials, structures, and management practices (NTZR). Springer
19. Faleschini F et al (2023) Sustainable mixes for 3D printing of earth-based constructions. Constr Build Mater 398:132496
20. Bhusal S, Sedghi R, Hojati M (2023) Evaluating the printability and rheological and mechanical properties of 3D-printed earthen mixes for carbon-neutral buildings. Sustainability 15(21):15617
21. Bhusal S, Uviña F, Hojati M (2022) Preliminary study on 3D printing of locally available earthen materials in New Mexico. In: Proceedings of the preliminary study on 3D printing of locally available earthen materials, Santa Fe, NM, USA, pp 23–25
22. Sahana C et al (2024) 3D printing with stabilized earth: material development and effect of carbon sequestration on engineering performance. Cement Concr Compos 152:105653

Rethinking Architectural and Structural Design in Shotcrete 3D Printing

Robin Dörrie(✉), Bartłomiej Sawicki, and Harald Kloft

Technische Universität Braunschweig, 38100 Braunschweig, Germany
r.doerrie@tu-braunschweig.de

Abstract. Achieving a net-zero future in construction requires new approaches for architectural and structural design to minimize material consumption, reduce waste, and enhance resource efficiency. Topology optimization combined with robotic additive manufacturing techniques offers significant potential for material-efficient structural design by integrating sustainable, high-performance materials and processes while maximizing load-bearing and functional capacities. This study presents conceptual exploration towards topology optimization algorithms based on Shotcrete 3D Printing (SC3DP) constraints. A comparison of the architectural landmark "House with One Wall" by Christian Kerez, and a version which is optimized utilizing a fabrication-informed topology optimization workflow, are evaluated. SC3DP enables the additive manufacturing of mass-minimized, reinforced, and geometrically complex concrete structures without any formwork. The research focuses quantitatively on material savings, and qualitatively on architectural forms. The results indicate that through topology optimization and digital fabrication up to 40% material reduction can be achieved, demonstrating the potential of SC3DP for resource-efficient construction. Nevertheless, the conceptual consideration underlines, that it needs to be further investigated on the potential for architectural design, whether the structural optimization should be the only goal, if the long-lasting impacts on the global scale is expected. Overall, the findings underscore the opportunities and threats when employing digital optimization tools regarding material savings especially in combination with novel AM methods. While topology-optimized structures may diverge from traditional design thinking, they hold immense potential to create sustainable, long-lasting structures while encouraging automated design workflows.

Keywords: Shotcrete 3D Printing · Additive Manufacturing · Topology Optimization · Structural Design · Digital Concrete

1 Introduction

Architectural design is the balance between fulfilling structural and functional requirements and searching for formal and spatial freedom. It embraces, translates, and shapes the technological possibilities, in order to reflect the societal and human needs. The role of architecture is by far not passive – in the past, it frequently sparked new discourses on contemporary questions, stimulating advances [1].

H. Beushausen et al. (Eds.): NTZR 2025, LNCE 873, pp. 473–484, 2026.
https://doi.org/10.1007/978-3-032-27698-8_43

Today, the central societal need is mitigating the climate crisis. The Architecture, Engineering, and Construction (AEC) industry is a significant contributor to the climate change [2–5]. But, as the built environment influences daily life profoundly impacting directly human health, behavior, and identity [6], the AEC sector is also the key to change our patterns of living, working, and travelling [7]. Therefore, it must become once more a leading actor in innovation, and drive the transition towards the carbon-neutral future.

This study focuses on the potential of Additive Manufacturing in Construction (AMC) which promises highly efficient building components and great material savings [8–12] in combination with digital tools for topology optimization and reduced material usage [13, 14]. The main goal of this publication is to discuss the limits of form optimization with the sole goal of material reduction, as constrained by the intersection to architectural design. It is performed on the example of an architectural landmark building 'House with One Wall' by Christian Kerez, whose central structural wall element is form-optimized for additive manufacturing with Shotcrete 3D Printing (SC3DP). The resulting structure is critically discussed from the point of view of its architectural design as compared to the reference one, in order to demonstrate the challenges of a purely technology-based approach to reduce the environmental footprint of reinforced concrete structures.

2 State of the Art

Recent prototypical projects from around the world showcase how digital-physical design-to-fabrication workflows [15] utilize digital design tools, additive manufacturing (AM), robotic fabrication and innovative materials to tackle the challenges of contemporary construction, including sustainability [16]. Three examples of structures which utilized different robotic fabrication methods are presented in Fig. 1.

The Server Hotel (see Fig. 1 a.) utilizes classical simple form to provide enclosed and obstacle-free space for server rooms [17–19]. The digital fabrication is revealed by design of façade. With undulating surfaces increasing the transverse stiffness, the required element thickness was reduced, saving up to 25% of material and hence reducing the carbon footprint. Nevertheless, similar forms are already known in the architecture of concrete, and therefore this example does not rise particular questions regarding its acceptance among architects. It is proven by the fact, that it represents a commercial rather than demonstrative project.

The second example, DFAB House built by NCCR Digital Fabrication to demonstrate the capabilities of developed digital fabrication methods, presents an interesting example in particular regarding its element 'Smart Slab' [20–22]. It combines topology optimization methods applied to complex geometrical situation of only one, central, and S-shaped wall, translated to physical world thanks to additive and subtractive digital fabrication of formwork. It is further enabled by high-performance cementitious fiber-reinforced material, leveraging the synergy of all process steps to achieve the unique and optimized form (see Fig. 1 b). Presenting unique architectural features, so radically different from flat ceilings we are used to nowadays, it rose some questions regarding the broad acceptance of this technological solution. As such, after more than a decade of the development, the 'Smart Slab' element remains a demonstrator rather than scalable and broadly implemented technology.

The last example is the 'Maison Fibre', a radical attempt to present the capabilities of coreless filament winding fabrication presented on Architectural Biennale di Venezia 2021 [23] (see Fig. 1 c.). Through use of glass, carbon and in the future natural fibers which are all materials of extremely good strength-to-weight ratio, as well as through topology optimization and robotic fabrication, the load-bearing structure of a multistory building was demonstrated. Respecting the realistic load requirements, the structural system was so efficient and lightweight that the units not only could be moved by hand, but they seemed do defy the physical laws. 'Maison Fibre' sparked discussion regarding the function of building materials and limitation of structural optimization to make the resulting buildings livable, as in this extreme case it did not provide a fundamental shelter function. Therefore, for now, it remains far away from impactful application in AEC.

a. b. c.

Fig. 1. a. Server Hotel, SVV Architekten/Mense Korte, Heidelberg, Germany, © Peri 3D Construction GmbH [18]; b. DFAB House, NCCR Digital Fabrication, Zurich, Switzerland, © Roman Keller [21]; c. 'Maison Fibre', ITECH Research Building 2021, ICD/ITKE, Stuttgart, Germany, © ICD/ITKE [23, 24].

The three abovementioned examples showcase that the combination of digital fabrication and topology optimization holds a great potential [20, 25]. Especially additive processes, particularly 3D printing techniques, enable the fabrication of highly complex geometries that are typically unfeasible or uneconomical using conventional construction methods. Through topology optimization, structures can be optimized to represent only a fraction of the original mass by still meeting the same loadbearing requirements [26]. However, due to the nature of the material and process, additive manufacturing of topology-optimized concrete elements still is limited by challenges such as the reliable integration of reinforcement [27, 28], control over material behavior during material application and curing [29], as well as ensuring structural continuity and code compliance.

Furthermore, a big challenge in utilizing topology optimization software for AM is the implementation of fabrication constraints, as these are of utmost importance to create realistic and manufacturable structures with the given process [30]. Software plugins for general design workflows like Ameba, Millipede, or Karamba, lack the ability to create optimizations based on characteristics of the fabrication processes. Thus, a standalone software has to be employed for the complex calculations. In this study the software nTop [31] was utilized as it provides the option to include material properties, manufacturing constraints and provides high controllability over the optimization process.

3 Methods

The following chapter explores the potential of combining AM with optimization algorithms. Specifically, an optimization approach is presented that uses the SC3DP process constraints [32–34] as a basis for generating optimized wall structures for the 'House with One Wall' by Christian Kerez.

The Architect describe the building like this: "A two-family house is a building with a wall that divides it into two halves. The dividing wall is the only wall in the entire house. It cannot be crossed anywhere. It has to fulfil functions conventionally assigned to several architectural elements. It is the loadbearing structure and the installation core, its folds define all of the rooms and it determines how the view from the entirely glazed building is divided between the two living units.

The simplicity of the concept, the reduction of the architecture to a single element, creates substantial dependence and, in turn, makes the building very complex. It is only through that dependence that the wall acquires a compelling and cogent character although, in itself, it can follow any chosen course.

The wall between the two units has folds in it so that it will not fall over, like folding a piece of paper so that it can stand on end. The folds are different from floor to floor. One room is concave, another convex. One room is open plan, another has sections. The shape of the wall is reversed in each of the flats – on one side a protruding bend, on the other a niche; on one side a wide space, on the other a narrow one…" [35]. Figure 2 shows the building and an isometric view of the floorplans.

In order to explore topology optimization algorithms based on SC3DP constraints, the "House with One Wall" is taken as a starting form. The fabrication constraints of the SC3DP process are implemented to investigate the potential material savings and underlaying changes in architecture and structure itself, which appear when changing the topology. As basis for the comparison the values shown in Table 1 are estimated based on the architects plans.

For the optimization process in the software nTop [31], the algorithm uses the Solid Isotropic Material with Penalization (SIMP) method, continuously optimizes the structure. Opposing this method the Bi-directional evolutionary structural optimization (BESO) method would discretize and iteratively add and delete material in a binary manner. Therefore SIMP method it is implemented in nTop and brings advantages in terms of a more detailed geometry as foundation for the print strategy planning, easier implementation of fabrication constraints as well as integration in digital workflows for the printing process. The setup targets minimum compliance under multiple load case scenarios. The design domain includes the volume of the house (approx. 18.5 m $\times$ 12.0 m $\times$ 10.5 m) and the single walls (approx. 18.5 m $\times$ 0.4 m $\times$ 2.65 m). The boundary conditions include fixed supports at the base of each wall, reflecting the monolithic nature

a.

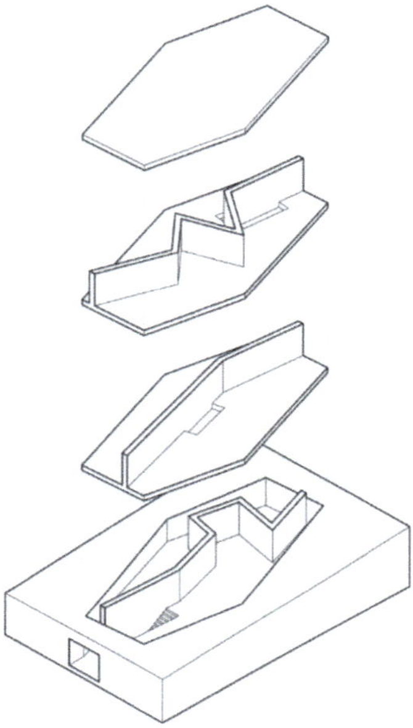

b.

Fig. 2. a. "House with One Wall" by Christian Kerez, ©Walter Mair; b. conceptual isometric visualization of the floor plan layouts, © Macromicro Studio

Table 1. Estimated values of "House with One Wall" for comparison

Concrete type*	Concrete vol. [m^3]	Reinforcement type*	Reinforcement mass [t]
C 30/37	247	B500B	33
Comp. strength 30 MPa/37 MPa		Yield strength 500 MPa	

* According to Eurocode 2 (EN 1992-1-1) [36]

of reinforced concrete. A uniformly distributed load on the top of each slab and vertical loads distributed uniformly on the sides of the building are applied according to the Eurocode load cases [36]. Live loads are disregarded due to the scale of the case study building and as simplification. The Material properties assumed Young's modulus $E = 25$ GPa, Poison's ratio $\upsilon = 0.2$ and density $\rho = 2100$ kg/m^3. As a stress constraint the compressive strength of Eurocode concrete class C30/37 was assumed and 2.9 MPa for tensile strength to check for the integration of reinforcement afterward. Furthermore, the optimization constraints were set up based on preliminary research regarding the design and process limitations of the SC3DP process [37–42]. The process constraints include limiting the minimal feature size to 80 mm, limiting the overhang to 55°. The optimization was then performed reaching convergence after 150 iterations. As post-processing steps a smoothening and remeshing was utilized.

4 Results

The following section presents the results of the topology optimization and the comparison to the reference structure, focusing on material efficiency, structural performance, and architectural implications. Figure 3 shows the optimization results.

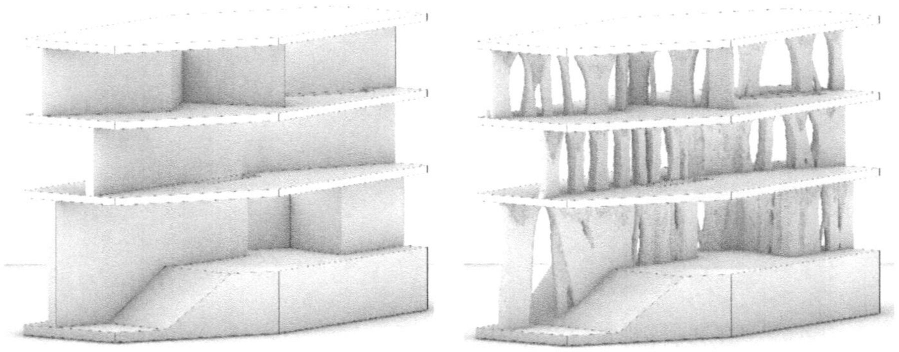

a. b.

Fig. 3. Resulting structure for the optimization approach a. Reference structure b. Topology optimized structure

In the optimized structure, the amount of reinforced concrete in walls is reduced by approx. 40%, from around 247 m^3 to 153 m^3, while keeping the same loadbearing capacity and load paths. The original separating wall structure is dissolved into various arches, tree-like forms, evolving from the floor slabs towards the ceiling. The reduced wall elements vary from thin column-like elements to rather elongated wall segments, providing the necessary stiffness in corner regions while long segments purely separating space are opened up. Additionally, the thickness of the wall segments vary, replacing the straight formwork surfaces. Overall the geometry reflects the SC3DP constraints such as printable overhang angles, self-supporting curvature and element thickness. The planar aesthetic of the original structure is replaced by a spatially dynamic and porous structure.

5 Discussion

The exploration focused on the optimization of structural elements to minimize material usage and improve the ecological footprint of reinforced concrete structures while utilizing novel AM techniques. Although this target-oriented but one-sided approach achieves a high value of material savings. It raises many questions regarding the function of the structure and overall architecture.

The structure of "House with One Wall" demonstrates the state of the art of utilization of systematic formwork. Flat surfaces, sharp edges and rectangular shapes define the living space. The design is influenced by contemporary construction methods and creates a fascinating space through the combination of these techniques and thoughtful design. However, the structure contains more material than structurally necessary.

When observing the topology optimized structure, three main questions arise. Firstly, the question of retaining the functions of each wall element should be discussed. The multidimensional range of requirements of a wall cannot be retained when optimizing the structures topology. The function of enclosing a space and zoning different areas in the building is lost due to the dissolvement. Furthermore, the ability of incorporating ducts and pipes in a structured manner is decreased due to non-uniform shapes. Only the loadbearing capabilities are retained and material efficiency is increased. To retain all features the structural element has to be cladded or encased by a second material. This can be an additional feature to integrate an insulating or sound absorbing material to improve the spatial climate, but in the same time complicating the fabrication or construction process and increasing the environmental footprint.

The results show increasing material efficiency and retaining structural performance when decreasing the amount of material. The decrease of material amount however also decreases the architectural integrity. To achieve a balance between design and function, implementation strategies should incorporate the secondary functions such as insulation, acoustics, and service integration through hybrid manufacturing or multi-material printing approaches. These approaches preserve architectural integrity while ensuring structural and ecological optimization.

Secondly, the potentials and limitations of the optimization strategy must be critically evaluated in terms of their applicability to real-world construction projects. In general, topology-optimized structural elements are challenging to realize using conventional construction methods. The production of complex geometries typically requires special formwork, which is both costly and resource-intensive in terms of material waste and labor. In contrast, AM techniques, particularly SC3DP, offer a promising alternative. These methods eliminate the need for traditional formwork and enable the integration of reinforcement and functional elements directly during the fabrication process. Moreover, the inherently digital nature of AM workflows aligns well with the complexity of optimized geometries, facilitating seamless integration into a broader digital ecosystem, including digital twins and tools such as Building Information Modeling (BIM), Fabrication Information Modeling (FIM), and data-driven quality control [43–45]. An additional advantage of this digital integration is the increased precision of robotic fabrication. With accurate digital planning and automation, traditional safety factors could potentially be reduced, leading to further material savings and improved overall efficiency. However, for applicable calculations also a variety of additional load scenarios should be regarded such as live loads or cantilevers. These can significantly influence the design and further manufacturing steps such as reinforcement integration or print strategy development.

Thirdly, the architectural design of such structures must be given to be implemented as viable and livable alternatives to traditional construction. The topology optimized structures contradict the known architectural values. Different studies have explored the acceptance of integrating new technologies into the construction practice [46–49]. Often new technologies or shapes are perceived as foreign or unattractive and therefore avoided. Additionally, they often do not align with regional construction methods. A compromise of utilizing local materials, construction methods and optimization based on these constraints could increase the acceptance. Also integrating qualitative human input into the optimization workflow, such as architectural constrains, spatial logic or shape preferences help balancing performance with perception. Furthermore, educating the broad public on the advantages can increase acceptance by stakeholders and lead to a faster implementation. Demonstrators, prototyping and visualizations can bridge the gap between the abstract computational spaces and provide spatial understanding of the new structures.

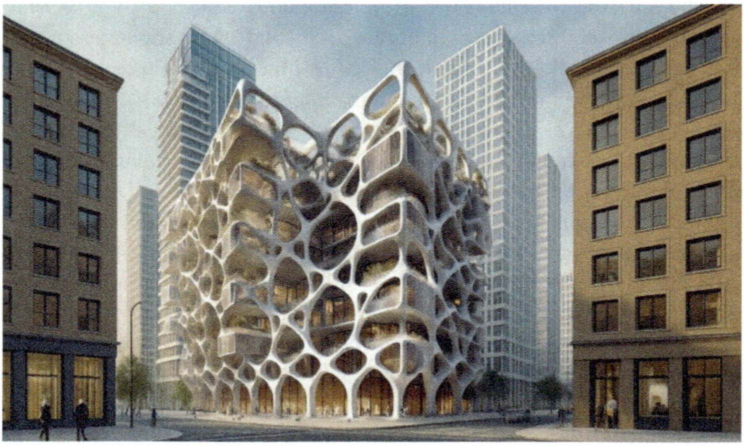

Fig. 4. Visualization of an optimized structure within a given urban context.

Finally, a rendering illustrates the potential appearance of a building incorporating optimized structural features. The visualization highlights the significant transformation of the spatial configuration and the resulting adaptions introduced by the topology optimization (see Fig. 4).

6 Conclusion and Outlook

Topology optimization in combination with additive manufacturing methods allows for the force-flow related application of material in complex structural systems for the most structurally-efficient solution, limiting the carbon footprint. Nevertheless, the question arises: where is the boundary between striving for the most efficient structural topology, and the architectural value of the achieved solution. Understanding of this concern is critical to introduce the optimized design and construction methods as scalable and impactful solutions for Net-Zero future. In this study it is critically evaluated how utilizing SC3DP constraints in combination with topology optimization improve an architectural landmark of concrete construction, resulting in material savings of up to 40%. From an architectural perspective, topology optimization applied to specific given and predefined geometries offers significant potential to save material by retaining the structural performance. The optimized structure is highly efficient, however lacks spatial clarity and functionality. As such, topology optimization should be strategically applied to elements with a clearly defined structural role.

Equally crucial is the consideration of architectural design: topology-optimized forms often deviate from familiar architectural languages and may face resistance if perceived as foreign or impractical. To address this, a balance must be struck between structural performance and perceptual clarity, by integrating local materials, architectural constraints, and human-centered design preferences. Public education, prototyping, and visualization can further bridge the gap between abstract computation and spatial understanding, fostering broader acceptance of these novel construction methods.

Funding. This research was funded by the Deutsche Forschungsgemeinschaft (DFG, German Research Foundation) – Projektnummer 414265976 – TRR 277. The authors thank the DFG for the support within the CRC/ Transregio 277 - Additive Manufacturing Construction. (Project A04).

References

1. Corbusier L (2007) Toward an architecture. Getty Research Institute, Los Angeles, CA
2. Hamilton I, Kennard H, Rapf O, Amorocho J, Steuwer S, Kockat J, Toth Z (2024) Global status report for buildings and construction-beyond foundations: mainstreaming sustainable solutions to cut emissions from the buildings sector
3. Camarasa C, Mata É, Navarro JPJ et al (2022) A global comparison of building decarbonization scenarios by 2050 towards 1.5–2 °C targets. Nat Commun 13(1):3077
4. Onat NC, Kucukvar M (2020) Carbon footprint of construction industry: a global review and supply chain analysis. Renew Sustain Energy Rev 124:109783
5. Ma M, Zhou N, Feng W, Yan J (2024) Challenges and opportunities in the global net-zero building sector. Cell Rep Sustain 1(8):100154
6. The sociology of architecture: constructing identities (2011). Liverpool University Press, Liverpool
7. Sobek W, Heinlein F (eds) (2023) Über die Randbedingungen des Zukünftigen. avedition, Stuttgart
8. de Schutter G, Lesage K, Mechtcherine V, Nerella VN, Habert G, Agusti-Juan I (2018) Vision of 3D printing with concrete—technical, economic and environmental potentials. Cem Concr Res 112:25–36
9. Gengnagel C, Baverel O, Betti G, Popescu M, Thomsen MR, Wurm J (eds) (2023) Towards radical regeneration. Springer International Publishing, Cham
10. Keating SJ, Leland JC, Cai L, Oxman N (2017) Toward site-specific and self-sufficient robotic fabrication on architectural scales. Sci Robot 2(5)
11. Bier H, Mostafavi S (2015) Structural optimization for materially informed design to robotic production processes. Am J Eng Appl Sci 8(4):549–555
12. Flatt RJ, Wangler T (2022) On sustainability and digital fabrication with concrete. Cem Concr Res 158:106837
13. Menna C, Mata-Falcón J, Bos FP et al (2020) Opportunities and challenges for structural engineering of digitally fabricated concrete. Cem Concr Res 133:106079
14. Hesselgren L, Kilian A, Malek S, Olsson K-G, Sorkine-Hornung O, Williams C (eds) (2018) AAG 2018—advances in architectural geometry 2018. Klein Publishing GmbH, Wien
15. Dörrie R, Gantner S, Amiri FS et al (2025) From digital to real: optimised and functionally integrated Shotcrete 3D printing elements for multi-storey structures. Buildings 15(9):1461

16. Nyamekye P, Lakshmanan R, Tepponen V, Westman S (2024) Sustainability aspects of additive manufacturing: leveraging resource efficiency via product design optimization and laser powder bed fusion. Heliyon 10(1):e23152
17. Placzek G, Schwerdtner P (2024) A global snapshot of 3D-printed buildings: uncovering robotic-oriented fabrication strategies. Buildings 14(11):3410
18. Project Website Peri 3D Construction. https://www.peri3dconstruction.com/wavehouse-heidelberg. Accessed May 2025
19. Project Website Heidelberg Materials. https://www.heidelberg-it.de/serverhotel-der-heidelberg-it-in-europas-groesstem-3d-gebaeude/. Accessed May 2025
20. Jipa A, Bürgin T, Dillenburger B (2023) 3D-printed formworks for glass fibre reinforced shotcrete. ETH Zurich
21. Homepage DFAB House. https://dfabhouse.ch. Accessed May 2025
22. Anzalone P, Del Signore M, Wit AJ (eds) (2018) Proceedings of the 38th annual conference of the association for computer aided design in architecture (ACADIA). ACADIA
23. Gil Pérez M, Früh N, La Magna R, Knippers J (2022) Integrative structural design of a timber-fibre hybrid building system fabricated through coreless filament winding: Maison Fibre. J Build Eng 49:104114
24. Project Website ITKE. https://www.itke.uni-stuttgart.de/research/built-projects/maison-fibre-2021/. Accessed May 2025
25. Brackett D, Ashcroft I, Hague R (2011) Topology optimization for additive manufacturing. University of Texas at Austin
26. Stoiber N, Kromoser B (2021) Topology optimization in concrete construction: a systematic review on numerical and experimental investigations. Struct Multidisc Optim 64(4):1725–1749
27. Amir O (2013) A topology optimization procedure for reinforced concrete structures. Comput Struct 114–115:46–58
28. Xia Y, Langelaar M, Hendriks MA (2020) A critical evaluation of topology optimization results for strut-and-tie modeling of reinforced concrete. Comput Aided Civ Eng 35(8):850–869
29. Vantyghem G, de Corte W, Shakour E, Amir O (2020) 3D printing of a post-tensioned concrete girder designed by topology optimization. Autom Constr 112:103084
30. Bi M, Tran P, Xia L, Ma G, Xie YM (2022) Topology optimization for 3D concrete printing with various manufacturing constraints. Addit Manuf 57:102982
31. nTopology Inc. (2025) nTop
32. Lindemann H, Gerbers R, Ibrahim S et al (2019) Development of a Shotcrete 3D-printing (SC3DP) technology for additive manufacturing of reinforced freeform concrete structures. In: Wangler T, Flatt RJ (eds) First RILEM international conference on concrete and digital fabrication—digital concrete 2018. Springer International Publishing, Cham, pp 287–298
33. Kloft H, Hack N, Sawicki B et al (2025) Digital fabrication strategies for sustainability. In: Różański A, Bui Q-B, Sadowski Ł, Tran MT (eds) Proceedings of the 4th international conference on sustainable development in civil, urban and transportation engineering. Springer Nature Singapore, Singapore, pp 35–44
34. Hack N, Kloft H (2020) Shotcrete 3D printing technology for the fabrication of slender fully reinforced freeform concrete elements with high surface quality: a real-scale demonstrator. In: Bos FP, Lucas SS, Wolfs RJ, Salet TA (eds) Second RILEM international conference on concrete and digital fabrication. Springer International Publishing, Cham, pp 1128–1137
35. Online Article—House with One Wall/Christian Kerez. https://www.archdaily.com/604014/house-with-one-wall-christian-kerez. Accessed May 2025
36. Fingerloos F, Hegger J, Zilch K (eds) (2016) Eurocode 2 für Deutschland: DIN EN 1992-1-1 Bemessung und Konstruktion von Stahlbeton- und Spannbetontragwerken - Teil 1-1:

Allgemeine Bemessungsregeln und Regeln für den Hochbau mit Nationalem Anhang: kommentierte und konsolidierte Fassung, 2nd edn. Beuth; Ernst & Sohn, Berlin, Wien, Zürich, Berlin

37. Kloft H, Krauss H-W, Hack N et al (2020) Influence of process parameters on the interlayer bond strength of concrete elements additive manufactured by Shotcrete 3D printing (SC3DP). Cem Concr Res 134:106078
38. Dörrie R, Kloft H (2022) Force flow compliant robotic path planning approach for reinforced concrete elements using SC3DP. In: Buswell R, Blanco A, Cavalaro S, Kinnell P (eds) Third RILEM international conference on concrete and digital fabrication. Springer International Publishing, Cham, pp 370–375
39. Dörrie R, David M, Freund N, Lowke D, Dröder K, Kloft H (2023) In-process integration of reinforcement for construction elements during Shotcrete 3D printing. Open Conf Proc, 3
40. Lachmayer L, Böhler D, Freund N, Mai I, Lowke D, Raatz A (2023) Modelling the influence of material and process parameters on Shotcrete 3D printed strands—cross-section adjustment for automatic robotic manufacturing. Autom Constr 145:104626
41. Dressler I, Freund N, Lowke D (2020) Control of strand properties produced with Shotcrete 3D printing by accelerator dosage and process parameters. In: Bos FP, Lucas SS, Wolfs RJ, Salet TA (eds) Second RILEM international conference on concrete and digital fabrication. Springer International Publishing, Cham, pp 42–52
42. David M, Freund N, Dröder K, Lowke D (2023) The effects of nozzle diameter and length on the resulting strand properties for Shotcrete 3D printing. Mater Struct 56(8)
43. Li C, Petzold F (2023) Towards informed design decision support of additive manufacturing in construction—the use of integrated knowledge in BIM-based architectural design. Archit Plann J (APJ) 28(3)
44. Slepicka M (2021) BIM für die Additive Fertigung im Bauwesen. In: Borrmann A, König M, Koch C, Beetz J (eds) Building information modeling. Springer Fachmedien Wiesbaden, Wiesbaden, pp 545–561
45. Slepicka M, Vilgertshofer S, Borrmann A (2021) Fabrication information modeling: closing the gap between building information modeling and digital fabrication. In: 38th international symposium on automation and robotics in construction, Dubai, UAE, Nov 02–04 2021
46. Mishra AS (2021) Social acceptance prediction model for generative architectural spaces in India. JoADRCUA 6(3):50–57
47. García López E, Heard C (2023) Social acceptance of a thermal architectural implementation proposal. Sustainability 15(5):4121
48. Zhou A, Thomaschke R, Wessels A, Glunz S, Speck T, Kiesel A (2024) (Not) in my city: an explorative study on social acceptance of photovoltaic installations on buildings. Technol Soc 79:102725
49. Pasetti Monizza G, Bendetti C, Matt DT (2018) Parametric and generative design techniques in mass-production environments as effective enablers of Industry 4.0 approaches in the building industry. Autom Constr 92:270–285

3D Concrete Printing in the Global South: Evaluating Its Role in Mitigating Africa's Housing Crisis

Luca Breseghello(✉)

Technical University of Denmark (DTU), Lautrupvang 15, 2750 Ballerup, Denmark
lucbe@dtu.dk

Abstract. This study investigates the potential of 3D concrete printing (3DCP) as a sustainable solution to the housing crisis in Africa and the Global South. Rapid urbanisation and population growth in these regions present significant challenges in meeting housing demands, with traditional construction methods failing to keep pace. 3DCP offers a promising alternative by enabling cost-effective, resource-efficient, and locally adaptable construction processes. The benefits of 3DCP, such as reduced material waste, design flexibility, and the incorporation of locally sourced materials, are explored in the context of addressing environmental and economic challenges related to imported building materials. Furthermore, integrating 3DCP with regional development goals presents opportunities for job creation, skill development, and local manufacturing of printers and materials. However, challenges remain in adopting this technology, including the need for tailored regulatory frameworks, technical training, and investment in infrastructure. The research also highlights the importance of policy development, economic incentives, and community engagement in facilitating widespread adoption. In conclusion, with strategic efforts in policy, education, and technological innovation, 3DCP has the potential to significantly contribute to solving Africa's housing crisis and advancing sustainable construction practices across the Global South.

Keywords: 3D Concrete Printing · African housing · Housing crisis · Sustainability · Resource-efficiency

1 Introduction

According to UN-Habitat [1], approximately 55% of the global population currently resides in urban areas. This figure is projected to rise to around 68% by 2050, adding roughly 2.2 billion urban inhabitants. 90% of this urban expansion will occur in cities and urban agglomerations in Asia and sub-Saharan Africa. Historical trends indicate that Africa's urbanization rate has increased from approximately 15% in 1960 to 40% in 2010, with projections suggesting a further increase to 60% by 2050 (Fig. 1). Consequently, the urban population in Africa is expected to triple over the next 50 years, thereby fundamentally transforming the region's demographic and infrastructural landscape. To accommodate the escalating demand driven by rapid population growth and

H. Beushausen et al. (Eds.): NTZR 2025, LNCE 873, pp. 485–496, 2026.
https://doi.org/10.1007/978-3-032-27698-8_44

urbanisation, it is estimated that at least 40 million new housing units will be required, with most of these developments concentrated in these regions [2]. The accelerated rate of urbanisation in Africa poses a substantial challenge to the development of affordable housing.

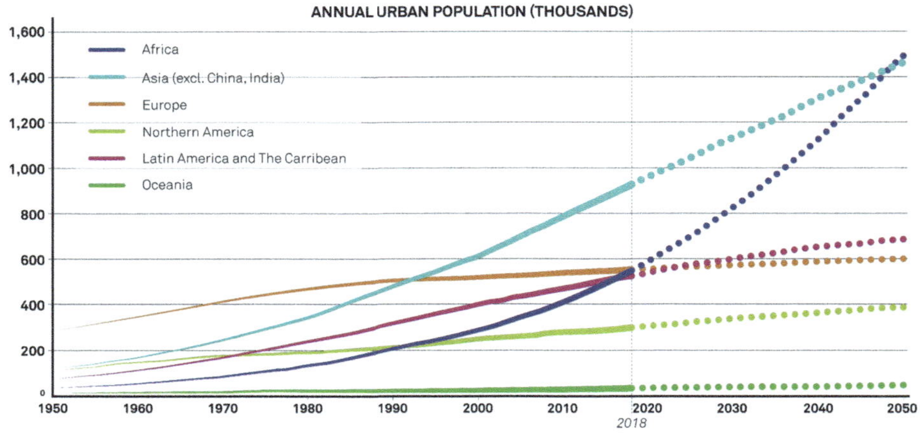

Fig. 1. Urban population per continent, data and projection (1950–2050). Data from [3].

Conventional construction methods have proven insufficient in addressing these challenges sustainably, widely contributing to the global climate crisis and poor living conditions. Environmentally unsustainable materials and production practices are coupled with high reliance on labour-intensive and inefficient processes, decreasing access to skilled workers and, consequently, lengthy construction timelines and high costs. In sub-Saharan Africa, these challenges are further fuelled by the negative continent's economic situation and by the absence of policies and regulatory frameworks, low awareness, knowledge and acceptance of the need for building standards and environmental policies by governing bodies across the continent. Moreover, reliance on labour-intensive practices, in the context of limited access to skilled labour and locally sourced sustainable materials, has been shown to increase construction timelines and costs by 30–50% in some studies [4].

Consequently, a multidisciplinary approach is essential to develop and integrate innovative construction strategies that enhance productivity and improve building processes' environmental, economic, and social sustainability. In addition to robust policy frameworks and economic support from governing institutions, the successful adaptation and implementation of sustainable, efficient, and locally optimised building technologies and materials are critical to mitigating the African housing crisis. 3D Concrete Printing (3DCP) is increasingly being explored globally as a promising technology to address construction challenges by shortening construction times, making construction sites safer, reducing material waste through formwork-free fabrication and optimising material usage through design engineering. Both academic and industrial sectors are intensifying efforts to advance 3DCP technologies, encompassing robotic motion systems, nozzles, pumps, materials, and design and engineering solutions to harness the full potential of additive manufacturing. However, a significant disparity exists between the Global North and Global South in terms of resources, focus, as well as research and industry outputs in this field. Most 3DCP developments are concentrated in the Global North, characterised by high technological standards and abundant resources. Nevertheless, a growing body of 3DCP academic work and industrial projects developed in the Global South exists.

1.1 Slums, Informal Settlements and Inadequate Housing

Understanding the scope of the crisis requires clarity on terminology related to housing deprivation. According to the United Nations, a *slum* household is defined as a group of individuals residing together in an urban setting who lack one or more of five essential conditions: structurally sound and durable housing, sufficient living area (no more than three people per room), access to safe water at an affordable price, adequate sanitation facilities, and secure tenure protecting against forced eviction [3]. Expanding on this, Sustainable Development Goal Indicator 11.1.1 broadens the focus to encompass *informal settlements*—dwellings constructed without legal claim to the land, necessary planning permissions, or adherence to formal building codes—and *inadequate housing*. The latter category includes housing that fails fundamental habitability criteria due to poor location, lack of safety, social exclusion, accessibility barriers, cultural inappropriateness, or unaffordability, often benchmarked as housing costs consuming over 30% of household income. Within this framework, *affordable housing* is understood as adequate shelter in quality and location, with costs that do not prevent occupants from meeting other basic living expenses or compromise their fundamental human rights.

1.2 Research Aim

This study provides an overview of the status of concrete construction in Africa, evaluating the viability of 3DCP as an efficient technology to sustainably mitigate the housing deficit in Africa and the Global South. Specifically, it seeks to assess the applicability of 3DCP in addressing the acute housing shortages prevalent in these regions, where traditional construction methodologies have proven inadequate in delivering affordable and scalable housing solutions. The investigation focuses on the sustainability dimensions of 3DCP, encompassing its environmental, economic, and social impacts. These facets are essential for understanding how 3DCP can contribute to sustainable development, particularly in areas characterised by resource constraints and rapid urbanisation. Furthermore, the research aims to identify the principal barriers impeding the widespread adoption of 3DCP in these regions. These challenges include technical issues such as the necessity for specialised training, the substantial initial capital investment required for 3D printing equipment, and economic and regulatory obstacles that must be addressed to enhance the technology's feasibility and scalability in these contexts.

2 Cement and Concrete Construction Development

Given the scale and urgency of housing and infrastructure needs across the continent, concrete construction is one of the most practical and scalable techniques. Concrete durability, adaptability to local conditions, and potential for industrialised production make it essential to meet Africa's growing and rapid demand for buildings and infrastructure. In the last two decades, the concrete industry in Africa has undergone a market transformation driven by economic shifts rather than regulatory reform or anti-competitive behaviour. Reference [6] points out that, until the 2010s, high cement prices and low consumption were the norm, primarily due to small national markets and limited production capacity. However, as several African economies experienced rapid growth, new cement plants, often larger and more efficient, entered the market, increasing supply and reducing marginal costs. This expansion led to a significant drop in cement prices between 2011 and 2017, the steepest of any continent. The evolution signals increased demand and a gradual industrial scaling-up, with implications for how concrete might support more affordable and rapid construction across the region. Figure 2 illustrates that, while Africa's cement production remains comparatively low, it is projected to experience a substantial increase by 2050, surpassing most regions in growth rate. This anticipated surge underscores the continent's accelerating demand for construction materials, reflecting broader trends in urbanisation, infrastructure development, and housing expansion.

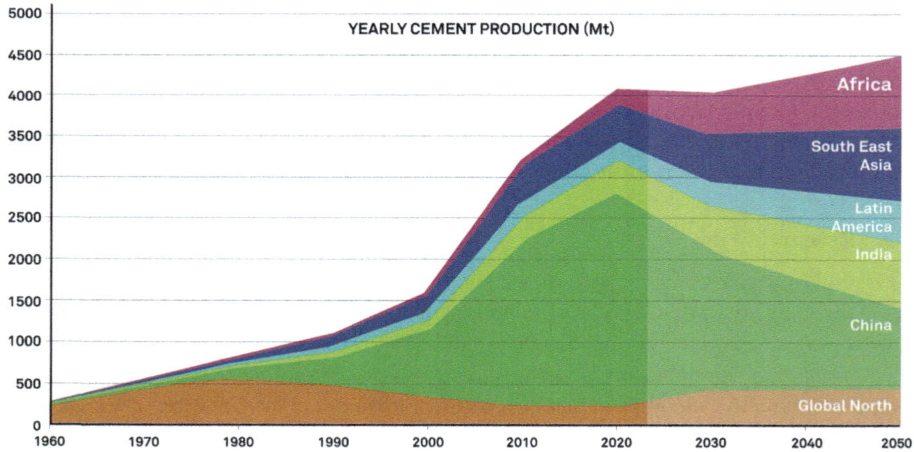

Fig. 2. Cement production from 1960 to 2050 (projection) (Adapted from [7])

While concrete remains the most viable material for addressing the continent's pressing housing and infrastructure demands, the cement industry still faces significant challenges, reflecting its immaturity [8]. Construction practices are often manual and decentralised, with on-site mixing still the norm due to the limited penetration of ready-mix and precast concrete systems. Climatic conditions - high temperatures and humidity - further challenge the application of standards developed for temperate climates.

Cement plant density remains low, leading to high transportation costs that inflate local prices and limit accessibility, especially in inland or poorly connected regions. This creates space for international suppliers, resulting in various imported cement types with inconsistent quality control and lower local development. The local production landscape is dominated by Ordinary Portland Cement (OPC) and a limited number of blended variants, with little availability of industrial by-products like fly ash to enhance performance or reduce carbon intensity. Although new plants are being commissioned across the continent, the pace of industrial scaling is still catching up to demand. Consequently, the cement sector in Africa remains fragmented and heavily dependent on geography, logistics, and international trade dynamics. This underscores the urgent need for local capacity building, regulatory harmonisation, and investment in infrastructure to unlock the full potential of concrete construction on the continent.

The limited correlation between cement prices and overall construction costs highlights the concrete industry's infancy on the continent [8]. As illustrated in the image (Fig. 3), this variability underscores the instability inherent in a sector still hampered by inadequate infrastructure, fragmented supply chains, and a lack of economies of scale. These factors contribute to a cost structure that is more volatile than in more mature markets, signalling a pressing need for sectoral consolidation and professionalisation as the industry develops.

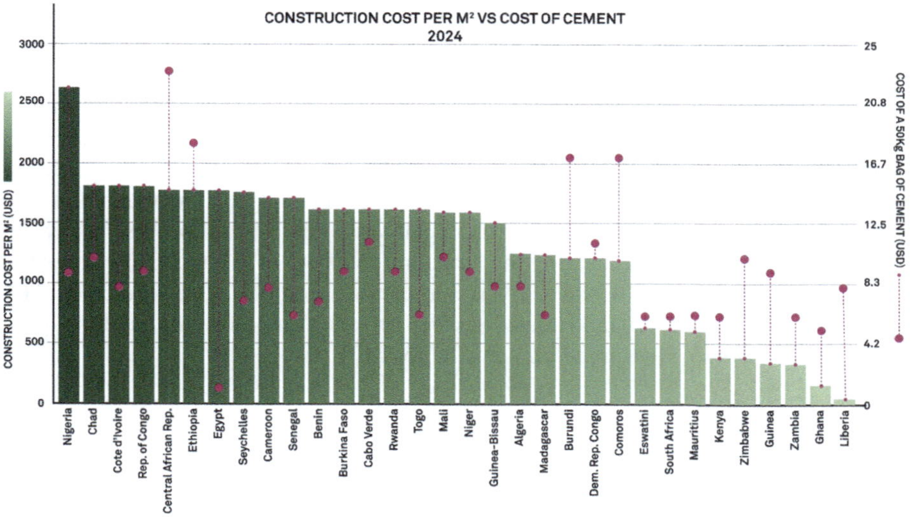

Fig. 3. Cost comparison of overall construction per square meter against the cost of a 50 kg bag of cement for African countries (according to availability). While construction costs are relatively consistent (bars), cement prices are highly variable (lines). (Adapted from [8])

A significant proportion of sub-Saharan Africa's future building stock remains unconstructed, and considering the growth of the region's concrete industry, a strategic opportunity exists to embed emission-reducing solutions and sustainability principles throughout the concrete construction supply and value chain [9]. This entails the adoption of innovative technologies, the utilisation of sustainable materials, and the implementation of design strategies that prioritise resource efficiency and environmental stewardship. To realise this potential, supportive policies and economic incentives are essential. Such an integrated approach can facilitate urbanisation that addresses housing and infrastructure needs while minimising material consumption and environmental impact.

3 Status of 3D Concrete Printing in Africa

3DCP encompasses advanced additive manufacturing techniques that fabricate concrete structures layer by layer, eliminating the need for traditional formwork and enhancing design flexibility. The primary method employed is extrusion-based printing, wherein a concrete mixture is precisely extruded through a nozzle to form successive layers, guided by digital instructions derived from a 3D model. This technique uses various systems, notably gantry-based setups and robotic arms. Gantry systems operate on fixed frames that move the printhead along Cartesian coordinates, facilitating large-scale constructions with high precision. Conversely, robotic arm systems offer multiple axes of movement, enabling the creation of complex geometries and intricate designs that are challenging to achieve with conventional methods [10]. 3DCP offers potential advantages, including the ability to produce complex geometries that are difficult to achieve with traditional techniques. Additionally, it reduces material waste through precise material placement and decreases reliance on manual labour via automation [11]. Other potential benefits, though not yet fully realised or proven, include faster construction times and cost savings by minimising formwork use and optimising material efficiency. However, the adoption of 3DCP faces several challenges. A current hurdle is developing printable concrete mixes with suitable rheological properties, such as workability and buildability, balancing them with environmental sustainability and process scalability. Another critical concern is ensuring the structural integrity of printed elements, particularly concerning bonding between layers and the incorporation of reinforcement [12]. Moreover, the absence of standardised testing protocols and design guidelines poses additional barriers to widespread implementation [13].

3.1 Research and Technology Development

The state-of-the-art 3DCP in Africa showcases an emerging field with significant potential to, at least partially, mitigate the continent's pressing housing deficits [14]. While still in its infancy, the number of small-scale buildings manufactured with 3DCP is growing in African nations, including South Africa, Kenya, Malawi and Angola. These initiatives range from pilot projects to the construction of the first 3D-printed houses and schools, involving universities, local companies, NGOs, and international partners. The key focus is leveraging 3DCP to deliver faster, affordable, and sustainable housing solutions. Barriers to widespread development exist in the upskilling of the workforce, the necessary upfront investment, and the continuous technological advancement of 3DCP hardware and materials.

Research Initiatives. Since 2016 [15], a team of researchers from the Department of Civil Engineering at Stellenbosch University, South Africa, has conducted research on materials and structural engineering for 3DCP. Their research focuses on creating printable materials, ranging from lightweight to high-performance concrete, and developing analytical models for printed structural elements. In 2020, the University of Johannesburg, South Africa 2020, started research and a pilot program to demonstrate the effectiveness of 3DCP for sustainable human settlements, working on studies on cost-benefit analysis and qualitative research on the availability of 3DCP materials and social-technical acceptability. In Egypt, preliminary investigations have been conducted into the applicability, impact, and constraints of 3DCP [17, 18]. While 3DCP shows promise for enhanced environmental and productivity outcomes, their qualitative analyses reveal that limited technical and economic resources, weak policy frameworks, and low stakeholder awareness significantly hinder its adoption. Moreover, they highlight the need for region-specific materials tailored to local climate, resource availability, and traditional practices.

Technology Development. The IROKO™ 3D Concrete Printing Machine was developed through a collaboration between 14Trees and PMSA, and it is designed for mobility, affordability, and reliability. It features a patented technology, namely LevelTrac™, for easy setup on various sites and is built to withstand harsh conditions, ensuring long-term usability. Local technology development is an important step towards independent, sustainable growth of the 3DCP construction sector in the African continent, with hardware, materials and knowledge adapted to the specific context. While 14Trees was founded in Africa, it has moved parts of its operation to Spain, intending to expand globally.

3.2 Notable Projects

While the number of research institutions in Africa focusing on 3DCP remains limited, the continent has witnessed a notable increase in the construction of 3D-printed buildings. This effort is largely driven by construction companies' interest in finding more efficient construction strategies and by technology companies, NGOs and international partners.

The chronological development of 3DCP projects in Africa reveals a transition from small-scale prototypes to increasingly complex, application-oriented structures. One of the earliest documented initiatives was BeMore3D's 53 m^2 prototype house constructed during the Solar Decathlon Africa in Benguerir, Morocco, in 2019, which tested the viability of extrusion-based 3DCP under North African environmental conditions. In 2021, 14Trees—a joint venture between Holcim and British International Investment—constructed a 3D-printed primary school in Malawi's Salima District, achieving complete wall fabrication within 18 h. Concurrently, Thinking Huts completed a 3D-printed school in Fianarantsoa, Madagascar, employing hybrid methods that integrated printed walls with conventional roof and finishing systems to facilitate community-based implementation. In 2022, 14Trees advanced the sustainability profile of 3DCP with a prototype house in Nairobi, Kenya, which became the first 3D-printed building globally to receive IFC EDGE Advanced certification, demonstrating a 42% reduction in operational energy consumption, a 24% reduction in water usage during construction, and a 69% reduction in embodied energy in materials, relative to baseline construction methods. Also in 2022, Power2Build completed a 53 m^2 residence in Angola using COBOD International's BOD2 gantry system, reporting improved print speed, labour efficiency, and potential cost savings. Simultaneously, 14Trees launched the Mvule Gardens project in Kilifi, Kenya, comprising 52 housing units within a regenerative development framework.

Expansion continued in Malawi by printing a prototype house in Lilongwe in 2023—reportedly completed in 12 h—and delivering a 3D-printed Child-Friendly Space in Phalombe District. Hassell Studio has announced a 2024 plan to construct the Hope Village Community Centre in Tanzania using 3DCP and locally sourced materials. More partnerships and companies adopting 3DCP are growing, such as Orascom 3D in Egypt. These developments reflect both the technical maturation of 3DCP in diverse African contexts and its strategic deployment to address region-specific infrastructure gaps, particularly in housing and education (Table 1).

Table 1. Overview of 3DCP projects developed in Africa. All projects can be found on the company's website.

Company	Location	Project/research	Year
BeMore	Ben Guerir, Morocco	53 m^2 temporary showcase house	2019
14Trees	Salima District, Malawi	3D-printed school	2021
14Trees, Thinking Huts	Fianarantsoa, Madagascar	3D-printed school	2021
Power2Build, COBOD	Luanda, Angola	3D-printed house	2021
14Trees	Nairobi, Kenya	1st IFC EDGE Advanced 3D-printed building in the world	2022
14Trees	Kilifi, Kenya	Mvule Gardens – 52-Unit Housing Project	2022
Power2Build	Angola	53 m^2 house	2022
CyBe Construction	Johannesburg, South Africa	UJ House	2022
14Trees	Lilongwe, Malawi	Prototype house	2023
14Trees	Phalombe District, Malawi	Child-friendly space (CBCC)	2023
Hassell Studio	Tanzania	Hope Village Community Building	2024

4 Barriers and Strategies for Sustainable Scaling of 3DCP in Africa

Despite the promising potential of 3DCP to broadly impact the construction industry in Africa, several significant barriers hinder its widespread adoption. High initial capital investments required for procuring 3D printing equipment and materials pose a substantial financial challenge, particularly in economies where construction budgets are already constrained. Additionally, the novelty of the technology results in a scarcity of skilled operators and a lack of standardised training programs, leading to a workforce unprepared for its implementation. The absence of comprehensive building codes and regulatory frameworks tailored to 3DCP further complicates its integration into existing construction practices. Moreover, traditional procurement processes, which favour conventional construction methods, often exclude innovative technologies like 3DCP, thereby limiting opportunities for its application. Reference [18] highlights these challenges, noting that the South African construction industry exhibits reluctance toward adopting innovative solutions, citing factors such as high capital costs, potential labour displacement, and a general lack of awareness and training opportunities for on-site employees. Without targeted interventions to address financial, educational, and regulatory obstacles, the adoption of 3DCP in Africa will remain limited.

4.1 Utilisation of Locally Sourced Material

In pursuing the sustainable scaling of 3DCP across Africa, using locally sourced materials is critical to address the continent's growing housing and infrastructure challenges. Africa's abundant natural resources, particularly clay deposits, offer significant potential for sustainable construction. These resources can be used to produce limestone calcined clay cements (LC3), limiting reliance on imported clinker, thereby decreasing associated carbon emissions and mitigating price volatility in cement markets [19]. However, using locally sourced materials in 3DCP requires ensuring consistent material quality, which necessitates the development of robust testing protocols and standardisation. Additionally, raw material extraction can have environmental consequences, as illegal mining contributes to riverbank erosion, habitat destruction, and local conflicts [20]. Similarly, poorly regulated limestone quarrying can degrade air and water quality. Africa's tradition of vernacular construction, utilising earth-based materials like adobe and rammed earth, offers valuable insights into sustainable building practices. Often considered low-performance, these materials are inherently sustainable and suitable for non-structural applications.

Furthermore, Africa's agricultural sector produces substantial biomass waste that can be converted into bio-based construction materials. Agricultural byproducts such as rice husks and coconut shells, which have pozzolanic properties, can enhance 3DCP by improving rheology and buildability [21].

In conclusion, integrating locally sourced and secondary materials into 3DCP provides a multi-dimensional opportunity to advance sustainable construction in Africa. This strategy can help meet the continent's housing needs while minimising environmental impact, promoting economic resilience, and preserving cultural heritage.

4.2 Resource, Climate and Socially Responsible Design

The design freedom at no cost inherent in 3DCP offers the possibility of designing and building housing and optimising environmental, economic and social costs, which is of paramount relevance in rapidly urbanising regions with a pressing and growing need for housing. Computational design methods allow the precise control of parameters such as wall thickness and internal void geometry, thereby reducing material usage while improving resilience and climate adaptation. Moreover, the same formal flexibility offers avenues to seamlessly integrate new structures into existing complex urban fabrics by designing buildings that complement established cities' spatial, cultural, and historical contexts. Parallelly, integrating the local manual workforce into these processes is of utmost importance, as well as devising hybrid construction processes where man and machine collaborate in the construction process, e.g., designing and 3D printing small-scale blocks that workers can manually place.

4.3 Policy Development and Economic Support

Developing appropriate policy frameworks and economic support mechanisms is crucial for the widespread adoption of 3DCP across Africa. A primary challenge in implementing 3DCP is the absence of tailored building codes and standards that address the distinct characteristics of this emerging technology. However, Africa's relatively low standardisation provides a unique opportunity to bypass the constraints of outdated regulatory frameworks and establish innovative, context-specific standards that align with regional needs and sustainability objectives. By proactively developing forward-thinking policies, Africa can set a global precedent for integrating 3DCP into sustainable urban development, ensuring safety and efficiency in its deployment.

Economic support through targeted government incentives, public-private partnerships, and promoting local manufacturing capabilities for hardware and knowledge related to 3DCP—including materials, technology, and required design and engineering expertise—can significantly reduce costs and cultivate local expertise. By fostering the development of region-specific policies and economic frameworks, 3DCP can be effectively scaled to address Africa's pressing housing and infrastructure demands, driving innovation, sustainable growth, and economic resilience.

4.4 Promoting Social Acceptance and Capacity Building

Addressing social barriers and fostering awareness is crucial to facilitate the widespread adoption of 3DCP in Africa. Implementing awareness campaigns, training programs, and community engagement initiatives can help build trust in the technology and encourage its acceptance. Additionally, promoting research and development focused on locally improving 3DCP technology, material properties, and construction processes is essential for empowering local industries and workforces. These efforts will not only enhance the technical capabilities of 3DCP but also ensure that the technology is tailored to meet regional needs and capacities, thereby driving both technological innovation and socio-economic development.

5 Conclusions

This study has explored the potential of 3DCP as an innovative solution to Africa's affordable housing crisis. As CAHF notes, "*We cannot rely on precedent in unprecedented times*" [8]. Embracing innovative construction technologies like 3DCP can become tools to address the housing crisis effectively and sustainably. While 3DCP offers advantages such as reduced construction time and material efficiency, its adoption is hindered by significant barriers, including high initial investment costs, lack of skilled labour, and inadequate regulatory frameworks. Addressing these challenges requires concerted efforts from governments, industry stakeholders, and educational institutions to develop supportive policies, invest in training programs, and create an enabling environment for innovation and for including and empowering local communities in steering the process. However, first and foremost, working with technology in sub-Saharan Africa means not only providing tools but also learning, and adapting them to the cultural, social and economic framework in which they are implemented.

References

1. UN-Habitat (2022) World cities report 2022: envisaging the future of cities. United Nations Human Settlements Programme, Nairobi, Kenya
2. Behr DM, Chen L, Goel A, Haider KT, Singh S, Zaman A (2021) Introducing the Adequate Housing Index (AHI): a new approach to estimate the adequate housing deficit within and across emerging economies (Policy Research Working Paper No. 9830). International Finance Corporation
3. United Nations Human Settlements Programme (UN-Habitat) (2006) The state of the world's cities 2006/2007: The millennium development goals and urban sustainability: 30 years of shaping the Habitat Agenda
4. Mahame C, Kikwasi GJ, Baruti MM (2023) Barriers to the effective selection of sustainable materials for residential building projects: a qualitative study. Sustainability 16(21):9526. https://doi.org/10.3390/su16219526
5. Leone F, Macchiavello R, Reed T (2025) The high and falling price of cement in Africa. Am Econ J Appl Econ 17(2):1–40
6. École Polytechnique Fédérale de Lausanne (EPFL) (2023) The future of construction with more sustainable cement. EPFL. https://www.epfl.ch/labs/lmc/the-future-of-construction-with-more-sustainable-cement/. Accessed 27 Feb 2025
7. Schmidt W, Tchetgnia Ngassam I, Breitschaft G, Virchow S (2018) Challenges of the growing African cement market—environmental issues, regulatory framework, and quality infrastructure requirements. In Diouri A, Boukhari A, Ait Brahim L, Bahi L, Khachani N, Saadi M, Aride J, Nounah A (eds) MATEC web of conferences, 149, 01014. Federal Institute for Materials Research and Testing (BAM), Berlin. https://doi.org/10.1051/matecconf/201814901014
8. Centre for Affordable Housing Finance in Africa (CAHF) (2024) Housing finance in Africa: a review of Africa's housing finance markets 2024. Retrieved from https://housingfinanceafrica.org
9. Schmidt W et al (2021) Sustainable circular value chains: From rural waste to feasible urban construction materials solutions. Dev Built Environ 6:100047. https://doi.org/10.1016/j.dibe.2021.100047
10. Buswell RA et al (2020) A process classification framework for defining and describing digital fabrication with concrete. Cem Concr Res 134:106068. https://doi.org/10.1016/j.cemconres.2020.106068
11. Hassan H, Rodriguez-Ubinas E, Al Tamimi A, Trepci E, Mansouri A, Almehairbi K (2024) Towards innovative and sustainable buildings: a comprehensive review of 3D printing in construction. Autom Constr 163:105417. https://doi.org/10.1016/j.autcon.2024.105417
12. Wolfs R (2023) The status quo of 3D concrete printing: are we there yet? RILEM Tech Lett
13. Vasilic K (2025). Standardization aspects of concrete 3D printing: state of the art, requirements and first steps towards standardization. RILEM Tech Lett 9:98–105. https://doi.org/10.21809/rilemtechlett.2024.201
14. Moghayedi A, Mahachi J, Lediga R, Mosiea T, Phalafala E (2024) Revolutionizing affordable housing in Africa: a comprehensive technical and sustainability study of 3D-printing technology. Sustain Cities Soc 105:105329. https://doi.org/10.1016/j.scs.2024.105329

15. Van Zijl GPAG, Paul SC, Tan MJ (2016) Properties of 3D printable concrete. In: Proceedings of the 2nd international conference on progress in additive manufacturing (Pro-AM 2016), pp 421–426
16. Geneidy O, Ismaeel WSE, Abbas A (2019) A critical review for applying three-dimensional concrete wall printing technology in Egypt. Archit Sci Rev 62(5):438–452. https://doi.org/10.1080/00038628.2019.1596066
17. El-din AA, Ibrahim M, El-Sayed Z (2021) Applying 3D printing as a new building technology: potentials and challenges in the Egyptian context. REAL CORP, Vienna. CORP—Competence Center of Urban and Regional Planning
18. Windapo A, Yende M, Kala O, Maboreke L, Matshidze L (2023) The adoption of 3D printing as a construction methodology in South Africa: what are the barriers? In: Haupt TC, Akinlolu M, Simpeh F, Amoah C, Armoed Z (eds) Construction in 5D: deconstruction, digitalization, disruption, disaster, development. Lecture Notes in Civil Engineering, vol 245. Springer, Cham
19. Scrivener KL, John VM, Gartner EM (2018) Eco-efficient cements: potential economically viable solutions for a low-CO_2 cement-based materials industry. Cem Concr Res 114:2–26. https://doi.org/10.1016/j.cemconres.2018.03.015
20. Valentini L (2023) Sustainable sourcing of raw materials for the built environment. Mater Today: Proc. https://doi.org/10.1016/j.matpr.2023.07.308
21. Pimentel Tinoco M, Gouvêa L, De Cássia Magalhães Martins K, Dias Toledo Filho R, Aurelio Mendoza Reales O (2023) The use of rice husk particles to adjust the rheological properties of 3D printable cementitious composites through water sorption. Constr Build Mater 365:130046. https://doi.org/10.1016/j.conbuildmat.2022.130046

3D Concrete Printing in Africa: A Kenyan Perspective

Ian M. Mararo(✉) and Adewumi J. Babafemi

Stellenbosch University, Stellenbosch 7599, South Africa
28213343@sun.ac.za

Abstract. Innovative construction methods in the housing sector have the potential to overcome the challenges posed by rising construction costs and stagnant productivity. By addressing these bottlenecks, such advancements can significantly enhance housing delivery and urbanisation, ultimately benefiting all stakeholders involved. In recent years, the construction industry has experienced significant innovation breakthroughs, largely driven by the advent of Industry 4.0 technologies, with 3D concrete printing (3DCP) standing out as a transformative advancement. Gathering expert opinions from the Kenyan construction sector, where the largest 3DCP affordable housing project in Africa is underway, this paper aims to capture the attitudes and perceptions of construction stakeholders on the use of 3DCP in Kenya. Online interviews were conducted with stakeholders both in industry and academia. The interviews revealed that economic and socio-cultural factors are the primary barriers to the adoption of 3DCP for affordable housing in Kenya. The main economic barrier cited was the high cost, while resistance to change was the dominant socio-cultural challenge. Addressing these challenges requires a multi-pronged, multi-stakeholder approach. To lower costs, a combination of research and development efforts, along with 3DCP-specific policies such as tax incentives, could be implemented. To overcome resistance to the technology, social networks emerged as the key strategy for raising awareness among construction stakeholders.

Keywords: 3D concrete printing · Stakeholders · Kenya · Perspective

1 Introduction

Innovations are viewed as tools that improve on a sector's previous results through changes that enhance project efficiency and effectiveness [1]. Unlike the manufacturing and energy sectors, which have achieved substantial productivity gains through innovations, the construction industry has lagged [2]. It is characterised by temporary project-based teams, fragmented supply chains, thin profit margins, over-reliance on low-skilled labour and lengthy development cycles [3, 4]. These inherent characteristics make the construction industry one of the least digitised industries globally [5]. Noticeably, the industry has experienced significant innovation breakthroughs in recent years, largely driven by the advent of Industry 4.0 technologies, with 3D concrete printing (3DCP)

H. Beushausen et al. (Eds.): NTZR 2025, LNCE 873, pp. 497–508, 2026.
https://doi.org/10.1007/978-3-032-27698-8_45

standing out as a transformative advancement [6]. Innovative construction methods, such as 3DCP, have the potential to overcome the challenges posed by rising construction costs and stagnant productivity [7].

Innovations such as 3DCP can really benefit the African housing construction sector. Currently, with an urban population of about 700 million, the region faces extreme difficulties in providing adequate housing for its urban dwellers [8]. Several affordable housing schemes have been implemented across Africa such as the Family homes Fund in Nigeria, the Affordable Housing Program in Kenya and the Comprehensive Housing Plan in South Africa however, they all share the same challenges of slow delivery process [9]. Improving construction productivity is especially critical for the region's housing sector, which faces acute shortages, particularly in Kenya, where the annual housing deficit is estimated at 200,000 units [10]. Slow delivery of projects has been attributed to the limited used of construction innovations coupled with the prevalent use of traditional construction methods [11–13]. 3DCP has the potential to revolutionise affordable housing delivery, addressing the challenges of high costs, poor quality and slow project execution.

3DCP uses specialised concrete to create monolithic structures without the need for formwork. This formwork-less approach to construction gives 3DCP the potential to enhance construction productivity in developing regions to simultaneously cope with the shortages of built infrastructure and urbanisation challenges [14]. Despite these benefits, 3DCP adoption has been relatively low [15]. While slow adoption is often attributed to industry-specific challenges, such as complex construction sites, fragmented supply chains, and temporary teams, other critical factors unrelated to the industry's inherent characteristics also play a role [16]. One such factor is the readiness of the target market to embrace and implement the innovation, a challenge evident in Africa's 3DCP adoption landscape [17].

The widespread adoption of 3DCP, like most construction innovations, relies on both financial and knowledge resources to facilitate its implementation [18]. Financial resources are critical for funding research and development to enhance innovation, while knowledge resources are essential for effectively communicating its functionality and benefits to gain stakeholder support [19]. Stakeholders—such as contractors, designers, clients, end-users, developers and governments—play a pivotal role in either driving or hindering the adoption of construction innovations [20]. The primary source of knowledge in construction lies with these individuals, as they hold the tacit expertise necessary to refine and adopt new construction innovations, thereby enhancing buildability [21, 22]. Beyond resolving technical challenges, promoting 3DCP adoption also relies on resolving non-technical factors such as supply chain management, standardisation and customer acceptance [23]. Both dimensions rely on strong support and cooperation from construction stakeholders; therefore, resistance from any party is likely to hinder the successful implementation of 3DCP in the industry [24]. Capturing and addressing these stakeholders' perspectives and concerns is crucial when introducing novel construction technologies to the market [25]. The benefits and challenges of 3DCP are not uniform across industry stakeholders, as each group plays a distinct role in the construction process [25, 26]. Each of these stakeholder groups not only holds diverse perceptions towards 3DCP technology but also share similarities, which may be shaped by the extent to which 3DCP technology becomes prevalent and integrated into the industry [27].

Compared to the developed nations, there exists a dearth of studies analysing 3DCP technology in developing countries, particularly regarding its perception amongst industry stakeholders in regions like Kenya. To address this gap, the study aims to investigate how 3DCP is perceived by Kenyan construction stakeholders. Kenya is an ideal case study as it hosts the continent's largest 3D-printed housing project, providing a unique opportunity to assess stakeholders' perception toward the technology. Additionally, Kenya has a strong policy push for affordable housing construction which creates a conducive environment for analysing the potential application of 3DCP for affordable housing construction.

2 Research Method

This research adopted a qualitative approach using semi-structured online interviews as the primary data collection instrument. To strengthen the interpretation of the stakeholders' perceptions, the findings were also compared with three peered reviewed studies published within the last five years namely: [29] which investigated the application of 3DCP for housing delivery in South Africa, [30] which explored the use of 3DCP for remote housing solutions in Australia, and [31], which investigated factors affecting 3D printing technology in the Indian construction sector. Table 1 highlights the justification for using each case study.

Table 1. Justification for using studies

Study	Justification for comparative analysis
[29]	Explores the use of 3DCP in a region with comparable socio-economic dynamics and housing challenges as Kenya
[31]	Investigates adoption of 3DCP in a region with a higher industrialised economy but with similar developing status as Kenya
[30]	Explores 3DCP use in remote areas, which often reflect conditions in developing regions in terms of limited infrastructure and resource constraints

Purposeful sampling was used to select interviewees, a technique frequently used in qualitative research to gather insightful information with limited resources [32, 33]. Participants had to be construction stakeholders in academia or industry with at least 10 years of professional experience. This threshold ensured participants were registered with relevant regulatory bodies, had experienced policy shifts due to Kenya's five-year political cycles, and had been involved across an entire project's lifecycle, enabling informed perspectives on 3DCP adoption. Identifying perception differences between industry and academia is essential for fostering collaboration and driving commercialisation of construction technologies [28]. Snowball sampling method was subsequently used to identify additional participants who met the criteria resulting in 11 interviews—five from academia and six from industry. This sample size was considered sufficient for qualitative research aligning with recommendations that six to ten interviews achieve meaningful insights [34].

The profile of the interviewees is presented in Table 2. It is important to note that interviewee P4 was included in the study (despite having less than 10 years of construction experience) because all their experience has been in 3DCP which proved beneficial to the study.

Table 2. Demographics of interviewees

Interviewee ID	Stakeholder category	Area of expertise	Construction experience
A1	Academic	Civil and structural engineering	14 years
A2	Academic	Structural and material engineering	13 years
A3	Academic	Structural engineering	28 years
A4	Academic	Architecture and green construction	13 years
A5	Academic	Civil engineering	16 years
P1	Industry	Quantity surveying and contract management	16 years
P2	Industry	Architecture and urban planning	25 years
P3	Industry	Building services engineering	22 years
P4	Industry	Materials engineering	5 years
P5	Industry	Civil and structural engineering	25 years
P6	Industry	Construction project management	12 years

3 Data Analysis

3.1 Awareness

All interviewees demonstrated awareness of 3DCP as shown in Fig. 1 however, the depth of understanding varied considerably and was linked to the stakeholders' practical experience with the technology. Stakeholders P4 and P6 exhibited the most knowledge evidenced by their practical experience with 3DCP:

"I would say I'm very familiar with the field. I've been involved in the field since it basically started, which I would say was around 2018 when it really started to kick off" – P4

"I have delivered more than 100,000 square meters of buildings including the biggest affordable housing project in Africa, located in Kenya. We recently delivered the first 3D concrete printed control room for a solar farm in the world, a total of 480 square metres.... This is my expertise" – P6

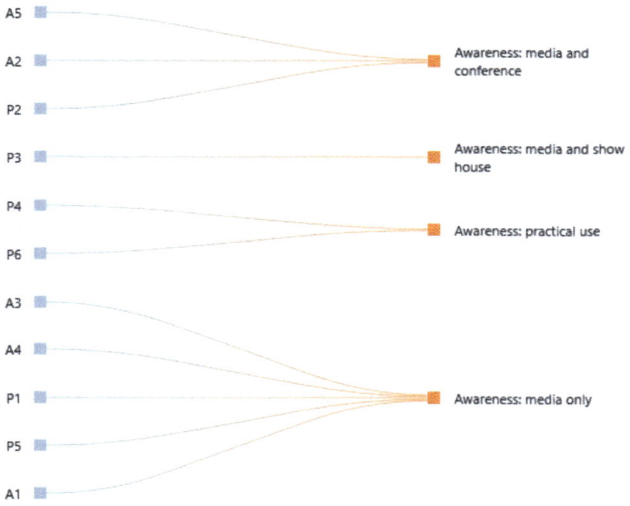

Fig. 1. Awareness of 3DCP

3.2 Readiness

The interviewees were asked if Kenya is ready to adopt 3DCP as a construction method for housing and Fig. 2 summarises the insights shared by the interviewees. Factors such as local capacity to produce the required concrete, end-user satisfaction, and global connectivity were mentioned as key indicators of Kenya's readiness to adopt 3DCP. However, when asked specifically about affordable housing, the support for 3DCP was conditional. In fact, the interviewees highlighted challenges more frequently than the benefits as illustrated in Table 3. Notably, economic and socio-cultural challenges were the most frequently mentioned challenges accounting for about 65% of all the barriers mentioned whereas economic benefits accounted for about 60% of all the benefits mentioned.

Table 3. 3DCP attributes as mentioned by interviewees

Benefits of 3DCP adoption		Barriers to 3DCP adoption	
Type	Mentions	Type	Mentions
Fast construction	8	Cost of technology	21
Guaranteed quality	8	Resistance to change	19
Less waste	3	Printer capacity	7

(continued)

Table 3. (*continued*)

Benefits of 3DCP adoption		Barriers to 3DCP adoption	
Type	Mentions	Type	Mentions
Reduced material use	3	Logistics	6
Client satisfaction	2	Reduced labour	4
Culturally aligned	2	Maintenance and repair	3
Reduced labour cost	1	Embodied energy	2
Safer	1	Lack of knowledge	1

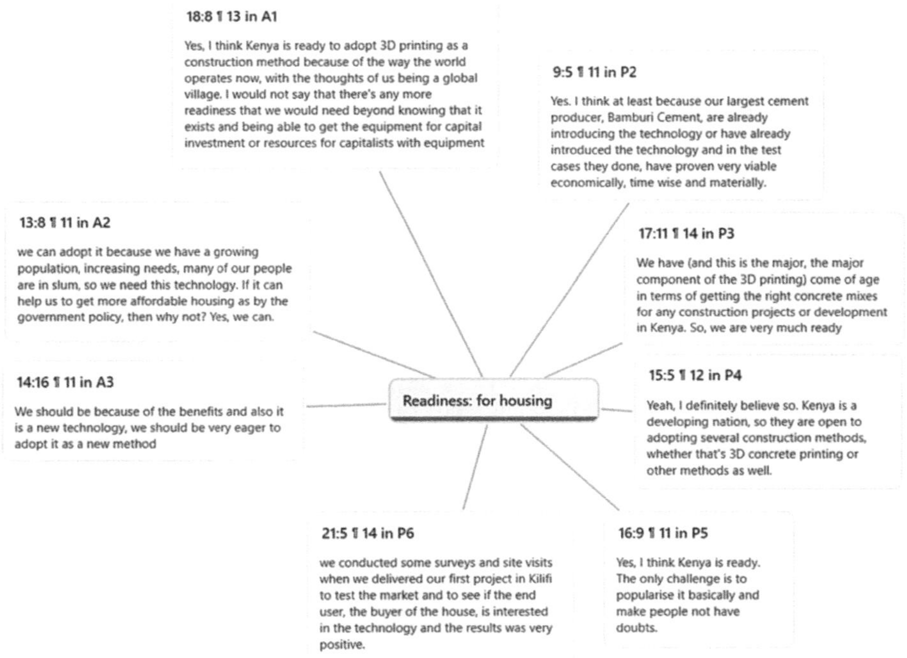

Fig. 2. Interviewees remarks on Kenya's readiness to adopt 3DCP

3.3 Adoption Strategies

The interviewees proposed five strategies to boost the adoption of 3DCP in Kenya: social networks, government policy, research and development and tax incentives, as shown in Fig. 3. The most frequently cited enablers were knowledge exchange through social networks and increased investment in research and development.

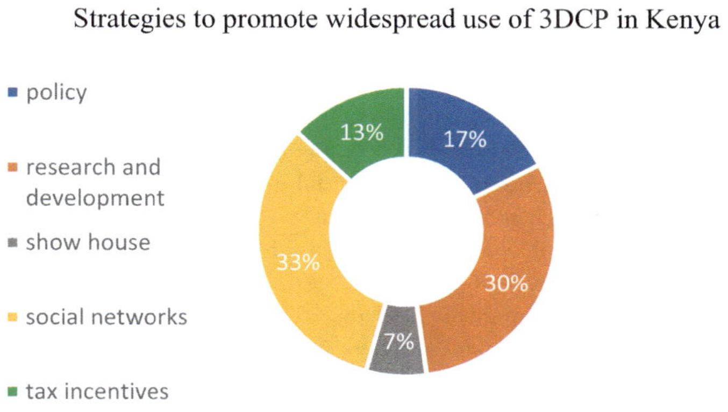

Fig. 3. Proposed adoption strategies

3.4 3DCP for Other Sectors

Beyond housing, the interviewees identified other sectors in which 3DCP could add value in Kenya as shown in Fig. 4. The education sector was the most frequently cited owing to the simplicity and uniformity of classroom structures:

"...classrooms and schools. Public schools. Those are very orthogonal, very dimensionally regular buildings, very simple." – P2

"Education sector. You have a massive program of school construction in Kenya, in everywhere in Africa, education is key, and the construction of schools is very important." – P6

The industrial sector, particularly data centres and warehouses, also emerged as a promising area for 3DCP applications:

"...niche industry like warehouses, data centres can also fit because these are sectors that are booming, we need a lot of data centres and warehouses." – P6

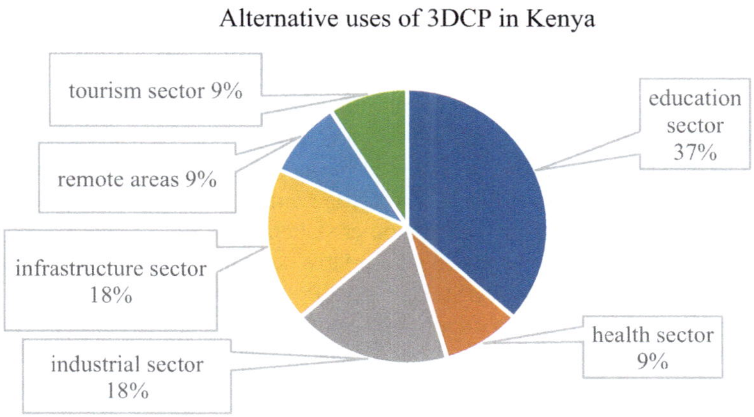

Fig. 4. Sectors that could benefit from 3DCP

4 Discussion

4.1 Perceived Attributes

Kenyan stakeholders perceived 3DCP as offering notable advantages over conventional methods in terms of speed, quality, and material efficiency. These aligned with global narratives: In India, 3DCP has been championed for time saving and design flexibility [31], and in Australia rapid construction was highly regarded in remote and disaster affected regions [30]. However, similar to Kenya, perceived cost savings in South Africa were negated by market immaturity and technical unfamiliarity [29]. In Kenya, the perceived high costs made the stakeholders question whether 3DCP aligns with the economic realities of low-income housing.

In Kenya, the stakeholders noted some alignment of 3DCP with traditional earthen construction however, resistance to abandon traditional construction practices and incompatibility of 3DCP with high-storey construction emerged as strong counterweights. In South Africa, 3DCP did not align with dominant contractor models of procurement however, 3DCP aligned with India's industrial modular housing ideas [29, 31]. Compatibility was challenged by climatic and geographic extremes in remote Australia [30]. The resistance to change by majority of the Kenyan stakeholders is evidence of a risk-averse sector which serves as a major barrier to 3DCP adoption in Kenya.

Kenya's need for specialists' skills to operate and maintain 3DCP technology aligns with the challenges faced in South Africa and India. Both studies revealed that lack of training and inadequate skilled personnel affected 3DCP widespread adoption. The study on Australia also revealed that despite integration of robotic systems in their construction sector, there was still need for progressing training skilled operators to facilitate scalability of the 3DCP technology [30]. Transitioning the abundant low skill labour in Kenya to highly skilled technicians is going to require significant investments in form of finances, time and ultimately employment opportunities. When compared to the perceived ease of use of traditional brick and mortar construction, 3DCP proves too complex for the current construction labour force.

Kenya boasts the biggest 3DCP housing development in Africa which acts as the pilot project for 3DCP in Kenya. Similarly, India and Australia have established trial sites that have served as successful deployments using 3DCP contractors L&T and WASP respectively [30, 31]. South Africa on the other hand, lack real world application of 3DCP which has limited the stakeholders' engagement with the technology [29]. The presence of a test project in Kenya coupled by media coverage on the same, shall be favourable towards increasing 3DCP knowledge amongst stakeholders, building trust and potentially encouraging buy-in from other stakeholders.

It can be inferred that documenting the relative advantages of the 3DCP projects in Kenya were deliberate actions, by Stakeholders P6 and P4, to promote 3DCP visibility which led to the construction of a solar power control room using 3DCP signalling further success with the technology in Kenya. Australia also leveraged the media to demonstrate the value of using 3DCP in crisis responses [30]. Observability was noticeably weak in South Africa due to lack of trial projects and limited knowledge dissemination.

In sum, Kenya's journey towards adopting 3DCP reflects a complex interplay of opportunities and constraints shaped by its perceived benefits, technical complexity, contextual fit, practical feasibility, and demonstrability within the local construction sector. While stakeholders recognise the relative advantages of 3DCP over traditional construction methods, compatibility and complexity remain the critical barriers that are limiting widespread adoption. Technical complexity is a critical barrier echoed in other markets as well namely: India and Australia, all of which are more developed than Kenya. Despite this, Kenya's 3DCP market is strengthened by the presence of the largest 3DCP housing project and growing the market shall heavily depend on addressing cost perceptions, expanding training programs, and maintaining momentum through continued demonstration and visibility efforts.

4.2 Promoting 3DCP Adoption

The interviewees proposed various strategies as shown in Fig. 3 to overcome existing barriers and promote adoption of 3DCP in Kenya. To address affordability, the primary challenge, the interviewees suggested investing in research and development initiatives to develop cost effective ways of locally producing the 3D concrete mix at scale and maintain the same quality standards. Additionally, they proposed implementing tax incentives for affordable housing projects using 3DCP to lower the costs associated with importation of 3DCP equipment. To tackle resistance to change, social networks emerged as the most frequently mentioned strategy. These networks facilitate knowledge exchange by creating platforms where stakeholders can publicise 3DCP technology to other industry stakeholders consequently leading to greater awareness of the technology. Such networks, formal or informal, could motivate designers to take an interest in the technology, be aware of 3DCP, and probably this will have a knock-on effect on future designs or other stakeholder interactions.

Research and development (R&D) initiatives can be said to have the greatest scope of influence by directly addressing technical, economic and human capital barriers. Despite this, R&D initiatives in developing regions suffer from lack of stable funding and lack of policy support. Therefore, a two-pronged approach of R&D initiatives backed with government policy would have the greatest influence in pushing 3DCP adoption in the Kenyan housing construction industry. These two strategies would effectively break down the significant barriers of cost and resistance to change that inhibit widespread adoption of 3DCP in Kenya.

5 Conclusion

The study revealed that despite the dearth of 3DCP projects in Kenya, construction stakeholders have a solid understanding of 3DCP technology and recognised its viability for application in Kenya. Among the perceived benefits, rapid construction and guaranteed quality received the highest ratings in support of 3DCP for affordable housing delivery. However, affordability and resistance to change were identified as the most significant barriers to act as adoption. These challenges can be said to be interconnected: the high cost deters stakeholders from adopting the technology, while resistance to its use limits widespread adoption, which in turn keeps costs high. Addressing these barriers requires a multi-pronged, multi-stakeholder approach. To lower costs, a combination of research and development efforts, along with 3DCP-specific policies such as tax incentives, could be implemented. To overcome resistance to the technology, social networks emerged as the key strategy for raising awareness among construction stakeholders.

This study captures the perception of local experts on the use of 3DCP in Kenya, contributing to the limited body of knowledge on 3DCP in African contexts. The study offers practical insights for policy makers, highlighting the importance of embedding 3DCP within affordable housing programmes by fostering an environment that supports innovation. At the same time, it provides guidance for private sector investors and donors, emphasising the value of initiating R&D initiatives that would demonstrate long-term value of 3DCP in the Kenyan housing sector. Importantly, the findings show that construction innovation adoption is not solely determined by technological merits but is also shaped by economic and socio-cultural factors of its environment.

This study is limited by its lack of quantitative analysis of performance metrics such as life-cycle costing and durability performance. The novelty of the technology in Kenya has resulted in 3DCP contractors in the region to be reluctant to share cost or material data necessary for comparative analysis. Furthermore, the study was restricted to housing applications which may not have captured the technology's full potential. Consequently, this study may form a part of future studies that employ mixed-method approaches that combine stakeholders' interviews and quantitative performance data of 3DCP projects. This study would also be suitable to conduct a longitudinal study tracking changes in 3DCP perception over time as adoption rates increase.

In conclusion, while 3DCP currently occupies a niche market in Kenya, it has the potential to penetrate the affordable housing sector over time—particularly if its initial capital costs decrease, which shall be a catalyst for its widespread adoption.

References

1. Lekan A, Aigbavboa C, Babatunde O, Olabosipo F, Christiana A (2022) Disruptive technological innovations in construction field and fourth industrial revolution intervention in the achievement of the sustainable development goal 9. Int J Constr Manag 22(14):2647–2658
2. Hasan A, Baroudi B, Elmualim A, Rameezdeen R (2018) Factors affecting construction productivity: a 30 year systematic review. Eng, Constr Archit Manag 25:916–937
3. Hassan H, Rodriguez-Ubinas E, Al Tamimi A, Trepci E, Mansouri A, Almehairbi K (2024) Towards innovative and sustainable buildings: a comprehensive review of 3D printing in construction. Autom Constr 163
4. Ninan J, Sergeeva N, Winch G (2022) Narrative shapes innovation: a study on multiple innovations in the UK construction industry. Constr Manag Econ 40(11–12):884–902
5. Regona M, Yigitcanlar T, Xia B, Li RYM (2022) Opportunities and adoption challenges of AI in the construction industry: a PRISMA review. J Open Innov: Technol, Market, Complexity 8(1)
6. Ambily PS, Kaliyavaradhan SK, Rajendran N (2024) Top challenges to widespread 3D concrete printing (3DCP) adoption—a review. Eur J Environ Civ Eng 28(2):300–328. https://doi.org/10.1080/19648189.2023.2213294
7. Zulu SL, Saad AM, Gledson B (2023) Exploring leaders' perceptions of the business case for digitalisation in the construction industry. Buildings 13(3)
8. World Green Building Council (WGBC) (2022) Sustainable and Affordable Housing: Spotlighting Action from Across the World Green Building Council Network [online] Available at: https://worldgbc.org/article/sustainable-and-affordable-housing/ [2025 Feb 10]

9. Moghayedi A, Mahachi J, Lediga R, Mosiea T, Phalafala E (2024) Revolutionizing affordable housing in Africa: a comprehensive technical and sustainability study of 3D-printing technology. Sustain Cities Soc 105329
10. Centre for Affordable Housing Finance in Africa (CAHF) (2023) Affordable Housing and Housing Finance in Kenya
11. Irungu MN, Diang'a S, Gwaya A (2022) Constraints facing the effective implementation of affordable housing programme in Kiambu County. East Afr J Eng 5(1):195–204
12. Mahachi J (2021) Innovative building technologies 4.0: fast-tracking housing delivery through 3D printing. S Afr J Sci 117(11–12)
13. Moghayedi A, Awuzie B, Omotayo T, Le Jeune K, Massyn M (2022) Appraising the nexus between influencers and sustainability-oriented innovation adoption in affordable housing projects. Sustain Dev 30(5):1117–1134
14. Qi L, Wong PSP, Holdsworth S (2024) Towards understanding the impact of the perceptions about construction innovation on advanced technology adoption. Int J Constr Manag 24(14):1521–1529
15. Peavey JB, Hudson E, Summy ZA, Violette J (2023) Cityscape 3D concrete printed houses: barriers to adoption and construction practices. Cityscape: A J Policy Dev Res 25
16. Duncan A, Kingi V, Brunsdon N (2018) Adopting new ways in the building and construction industry (BRANZ Study Report SR406) [Internet]. [cited 2024 Nov 21]. Available from: https://www.branz.co.nz/pubs/research-reports/sr406/
17. Mimoun L, Paris H, Lez F, Torres T, Sobande F (2017) When high failure, risky technology leads to market expansion: the case of the fertility services market [Internet]. Vol. 45. Association for Consumer Research. Available from: http://www.acrwebsite.org/volumes/1023813/volumes/v45/NA-45
18. Ozorhon B, Oral K (2017) Drivers of innovation in construction projects. J Constr Eng Manag 143(4)
19. Chen X, Chang-Richards A, Ling FYY, Yiu TW, Pelosi A, Yang N (2023) Developing a readiness model and a self-assessment tool for adopting digital technologies in construction organizations. Build Res Inf 51(3):241–256
20. Shibeika A, Harty C (2015) Diffusion of digital innovation in construction: a case study of a UK engineering firm. Constr Manag Econ 33(5–6):453–466
21. Wimalaratne PLI, Kulatunga U, Gajendran T (2023) 'Buildability' in the digital age: a phenomenological discourse of industry practitioners' perceptions. Buildings 13(11)
22. Besklubova S, Skibniewski MJ, Zhang X (2021) Factors affecting 3D printing technology adaptation in construction. J Constr Eng Manag 147(5)
23. Kanyilmaz A, Demir AG, Chierici M, Berto F, Gardner L, Kandukuri SY et al (2022) Role of metal 3D printing to increase quality and resource-efficiency in the construction sector. Addit Manuf [Internet] 50:102541. Available from: https://www.sciencedirect.com/science/article/pii/S2214860421006886
24. Zhang J, Wang J, Dong S, Yu X, Han B (2019) A review of the current progress and application of 3D printed concrete. Compos Part A Appl Sci Manuf [Internet]. 125:105533. Available from: https://www.sciencedirect.com/science/article/pii/S1359835X19302829
25. Wang QC, Yu SN, Chen ZX, Weng YW, Xue J, Liu X (2023) Promoting additive construction in fast-developing areas: an analysis of policies and stakeholder perspectives. Dev Built Environ 1:16
26. Anjum T, Dongre P, Misbah F, Nanyam VPSN (2017) Purview of 3DP in the Indian built environment sector. Procedia Eng [Internet]. 196:228–235. Available from: https://www.sciencedirect.com/science/article/pii/S1877705817330539
27. Chang R, Antwi-Afari M (2023) Critical success factors for implementing 3D printing technology in construction projects: academics and construction practitioners' perspectives. Constr Innov

28. Jang Y, Kim K, Leite F, Ayer S, Cho YK (2021) Identifying the perception differences of emerging construction-related technologies between industry and academia to enable high levels of collaboration. J Constr Eng Manag 147(10)
29. Aghimien D, Aigbavboa C, Aghimien L, Thwala WD, Ndlovu L (2020) Making a case for 3D printing for housing delivery in South Africa. Int J Hous Markets Anal 13(4):565–581
30. Bazli M, Ashrafi H, Rajabipour A, Kutay C (2023) 3D printing for remote housing: benefits and challenges. Autom Constr 148
31. Minde P, Patil J (2025) A comprehensive review of factors affecting 3D printing technology adaptation in the Indian construction sector. J Build Pathol Rehabil 10
32. Creswell JW, Creswell JD (2018) Research design: qualitative, quantitative, and mixed method approaches, 5th edn. Sage Publications, Thousand Oaks, CA
33. Palinkas L, Horwitz S, Green C, Wisdom J, Duan N, Hoagwood K (2013) Purposeful sampling for qualitative data collection and analysis in mixed method implementation research. Adm Policy Ment Health 42
34. Bekele WB, Ago FY (2022) Sample size for interview in qualitative research in social sciences: a guide to novice researchers. Res Educ Policy Manag 4(1):42–50

Resource Efficiency of Shotcrete 3D Printed Slabs Through Material, Process, and Form

Bartłomiej Sawicki(✉), Hannes Eichler, and Harald Kloft

Institute of Structural Design (ITE), Technische Universität Braunschweig, Braunschweig 38-106, Germany
b.sawicki@tu-braunschweig.de

Abstract. Digital fabrication with concrete is frequently promoted as a key technological solution to the sustainability crisis of the construction industry. Simultaneously, it is criticized for its large net material carbon footprint. Thus, the environmental impact should be assessed for a complete structural solution, representing three mutually dependent opportunities for resource efficiency: material, processing, and structural form. This paper presents a critical analysis of the resource efficiency of Shotcrete 3D Printing (SC3DP) slab through two parameters: mass and Global Warming Potential (GWP). Four scenarios are considered, utilising: (1) current state-of-the-art cast concrete, (2) reinforced SC3DP, (3) post-tensioned SC3DP, and (4) hybrid approach with cast plate and SC3DP ribs. Each design respects the opportunities and limitations of the given fabrication process, as based on the knowledge available from the previous research. Furthermore, two alternatives for SC3DP material concrete are considered, presenting a holistic approach. The results demonstrate that for the overly GWP-intensive SC3DP material, despite optimised form and lower mass, the slabs (2) and (3) have similar footprint as the reference, while the hybrid solution (4) offers reduction of the footprint by 20%–40%. However, with the use of the material tailored for the SC3DP process, all digitally-fabricated alternatives (2)–(4) demonstrate quite similar carbon footprint, being around 50% lower than the reference solution, but option (4) has higher mass, which would affect the resource efficiency of the whole building. As such, this research demonstrates the importance of consideration of material, process, and structural form in a holistic manner in order to provide resource-efficient design and fabrication alternatives for Net-Zero construction.

Keywords: Additive Manufacturing in Construction · LCA · sustainability

1 Introduction

In order to make the Net-Zero vision realisable, the construction industry needs to bring back the resource efficiency [1]. The resource efficiency, i.e., building more with less, can be achieved by lowering the net material footprint, as well as by reducing the material quantity to achieve a given (structural) function [2]. It was a common practice before industrial revolutions, and only with advent of (monetary) cheap materials and (monetary) expensive labour was lost for the sake of process optimisation. However, if

H. Beushausen et al. (Eds.): NTZR 2025, LNCE 873, pp. 509–521, 2026.
https://doi.org/10.1007/978-3-032-27698-8_46

we, as society, are aiming to achieve net zero emissions and respect the boundaries of Earth in terms of resource availability, we need to start considering environmental costs as equally important as monetary costs.

The current paper discusses the resource efficiency represented by material mass and Global Warming Potential (GWP), per square meter of reinforced concrete slabs. As slabs represent the largest part of mass of reinforced concrete buildings, and define dimensions of other structural elements [3], they seem most pertinent to discuss the resource efficiency. This study is embedded in the ongoing discussion regarding the strategies for reduction of carbon intensity of construction [4] and takes over the fundamental design assumptions from similar studies [3], but discusses them in the context of increasing the resource efficiency by using the new processing method: additive manufacturing (AM) with Shotcrete 3D Printing (SC3DP).

To compare the design and fabrication strategies, four types of reinforced concrete slabs with spans of 6 m, 9 m, and 12 m are compared: (a) flat reinforced cast concrete slab, (b) ribbed SC3DP reinforced concrete slab, (c) post-tensioned SC3DP ribbed slab, and d) hybrid solution with cast reinforced concrete plate and reinforced SC3DP ribs. For each design the resource efficiency is analysed in terms of the required material quantity, as well as the associated cradle-to-gate environmental footprint based on a Life Cycle Assessment (LCA). As the focus is put on the design for fabrication, other relevant strategies for resource efficiency, such as extraction method, maintenance of structures, reuse, or transportation and assembly, are not discussed. Instead, the mutually interrelated strategies for resource efficiency through optimization of form, process, and material, as enabled by digital fabrication, are discussed.

2 SC3DP—State of the Art

The SC3DP process is based on thick-flow wet concrete spraying. Thanks to digital control of the process parameters, including: the air volume flow, the concrete volume flow, the nozzle traverse speed and distance [5], as well as an optimized nozzle geometry [6], SC3DP allows for layer-by-layer dynamic printing of concrete elements without formwork [7, 8]. This section briefly discusses the previous experimental research, being the foundation of the theoretical consideration presented in this paper.

2.1 Resource Efficiency of SC3DP Through Form, Process, and Material

Resource efficiency through form utilises the flexibility of SC3DP for fabrication of tailored structural forms at no additional cost [9]. This opens-up potentials for meeting structural and additional demands with lowest material quantities by efficiently guiding the flow of forces with shapes [10] and layouts of reinforcement [11] to achieve the required structural efficiency and performance [2].

The external outline, or shape, of an element liberated from the limitations of formwork can draw from well-known principles to guide the flow of forces. One such example is a column capital, providing a base to a structure it supports without excessive concentration of stresses. This type of detail is foregone in modern concrete construction due to cost of formwork, and usually replaced with punching reinforcement of slabs. Importantly, the SC3DP column capitals, contrary to their antique monolithic counterparts, do not bring additional mass, as their thickness is constant, forming a hollow space thanks to flexibility of fabrication process with inclined layers [12].

The structural form stems not only from the external shape, but also from the, usually internal, reinforcement alignment. It can be integrated as a complex layout in a simple shape as interlayer reinforcement, for example a set of bent-up rebars providing both bending and shear resistance [13, 14]. The reinforcement alignment can be also digitally guided by optimal concrete layer printing, forming non-planar surfaces onto which the reinforcement can be deposited to comply with force flow [15]. Yet another method is green-state milling, where reinforcement is integrated into grooves formed in stable but not yet hardened concrete along the principle stress lines [16].

As such, the digital methods bring unique opportunity to reexplore the optimized forms known from the nature, as well as from the past of architecture and engineering. With no additional fabrication cost, the reduced environmental footprint can be brought by resource efficiency enabled by structural form.

Resource efficiency through process is understood as maximum production output with minimum use of auxiliary resources. It characterises serial production, and in the past allowed for scalability. This approach is clearly visible in reinforced concrete, where the efficiency through form was lost as a trade-off against high labour costs due to the introduction of systemic formwork [17] allowing for rapid fabrication of planar elements such as flat slabs, at a cost of increased material consumption.

Taking the best from traditional and digital fabrication leads to hybrid approaches. Here, the more complex, and sometimes more resource-intensive AM can be used to place materials precisely in critical areas, while conventional production processes can provide the gross volume of an element. For certain design cases this can lead to the most efficient use of resources, minimizing the environmental footprint [18, 19].

Such an example is hybrid fabrication of ribbed slab using SC3DP [12]. The flat and volumetric plate can be made with cast concrete, with subsequent addition of material by fresh-on-fresh spraying where structurally required. In this way Nervi-like efficient structural form can be achieved, with the most efficient processing (Fig. 1).

Fig. 1. Left: form-optimized column capital enabled through SC3DP; Centre: casting of hybrid slab deck; Right: SC3DP or ribs on previously cast deck of hybrid-AM slab.

2.2 Structural Performance and Material Properties of SC3DP

The mechanical properties of SC3DP material were investigated mostly in terms of compressive and flexural strength. The compressive strength along and across layers was respectively 71.1 MPa and 68.1 MPa, with cast reference reaching 56.9 MPa. The flexural strength with horizontal and vertical alignment was 7.6 MPa and 5.9 MPa, with cast reference of 7.4 MPa [20]. Those results vary depending on fabrication parameters, e.g., nozzle design [6] or accelerator dosage [5]. However for a given set of parameters, the coefficient of variation of results is well below 10% [21]. The above results, although limited, present clearly that SC3DP concrete is nearly anisotropic.

Likewise, the reinforcement bond varies depending on processing and integration method. For steel ribbed reinforcement, the bond strength is in the range of 22.8 MPa–26.2 MPa for interlayer reinforcement [22], while reaching only 3.8 MPa–9.7 MPa for green-state milling [16]. In case of short rebars insertion, average bond strength of 9.8 MPa and 31.8 MPa was found respectively without and with rebar vibration [23]. The above results demonstrate clearly that the bond strength varies largely depending on method of reinforcement integration which is closely linked to the structural form [11]. The bond depends also on the material and processing parameters. Overall, the bond strength varies from comparable to poor quality, up to surpassing the good quality bond in cast concrete of a comparable compressive strength.

There exists only limited literature regarding full-scale structural testing of SC3DP reinforced concrete. The first study considered two beams of 2 m span with green-state milling-integrated steel ribbed bars, tested under four-point bending [16]. The second study [14] considered shear and bending resistance of 3 m span beams with interlayer integration of steel reinforcement bars. In both cases the structural response could be precisely foreseen using existing models for concrete.

Overall, the existing concrete calculation models seem to be applicable to SC3DP elements. Special attention needs to be paid to fabrication parameters and methods, as well as the detailing. Furthermore, thanks to the processing, the material properties achieved with the same material are similar or superior to the cast concrete.

3 Structural Design

The material consumption and carbon footprint are largely defined by the fundamental assumptions regarding structural requirements. In order to assure the basic comparability with previous research, the design scenario from [4] is taken.

3.1 Design Boundary Conditions and Loading

The design scenario [4] is a single-direction slab with a span of 6 m, 9 m, and 12 m, with live load 2.4 kN/m^2 and imposed dead load 1.0 kN/m^2. The partial safety factors as imposed by Eurocode are used, i.e. $\gamma_G = 1.35$, $\gamma_Q = 1.5$ for dead and live load respectively. A stripe of width $b = 1$ m is considered, i.e., reflecting the 1 m spacing of the principal beams. It was verified that in all cases the full width of slab can be taken as effective width. The Eurocode cast concrete class C35/45 is taken comparable to concrete in [3,

4]. Steel with yielding strength of 500 MPa is assumed for sake of comparability with the European LCA, being slightly higher than 415 MPa assumed in [4].

Verification is performed only at the Ultimate Limit State (ULS), encompassing bending and shear resistance, and assuring failure through reinforcement yielding before concrete crushing. The Serviceability Limit State (SLS), except of concrete deck, is not considered. The deflection should be calculated taking into account variable static height. The crack opening of the rib is expected not to be decisive thanks to low concrete cover and relatively small rebar diameters. The deck reinforcement is governed by the exposure class XC1 resulting in maximum crack width $w_k = 0{,}4$ mm. The early age tensile strength $f_{ct,eff} = 0.65 f_{ctm}$ and $f_{ct,eff} = 0.8 f_{ctm}$ are assumed respectively for cast and SC3DP concrete according to EN 1992-1-1 and [24].

3.2 Material Strength

The partial safety factors regarding material strength are used as per in-situ construction according to EN 1992-1-1. For cast concrete it is respectively for steel and concrete $\gamma_S = 1.15$, and $\gamma_C = 1.5$. In case of SC3DP elements a higher fabrication precision regarding material and geometry can be taken into account. Based on previous experience and code provisions, the partial safety factors for SC3DP elements are taken as $\gamma_S = 1.1$, and $\gamma_C = 1.3$. This is extensively discussed below.

The concrete cover in all cases is designed for exposure class XC1 – concrete inside buildings with low air humidity, resulting in environmental requirement $c_{min,dur} = 10$ mm. Therefore in all cases the bond requirement prevails, with $c_{min,b} = \varnothing$, where $\varnothing$ is the diameter of rebar or tendon duct. It is further increased by $\Delta c_{dev} = 10$ mm for cast concrete and $\Delta c_{dev} = 5$ mm for SC3DP due to higher precision, respectively.

Partial safety factors for SC3DP are modified as per Annex A, EN 1992-1-1. If the reduced geometric deviations are assured (clause A.2.1), the partial safety factor for steel can be reduced to $\gamma_S = 1.1$. In SC3DP it is justified by the fact that after placement of reinforcement, concrete is applied in the manner that no displacement can be imposed on reinforcement, and thus the final position of reinforcement is known. Furthermore, if thanks to the quality control the coefficient of variation of the concrete strength does not exceed 10% (A.2.1) and if the assessment is based on concrete strength in finished structure including the fabrication process (A.2.3), the partial safety factor for concrete can be reduced to $\gamma_C = 1.3$. This should be usually considered with the conversion factor $\eta = 0.85$; however, in the current case, η is considered in calculation of design compressive strength of SC3DP material, as discussed below.

Design compressive strength of SC3DP is determined according to EN 1990 annex D 'Design assisted by testing'. Such a derivation is performed for a single material property—the compressive strength—on basis of clause D7.2, and the previous material and full-scale testing explained in detail in [14]. Two SC3DP beams were fabricated on two separate days, but using the same concrete dry-mix and SC3DP system. Together with each beam an assisting element (1 m × 0.15 m × 0.11 m) was produced and cured. From each, five cylindrical cores $d = 72$ mm and $h/d \approx 2$ were extracted and tested in compression together with beams. The results are presented in the Table 1.

Utilizing the annex D with $\gamma_c = 1.3$, the characteristic and design compressive strength of SC3DP concrete are found respectively as $f_{ck,SC3DP} = 52.8$ MPa and $f_{cd,SC3DP} = 40.6$ MPa. The obtained $f_{ck,SC3DP}$ is only 1.2% lower than the minimum core specimen value, and only 3.4% lower than the mean, demonstrating excellent coherency not achievable for cast concrete. The elasticity modulus is $E_{C,SC3DP} = 26$ GPa [14]

Table 1. Compressive testing results of specimens cored from two elements fabricated on two different dates using SC3DP.

Fabrication date	Core specimen compressive strength [MPa]					Mean [MPa]	STDV [MPa]	CoV
	1	**2**	**3**	**4**	**5**			
02.08.2024	55.7	53.5	54.1	53.4	53.4	54	1	0.018
05.08.2024	55.7	55.7	55	55.1	54.9	55.3	0.39	0
Overall results from all cores						**54.6**	**0.96**	**0.017**

3.3 Cast Concrete Solution (Flat)

A cast concrete flat slab with two reinforcement layers serves as the comparative reference for the SC3DP alternatives. Slab height and bending reinforcement are selected to meet the ULS requirements. Minimum reinforcement for crack control is considered. Spacers, typically used between the top and bottom layers, are disregarded. The shear resistance comes from concrete without additional shear reinforcement. The results of the design for different spans are presented in Fig. 2.

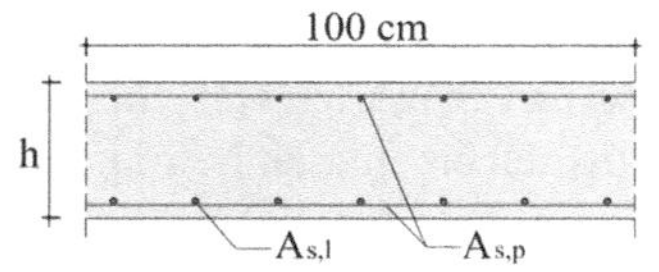

Span		6.0 m	9.0 m	12.0 m
$A_{s,l}$	[type]	ø10mm / 9cm	ø12mm / 7.5cm	ø16mm / 8.5cm
$A_{s,V}$	[type]	-	-	-
$A_{s,p}$	[type]	ø10mm / 15cm	ø10mm / 12.5cm	ø10mm / 10cm
h_{min}	[cm]	16	28	36
Mass	[kg/m^2]	400	700	900

Fig. 2. Geometry and reinforcement of flat slab.

3.4 Reinforced Fully SC3DP (SC3DP)

The element is produced upside-down. First, SC3DP plate is manufactured with interlayer steel mat. Subsequently, the central part of rib is SC3DP manufactured fresh-on-fresh, with green-state-milling integration of pre-bent shear reinforcement. Subsequently, the longitudinal reinforcement is placed directly on the core of the rib, following its curvature. Finally, the cover concrete is applied with digital shotcrete. The design is presented in Fig. 3.

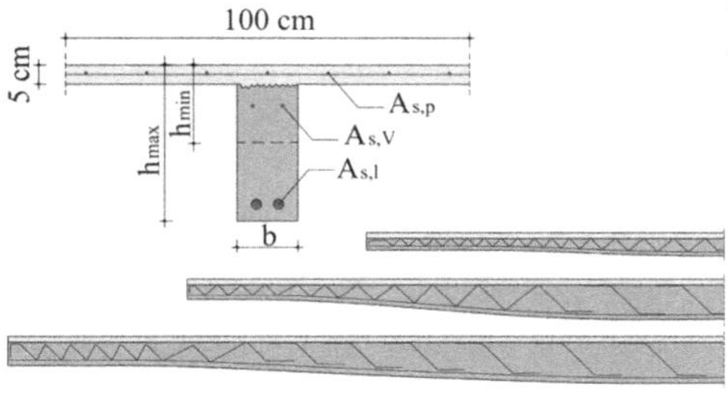

Span		6.0 m	9.0 m	12.0 m
$A_{s,l}$	[type]	2xø16mm	2xø20mm	2xø25mm
$A_{s,V}$	[type]	2xø6mm	2xø6mm	2xø6mm
$A_{s,p}$	[type]	Q257	Q257	Q257
h_{min}	[cm]	15	18	25
h_{max}	[cm]	20	35	40
b	[cm]	11	13	15
Mass	[kg/m^2]	165	215	241

Fig. 3. Geometry and reinforcement of SC3DP slab.

The longitudinal reinforcement is verified every 1 m for bending at ULS and interpolated in-between. No additional detailing is considered near supports, as the longitudinal reinforcement is not fully activated there, and the bond of interlayer reinforcement in SC3DP is comparable or higher than in cast concrete [22]. The shear reinforcement is designed according to EN 1992-1-1 clause 6.2.3 (4) with inclination of 45° under assumption of shear crack angle $\theta \approx 20°$. Towards the mid-span the geometric requirement of minimum one shear reinforcement bar passing the crack prevails. Verification is also performed for $\theta \approx 45°$ crack at distance d from the support. Bond strength $\tau =$ *3.5* MPa, the lower bound of available results [16], is taken. For $\varnothing = 6$ mm the required anchorage length is $l_0 = 212$ mm $\approx 35\varnothing$, equal to Eurocode recommendation for low-quality bond. Continuous truss-like bar is used were more favourable due to anchorage length. The mandrel radius is taken as per Eurocode, i.e., $4\varnothing$. The slab reinforcement is a single layer Q257 steel mat ($\varnothing$ 7.0 mm/150 mm) due to the cracking requirements and transversal strength under distributed design load. Thanks to use of lower partial safety factors the total mass is reduced by around 9%.

3.5 Post-Tensioned SC3DP (SC3DP PS)

The fabrication process is similar as in Sect. 3.4. To facilitate the installation of post-tensioning duct, the geometry of layers produces a surface on which the duct can be placed, similar to [15]. The resulting geometry is presented in Fig. 4.

A single-duct post-tensioning system with number and size of strands selected from catalogue of a commercial provider is used. The ULS design is performed assuming the full resistance of strands, resulting in $P_{md} = P_{mk} / \gamma_s$, with $\gamma_s = 1.15$. The verification at post-tensioning application is performed assuming fully developed strength of concrete, characteristic self-weight of the element as a simply-supported beam, and activation of the full characteristic strength of prestressing, i.e., with force P_{mk}. The compressive stress in concrete cannot exceed $0.6 f_{cd}$. The minor tensile stresses in the plate are carried by steel mat reinforcement. As the upper flange will be under compression at SLS, which is not verified here, no crack control reinforcement is needed. As the structural reinforcement in transversal direction, as well as to compensate for tensile stresses at post-tensioning, a single layer of Q188 steel mat reinforcement ($\varnothing$ 6.0 mm/150 mm) is used. No shear reinforcement is required, as shear is carried by the vertical component of P_{md} prestressing force in inclined tendons, as well as by concrete. The precise control of the vertical component at tendon inclination 1°–5° would be challenging even with SC3DP; it is however compensated by low f_{td} = 0.5 MPa. The shear resistance of concrete is calculated according to EN 1992-1-1. Thanks to lower partial safety factors the total mass is reduced by around 15%.

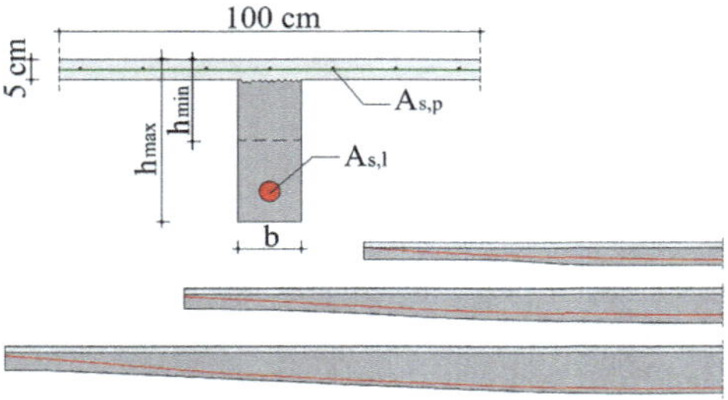

Span		6.0 m	9.0 m	12.0 m
$A_{s,l}$	[type]	VSL 6-1	VSL 5-3	VSL 5-4
$A_{s,V}$	[type]	-	-	-
$A_{s,p}$	[type]	Q188	Q188	Q188
h_{min}	[cm]	15	17.5	20
h_{max}	[cm]	20	30	40
b	[cm]	10	13.5	15
Mass	[kg/m²]	160	196	232

Fig. 4. Geometry and reinforcement of SC3DP PS slab.

3.6 Hybrid Cast and SC3DP (Hybrid)

The fabrication process considers firstly casting a reinforced plate, including protruding u-shaped stirrups. The interface between plate and rib is finished in a "rough" quality, according to EN 1992-1-1. Subsequently the ribs are SC3DP fabricated until formation of parabolic surface for interlayer integration of longitudinal reinforcement. Next, the stirrups are inserted as vibrated short rebars [23, 25] with lap splicing to the previously installed stirrups. Finally, the remaining layers of concrete are applied.

The 8 cm thick plates are reinforced with two layers of welded wire mesh, type Q188, covering the minimum reinforcement for crack control which is considered as a relevant criterion, since slabs and ribs are fabricated at different times and can hinder each other in their individual contraction. The plates' structural demands to span orthogonally between the ribs are covered by the minimum reinforcement. Stirrups are selected and staggered according to the required shear resistance of the slabs but also to ensure shear resistance of the interface between plate and ribs Fig. 5.

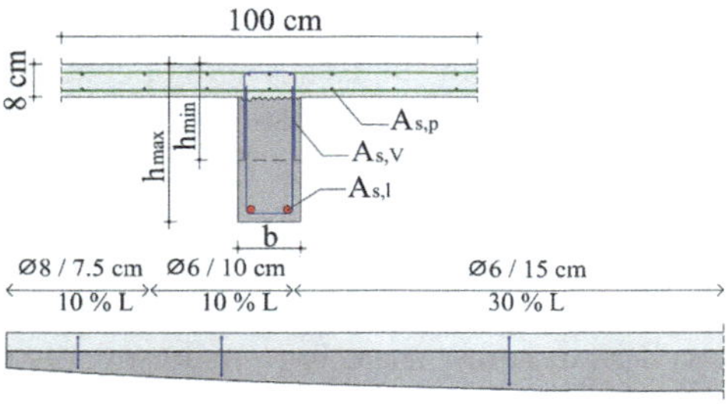

Span		6.0 m	9.0 m	12.0 m
$A_{s,l}$	[type]	2xø16mm	2xø20mm	2xø25mm
$A_{s,V}$	[type]	According to drawing		
$A_{s,p}$	[type]	2xQ188	2xQ188	2xQ188
h_{min}	[cm]	15	18	20
h_{max}	[cm]	25	35	44
b	[cm]	11	13	15
Mass	[kg/m²]	239	270	307

Fig. 5. Geometry and reinforcement of hybrid slab.

4 LCA Assumptions

A cradle-to-gate LCA is executed to compare the different design and fabrication strategies with respect to the GWP in the production stage (life cycle stages A1-A3 according to EN 15804). To quantify the emissions for conventional concrete and reinforcement, LCA assumptions from the German database Oekobaudat.de are used [26]. For SC3DP material, there is only limited data available, with the results of the study being significantly influenced by the CO_2 emissions of the shotcrete mortar. In order to demonstrate a potential range of SC3DP, two LCA scenarios were analyzed: A) a product-specific assumption for a commercially available material [27] representing the upper bound with extremally high CO_2 intensity; and B) CO_2-reduced material, based on [28], as more realistic. All LCA assumptions used are summarized in Table 2.

Table 2. Life cycle analysis assumptions and databases.

Material	Database/EPD	GWP_{tot} (A1-A3)
Cast concrete (German average)	Oekobaudat [26]: concrete C35/45 (InformationsZentrum Beton GmbH)	220 $kgCO_{2eq}/m^3$
CO_2 intensive SC3DP concrete A) (EPD)	MC-Bauchemie Müller GmbH & Co. [27]: Nafufill RM 40	$\approx$ 900 $kgCO_{2eq}/m^3$
CO_2 reduced SC3DP concrete B) (literature)	Mobbot SA [28] proprietary mix	$\approx$ 350 $kgCO_{2eq}/m^3$
Reinforcement (German average)	Oekobaudat [26]: reinforcement steel wire (Sphera Solutions GmbH)	0.65 $kgCO_{2eq}/kg$
Prestressing wire (EPD)	Oekobaudat [26]: prestressing wire and strand (Voestalpine AG)	2.9 $kgCO_{2eq}/kg$

5 Results

The GWP of each design and fabrication method is presented in Fig. 6. Even with assumption of extremally high net environmental footprint of the shotcrete material, the fully SC3DP solutions lay in the vicinity of the flat slab. For this assumption the hybrid solution is the most favourable, as most of the compressive concrete volume is provided by relatively low carbon-intensive cast concrete, while the material of higher intensity is strategically placed in optimised ribs. The results would be even more favourable for the hybrid solution considering its relatively small mass, allowing for reduction of the subsequent structural elements, such as columns and foundations. Therefore, even radically more resource-intensive material can provide favourable solution thanks to the resource efficiency achieved through form and digital processing.

If a realistic, low-intensity shotcrete material is assumed, the three SC3DP strategies are similar and radically better than the flat slab, representing the combined effect of resource efficiency through form, process, and material. Should the complete building be considered [3], the fully SC3DP options could be better than the hybrid one, as their mass is around 30% smaller, especially for 6 m span, thanks to thinner deck achieved by higher strength of material, lower safety factors, and one-step fabrication process.

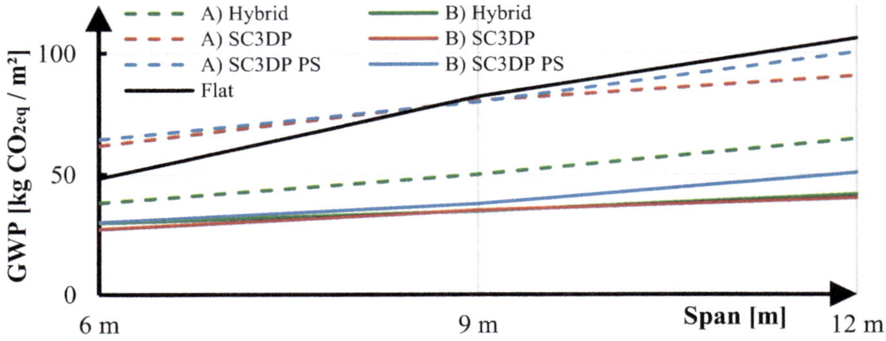

Fig. 6. Total GWP for different systems (see section titles for abbreviation) and spans: scenario A) for carbon-intensive SC3DP concrete and scenario B) for CO2-reduced SC3DP concrete.

6 Discussion and Conclusions

This study discusses the resource efficiency of reinforced Shotcrete 3D Printed slabs of 6 m, 9 m, and 12 m span in terms of environmental footprint and mass. The efficiency is achieved thanks to synergetic optimisation of structural form, fabrication process, and material. Four design and fabrication strategies are considered.

The results demonstrate that when the efficiency is achieved only by form and process, the hybrid cast-SC3DP solution is the most competitive, even with strategic placement of an extremally environmentally intensive material. When shotcrete material with a small environmental footprint is considered, the fully SC3DP solutions are the best in terms of both mass and GWP through the synergy of: (a) optimised structural form in terms of external shape and internal reinforcement arrangement, achievable with digital fabrication; (b) reduction of safety factors due to digitally controlled and precise fabrication process; and (c) use of resource-efficient material tailored to the processing.

The reduction of resource consumption expressed through mass would render the fully SC3DP solutions even more favourable, should the complete structural system including columns and foundations, as well as the construction process, be considered. Likewise, to assess monetary costs, the complete value chain needs to be considered [9]. This is not performed for the sake of brevity and to avoid speculation.

Importantly, the results of this study demonstrate the weakness of existing LCA methods, if they are about to serve as metrics to select the most resource-efficient design and fabrication strategy. This well-known and broadly discussed issue [29] is further exacerbated in case of the digital fabrication, which unites material, process, and form in the new digital-real design-to-fabrication workflows [30]. Hence, new holistic and scalable design and assessment methods need to be developed if we are about to unlock the full potential of the Additive Manufacturing in Construction.

Nevertheless, it can be concluded that the digital fabrication methods allow for increase of resource efficiency through simultaneous optimisation of material, process, and form, paving the way towards Net-Zero construction.

References

1. Allwood JM, Ashby MF, Gutowski TG, Worrell E (2011) Material efficiency: a white paper. Resour Conserv Recycl 55:362–381. https://doi.org/10.1016/j.resconrec.2010.11.002
2. Gebhard L, Mata-Falcón J, Ammann R et al (2024) Enhancing structural efficiency with digital concrete – principles, opportunities and case studies. Cem Concr Res 185:107645. https://doi.org/10.1016/j.cemconres.2024.107645
3. Ismail MA, Mueller CT (2021) Minimizing embodied energy of reinforced concrete floor systems in developing countries through shape optimization. Eng Struct 246:112955. https://doi.org/10.1016/j.engstruct.2021.112955
4. Feickert K, Mueller CT (2025) Policy and design levers for minimizing embodied carbon in United States buildings: a quantitative comparison of current and proposed strategies. Build Environ 270:112485. https://doi.org/10.1016/j.buildenv.2024.112485
5. Lachmayer L, Böhler D, Freund N et al (2023) Modelling the influence of material and process parameters on shotcrete 3D printed strands—cross-section adjustment for automatic robotic manufacturing. Autom Constr 145:104626. https://doi.org/10.1016/j.autcon.2022.104626
6. David M, Freund N, Dröder K, Lowke D (2023) The effects of nozzle diameter and length on the resulting strand properties for shotcrete 3D printing. Mater Struct 56:157. https://doi.org/10.1617/s11527-023-02246-1
7. Lindemann H, Gerbers R, Ibrahim S et al (2018) Development of a shotcrete 3D-printing (SC3DP) Technology for Additive Manufacturing of reinforced freeform concrete structures. In: Digital Concrete 2018. Springer, Cham
8. Neudecker S, Bruns C, Gerbers R et al (2016) A new robotic spray Technology for Generative Manufacturing of complex concrete structures without formwork. Procedia CIRP 43:333–338. https://doi.org/10.1016/j.procir.2016.02.107
9. Lowke D, Kloft H, Hack N et al (2024) Potenzial additiver Fertigung und Performance im Verfahrensvergleich zu konventioneller Fertigung im Stahlbetonbau. Deutscher Beton-und Bautechnikverein (DBV)
10. Lowke D, Anton A, Buswell R et al (2024) Digital fabrication with concrete beyond horizontal planar layers. Cem Concr Res 186:107663. https://doi.org/10.1016/j.cemconres.2024.107663
11. Kloft H, Sawicki B, Bos F et al (2024) Interaction of reinforcement, process, and form in digital fabrication with concrete. Cem Concr Res 186:107640. https://doi.org/10.1016/j.cemconres.2024.107640
12. Kloft H, Hack N, Mainka J et al (2019) Additive Fertigung im Bauwesen: erste 3-D-gedruckte und bewehrte Betonbauteile im Shotcrete-3-D-Printing-Verfahren (SC3DP). Bautechnik 96:929–938. https://doi.org/10.1002/bate.201900094
13. Dörrie R, Freund N, Herrmann E et al (2023) Automated force-flow-oriented reinforcement integration for shotcrete 3D printing. Autom Constr 155:105075. https://doi.org/10.1016/j.autcon.2023.105075
14. Sawicki B, Chen Z, Leusmann T, Kloft H (2025) Digitally fabricated reinforced concrete beam: design, testing, and comparison against conventional cast reinforced concrete solution. Engineering Structures Submitted:

15. Dörrie R, Kloft H (2022) Force flow compliant robotic path planning approach for reinforced concrete elements using SC3DP. In: Buswell R, Blanco A, Cavalaro S, Kinnell P (eds) Third RILEM international conference on concrete and digital fabrication. Springer International Publishing, Cham, pp 370–375
16. Dörrie R, Kloft H, Sawicki B et al (2024) Automated reinforcement integration in shotcrete 3D printing through green state milling. In: 4th RILEM international conference on concrete and digital fabrication, Munich, Germany
17. Forty A (2016) Concrete and culture: a material history, Reprint edition. Reaktion Books, London
18. Eichler H, Sawicki B, Kloft H (2025) Quantification of potential of hybrid wire and arc additively manufactured I-beams for resource-efficient and sustainable steel structures. In: Różański A, Bui Q-B, Sadowski Ł, Tran MT (eds) Proceedings of the 4th international conference on sustainable development in civil. Urban and Transportation Engineering. Springer Nature, Singapore, pp 411–419
19. Eichler H, Sawicki B, Kloft H (2025) Hybrid wire and arc additive manufactured I-beams—assessment of material efficiency and carbon footprint. Archives of Civil and Mechanical Engineering Submitted
20. Kloft H, Krauss H-W, Hack N et al (2020) Influence of process parameters on the interlayer bond strength of concrete elements additive manufactured by shotcrete 3D printing (SC3DP). Cem Concr Res 134:106078. https://doi.org/10.1016/j.cemconres.2020.106078
21. Dressler I, Freund N, Lowke D (2020) The effect of accelerator dosage on fresh concrete properties and on interlayer strength in shotcrete 3D printing. Materials 13:374. https://doi.org/10.3390/ma13020374
22. Freund N, Lowke D (2022) Interlayer reinforcement in shotcrete-3D-printing : the effect of accelerator dosage on the resulting bond behavior of integrated reinforcement bars. Open Conference Proceedings 1:83–95. https://doi.org/10.52825/ocp.v1i.72
23. Freund N, David M, Dröder K, Lowke D (2024) Vibrated short rebar insertion—the effect of integration time on the resulting bond quality. In: Fourth RILEM international conference on concrete and digital fabrication. Springer, Cham, pp 327–334
24. DBV (2016) Merkblatt—Begrenzung der Rissbildung im Stahlbeton- und Spannbetonbau (Leaflet—Crack control in reinforced and prestressed concrete structures)
25. Freund N, Dressler I, Lowke D (2020) Studying the bond properties of vertical integrated short reinforcement in the shotcrete 3D printing process. In: Bos FP, Lucas SS, Wolfs RJM, Salet TAM (eds) Second RILEM international conference on concrete and digital fabrication. Springer International Publishing, Cham, pp 612–621
26. Bundesministerium für Wohnen, Stadtentwicklung und Bauwesen (2023) ÖKOBAUDAT
27. MC-Bauchemie Müller GmbH & Co. KG (2023) Declaration of Conformity with a Model EPD Nafufill RM:40
28. Favier A, Petit A (2022) Strategies for reducing the environmental footprint of additive manufacturing via sprayed concrete. In: Buswell R, Blanco A, Cavalaro S, Kinnell P (eds) Third RILEM international conference on concrete and digital fabrication. Springer International Publishing, Cham, pp 105–110
29. Röck M, Sørensen A, Tozan B, et al (2022) Towards embodied carbon benchmarks for buildings in Europe-#2 setting the baseline: a bottom-up approach. https://doi.org/doi.org/10.5281/zenodo.5895051
30. Dörrie R, Gantner S, Amiri FS et al (2025) From digital to real: optimised and functionally integrated shotcrete 3D printing elements for multi-storey structures. Buildings 15:1461. https://doi.org/10.3390/buildings15091461

Advanced 3D Concrete Printing (3DCP) with Low-Carbon Calcium Sulfoaluminate (CSA) Cement Blends

Ferdinand Senf(✉), Marvin Abstoß, Henrik Funke, and Sandra Gelbrich

Chemnitz University of Technology, 09126 Chemnitz, Germany
ferdinand.senf@mb.tu-chemnitz.de

Abstract. Ecological demands are driving clinker reduction in construction, which may adversely affect material properties and delay production. The partial replacement of clinker by calcium sulfoaluminate (CSA) cement in blended systems based on blast furnace or pozzolanic cement offers advantages such as accelerated setting while simultaneously reducing CO_2 emissions. Due to the use of the new binder composition in 3D concrete printing (3DCP), additional efficiency can be achieved. From a materials engineering perspective, a controlled retardation of setting is necessary to enable the integration of CSA cement-based mixtures into the 3DCP process. At high levels of clinker substitution with latent hydraulic or pozzolanic binders, low amounts of citric acid used as setting time retarder lead to losses in mechanical strength, as these binders develop little to no strength in the absence of Portlandite. Therefore, a partial clinker replacement of 51 wt.% through 30 wt.% CSA cement with 21 wt.% Supplementary Cementitious Materials, such as ground granulated blast furnace slag or metakaolin, along with 0.76 wt.% citric acid to achieve a 90 min processing window has proven technically viable with respect to strength development. The investigated ternary binder systems demonstrate a balanced interaction without adverse synergistic effects, even in the presence of setting time retarders. Furthermore, the developed mixture exhibits high compatibility with 3DCP processes and ensures stable and continuous extrusion. Mechanical strength is not negatively affected by the 3DCP process.

Keywords: Additive Manufacturing · Calcium Sulfoaluminate Cement · 3D Concrete Printing

1 Introduction

The European Green Deal establishes a framework for achieving climate neutrality within the European Union by 2050. The construction sector, which accounts for a share of global CO_2 emissions, offers considerable potential for reduction emissions across all stages of the value chain, including clinker production, cement manufacturing, concrete formulation, construction processes, and (re)carbonation of concrete structures [1].

H. Beushausen et al. (Eds.): NTZR 2025, LNCE 873, pp. 522–534, 2026.
https://doi.org/10.1007/978-3-032-27698-8_47

This research investigates two complementary strategies to reduce CO_2 emissions associated with concrete production and application. The first strategy involves the development of novel cementitious systems with reduced clinker content. By partially replacing ordinary Portland cement (OPC) with alternative binders such as Calcium sulfoaluminate (CSA) cement and various Supplementary Cementitious Materials (SCMs), the clinker content in the final concrete mixtures can be reduced by 51 wt.%. This approach directly addresses the emissions-intensive nature of clinker production and contributes to lowering the embodied carbon of the material.

The second strategy focuses on the integration of these low-clinker concrete mixtures into the 3D concrete printing (3DCP) process. As a digitally controlled layer-by-layer fabrication process, 3DCP enables highly efficient use of materials by allowing for precise deposition of concrete and the design of structurally optimized geometries, such as hollow sections. Furthermore, 3DCP enables construction without the need for formwork or support structures, which reduces material waste. The automated process also offers potential savings in construction time, energy use, and labor input, making it an effective method for improving both environmental and economic performance in the construction industry.

By combining advancements in low-emission cement materials with innovative digital construction technologies, the presented study contributes to the development of more sustainable building practices and supports the long-term climate objectives set forth by the European Green Deal.

2 Materials and Methods

2.1 Materials

To address the environmental impact of conventional binder systems, this study utilizes a ye'elimite-rich CSA cement, which emits around 30% less CO_2 than OPC due to its lower limestone requirement and reduced clinkerization temperature [2]. This cement, containing approximately 50 wt.% ye'elimite and 30 wt.% belite, serves as the basis for various blended systems. Table 1 provides an overview of the material composition of the 3DCP mixture used for all subsequent experiments.

To further reduce the environmental footprint, the CSA cement is combined with standardized cements according to DIN EN 197–1 [3], namely blast-furnace cements (CEM III/A 52.5 L-LH/SR/LA and CEM III/B 42.5 N-LH/SR/LA), pozzolan cement (CEM IV/A (P) 42.5 R), and Portland cement (CEM I 52.5 N), which serves as a reference for comparison reasons. To improve the hydration balance between CEM I and CSA cement in the ternary systems, various supplementary cementitious materials (SCMs) are incorporated. These include ground granulated blast furnace slag (GGBFS, Blaine specific surface area approx. 5000 cm^2/g [4]), fly ash (FA, approx. 3800 cm^2/g Blaine [5, 6]), metakaolin (MK, approx. 20000 cm^2/g Blaine [7]), and limestone powder (LP, approx. 6500 cm^2/g Blaine [8]). To control the setting behavior, citric acid monohydrate is used as a set retarder.

Table 1. Material composition for 3D concrete printing.

Component	[kg/m^3]
Binder	571
Quartz sand 0/2 mm	909
Limestone powder	317
Methylcellulose based Stabilizer	2.8
PCE based Superplasticizer	2.9
Setting time retarder	4.3
w/b	0.50

2.2 Methods

Compressive Strength. The compressive strength of the developed concrete mixtures is measured according to DIN EN 12390–3:2019 [9, 10].

Bending Tensile Strength. The 4-point bending tensile strength of the developed concrete mixtures is measured according to DIN EN 196–1:2016 [9, 11].

3D Concrete Printing Process. The 3DCP process was executed with a KUKA KR90 R2700 six-axis industrial robot, equipped with a PFT Bolero 230 V screw pump extrude for continuous material deposition. All elements are printed using a 30% pumping rate, corresponding to a printing speed of 0.06 m/s and a 20 mm nozzle diameter. Specimens are printed in layers and tested for directional compressive and 4-pt.-bending tensile strength. All mechanical tests are conducted according to DIN EN 12390–3:2019 and DIN EN 196–1:2016 [9–11]. Geometric modelling, toolpath generation, and process control are carried out using Grasshopper, a parametric design tool integrated as an add-on into the Rhinoceros 3D CAD environment [12, 13].

3 Results and Discussion

3.1 Material Engineering

CSA Cement Blends with Other Cement Types. The material performance of CSA cement is influenced by its combination with other types of cement, particularly through the targeted selection of pozzolanic and latent-hydraulic supplementary materials. The interaction with organic set retarders gives rise to a complex system of effects that substantially governs both the hydration kinetics and the development of early and final strengths. Especially in purely material-based approaches for 3DCP, a controlled delay in the setting of CSA cement is essential to ensure a sufficient time window for the transport and extrusion of fresh concrete. Against this background, the question arises as to which conventional cement types can be combined with CSA cement to create a technically viable and durable mixture.

Figure 1 presents the compressive strength development of binder systems containing 30 wt.% CSA combined with 70 wt.% of various cement types: CEM I 52.5 N, CEM III/A 52.5 L, CEM III/B 42.5 N, and CEM IV/A (P) 42.5 R. Each mixture is modified with an addition of 0.76 wt.% citric acid. For comparison, the reference mixture with 100 wt.% CEM I and without citric acid or CSA cement has a compressive strength of 61.7 MPa, while the mixture consisting of 100 wt.% CSA cement with 0.76 wt.% citric acid reaches 65.7 MPa, according to [14].

In Fig. 1, it can be observed that the hydration of CEM I is not notably affected by the presence of citric acid. In contrast, mixtures containing CEM III/A (in this case with 59.7 wt.% GGBFS) exhibit a reduced strength development when combined with CSA and citric acid. This inhibitory effect is further pronounced in the CEM III/B-based system, which contains 69.8 wt.% GGBFS. The use of the pozzolanic cement CEM IV/A (containing 16.8 wt.% trass) results in a comparatively smaller reduction in strength, likely due to the lower content of SCMs.

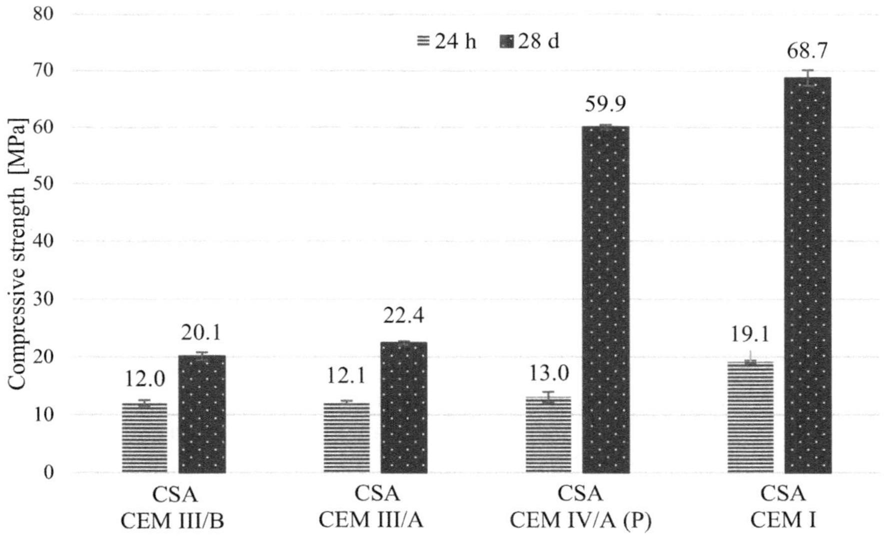

Fig. 1. Compressive strength of binder systems comprising 30 wt.% CSA cement and 70 wt.% of alternative cement types.

CSA-CEM III/B Blends with Varying Citric Acid Dosages. Building on these findings, the study examines how a reduction in citric acid content and the associated shortening of the processing window affects the strength development of CSA-CEM III/B mixtures.

For a mixture containing 30 wt.% CSA cement, a dosage of 0.76 wt.% citric acid is required to achieve a processing window of 90 min. For a processing time of 60 min, 0.60 wt.% is sufficient, while 0.44 wt.% is required for 30 min. At lower CSA cement contents, the necessary dosage of retarder decreases [14]. Across all tested mixture ratios, a trend is observed: as the retarder dosage decreases and the processing window shortens, the compressive strength after 28 days increases. Nevertheless, the results in Fig. 2 clearly demonstrate that even small amounts of retarders have a significant impact on the mechanical properties of CSA-CEM III/B mixtures.

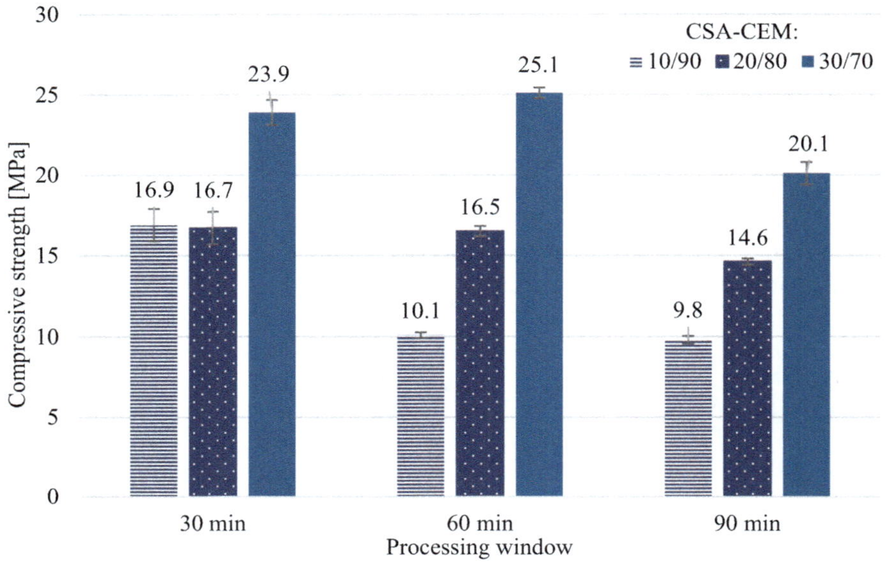

Fig. 2. Compressive strength development of various CSA + CEM III/B blends including different dosages of set retarders.

Blending CSA Cement with SCMs. In the following section, the ratio between Portland cement clinker and SCMs is systematically adjusted while maintaining a constant CSA cement content of 30 wt.%. To determine the maximum SCM replacement level at which the 28-day compressive strength remains unaffected by the presence of citric acid, various SCMs are examined and compared. These include GGBFS as a latent hydraulic binder, MK and FA as pozzolanic materials, and LP as an inert filler for reference.

As illustrated in Fig. 3, a distinct reduction in compressive strength is observed between 21 wt.% and 28 wt.% SCM replacement. For GGBFS, the strength at 21 wt.% replacement shows a moderate decrease of 13.1% compared to the reference mixture containing 100 wt.% Portland cement clinker. However, at 28 wt.% GGBFS, the reduction increases significantly to 36.4%. A similar trend is seen with MK, where the compressive strength drops from a 4.1% reduction at 21 wt.% to a 51.4% reduction at 28 wt.%. FA exhibits a more pronounced loss in strength even at 21 wt.% (27.5% reduction), which further increases to 39.0% at 28 wt.% replacement.

The role of GGBFS, MK, and FA in the hydration process is further highlighted when compared to LP, which serves as an inert filler. Despite having a higher Blaine fineness (6500 cm^2/g) than GGBFS (5000 cm^2/g) and FA (3800 cm^2/g), LP results in notably lower compressive strength, underscoring the contribution of the other SCMs to the hydraulic reaction.

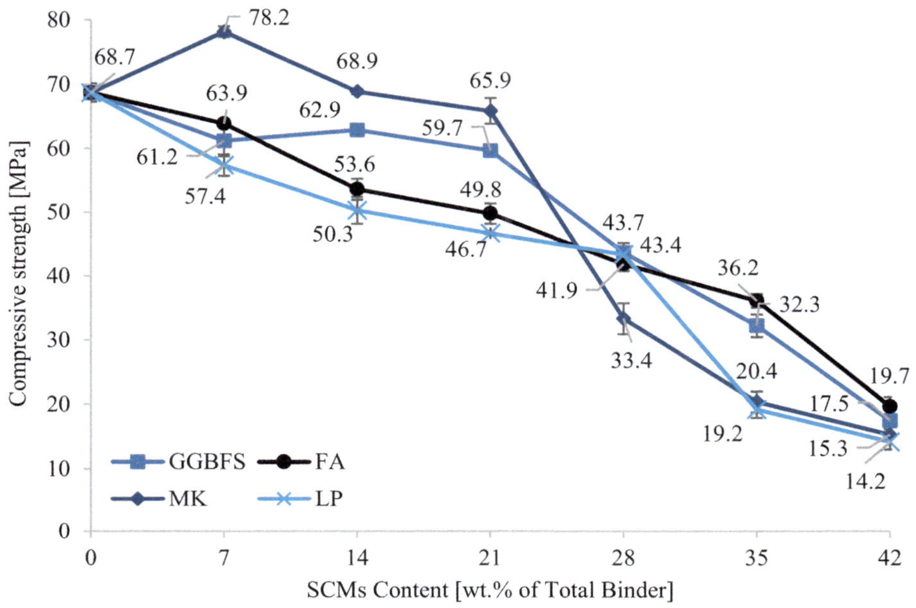

Fig. 3. Compressive strength development of CSA, CEM I and various SCMs.

Early Hydration of CSA Cement Blends. To highlight the potential of CSA cement for applications such as 3DCP, Fig. 4 presents the 4-point bending tensile strength development during the first 8 h of hydration. Four binder systems are compared, based on the concrete composition in Table 1. The only difference among the mixtures lies in the binder type: one uses CEM I, another CEM III/A, while the remaining two are ternary systems consisting of 30 wt.% CSA cement, 49 wt.% CEM I, and 21 wt.% GGBFS, as previously identified in Fig. 3. One of the CSA cement-based mixtures includes 0.76 wt.% citric acid as retarder, while the other contains no retarder.

The CEM III/A-based mixture exhibits its first measurable bending tensile strength (0.3 MPa) after 5 h of hydration, whereas the CEM I mixture reaches the same level one hour earlier, after 4 h. This reflects delayed early strength development associated with reduced clinker content (from 100 wt.% in CEM I to approximately 40 wt.% in CEM III/A). Measurable strength values are recorded once the applied force exceeds 120 N, corresponding to a minimum detectable bending tensile strength of approximately 0.25 MPa.

In contrast, the CSA cement binders demonstrate accelerated strength development. The mixture without citric acid begins gaining strength immediately after water addition, while the mixture containing 0.76 wt.% citric acid shows a delayed strength onset that aligns with the designed 90 min open time.

This early reactivity of CSA cement-based systems presents a clear advantage for 3DCP, particularly in the production of precast elements in rapid succession. Unlike pozzolanic or blast furnace cements, which typically require extended setting times due to their low clinker content, CSA cement-containing binders support rapid strength gain and enhanced buildability. These advantages are further explored in the following section, which addresses their implication for structural stability during 3DCP.

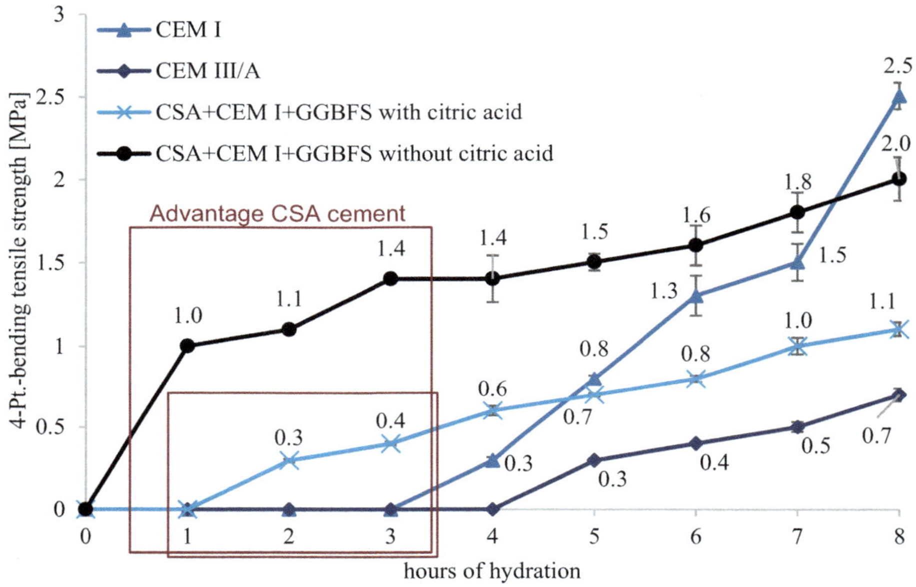

Fig. 4. Early age hydration of CSA cement blends.

3.2 3D Concrete Printing

Anisotropic Strength Analysis of Printed Specimens. Building upon the demonstrated early strength development of the CSA cement-based binder system discussed in the previous section, the following chapter investigates its performance in the context of 3DCP. The mixture used therefore is based on the composition provided in Table 1, consisting of 30 wt.% CSA cement, 49 wt.% CEM I, and 21 wt.% GGBFS, with a citric acid dosage of 0.76 wt.% to control setting time (see Figs. 3 and 4).

Figure 5 shows the two printed geometries used for mechanical testing: a plate specimen measuring 400 × 400 × 60 mm^3 and a beam specimen of 600 × 150 × 120 mm^3. From these printed components, smaller test specimens are extracted for mechanical evaluation. Specifically, 40 × 40 × 160 mm^3 prisms and 100 × 100 × 100 mm^3 cubes are cut to assess direction-dependent compressive strength and 4-point bending tensile strength after 28 days, as presented in Figs. 6 and 7.

Fig. 5. Printed test specimens for anisotropic strength analysis.

The 3DCP process of the specimens was completed reliably and without complications. The material exhibited the appropriate consistency, ensuring both good pumpability and stable layer-by-layer build up. All four components (one plate and three beams) were successfully printed within 60 min from the time of water addition, confirming that the 90 min processing window, adjusted by citric acid, was sufficient for the entire printing operation.

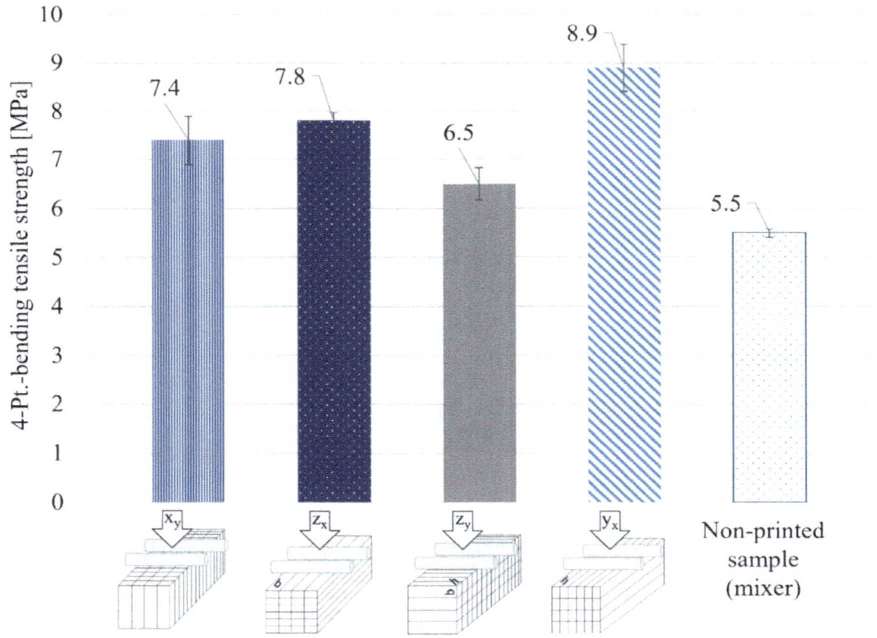

Fig. 6. Directional 4-Pt.-bending tensile strength from printed test specimens.

The anisotropic 4-Pt.-bending tensile strength results are shown in Fig. 6. A distinction can be observed between loading applied parallel to the printed layers (xy and zy) and loading applied perpendicular to the layers (yx and zx). The tensile strengths in the yx- and zy-direction (8.9 MPa and 7.8 MPa) are higher than those in the xy- and zy-directions, which reach 6.5 MPa and 7.4 MPa.

When compared to the corresponding reference specimen produced without 3DCP, which exhibits a tensile strength of 5.5 MPa, the yx-direction in the printed specimen demonstrates a strength increase of approximately 61.8%. This suggests a favorable influence of the printing process and orientation on the mechanical performance, particularly in the direction perpendicular to the layer interface.

The corresponding compressive strength results are illustrated in Fig. 7. In this context, the directional dependence is less pronounced. The compressive strengths in the x-,y-, and z-directions are measured at 66.4 MPa. 67.3 MPa and 65.9 MPa, respectively, indicating near-isotropic behavior. Notably, all printed directions show an increase of approximately 10% in compressive strength compared to the non-printed reference specimen, which reached 59.7 MPa.

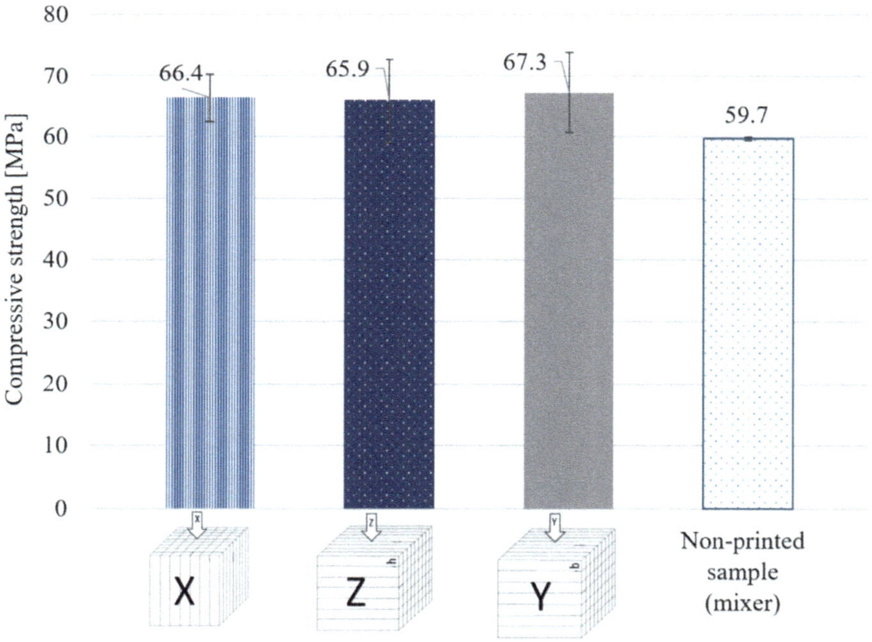

Fig. 7. Directional compressive strength from printed test specimens.

All in all, these findings underline the potential of the CSA cement-blended mixture for the 3DCP process not only in enabling geometric freedom but also in achieving mechanical performance equal to or superior to that of conventionally mixed counterparts.

Modeling and Additive Manufacturing of the Precast Concrete Elements. Building on the previously analyzed directional strength behavior, reference elements were printed to validate the material's suitability for more architectural applications. The components, pilasters (lesenes) for façade integration, are illustrated in the digital model and while 3DCP in Fig. 8.

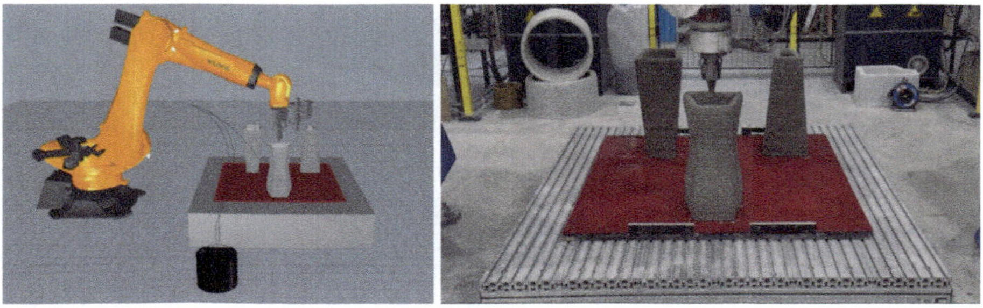

Fig. 8. Modeling and 3DCP of the reference elements.

For these precast concrete elements, the printing parameters such as speed (0,06 m/s) and layer width (20 mm) are kept consistent with those used in the test specimens, while the layer height is reduced from 10 mm to 5 mm to achieve a more refined surface appearance.

To further investigate the geometric limitations a structural response of the printable material, three lesenes with varying shapes and corresponding overhangs are extruded.

Lesene A features a trapezoidal base with side lengths of $a_1 = 200$ mm, $b_1 = 245$ mm and $c_1 = 70$ mm (opposite side of a_1). It exhibits an inward overhang of 50 mm on three sides over the full height of 500 mm ($= h$), corresponding to an overhang ratio of 10%. Lesene B serves as the inverse geometry to Lesene A, also with a trapezoidal base ($a_2 =$ 110 mm, $b_2 = 200$ mm and $c_2 = 70$ mm) and shows an outward overhang of 50 mm on three sides. Lesene C represents a hybrid design, combining the characteristics of both A and B. It includes an inward overhang over the lower portion of the height, transitioning into an outward overhang in the upper portion. Both the inward and outward section have an overhang of 12% on three sides. Each of the three lesenes features one vertical, flat face to enable attachment to the façade (Fig. 9).

All printed reference elements remained stable throughout the process, with no signs of structural failure. The CSA-based mixture proved capable of supporting overhanging structures and hollow sections within the tested geometry.

Fig. 9. Printed reference elements with dimensions.

4 Conclusions

This paper explored the use of CSA cement in combination with various conventional cements and SCMs to reduce the clinker content, and thus the CO_2 footprint, of binder systems. In parallel, the interaction between CSA cement-based systems and the set retarder citric acid was systematically investigated, with the aim of tailoring these binders for application in 3DCP. The targeted use of CSA cement enables not only low-carbon material design but also takes advantages of 3DCP's inherent efficiency in precise material placement and structurally optimized geometries such as hollow sections, further enhancing sustainability.

Key findings from the materials analysis can be summarized as follows:

1. CSA cement in combination with conventional cements such as CEM III/A and CEM III/B exhibited significant strength reductions even at low citric acid dosages, indicating compatibility issues when combined with high latent hydraulic components.
2. A ternary binder composition comprising 30 wt.% CSA cement, 21 wt.% GGBFS, and 49 wt.% CEM I proved effective, reducing the clinker content substantially while maintaining a well-balanced hydration profile.
3. These ternary systems demonstrated superior early strength development compared to pure CEM I or CEM III/A binders, both with and without the use of citric acid. This performance makes them particularly well-suited for 3DCP applications, where early build-up and open time must be carefully controlled.

The selected binder system proved suitable for 3DCP, achieving mechanical properties equal to or exceeding those of conventionally mixed materials. The successful printing of reference façade elements (lesenes) demonstrated the potential for material savings through hollow geometries and eliminated the need for formwork, highlighting the key advantages of digitally controlled additive manufacturing.

In summary, this work supports the ambitions of the European Green Deal by addressing CO_2 reduction both materially through clinker-reduced CSA cement-based binder systems and technologically through resource-optimized 3DCP, enabling more sustainable construction practices.

Acknowledgements. The authors would like to thank the sponsors for the financial support provided by the Department Bank of Saxony (SAB) via the European Regional Development Fond (ERDF).

References

1. CEMBUREAU – The European Cement Association (2020) Cementing the European green deal: executive summary of the 2050 carbon neutrality roadmap [PDF]. CEMBUREAU.
2. Verein Deutscher Zementwerke (VDZ) (2020) Dekarbonisierung von Zement und Beton–Minderungspfade und Handlungsstrategien. Retrieved from https://www.vdz-online.de/fileadmin/wissensportal/publikationen/zementindustrie/VDZ-Studie_Dekarbonisierung_von_Zement_und_Beton.pdf. Accessed 25 July 2025
3. DIN EN 197-1:2011–11. Cement—part 1: composition, specifications and conformity criteria for common cements—German version EN 197–1:2011
4. HeidelbergCement AG (2016) ZKG-Handbuch Zementanlagenbau 2016/2017: Leitfaden für Einsteiger in die Zementindustrie. Bauverlag BV GmbH. ISBN 978–3–7625-3678-9.
5. BauMineral GmbH (2020) Product data sheet: EFA-Füller® D4 Blaine 3800 [Fly ash material specification]. Uniper Kraftwerke GmbH. https://www.baumineral.de/
6. DIN EN 450–1:2012–10. Fly ash for concrete—part 1: definition, specifications and conformity criteria—German version EN 450–1:2012
7. Astra Technologia Betonu (2020) Technical data sheet: Astra MK-40—mineral activator for hydraulic binders. https://www.astra-polska.com
8. Deutsches Institut für Bautechnik (2020) European technical assessment: sh_stoneash, sh_easyflow, sh_compact (ETA-20/0231). Calcitwerke Schön + Hippelein GmbH & Co, KG. https://www.dibt.de
9. DIN EN 12390–2. Testing hardened concrete—part 2: making and curing specimens for strength tests; German version EN 12390–2:2019
10. DIN EN 12390–3:2019–10. Testing hardened concrete—part 3: compressive strength of test specimens—German version EN 12390–3:2019.
11. DIN EN 196–1:2016–11. Methods of testing cement—part 1: determination of strength—German version EN 196–1:2016
12. Abstoß M, Funke H, Gelbrich S (2025) Combined additive manufacturing process for the production of sustainable concrete modules for the construction of a tiny house. In: MATEC web of conferences, vol 409. EDP Sciences, p 06001
13. Lindner M, Gliniorz R, Funke H, Gelbrich S (2024) Anisotropic properties of additively manufactured concrete elements. Appl Res 3(3):e202300035

14. Senf F, Javed R, Funke H, Gelbrich S (2025) Enhanced environmental sustainability of 3D-printed concrete mixtures based on calcium sulfoaluminate (CSA) and blast furnace cement (CEM III/B). In: MATEC web of conferences, vol 409. EDP Sciences, p 06004

A Computational Framework for Topological Design Optimization of 3D-Printed Hempcrete Walls: An AI-Driven Approach to Minimize Global Warming Potential and Enhance Thermal Performance

Yasaman Norouzi(✉) and Manish K. Dixit

Department of Construction Science, Texas A&M University, College Station, USA
{Jasmine.norouzi,Mdixit}@tamu.edu

Abstract. Emerging 3D printing technologies in construction create new opportunities for novel wall designs that balance operational and embodied impact performance to deliver net-zero energy buildings. Although much research has focused on building geometry optimization and alternative material selection, a significant gap remains in minimizing the total global warming potential (GWP) within a single material framework. In this study, we propose a data-driven approach leveraging machine learning (ML) models to first predict and then optimize the GWP of buildings with 3D-printed external walls. Using parametric modeling in Rhino with the Grasshopper plugin, we computationally generate diverse geometries of 3D-printed lightweight concrete wall sections. We then employ two additional scripts to calculate thermal conductivity and simulate the building's energy consumption by iteratively altering external wall cross-sections. Additionally, the volume of each wall section is extracted for embodied impact calculations. After incorporating GWP data into our initial dataset, an ML algorithm is trained and tested to predict the embodied and operational impacts of buildings with 3D-printed walls. A second ML algorithm is then developed and employed to optimize the wall cross-section geometries for enhanced embodied and operational performance. The goal is to identify design configurations that minimize total GWP by modifying wall cross-section geometries, which affect both thermal conductivity and embodied emissions. The presented dual-model framework accelerates the design process and offers a streamlined path toward developing zero-carbon or carbon-negative 3D-printed wall systems.

Keywords: Building energy optimization · Global warming potential (GWP) · 3D print in construction · Hempcrete · Machine learning prediction model

1 Introduction

The construction sector is undergoing a significant transformation driven by advancements in 3D printing technology, which enables innovative and more sustainable building solutions [1, 2]. Construction and building operations contribute to roughly 33% of

H. Beushausen et al. (Eds.): NTZR 2025, LNCE 873, pp. 535–546, 2026.
https://doi.org/10.1007/978-3-032-27698-8_48

global greenhouse gas (GHG) emissions and consume about 40% of total annual energy use [3, 4]. A major portion of this energy is attributed to operational energy (OE), which includes heating, cooling, lighting, and the use of appliances and equipment.

Reducing OE is therefore critical to achieving sustainability goals in the built environment [5]. However, alongside operational energy, embodied energy—the energy required to produce, transport, and assemble building materials—also contributes significantly to a building's environmental impact [6, 7]. From raw material extraction through manufacturing and transportation, construction materials consume substantial primary energy inputs [8–11]. This total energy expenditure, known as embodied energy, plays a vital role in assessing building sustainability [6, 12].

A widely used metric for evaluating these environmental impacts is Global Warming Potential (GWP), which measures the heat trapped by greenhouse gases relative to CO_2. Achieving meaningful reductions in a building's GWP necessitates a dual focus on both embodied and operational emissions [13]. One effective strategy for reducing embodied impacts is the adoption of bio-based materials, which generally exhibit lower carbon footprints than conventional materials like steel and concrete [14]. Among these, hempcrete—a bio-composite composed of hemp shiv and a lime-based binder—shows strong potential due to its low or even negative carbon footprint. Hemp, being a fast-growing crop, actively sequesters CO_2 from the atmosphere during growth [14], making it particularly well-suited for sustainable applications such as 3D-printed construction, where it could replace high-emission materials like concrete.

The use of 3D printing in construction—commonly referred to as additive manufacturing—further enhances sustainability by offering precise geometric control, improving material efficiency, and reducing construction waste [15, 16]. Despite its potential, most current research in this area focuses on material-level or whole-building optimizations, overlooking the potential for improving environmental performance by refining wall section geometry. Additionally, traditional simulation tools for evaluating thermal conductivity and building energy use are computationally intensive and often slow, limiting the feasibility of iterative design exploration.

This study addresses these gaps by developing an AI-Driven methodology to predict and optimize the thermal performance and total GWP of 3D-printed wall sections made from a fixed hempcrete mix. Using parametric modeling in Rhino Grasshopper, we generate a wide range of wall geometries—such as zigzag and honeycomb patterns—designed to systematically vary the distribution of solid material and air gaps, which directly affect both thermal performance and material usage.

Two core assumptions underpin this research:

- A fixed hempcrete mix, validated in prior studies, is adopted to ensure printability and eliminate the need for physical experimentation. Instead, computational tools (THERM and EnergyPlus) are used to simulate thermal behavior and energy performance. The generated results will be further validated using physical 3D printing tests conducted in the Texas A&M University lab, as part of the HESTA project funded by the U.S. Department of Energy's ARPA-E program.
- Given hempcrete's non-load-bearing nature [17], the structural function is assumed to be provided by external framing systems (e.g., wooden studs), like the approach in Sharma et al. [18]. Thus, strength is not considered in the optimization process.

By integrating bio-based materials with computational design and machine learning, this research contributes to the development of advanced methodologies for minimizing building GWP and promoting scalable, low-carbon construction solutions.

2 Literature Review

2.1 Bio-Based Materials in Construction: Hempcrete

The construction industry's growing emphasis on sustainability has led to increasing attention on bio-based materials due to their environmental advantages. Among these alternatives, hempcrete has emerged as a notable material because of its carbon sequestration capability, thermal insulation performance, and comparatively low embodied emissions when measured against conventional materials like concrete [19]. Hempcrete is produced by combining hemp shiv, derived from the stalks of the Cannabis sativa plant, with a lime-based binder [20, 21]. This composite is primarily used in non-load-bearing applications, especially for building envelopes and insulation layers. Numerous studies have confirmed hempcrete's potential to store atmospheric CO_2, helping reduce the overall environmental burden throughout a building's lifecycle [14].

Although hempcrete's mechanical strength often falls short for load bearing uses [22], its renewable nature and fast growth rate offer a compelling alternative to traditional materials, particularly in contexts where environmental performance and carbon mitigation are priorities [23].

2.2 Additive Manufacturing (3D Printing) with Hempcrete in Construction

3D printing, or additive manufacturing, is transforming the construction sector by enabling precision fabrication of complex building geometries and reducing waste through material-efficient design [24]. When paired with sustainable materials like hempcrete, additive manufacturing can lower environmental impacts while enhancing productivity [25]. Research in this domain has so far concentrated on cost-effectiveness, environmental evaluation, lifecycle performance, and safety concerns related to digital fabrication techniques [25].

The unique ability of 3D printing to control layered geometry facilitates the design of multi-functional wall systems. For instance, previous work has shown that a dense outer shell made from printable hempcrete can be combined with a less dense infill for insulation, allowing both thermal and structural goals to be addressed simultaneously. This technique has been shown to sequester between 1.2–16.7 kg CO_2eq/m^2, depending on mix composition and application strategy [14]. Sinka et al. [26] extended this research by exploring various fast-setting binders in combination with hemp shives, demonstrating the environmental viability of 3D-printed hempcrete composites as sustainable replacements for more carbon-intensive alternatives.

2.3 Optimization of Wall Designs for Environmental Performance

Wall geometry plays a crucial role in influencing thermal behavior, energy efficiency, and material consumption in buildings. Prior studies have highlighted how intentional geometric modifications—like embedded air pockets—can disrupt thermal bridges and reduce heat transfer, leading to better insulation and lower energy use [27, 28]. Alzahrani et al. [29], for instance, demonstrated how specific infill designs within 3D-printed concrete walls can reduce thermal conductivity without increasing material volume. Briels et al. [30] showed that cellular configurations can boost thermal resistance by up to 24%, indicating a strong correlation between geometry and performance.

Other studies, such as those by Dielemans et al. [31], further support the idea that cellular structures can serve dual roles, simultaneously addressing thermal and structural requirements, which also leads to reductions in embodied energy. Mahadevan et al. [32] emphasized that refined wall geometry not only improves operational performance but can also help lower Global Warming Potential (GWP). Similarly, Antony et al. [33] explored the fabrication of hemp/PLA honeycomb sandwich structures through FDM printing. These designs were shown to be lightweight, mechanically sound, and biodegradable, offering environmentally conscious options for structural applications.

These findings collectively underscore the critical role of geometric optimization in 3D-printed wall design. They also highlight how AI-driven modeling and simulation can enhance this process by accelerating the discovery of low-GWP geometries without relying solely on trial-and-error or conventional simulation tools [34].

3 Research Motivation and Focus

Although there has been meaningful progress in developing sustainable construction materials such as hempcrete [35–37] and in optimizing building-scale performance parameters [38, 39], limited research has addressed the role of wall section geometry in reducing the total Global Warming Potential (GWP) of buildings constructed with 3D-printed walls [40]. Much of the existing literature emphasizes material selection [14, 21] or macro-scale optimization strategies [29, 30, 38, 39, 41], while the influence of micro-scale geometrical and topological variation in wall sections remains underexplored.

In particular, the potential of using machine learning (ML) to predict and optimize both thermal performance and total GWP—including embodied and operational impacts—of various 3D-printed wall designs has not yet been systematically investigated [29, 40]. Conventional simulation-based approaches are not only computationally intensive but also inefficient for high-resolution geometric iteration, often limiting their applicability in iterative design workflows [42, 43]. This creates a critical need for a computationally efficient, data-driven methodology that can identify optimal wall configurations by learning from simulation data and predicting performance without repeating resource-intensive analyses.

This research responds to that gap by investigating three central questions:

- How do variations in geometric and topological configurations of 3D-printed hempcrete walls affect thermal conductivity and building energy consumption?
- Can machine learning algorithms reliably predict thermal conductivity and total GWP across a wide range of geometric wall designs?
- How effectively can AI-driven optimization frameworks identify wall geometries that minimize total GWP, considering both operational and embodied environmental impacts?

This study aims to bridge the methodological divide between design geometry and performance optimization, offering a scalable, intelligent solution to improve environmental outcomes in 3D-printed, bio-based construction systems.

4 Methodology

This study presents a structured methodology integrating parametric design, thermal and energy simulation, life cycle impact analysis, and machine learning to optimize the geometry of 3D-printed hempcrete walls to predict and optimize the Global Warming Potential (GWP). The methodology consists of five main stages as shown in Fig. 1:

1. Generation of diverse wall geometries
2. Calculation of thermal conductivity
3. Simulation of building energy consumption and assessment of embodied environmental impacts
4. Development and evaluation of machine learning models
5. AI-driven optimization of wall designs

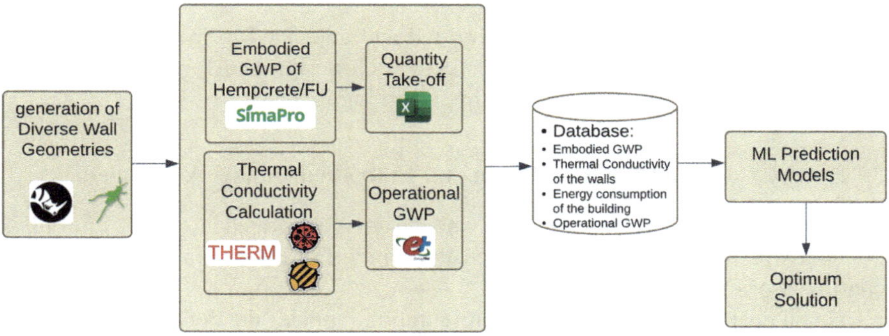

Fig. 1. Research methodology workflow

4.1 Parametric Generation of Wall Geometries

A broad range of wall section configurations is generated using Grasshopper, the visual programming interface in Rhinoceros 3D. Three geometric typologies—zigzag, honeycomb, and Voronoi—are selected for their adaptability to 3D printing and their relevance in prior research [14, 26, 29, 33]. Each design varies in pattern density, wall thickness distribution, and material-to-air ratio, while maintaining a constant total wall thickness to isolate the impact of geometry on thermal performance.

These geometries are algorithmically iterated to construct a comprehensive dataset of wall section designs, enabling controlled comparisons across variations in shape and topology (Fig. 2).

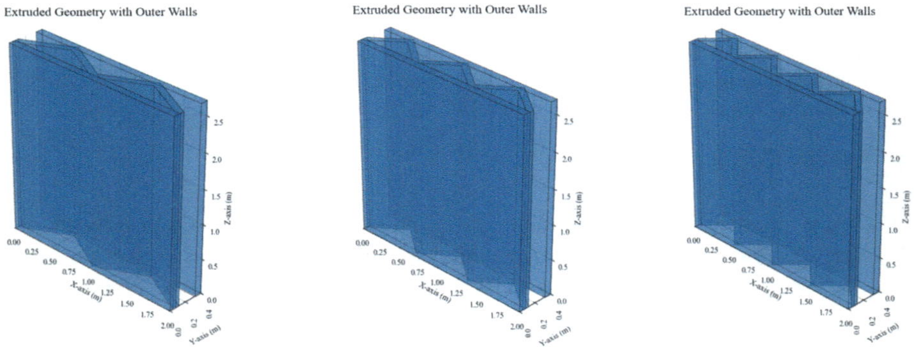

Fig. 2. Examples of different wall section configurations

4.2 Thermal Conductivity Analysis

THERM, a validated 2D finite element software tool for steady-state heat transfer modeling [44], is used to calculate the thermal conductivity of each wall configuration. Each geometry is translated into THERM-compatible input, where boundary conditions are standardized across simulations (Fig. 1).

Thermal conductivity results are exported and appended to the geometry dataset. This step quantifies the thermal resistance of each design and forms a key input for subsequent energy simulations and machine learning models.

4.3 Energy Consumption Simulation and Embodied Impact Assessment

To assess operational performance, each wall design is integrated into a baseline building model—a small office prototype located in Tampa, Florida—obtained from the U.S. Department of Energy [45].

Using a custom script, each wall configuration replaces the default exterior wall in the EnergyPlus simulation engine. The internal layout, mechanical systems, and climate conditions remain fixed to isolate the influence of wall geometry. Annual heating and cooling loads are extracted from each simulation, representing the operational energy use associated with each design. The embodied GWP of each wall configuration is evaluated using SimaPro version 9.6.0.1, paired with the North American DATASMART LCI database [14, 25]. Hempcrete quantities are calculated from wall volumes, and impacts are assessed using a cradle-to-gate boundary. Since the material mix ratio is fixed across all geometries, differences in embodied GWP arise solely from volumetric and surface design changes. This enables a direct evaluation of how geometric variation affects embodied carbon outcomes.

4.4 Machine Learning Model Development

To develop robust predictive models for estimating thermal conductivity, operational energy use, and total GWP based on wall geometry parameters, appropriate machine learning algorithms will be meticulously selected. This selection will be guided by the complexity of the relationships observed between inputs and outputs, the specific characteristics of the generated dataset, and the nature of the prediction task. Model performance will be rigorously evaluated using standard metrics such as Mean Squared Error (MSE) to quantify the average squared difference between predicted and actual values, and R-squared (R^2) to indicate the proportion of the variance in the dependent variable that is predictable from the independent variables. This robust evaluation will ensure the selection of the most accurate and reliable models for practical application, significantly reducing the need for repeated simulation workflows and thereby accelerating the design iteration process. The dataset will be divided into training (70%), testing (20%), and validation (10%) subsets to ensure robust model development (Fig. 3).

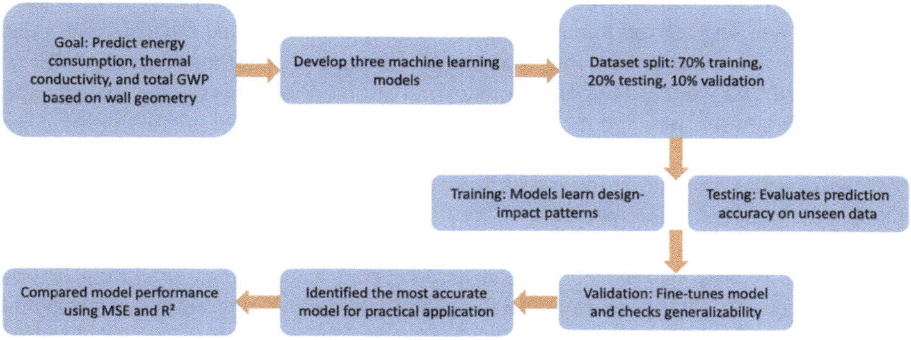

Fig. 3. Machine learning prediction models workflow

4.5 AI-Driven Optimization of Wall Geometries

Using the best-performing prediction model, an optimization algorithm is implemented to identify wall geometries that minimize total GWP, combining both embodied and operational impacts. The optimization process is constrained by factors essential for practical application, including printability, fixed wall thickness, and non-load-bearing material assumptions. Printability is addressed by operating under the core assumption of a fixed hempcrete mix ratio, which has been validated for printability in prior studies, specifically through the HESTA project, Texas A&M University. This fixed material composition ensures consistency in embodied impact calculations and eliminates the need for extensive physical material experimentation within this computational study. Furthermore, the methodology's results will be cross-referenced with physical 3D printing tests conducted in the Texas A&M University HESTA lab, as part of the U.S. Department of Energy's ARPA-E program, for practical validation.

5 Discussion

The methodology presented in this study demonstrates a robust and integrated approach to optimizing the environmental performance of 3D-printed hempcrete walls. By combining parametric geometry generation, validated simulation tools, and machine learning models, the framework enables scalable, data-driven evaluation and optimization of wall section designs with respect to thermal conductivity and total Global Warming Potential (GWP). One of the key strengths of this approach is its ability to streamline the traditionally time-consuming iterative design process. Through the integration of THERM for thermal analysis and EnergyPlus for operational energy simulation, the framework produces high-fidelity performance data across a large design space. Machine learning models trained on this data replace the need for repeated simulations in later design stages, enabling rapid performance prediction with minimal computational overhead.

Another significant advantage is the isolation of geometry as the sole variable, made possible by fixing the material mix ratio and maintaining constant wall thickness. This allows for a clear, unambiguous evaluation of how geometric and topological design decisions alone impact both operational and embodied environmental performance. Moreover, by leveraging a fixed hempcrete composition with validated environmental data, the framework ensures consistency in embodied impact calculations using SimaPro and standardized LCI sources. To complement simulation-based analysis, real-world experimental validation is planned through 3D printing tests conducted in collaboration with the Texas A&M HESTA lab, as part of the ARPA-E research initiative. This validation step will further strengthen the credibility of simulation results and help calibrate the models for practical implementation.

Despite these strengths, the methodology is subject to several limitations. First, the analysis focuses exclusively on non-load-bearing wall designs, assuming external structural support. This is appropriate for hempcrete, which inherently lacks structural capacity. However, this exclusion limits the framework's applicability to structural performance analysis. In future work, this methodology could be extended to incorporate structural considerations for alternative materials with load-bearing capabilities, thereby broadening its utility for integrated building systems. Second, the scope is limited to a single climate zone (Florida, TX), which may affect the generalizability of operational results across diverse environmental contexts. Future studies should extend the framework to account for different climates and thermal zoning requirements.

Furthermore, while the current implementation explores a defined range of geometric typologies, the optimization process could be enhanced by incorporating multi-objective algorithms and additional constraints, such as 3D printability, cost, or construction time. There is also potential for incorporating real-world experimental validation, particularly to refine thermal bridging effects and construction tolerances not fully captured in simulation.

Overall, this methodology offers a replicable and adaptable framework that can accelerate sustainable material design by bridging simulation, data analytics, and optimization. It contributes to broader efforts to reduce carbon emissions in the built environment while paving the way for smarter, geometry-informed design strategies in 3D-printed construction.

6 Conclusion

This study introduced a data-driven methodology to optimize the thermal and environmental GWP of 3D-printed hempcrete wall sections by systematically varying geometry while maintaining a fixed material composition. The proposed framework leverages parametric modeling, energy and thermal simulations, and machine learning to predict and minimize total Global Warming Potential (GWP), incorporating both embodied and operational components.

The core contribution of this work lies in its ability to transform a traditionally time- and computation-intensive process into a scalable, ML-assisted design workflow. By predicting performance outcomes based solely on geometric inputs, the approach facilitates faster, more informed decision-making during early-stage design.

Beyond methodological innovation, the study supports broader goals of sustainable development by offering practical tools for reducing carbon emissions in the building sector. Upon the completion of the AI-driven optimization, a rigorous quantitative comparison will be conducted between the resulting optimized wall geometries and the initial baseline designs. This will precisely quantify the improvements in both operational and embodied environmental performance, thereby strengthening the practical relevance of our findings and illustrating a clear path towards low-carbon construction solutions. This paper contributes to the advancement of low-carbon construction practices and aligns with global climate objectives, particularly in the context of SDG 9 (Industry, Innovation, and Infrastructure), SDG 11 (Sustainable Cities and Communities), and SDG 13 (Climate Action).

Future work will build on this framework by integrating structural performance constraints, expanding to diverse climate zones, and exploring multi-objective optimization that includes cost, printability, and construction logistics.

References

1. Sakin M, Kiroglu Y (2017) 3D printing of buildings: construction of the sustainable houses of the future by BIM. Energy Procedia 134:702–711. https://doi.org/10.1016/J.EGYPRO.2017.09.562
2. Adaloudis M, Roca J (2021) Sustainability tradeoffs in the adoption of 3D concrete printing in the construction industry. J Clean Prod Adv Online Publ. https://doi.org/10.1016/J.JCLEPRO.2021.127201
3. Yan H, Shen Q, Fan LC, Wang Y, Zhang L (2010) Greenhouse gas emissions in building construction: a case study of one Peking in Hong Kong. Build Environ 45(4):949–955. https://doi.org/10.1016/j.buildenv.2009.09.014
4. Kisku N, Joshi H, Ansari M, Panda SK, Nayak S, Dutta SC (2017) A critical review and assessment for usage of recycled aggregate as sustainable construction material. Constr Build Mater 131:721–740. https://doi.org/10.1016/j.conbuildmat.2016.11.029
5. Asim N et al (2022) Sustainability of heating, ventilation and air-conditioning (HVAC) systems in buildings—an overview. Int J Environ Res Public Health 19(2):1016. https://doi.org/10.3390/ijerph19021016
6. Dixit MK, Culp CH, Fernandez-Solis JL (2015) Embodied energy of construction materials: integrating human and capital energy into an IO-based hybrid model. Environ Sci Technol 49(3):1936–1945. https://doi.org/10.1021/es503896v
7. Praseeda KI, Reddy BVV, Mani M (2016) Embodied and operational energy of urban residential buildings in India. Energ Buildings 110:211–219. https://doi.org/10.1016/j.enbuild.2015.09.072
8. Dixit MK (2017) Life cycle embodied energy analysis of residential buildings: a review of literature to investigate embodied energy parameters. Renew Sust Energ Rev 79:390–413. https://doi.org/10.1016/j.rser.2017.05.051

9. Marszal AJ et al (2011) Zero energy building – a review of definitions and calculation methodologies. Energ Buildings 43(4):971–979. https://doi.org/10.1016/j.enbuild.2010.12.022
10. Dixit MK (2017) Embodied energy and cost of building materials: correlation analysis. Build Res & Inf 45(5):508–523. https://doi.org/10.1080/09613218.2016.1191760
11. Monteiro H, Fernández JE, Freire F (2016) Comparative life-cycle energy analysis of a new and an existing house: the significance of occupant's habits, building systems and embodied energy. Sustain Cities Soc 26:507–518. https://doi.org/10.1016/j.scs.2016.06.002
12. Stephan A, Stephan L (2016) Life cycle energy and cost analysis of embodied, operational and user-transport energy reduction measures for residential buildings. Appl Energy 161:445–464. https://doi.org/10.1016/j.apenergy.2015.10.023
13. Ibn-Mohammed T, Greenough R, Taylor S, Ozawa-Meida L, Acquaye A (2013) Operational vs. embodied emissions in buildings—a review of current trends. Energ Buildings 66:232–245. https://doi.org/10.1016/j.enbuild.2013.07.026
14. Sinka M, Spurina E, Korjakins A, Bajare D (2022) Hempcrete – CO_2 neutral wall solutions for 3D printing. Environ Clim Technol 26(1):742–753. https://doi.org/10.2478/rtuect-2022-0057
15. Hager I, Golonka A, Putanowicz R (2016) 3D printing of buildings and building components as the future of sustainable construction. Procedia Eng 151:292–299. https://doi.org/10.1016/J.PROENG.2016.07.357
16. Kontovourkis O (2021) 3D printing technology within a regenerative construction framework. Future City. https://doi.org/10.1007/978-3-030-71819-0_13
17. Pretot S, Collet F, Garnier C (2014) Life cycle assessment of a hemp concrete wall: impact of thickness and coating. Build Environ 72:223–231. https://doi.org/10.1016/j.buildenv.2013.11.010
18. Sharma S, El Tahlawi M, Delavar MA, Sideris P (2025) Structural design methodology for low-rise 3D printed concrete (3DPC) buildings subjected to non-seismic loading: description, application and validation. J Build Eng 105:112200
19. Arrigoni A, Pelosato R, Melià P, Ruggieri G, Sabbadini S, Dotelli G (2017) Life cycle assessment of natural building materials: the role of carbonation, mixture components and transport in the environmental impacts of hempcrete blocks. J Clean Prod 149:1051–1061. https://doi.org/10.1016/j.jclepro.2017.02.161
20. Jami T, Karade SR, Singh LP (2019) A review of the properties of hemp concrete for green building applications. J Clean Prod 239, Article 117852. https://doi.org/10.1016/j.jclepro.2019.117852
21. Birjukovs M, Sinka M, Jakovics A, Bajare D (2022) Combined in situ and in silico validation of a material model for hempcrete. Constr Build Mater 321, Article 126051. https://doi.org/10.1016/j.conbuildmat.2021.126051
22. De Bruijn PB, Jeppsson K-H, Sandin K, Nilsson C (2009) Mechanical properties of lime–hemp concrete containing shives and fibres. Biosyst Eng 103(4):474–479. https://doi.org/10.1016/j.biosystemseng.2009.02.005
23. Rehman M et al (2021) Evaluation of hemp (Cannabis sativa L.) as an industrial crop: a review. Environ Sci Pollut Res 28(38):52832–52843. https://doi.org/10.1007/s11356-021-16264-5
24. Das AK, Agar DA, Rudolfsson M, Larsson SH (2021) A review on wood powders in 3D printing: processes, properties and potential applications. J Mater Res Technol 15:241–255. https://doi.org/10.1016/j.jmrt.2021.07.110
25. Mognol P, Lepicart D, Perry N (2006) Rapid prototyping: energy and environment in the spotlight. Rapid Prototyp J 12(1):26–34. https://doi.org/10.1108/13552540610637246
26. Sinka M, Zorica J, Bajare D, Sahmenko G, Korjakins A (2020) Fast setting binders for application in 3D printing of bio-based building materials. Sustainability 12(21), Article 8838. https://doi.org/10.3390/su12218838

27. Marais H, Christen H, Cho S, Villiers W, Zijl G (2021) Computational assessment of thermal performance of 3D printed concrete wall structures with cavities. J Build Eng 41:102431. https://doi.org/10.1016/J.JOBE.2021.102431
28. Suntharalingam T et al (2021) Energy performance of 3D-printed concrete walls: a numerical study. Buildings. https://doi.org/10.3390/buildings11100432
29. Alzahrani A, Alghamdi A, Basalah A (2022) Computational optimization of 3D-printed concrete walls for improved building thermal performance. Buildings. https://doi.org/10.3390/buildings12122267
30. Briels D et al (2022) Thermal optimization of additively manufactured lightweight Concrete Wall elements with internal cellular structure through simulations and measurements. Buildings. https://doi.org/10.3390/buildings12071023
31. Dielemans G, Briels D, Jaugstetter F, Henke K, Dörfler K (2021) Additive Manufacturing of Thermally Enhanced Lightweight Concrete Wall Elements with Closed Cellular Structures. 9:59–72. https://doi.org/10.7480/JFDE.2021.1.5418
32. Mahadevan M, Francis A, Thomas A (2020) A simulation-based investigation of sustainability aspects of 3D printed structures. J Build Eng 32:101735. https://doi.org/10.1016/j.jobe.2020.101735
33. Antony S, Cherouat A, Montay G (2020) Fabrication and characterization of hemp fibre based 3D printed honeycomb sandwich structure by FDM process. Appl Compos Mater 27(6):935–953. https://doi.org/10.1007/s10443-020-09837-z
34. Alhumayani H, Gomaa M, Soebarto V, Jabi W (2020) Environmental assessment of large-scale 3D printing in construction: a comparative study between cob and concrete. J Clean Prod. https://doi.org/10.1016/j.jclepro.2020.122463
35. Muhit IB, Omairey EL, Pashakolaie VG (2024) A holistic sustainability overview of hemp as building and highway construction materials. Build Environ 111470
36. Asghari N, Memari AM (2024) State of the art review of attributes and mechanical properties of hempcrete. Biomass 4(1):65–91
37. Jellen AC, Memari AM (2025) State-of-the-art review of Hempcrete for residential building construction. Designs 9(2):44
38. Lu J, Luo X, Cao X (2024) Research on geometry optimization of park office buildings with the goal of zero energy. Energy 306:132179
39. Salata F, Ciardiello A, Dell'Olmo J, Ciancio V, Ferrero M, Rosso F (2024) Geometry optimization in the schematic design phase of low-energy buildings for all European climates through genetic algorithms. Sustain Cities Soc 112:105639
40. Hassan H, Rodriguez-Ubinas E, Al Tamimi A, Trepci E, Mansouri A, Almehairbi K (2024) Towards innovative and sustainable buildings: a comprehensive review of 3D printing in construction. Autom Constr 163:105417
41. Jipa A, Dillenburger B (2021) 3D printed formwork for concrete: state-of-the-art, opportunities, challenges, and applications. 3D Print Addit Manuf 9:84–107. https://doi.org/10.1089/3dp.2021.0024
42. Liu Y, Zhao T, Ju W, Shi S (2017) Materials discovery and design using machine learning. J Mater 3(3):159–177
43. Qu K et al (2023) Optimal design of building integrated energy systems by combining two-phase optimization and a data-driven model. Energ Buildings 295:113304
44. Lawrence Berkeley National Laboratory (2022) THERM. Retrieved 29 Mar 2025, from https://windows.lbl.gov/software-tools#therm-heading
45. U.S. Department of Energy (n.d.) Prototype Building Models. (online) Available at https://www.energycodes.gov/prototype-building-models

Towards the Integration of Additive Manufacturing into a Cyber-Physical Construction System

Gerrit Placzek[1(✉)], Julian Tesche[2], Karam Mawas[3], Jan Thormählen[1], Patrick Schwerdtner[1], Norman Hack[2], Mehdi Maboudi[3], and Markus Gerke[3]

[1] Institute for Construction Engineering and Management, Technische Universität Braunschweig, Braunschweig, Germany
g.placzek@tu-braunschweig.de

[2] Institute of Structural Design, Technische Universität Braunschweig, Braunschweig, Germany

[3] Institute of Geodesy and Photogrammetry, Technische Universität Braunschweig, Braunschweig, Germany

Abstract. As one of the most important sectors of the global economy, the construction industry has continuously sought to optimize its processes. However, construction remains characterized by a highly fragmented value chain, low productivity, and limited digitalisation. Introducing Cyber-Physical Construction Systems offers a way to address these challenges by enabling a more flexible, automated, and collaborative approach. In this paper, a cyber-physical design-to-fabrication workflow was developed using terrestrial laser scanning as a sensing tool, augmented reality as a design interface, and an extrusion concrete printer as an actuator. As a result of the cyber-physical construction workflow, two benches were digitally fabricated as a case study by combining reclaimed bricks with 3D concrete-printed components. The developed workflow enables designers to use augmented reality to generate an adapted path for 3D-printed structures. The study highlights the potential of Cyber-Physical Construction Systems to improve automation and efficiency while fostering human-robot collaboration through human-in-the-loop approaches. However, the case study revealed key challenges, including gaps in automation and scalability, that must be addressed for broader implementation. Future work should focus on refining the integration of sensing technologies, improving communication between digital and physical systems, and optimizing collaborative workflows to leverage the benefits of Cyber-Physical Construction Systems in construction fully.

Keywords: Cyber-Physical Systems · 3D Concrete Printing · Augmented Reality · Laser Scanning · Digital Fabrication

1 Introduction

The construction industry, as one of the most important sectors of the global economy, is largely characterised by low productivity, limited digitalisation and a highly fragmented value chain. The combination of labour-intensive manual tasks, an ageing workforce, and

H. Beushausen et al. (Eds.): NTZR 2025, LNCE 873, pp. 547–558, 2026.
https://doi.org/10.1007/978-3-032-27698-8_49

a declining availability of skilled labour highlights the growing need to integrate digital fabrication processes. Additionally, the physical and the digital worlds are increasingly connected and interacting with each other. In the context of industrial production, we now commonly speak of Industry 4.0, Internet-of-Things, or cyber-physical systems (CPS). The first genuinely digital construction technologies applied in construction practice, such as depositing processes in 3D concrete printing (3DCP), are based on the principles of Industry 3.0 [3]. Since there is no communication between the machines and processes, no digital continuous and bidirectional data flow is guaranteed [4]. In this case, data from the 3D model is sent to the printer, and then printed *blindly.* In research, on the other hand, the first cyber-physical Additive Manufacturing (AM) systems for construction are being investigated in which, for example, the actual as-is condition is fed back into a 3D model by utilizing a 3D laser scanner to adapt the subsequent printing paths accordingly [5]. However, these CPSs are yet limited to individual production steps and are not embedded into the overall construction process, which consists of many individual components and production steps in a sequential or synchronized manner [6].

To research seamless, digitally continuous construction processes from design to realization across all facets, and not be limited to individual, singular production systems, the Digital Construction Site (DCS) has been installed at TU Braunschweig. We refer to the DCS as our Cyber-Physical Construction System (CPCS), that combines the following: a stationary, large-scale 3D concrete printer, multiple mobile robots on caterpillar tracks, a Digital Engineering Center (DEC) for VR/AR applications, drones, a variety of reality capture sensors (e.g., 3D scanning), and a system tracker (see Fig. 1). In this paper, we present a preliminary case study in which the following key steps are executed manually or semi-manually: (i) 3D concrete printing, (ii) toolpath generation, (iii) laser scanning, (iv) quality control, and (v) AR-based form finding are executed manually or semi-manually. Our goal is to highlight the fragmented processes and potential failure points in the current setup. These insights will guide the next phase of the CPCS (and our future work), in which we will:

1. bridge gaps between individual actuators and software modules.
2. automate the remaining manual workflows, and
3. establish real-time data integration across all hardware.

A fully realized CPCS will ultimately enable continuous digital additive manufacturing strategies along the entire on-site production chain, overcoming the limitations of scaled-down in-lab experiments and demonstrating design-to-fabrication continuity.

The paper is structured as follows: Sect. 2 provides an overview of the key components of a CPS and its potential for integration into construction. Moreover, a brief overview of the current state of the art in design-to-fabrication workflows is given. In Sect. 3, a case study is designed to explore cyber-physical design-to-fabrication workflows. Section 4 shows the results of the workflow used to produce a hybrid bench consisting of reclaimed bricks and 3D-printed components, followed by a discussion about future research in Sect. 5. Finally, Sect. 6 draws conclusions.

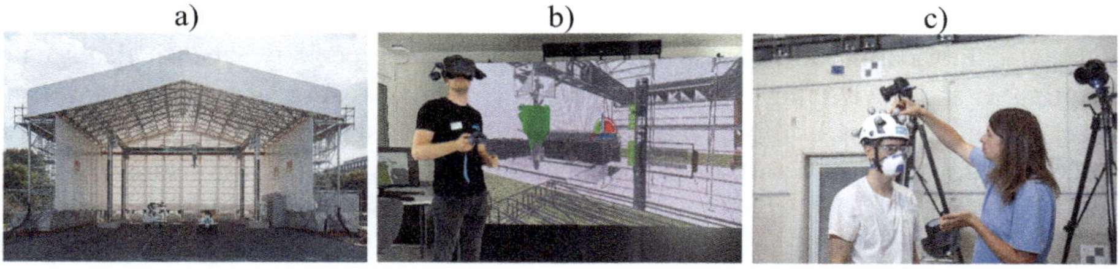

Fig. 1. Digital Construction site: a) stationary and mobile 3D construction printer (© Tjark Spille); b) virtual reality powerwall at the digital engineering center c) VICON scanning and tracking system.

2 Background

2.1 Cyber-Physical Systems in Construction

With the introduction of CPS in architecture and construction, the challenges emerging from traditional sequential and fragmented workflows in the architecture, engineering, and construction (AEC) industries can be tackled. In addition, CPSs aim for a cost-effective production with high efficiency and increased flexibility [7]. Tao. et al. (2019) describe CPSs as "multidimensional and complex systems that integrate the cyber world and the dynamic physical world" [1]. Furthermore, in the context of construction, CPSs consist of two key elements: firstly, the "physical to cyber-bridge", where sensors are used to acquire information about the building components and secondly, the "cyber to physical-bridge", where the gathered information is used to adapt to changes in the process [2].

However, there a many important differences when comparing construction to industrial production [6]. Since the construction site is mostly an unstructured, constantly changing environment with dynamic processes, standard automation procedures will not be applicable and require an adapted "Cyber-Physical Construction System" [6]. Using the specific term *CPCS* distinguishes it from the broader term *CPS*, which applies to other industries. While the construction industry is still highly labour-intensive, a CPCS must not only account humans as a crucial component. By actively using *human-in-the-loop* approaches, collaborative workflows are created where humans and robots can contribute their respective skills. What is more, the emergence of CPS comes along with a digital transformation that could not only speed up the automation on construction sites, but also allow CPS-enabled prefabrication.

The key elements of a CPS—and thereby also of a CPCS—to facilitate bidirectional coordination between the virtual models (cyber world) and their physical representations (physical world) are different *layers,* such as sensing, communication, application and actuation layer are considered [8]: While sensing devices could capture the reality, they could be used to compute the deviation from design, track the motion of on-site personnel and locate materials and equipment and facilitate time-stamped progress monitoring. The data obtained from the sensing layer is converted by the communication layer and will be analyzed by the application layer. Robotics and (digital-assisted) human workers as part of the actuation layer execute the instructions provided by the application layer.

CPCS offers fabrication flexibility and efficiency due to environment-aware design and automated fabrication and quality monitoring at once, especially when it deals with adapting to and developing existing structures [6]. For existing structures, significant effort is often required, as most available 2D plans must first be digitized, and automated processes are often only partially applicable. Digital design and fabrication have the potential to significantly reduce construction costs, resource consumption, and CO_2 emissions by integrating reused building components into newly conceived forms and functions. This approach enables the sustainable transformation of traditional linear construction into circular construction [2].

2.2 State of the Art in Design-to-Fabrication-Workflows for Existing Contexts

The following brief literature review demonstrates how cyber-physical construction processes, such as environmental-aware, augmented reality-based design to robotically fabrication processes are offering significant improvements in customized design, fabrication productivity and precision.

Naboni et al. (2022) hypothesized that equipping printing machines with sensing devices and adaptive design algorithms could extend the range and effectiveness of 3DCP [10]. Therefore, they developed a real-time and environment-aware robotic 3DCP process, called SENS-ENV. By using a depth camera, an unknown and unstructured ground was scanned, followed by an autonomously generated tailored design and toolpath. During fabrication, the print path was dynamically updated when defects or changes in the environment occurred. Similarly, Dielemans et al. (2024) developed a design-to-fabrication workflow for the in-situ repair of existing building structures, in which an incomplete and partially demolished brick structure was first scanned with a depth camera. Thereafter, a static print path was generated based on the 3D scans [11]. Then, also due to workshop constraints, a mobile robot printed the missing part with clay. Two different printing paths were also created to explore further design possibilities within the specified brick dimensions. In both approaches, the aim was to keep the print width consistent throughout the manufacturing process.

Zhan et al. (2022) showed a promising approach to actively change the printing width during fabrication by controlling the robot speed [12]. This has increased print resolution and improved structural integrity. Again, the toolpath generation algorithm is one of the main keys to exploring innovative designs.

Yet existing approaches have focused only on the ease of fabrication with the given robotic system, excluding the possibility of human interaction in the design or fabrication process. Therefore, human-machine collaboration seems favorable. Mittenberger et al. (2022) developed an Interactive Robotic Plastering (IRoP) system, enabling designers and skilled workers to intuitively engage with an in-situ robotic plastering process through a projection-based augmented reality digital-physical workflow [13]. The results indicated that collaborative human-machine workflows can foster a socially sustainable integration of robotic construction processes.

Despite existing and sophisticated approaches, further research is needed to take a holistic view of CPS in the built environment, from design through engineering to construction. In particular, the use of 3DCP in the context of existing buildings, where people are not just passive observers but are actively involved in the design-to-fabrication process, remains largely unexplored. To further investigate the potentials of CPCS in this context, the research project "Integration of additive manufacturing into a cyber-physical construction system" of the Collaborative Research Centre "Additive Manufacturing in Construction" was funded from the German Research Foundation (DFG—project number 414265976—TRR 277) [9]. Within this research project, cyber-physical design and fabrication strategies for AM will be explored based on case studies from the perspectives of structural design, geodesy and photogrammetry as well as construction management.

3 Case Study: General Workflow and Experimental Setup

This first case study at the DCS was conducted with a workshop at the TU Braunschweig, prior to Digital Concrete 2024. Eight international students and the authors of this paper developed and evaluated novel fabrication scenarios (case studies) for the use of CPS on digital construction sites. The aim of the case study present here was multifaced: It was partitially pre-designed from the authors to explore the potential benefits of CPS in construction, as seen from the perspectives of structural design, geodesy and photogrammetry as well as construction management to identify technical challenges as a basis for future improvements of our research infrastructure. It also served as a teaching material, with its scope tailored to fit a two-day workshop format for PhD students, and setting the boundaries for design and fabrication complexity. Generally, the developed design-to-fabrication workflow can be summarized as followed (see Fig. 2): A brick structure built from non-planar, 140-year-old reclaimed bricks will be scanned to enable an automated script to design a 3D-printed concrete bench on top of them.

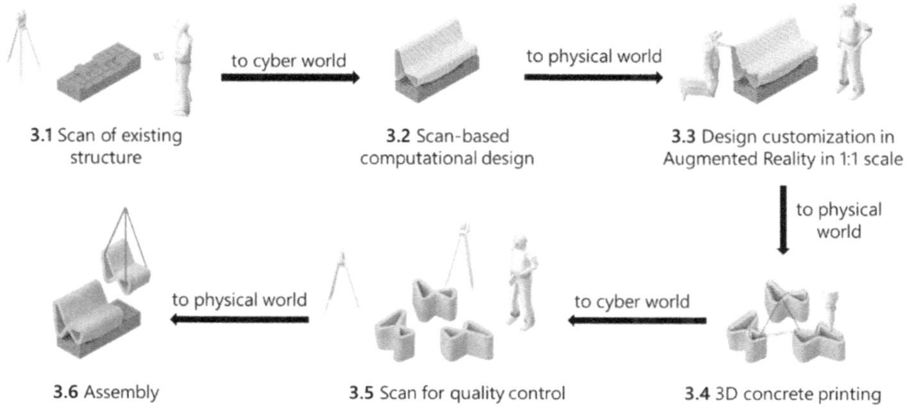

Fig. 2. Digital-to-fabrication workflow for the case study

Designers can adjust the geometry and path planning of the bench using an augmented reality app. Once finalized, the bench components are sliced, printed on a construction-scale concrete 3D printer and rescanned for quality control. After checking tolerances, the bench is assembled onto the brick structure. Although the bench as such might not be of high relevance for practical application, this case study was designed to demonstrate a CPCS workflow that integrates physical-to-digital-to-physical transitions through the use of scans, highlighting the potential of CPCS.

3.1 Scan of Existing Structure

In this case study, a three-stage approach for reality capture during the printing processes for quality control (QC) using a terrestrial laser scanner (TLS) was proposed. The three stages are as follows: (i) pre-processing, (ii) after printing, and (iii) after assembly. In the pre-processing stage, TLS captures the existing structure's geometry to facilitate design adaptation. After printing, the scanner provides feedback on the printed object, integrating real-world data into the digital world to refine the printed object for any potential artifacts that might exist. After the curing and assembly of the printed object, a subsequent cycle of quality control and data capture is initiated to facilitate a comparison between the as-built object and the original design. The scanning after assembly (iii) becomes the starting point of the process again (i). These stages synchronize the digital and physical worlds, enabling comprehensive analysis throughout the printing cycle.

3.2 Computational Design

In the next step, the captured point cloud of the brick structure was manually imported into a Rhino Grasshopper environment. There, it was meshed to simulate two curves using Rhino Kangaroo, which drape like ropes over the irregular and stepped brick structure (see Fig. 3a). These curves defined the future contact surfaces of the concrete components on the bricks. Since the concrete components were to be rotated after printing, the maximum slopes of these curves are limited by an overhang angle of 15°, which is the maximum that can be printed with the current setup.

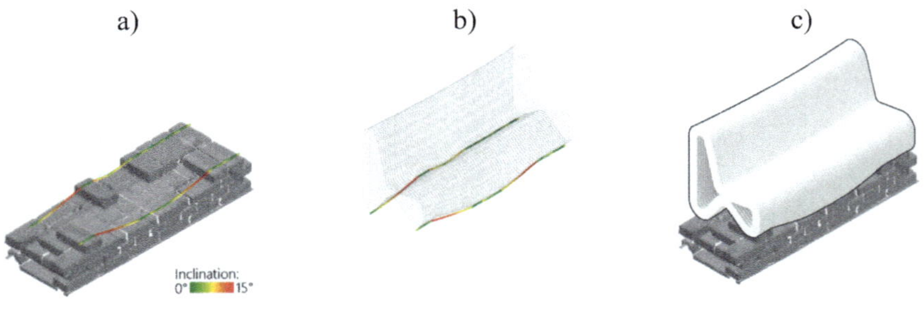

Fig. 3. Path planning based on scan: a) definition of interface curves; b) path planning based on interface curves; c) print prediction for augmented reality modifications

Although the drape approach does not maximize the contact areas between the concrete and the bricks, it was preferred for aesthetic reasons, as it gives the bottom part of the concrete pieces a textile-like appearance (see Fig. 3b). Based on these two contact curves, the geometry respectively path planning for a generic bench was generated, which was used as a base for further modification (see Fig. 3c).

3.3 Design Customization with Augmented Reality

This automatically generated geometry was then refined by a designer using AR with the Microsoft HoloLens 2 head-mounted display to meet subjective requirements for aesthetics or ergonomics. To achieve this, the geometry was transmitted from Grasshopper to a custom HoloLens 2 Unity-based application at runtime. Within the AR environment, the geometry was evaluated as a hologram and iteratively modified using control points until the desired shape was achieved. Bilateral communication was implemented between Rhino Grasshopper and the HoloLens 2 using the User Datagram Protocol (UDP), a lightweight communication protocol designed for fast and efficient data exchange. This setup allowed the HoloLens 2 to act solely as a visualization and interaction tool, with all computational tasks handled within Grasshopper. This was carried out at a 1:1 scale and could be done either in situ on the physical brick structure or using a holographic representation of the scanned brick structure (see Fig. 4a). The main advantage of this approach is that it provides immediate feedback within the construction environment, allowing real-time adjustments to be made to enhance the design process.

3.4 3D Concrete Printing

Once a satisfactory geometry was found in AR, it was prepared for 3D printing. The first step in Grasshopper was to divide the object (DC-Bench) into three parts, which serves three purposes: to minimize the overall printing time, to minimize the risk of the material collapsing during printing due to the printing height, and to make the bench assembly more convenient. After this segmentation, the centerline geometries of the individual components of the bench are exported from Grasshopper as a mesh in a STEP file (.stp). This file is then loaded into the COBOD BOD 2 slicer software, named "COBOD Slice".

Special considerations must be considered regarding layer height, nozzle dimensions, robot speed, extrusion rate and layer cycle time. The printing system used was a COBOD BOD 2 with a total build volume of 9.57 m $\times$ 6.47 m $\times$ 6.02 m. However, due to the small print volume, two different designs were printed at the same time, using only half of the printer's working area capacity. The material delivery system, consisting of a silo, mobile batching plant and progressive cavity pump, purchased along with the COBOD BOD 2, mixes and delivers the material to the print head. A 50 $\times$ 40 mm trapezoidal nozzle is used for extrusion (see Fig. 4b). TectorPrint 834, a dry mortar compound developed by Holcim specifically for the requirements of 3D concrete printing, was used as the print material.

3.5 Quality Control Before and After Assembly

For the two final steps of reality capture, namely (ii) after printing and (iii) after assembly, a TLS was used again to scan the printed object. Several capturing points were used to capture the entire object. To analyse the scanned data, the registration between the as-designed model and its point cloud replica should be performed. The process was done using the ICP (iterative closest point) method after a manual pre-registration [14]. Then, the cloud-to-mesh (C2M) distance between the point cloud data (PCD) and the as-designed model was computed. As a result, each point in the PCD, got assigned a distance to the model, providing directly a quality measure for the printing process.

3.6 Assembly

The last step remaining was the assembly. Once the concrete components had hardened (4 weeks), they were carefully lifted onto the bricks by hand and aligned correctly. As mentioned in the section before, the assembled bench was then scanned once again (see Fig. 4c).

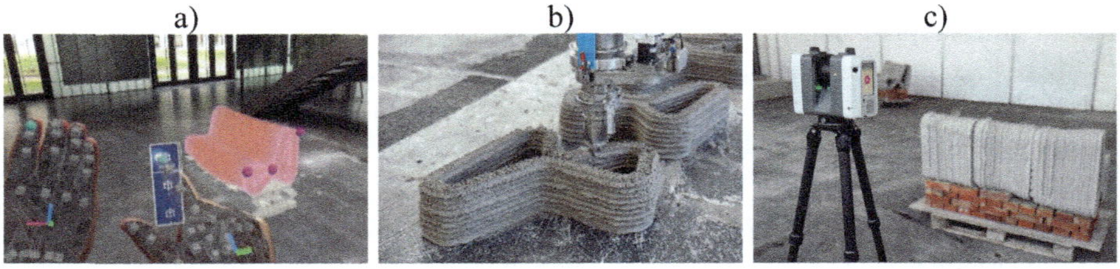

Fig. 4. Impressions from design-to-fabrication-workflow: a) design customization with AR; b) 3D concrete printing; c) Quality control after assembly

4 Result

As the first experiment at the DCS, a cyber-physical design-to-fabrication workflow was successfully executed as part of the Digital Concrete Workshop 2024, resulting in two hybrid benches 1350 × 510 × 900 mm^3 (l × b × h) (see Fig. 5d). This workflow successfully used reality capture, augmented reality and 3D concrete printing within the context of an existing built environment.

The printing process took about one hour for all six components (two different DC-Bench designs), including a 0.5 m tall wall of 2.0 m length and a lettering of the DC 2024. Each component weighs around 100 kg. The precise assembly of the parts on the brick structure proved to be difficult, as there was no visual indication of the correct positioning. Care was also needed as the brick structure is merely laid and not fixed with mortar. The fact that the bench components remain unattached might influence load-bearing capacity and durability, though this aspect was beyond the scope of this paper.

For proper analysis of point cloud data, a noise removal filter has been applied to remove existing outliers that may exist due to various reasons and especially for mixed pixels, which occur in laser scanning around sharp edges [15]. The noise removal function is selected to search for the six neighbour's mean distance from the PCD with a confidence threshold of 99.73%.

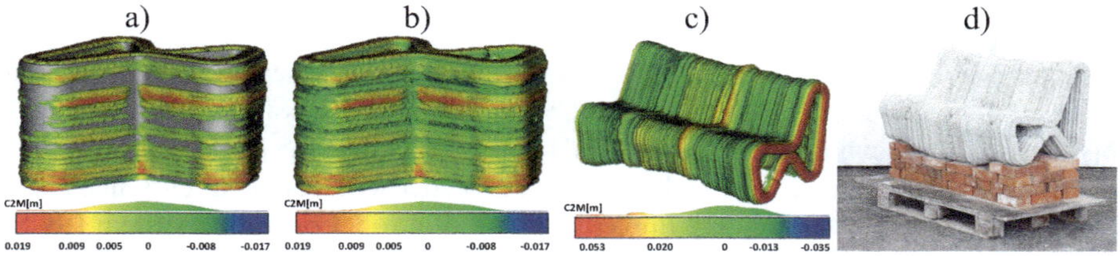

Fig. 5. Results from quality control: a) ICP alignment results between the PCD and the as-designed model. The PCD is colorized based on the C2M. b) C2M deviation map between as-printed and as-designed model; c) C2M deviation map results for the entire bench; d) Final DC-Bench No. 1

Then, the alignment between the designed model and the PCD is performed based on the ICP algorithm. The results indicate that ICP distributes the error over the whole object (see Fig. 5a). A direct co-registration based on an absolute coordinate frame definition could be used to avoid such effects, but due to time constraints it was not possible in this use case experiment [16]. This is a very common behaviour of ICP as it tends to get stuck in local minima. Later, C2M is computed between the PCD and the as-designed model (see Fig. 5b).

Red coloured areas from Fig. 5b shows that these areas are above the designed model, which means that a subtractive process must be considered to mill the excess material. However, the blue area, which indicates the minus distance, refers to the areas that require additional material. Nevertheless, the results of C2M depend on the results of the alignment process. As it can be clearly seen in Fig. 5a, the model is in the center of the PCD. This indicates that any deformation in the process will affect the results of what happened before the deformation as well as after. This is because the results are only relative and have no absolute reference to compare to. The analyses and results are getting even more pronounced, when we compare the entire bench with the model. This comparison reveals different magnitudes of distance and areas that need to be corrected, just from looking at one part (see Fig. 5c).

5 Discussion

The case study highlights the challenges and opportunities of using cyber-physical construction processes. It was not only a milestone in learning about CPCS at the new research infrastructure DCS, but was also part of its launch and putting the DCS in operation. However, both the realized designed-to-fabrication-workflow and the current version of the DCS represents only partial steps toward a fully CPCS. The cyber-physical workflow developed in this case study is currently limited to the component scale, with automation gaps within the digital process chain. Therefore, the key findings are discussed below, structured along two main directions: (1) gaps in automation and scalability and (2) lessons learned and future research outlook.

5.1 Gaps in Automation and Scalability

As stated previously, the constraints of the workshop limited the scale and complexity of the case study. Although the efficiency of CPCS results from a continuous data flow, in this case study the cyber and physical worlds are not fully integrated and require manual data transfers. Therefore, a technical framework is needed to integrate the current isolated systems, for example through the implementation of a robot operating system (ROS). However, it is not only the design process itself that lacks of a seamless data flow; the assembly process also remained highly manual without any direct feedback regarding the correct component positioning. Due to the weight of concrete components, sophisticated robotic solutions are needed, capable of lifting and rotating large or heavy components. In the context of human-in-the-loop approaches, an alternative solution could be the use of augmented reality holograms that provide real-time visual guidance and feedback once the component is correctly positioned.

So far, the case study was only conducted using a bench as an example, which results in a current lack of scalability. However, the developed workflow itself is scalable and could be applied to more challenging environments, which should be tested in a next step.

5.2 Lessons Learned and Future Research Outlook

Independent of the aforementioned gaps, future research outlooks can be derived from the development and execution of the case study as lessons learned:

- The design for the DC-Bench was generated using a cloth simulation, facilitating rapid prototyping of aesthetics and ergonomic design suitable for the workshop. Further research should target more efficient strategies involving structural and resource optimization.
- The overhang stability (top layer to the previous one) was checked during the design phase. However, a more comprehensive prognosis of critical fabrication parameters, such as printing speed and layer time, could be implemented as a simulation and thereby part of the path planning generation process.
- During the fabrication of the components of the benches, the batch mixes were adjusted manually and based on the mixing operator's experiences, leaving room for automation and real-time control, but also non-uniform layers. Future research should include a real-time quality control of the layers during fabrication to address potential issues automatically and on the fly.
- Since concrete 3D printing is prone to high temperatures causing uncontrolled pre-hydration, a cyber-physical construction processes seems predestined for an integration of an automated material quality control loops.
- Quality control and reality capturing was conducted solely on the finished components, without integrating process monitoring. Future implementations could include quality control to continuously monitor during the printing process.
- Relying on ICP can result in an averaged registration error across the entire object. This can lead to issues, such as getting stuck in local minima, that directly affect the results of C2M distances. Consequently, it becomes challenging to identify areas requiring material addition or subtraction, which leads to complications in surface refinement and increases processing time.

6 Conclusion

This paper presented the concept of a cyber-physical design-to-fabrication workflow developed for Digital Concrete 2024 Workshop in Braunschweig. The experiment served as a first case study and marked an important milestone, as it lays the foundation for future research at the newly established DCS. While the current implementation remains at an early stage, this study has revealed key research areas for further development, especially in terms of automating data flow and enhancing robotic support for material handling or human-in-the-loop approaches.

Funding and Acknowledgment. This research was funded by the Deutsche Forschungsgemeinschaft (DFG, German Research Foundation)—TRR 277/2 2024—Project number 414265976 (Subproject C06). The authors thank the DFG for the support within the CRC/TRR 277—Additive Manufacturing in Construction. This research was part of the Digital Concrete Workshop in Braunschweig, Germany (Sept., 01–03) prior to the Digital Concrete 2024 conference in Munich, Germany (Sept., 04–06). We would like to thank all participants for their contribution within workshop B "Cyber-Physical Construction @ Digital Construction Site". A special thank goes to PERI 3D Construction for their support.

References

1. Tao F, Qi Q, Wang L, Nee A (2019) Digital twins and cyber-physical systems toward smart manufacturing and industry 4.0: correlation and comparison. Engineering 5:653–661. https://doi.org/10.1016/j.eng.2019.01.014
2. Vasey L, Menges A (2020) Potentials of cyber-physical systems in architecture and cosntruction. In: Sawhney A, Riley M, Irizarry J (eds) Construction 4.0: an innovation platform for the built environment. Routledge, pp 91–112
3. Sawhney A, Riley M, Irizarry J (2020) Construction 4.0. Introduction and overview. In: Sawhney A, Riley M, Irizarry J (eds) Construction 4.0: an innovation platform for the built environment. Routledge
4. Slepicka M, Borrmann A (2024) Fabrication information modeling for closed-loop design and quality improvement in additive manufacturing for construction. Autom Constr 168:105792. https://doi.org/10.1016/j.autcon.2024.105792
5. Hack N, Jantzen C, Brohmann L, Gerke M, Mawas K, Maboudi M (2022) A closed-loop workflow for quality inspection and integrated post-processing of 3D-printed concrete elements. In: Buswell R, Blanco A, Cavalaro S, Kinnell P (eds) Third RILEM international conference on concrete and digital fabrication. RILEM bookseries, vol 37. Springer International Publishing, Cham, pp 364–369. https://doi.org/10.1007/978-3-031-06116-5_54
6. Naboni R (2022) Cyber-physical construction and computational manufacturing. In: Bolpagni M, Gavina R, Ribeiro D (eds) Industry 4.0 for the built environment. Structural integrity, vol 20. Springer International Publishing, Cham, pp 515–540. https://doi.org/10.1007/978-3-030-82430-3_22
7. Kantaros A, Ganetsos T (2024) Integration of cyber-physical systems, digital twins and 3D printing in advanced manufacturing: a synergistic approach. Am J Eng Appl Sci 17:1–22. https://doi.org/10.3844/ajeassp.2024.1.22
8. Anumba CJ, Akanmu A, Yuan X, Kan C (2021) Cyber—physical systems development for construction applications. Front Eng Manag 8:72–87. https://doi.org/10.1007/s42524-020-0130-4

9. Additive Manufacturing in Construction (AMC) (2024) TRR277: Project C 06—additive manufacturing in construction (AMC) TRR277. https://amc-trr277.de/projects/project-area-c/focus-area-c-06/
10. Naboni R, Breseghello L, Sanin S (2022) Environment-aware 3D concrete printing through robot-vision. In: Pak B, Wurzer G, Stouffs R (eds) Proceedings of the 40th international conference on education and research in computer aided architectural design in Europe (eCAADe) [Vol 2]. eCAADe proceedings. eCAADe, pp 409–418. https://doi.org/10.52842/conf.ecaade.2022.2.409
11. Dielemans G, Lachmayer L, Khader N, Hack N, Raatz A, Dörfler K (2024) Robotic repair: in-place 3D printing for repair of building components using a mobile robot. In: Tan MJ, Li M, Tay YWD, Wong TN, Bartolo P (eds) Construction 3D printing. springer tracts in additive manufacturing. Springer Nature Switzerland, Cham, pp 156–164. https://doi.org/10.1007/978-3-031-64269-2_20
12. Zhan Q, Wu H, Zhang L, Yuan PF, Gao T (2021) 3D concrete printing with variable width filament. In: Stojaković V, Tepavčević B (eds) Proceedings of the 39th international conference on education and research in computer aided architectural design in Europe (eCAADe). eCAADe proceedings. eCAADe, pp 153–160. https://doi.org/10.52842/conf.ecaade.2021.2.153
13. Mitterberger D, Ercan Jenny S, Vasey L, Lloret-Fritschi E, Aejmelaeus-Lindström P, Gramazio F, Kohler M (2022) Interactive robotic plastering: augmented interactive design and fabrication for on-site robotic plastering. In: Barbosa S, Lampe C, Appert C, Shamma DA, Drucker S, Williamson J, Yatani K (eds) CHI conference on human factors in computing systems. ACM, New York, NY, USA, pp 1–18. https://doi.org/10.1145/3491102.3501842
14. Besl PJ, McKay ND (1992) A method for registration of 3-D shapes. IEEE Trans Pattern Anal Mach Intell 14:239–256. https://doi.org/10.1109/34.121791
15. Wang Q, Kim M-K, Cheng JC, Sohn H (2016) Automated quality assessment of precast concrete elements with geometry irregularities using terrestrial laser scanning. Autom Constr 68:170–182. https://doi.org/10.1016/j.autcon.2016.03.014
16. Mawas K, Maboudi M, Gerke M (2022) Automatic geometric inspection in digital fabrication. Int Arch Photogramm Remote Sens Spatial Inf Sci XLIII-B1:459–466. https://doi.org/10.5194/isprs-archives-XLIII-B1-2022-459-2022

Sustainable Precast Concrete Construction Using Hybrid Additive Manufacturing: 3D Concrete Printing and Shotcrete with Low-Carbon Concrete

Marvin Abstoss(✉), Ferdinand Senf, Ralf Gliniorz, Henrik Funke, and Sandra Gelbrich

Department of Lightweight Structures and Polymer Technology, Chemnitz University of Technology, 09107 Chemnitz, Germany
marvin.abstoss@mb.tu-chemnitz.de

Abstract. In this research, a low carbon concrete was developed using blast furnace (BFS) cement. The properties of the material have been optimized to enable the integration of a hybrid additive manufacturing process consisting of robot-assisted 3D concrete printing (3DCP) and shotcrete. Using a maximum grain size of 5.6 mm in both process steps reduces the cement content and consequently lowers the overall carbon footprint. The mixture is used to produce precast concrete modules for tiny house construction. Implementing a multitool nozzle system simplifies the process and eliminates the need for tool changes. Milling or brushing techniques are applied to the modules to create suitable joints and ensure precise measurements. The milling waste was used in the project to partially substitute the aggregate fraction in the concrete mixture, thereby improving the ecological footprint by creating a circular economy. The modules are produced with a force-flow-optimized design. Through simulation and calculation, it was possible to minimize wall thickness and integrate local reinforcement structures in line with the force flow. These structures create internal cavities within the module that are filled with an insulating calcium sulfoaluminate (CSA) cement mixture. This cement enables rapid setting and early stiffening. This allows for the rapid formation and subsequent closure of the layer using either 3DCP or shotcrete. The use of CSA cement is essential in this context, as long waiting times would hinder the production process. Furthermore, using CSA cement improves the carbon footprint due to lower calcination temperatures compared to ordinary Portland cement (OPC).

Keywords: Additive Manufacturing · 3D Concrete Printing · Shotcrete

1 Introduction

Record levels of greenhouse gases are being emitted, and the targets set out in the Paris Climate Agreement are being missed. The construction industry now generates 37 percent of global CO_2 emissions [1].

H. Beushausen et al. (Eds.): NTZR 2025, LNCE 873, pp. 559–571, 2026.
https://doi.org/10.1007/978-3-032-27698-8_50

In order to meet future sector targets, one approach is to replace ordinary Portland cement (OPC—also defined as CEM I according to European standards) with more climate-friendly alternatives and reduce the proportion of cement in concrete. Another way is to use materials efficiently and in a power-flow-oriented way, resulting in weight and raw material savings and reduced industry carbon emissions. The ability to address rising carbon emissions and a worsening shortage of skilled workers makes additive manufacturing processes, such as 3DCP and shotcrete, essential for the future of the construction industry. These processes are characterized by a high degree of design freedom, which can lead to material savings through the optimized design of components according to the flow of forces. This type of production offers a high degree of automation, which will counteract the future shortage of skilled labor. A further tendency in the building trade is the functionalization of concrete components, e.g. the integration of heating elements, LEDs for direct or indirect lighting, and other electrified functionalities, like touch sensors [2]. The digital process chain of automated production processes simplifies the planning and implementation of this functionalization. Subsequent automated mechanical processing in the fresh state of the concrete makes it possible to integrate precisely fitting structures. The resulting automated process chain is a step towards the digitalization and automation in the construction industry.

Waste materials are produced during the machining process. These should be reused in the process to realize a material cycle. Classification requires characterization of the grain sizes and the influence on the mechanical parameters of the concrete.

The research focuses on producing modular tiny house elements as precast concrete components. These elements are produced using a hybrid additive manufacturing process. The process involves 3DCP and shotcrete of mineral-bonded materials. These modules can be planned digitally in advance and converted into production programs using parametric programming. Specific functionalization and localized post-processing (e.g. milling) can be programmed and executed [3, 4]. Automated production also enables mineral-bonded insulation to be integrated into the hybrid process. The hybrid additive manufacturing process enables force-flow-orientated design. Furthermore, this research contributes to solving two major problems: carbon emissions and the shortage of skilled labor.

2 Material and Methods

2.1 Material Development and Investigations

In this work, a concrete formulation was developed with low carbon emissions from the raw materials used for hybrid additive manufacturing. The BFS cement used is defined as CEM III/B 42.5 N according to European standards. Another key aspect is the use of a maximum grain size of 5.6 mm for the hybrid additive manufacturing process, which reduces the amount of cement in the concrete even more. A mineral-bonded insulating material has also been developed combining the BSF cement with CSA cement. In order to determine the environmental impact, the CO_2 emissions for the developed cement mixture were calculated in comparison to OPC, and the overall carbon neutrality was assessed by integrating the carbon-positive values of the biochar from EcoLocked into the insulating material [5].

The Haegermann Flow Table, in accordance with DIN EN 1015–3 [6], was used to determine the flowability. According to DIN EN 12390-16:2019 [7], the shrinkage of the concrete was determined by measuring the change in length using the shrinkage channel. Determination of the air content and the bulk density of the fresh concrete was done using an air content tester acc. to DIN EN 12350-6 and DIN EN 12350-7 [8, 9]. According to DIN EN 12390-2 [10], all test specimens were stored for the purpose of determining their compressive and 4-point bending tensile strength (4PBTS). Compressive strength testing was performed on six specially manufactured test specimens with an edge length of 150 mm, in accordance with DIN EN 12390-3, using the ToniNORM (3000 kN load frame) from Toni Technik [11]. Determination of the 4PBTS was conducted on ten specimens with dimensions of 160 × 40 × 40 mm^3 (L × W × H according to DIN EN 196-1) in accordance with DIN EN12390-5 [12, 13]. The support span was 120 mm and the loading rate was 50 N/s.

To evaluate the concrete's durability against freeze-thaw cycles, the CDF test (Capillary Suction, De-icing Agent and Freeze-Thaw Test) was used [14]. The test followed the BAW (Bundesanstalt für Wasserbau—Federal Waterways Engineering and Research Institute) criterion, which specifies a maximum mass loss of 1500 g/m^2 after a defined number of freeze-thaw cycles [15]. This ensures the concrete is resistant to environmental influences in frost zones and meets the necessary standards for precast elements.

The tests to determine the particle size distribution were carried out using a Retsch flat sieve shaker. The tests were executed in accordance with DIN EN 12620:2008-07 to ensure high accuracy and reproducibility of the results for the material samples in question [16]. The "Bettersizer S3 Plus", which combines laser diffraction and dynamic image analysis in a single device, was used to analyze the small-grain fractions (< 1 mm) more precisely. The tests were carried out in acc. With ISO 13320 [17].

2.2 Technology for the Hybrid Additive Manufacturing Process

The 3DCP and shotcrete processes were carried out using a KUKA KR 90 6-axis industrial robot, as shown in Fig. 1.

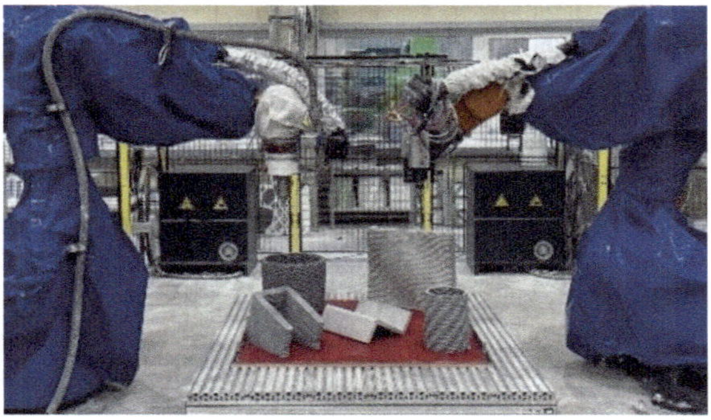

Fig. 1. Robots for hybrid additive manufacturing process.

A screw extruder pump was used to convey the concrete. Specially developed nozzles can be attached to the robot, e.g. for the forced mixing of additives fed to the nozzle, or the multitool nozzle for combined printing/shotcrete technology [18, 19]. The second robot can be used for tasks such as inserting reinforcement or processing concrete. Different tools, like milling heads or brushes, were used to process the fresh concrete. Milling tests were carried out in the fresh state using the second KUKA KR 90 robot. The two robots can work in parallel if required using the team application. The synchronised coordinate systems allow for the precise post-processing of the elements.

2.3 Planning and Programming of Robot Paths

Rhinoceros 3D and Grasshopper software were used to control the KUKA robots. Rhinoceros 3D is an efficient software for computer-aided 3D modelling (CAD) and computer-aided design. Grasshopper is a visual programming language and environment that is fully integrated into Rhinoceros 3D. It is a graphical algorithm editor that allows users to create complex parametric models and generative design without having to write traditional code. The 3DCP, shotcrete and element processing programs essentially consist of the following steps: creation or preparation of the target geometry, creation of the path curves, and generation of the robot commands.

3 Results and Discussion

3.1 Material Properties and Carbon Emissions

Table 1 contains the qualitative and quantitative compositions of the concrete used for printing and spraying. The mixture contains a water/cement ratio of 0.42.

Table 1. Composition of concrete

Component	Proportion [kg/m^3]
CEM III/B 42.5 N	590
Quartz sand 0–2 mm	570
Quartz gravel 0–5.6 mm	550
Mineral powder	200
Stabiliser	2.1
Water	240
Shrinkage reducer	3.5
Superplasticiser	4.5

The superplasticiser reduced the required water content and, when used with the stabiliser, ensured a more consistent mix that is less likely to separate.

The slump flow was found to be in a range between 13 and 14 cm, which was an acceptable value for estimating the material's printability and sprayability.

CO_2 emissions were reduced by around 66% by using BFS cement instead of OPC (OPC: 529 kg vs. BFS: 180 kg) [20, 21]. By using an aggregate with a maximum grain size of 5.6 mm, a relatively low cement content of 590 kg/m^3 can be used with a water-cement ratio of 0.42, thereby reducing CO_2 emissions.

Furthermore, the compressive and 4PBTS were tested after one, two and 28 days, and the results were compared with those of a similar concrete made with OPC (see Figs. 2 and 3). In practice, the elements are to be demolded after a maximum of two days, making the strength at this point more critical.

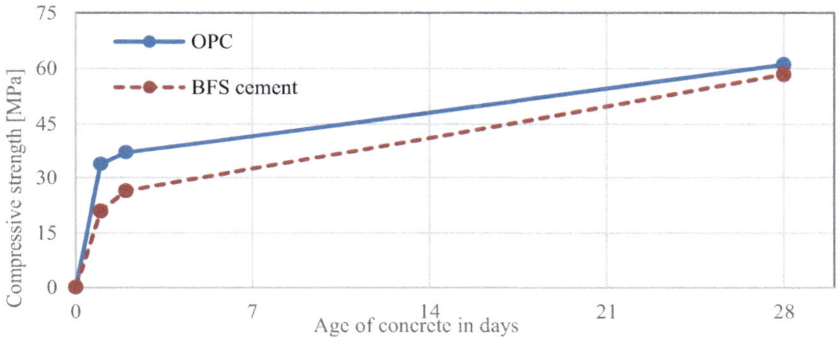

Fig. 2. Compressive strength in comparison with OPC concrete.

Based on the compressive strength, the concrete mixture with a lower clinker content could be determined in comparison to a similar concrete mixture with OPC in the early hydration stages. A typical effect was seen in the strength development of ground BFS cement compared to OPC. Initially, OPC provides faster early strength gain, but the latent hydraulic reaction of BFS contributes to a more gradual and sustained strength increase over time. This long-term strength gain causes the BFS mixture to eventually catch up to, and even surpass, the strength of the OPC mixture after about 28 days.

A similar trend was seen for the 4PBTS. After 28 days, there was an increase in the strength of the BFS mixture beyond the strength of the OPC mixture.

Based on these results, it was concluded that using a BFS cement would also be successful in precast concrete plants where short shuttering times of less than one day are required. Sufficient strength for demoulding components was achieved.

In the CDF test, which was used to evaluate the durability of concrete against freeze-thaw cycles, weathering of 1149 g/m^2 was observed after 28 cycles (see Fig. 4).

It has been determined that the freeze-thaw resistance according to the BAW criterion has been passed [15]. This ensures that the spraying and pressure process, which manufacturers the precast concrete elements, makes them resistant to environmental influences in frost zones.

Gaps were created in the concrete modules due to the force-flow-oriented reinforcement of the elements. These gaps were then filled with a mineral-bound insulating

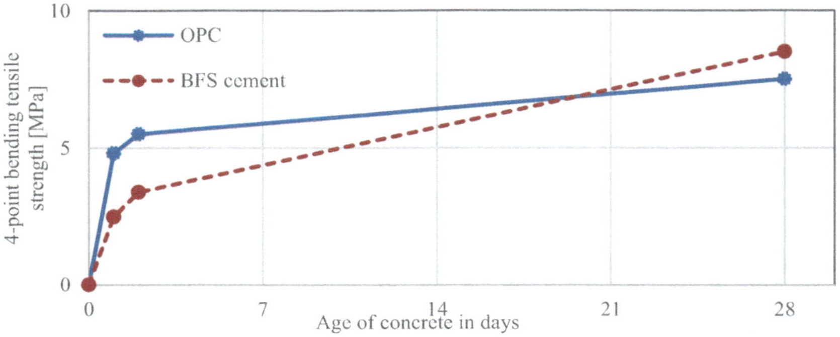

Fig. 3. 4-point bending tensile strength in comparison with OPC concrete.

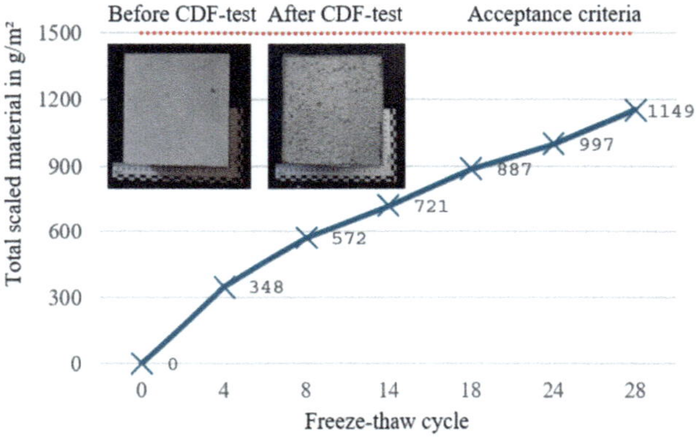

Fig. 4. Progression of weathering in the CDF test.

material. To this end, tests were carried out with various filling materials, including expanded glass. However, the best material for automatic conveying was achieved by mixing water, an air-entraining agent, BFS and CSA cement.

Combining BFS with CSA cement enabled the mixture to set and stiffen quickly while emitting low levels of carbon. This allowed the layer to be formed and closed quickly using either 3DCP or shotcrete. Using CSA cement was essential in this context as long waiting times would have hindered the production process. Furthermore, using CSA cement improves the carbon footprint due to lower calcination temperatures compared to OPC. This mortar mixture is both cost-effective and stable. The quantity of air-entraining agent used determined the fresh concrete density (0.4–0.6 kg/l). The integration of biochar from EcoLocked allows for the modules to achieve CO_2 neutrality; however, this increases the density of the insulating mixture to 1 kg/l, thereby reducing its thermal insulation performance. Depending on the module's application, this trade-off is deemed acceptable.

3.2 Hybrid Additive Manufacturing Process

Characterization of Parameters. The research determined the limits of 3DCP parameters, correlating printing speed of 60–240 mm/s with material feed rates of 1–4 l/min to achieve specific layer widths of 20–40 mm and heights of 5–20 mm. For shotcrete application, parameter variations were explored to define process limits. A constant air pressure of 5.5 bar was used. To ensure surface quality and avoid voids, a maximum lateral path distance of 60 mm was maintained, with a nozzle distance of 150–160 mm. Spraying speed varied from 120–240 mm/s with feed rates of 2–4 l/min.

The optimized parameters successfully prevented common defects in concrete additive manufacturing, including segregation, blue coloration, and surface irregularities such as mountain and valley formations (see Fig. 5). The quality of the resulting elements was quantitatively assessed via 3D scanning (see Fig. 5B).

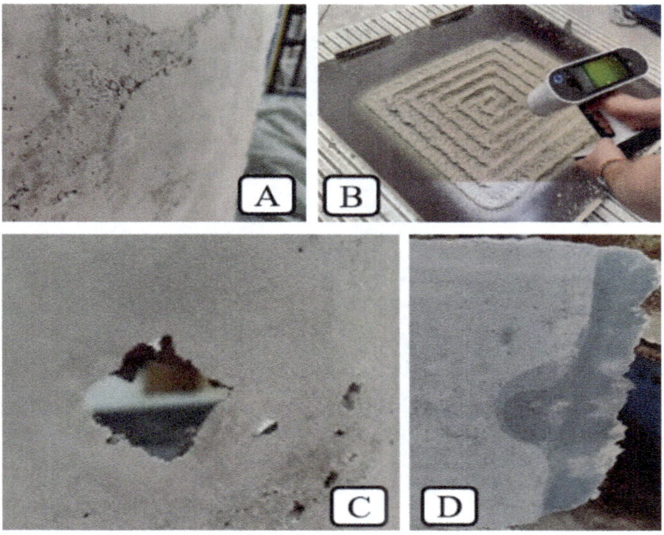

Fig. 5. Material defects in additive manufacturing: segregation phenomena (A), mountain-valley (B), defects (C) and blue coloring (D).

Applying the validated parameters enabled shotcrete to produce a high-quality and even surface (see Fig. 6). This was necessary for the subsequent 3DCP process step.

Application of hybrid additive Process in Practice: Production of a Module. The work involved combining the individual manufacturing processes of 3DCP and shotcrete and validating them using a small-scale wall module measuring 2 x 1 m2 (see Fig. 7). The multitool nozzle was used for production in order to avoid tool changes between the individual production steps and to avoid time-consuming cleaning processes for the pipe and nozzle adapters.

In a first production step, formwork edges were manufactured using the 3DCP process. The surface was then sprayed. After that, a reinforcing structure was applied to the

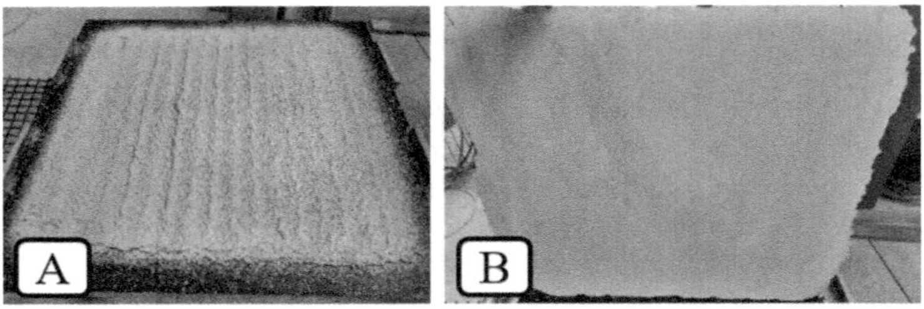

Fig. 6. Defect-free front (A) and back (B) sides due to spray processing.

Fig. 7. Steps to produce a module with hybrid additive manufacturing: printing edge and spraying surface (A), printing reinforcement structure (B), filling gaps with insulation mixture (C) and closing surface via printing (D).

flat surface. This structure was calculated in advance according to the dead, wind, and snow loads corresponding to zone two in central Germany (depending on altitude), again using the 3DCP process. This was followed by the automated application of the insulating mixture, which itself formed a flat surface with the reinforcing structure thanks to its flowability.

Finally, another 3DCP process step was carried out on this closed surface to close the module. Closing the structure with a further layer of shotcrete would also be possible and will be investigated in the further course of the research project.

3.3 Post-processing of Fresh Concrete and Recycling

Characterization of Parameters. To characterize the process parameters for post-processing in the fresh state of the concrete, various tests were carried out (see Fig. 8).

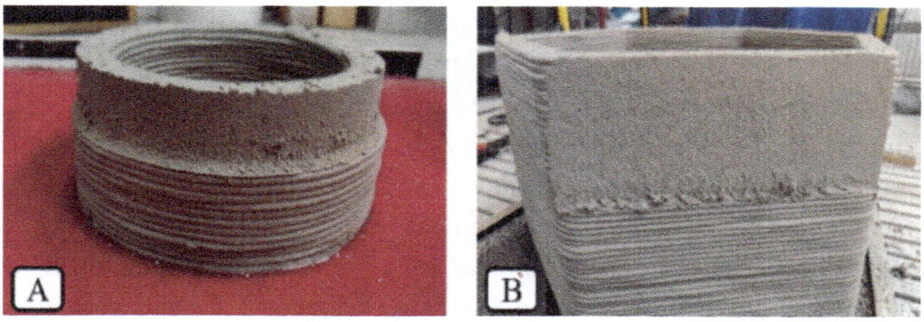

Fig. 8. Milling (A) and brushing (B) results.

The best results for milling were achieved with a waiting time of four hours or more after the test specimens were printed. After six hours, the concrete began to harden, which led to higher tool wear. Milling within four hours of printing led to a smeared surface rather than defined cutting. For brushing, however, a shorter waiting time (between 2.5 and 3.5 h) was preferable.

When milling, the infeed was selected according to the milling cutter used. A recommended value was half the diameter of the milling cutter or less, and an infeed of more than 10 mm was not advisable. In contrast, an infeed of up to 5 mm was suitable for brushing. At higher values, the brushes would penetrate too far into the concrete, which might cause the spindle to stop.

The feed rate and speed were selected for both processes, taking the above parameters into account. Successful printing parameters were a feed rate of 120–240 mm/s and a speed of 9,000–12,000 rpm. For brushing, lower feed rates (up to 60 mm/s) and speeds (up to 6,000 rpm) were suitable.

Examination and Reuse of the Removed Milling Material. The milling waste produced during the milling process was to be reused in the base material. therefore, for milling tests 3, 4 and 5 h after 3DCP, grading curves for the milling material were set up and compared (see Fig. 9).

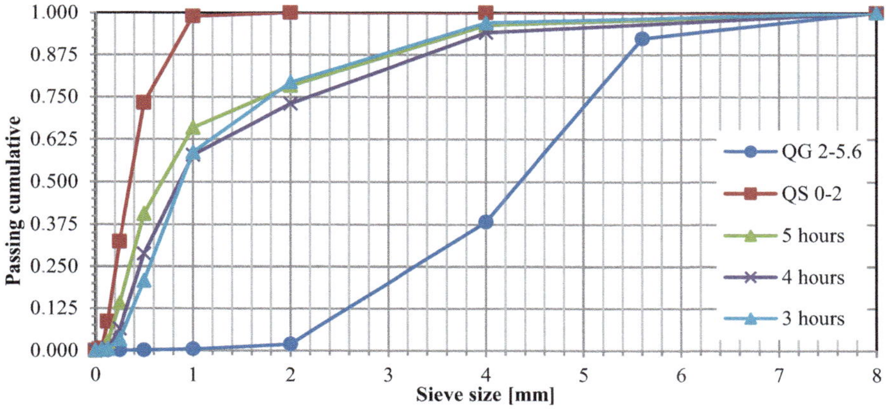

Fig. 9. Sieve lines for quartz sand, quartz gravel and milling material.

The longer the time between the 3DCP of the material and the milling process, the finer the waste material tended to become. As the grading curves differed greatly from those of the starting materials (quartz sand 0–2 mm and quartz gravel 2–5.6 mm), the milled material was sieved into its individual fractions. These fractions were then combined according to the grading curves of the sand and gravel to produce a substitute material as similar as possible to the original.

Subsequently, the finest milled material (5 h) was examined in more detail using laser diffraction and dynamic image analysis in the grain size range 0–3000 mm (see Fig. 10, bottom). Furthermore, the original mineral powder used was examined with the same parameters in the grain size range 0–1000 mm (see Fig. 10, top).

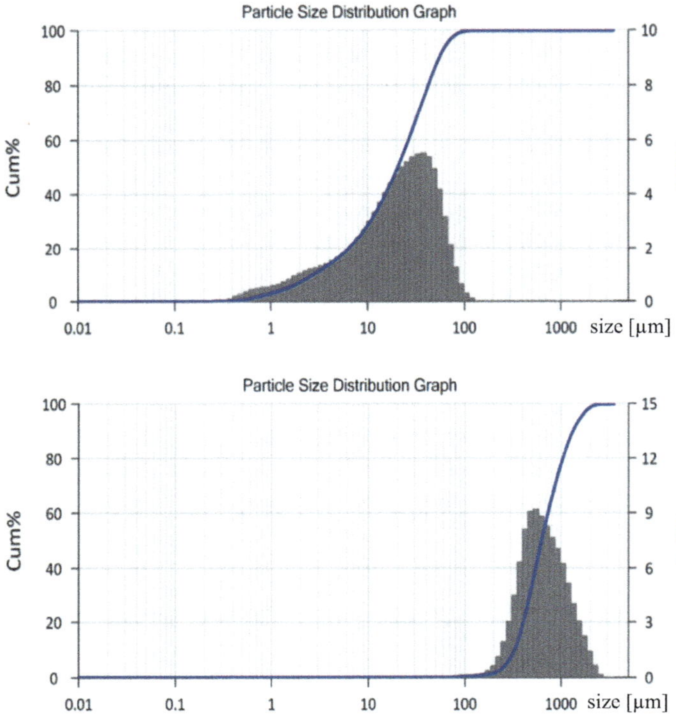

Table

Diam um	Percent
0.500	0.46
1.000	2.95
2.000	7.12
5.000	16.64
10.00	28.54
20.00	49.24
45.00	82.77
75.00	97.53
100.0	99.63
200.0	100.00

Table

Diam um	Percent
500.0	33.82
600.0	46.96
700.0	57.53
800.0	65.88
900.0	72.66
1000	78.27
1500	93.64
2000	98.71
2500	99.90
3000	100.00

Fig. 10. Sieve curves for milling material "5 h" (bottom) and mineral powder (top).

Due to the insufficient proportion of very fine grains in the milled material, it was not possible to substitute the mineral powder for research purposes or to use it in practice.

Test specimens were then produced for compressive strength and 4PBTS using the substitute sand 0–2 mm and gravel 2–5.6 mm (see Figs. 11 and 12). During this process, 10 wt.%, 20 wt.%, 30 wt.% and 40 wt.% of the original sand or gravel was replaced with the milled and replicated material.

As the substitution rate increased, the compressive strength of the mixtures decreased slightly, both for replacement in sand and in gravel. No trend was identified in 4PBTS

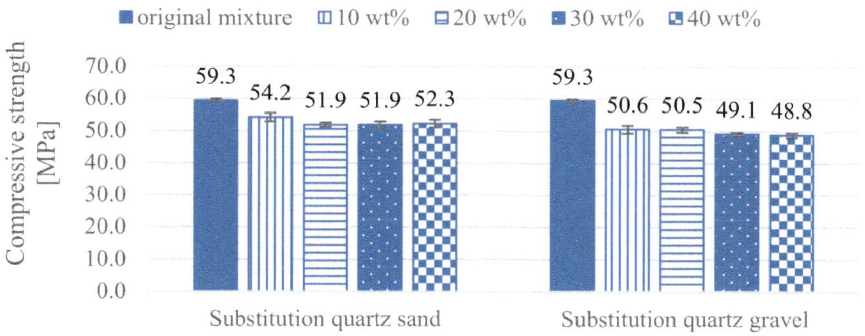

Fig. 11. Compressive strength through substitution of quartz sand and gravel

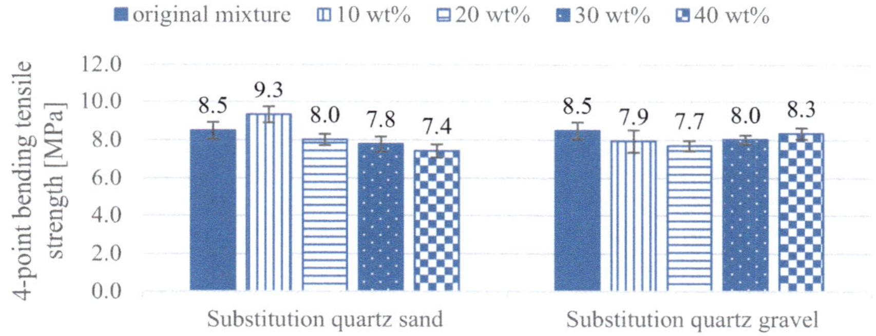

Fig. 12. 4-point bending tensile strength through substitution of quartz sand and gravel

with an increasing substitution rate. When considering the standard deviation, there were variations that fall within the tolerance range. However, both the compressive strength and the 4PBTS remained at a high level for all substitution rates and were therefore also be used for the hybrid additive manufacturing.

The spreading dimension was also measured during the production of the mixtures in order to characterize their suitability for the 3DCP and spraying process. All mixtures were measured with a spreading dimension between 13 and 14 cm and were therefore suitable for hybrid additive manufacturing.

4 Conclusions

This paper demonstrates the successful development of a hybrid additive manufacturing process for the sustainable 3DCP, shotcrete and post-processing. A new low carbon concrete mixture containing BFS cement with a maximum grain size of 5.6 mm was developed and characterised. Furthermore, milling waste from the post-processing stage could be used as an alternative material in the initial mixture. This is a significant step forward for the circular economy and reduces waste and the need for raw materials, further reducing the carbon footprint. The new hybrid technology allows for the efficient, digitalised production of complex modules without the need for formwork. These modules can then be used on a large scale for constructing demonstrator houses.

Acknowledgements. The authors would like to thank the sponsors for the financial support provided by the Department Bank of Saxony (SAB).

References

1. United Nations Environment Program (2022) Global status report for buildings and construction, efficient and resilient buildings and construction sector. Nairobi
2. Gliniorz R (2024) Multi-functional integration in clay-cardboard sandwich structure with semi-automated manufacturing. Springer Nature Switzerland AG 2025 ECCE, pp 130–143. ISBN 978-3-031-73775-6
3. Gelbrich S, Abstoß M, Funke H (2023) Complex concrete elements for the production of recyclable formwork by extrusion and milling. ce/papers 6(6):729–734. https://doi.org/10.1002/cepa.2815
4. Abstoß M, Rudolph E, Funke H, Gelbrich S (2021) 3D printing and milling of complex concrete elements for the production of casting resin molds. In: CPT worldwide—construction printing technology. CPT Worldwide, Nr., S.14-20. ISSN 2629-1592
5. EcoLocked GmbH: EcoLocked materials—products. https://www.ecolocked.com/en-us/products. Accessed 3 Sept 2025
6. DIN EN 1015-3 methods of test for mortar for masonry—part 3: determination of consistence of fresh mortar (by flow table); German version EN 1015-3:1999+A1:2004+A2:2006
7. DIN EN 12390-16 testing hardened concrete—part 16: determination of the shrinkage of concrete; German version EN 12390-16:2019
8. DIN EN 12350-6 testing fresh concrete—part 6: density; German version EN 12350-6:2019
9. DIN EN 12350-7 testing fresh concrete—part 7: air content—pressure methods; German version EN 12350-7:2019 + AC:2022
10. DIN EN 12390-2 testing hardened concrete—part 2: making and curing specimens for strength tests; German version EN 12390-2:2019
11. DIN EN 12390-3 testing hardened concrete—part 3: compressive strength of test specimens; German version EN 12390-3:2019
12. DIN EN 196-1 methods of testing cement—part 1: determination of strength; German version EN 196-1:2016
13. DIN EN 12390-5:2019-10 testing hardened concrete—part 5: flexural strength of test specimens; German version EN
14. DIN CEN/TS 12390-9:2017-05 testing hardened concrete—part 9: freeze-thaw resistance with de-icing salts—scaling
15. Merkblatt „Frostprüfung von Beton“ (BAW Merkblatt „Frostprüfung“) Dezember 2004; Bundesanstalt für Wasserbau; Eigenverlag; Karlsruhe, Hamburg, Ilmenau 2004
16. DIN EN 12620:2008-07: aggregates for concrete; German version EN 12620:2002+A1:2008
17. ISO 13320: Particle size analysis—Laser diffraction methods
18. Lindner M, Scharf-Wildenhain R, Gliniorz R, Vanselow K, Funke H, Gelbrich S, Kroll L (2021) Kalibrierverfahren für konstante Materialstränge bei robotergestützter Betonextrusion. Beton- und Stahlbetonbau 116, Sonderheft Schneller Bauen, 116, Ernst & Sohn 2021, pp 42–47. https://doi.org/10.1002/best.202100051. ISSN 1437-1006

19. Lindner M, Mandel K, Gelbrich S, Funke H, Kroll L (2021) Kontinuierliche Fertigung von dünnwandigen faserverstärkten Präzisions-Betonelementen für freigeformte Modulbauweisen (KoBeMo). BetonWerk International Nr. 1, 2021, pp 28–29. ISSN 1437-9023. https://www.cpi-worldwide.com/images/cpi/flips/2021/1/en/index.html?#p=28
20. Verein Deutscher Zementwerke e.V: Umwelt-Produktdeklaration - Portlandzement (CEM I), EPD-VDZ-20220153-IAG1-DE, 08.06.2022
21. Verein Deutscher Zementwerke e.V: Umwelt-Produktdeklaration - Hochofenzement (CEM III/A), EPD-VDZ-20220154-IAG1-DE, 09.06.2022

Performance of Deficient Reinforced Concrete Columns Strengthened with Jute Fiber-Reinforced Polymer and Wheat Fiber-Reinforced Polymer Composites

Kusum Saini(✉), Umakant Pandey, Karanpunit Singh, and Vasant Matsagar

Department of Civil and Environmental Engineering, Indian Institute of Technology (IIT) Delhi, Hauz Khas, New Delhi 110016, India
kusum.saini@iitdalumni.com, pandey.umakant0012@gmail.com, kpmann04@gmail.com, matsagar@civil.iitd.ac.in

Abstract. At the present time, with increasing climate change impacts and rising air pollution, exploring sustainable alternatives to carbon-intensive construction materials (CCM) are imperative. Natural fiber-reinforced polymer (NFRP) composites offer significant potential as substitutes to CCM for structural applications, such as building units, reinforcement, and strengthening in civil infrastructure. Therefore, it is essential to evaluate mechanical properties of NFRP and assess their performance for aforementioned applications. This study investigates the influence of these natural fiber-reinforced polymer (NFRP) composites on the lateral load-carrying capacity of deficient reinforced concrete columns. The biaxial jute long fiber-reinforced polymer (JFRP) and random wheat short fiber-reinforced polymer (WFRP) composites have been considered for strengthening. The tensile behavior of these composites is experimentally assessed in accordance with ASTM D3039 standard. For JFRP and WFRP composites, the dominant tensile failure modes are identified as lateral splitting and angled explosive failure, respectively. Finite element models are developed for both unstrengthened and NFRP-strengthened reinforced concrete (RC) columns. The concrete is modeled using the concrete damage plasticity (CDP) model, while the steel reinforcement bars and ties are represented using elastic-plastic material model. This comparative analysis demonstrates the effectiveness of JFRP and WFRP composites in enhancing the lateral load-carrying capacity approximately by 10.4% to 4.1%, respectively. Furthermore, the study highlights the potential of NFRP composites as a sustainable solution for enhancing structural performance while significantly reducing the carbon footprint of conventional strengthening practices, addressing key challenges in both construction and environmental sustainability.

Keywords: concrete damage plasticity · jute fiber-reinforced polymer · natural fiber · reinforced concrete column · strengthening · tensile behavior · wheat fiber-reinforced polymer

H. Beushausen et al. (Eds.): NTZR 2025, LNCE 873, pp. 572–582, 2026.
https://doi.org/10.1007/978-3-032-27698-8_51

1 Introduction

The applications of fiber-reinforced polymer (FRP) composites are continuously increasing in the construction industry either in the form of reinforcement or strengthening. They have various advantages, such as being lightweight, having high insulating properties, and being less carbon-intensive than conventional materials. Recent frequent natural disasters, e.g., 2025 Myanmar earthquake and the 2024 Noto earthquake, resulted into huge loss of structures and lives, showing the necessity of strengthening to improve the performance of deficient structures. The steel and synthetic fiber-reinforced polymer (FRP) composites are commonly used for strengthening of structures. The ferrocement composites have been implemented effectively to strengthen concrete columns [1]. Some researchers showed the effective use of glass fiber-reinforced polymer (GFRP) [2] and carbon fiber-reinforced polymer (CFRP) [3, 4] for strengthening concrete. However, due to the high carbon footprint of synthetic FRP composites, there is a need to propose alternative natural FRP composites (bio-composites) for structures.

Bio-composites have the potential to be implemented for strengthening repair, and retrofitting of deficient structures [5–7]. There are various commonly used natural fibers, e.g., bamboo, banana, coir, hemp, jute, roselle, and sisal that are widely used in bio-composites for structural applications [8, 9], demonstrating high tensile strength, making them effective alternative for reinforcement and strengthening structures. Currently, there are many countries struggling to resolve the agricultural waste management problem. Approximately, 3.5% of greenhouse gas (GHG) emission is led by crop burning globally [10]. Therefore, the utilization of this agricultural waste in the construction industry needs to be encouraged [11].

Several researchers [5, 6, 12–14] have explored the use of natural fiber-based FRP composites, such as jute, sisal, and flax, for structural strengthening applications. Jute fiber-reinforced polymer (JFRP) composites have been widely utilized in various forms, including laminates, fabrics, and ropes. Yooprasertchai et al. [12] reported an approximate 10% improvement in the ductility of reinforced concrete (RC) columns strengthened with JFRP under lateral cyclic loading, with performance comparable to that of carbon FRP (CFRP). Notable enhancements in shear strength and flexural capacity were observed in pre-damaged and undamaged beams when strengthened using JFRP and sisal FRP, respectively [13, 14]. Wahab et al. [15] and Varma et al. [16] demonstrated that hybridizing JFRP with synthetic or basalt fibers significantly improved the compressive strength of concrete cylinders, surpassing the performance of composites strengthened with individual fiber-based composites.

Additionally, JFRP composites enhanced the compressive strength of fire-damaged concrete by a factor of 3.69 [17]. The JFRP composites have been proven effective in improving the load-carrying capacity of structural members. However, no study has investigated the structural performance of concrete structures strengthened with agricultural wastes such as rice straw, wheat straw, or sugarcane bagasse-reinforced polymer composites. This study addresses this gap by examining the behavior of both biaxial JFRP composites and random short wheat straw fiber-reinforced polymer (WFRP) composites for potential use in strengthening of RC columns.

The study presents that if structural members are deficient in the strength to bear the applied loads, strengthening with natural fiber-reinforcement polymer (NFRP) composites can be used efficiently to enhance its performance. The main objectives of this study are (a) analyzing the tensile behavior of jute fiber-reinforced polymer (JFRP) and wheat fiber-reinforced polymer (WFRP) composites and (b) assessing the performance of deficient reinforced concrete columns strengthened with JFRP and WFRP with respect to unstrengthened column.

2 Numerical Modeling

Finite element (FE) models of square reinforced concrete (RC) columns unstrengthened and strengthened with jute fiber-reinforced polymer (JFRP) and wheat fiber-reinforced polymer (WFRP) are developed in Abaqus® [18]. The length and width of RC columns are taken as 1800 mm and 250 mm, respectively. The NFRP composites of 500 mm length and 5 mm thickness are applied for strengthening RC column at the top and bottom of columns, locations where the highest stress concentrations are obtained (see Fig. 1). Wrapping the entire column along the length has shown similar load carrying capacity as considered configurations for strengthening. Therefore, for a detailed performance analysis, the strengthening configuration used is presented in Fig. 1. Here the concrete is modeled with three-dimensional (3D), C3D8R elements, NFRP with shell elements, SC8R elements, and steel reinforcement and tie with truss elements, T3D2. Here, the steel reinforcement and steel ties are considered embedded in the concrete, and the bond between NFRP and RC column is considered perfectly bonded. An axial load of 500 kN and a lateral incremental dynamic displacement are applied at the top of the RC columns in the Y-direction (at the mid-node, assigned as the representative node for all surface nodes, see Fig. 1). The translational and rotational movements of the nodes at the top of the column are constrained in all other directions. An efficient progressive mesh size of 2.5 to 20 mm is considered in the FE models to capture failure perfectly (smaller mesh size in the high probable, high stress concentration area).

2.1 Material Properties

The concrete columns strengthened with JFRP and WFRP have been presented in this study. The concrete has been modeled using the concrete damage plasticity (CDP) model. Here, tensile and compressive stress-strain curves of the concrete are considered with respect to the ultimate tensile stress of 3.5 MPa and compressive stress of 30 MPa, respectively [19] (see in Table 1). The concrete columns are reinforced with four steel reinforcements bars of 10 mm diameter and ties 6 mm diameter at 200 mm spacing. The reinforcement is modeled with elastic-plastic behavior respective to 381 MPa yield strength and 604 MPa ultimate strength [12].

2.2 Tensile Behavior of JFRP and WFRP Composites

The JFRP composite laminates are made with jute fabric layups in epoxy resin using the compression molding technique. Whereas, WFRP composite laminates are made with wheat straw randomly spread in epoxy resin compressed manually. The wheat straws are treated with 1% NaOH, followed by cleaning and drying before molding (using hand lay-up and compression). The JFRP and WFRP composites are utilized for strengthening RC column; where the composites are modeled as per the tensile characteristics obtained experimentally, as shown in Fig. 2.

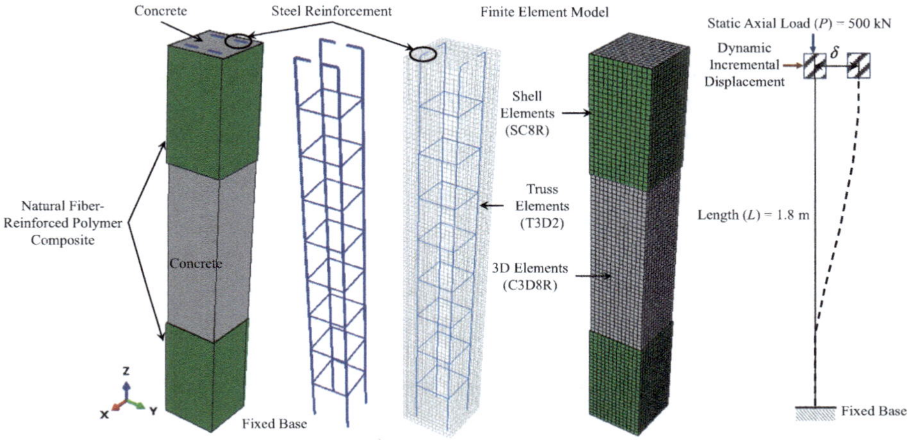

Fig. 1. The details of steel reinforced concrete column strengthened with natural fiber-reinforced polymer composites.

Table 1. Material properties of concrete adopted in the study [16].

S. No.	Properties	Concrete
1	Modulus of elasticity (MPa)	27691.47
2	Poisson's ratio	0.2
3	Density (kg/m^3)	2400
4	Compressive strength (MPa)	30
5	Tensile strength (MPa)	3.53
6	Dilation angle (°)	30
7	Eccentricity	0.1
8	σ_{b0}/σ_{c0}	1.16
9	K	2/3
10	Viscosity parameter	0.0001

Here, K is the ratio of the second stress invariant on the tensile meridian to that on the compressive meridian at the initial yield and σ_{b0}/σ_{c0} represents the ratio of initial equibiaxial compressive yield stress and initial uniaxial compressive yield stress.

In JFRP, the major failures are observed with respect to lateral failure in the gage area at the top and middle locations (LGT and LGM) as per ASTM D3039/D3039M-17 [20]. In WFRP, the major failure patterns are obtained with respect to angled and explosive failure in the gage area at the top and middle locations (AGT/XGT and AGM/XGM). The random spread of fibers in WFRP composites provides additional ductility in tension. In contrast, a brittle failure is obtained in JFRP composites while fracture in tension, see Figs. 2 and 3. In this study, high variation in the tensile behavior of WFRP composite samples is observed due to high variation in the manufacturing process (due to human errors) and non-uniformity in the fibers. Whereas, in JFRP composites, the continuous fibers are used biaxially with four laminates. The total thickness of JFRP and WFRP composites is taken as 5 mm for the numerical study. In Fig. 3, the stress-strain curves of JFRP and WFRP composites are presented. The average elastic modulus and peak tensile strength of JFRP and WFRP composites are taken based on Fig. 3(a) and (b), respectively. Here, high variation for WFRP composites as compared to JFRP composites is observed, due to high uncertainty in the manufacturing process. Accordingly, based on stress-strain curves, linear and quadratic tensile behaviors are proposed for both JFRP and WFRP composite laminates, respectively (see Fig. 3).

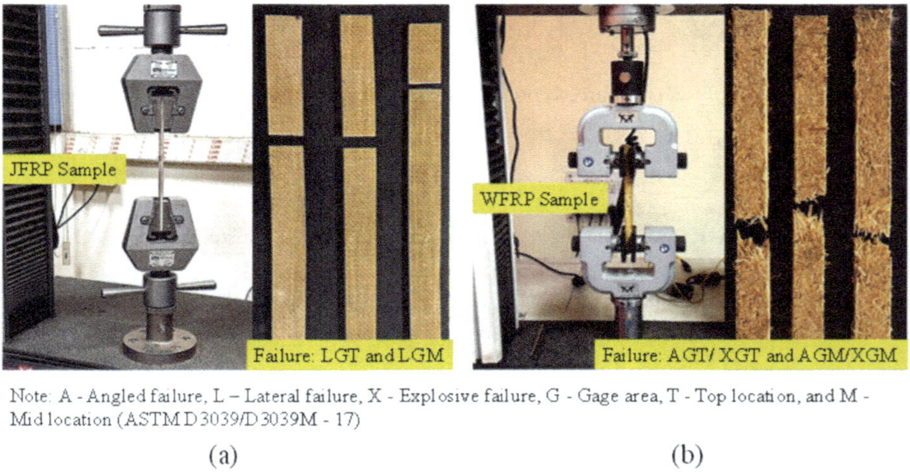

Fig. 2. Tensile testing and failure of (a) jute fiber-reinforced polymer (JFRP) and (b) wheat fiber-reinforced polymer (WFRP) composite laminates as per ASTM D3039/D3039M—17.

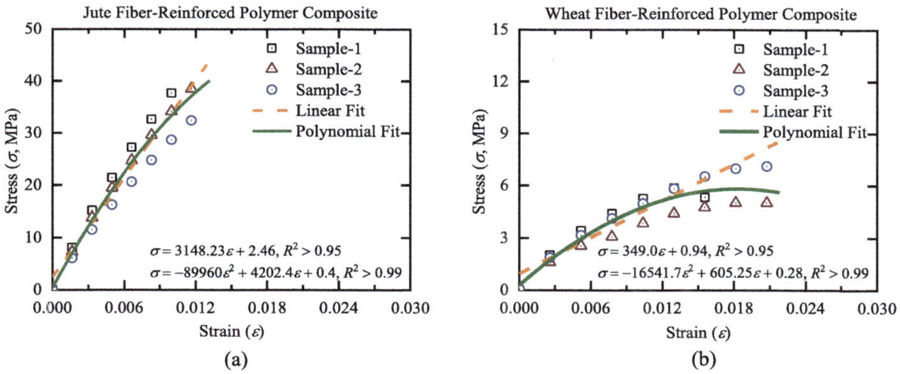

Fig. 3. Stress-strain curves of (a) jute fiber-reinforced polymer (JFRP) composites and (b) wheat fiber-reinforced polymer (WFRP) composite laminates.

3 Results and Discussion

The response of RC columns unstrengthened and strengthened with JFRP and WFRP composites is reported under axial static load and lateral dynamic load, applied at the top of columns. The response of these columns is investigated in terms of stress and strain time histories, force-deflection curve, and damage dissipation energy (energy due to damage in concrete) time histories. The peak response of RC columns unstrengthened and strengthened with JFRP and WFRP composites in terms of peak lateral loads, respective lateral deflections, and respective drifts, measured at point where axial and lateral deflection are applied, presented in Table 2. In the case of unstrengthened column, the peak lateral load 68.50 kN is observed with respect to 13.52 mm lateral deflection. Whereas, with respect to same load, 24.36% and 14.84% lesser deflections are obtained for the columns strengthened with JFRP and WFRP composites, respectively. A 15.0% and insignificant increment in the damage dissipation energy is reported due to strengthening with JFRP and WFRP composites, respectively (see Fig. 4(b)).

The failure contours are presented corresponding to same load as peak load capacity of unstrengthened column are observed (Figs. 5, 6 and 7). The incremental displacement on the RC columns results into tensile failure starts from the corner in concrete and followed by yield in main rebars at the top and bottom, presented in Figs. 5 and 6. Here, the highest strain generates in the reinforcements of unstrengthened RC column followed by columns strengthened by WFRP and JFRP composites. At a load of 68.50 kN, the principal stress in the main steel reinforcement bars is reduced by approximately 4.0% and 2.0% with JFRP and WFRP composites strengthening, respectively. Overall, a reduction of 17.5% in peak principal stress is observed in the steel reinforcement bars due to strengthening. From Fig. 7, it is observed that the dominating failure is governed by shear cracks followed by concrete crushing. The major failure patterns are observed as horizontal cracks followed by vertical cracks.

In addition, the failure patterns of the columns at same load are presented, in terms of principal strains, and damage in concrete based on the CDP model. It is observed that due to strengthening of the RC column with JFRP and WFRP composites, the peak load carrying capacity is enhanced by 10.4% and 4.1%, and respective drift by 24.5% and 7.1%, respectively (c.f., Fig. 4(a) and Table 2). At 80% of peak load, the post-peak drifts are observed 18.64% and 5.93% higher than that of the unstrengthened column, representing higher ductility. At same applied load, 68.50 kN, the RC columns strengthened with JFRP and WFRP composites have lesser lateral drift ratio by 24.4% and 14.91%, respectively. This shows higher load resistance of strengthened columns.

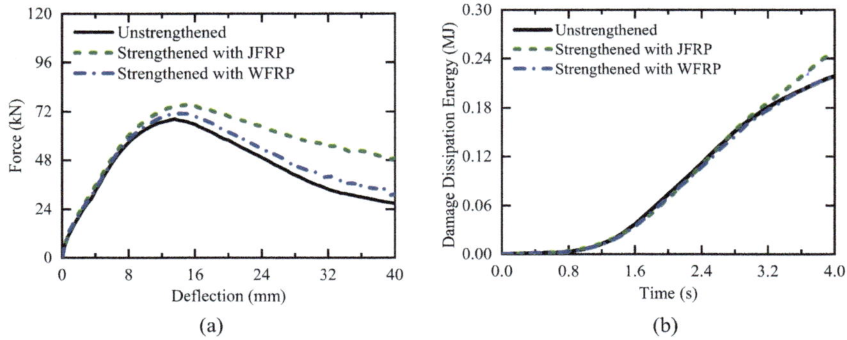

Fig. 4. Response of reinforced concrete columns unstrengthened and strengthened with JFRP and WFRP composites: (a) lateral force-displacement curves and (b) damage dissipation energy time histories.

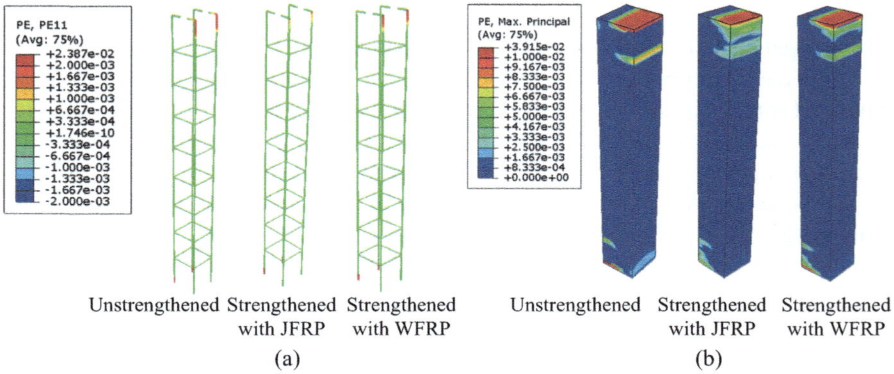

Fig. 5. Strain in reinforced concrete columns unstrengthened and strengthened with JFRP and WFRP composites: (a) principal strain in steel reinforcement bars and ties and (b) maximum principal strain in concrete corresponding to 68.50 kN in columns.

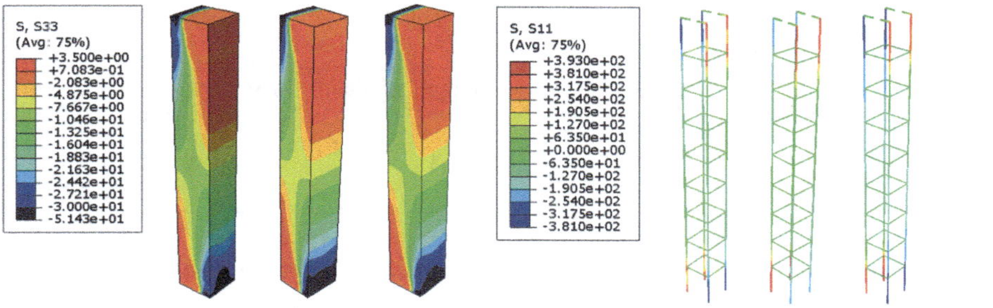

Unstrengthened Strengthened with JFRP Strengthened with WFRP

(a)

Unstrengthened Strengthened with JFRP Strengthened with WFRP

(b)

Fig. 6. Stress in reinforced concrete columns unstrengthened and strengthened with JFRP and WFRP composites: (a) principal stress along the column and (b) principal stress in steel reinforcement bars ad ties corresponding to 68.50 kN in columns.

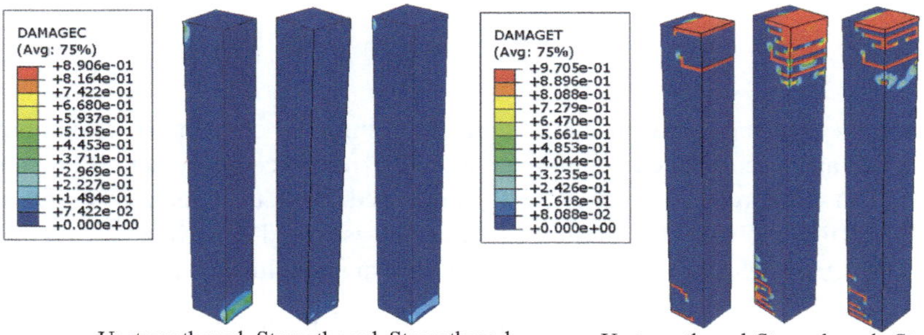

Unstrengthened Strengthened with JFRP Strengthened with WFRP

(a)

Unstrengthened Strengthened with JFRP Strengthened with WFRP

(b)

Fig. 7. Damage in reinforced concrete columns unstrengthened and strengthened with JFRP and WFRP composites: (a) in compression and (b) in tension corresponding to 68.5 kN in columns.

Table 2. Response of the building unstrengthened and strengthened with JFRP and WFRP composites subjected to lateral dynamic load.

Response	Unstrengthened	Strengthened with JFRP	Strengthened with WFRP
Lateral peak load (kN)	68.50	75.60	71.25
Drift at peak load (%)	0.751	0.935	0.804
Post-peak drift at 80% of peak load (%)	1.180	1.400	1.250

(continued)

Table 2. *(continued)*

Response	Unstrengthened	Strengthened with JFRP	Strengthened with WFRP
Lateral drift at 68.5 kN (%)	0.751	0.568	0.639

The study proposes the use of NFRP composites for strengthening of structural members as an alternative to high-carbon-intensive conventional materials. In the future, the bond between NFRP and concrete can be modeled to investigate delamination effects in strengthened RC columns. Additionally, the durability of NFRP composites used in civil infrastructure can be explored more comprehensively under real-life loading and environmental conditions. Additionally, modeling the beam-column joint and the column foundation will provide a deeper understanding of the impact of NFRP composite strengthening. The findings contribute to the development of green materials for civil infrastructure, offering a pathway to reduce carbon emissions and meet the United Nations Sustainable Development Goals (SDGs).

4 Conclusions

In this study, the dynamic performance of deficient reinforced concrete (RC) columns unstrengthened and strengthened with biaxial jute fiber-reinforced polymer (JFRP) and short wheat fiber-reinforced polymer (WFRP) is assessed. Here, the tensile properties of JFRP and WFRP composites for strengthening are investigated experimentally, as per ASTM D3039/D3039M—17. Accordingly, the following conclusions are drawn.

1. The dominating tensile failure modes in JFRP and WFRP composites are lateral and angled explosive failures, respectively.
2. The lateral load resistance of deficient RC columns can be enhanced by strengthening with JFRP and WFRP composites.
3. Strengthening RC columns with JFRP and WFRP composites of equal thickness enhances the maximum load-carrying capacity by 10.4% and 4.07%, respectively, while reducing lateral drift by 24.4% and 14.9% compared to unstrengthened columns at the same peak load.
4. The peak principal stresses in the main steel reinforcement bars of RC columns reduce by 17.5% due to strengthening with JFRP and WFRP composites.

Acknowledgement. The first author gratefully acknowledges the financial support received from The Batch of 1969 Innovation Fellow Award and Industrial Research and Development (IRD) Student Startup Scheme by Indian Institute of Technology (IIT) Delhi. The opinions expressed herein are those of the authors and not necessarily of the funding agency.

References

1. Kaish ABMA, Jamil M, Raman SN, Zain MFM, Nahar L (2018) Ferrocement composites for strengthening of concrete columns: a review. Constr Build Mater 160:326–340. https://doi.org/10.1016/j.conbuildmat.2017.11.054
2. Yoddumrong P, Rodsin K, Katawaethwarag S (2020) Seismic strengthening of low-strength RC concrete columns using low-cost glass fiber reinforced polymers (GFRPs). Case Stud Constr Mater 13:e00383. https://doi.org/10.1016/j.cscm.2020.e00383
3. Ozcan O, Binici B, Ozcebe G (2010) Seismic strengthening of rectangular reinforced concrete columns using fiber reinforced polymers. Eng Struct 32(4):964–973. https://doi.org/10.1016/j.engstruct.2009.12.021
4. Juntanalikit P, Jirawattanasomkul T, Pimanmas A (2016) Experimental and numerical study of strengthening non-ductile RC columns with and without lap splice by carbon fiber reinforced polymer (CFRP) jacketing. Eng Struct 125:400–418. https://doi.org/10.1016/j.engstruct.2016.07.019
5. Nwankwo CO, Mahachi J, Olukanni DO, Musonda I (2023) Natural fibres and biopolymers in FRP composites for strengthening concrete structures: a mixed review. Constr Build Mater 363:129661. https://doi.org/10.1016/j.conbuildmat.2022.129661
6. Kushwaha M, Saini K, Matsagar V (2024) Strengthening brick masonry structures with natural fiber elements for enhancing earthquake resistance. In: Kioumarsi M, Shafei B (eds) The 1st international conference on net-zero built environment. NTZR 2024. Lecture notes in civil engineering. Springer, Cham, p 237. https://doi.org/10.1007/978-3-031-69626-8_33
7. Khan AA, Saini K, Matsagar VA (2025) Performance of concrete structures strengthened with jute rope-reinforced polymer composites. Engineering Mechanics Institute (EMI) Conference 2025, California, United States of America (USA)
8. Saini K, Matsagar VA, Kodur VR (2024) Recent advances in the use of natural fibers in civil engineering structures. Constr Build Mater 411:134364. https://doi.org/10.1016/j.conbuildmat.2023.134364
9. Saini K, Saverio S, Matsagar V (2024) Flexural behavior of natural fiber-reinforced foamed concrete beams. Arch Struct Constr 4:157–172. https://doi.org/10.1007/s44150-024-00114-2
10. Ritchie H (2020) Sector by sector: where do global greenhouse gas emissions come from? Published online at OurWorldinData.org. https://ourworldindata.org/ghgemissions-by-sector [Online Resource]
11. Jagtap G, Saini K, Matsagar V, Dey S, Kodur V (2024) Greenhouse gas emissions and life-cycle assessment of biocomposites from agro-residues for sustainable infrastructure development and climate change mitigation. In: Civil Engineering innovations for sustainable communities with net zero targets. Taylor & Francis, CRC Press, pp 138–158. https://doi.org/10.1201/9781032686899-9
12. Yooprasertchai E, Wiwatrojanagul P, Pimanmas A (2022) A use of natural sisal and jute fiber composites for seismic retrofitting of nonductile rectangular reinforced concrete columns. J Build Eng 52:104521. https://doi.org/10.1016/j.jobe.2022.104521
13. Jirawattanasomkul T, Likitlersuang S, Wuttiwannasak N, Ueda T, Zhang D, Shono M (2020) Structural behaviour of pre-damaged reinforced concrete beams strengthened with natural fibre reinforced polymer composites. Compos Struct 244:112309. https://doi.org/10.1016/j.compstruct.2020.112309
14. Yinh S et al (2021) Strengthening effect of natural fiber reinforced polymer composites (NFRP) on concrete. Case Stud Constr Mater 15:e00653. https://doi.org/10.1016/j.cscm.2021.e00653
15. Wahab N, Srinophakun P, Hussain Q, Chaimahawan P (2019) Performance of concrete confined with a jute-polyester hybrid fiber reinforced polymer composite: a novel strengthening technique. Fibers 7(8):72. https://doi.org/10.3390/fib7080072

16. Varma DA, Joseph L, Madhavan MK, Jayanarayanan K, Pegoretti A (2024) Strength, durability and finite element analysis of hybrid jute/basalt fiber reinforced polymer confined concrete column under axial compression. Results Eng 22:102281. https://doi.org/10.1016/j.rineng.2024.102281
17. Kunawisarut A et al (2025) Strengthening fire-damaged concrete with jute fiber-reinforced biopolymer: experimental investigation and sustainability assessment. Constr Build Mater 474:141081. https://doi.org/10.1016/j.conbuildmat.2025.141081
18. Abaqus®: Finite Element Software. Dassault Systèmes Simulia Corp.
19. Qasem M, Hasan M, Muhamad R, Chin C-L, Alanazi N (2025) Generalised calibration and optimization of concrete damage plasticity model for finite element simulation of cracked reinforced concrete structures. Results Eng 25:103905. https://doi.org/10.1016/j.rineng.2024.103905
20. ASTM D3039/D3039M—17 (2017) Standard test method for tensile properties of polymer matrix composite materials. American Society of Testing and Materials (ASTM) International, West Conshohocken, United States

Laboratory Evaluation of the Shape Effect of Reinforced Concrete Columns on Axial Performance Under Electromagnetic Field

Behdad Abbaspour[1(✉)], Ali Kheyroddin[1], Omid Rezaifar[1], and Mahdi Kioumarsi[2]

[1] Department of Structural Engineering, Semnan University, 5th km of Damghan Road Semnān, Semnān, Semnan, Iran
abbaspourb@semnan.ac.ir
[2] Department of Built Environment, Oslo Metropolitan University, Oslo, Norway

Abstract. This study examines the influence of geometric shape and electromagnetic field (EMF) conditions on the axial performance of reinforced concrete columns. Two types of specimens, with dimensions of 75 × 15 × 15 cm (cuboidal) and 15 × 75 cm (cylindrical), were experimentally evaluated under three levels of magnetic field intensity (0, 0.35, and 0.7 T) and three levels of electric current (0, 24, and 36 A). The experiments were designed to quantitatively assess changes in axial strength as a function of geometric shape and electromagnetic parameters. The results revealed that applying a magnetic field up to 0.35 T improved the axial strength of both specimen types. However, a significant reduction in axial strength was observed at higher magnetic field intensities (0.7 T). This reduction was more pronounced in cuboidal specimens due to non-uniform stress distribution and strain concentration at the corners. On the other hand, applying electric current at all levels consistently reduced the axial strength, particularly in cuboidal specimens, which exhibited more unstable behavior under the influence of the electric field. The analysis of geometric shape effects showed that cylindrical specimens demonstrated better resistance to magnetic and electric fields due to their more uniform stress distribution and the absence of sharp corners. In contrast, cuboidal specimens exhibited greater performance degradation due to stress concentration and corner vulnerability.

Keywords: Reinforced concrete column · axial load · electromagnetic field · electric current · magnetic field

1 Introduction

The structural integrity and load-bearing capacity of reinforced concrete (RC) columns are pivotal in the design and performance of modern infrastructures. Traditionally, the cross-sectional geometry of these columns has been recognized as a critical determinant of their axial and lateral load responses. Innovative configurations, such as L-shaped, T-shaped, and cross-shaped sections, have been explored to enhance both the aesthetic

H. Beushausen et al. (Eds.): NTZR 2025, LNCE 873, pp. 583–595, 2026.
https://doi.org/10.1007/978-3-032-27698-8_52

appeal and structural efficiency of buildings. Rahaman et al. [1] conducted a comparative analysis revealing that buildings with specially shaped columns exhibit superior lateral load resistance compared to those with conventional rectangular columns.

In a pioneering study, Abavisani et al. [2] introduced an innovative approach to real-time structural behavior control of RC beams through the application of alternating magnetic fields (AMFs) and alternating current (AC) electricity. Their experiments demonstrated that applying a 0.5 T AMF to small-scale RC beams during both the fresh and hardened phases influenced the beams' structural properties, including load-bearing capacity, deflection, bending stiffness, and ductility. Furthermore, the introduction of AC with varying intensities up to 36 A to the steel reinforcement system revealed a cubic relationship between current intensity and beam ductility, suggesting a potential method for enhancing the ductility of RC beams through electrical means.

Extending this concept to RC columns, Rezaifar et al. [3] explored the feasibility of improving compressive behavior and controlling structural responses in real-time by leveraging the interaction between AMFs and AC electricity. Their laboratory experiments involved applying a 0.5 T AMF with a frequency of 50 Hz externally to scaled-down RC columns and/or running AC of intensities up to 36 A through the reinforcing web. The specialized test setup allowed for simultaneous application of AMF, AC, and uniaxial compression, revealing that such electromagnetic interventions could modulate the compressive strength and deformation characteristics of RC columns.

Further investigations by Hajforoush et al. [4] examined the engineering properties of steel fiber-reinforced concrete (SFRC) exposed to homogeneous magnetic fields. Their findings indicated that magnetic field exposure influenced the orientation of steel fibers within the concrete matrix, leading to enhancements in compressive strength and ductility. This suggests that magnetic fields can be strategically employed to optimize the internal microstructure of SFRC, thereby improving its mechanical performance.

Early investigations by Buyukozturk [5] demonstrated that the electromagnetic properties of hardened concrete could be measured over a microwave frequency range, providing insights into the material's behavior and its interaction with electromagnetic waves. Subsequent studies have delved deeper into the effects of EMFs on concrete's microstructure and mechanical performance. For instance, Carrera et al. [6] explored how the orientation of steel fibers within high-performance concrete could be influenced by electromagnetic fields, leading to notable improvements in compressive strength. Similarly, Abavisani et al. [7] investigated the impact of alternating magnetic fields on fine-aggregate steel chip–reinforced concrete, observing enhancements in mechanical properties.

Despite various advancements in understanding the behavior of reinforced concrete (RC) columns under electromagnetic fields, a comprehensive evaluation of how different cross-sectional shapes influence axial performance in such environments remains limited. While previous studies have predominantly focused on material types, reinforcement techniques, or electromagnetic field parameters, the specific role of column geometry has received insufficient attention. This indicates a significant research gap regarding the interaction between cross-sectional shape and axial performance under electromagnetic exposure. Addressing this gap is crucial, especially considering the growing presence of electromagnetic fields in modern urban environments due to the proliferation of electronic devices and communication infrastructure. Therefore, this study aims to systematically investigate the axial behavior of RC columns with varying geometric configurations under the influence of electromagnetic fields. By integrating experimental assessments with theoretical analysis, the research endeavors to clarify the interplay between column geometry and electromagnetic exposure, ultimately contributing to improved design strategies for RC structures in electromagnetically active settings.

2 Experimental Program

2.1 Specimen Design and Geometric Configurations

To systematically evaluate the axial behavior of reinforced concrete (RC) columns under electromagnetic field (EMF) conditions, a total of 18 column specimens were fabricated with two distinct cross-sectional geometries: cuboidal and cylindrical. The geometric dimensions of the specimens were carefully selected to ensure comparability of height and cross-sectional area while reflecting common column configurations used in structural applications.

The cuboidal specimens measured 75 cm in height and had a rectangular cross-section of 15 × 15 cm. In contrast, the cylindrical specimens had the same height of 75 cm with a diameter of 15 cm. These dimensions were chosen to maintain a consistent slenderness ratio and axial load path across all samples, facilitating a reliable comparison of axial performance under varying test conditions (Fig. 1).

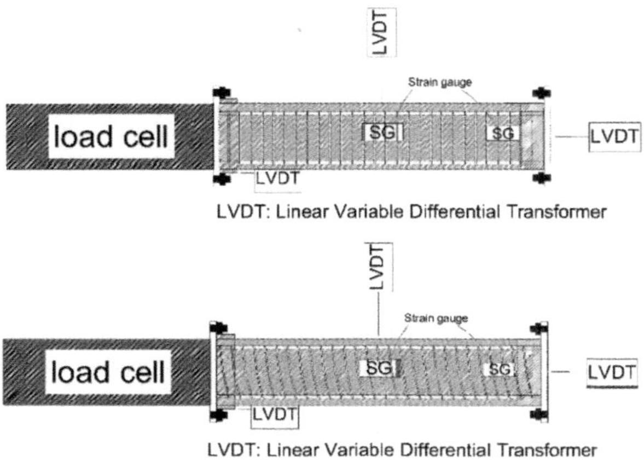

Fig. 1. Experimental test setup illustrating the instrumentation layout and reinforcement configuration of concrete specimens. The specimens are equipped with linear variable differential transformers (LVDTs) at both ends and at mid-span to accurately capture displacement. Strain gauges (SG) are symmetrically bonded on the surface to monitor longitudinal strain distribution. The load application was performed using a calibrated load cell to ensure precise axial loading.

Each specimen was internally reinforced using a longitudinal steel bar configuration designed according to conventional structural detailing methods. The cubic columns were reinforced with four longitudinal-shaped bars (φ8 mm) located at the corners and tied together with 3 mm diameter closed stirrups spaced 2.5 cm apart. For the cylindrical specimens, the reinforcement consisted of four symmetrical longitudinal bars (φ8 mm) with circular stirrups (3 mm) spaced 2.5 cm apart along the height. This reinforcement design ensures equivalent confinement levels in the two geometries.

The concrete mix used for all specimens was designed for a target compressive strength of approximately 20 MPa, ensuring uniform material properties across the test matrix. The concrete was mixed, cast, and cured under identical laboratory conditions to minimize variability. All specimens were demolded after 24 h and cured in water for 28 days before testing.

The experimental matrix incorporated three levels of magnetic field intensity (0, 0.35, and 0.7 T) and three levels of direct electric current (0, 24, and 36 A), resulting in a comprehensive factorial combination of geometric and electromagnetic variables. The application of the electromagnetic field was conducted using a custom-designed EMF chamber with adjustable field intensity and current flow, allowing for consistent and controlled exposure of specimens during testing.

This experimental design enabled a focused analysis of how the interaction between geometric shape and electromagnetic parameters influences the axial strength and failure characteristics of RC columns. The next sections detail the test setup, instrumentation, and loading protocol used to obtain quantitative data.

2.2 Material Properties

All materials employed in this experimental program conformed to the relevant ASTM standards. Ordinary Portland Cement (Type II), complying with ASTM C150 [8], was used as the primary binding material. The fine aggregate consisted of natural river sand with a specific gravity of 2.62 and a fineness modulus of 2.7. Crushed coarse aggregate with a nominal maximum size of 10 mm and a specific gravity of 2.68 was utilized. Tap water was used for both mixing and curing purposes [9]. The detailed properties of the materials are presented in Table 1.

Concrete was mixed and cast in standard molds, and all specimens were moist-cured at 23 ± 2 °C with relative humidity above 95% for 28 days. The mechanical properties of the concrete, including compressive and tensile strength, were determined using standard cylindrical samples (150 mm × 300 mm), tested in accordance with ASTM C39 [10] and ASTM C496 [11], respectively.

The electromagnetic field (EMF) was applied using a laboratory-generated coil system capable of producing uniform magnetic fields ranging from 0 to 7 T (T). The influence of EMF was studied on reinforced concrete columns with different cross-sectional shapes (circular, square, and rectangular) to evaluate the axial performance variations under magnetic stimulation.

The longitudinal reinforcement consisted of Φ8 steel bars with a yield strength of **400 MPa, and the lateral confinement was provided using** 3 mm diameter binding wire with a yield strength of 240 MPa.

Table 1. Mix design proportions for C20 concrete

Mix design	Gravel (kg/m^3)	Sand (kg/m^3)	Cement (kg/m^3)	Water (kg/m^3)	Density (kg/m^3)
C20	1213	595	300	186	2294

2.3 Electromagnetic Field Setup

To investigate the effect of electromagnetic fields on the axial performance of reinforced concrete (RC) columns, a controlled electromagnetic field was applied during the experimental procedure. The setup consisted of two main components:

(A) A magnetic flux generator, which was responsible for producing a stable and uniform magnetic field; and
(B) An electric current controller, designed to induce and regulate the electric current through the reinforcing bars.

Each component was operated independently to ensure precise control over both the intensity and the uniformity of the applied field.

(A) **Magnetic Flux Generator**

A uniform magnetic field was generated using a custom-designed Helmholtz coil system, comprising two parallel copper coils with 3000 turns of 2.5 mm insulated wire, mounted on a non-conductive frame. Powered by a programmable AC supply (10 A, 30 V), the system allowed precise adjustment of coil spacing to ensure field uniformity across the specimen. Magnetic field intensity, ranging from 0.35 to 0.7 T, was continuously monitored and calibrated using a Gaussmeter to maintain consistent exposure throughout the experiments [12].

(B) **Electric Current Control Device**

To investigate the influence of electric current on reinforced concrete columns, a dedicated control system comprising a precision variable AC power supply (0–40 A) was employed. The current was applied directly to the longitudinal steel bars via insulated copper terminals at both ends, ensuring full electrical isolation from the concrete. Current intensity and duration were systematically varied according to the experimental matrix. Real-time monitoring was conducted using a digital multimeter and clamp meter, while all electrical and magnetic parameters were recorded through a data acquisition system under stable laboratory conditions [13].

2.4 Test Procedure

To investigate the axial behavior of RC columns under an electromagnetic field, specimens were subjected to monotonic axial compression using a 2000 kN servo-controlled UTM at a displacement rate of 0.5 mm/min until failure. Axial load was applied concentrically to ensure uniform stress distribution. A digital acquisition system recorded load-displacement data in real-time. LVDTs and electrical strain gauges, including those placed on reinforcement bars at critical and mid-height locations, enabled precise strain measurements (Fig. 2). A custom electromagnetic coil, powered by an AC supply and monitored via Gaussmeter, established the controlled field environment. Data were captured at 10 Hz and post-processed to assess axial behavior and peak load response under coupled mechanical and electromagnetic loading [14].

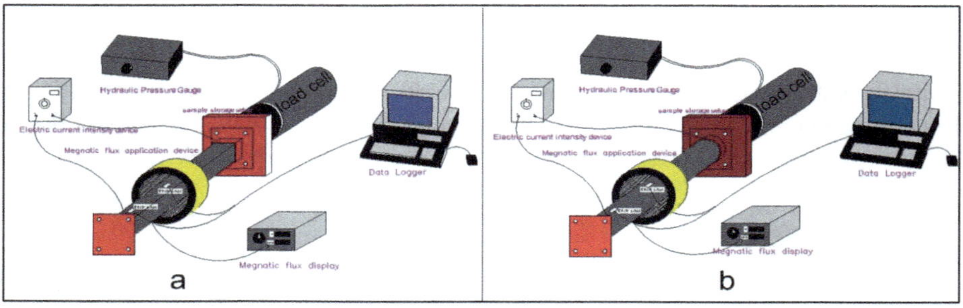

Fig. 2. Experimental setup for applying combined magnetic and electric fields during axial compression testing of concrete specimens. (a) Setup configuration for cubic concrete specimens, showing the arrangement of the magnetic flux applicator, electric current intensity device, hydraulic loading system, and data acquisition components. (b) Setup configuration for cylindrical concrete specimens under identical electromagnetic and mechanical conditions. Both setups utilize a load cell for force measurement and a data logger for synchronized recording of mechanical and electromagnetic parameters during the test.

3 Results and Discussion

The changes in axial performance under the electromagnetic field may result from localized heating and microstructural effects, such as increased porosity or microcracks. The EM field can also influence ionic movement, potentially affecting hydration or causing electrochemical reactions at the steel–concrete interface [15, 16].

The axial load–displacement behavior of the tested concrete specimens under uniaxial compression is presented and analyzed in this section. The results revealed significant variations in load-bearing capacity and deformation characteristics based on the cross-sectional shape, type of confinement, and presence of the electromagnetic field.

In general, specimens exhibited an initial linear response, followed by a nonlinear region leading up to the peak load, after which a gradual or sudden drop in load-carrying capacity was observed, depending on the level of confinement. Cylindrical specimens demonstrated a more stable post-peak behavior compared to cubic ones, which tended to fail more abruptly.

The experimental results indicate that the axial behavior of cylindrical reinforced concrete columns is notably affected by variations in magnetic flux density and electric current intensity. As shown in Table 2, when no magnetic field was applied (M0), increasing the electric current from 0 to 24 A led to an increase in the maximum axial load (Fmax) from 308 to 348.5 kN. This suggests that moderate current exposure may enhance load-bearing capacity Fig. 3.

Table 2. Experimental results of cylindrical reinforced concrete columns under different magnetic flux densities and electric currents

Specimen ID	Magnetic flux (T)	Current intensity (A)	Longitudinal displacement (mm)	Fmax (kN)	kmax (kN/mm)
C-N-M0-E0	0	0	41.42	308	18.85
C-N-M0-E1	0	24	36.74	348.5	17.03
C-N-M0-E2	0	36	40.65	275	13.78
C-N-M1-E0	0.35	0	38.97	352	20.47
C-N-M1-E1	0.35	24	33.55	358.7	21.38
C-N-M1-E2	0.35	36	34.61	283.5	19.75
C-N-M2-E0	0.7	0	30.93	254.5	15.64
C-N-M2-E1	0.7	24	34.7	252.5	12.06
C-N-M2-E2	0.7	36	41.5	224.5	19.73

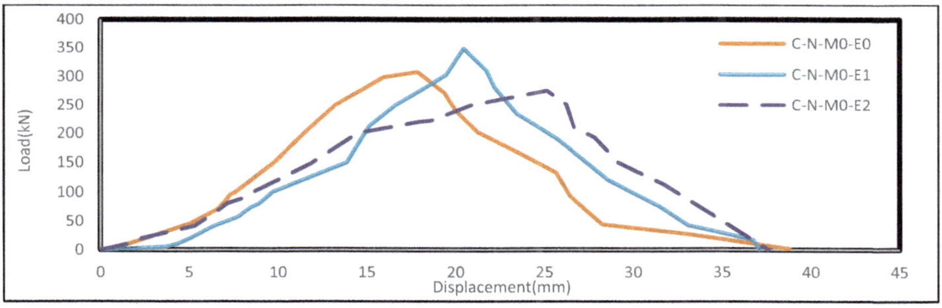

Fig. 3. Presents the axial load-displacement curves of cylindrical specimens in the absence of a magnetic field, reflecting their baseline structural behavior.

However, a further increase to 36 A caused a drop in Fmax to 275 kN, indicating that excessive current may begin to degrade the structural performance.

A similar trend was observed under a magnetic flux density of 0.35 T (M1). The specimen subjected to 24 A current (C-N-M1-E1) reached the highest recorded Fmax of 358.7 kN, while the application of 36 A current (C-N-M1-E2) resulted in a decreased strength of 283.5 kN. Stiffness values (kmax) followed a comparable pattern, with the highest stiffness (21.38 kN/mm) observed under moderate magnetic and electrical conditions (Fig. 4).

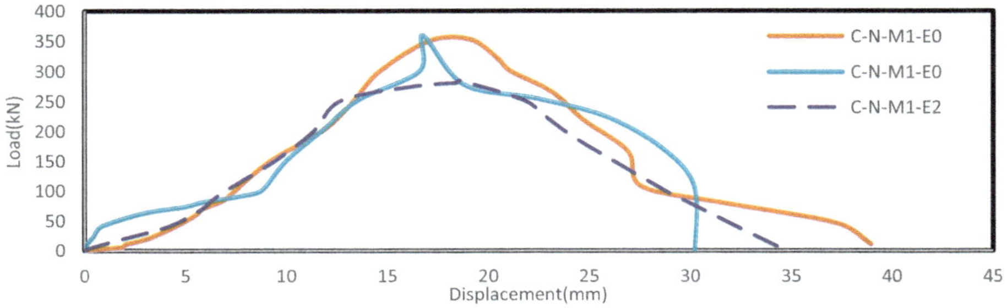

Fig. 4. Axial load–displacement curves of cylindrical specimens subjected to a 0.35 T magnetic field

In contrast, at a higher magnetic flux of 0.7 T (M2), all specimens showed a general decline in both load capacity and stiffness. Notably, specimen C-N-M2-E2 experienced the lowest Fmax (224.5 kN), despite displaying one of the highest displacements, indicating considerable softening due to strong electromagnetic exposure (Fig. 5).

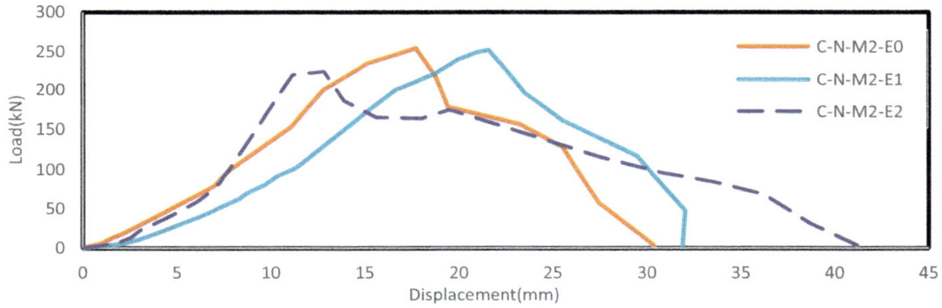

Fig. 5. Axial load–displacement curves of cylindrical specimens subjected to a 0.7 T magnetic field.

The results obtained from reinforced concrete cubic columns subjected to different levels of magnetic flux density and electric current are summarized in the following discussion. The findings reveal that both magnetic and electrical stimuli significantly influence the structural performance of the specimens. As shown in Table 3, specimen S-N-M0-E1, without a magnetic field but exposed to 24 A current, achieved the highest peak load (312.5 kN) and stiffness (40.31 kN/mm) among the control group, indicating the positive impact of moderate electrical stimulation Fig. 6.

Table 3. Experimental results of cubic reinforced concrete columns under different magnetic flux densities and electric currents

Specimen ID	Magnetic flux (T)	Current intensity (A)	Longitudinal displacement (mm)	Fmax (kN)	kmax (kN/mm)
S-N-M0-E0	0.0	0	35.31	290.0	24.83
S-N-M0-E1	0.0	24	30.99	312.5	40.31
S-N-M0-E2	0.0	36	44.25	270.0	17.24
S-N-M1-E0	0.35	0	33.67	376.0	27.6
S-N-M1-E1	0.35	24	22.39	356.5	44.3
S-N-M1-E2	0.35	36	32.26	319.5	30.97
S-N-M2-E0	0.7	0	20.77	229.0	17.98
S-N-M2-E1	0.7	24	29.59	222.0	14.78
S-N-M2-E2	0.7	36	30.41	211.0	12.98

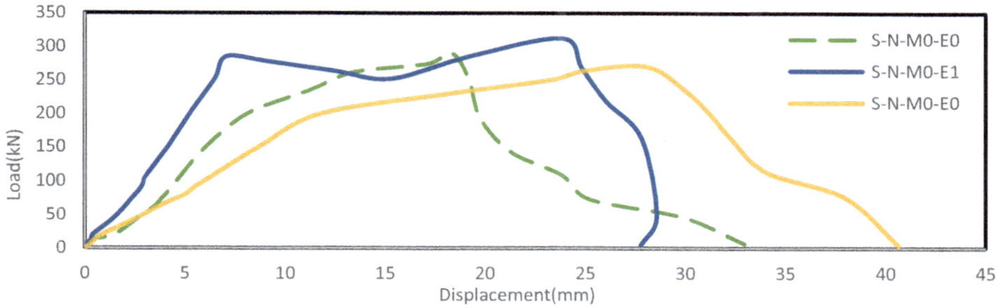

Fig. 6. Presents the axial load-displacement curves of cubic specimens in the absence of a magnetic field, reflecting their baseline structural behavior

Under a magnetic flux of 0.35 T, S-N-M1-E0 reached the highest load capacity (376.0 kN), highlighting the effectiveness of magnetic stimulation alone. However, when magnetic flux was combined with electric current (S-N-M1-E1 and S-N-M1-E2), a reduction in peak load was observed, suggesting a potential interaction effect (Fig. 7).

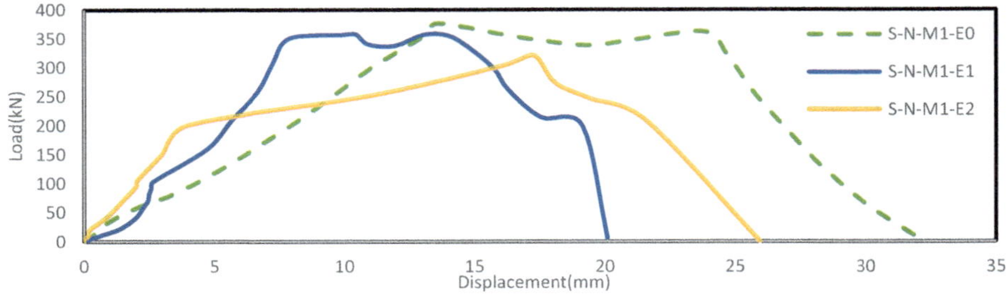

Fig. 7. Axial load–displacement curves of cubic specimens subjected to a 0.35 T magnetic field

At the highest magnetic flux level (0.7 T), increasing current intensity correlated with a noticeable drop in both load capacity and stiffness, with S-N-M2-E2 showing the lowest stiffness (12.98 kN/mm) (Fig. 8). These results emphasize the complex and sometimes opposing effects of electromagnetic fields on the mechanical behavior of reinforced concrete.

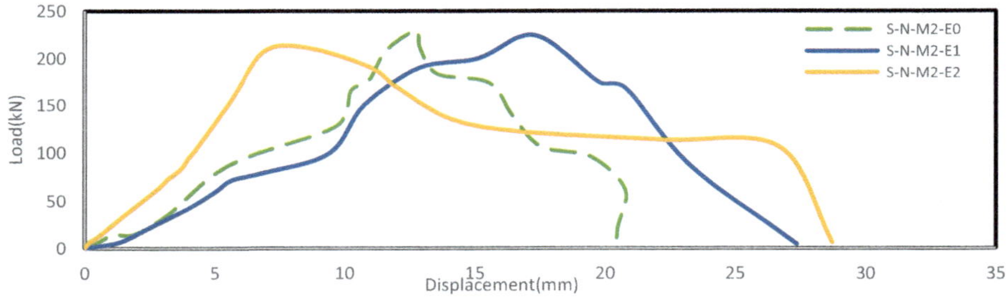

Fig. 8. Axial load–displacement curves of cubic specimens subjected to a 0.7 T magnetic field

The differences in failure mechanisms between cylindrical and square concrete columns are mainly due to their geometric characteristics. Cylindrical columns showed more uniform stress distribution and failed gradually through compressive crushing. In contrast, square columns experienced stress concentration at the corners, leading to earlier diagonal cracking and more brittle failure. The electromagnetic field further intensified these effects, especially in square columns, likely due to localized stress amplification and its interaction with the steel reinforcement (Fig. 9).

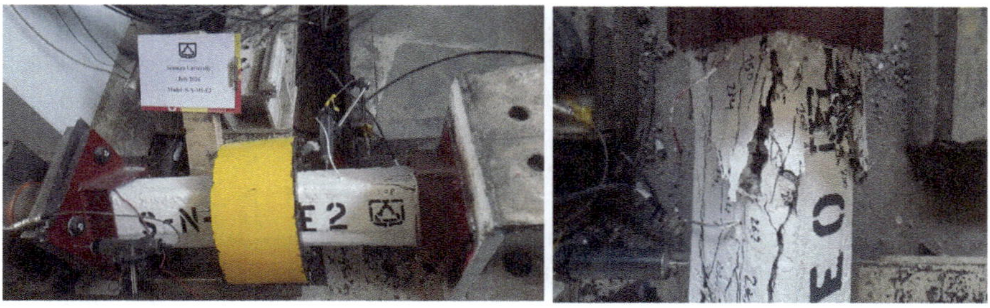

Fig. 9. Failure patterns of reinforced concrete column specimens under axial loading

4 Conclusion

This research provided a comprehensive experimental evaluation of the axial performance of reinforced concrete columns with different cross-sectional shapes (cylindrical and cubic) under the influence of electromagnetic fields, including varying magnetic flux densities and electric current intensities. The study aimed to identify how the combination of geometric configuration and electromagnetic stimulation affects the mechanical behavior, strength, stiffness, and ductility of reinforced concrete under uniaxial compression.

The findings clearly demonstrate that the shape of the concrete column plays a crucial role in its structural response. Cylindrical specimens consistently exhibited more stable post-peak behavior and better ductility, while cubic specimens tended to fail more abruptly with lower deformation capacity, reflecting their inherently more brittle nature. This highlights the importance of cross-sectional geometry in enhancing structural resilience under complex loading conditions.

Moreover, the influence of electromagnetic fields was found to be nonlinear and highly dependent on intensity. The application of moderate electric current (24 A), particularly in the absence or presence of a low magnetic field (0.35 T), generally improved both load-bearing capacity and energy absorption characteristics. For instance, under 0.35 T magnetic flux, cylindrical specimens achieved the highest peak load and stiffness when combined with 24 A current, indicating a synergistic effect at optimal levels of stimulation.

However, higher levels of electric current (36 A) or magnetic flux density (0.7 T), especially when applied simultaneously, led to a noticeable decline in axial strength and stiffness, particularly in cubic specimens. This degradation is attributed to possible microstructural damage, internal heating, and unfavorable electrochemical interactions induced by excessive electromagnetic exposure.

Notably, cylindrical specimens under optimized electromagnetic conditions showed a better balance between strength and deformability, making them more suitable for applications where both high capacity and post-peak ductility are critical.

In conclusion, the results of this study underscore the dual influence of geometry and electromagnetic parameters on the axial behavior of reinforced concrete columns. While moderate electromagnetic stimulation can be strategically used to enhance mechanical performance, particularly in cylindrical configurations, excessive levels may lead to adverse effects. These findings provide important insights for the future development of smart structural systems and non-destructive enhancement techniques, where electromagnetic fields are employed to tailor the performance of concrete elements in structural engineering applications. Further investigations on long-term durability, microstructural changes, and scalability to real-world conditions are recommended to extend the applicability of this method.

Beyond the mechanical performance, the comparison between cubic and cylindrical columns also suggests potential implications for material efficiency. For columns designed to withstand equivalent axial loads, cylindrical specimens—due to their superior load-bearing capacity and ductility—may allow for a reduction in cross-sectional area or reinforcement requirements while maintaining structural performance. This implies that adopting cylindrical configurations in design could contribute to a more efficient use of concrete and steel materials. Such geometrical optimization aligns with the broader goals of sustainable construction and structural design strategies aimed at minimizing resource consumption without compromising safety or durability.

Electromagnetic techniques can enhance durability and structural performance with less material use and minimal intervention. These benefits support net-zero goals by promoting sustainable and low-carbon construction methods.

These results can assist in the design of more intelligent, efficient, and sustainable structural systems, particularly in environments where exposure to electromagnetic fields is expected.

References

1. Rahaman A, Anik AM, Serker NHMK (2018) Effect of special shaped column on lateral load resistance capacity of reinforced concrete (RC) building. Am J Civ Eng 6(5):147–153
2. Abavisani I, Rezaifar O, Kheyroddin A (2017) Magneto-electric control of scaled-down reinforced concrete beams. ACI Struct J 114(1):233–244
3. Rezaifar O, Abavisani I, Kheyroddin A (2017) Magneto-electric active control of scaled-down reinforced concrete columns. ACI Struct J 114(5):1351–1362
4. Hajforoush M, Kheyroddin A, Rezaifar O (2020) Investigation of engineering properties of steel fiber reinforced concrete exposed to homogeneous magnetic field. Constr Build Mater 252:119064
5. Buyukozturk O (1997) Electromagnetic properties of concrete and their significance in nondestructive testing. Transp Res Rec 1574(1):10–17
6. Carrera K, Künzel K, Konrád P (2022) Mechanical properties of high-performance concrete with steel fibers oriented by an electromagnetic field. J Mater Civ Eng 34(9):04022265
7. Abavisani I, Rezaifar O, Kheyroddin A (2018) Alternating magnetic field effect on fine-aggregate steel chip–reinforced concrete properties. J Mater Civ Eng 30(6):04018087. https://doi.org/10.1061/(ASCE)MT.1943-5533.0002267
8. ASTM International (2022) Standard specification for Portland cement. ASTM C150/C150M-22, West Conshohocken, PA. https://doi.org/10.1520/C0150_C0150M-22
9. Badarloo B, Kari A, Jafari F (2018) Experimental and numerical study to determine the relationship between tensile strength and compressive strength of concrete. Int J Civ Eng 16(12):1655–1665
10. ASTM International (2023) Standard test method for compressive strength of cylindrical concrete specimens. ASTM C39/C39M-23, West Conshohocken, PA

11. ASTM International (2017) Standard test method for splitting tensile strength of cylindrical concrete specimens. ASTM C496/C496M-17, West Conshohocken, PA
12. Hill RM, Lassiter DS (1996) Design of Helmholtz coils for uniform magnetic field generation. Rev Sci Instrum 67(9):3233–3237
13. Aghajani A, Saatchi A, Golzari MA, Reisi K (2014) Investigation of the effects of high-voltage AC stray current on concrete using electrochemical impedance spectroscopy. Transp Res J 11(2):117–133
14. Rezaifar O, Ganji R (2013) Investigating the effect of magnetic field on behavioral properties of concrete. In: 2nd National conference on concrete industry. Semnan University, Iran
15. Lee H, Jinho Park JKS, Yang H (2019) Electromagnetic shielding performance of carbon black mixed concrete with Zn–Al metal thermal spray coating. Materials 12(17)
16. Zhang Y, Liu Q, Cai H (2021) A fully coupled electro-chemo-mechanical phase field method to simulate corrosion mechanisms in reinforced concrete. Corros Sci 200:110–123

Investigation of Longitudinal Reinforcement Anchorage in Reinforced Concrete Beams Using Mechanical Splice (Coupler) in the Joint with a Steel Section Encased in Reinforced Concrete Column (SRC Composite Column)

Seyedeh Marzieh Qiyami Taklymi[1](✉), Ali Kheyroddin[1], Omid Rezaifar[1], and Mahdi Kioumarsi[2]

[1] Structural Engineering, Department of Civil Engineering, Semnan University, Semnan, Iran
{m.qiyami,kheyroddin,orezayfar}@semnan.ac.ir
[2] Department of Built Environment, Oslo Metropolitan University, Oslo, Norway
Mahdi.kioumarsi@oslomet.no

Abstract. In recent decades, the use of composite steel–concrete columns has increased in tall structures due to their numerous advantages. The choice of construction materials significantly influences the properties of each structure, making it one of the most critical factors. Composite columns, due to the presence of a steel section encased in reinforced concrete (SRC), exhibit increased confinement, flexural resistance, ductility, stiffness, and energy absorption compared to reinforced concrete columns.

This study investigates the effects of incorporating a steel transitional part (TP) in the reinforced concrete beam (RCB) and a steel section encased in the reinforced concrete column (SRCC) on the performance of specimens. In this study, three experimental specimens were examined. The SRCC-RCB-TP-Coupler specimen demonstrated notable improvements in mechanical properties, including an 18% and 9% increase in load-bearing capacity compared to RCC-RCB and SRCC-RCB specimens, respectively. Stiffness also increased significantly, by 33.5% compared to SRCC-RCB and by over 100% compared to RCC-RCB. Additionally, the use of the transitional part improved ductility, with a 16% increase compared to RCC-RCB and an 8% increase compared to SRCC-RCB specimens. The SRCC-RCB specimen also exhibited an 8% increase in ductility compared to RCC-RCB. These results indicate that encasing an I-shaped steel section in the reinforced concrete column and utilizing a steel transitional part in the beam–column joint enhances load-bearing capacity, stiffness, and ductility. This configuration improves force distribution and extends the plastic hinge length in the beam, enhancing the specimen's performance under load.

Keywords: Composite Column · Transitional Part (TP) · Mechanical Splice (Coupler) · Stiffness · Ductility

H. Beushausen et al. (Eds.): NTZR 2025, LNCE 873, pp. 596–608, 2026.
https://doi.org/10.1007/978-3-032-27698-8_53

1 Introduction

The connections between beams and columns play a critical role in the seismic performance of structures, including reinforced concrete (RC) ones. To prevent premature connection failure during earthquakes, the strong column-weak beam principle is often employed. In structures with Steel Reinforced Concrete (SRC) columns and beams, this principle helps maintain the integrity of connections. SRC members, which consist of steel sections encased in reinforced concrete, provide high load-bearing capacity, stiffness, and significant ductility. Due to these advantages, SRC columns have become increasingly popular in modern construction, especially in tall buildings, as they offer substantial energy absorption capabilities while addressing architectural and economic constraints [1]. The combination of steel and reinforced concrete sections improves load-bearing capacity and deformation behavior. SRC (Steel-Reinforced Concrete) members, due to their seismic performance, are used in tall buildings in earthquake-prone regions [2]. The steel section enhances resistance to compressive buckling, while the concrete cover prevents corrosion. These columns are beneficial in earthquake scenarios because of their higher load capacity, energy absorption, and deformability [3, 4]. Confinement in RC columns increases lateral deformation capacity, and in SRC columns, the internal steel section further contributes to this capacity. Additionally, reinforced concrete reduces local buckling in the steel section of SRC columns [5].

Rong et al. [6] investigated the seismic performance of S-RC-SRC composite connections, revealing that flexural and shear failures occurred in the connection zone. The reinforcement ratio was identified as a key factor influencing the failure mode. They recommended incorporating hoop beams in the connection zone to improve load-carrying capacity and ductility. Li et al. [7] studied the axial compressive resistance of composite steel T-shaped columns with minimal reinforcement in concrete. Their findings indicated that low reinforcement and thin concrete cover could prevent instability under axial compression. Furthermore, they suggested using non-combustible boards, closely spaced transverse reinforcements, and reducing their spacing to prevent cracking and spalling of the concrete cover. Jin et al. investigated the cyclic lateral behavior of SRC-RC composite columns under lateral bending, showing that their lateral resistance is higher than that of RC columns [8]. Gautham et al. studied the seismic performance of soft-story reinforced concrete frames with SRC columns using an energy-based approach. Their results demonstrated that SRC columns reduce the seismic vulnerability of RC frames in soft-story configurations [9].

Studies on H-shaped steel-section SRC columns under lateral impact loading [10] show that increasing impact velocity leads to higher localized shear failure and a greater percentage of energy absorption by the steel section, improving structural impact resistance. Pan et al. proposed a ring beam connection for linking an RC beam with an SRC column, with its behavior being influenced by factors such as flexural and shear behavior of the RC column and beam, rebar development length, shear transfer between steel and concrete, and concrete confinement level [11].

Despite numerous studies on the joint between steel beams and SRC columns, limited research has focused on the joint between SRC columns and reinforced concrete beams. The primary objective of this study is to investigate the behavior of the connection between SRC columns and reinforced concrete beams under combined lateral cyclic loading and axial force, along with the anchorage of the longitudinal reinforcements of the beam using mechanical methods (couplers) and the use of a transitional part in the joint region between the beam and column. Previous studies have primarily observed the passage of longitudinal reinforcements through the flange section of the SRC column. However, no research has explored the passage through the web of the steel section or the anchorage using hooks, mechanical splices (couplers), or transitional parts (TP) to enhance joint performance. In this study, the longitudinal reinforcements of the reinforced concrete beam pass through the web of the steel section, with anchorage either inside the column or behind the steel section using a coupler, and the beam is of the discontinuous and unconfined type according to the ACI318-19 standard. Key parameters investigated include load-carrying capacity, ductility, stiffness, and energy absorption under lateral cyclic loading.

2 Experimental Program

An experimental investigation was conducted on the seismic performance of a steel section encased in reinforced concrete column and a reinforced concrete beam, along with the longitudinal reinforcement of the beam being anchored through mechanical slice (coupler) and the use of a transition part at the beam-to-column joint. The details of the laboratory test specimens will be presented in the following sections.

2.1 Experimental Specimens

In this study, three specimens of lateral beam-column connections were fabricated, tested, and evaluated. The geometric details of the experimental specimens, including dimensions and cross-sections, are shown in Fig. 1. The specimens were categorized based on the type of column, either reinforced concrete (RC) or steel encased in reinforced concrete (SRC), and were named RCC-RCB, SRCC-RCB, and SRCC-RCB-TP-Coupler, respectively. Here, "RC" refers to reinforced concrete, "SRC" represents steel encased in reinforced concrete, "TP" stands for transitional part, "Coupler" indicates the anchorage of longitudinal beam reinforcements using mechanical splices, "C" refers to the column, and "B" refers to the beam. The RCC-RCB specimen represents a joint with a reinforced concrete column and a reinforced concrete beam. The SRCC-RCB specimen represents a joint with a steel column encased in reinforced concrete and a reinforced concrete beam. Finally, the SRCC-RCB-TP-Coupler specimen represents a steel section encased in reinforced concrete, with a reinforced concrete beam and a transitional part in the beam-to-column joint region. The longitudinal reinforcements of this specimen are anchored using mechanical splices (Couplers). The geometric details and design of the RCC-RCB, SRCC-RCB, and SRCC-RCB-TP-Coupler experimental specimens were carried out according to the ACI 318-2019 standard. The longitudinal reinforcement bars of the beam and column have a diameter of 14 mm, and the transverse reinforcement bars used for both elements have a diameter of 10 mm. The steel percentage used in the columns is 1%, and in the beams, it is 2%. Based on the design and dimensions of the columns, the critical region length was determined, and the spacing of 10 mm diameter ties was set at 85 mm intervals, following ACI 318-2019 provisions. In the beams, tie spacing was designed at 70 mm intervals in the critical region (600 mm from the joint edge), and 120 mm intervals beyond the critical region.

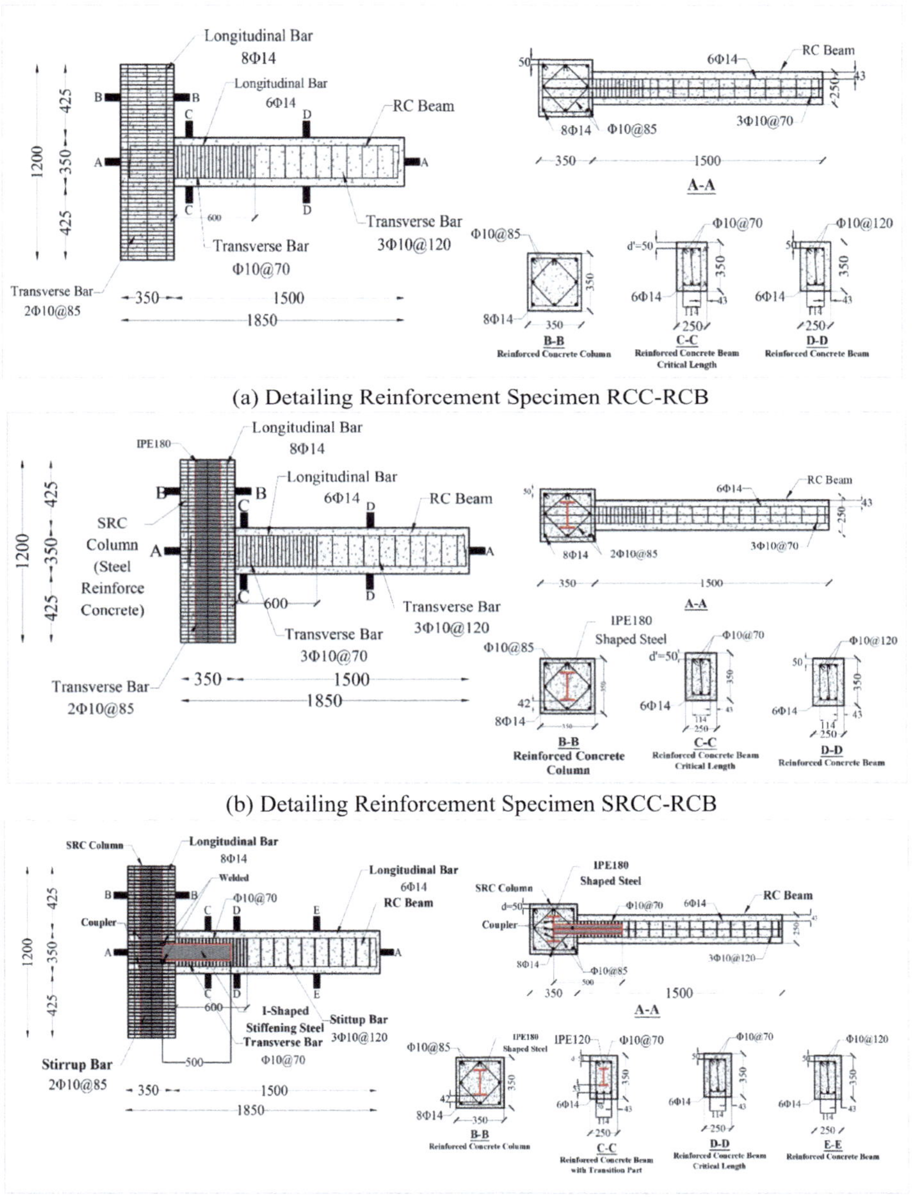

(a) Detailing Reinforcement Specimen RCC-RCB

(b) Detailing Reinforcement Specimen SRCC-RCB

(c) Detailing Reinforcement Specimen SRCC-RCB-TP-Coupler

Fig. 1. Detailing specimens

Due to experimental conditions and available hydraulic jacks, a constant axial compressive load of 400 kN was applied to both specimens, approximately 0.8% of the column's axial capacity. The specimens were designed based on the strong column-weak beam principle to ensure plastic hinge formation in the beams and prevent connection or column failure. The SRC column section is made of an IPE180 steel profile with dimensions of 91 mm for the top flange width, 180 mm for the web height, and 1200 mm in length. The TP (Transition Part) section uses an IPE120 profile with a flange width of 64 mm, a height of 120 mm, and a length of 500 mm. To allow the longitudinal reinforcement bars of the beam to pass through the steel section, holes with a diameter of 25 mm are made in the web of the SRC column. No stiffeners are used in the web or flange to assess their effect on the load-bearing capacity. The study examines the impact of the I-shaped steel section and the use of mechanical splices (couplers) on plastic hinge rotation, crack propagation, and load-carrying capacity. The longitudinal reinforcement bars of the beam are anchored with 90-degree hooks following the ACI 318-2019 standard, while the SRCC-RCB-TP-Coupler specimen uses a coupler for reinforcement anchorage. Shear studs, compliant with BS EN-ISO 13918, are installed on the flanges and web of the SRC and TP steel sections at intervals of 150 mm and 125 mm, respectively (Fig. 2).

(a) (b) (c)

Fig. 2. Details longitudinal reinforcement: (a) Anchorage length, (b) longitudinal reinforcement of the beam, (c) detailing mechanical splice (Coupler)

2.2 Material Properties

The concrete mix was designed per ACI211-91 using Type II Portland cement, crushed sand (9/5 mm), and washed sand. Cylindrical specimens (20 × 10 cm) were cured for 28 days and tested for compressive strength using a Universal Testing Machine (ASTM-C39-86). The average compressive strength was 32 MPa.

2.3 Experimental Setup

Figure 3 illustrates the test setup and loading procedure. A 400 kN axial compressive load was applied to all specimens, and lateral cyclic loading was applied 125 mm from the beam end. To prevent out-of-plane movement, restraining beams were installed 750 mm from the connection point and secured to the laboratory frame. Cyclic loading was applied to the beam end using a 1000 kN hydraulic jack and a 500 kN load cell. Axial loading on the columns was similarly applied with a 1000 kN hydraulic jack and a 500 kN load cell. Four uniaxial strain gauges were installed on the intermediate reinforcement bars in the critical region of the beam, spaced 50 mm apart, and positioned 50 mm from the joint point. The bars were smoothed, and the gauges were bonded with special adhesive. Their proper functioning was verified using an ohmmeter, ensuring a resistance of 120 Ohms before and after concrete pouring.

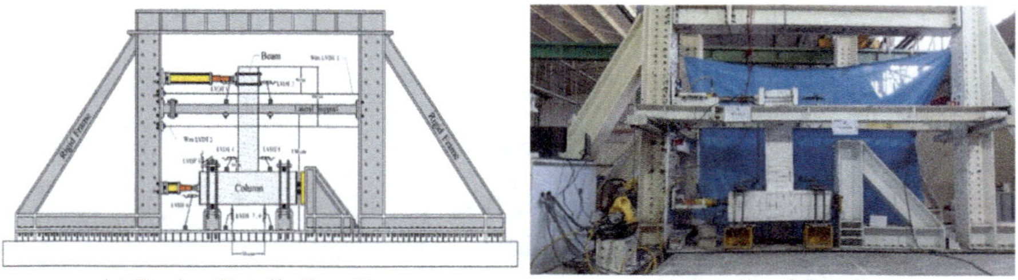

(a) Design Details Test Setup (b) Details of Test Setup

Fig. 3. Details test setup

The RC and SRC columns were subjected to a constant axial compressive force of 400 kN, with loading conducted per ACI-374.1–05 [12]. The test continued until the load-carrying capacity decreased by 15% or until longitudinal reinforcements fractured, as shown in Fig. 4.

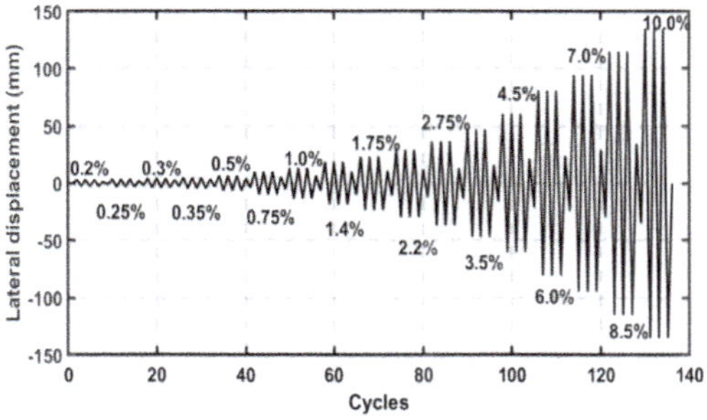

Fig. 4. Loading protocol [12]

3 Experimental Results and Discussion

The parameters examined in this study include the mode of crack initiation and propagation, failure mode, strength capacity (load-carrying capacity), hysteresis curve, ductility, energy dissipation, and stiffness curve of each specimen under loading.

3.1 Crack Pattern, Failure Mode, and Hysteresis Curve

The RCC-RCB specimen consists of a reinforced concrete column (RCC) and a beam (RCB). Cracks initiated within 100 mm of the connection at a drift of 0.2% under loads of 13 kN (compressive) and 20.9 kN (tensile), eventually leading to flexural cracks. At a

2.2% drift, the beam's longitudinal reinforcement yielded, and plastic hinge formation occurred (see Fig. 5). The specimen exhibited nearly symmetric hysteresis loops (see Fig. 8a), indicating stable energy dissipation and resistance to degradation. The ultimate tensile and compressive forces were 45.7 kN and – 43.5 kN, with stiffness gradually reducing due to crack propagation near the joint while maintaining stable cyclic behavior characteristic of ductile structures under repeated loading.

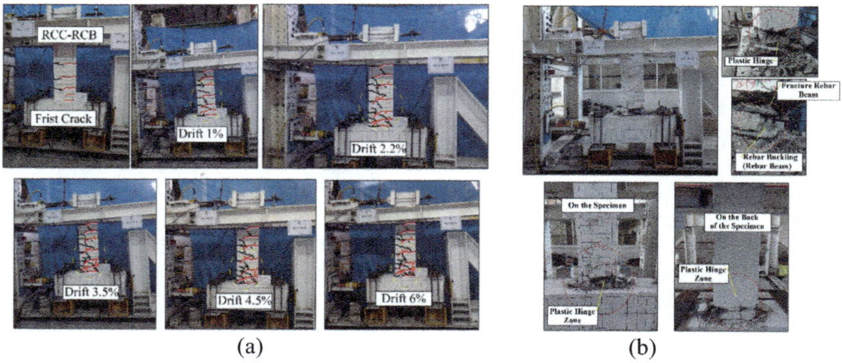

Fig. 5. (a) Crack pattern, (b) ultimate failure RCC-RCB specimen

The SRCC-RCB specimen, composed of a steel-reinforced concrete column (SRCC) and a reinforced concrete beam (RCB), showed crack initiation at a drift of 0.2%, with compressive and tensile loads of 16.5 kN and 13 kN, respectively. Surface cracks extended into flexural cracks in the beam-column joint (see Fig. 6). The hysteresis curve in Fig. 8-b shows symmetric loops, indicating stable energy dissipation. The specimen's maximum load-carrying capacity was 47.3 kN (tensile) and – 48.7 kN (compressive). The SRCC-RCB specimen exhibited about 3.5% higher load-bearing capacity in tension and 12% in compression compared to the RCC-RCB specimen, with stable cyclic behavior despite reduced stiffness due to crack propagation.

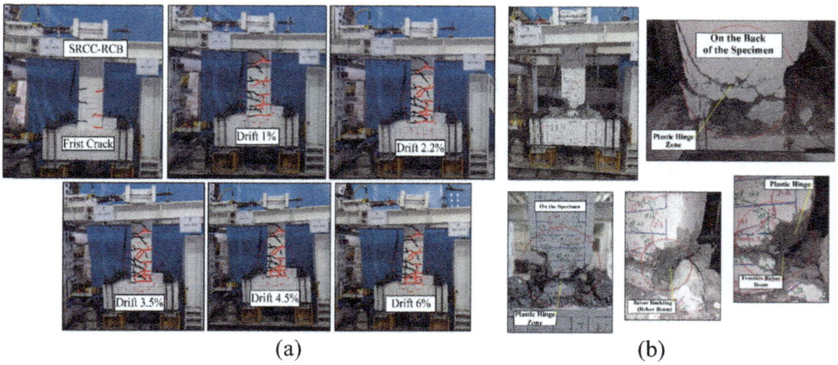

Fig. 6. (a) Crack pattern, (b) ultimate failure SRCC-RCB specimen

The SRCC-RCB-TP-Coupler specimen consists of a steel-reinforced concrete column (SRCC) and a reinforced concrete beam (RCB), along with a transitional part (TP) at the beam-column joint. A mechanical splice (coupler) anchors the longitudinal reinforcement of the beam. Cracks appeared near the connection, with initial surface cracks at 0.2% drift, under compressive and tensile forces of – 14.2 and 16.7 kN. These cracks propagated, leading to flexural cracks in the joint (see Fig. 7). The hysteresis diagram shows stable energy dissipation, with maximum tensile and compressive forces of 52.6 kN and – 52.4 kN, respectively, indicating the ultimate load-carrying capacity (see Fig. 8-c).

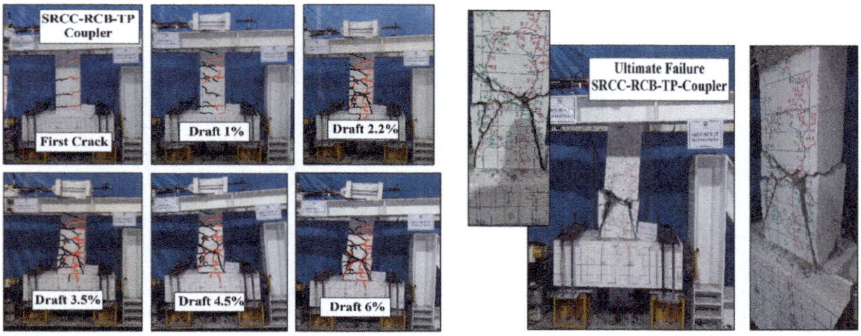

(a) (b)

Fig. 7. (a) Crack pattern, (b) ultimate failure SRCC-RCB-TP-coupler specimen

The gradual reduction in stiffness during subsequent cycles is due to flexural cracks near the beam-column joint. Despite this, the specimen maintains stable cyclic behavior, efficiently dissipating energy during loading and unloading, characteristic of ductile structures. The load-bearing capacity of the SRCC-RCB-TP-Coupler specimen increased by 11% in tension and 8% in compression compared to the SRCC-RCB specimen, and by 15% in tension and 20.1% in compression compared to the RCC-RCB specimen. Furthermore, the plastic hinge length has increased in the SRCC-RCB and SRCC-RCB-TP-Coupler specimens compared to the RCC-RCB specimen. In the SRCC-RCB and SRCC-RCB-TP-Coupler specimens, flexural cracks propagate along the beam and are not confined to only 0.5 times the length of the beam.

Figure 8 shows the hysteresis curves and backbone comparison of the RCC-RCB, SRCC-RCB, and SRCC-RCB-TP-Coupler specimens, revealing that the SRCC-RCB-TP-Coupler has a higher load-bearing capacity and initial stiffness. These curves help determine key parameters like maximum load capacity, yield capacity, ductility, stiffness, and absorbed energy.

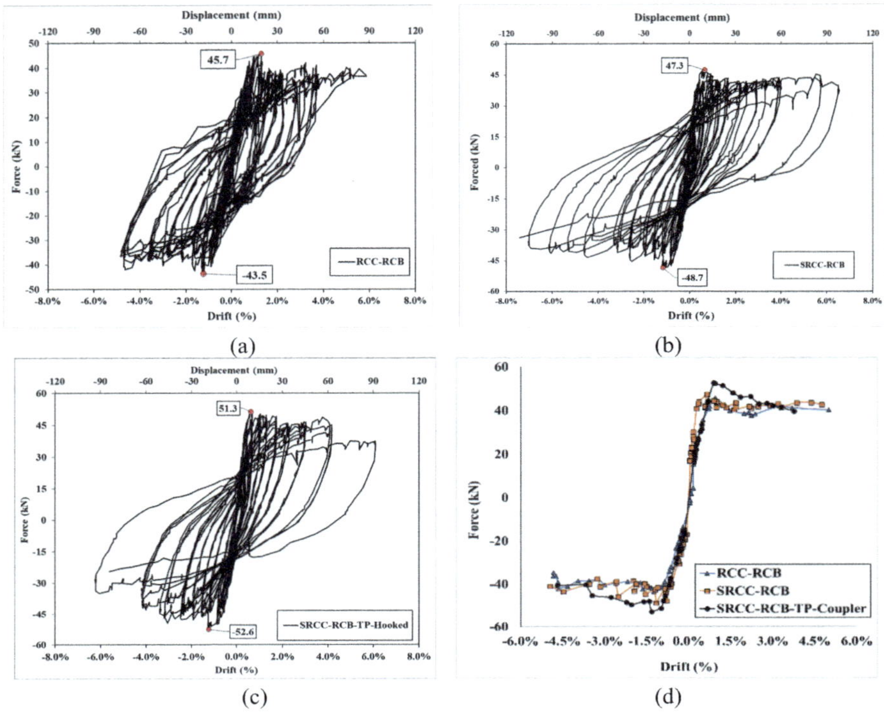

Fig. 8. Hysteresis curve: (a) RCC-RCB, (b) SRCC-RCB, (c) SRCC-RCBTP-Coupler, (d) Push-over specimens

Lateral ductility is the ratio of maximum displacement to displacement at the yield point (Eq. 1), with maximum displacement defined as the point where the force is 85% of the yield force, as determined from the slope of the elastic region in the backbone curve (see Fig. 9) [13]. Where Δ_u represents the maximum displacement and Δ_y denotes the displacement at the yield point (Table 1).

$$\mu = \frac{\Delta_u}{\Delta_y} \tag{1}$$

Lateral ductility values for the RCC-RCB, SRCC-RCB, and SRCC-RCB-TP-Coupler specimens are 6.91, 7.45, and 8.04, respectively. The SRCC-RCB-TP-Coupler specimen shows a 16 and 8% increase in ductility compared to the RCC-RCB and SRCC-RCB specimens, while the SRCC-RCB specimen shows a 8% increase compared to RCC-RCB. These increases in ductility are attributed to the use of a steel-reinforced column, a transitional part at the beam-column joint, and a mechanical splice for the beam's longitudinal reinforcements.

Table 1. Ductility specimen

Specimen	Δ_y (Tension)	Δ_u (Tension)	μ (+)	Δ_y (Compression)	Δ_u (Compression)	μ (-)	μ_{avg}
RCC-RCB	10.2	88	8.63	– 14	– 72.6	5.18	6.91
SRCC-RCB	9.4	34.63	3.68	– 12.04	– 90.8	7.54	7.45
SRCC-RCB-TP-Coupler	8.1	71.1	8.78	– 8.1	– 69.4	7.31	8.04

The area under the hysteresis curve represents the dissipated energy, indicating the ductility and energy absorption capacity of the element, which enhances its seismic performance. The energy dissipated in each cycle is calculated, with the average energy over three cycles being considered. A cumulative plot of energy dissipated versus lateral drift is shown, up to a drift of 4.5% (see Fig. 9). In this equation, the areas of CDA and ABC correspond to the shaded areas. The areas of ODF and OBE also correspond to the triangles formed by the idealized hysteresis curve.

$$E_{Dissipation} = \frac{S_{CDA} + S_{ABC}}{S_{ODF} + S_{OBE}} \quad (2)$$

$$E_{Dissipation} = \frac{S_{CDA} + S_{ABC}}{S_{ODF} + S_{OBE}} \quad (2)$$

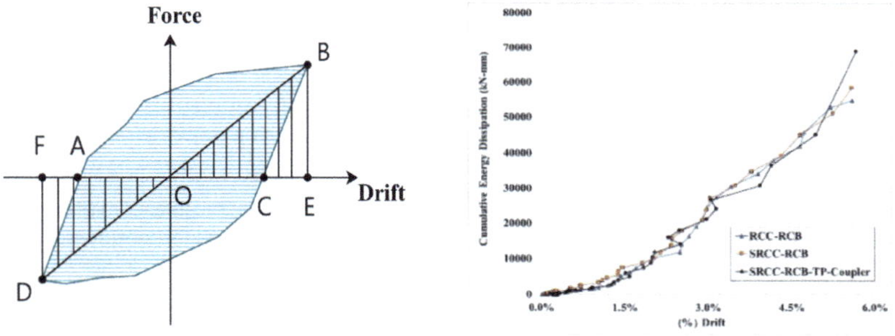

(a) (b)

Fig. 9. Energy dissipation: (a) Calculation of the area under the hysteresis curve, (b) cumulative energy dissipation

The stiffness for each cycle of the specimen is calculated using Eq. 3, and the stiffness-lateral drift curve is plotted. In this equation, $+F_i$ and $+F_i$ represent the maximum lateral force in each cycle under compression and tension, respectively. $+\Delta_i$, and $-\Delta_i$ illustrate the corresponding displacements to maximum load for each cycle of compression and tension, respectively. The equation defines the maximum lateral force in each cycle for compression and tension, along with the corresponding displacements for each load. Stiffness is derived from the pushover (backbone) analysis and represents maximum stiffness in both loading directions. Figure 10 shows the stiffness degradation under both compressive and tensile loads, revealing a gradual reduction due to the development and propagation of cracks in the specimens [14].

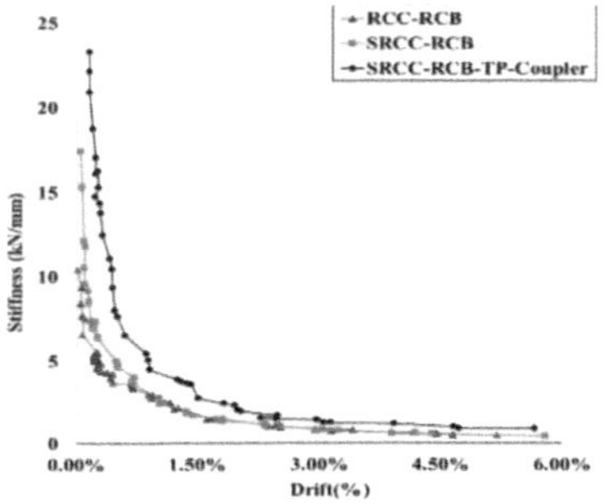

Fig. 10. Stiffness

The stiffness comparison shows that the RCC-RCB specimen has an initial stiffness of 10.40, with a lower rate of stiffness reduction. The SRCC-RCB specimen has a 68% higher initial stiffness (17.44), while the SRCC-RCB-TP-Coupler specimen has 33.5% more stiffness (23) than SRCC-RCB, and over 100% more than RCC-RCB.

$$K_e = \frac{\sum_{i=1}^{n} F_i^j}{\sum_{i=1}^{n} \Delta_i^j} = \frac{|+F_i| + |-F_i|}{|+\Delta_i| + |-\Delta_i|} \quad (3)$$

4 Conclusion

Three specimens were tested: a reinforced concrete beam and column, an SRC column with a reinforced concrete beam, and an SRC column with a reinforced concrete beam and a steel transitional section at the beam-column connection. This study examines the effects of using an SRC column with an I-shaped steel section, a steel transitional part, and mechanical splices (couplers) for restraining the beam's longitudinal reinforcements. The study evaluates the impact of these methods on the specimens' performance:

- The SRCC-RCB-TP-Coupler specimen shows superior seismic performance compared to the SRCC-RCB and RCC-RCB specimens, with higher load-bearing capacity, stiffness, ductility, and energy dissipation. The SRCC-RCB specimen also demonstrates better seismic performance than the RCC-RCB specimen, offering increased stiffness and load capacity. The I-shaped steel section in the reinforced concrete column enhances lateral resistance.
- In both the SRCC-RCB and SRCC-RCB-TP-Coupler specimens, cracks propagated along the entire length of the beam, while in the RCC-RCB specimen, cracks did not spread throughout the entire length. Additionally, in both the SRCC-RCB and SRCC-RCB-TP-Coupler specimens, the damage and failure occurred away from the connection zone and were confined to the beam. The plastic hinge length in these specimens was greater than in the RCC-RCB specimen.
- The SRCC-RCB and SRCC-RCB-TP-Coupler specimens show increased stiffness compared to the RCC-RCB specimen. The ductility of the SRCC-RCB-TP-Coupler has improved by 16% and 8% compared to the RCC-RCB and SRCC-RCB, respectively. The SRCC-RCB specimen shows a 8% increase in ductility compared to the

RCC-RCB. These improvements are attributed to the use of a steel transitional section at the beam-column joint and the restraining of the beam's longitudinal reinforcements with a coupler.

- The SRCC-RCB-TP-Coupler specimen, with its steel transitional section and mechanical splice (coupler) anchoring, demonstrates higher energy absorption and stiffness. Its initial stiffness of 23 is 33.5% higher than that of the SRCC-RCB specimen (17.44) and significantly exceeds the stiffness of the RCC-RCB specimen (10.4), highlighting the significant impact of these parameters.
- Although sustainability was not the primary focus of this study, the proposed detailing strategies may indirectly support sustainable or net-zero construction by improving the service life and reducing long-term maintenance needs. Enhanced durability and ease of retrofitting can minimize resource consumption over the structure's lifespan.

Acknowledgements. This research was funded by Semnan University, research grant No. 1403993.

References

1. El-Tawil S, Deierlein GG (1999) Strength and ductility of concrete encased composite columns. J Struct Eng 125:1009–1019
2. Wang G-Y, Zhang C, Xu J, Zhang D-M (2020) Post-fire seismic performance of SRC beam to SRC column frames. Structures 25:323–334
3. Mirza SA, Lacroix EA (2004) Comparative strength analyses of concrete-encased steel composite columns. J Struct Eng 130:1941–1953
4. Chen C-C, Chen C-C, Shen J-H (2018) Effects of steel-to-member depth ratio and axial load on flexural ductility of concrete-encased steel composite columns. Eng Struct 155:157–166
5. Shanmugam NE, Lakshmi B (2001) State of the art report on steel–concrete composite columns. J Constr Steel Res 57:1041–1080
6. Rong B, Sun J, Xu M, Zhang R, Sun Y, Zhang W, Zhang W (2021) Experimental and numerical research on seismic performance of S-RC-SRC composite frame. J Build Eng 43:103119
7. Li C-H, Liu XC, Zhang A-L, Chen X, Zhang D, Ren Y-Z, Zhang J, Li Z (2024) Axial compressive strength of concrete encased extended T-shaped steel composite columns with less reinforcement. J Build Eng 82:108207
8. Jain A, Sahoo DR, Jain AK (2024) Lateral cyclic behavior of SRC-RC hybrid columns bending around strong and weak axes: effects of embedment lengths. J Perform Constr Facil 38:4023062
9. Gautham A, Sahoo DR (2022) Seismic collapse performance of soft-story RC frames with SRC columns designed using the energy-based method. J Perform Constr Facil 36:4022062
10. Wei J, Xue J, Hu Z, Qi L, Xu J (2024) Experimental study on dynamic shear-behavior of H-shaped SRC columns under lateral impact loading. J Constr Steel Res 213:108396
11. Pan P, Lin X, Lam A, Chen H, Ye L (2014) Monotonic loading tests of ring-beam connections for steel reinforced concrete columns and RC beams. J Struct Eng 140:4013092
12. ACI Committee (2005) Acceptance criteria for moment frames based on structural testing and commentary. ACI 374:1–5
13. Park R (1989) Evaluation of ductility of structures and structural assemblages from laboratory testing. Bull New Zeal Soc Earthq Eng 22:155–166

14. Zhang J, Pei Z, Rong X (2022) Experimental seismic study of an innovative precast steel–concrete composite beam–column joint. Soil Dyn Earthq Eng 161:107420

Retrofitting Office Buildings in Norway: Challenges and Innovations

Emma Zheng Liang(✉) and Arne Ronny Sannerud

Høyskolen for Yrkesfag, Gjøvik, Norway
{emma.liang,arne.ronny.sannerud}@hfy.no

Abstract. Norway aims to reduce greenhouse gas emissions by 55% by 2030 compared to 1990 levels, and to transition to a low-emission society by 2050. Retrofitting existing office buildings plays a crucial role in achieving these targets these targets by reducing energy use and carbon footprints. However, there is limited research that systematically examines how to coordinate different systems to reduce carbon emissions particularly in areas such as building processes for controlling climate gas emissions related to energy efficiency, integration of technical installations, and material reuse.

This study explores the challenges and innovative solutions involved in upgrading office building to meet sustainability goals, with a focus on the process quality of system integration. Using a case study approach, it identifies best practices and potential barriers. The findings offer practical insights to support and accelerate Norway's transition to a low carbon built environment.

Keywords: Retrofitting Office Buildings · energy efficiency · building processes · process quality

1 Introduction

According to a report by the Intergovernmental Panel on Climate Change (IPCC), the United Nations climate science body in October 2018, concluded that the impacts and risks associated with global warming are significantly greater at 2 °C than to 1.5 °C. To limit global temperature rise to 1.5 °C, the report calls for rapid reductions in CO_2 emissions by 2030 [1]. In response, Norway has set a target to reduce emissions by at least 55% by 2030 compared to 1990 levels. This commitment is set in the Norwegian Climate Change Act, which establishes national climate targets for 2030 and 2050, and serves as the legal foundation for climate policy. The Act further aims to facilitate the transition to a low emission society, with the aim of achieving a 90–95% reduction in emissions by 2050 compared to 1990 levels [2]. Gholipour et al. [3] pointed out the building industry is often referred to as the '40% sector', but it more precisely accounts for around 38% of global CO_2 emissions and 35% of total energy consumption, with buildings alone accounting for around 55% of electricity use according to United Nations Environment Programme in 2020 [4]. In Norway it demonstrates a different emission profile compared with other European countries which CO_2 emissions are associated with operational energy

H. Beushausen et al. (Eds.): NTZR 2025, LNCE 873, pp. 609–617, 2026.
https://doi.org/10.1007/978-3-032-27698-8_54

use. Norwegian buildings have relatively lower operational emissions due to its near 100% renewable electricity production which primarily from hydropower compared to other European [5]. Consequently, embodied emissions from materials such as concrete, steel, and insulation now represent the dominant source of life cycle CO_2 emissions in Norwegian buildings [6]. This again emphasize the importance for decarbonizing buildings in retrofitting of existing buildings with minimizing material-related emissions. To achieve the goal of decarbonizing both material-related and operational emissions, effective and interdisciplinary collaboration is essential throughout the entire building process. The quality and integration of processes across the project lifecycle play an important role to reach the goal of reducing CO_2 emissions.

This paper explores the challenges and innovations associated with retrofitting an existing office building with a particular focus on the process quality throughout the project. By combining findings from the case study, the paper aims to provide practical insights into how process quality can support effective retrofitting projects. The structure of the paper is as follows: Sect. 2 provides background information on the context and significance of retrofitting office buildings in Norway. Section 3 presents the selected case study, highlighting critical process steps and quality factors essential for project outcomes. Finally, Sect. 4 discusses key findings, identifies practical barriers encountered in the field, and concludes with implications for future research and practice.

2 Retrofitting Office Buildings in Norway

The United Nations Environment Programme (UNEP) 2020 Global Status Report for Buildings and Construction highlights retrofitting existing buildings as a key strategy to reduce energy use and emissions in the sector, which accounts for 37% of global energy-related emissions. Retrofitting through for example better insulation, efficient windows, reduce material using and modern Heating, ventilation, and air conditioning (HVAC) systems contributes to support the transitioning to a zero-emission, energy efficiency, and resilient built environment [4]. The latest version of the Norwegian building regulations (TEK17), which applies to both for new construction and renovation projects, mandates the calculation of greenhouse gas emissions in accordance with the Norwegian Standard NS 3720:2018. This standard provides a structured methodology for assessing the climate impact of buildings over their life cycle, including materials, energy use, and end-of-life scenarios [7].

In Norway, the carbon emission profile of buildings contrasts notably with that of many other European countries. This is largely due to Norway's electricity production, which is renewable predominantly sourced from hydropower. According to the International Energy Agency (IEA), approximately 92% of Norway's electricity generation in 2020 came from hydropower, resulting in one of the lowest carbon intensities for operational energy use in Europe [8]. Norwegian buildings exhibit significantly lower operational emissions compared to buildings in countries with more carbon-intensive energy supplies. This shift in emissions profile means that embodied carbon emissions associated with the extraction, manufacturing, and transport of construction materials constitutes a larger share of the total life-cycle emissions. Recent analyses show that in countries like Norway, Sweden, and Canada, the distribution of embodied versus operational emissions in new buildings is approaching a 50/50 ratio [9]. This highlights the critical importance of retrofitting strategies must therefore go beyond traditional energy upgrades and include circular design principles, such as reuse of existing structural components, integration of low-carbon materials, and implementation of Environmental Product Declarations (EPDs). These strategies are aligned with international frameworks, such as the EU taxonomy and the Paris Agreement, which emphasize life-cycle-based emissions reductions [10].

Despite increasing awareness of the importance of retrofitting importance, retrofit projects in Norway often encounter several barriers. These include technical challenges related to an aging building stock, the absence of comprehensive digital documentation, and incompatibilities between old systems and modern technologies. Furthermore, as the building industry undergoes a green transition, traditional professionals require updated knowledge and ongoing learning to adapt to new sustainability-driven practices. Economic and 'old project management' constraints also present challenges, including fragmented project delivery methods [11]. Addressing these obstacles calls for an effective and structured approach that fosters better coordination among project stakeholders and integrates sustainability knowledge throughout the process.

Innovative solutions are beginning to emerge in response to these challenges, such as BREEAM-NOR certification are often used to guide sustainable decision-making throughout the renovation process. Case study from Norway illustrate how public building owners are leveraging cross-disciplinary collaboration to reduce emissions [12].

3 Case Study: Office Project in the Centrum of Oslo

This study explores the challenges and innovative solutions involved in retrofitting an existing office building with a particular focus on the process quality throughout the project. While there is increasing interest in reducing emissions from the built environment, few studies have systematically examined how different technical systems are coordinated during the retrofit process to reduce greenhouse gas emissions. This case study contributes by looking into how energy efficiency measures, integration of technical installations, and material reuse are handled in practice, with specific attention to timing and process flow of real building project.

The analysis is based on document reviews, climate gas calculations, and a life-cycle carbon assessment from a real building project. The findings show that early clarification of system integration, especially linking material using, energy supply, control systems, HVAC and monitoring contributed to clear and traceable carbon outcomes. Reuse of structural elements, such as steel and facade components, also led to measurable reductions in embodied emissions. In addition, the use of a project-specific GHG budget and regular monitoring helped ensure that all disciplines followed up on the carbon targets throughout the process.

Although the study is limited to one case, the detailed process analysis gives practical and transferable insights. It demonstrates that material reuse can be both economically viable and environmentally effective, but only if supported by a structured process starting with early target-setting and clear decision-making. These results can support future projects aiming to improve climate performance and point to the need for further research to follow up across a broader selection of retrofit cases.

In this study, "process quality of system integration" refers to how well the technical systems, such as material use, energy supply, HVAC, automation, and monitoring are coordinated throughout the project phases. It is not just about the final performance outcome, but also about how the coordination is planned, followed up, and adjusted during the process. This includes when key decisions are made, how responsibilities are shared, and whether the systems are compatible both technically and operationally.

In this case study, we looked at how timing, role distribution, and technical alignment influenced the project's ability to reduce greenhouse gas emissions. This also includes how well the project team worked across disciplines to align decisions with the carbon budget and reuse goals.

While the concept is not fully developed in existing literature, this paper uses the case to show why the integration process matters, and how it can affect both energy performance and carbon outcomes. The observations in this project can help to better understand what kind of coordination is needed to reach sustainability targets in retrofit project.

3.1 An Introduction to the Project

Located in central Oslo, the retrofitting office project is a 3400 m^2, six-story building currently undergoing extensive refurbishment to accommodate new tenants. The project owners have set ambitious targets for material reuse and circularity, aiming to minimize environmental impact. To achieve these goals, the project team has prioritized sustainable design and construction practices, targeting BREEAM-NOR [26] v6.0 Very Good certification and compliance with FutureBuilt [28] criteria for circular buildings (Fig. 1).

Upon completion in March 2024, the project successfully reused approximately 50 metric tons of materials excluding the main structural components of the existing building that were retained, such as the load-bearing structure, roof, facade, and floor slabs. The reuse of these major elements was the primary factor in achieving a reduction of over 60% in CO_2 emissions from building materials compared to the construction of a new office building. Decisions regarding material and product reuse were based not only on their potential to reduce emissions but also on cost and effort required.

4 Analysis: The Building Process for the Case Study

It is worth noting that certain BREEAM-NOR credits cannot be achieved unless they are addressed in accordance with the specific project stages. Each relevant BREEAM-NOR issue specifies the corresponding Bygg21 Fasenorm [29] step in which it must be considered to ensure compliance and eligibility.

Following the structured process defined in BREEAM-NOR, this retrofit office project was developed across the following sequential stages: strategic definition, preparation and brief, concept design, technical design, construction, handover and close-out, and in use (Fig. 2). This alignment ensured that sustainability requirements including material reuse, emission accounting, and verification were systematically integrated and evaluated within the appropriate phase of delivery.

Fig. 1. Case study: project in the centrum of Oslo

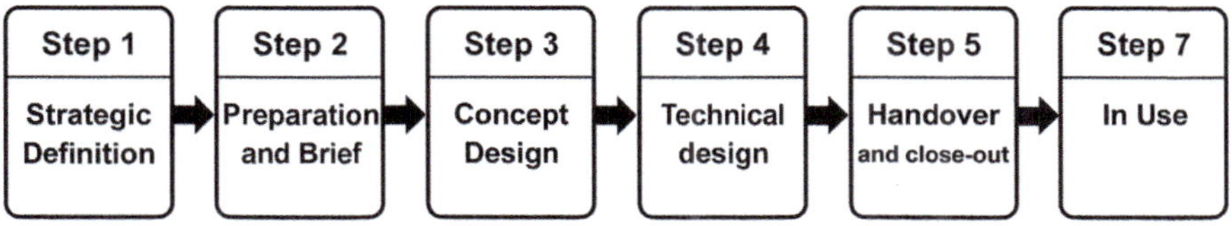

Fig. 2. REEAM-NOR assessment stages in relation to the project work stage

From the initial stages, the project team established a comprehensive climate gas—GHG budget, calculated in accordance with NS 3720:2018 [13]. This budget encompassed emissions from the building's full life cycle including material production, construction, operation, and end-of-life and was finalized by the schematic design phase (step 3), aligning with BREEAM-NOR v6.0 standards [14].

The comprehensive greenhouse gas account that calculates emissions related to the following modules during the life of the building according to NS 3720:2018:

- Production of materials, A1–A3
- Transport of materials to the construction site, A4
- Energy use on the construction site, A5
- Future replacements, B4
- Energy consumption in operations, B6

- Transport in operations, B8
- End of life cycle, C1–C4.

Establishing this budget early enabled transparent benchmarking of sustainability goals and informed decision-making throughout the project.

The GHG budget served as the foundation for developing targeted sustainability strategies. It transformed climate goals into measurable and actionable design parameters. Early target-setting also clarified roles and responsibilities across the design, engineering, and construction disciplines, fostering collaboration around a shared carbon framework. Moreover, it enabled trade-off assessments between emissions, cost, and performance, ensuring that environmental ambitions remained feasible within project constraints [15].

During the concept phase, the team conducted early GHG calculations to evaluate alternative design approaches. These life-cycle assessments (LCAs) compared different structural systems, envelope materials, and foundation solutions. For instance, retaining rather than replacing the existing foundations was shown to yield significant emission savings over the building's service life. All calculations adhered to NS 3720:2018, enabling consistent comparison and evidence-based decision-making [16].

As the project transitioned into technical design, the initial GHG principles were formalized into concrete specifications. These included maximum allowable CO_{2e} per material unit and mandatory use of third-party Environmental Product Declarations (EPDs) for critical materials [17]. Procurement documents were aligned accordingly, embedding carbon performance requirements into contractor responsibilities and supplier selection criteria [18].

Throughout construction, strict compliance with the climate strategy was maintained. The contractor implemented a detailed material sourcing plan, rooted in the early-stage carbon targets. This plan outlined specific thresholds for emissions, priorities for reuse, and documentation requirements. Crucially, major structural components such as the load-bearing system, roof, facade, and slabs were retained, substantially lowering embodied emissions [19].

Approximately 50 metric tons of additional materials were reused beyond the retained structure. Although smaller in quantity, these materials supported the project's broader circular economy goals. The reuse of primary structural elements alone accounted for most emissions reductions, enabling the project to exceed its goal of a 60% reduction in embodied GHG emissions compared to a new-build baseline [20]. This outcome illustrates the central role of structural reuse in climate-smart retrofit strategies.

Following completion, a GHG performance review was conducted using as-built data to reassess emissions from materials, transport, and site activities. This post-construction analysis, conducted in line with NS 3720:2018, verified whether the project had met its original targets. Any deviations were documented and reported to the client, accompanied by explanations and mitigation efforts, reinforcing a culture of accountability [21].

This structured, target-driven process significantly improved decision transparency and environmental quality assurance. By embedding GHG monitoring throughout each phase, the project avoided carbon lock-in and maintained alignment with evolving goals. The early carbon budget also served as a tangible tool to balance sustainability with economic and technical considerations, ensuring that climate objectives were practical and achievable [22].

The approach outlined in this project provides a replicable framework for reducing carbon emissions in building retrofits. It highlights the importance of integrating life-cycle carbon budgeting, performance-based specifications, and post-construction verification into mainstream project workflows. Particularly in dense urban environments, where reuse potential is high, this method offers both environmental and economic advantages [23].

CO_2 emissions remain a critical metric in evaluating the climate impact of construction projects. This retrofit project indicate a total of 97,293 kg CO_{2eq} within the boundaries set by TEK17, encompassing stages A1-A3 (material production), A4 (transport to site), and B4 (material replacement over a 60-year life span) [24]. With a gross floor area of 3400 m^2, this translates to 0.48 kg CO_{2eq}/m^2/year. This is significantly lower than BREEAM-NOR v6.0's reference value of 6.8 kg CO_{2eq}/m^2/year, equating to an estimated 93% reduction as of November 2023 [14]. The CO_2 accounting was performed by Rambøll Norway using the OneClick LCA tool, which aligns with FutureBuilt's methodology for reused materials. According to this method, 20% of emissions from A1–A3 are attributed to a reused material when replacing new materials of the same type [25].

Beyond measurable carbon reductions, the project generated valuable knowledge for the building sector. As expectations and client demands for low emission buildings increase, this project exemplifies how process quality and early carbon integration can deliver tangible climate outcomes. The main contributor to this substantial emissions reduction was the retention of the existing load-bearing structure and envelope. While material reuse played an important role in reducing emissions, it was not the sole determining factor in achieving the project's ambitious CO_2 targets. The case also demonstrates the importance of tracking emission and material selection early in the process.

5 Conclusion

Through analysis of reuse in the building industry and case study, retrofit project case study in Norway, it can be funded that material reuse is gradually gaining notice. This example demonstrates that significant reuse can be both economically viable and environmentally impactful. The project has shown that with structured planning, early carbon target setting, and integration of reuse criteria across all design and construction phases, meaningful climate benefits can be achieved without compromising functional or financial feasibility.

This study aimed to explore how carbon reduction targets can be integrated into the retrofit process of an existing Norwegian office building, with a specific focus on process quality and system coordination. The project followed the defined BREEAM-NOR structure from early phase to completion and applied a project-specific GHG budget based on NS 3720:2018 to support decision-making throughout.

The main finding is that early establishment of a GHG budget, combined with continuous monitoring and integration across all project stages, was critical for achieving the project's ambitious climate goals. The project documented a 93% reduction in embodied emissions compared to a new-build baseline, mainly due to the retention of the existing structure and reuse of key materials.

Although this is a single case study, the insights are relevant for similar Norwegian projects, especially in dense urban areas where reuse potential is high. Future research should include more cases to better understand which factors support or limit implementation across different contexts. Scaling up reuse and emissions tracking will also depend on developing better infrastructure for storage, testing, and trading of reclaimed materials.

This case shows that ambitious climate goals are possible in retrofit projects, as long as carbon performance is integrated from the beginning and followed up with real decisions, not just intentions. The adoption of life-cycle-based GHG budgets and consistent monitoring contributed substantially to process quality. It enabled precise tracking of sustainability performance and ensured alignment with established targets. Notably, the success of the project relied not only on material reuse itself but also on a process quality that supported transparent decision-making and accountability.

While TEK17 [26] has taken an important step by requiring GHG calculations for larger new buildings, the current regulation only asks for submit a GHG report based on NS 3720:2018. Without structured follow-up, early ambitions risk being lost along the way. To ensure that documented emissions reflect actual outcomes, it would more effective be revised to include process-based requirements such as stage-by-stage carbon budgeting, verification of key decisions, and final performance checks at project close-out. Integrating these elements into regulation would strengthen accountability, close the gap between intention and result, and support Norway's broader climate goals in the building sector.

It is important to note that the research is still in its early stages. Conducting a more comprehensive survey in the future to accurately track greenhouse gas emissions across different stages would provide more precise and meaningful insights.

References

1. IPCC special report on the impacts of global warming. https://www.ipcc.ch/sr15/
2. The official site of Norwegian climate change and policy. https://www.regjeringen.no/no/tema/klima-og-miljo/innsiktsartikler-klima-miljo/klimaendringer-og-norsk-klimapolitikk/id2636812/
3. Gholipour HF, Arjomandi A, Yam S (2022) Green property finance and CO_2 emissions in the building industry. Glob Finance J 51. https://doi.org/10.1016/j.gfj.2021.100696
4. United Nations Environment Programme (2020) 2020 Global status report for buildings and construction: towards a zero-emission, efficient and resilient buildings and construction sector. Tilgjengelig fra: https://globalabc.org/sites/default/files/inlinefiles/2020%20Buildings%20GSR_FULL%20REPORT.pdf. 3 Apr 2022
5. Statistics Norway (2021) Energy balance and energy use in buildings, 2020. SSB, Oslo
6. Miljødirektoratet (2020). Klimagassutslipp fra byggeplasser og materialbruk i byggsektoren. Norwegian Environment Agency, Oslo
7. Norwegian Building Regulations (TEK17). https://www.dibk.no/regelverk/byggteknisk-forskrift-tek17/17/17-1

8. IEA (2022) Norway 2022. IEA, Paris. https://www.iea.org/reports/norway-2022, Licence: CC BY 4.0
9. Embodied carbon vs. operational carbon. https://oneclicklca.com/en/resources/articles/embodied-carbon-vs-operational-carbon. 10 Feb 2025
10. European Commission (2021) EU taxonomy for sustainable activities. Brussels
11. Arge K, Hjelseth E (2020) Barriers to sustainable retrofit in the Norwegian context. J Green Build 15(4):89–105
12. Statsbygg (2021) Miljøstrategi for Statsbyggs eksisterende byggportefølje. Oslo
13. Standard Norge (2018) NS 3720:2018—Methods for greenhouse gas calculations for buildings. Standards Norway, Oslo
14. Grønn Byggallianse (2021) BREEAM-NOR v6.0 manual. Norwegian Green Building Council, Oslo
15. Pomponi F, Moncaster A (2016) Embodied carbon mitigation and reduction in the built environment—what does the evidence say? J Environ Manage 181:687–700
16. Passer A, Kreiner H, Maydl P (2012) Assessment of the environmental performance of buildings: a critical evaluation of the influence of technical building equipment on residential buildings. Int J Life Cycle Assess 17(9):1116–1130
17. Säynäjoki A, Heinonen J, Junnila S (2012) A scenario analysis of the life cycle greenhouse gas emissions of a new residential area. Environ Res Lett 7(3):034037
18. Giesekam J, Barrett JR, Taylor P (2016) Construction sector views on low carbon building materials. Build Res Inf 44(4):423–444
19. Eberhardt LCM, Birkved M, Birgisdóttir H (2020) Building design and construction strategies for a circular economy. Archit Eng Des Manag 16(1):16–31
20. Cederquist K, Wallbaum H, Baudire H (2021) Life cycle assessment of building refurbishment: a comparative analysis of scenarios. Energy Build 231:110566
21. Hollberg A, Ruth J (2016) LCA in architectural design—a parametric approach. Int J Life Cycle Assess 21:943–960
22. Fufa SM, Skaar C, Haavik TK (2022) Integrated emission budget planning for low-carbon retrofitting of existing buildings. Energy Build 264:112079
23. Andersen CE, Bjørberg S (2019) Strategies for achieving low-carbon renovation in Norwegian office buildings. In: Proceedings of SBE19. Graz, Austria
24. Direktoratet for byggkvalitet (2017) Byggteknisk forskrift (TEK17). DIBK, Oslo, Norway. Available at: https://dibk.no
25. FutureBuilt (2020) Klimagassregnskap for bygg – Metodebeskrivelse. FutureBuilt, Oslo
26. Norway's Technical Building Regulations (TEK17) § 17-1: Greenhouse gas account. https://www.dibk.no/regelverk/byggteknisk-forskrift-tek17/17/17-1?_t_q=KLIMAGASS. Last accessed 10 Sep 2025

Professionals' Awareness and Application of Industry 4.0 Technologies in Nigeria

Olatunji Joseph Oladiran[1](✉), Olusegun A. Oguntona[1], and Mohammed Elelu-Bashir[2]

[1] Faculty of Engineering, Built Environment and Information Technology, Walter Sisulu University, Mthatha, South Africa
tungybox2000@gmail.com, ooguntona@wsu.ac.za

[2] Department of Building, University of Lagos, Lagos, Nigeria

Abstract. Industry 4.0 aims to create advanced building sites, virtual environments, and simulated storage systems that allow contractors to manage and analyze data from various stages of construction and from clients after the building project is complete. The objectives are to determine the awareness of Nigerian construction professionals of Industry 4.0 technologies; and to evaluate the extent to which the technologies are used in the construction industry in Nigeria. Cross sectional survey research was adopted to carry out the study in Lagos State. The population of the study is construction professionals in Lagos State, Nigeria. Convenience sampling technique was employed to select a sample of 110 respondents. Frequency, percentage, mean and rank were used to analyze the data. The finding reveals that construction professionals are somewhat aware but rarely used Industry 4.0 technologies. The study concludes that there is variability in the awareness levels across different Industry 4.0 technologies, indicating potential areas for further education and communication to enhance professionals' understanding of these technologies. It is recommended that the awareness of Industry 4.0 technologies should be improved. This can be done with adequate education and campaign of the technologies.

Keywords: Application · Awareness · Industry 4.0 · Innovation · Technology

1 Introduction

A strong construction industry is crucial for a country's overall economic growth, as it contributes to meeting important societal needs like providing housing, building necessary infrastructure, and creating employment opportunities [1]. Construction is one of the world's largest industries [2]. [3] asserts that the construction industry is a major contributor to the global economy, accounting for 10% of the world's GDP. As it is needed for nearly all types of services, every nation, including Nigeria, has a robust construction sector. Given its significant impact on the economy and resources, it is essential to plan and execute actions in this industry with great care. One of the biggest issues the construction sector faces is the failure of projects to be completed within their allocated budget and timeframe [4]. There is a strong need to improve productivity and lower costs

H. Beushausen et al. (Eds.): NTZR 2025, LNCE 873, pp. 618–629, 2026.
https://doi.org/10.1007/978-3-032-27698-8_55

in the construction industry [5, 6]. The growing demand for affordable housing, social infrastructure, and transportation, driven by population growth and economic expansion, has increased the importance of raising productivity as it is a key indicator of a country's quality of life and potential for future growth [7]. Industry 4.0 has brought tremendous improvement to construction [8, 9]. The technology is characterized by a shift towards digitization and the use of smart production methods. Critical elements of construction 4.0 include cyber-physical systems (CPS), the Internet of things (IoT), the Internet of services (IoS), artificial intelligence (AI), big data, and smart production, which can help to improve the operation of the construction industry. Integrating elements of industrial revolution 4.0 into construction projects involves using IoT to combine data from multiple sources, and implementing advanced technology such as laser scanners, drones, and 3D printers, with the goal of improving the management of construction projects throughout the design, construction, and in-use stages, resulting in more sustainable and safe outcomes [10]. Hence, these technologies can be employed to minimize a number of challenges confronting the Nigerian construction sector. Researches are currently ongoing in Nigeria on this concept but there is still a lot of vacuum to be filled. Thus, the aim of this study is to investigate the awareness and application of Industry 4.0 in Nigeria. The objectives include to determine the awareness of Nigerian construction professionals of Industry 4.0 technologies to evaluate the extent to which these technologies are used. The significance of this study lies in the improvement of the construction industry's productivity via technology-driven processes.

2 Literature Review

The application of construction 4.0 technologies in the construction industry can be categorized into three, including smart factory, simulation and modelling, and digitization and visualization [11].

2.1 Smart Factory

[11] uses an image of a "Smart Factory" to demonstrate the use of automation technology. [9] explain that a smart environment is equipped with various technical devices such as sensors, readers, and computers that are seamless and easy to interact with for the users. An example is a room with automatic lighting that activates when the light intensity is low and people are present. The cluster suggests certain construction 4.0 technologies that can be beneficial in the construction industry, including:

Cyber-Physical-Systems (CPS). [6] defines CPS as large-scale systems that are geographically distributed, composed of different elements and are critical for life, in which embedded devices such as sensors and actuators are networked to detect, monitor, track and regulate the physical environment. These systems enable a shift from virtual to real-world scenarios. By using sensors, the cyber world collects data and creates a platform for sharing it, which facilitates advancements in construction process management, as-built documentation, and sustainable building techniques [12]. [11] states that they are used in construction to validate project prototypes and to coordinate the use of virtual models with actual construction.

Radio-Frequency-Identification (RFID). Many industries, services, scientific, and social spheres have a need for automated element identification and data collection without human input [6]. In response, multiple technologies have been developed and implemented, one of which is radio frequency identification (RFID). RFID uses electromagnetic impulses to transmit data [12]. The system includes a reader and a tag. The magnetic field produced by the reader activates the tag when it is in close proximity or within the range of the reader. RFID solutions can be used to automate the construction process, track and manage construction assets such as tools, materials, and equipment [11]. RFID can also be used for inventory management and theft prevention. By enabling equipment and machines to automatically transmit performance data to engineers, RFID may help the construction sector move away from reactive maintenance [10].

The Internet of Things and Services (IoT and IoS). Another key concept in construction 4.0 that drives the creation of virtual networks for smart manufacturing environments is the Internet of Things (IoT) and Internet of Things Services (IoS) [6]. IoT is a network of physical objects that are digitally connected to sense, monitor and interact within a company and between the company and its supply chain, which enables agility, visibility, tracking, and information sharing to facilitate timely planning, control and coordination of supply chain processes [13]. While the IoS is a new business model that will transform how services are provided and enable higher value creation as the IoT advances. IoS involves offering services with the aid of internet service [14]. IoT and IoS in construction create virtual networks to support a smart factory environment and help move away from reactive maintenance to predictive maintenance, where machines can be repaired before damage occurs [11]. Applying IoT technology in oil and gas extraction, mining, and building sites improves health and safety conditions, monitors the condition of equipment, and optimizes operations [6]. IoT implementation in the construction industry involves embedded sensors on real-world objects like cars, robots and building components that can connect to the internet. Equipment with sensors, such as cranes, dozers and loaders, transmit performance data to engineers via the internet for analysis [15]. Engineers can anticipate and prevent impending equipment breakdowns based on data received.

Additive Manufacturing. Additive Manufacturing is a relatively new technology that can be used for modularization [6]. The main advantage of this technology is the ability to create building components based on digital models, such as CAD data. [11] highlights that using additive manufacturing allows for the automation of even the most complex 3-D printed architectural components. 3D printers can directly read specifications for component parts from a digital model and print them layer by layer until the elements are completed, without the need for additional labor. [16] suggest that using 3D printing may reduce the overall cost of concrete in construction projects by 35–60% by eliminating the need for a framework.

Robotics. Robotics can be used to automate the assembly of steel beams in construction, replacing human labor [11]. Robotics has been adapted for use in the construction industry to address various challenges, such as planning routes for material transportation and automating operations at construction sites [9]. Robotics is being used for tasks such as examining civil structures, assembling wooden structure elements, installing steel, and building brick or concrete blocks, among others.

2.2 Simulation and Modelling

This section discusses the implementation of construction 4.0 and the role of simulation and modeling in it. Construction projects are unique and challenging in nature, and are heavily impacted by external factors such as changes in supply, unpredictable weather, and employee performance [6]. The use of second-category technology, such as simulation tools and models, can help improve project design and more efficiently manage the construction process because of the complexity of building projects [10]. This cluster primarily focuses on simulations that can be used to analyze the impact of external factors, such as weather, labor performance, and supply fluctuations, on building projects [17]. Additionally, simulations can be used to improve the design of building projects. Technologies commonly used in this cluster include project management, frameworks for resource planning, and simulation tools. Examples of some of these tools include:

Augmented/Virtual Reality (AR/VR). Virtual reality is an experience where a person is surrounded by a computer-generated three-dimensional depiction that allows them to move around, view from different perspectives, capture and manipulate it, while augmented reality is a technology that superimposes a computer-generated image on top of a user's view of the real-world [13]. Virtual reality (VR) is an artificial environment created by computer technology and software. Users can interact and become fully immersed in a 3D world through this technology [8]. It can create additional experiences that are similar to reality by stimulating multiple senses such as vision, hearing, touch and even smell in this artificial environment. Virtual reality, which is a precise virtual model of the project, represents the next step for 3D modeling in the construction industry [6]. The main difference between 3D modeling and virtual reality is that the latter allows users to physically move around inside the project, which can be used as a tool for providing risk-free safety training [11]. However, barriers to the widespread use of VR include low-cost computing power and accessibility to content [18].

Building Information Modelling. Building Information Modeling (BIM) is the most widely recognized and prevalent concept within construction 4.0 [19]. BIM is an integrated approach that significantly improves project understanding and provides predictable results, according to Autodesk. This visibility allows all stakeholders to stay organized, improve accuracy, reduce waste, and make informed decisions earlier in the process, which contributes to the success of the project [6]. BIM digitally represents a facility's structural and operational elements throughout project life cycle [20]. By providing relevant data such as cost estimation, planned project, material inventory, and technical data on construction components such as geometry and spatial relations, BIM is a shared knowledge resource that offers a reliable foundation for decisions throughout its life cycle, allowing project partners to work together effectively throughout the entire life cycle. Creating a virtual model of a building and using information such as project schedule, cost estimates, and material inventories as input, BIM also makes it possible to design and manage construction projects [11].

2.3 Digitization and Virtualization

This topic discusses the significance of digitalization technologies in the construction industry. The fourth industrial revolution generates a large amount of digital data that cannot be managed using traditional methods, therefore, specific technologies and concepts are required to manage the digital environment and value chain [5]. A few examples of these include:

Big Data. Big data refers to techniques for analyzing, extracting information from, or otherwise handling large and complex data sets that are too big for traditional data processing software. As technology advances, the world is being flooded with new data, and by utilizing statistical and algorithmic techniques, scientists have the potential to uncover valuable insights that can improve various fields [6]. Big data can be characterized by the "3Vs" - volume, velocity and variety [13]. Big data generates more data than was ever stored on the internet 20 years ago in terms of volume and velocity, and it also includes a variety of structured and unstructured data, with unstructured data being more prevalent. [6] relates these three Vs to construction as follows: Volume - a large amount of data including design data, planning data, contracts data, and others; variety - different types of data format (text, audio, video, graphs) can be found in construction, such as software formats like DWG, DGN, RVT and so on; and velocity - the dynamic nature of the construction industry and various sources of data result in a continuous flow of data. A big data solution can be used to combine all the data from devices and personnel as well as external data and make it accessible to all parties involved in a building project. The collection of historical big data allows for reducing uncertainty when making decisions [11]. Big data is mostly utilized in urban planning and management in relation to construction 4.0, where it is used to evaluate factors such as the impact of social factors on urban growth and "smart cities," which aim to improve the quality of life for citizens [9].

Cloud Computing. Cloud computing allows users to access a shared pool of personalized resources via the internet [26]. It involves outsourcing data processing and storage to a remote location, enabling multiple users to utilize the service without the need for individual licenses. This technology is widely used in the construction industry as it promotes the integration of tasks in BIM systems. [27] posit that cloud-based BIM systems can facilitate collaboration among project participants and be used in various aspects of building construction, such as creating sustainable structures and managing energy usage. Cloud computing also offers integrated services that can be accessed online, such as cross-company collaboration on building sites.

Mobile Computing. The construction industry is known for its high volume of data and its performance depends on the availability of accurate and current data. To keep stakeholders informed and to help them make wise decisions, the generated data needs to be processed, saved, and provided to them. These and other factors put pressure on the construction sector to digitize data flow and supply. Utilizing mobile computing and handheld computers gives the stakeholders access to precise, dependable, and real-time data as needed [28]. The usage of mobile devices to promote collaboration and communication during the construction process is referred to as mobile computing.

Another tool that can help the construction industry be more productive in terms of communication, hiring, project management, and client networking is social media [29].

3 Research Method

This study employs a qualitative methodology to investigate construction 4.0 technologies, their awareness and adoption in Lagos, Nigeria. The population of the study was construction professionals, including architects, builders, engineers, quantity surveyors, planners and project managers. Purposive sampling technique was used to select a sample of 100 professionals, determined by using Cochran formula for an infinite population. A questionnaire was designed to elicit information from the respondents. The questionnaire sought the respondents and organizations information; and the awareness and adoption of nine construction 4.0 technologies coined from literature using Likert scales. Frequency, percentage, mean and rank were employed to analyze the data.

4 Findings and Discussions

4.1 Demographic of Respondents

The demographic characteristics of the respondents are presented in Table 1.

Table 1 presents the demographic information of respondents who participated in the survey within the context of the construction 4.0. This information offers valuable insights into the demographic composition of the respondents, including their professional roles, educational qualifications, academic backgrounds, professional affiliations, status, and years of experience in the construction field.

4.2 Awareness of Construction Professionals of 4.0 Technologies in the Construction Industry in Nigeria

Table 2 represent a survey of respondents' perceptions regarding their level of awareness of various construction 4.0 technologies. Each concept represents a different aspect of technology that may have an impact on the construction industry. The awareness levels are measured on a scale ranging from "Not at all aware" to "Extremely aware," and the mean scores provide an average awareness level for each concept. The criterion for the selection of the technologies for awareness is any technologies with a 2.5 and above MS in Table 2. Thus, the respondents are somewhat aware of all the technologies investigated. However, here is the breakdown of the table's key points:

Big Data Analysis. The concept of using data analysis to extract insights from construction-related data appears to be relatively well understood by respondents, with the highest mean score of 3.407. This suggests that respondents are familiar with the need of analysing large volumes of data to inform decision-making in construction.

3D Printing. Respondents show awareness of 3D printing in construction, with a mean score of 3.3721. This implies they recognize the potential of additive manufacturing techniques to create three-dimensional objects in construction.

Table 1. Background information of respondents

Background information	Frequency	Percent
Designation of the respondent		
Project manager	9	10.47
Architect	10	11.63
Builder	4	4.65
Civil Engineer	27	31.40
Electrical Engineer	5	5.81
Mechanical Engineer	8	9.30
Quantity Surveyor	20	23.26
Missing	3	3.49
Total	86	100.0
Highest academic qualification		
HND	9	10.47
B. Sc./B. Tech	45	52.33
M. Sc./MBA	24	27.91
PhD	3	3.49
Missing	5	5.81
Total	86	100.0
Project manager	3	3.49
Architect	10	11.63
Builder	9	10.47
Civil Engineer	28	32.56
Electrical Engineer	5	5.81
Mechanical Engineer	10	11.63
Quantity Surveyor	21	24.42
Total	86	100.0
Professional affiliation		
NIA	10	11.63
NIOB	4	4.65
NIQS	24	27.91
NSE	39	45.35
NA	3	3.49
PMP	6	6.98

(*continued*)

Table 1. *(continued)*

Background information	Frequency	Percent
Total	86	100.0
Professional status		
Graduate member	22	25.58
Corporate member	57	66.28
NA	3	3.49
Student	4	4.65
Total	86	100.00
Years of experience in construction		
1–5 years	24	27.91
6–10 years	37	43.02
11–15 years	21	24.42
16–20 years	4	4.65
Total	86	100.00

Table 2. Level of awareness of construction workers regarding construction 4.0 technologies in the construction industry in Nigeria

Industry 4.0 technologies	1	2	3	4	5	MS	Rank
Big Data Analysis:	2	18	26	23	17	3.41	1
3D Printing	4	23	13	29	17	3.37	2
Cloud Computing	1	22	31	21	11	3.22	3
Simulation:	7	26	14	20	19	3.20	4
IoT	6	28	20	11	21	3.15	5
Machine Automation	2	24	30	22	8	3.11	6
Augmented Reality	8	28	13	22	15	3.09	7
Cybersecurity:	10	27	18	11	20	3.05	8
Artificial Intelligence:	2	33	20	23	8	3.02	9

Note 1 represents Not Aware; 2 represents Slightly Aware; 3 represents Somewhat Aware; 4 represents Moderately Aware; 5 represents Extremely Aware; and MS represents Mean Score

Cloud Computing. Respondents seem to be relatively aware of cloud computing's relevance to construction, as indicated by the mean score of 3.2209. This suggests that the idea of using remote servers to store and manage construction-related data is moderately understood.

Simulation. Similar to AR, respondents are somewhat aware of using computer simulations in construction (mean score of 3.2093). This involves creating virtual models to plan and visualize construction projects.

IoT (Internet of Things). The respondents show some level of awareness for this concept, with a mean score of 3.1512. This suggests that respondents have some awareness of IoT's potential applications in construction, such as using sensors and connectivity to gather and exchange data.

Machine Automation. Respondents exhibit a somewhat level of awareness of machine automation, with a mean score of 3.1163. This indicates that the concept of using programmable machines in construction processes is also familiar to the respondents.

Augmented Reality (AR). The respondents' awareness of AR's potential in construction has a mean score of 3.093. This implies that while they have some understanding of AR's application in creating enhanced digital overlays on the real world, there is room for increased awareness.

Cybersecurity. While respondents display awareness of the importance of cybersecurity in construction (mean score of 3.0465), this could indicate room for further education about the risks and measures to protect construction-related digital assets.

Artificial Intelligence (AI). The respondents' awareness of AI in construction is also similar to others with a mean score of 3.0233. While still above the midpoint of the awareness scale, this suggests that respondents might have somewhat less familiarity with AI's role in construction-related tasks.

Comparing these technologies, it becomes apparent that while somewhat awareness prevails across the board, nuances exist in the degree of understanding. Big data analytics and additive manufacturing register comparatively higher levels of awareness, possibly due to their prominent roles in the modernization of industries. Other concepts record similar level of awareness, which underscores the industry's growing awareness of industry 4.0. The finding of this study disagree with [21], which confirmed low awareness of Industry 4.0 in South India construction sector. However, the finding of this study is in congruence with [22], which discovered overall moderate awareness of railway 4.0 technologies; participants' level of awareness on 85% of all the components of railway 4.0 technologies revealed a mean score above average, though the most familiar technologies were building information modelling (BIM) and IoT, unlike big data and 3D printing in this present study.

4.3 Utilization of Industry 4.0 Related Technologies in the Nigerian Construction Industry

Table 3 sheds light on the frequency of usage of the Industry 4.0 technologies. Artificial Intelligence is the most frequently used technology based on the opinion of the respondents with a mean score 2.849. This is followed in descending order IoT, Big data, Cloud computing, Machine automation and so on; with the least as Simulation (2.000).

The baseline for the selection of the technologies for utilization is any technologies with a 2.5 and above MS in Table 3. Implicitly, the frequency of utilization of Industry

Table 3. Utilization of construction 4.0 technologies in the Nigerian construction industry

	1	2	3	4	5	MS	Rank
Artificial intelligence	10	29	20	18	9	2.84	1
Internet of things	7	43	14	4	18	2.80	2
Big data analytics	15	22	29	10	10	2.74	3
Cloud computing	25	30	14	12	5	2.33	4
Machine Automation	22	33	18	8	5	2.31	5
Cyber Security	15	44	17	10	0	2.26	6
Augmented Reality	29	24	24	9	0	2.15	7
3D Printing	27	34	12	13	0	2.13	8
Simulation	33	33	7	13	0	2.00	9

Note 1 represents Never; 2 represents Rarely; 3 represents Sometimes; 4 represents Often; 5 represents Always; and MS represents Mean Score

4.0 technologies in the Nigerian construction industry appears to be rare, except for AI, IoT and Big Data that meets the set baseline. While some technologies like IoT and AI show a higher degree of utilization, others like advanced robotics and additive manufacturing exhibit lower frequency of adoption. The overall mean scores indicate that the industry is gradually incorporating these technologies, but there is still room for further integration and advancement. This finding aligns with [22], which posit slow rate of adoption of Industry 4.0 in the construction sector in comparison with other sectors like health and manufacturing, although the scenario in the developed countries is better than those of developing countries. Similarly, [23] also opine dull adoption of Industry 4.0 in the construction industry, nonetheless, the developed countries has better progress than the developing in this regard. [24] is also in tandem with the finding of this study, which shows many companies are not utilizing Industry 4.0 presently. Additionally, this finding in some sort align with [25], which revealed that the construction industry was missing among the top three industries that employ Industry 4.0 technologies in Turkey.

5 Conclusions

Industry 4.0 technologies were investigated empirically for awareness and utilization among construction professionals in Nigeria. The following conclusions are drawn based on the findings of the study:

There is variability in the awareness levels across different Industry 4.0 technologies, indicating potential areas for further education and communication to enhance professionals' understanding of these technologies.

Industry 4.0 technologies are rarely utilized in the Nigerian construction industry. The implication is that the industry is not technology-driven adequately.

6 Recommendations

Based on the findings of the study, the following recommendations are made:

The awareness of Industry 4.0 technologies should be improved. This can be done with adequate education and campaign of the technologies.

The utilization of Industry 4.0 technologies should more frequent on construction projects. This can be enforced by policymakers, practitioners, and educators collaboratively in the Nigerian construction sector.

References

1. Danso, F.O.: Occupational health and safety issues involving casual workers on building construction sites in Ghana: a Kumasi study. In: Kwame Nkrumah University of Science and Technology (2010)
2. Zahradnik, T.: Comparison of requirements for occupational health and safety on the construction site for the Czech Republic and Norway. In: Czech Technical University (2017)
3. Munano, A.N.: Preconstruction planning: exploring the factors that influence timeliness of project completion for public sector buildings in Kenya. In: Jomo Kenyatta University (2012)
4. Babuong C (2016) Investigating pre-construction planning in the construction industry of Kenya: practitioners perspective. Int. J. Soft Comput. Eng. 6(1):38–42
5. Chowdhury, T., Adafin, J., Wilkinson, S.: Review of digital technologies to improve productivity of New Zealand construction industry. J. Inf. Technol. Constr. **24**, 569–587 (2019). https://doi.org/10.36680/j.itcon.2019.032
6. Ebrahimi, S.: Awareness of the Turkish construction industry towards industry 4.0 technologies and concepts. In: Middle East Technical University (2019)
7. Turner, C.J., Oyekan, J., Stergioulas, L., Griffin, D.: Utilizing industry 4.0 on the construction site: challenges and opportunities. IEEE Trans. Ind. Inform. **20**(2652), 1–13 (2020). https://doi.org/10.1109/TII.2020.3002197
8. Maskuriy, R., Selamat, A., Maresova, P., Krejcar, O.: Industry 4.0 for the construction industry: review of management perspective. Economies **7**(68), 1–14 (2019). https://doi.org/10.3390/economies7030068
9. Forcael, E., Ferrari, I., Opazo-Vega, A., Pulido-Arcas, J.A.: Construction 4.0: a literature review. Sustainability **19**(9755), 1–28 (2020)
10. Osunsanmi, T.O., Aigbavboa, C., Oke, A.: Construction 4.0: the future of the construction industry in South Africa. Int. J. Civ. Environ. Eng. **12**(3), 206–212 (2018)
11. Yakovyna, A.: Industry 4.0 transition strategies for Ukrainian construction SMEs. In: University of Twente, vol. 4 (2020)
12. Nilsen, S., Nyberg, E.: The adoption of Industry 4.0- technologies in manufacturing—a multiple case study. In: KTH Industrial Engineering and Management (2016)
13. Olsson, J.G., Yuanjing, X.: Industry 4.0 adoption in the manufacturing process. In: Linnaeus University (2019)
14. Pérez, D., Alarcón, F., Boza, A.: Industry 4.0: a classification scheme. In: International Joint Conference—CIO-ICIEOM-IIE-AIM (IJC 2016), July 13–15, San Sebastián, Spain (2016)
15. Lau SE et al (2019) Revolutionizing the future of the construction industry: Strategizing and redefining challenges. WIT Trans. Built Environ. 192:105–115. https://doi.org/10.2495/BIM190101
16. Lloret, E., Shahab, A.R., Linus, M., Flatt. R.J., Gramazio, F., Matthias Kohler, M. Langenberg, S.: Complex concrete structures: merging existing casting techniques with digital fabrication. Comput.-Aided Des. **60**, 40–49 (2015)

17. Alaloul, W.S., Liew, M.S., Amila, N., Abdullah, W., Kennedy, I.B.: Industrial Revolution 4.0 in the construction industry: challenges and opportunities for stakeholders. Ain Shams Eng. J. **11**(1), 225–230 (2020). https://doi.org/10.1016/j.asej.2019.08.010
18. Bardi, J.: What is Virtual Reality? VR Definition and Examples. Retrieved June 14, 2019, from https://www.marxentlabs.com/what-is-virtual-reality/ (2019)
19. Oesterreich, T.D., Teuteberg, F.: Understanding the implications of digitization and automation in the context of industry 4.0: a triangulation approach and elements of a research agenda for the construction industry. Comput. Ind. **83**, 121–139 (2016)
20. Nguyen, T.D., Adhikari, S.: The role of BIM in integrating digital twin in building construction: a literature review. Sustainability **15**(13), 10462 (2023). https://doi.org/10.3390/su151310462
21. Safar, L., Sopko, J., Dancakova, D., Woschank, M.: Industry 4.0—awareness in South South India. Sustainability **12**(8), 3207 (2020). https://doi.org/10.3390/su12083207
22. Awodele, I.A., Mewomo, M.C., Municio, A.M.G., Chan, A.P.C., Darko, A., Taiwo, R., Olatunde, N.A., Eze, E.C., Awodele, O.A.: Awareness, adoption readiness and challenges of railway 4.0 technologies in a developing economy. Heliyon **10**, e25934 (2024)
23. Sepasgozar, S.M.E.: Digital technology sutilisation decisions for facilitating the implementation of industry 4.0 technologies. Constr. Innov. **21**(3), 476–489 (2021)
24. McDermott, O., Nelson, S., Antony, J., Sony, M.: Industry 4.0 readiness in west of Ireland small and medium and micro enterprises–an exploratory study. Qual. Manag. J. **30**(2), 105–120 (2023). https://doi.org/10.1080/10686967.2023.2171325
25. Sarı, T., Güleş, H.K., Yigitol, B.: Awareness and readiness of industry 4.0: the case of Turkish manufacturing industry. Adv. Prod. Eng. Manag. **15**(1), 57–68 (2020)
26. Mittai, A.: Cloud computing. Retrieved on 12 th June, 2025 from: https://www.researchgate.net/publication/392400831_Cloud_computing (2025)
27. Onungwa I, Uduma-Olugu N, Shelden DR (2021) Cloud BIM technology as a means of collaboration and project integration in smart cities. SAGE Open **11**(3):1–9
28. Mercom.: Mobile computing uses in different industries. Retrieved on 12th June, 2025 from: https://www.mercomrepair.com/mobile-computing-uses-in-different-industries (2024)
29. Silvero, M., Renukappa, S., Suresh, S., Donastorg, A.: Mobile computing in the construction industry: main challenges and solutions. In: Leadership, Innovation and Entrepreneurship as Driving Forces of the Global Economy (2016). https://doi.org/10.1007/978-3-319-43434-6_8

GPSR Compliance

The European Union's (EU) General Product Safety Regulation (GPSR) is a set of rules that requires consumer products to be safe and our obligations to ensure this.

If you have any concerns about our products, you can contact us on ProductSafety@springernature.com

In case Publisher is established outside the EU, the EU authorized representative is:

Springer Nature Customer Service Center GmbH
Europaplatz 3
69115 Heidelberg, Germany

Batch number: 10844685

Printed by Printforce, the Netherlands

Lecture Notes in Civil Engineering 873

Lecture Notes in Civil Engineering (LNCE) publishes the latest developments in Civil Engineering—quickly, informally and in top quality. Though original research reported in proceedings and post-proceedings represents the core of LNCE, edited volumes of exceptionally high quality and interest may also be considered for publication. Volumes published in LNCE embrace all aspects and subfields of, as well as new challenges in, Civil Engineering. Topics in the series include:

- Construction and Structural Mechanics
- Building Materials
- Concrete, Steel and Timber Structures
- Geotechnical Engineering
- Earthquake Engineering
- Coastal Engineering
- Ocean and Offshore Engineering; Ships and Floating Structures
- Hydraulics, Hydrology and Water Resources Engineering
- Environmental Engineering and Sustainability
- Structural Health and Monitoring
- Surveying and Geographical Information Systems
- Indoor Environments
- Transportation and Traffic
- Risk Analysis
- Safety and Security

To submit a proposal or request further information, please contact the appropriate Springer Editor:

- Pierpaolo Riva at pierpaolo.riva@springer.com (Europe and Americas);
- Swati Meherishi at swati.meherishi@springer.com (Asia—except China, Australia, and New Zealand);
- Wayne Hu at wayne.hu@springer.com (China).

All books in the series now indexed by Scopus and EI Compendex database!

Hans Beushausen · Mahdi Kioumarsi ·
Behrouz Shafei
Editors

Proceedings of the 2nd International Conference on Net-Zero Built Environment

Innovations in Materials, Structures, and Management Practices (Net-Zero Future 2025)

Set 2

Springer

Editors
Hans Beushausen
Department of Civil Engineering
University of Cape Town
Cape Town, South Africa

Mahdi Kioumarsi
Department of Built Environment
Oslo Metropolitan University
Oslo, Norway

Behrouz Shafei
Department of Civil, Construction
and Environmental Engineering
Iowa State University
Ames, IA, USA

ISSN 2366-2557 ISSN 2366-2565 (electronic)
Lecture Notes in Civil Engineering
ISBN 978-3-032-27697-1 ISBN 978-3-032-27698-8 (eBook)
https://doi.org/10.1007/978-3-032-27698-8

This work was supported by the University of Cape Town.

This Springer imprint is published by the registered company Springer Nature Switzerland AG
The registered company address is: Gewerbestrasse 11, 6330 Cham, Switzerland

Preface

The *Proceedings of the 2nd International Conference on Net-Zero Built Environment: Innovations in Materials, Structures, and Management Practices (Net-Zero Future 2025)* took place in Cape Town, South Africa, from 5 to 7 November 2025, building on the success of the inaugural conference held in Oslo, Norway, in 2024. The conference formed part of the Net-Zero Future initiative, supported by the Research Council of Norway and the Norwegian Directorate for Higher Education and Skills. Through this initiative, a global partnership was established among Norway, the USA, Germany, South Africa, and India, enabling collaborative research and educational exchanges aimed at moving toward a net-zero built environment. The project aligns directly with several United Nation's Sustainable Development Goals, including those related to quality education, innovation, sustainable cities, climate action, and international cooperation.

Growing concerns over the deterioration of the built environment under escalating stressors, alongside the environmental burden associated with its expansion, have highlighted the urgent need to rethink conventional approaches to infrastructure development. Civil engineering, which is central to the planning, construction, and stewardship of infrastructure, plays a pivotal role in this transformation. The Net-Zero Built Environment conference series serves as a unique platform to examine the net-zero imperative through the lens of civil engineering and several related fields, underscoring the necessity of dramatically reducing the carbon footprint throughout the world.

Achieving net-zero in the built environment encompasses the implementation of new insights, practices, and policies to design, deliver, operate, and eventually retire physical infrastructure components with no net carbon emissions over their entire life cycle. Meeting this ultimate goal demands integrated strategies spanning material selection, structural innovation, and modern management practices. These three areas were therefore established as the core themes of the conference series. The Cape Town event drew substantial participation from academia, industry, and the public sector, with delegates sharing the state-of-the-art and practice across all three thematic areas.

Following a thorough peer-review process carried out by members of the conference's international scientific committee together with additional expert reviewers, 111 full-length papers were selected for inclusion in the conference proceedings. These contributions reflect a broad spectrum of authors and research teams from across the world, each offering distinct perspectives and new insights into pathways toward a net-zero built environment. Aligned with this overarching ambition, the papers showcase recent advances in: (i) low-carbon and carbon-neutral materials and manufacturing methods, (ii) robotic and automated construction techniques that minimize formwork and on-site effort, (iii) innovative structural concepts and detailing strategies that maximize performance while reducing material demand, (iv) latest condition assessment and structural health-monitoring approaches, and (v) progressive life-cycle assessment methods and infrastructure management frameworks.

We acknowledge that reaching net-zero is more than a technical or policy target; It is a ethical responsibility to safeguard a sustainable future. Lowering carbon emissions contributes to minimizing environmental stressors, mitigating the frequency and intensity of extreme weather events, and preserving fragile ecosystems. All contributors to the enclosed proceedings share a commitment to advancing the materials, structures, and management practices that benefit communities worldwide. Through the dissemination of emerging ideas and innovations, along with lessons learned from the past, the themes explored in this book provide meaningful contributions to the net-zero agenda. By equipping researchers, engineers, decision-makers, and the public with relevant tools and knowledge base, we aim to support collective progress toward a global net-zero future.

Hans Beushausen
Mahdi Kioumarsi
Behrouz Shafei

Contents

An Innovative Solution to Repair and Restore Deteriorating Concrete Girder Ends

Amirhosein Vakili[1], Weizhuo Shi[2], and Behrouz Shafei[1](✉)

[1] Iowa State University, Ames, IA, USA
shafei@iastate.edu
[2] University of Nottingham, Ningbo, China

Abstract. One of the foremost issues associated with the utilization of precast prestressed concrete beam (PPCB) bridges pertains to the occurrence of cracking and spalling in the concrete near the beam ends, where the demand due to environmental stressors is at the highest. This phenomenon is primarily attributed to the infiltration of water and deicing chemicals, leading to the exposure and subsequent corrosion of the reinforcement and prestressing strands within the beams. Allowing this deterioration to continue will reduce beam capacity, and may threaten bridge integrity, safety, and durability. This issue was addressed in the current study by testing ultra-high performance concrete (UHPC) for strengthening and repairing prestressed concrete beam ends. The repair method involves removing the corroded concrete from the beam web and bottom flange, then encasing the beam end with UHPC. This repair can be regarded as a conventional bridge repair method because it is substantially less expensive than replacing a beam. To investigate this solution, artificially damaged prestressed concrete beams (both unrepaired and repaired with UHPC) were subjected to full-scale experimental tests. Observations indicated that the UHPC exhibited exceptional performance as an end patch. The study resulted in a comprehensive set of suggestions pertaining to optimal repair techniques and suitable retrofit materials for the rehabilitation of prestressed concrete beam ends.

Keywords: Precast Prestressed Concrete Beam · Corroded Girder · Bridge Repair · Experimental Study

1 Introduction

The increasing demand for structures that possess enhanced strength, improved efficiency, and extended durability has led to the growing popularity of precast prestressed concrete beam (PPCB) bridges within the engineering community [1, 2]. Prestressed concrete bridges have been utilized in the United States since their initial implementation in 1949. Over the past decade, these bridges have constituted a substantial proportion of all bridges constructed in the US, which is roughly 22% [3]. In this category of bridges, however, the transfer of prestressing pressure from the steel strands to the concrete induces significant strains within the beam, particularly at its ends. The occurrence of

H. Beushausen et al. (Eds.): NTZR 2025, LNCE 873, pp. 630–640, 2026.
https://doi.org/10.1007/978-3-032-27698-8_56

significant strains frequently results in the development of cracks at the ends of the beam [4]. The implementation of appropriate steel reinforcing design can effectively limit the width of these cracks to an acceptable level. Nevertheless, it is important to note that despite such measures, the presence of cracks still permits the penetration of water and chlorides into the ends of the beam [5]. The presence of cracked beam ends that are susceptible to moisture is a significant concern [6, 7], especially when located beneath expansion joints that are prone to leakage (Fig. 1).

Fig. 1. Severely deteriorated prestressed concrete girder in a bridge in Iowa.

To assess the behavior of the corroded concrete girders, several experimental and numerical investigations have been conducted. Among the relevant studies, Recupero & Spinella [8] investigate the influence of tendon corrosion on the response behavior of corroded prestressed concrete beams subjected to transversal load. These experimental studies were conducted on beams with artificially induced defects and varying degrees of tendon corrosion. The corroded beams showed a significant reduction in load-bearing capacity compared to control specimens with non-corroded tendons. Belletti et al. [9] proposed a numerical approach for modeling the effects of nonuniform corrosion in prestressed concrete structures using nonlinear finite element analysis. The study considered two modeling strategies: the smeared approach with perfect bond between concrete and strands and the discrete approach considering a deteriorated bond–slip relationship. The comparison between corroded and uncorroded beams demonstrated the model's potential in predicting ultimate load, displacement, and failure mode. Liu et al. [10] addressed the critical issue of fatigue behavior in corroded prestressed concrete (PC) beams, crucial for understanding the potential failure of aging concrete bridges. Thirteen PC beams underwent accelerated corrosion before high-cycle fatigue testing, resulting in fatigue fractures of corroded tensile steel bars at minimum cross-sections with severe corrosion pits.

In an attempt to address the mentioned structural performance concerns, a number of repair methods have been created and tested. Materials with particular qualities are preferred for the greatest long-term outcomes when repairing concrete damaged by corrosion. Investigations into the behavior of damaged reinforced concrete beams strengthened with ultra-high performance concrete layer show that these beams, when compared to unstrengthened reinforced concrete beams, exhibit monolithic behavior without interfacial debonding, leading to significantly higher ultimate loads and cracking [11–14]. Zhu et al. [15] employs finite element analysis to simulate the structural response of pre-damaged reinforced concrete beams strengthened with reinforced UHPC layers. Their findings offer insights for designing and constructing pre-damaged RC beams reinforced with UHPC layers to extend service life and reduce maintenance costs. Zhang et al. [16] presented an innovative method that uses a layer of pretensioned prestressed ultra-high-performance concrete (P-UHPC) to improve the flexural performance of pre-damaged reinforced concrete beams. The flexural capacity and fracture resistance were evaluated using a four-point bending test on two P-UHPC strengthened beams, one reinforced UHPC layer strengthened beam, and one unstrengthened RC control beam. The P-UHPC layer exhibited a typical bending failure mode with one main crack at the ultimate state and densely distributed micro-cracks in the UHPC layer.

This study investigated novel alternatives for the repair and retrofitting of prestressed concrete beam ends that have sustained damage over time due to beam-end deterioration. The primary objectives were to restore the beam's entire load-bearing capacity and prolong the operational lifespan of PPCB bridges. For this purpose, UHPC was selected as the repair material of choice. The repair solution was evaluated to understand the load transfer mechanisms and quantify reductions in the impact of excessive loading on the corroded web and flange. This was accomplished by running a comprehensive experimental program on different portions of a prestressed concrete bridge girder. Each beam web had a damaged region with a height of approximately 1 ft (0.30 m), while the bottom flange exhibited a damaged length of 2 ft (0.6 m). By using a variety of response characteristics, including concrete crack patterns and structural capacity, the study conducted detailed analyses to determine the effects of using UHPC to repair the corroded areas of the concrete beams.

2 Full-Scale Beam Tests

This section provides extensive details regarding the test specimens, the laboratory test setup, and the experimental results. The beam utilized for experimentation was a bulb-tee C-shaped beam from the Iowa Department of Transportation, with a length of 115 feet (35 m) (referred to as the BTC115 beam) (Fig. 2). The BTC115 beam was cut using a wire saw technique. Two end pieces measuring two feet and five inches in length were cut off from the beam. To mitigate the impact of variations in tendon profile, a symmetrical cut was performed on the remaining beam. To ensure the beams would experience shear failure rather than bending, each piece was precisely cut to a length of 11 feet (3.35 m). Two samples, one of them corroded and another repaired with UHPC, were subjected to full-scale experimental tests.

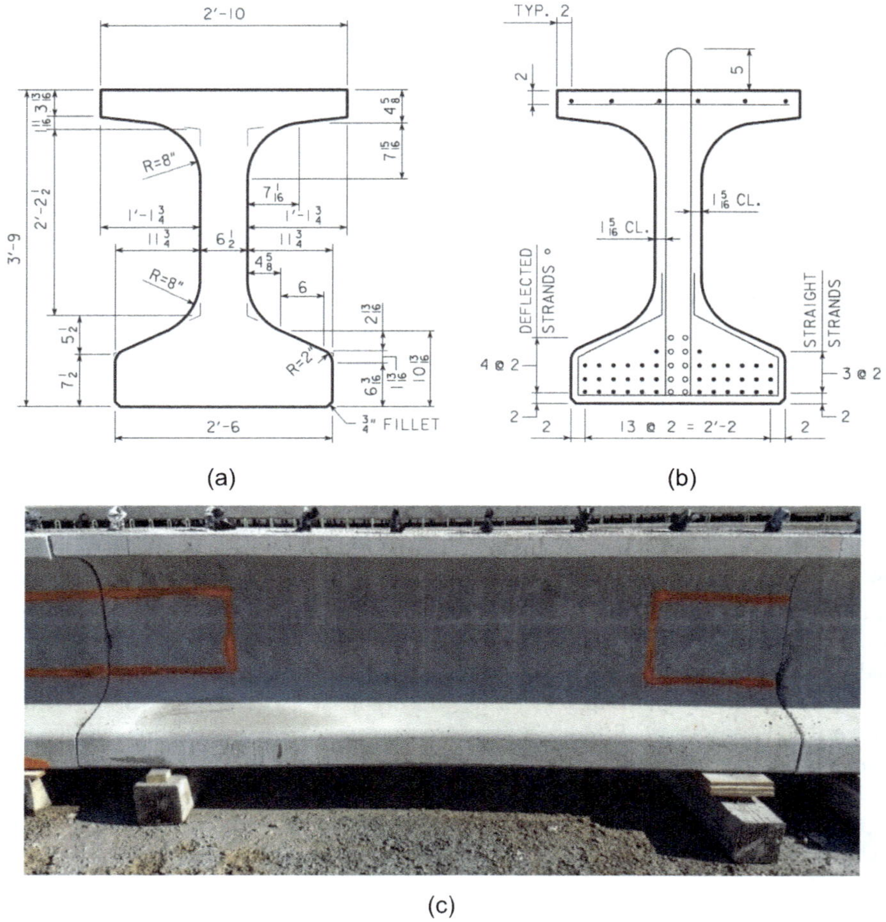

Fig. 2. BTC115 beam properties: (a) cross-section; (b) midspan reinforcement layout; and (c) cut beam segment (1 ft. = 0.30 m; 1 in. = 2.54 cm).

2.1 Deterioration Simulation

To examine the effects of deterioration at the beam ends of bridge girders, beams were intentionally damaged to replicate the degradation caused by corrosion. A major focus of the investigation was evaluating damage to the beam webs because the research focused on the shear capacity of the girders. The removal of concrete from the beam web region led to a reduction of roughly 30% in the width of the beam web. Each beam web had a damaged region characterized by a height of roughly 1 ft (0.3 m). The length of this damaged area is 1 ft (0.3 m). The aforementioned procedure was replicated on both surfaces of the web as well as on each end of the specimen. The utilization of simulated damage facilitated the consistent replication of damaged regions across all specimens, hence enabling their systematic testing and subsequent comparison.

2.2 Patching Material and Plan

To achieve the long-term efficacy of concrete repair in the face of corrosion-induced damage, the selection of materials with specialized properties becomes paramount. Two main factors, both necessary to achieve the best results, influenced the selection of alternate patching materials in this investigation. Firstly, the requirement for a sufficient bond strength was identified as critical in realizing a resulting structure that exhibits monolithic behavior. The attainment of an effective bond is imperative as it directly influences the structural integrity. Second, materials with small crack widths and low permeability were used in order to reduce continuous damage caused by corrosive environments. In order to strengthen the repaired concrete against the corrosive effects of environmental elements, this strategic decision attempts to obstruct the ongoing penetration of chloride-laden deck runoff. It is essential to consider these specific qualities in addition to the requirement that the patching material of choice have excellent mechanical qualities. Among the essential characteristics that add to the overall efficacy and endurance of the concrete restoration are strength, ductility, and durability. In the field of concrete rehabilitation, the synergistic combination of these qualities guarantees the structural integrity of the repaired concrete while also providing protection against future deterioration. This promotes long-term solutions.

To achieve the mentioned properties, UHPC was used for the repaired material [17–20]. To evaluate the bond performance of the patching materials chosen for use in beam-end repairs, numerous small-scale tests were carried out. The bond strength was assessed by using the splitting tensile strength test for tension specimens and the slant shear strength test for shear specimens. Shi and Shafei [21] provide extensive details about materials, test specimen preparation, and laboratory test setup. The full-scale part of this study involved testing two specimens cut from the BTC 115 beam. Upon the replication of deterioration-induced damage, the first beam received no repair or patching and the second one received UHPC patching.

2.3 Test Setup and Instrumentation

Experimental testing was conducted in order to verify that the beam ends experienced the highest shear stresses and that shear would be the predominant mode of failure. In this research, beam segments were subjected to four-point bending tests to evaluate their shear reaction. The experimental configuration is depicted in Fig. 3. The beam segments were simply supported with a hinge and a roller. The distance between supports, measured from center to center, was 10 feet (3.05 m). The arrangement described results in a reduced ratio of span to depth, hence promoting the formation of shear cracks. Two-point loads were positioned at a distance of 3 feet (0.91 m) from each other, with symmetrical placement relative to the center of the beam span. Two steel plates measuring 10 inches by 10 inches (254 X 254 mm) were positioned on the upper surface of the beam in order to evenly distribute the load onto the beam. Two hydraulic actuators were used to apply each point load, and 200-kip (890 kN) load cells were used to measure the results. In the initial beam test, the load was applied in increments of 25 kips (111 kN) until the actuators' capacity was reached. The occurrence of cracking on the surface of the beam was monitored at intervals of 100 kips (445 kN). Following the initial loading, the beam was subjected to the maximum load and subsequently unloaded for three further cycles. The process of cracking and the subsequent propagation of cracks were monitored for each subsequent maximum loading event. Following the end of the initial beam test, it was decided to increase the load increment from 25 kips (111 kN) to 50 kips (222 kN) for the subsequent loading steps.

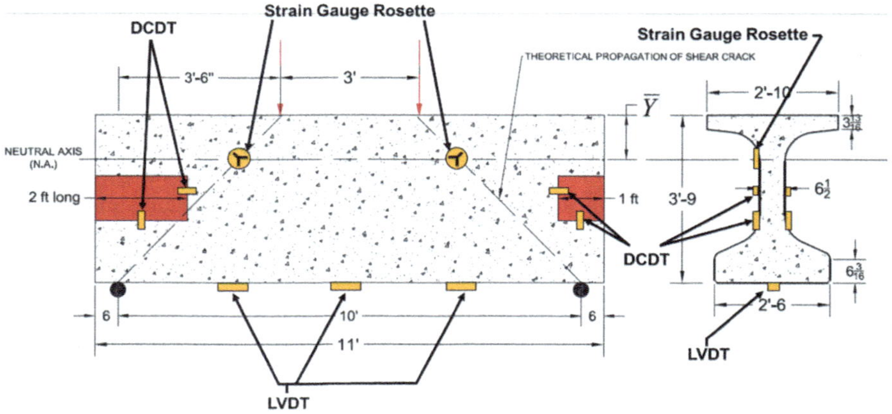

Fig. 3. Four-point loading laboratory test set up (1 ft. = 0.30 m; 1 in. = 2.54 cm).

3 Experimental Results and Discussion

To determine when the patch delaminated from the girder, displacement gauges were positioned on each patch and used to record the distance between the patch and the girder. Every displacement gauge reading is less than 0.01 inches (0.25 mm). The data acquired by the sensors employed in this experiment do not exceed the margins of error. This indicates that there was no separation between the girder substrate and the beam web patches. In full-scale testing, UHPC patches showed satisfactory bonding performance.

The results from the first loading cycle are included in this section for the purpose of comparing the beam specimens. In every test, cracking happened during the initial loading cycle. Figure 4 shows a plot of load versus midspan displacement for each specimen. The elastic stiffness values for the corroded model is 2877 kips/in (504 kN/mm) and for UHPC repaired model is 4360 kips/in (764 kN/mm). It is noteworthy that an increase in elastic stiffness is evident in the UHPC repaired model. The utilization of high-strength material and the increased cross-sectional area contributed to the improvement that was noticed. This combination shows how the material helps to improve the girders' elastic stiffness by allowing the reinforced girder's tensile zone to persevere more of tensile stress before breaking. Following the elastic stage, characterized by visible cracking in the experiment, the stiffness of the corroded girder gradually diminishes, causing a deviation from the curves of the repaired model. Conversely, the stiffness of UHPC repaired girder remains nearly constant, exhibiting linear behavior with the applied load during this stage. This resilience is attributed to the ability of UHPC to maintain high tensile strength post-cracking, thereby contributing to the preservation of the strengthened beam's stiffness. In the final loading stage, girders subjected to patching repair exhibit lower deflections, with UHPC demonstrating the most pronounced reduction in deflection values.

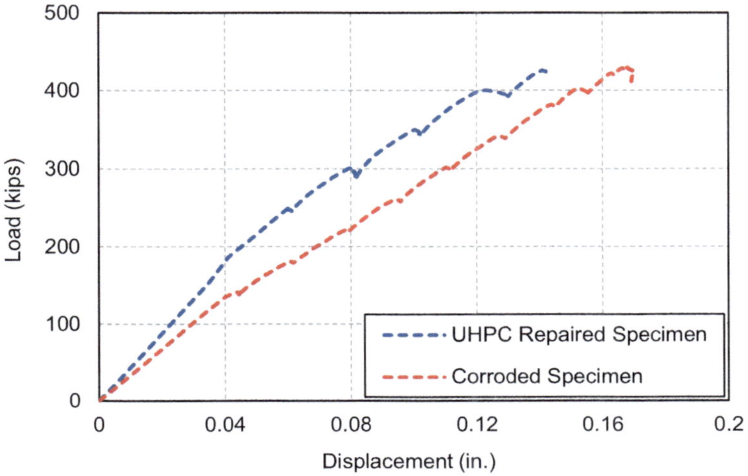

Fig. 4. Load versus midspan displacement curve.

The corroded beam experienced failure. This failure event happened in the second load cycle, reaching a peak load of 400.6 kips (1780 kN). The failure modes of the model belong to a typical shear failure, which is characterized by the diagonal shear cracking. The onset of cracking was initially observed within the load range of 250 kips (1112 kN) to 300 kips (1334 kN). Subsequently, during the second loading cycle, a brittle failure event transpired at a peak load of 400.6 kips (1780 kN). Notably, this failure was characterized by significant shear cracking in the web, particularly above the designated theoretical shear cracking path. Figure 5 illustrates the distribution of cracks in various end parts of the tested girders after failure. Throughout the experiment, multiple primary cracks were developed concurrently towards the bottom flange of the girder, resulting in the bursting of the bottom flange. It should be mentioned that the prestressed strands underwent a contraction of approximately 1.25 inches (32 mm) into the beam subsequent to the bursting failure within the bottom flange. This contraction serves as an indication that residual prestressing forces persisted within the strands even after the original full-length beam was cut into the shorter test specimens.

In the UHPC repaired specimen, the first obvious cracks appeared in the load range of 300–350 kips (1334–1557 kN). The cracks occurred in the pre-existing concrete outside of the designated repair area and extended at a 40° angle from the horizontal over the upper web. Additionally, the crack was seen along the anticipated direction as well as above the predicted theoretical shear path. A few pre-existing cracks that had developed as a result of pretensioned steel cable cutting expanded horizontally under loads greater than 300 kips. Cracking spread around 1–2 inches (25.4 mm to 50.8 mm) during the second and third load cycles, and very little to nothing more during the last loading stage. As reflected in Fig. 5, the UHPC patching surface on both specimens, demonstrating an effective bond between the substrate and the patching material. Furthermore, when loads above 300 kips (1334 kN) were applied to UHPC repaired specimen, cracks appeared on the bottom flange at the interface between the UHPC patch and the concrete substrate. Crucially, the strong bonding of the UHPC interface in the repaired models was demonstrated by the fact that, despite the presence of surface cracks, there was no separation during the entire loading procedure.

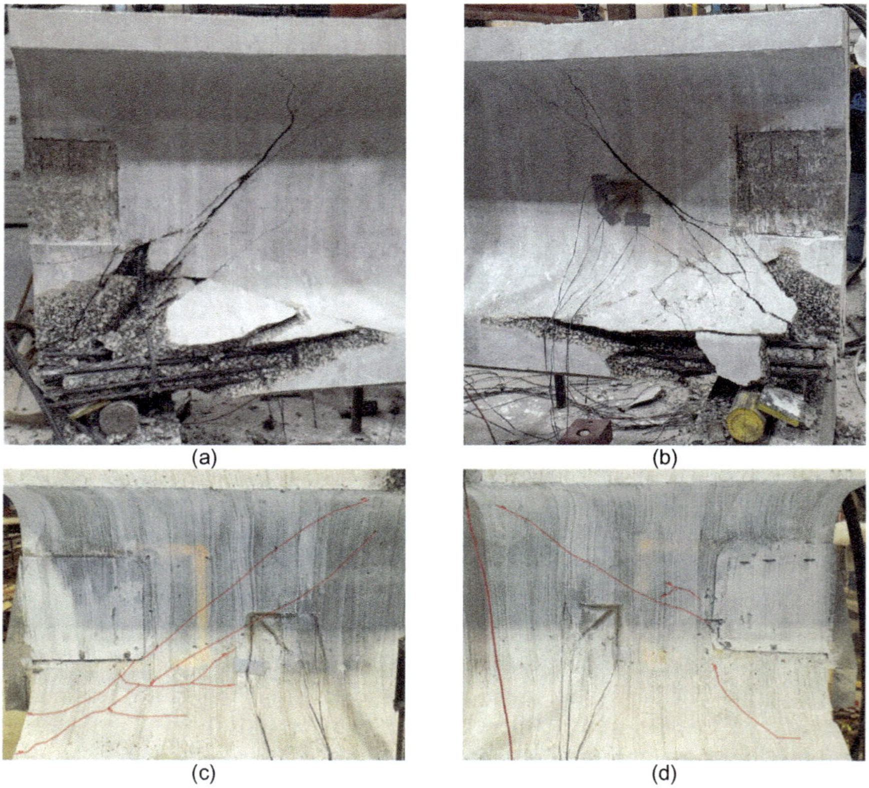

Fig. 5. Crack pattern and failure of specimens: (a) corroded left side; (b) corroded right side; (c) repaired left side; and (d) repaired right side.

The observed cracking outside the UHPC repair zone suggests that the repair effectively restored and even enhanced the capacity of the damaged region, causing stress redistribution toward adjacent unrepaired concrete. In the repaired specimen, the UHPC maintained its integrity without delamination, indicating that the interface bond strength exceeded the tensile capacity of the surrounding substrate concrete. Potential detailing adjustments to further shift failure away from critical interfaces can include extending the UHPC encasement beyond the damaged zone to provide a transition length for stress transfer and tapering the patch thickness to minimize abrupt stiffness changes.

These measures can further reduce localized stress concentrations and enhance overall performance under high shear demands.

4 Conclusions

This study delves into the behavior of prestressed bulb-tee C-shaped beams with corrosion-induced damage at their ends. Experimental test of the girders was conducted using UHPC material. A four-point test was performed on damaged and strengthened specimen to assess crack patterns, load-bearing capacity, and the bonding performance of the retrofitting materials. UHPC material exhibited satisfactory bond strength with the damaged surfaces. Horizontal and vertical LVDTs positioned on each part of the patch indicated minimal displacement (below 0.01 in), suggesting that the beam web patches did not separate from the girder substrate. This underscores the adequacy of bond strength, rendering both materials suitable for use as patch repair materials. In the damage model, the diagonal shear cracking failure event occurred during the second load cycle, reaching a peak load of 400.6 kips (1780 kN). Throughout the progressive failure process in the experiment, a simultaneous development of multiple primary cracks was observed, converging towards the bottom flange of the girder. This cumulative cracking phenomenon ultimately led to the rupture and bursting of the bottom flange. Upon completion of the monotonic loads, the repaired specimen maintained its integrity and performance with no failure. In the repaired specimen, minor instances of cracking were observed both above the prescribed theoretical shear path and along the anticipated direction. This observation suggests that the patching technique exhibited commendable behavior during both loading and unloading phases.

Acknowledgements. The authors would like to acknowledge the support provided by the Iowa Department of Transportation and the Iowa Highway Research Board. Additionally, the authors would like to recognize Quin Rogers, Doug Wood, and Owen Steffens for their assistance with laboratory tests.

References

1. Shafei, B., Phares, B., Shi, W.: Beam End Repair for Prestressed Concrete Beams. Iowa State University (2020)
2. Hosteng, T., Abu-Hawash, A., Port, G., Phares, B.: Laboratory Investigation of Concrete Beam-End Treatments. Iowa State University (2015)
3. Wu, N.C., Chase, S.: An Exploratory Data Analysis of National Bridge Inventory. Mid-Atlantic Universities Transportation Center (2010)
4. Tadros, M.K., Badie, S.S., Tuan, C.Y.: Evaluation and Repair Procedures for Precast/Prestressed Concrete Girders with Longitudinal Cracking in the Web, vol. 654. Transportation Research Board (2010)
5. Kannel, J., French, C.E., Stolarski, H.K.: Release methodology of strands to reduce end cracking in pretensioned concrete girders. PCI J. **42**(1) (1997)
6. DeJong, A., Shi, W., Shafei, B., Hosteng, T.: Integral abutment connections with grouted reinforcing bar couplers and ultra-high performance concrete. J. Bridg. Eng. **26**(8), 04021042, 1–15 (2021)

7. Karim R, Shafei B (2021) Performance of fiber-reinforced concrete link slabs with embedded steel and GFRP rebars. Eng. Struct. 229:111590
8. Recupero A, Spinella N (2019) Experimental tests on corroded prestressed concrete beams subjected to transverse load. Struct. Concr. 20(6):2220–2229
9. Belletti B, Vecchi F, Bandini C, Andrade C, Montero JS (2020) Numerical evaluation of the corrosion effects in prestressed concrete beams without shear reinforcement. Struct. Concr. 21(5):1794–1809
10. Liu X, Zhang W, Gu X, Ye Z (2023) Assessment of fatigue life for corroded prestressed concrete beams subjected to high-cycle fatigue loading. J. Struct. Eng. 149(2):04022242
11. Zhang Y, Li X, Zhu Y, Shao X (2020) Experimental study on flexural behavior of damaged reinforced concrete (RC) beam strengthened by toughness-improved ultra-high performance concrete (UHPC) layer. Compos. B Eng. 186:107834
12. Al-Huri MA, Ahmad S, Al-Osta MA, Al-Gadhib AH, Kharma KM (2023) Performance of corroded RC beams strengthened in flexure using UHPC: effect of configuration and thickness of the UHPC layers. Eng Struct 292:116519
13. Ahmad S, Sharif AM, Al-Osta MA, Al-Zahrani MM, Sharif AM, Al-Huri MA (2023) Flexural performance of pre-damaged RC beams strengthened with different configurations of UHPFRC layer–experimental and analytical investigation. Structures 48:1772–1787
14. Al-Osta MA, Kharma KM, Ahmad S, Maslehuddin M, Al-Huri M, Khalid H (2023) Strategies for strengthening of corroded reinforced concrete beams using CFRP laminates and UHPC jacketing. Struct. Concr. 24(1):1546–1571
15. Zhu Y, Zhang Y, Li X, Chen G (2021) Finite element model to predict structural response of predamaged RC beams reinforced by toughness-improved UHPC under unloading status. Eng Struct 235:112019
16. Zhang Y, Huang S, Liu Y, Fan W, Shao X (2023) Flexural behavior of damaged RC beams strengthened with prestressed UHPC layer. Eng. Struct. 283:115806
17. Qureshi, M.H., Nawaz, S.R., Vakili, A., Kioumarsi, M., Shafei, B.: Assessment of the structural potential of nonconventional material alternatives in shear walls. In: The International Conference on Net-Zero Built Environment: Innovations in Materials, Structures, and Management Practices. Springer Nature (2024)
18. Karim R, Shafei B (2021) Flexural response characteristics of ultra-high performance concrete made with steel microfibers and macrofibers. Struct. Concr. 22(6):3476–3490
19. Karim R, Najimi M, Shafei B (2019) Assessment of transport properties, volume stability, and frost resistance of non-proprietary ultra-high performance concrete. Constr. Build. Mater. 227:117031
20. Kazemian M, Shafei B (2022) Internal curing capabilities of natural zeolite to improve the hydration of ultra-high performance concrete. Constr. Build. Mater. 340:127452
21. Shi W, Shafei B (2021) Bond characteristics between conventional concrete and six high-performance patching materials. Constr. Build. Mater. 308:124898

Parametric Investigation of Cellular Steel Beams

Themistoklis Tsalkatidis(✉) and Abdimalik Sheikh Muhumed

Norwegian University of Life Sciences, 1433 Ås, Akershus, Norway
{themistoklis.tsalkatidis,abdimalik.sheikh.muhumed}@nmbu.no

Abstract. Cellular beams are beams with web openings. Their construction material is typically steel, and they are specifically made from either hot rolled steel profiles or steel plating. The main advantage of cellular beams over solid ones, besides the obvious weight decrease due to the use of less material, is that they allow the passing of HVAC equipment through their perforations, thus resulting in a significant reduction in the total floor height of a building. The current paper investigates the effect of three parameters on the structural behavior of steel cellular beams and focuses on the shear capacity of the web-post that is considered a key performance indicator. These parameters are the diameter of the web opening, the thickness of the flanges and the thickness of the web of the steel beam. The reference numerical model consists of a universal steel beam section with dimensions UB457 × 152 × 52. The top and bottom flange thickness are both equal to 10.9 mm, the web thickness is set equal to 7.6 mm and the web opening diameter is set equal to 315 mm. The reference numerical model is validated against experimental results from international literature [1]. The numerical study is performed by constructing three-dimensional numerical models using ANSYS software package. A total number of twelve case studies are examined. The numerical analysis results indicate that a combination of top and bottom flange thicknesses equal to 10.9 mm, a web thickness equal to 11.4 mm and a web-opening diameter equal to 315 mm has been proved to be the best performance wise. The shear capacity of the web-post increased by 45.5% in respect to the initial reference value when this combination is used. Overall, cellular beams provide a sustainable design option with less material usage and good structural performance.

Keywords: Cellular Steel Beams · Numerical Investigation · Sustainable Construction Elements

1 Introduction

According to data collected and distributed from several sources, the construction sector is one of the leading culprits of releasing carbon dioxide emissions due to energy consumption and material storage, thus contributing to global climate change [2–5]. Therefore, the responsible approach of society must be to act and to make a shift towards a more sustainable built environment. This shift can be achieved by selecting and enforcing effective construction designs and systems, and it is in line with the UN Sustainable

H. Beushausen et al. (Eds.): NTZR 2025, LNCE 873, pp. 641–649, 2026.
https://doi.org/10.1007/978-3-032-27698-8_57

Development Goals (SDGs), and more specifically with SDGs 7, 9 and 11 [6], and their described topics. A key parameter in achieving a sustainable construction solution is to reduce the amount of material used when designing such a system. The concept in this case is to minimize any excess material and to keep only the effective one [7]. Cellular steel beams are structural elements that are inherently constructed with significant less material over traditional solid steel beams [8], thus making them an attractive alternative for sustainable construction elements.

Moreover, several research studies on cellular steel beams have confirmed their noteworthy structural properties which can be attributed to their optimized design. Chung et al. [9] examined the effect of the Vierendeel mechanism on steel beams with web openings and discovered that its effect is more significant in beams having circular over rectangular openings. The reason is that in the first case, shear yielding is the dominant failure mechanism whereas in the latter is bending. Erdal and Saka [10] investigated both experimentally and numerically twelve cellular steel beams with different geometry, but they were subjected to the same support and loading conditions. The twelve beam specimens experienced different failure modes, mainly web-post buckling and Vierendeel bending, because of the changes made in their geometric properties. Boissonnade et al. [11] conducted three bending tests and a parametric numerical analysis on cellular steel girders with a focus on examining the failure mode of the specimens in relation to local and global instability. Their research findings underline the overconservative results when using the current design rules of Eurocode 3 thus resulting in proposing a new set of design formulas to reduce this tendency. Zewudie et al. [12] investigated the buckling performance of circular arched cellular steel beams and discovered that in this configuration the shear capacity of the web-post is overestimated when using the same design approach as non-arched beam and this effect is more prominent in steel buildings that are low rise and have long spans.

The aim of the current paper is to investigate the effect of three parameters on the structural behavior of cellular steel beams, i.e. steel beams with circular web openings. The examined parameters are the diameter of the web opening, the thickness of the two flanges and the thickness of the web of the cellular beam. Twelve different parametric combinations are explored. The shear strength of the web-post is used as a key performance indicator of the modified cellular steel beam in each case. The reference numerical model has been created according to the tested specimen by Tsavdaridis and D'Mello [1], and it is validated against their experimental results [1, 8]. Tsavdaridis and D'Mello studied the effect of web spacings and web openings on the web-post behaviour of beams with web openings. Circular, hexagonal and elongated web openings were examined. The analysis results indicated that the hexagonal and elongated web openings improve the structural performance of perforated against cellular beams, especially when the focus is placed on the web-post buckling.

2 Analysis

The Finite Element Method (F.E.M.) is the method selected to treat both the physical and the numerical problem. F.E.M is an effective technique that solves complex engineering problems by converting them into a group of simplified problems that are easier to solve

[13, 14]. The addition of these problems is equivalent to the original problem and the solution is significantly accurate. The method utilizes user-selected and induced convergence criteria to achieve a converged solution. The selected convergence criteria and their corresponding root-finding algorithms rely upon the problem under consideration [13, 14].

The presented inquiry starts with the creation of a reference model of cellular steel beam. The numerical results of the reference simulation are compared and validated against the experimental results resulting in the calibration of the finite element model. The numerical model employs both geometric and contact nonlinearities, and the stiffness matrix is updated at each load step. The Newton-Raphson incremental-iterative algorithm is selected to solve the model due to its robust characteristics [14, 15]. Furthermore, the second stage of the study is to perform new numerical analyses on modified cellular steel beams in respect to the reference model. The focus is placed on three key parameters that are expected to have a significant effect on the structural performance of the cellular beam. These parameters are the diameter of the web opening, the thickness of the steel flanges and the thickness of the web of the steel beam. The scope of these analyses is to quantify the effect of the aforementioned parameters and to provide an insight into the vulnerable parts of the cellular beam [1, 8].

2.1 Reference Case

The numerical analysis is conducted using ANSYS 2019 Academic R3 version software package [15]. The reference numerical model consists of a universal steel beam section with dimensions UB457 × 152 × 52. The top and bottom flange thickness are both equal to 10.9 mm, the web thickness is set equal to 7.6 mm and the web opening diameter is set equal to 315 mm in the reference model as described in the test campaign of Tsavdaridis and D'Mello [1]. The length of the examined beam is 1700 mm and spacing distance of the circular web perforations is 409.5 mm i.e. 1.3 times the diameter of the web opening (s equals to 1.3*do) resulting in an opening to depth ratio of 0.7. The cellular beam is pin-supported, and it is loaded by a vertical point load placed at mid-span. At each support a stiffener having 10.9 mm of thickness is used and at the point of load application a stiffener having 20 mm of thickness is placed to avoid local buckling [1, 8]. The reference model of the cellular steel beam is shown in Fig. 1.

The constructed numerical simulation is three-dimensional. For the steel beam, SHELL181 element is selected as the element type from the software library. This structural element is a shell element that has four nodes with six translations and six rotations degrees of freedom at each node [15]. Steel S355(Fe510) is the material used in the analysis which has a yield strength of 355 MPa and an ultimate strength of 510 MPa. The modulus of elasticity of steel is 210 GPa [1, 8].

Due to modelling requirements, surface-to-surface contact elements are used to connect the top and bottom half of the cellular beam to simulate the re-welding fabrication phase of the cellular beams, at mid-height of the steel profile. These elements overlay the existing shell elements used in steel beam. ANSYS software offers many options to ensure contact both in a variety of contact conditions and contact elements. However, modelling contact between bodies is a complex and delicate task since it depends on many factors such as relative stiffness and contact laws [16]. CONTA175 and TARGE170

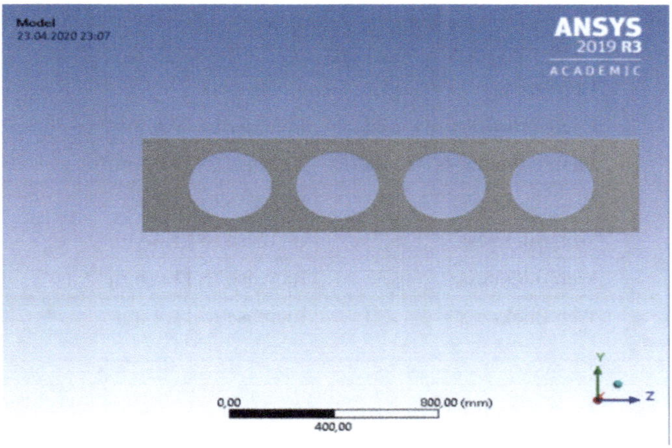

Fig. 1. The reference cellular steel beam under study.

are selected as contact elements to establish the interface contact in this case and the 'bonded always' contact condition is applied due to this condition found closer to the physical problem [8, 15].

2.2 Parametric Investigation

The next stage of the research study, after the calibration of the reference model (case 1a), is to conduct a parametric study with a focus on the three parameters i.e. the diameter of the web openings (cases 1b, 1c and 1d), the thickness of the steel flanges (cases 2a, 2b, 2c and 2d) and the thickness of the web of the steel section (cases 3a, 3b, 3c and 3d). Cases with a subscript have the original diameter of web openings, case with b subscript have web openings diameter of 157.5 mm and cases with c subscript have web openings diameter of 105 mm. Case studies with d subscript do not have any web openings. A total number of twelve numerical models are constructed, four models for each examined parameter. The twelve case studies are presented in Table 1.

Table 1. Examined case studies during parametric numerical study.

Parameter	Case	Comments
Opening diameter	1a	Diameter of 315 mm
Opening diameter	1b	Decrease to 157.5mm
Opening diameter	1c	Decrease to 105 mm
Opening diameter	1d	Decrease to 0 mm
Flange thickness	2a	Increase to 19.6 mm
Flange thickness	2b	Increase to 19.6 mm

(continued)

Table 1. *(continued)*

Parameter	Case	Comments
Flange thickness	2c	Increase to 19.6 mm
Flange thickness	2d	Increase to 19.6 mm
Web thickness	3a	Increase to 11.4 mm
Web thickness	3b	Increase to 11.4 mm
Web thickness	3c	Increase to 11.4 mm
Web thickness	3d	Increase to 11.4 mm

3 Results and Discussion

The numerical findings of the case studies are depicted in the following Figures and Tables. Focus is placed on the developed von Mises stress, the web-post shear strength and the maximum deflections at mid-span of the cellular steel beams. Figure 2 presents the distribution of the von Mises stresses found in the reference case and Fig. 3 shows the corresponding maximum deflection at mid-span. The maximum value of the deflection at mid-span for the reference model is 1.3 mm.

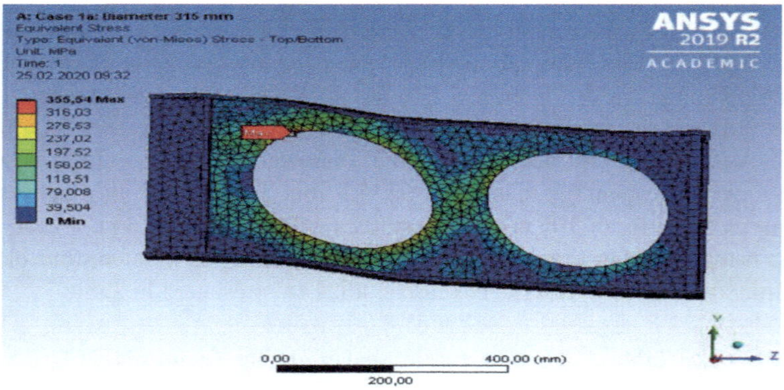

Fig. 2. Von Mises stress distribution of the reference model.

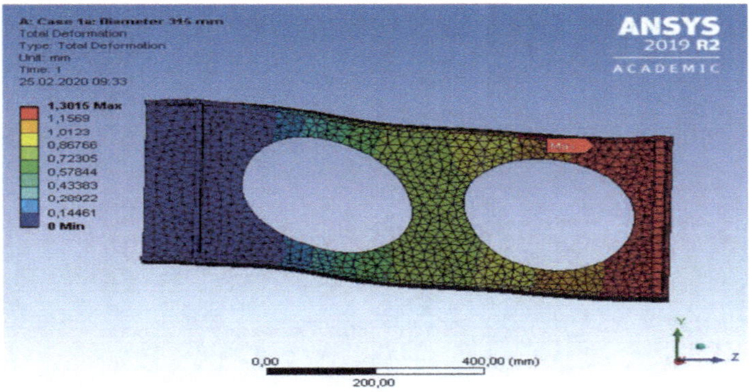

Fig. 3. Maximum deflection of the reference model.

After examination of the numerical results, the failure mode of the cellular beam is web-post buckling whereas the failure mode of the beam with no openings is buckling of the top flange. The shear capacity of the web-post depends on the geometry of the steel beam and the diameter of its web openings. It is calculated as equal to 255 kN for the reference model which is very close to 250 kN that it is measured during the test [1]. Table 2 presents all the values of the shear strength of web-post and the maximum deflections at mid-span for all cellular beam configurations examined.

Table 2. Shear capacity of web-post of cellular beams.

Case	Web-post shear capacity (kN)	Maximum deflection at mid-span (mm)
1a	255	1.3
1b	324.5	0.88
1c	343	0.8
2a	266	1.2
2b	331	0.76
2c	354	0.66
3a	371	1.4
3b	456	0.97
3c	473.5	0.85

As shown in Figs. 4 and 5, the shape of the von Mises stress distribution in cases 1b and 1c changes significantly from the reference case 1a (see Fig. 2). This change can be attributed to the decrease in the opening diameter and the increase in the number of web perforations for cases 1b and 1c. The diameters of the circular web openings in cases 1b and 1c are reduced by 50% and 67% in respect to case 1a whereas the number of web openings is increased from 3 to 8 and 12 respectively [8]. The scope is to maintain the spacing of 1.3*do that it is used in the experimental analysis [1]. The number of the openings remains the same in all cases of perforated beam. When examining the other two parameters (cases 2 and 3), the same notion is employed as in cases 1.

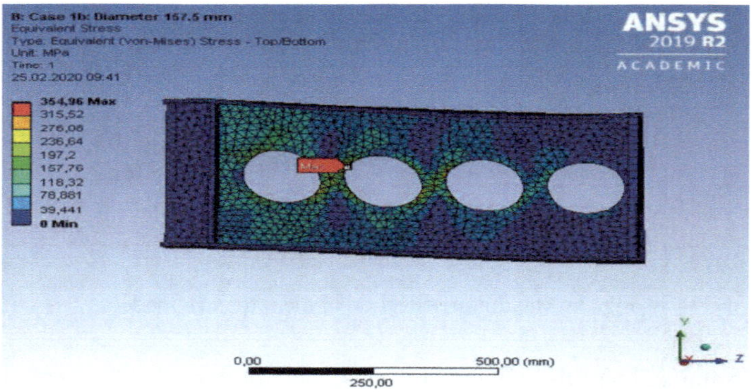

Fig. 4. Von Mises stress distribution of case 1b.

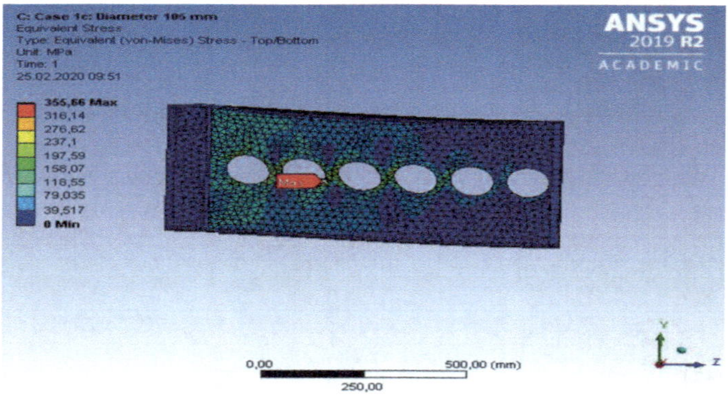

Fig. 5. Von Mises stress distribution of case 1c.

Figure 6 shows the smaller mid-span deflection calculated in a cellular beam configuration during the parametric analysis. This belongs to case 2c, and it is found to be equal to 0.66 mm.

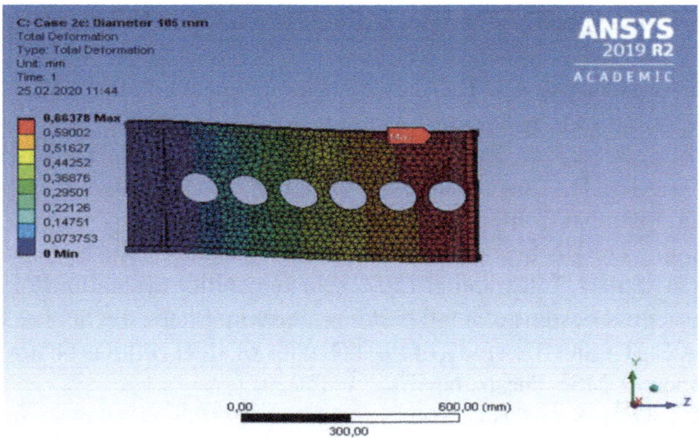

Fig. 6. Maximum deflection at mid-span of case 2c.

4 Conclusions

The current research study presents new findings from a numerical analysis performed to a sustainable structural element such as the cellular steel beam that can be used for future projects. This can be mostly attributed to conducted parametric analysis which can be cooperative in the optimization of the cellular beam so that the amount of material used in these elements can be significantly decreased. As understood, this is an effective and sustainable approach when designing structural elements and systems in general. Moreover, as expected, cellular beams with smaller openings areas are found to have less deflection and to have higher capacity before failure due to having a higher surface of effective areas. The increase in flange thickness contributes mostly to the decrease in mid-span deflection (minimum deflection value found in case 2c) whereas the increase in web thickness affects mostly the web-post shear capacity (maximum shear strength value found in case 3c). This is obvious when comparing cases 2 and 3 to cases 1, with the same diameters of web openings. Therefore, if having a high web-post shear strength is of primal importance then the web thickness is to be increased while if keeping mid-span deflection small is prioritized then the flange thickness is to be increased. The use of stiffeners in the cellular beam is found to be beneficial for the prevention of Vierendeel bending failure which is quite common in T-sections, and for that reason it is highly recommended. The stress concentrations around the circular web openings of the cellular beam tend to weaken the cross-section of the beam by reducing the effective area of the beam and must be considered carefully during analysis.

References

1. Tsavdaridis KD, D'Mello C (2011) Web buckling study of the behaviour and strength of perforated steel beams with different novel web opening shapes. J. Constr. Steel Res. 67(10):1605–1620
2. Keskin FS, Martinez-Vazquez P, Baniotopoulos C (2021) An integrated method to evaluate sustainability for vulnerable buildings addressing life cycle embodied impacts and resource use. Sustainability 13:10204

3. Sun Y, Yan C, Xing H (2024) Can green buildings reduce carbon dioxide emissions? Energy 312:133613
4. Khozema, A.A., Mardiana A.I., P., Yusri, Y.: Issues, impacts, and mitigation of carbon dioxide emissions in the building sector. Sustainability **12**, 7427 (2020)
5. Rizvanolli, B.V., Sáez, P.V., Rio Merino, M., Kontovourkis, O., Tsalkatidis, T., Peretti, G., Galán, A., Friedrich, D.: Current status of sustainable construction in Europe. In: Hajrizi, E. (ed.) UBT International Conference, Pristina (2019)
6. United, Nations.: The UN Sustainable Development Goals, United Nations, New York (2015)
7. Shishegaran A, Safari S, Karami B (2022) Sustainability evaluation for selecting the best optimized structural designs of a tall building. Sustain. Mater. Technol. 133:e00482
8. Muhumed, A.S.: Parametric study of the behavior of steel cellular beams with various web opening diameters. MSc Thesis, NMBU, Ås (2020)
9. Chung KF, Liu TCH, Ko ACH (2001) Investigation on Vierendeel mechanism in steel beams with circular web openings. J. Constr. Steel Res. 57(5):467–490
10. Erdal F, Saka MP (2013) Ultimate load carrying capacity of optimally designed steel cellular beams. J. Constr. Steel Res. 80:355–368
11. Boissonnade N, Nseir J, Somja H (2024) Experimental and numerical investigations towards the lateral torsional buckling of cellular steel beams. Thin-Walled Struct. 195:101388
12. Zewudie BB, Zerfu K, Agon EC (2024) Numerical investigation of elastic-plastic buckling performance of circular arched cellular steel beam using nonlinear finite element analysis method. Heliyon 10:e25292
13. Bi Z (2017) Finite Element Analysis Applications: A Systematic and Practical Approach, 1st edn. Academic Press, Cambridge MA
14. Bathe K-J (2006) Finite Element Procedures, 2nd edn. Prentice Hall, Upper Saddle River NJ
15. ANSYS Inc.: Ansys Manual v.2019, Canonsburg PA (2019)
16. Tsalkatidis T (2014) Numerical simulation and analytical study of glulam timber beams. Int. J. Eng. Technol. 3(2):129–136

Evaluation of Curved End Shear Wall (CESW) in Reinforced Concrete Tall Buildings by Seismic and Wind Loads

Ali Kheyroddin[1], Mehran Akhavan Salmassi[2](✉), and Mahdi Kioumarsi[3]

[1] Faculty of Civil Engineering, Semnan University, Semnan, Iran
kheyroddin@semnan.ac.ir
[2] Department of Civil Engineering, Seismic Geotechnical, and High-Performance Concrete Research Center, Semnan Branch, Islamic Azad University, Semnan, Iran
m.akhavansalmassi@iau.ir
[3] Department of Built Environment, OsloMet—Oslo Metropolitan University, Oslo, Norway
mahdi.kioumarsi@oslomet.no

Abstract. Architectural requirements affect structural design investigations. In addition, seismic and wind loads have an essential effect on tall buildings. On the other hand, vortex shedding usually occurs in slender cylindrical structures and some tall buildings with flat bodies. Sometimes, the phenomena mentioned above cause resonance occurrence in tall buildings. This study aimed to evaluate the curved end shear wall (CESW) in two 20-story reinforced concrete tall buildings with and without CESW. The curved end shear wall connected the end of shear walls in all stories, reducing the stresses at the end. The research cases were modeled by using ETABS software. The results of the seismic load showed that the maximum story drift decreased by 34% by CESW. Also, the first period showed a 40% reduction in tall buildings with CESW. The finding related to wind load indicated that the critical wind speed for vortex shedding declined by 17% in tall buildings with CESW. The curved end shear wall revealed suitable performance in reinforced concrete tall buildings, decreasing subjected seismic and wind loads.

Keywords: End shear walls · Tall buildings · Vortex shedding · Seismic load · Wind load · Reinforced concrete structures

1 Introduction

Architectural configurations and lateral load-bearing mechanisms profoundly influence how high-rise buildings respond to external forces such as seismic activity and wind. In this regard, the curved end shear walls (CESW) were suggested for improving tall buildings. CESW elements span vertically, linking shear wall segments at each story level. Prior studies have examined various lateral resistance strategies in tall structural design.

In a 30-story building with end shear walls and a helipad, the skewness and kurtosis coefficients were found to decrease by 18% and 70%, respectively. These findings underscore the significant role of end shear walls in improving the performance of reinforced

H. Beushausen et al. (Eds.): NTZR 2025, LNCE 873, pp. 650–661, 2026.
https://doi.org/10.1007/978-3-032-27698-8_58

concrete tall buildings with helipads [1]. Using nonlinear time history analysis, the performance of two tall buildings with end shear walls was improved significantly. The kurtosis coefficient increased by 85%, while the skewness coefficient in the X direction decreased by 75%. Additionally, end shear walls contributed to a reduction in residual drift by up to 22%, demonstrating their substantial effect on the buildings' lateral systems [2]. Chaos theory was employed to evaluate the dynamic complexity of structures incorporating end shear walls, showing that the presence of end shear walls in a 30-story structure effectively mitigated the rise of chaotic patterns in graphs such as Chebyshev, Gaussian, and logistic mappings. These findings highlight the critical importance of end shear walls in stabilizing the lateral load systems of such buildings [3]. An assessment of reinforced concrete tall buildings with end shear walls, subjected to sequences far from the fault, revealed notable findings. Quantile-quantile (Q-Q) plots revealed nearly a 20% contraction in the typical statistical spread across both primary horizontal axes for thirty-story structures incorporating end shear walls. Additionally, the kurtosis coefficient dropped by 61% in the X direction and 92% in the Y direction. By minimizing the dispersion of inter-story drift data, end shear walls significantly enhanced confinement effects and bolstered seismic performance [4].

The wind-induced response control model for high-rise buildings, developed using a resizing method, demonstrated its effectiveness in enhancing structural stiffness. By employing the resizing algorithm, the natural frequency of the building was increased, contributing to improved vibration control. This design approach was applied to mitigate wind-induced vibrations in a real 37-story building during its early structural design phase [5]. An analytical investigation into the across-wind response of cylinders caused by vortex shedding revealed that the chimney's behavior is highly influenced by the selection of the critical diameter, which plays a key role in determining critical wind velocity as well as the lift and aerodynamic coefficients. Furthermore, the study highlighted the significance of the lock-in condition in the second mode, particularly concerning bending stresses [6]. The analysis of earthquake and wind load effects on existing reinforced concrete (RC) minarets in Northern Cyprus indicated that static wind load calculations underestimate both deflections and base reactions. As a result, the use of static wind load should be avoided when designing RC minarets to ensure more accurate and reliable structural performance evaluations [7]. The modeling of vortex-shedding effects on the stochastic response of tall buildings under non-synoptic wind conditions introduced the wavelet-Galerkin (WG) approximation method. This approach demonstrated high potential and efficiency in addressing a variety of stochastic dynamic scenarios and solving related equations of motion. It effectively accounted for factors such as vortex shedding effects, intricate wind loads, coupling dynamics, nonlinearity, and non-stationarity [8].

The prediction of dynamic loads and responses of a tall building under tornado wind conditions, using numerical analysis and experimental validation for peak accelerations, demonstrated strong agreement between the numerical and experimental results, with errors ranging from 7.5 to 16%. The study highlights the critical need for precise definitions of wind field and structural parameters to accurately predict dynamic responses. This method significantly improves the capability of wind tunnel facilities to analyze the effects of non-stationary tornado winds on structural loads and responses [9]. The evaluation of seismic performance for reinforced concrete (RC) shear walls, with and without damped outriggers and controlled rocking, demonstrated notable enhancements in structural behavior. The study found that the performance of traditional RC shear wall systems could be significantly improved by implementing one of the following: (a) damped outrigger at the roof level, (b) controlled rocking at the base level, or the innovative controlled outriggered rocking wall (CORW) system [10].

2 Modeling of Buildings

This research analyzed the seismic and wind behavior of CESW in reinforced concrete high-rise normal distribution buildings by modeling and evaluating two 20-story structures using ETABS software for linear static analysis. Each level featured CESW components joining the shear wall terminals, creating a continuous lateral support mechanism. The first model featured a reinforced concrete high-rise building without curved end shear walls, while the second included them. The study accounted for dead and live loads of 170 kg/m^2 and 200 kg/m^2, respectively, along with a Concrete compressive strength (fc) of 50 MPa, Poisson's ratio (υ) of 0.15, and steel yield strength (fy) of 400 N/mm^2. Both models adhered to the ASCE7–16 code [11] and incorporated three-dimensional concrete moment frames with regular plans, 4-m spans, and 4-m story heights at the same site location (D type). Also, analysis methods and loading combinations considered by ASCE7–16 code. Each story included a core-area opening of 1.80 $\times$ 2.40 m. Also, the cantilever roof in the high rise building with a CESW is considered to be 1.35 m.

Figure 1 depicts the three-dimensional view and layout of a high-rise reinforced concrete building without curved end shear walls. This structure comprises 20 story levels, a reinforced concrete core, and shear walls. In contrast, Fig. 2 shows the three-dimensional view and design of a high-rise reinforced concrete building incorporating curved end shear walls (CESW). A curved end shear wall, characterized by its curved design, connects the ends of all shear walls across every story level. This building includes 20 story levels, a reinforced concrete core, shear walls, and additional reinforced concrete curved end shear walls.

Table 1 presents the specifications of the element sections, providing the details of the reinforced concrete sections for columns and beams. It also includes the dimensions of the shear walls, curved end shear walls, concrete core, and concrete slabs.

Table 2 outlines the specifications of the tall structure models, designated T1 for the tall structure without curved end shear walls and T2 for the tall structure with curved end shear walls.

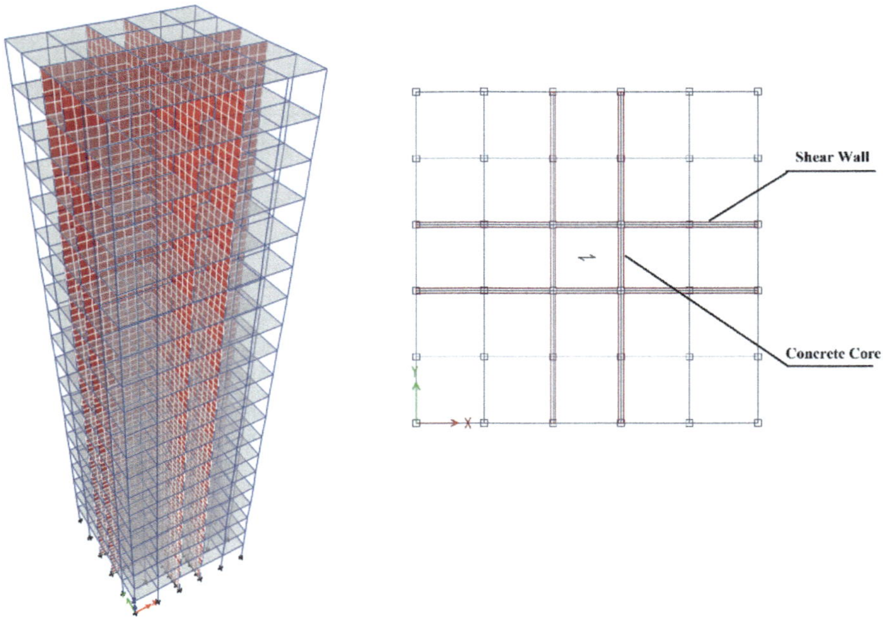

Fig. 1. Three-dimensional view and plan of the high-rise building without curved end shear walls (T1).

Table 2. The specifications of the tall structure models.

Model	Explanation	No. of stories	Total height (m)	Plan dimensions (m × m)
T1	Tall building with reinforced concrete moment frame—without curved end shear walls	20	80	20 × 20
T2	Tall building with reinforced concrete moment frame—with curved end shear walls (CESW)	20	80	22.70 × 22.70

3 Results and Discussions

Figure 3 illustrates the displacement patterns of the T1 and T2 20-story high-rise reinforced concrete buildings, both of which show a predictable increase in displacement as the height rises. However, the displacements observed in the T1 building are significantly larger compared to those in the T2 building. Specifically, the maximum displacement

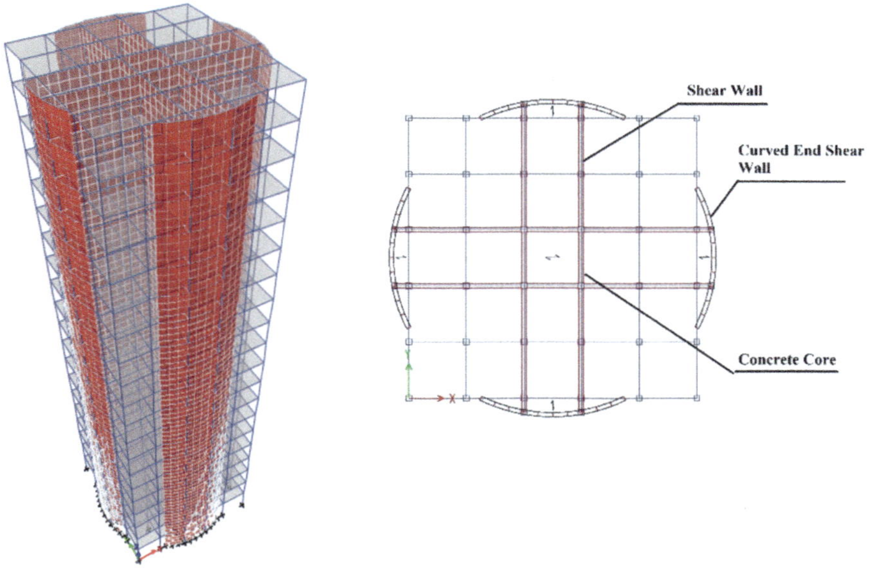

Fig. 2. Three-dimensional view and plan of the high-rise building with curved end shear walls (CESW) (T2).

Table 1. Specifications of the sections.

Label	Dimension
Columns	500*500 mm (Stories 1–10), 400*400 mm (Stories 11–20),
Beams	400*400 mm
Curved end shear walls	300 mm
Shear walls	300 mm
Concrete core shear wall	300 mm
Concrete slabs	150 mm

at the top of the structures measures 50.282 mm for T1 and 35.777 mm for T2. This reduction in displacement in the T2 building highlights the effectiveness of curved end shear walls in enhancing structural stability and mitigating displacement.

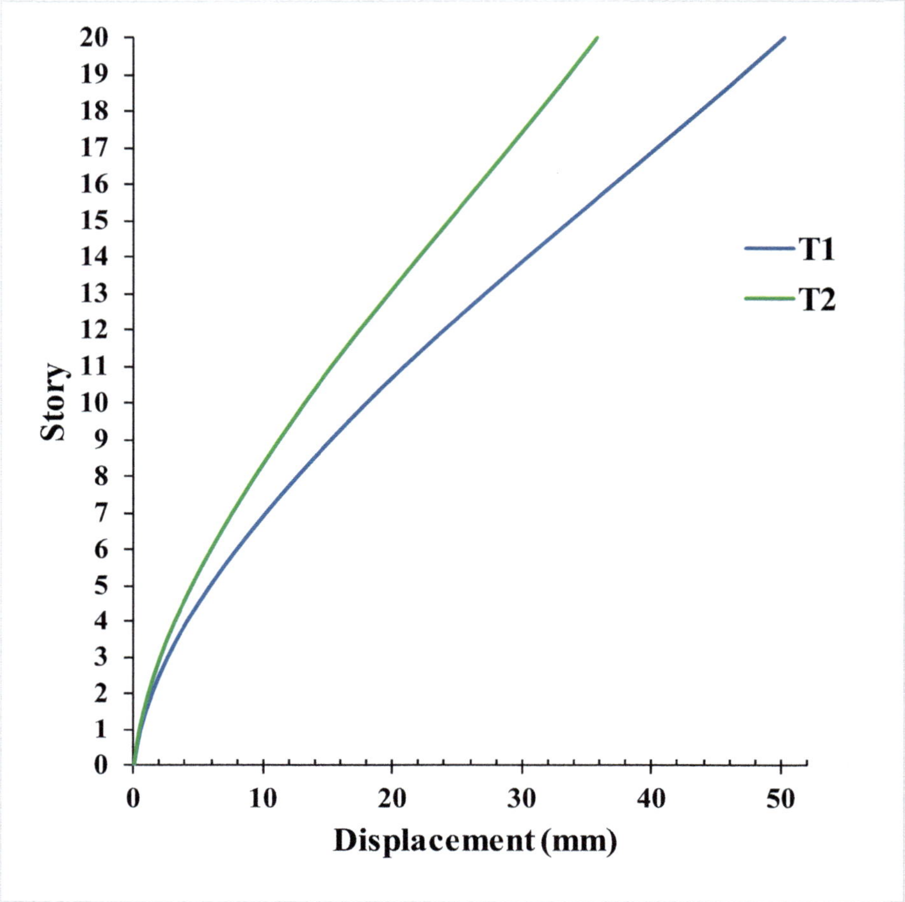

Fig. 3. The displacement of T1 and T2 buildings.

Figure 4 depicts the drift ratio (%) across all story levels for the T1 and T2 buildings. The incorporation of curved end shear walls in the T2 building led to a significant reduction in the drift ratio. Specifically, the maximum drift ratios were recorded as 0.1016% at the 14th story for the T1 structure and 0.0657% at the 15th story for the T2 structure. These results underscore the effectiveness of curved end shear walls in minimizing drift and enhancing structural performance

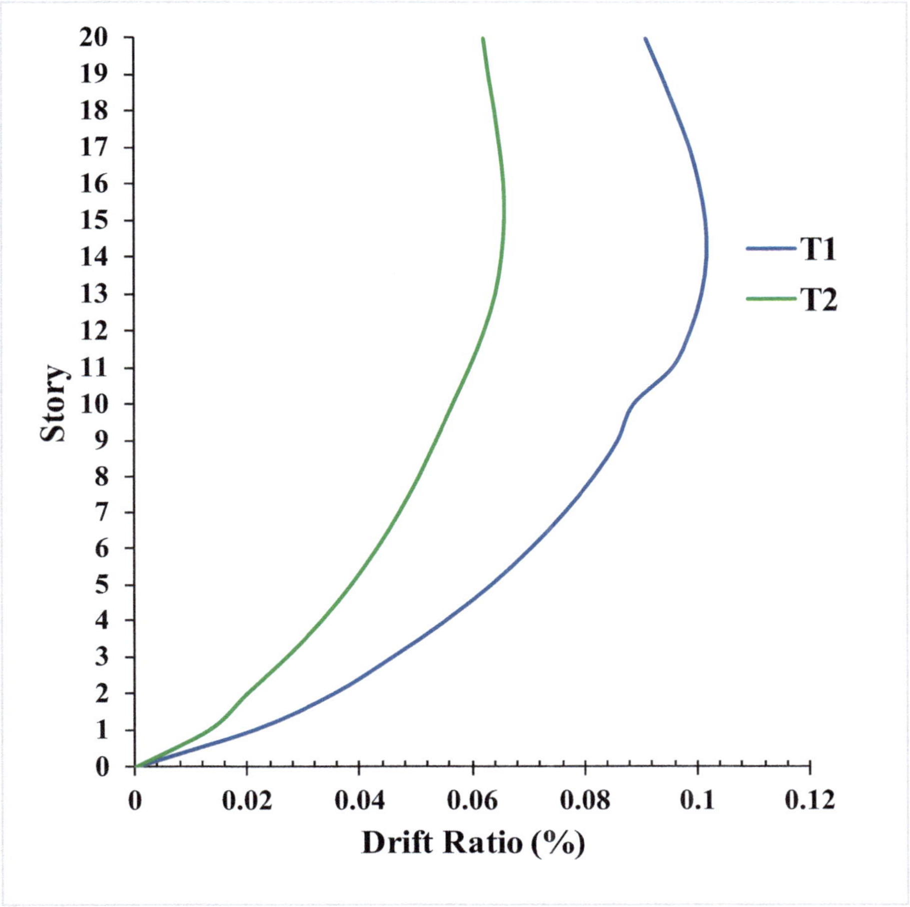

Fig. 4. The drift ratio (%) of T1 and T2 buildings.

Figure 4 illustrates the drift ratio (%) throughout the story levels of the T1 and T2 buildings. The addition of curved end shear walls in the T2 building resulted in a substantial decrease in the drift ratio. The highest drift ratios were observed as 0.1016% at the 14th story for the T1 structure and 0.0657% at the 15th story for the T2 structure. These findings highlight the efficiency of curved end shear walls in reducing drift and improving overall structural performance.

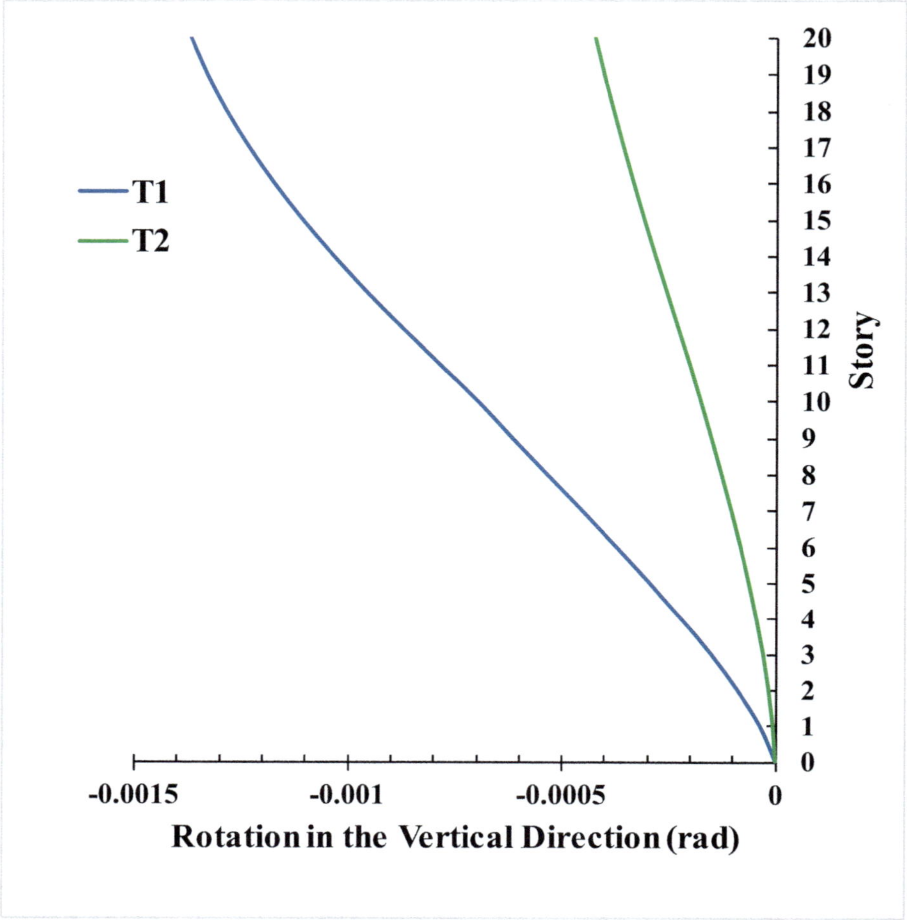

Fig. 5. The rotation in the vertical direction for the T1 and T2 buildings.

Rotation in the vertical direction plays a critical role in the seismic performance of tall structures. According to Fig. 5, the vertical rotation for the T1 and T2 20-story high-rise buildings increases as the story level rises. The analysis reveals maximum rotation in the vertical rotations of – 0.001368 rad for T1 and – 0.000426 rad for T2. These findings demonstrate that the incorporation of curved end shear walls greatly minimized rotation in the vertical rotation in the T2 high-rise building.

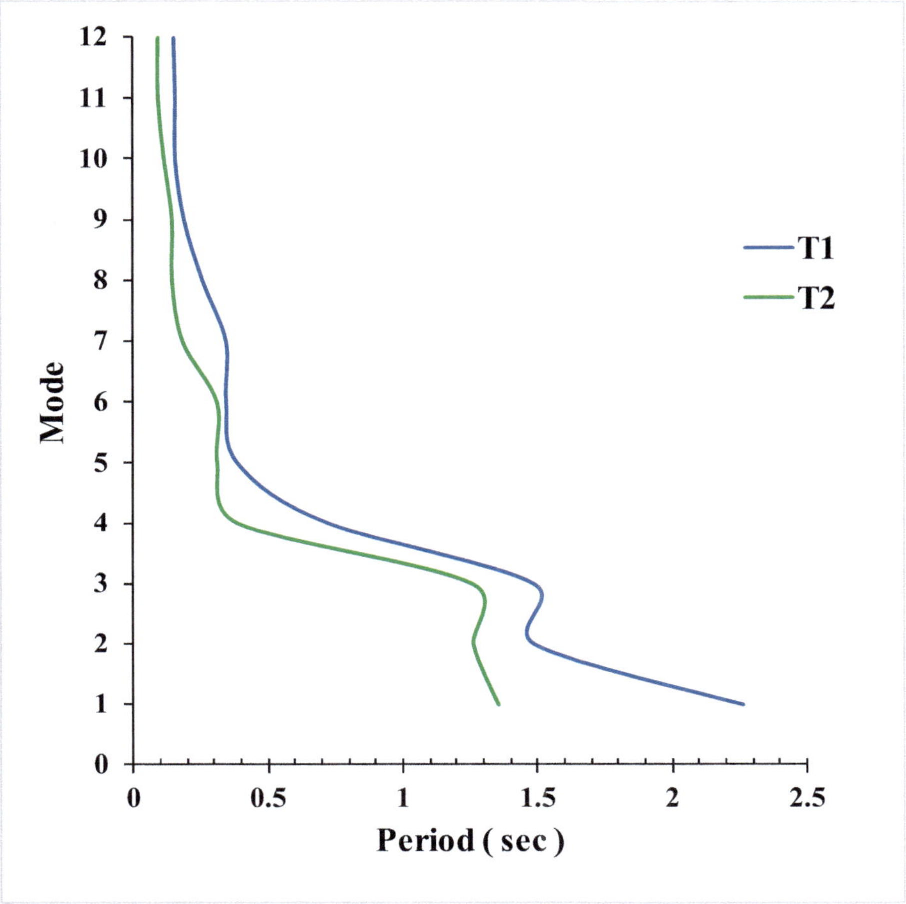

Fig. 6. The periods for the first 12 modes of the T1 and T2 buildings.

The fundamental period plays a pivotal role in the design and analysis of tall buildings. Fig. 6 illustrates the periods corresponding to the first 12 modes for the T1 and T2 buildings. As expected, the period decreases with an increase in mode number. The fundamental periods were measured at 2.265 and 1.365 s for the T1 and T2 buildings, respectively. This significant variation in the fundamental period is primarily attributed to the influence of the curved end shear walls, particularly in the first (fundamental) period.

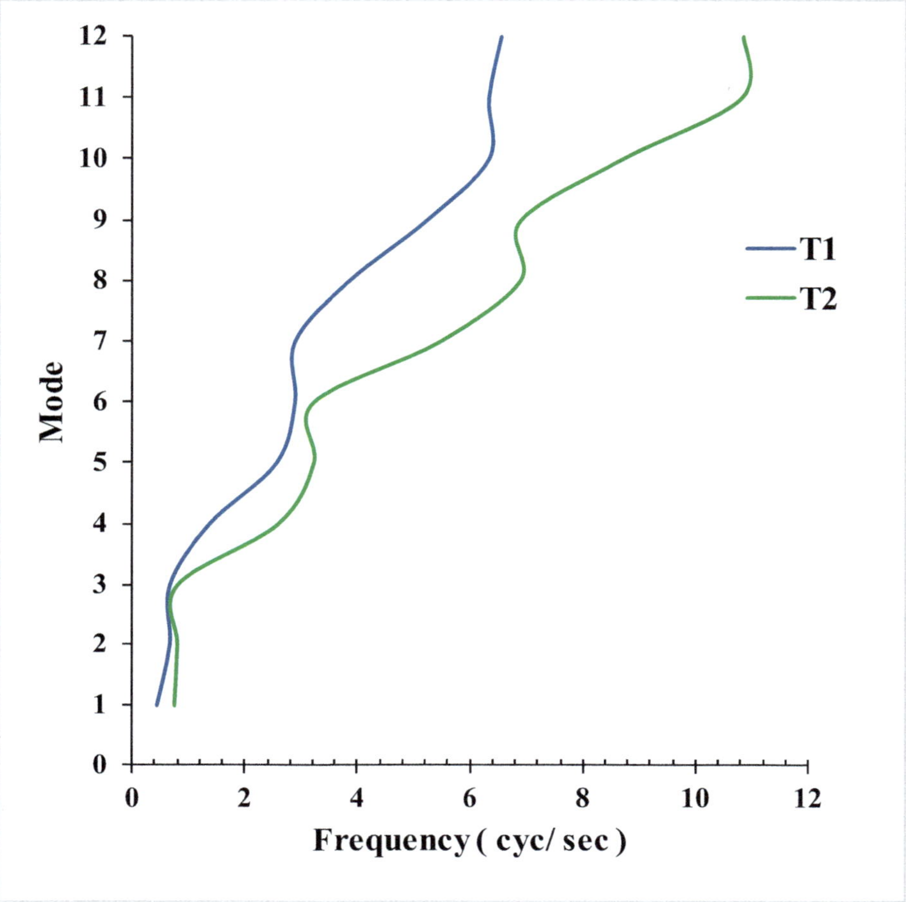

Fig. 7. The frequency for the first 12 modes of the T1 and T2 buildings.

The frequency is a crucial factor in the design and evaluation of tall structures. Figure 7 demonstrates the frequency associated with the first 12 modes for the T1 and T2 buildings. As anticipated, the period reduces as the mode number increases. The frequency of the first mode recorded was 0.441 cycles per second for the T1 building and 0.738 cycles per second for the T2 building. This substantial difference in frequency is largely due to the impact of the curved end shear walls, especially on the frequency behavior.

On the other hand, the vortex shedding in tall structures arises when alternating vortices detach perpendicularly to the wind flow, inducing oscillations. If the vortex shedding frequency aligns with the natural frequency of the structure or its components in the direction perpendicular to the wind, resonance can occur. This phenomenon poses a risk of structural fatigue, potentially compromising the integrity of the structure over time.

The critical wind speed for vortex shedding can be determined using Eq. 1

$$V_{HC} = (f_{ni}\ w)/S \tag{1}$$

$$W = \left(\sum h_i w_i\right)/\left(\sum h_i\right) \tag{2}$$

In this context, W represents the effective width of the structure or building in the direction of wind flow, and f_{ni} refers to the natural frequency of the structure in the desired mode, perpendicular to the wind flow. The Strouhal number (St), a key dimensionless parameter in fluid dynamics, effectively characterizes vortex shedding phenomena. This number is crucial in wind engineering as it helps predict vortex-induced vibrations, which can impact the stability and comfort of tall structures. The Square buildings experience stronger vortex shedding, leading to higher wind-induced vibrations. Circular buildings have a more stable aerodynamic profile, reducing vortex-induced oscillations. The Strouhal number for square buildings is lower than for circular buildings, meaning vortex shedding occurs at a different frequency. The Strouhal number typically has a value of approximately 0.2 for structures with circular plans and around 0.1 for those with square or rectangular plans in tall buildings, reflecting the influence of geometric configurations on the shedding frequency [12].

Also, The S_{T1}, S_{T2}, W_{T1}, W_{T2}, f_{ni}- $_{T1}$ and f_{ni}- $_{T2}$ are considered as follows:

$$S_{T1} = 0.1 \quad S_{T2} = 0.2, \quad W_{T1} = W_{T2}$$
$$f_{ni-T1} = 0.441 \text{cyc/s} \quad f_{ni-T2} = 0.738 \text{ cyc/s}$$

The critical wind speed for the vortex shedding ratio corresponding to T2 and T1 is obtained:

$$(V_{HC-T2}/V_{HC-T1}) = 0.83$$

The mentioned ratio presented a suitable performance of CESW in tall buildings under wind load.

4 Conclusion

This study used ETABS software to model and evaluate two 20-story reinforced concrete structures. The first model, T1, represented a high-rise building without curved end shear walls, while the second model, T2, included curved end shear walls (CESW). The analysis revealed several benefits of integrating CESW:

- Maximum story drift decreased by 34% in the T2 structure compared to T1.
- The maximum (top) horizontal displacement was reduced by 29% in T2.
- Absolute maximum rotation in the vertical direction decreased by 69% in the T2 building.
- The critical wind speed for vortex shedding dropped by 17% in tall buildings featuring curved end shear walls.

These findings highlight the role of curved end shear walls in significantly improving the seismic performance of high-rise reinforced concrete tall buildings.

References

1. Kheyroddin A, Akhavan Salmassi M, Kioumarsi M (2024) The impact of end shear walls on the behavior of reinforced concrete tall buildings with helipad for personal transportation. Iran J Sci Technol Trans Civ Eng:1–14. https://doi.org/10.1007/s40996-024-01574-8
2. Akhavan Salmassi M, Kheyroddin A, Hemmati A (2024) Enhancement of the performance of two tall buildings with end shear walls using nonlinear time history analysis: a case study. Iran J Sci Technol Trans Civ Eng 48(4):2019–2029. https://doi.org/10.1007/s40996-023-01289-2
3. Akhavan Salmassi M, Kheyroddin A, Hemmati A (2024) Evaluating the nonlinear behavior of tall buildings with end shear walls through chaos theory. Innov Infrastruct Solut 9:78. https://doi.org/10.1007/s41062-024-01385-w
4. Akhavan Salmassi M, Kheyroddin A, Hemmati A (2023) Evaluation of reinforced concrete tall buildings with end shear walls subjected to sequences far from the fault. Scientia Iranica. https://doi.org/10.24200/sci.2023.60961.7078
5. Choi SW, Seo JH, Lee HM, Kim Y, Park HS (2015) Wind-induced response control model for high-rise buildings based on resizing method. J Civ Eng Manag 21(2):239–247. https://doi.org/10.3846/13923730.2013.802742
6. Datta TK, Jain AK (1987) An analytical study of the across-wind response of cylinders due to vortex shedding. Eng Struct 9(1):27–31. https://doi.org/10.1016/0141-0296(87)90037-X
7. Reşatoğlu R, Mirata T, Karaker L (2018) Earthquake and wind load effects on existing RC minarets in North Cyprus. Int J Eng Technol 7(4):3074–308523. https://doi.org/10.14419/ijet.v7i4.16598
8. Le T-H, Caracoglia L (2016) Modeling vortex-shedding effects for the stochastic response of tall buildings in non-synoptic winds. J Fluids Struct 61:461–491. https://doi.org/10.1016/j.jfluidstructs.2015.12.006
9. Tiwari ND, Sarkar PP (2025) Prediction of dynamic loads and responses of a tall building in tornado wind: numerical analysis and experimental validation. Eng Struct 326:119510. https://doi.org/10.1016/j.engstruct.2024.119510
10. Tobber L, Yang TY, Najam FA (2025) Seismic performance of reinforced concrete (RC) shear walls with and without damped outriggers and controlled rocking. Eng Struct 330:119899. https://doi.org/10.1016/j.engstruct.2025.119899
11. ASCE 7-16 (2016) Minimum design loads and associated criteria for buildings and other structures. American Society of Civil Engineers (ASCE). https://doi.org/10.1061/9780784414248
12. Gu M, Quan Y (2004) Across-wind loads of typical tall buildings. J Wind Eng Ind Aerodyn 92(13):1147–1165. https://doi.org/10.1016/j.jweia.2004.06.004

Assessing Coastal Bridge Vulnerability Under Intensifying Hurricane Loads

Ravi Shastri[1,2], Samit Ray-Chaudhuri[1], and Behrouz Shafei[2(✉)]

[1] Indian Institute of Technology Kanpur, Kanpur, Uttar Pradesh 208016, India
[2] Iowa State University of Science and Technology, Ames, IA 50011, USA
shafei@iastate.edu

Abstract. Increasing frequency and intensity of hurricanes pose serious hazards to the aging coastal bridges. A high-intensity hurricane generates a heightened storm surge, severe wind forces with steepened dynamicity, and substantial wave impacts compared to what anticipated for the bridge when it was initially designed. Conventional design philosophy considers the demand uncertainty by assigning load factors to the hurricane forces and uses the obtained deterministic value for structural design. A sufficiently high value of load factor is considered to ensure structural safety against the chaotic nature of the hurricane phenomenon. Although these load factors provide a reasonable margin of protection against the design demand, or at times even at higher levels, they fail to quantify the structural performance, even for the level for which it is designed. The lack of clarity over the performance assessment sometimes leads to overestimating demand forces for the design level and uncertainty at the higher design levels. The overestimation leads to an unnecessary increase in material usage, adding to the structure's carbon footprint. Besides, an intensified hurricane load can create a challenging situation even for the overestimated load factor, as no information about the exact performance behavior is available for assessment. Adding to this, the capacity degradation of an aging bridge would generate new questions about structural safety. In response to the outlined aspects, this study presents an analysis framework to evaluate bridge vulnerability under intensifying hurricane scenarios using performance-based assessment. This work uses a representative coastal bridge to examine the structural fragility. The findings emphasize the need for scenario-specific, failure-mode-driven evaluations and support scalable, material-efficient adaptation strategies consistent with the net-zero future goals.

Keywords: Bridges · Hurricane · Performance assessment · Fragility surface

1 Introduction

According to the United States Environmental Protection Agency, an increase in sea surface temperatures over the years has led to an increased accumulated cyclone energy (ACE) index, making it move above normal with a rise in frequency over the past decade. In the Atlantic, where hurricanes dominate ACE, which implies a future rise in hurricane intensity under continued warming. The National Oceanic and Atmospheric

H. Beushausen et al. (Eds.): NTZR 2025, LNCE 873, pp. 662–670, 2026.
https://doi.org/10.1007/978-3-032-27698-8_59

Administration's Geophysical Fluid Dynamics Laboratory [1] model simulations also predict an increase of about 14% globally for rainfall rates of a storm for a rise of 2° in global temperatures. Peer-reviewed research articles also projected an increased U.S. coastal hurricane risk with an impact on storm surge inundations due to a changing climate.

The increasing risk associated with hurricane occurrences and related events mandates re-evaluating structural vulnerability, specifically for aging coastal bridges, to provide a resilient bridge infrastructure system. Several studies [2–10] have been conducted in the past that evaluate damage to bridges and other coastal transportation networks due to hurricanes and related events, highlighting the importance of multiple intensity measures coming into play for these events. Recent studies have started focusing on adaptive methodologies of climate-resilient design to ensure safety against intensifying natural events. Ayyub [11] characterized extreme events like precipitation and flooding in multi-hazard scenarios. The study also provided adaptive design and risk management frameworks followed by life-cycle cost analyses. Asam et al. [12] highlighted the rationale behind requiring a transformative approach to respond to extreme events. The referenced study's focus was primarily on transportation infrastructure, suggesting steps to make the transportation system adaptive and robust.

Contemporary design guidelines [13, 14] suggest a systematic method for force estimation due to hurricane events and constant load factors to account for uncertainty in the event intensity. A deterministic estimation of design demand relies primarily on historical hazard assumptions, which may be insufficient to ensure safety given the increasing frequency and intensity of extreme events. This approach may also fail to predict structural performance, especially for aging bridge infrastructure accurately.

Considering this, a probabilistic estimation of bridge stability under these forces is necessary. This study models a typical bridge for a site in the coastal regions of Puerto Rico. It estimates its failure probabilities under increasing hurricane intensities, modeled using changing storm surge depths (d_s) and 1-min sustained wind speed at 10 m above ground ($U_{1-\text{min}}$). Exceedance of anchor bolt capacity upon deck unseating is considered the failure criterion. The chosen intensity measures (IMs) and damage criterion develop a fragility surface for the static force values determined following [14]. Probabilistic estimation of performance would help select the highly vulnerable bridges of the region and subsequently plan retrofit strategies to increase structural resilience with the least material usage.

2 Hurricane-Induced Bridge Failure Mechanisms

Numerous coastal bridges have been damaged due to hurricane events in the past [15]. Padgett et al. [16] reports the damaged superstructure that included deck unseating and failure of parapets due to Hurricane Katrina. Studies, such as [17–21], documented other failure modes such as substructure damage, damage to mechanical and electrical systems, pile damage, and scour damage. Of all the damage types, deck unseating was the most common observed damage listed in the literature. So, deck unseating was selected for this study, as the main criterion for damage representation. Since a deck is connected to the pier cap in a bridge, deck unseating is directly linked to the connection strength between the pier cap and the deck.

Typically, the pier caps and decks of a concrete bridge are connected using elastomeric bearings and anchor bolts in parallel [2, 9]. While elastomeric bearings primarily resist compressive forces with a small displacement and rotation allowance, an anchor bolt is designed to withstand the uplift forces, and its rigidity is very high compared to elastomeric bearings. Thus, anchor bolts contribute to most of the connection strength between a pier cap and deck, which implies that a connection failure could be directly attributed to anchor bolt capacity for a conservative estimation.

To understand the failure mechanism mentioned above, it is essential to understand the forces imposed due to a hurricane in addition to structural configuration information. Past studies and code provisions [2, 9, 14] suggest simplifying the hurricane-imposed demand through three types of forces: (i) horizontal, (ii) vertical, and (iii) slamming forces.

3 Methodology

A vulnerability study, popularly represented by a fragility function, encapsulates estimating the probability of structural failure under an applied set of forces. Identifying a failure or non-failure instance is done by comparing the demand and capacity values. The cases where demand exceeds the structural capacity are classified as failure, else a non-failure instance. Demand and capacity estimation involve following a methodological and systematic procedure, the details of which are presented in the upcoming subsections.

3.1 Bridge Description

A typical two-span concrete bridge, supported by continuous supports mid-span and abutments at the ends, is modeled to perform the vulnerability study for this work. A 23-m-wide bridge slab is simulated using closely spaced beams spanning in the longitudinal direction. Pier cap is modeled using beams spanning in the transverse direction. Both the slab and pier caps are assumed to behave linearly elastic. Non-linearity is included in the bridge only through the fiber section for the pier cross-section. The bridge configuration is adopted from the work by Alipour et al. [22].

The supported abutment ends are modeled using simple elastic springs in transverse and longitudinal directions; the stiffness values are taken from the California Department of Transportation [23]. The vertical supports are simulated using a spring of high rigidity to mimic the rigid element behavior. Slab beams to the pier cap are connected using parallel anchor bolts and an elastomeric bearing. Anchor bolts are made of nonlinear elastic steel material. Fig. 1 shows the simplified configuration for the bridge. The bridge is designed to provide adequate functional safety against the hurricane category I, for the criteria listed in [13].

3.2 Load Model

As discussed in Sect. 2, the hurricane forces are modeled using three force types: horizontal forces acting in each of the longitudinal and transverse directions of the bridge,

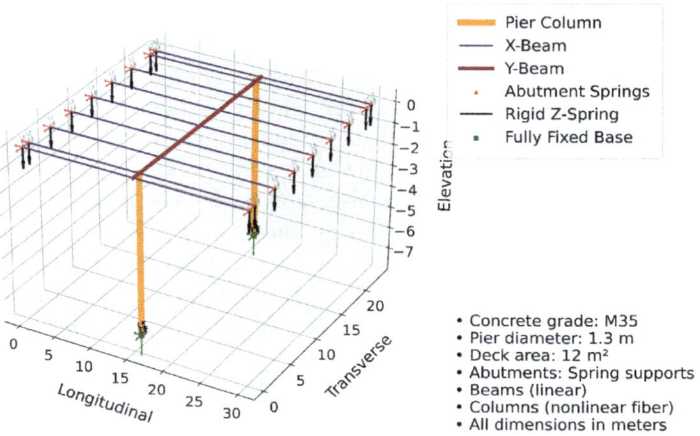

Fig. 1. The bridge model developed for the current study.

and a combination of vertical and slamming forces acting in the vertical direction. These forces are distributed throughout the bridge deck and applied as nodal loads in the respective directions. The mathematical formulation for these forces is given in [14]. The force values thus estimated are deterministic, as these values are unique for the given input parameters. Since the input is assumed to be constant for a chosen hurricane category, so are the obtained forces. However, a vulnerability study requires studying the probabilistic comparison of demand and capacity values. For the demand side, the variability could be included in analyses by presuming the parameter values to lie within a range instead of a single deterministic value, thus yielding a set of force values instead of a single force value.

Hurricane intensities can be represented using multiple parameters, such as $U_{1-\mathrm{min}}$, d_s, significant wave height (H_s), and peak spectral period (T_p). For this study, d_s and $U_{1-\mathrm{min}}$ are the IMs under consideration. The IM values for different hurricane categories help estimate the relevant parameters to calculate force values, as described in [14]. Uncertainty in demand estimation is introduced by selecting a set of values for the parameters, d_s, H_{max}, and T_p, instead of choosing single values, as done conventionally for a deterministic force estimation method. For this study, 50 samples for each d_s and H_{max}, along with 30 samples of T_p, are selected, thus making a total of 1500 combination of force values for each hurricane category. A step-by-step procedure for the uncertainty quantification of demand values is discussed in [9].

Calculation of T_p involves estimating roots to a conditional probability equation, which does not yield real roots for a few instances, hence, an attempt to sample 1500 values, results in a value less than 1500. The sampled values for the hurricane forces are shown in Fig. 2, to visualize the range of variation of forces for the chosen hurricane category. The plot shown in Fig. 2 is estimated for hurricane category I.

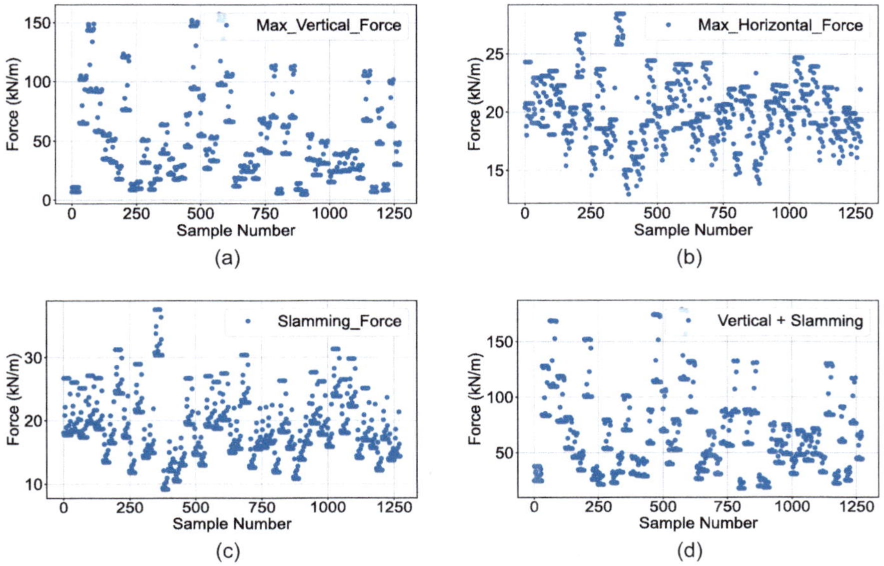

Fig. 2. Sampled values for (a) vertical, (b) horizontal, (c) slamming, and (d) total vertical force.

3.3 Failure Criterion

Once the numerical model for the concerned structure and demand force estimation is ready, the next step is to perform structural analyses to obtain structural responses. The estimated response values are then compared to the capacity values to separate failure and non-failure cases. A structure can have multiple response parameters, say, member forces, deformations, joint rotation, etc. As discussed in Sect. 2, this work involves deck unseating as the sole failure criterion, which is modeled by the deck-pier cap connection failure, further implied by a failed anchor bolt at any connection.

Instances of exceedance of axial forces in the anchor bolt above the threshold tension capacity, or shear force exceeding the shear capacity, are marked as failure; otherwise, a non-failure case. The connection capacity between concrete and anchor bolts is determined as the minimum values corresponding to different failure modes for each failure type: tension and shear. The formulations for these capacity values are taken from [24].

3.4 Fragility Surface Development

A ratio of failure instances to the total sample of force values for any hurricane category yields the failure probability for the selected category. The probability values are obtained for each category discussed in the Saffir-Simpson scale. In addition, increasing the IM values following a linear trend assumes five more category values. Once the failure probability values (P_f) for all 10 categories are estimated, these values are plotted against each of the two IMs. Lognormal functions are fitted to the obtained data points using the maximum likelihood estimator.

Since hurricane intensity is represented through two IMs, a fragility surface is fitted to the probability data by considering the two IMs independent. Development of the fragility surface is through the maximum likelihood estimator, as performed separately for the two IMs.

4 Results and Observations

The computed uplift forces, i.e., a combination of vertical and slamming forces, are compared against developed axial forces in the anchor bolts to distinguish the failure and non-failure cases. Developed shear forces in the bolts were significantly less than the shear capacity of the connection strength; hence, all the failed cases were due only to tensile forces. Fig. 3 compares these force values for a category V hurricane. P_f is obtained for each hurricane category as a ratio of failed cases to the total load instances for the selected hurricane category. P_f calculation is followed by fitting fragility curves and surfaces for individual and a combination of IMs.

The maximum likelihood estimator obtains the mean (μ) and standard deviation (σ) values for the two IMs separately. These parameters obtain fragility curves for d_s and $U_{1\text{min}}$, shown in Fig. 4. Probability values are further used to develop the fragility surface, assuming the two IMs are independent.

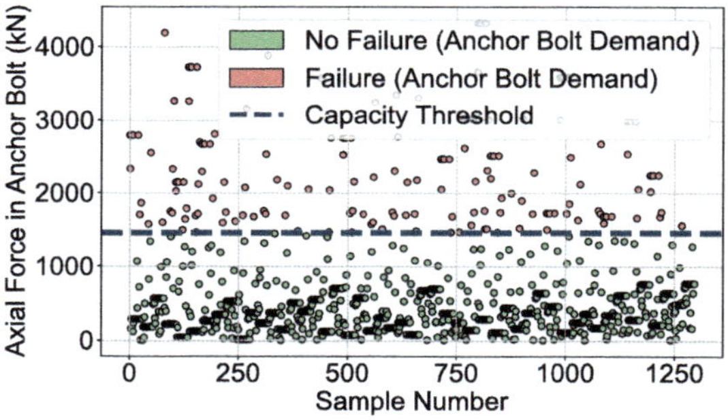

Fig. 3. Demand force versus anchor bolt-concrete connection capacity.

Table 1. Fitted statistical parameters for both the intensity measures, d_s and $U_{1\text{min}}$.

Statistical parameters	IM_1 (storm surge depth, d_s) (in m)	IM_2 (wind speed, $U_{1\text{min}}$) (in m/s)
Mean (μ)	7.48	101.68
Logarithmic standard deviation (σ)	0.32	0.24

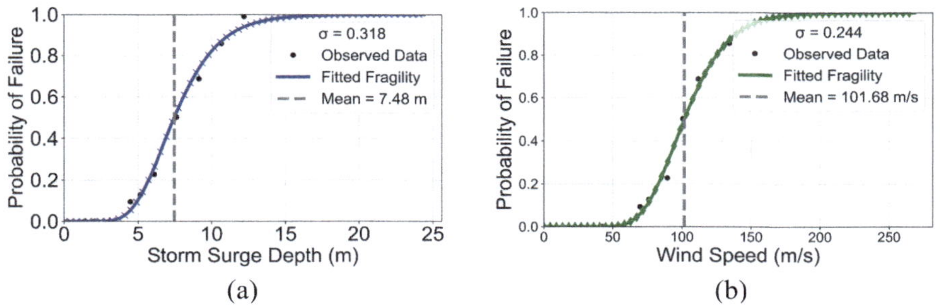

Fig. 4. Fragility plot for (a) storm surge depth d_s, and (b) wind speed $U_{1\min}$.

Fig. 5 shows the plotted fragility surface. The fragility function parameters for the two IMs are listed in Table 1. Results from the fragility analyses indicate that the bridges show instances of failure for higher hurricane categories, i.e., IV and V, where d_s values range from 3.0 to 5.2 m for regions in Puerto Rico, as obtained from SLOSH MOM [25]. Fig. 4a shows that the probability of failure is 5–10%.

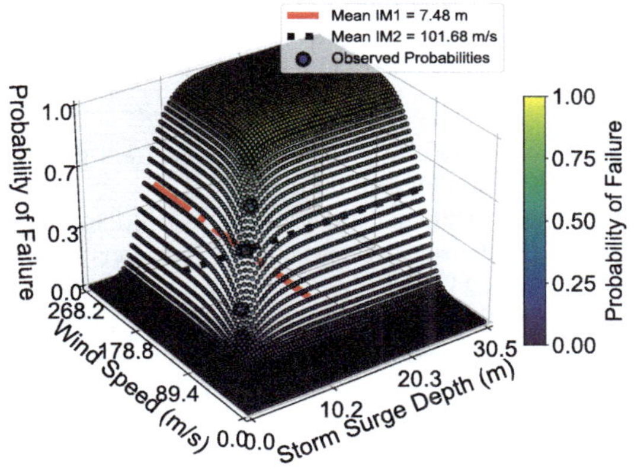

Fig. 5. Fragility surface plot for hurricane-induced forces.

A qualitative comparison of Hurricane Ian's damage [26] that made landfall as a category IV hurricane with d_s as high as 4.26 m witnessed multiple bridge damage instances, including in Sanibel Causeway, New Pass, and Big Hickory Pass. These three bridges were rendered inoperative, disrupting connectivity to barrier islands as mentioned in [26]. Such real-world damage instances further confirm the accuracy of the predictions obtained through the current study. The overall situation also highlights the need for performance-based approaches instead of relying only on code-specified loads.

5 Conclusions

Considering uncertainties with the estimation of d_s, $H_{\max}$, and T_p values leads to significant variation in the force values. Vertical and slamming forces exhibit a larger variation in force values than horizontal forces. The probable reason behind such behavior is their direct dependence on storm surge and wind speed for the vertical and slamming forces. While the values are scattered throughout the plot for horizontal forces, stating that the values are somewhat uniformly distributed, the trend differs for vertical and slamming forces, where occasional rises in values are seen. A plausible explanation for this trend could be the dynamism of wind and wave height variation.

From the structural analyses for changing intensity of hurricane forces, it can be noted that the probability values are 0 for the first three hurricane categories and increase with further magnified intensities. Since the bridge was designed to resist the demand forces for hurricane category I, it remained safe under that demand level even with an increased uncertainty. Furthermore, the bridge's survival under Category II and III loading conditions suggests that the 1.75 load factor prescribed to the wave forces may be overly conservative, leading to inefficient and excessive use of materials. For judicial material usage, a need to examine performance assessment through a fragility analysis is a viable option.

References

1. National Oceanic and Atmospheric Administration. Global warming and hurricanes. https://www.gfdl.noaa.gov/global-warming-and-hurricanes/
2. Saeidpour A, Chorzepa MG, Christian J, Durham S (2018) Parameterized fragility assessment of bridges subjected to hurricane events using metamodels and multiple environmental parameters. J Infrastruct Syst 24
3. Torabi T, Shafei B, Alipour A (2025) intensity measures for flood hazards in fragility assessments of bridges. In: Lecture notes in civil engineering. Springer Nature, pp 1539–1547
4. Twumasi-Boakye R, Sobanjo JO (2018) Resilience of regional transportation networks subjected to hazard-induced bridge damages. J Transp Eng Part A: Syst 144
5. Alipour A, Shafei B, Shinozuka M (2013) Capacity loss evaluation of reinforced concrete bridges located in extreme chloride-laden environments. J Struct Infrastruct Eng 9(1):8–27
6. Testa AC, Furtado MN, Alipour A (2015) Resilience of coastal transportation networks faced with extreme climatic events. Transp Res Rec 2532:29–36
7. Alipour A, Shafei B, Shinozuka M (2013) Reliability-based calibration of load and resistance factors for design of RC bridges under multiple extreme events: scour and earthquake. J Bridg Eng 18(5):362–371
8. Alipour A (2016) Post-extreme event damage assessment and response for highway bridges: a synthesis of highway practice. Transportation Research Board, Washington, D.C.
9. Saeidpour A, Chorzepa MG, Christian J, Durham S (2019) Probabilistic hurricane risk analysis of coastal bridges incorporating extreme wave statistics. Eng Struct 182:379–390
10. Amini K, Padgett JE (2023) Probabilistic risk assessment of hurricane-induced debris impacts on coastal transportation infrastructure. Reliab Eng Syst Saf 240:109579
11. Ayyub BM (2018) Climate-resilient infrastructure: Adaptive design and risk management. American Society of Civil Engineers

12. Asam S, Bhat C, Dix B, Bauer J, Gopalakrishna D (2015) Climate adaptation guide for transportation systems management, operations, and maintenance. Federal Highway Administration, Washington, DC
13. AASHTO (2020) LRFD bridge design specifications. American Association of State Highway and Transportation Officials, Washington, D.C.
14. AASHTO (2008) Guide specifications for bridges vulnerable to coastal storms. American Association of State Highway and Transportation Officials, Washington, D.C.
15. Chorzepa MG, Saeidpour A, Christian JK, Durham SA (2016) Hurricane vulnerability of coastal bridges using multiple environmental parameters. Int J Saf Secur Eng 6:10–18
16. Padgett J et al (2008) Bridge damage and repair costs from hurricane Katrina. J Perform Constr Facil 22:94–100
17. Okeil AM, Cai CS (2008) Survey of short-and medium-span bridge damage induced by hurricane Katrina. J Perform Constr Facil 22:373–380
18. Xiao H, Huang W, Chen Q (2010) Effects of submersion depth on wave uplift force acting on Biloxi Bay bridge decks during hurricane Katrina. Comput Fluids 39:1390–1400
19. Lamb R, Aspinall W, Odbert H, Wagener T (2017) Vulnerability of bridges to scour: insights from an international expert elicitation workshop. Nat Hazards Earth Syst Sci 17:1393–1409
20. Fan G et al (2025) Study on scour simulation and boundary condition conversion technology for a shallow foundation bridge. Sci Rep 15
21. Oppong K, Saini D, Shafei B (2020) Vulnerability assessment of bridge piers damaged in barge collision to subsequent hurricane events. J Bridg Eng 25
22. Alipour A, Shafei B, Shinozuka M (2011) Performance evaluation of deteriorating highway bridges located in high seismic areas. J Bridg Eng 16:597–611
23. California Department of Transportation. Caltrans. https://dot.ca.gov/
24. ACI Committee 318 (2014) Building code requirements for structural concrete (ACI 318–14) and commentary (ACI 318R-14). American Concrete Institute, Farmington Hills, MI
25. NOAA (2025) SLOSH maximum of the maximum (MOM) storm surge map for Puerto Rico. National Oceanic and Atmospheric Administration, Miami, FL
26. Lee County Emergency Management (2023) Hurricane Ian after-action report. Lee County, Fort Myers, FL

Analysis of Tornado Loads on Wind Turbine Structural Performance: Implications for Sustainable Design and Resilience

Muhammad Usama Aslam and Alice Alipour(✉)

Department of Civil, Construction, and Environmental Engineering, Iowa State University, Ames, IA, USA
alipour@iastate.edu

Abstract. The increasing adoption of renewable energy, especially wind power, plays a critical role in the global transition toward a net-zero carbon future. However, with the taller and more slender wind turbine technologies and deployment in high wind regions comes the potential for damaging winds impacting these structures. In fact, tornadoes have resulted in multiple damage to wind turbines in the past few years. This paper presents a comparative analysis of five tornado models on wind turbine structural performance. The study focuses on evaluating the effects of these tornado models on wind turbine resilience using finite element modeling. This research emphasizes the importance of accounting for severe weather events in wind turbine structural assessments by analyzing how different tornado models impact wind turbine design under extreme conditions. The results highlight how tornado-induced loads influence turbine stability and can inform more resilient design practices. As the adoption of wind energy continues to rise, understanding the effects of extreme weather on turbine structures is crucial for ensuring the long-term sustainability and reliability of wind energy systems. The findings have implications for the long-term sustainability of wind energy systems, particularly in regions prone to extreme weather events.

Keywords: Tornado loads · Wind turbine resilience · Finite element analysis · Structural performance · Sustainable energy systems

1 Introduction

Net-zero emission targets require a worldwide transition to clean energy technology, and wind power has been a central pillar of that process. Its ability to scale, minimal operational emissions, and quick technological development make it a key performer in global decarbonization. Wind power is one of the mature and fast-developing solutions that has a scalable and sustainable way of replacing fossil fuels. The globally installed wind energy capacity increased from fewer than 100 GW of power in 2000 to over 1,000 GW in 2023, demonstrating its crucial contribution to the clean energy revolution [1]. Yet, the dependability of wind power equipment is constantly tested by the increasing frequency and intensity of violent climatic occurrences, including tornadoes. Though

H. Beushausen et al. (Eds.): NTZR 2025, LNCE 873, pp. 671–683, 2026.
https://doi.org/10.1007/978-3-032-27698-8_60

standard wind design codes like IEC 61400 contain elaborate descriptions of operational and storm winds, they tend to fail when it comes to dealing with the highly dynamic and turbulent conditions created by tornadoes. This design deficiency is of paramount importance, particularly in areas like the central and southeastern part of the USA, where past statistics between the years 1950 and 2019 indicate concentrated activity of violent EF4 and EF5 occurrences [2]. The geographical overlap between tornado-prone regions and active wind energy development zones heightens the urgency to re-evaluate wind turbine structural resilience under such conditions. Tornado wind fields are characterized by non-stationary, non-Gaussian flow behavior, often involving embedded vortices, abrupt direction reversals, and intense vertical components. These features introduce aerodynamic instabilities and transient load reversals that differ significantly from synoptic wind patterns typically considered in structural wind engineering. Unlike hurricane or straight-line winds, tornadoes impose localized pressure spikes and complex uplift profiles that can exceed standard design thresholds and trigger cascading failure mechanisms in turbine towers, nacelles, and blades.

Addressing this challenge requires a dual focus on energy resilience and structural engineering. Qudaisat and Alipour [3] highlighted the role of distributed energy resources in enhancing infrastructure resilience to extreme events. Aslam and Alipour [4] analyzed wind turbine failures in the 2024 Greenfield tornadoes, identifying key damage patterns and emphasizing the need for improved tornado-resistant design of utility-scale turbines. Alipour et al. [5] developed a comprehensive analytical model to characterize tornado-induced loads on lattice structures, revealing that conventional wind load approximations substantially underestimate the magnitude and complexity of these forces, particularly due to the influence of tornado swirl ratios and transient pressure gradients. From a climatological standpoint, Romanic et al. [6] conducted a detailed analysis of tornado and wind climatology across tornado alley, emphasizing spatial and temporal variability. The study integrated meteorological data with wind resource modeling to support strategic planning for renewable energy infrastructure. Broader energy resilience strategies also intersect with infrastructure design. Younesi et al. [7] proposed methods for enhancing the resilience of zero-carbon energy communities by utilizing network reconfiguration and load-carrying metrics, thus emphasizing the interconnected nature of structural and energy resilience. From a structural enhancement perspective, Faharidine et al. [8] demonstrated improved buckling resistance and stability in isotrussed towers subjected to high wind loads, a valuable precedent for improving wind turbine tower designs. In the context of digital energy forecasting, Elmousalami et al. [9] highlighted the importance of integrating AI-driven forecasting models to improve decision-making and operational reliability in zero-carbon wind energy systems. These approaches reinforce the role of smart predictive analytics in enhancing turbine performance and environmental resilience. In terms of broader policy and planning, Goyal et al. [10] emphasized the necessity of embedding resilience into wind energy expansion strategies to meet carbon neutrality goals without compromising system reliability. Tornadoes' directional volatility further complicates structural design. Kwon and Kareem [11] proposed a gust-front factor to adjust design parameters for sudden shifts in wind vector orientation conditions, particularly relevant in tornadic flow fields. On the economic and sustainability front, Adhikari et al. [12] demonstrated that integrating tornado-resistant features into residential buildings results in long-term lifecycle savings and lower carbon emissions, reinforcing the case for resilience-driven design. Nationally, NIST [13] has outlined critical research gaps and called for new design codes, experimental validation, and policy initiatives to mitigate tornado-related risks across critical infrastructure.

Despite meaningful advancements in wind energy research and resilience planning for hurricanes and earthquakes, the structural response of wind turbines under tornado-induced loading remains insufficiently addressed. Prior studies have predominantly relied on idealized or simplified loading assumptions, often neglecting the non-stationary, multi-vortex, and highly transient nature of real tornado wind fields. Notably, empirical tornado models that capture the spatial and temporal complexity of observed tornado events have yet to be directly integrated into finite element models for wind turbine analysis. This research aims to close that critical gap by incorporating empirically derived tornado wind profiles into a high-fidelity FEM framework. This approach offers a more realistic evaluation of structural demands and potential failure mechanisms, thereby advancing the resilience and reliability of wind energy infrastructure in tornado-prone regions. Closing this gap is essential to securing the long-term performance of wind energy infrastructure and ensuring its contribution to a sustainable, net-zero future.

2 Review of Empirical and Analytical Tornado Models

Tornado-induced loads pose a significant challenge for structural engineers due to their highly localized intensity, rapidly changing direction, and lack of consistent forecasting data. Although numerous theoretical and empirical models have been developed to describe tornado dynamics, their integration into structural engineering applications, particularly in finite element modeling (FEM) of wind turbine systems, has been limited. These models aim to capture the spatially and temporally complex characteristics of tornado vortices, including their unique pressure distributions and velocity gradients. However, the use of such models in simulating tornado impacts on wind turbine structures remains relatively rare in the literature. While prior research has made strides in characterizing tornado flows, few studies have systematically embedded these models within detailed FEM frameworks for engineering evaluation. The need for robust evaluation frameworks has been emphasized across various structural domains, as demonstrated by Aslam et al. [14] This study contributes to filling that gap by implementing a range of tornado models into FEM simulations, enabling a comparative assessment of their structural implications on wind turbine installations.

Rankine [15] introduced a basic vortex flow model with a solid-body core and outer irrotational region, but it neglects viscous and vertical effects. Bjerknes [16] advanced pressure-based models for cyclonic tangential velocities, limited by the exclusion of radial and vertical flows. Burgers [17] incorporated eddy viscosity into smooth velocity transitions, though both remained limited by steady-state and axisymmetric. Sullivan [18] applied the Navier-Stokes equations in a two-cell vortex model, capturing downdrafts and inflow but lacking transient dynamics. Fujita [19] used empirical observations to model vortex structure, aiding intensity classification but oversimplifying turbulence. Kuo [20] added 3D and boundary-layer effects, improving near-ground accuracy under steady-state assumptions. Vatistas et al. [21] offered closed-form empirical solutions for tangential velocity but remained 2D. Wood and White [22] refined curve fitting for tangential winds in parametric models, still limited by axisymmetric. Baker and Sterling [23] integrated all wind components for debris and load prediction, advancing tornado-resilient design. Kim and Matsui [24] compared models for structural applications, highlighting the need for context-specific selection.

While a wide range of tornado vortex models has been proposed, ranging from simplified analytical formulations to complex empirical and computational frameworks, few have been systematically implemented within finite element simulations for structural analysis. Many existing models either lack the dimensional fidelity required for structural loading assessments or depend on assumptions that limit their applicability in dynamic, real-world tornado scenarios. To address this gap, the present study selects five representative models: two developed by Baker, along with the Fujita, Kuo-Wen, and Burger–Rott models. These were chosen for their three-dimensional formulation, historical and scientific relevance, and suitability for integration into finite element modeling environments. Together, they offer a well-rounded comparison of both empirical and theoretical perspectives on tornado wind fields, enabling a robust evaluation of their effects on wind turbine structural response. The detailed formulations for the radial and tangential velocities of these five tornado models are summarized in Table 1. Further details on specific expressions and their implementation can be found in the referenced papers.

Table 1. Summary of numerical expressions for analytical and empirical tornado models

Model	Radial velocity (U)	Tangential velocity (V)
Baker Model [25]	$U = U_{\max}\frac{-4\left(\frac{r}{r_m}\right)\left(\frac{z}{z\mathrm{m}}\right)}{\left\{1+\left(\frac{r}{r\mathrm{m}}\right)^2\right\}\left\{1+\left(\frac{z}{z\mathrm{m}}\right)^2\right\}}$	$V = U_{\max}\frac{K_B\left(\frac{r}{r_m}\right)\ln\left\{1+\left(\frac{z}{z_m}\right)^2\right\}}{1+\left(\frac{r}{r_m}\right)^2}$
Fujita Model	$U = V \cdot \tan\alpha$	$V = V_{\max}F(r)F(z)$
Kuo-Wen Mode	$U_o = 0$ $U_i =$ $V_o\{0.672\exp(-\pi\eta_{KW})\sin((b+$ $1)\pi\eta_{KW})\}$	$V_o =$ $1.4\left(\frac{V_{o,\max}}{r/r_\infty}\right)\left\{1-\exp\left(-1.256\left(\frac{r}{r_\infty}\right)^2\right)\right\}$ $V_i = V_o\{1-$ $\exp(-\pi\eta_{KW})\cos(2b\pi\eta_{KW})\}$
Burger-Rott Model	$\overleftharpoon{U} = ar$	$\overline{V} =$ $\left(\frac{\overline{V}_{\max}}{K_{BR2}}\right)\left(\frac{R}{r}\right)\left\{1-\exp\left[-K_{BR1}(r/R)^2\right]\right\}$
Baker Model [26]	$\overline{U} = -K\frac{\overline{r}^{\gamma-1}}{(1+\overline{r}^2)\gamma/2}\frac{\overline{z}^{(\gamma-2)/2}(1-\overline{z}^2)}{(1+\overline{z}^2)(\gamma+2)/2}$	$\overline{V} = \frac{4\overline{rz}}{(1+\overline{r}^2)(1+\overline{z}^2)}$

2.1 Comparative Evaluation of Radial and Tangential Velocity Profiles in Tornado Vortex Models

A comparative evaluation of five established tornado vortex models, Baker [25, 26], Burgers-Rott, Fujita, and Kuo-Wen, was conducted to assess their representation of wind velocity components along both the vertical and radial domains. Fig. 1a illustrates the normalized radial velocity profiles as a function of height. The Baker [25] model exhibits a nonlinear inflow pattern that intensifies with height and peaks around the mid-level of the domain, characteristic of energy-conserving vortex formulations. The Kuo-Wen model, by contrast, presents a complex inflow structure influenced by near-surface boundary layer effects, including radial flow reversal. The Fujita model captures concentrated inflow near the surface, while the Burgers-Rott model demonstrates a steady, monotonic decrease in radial inflow velocity with height, reflecting its laminar, steady-state nature and the absence of vertical complexity. Tangential velocity distributions with height are shown in Fig. 1b. The Baker and Burgers-Rott models produce nearly linear vertical growth, consistent with idealized axisymmetric vortex structures. Meanwhile, the Fujita and Kuo-Wen models capture rapid increases in tangential velocity within the lower boundary layer, followed by stabilization aloft, aligning more closely with observed low-level intensifications in tornadic wind fields. Normalized radial velocity profiles are presented in Fig. 2a,b, showing variations in radial inflow behavior. The Baker models show maximum inflow near the core with smooth decay, while the Fujita and Kuo-Wen models exhibit more abrupt, localized inflow closer to the vortex center. The Burgers-Rott model shows a linear decline across the radial domain. Tangential velocity profiles in Fig. 2b exhibit a near-linear increase in the core region, followed by a gradual decay with distance. Baker and Burgers-Rott models produce broader distributions, while Fujita and Kuo-Wen models predict sharper peaks, indicating a more compact and intense vortex structure.

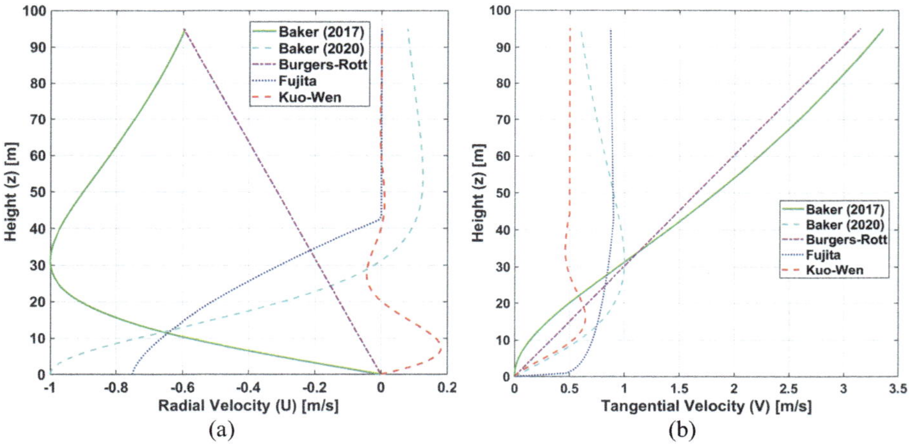

Fig. 1. Velocity profiles of tornado models along height (a) radial velocity (b) tangential velocity

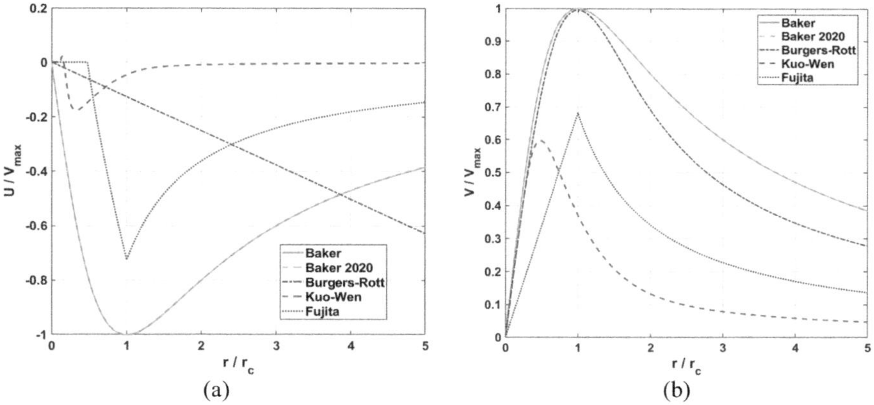

Fig. 2. Velocity profiles of tornado models along radial distance (a) radial velocity (b) tangential Velocity

2.2 Transient Tornado Wind Loads on Tower

Accurate estimation of tornado-induced wind loads on wind turbine towers requires a clear understanding of the velocity components within a tornado wind field. Tornado wind loads are composed of three primary components: radial, tangential, and vertical velocities. Among these, the radial velocity, directed toward or away from the tornado core, exerts lateral pressure on the tower, particularly when aligned with the tornado's translational motion. The tangential velocity, resulting from the vortex's rotational circulation, induces lateral and torsional forces significantly influence the tower's structural response. The vertical velocity component, while present in tornado flows, is not considered in this study due to its relatively minor contribution to the overall loading on the tower, as the pole base is fixed, making the vertical reaction almost negligible. In addition to the velocity components, tornado-induced load characteristics are influenced by spatial variability in vortex intensity, the evolving tornado diameter, and the relative strike angle of the tornado's path with respect to the tower. Fig. 3 illustrates the tornado's translational trajectory and its resolved velocity components with respect to a stationary wind turbine tower. In this study, the tornado parameters were assumed to follow a straight-line translational path with a velocity $V_T = 40$ m/s, originating at $(x_0, z_0) = (-400, \pm 100)$ m. The maximum radial distance r_m is kept 100 m to capture the tornado's wind field. These parameters were assumed based on wind speeds from the Greenfield tornado [27], where the maximum recorded wind speeds reached 82.7 m/s. The velocity components are resolved into two main directions relative to the wind turbine's position: the radial velocity V_{RD} and the tangential velocity V_{TN}. The tornado translates across the ground with a speed V_T, and the wind velocities at the tower are resolved based on a combination of both the radial and tangential components, relative to the angle θ between the tornado's radial direction and its path of motion.

The translational velocity of the tornado, denoted as $V_{T,}$ is the resultant velocity of the tornado's motion in both the x and z directions. The total translational speed is given by Eq. 1:

$$V_T = \sqrt{V_x^2 + V_z^2} \tag{1}$$

Where, V_x is the velocity component in the x-direction, defined as $V_x = \frac{dx}{dt}$, describing the tornado's motion along its path. And V_z is the velocity component in the z-direction, defined as $V_z = \frac{dz}{dt}$, representing the tornado's motion perpendicular to its path.

The wind velocities in the global coordinate directions are calculated using the following expressions, Eqs. 2 and 3 [28]:

$$\mathrm{V_{ix}} = (\mathrm{V_{RD}})_\mathrm{i}(\cos\theta) - (\mathrm{V_{TN}})_\mathrm{i}(\sin\theta) \tag{2}$$

$$\mathrm{V_{iz}} = (\mathrm{V_{RD}})_\mathrm{i}(\sin\theta) + (\mathrm{V_{TN}})_\mathrm{i}(\cos\theta) \tag{3}$$

The resulting wind load on the tower is evaluated using the drag force Eq. 4:

$$\mathrm{F_{ij}} = 0.5\ \mathrm{C_d}\ \rho\ \mathrm{V_{ij}^2}\ \mathrm{A} \tag{4}$$

where, $C_\mathrm{d} = 1.2$ is the drag coefficient for a cylindrical tower, $\rho = 1.225$ kg/m^3 is the air density, A is the projected area of the tower segment, and V_ij is the effective wind velocity acting on element ith element along the jth global coordinate direction.

The size of a tornado strongly influences the likelihood of a wind turbine experiencing tornado loads. A larger radius expands the area affected by strong winds, increasing the risk to turbines near the tornado path. Turbines closer to the core or within high-intensity zones face greater loads. The turbine's position, determined by the radial distance r(t) and angle θ relative to the tornado's path, directly affects both the magnitude and direction of the applied forces.

3 Numerical Simulation and Structural Behavior of Wind Turbine Tower

A finite element model (FEM) of the Vestas V110–2.0 MW [29] wind turbine tower was created using ABAQUS software to study its structural behavior under extreme wind conditions. As can be seen in Fig. 4, the tower, modeled as a 95-m-tall steel structure, was discretized using linear reduced-integration shell elements (S4R). Shell thickness varied from 30 mm at the base to 15 mm at the top, closely matching realistic construction practices. Structural steel (grade S355) with an elastic modulus of 200,000 MPa, a yield strength of 355 MPa, and a density of 7850 kg/m^3 was assigned. The elastic-plastic constitutive law with isotropic hardening at 0.1% of the elastic modulus was used to simulate accurate stress-strain behavior, particularly near critical connections. The base of the tower was modeled as a rigid, fixed boundary to replicate its connection to the foundation, ensuring accurate boundary conditions. Mesh refinement was strategically applied along the tower to accurately capture stress concentrations and structural response without compromising computational efficiency. This model balanced detailed

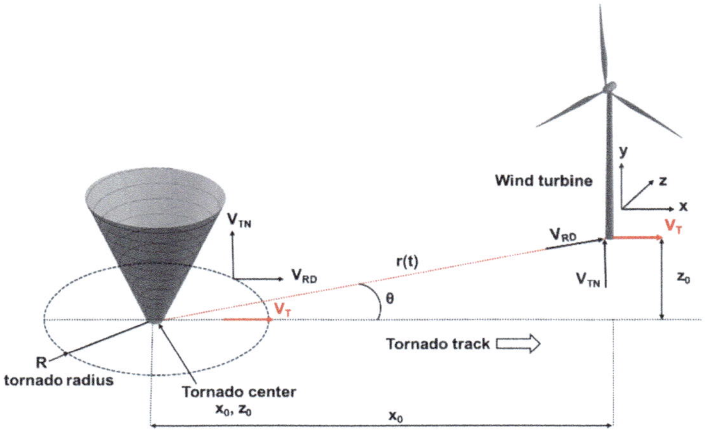

Fig. 3. Visualization of tornado path and directional wind velocity components acting on the wind turbine.

geometric representation and computational practicality, effectively analyzing deformation and stress distributions under extreme tornado-induced aerodynamic loads. As shown in Fig. 5, a modal analysis was performed to determine the dynamic characteristics of the tower. The tower was represented as a pole structure with lumped masses representing the blades and nacelle, facilitating simplified modeling without sacrificing accuracy. The analysis identified natural frequencies of 0.147, 0.148, and 4.47 Hz for the first three modes, corresponding to vertical and horizontal bending modes critical for understanding dynamic wind responses.

3.1 Failure Analysis and Comparative Force Evaluation of Tornado Wind Models

Finite element simulations were conducted to analyze the progressive structural failure of a wind turbine tower subjected to severe tornado-induced loading, as depicted in Fig. 6 represents the failure progression specifically under the Baker [25] tornado wind model. Although several simulations were conducted for all five models, including Kuo-wen, Fujita, Burgers-Rott, and Baker [26], only the Baker [25] model demonstrated plastic hinge formation at a realistic threshold of wind speed (above 50 m/s). In contrast, other models predicted initial deformation at significantly higher wind speeds, some only showing plastic hinge formation within a range of 130–240 m/s, which is highly unrealistic and does not reflect actual structural behavior under known tornado conditions. This realistic response reinforces the Baker [25] model's suitability for evaluating actual tornado-induced structural behavior. The selected wind speeds for these simulations were based on conditions observed during the Greenfield tornado event, which reached Enhanced Fujita Scale level EF-4, thus justifying the use of extreme wind speeds in the analysis. Figure 6 shows the stress distribution and deformation response of the tower under a tornado. Initially, at wind speeds up to 40 m/s, the structure remained

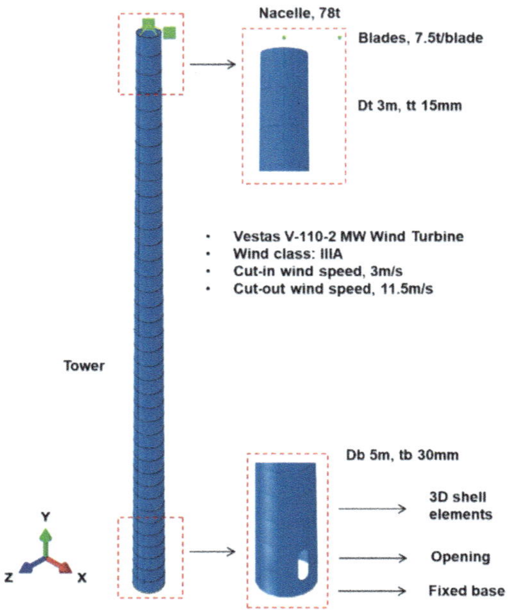

Fig. 4. Finite element model using lumped masses.

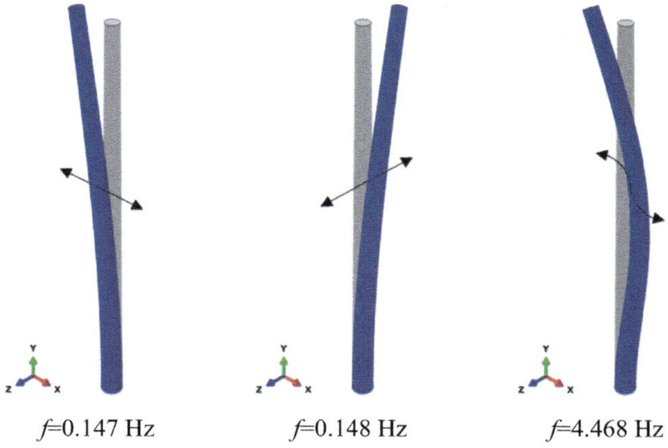

Fig. 5. Modal analysis of the wind turbine.

stable, with minimal deformation and stress levels well within permissible material limits. Von Mises stress distributions indicated uniform, manageable stress concentrations, demonstrating the tower's initial structural integrity. As wind speeds exceeded 50 m/s, structural behavior changed markedly. Localized plastic yielding appeared near geometric discontinuities, with stress levels surpassing the yield strength, leading to the formation of a single plastic hinge at approximately 9 m above the base. A nonlinear dynamic implicit analysis using the Newton-Raphson method effectively captured the progression from initial yielding to full collapse. As loading intensified, global buckling and large lateral deflections emerged, with rapid stress redistribution driving failure propagation. Collapse was ultimately marked by severe bending, buckling, and rotations, with stress concentrations at hinge zone underscoring the interaction between localized yielding and global instability under tornadic loading.

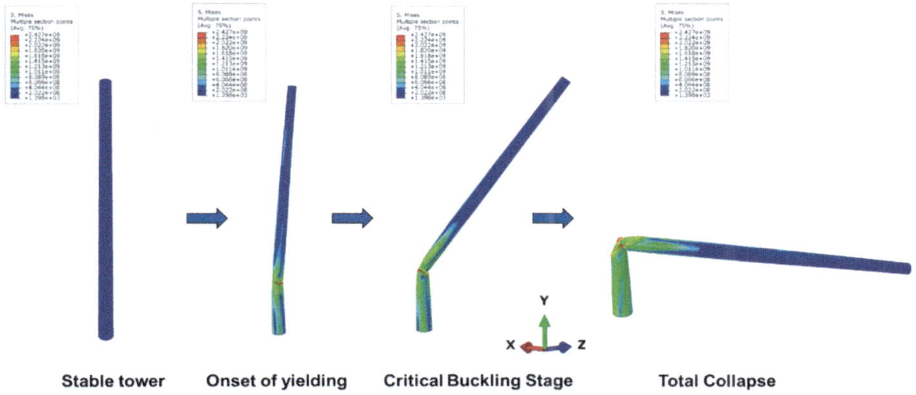

Fig. 6. Progressive failure stages of the wind turbine tower under tornado-induced loads

Comparative evaluations of reaction forces (F_X and F_Z), as illustrated in Fig. 7a,b, were conducted using five established tornado wind field models: Baker [25, 26], Burgers-Rott, Fujita, and Kuo-wen. Notably, up to a wind speed of 40 m/s, the Baker [25] model predicted significantly higher reaction forces than other models, accurately capturing the true nature of severe damage potential in real-world scenarios where structures like wind turbines experience substantial forces and structural impairment at high wind speeds. In contrast, other models predicted relatively modest reaction forces at the same wind speed, highlighting the conservative and realistic nature of the Baker model under these extreme conditions. This observation underscores the critical importance of selecting an appropriate tornado model for accurately predicting structural responses and ensuring adequate design resilience.

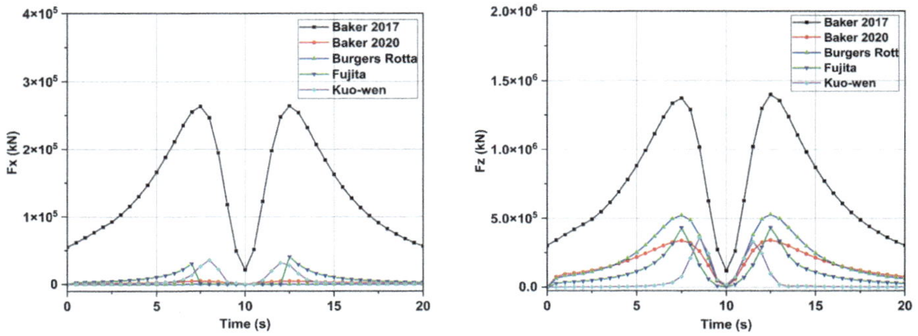

(a) (b)

Fig. 7. Comparison of reaction forces for tornado wind field models, (a) F_X, (b) F_Z

4 Conclusions

As the global push toward net-zero emissions accelerates, the reliability of renewable energy infrastructure under extreme weather conditions becomes increasingly critical. This study investigates the structural failure mechanisms of wind turbines subjected to tornado-induced loads by integrating empirical tornado wind field models with finite element analysis (FEM). Through simulations using five distinct tornado models the research highlights substantial variation in structural response, emphasizing the importance of accurate wind field representation in design standards.

Among the models analyzed, the Baker [25] model most effectively replicated observed real-world failure mechanisms, affirming its suitability for structural assessments in tornado-prone regions. Modal analysis identified key natural frequencies of the turbine tower corresponding to tower bending and deformation modes critical to understanding dynamic response.

By informing the design of more robust wind turbine structures, this work supports the development of climate-resilient renewable energy systems. It contributes directly to the broader goal of a sustainable, net-zero carbon future by enhancing the longevity and reliability of wind power infrastructure in the face of intensifying weather extremes.

Acknowledgements. This research was primarily supported by the U.S. National Science Foundation (NSF) under research grant 2330150. Their support is gratefully acknowledged.

References

1. IRENA (2024) Renewable capacity statistics 2024. International Renewable Energy Agency
2. NOAA (2020) Tornado Database 1950–2019. National Weather Service Storm Prediction Center
3. Qudaisat M, Alipour A (2025) Future-proofing energy infrastructure resilience with distributed energy resources. Dev Built Environ 23:100696

4. Aslam MU, Alipour A (2025) Structural failure analysis of wind turbines to tornadoes in greenfield tornadoes of 2024. Eng Fail Anal. https://doi.org/10.1016/j.engfailanal.2025.110387
5. Alipour A, Sarkar P, Dikshit S, Razavi A, Jafari M (2020) Analytical approach to characterize tornado-induced loads on lattice structures. J Struct Eng 146(6):04020108
6. Romanic D, Parvu D, Refan M, Hangan H (2018) Wind and tornado climatologies and wind resource modelling for a modern development situated in "tornado alley". Renew Energy 115:97–112
7. Younesi A, Wang Z, Siano P (2024) Enhancing the resilience of zero-carbon energy communities: leveraging network reconfiguration and effective load carrying capability quantification. J Clean Prod 434:139794
8. Faharidine M, Aslam MU, Choufaikat MM (2024) Improving stability and buckling resistance of self-supporting isotrussed telecommunication tower under wind load. J Rehabil Civ Eng:54–68. https://doi.org/10.22075/jrce.2024.33439.2013
9. Elmousalami H, Alnaser AA, Hui FKP (2025) Sustainable AI-driven wind energy forecasting: advancing zero-carbon cities and environmental computation. Artif Intell Rev 58. https://doi.org/10.1007/s10462-025-11191-0
10. Goyal V et al (2024) Role of wind energy in achieving global carbon neutrality: challenges and opportunities. E3S Web Conf 591:02002
11. Kwon D, Kareem A (2009) Gust-front factor: A new framework for wind load effects on structures. J Struct Eng:5. https://doi.org/10.1061/(asce)st.1943-541x.0000010
12. Adhikari P, Mahmoud H, Xie A, Simonen K, Ellingwood B (2020) Life-cycle cost and carbon footprint analysis for light-framed residential buildings subjected to tornado hazard. J Build Eng 32:101657
13. NIST (2024) State of the art and research needs in Design for Tornadoes. National Institute of Standards and Technology. https://www.nist.gov/publications/state-art-and-research-needs-design-tornadoes-structural-engineering-institute-workshop
14. Aslam MU, Mahamoudou F, Ahmed B, Suliman M (2024) Reliability analysis of slope based on Fellenius's and Bishop's method. Tech J Univ Eng Technol Taxila
15. Rankine (1858) A manual of applied mechanics.pdf
16. Bjerknes V (1914) Meteorology as an exact science 1. Mon Weather Rev 42:11–14. https://doi.org/10.1175/1520-0493(1914)42<11:maaes>2.0.co;2
17. Burgers JM (1948) A mathematical model illustrating the theory of turbulence. Adv Appl Mech 1:171–199. https://doi.org/10.1016/S0065-2156(08)70100-5
18. Sullivan RD (1959) A two-cell vortex solution of the Navier-stokes equations. J Aerosp Sci 26:767–768. https://doi.org/10.2514/8.8303
19. Fujita TT h (1978) Workbook of tornadoes and high winds for engineering applications, p 144
20. Kuo HL (1971) Axisymmetric flows in the boundary layer of a maintained vortex. J Atmos Sci 28:20–41. https://doi.org/10.1175/1520-0469(1971)028<0020:afitbl>2.0.co;2
21. Vatistas GH, Kozel V, Mih WC (1991) A simpler model for concentrated vortices. Exp Fluids 11:73–76. https://doi.org/10.1007/BF00198434
22. Wood VT, White LW (2013) A parametric wind-pressure relationship for rankine versus non-rankine cyclostrophic vortices. J Atmos Ocean Technol 30:2850–2867. https://doi.org/10.1175/JTECH-D-13-00041.1
23. Baker CJ, Sterling M (2018) A conceptual model for wind and debris impact loading of structures due to tornadoes. J Wind Eng Ind Aerodyn 175:283–291. https://doi.org/10.1016/j.jweia.2017.11.029
24. Kim YC, Matsui M (2017) Analytical and empirical models of tornado vortices: A comparative study. J Wind Eng Ind Aerodyn 171:230–247. https://doi.org/10.1016/j.jweia.2017.10.009

25. Baker CJ (2017) Modelling wind fields and debris flight in tornadoes. J Wind Eng and Ind Aerodyn 168:312–321. https://doi.org/10.1016/j.jweia.2017.06.017
26. Baker C, Sterling M, Jesson M (2020) The lodging of crops by tornadoes. J Theor Biol 500:110309. https://doi.org/10.1016/j.jtbi.2020.110309
27. Wikipedia (2024) 2024 Greenfield tornado. Available at: https://en.wikipedia.org/wiki/2024_Greenfield_tornado
28. AbuGazia M, El Damatty AA, Dai K, Lu W, Ibrahim A. Numerical model for analysis of wind turbines under tornadoes. Eng Struct 2020; 223. doi:https://doi.org/10.1016/j.engstruct.2020.111157.
29. U.S. Department of Energy (2021) Wind turbine database. U.S. Department of Energy. Retrieved from https://energy.usgs.gov/uswtdb/viewer

Life Cycle Fatigue of Transmission Lines under Wind

Pooria Mazaheri and Alice Alipour(✉)

Department of Civil Construction and Environmental Engineering, Iowa State University, Ames, IA, United States
alipour@iastate.edu

Abstract. Transmission towers experience persistent wind loading throughout their service life, leading to cumulative fatigue damage, particularly at structural joints where stress concentrations occur. This study introduces a fatigue-based framework for assessing the long-term performance of transmission towers by directly analyzing stress cycles in joint regions under realistic wind excitation. Using a validated finite element model of a steel lattice transmission line system carrying three conductors, subjected to direction-dependent wind fields, dynamic structural responses are simulated and time-varying stress histories at critical connection points are extracted. Fatigue damage is estimated using the Rainflow counting method and Palmgren-Miner rule, incorporating appropriate stress concentration factors to reflect local joint behavior. The resulting fatigue indices are used to evaluate the evolution of vulnerability over time, identify critical joint locations, and construct lifetime fragility relationships. To complement the fatigue assessment, a risk-based inspection planning strategy is adopted to minimize expected life-cycle costs while maintaining structural reliability. By integrating probabilistic fatigue predictions with inspection intervals and conditional failure consequences, the approach enables informed maintenance scheduling under uncertainty. This methodology supports targeted assessment of fatigue-prone details and contributes to condition-based maintenance practices aimed at extending the reliable service life of transmission infrastructure.

Keywords: Life-cycle assessment · Wind-Induced Fatigue · Finite Element Analysis · Risk-Based Inspection

1 Introduction

Transmission towers must withstand fluctuating loads over decades, with wind being the most persistent and unpredictable stressor. While design codes focus on extreme events, they often neglect fatigue from daily wind cycles, a key concern for steel lattice towers where joint details like bolt holes, gusset edges, and attachments face localized high-cycle fatigue due to stress concentrations. Fatigue damage is thus highly localized, not uniformly distributed across the structure. Wang [1] and Qarib et al. [2] have highlighted the contribution of wind-induced and vortex-induced vibrations to significant fatigue accumulation in slender transmission components, especially under resonant conditions.

H. Beushausen et al. (Eds.): NTZR 2025, LNCE 873, pp. 684–697, 2026.
https://doi.org/10.1007/978-3-032-27698-8_61

Repetto and Solari [3] emphasized the importance of directional wind fields in intensifying local stress amplitudes, particularly in towers aligned with prevailing wind directions. These effects can exacerbate fatigue at structurally sensitive points, reinforcing the need for detailed stress analysis at local joints under realistic wind patterns. Despite the structural redundancy and high static safety margins in tower design, fatigue failures have been observed in real-world settings, often manifesting as progressive cracking in welded or bolted connections [4]. To support accurate fatigue life prediction, recent studies have focused on local stress concentration effects and fatigue class identification. Muñiz-Calvente et al. [5] compared time-domain and spectral approaches, concluding that time-domain methods based on Rainflow counting and Palmgren-Miner accumulation offer better alignment with random wind loading. Ribeiro et al. [6] developed a practical, full-cycle fatigue analysis for guyed lattice towers, using a 120-m tower and demonstrating how connection regions between modules serve as fatigue-critical locations. Li et al. [7] similarly employed finite element simulations and stress concentration factor (SCF) adjustments to quantify local damage in wind-sensitive tower joints. Nonlinear finite element modeling combined with stochastic wind simulation [8] allows for time-varying stress histories, as demonstrated by Tsai and Alipour [9]. Eurocode design standard for steel structures [10] provides the required fatigue classifications, while supplemental S-N curves and SCFs [11] guide detail-specific specifications for bolted connections. Realistic modeling has also considered loosening effects in bolted joints [12] and the role of residual stresses [13]. The integration of SCFs into detailed finite element modeling is essential to avoid non-conservative fatigue life estimates. Additionally, as Martindale and Wirsching [14] have shown, probabilistic modeling helps characterize uncertainty in loading and material behavior, making it crucial in risk-based frameworks. The inspection and maintenance side of fatigue management has also evolved. Risk-based inspection (RBI) strategies, as discussed by Mendoza et al. [15], provide probabilistic and cost-optimized decision frameworks. These approaches assess inspection timing and extent based on expected damage accumulation and failure probabilities.

While fatigue in steel structures is well-studied, its application to transmission towers, especially at the system level under realistic wind, is limited. Past work often overlooks joint behavior and hotspot stress. This study introduces a two-phase framework combining fatigue life modeling with RBI planning. Using a validated FE model, joint stress histories are extracted, scaled with SCFs, and used to estimate damage via Rainflow counting and Miner's rule. Results identify vulnerable joints across wind directions, feeding into a probabilistic inspection model that optimizes intervals to reduce life-cycle cost. The case study highlights how fatigue-informed maps enable targeted, cost-effective inspections.

2 Methodology

This section outlines the proposed two-phase framework, consisting of (1) fatigue damage modeling under wind excitation and (2) a probabilistic inspection planning approach based on RBI. As illustrated in Fig. 1, the framework integrates dynamic stress histories obtained from finite element simulations with stochastic fatigue analysis to identify critical joints and inform optimal maintenance schedules. Phase one is detailed in the following subsections and phase two is detailed in Sect. 3.

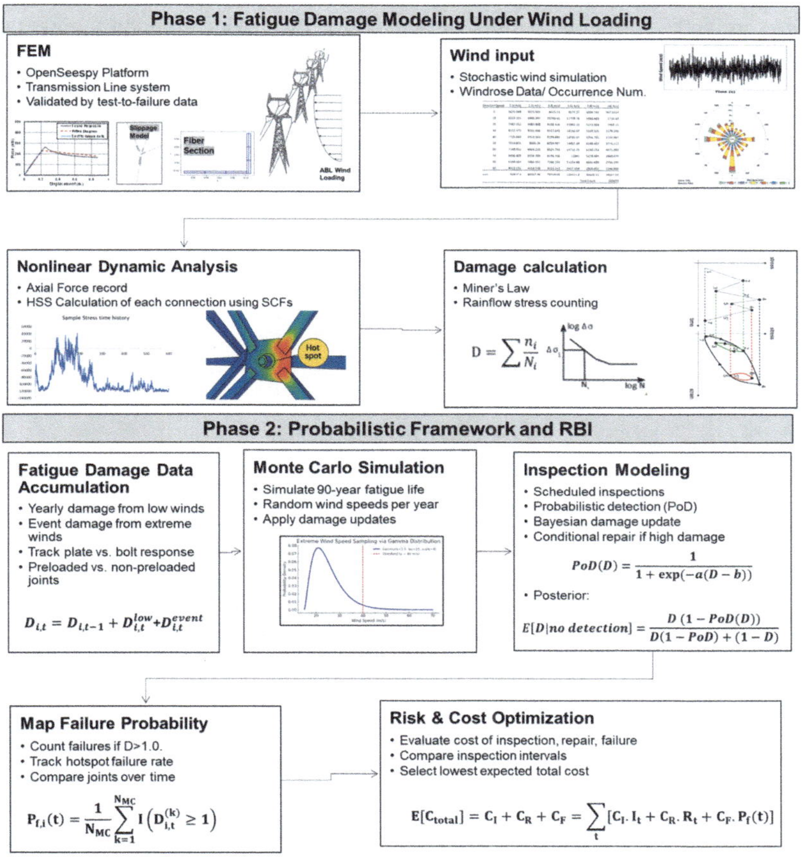

Fig. 1. Two-phase framework for fatigue modeling and risk-based inspection planning

2.1 Finite Element Model

The modeled structure is a 500 kV lattice transmission tower, 44.5 m tall, carrying three-phase conductors and two ground wires. The geometry, including legs, braces, and framing components, was developed in Python using OpenSeesPy based on construction drawings. High-strength A572 Gr.50 was assigned to legs and A36 to braces. Members were modeled with bilinear stress–strain behavior using fiber sections, displacement-based beam-column elements, and appropriate geometric transformations. Mass distribution was assigned based on steel density definitions within OpenSees. A schematic view of the transmission line system is presented in Fig. 2.

Slippage in bolted joints was captured using zero-length hysteretic elements calibrated from test data. Initial geometric imperfections (0.075% of member length) were applied at element ends. The conductor system was modeled with 300 m truss spans using a corotational formulation to capture large deflections. Conductors were prestressed to simulate sag, and 3.5 m elastic links were used for insulators. Boundary and coupling effects were included to ensure realistic tower-line interaction under wind loading.

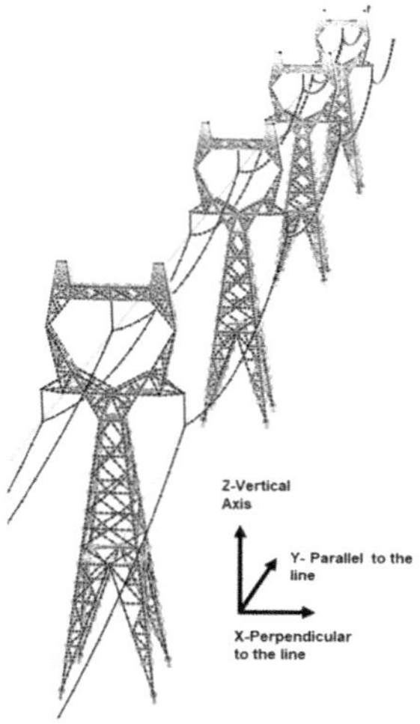

Fig. 2. Schematic view of the FEM (produced using BraineryWiz [16])

2.2 Wind field characterization and dynamic loading

To model realistic high-cycle, low-amplitude wind-induced fatigue on the transmission tower system, wind data were extracted for a 50-year historical period from the Quad Cities weather station (DVN) in Davenport, Iowa. Ten-minute wind speed and direction records were obtained from the Iowa Environmental Mesonet [17], and the frequency of occurrence for each wind speed/direction pair was averaged to represent an annual wind loading spectrum. The resulting wind occurrence data were used to define loading probabilities across wind scenarios. The wind rose diagram in Fig. 3 illustrates the dominant wind directions, with prevailing winds from the south and west sectors. For fatigue analysis, each wind scenario (defined by speed and direction) was assigned a 10-min stochastic wind time series using the stochastic simulation method by Deodatis [8]. These 10-min simulations match the record resolution and were used directly to estimate fatigue damage without further scaling. However, for extreme wind event simulations, a duration of one hour was assumed to capture the cumulative effect under rare high-load conditions. This assumption follows the findings of Li and Kopp [18], who concluded that a one-hour wind duration remains a valid representation for extreme synoptic events across most regions in North America.

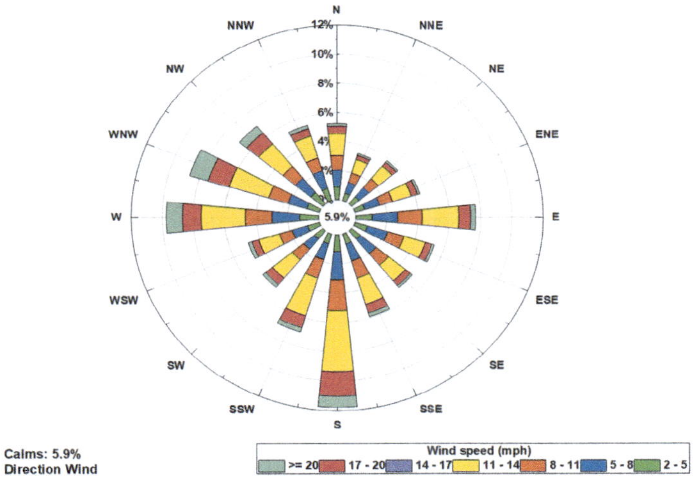

Fig. 3. Wind rose diagram for Davenport, IA (DVN station), showing wind direction and speed distribution based on a 50-year historical dataset

The wind field is modeled as a stationary, direction-dependent boundary layer flow using the atmospheric boundary layer formulation as per Eq. (1).

$$U(z) = U_{10m}(z/10)^{0.15} \tag{1}$$

where U_{10} is the wind speed at the 10 m reference height. Horizontal coherence is assumed; vertical turbulence follows the Von Karman spectrum. Fig. 4 shows a sample realization of the 10-min wind time series. For each tower component and conductor element, wind-induced drag forces are computed as a function of time and height:

$$F(y, z, t) = \frac{1}{2} \rho\, k_z k_{zt} U(y, z, t)^2 G\, C_f A_{x,y} \tag{2}$$

where $\rho = 1.23$ kg/m^3 is the air density, C_f is the drag coefficient (assumed as 1.0 for cables and directionally calibrated for tower panels from work done by Alipour et al. [19]), and $A_{x,y}$ is the projected area. The coefficients k_z, k_{zt}, and G follow ASCE Manual of Practice No. 74 guidance for transmission tower analysis. For extreme wind scenarios, a duration scaling factor of 6 was applied to the 10-min fatigue simulation results, consistent with Li and Kopp [18]. This extrapolated response was used in conjunction with the rainflow counting method [20] to estimate cumulative fatigue damage.

2.3 Nonlinear dynamic analysis

Nonlinear dynamic analysis was used to assess the tower's wind response. For each wind speed and direction, a stochastic time series was applied to the OpenSees model. A 4% damping ratio was used for the tower and supports, and 10% for conductors, per ASCE MOP No. 74. Axial force histories were extracted, shear and moment effects were negligible. Nominal stresses from axial forces were scaled by SCFs based on joint

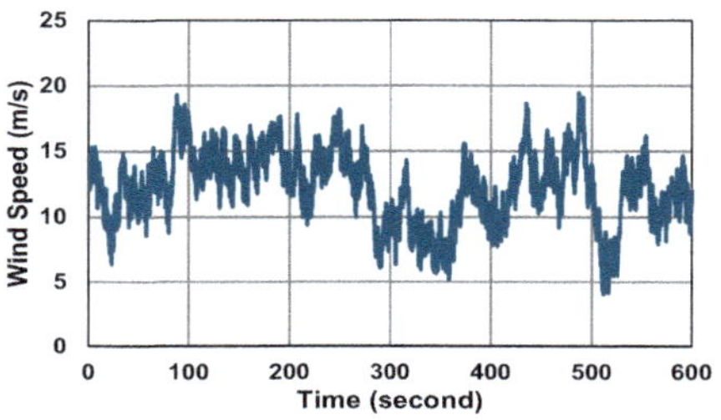

Fig. 4. Sample 10-min stochastic wind speed time series generated using the spectral simulation method for a mean wind speed of 10 m/s

geometry to compute hotspot stresses. As shown in Fig. 5, stress ranges were higher in the crossarm than the leg, indicating stronger wind-induced vibrations at elevated, less-restrained members.

2.4 Fatigue Damage Assessment

The fatigue life of transmission tower joints was assessed using an SCF-based cycle-counting method that integrates Eurocode S–N curves and Miner's rule. Joint-specific SCFs were assigned based on connection details and bolt configuration (Table 1), with stress cycles derived from climate-informed wind scenarios. This setup ensures accurate stress amplification and fatigue estimation across joint types.

Table 1. Number of bolts per connection type

Connection type	Number of bolts	Connection type	Number of bolts
Leg splice (vertical chord)	4 bolts	Top-arm splice	3–4 bolts
Main horizontal bracing (diagonal to chord)	2 bolts	Base gusset or foundation brace	2 bolts
X-bracing intersection (mid-panel diagonals)	1 bolt	Mid-span transverse bracing	1 bolt
Bracing (diagonal to transverse brace)	2 bolts	Internal cage/diagonal bracing	1 bolt

SCFs accounted for joint geometry and load paths, set to 1.0 for preloaded connections, and computed for non-preloaded cases using Maljaars and Euler's formulations [11]. To adjust local stress ranges for fatigue damage, SCFs were calculated based on connection load transfer: pin loading, bypass loading, or a combination of both (Eqs. 3–6). Here, d_0 is the bolt hole diameter, w the plate width, and k the number of bolt rows across the width, used to weigh pin and bypass effects. A calibration constant $c_I = 2.55$ scales the SCF to align with Eurocode fatigue classes.

$$K_t^{pin} = 12.88 - 52.71(d_0/w) + 89.76(d_0/w)^2 - 51.67(d_0/w)^3 \tag{3}$$

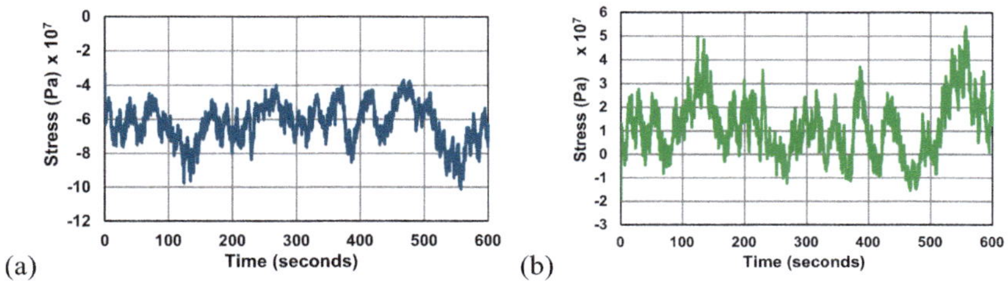

Fig. 5. Sample axial stress time histories extracted from the transmission tower model under wind loading: (a) stress response at an element in the crossarm; (b) stress response at a leg element

$$K_t^{byp} = 2 + 0.284(1 - d_0/w) - 0.6(1 - d_0/w)^2 + 1.32(1 - d_0/w)^3 \tag{4}$$

$$K_t = 1/k \times K_t^{pin} + (k - 1)/k \times K_t^{byp} \tag{5}$$

$$SCF = K_t/c_1, \, with \; c_1 = 2.55, \, and \; SCF \geq 1.0 \tag{6}$$

Fatigue life estimation was performed using Eurocode-classified S-N curves [10], with fatigue resistance parameters selected based on connection type. For plates in non-preloaded connections, a FAT class of 90 was used, corresponding to log $a = 16.072$ and slope parameter $m = 5$. For preloaded bolts, the FAT112 curve was adopted, with log $a = 15.87$ and the same slope $m = 5$ [11]. The number of cycles to failure for a given stress range, $\Delta\sigma$, was calculated using Basquin's law, expressed in Eq. (7).

$$\log \mathrm{N} = \log\ a - m.\log\ \Delta\sigma \tag{7}$$

For each joint, rainflow counting was applied to the SCF-amplified stress histories to extract individual stress ranges and their corresponding number of cycles. Plate stresses were scaled directly by the computed SCF values, while bolt stresses were derived from the axial force using the relation in Eq. (8).

$$\sigma_{\mathrm{bolt}} = \mathrm{F}/\mathrm{n}_{\mathrm{bolt}}.\mathrm{A}_{\mathrm{bolt}} \tag{8}$$

where F is the axial force in the member, n_{bolt} is the number of bolts per joint, and A_{bolt} is the tensile stress area of a single bolt. Fatigue damage accumulation was calculated using Miner's rule, summing the damage contributions from all identified stress cycles as:

$$D = \sum (n_i/N_i) = \sum \left(n_i/\left(a.(\Delta\sigma_i)^{-m}\right) \right) \tag{9}$$

where n_i is the number of cycles at stress range $\Delta\sigma_i$, and N_i is the corresponding number of cycles to failure derived from the S–N curve.

3 Probabilistic Fatigue Framework and RBI Strategy

This phase integrates fatigue damage modeling, probabilistic simulation, inspection planning, and economic optimization into a unified framework for transmission tower joints.

3.1 Fatigue Damage Modeling

Let $D_{i,t}$ denote the cumulative fatigue damage at joint i in year t. The total damage evolves annually based on the previous year's accumulated damage and new increments due to both regular and extreme wind events:

$$D_{i,t} = D_{i,t-1} + D^{low}{}_{i,t} + D^{event}{}_{i,t} \tag{10}$$

where the $D^{low}{}_{i,t}$ is the fatigue damage from daily wind fluctuations, $D^{event}_{i,t}$ is the damage from extreme wind events. To compute $D^{low}_{i,t}$ each year's wind load is decomposed into multiple wind speed–direction combinations indexed by $j \in [1, N_w]$. For each wind case, fatigue damage is evaluated using stress histories derived from finite element simulations and processed via rainflow counting and Miner's rule:

$$D^{low}{}_{i,t} = \sum f_i.\Delta D_{i,j}; for\, j \in [1, N_W] \tag{11}$$

where N_w is the number of wind cases, f_j is the normalized frequency of occurrence of wind case j to the yearly number of occurrences, and $\Delta D_{i,j}$ damage increment from case j. Event-based fatigue damage for each joint was evaluated using a duration-adjusted form of Miner's rule. For a given extreme wind realization, the damage increment was calculated as:

$$D^{event}_{i,t} = \delta.\sum_k \frac{n_k}{a.\Delta\sigma_k{}^{-m}} \tag{12}$$

Where, δ is the duration factor for one-hour exposure, $\Delta\sigma_k$ the stress range, and a, m are S–N curve parameters (see Sect. 2.4). This formulation was used to precompute joint-level fatigue damage across wind speeds and directions. Extreme wind-induced damage was captured via Monte Carlo Simulation (MCS) with one million realizations of annual high wind events, sampled from a Gamma distribution ($\alpha = 2.5$, $\theta = 4$, $\mu = 15$). Each year sampled wind speeds were matched to precomputed structural response cases (15–50 m/s, all directions), and the resulting damage was added to the cumulative total. If wind exceeded 40 m/s, a threshold where prior simulations showed higher stresses than $0.7\,Fy$ at a joint [10], preload loss was assumed, triggering a different damage path. This two-path model (preloaded vs. non-preloaded) captures evolving joint behavior over time.

3.2 Joint-Level Failure Probability

The annual failure probability of each joint was computed by tracking the proportion of MCSs in which accumulated fatigue damage exceeded the failure threshold of 1.0. For joint i in year t, this probability was calculated using Eq. (13):

$$P_{f,i}(t) = \frac{1}{N_{MC}} \sum_{k=1}^{N_{MC}} I\left(D^{(k)}_{i,t} \geq 1.0\right) \tag{13}$$

Where N_{MC} is the number of MCSs, and the indicator function $I(\cdot)$ flags realizations that cross the damage limit. This process was performed independently for plates and bolts using their respective SCFs and S-N parameters. The result is a joint-level, time-dependent probability of fatigue failure across the simulated service life.

3.3 RBI and Cost Optimization

Inspections are key to managing joint-level fatigue in transmission towers under variable wind loads. Here, we implement a risk-based inspection strategy within the probabilistic fatigue framework. Scheduled inspections at 5, 10, and 15-year intervals were evaluated using a logistic PoD function (Eq. 14) applied to joint fatigue histories to assess detection and repair benefits.

$$PoD(D_{i,t}) = 1/\left(1 + \exp\left(-a\left(D_{i,t-b}\right)\right)\right) \tag{14}$$

The parameters of this sigmoidal function were calibrated based on the interpretation of Miner's damage ranges, summarized in Table 2. Parameter $b = 0.6$ corresponds to the threshold between the degradation zone and the high-risk zone, where microcrack initiation becomes likely according to standards such as Eurocode [10], the FHWA Steel Bridge Design Handbook [21], and studies such as Mendoza et al. [15]. Parameter $a = 20$ was selected to ensure a steep transition from low detectability ($PoD \approx 0.002$ at $D = 0.3$) to high detectability ($PoD \approx 0.92$ at $D = 0.8$), reflecting how field inspections become significantly more effective as structural degradation progresses.

Table 2. Interpretation of Miner's Fatigue damage and expected structural behavior

Miner's damage (D)	Interpretation	Miner's damage (D)	Interpretation
$0 < D < 0.3$	Safe / No observable damage	$0.6 \leq D < 1.0$	High-risk zone / pre-crack
$0.3 \leq D < 0.6$	Degradation zone	$D = 1.0$	End of fatigue life

Once an inspection occurs, if no damage is detected, a Bayesian posterior update is applied to adjust the joint's damage estimate as:

$$E\left[D_{i,t}|No\ Detection\right] = \frac{D_{i,t}\ (1 - PoD(D_{i,t}))}{D_{i,t}\left(1 - PoD\left(D_{i,t}\right) + (1 - D_{i,t}\right)} \tag{15}$$

If the updated estimate exceeds a repair threshold $D_{repair} = 0.3$, the joint is assumed to be repaired (i.e., damage reset to zero and preload restored). Otherwise, the joint remains in its degraded state. This threshold aligns with the upper bound of the "safe" zone in Table 2, beyond which preload degradation and bolt slip are likely. To evaluate the lifecycle efficiency of different inspection plans, the expected cost was computed over a 90-year period as:

$$E[C(\Delta t)] = \sum_{t=1}^{T}\left[C_I.I_{inspect}(t) + C_R.E\left[Repairs(t)\right] + C_F.\sum_i P_{f,i}(t)\right] \tag{16}$$

Here, C_I is the cost per inspection, C_R is the cost of repair per joint, and C_F represents the consequence of failure. The indicator $I_{inspect}(t)$ is 1 when inspection occurs, and

$\sum P_{f,i}(t)$ captures cumulative failure probabilities across all joints in year t. By comparing outcomes across different inspection frequencies, the framework identifies the inspection interval that yields the lowest total lifecycle cost while maintaining structural reliability.

4 Results

This section summarizes the fatigue and inspection results for the 500 kV transmission line system, highlighting vulnerable joints, evaluating inspection strategies, and quantifying cost-risk tradeoffs over a 90-year service life.

4.1 Joint Vulnerability and Fatigue Hotspots

Fatigue simulations across wind scenarios consistently flagged upper joints near the crossarm base, especially at bracing intersections, as early damage hotspots (Fig. 6). Fatigue damage estimates, summarized in Table 3, show that for joints initially modeled as preloaded, the accumulated damage (e.g., $D = 9.91 \times 10^{-5}$ for joint 758–284) remains low. However, once preload is lost, after an extreme wind event or damage accumulation in years over regular wind conditions, those same joints exhibit an order-of-magnitude increase in damage (e.g., D = 0.00111 for the same joint). These hotspots were flagged consistently across all MCSs, confirming that their vulnerability stems from deterministic geometric and loading conditions rather than random fluctuations. As such, they represent prime candidates for targeted condition monitoring and inspection prioritization throughout the structure's service life.

Table 3. Fatigue-prone locations identified by joint tag and element number

Element number-joint tag		Location	$D_{preloaded}^{*}$	$D_{non\text{-}preloaded}^{*}$
761–285	758–284	Diagonal Bracing at Tower Parallel to the Line Face	9.91×10^{-05}	0.00111
761–293	758–290			
760–292	757–290	Horizontal Bracing at Tower Parallel to the Line Face	8.94×10^{-05}	0.00092
760–295	757–294			
769–290	770–291	Horizontal Bracing at Tower Perpendicular to the line Face	8.27×10^{-05}	0.00085
769–298	770–298			

4.2 Fatigue Failure Probability

As outlined in Sect. 3.1, fatigue damage accumulates annually through two main pathways: (1) low-amplitude, high-cycle stress cycles driven by daily wind fluctuations, and (2) rare but high-impact damage from extreme wind events. Using MCSs over a 90-year horizon, joint-level damage histories were computed and compared against a failure threshold of $D = 1.0$ to determine annual failure probabilities. Fig. 7a presents the evolution of plate failure probability over time in the absence of inspection. The average probability steadily increases, reaching approximately 80% by the year 90, emphasizing the long-term vulnerability of transmission tower joints under persistent wind exposure. Notably, fatigue risk is not uniformly distributed across all joints.

Connections in diagonal bracing elements parallel to the transmission line (e.g., 758–284, 759–284) emerge as the most fatigue-prone. These joints are aligned with the dominant wind direction and experience amplified stress variations due to their inclined geometry and directional wind loads. In contrast, horizontal bracing perpendicular to the line (e.g., 769–290, 770–291) shows the lowest failure probabilities, benefitting from more favorable load paths and reduced cyclic stress exposure.

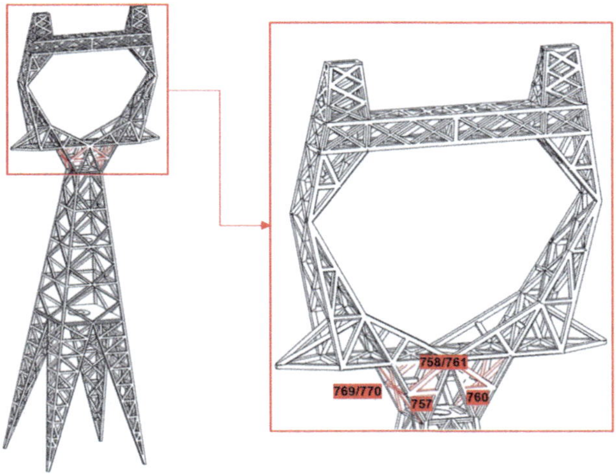

Fig. 6. Visual identification of fatigue-sensitive joint locations on the tower structure

4.3 RBI Strategy and Cost Implications

RBI begins by addressing where, when, and what to inspect. Fatigue-prone joints were pre-identified based on wind-induced damage. Inspection intervals of 5, 10, and 15 years were evaluated using the risk-based method in Sect. 3.3. A sigmoid PoD function, calibrated to Miner's damage, modeled detection sensitivity. Bayesian updating was applied when inspections found no damage, and joints with posterior damage above 0.3 were assumed repaired. This threshold marks the transition from safe to degraded behavior. Inspections focused on plate components, as bolts showed negligible damage. Costs were set at \$10 k per inspection, \$20 k per repair, and \$400 k [22], per failure, held constant across scenarios. Life-cycle costs were then calculated over 90 years, including inspection, repair, and failure expenses. Fig. 7b illustrates the cumulative risk (in USD) under each inspection plan. Without any inspections, expected failure costs rise sharply, exceeding \$350 k by the year 90. Incorporating inspections markedly reduces risk, particularly for 5-year and 10-year intervals, by regularly catching and addressing damage before it progresses. The 15-year strategy, while better than no inspection, shows sharper post-inspection spikes and overall higher costs. These trends show the value of routine monitoring and early intervention to control long-term failure risk in fatigue-sensitive components.

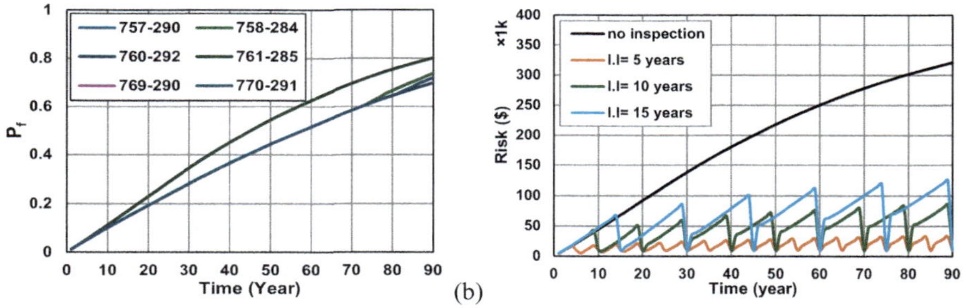

(a) (b)

Fig. 7. (a) Time-dependent joint-level failure probability (P_f) for selected plate connections over a 90-year service life (legend: Element No.—Joint Tag); (b) Time-dependent risk accumulation under different inspection intervals for plate components

5 Conclusions

This study proposed a two-phase fatigue and inspection framework for wind-exposed transmission towers. A validated FEM of a 500 kV lattice tower was subjected to 60 wind scenarios, generating joint-level stress histories processed via rainflow counting and Miner's rule. Fatigue damage was tracked over 90 years with 1 million MCSs, incorporating SCFs and preload loss. Diagonal bracing joints (e.g., 758–284) showed 10 times more damage than perpendicular bracing (e.g., 770–291), with failure probabilities exceeding 80% without inspection. A risk-based inspection strategy, using probabilistic PoD and Bayesian updates, substantially reduced lifecycle risk relative to the no-inspection case, with shorter inspection intervals providing stronger control of accumulated risk and post-inspection cost spikes.

References

1. Wang Q, Li Z, Wang F, Li H (2023) Fatigue life prediction of power transmission towers based on nonlinear finite element analysis. Appl Math Nonlinear Sci 9:00630
2. Qarib H, Jeddi AB, Hur J, Shafieezadeh A (2022) Development of a Vortex-Induced Vibration Analysis Process and Evaluation of Fatigue Damage Risk. In: Electrical Transmission and Substation Structures 2022. American Society of Civil Engineers Reston, VA, pp 524–535
3. Pia RM, Giovanni S (2004) Directional Wind-Induced Fatigue of Slender Vertical Structures. J Struct Eng 130:1032–1040. https://doi.org/10.1061/(ASCE)0733-9445(2004)130:7(1032)
4. Lake R (1998) Investigation into the fatigue life of a flexible lattice steel transmission tower

5. Muñiz-Calvente M, Álvarez-Vázquez A, Pelayo F, Aenlle M, García-Fernández N, Lamela-Rey MJ (2022) A comparative review of time- and frequency-domain methods for fatigue damage assessment. Int J Fatigue 163:107069. https://doi.org/10.1016/j.ijfatigue.2022.107069
6. Ribeiro D, Bragança C, Montenegro PA, Carvalho H, Costa B, Marques F (2022) Wind-induced fatigue analysis of high-rise guyed lattice steel towers. Structure 36:719–734
7. Li J, Wang B, Wang T, Li Z (2023) Nonlinear finite element analysis and fatigue damage assessment of wind-induced vibration for the tension cable-supported power transmission structure. Buildings 13:2924
8. Deodatis G (1996) Non-stationary stochastic vector processes: seismic ground motion applications. Probabilistic Eng Mech 11:149–167. https://doi.org/10.1016/0266-8920(96)00007-0
9. Tsai L-W, Alipour A (2020) Assessment of fatigue life and reliability of high-mast luminaire structures. J Constr Steel Res 170:106066
10. de Normalisation CE. Eurocode 3: design of steel structures– part 1–10: material toughness and through-thickness properties. EN 1993-1-10, Comité Européen de Normalisation, Brussels; 2005.
11. Maljaars J, Euler M (2021) Fatigue SN curves of bolts and bolted connections for application in civil engineering structures. Int J Fatigue 151:106355
12. Wenqiang J, Ze M, Tong W, Qing Z, Lanqi Y, Wei Z (2025) Loosening process of bolted connection in transmission tower under transverse cyclic load. Structure 73:108399
13. Qu W, Wang M, Zhou Q (2020) Effect of welding residual stress on along-wind fatigue life of welded joints in guyed mast. Shock Vib 2020:6467354
14. Martindale SG, Wirsching PH (1983) Reliability-based progressive fatigue collapse. J Struct Eng 109:1792–1811
15. Mendoza J, Bismut E, Straub D, Köhler J (2021) Risk-based fatigue design considering inspections and maintenance. ASCE ASME J Risk Uncertain Eng Syst Civ Eng 7:04020055
16. Welcome to BraineryWiz's documentation!—BraineryWiz documentation 2024
17. daryl herzmann akrherz@iastate.edu. Iowa Environmental Mesonet 2019
18. Li SH, Kopp GA (2023) Duration of extreme synoptic wind speeds for North America in a changing climate and its engineering implementation. Struct Saf 104:102349
19. Alipour A, Sarkar P, Dikshit S, Razavi A, Jafari M (2020) Analytical Approach to Characterize Tornado-Induced Loads on Lattice Structures. J Struct Eng:146. https://doi.org/10.1061/(asce)st.1943-541x.0002660
20. Musallam M, Johnson CM (2012) An efficient implementation of the rainflow counting algorithm for life consumption estimation. IEEE Trans Reliab 61:978–986
21. Mertz DR (2012) Steel bridge design handbook: design for fatigue. United States. Federal Highway Administration, Office of Bridge Technology
22. MISO (2010) Midcontinent Independent System Operator I. Home

Service-Life Performance of Base-Isolated Hospital Building Under Multiple Hazards

Kunal Dahiya[1], Sarranya Banerjee[2(✉)], and Vasant Matsagar[2]

[1] Central Public Works Department (CPWD), New Delhi 110011, India
[2] Indian Institute of Technology (IIT) Delhi, New Delhi 110016, India
sarranya.banerjee@civil.iitd.ac.in

Abstract. Base isolation is an attractive technique for protection of structures against earthquakes. Especially, lifeline structures such as hospitals need to remain operational during and after natural disasters like earthquakes or cyclones. However, various environmental factors gradually degrade the performance of the base-isolated buildings in the form of deterioration of the material properties utilized in the base isolation as well as in the primary structure itself. Therefore, the present study aims to investigate the service-life performance of a base-isolated hospital building situated in New Delhi, India under multiple hazards. A detailed numerical study is conducted wherein the lead rubber bearing (LRB)-isolated hospital building is subjected to nonconcurrent earthquake and wind excitations. The effect of chloride-induced degradation of the structural elements and the oxidation-based degradation of the rubber in the LRBs is investigated over the service-life of the building. Interestingly, due to the addition of the LRBs, the hospital building could attain operational performance under earthquakes even with reduced structural member sizes, which in turn diminishes the carbon footprint of the entire project. Therefore, this study further compares the multi-hazard service-life performance of a regular base-isolated structure against a base-isolated structure with reduced carbon footprint.

Keywords: base isolation · hospital · multi-hazard · service-life performance · vibration control

1 Introduction

Lifeline structures are required to be designed with specific emphasis on their service-life performance. For instance, a hospital building should be functional even during and after earthquakes (Banerjee and Matsagar 2023a). Therefore, vibration control mechanisms are extensively utilized to enable continued functionality in lifeline structures under multiple hazards (Banerjee et al. 2025). Base isolators are effective vibration control devices against earthquakes (Banerjee and Matsagar 2024). There are various types of base isolation systems, such as elastomeric systems, friction-based systems, and rolling-type devices. Lead rubber bearing (LRB) is one of the most commonly utilized base isolator, wherein alternate layers of rubber and steel are stacked up with a lead plug

H. Beushausen et al. (Eds.): NTZR 2025, LNCE 873, pp. 698–710, 2026.
https://doi.org/10.1007/978-3-032-27698-8_62

centrally located in the device (Kelly 1986). The rubber provides the flexibility in shear and the lead plug increases the damping of the isolator. In this study, LRBs are utilized to isolate the hospital building from the damaging effects of the earthquakes.

However, the service-life performance of structures is dependent on various other factors; especially, elastomeric isolators like LRBs, they become stiffer with time due to thermal oxidation (Itoh and Gu 2009; Mazza 2019). Hence, the isolation time period reduces with ageing and the structural responses increase. Further, commonly found elements in the atmosphere such as water, oxygen, carbon dioxide, or chloride ions cause severe degradation in reinforced concrete (RC) members. Essentially, these factors cause corrosion in the reinforcing steel, which increases the volume of steel and cause further cracks in the concrete, which then increases the exposure of the steel to more of these degrading elements (Bhaskar et al. 2022). Essentially, the corrosion in steel degrades the elastic modulus of the structure. The effect of the degradation of the base isolators or the structural members are studied independently in literature, however the combined deterioration effect on the performance of the base-isolated structure is not well-explored, especially under multi-hazard scenarios.

Therefore, the objectives of this study are: (i) to estimate the effect of the chloride-induced degradation of reinforced concrete members and oxidation-based degradation of LRBs; and (ii) to obtain the service-life performance of a base-isolated hospital building under multiple hazards involving nonconcurrent earthquake and wind excitations.

2 Modeling of Base-Isolated Hospital Building

A base-isolated hospital building in Dwarka, New Delhi, India is investigated in this study. The reinforced concrete hospital building has eight stories. The natural time period of the building is 0.37 s. The exact details on the plan and structural elements of the hospital building are reported in Banerjee and Matsagar (2023b), thus not repeated herein. In this study, a simplified plan of the same hospital building is considered with equal column sizes. This simplification is undertaken to further utilize the same pattern of the building with reduced column sizes and base isolation.

Figure 1 shows the simplified floor plan of the considered hospital building with (a) Plan A: 1200 × 800 mm columns (close to actual sizes) and (b) Plan B: 600 × 600 mm columns (reduced sizes). The eight-story hospital building has a story height of 3.6 m for each story. The 28-day characteristic compressive strength of concrete is 30 MPa. Inside the hospital buildings, partitions consist of 100 mm thick fly-ash brick walls, and the floor finish measures 75 mm in thickness. Consideration is given to 25% of the 2 kN/m^2 imposed load on each floor during earthquake events [in compliance with IS 875 (Part 2), 1987].

Now, utilizing the two building plans, Plan A and Plan B, three different buildings are analyzed in this study, Building A, Building B, and Building C (as shown in Fig. 2). Building A represents the actual uncontrolled hospital building with Plan A, i.e., with column sizes of 1200 × 800 mm. Further, Building B is the base-isolated hospital building with Plan A, i.e., 1200 × 800 mm columns. Essentially, only the base isolation system is added to Building A to convert it into Building B. Finally, Building C is a base-isolated building with Plan B, i.e., 600 × 600 mm columns. In Building C the floor beams and slabs are designed with the same principles as Building A and Building B, however the columns have reduced sizes. This reduction in the column sizes is possible because of the base isolation of the building. Interestingly, due to the base isolation, the design shear forces on the superstructure have reduced significantly, which in turn diminished the size of the columns. Importantly, the reduction in the sizes of the columns result in a decrease of 518.4 m^3 of concrete. This is a huge saving in terms of both, (a) cost of construction and (b) emission of greenhouse gases due to the construction. Essentially, the reduction in reinforced concrete saves 181.44 ton of CO_2-equivalent greenhouse gas emissions (Hammond et al. 2011).

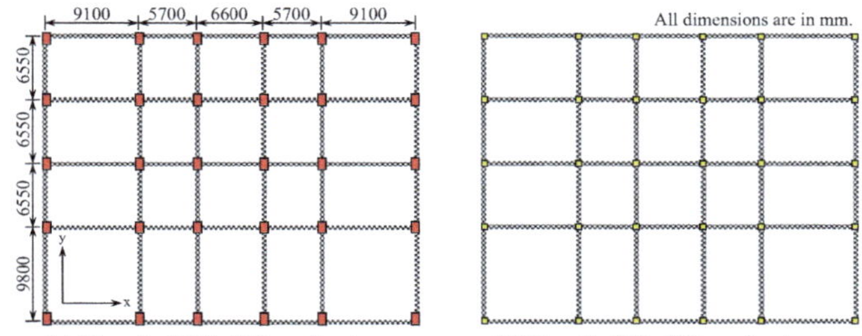

Fig. 1. Simplified floor plan of the considered hospital building with (a) 1200 × 800 mm columns and (b) 600 × 600 mm columns.

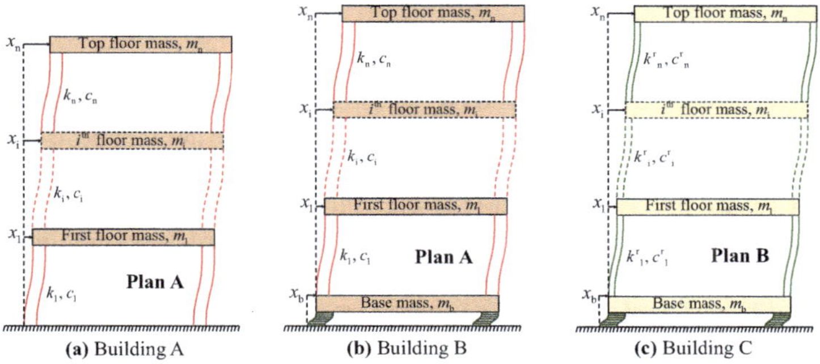

Fig. 2. Schematic diagrams of the three buildings investigated in this study.

To analyze the performance of the three buildings, Building A, Building B, and Building C, under the multi-hazard scenarios of earthquakes and winds, the buildings are simplified into multi-degree-of-freedom (MDOF) shear building models. The masses of the beams, floors, and columns are lumped at the floor levels. The equation of motion for the uncontrolled MDOF shear building model, i.e., for Building A is given as:

$$[M]\{\ddot{X}(t)\} + [C]\{\dot{X}(t)\} + [K]\{X(t)\} = F(t), \tag{1}$$

where, $[M]$, $[C]$, and $[K]$ are the mass, stiffness, and damping matrix, respectively, of the uncontrolled structure, i.e., Building A. Further, $\{X(t)\}$, $\{\dot{X}(t)\}$, and $\{\ddot{X}(t)\}$ are the displacement, velocity, and acceleration response vectors, respectively. Here, $F(t)$ is the vector containing the earthquake-induced forces or the wind-imparted forces on the various degrees-of-freedom (DOFs) of the structure. For the earthquake-induced forces, the expression of $F(t)$ is obtained as:

$$F(t) = [M]\{r\}\ddot{X}_g(t), \tag{2}$$

where, $\{r\}$ is the influence vector with the value of unity for all the DOFs. Further, $\ddot{X}_g(t)$ is the ground acceleration acting at the particular site. The damping ratio of the uncontrolled structure is considered as 2%. For the base-isolated buildings, i.e., Building B and Building C, the governing equation of motion is:

$$[M_{\mathrm{b}}]\{\ddot{X}_b(t)\} + [C_{\mathrm{b}}]\{\dot{X}_b(t)\} + [K_{\mathrm{b}}]\{X_{\mathrm{b}}(t)\} = F_{\mathrm{b}}(t), \tag{3}$$

where, $[M_{\mathrm{b}}]$, $[C_{\mathrm{b}}]$, and $[K_{\mathrm{b}}]$ are the mass, stiffness, and damping matrix, respectively, of the base-isolated structure, i.e., Building B or Building C. Further, $\{X_{\mathrm{b}}(t)\}$, $\{\dot{X}_b(t)\}$, and $\{\ddot{X}_b(t)\}$ are the displacement, velocity, and acceleration response vectors of the base-isolated structure, respectively. Here, $F_{\mathrm{b}}(t)$ is the vector containing the earthquake-induced forces or the wind-imparted forces on the various degrees-of-freedom (DOFs) of the base-isolated structure. Essentially, due to the addition of base isolation, a new DOF is added, which represents the mass, stiffness, and damping of the base-isolation layer.

In this study, lead rubber bearings (LRBs) are utilized to isolate the considered hospital building from the ill-effects of the earthquakes. The design isolation time period for both Building B and Building C is considered as 4 s. Herein, the LRBs are modeled simply using their effective stiffness (k_{b}) and effective damping (c_{b}) parameters. The damping ratio of the LRBs is assumed as 15% and the stiffness of the LRBs is evaluated from the design isolation period. Therefore, the natural time periods for Building B and Building C are 4 s, whereas the natural time period of the Building A is obtained from modal analysis as 0.38 s (which is close to the actual hospital building). Now, Eqs. 1 and 3 are solved using numerical integration for the various structural configurations and loading scenarios. Further, the service-life degradation of the uncontrolled and base-isolated buildings is considered in the subsequent section.

3 Service-Life Degradation of Base-Isolated Structure

Along the service-life of a structure, various environmental factors contribute towards degradation of its mechanical properties. Therefore, it is necessary to analyze the effect of this degradation in the performance of the structure subjected to different loading scenarios. In this section, two types of degradation are studied in the following subsections.

3.1 Degradation of Lead Rubber Bearings Due to Ageing

The mechanical characteristics of lead rubber bearings undergo changes over time due to the aging of the rubber caused by external factors. One of the major factors contributing towards deterioration of the elastomeric isolators is thermal oxidation of rubber. The time-dependent degradation of the elastomeric bearings was analyzed and modeled based on accelerated thermal oxidation tests conducted by Itoh et al. (2006), and later by Itoh and Gu (2009). They have concluded that the effective stiffness of each rubber bearing at any time during its service-life for any temperature at side is given as:

$$k_b(t_d) = k_{b,0}(1 + k_s \Delta f_s), \tag{4}$$

where, $k_b(t_d)$ is the degraded stiffness of the LRB after t_d years of construction. Further, $k_{b,0}$ is the initial effective stiffness of the LRB, Δf_s and k_s are given as:

$$\Delta f_s = 0.066(t_{ref})^{0.515} \tag{5}$$

$$k_s = 2d_c(a + b - d_c)/3ab, \tag{6}$$

respectively. Here, t_{ref} is the equivalent reference time for the accelerated oxidation in the laboratory. Further, a and b are the longitudinal and transverse dimensions of the LRBs. For the present study, LRBs with side lengths, a and b, equal to 400 mm is considered. The critical depth, d_c, is given by:

$$d_c = \alpha e^{\beta/T}, \tag{7}$$

where, $\alpha(= 0.0008\text{mm})$ and $\beta\left(= 3.31 \times 10^{-3}\right)$ are coefficients obtained from the accelerated thermal oxidation tests. Here, T is the average temperature at site, equal to to 303 K (30 °C), assumed for the present study. To obtain the equivalent reference time, t_{ref}, based on the actual time passed, t_d, the Arrhenius methodology could be utilized:

$$\ln\left(\frac{t_{ref}}{t_d}\right) = \frac{E_a}{R}\left(\frac{1}{T_{ref}} - \frac{1}{T}\right). \tag{8}$$

Here, $E_a\left(= 9.49 \times 10^4 \text{ J/mol}\right)$ is the activation energy for rubber, and $R(= 8.314\text{J/mol K})$ is the gaseous constant. Further, T_{ref} is the reference temperature at the laboratory, which is equal to 343 K (Itoh et al. 2006). With this, the degraded stiffness, $k_b(t_d)$, could be obtained after any number of years of service-life. It is to be noted that deterioration in the damping of the LRB is not considered due to thermal oxidation. Essentially, due to the thermal oxidation of rubber, the stiffness of the LRBs increases, thus reducing its isolation capabilities. Hence, using Eq. 4, the stiffness of the isolation system after any number of years is obtained and incorporated in the simulations.

3.2 Chloride-Induced Degradation of Reinforced Concrete

Chloride-induced corrosion plays a key role in the degradation of reinforced concrete (RC) structures. Essentially, various corrosion-inducing elements form rust around the reinforcing bars inside the concrete. Due to rust, the volume of reinforcement increases, which forms further cracks in the concrete. The cracks allow more corrosive elements to reach the reinforcing bars and continually form rust in the reinforcing bars, thus reducing the strength of the RC members. The effect of chloride-induced corrosion in the RC members is quantified by the reduction of the compressive strength of the concrete as (Vecchio and Collins 1986):

$$f_{\mathrm{ck}}^{*} = f_{\mathrm{ck}}\left(\frac{1}{1 + k_{\mathrm{r}}(\varepsilon_1/\varepsilon_{\mathrm{c0}})}\right), \tag{9}$$

where, $k_{\mathrm{r}}(= 0.1)$ is the coefficient representing the roughness and diameter of the reinforcing bars (Coronelli and Gambarova 2004). Further, $\varepsilon_{\mathrm{c0}}$ $(= 0.002)$ is the strain at the peak compressive strength, f_{ck}; and ε_1 is the tensile strength in the cracked concrete evaluated as:

$$\varepsilon_1 = (b_{\mathrm{f}} - b_0)/b_0. \tag{10}$$

Here, b_0 and b_{f} are the section widths before and after the corrosion-induced cracking in the section, respectively. The difference in the section widths after and before the corrosion can be estimated as:

$$b_{\mathrm{f}} - b_0 = n_{\mathrm{bars}} w_{\mathrm{c}}, \tag{11}$$

where, n_{bars} is the number of reinforcing bars in each layer. In this study, n_{bars} is considered as 4. Further, w_{c} is the crack width at a certain level of corrosion after a certain duration, expressed as:

$$w_{\mathrm{c}} = 2\pi(v_{\mathrm{rs}} - 1)d_{\mathrm{e}}. \tag{12}$$

Here, $v_{\mathrm{rs}}(= 2)$ represents the expansion of the reinforcing bars with respect to the undamaged material (Molina et al. 1993). Further, d_{e} is the corrosion penetration depth, calculated as:

$$d_{\mathrm{e}} = 0.0115 i_{\mathrm{co}} t. \tag{13}$$

Here, i_{co} is the corrosion current density in cm/year; and t is the time since corrosion initiated. Here, i_{co} is considered as 2 cm/year based on practical estimates for building structures (Bhaskar et al. 2022). With this, the variation in the characteristic compressive strength of the concrete can be estimated using Eq. 9. Further, the elastic modulus of the concrete can be estimated as:

$$E_{\mathrm{c}} = 2.1 \times 10^5\left(\frac{\gamma}{2.3}\right)^{1.5}\left(\frac{f_{\mathrm{ck}}}{200}\right)^{0.5}. \tag{14}$$

Here, $\gamma\left(= 2.5\mathrm{ton/m}^3\right)$ is the density of concrete. Therefore, utilizing Eq. 14, the degradation in the elastic modulus of the concrete can be modeled, using which the degraded performance of the structure is evaluated.

4 Numerical Study

The performance of the three reinforced concrete (RC) buildings described in Fig. 2 is assessed under the multi-hazard scenario of nonconcurrent earthquakes and winds. Further, the variation in the structural performance due to the degradation of its properties is emphasized.

4.1 Performance Under Earthquakes

Seven different earthquake records are selected for the performance assessment as shown in Table 1. The selected earthquake records have a wide variety of peak ground acceleration (PGA) ranging from $0.16g$ to $0.89g$. Further, the selected earthquake records are scaled corresponding to the site-specific earthquake at New Delhi, India. IS 1893 (Part 1), 2016 is utilized to obtain the site-specific target earthquake spectrum. The importance factor is considered to be 1.5 for the hospital building and medium-type soil condition is assumed. For converting the response spectrum to maximum considered earthquake (MCE) scenario, the response spectrum is multiplied by a factor of 2 (Banerjee and Matsagar 2023b). This is necessary since the response spectrum provided in the standard is for design basis earthquake (DBE), however the current base-isolated structures are tested under MCE conditions. With this, the recorded earthquakes are converted to site-specific design earthquakes using the software, SeismoMatch (2022). The site-specific earthquakes are now utilized for further analysis of the various building structures.

Table 1. Details of selected recorded earthquake excitations.

Notation	Earthquake name	Event year	Recording station	Component	PGA (g)	Duration (s)
EQ1	Koyna Dam #02	1967	Koyna Dam, Shear Zone Gallery	Long	0.160	10.70
EQ2	Uttarkashi	1991	Uttarkashi	N15W	0.237	39.92
EQ3	Gazli	1976	Karakyr Point	EW	0.706	13.02
EQ4	Ardal	1977	Naghan 1	Long	0.890	11.46
EQ5	Mammoth Lakes	1980	Convict Creek	180	0.392	65.00
EQ6	Erzincan	1992	Erzincan-Meteorologij Mudurlugu	N279	0.502	20.74
EQ7	Friuli	1976	Tolmezzo-Diga Ambiesta	NS	0.350	36.53

Figure 3 shows the time-history of the top floor absolute acceleration responses of the various buildings under the EQ6 earthquake, at (a) 0 year, (b) 25 year, (c) 50 year, and (d) 100 year of their service life. Evidently, the addition of the LRB-based isolation system (Building B) to the uncontrolled building (Building A) has reduced the top floor acceleration response of the structure. Additionally, Building C has performed similar to the Building B in terms of acceleration response of the structure. Interestingly, both Building B and Building C have resulted in 92% and 90% reduction in the top floor acceleration response of the hospital building, respectively, at the virgin state (0 year). Further, with the passage of time, considering the degradation in the buildings due to chloride-induced corrosion as well as thermal oxidation of rubber, the variation in the time-history responses of the various buildings are observed in Fig. 3. Due to the chloride-induced degradation of the structure, the response of Building A has deteriorated with time (structural stiffness is reduced), showing worst response after 50 years. Similarly, due to the thermal oxidation of rubber (increase in stiffness of the isolators) and chloride-induced degradation of the primary structural members, the responses of the base-isolated buildings (Building B and C) have also deteriorated. Importantly, the structural responses of Building C, with lesser sizes of columns (i.e., less carbon footprint) are slightly higher than Building B where the base-isolated building has the same sizes of columns as the primary uncontrolled structure. This difference in the performance of Building B and Building C is more pronounced with the degradation of the structural parameters. Nevertheless, even with the degraded performance, the response of Building C is much lesser than the response of Building A.

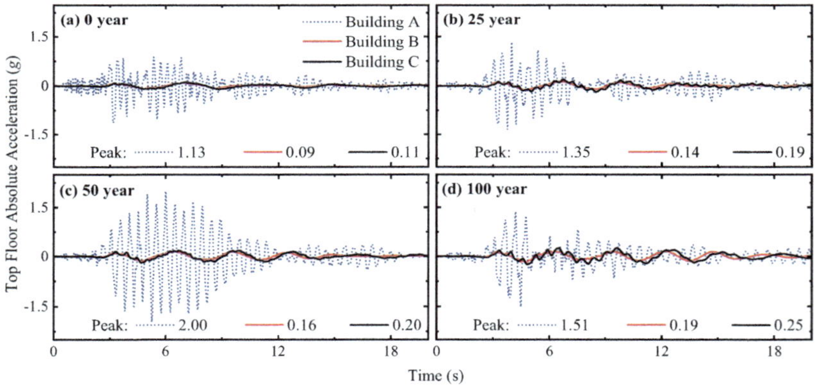

Fig. 3. Time-history of top floor absolute acceleration response of the various buildings under the EQ6 earthquake, at (a) 0 year, (b) 25 year, (c) 50 year, and (d) 100 year of their service life.

Figure 4 shows that time-history top floor absolute acceleration response of Building B and Building C considering degradation in the structural parameters, with distinction between the degradation in the isolator due to thermal oxidation and the complete degradation considering both thermal oxidation of rubber and chloride-induced degradation of structural RC members. With the passage of time, thermal oxidation makes the LRBs stiffer, hence there is a distinct reduction in the time-period of the isolated structure response, as observed in Fig. 4. Both for Building B and C, there is a constant increase in the top floor acceleration response of the structure with the passage of time. Interestingly, in Fig. 4b and d the effect of chloride-induced corrosion of the RC members is evident. Due to the degradation of the structural members, the time period of the primary structure increases, hence the degraded response of the base-isolated structure is dominated by the superstructure response. As expected, here too the affect is more pronounced in Building C, due to reduced member sizes.

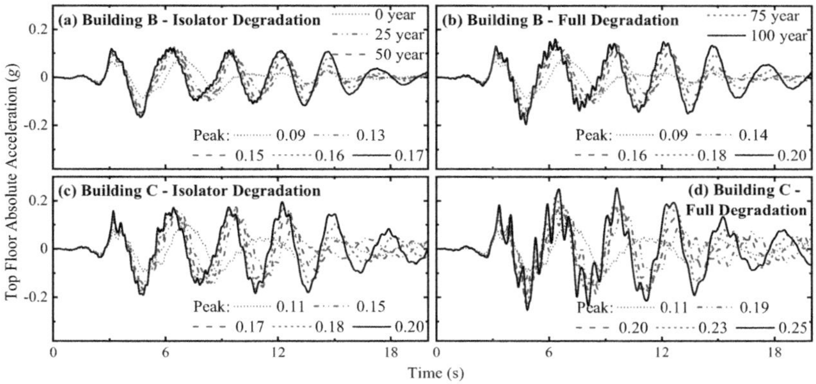

Fig. 4. Time-history of top floor absolute acceleration response of (a) Building B with only isolator degradation, (b) Building B with full (isolator and structural) degradation, (c) Building C with only isolator degradation, and (d) Building C with full (isolator and structural) degradation, at various stages of their service life.

Figure 5 shows important results on the peak top floor absolute acceleration response of the structures under all earthquakes. Now, to ensure the the hospital building is continuously serviceable even during and after the earthquakes, it is necessary to limit the structural vibrations to tolerable limits for its inhabitants. An acceleration of $0.15g$ is selected as the limit for tolerance amongst the occupants of the buildings (Griffis 1993). As observed from the figure, for Building B, only for three cases, the response has exceeded the tolerable limits that too after 50 years. However, for Building C, all the earthquakes have caused the exceedance of the tolerable limits. In fact, for three cases the tolerable limits are exceeded within just 25 years. This shows that although the design with lesser member sizes is able to reduce the earthquake-induced vibrations, it is not as effective as the building with larger member sizes in the long run.

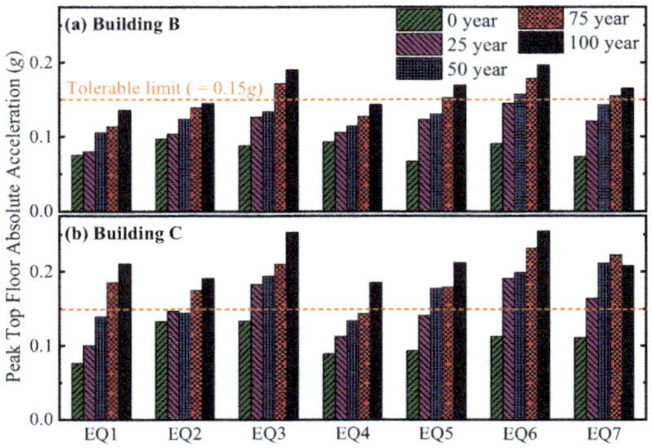

Fig. 5. Peak top floor absolute acceleration response of (a) Building B and (b) Building C under all the earthquakes, at various stages of their service life.

4.2 Performance Under Winds

The wind response of the building structures are evaluated using the site specific wind conditions at New Delhi, India. The basic wind speed is 47 m/s from the Indian Standard. The wind velocity distribution over the height of the buildings and the specific wind forces are obtained following the procedure outlined in Banerjee and Matsagar (2024). Essentially, the fluctuating wind velocities are obtained from the NatHaz online wind simulator (Kwon and Kareem 2006), and the final wind force at the ith story of the structure is evaluated as:

$$F_{\text{i}}(t) = 0.5\rho C_{\text{d}} A_{\text{i}}[U_{\text{i}} + u_{\text{i}}(t)]^2, \tag{15}$$

where, ρ and C_{d} are the density of the air and the drag coefficient which are equal to 1.2kg/m^3 and 1.3, respectively. Further, A_{i} is the tributary area of the ith floor, U_{i} is the mean wind speed, and $u_{\text{i}}(t)$ is the fluctuating wind speed. A 15-min duration for the wind force is considered.

Figure 6 shows the time history top floor acceleration response of the various buildings. As expected, the response of the non-isolated structure, Building A, is significantly lesser than the response of the base-isolated structures, Building B and C. Essentially, the base isolators lengthen the time period of the structures, making it more vulnerable to wind. Interestingly, the response of Building C is even considerably higher than the response of Building B. This is due to the lesser stiffness of the superstructure of Building C attributed to the smaller member sizes. Moreover, it is found that the structural and isolator degradation does not have a significant effect on the response of the various buildings under winds.

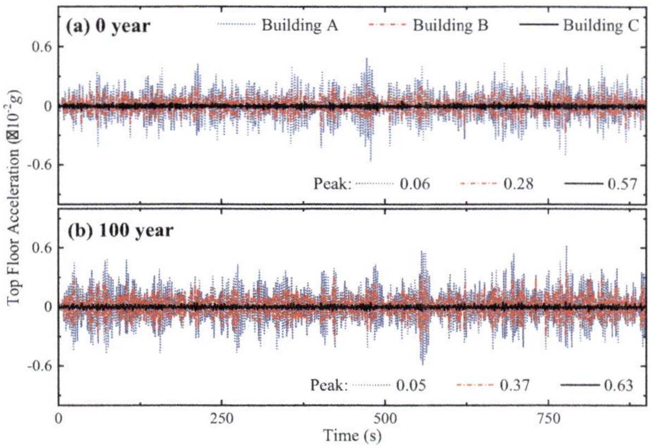

Fig. 6. Time-history of top floor acceleration response of the various buildings under the wind excitation, at (**a**) 0 year and (**b**) 100 year of their service life.

Figure 7 shows the peak top floor acceleration response of the various buildings. Here too, it is observed that Building A has the least response followed by Building B and then Building C. Interestingly, the response of Building C is found to be always over the perceptible limit (0.005g) of the inhabitants (Griffis 1993). Thus, reducing the size of the members results in a negative performance of the building under wind excitations.

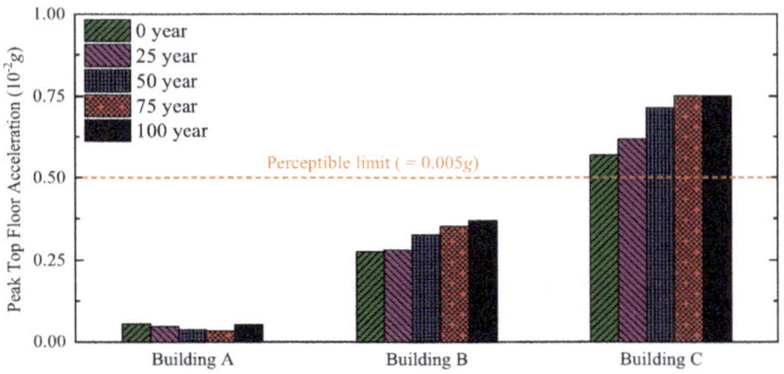

Fig. 7. Peak top floor acceleration response of the various buildings under the wind excitation at various stages of their service life.

5 Conclusions

The service-life performance of a base-isolated hospital building in New Delhi, India is assessed under multi-hazard scenario of nonconcurrent earthquakes and winds. Three buildings are compared: Building A, non-isolated building; Building B, base-isolated building; and, Building C, base-isolated building with reduced sizes of structural members. Two types of service-life degradation are considered: thermal oxidation of elastomeric isolators and chloride-induced corrosion of reinforced concrete members. The responses of the various buildings under nonconcurrent earthquakes and winds are obtained.

Results show that the LRB-based isolation system could effectively reduce the acceleration response of the hospital building by about 90%. Irrespective of whether the structural members are bulkier or lean, considerable reduction in the structural responses is observed due to base isolation. However, with the passage of time, the performance of the uncontrolled as well as the base-isolated structures deteriorated. The effect of structural degradation was more pronounced on Building C (with reduced member sizes). Under all the earthquakes, the acceleration response of Building B exceeded the tolerable limits only three time, that too after 50 years degradation. However, Building C resulted in exceedance of tolerance limits of structural acceleration from as early as 25 years. Subsequently, under site-specific wind excitations, Building A and Building B always produced lesser acceleration response than the perceptible limits of its inhabitants. On the other hand, the acceleration response of Building C exceeded the perceptible limits all the time during its service-life. Hence, it is observed that although a lesser carbon-intensive design of the hospital building is possible with the help of base isolation, its performance is slightly detrimental than the performance of the original building with base-isolation, under the multiple hazards. Although there is a saving of 181.44 ton of CO_2-equivalent greenhouse gas emissions in Building C, its performance is found to be slightly poor when compared to Building B, which could affect in additional greenhouse gas emissions in maintenance or repair during its service-life. Thus, designers should keep this factor in mind while deciding on the optimal design of base-isolated structures.

References

1. Banerjee S, Matsagar VA (2023a) Earthquake response control of hospital building using unbonded fibre-reinforced elastomeric isolators" in Coalition for Disaster Resilient Infrastructure (CDRI) 2022 conference proceedings. Sustain Resilient Infrastruct 8:85–92
2. Banerjee S, Matsagar V (2023b) Hybrid vibration control of hospital buildings against earthquake excitations using unbonded fiber-reinforced elastomeric isolator and tuned mass damper. Buildings 13(7):1724
3. Banerjee S, Matsagar VA (2024) A trilinear hysteretic model and design procedure for unbonded fiber-reinforced elastomeric isolators. Eng Struct 304:117616
4. Banerjee S, Saito T, Matsagar V (2024) Bidirectional performance assessment of UFREI-isolated structures with single and multiple tuned mass dampers under earthquakes and winds. Soil Dyn Earthq Eng 180:108618
5. Banerjee S, Saito T, Matsagar V (2025) Robust optimum design for hybrid vibration control of structures under multihazard scenarios of earthquakes and winds. ASCE-ASME J Risk Uncertain Eng Syst, Part A: Civ Eng 11(1):04025002
6. Bhaskar P, Baheti A, Matsagar V (2022) Service-life damage assessment of a reinforced concrete structure under multi-hazard seismic and wind actions. Trans Indian Natl Acad Eng 7(3):1017–1031
7. Coronelli D, Gambarova P (2004) Structural assessment of corroded reinforced concrete beams: modeling guidelines. J Struct Eng 130(8):1214–1224
8. Griffis LG (1993) Serviceability limit states under wind load. Eng J Am Inst Steel Constr 30(1):1–6
9. Hammond G, Jones C, Lowrie EF, Tse P (2011) Embodied carbon. The inventory of carbon and energy (ICE). Version (2.0), vol 16
10. IS 875 (Part 2): Design loads (other than earthquake) for buildings and structures—Code of practice: Part 2 imposed loads. Bureau of Indian Standards, New Delhi, India (1987)

11. IS 1893 (Part 1): Criteria for earthquake resistant design of structures—Code of Practice: Part 1 General provisions and buildings. Bureau of Indian Standards, New Delhi, India (2016)
12. Itoh Y, Gu H, Satoh K, Kutsuna Y (2006) Experimental investigation on ageing behaviors of rubbers used for bridge bearings. Structural Eng Earthquake Eng JSCE 23(1):17s–31s
13. Itoh Y, Gu HS (2009) Prediction of aging characteristics in natural rubber bearings used in bridges. J Bridg Eng 14(2):122–128
14. Kelly JM (1986) Aseismic base isolation: review and bibliography. Soil Dyn Earthq Eng 5(4):202–216
15. Kwon D, Kareem A (2006) NatHaz on-line wind simulator (NOWS): simulation of Gaussian multivariate wind fields. NatHaz Modeling Laboratory Report, University of Notre Dame. http://windsim.ce.nd.edu/
16. Mazza F (2019) Effects of the long-term behaviour of isolation devices on the seismic response of base-isolated buildings. Struct Control Health Monit 26(4):e2331
17. Molina FJ, Alonso C, Andrade C (1993) Cover cracking as a function of rebar corrosion: Part 2—numerical model. Mater Struct 26:532–548
18. SeismoMatch Software (Version 2022), https://seismosoft.com/products/seismomatch/, last accessed 2025/04/29
19. Vecchio FJ, Collins MP (1986) The modified compression-field theory for reinforced concrete elements subjected to shear. ACI J 83(2):219–231

Extending the Service Life of Bridge Structures Through Improved End Joint Details

Shadi Azad and Behrouz Shafei(✉)

Iowa State University, Ames, IA 50011, USA
shafei@iastate.edu

Abstract. This study explores improved bridge end details to enhance the service life of integral and semi-integral abutment bridges. The research involved comprehensive inspections of nine bridge structures, focusing on approach slabs, abutment connections, and joint response characteristics. Field observations revealed significant performance issues, including preformed tied approach joints expanding beyond design specifications and approach slabs shifting away from tied connections. Three-dimensional finite element models were developed to analyze approach slab response to bridge thermal movements. The numerical analysis demonstrated that skew angles dramatically alter stress distributions in tie bars across approach joints, with forces concentrating at acute approach slab corners. These findings provided essential guidance for optimizing bridge end details to extend the service life of bridge structures, thereby reducing maintenance requirements and replacement frequencies. Implementing these improved connection details supports sustainability goals by decreasing material consumption and embodied carbon associated with premature rehabilitation needs, contributing to resource-efficient transportation infrastructure aligned with net-zero objectives.

Keywords: Approach slabs · tied connections · integral abutments · thermal movement · skew effects

1 Introduction

Transportation infrastructure represents a significant contributor to global carbon emissions, both through direct operational impacts and embodied carbon in construction materials [1]. Bridge structures, as critical components of transportation networks, present unique sustainability challenges due to their material-intensive nature and specialized maintenance requirements [2]. The movement toward net-zero infrastructure [3] necessitates a comprehensive reevaluation of conventional bridge design approaches, with particular emphasis on components that frequently require maintenance or premature replacement [4, 5]. Jointless bridge designs have emerged as a promising approach to enhance transportation network sustainability by eliminating traditional expansion joints, which have historically represented significant maintenance liabilities and failure points in bridge systems [6, 7]. By transitioning to integral and semi-integral abutment configurations, transportation departments aim to extend service life while

H. Beushausen et al. (Eds.): NTZR 2025, LNCE 873, pp. 711–720, 2026.
https://doi.org/10.1007/978-3-032-27698-8_63

reducing maintenance interventions that contribute to life-cycle carbon emissions [8]. These designs have demonstrated particular value in regions experiencing extreme weather conditions, where temperature fluctuations and freeze-thaw cycles accelerate joint deterioration [9].

The sustainability implications of bridge end details extend beyond immediate construction considerations. As highlighted by Ali et al. [10], premature deterioration of bridge slabs contributes significantly to unplanned maintenance activities, with associated environmental impacts through construction equipment emissions, traffic disruptions, and material production. DeJong et al. [11] demonstrated that grouted rebar couplers embedded in ultra-high-performance concrete at the abutment significantly improve thermal load transfer and eliminate the need for traditional expansion joints, thereby reducing corrosion and maintenance demands. Full-scale tests and subsequent simulations [12] confirmed that integral abutment details using precast pile couplers can effectively accommodate cyclic thermal expansion while simplifying reinforcement layout. Karim and Shafei [13] showed that fiber-reinforced concrete link slabs can reduce joint deterioration, suggesting improved durability for deck-to-end connections. It was further observed that partial-depth link slabs, when designed with appropriate end-fixity conditions, provide effective retrofit solutions for joints, thereby extending service life [14].

The connection between approach slabs and bridge decks represents a critical interface that directly impacts sustainable performance objectives. The disparity between controlled testing environments and actual service conditions highlights the need for design approaches that better account for long-term material behavior under service loads. Recent advances in computational modeling techniques have enabled more comprehensive analysis of bridge end details under thermal loading conditions. Finite element studies by Khasawneh [15] demonstrated that soil confinement characteristics significantly influence slab movement patterns, particularly in skewed configurations where asymmetric pressure distributions develop. These findings suggest that current design methodologies may substantially underestimate the complex stress states experienced by connection systems during seasonal temperature fluctuations.

From a sustainability perspective, optimized bridge end details represent a critical opportunity to reduce life-cycle carbon impacts through extended service intervals and reduced material consumption. Research by Khatami et al. [16] established quantitative relationships between bridge component durability and environmental impact metrics, demonstrating that doubling the service life of bridge slabs could reduce lifetime carbon emissions by approximately 33% for typical highway overpasses. The integration of improved design methodologies with enhanced construction practices offers significant potential for advancing net-zero objectives in transportation infrastructure. As demonstrated by Mendoza-Lugo et al. [17], incremental improvements in bridge component performance can yield substantial cumulative benefits across transportation networks when systematically implemented through updated design standards. These modifications represent relatively low-cost interventions with disproportionately high sustainability returns through extended infrastructure service life and optimized maintenance scheduling.

In this study, the in-service performance of bridge end details was systematically evaluated through field inspections of nine bridge structures with semi-integral abutments and tied approach slabs. Comprehensive data collection focused on joint and approach slab issues to identify recurring performance deficiencies. Three-dimensional finite element models were developed using ABAQUS/CAE to analyze the mechanical behavior of tied approach slabs under thermal loading conditions, with particular emphasis on the effects of skew geometry on stress distribution patterns. The comparative analysis between free and tied configurations quantified the complex multi-directional forces experienced during thermal cycles, establishing relationships between design parameters and observed deterioration mechanisms. The results provide the bridge engineering community with recommendations regarding connection details for improved sustainability outcomes through extended service life, reduced maintenance requirements, and decreased embodied carbon, directly supporting net-zero objectives by minimizing the environmental impacts associated with premature rehabilitation activities.

2 Inspection of Bridge End Details and Joint Performance

A comprehensive field inspection program was conducted to evaluate the in-service performance of jointless bridges with particular emphasis on approach slab connections and joint behavior. Nine bridge structures across Iowa were systematically selected based on geometric parameters including length, width, skew angle, abutment type, and approach slab connection details. The inspection methodology focused on identifying performance deficiencies that could impact service life and sustainability metrics, concentrating on sleeper slabs, approach slabs, deck joints, embankments, abutment faces, and bearing conditions. The selected bridges represented various configurations and service conditions, with inspection sites including both newly constructed semi-integral abutment bridges and structures retrofitted from conventional to semi-integral designs. The inspection program specifically targeted bridges with tied approach slabs to assess their effectiveness in maintaining joint integrity through thermal cycles. Ambient temperatures during inspections ranged from 15.5 to 26.7 °C, providing a consistent baseline for comparative analysis.

Field measurements of joint widths were recorded at both the deck-to-approach slab interface (Joint 1) and the approach slab-to-sleeper slab or transition slab interface (Joint 2) for comparison against design specifications (Table 1). The investigations revealed several recurring performance issues that have significant implications for bridge maintenance requirements. Most notably, tied approach connections demonstrated joint expansion well beyond design parameters. Specifically, preformed tied approach joints designed for 6.4 mm widths were observed with actual measurements up to 41.3 mm, representing a significant increase from specified dimensions. This substantial deviation indicates failure of the tie bar system to maintain the intended connection between the bridge structure and approach slab.

Concurrent with tied joint expansion, the field data showed corresponding decreases in joint widths at sleeper slab interfaces, providing clear evidence of approach slab migration away from the bridge deck. This bidirectional joint response demonstrates that the tied connections are not effectively distributing thermal movement as intended in the

Table 1. Original bridge design and field investigations joint widths.

Bridge name	Length (m)	Skew angle	Year built	Joint 1		Joint 2	
				Design (mm)	Measured (mm)	Design (mm)	Measured (mm)
Polk 1215	101	2.3°	2016	25.4	50.8	76.2	38.1
Pottawattamie 111	62	0°	2011	50.8	38.1	–	–
Cass 113	37	0°	2013	–	47.6	–	6.35
Cass 213	68	7.75°	1965	6.4	12.7	63.5	50.8
Jasper 310N	71	6.25°	1962	6.4	22.2	88.9	69.9
Jasper 310S	71	6.25°	1962	6.4	19.1	88.9	47.6
Bremer 208N	49	2.25°	1994	6.4	41.3	88.9	63.5
Bremer 208S	49	2.25°	1994	6.4	25.4	88.9	44.5
Blackhawk 108	40	0°	2010	6.4	25.4	76.2	73.0

design. The vertical bar configuration employed in most of the inspected bridges appears inadequate for long-term performance despite favorable results in previous short-term studies. Inadequate drainage and resulting embankment deterioration emerged as significant contributing factors to premature performance degradation. Multiple instances of substantial voids were documented at the interface between abutment faces and embankments, with measured depths exceeding 55.88 cm. These void formations appear to result from a combination of ineffective surface water management and soil erosion patterns. Concrete slope protection exhibited severe displacement with gaps between protection panels and vertical settlement evidenced by displacement markers on the abutment face. The inspection of the bridges also revealed critical drainage system failures, with embankment drains either partially or completely obstructed (Fig. 1). These drainage deficiencies directly contribute to accelerated deterioration of structural components through increased moisture exposure, especially under freeze-thaw cycles.

Fig. 1. An example of blocked embankment drainage.

Analysis of the compiled inspection data revealed correlations between bridge geometric parameters and observed performance patterns. Bridges with higher skew angles demonstrated more pronounced joint deterioration, suggesting that skew geometry introduces asymmetric force distribution in approach slab connections. The comparative performance between newly constructed semi-integral abutments and retrofitted structures indicated that original design implementation generally outperformed conversion projects, though exceptions were noted.

The measured joint widths demonstrated a clear pattern of tied approach joint failure across multiple structural configurations. This consistent finding suggests a fundamental limitation in current tied approach design methodologies rather than isolated construction or maintenance deficiencies. The documented performance deficiencies have direct implications for bridge service life and environmental sustainability objectives. Each identified failure mode contributes to accelerated deterioration cycles that necessitate earlier intervention and rehabilitation. The progression from minor joint expansion to significant approach slab displacement ultimately leads to ride quality degradation, increased dynamic loading, and potential structural damage requiring resource-intensive repairs. From a sustainability perspective, these premature failures represent substantial embodied carbon implications. The production of concrete and reinforcing steel for rehabilitation or replacement activities contributes significantly to the transportation sector's carbon footprint. By addressing the root causes of joint connection deterioration, significant reductions in life-cycle embodied carbon can be achieved through extended service intervals and reduced material consumption.

The comparative analysis of bridges with different design ages also provides valuable insights into performance evolution over time. Structures retrofitted from conventional to semi-integral configurations exhibited distinct deterioration patterns from those initially constructed as semi-integral systems, suggesting that design optimization must consider both initial construction and potential future adaptation requirements. The field investigation findings strongly indicate that current tied approach connection details require fundamental reconsideration to achieve intended performance objectives. The consistent failure of vertical tie bar systems to maintain joint integrity suggests that alternative connection geometries or reinforcement configurations may be necessary. The migration of approach slabs away from bridge decks despite tie bar presence indicates that thermal movement forces exceed the restraint capacity of current connection details.

These observations provided critical direction for the subsequent finite element modeling phase of this research, particularly regarding the need to analyze stress distribution patterns within approach slabs and connection systems under thermal loading scenarios. By addressing the identified performance limitations through improved design details, significant advancements in bridge sustainability can be achieved through extended service life, reduced maintenance requirements, and decreased material consumption over infrastructure lifecycles. These improvements directly support net-zero objectives by minimizing the embodied carbon associated with premature rehabilitation activities.

3 Investigation of Bridge End Response

To analyze the mechanical behavior of tied approach slabs under thermal loads, a comprehensive three-dimensional finite element model was developed using ABAQUS/CAE software. The modeling approach focused specifically on the complex interaction between the approach slab and bridge abutments during thermal cycles, with particular emphasis on the performance of tie bar connections in a semi-integral abutment configuration. Rather than attempting to simulate the entire bridge structure, the model strategically isolated critical components including the approach slab, connection region, supporting soil, and the abutment section to enable detailed analysis of joint performance metrics. The finite element analysis framework was designed to simulate realistic bridge thermal movement conditions by applying calculated displacements to the abutment structure while simultaneously incorporating corresponding temperature change to connected elements. This dual-loading approach effectively captured the complex response mechanisms of the approach slab subjected to bridge expansion and contraction cycles.

The model geometry was based on a representative bridge, featuring an 8.53 m wide approach slab with 25.40 cm thickness in a trapezoidal configuration (4.57 to 13.11 m). The soil matrix was modeled as a cohesive block in a trapezoidal shape similar to that of the approach slab with a depth of 61 cm. The abutment dimensions included a width of 1.19 m and a depth of 84 cm. Material properties were specified to reflect standard Iowa Department of Transportation (DOT) construction specifications, with concrete compressive strength of 27.58 MPa, Young's modulus of 24856 MPa, Poisson's ratio of 0.2, and thermal expansion coefficient of 7.4×10^{-6} /°C, applied to approach slabs and abutment structures. The soil matrix was modeled as modified subbase material with properties specified by the Iowa DOT, featuring a compacted unit weight of 2243 kg/m^3. An elastic modulus of 137.90 MPa was selected based on literature values for compacted gravel, with a corresponding Poisson's ratio of 0.35. The steel material featured a density of 7850 kg/m^3, Young's modulus of 2×10^5 MPa, Poisson's ratio of 0.3, and thermal expansion coefficient of 11.7×10^{-6} /°C. No. 25 rebars spaced at 37.6 cm were modeled using wire elements to tie the approach slab to the abutment. Concrete-to-concrete interfaces utilized a coefficient of 0.60, while concrete-to-soil contacts were modeled with a coefficient of 0.55 representing interaction with gravel or coarse sand substrates. Three-dimensional solid C3D8R elements were selected for concrete and soil representation to facilitate accurate simulation of frictional contact interfaces. Structural reinforcement was modeled using embedded element techniques, with SFM3D4 surface elements representing distributed reinforcement within concrete host elements. A mesh size of 63.5 mm was adopted based on mesh sensitivity analysis.

Four distinct analysis cases were developed to comprehensively investigate approach slab behavior: two cases examining free (non-tied) approach slab response under expansion and contraction, and two parallel cases analyzing tied connection performance under the same loading conditions. Temperature changes of ± 100°F (37.8 °C) were applied to concrete elements to simulate seasonal extremes, with corresponding abutment displacements (54.9 mm) applied to reproduce bridge movement effects. Boundary conditions included fixed soil base in three translational directions, lateral soil restraints, and vertical abutment support. To ensure accurate representation of thermal response characteristics, extensive model calibration was performed using field instrumentation data. The finite element model was validated against field monitoring data. Figure 2 demonstrates the agreement between the simulated and measured strain-temperature relationships.

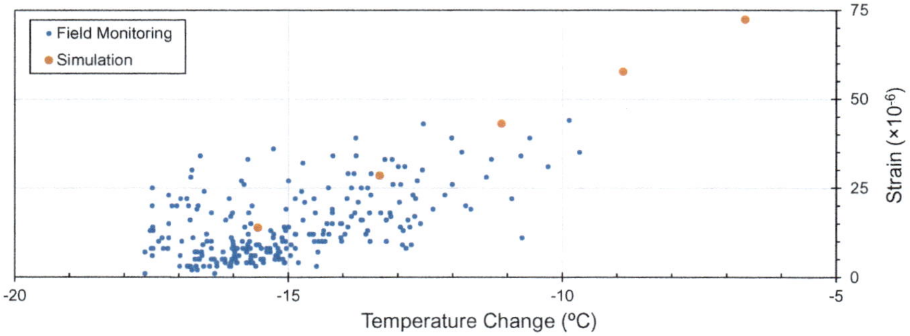

Fig. 2. Strain-temperature comparison between field measurements and simulation.

The numerical analysis showed distinct movement characteristics between free and tied approach slab configurations. The free configuration demonstrated movement patterns parallel to the bridge transverse axis. In contrast, the tied configuration showed displacement contours almost parallel to the abutment-approach slab joint, potentially due to transverse displacement of the bridge under thermal loads. Also, the tied configuration exhibited significant longitudinal displacements compared to the free cases as shown in Fig. 3. In the tied connection model, expansion loading produced substantial transverse displacement accompanying the longitudinal movement. This coupled displacement pattern confirms that skewed bridges subject tied approach slabs to complex multi-directional movements that significantly alter stress distribution characteristics compared to non-skewed configurations.

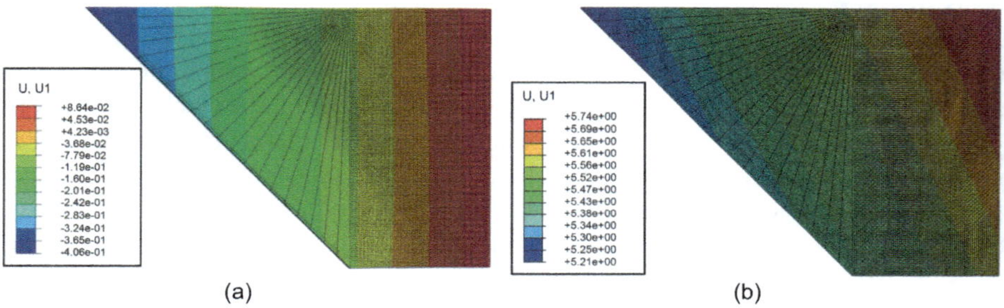

Fig. 3. Longitudinal displacement contours of the approach slab under expansion: (a) free approach slab, and (b) tied approach slab (cm).

Contact pressure analysis revealed critical differences between expansion and contraction loading scenarios in the tied connection model. Under contraction, contact pressures maintained relatively uniform distributions across the slab surface. However, expansion loading generated highly concentrated pressure zones near the approach slab-abutment interface, with particularly intense concentrations at the acute approach slab corner corresponding to the obtuse bridge corner. These pressure concentrations directly translate to uneven frictional resistance forces that introduce additional stress into the connection system.

Shear Force-longitudinal displacement analysis further demonstrated the asymmetric response characteristics of skewed approach slabs. During contraction, the system exhibited classical friction behavior with force plateauing at approximately 347 kN after initial movement. Expansion loading, however, showed progressive force increases reaching 440 kN, attributed to complex soil-structure interaction effects near the abutment interface. Most notably, transverse forces reached magnitudes comparable to longitudinal forces during expansion cycles, indicating that skewed configurations subject tie bar systems to significant multi-axial loading conditions not typically accounted for in conventional design approaches.

The numerical analysis findings provide critical insights into performance limitations of current tied approach slab systems, particularly in skewed bridge configurations. The concentrated stress patterns and complex multi-directional loading conditions identified through finite element modeling help explain the premature deterioration observed during field investigations. By quantifying these effects, the research establishes a scientific basis for developing improved connection details that can better accommodate the thermal movement characteristics of integral and semi-integral abutment bridges.

From a sustainability perspective, these findings demonstrate that current design approaches may significantly underestimate the complex stress states developed in skewed bridge configurations, particularly at the acute corners where field inspections revealed the most pronounced deterioration patterns. The numerical models suggest that optimized connection details must account for both longitudinal and transverse movement components to achieve the intended service life, particularly in bridges with skew angles exceeding 30 degrees. The comparison between tied and non-tied configurations clearly demonstrates that while tied connections offer theoretical advantages for preventing water infiltration, their implementation in skewed bridges introduces complex loading conditions that must be addressed through improved reinforcement configurations and connection details to fully realize their potential service life benefits. Enhancing bridge-end components can extend service life by 10–20 years [11, 12]. Extending service intervals by even 5–10 years has been shown to reduce life-cycle carbon emissions by approximately 9–17% [18], contributing to measurable carbon footprint and material savings aligned with net-zero objectives.

4 Conclusions

This research provides comprehensive insight into bridge end detail performance through integrated field investigations and finite element analysis. Field investigations of nine bridges revealed substantial performance deficiencies in preformed tied approach connections, with joint widths expanding significantly beyond design specifications. This coincided with approach slab migration, void formation at abutment-embankment interfaces, drainage failures, and concrete deterioration, particularly in skewed configurations. Numerical analysis demonstrated that skew geometry fundamentally alters load distribution patterns, with skewed approach slabs experiencing substantial transverse displacements (as high as 81% of longitudinal movement) during thermal cycles. Contact pressures become highly concentrated at acute corners, and conventional vertical tie bar systems prove inadequate for managing complex stress distributions in skewed configurations. These findings establish clear need for optimizing bridge end details, particularly for tie bar designs that address multi-directional force patterns. Recommendations include asymmetric reinforcement patterns at acute corners, non-uniform tie bar spacing, and design standards accounting for transverse movement in skewed configurations. Improved design detailing offers quantifiable sustainability benefits through extended service life, reduced material consumption, decreased construction carbon emissions, enhanced infrastructure resilience, and optimized resource allocation across transportation networks.

Acknowledgement. The research study, results of which reported in this paper, was sponsored by the Iowa Department of Transportation and Iowa Highway Research Board. The authors would like to thank the sponsors for this support. Opinions, findings, and conclusions expressed in this paper are of the authors and do not necessarily reflect those of the sponsors.

References

1. Lee SH, An LS, Kim HK (2024) Life-cycle environmental impact assessment of bridge designs considering maintenance strategy. In: Bridge maintenance, safety, management, digitalization and sustainability, pp 580–585
2. Torabi T, Shafei B, Alipour A (2024) Intensity measures for flood hazards in fragility assessments of bridges. In: The international conference on net-zero built environment: innovations in materials, structures, and management practices, pp 1539–1547
3. Bozorgmehr Nia S, Shafei B (2025) Carbon footprint reduction through repurposing industrial solid wastes into sustainable construction materials: a state-of-the-art review. J Clean Responsible Consum 100310
4. Azad S, Shafei B (2024) Performance evaluation of glass fiber-reinforced polymer (GFRP) bars in bridge decks. In: The international conference on net-zero built environment: innovations in materials, structures, and management practices, pp 1361–1370
5. Pang B, Yang P, Wang Y, Kendall A, Xie H, Zhang Y (2015) Life cycle environmental impact assessment of a bridge with different strengthening schemes. Int J Life Cycle Assess 20:1300–1311
6. Azad S, Shafei B (2024) Performance evaluation of semi-integral abutment bridge ends based on approach slab details. J Bridg Eng 30(3):04024117
7. Shafei B, Phares B, Shi W, Azad S (2023) Increase service life at bridge ends through improved abutment and approach slab details and water management practices, Report No. IHRB Projects TR-722 and TR-739, Ames, IA
8. Hassiotis S, Khodair Y, Roman E, Dehne Y (2006) Evaluation of integral abutments, Report No. FHWA-NJ-2005-025, Washington, DC
9. Aktan H, Attanayake U, Ulku E (2008) Combining link slab, deck sliding over backwall, and revising bearings, Report No. RC-1514, Kalamazoo, MI
10. Ali MS, Aslam MS, Mirza MS (2016) A sustainability assessment framework for bridges–a case study: Victoria and Champlain Bridges, Montreal. Struct Infrastruct Eng 12(11):1381–1394

11. DeJong A, Shi W, Shafei B, Hosteng T (2021) Integral abutment connections with grouted reinforcing bar couplers and ultra-high-performance concrete. J Bridg Eng 26(8):04021042
12. DeJong A, Shi W, Shafei B, Hosteng T (2021) Laboratory and numerical investigation of integral abutment connections with pile couplers. Eng Struct 248:113159
13. Karim R, Shafei B (2021) Performance of fiber-reinforced concrete link slabs with embedded steel and GFRP rebars. Eng Struct 229:111590
14. Karim R, Shafei B (2022) Addition of partial-depth link slabs to bridge structures: role of support conditions. J Bridg Eng 27(7):04022049
15. Khasawneh YA (2014) Soil structure interaction of integral abutment bridges. Doctoral Dissertation, Purdue University, Indiana
16. Khatami D, Shafei B, Bektas B (2023) Data-assisted prediction of deterioration of reinforced concrete bridges using physics-based models. J Infrastruct Syst 29(2):05023003
17. Mendoza-Lugo MA, Nogal M, Morales-Nápoles O (2024) Network-level optimisation approach for bridge interventions scheduling. Struct Infrastruct Eng 1–17
18. Raeisi F, Algohi B, Mufti A, Thomson DJ (2021) Reducing carbon dioxide emissions through structural health monitoring of bridges. J Civ Struct Heal Monit 11(3):679–689

Assessing Life Cycle Calculation Tools for Railway Bridges: A Comparative Study Using a Full-Scale Laboratory Testbed

Daria Sas[1], Lisa Persson[2], Peter Simonsson[2], Jens Eliasson[3], Vedad Coric[4(✉)], Chao Wang[4], and Gabriel Sas[4]

[1] iTid Tarinfo AB, 97339 Luleå, Sweden
daria.sas@itid.se
[2] Ramboll Sweden AB, 11851 Stockholm, Sweden
{lisa.persson,peter.simonsson}@ramboll.se
[3] ThingWave AB, 97346 Luleå, Sweden
jens.eliasson@thingwave.com
[4] Luleå University of Technology, 97187 Luleå, Sweden
{vedad.coric,chao.wang,gabriel.sas}@ltu.se

Abstract. Sweden's extensive network of over 4,000 railway bridges calls for innovative, data-driven strategies to ensure structural reliability, extend service life, and to meet evolving sustainability goals and increasing load demands. As part of a larger initiative, this study focuses on a comparative evaluation of three widely used life cycle calculation tools. We examined Trafikverket's Klimatkalkyl (employed by Swedish government agencies), OneClick LCA (prominent in industrial practice), and ORIS Material Intelligence (a commonly applied approach). This comparative study is based on a unique full-scale replica of a 1970s reinforced concrete trough railway bridge, housed indoors at Luleå University of Technology in Luleå and equipped with over 100 sensors (including Fiber Optic Sensing (FOS) and Strain Gauges). This testbed generates up to 2 TB of data daily, serving as a robust foundation for structural health monitoring, life cycle and life costs assessments. By analysing the same bridge through various tools, we highlight how their differing algorithms may lead to variations in environmental impact estimates, resource optimization strategies as well as life cycle cost projections in the future. This study aims to determine the most suitable tool for guiding sustainable infrastructure decisions. It also aims to illustrate how these insights feed into the overarching initiative of a holistic bridge management platform, intended to provide a comprehensive, real-time overview of bridge conditions at both the asset and network levels. At last, this work supports Sweden's broader transition to circular, climate-resilient infrastructure, equipping decision-makers with the comparative knowledge needed to optimize bridge longevity, minimize environmental impacts.

Keywords: Life Cycle Assessment (LCA) · comparative study · tools · Structural Health Monitoring (SHM) · Sustainable Infrastructure · Railway Bridges · Full-Scale Laboratory Test

H. Beushausen et al. (Eds.): NTZR 2025, LNCE 873, pp. 721–732, 2026.
https://doi.org/10.1007/978-3-032-27698-8_64

1 Introduction

As Sweden advances toward Net Zero, railway bridges are crucial for low-carbon mobility. Sweden has approximately 4,000 railway bridges. Concrete bridges constitute about 85% of these bridges, while about 46% (1,673 bridges) are over 50 years old. To address the issue of ageing infrastructure, the government has announced plans to invest 110 billion euros in the national transport system. This significant investment aims to refurbish and upgrade the country's transportation network. Furthermore, the infrastructure budget is projected to increase even more in upcoming years [1]. According to the Circular Gap Report 2022 [2], only 3% of Sweden's economy operates on a circular basis, with housing and infrastructure consuming 47% of the materials. Increasing circularity in these sectors is essential. Conducting bridge life-cycle assessment (LCAs) effectively supports circular economy (CE) principles by reducing material demand, minimise waste, and the environmental footprint of bridges. Thus, this offers cost-saving opportunities and assists in the transition to sustainable, climate-resilient infrastructure [2, 3].

Despite growing interest, progress in applying circular construction principles to bridges remains limited [4, 5]. Aging post-WW2 railway bridges, often with 50-to-70-year design lives, face three strategic options: demolish and replace, strengthen and use, or continue use. Decisions typically rely on code-compliant calculations and inspections. However, emerging research, particularly utilizing structural health monitoring (SHM), suggests that many bridges possess hidden load-carrying capacity that could safely extend their lifespan beyond their original design, thereby shifting the focus from reactive to proactive, predictive maintenance strategies [6].

At the same time, LCA research on railway bridges has matured. Most recent studies comply with ISO 14040/44 and EN 15804 + A2, adopting cradle-to-grave system boundaries that convers material production, construction, maintenance, and end-of-life stages [7–10]. Methodological advances include comparative analyses of design-options (e.g., ballasted vs. slab track superstructures [11]) and evaluation of novel material solutions (e.g., UHPFRC deck overlays, [9], or zero-cement, aramid-tendon bridges [12]). These studies consistently identify use-phase maintenance intensity and end-of-life steel recycling as critical determinants of total greenhouse gas emissions [7, 8]. While data driven frameworks combining SHM and LCA have been proposed in general bridges, their specific application to railway bridges is still developing.

A critical gap persists in the literature: a lack of systematic head-to-head comparative evaluation of the digital infrastructure LCA platforms. Therefore, this study presents a comparative evaluation of three LCA tools used to assess railway bridges: (1) Trafikverket's Klimatkalkyl (national policy tool); (2) OneClick LCA (industry standard); (3) ORIS Material Intelligence (emerging material-performance platform) using a single case study bridge. The analysis assumes construction of a new superstructure, substructure, and foundations. To analyse the case study bridge within the context of its full life cycle, a suitable location for such a bridge was found, and a hypothetical scenario developed. For this purpose, the Iron Ore Line (Malmbanan) in northern Sweden was chosen (see Fig. 3 for map), which is owned and managed by the Swedish Transport Administration (Trafikverket).

The Iron Ore line, spanning from Narvik in Norway to Luleå on the Swedish coast, is approximately 470 km long and is known as one of the busiest iron ore lines globally.

This railway line holds immense significance in the Swedish economy, facilitating the transportation of substantial quantities of iron ore from the mines in Kiruna and Gällivare to the ports of Luleå and Narvik. A significant portion of the Iron Ore Line runs north of the Arctic Circle, making it not only a key part of Sweden's infrastructure but also a unique engineering environment due to the extreme environmental conditions [13]. By benchmarking the three tools against this case study, the analysis enables comparison of embodied carbon results by revealing algorithmic differences, the study highlights how tool choice influences environmental impact estimates, resource-optimisation strategies, and by extension, future life-cycle cost projection.

2 Material and Methods

2.1 Life Cycle Assessment and Its Modules

LCA is a standardized methodology designed to evaluate the environmental impact of a product throughout its life cycle, as defined by ISO 14040:2006 [14]. While LCA primarily focuses on sustainability from an environmental perspective, it excludes other sustainability dimensions such as economic and social impacts. Conducting a detailed environmental assessment of bridges poses significant challenges due to the inherent complexities of both environmental factors and the structural specificity of bridges [15]. To address these complexities, standardized life cycle modules and system boundaries for carbon assessment, provide a structured approach for evaluating environmental impacts.

The life cycle module presented in the figure above is structured as follows: A0 represents the pre-design period, which includes activities such as surveys, field trips to construction sites, and necessary measurements before building. These tasks typically have minimal impact on building assessments and may be excluded from the life cycle analysis process, although they can significantly reduce carbon emissions in stages A1-A5 in new construction projects.

Stages A1-A5 encompass upfront carbon emissions in product and construction, known as the product stage module. A1 involves the impact from recycled materials or raw material extraction for manufacturing or processing for use in construction. A2 covers the transport of raw materials to factories, while A3 focuses on the manufacturing of new materials. A4 involves the transport of materials from the manufacturer to the building site, which requires manual calculation. A5 includes the construction and installation process, incorporating waste generation.

The use stage module, B1-B9, assesses climate impact and energy use during the structure's in-use period, such as maintenance and operations, and includes the re-absorption of carbon dioxide.

The end-of-life module, C, examines the interaction of specific environmental conditions with the socio-economic system rather than focusing on mere local extreme values of climatological parameters.

Finally, phase D aims to assess the benefits and loads resulting from the net flow of secondary fuels or materials exiting the product system, where circular economy principles can be applied to reduce emissions in stages A1-A5, considering benefits and loads beyond the construction work's life.

2.2 Life Cycle Calculations Tools

We benchmarked three web LCA platforms widely used: OneClick LCA (module Infrastructure Carbon EN 17472), Klimatkalkyl version 8 (Trafikverket's national climate-calculation tool) and ORIS material intelligence.

Aligned with ISO 14040/44 and the construction-production provisions of EN 15804 + A2 (European standard for environmental product declaration, EPDs) [14], we analyse each platform through 3 interacting "*engine layers*":

(i) **system boundary** – which EN 15804 modules are modelled.
(ii) **data provenance** – primary databases and default factors.
(iii) **algorithmic routines** – e.g. carbonation credit, maintenance schedules.

Table 1 shows how these layers are implemented and why they matter; the contrasts explain the ≈ 30% spread in embodied-carbon results discussed in later sections [16–18].

Table 1. Comparison of Engine layers across Klimatkalkyl, One Click, and ORIS LCA Software.

Engine layer	Klimatkalkyl	One Click	ORIS
System boundary	Covers A1–A5 + key B/C; omits traffic emissions	Defaults to A1–A3; optional A4–D	Runs A1–A5; built-in road-traffic model
Data provenance	Generic + Swedish EPDs	> 300 k EPDs (Ecoinvent + GaBi) + generic	Ecoinvent v3.10 + ~ 35 k geo-tagged datasets
Algorithmic routines	No carbonation credit; fixed transport	Implements EN 16757 carbonation and editable transport	Mix-design optimiser linked to full transport & maintenance

All three tools follow the ISO logic but differ in databases, licence tiers and module coverage. We therefore used the research/extended licence for each tool and disabled any feature that would break boundary alignment (e.g. One Click LCA's A0 pre-construction module), keeping every run within A1–A5 + C1–C3 [19].

To ensure comparability we supplied the same bill-of-quantities (BoQ) from the case bridge and fixed transport at 300 km rail + 100 km road (typical for the Iron-Ore Line). When primary data were missing, One Click LCA and ORIS automatically uplift cradle-to-gate factors, while Klimatkalkyl still uses a 2015 Swedish reference year (with an optional 2020 set in v 8) [16]. Both approaches introduce a conservative safety buffer.

BoQ, transport and construction information were entered under the five query groups shared by all tools: Earthworks, Construction Materials, Construction-Site Operations, Use Phase, and Emissions & Removals, ensuring like-for-like comparison across EN 15804 modules A1–A5 and C1–C3.

2.3 Calculations Assumptions and Limitations

The comparative LCA is based on a three-week construction window at a temporary casting yard in Koskivaara, Sweden. The finished deck is then shunted 40 km by rail to its permanent position on the Iron Ore line. A 120-year service life is assumed, and the system boundary spans EN-15804 + A2 modules A1-A5, B1, B5, C1-C4); use-phase traffic loads (B6 energy and B7 water) are excluded since they are identical across design options.

Concrete, ballast and aluminium components are delivered from Gällivare (60km by road); while reinforcing steel, formwork materials and processional services originates from Luleå (150 km by road). Site staff are housed in a temporary barrack for the three-week period.

Table 2 list the material inputs for the reference bridge, complied with contractor submittals and expert estimates. Where a quantity was supplied in kilograms or meters, it was converted to the unit required by each LCA engine. Densities and length-to-mass factors follow the respective environmental-product declarations (EPDs) for each material.

All assumptions are based on expert opinions by engineers involved in ongoing research at LTU. The inventory therefore represents a best-estimate baseline, but results remain sensitive to (i) future change in material EPDs, (ii) on-site fuel efficiency, and (iii) rail versus road modal splits, uncertainties that are examined in the sensitivity analysis.

2.4 Case Study

A full-scale reinforced concrete through railway bridge is located inside the structural engineering lab at Luleå University of Technology (LTU). The superstructure was cast with C40/50 concrete and Ks40 reinforcement (approx. 400 MPa yield strength) in 2023 just outside the LTU lab. The track comprises UIC60 rails and 12 prestressed concrete sleepers, embedded in 270 mm of Class-I ballast. Table 3 summarises the technical information about the superstructure.

Figure 1 shows the case bridge under construction, and the testbed in the lab with bridge loaded with rails, ballast, and sensors. Figure 2 illustrates a cross section of a concrete slab and a typical geometry of a bridge.

This particular through railway bridge was selected as a case study due to the comprehensive understanding and data available on the materials used and the operations performed during its construction.

Figure 3 illustrates map of northern Sweden, with the Iron Ore Line highlighted in white. The route follows a path from Luleå through Gällivare, Kiruna, and continues westward to the port of Narvik in Norway.

A 6m reinforced concrete trough bridge located in Lautajoki, approximately 200 km north of Luleå, provides an in-service analogue. This bridge slab was replaced in 1970 and operates on the original abutment [20]. For the presented study we assume for methodology comparison only, that the Trafikverket commissions a new parallel line on the same corridor and adopts the LTU through bridge designed for all $\leq$ 7m spans. This hypothetical but plausible scenario defines the boundary conditions (traffic class 30-tonne axel load, -40 ° C design temperature) used in the comparative LCA tool run.

Table 2. Summary of quantities

Description	Material	Quantities	Unit	Source of information
Superstructure				
Bridge deck, cast in situ	Concrete C40	13.2	[m^3]	Contactor
	Reinforcement Ks40	2.8	[ton]	Contactor
Formwork				
	Timber	17.2	[m^2]	Contactor
	Nails	2	[kg]	Contactor
	Reinforcement Ks40	6	[kg]	Contactor
	Form oil	17	[kg]	Contactor
	Timber	17.2	[m^2]	Contactor
Railway				
	Sleepers	3700	[kg]	Contactor
	Rails track UIC60	820	[kg]	Contactor
	Ballast class I	17400	[kg]	Contactor
	Railing safety barriers in steel	1120	[kg]	Assumption
	Aluminium cable protection	66	[kg]	Assumption
Substructure				
Abutments, cast in situ	Concrete C40	17,2	[m^3]	Assumption
	Reinforcement Ks40	2	[ton]	Assumption
	Piles, pre-casted	140	[m]	Assumption
Formwork				
	Timber	17,2	[m^2]	Assumption
	Nails	2	[kg]	Assumption
	Reinforcement Ks40	6	[kg]	Assumption
	Form oil	17	[kg]	Assumption

3 Results

Lifecycle carbon assessment tools are essential for evaluating the environmental impact of construction projects. This analysis compares three prominent LCA tools: OneClick LCA, Klimatkalkyl and ORIS, focusing on their capabilities and limitations in the context of material and work processes. The overall difference across life cycle modules is shown in Fig. 4.

The three platforms also differ markedly in workflow transparency and modelling depth. OneClick LCA offers the richest catalogue of materials and plant, lets users

Table 3. Technical Information of superstructure

Description	Quantities	Unit
Bridge		
Length	7.1	[m]
Width	4.2	[m]
Height	1.02	[m]
Bridge total weight	29500	[kg]
Ballas, rails and sleepers		
Ballast weight	11	[m^3]
Rails weight (the two of them)	842	[kg]
12 sleepers' weight	3700	[kg]

Fig. 1. The bridge under construction (left) and testbed at LTU (right)

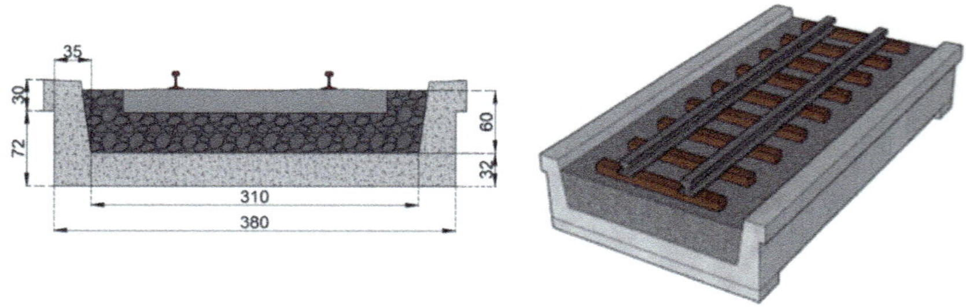

Fig. 2. A diagram of a cross section of a concrete slab (left) and through bridge typical geometry (right)

bundle items with their associated site processes, and displays each item's share of total emissions directly in the results dashboard. These features speed hotspot identification, yet they also raise the risk of double counting when the same material is entered both as

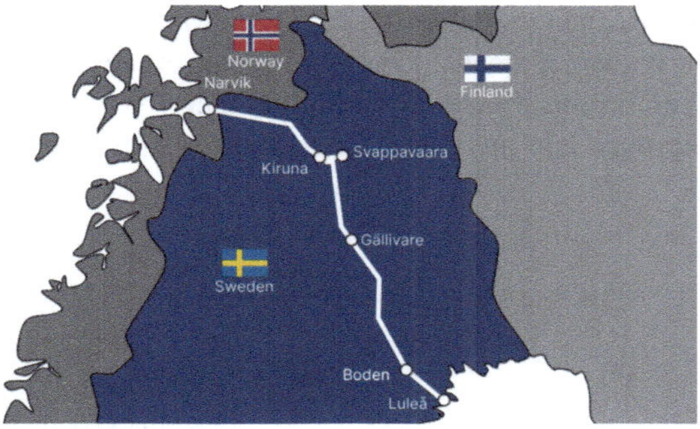

Fig. 3. Map of the Iron Ore Line in Northen Sweden

a removal item and as new work. OneClick LCA is the only engine that automatically credits concrete carbonation over the service life, delivering a negative D-module that can trim totals by roughly 6-ton CO_2e for this bridge.

Klimatkalkyl v8 matches OneClick on database breadth but embeds several layers of default components inside each construction element, so inexperienced users can by mistake duplicate materials or machinery. Material lifespans are hard-wired in the background model, changing them require manual overrides in a sperate CSV file, and the engine does not model carbonation or other carbon-removal mechanisms.

ORIS Material Intelligence currently limits calculations to A1-A5 and uses a fixed bill-of-material (BoM), which virtually eliminates double-counting but prevents nuanced work-process inputs. Material quantities are accepted only in tonnes or cubic meters, and the user cannot yet assign spill factors, adjust component lifespans, or claim carbonation credits.

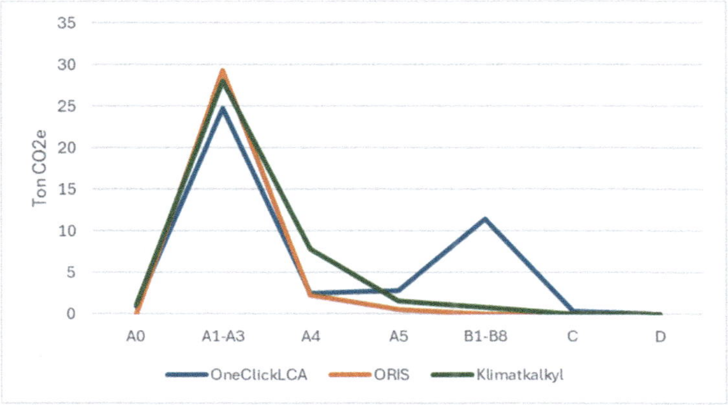

Fig. 4. Comparative results for each platform for each LCA module

These observations explain much of the ~ 30% spread in total emissions (Table 4 and illustrated in Fig. 4). OneClick LCA's granular work-process modelling and carbonation credit push its gross total higher than ORIS but lower once module D is included; Klimatkalkyl sits in between, elevated mainly by conservative freight factors and the absence of negative emissions. Practitioners should therefore declare database versions, maintenance assumptions, and any carbon-removal algorithms when comparing LCA outputs across tools.

Table 4. Results from climate calculation tools

Module	Tools			Comment
	OneClickLCA	ORIS	Klimatkalkyl	
A0 – pre-construction*	0.914	–	1.091	
A1-A3 – production stage	24.765	29.33	28.020	
A4 – transport to site	2.425	2.21	7.788	
A5 – construction-stage processes	2.765	0.55	1.497	
B1-B8 – use-phase maintenance**	11.481	–	0.766	
C – end-of-life	0.327	–	–	
D – beyond system boundary	– 5920***	–	–	*Excluded from totals
Total:	42.6-ton CO_2e	32.1-ton CO_2e	39.1-ton CO_2e	

Table 5 shows percentage differences between the results obtained from three different tools (OneClick LCA, ORIS, and Klimatkalkyl) for various modules. The largest percentage difference was observed between OneClick LCA and Klimatkalkyl for module B1-B8 (175.06%). Significant differences were also noted for module A5 between OneClick LCA and ORIS (133.66%) and between ORIS and Klimatkalkyl (92.54%).

Table 5. Differences between the tools within the various modules

Module	Tools		
	OneClickLCA vs Klimatkalkyl	OneClickLCA vs ORIS	ORIS vs Klimatkalkyl
A0 – pre-construction*	17.67%	0%	0%
A1-A3 – production stage	0%	16.88%	4.57%
A4 – transport to site	0%	9.28%	111.44%
A5 – construction-stage processes	0%	133.66%	92.54%
B1-B8 – use-phase maintenance**	175.06%	0%	0%

* A0 is available only in OneClick LCA and Klimatkalkyl; it was excluded from the cross-tool total
** Use-phase traffic energy (B6) and water (B7) are identical across options and therefore excluded
*** Negative values reflects carbonation credit in concrete (automatically modelling in OneClick LCA)

4 Discussions and Future Work

Among the three evaluated tools, OneClick LCA is the most versatile platform for Swedish railway-bridge projects because it (i) covers the full EN15804 + A2 life-cycle boundary, including A0 pre-construction, all B-modules, and C/D end-of-life phases; (ii) allows user-defined maintenance schedules and spill factors; and (iii) credits carbon-removal mechanisms such as concrete carbonisation. These features enable a circular economy perspective that is difficult to replicate in Klimatkalkyl v8 or ORIS, whose scopes stop at A5 and whose databases lack negative-emission pathways. The flexibility, however, comes at the cost of greater modelling responsibility. Users must guard against double-counting when demolition items and replacement materials are entered simultaneously, and they must validate any automatically generated carbonation credit against project-specific exposure conditions.

This divergence between tools is not just academic. Trafikverket's proposed carbon cap for new railway structures (400 kg CO_2e per m^2) means that a 30% spread in A1–A5 totals can shift a bridge from "compliant" to "non-compliant." In Table 4, the 7.8-ton vs. 2.3-ton delta in A4 transport alone translates to a swing of ~ 180 kg CO_2e/m^2 for the 7 m span case bridge.

The largest numerical divergence (Table 4) is in A4 (transport). All tools allow editing haulage distances but lock vehicle emission factors. Klimatkalkyl's conservative Swedish diesel factor (~2 × European default) explains its 7.8-ton CO_2e vs. ~ 2.3-ton CO_2e for others. A similar discrepancy exists in B1–B5: OneClick LCA models rail maintenance per EN17472 TC-3; other tools require manual input, leading to near-zero impacts if unaddressed. Transparency on default factors and maintenance algorithms is critical when LCA outcomes inform procurement or permitting. While limited to a single case, the approach offers a transferable framework for comparative assessment of LCA tools in other infrastructure domains.

OneClick LCA includes an optional life-cycle-cost (LCC) module that can ingest the same BoQ and maintenance plan for the carbon model. Integrating this cost dimension with the carbon baseline is a logical next step for ageing assets such as Sweden's post-war concrete-bridge stock, where budget constraints compete with decarbonisation targets. Future research should therefore (i) benchmark OneClick's LCC outputs against dedicated cost-analysis tools, and (ii) test how joint carbon-cost optimisation alters retrofit versus replacement decisions for bridges approaching the end of their nominal 70-year design life.

Acknowledgments. This study was financially supported by the Swedish Transport Administration (Trafikverket); Vinnova (Strategic Innovation Program InfraSweden's call "*Increased circularity for the transport infrastructure*"); and the Swedish Agency for Economic and Regional Growth (Tillväxtverket, initiative "*Digitalisering som främjar omställning*").

References

1. Framtidsinvesteringar och tillväxtreformer för ett rikare och tryggare Sverige. Regeringskansliet. https://www.regeringen.se/pressmeddelanden/2024/09/framtidsinvesteringar-och-tillvaxtreformer-for-ett-rikare-och-tryggare-sverige/. Accessed 2025
2. Circle E (2022) The circularity gap report 2022. https://www.circularity-gap.world/2022
3. Ellen MacArthur F (2013) Towards the circular economy vol 1: an economic and business rationale for an accelerated transition
4. Ossio F, Salinas C, Hernández H (2023) Circular economy in the built environment: A systematic literature review and definition of the circular construction concept. J Clean Prod 414:137738. https://doi.org/10.1016/j.jclepro.2023.137738
5. Ottosen LM (2024) Innovations in buildings and bridge engineering associated with circular economy. In: Proceedings of the 12th International Conference on Bridge Maintenance, Safety and Management
6. Kralovec C, Schagerl M (2020) Review of structural health monitoring methods regarding a multi-sensor approach for damage assessment of metal and composite structures. Sensors 20(3):826. https://www.mdpi.com/1424-8220/20/3/826
7. Du G, Karoumi R (2013) Life cycle assessment of a railway bridge: comparison of two superstructure designs. Struct Infrastruct Eng 9(11):1149–1160. https://doi.org/10.1080/15732479.2012.670250
8. Al-Gburi M, Gonzalez-Libreros J, Sas G, Nilsson M (2022) Quantifying the environmental impact of railway bridges using life cycle assessment: a case study. In: IABSE, Prague, 2022: IABSE, pp 1796–1803. https://doi.org/10.2749/prague.2022.1796. https://structurae.net/en/literature/id/10676617
9. Bizjak KF, Šajna A, Slanc K, Knez F (2016) Environmental life cycle assessment of railway bridge materials using UHPFRC. Mater Geoenviron 63(4):183–198. https://doi.org/10.1515/rmzmag-2016-0017
10. Zerin AI, Kasuga A (2023) Life cycle assessment of next generation non-metallic bridges. Struct Concr 24(2):1914–1931. https://doi.org/10.1002/suco.202200852

11. Blengini GA (2009) Life cycle of buildings, demolition and recycling potential: a case study in Turin, Italy. Build Environ 44(2):319–330. https://doi.org/10.1016/j.buildenv.2008.03.007
12. Bizjak KF, Knez F, Lenart S, Slanc K (2017) Life-cycle assessment and repair of the railway transition zones of an existing bridge using geocomposite materials. Struct Infrastruct Eng 13(3):331–344. https://doi.org/10.1080/15732479.2016.1158288
13. J. International Railway (2024) Sweden to upgrade Ore Line from Luleå to Narvik (ed)
14. S. International Organization (2006) ISO 14040:2006 Environmental management—Life cycle assessment—Principles and framework. https://www.iso.org/standard/37456.html
15. Du G (2015) Life cycle assessment of bridges, model development and case studies. https://kth.diva-portal.org/smash/record.jsf?pid=diva2:793949
16. Trafikverket (2024). Klimatkalkyl – Beräkning av infrastrukturens klimatpåverkan och energianvändning, version 8.0
17. One Click LCA. How One Click LCA works with data. https://oneclicklca.com/help/data-quality. Accessed
18. Oris and Ramboll. ORIS × Ramboll FuelSave partnership press release. https://oris.com/press/fuelsave-partnership. Accessed
19. OC LCA. General infrastructure LCA guidance. One Click LCA. https://help.oneclicklca.com/en/articles/275779-general-infrastructure-lca-guidance. Accessed 11 May 2025
20. Paulsson B, et al (1996) 30 ton på Malmbanan. Rapport 3.3 Infrastruktur. https://ltu.diva-portal.org/smash/get/diva2:1888895/FULLTEXT01.pdf

A Novel Co-disposal Stone Column Reinforcement Technique for Enhanced Stability of Copper Mine Tailings Dam

Akshay Kumar Singh and Anumita Mishra(✉)

Department of Civil Engineering, Indian Institute of Technology Roorkee, Roorkee, India
{ak_singh,anumita.mishra}@ce.iitr.ac.in

Abstract. Mine tailings dams represent a significant hazard in the mining industry, with failure rates occurring at an alarming frequency of 2–3 incidents annually worldwide. These failures pose severe environmental and geotechnical risks, particularly concerning the stability of impoundments constructed using the upstream method, which remains prevalent despite its known vulnerabilities. The critical need for enhanced stability measures in copper mine tailings (CMT) dam construction has become increasingly urgent, especially given the growing scale of mining operations and the corresponding volume of stored tailings material. This study addresses this pressing challenge by evaluating a novel co-disposal technique incorporating waste rock inclusions (WRI) in the form of stone columns within a CMT impoundment. A numerical analysis was conducted using 3D axisymmetric models converted to 2D plane strain configurations via established literature methods, simulating stone columns with radii ranging from 0.5 to 10 m and center-to-center spacings of 10 to 40 m. A subsequent nonlinear regression analysis was performed to model the Factor of Safety (FOS) as a function of column radius (r) and spacing (s), incorporating interaction ($r \times s$) and nonlinear terms ($r \times s^2$). The regression model achieved exceptional predictive accuracy ($R^2 = 0.96$) and generalizability ($R^2{}_{CV} = 0.88$), identifying radius as the dominant stability driver ($\beta = 0.71$, $p < 1e{-}3$) and revealing critical diminishing returns at wider spacings ($\beta = -0.08$, $p < 1e{-}3$). Results demonstrated that optimized stone column configurations increased FOS by up to 70% compared to conventional designs, with closer spacing and larger radii synergistically enhancing stability. These findings provide actionable guidelines for balancing material efficiency and geotechnical performance, offering a data-driven framework to improve tailings dam resilience while addressing a persistent industry challenge.

Keywords: Tailings Dam Stability · Waste Rock Inclusions · Stone Column Reinforcement · Finite Element Analysis

1 Introduction

The mining industry produces large amounts of waste, requiring tailings dams for safe storage. These dams are vital for environmental protection, as failures can cause ecological harm, economic losses, and human casualties [1]. Growing global demand for

H. Beushausen et al. (Eds.): NTZR 2025, LNCE 873, pp. 733–745, 2026.
https://doi.org/10.1007/978-3-032-27698-8_65

minerals has increased mining activities, producing vast amounts of water-rich, chemically laden tailings. The copper industry alone generates over 20 million metric tonnes annually, with around 3,500 active tailings dams storing slurry worldwide [2]. However, the structural integrity of these dams remains a pressing concern, as evidenced by several high-profile failures, including the Mount Polley dam (Canada), Padcal mine (Philippines), Aitik mine (Sweden), and Niujiaolong tailings dam (China) [3]. The main causes of tailings dam failures include slope instability, seismic actions, overtopping, structural inadequacies, seepage forces, foundation failure, and erosion, highlighting the critical need for improved stabilization techniques.

Tailings dam stability is vital to prevent catastrophic failures. Traditional methods like ripraps and chemical additives combat erosion and boost slope strength. Upstream tailings (USM) dams face unique issues like pore pressure buildup during construction. Finite element modeling shows strategic rockfill placement enhances stability by reducing pore pressures [4]. Optimization techniques reduce rockfill volume while ensuring dam safety through staged consolidation and monitoring. Ground improvement methods like stone (gravel) columns and geosynthetic reinforcement enhance stability by accelerating consolidation, dissipating pore pressure, and increasing shear strength. Studies show gravel columns improve bearing capacity by up to 2.28 times and limit displacements. Waste rock inclusions further boost static/seismic stability and aid pore pressure dissipation post-earthquakes [5]. Geogrid reinforcement in USM nearly doubles safety factors, reducing failure risks in both drained and undrained conditions, as proven by case studies and numerical analyses [6, 7]. Geogrid reinforcement effectiveness depends mainly on layer length and spacing, with stiffness being less critical. Properly designed multi-layered systems significantly enhance slope stability by increasing resisting moments [8]. However, research on optimizing stone column parameters (diameter, spacing) for pore pressure control in copper mine tailings (CMT) dams remains limited. This gap underscores the need for further investigation into the combined effects of stone column geometry and arrangement on the geotechnical performance of tailings storage facilities.

In light of the above, this study investigates the stability of CMT dams reinforced with stone columns through numerical analysis using the finite element software PLAXIS. The research employs 3D axisymmetric models converted to 2D plane strain configurations using established literature methods, evaluating the influence of stone column radii (0.5–10 m) and center-to-center spacings (10–40 m) on dam stability. A nonlinear regression analysis is subsequently performed to model the Factor of Safety (FOS) as a function of these parameters, incorporating interaction ($r \times s$) and nonlinear ($r \times s^2$) terms to quantify their synergistic effects. The derived model identifies statistically significant predictors, achieving high predictive accuracy ($R^2 = 0.96$) and robust generalizability ($R^2_{CV} = 0.88$). By linking geometric configurations to stability outcomes, this research provides actionable guidelines for optimizing stone column designs, advancing safer and more cost-effective practices for tailings dam construction in geotechnically challenging environments.

2 Materials and Methods

2.1 Study Site: Malanjkhand Copper Mine Tailings Dam

The Malanjkhand Copper Mine (22°0′54″N, 80°43′20″E), India's largest copper operation, features an upstream-method dam storing 2.2 million tonnes/year of fine tailings in seismic zone III. For this study, partially dry CMT specimens were collected at 1m depth. Laboratory tests (sieve analysis, compaction, permeability, and direct shear tests) characterized tailings properties are shown in Table 1 and Fig. 1, while bedrock and clay layer data are referenced literature [9]. Results inform ground improvement strategies for such high-moisture, stability-challenged tailings facilities.

Table 1. Material parameters for the model

Layer	$\gamma_{dry}\left(\frac{kN}{m^3}\right)$	$\gamma_{sat}\left(\frac{kN}{m^3}\right)$	μ	E_{ref} (kPa)	ϕ	c(kPa)
CMT	15.6	17.5	0.30	2348	26.40	1
Intermediate tailings	15.6	17.6	0.30	3348	26.40	1
Clay[a]	17.1	19.5	0.34	7520	11.90	27.5
Granite[a]	25	25	0.22	12×10^6	40.00	1000
Stone column	19	20	0.3	30000	40	5

[a]Xu and Wang[9]

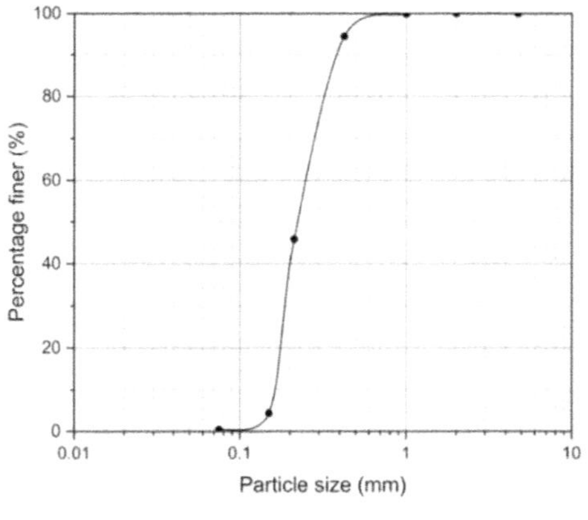

Fig. 1. Particle size distribution curve for CMT samples from Malanjkhand, India

2.2 Numerical Modelling

The numerical analysis in this study was performed using PLAXIS, with a stepwise validation approach to ensure the reliability of the modelling methodology and its applicability to the Malanjkhand copper mine tailings dam.

Validation of Upstream Tailings Dam Model

The upstream tailings dam model was validated against finite element approaches by Singh and Mishra [4] and Psarropoulos and Tsompanakis [10], simulating staged construction with Mohr-Coulomb material behaviour. Mesh refinement focused on failure-prone zones, with fixed-base and roller lateral boundaries to restrict horizontal movement. The factor of safety (FOS) was calculated using the shear strength reduction technique at each stage. Results (see Table 2 and Fig. 2) align with Psarropoulos and Tsompanakis [10], confirming the model's accuracy in capturing stability evolution during sequential raises, critical for assessing upstream dam safety.

Table 2. Validation of the upstream method in the factor of safety value

Construction stage	1	2	3	4	5	6
Reported study	3.09	3.12	1.67	1.77	1.26	1.15
Present study	3.07	3.1	1.7	1.8	1.3	1.2

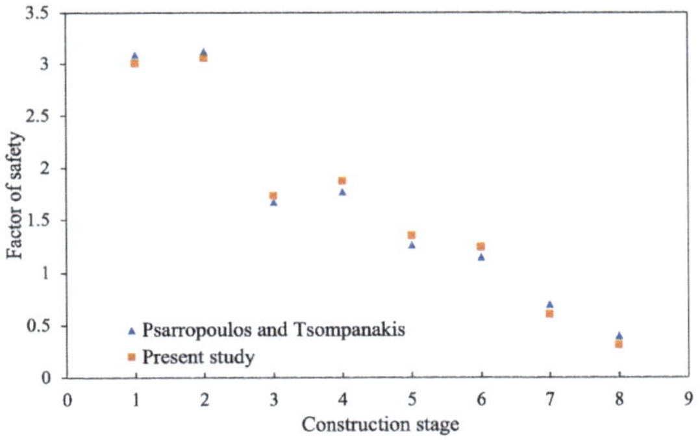

Fig. 2. Comparison of stage-wise FOS for USM in the present study and Psarropoulos and Tsompanakis [10].

Validation of Stone Column Inclusion

The stone column modelling was validated using Tan et al. [11] methodology, which converts 3D column arrangements to 2D plane-strain models through two approaches: permeability-matching for flow path accuracy and area equivalence for geometric consistency. Both stone columns and surrounding soil were modelled using elastic and elasto-plastic constitutive laws, with fixed-base and horizontally restrained boundary conditions to represent infinite ground extent. Validation involved comparing plane-strain results against axisymmetric simulations and field data, confirming the method's reliability for efficient 2D finite element analysis of column-reinforced ground. The geometry of the model is shown in Fig. 3. Comparison of settlement for embankment with stone column in the present study and Tan et al. [11] is shown in Fig. 4.

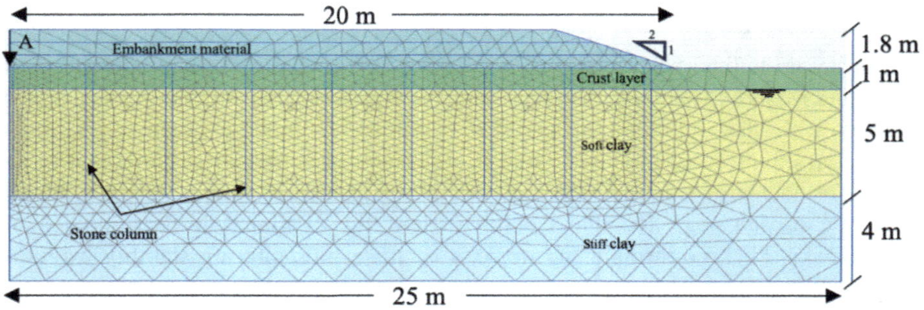

Fig. 3. Discretized mesh of the embankment with stone column

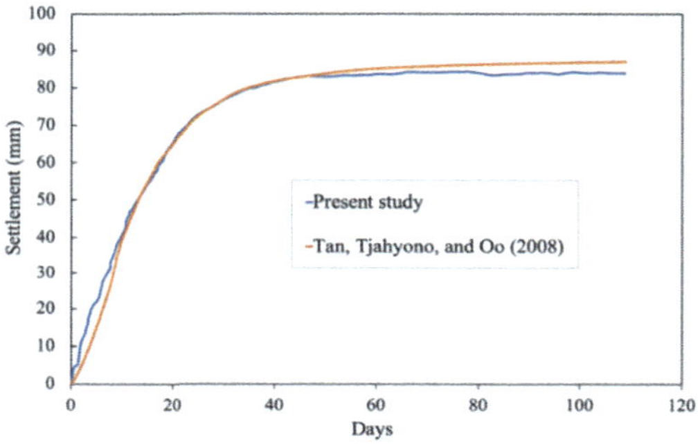

Fig. 4. Comparison of settlement for embankment with stone column in the present study and Tan et al. [11].

Modelling of Malanjkhand Copper Mine Tailings Dam with Stone Columns

Following validation, the Malanjkhand copper mine tailings dam was modelled using the geometry and construction details from Singh and Mishra [4]. The stage-wise construction of the dam was considered, and the raising of each stage was 10 m. A height of 30 m could be achieved by this method, beyond which the FOS of the dam reduced to undesirable values. To enhance overall dam safety, stone columns were incorporated beneath the embankment. Discretized mesh of the USM in the present study, as shown in Fig. 5. For this purpose, the embankment length was discretized according to the spacing of the stone columns. To optimize material usage, initial analyses were performed with a wider spacing of 40 m, which was subsequently reduced to 20 m and 10 m to systematically assess the influence of column density on stability. The width of the stone columns to be incorporated in the model was determined using the expression proposed by Tan et al. [11], facilitating the conversion of the three-dimensional column arrangement to an equivalent two-dimensional plane strain model suitable for PLAXIS analysis (see

Fig. 6 and Eqs. 1 and 2).

$$\frac{r_{ax}}{R_{ax}} = \sqrt{\frac{r_{ps}}{R_{ps}}} \tag{1}$$

$$R_{ax} = 1.13R_{ps} \tag{2}$$

where r_{ax} and R_{ax} are the radii of the stone column, and the drainage zone in the axisymmetric geometry and r_{ps} and R_{ps} are the respective radii in an equivalent plain strain geometry. The R_{ps} and the r_{ps} used in the current study and their corresponding axisymmetric counterparts are given in Table 3.

Table 3. Values of r_{ax} for various combinations of r_{ps} and R_{ps}

Width in model, m $(2 \times r_{ps})$	Spacing used in model $(2 \times R_{ps})$, m		
	10	20	40
1	1.3	1.8	2.5
1.5	1.8	2.5	3.6
2	2.2	3.1	4.4
5	2.5	3.6	5.1
10	4.0	5.7	8.0

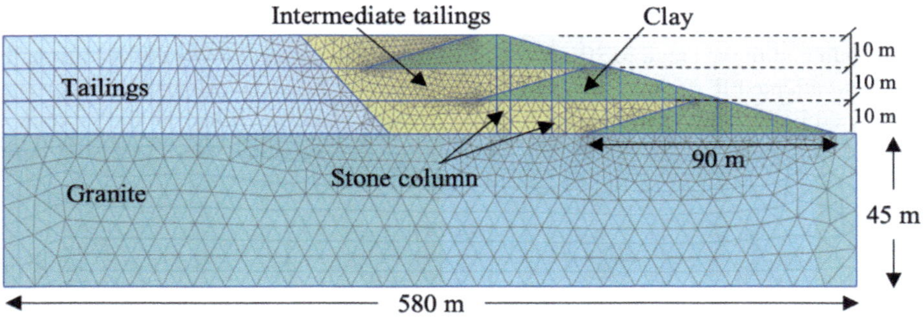

Fig. 5. Discretized mesh of the USM with stone column in the present study

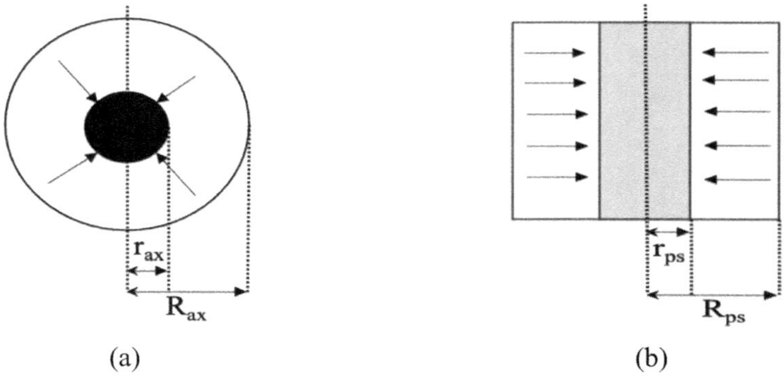

Fig. 6. Cross sections of unit-cell stone column in (a) axisymmetric geometry and (b) plane-strain conversions

The inclusion of stone columns is a well-established ground improvement technique, known to increase shear strength, accelerate consolidation, and provide vertical drainage, thereby reducing the risk of liquefaction and excessive settlement in tailings dams. The parametric evaluation of stone column and drainage zone radii allowed for the identification of configurations that maintain the FOS above recommended thresholds, enabling safer staged construction of higher embankments at Malanjkhand.

Statistical analysis

A nonlinear regression analysis was conducted to predict the FOS of embankments reinforced with stone columns, incorporating radius (r_{ax}), zone of influence (R_{ax}), (denoted as r and s henceforth) and their interaction terms. To assess the model's generalization capability, the 5-fold cross-validation was conducted, where the model was trained on 4 subsets and tested on the remaining one, for each fold. The process was repeated 5 times, each time with a different fold as the test set. The mean squared error (MSE) for each fold and the average MSE across all folds (MSE_{CV}) were reported. Additionally, a cross-validated R-squared (R^2_{CV}) was derived to evaluate the proportion of variance explained by the model. Further, adjusted R^2 value, which is based on the number of predictors, penalizing the addition of non-informative variables, was calculated based on Eq. (3)

$$R^2_{adj} = 1 - \left(\frac{\left(1 - R^2\right)(n-1)}{n-p-1} \right) \tag{3}$$

where n is the number of observations and p is the number of predictors.

3 Results and Discussions

Numerical simulations of the Malanjkhand copper mine tailings dam reinforced with stone columns demonstrate the significant influence of stone column diameter and spacing on dam stability. The factor of safety was used as the primary metric to assess the stability of each configuration. The results of the various analyses have been tabulated in Table 4.

Table 4. Variation of FOS with different spacings and radii

Width in model, m $(2 \times r_{ps})$	Spacing used in model $(2 \times R_{ps})$, m		
	10	20	40
1	1.92	1.88	1.86
1.5	2.01	1.92	1.87
2	2.08	1.97	1.89
5	2.20	2.05	1.92
10	2.66	2.29	2.04

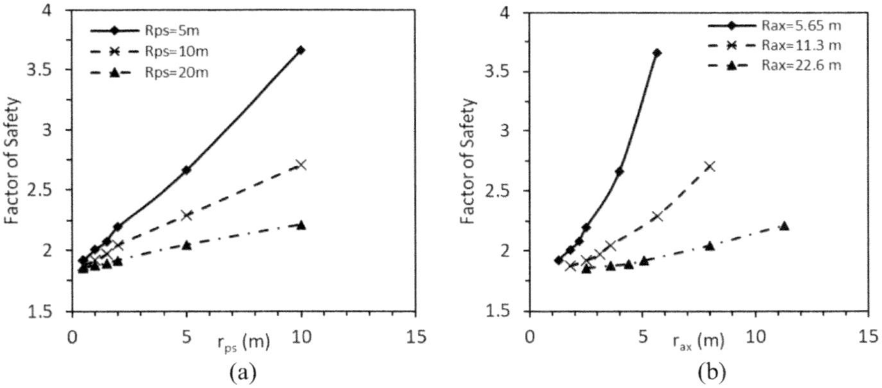

Fig. 7. (a) Parametric analysis of stone column spacing and radius in plane strain case, (b) Corresponding parametric analysis for axisymmetric values of radius and spacing.

Figures 7 (a) and (b) demonstrate that the stone column radius and spacing affect FOS in both plane strain and axisymmetric models. Results show FOS increases with larger diameters across all spacings, consistent with geotechnical principles where greater area replacement improves load transfer, reduces settlement, and enhances shear strength. The bearing capacity proves most sensitive to diameter at closer spacings, maximizing soil-column interaction. Reduced spacing also boosts stability by increasing composite stiffness and pore pressure dissipation, though benefits diminish when spacing falls below twice the diameter. Notably, plane strain shows linear trends while axisymmetric models exhibit nonlinear behaviour, particularly at smaller radii. The most significant FOS gains occur at tight spacings with larger radii; at wider spacings, even substantial radius increases yield marginal improvements. This radius-spacing interaction highlights the need for balanced parameter optimization to achieve cost-effective stability, considering both technical performance and practical constraints like constructability and materials.

3.1 Statistical Analysis Results

Initially, linear (r, s), quadratic (r^2, s^2) and interaction (r × s) terms were included in the regression analysis to capture potential curvature in the relationships, but the quadratic terms proved to be non-significant and were subsequently removed from the refined model (r^2: $\beta = 0.00$, $p = 0.90$; s^2: $\beta = 1e{-}3$, $p = 1.4e{-}2$). The final model performance metrics have been summarized in Table 5, while the regression analysis coefficients are presented in Table 6.

$$FOS = 1.202 + 0.709r + 0.027s - 0.077rs + 0.002\ rs^2 \quad (4).$$

Table 5. Model Predictive performance metrics

Metric	Value
R- squared	0.96
Adjusted R-squared	0.95
Root mean squared error	0.09
Cross-validation MSE	0.02
Cross-validation R-squared	0.88
F-statistic	88
F-statistic p-value	2.81×10^{-9}

Table 6. Details of regression coefficients

Term	Estimate	Std. Error	t-statistic	p-value	Inference
Intercept	1.20	0.11	11.31	4.23e–8	Baseline FOS when r = 0 and s = 0
r	0.71	5.2e–2	13.76	3.99e–9	Strong positive effect, highly significant
s	0.02	8e–3	3.39	0.004	Small but significant positive effect
$r \times s$	– 0.07	8e–3	– 10.21	1.41e–7	Significant negative interaction effect
$r \times s^2$	2.e–3	3e–4	8.56	1.06e–6	Small but significant positive interaction

The model demonstrates exceptional predictive power, explaining 96.4% of the variance in FOS ($R^2 = 0.96$, $p < 1e{-}3$). A high adjusted R^2 value ($R^2_{adj} = 0.95$) confirmed the model's reliability after penalizing overfitting. Cross-validation metrics further validated the model's generalizability, with a cross-validated R-squared of 0.88 and a mean squared error (MSE) of 0.02.

The lack of significance for r^2 ($p = 0.90$) indicates an effectively linear relationship between radius and FOS, with no evidence of diminishing returns or acceleration in gains, within the tested range. While s^2 initially showed marginal significance ($p = 1.3e{-}2$, $p = 1.4e{-}2$), its exclusion in the refined model did not degrade predictive power (R^2 remained stable at 0.96). This suggests that spacing influences FOS primarily through linear and interaction effects (e.g., $r \times s$ and $r \times s^2$) rather than intrinsic curvature. Removing r^2 and s^2 improved parsimony without sacrificing accuracy, as evidenced by minimal change in R^2 ($0.97 \rightarrow 0.96$) and improved cross-validation performance (R^2_{CV} increased from 0.82 to 0.88).

A strong positive coefficient for r ($\beta = 0.70$, $p < 1e{-}3$) highlights that increased column radius directly enhances stability. This aligns with geotechnical principles, where larger columns enhance load-bearing capacity and shear resistance. A 1-unit increase in r increases FOS by 0.70, making it the most influential design parameter.

A positive coefficient for spacing, s ($\beta = 0.02$, $p = 0.01$), suggests moderate spacing improves stability up to a threshold. This suggests that a moderate spacing (e.g., $s = 15$–25) improves stability by optimizing stress distribution, beyond which, the negative interaction term ($r \times s$: $\beta = -0.07$) dominates, reducing gains from larger radii.

The $r \times s^2$ term ($\beta = 0.002$) captures a critical nonlinearity, implying that larger radii amplify stability improvements at intermediate spacings. For example, increasing r from 5 to 10 at $s = 30$ improves FOS by 3.8 (from 2.29 to 6.09), whereas the same increase at $s = 40$ yields only 1.7 (from 2.04 to 3.74).

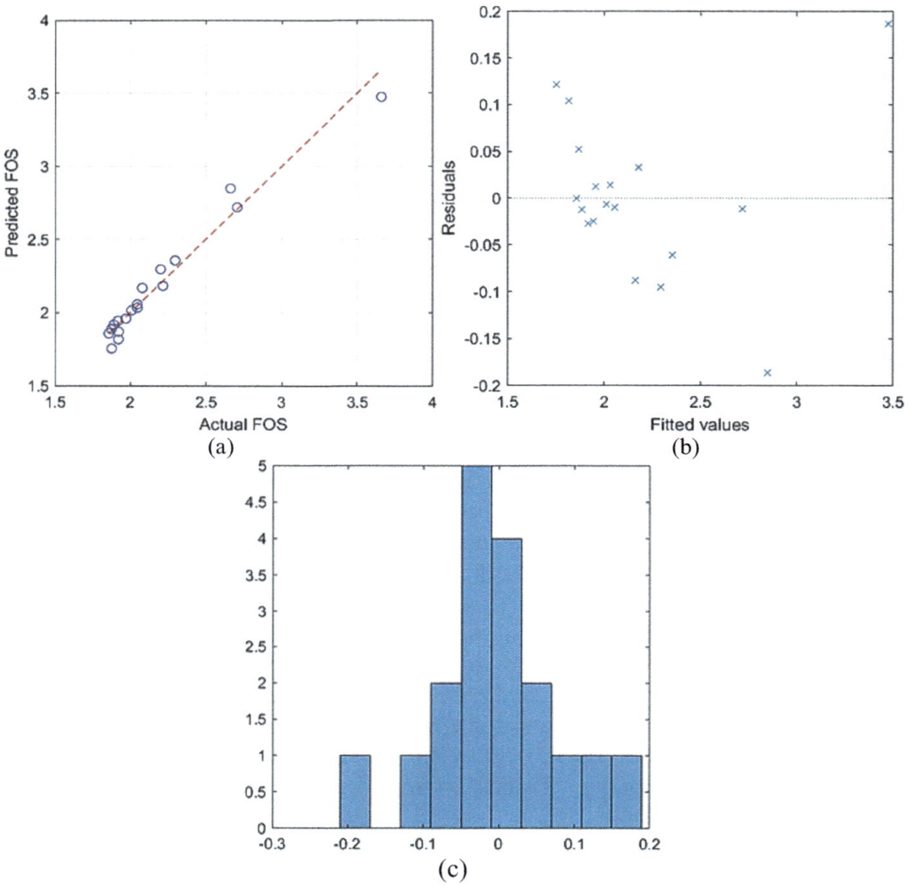

Fig. 8. (a) Actual and predicted values for FOS, (b) Residual and fitted values, (c) Residual distribution.

The residual analysis for this nonlinear regression model provides valuable insights into model adequacy and assumptions. The Predicted vs. Actual FOS plot (Fig. 8 (a)) shows a tight clustering of points around the diagonal, suggesting strong agreement between the model's predictions and the actual values, with only a few moderate deviations for higher FOS values. The residuals vs. fitted values plot (Fig. 8 (b)) demonstrates that the residuals are generally randomly scattered around zero, indicating the model captures the main structure of the data without major systematic bias. However, a slight funnel shape is observed, with residual variance increasing at higher fitted values, suggesting mild heteroscedasticity. This means that while the model's predictions are largely unbiased, the error variance is not perfectly constant across all levels of the predicted FOS. The histogram of residuals (Fig. 8 (c)) reveals an approximately symmetric, bell-shaped distribution centered near zero, with only minor left skewness and no significant outliers. This supports the assumption of normality, which underpins reliable inference for p-values and confidence intervals. Overall, the residuals indicate a robust model fit, with only minor concerns about non-constant variance.

4 Conclusions

This study investigated the stability of tailings dams reinforced with stone columns, focusing on the Malanjkhand copper mine case, through numerical simulations and nonlinear regression modelling. A refined nonlinear regression model, excluding insignificant quadratic terms, achieved exceptional predictive accuracy ($R^2 = 0.96$) and generalizability $R^2{}_{CV} = 0.88$), highlighting the dominance of radius, spacing, and their nonlinear interactions in governing stability. The dominance of radius as a predictor aligns with geotechnical principles, where larger columns enhance load-bearing capacity and shear resistance. Conversely, the negative coefficient for spacing (s) underscores the trade-off between material efficiency and stability, as wider spacing reduces lateral confinement and stress redistribution. The following practical design insights can be obtained from the study:

1. Radius of the stone column can be prioritized over the spacing, as larger stone column radii ($r \geq 5$) yield the most substantial stability gains, with a 1-unit increase in radius boosting FOS by 0.70.
2. Moderate spacing ($s = 5$–25m) maximizes the composite effect of columns and soil, while higher spacing negates the benefits of larger radii.
3. The quadratic interaction effect($r \times s^2$) can be leveraged to amplify stability at intermediate spacings, but overdesigning with excessively small spacings (e.g., $s < 10$) must be avoided, where marginal benefits plateau.

The co-disposal technique and optimization framework presented here offer promising opportunities for application across diverse mining operations, including gold and iron ore tailings impoundments. For gold tailings, which often exhibit similar fine-grained characteristics to copper tailings, the stone column approach can effectively address low permeability and poor drainage conditions. In iron ore tailings, particularly those with high clay content and susceptibility to liquefaction, the enhanced drainage pathways and load transfer mechanisms can significantly improve long-term performance. It should also be noted that while this research focuses on geotechnical stability, the performance of the waste rocks in chemically active environments needs to be studied, particularly regarding chemical interactions between tailings, waste rock, and groundwater. Future investigations could examine the selection of chemically resistant waste rock materials, protective measures such as encapsulation, and the development of hybrid systems that combine structural enhancement with environmental protection for sustainable tailings management.

References

1. Rico M, Benito G, Salgueiro AR, Díez-Herrero A, Pereira HG (2008) Reported tailings dam failures: a review of the European incidents in the worldwide context. J Hazard Mater 152(2):846–852. https://doi.org/10.1016/j.jhazmat.2007.07.050

2. Bussière B, Colloquium B (2007) Hydrogeotechnical properties of hard rock tailings from metal mines and emerging geoenvironmental disposal approaches. Can Geotech J 44(9):1019–1052. https://doi.org/10.1139/T07-040
3. Piciullo L, Storrøsten EB, Liu Z, Nadim F, Lacasse S (2022) A new look at the statistics of tailings dam failures. Eng Geol 303(2):1–13. https://doi.org/10.1016/j.enggeo.2022.106657
4. Singh AK, Mishra A (2023) Stability of copper mine tailings dam for upstream, downstream, and centreline construction methods: numerical analysis. In: Dimitrios Z, Tugce B (eds) 9th International congress on environmental geotechnics 2023, vol 8, pp 1–10. https://doi.org/10.53243/ICEG2023-261
5. Ferdosi B, James M, Aubertin M (2015) Effect of waste rock inclusions on the seismic stability of an upstream raised tailings impoundment: a numerical investigation. Can Geotech J 52(12):1930–1944. https://doi.org/10.1139/cgj-2014-0447
6. Koerner RM, Soong TY (2001) Geosynthetic reinforced segmental retaining walls. Geotext Geomembr 19(6):359–386. https://doi.org/10.1016/S0266-1144(01)00012-7
7. Sasaki Y, Kano S, Tsuji T (2004) Embankment reinforcement by geogrid to reduce its settlement during earthquakes. In: 13th World conference on earthquake engineering 2004, vol 28, pp 1–15
8. Mudenge ST, Kalumba D (2023) Reliability slope stability analysis of geosynthetics reinforced tailings dams: a platinum mine case study. In: Dimitrios Z, Tugce B (eds) 9th International congress on environmental geotechnics 2023, vol 8, pp 1–9
9. Xu B, Wang Y (2015) Stability analysis of the Lingshan gold mine tailings dam under conditions of a raised dam height. Bull Eng Geol Environ 74(1):151–161. https://doi.org/10.1007/s10064-014-0602-z
10. Psarropoulos PN, Tsompanakis Y (2015) Stability of tailings dams under static and seismic loading. Can Geotech J 45(5):663–675. https://doi.org/10.1139/T08-014
11. Tan SA, Tjahyono S, Oo KK (2008) Simplified plane-strain modelling of stone-column reinforced ground. J Geotech Geoenviron Eng 134(2):185–194. https://doi.org/10.1061/(ASCE)1090-0241(2008)134:2(185)

Investigation of Crop Residues for Thermal and Acoustic Panel Production

Merve Aslan Kadız[1], Gülşah Balamut Arslan[1], Paulina Faria[2], Hülya Kalaycıoğlu[3], Dino Clarke[4](✉), and Salise Oktay[1]

[1] Kastamonu Entegre Ağaç San. Tic. A.Ş, R&D Center, İstanbul, Türkiye
{merve.aslan,gulsah.balamut,salise.oktay}@keas.com.tr
[2] Department of Civil Engineering, CERIS, NOVA University Lisbon, Caparica, Portugal
paulina.faria@fct.unl.pt
[3] Faculty of Forestry, Karadeniz Technical University, Trabzon, Türkiye
khulya@ktu.edu.tr
[4] NTE Company (Pty) Ltd, Pietermaritzburg, South Africa
dino@nte.co.za

Abstract. The aim of this study is to protect forest resources and investigate the potential of using agricultural waste, namely crop residues, to produce thermal and acoustic particleboards. Agricultural resources commonly grown in Turkey, and corresponding wastes and byproducts, were examined. Corn stalk, corn cob and hemp hurds were selected to produce light-weight composite boards to be used as acoustic building panels. Experimental sets were prepared at 100% corn stalk, 100% corn cob, 100% hemp hurds and 100% wood. Urea-formaldehyde and tannin-based resins were selected as binders. The panel density was targeted at 400 ± 50 kg/m^3. Panel prototypes samples were prepared with a laboratory scale hot press. These samples were tested for their physical and mechanical properties according to European standards, including durability, thermal conductivity and the sound absorption coefficient. The results obtained show promising and competitive results when compared to literature and commercial boards, justifying further studies and boards scale-up.

Keywords: Biomass Composite · Lignocellulosic Materials · Insulation

1 Introduction

The current global challenges related to carbon emissions and climate change have drawn significant attention to green building concepts. These buildings emphasize the use of renewable natural resources to reduce overall carbon emissions and energy consumption. In residential, educational, and workplace buildings, noise control and acoustic comfort, thermal insulation and thermal comfort together with energy efficiency, and overall indoor environmental quality have become major societal concerns. Consequently, the demand for both acoustic and thermal insulation boards and solutions that are cost-effective, renewable, durable, healthy, and environmentally friendly is steadily increasing. This has led to a surge in research on alternative raw materials and the reuse

H. Beushausen et al. (Eds.): NTZR 2025, LNCE 873, pp. 746–758, 2026.
https://doi.org/10.1007/978-3-032-27698-8_66

and recycling of waste in the production of building insulation products, in line with sustainable development goals [1–3].

The growing demand for natural materials in the construction sector has increased pressure on forest resources, contributing to deforestation and biodiversity loss. To mitigate these impacts, it has become essential to explore alternative lignocellulosic sources—particularly non-wood materials and agricultural wastes and byproducts, namely those abundantly available in countries like Turkey, such as corn stalks, corn cobs, and hemp hurds [4].

Research shows that non-wood lignocellulosic materials can be effectively utilized in the production of particleboards and insulation products, offering comparable or even superior performance to conventional wood-based alternatives [1, 2, 5]. Moreover, the valorization of such residues supports circular bioeconomy principles by converting waste streams into value-added materials and reducing environmental footprints [4, 6, 7].

Numerous studies have evaluated the potential of agro-forestry industrial wastes in insulation board production, while efforts continue to develop more environmentally friendly adhesives such as tannin-based or low-emission urea-formaldehyde resins [8, 9]. Given the innovative nature of these materials, new testing methodologies have also been developed and refined [2, 3].

Given the significant environmental footprint of the building industry, greener alternatives, such as bio-based particleboards, can play a key role in reducing that impact. In particular, agricultural fibers like corn stalks have shown promising acoustic and thermal properties, making them suitable for insulation applications [10]. By incorporating these materials into construction products, it is possible to meet evolving technical standards while fostering sustainable resource utilization.

In this study, instead of obtaining wood from trees that grow in about 30–40 years, it is aimed to use agricultural wastes obtained annually for producing bio-based particleboards with promising acoustic and thermal properties, underscoring their potential as sustainable building materials and affirming the role of non-wood lignocellulosic resources in minimizing the environmental impact of the construction sector. Beyond their renewable resource potential, agricultural wastes represent a significant waste stream that is often underutilized or inadequately managed, leading to environmental problems such as open-field burning, greenhouse gas emissions, and soil nutrient depletion. By incorporating these residues into building materials, it is possible to mitigate such environmental impacts, reduce disposal costs, and decrease the reliance on landfilling or incineration. This valorization approach aligns with circular bioeconomy and waste hierarchy principles, in which agricultural waste is transformed into value-added construction materials, contributing simultaneously to waste minimization, resource efficiency, and climate change mitigation.

Due to the impact of the building industry, and the importance the use of greener particleboards can have on reducing that impact, the aim of this study is to protect forest resources and investigate the potential of using agricultural waste, namely crop residues, to produce thermal and acoustic particleboards.

2 Materials and Methods

2.1 Materials

The feasibility of using crop residues as an alternative to wood in particleboard manufacturing was evaluated. Raw material selection was based on the prevalence and availability of agricultural waste in Türkiye. Corn, as one of the most extensively cultivated crops in the country [11], was identified as a key lignocellulosic source; therefore, corn stalks and corn cobs were incorporated into the experimental design. Hemp hurds were also included, as hemp is a renewable natural low-density resource with good mechanical properties, and can be found as a byproduct of the textile industry. Therefore, its use aims to enhance both the thermal and mechanical performance of the panels. Additionally, this approach supports industrial symbiosis by valorizing cross-sectoral residues and contributes to the environmental sustainability of the bio-based boards.

The three lignocellulosic raw materials - corn stalks, corn cobs, and hemp hurds - were used to produce lightweight composite panels intended for acoustic and thermal insulation applications. All materials were obtained from agricultural production areas across Türkiye. Before panel fabrication, the residues were mechanically processed and classified to a standardized particle size suitable for laboratory-scale board production. The dimensions of the particles used are in the range of 2.1–0.7 mm sieve. Four experimental groups were prepared, each composed of a single raw material: 100% corn stalk, 100% corn cob, 100% hemp hurds, and 100% wood particles, the latter as control group for comparison (Fig. 1).

Fig. 1. From left to right, cut wood particles and hemp hurds, corn cob before preparation and cut corn stalk.

Two binders were utilized: a urea-formaldehyde (UF) glue, representing conventional synthetic binders (Kastamonu Entegre, Türkiye), and a tannin-based bio-binder, sourced from NTE / Bondtite (South Africa). The selection of the binder systems was based on their relevance to both traditional and bio-based particleboard technologies, with particular attention paid to emission profiles and environmental compatibility. Their properties are presented in Table 1.

2.2 Production of the Particleboard Samples

The target density for all boards was set at 400 ± 50 kg/m^3, suitable for acoustic and thermal insulation applications. Laboratory-scale prototypes were produced using a 500 $\times$ 500 mm^2 hot press, with a nominal panel thickness of 16 mm. The conventional UF resin was used for reference board production, reflecting common industrial practice. To evaluate a more environmentally sustainable alternative, the tannin-based bio-resin was also employed. For both resin types, the adhesive was applied at a loading level of 12% based on oven-dry particle weight. A laboratory-scale spray gluing system was operated to apply the glue uniformly to the particles prior to mat formation and hot pressing. In this context, particleboard prototype production stages are given in Fig. 2.

Table 1. Properties of the adhesives used.

Properties	UF (1.22 mol)	Tannin
Solid (%)	65 ± 2	47
Density (g/cm^3)	1270–1295	1215
pH (25 °C)	7.8–9	6.47
Viscosity (cP 20 °C)	100–600	200
Gel point (sn, 100 °C)	35–45	110

Fig. 2. Particleboard prototype production stages

Experimental design for the production of the particleboards is given in Table 2. Press parameters applied in particleboard production are given in Table 3.

Table 2. Particleboard prototypes experimental design.

Board type	Lignocellulosic material	Adhesive type	Adhesive content (%, wt)
WU	Wood	UF	12
WT	Wood	Tannin	
HU	Hemp hurds	UF	
HT	Hemp hurds	Tannin	
CSU	Corn stalk	UF	
CST	Corn stalk	Tannin	

(continued)

Table 2. (*continued*)

Board type	Lignocellulosic material	Adhesive type	Adhesive content (%, wt)
CCU	Corn cob	UF	
CCT	Corn cob	Tannin	

Table 3. Hot- press parameters for the particleboards prototype production.

Temperature	Pressure	Pressure time
215 °C	80 bar	4:05 (min)

2.3 Test Methods and Samples

Test samples were prepared and tested according to EN standards. For some tests, additionally to the innovative particleboards developed, samples of commercial particleboard and MDF were also tested for comparison. The samples were conditioned at 60 ± 5% relative humidity and at a temperature of 23 ± 2 °C prior to tests, until constant mass. The prototypes were tested physically and mechanically according to specification for particleboards, defined by EN 312 [12]. The average value of five samples were tested for each property, except when something contrary is mentioned. The following tests were performed:

- Density, based on EN 323 [13];
- Thermal conductivity, using a portable ISOMET 2114 with an IPS1105 surface probe with 60 mm diameter and measurement range 0.04–0.3 W/(m K); the measurements were performed with the samples in controlled laboratory conditions of temperature (T) and relative humidity (RH): lower RH, at 18.3 ± 0.8 °C and 47.6 ± 2.9%, and higher RH, at 18.5 ± 0.6 °C and 67.6 ± 5.4%; three measurements were made per particleboard type; the average and standard deviation are considered (Fig. 3a);
- Sound absorption coefficient using the impedance tube - transfer function method, based on ISO 10534-2 [14]; this test was just performed with the particleboards produced with tannin glue;
- Internal bond, based on EN 319 [15];
- Bending strength and Elasticity modulus, based on EN 310 [16];
- Surface soundness, based on EN 311 [17].

The samples were produced as boards with circa 20 cm × 20 cm × 2 cm. They were used for non-destructive tests and then cut into smaller specimens for the different tests. The samples tested for acoustic properties are circular (Fig. 3b), with 99 mm diameter and 16 mm thickness.

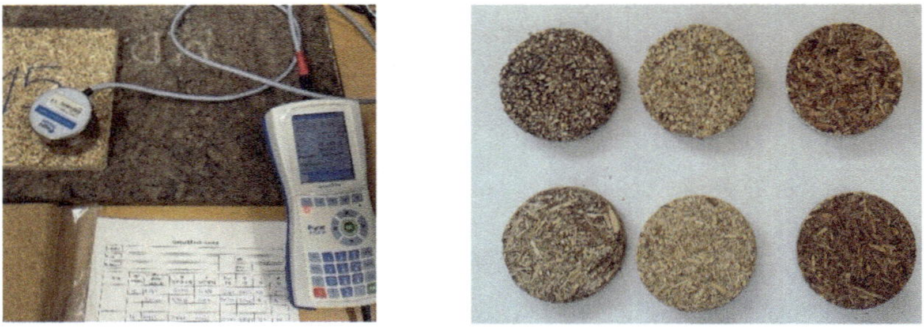

Fig. 3. A particle board being tested for thermal conductivity, with the ISOMET equipment and on top of expanded cork board to insulate from the table, and circular samples for the acoustic test.

3 Results and Discussion

3.1 Physical Properties

The density of the boards is given in Fig. 4. Results were obtained within the target value range of 400 ± 50 kg/m^3 and variations are low between all tested particleboards.

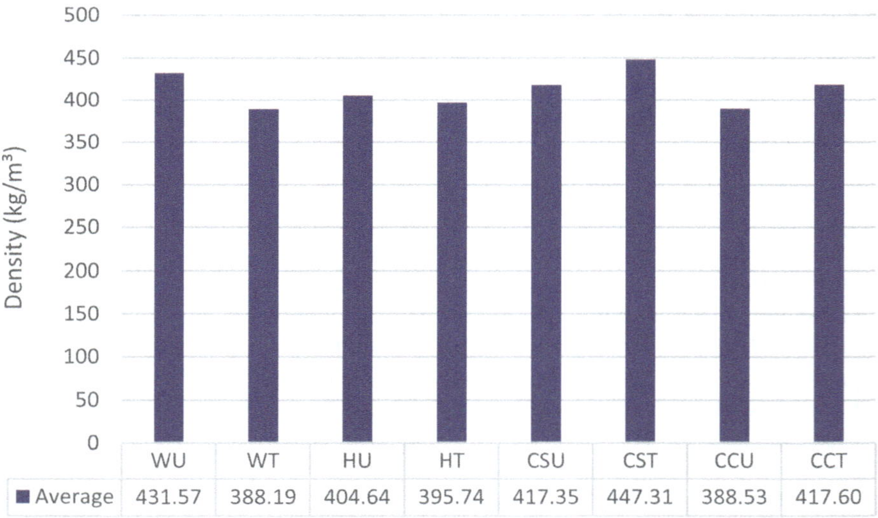

Fig. 4. Test results of density of the particleboards with wood (control), hemp, corn stalk and cob, with UF and tannin glues.

The difference in densities between UF- and tannin-bonded panels can be explained by the combined effects of adhesive viscosity and the porous structure of the lignocellulosic particles. In wood-based panels, the UF adhesive—with lower viscosity—penetrates more effectively into the relatively uniform and compact wood particles, enabling better particle compaction during pressing and resulting in higher density. In contrast, the agricultural residues, such as corn stalk and cob, have a more porous and heterogeneous structure, which allows UF resin to penetrate excessively and unevenly, reducing mat stability and compaction during pressing and thus leading to lower density. The tannin-based adhesive, having higher viscosity, tends to remain more at the surface of these porous particles, improving mat cohesion and compaction, and therefore resulting in higher final panel density. Additionally, slight deviations related to laboratory-scale hot-press equipment—including temperature distribution and pressure homogeneity—may have contributed to the observed variations in panel density.

The thermal conductivity (average and standard deviation) is presented in Table 4, for lower and higher RH. Additionally to the innovative particleboards developed, commercial particleboard and MDF were also tested for comparison.

Table 4. Thermal conductivity (average and standard deviation) for commercial particleboard, MDF and the developed boards.

Board	Thermal conductivity [W/(m K)]			
	Lower RH		Higher RH	
	Ave.	Stdv.	Ave.	Stdv.
Ref PB	0.116	0.002	0.121	0.002
Ref MDF	0.110	0.001	0.114	0.001
WU	0.074	0.003	0.076	0.002
WT	0.075	0.003	0.078	0.002
HU	0.073	0.004	0.074	0.004
HT	0.088	0.002	0.089	0.002
CSU	0.075	0.001	0.078	0.002
CST	0.086	0.003	0.089	0.002
CCS	0.076	0.002	0.078	0.001
CCT	0.088	0.001	0.091	0.000

Results show that all tested boards present very low standard deviation. As expected, the thermal conductivity when the boards are conditioned at lower RH is lower than at higher RH, although for some boards the difference is very low and within the standard deviation. The control wood-UF board has a thermal conductivity very similar to the wood-tannin board, both at lower and higher RH. Comparing the boards with corn cob, corn stack and hemp, the boards with UF have lower thermal conductivity than the similar boards with the tannin glue. However, and as intended, all the innovative particleboards have significantly lower thermal conductivity than the commercial MDF and particleboards. Analysing available literature, Abobakr et al. [18], testing particleboards produced with rice husk and epoxy resins achieved thermal conductivity values ranging from 0.064 to 0.076 W/(m K). Faria et al. [19], testing particleboards produced with bean straw and rubberwood particles glued with vegetable polyurethane adhesive registered thermal conductivity ranging from 0.097–0.106 W/(m K). Essid et al. [20], testing flax shives particleboards with lignin and epoxy resin glues, reached thermal conductivities of 0.074–0.081 W/(m K). Comparing with literature and commercial boards, it can be concluded that the boards developed in the present study are competitive and can have a positive effect on energy efficiency.

Noise is an important problem in residential and workplaces in urban areas. The demand for acoustic solutions which are cost-effective, renewable, eco-friendly in nature is increasing. Figure 5 shows the graph of the sound absorption coefficient measurements of the boards produced with the four biomasses and tannin glue. The acoustic tests were conducted using the same particleboard specimens that were characterized for thermal and mechanical properties. Therefore, the thickness and density values of the test specimens correspond to those provided earlier in the Materials and Methods and Results sections. Specifically, the circular samples used for the acoustic tests had a diameter of 99 mm and a thickness of 16 mm, with density values within the target range of 400 $\pm$ 50 kg/m^3 as shown in Fig. 4.

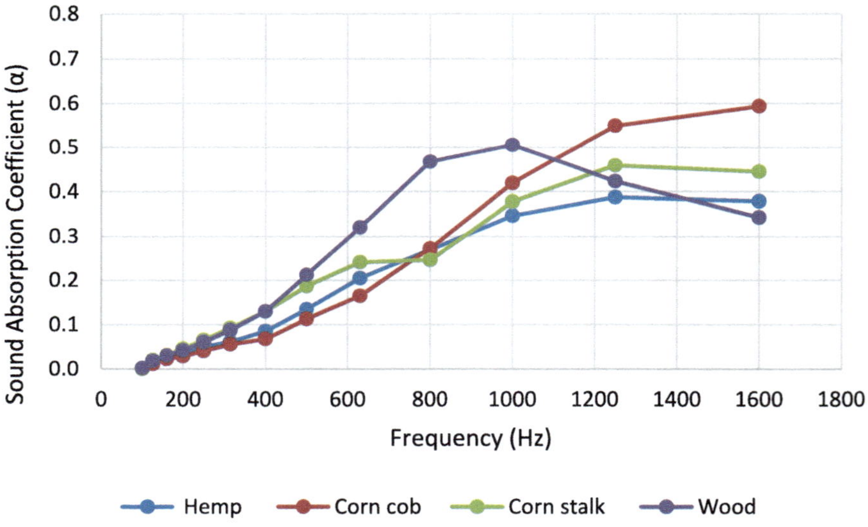

Fig. 5. Sound absorption coefficient in function of the frequency for the boards with hemp, corn cob, corn stalk and wood with tannin glue.

It can be seen that, for frequencies above 1100 Hz, the best result belongs to the corn cob-based boards, followed by the hemp-based ones, both better than the control wood-based board. The latter is better for frequencies between 500 Hz and 1100 Hz.

3.2 Mechanical Properties and Durability

The results obtained from the internal bond tests of the boards are given in Fig. 6. It can be observed that the hemp boards, both with UF and tannin glues, present the higher internal bond. The corn cob boards present intermediate values, while the boards with corn cob are the lowest ones. The tannin significantly reduces the internal bond of the control wood particle board and the hemp-based board. But the tannin glue slightly increases the internal bond for the corn stalk board and particularly the corn cob board, which is clearly positive.

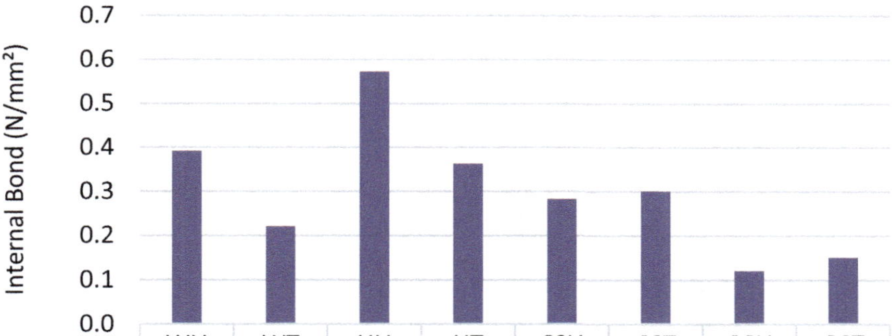

	WU	WT	HU	HT	CSU	CST	CCU	CCT
■ Average	0.39	0.22	0.57	0.36	0.28	0.30	0.12	0.15

Fig. 6. Test results of internal bond of the particleboards with wood (control), hemp, corn stalk and cob, with UF and tannin glues.

The results obtained from the bending strength tests of the boards are given in Fig. 7. It can be observed that the bending is higher for the hemp-based boards, followed by the corn stalk ones, the latter also competitive in comparison to the control wood-based boards. Again, the corn cob boards have the lowest results of bending strength. The tannin glue significantly reduces the bending capacity of the control wood particleboard, but not so much for the hemp-based and even less for the corn stalk board. The tannin glue is beneficial for the corn cob board, which is in accordance with the internal bond results.

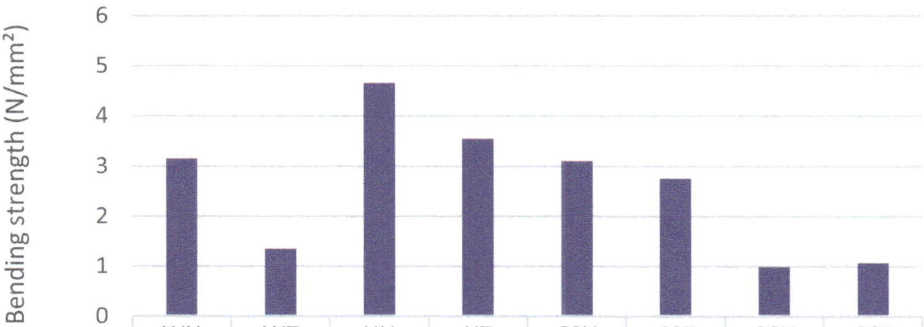

	WU	WT	HU	HT	CSU	CST	CCU	CCT
■ Average	3.15	1.34	4.65	3.54	3.10	2.74	0.98	1.06

Fig. 7. Test results of bending strength of the particleboards with wood (control), hemp, corn stalk and cob, with UF and tannin glues.

In Fig. 8, elasticity modulus test results are given. Accordingly, the test results of hemp-containing boards with both glues are higher than those of boards produced from other lignocellulosic sources, while the ones with corn cob present the lowest values. Again, the corn stalk boards present competitive performance when compared to the control wood-based boards.

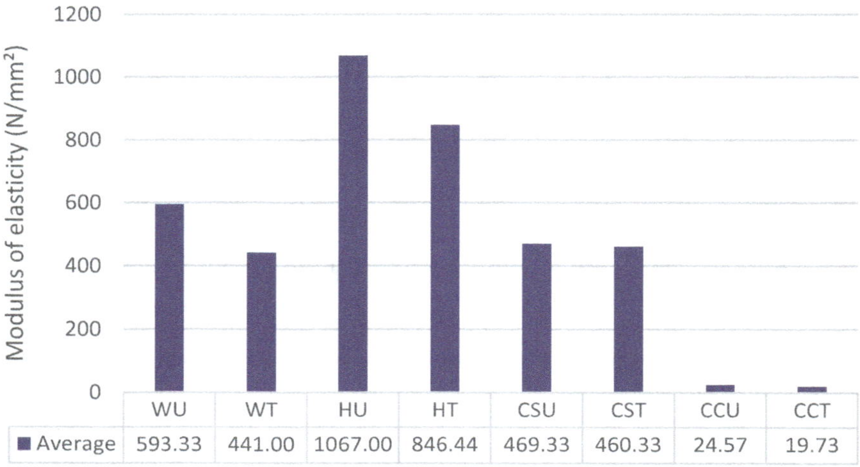

Fig. 8. Test results of modulus of elasticity of the particleboards with wood (control), hemp, corn stalk and cob, with UF and tannin glues.

Figure 9 shows the results of the surface durability tests. According to these results, the durability of hemp-containing boards is always higher than those of other tested particleboards. There is a decrease of strength when using tannin glue instead of the UF one, but that is higher for the control wood-based boards and lower for the hemp and corn stalk boards. The tanning glue seems very positive for the corn cob board, showing a high durability in comparison to the UF when using this biomass.

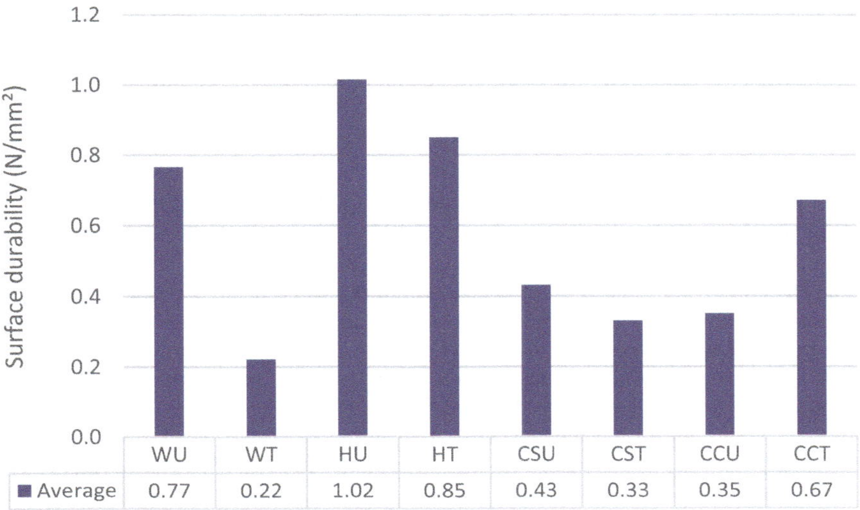

Fig. 9. Test results of surface durability of the particleboards with wood (control), hemp, corn stalk and cob, with UF and tannin glues.

4 Conclusions

It is important to reduce harvest pressure in face of increasing wood demands. The aim of this study is to protect forest resources and investigate the potential of using agricultural waste in the insulation panel industry instead of fossil-based materials.

Agricultural resources and waste byproducts commonly grown in Turkey were examined. Corn stalk, corn cob and hemp hurds were selected to produce lightweight composite boards to be used as acoustic and thermal insulation panels. Experimental sets were prepared at 100% corn stalk, 100% corn cob, 100% hemp hurds, and 100% wood as control for comparison. Urea Formaldehyde and tannin-based resins were selected as binders. The boards density was targeted at 400 ± 50 kg/m^3 and prototypes were prepared with a laboratory scale hot press, and tested for their physical and mechanical properties according to European Norms, including the sound absorption coefficient and thermal conductivity.

As a result of the tests, it was seen that the chosen crop residues have potential to be used as a substitute for wood for particleboards production. For thermal conductivity all the crops tested were viable and with lower values in comparison to commercial particleboards and MDF. The most positive results in terms of acoustic tests were obtained with corn cob-based particleboards.

Results justify further investigations. Therefore, for complementary characterization and numerical simulation of the effect these panels can have when used indoors, additional experimental sets are made using tannin-based resins made of lignocellulosic residues. By integrating agricultural waste into durable construction materials, this approach helps to reduce the environmental burden of residue accumulation, including minimizing greenhouse gas emissions from open-field burning and reducing soil nutrient loss. The carbon present in lignocellulosic biomass can be effectively sequestered within the panels, acting as a carbon sink during their service life. Additionally, the replacement of petrochemical-based adhesives with bio-based binders such as tannin resins lowers the emission of formaldehyde and other volatile organic compounds, thus improving indoor air quality and reducing toxicity. Beyond material efficiency, promoting hemp cultivation contributes to biodiversity, improves soil regeneration, and supports low-impact agricultural systems. These synergies address multiple United Nations Sustainable Development Goals (SDGs), including responsible consumption, climate action, and sustainable communities.

Acknowledgments. This work is supported within the scope of the EUROGIA 2030 project Morewood, ID number EU2030–19-2. Moreover the authors would like to thank the Scientific and Technical Research Council of Turkey (TUBITAK) 2244 Industrial PhD Program-Karadeniz Technical University-Kastamonu Entegre Ağaç San. Tic. A.Ş, Project No: 118C101, the Portuguese Agency for Innovation (ANI) for funding COMPETE2030-FEDER-00365700, operation 12706, and Tânia Santos for helping with the experimental campaign.

References

1. Barbu MC, Réh R, Irle M (2014) Non-wood lignocellulosic composites. BioResources 9(1):7–23
2. Cintura D, Palma A, Fiore V, Valenza A (2021) Agro-industrial wastes as building insulation materials: a review and challenges for Euro-Mediterranean countries. Industrial Crops and Product 171:113833
3. Ingrao C, Messineo A, Beltramo R, Yazan DM, Huisingh D, Siracusa V (2014) Recycled PET fibre-based panels for building thermal insulation: environmental impact and improvement potential assessment. Sci Total Environ 493:914–929
4. Maisarah T, Ahmad Z, Ismail H (2016) A review of natural fibers and processing operations for the production of binderless boards. BioResources 11(2):5600–5617
5. Ndububa EE, Nwobodo D (2015) Particleboard from fonio husk bonded using gum Arabic as a resin glue. Int J Innov Sci Eng Technol Res 11(1):97–110
6. Pędzik M, Rogoziński T, Auriga R (2022) Particleboard from agricultural biomass and recycled wood waste: a review. Materials 15(1):1–18
7. Pędzik M, Janiszewska D, Rogoziński T (2021) Alternative lignocellulosic raw materials in particleboard production: a review. Ind Crops Prod 174:114162
8. Sugahara K, Takahashi M (2019) High-density particleboard made from agro-industrial waste and castor oil-based polyurethane adhesive. BioResources 14(3):5162–5170
9. Tajuddin M, Ahmad Z, Ismail H (2016) A review of natural fibers and processing operations for the production of binderless boards. BioResources 11(2):5600–5617
10. Zhang Y, Li X, Ren H, Liu X (2017) Preparation and characterization of lightweight sound-absorbing boards made from corn stalk and urea-formaldehyde resin. Constr Build Mater 146:636–643
11. TurkStat (2022) Crop production statistics. Ministry of Agriculture and Forestry
12. EN 312 (2010) Particleboards—Specifications. CEN
13. EN 323 (1993) Wood-based panels—Determination of density. CEN
14. ISO 10534-2 (2023) Acoustics—Determination of acoustic properties in impedance tubes. Part 2: two-microphone technique for normal sound absorption coefficient and normal surface impedance. ISO

15. EN 319 (1993) Particleboards and fibreboards—Determination of tensile strength perpendicular to the plane of the board. CEN
16. EN 310 (1993) Wood-based panels—Determination of modulus of elasticity in bending and of bending strength. CEN
17. EN 311 (2002) Wood-based panels—Surface soundness—Test method. CEN
18. Abobakr H, Kakou CA, Bensalah MO, Bouhfid R, Qaiss AK, Raji M (2025) Development of lightweight insulation particleboard with superior thermo-mechanical and acoustic properties based on rice husks. Journal of Building Engineering 104:112289
19. Faria DL, Gonçalves FG, Maffioletti FD, Scatolino MV, Soriano J, Protásio TP, Martinez Lopez Y, Benigno Paes J, Mendes LM, Guimarães Junior JB, Tonoli GHD (2024) Particleboards based on agricultural and agroforestry wastes glued with vegetal polyurethane adhesive: an efficient and eco-friendly alternative. Ind Crops Prod 214:118540
20. Essid S, Hegde VJ, Mahieu A, Bizet L, Leblanc N, Saouab A (2021) Comparison of the properties of flax shives based particleboards prepared using binders of bio-based lignin and partially bio-based epoxy resin. Int J Adhes Adhes 109:102915

Experimental Investigation of Flexural Performance and Ductility Enhancement in Glued Laminated Bamboo (Glubam) Beams: Effects of Steel Fuse Bars and Spliced Toothed Metal Plate Connections

Mohamed Adel[1,2](✉), Weichang Guo[3], Hongyi Pan[3], Bibin Wang[4], and Yan Xiao[1]

[1] Zhejiang University—University of Illinois at Urbana-Champaign Institute, Haining 314400, Zhejiang, People's Republic of China
mohamedadel@eng.cu.edu.eg

[2] Structural Engineering Department, Faculty of Engineering, Cairo University, Giza 12613, Egypt

[3] College of Civil Engineering and Architecture, Zhejiang University, Hangzhou 310058, Zhejiang, People's Republic of China

[4] Zhejiang Provincial Science and Technology Cooperation Center, Hangzhou 310000, Zhejiang, People's Republic of China

Abstract. Glued laminated bamboo (glubam) is an innovative engineered material that leverages bamboo's high strength and rapid growth cycle, offering a sustainable alternative for modern construction. While demonstrating considerable potential, glubam's widespread adoption in large-span structures faces challenges, including brittle flexural failure modes, limited ductility, and practical constraints on construction lengths. This study investigates two distinct approaches to enhance the flexural performance of thin-strip glubam beams: (1) incorporation of steel fuse bars and (2) implementation of spliced toothed metal plate connections (MPCs). The experimental investigation comprised two comprehensive test series. The first series examined pre-cracked glubam beams under four-point bending, comparing specimens with and without steel fuse bars across three distinct configurations. Results demonstrated that fuse bar reinforcement could increase ductility by up to 2.5 times, though this improvement was accompanied by a 45% reduction in load-bearing capacity. The second test series evaluated spliced beam configurations using MPCs, analyzing performance variations between beams with splices at the inner layer, outer layers, and those un-spliced. Spliced specimens maintained 65% of the ultimate load capacity observed in un-spliced beams, with only marginal stiffness reductions. Notably, spliced configurations exhibited comparable failure modes and enhanced ductility characteristics, except in cases of MPC rupture. These findings provide critical insights into ductility enhancement strategies for glubam beams, offering practical solutions for extending their application in large-span construction while maintaining structural reliability.

Keywords: Glubam · Flexural behavior · Steel fuse bars · Toothed metal plate connections · Ductility enhancement

H. Beushausen et al. (Eds.): NTZR 2025, LNCE 873, pp. 759–770, 2026.
https://doi.org/10.1007/978-3-032-27698-8_67

1 Introduction

The increasing adoption of bamboo and timber as sustainable, bio-based construction materials is driven by their lightweight nature, superior mechanical properties, and environmental benefits [1]. Bamboo, in particular, offers distinct advantages, including remarkable rapid growth rates and high strength-to-weight ratio, positioning it as an environmentally favorable solution that can contribute to carbon emission reduction and sustainable forest management [2]. Among engineered bamboo products, glued laminated bamboo (glubam) has emerged as a promising alternative to traditional timber materials, especially in bamboo-abundant regions such as China [2–5]. Developed through advanced manufacturing processes involving hot- and cold-pressing techniques, glubam exhibits consistent mechanical properties suitable for demanding structural applications [1–3].

Glubam boards are typically categorized into two primary types: thick-strip (unidirectional) and thin-strip (multidirectional) laminates. The thin-strip board demonstrates superior performance characteristics, attributable to its optimized fiber orientation incorporating transversal fibers [2, 4, 5]. Extensive comparative studies have revealed that glubam outperforms conventional glued laminated timber (glulam) in strength and durability characteristics [2, 5–10]. However, a persistent limitation affecting glubam's structural application is its tendency toward brittle flexural failure, which significantly constrains both ductility and overall structural robustness [11–14].

Recent research has explored various reinforcement strategies to address these limitations, including steel reinforcement, fiber-reinforced polymers (FRP), and glass fiber-reinforced polymers (GFRP). These approaches have demonstrated measurable improvements in stiffness, load capacity, and ductility when applied to glulam beams [15–17]. Particularly noteworthy is the emerging application of fuse bars – high-yield threaded rods featuring strategically reduced cross-sections–which have shown considerable enhancement in both ductility and energy dissipation in timber and concrete structural members [18, 19]. Despite glubam's superior material properties, research specifically addressing fuse bar reinforcement in glubam beams remains limited. Investigating this area is essential to enhance glubam's structural capabilities and expand its practical applications in modern engineering.

Developments in beam splicing techniques, particularly those employing finger joints or metal plate connections (MPCs), offer practical solutions to overcome dimensional and transportation limitations inherent in bamboo-based structural systems. While MPC technology has been extensively researched and implemented in timber construction [20, 21], its adaptation for glubam structures represents a more recent innovation. Preliminary studies have validated the fundamental feasibility of MPC applications in glubam, demonstrating competitive strength performance and acceptable failure modes [22, 23]. However, a comprehensive investigation of bending behavior in spliced glubam beam configurations remains necessary to ensure their reliable performance in full-scale engineering applications.

The present study addresses these critical research gaps through a systematic experimental investigation of two innovative approaches for enhancing thin-strip glubam beam performance. The research methodology encompasses two distinct experimental series:

The first evaluates the effectiveness of steel fuse bars in improving the flexural ductility of pre-cracked glubam beams through four-point bending tests, examining multiple pre-crack configurations and fuse bar implementations. The second series investigates the structural performance of spliced glubam beams utilizing MPC at the inner-layer and outer-layers compared with un-spliced beam configurations. Comprehensive performance evaluation includes analysis of stiffness characteristics, failure mechanisms, strain distribution patterns, and ductility parameters to thoroughly assess the efficacy of these reinforcement techniques.

2 Experimental Program

The experimental study consisted of two test series, each involving thin-strip glubam beams subjected to four-point flexural loading until failure. Series-1 examined ductility enhancement through steel fuse bar incorporation in pre-cracked beams, while Series-2 assessed the structural viability of spliced configurations using MPCs.

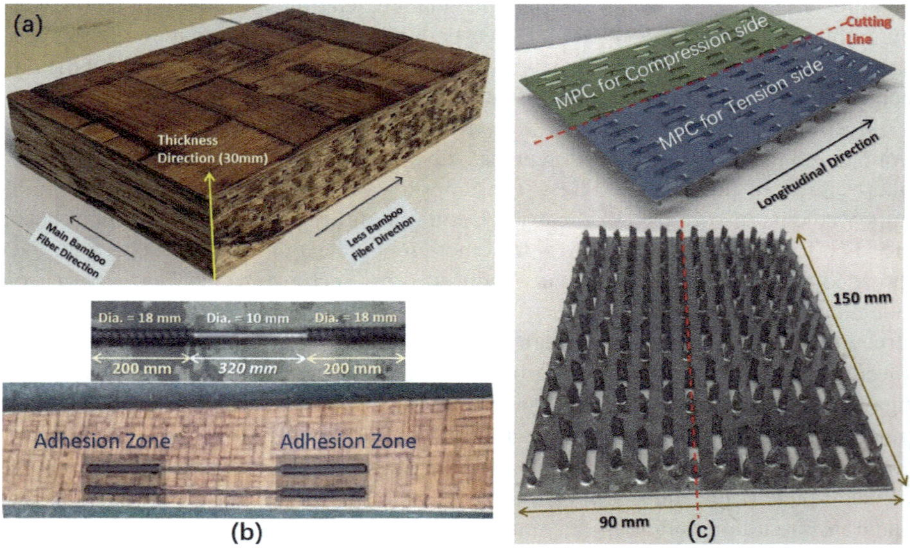

Fig. 1. (**a**) Thin-strip glubam board; (**b**) fuse steel bars configuration and installation; **c** toothed metal plate connection (MPC) layout.

2.1 Materials and Specimen Fabrication

All test specimens were fabricated through a cold lamination process, where thin-strip glubam boards were bonded with phenolic resin adhesion. As illustrated in Fig. 1(a), the glubam boards featured a bidirectional lamination pattern of 2 mm-thick bamboo strips produced through hot-pressing. The material exhibited an average tensile strength of 80 MPa, a compressive strength of 40 MPa, and a mean density of 862 kg/m^3.

For the first series, hot-rolled steel fuse bars (grade HRB450) were employed, featuring a strategically reduced diameter (10 mm) at the mid-section to facilitate early controlled yielding within this section (Fig. 1(b)). These bars demonstrated characteristic yield and tensile strengths of 450 MPa and 640 MPa, respectively. To ensure secure bonding and effective load transfer, the larger diameter ends (18 mm) of the fuse bars were embedded into the glubam boards using AB epoxy adhesive.

The fabrication of Series-1 specimens involved the cold lamination of four thin-strip glubam boards, producing beams with a total length of 3000 mm and a cross-section of 120 × 223 mm. To promote the controlled yielding of the fuse bars prior to the beam's brittle failure under flexural loading, pre-cracks were introduced at mid-span. These pre-cracks, manually cut to a depth of half the beam section, with configurations including single and triple pre-cracks (Fig. 2(a)), simulate initial damage and were intended to trigger the yielding of the fuse bars, thus enhancing ductility and energy dissipation. In the reinforced specimens, four fuse bars were installed in pre-cut grooves at two levels along the beam height to enhance ductility and energy dissipation (Figs. 1(b) and 2(a)).

For the second series, the specimens incorporated galvanized toothed metal plate connections (MPCs), characterized by a tooth density of 1.3 teeth/cm^2 and a thickness of 1.04 mm (Fig. 1(c)). The MPCs were strategically segmented to optimize load distribution, with the sect. (45 × 150 mm) placed in the tensile stress zone and the compression zone.

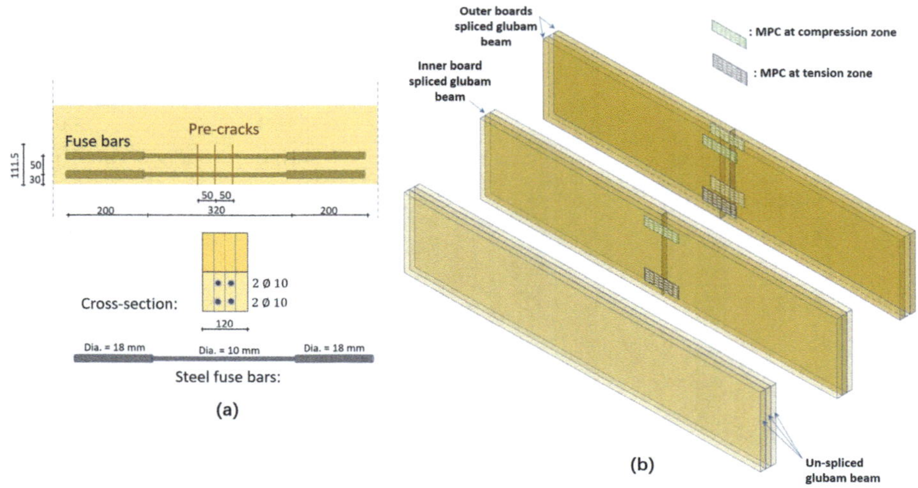

Fig. 2. Schematic detailing of: (**a**) thin-strip glubam beams with steel fuse bars (Series-1); and (**b**) spliced thin-strip glubam beams (Series-2); dimensions in mm.

The Series-2 specimens were fabricated by cold-laminating three thin-strip glubam boards, either spliced or un-spliced, resulting in beams measuring 3200 mm in length with a cross-section of 90 × 300 mm. Three different configurations were evaluated:

un-spliced beams, beams with splices in the inner board, and beams with splices in the outer-boards. The spliced beams featured joints at 1600 mm intervals, with MPCs positioned 20 mm from the board edges at both the compression and tension zones, as depicted in Fig. 2(b). To ensure precision and structural integrity, the splicing process was conducted using a QZ-100kN hydraulic press.

2.2 Testing Method

Table 1 summarizes the characteristics of the tested specimens, comprising six beams in each series, with distinct configurations. All beams were subjected to four-point bending tests following the GB/T 50329–2012 standard, as shown in Fig. 3. The loading was applied using a 300 kN hydraulic testing machine at a controlled rate of 6.0 kN/min till failure. The first series of beams had a simply supported span of 2700 mm, whereas the second series featured a span of 3000 mm.

Table 1. Characteristic and flexural loading results of thin-strip glubam beams.

Series	Beam ID	Size	No. of Pre-cracks	Fuse bars	No. of spliced boards	MOE (GPa)	Ultimate load (kN)	Ductility ratio
1	GB1-NC-1	120 × 223 × 3000	–	–	–	9.62	161.0	1.60
	GB1-NC-2		–	–	–	9.40	132.0	1.40
	GB1-1C-1		1	4Ø10	–	9.60	80.0	3.70
	GB1-1C-2		1	4Ø10	–	9.30	76.0	3.60
	GB1-3C-1		3	4Ø10	–	9.15	74.0	3.50
	GB1-3C-2		3	4Ø10	–	9.00	75.0	3.00
2	GB2-NS-1	90 × 300 × 3200	–	–	–	10.8	144.0	1.50
	GB2-NS-2		–	–	–	10.1	122.0	1.10
	GB2-1S-1		–	–	1 (inner)	8.70	86.0	2.20
	GB2-1S-2		–	–	1 (inner)	8.90	89.0	2.10
	GB2-2S-1		–	–	3 (outer)	8.15	80.0	3.50
	GB2-2S-2		–	–	3 (outer)	8.20	71.0	4.50

To capture the structural response, Linear Voltage Displacement Transducers (LVDTs) were positioned at the mid-span, loading points, and supports to record displacement levels throughout the loading process. Additionally, strain measurements were taken using 5 mm strain gauges placed at mid-span across the depth of the glubam beams (Fig. 3). Strain gauges were also affixed along the steel fuse bars in Series-1 specimens and the MPCs in Series-2 specimens to monitor their deformation behavior under flexural loading.

3 Results and Discussion

Table 1 also summarizes the results of the bending tests conducted on both series of thin-strip glubam beams. The flexural performance of the tested specimens was evaluated by key parameters, including the load–mid-span deflection response, modulus of elasticity (MOE), ultimate load (Pmax), calculated ductility ratio, failure modes, and strain distributions at mid-span.

Fig. 3. Actual four-point flexural loading conditions of Series-1.

3.1 Flexural Behavior

Fig. 4 illustrates the load–mid-span deflection curves for both series of thin-strip glubam beams, providing a detailed overview of their flexural performance, including their load-bearing capacity and deformation characteristics. The modulus of elasticity (MOE) and the ductility ratio were calculated following the CEN standard and are summarized in Table 1. The MOE was derived from the slope of a reference line (Line-1) connecting two key points on the load–mid-span deflection curve, corresponding to 20 and 40% of the ultimate load. The ductility ratio (μ) was computed as the ratio of the ultimate deflection (Δ_u) to the yielding deflection (Δ_y), both extracted from the load–mid-span deflection curves. The yielding point was determined as the intersection of Line-1 and a tangent line (Line-2), where Line-2 was constructed with a slope equal to one-sixth of Line-1's slope, thereby defining both the yielding load and corresponding deflection.

For the first series, results revealed minimal variation in the MOE among glubam beams with and without fuse bars (Table 1 and Fig. 4), indicating that the presence of the fuse bars mitigated the effects of pre-cracking. However, in series-2 specimens, the MOE decreased by 16% and 21.8% as the spliced boards increased from one to two, respectively, attributed to structural discontinuities that reduce the overall stiffness.

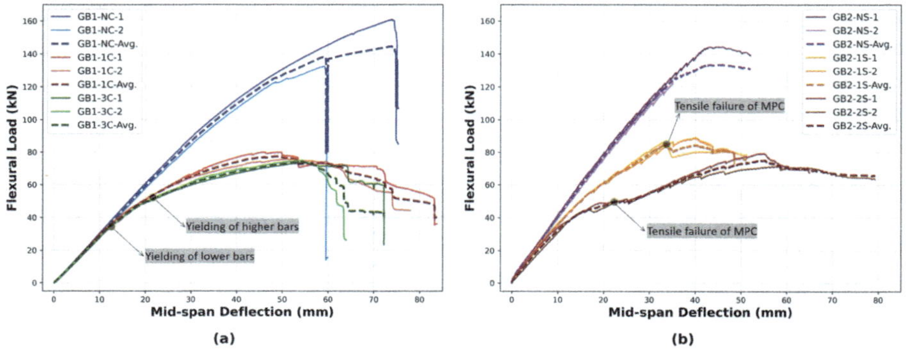

Fig. 4. Flexural response of thin-strip glubam beams: (**a**) with steel fuse bars (Series-1); (**b**) spliced with MPC (Series-2).

The influence of the fuse bars in Series-1 specimens became more pronounced as the applied load increased, promoting a ductile behavior with failure occurring at larger deflection levels compared to the sudden brittle failure in beams without bars (Fig. 4(a)). The results also indicated that the fuse bars yielded in both layers at an early stage before reaching the ultimate load, as shown in Fig. 4(a), underscoring their role in enhancing the overall ductile behavior of the beams. The group of beams with a single pre-crack exhibited the highest ductility ratio – approximately 2.5 times greater than specimens without fuse bars – allowing progressive yielding of the bars and improved deformability of the beams.

In Series-2, the spliced beams demonstrated notable ductile flexural behavior with greater deformability in beams where the outer boards were spliced, alongside earlier fracture of the MPCs in the tensile stress zone (Fig. 4(b)). The beams with outer board splicing exhibited the highest ductility ratio enhancement (3.1 times greater than control specimens), with rupture failure of the tensile MPCs near the ultimate flexural load (Table 1 and Fig. 4(b)). The results highlight the role of the MPCs in improving the flexural behavior of the spliced beams by enhancing ductility and maintaining acceptable stiffness, even with early rupture of the MPC in the tensile zone. These findings suggest that MPCs would be a perfect solution for the large-span glubam structures.

Finally, it is worth noting that the ultimate flexural load for both beam series decreased when incorporating either fuse bars or splicing with MPCs (Table 1 and Fig. 4). Optimizing the geometry, position, orientation, and material properties of the fuse bars and MPCs could further enhance the load-bearing capacity of thin-strip glubam beams, underscoring the need for further investigations.

3.2 Failure Mode

Fig. 5 shows the failure modes observed for both series of thin-strip glubam beams under flexural loading. In series-1, a distinct change in the failure mode was recorded from a brittle tensile failure in beams without fuse bars to a ductile compression failure with noticeable out-of-plane deformation in beams with fuse bars (Fig. 5(a)). This shift could be attributed to the contribution of fuse bars in enhancing the beam's deformability, as their yielding behavior prevented sudden failure. No signs of debonding were observed between the fuse bars, epoxy, and glubam boards, indicating a well-integrated system.

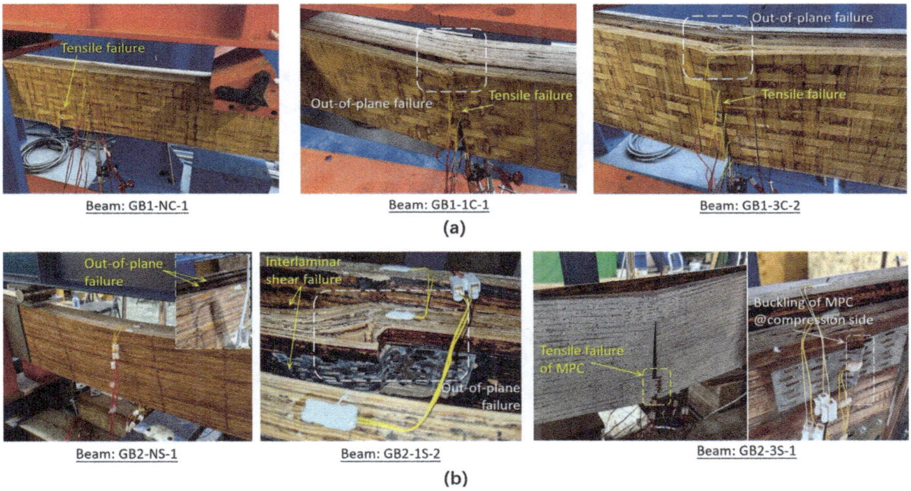

Fig. 5. Flexural failure of thin-strip glubam beams within: (**a**) Series-1; (**b**) Series-2.

In series-2, failure was characterized by a tensile rupture of the MPC on the tension side, followed by buckling for the MPC on the compression side. This was accompanied by interlaminar shear failure, as depicted in Fig. 5(a). The observed failure mechanisms highlight the structural role of the MPCs in influencing the flexural response of spliced beams.

3.3 Strain Distribution

Fig. 6 presents the strain distribution along the mid-span section of the thin-strip glubam beams at various load levels up to failure, highlighting the strain evolution in glubam, the fuse bars (Series-1), and the MPCs (Series-2). In Series-1, the pre-cracked beams with fuse bars exhibited a nonlinear strain distribution at higher load levels, indicating the yielding of fuse bars and a noticeable shift of the neutral axis towards the compression zone (Fig. 6(a)). This shift suggests significant plastic deformation in both the tensile and compression zones, contributing to the beam's increased deformability and enhanced ductility.

For the spliced glubam beams in Series-2, the tensile stress was primarily resisted by the MPCs till complete rupture at higher load levels. Beyond this point, the glubam experienced zero tensile strain at the mid-span due to separation, as shown in Fig. 6(b). In the compression zone, the glubam boards mainly sustained the compressive stress, with a minor contribution from the MPCs, which eventually led to compression buckling of the MPCs. Despite this failure mechanism, the MPC splicing technique demonstrated its effectiveness in promoting a ductile flexural response in the glubam beams.

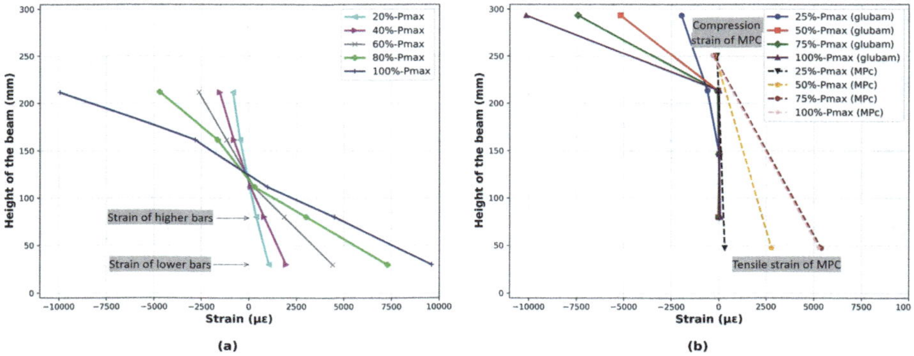

Fig. 6. Strain distribution across mid-span section of thin-strip glubam beams for: (**a**) Series-1 (beam: GB1-3C-2); (**b**) Series-2 (beam: GB2-2S-1).

4 Global Implications and Future Potential

While this investigation has focused on glubam applications in China, the findings are highly relevant to regions such as Africa, where bamboo grows abundantly yet remains underutilized in engineered construction. With many indigenous bamboo species available across Africa, glubam presents a promising solution for affordable, low-carbon housing. Nonetheless, challenges remain, particularly in terms of technological infrastructure, skilled labor, and the initial investment required for glubam processing. Future initiatives should focus on developing locally adapted manufacturing techniques to enhance cost-efficiency and expand the structural use of bamboo in resource-constrained settings.

Demonstration buildings utilizing glubam have already been successfully constructed in Kenya and Uganda [2], providing valuable proof of concept for the material's international applicability. Although these projects relied on prefabricated glubam components shipped from China, they offer a valuable precedent and indicate the potential to establish localized bamboo processing facilities in bamboo-rich African regions, enabling the construction of sustainable structures using locally sourced materials.

Beyond structural performance, the long-term durability remains a critical aspect of glubam's practical implementation, especially concerning connection techniques like MPCs under various environmental exposures. Although phenolic resin bonding provides improved moisture resistance, further research is needed to address potential joint deterioration and creep behavior under sustained loading to ensure reliable performance across different climatic conditions.

From a sustainability perspective, glubam offers compelling advantages as a rapidly renewable material with low embodied energy. The reinforcement strategies examined in this study further validate its structural viability, supporting its potential as an environmentally responsible alternative to conventional construction materials. These combined benefits position glubam as a promising solution that meets both engineering performance requirements and sustainability objectives across diverse global contexts.

5 Conclusion

This study experimentally investigated the flexural performance and ductility enhancement of thin-strip glued laminated bamboo (glubam) beams through two reinforcement strategies: (1) the incorporation of steel fuse bars in pre-cracked beams and (2) the use of spliced toothed metal plate connections (MPCs) to accommodate construction constraints. The key findings are summarized as follows:

- **Steel Fuse Bar Reinforcement:** The integration of fuse bars in pre-cracked beams significantly improved ductility, with a maximum increase of 2.5 times compared with unreinforced specimens. However, this reinforcement strategy reduced the ultimate load-bearing capacity by approximately 45%, highlighting a trade-off between strength and ductility. The results further indicate that fuse bars effectively mitigated the adverse effects of pre-cracking by promoting progressive yielding and controlled energy dissipation.
- **Performance of Spliced Beams with MPCs:** Beams spliced using toothed metal plate connections retained approximately 65% of the ultimate load capacity of un-spliced beams, with only marginal reductions in stiffness. The failure modes of spliced beams were comparable to those of un-spliced beams, except for the MPC rupture in the tension zone. Additionally, spliced beams, particularly those with outer board splicing, demonstrated enhanced deformability and improved ductility by 3.1 times compared with un-spliced beams.
- **Implications for Structural Applications:** These findings suggest that both reinforcement strategies provide viable solutions for enhancing the structural performance of glubam beams. Steel fuse bars are particularly beneficial for applications requiring improved ductility, while spliced MPC connections offer a practical approach to overcoming length limitations without severely compromising mechanical performance.
- **Future Research Directions:** Further investigations should focus on optimizing the configuration of fuse bars and MPCs to balance strength reduction and ductility enhancement.

These insights contribute to the broader application of glubam in long-span and modular construction systems, reinforcing its potential as a sustainable alternative to traditional timber materials.

References

1. Xiao Y, Yang RZ, Shan B (2013) Production, environmental impact and mechanical properties of glubam. Constr Build Mater 44:765–773
2. Xiao Y (2022) Engineered bamboo structures. CRC Press, Taylor and Francis Group
3. Xiao Y, Inoue M, Paudel SK (eds) (2008) Modern bamboo structures: proceedings of the first international conference. CRC Press
4. Li Z, He XZ, Cai ZM, Wang R, Xiao Y (2021) Mechanical properties of engineered bamboo boards for glubam structures. J Mater Civ Eng 33(5):04021058
5. Xiao Y, Wu Y, Li J, Yang RZ (2017) An experimental study on shear strength of glubam. Constr Build Mater 150:490–500
6. Chen CQ, Zhou SC, Lai ST, Xu MD, Xiao Y (2023) Impact toughness and quasi-static bending strength of glubam. J Mater Civ Eng 35(9):04023322
7. Chen CQ et al (2023) Compressive strength degradation of engineered bamboo subjected to fungal attack. npj Mater Degrad 7(1):92
8. He ZW, Xiao Y (2020) Experimental study on axial pull-out behavior of steel rebars glued-in glubam. J Mater Civ Eng 32(3):04020021
9. Xiao Y, Cai H, Dong SY (2021) A pilot study on cross-laminated bamboo and timber beams. J Struct Eng 147(4):06021002
10. Xiao Y, Zhou Q, Shan B (2010) Design and construction of modern bamboo bridges. J Bridg Eng 15(5):533–541
11. Wen J, Xiao Y (2023) The flexural behavior of cross laminated bamboo and timber (CLBT) and cross laminated timber (CLT) beams. Constr Build Mater 408:133739
12. Li H et al (2024) Comparative investigation on beam-to-beam connection with different types of connecting plates. Eng Struct 306:117790
13. Li TY et al (2023) Bending performance of nail-laminated bamboo-timber panels made with glubam and fast-grown plantation Chinese fir. Constr Build Mater 384:131425
14. Tomasi R, Parisi MA, Piazza M (2009) Ductile design of glued-laminated timber beams. Pract Period Struct Des Constr 14(3):113–122
15. Yang H, Liu W, Lu W, Zhu S, Geng Q (2016) Flexural behavior of FRP and steel reinforced glulam beams: experimental and theoretical evaluation. Constr Build Mater 106:550–563
16. Xu Q, Chen L, Harries KA, Zhang F, Wang Z, Chen X (2016) Experimental study and numerical simulation of long-term behavior of timber beams strengthened with near surface mounted CFRP bars. Mater Struct 50(1):45
17. Gentile C, Svecova D, Rizkalla SH (2002) Timber beams strengthened with GFRP bars: development and applications. J Compos Constr 6(1):11–20
18. Hamid NH, Mander JB (2014) Damage avoidance design for buildings. KSCE J Civ Eng 18(2):541–548
19. Cheng C-T, Mander JB (1997) Seismic design of bridge columns based on control and repairability of damage. National Center for Earthquake Engineering Research, Technical Report No. NCEER-97-0013
20. Vafadar F, Jaaranen J, Fink G (2024) Experimental stiffness investigation of finger joints in glued laminated timber beams using digital image correlation. Constr Build Mater 438:137095
21. Ling T, Mohrmann S, Li H, Bao N, Gaff M, Lorenzo R (2022) Review on research progress of metal-plate-connected wood joints. J Build Eng 59:105056
22. Wu JM, Xiao Y, Elise S (2016) Research on ultimate strength of toothed plate connected glubam joints. Ind Constr 46:118–123
23. Shan B, Deng JY, Wu JM, Peng Q, Li TY, Xiao Y (2022) Mechanical performance of toothed metal-plate connected glubam joints. J Mater Civ Eng 34:04022328

Bond Performance of Bamboo Strips in Bio-Concretes: Experimental and Analytical Investigation

Nathalia Andrade da Silva[1](✉), Marco Pepe[2], Enzo Martinelli[2], and Romildo Dias Toledo Filho[1]

[1] Federal University of Rio de Janeiro, Rio de Janeiro, RJ 21945-970, Brazil
nathalia.silva@coc.ufrj.br

[2] Department of Civil Engineering, University of Salerno, Via Giovanni Paollo II, 132 Fisciano, SA, Italy

Abstract. The search for sustainable, low-carbon construction materials has drawn attention to bamboo due to its renewability, fast growth, and favorable mechanical properties. This study investigates the bond behavior of bamboo strips used as reinforcement in bio-concretes, focusing on the influence of surface treatments with castor oil resin and sand. Pull-out tests were conducted to evaluate bond strength and fracture energy, and a simplified bilinear bond-slip model was applied for analytical validation. The results demonstrated that the resin-sand treatment increased the bond strength by up to 3.9 times compared to untreated bamboo, significantly improving interfacial performance. The analytical model showed good agreement with the experimental curves, especially for the reference condition (REF), and reasonably captured the behavior of treated samples (CR and CRS), despite their higher variability. The study confirms the potential of castor oil resin as an effective, bio-based treatment for improving the interface between bamboo and cementitious matrices, contributing to the development of more sustainable reinforcement systems.

Keywords: Bamboo reinforcement · Bio-concrete · Bond behavior · Pull-out test

1 Introduction

The construction sector is one of the main contributors to global greenhouse gas (GHG) emissions, driving the search for low-carbon materials and more sustainable solutions. In this context, bamboo has emerged as a promising material due to its renewability, fast growth, low carbon footprint, and its ability to sequester CO_2 during its life cycle [1–4]. In addition to its environmental advantages, bamboo also presents favorable mechanical properties, which has motivated research into its use as reinforcement in concrete elements [5–7].

The use of bamboo in bio-concretes, concretes incorporating materials of biological origin, is particularly attractive from both structural and environmental perspectives [1, 4, 8, 9]. However, for bamboo to be effectively applied as reinforcement, it is essential

H. Beushausen et al. (Eds.): NTZR 2025, LNCE 873, pp. 771–782, 2026.
https://doi.org/10.1007/978-3-032-27698-8_68

to ensure adequate bonding between the bamboo and the cementitious matrix, as bond strength is crucial for the efficient transfer of stress between materials.

To investigate this interaction, pull-out tests are widely used, allowing for the quantification of bond strength and characterization of the bond-slip relationship at the interface. These tests are especially relevant in the case of natural fibers, such as jute, sisal, and curauá, whose hygroscopic properties can lead to volumetric changes and interfacial stresses, causing debonding and microcracking [10, 11]. The same issues apply to bamboo, requiring specific treatments to mitigate water absorption effects.

Several studies have explored surface treatments to enhance bamboo-to-concrete bonding [6, 12–14], including epoxy resins with sand, industrial adhesives, surface grooving, wire wrapping, and mechanical clamps. These methods aim to increase surface roughness and promote better mechanical anchorage. However, the literature still lacks investigations on the use of castor oil-based resins as a surface treatment for bamboo strips used as structural reinforcement in bio-concretes.

In this context, the present study aims to evaluate the bond behavior between bamboo strips and a bio-concrete matrix, through pull-out tests and analytical simulation using a bond-slip model, with a focus on the influence of castor oil resin and sand treatments in improving the bamboo–matrix interface.

2 Materials and Methods

2.1 Materials

Bamboo Strips

For the production of bamboo strips, culms of the *Phyllostachys pubescens* species aged 3 to 5 years, sourced from the state of São Paulo, Brazil, were utilized. Culms from the intermediate region were selected. These culms were initially halved using a circular saw, followed by the division into rough strips, which were subsequently flattened to achieve a rectangular cross-section. The resulting strips have dimensions of 6 mm in width and 6 mm in thickness (Fig. 1).

Fig. 1. Bamboo strips.

Bio-Aggregates

The bamboo bio-aggregates were produced using waste materials sourced from companies located in the state of Rio de Janeiro, Brazil. The utilized waste materials include the following bamboo species: *Dendrocalamus asper*, *Bambusa Chungii*, *Phyllostachys pubescens*, *Guadua Chacoensis*, and *Phyllostachys aurea*.

The production of the bio-aggregates followed the steps bellow: the residues were received in the laboratory in the form of culms; the first processing stage consisted of shredding the residues in a two-shaft shredder, resulting in a coarse aggregate. Since the goal was to obtain a fine aggregate, an additional grinding process was performed using a knife mill, thus producing a bio-aggregate with smaller particles.

After the grinding process, the bio-aggregates are sieved using a mechanical sieve with a 5.60 mm mesh. All the passing material is used, while the retained residues are returned to the knife mill for a second grinding. A second sieving step is then carried out to ensure that all the processed material is utilized.

Another essential step concerning the bio-aggregates is the washing process to remove water-soluble extractives. The bio-aggregates are immersed in water heated to 80 °C for 1 h, using a biomass-to-water ratio of 1:2.5, as recommended. Afterward, the residues are air-dried, and once dry, the homogenization process is carried out using the elongated pile method.

The physicochemical properties of the material are as follows: water absorption of 65%, density of 590 kg/m^3, and moisture content of 15%.

Cementitious Materials and Additives

The cementitious matrix is composed of Brazilian Portland cement type CPV-ARI, supplied by Holcim, along with supplementary cementitious materials: fly ash from the company Pozofly and metakaolin from Metacaulim. The specific densities of the cement, fly ash, and metakaolin were 3.04 g/cm^3, 2.16 g/cm^3, and 2.66 g/cm^3, respectively.

The admixtures used were Glenium 51 superplasticizer, calcium chloride, and a Viscosity Modifying Agent (VMA), with the aim of accelerating the setting process, improving rheology, and ensuring workability.

2.2 Experimental Procedures

Bamboo Treatments

For the pull-out test, the bamboo strips were treated with Imperveg® UG 132 A resin, a plant-based polyurethane derived from castor oil. In the case of strips treated with both resin and sand, the sand used passed through a 1.68 mm sieve. The different treatment conditions applied in this study are presented and identified in Fig. 2. For the resin application, a single coat was applied to each strip. For the resin-sand treatment, manual sand spraying was performed one minute after the resin was applied. The resin was then allowed to dry for 24 h.

Production and Characterization of Bamboo-Based Bio-Concrete: Physical and Mechanical Properties

The bamboo bio-concrete (BBC) was prepared using a mixture composed of 35% bamboo bio-aggregate (BB), 40% Portland cement, 30% metakaolin, and 30% fly ash, according to the proportion proposed by Aguiar et al. [15]. In addition, 2% calcium chloride, 0.6% superplasticizer, and 0.15% viscosity modifying agent (VMA) were added, following the methodology of Andreola et al. [16]. The total water content is the sum of

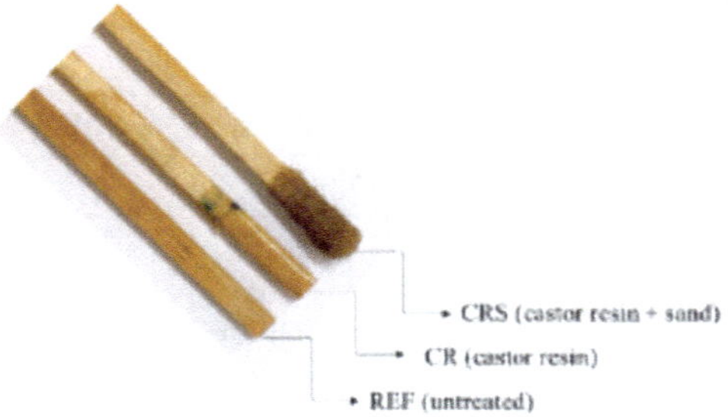

Fig. 2. Treated bamboo strips.

the water-to-binder ratio of 0.35 and the compensation water, which corresponds to the 65% absorption capacity of the bamboo bio-aggregate. The quantity of each material used is presented in Table 1.

Table 1. Mix design (kg/m^3).

Bamboo	Portanld cement	Metakaolin	Fly ash	Water	Superplasticizer	VMA	Calcium chloride
206.50	354.73	266.04	266.04	444.61	5.32	1.33	17.73

To prepare the bamboo-based bio-concrete mixture, a 5-l mixer was used. Initially, the cementitious materials were mixed together with the bamboo bio-aggregate for 1 min. Then, the mixing water previously combined with the superplasticizer (SP) was added, and the mixture continued for another 2 min. After that, the compensation water containing calcium chloride (CC) was incorporated and mixed for an additional 2 min. Finally, the VMA admixture was added, and the mixing process proceeded for 2 more minutes, totaling 7 min of mixing time.

The flow table test was performed on the fresh state of the bio-concrete in accordance with NBR 13276 [17].

After 28 days of curing, the apparent density, compressive strength, and elastic modulus of the bio-concrete were determined. To evaluate the apparent density, the samples were weighed and their dimensions measured using a caliper. The density was calculated by dividing the mass by the volume of each specimen. The compressive strength and elastic modulus were obtained in accordance with the Brazilian standards ABNT NBR 5739 [18] and ABNT NBR 8522 [19], respectively. Compression tests were performed on cylindrical specimens with 5 cm in diameter and 10 cm in height, using a loading rate of 0.3 mm/min. A total of five specimens were tested. These results are presented in Table 2.

Pull-out Tests

For the production of the specimens used in the pull-out test, cubic molds measuring

Table 2. Physical and mechanical properties of bamboo-based bio-concrete (coefficient of variation in parentheses).

Spread (mm)	Bulk density (kg/m^3)	Compressive strength (MPa)	Elastic modulus (GPa)
230	1251.29 (0.3%)	9.75 (7.0%)	3.5 (9.0%)

50 × 50 mm were used. The treated bamboo strips were centrally glued inside the molds (Fig. 3a), and then the bamboo bio-concrete was cast (Fig. 3b). The entire assembly was vibrated for 30 s. After 24 h, the specimens were demolded (Fig. 3c) and cured at a temperature of 22 ± 3 °C and a relative humidity of 55 ± 5%. Four specimens were tested for each variable, and the procedure followed the methodology described in reference [14].

(a) (b) (c)

Fig. 3. Pull-out test: (**a**) mold with bamboo strips glued and centered; (**b**) casting of bamboo bio-concrete into the molds; and (**c**) specimen after 28 days of curing.

After 28 days of casting, the specimens were subjected to the pull-out test using a Shimadzu AG-X testing machine with a capacity of 100 kN and a loading rate of 1 mm/min, as illustrated in Fig. 4.

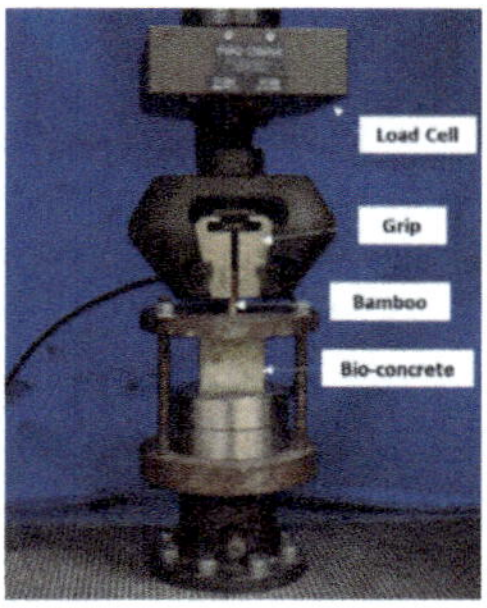

Fig. 4. Pull-out test setup.

Analytical Model of the Bamboo–Bio-Concrete Interface with Inverse Identification of the Bond Law

The bond behavior between the bamboo strip and the bio-concrete matrix was modeled using a continuous interface formulation. This approach was primarily based on the formulation proposed by [10], which was originally developed for natural fibers. Additional considerations and adaptations were made based on the experimental insights presented in [20]. The following assumptions were adopted:

- The bamboo is assumed to behave as a linear elastic material, with modulus of elasticity E_f, cross-sectional are A_f and perimeter P_f;
- The bio-concrete matrix is considered perfectly rigid.

The interaction between the bamboo and the matrix is described by a generalized bilinear bond-slip law, defined in Eq. 1 and illustrated in Fig. 5.

$$\mathrm{T(s)} = \left\{ \begin{matrix} \mathrm{k_{el}}.s \ \mathit{if} \ |\mathrm{s}| \le \mathrm{s_{el}} \\ \tau_\mathrm{r} - \mathrm{k_{in}}.(\mathrm{s} - \mathrm{s_{el}}) \ \mathit{if} \ \mathrm{s_{el}} < |\mathrm{s}| \le \mathrm{s_u} \\ 0 \ \mathit{if} \ |\mathrm{s}| > \mathrm{s_u} \end{matrix} \right\} \tag{1}$$

where $k_{el} = \tau_{max}/s_{el}$ is the slip modulus of the elastic branch ending at a slip value s_{el} with a corresponding bond stress τ_{max}, τ_r is the residual bond stress, and k_{in} is the post-peak slip modulus (strictly positive), resulting in a linear variation of stresses from τ_r to τ_u, the latter being achieved for a slip s_u.

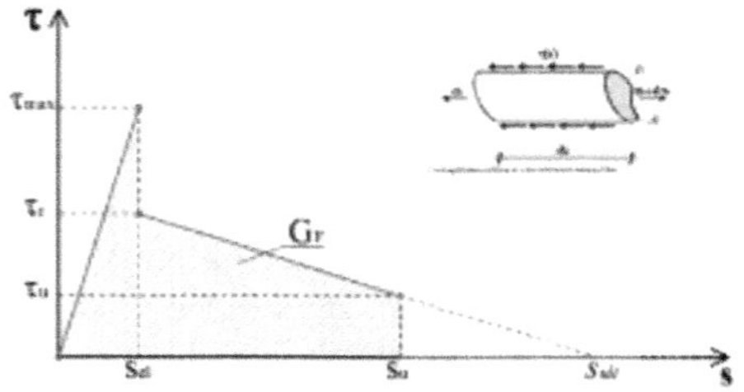

Fig. 5. Bilinear bond-slip relationship and equilibrium of a segment fiber element [11].

The distribution of slip along the embedded length of the strip, *s(z)*, is obtained from the differential equation resulting from the axial equilibrium of the strip and the compatibility of deformations:

$$\frac{\mathrm{d^2s}}{\mathrm{dz^2}} = \frac{\mathrm{P_f}}{\mathrm{E_f A_f}}.\tau\ (\mathrm{s}) \tag{2}$$

The piecewise solution of this equation allows the simulation of different stages of the pull-out test (elastic, softening, and debonding). In the purely elastic regime, the force applied at the loaded end of the fiber is given by:

$$\mathrm{F_0} = \mathrm{E_f A_f a_{el}.s_0}.\tanh(\mathrm{a_{el}L}),\ \mathrm{com}\ \mathrm{a_{el}} = \sqrt{\frac{\mathrm{P_f k_{el}}}{\mathrm{E_f A_f}}} \tag{3}$$

where s_0 is the slip at the loaded end and L is the embedded length.

To determine the parameters $[s_{el}\ \ s_u\ \tau_{max}\ \tau_r\ \tau_u]$ that define the bond-slip law shown in Fig. 5, an inverse identification process was employed. This was done by minimizing the objective function:

$$\Delta(q) = \sum_{i=1}^{n} \left[F_{sim}(s_i; q) - F_{exp}(s_i)\right]^2 \quad (4)$$

where **q** represents the parameter vector, F_{sim} is the force simulated from the analytical, and *Fexp* is the experimentally measured force.

3 Results and Discussion

3.1 Pull-Ou Tests

The pull-out tests were conducted using three specimens for each evaluated condition. Fig. 6 presents the average curves obtained for each group, while the numerical results are summarized in Table 3.

Analyzing the curves, it is observed that, in the initial phase, there is a linear relationship between the applied force and the displacement. This behavior persists until the bond stress reaches its maximum value, marking the onset of debonding, with the rupture of the chemical bond.

After the peak, the curve enters the post-peak phase, characterized by a drop in the pull-out force. At this stage, both chemical and mechanical adhesion have been overcome, and the resistance to extraction becomes governed primarily by friction between the surface of the strip and the matrix. This frictional interaction is responsible for the presence of residual force along the slip, evidenced by a curve with relatively constant or gradually decreasing values.

The frictional behavior is related to the surface roughness of the strip and the presence of solid particles, especially in the samples with surface modification (CR and CRS).

Compared to the reference condition (REF), which consisted of untreated bamboo strips, the use of castor oil-based resin (CR) resulted in a 119% increase in the maximum pull-out force (F_{max}). When castor resin was combined with sand (CRS), the increase was even more significant, reaching 276%, which corresponds to a force approximately 3.76 times greater than that of the REF group.

This improvement directly affected the maximum bond stress (τ_{max}), which increased by 120% for CR and 279% for CRS compared to REF. Notably (τ_{max}), nearly quadrupled, rising from 0.39 MPa (REF) to 1.48 MPa (CRS).

These results indicate a significant enhancement in the bond performance between the bamboo strip and the cementitious matrix, demonstrating that the load transfer mechanisms at the interface were substantially improved by the applied treatments, particularly in the CRS condition.

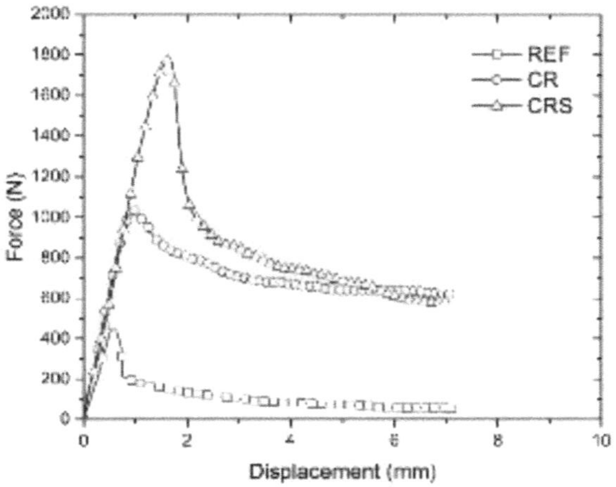

Fig. 6. Typical curves of the pull-out test.

Table 3. Experimental results of the pull-out test (coefficient of variation in parentheses).

	F_{max} (N)	s_u (mm)	τ_{max} (MPa)
REF	472.67 (1.6%)	9.44	0.39 (1.6%)
CR	1034.98 (8.7%)	18.96	0.86 (8.7%)
CRS	1778.97 (10.3%)	37.18	1.48 (10.3%)

3.2 Analytical Model

By applying the analytical model to the experimental results shown in Fig. 7, the curves presented in Figs. 7, 8, and 9 were obtained, along with the data reported in Table 4. These figures demonstrate that the simplified bond-slip law adopted (Fig. 6), together with the inverse identification procedure, was able to accurately reproduce the mechanical behavior observed in the experiments. The simulated results show good agreement with the global force-slip response of the tested specimens, enabling the effective identification of the corresponding bond-slip laws.

Based on the results, it is possible to evaluate key mechanical parameters such as the fracture energy (G_F) and the initial interface stiffness (k_{el}). The REF condition, using untreated strips, showed the lowest G_F value (1.10 N/mm) due to poor adhesion and minimal energy dissipation. In contrast, the CR and CRS conditions, which included castor oil resin and sand addition, exhibited higher adhesion, greater slip, and residual force, resulting in significantly higher G_F values, with CRS reaching the highest (19.10 N/mm). These findings demonstrate that surface modifications greatly enhance the energy dissipation capacity of the interface. Furthermore, the higher k_{el} values in CR and CRS samples indicate stiffer interfaces with more efficient load transfer.

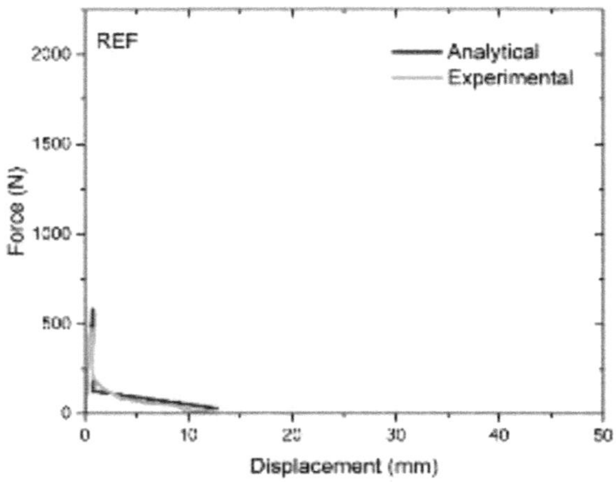

Fig. 7. Experimental and analytical results for REF.

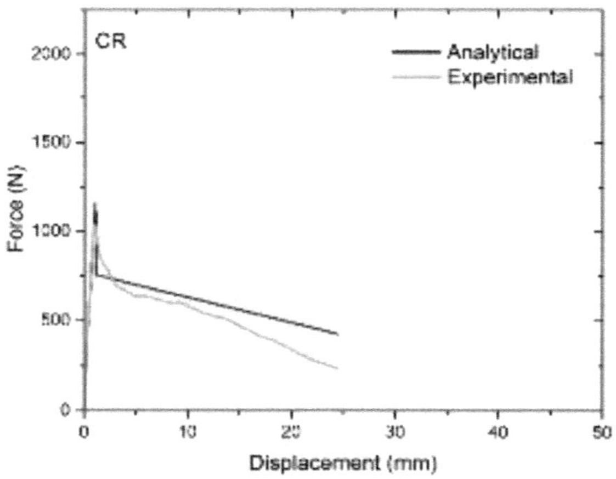

Fig. 8. Experimental and analytical results for CR.

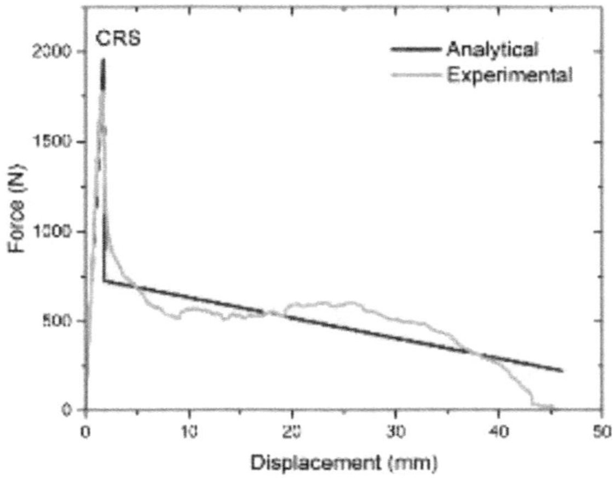

Fig. 9. Experimental and analytical results for CRS.

Table 4. Bond-Slip law for bamboo in bamboo bio-concrete.

	s_e (mm)	s_u (mm)	τ_{max} (MPa)	τ_r (MPa)	τ_u (MPa)	G_F (N/mm)	k_{el} (MPa/mm)
REF	0.64	12.78	0.44	0.12	0.04	1.10	0.69
CR	1.04	24.58	1.01	0.63	0.35	12.13	0.97
CRS	1.75	46.14	1.74	0.61	0.18	19.10	0.99

4 Conclusion

This study evaluated the bond behavior between bamboo strips and bamboo-based bio-concrete matrices, using pull-out tests and analytical simulations based on a simplified bond-slip model. The main findings are summarized as follows:

- The pull-out test showed that surface treatment with castor oil resin and sand increased the bond strength by up to 3.9 times compared to untreated strips, highlighting the importance of interface modification to enhance the system's performance.
- The comparison between experimental and analytical results confirmed the effectiveness of the adopted bond-slip model in reproducing the mechanical behavior observed in the tests.
- For the reference condition (REF), the analytical curve showed excellent agreement with the experimental data, especially in the initial loading and post-peak phases, which reflects lower energy dissipation and minimal interfacial variability.
- For the modified conditions (CR and CRS), although some local discrepancies were observed, particularly in residual force levels and displacement—the simplified bilinear bond-slip law still provided a good approximation of the global force-slip response. These differences are attributed to greater surface variability and frictional mechanisms at the treated interfaces.

- The accuracy of the analytical model was further supported by the strong correlation between the initial stiffness parameter (k_{el}) and the fracture energy (G_F), especially in the CRS condition, which exhibited the highest force and energy values, despite its more complex interfacial behavior.

Acknowledgments. The authors gratefully acknowledge the financial support of the Brazilian institution CAPES (Coordenação de Aperfeiçoamento de Pessoal de Nível Superior) – Finance Code 001. In addition, we gratefully acknowledge the financial support from the European Commission through the MSCA under GA ID: 101086440, which has enabled the realization of the BEST (Bio-based Energy-efficient Materials and Structures for Tomorrow) project: https://doi.org/10.3030/101086440. This work reflects the views and opinions of the authors only, and neither the European Union nor the granting authority can be held responsible for any use that may be made of the information contained herein.

References

1. da Silva NA, Andrade N et al (2024) Bamboo bio-concretes: a literature review of 8 years of laboratory researches at NUMATS/UFRJ. Procedia Struct Integr 64:1460–1467
2. Gan J et al (2022) Life cycle assessment of bamboo products: review and harmonization. Sci Total Environ 849:157937
3. Caldas LR et al (2020) Bamboo bio-concrete as an alternative for buildings' climate change mitigation and adaptation. Constr Build Mater 263:120652
4. Gangorra LCC et al (2024) Análise das propriedades físico-mecânicas de bioconcretos de bambu produzidos com partículas finas em substituição parcial aos bio-agregados. Matéria (Rio de Janeiro) 29(4):e20240364
5. Gunasti A (2023) Experimental evaluation of axial compression performance of precast panels from bamboo-reinforced concrete. Appl Eng Sci 16:100155
6. Widayanto E, Wiswamitra KA, Subchan BR (2025) Flexural behaviour of bamboo-reinforced concrete beams soaked in salt solution and molasse treatments and wrapped in fibreglass tape: an experimental and numerical study. Adv Bamboo Sci 11:100147
7. Karakus-Zambak O, Celik OC (2025) Behavior of bamboo reinforced concrete (BRC) beams under monotonic and dynamic loads. Constr Build Mater 458:139683
8. Araujo A et al (2022) Potential of earth-based bamboo bio-concrete in the search for circular and net-zero carbon solutions to construction industry. In: IOP conference series: earth and environmental science, vol 1122. IOP Publishing
9. Andreola VM et al (2023) Biological durability of bamboo bio-concretes. In: International conference on bio-based building materials. Springer, Cham
10. Ferreira SR et al (2016) Inverse identification of the bond behavior for jute fibers in cementitious matrix. Compos Part B 95:440–452
11. Ferreira SR et al (2018) Influence of natural fibers characteristics on the interface mechanics with cement based matrices. Compos Part B 140:183–196
12. Agarwal A, Nanda B, Maity D (2014) Experimental investigation on chemically treated bamboo reinforced concrete beams and columns. Constr Build Mater 71:610–617
13. Bala A, Gupta S (2023) Engineered bamboo and bamboo-reinforced concrete elements as sustainable building materials: a review. Constr Build Mater 394:132116
14. da Silva NA et al (2019) Analysis of bond-behavior of bamboo strips and bamboo bio-concrete. In: Non-conventional materials and technologies (NOCMAT), Nairobi, Kenya

15. Aguiar ALD et al (2024) Experimental investigation on the fire performance of wood bio-concrete using cone calorimeter. Fire Saf J 148:104225
16. Andreola VM, Hasparyk NP, Toledo Filho RD (2024) Effect of natural weathering on the mechanical strength of bamboo bio-concrete. Buildings 14(11)
17. Associação Brasileira de Normas Técnicas (ABNT) (2005) NBR 13276: Argamassa para assentamento e revestimento de paredes e tetos—Preparo da mistura e determinação do índice de consistência. Rio de Janeiro
18. Associação Brasileira de Normas Técnicas (ABNT) (2018) NBR 5739: Concreto—Ensaio de compressão de corpos de prova cilíndricos. Rio de Janeiro
19. Associação Brasileira de Normas Técnicas (ABNT) (2021) NBR 8522-1: Concreto endurecido—Determinação dos módulos. Rio de Janeiro
20. Caggiano A, Martinelli E, Faella C (2012) A fully-analytical approach for modelling the response of FRP plates bonded to a brittle substrate. Int J Solids Struct 49(17):2291–2300

Assessment of Economical Foundation Design for Ancillary Structures Based on Field Data

Mahtab Goudarzi and Alice Alipour(✉)

Department of Civil, Construction, and Environmental Engineering, Iowa State University, Ames, IA, USA
alipour@iastate.edu

Abstract. Wind loads prescribed for the design of cantilevered traffic sign structures are frequently conservatively high, resulting in overdesigned drilled shaft foundations that are neither cost-efficient nor sustainable. This study investigates the problem by directly quantifying the moments and torsional demands transferred from the superstructure to the foundation and subsequently applying these measurements to optimize foundation design. Two full-scale cantilevered sign structures were instrumented with rosette strain gauges and accelerometers to capture in-situ structural response under service conditions. Complementary static pull tests and free-vibration tests were performed to characterize load–deformation behavior and dynamic properties. The field data were used to calibrate and validate numerical structural models by comparing predicted responses with measured responses. Once validated, the models were subjected to synthetically generated extreme wind loads to evaluate load transfer mechanisms during severe wind events. Results demonstrate that reduced foundation sizes are sufficient to resist wind-induced demands while satisfying safety and serviceability criteria. These findings suggest that current design methodologies, which rely on highly conservative load and resistance factors, can lead to systematic overdesign. By integrating full-scale testing, advanced instrumentation, and calibrated numerical modeling, this study establishes a validated analytical framework for the rational design of foundations supporting cantilevered traffic sign structures. The proposed approach enables more efficient use of materials, reduces construction complexity, and lowers overall project costs without compromising structural reliability.

Keywords: Cantilevered traffic sign structure · Wind loading · Load transfer · Bending moment · Torsion · Long-term structural health monitoring

1 Introduction

Traffic sign structures are essential roadway assets that must withstand various dynamic loads [1]. Among them, cantilevered traffic sign structures are widely used on highways due to their simplicity and cost-effectiveness [2]. The increasing intensity and variability of environmental conditions highlight the importance of designing transportation infrastructure that can withstand a wide range of extreme loadings [3–5]. In this regard, multiple studies emphasize that wind can affect their design and performance. Wind

H. Beushausen et al. (Eds.): NTZR 2025, LNCE 873, pp. 783–795, 2026.
https://doi.org/10.1007/978-3-032-27698-8_69

loads on slender structures can induce not only large bending moments but also significant torsional forces, especially in structures with asymmetric configurations [6]. These combined bending and torsional demands are ultimately transferred into the foundation, typically a drilled shaft, which must resist the resulting overturning loads. In fact, deep foundations for sign support systems are often governed by extreme wind load combinations, requiring substantial size and strength to prevent failure [7]. While this ensures safety under rare events (design wind speeds can reach 115–170 mph (52–76 m/s) in some regions [7]), it can lead to overly overdesigned foundations that are not economical. With over a million linear feet of drilled shafts constructed annually in the U.S. at a cost of billions of dollars there is a strong incentive to optimize foundation designs by aligning them with actual demand [8]. Based on the provided explanation, researchers have investigated wind effects on traffic sign structures through both experimental and numerical studies. Experimentally, wind tunnel tests and full-scale field tests have been used to study wind-induced forces and vibrations [9]. According to Zuo et al. [10], cantilevered mast arm traffic signal structures are vulnerable to significant vibrations under specific wind conditions. Such a situation raises concerns regarding structural endurance and performance. In addition, this result highlights how field testing might reveal essential wind-induced behavior. Tsai and Alipour [11], through long-term monitoring of a traffic signal structure in Ames, Iowa, further highlighted the significance of investigating wind-induced vibrations. In another study performed by Paiva and Barros [12], it is demonstrated that cantilevered sign structures are subject to wind and gravity loads at the same time. Elements in such a structure might experience complicated stress distributions as a result of combined loading during biaxial bending and torsion. Specifically, vertical and horizontal elements function as beam-columns when subjected to bending moments, torsional stresses, and axial loads, requiring a comprehensive understanding of the wind load effect. Numerical studies have also provided valuable insights into the complex interactions of traffic sign structures under the wind loads. In this regard, Jafari et al. [13] demonstrated the need for effective modeling methodologies to mitigate wind-induced vibrations in traffic signal structures without compromising structural safety. These investigations have advanced knowledge of superstructure behavior, but limited studies have addressed the foundation's contribution to wind load resistance. In recent years, studies have increasingly examined foundation response and the combined behavior of superstructure and foundation systems under wind loading. Studies have shown that typical foundation design methods consider lateral and torsional loads separately, even though cantilevered signs apply both loads at the same time because of their L-shaped geometry [7]. McVay et al. [14] found that analyzing foundation stability for torsion and lateral loads separately is unsafe, as this approach may ignore their interaction effects. Current design guidelines lack specific provisions for combined lateral–torsional loading on deep foundations supporting signs [7]. Finite-element analyses of foundations under simultaneous bending and torque are exceedingly rare, and most available methods do not capture torsional behavior [7, 8]. To bridge this gap, some researchers have performed full-scale field tests on sign support foundations. For example, Boeckmann et al. [15] instrumented and laterally loaded pile-supported sign foundations to investigate load transfer under simulated wind, finding that design methods which consider only axial or uplift capacity can be overly conservative. Li et al. [16]

performed full-scale torsional load transfer tests on drilled shaft foundations and developed methodology to quantify the relationship between applied torque and rotational displacement, evaluate unit side resistance profiles, and estimate total torsional resistance. Their study demonstrated that torsional resistance is influenced by both soil type and depth, and that ignoring torsional contribution in design can underestimate foundation capacity. Such research highlights the importance of directly measuring torsion in real foundations to calibrate design methods. Furthermore, site investigations often reveal actual soil properties that exceed the conservative values assumed in design, suggesting many foundations have higher capacity than needed [16]. This disparity between design assumptions and reality reinforces the need for field validation and potential recalibration of foundation design criteria. Despite previous research, there is still a lack of full-scale evidence on how wind-induced moments, and torsion are transferred in cantilevered traffic sign structures, particularly in ways that could support more economical and sustainable foundation designs. Based on the aforementioned explanation, the current study investigates two full-scale cantilevered traffic sign structures, each supported by drilled shaft [18]. Through static pull tests, free-vibration tests, and synthetic wind simulations, the research quantifies the bending moments and torsional loads that wind creates in the superstructure and passes to the foundation top. By applying synthetically generated high wind loads to the validated structural model, the research demonstrates that smaller foundation systems can still safely resist these demands, providing field-based evidence to challenge overly conservative design assumptions and support more economical, sustainable foundation designs. Figure 1 shows the structures.

(a) (b)

Fig. 1. **a** Structure 1 on granular soil and, **b** Structure 2 on cohesive soil

2 Methodology

Two full-scale structures located in Madison, Wisconsin, were selected for testing. The site-specific subsurface conditions were characterized through geotechnical borings and laboratory tests to inform the soil characterizations and interpretation of field measurements. Lab results revealed contrasting foundation conditions that informed both testing and modeling. Structure 1 was underlain predominantly by granular soils, with a surface layer of lean clay (UCS 1.5–3 tsf [144–287 kPa]) over clayey sand, silty sand, and clayey sand with gravel and cobbles. Structure 2 was located on a more cohesive profile, beginning with lean clay underlain by a layer of poorly graded sand with gravel, lean clay, silty sand with gravel, and clayey sand with gravel and cobbles. Moisture contents ranged from 11.6–15.3% in the granular soils at structure 1 to as high as 40.9% in the surficial fill at structure 2, with measured moist unit weights between 120 and 149 pcf (19.0–23.4 kN/m^3) for structure 1 and 125 to 146 pcf (19.6–22.9 kN/m^3) for structure 2. In both sites, granular layers exhibited friction angles of 29–36°, exceeding the design assumption of 24°, while cohesive soils at structure 1 had higher cohesion (2000 psf [96 kPa]) than assumed and those at structure 2 matched the assumed 750 psf (36 kPa). These findings indicate that both sites possess foundation resistance equal to or greater than the original design assumptions.

Following the geotechnical investigations, structures were constructed using drilled shaft foundations designed in accordance with AASHTO LRFD Bridge Design Specifications [17] for a wind velocity of 115 mph (ultimate). Both structures utilized the same foundation with a 48-in. (1.22 m) diameter drilled shaft embedded 28 ft. (8.53 m) into the ground. The concrete achieved compressive strengths of 5762 psi (39.7 MPa) at 7 days and 7032 psi (48.5 MPa) at 14 days, well above the design assumption of 3500 psi (24.1 MPa). Installation for both structures followed the same procedure: drilling with an excavator-mounted auger and temporary casing to maintain borehole stability, lowering the reinforcement cage into position, and placing concrete. After the foundations were completed and the steel superstructures constructed, essential sensors were installed to obtain structural response. These included a load cell to directly measure applied forces during static pull tests and rosette strain gagues 12 in. Above the base of the structure to measure moments, torque, and shear in multiple directions. In addition, accelerometers were mounted at the mast arm tip to record structural motion in three orthogonal directions, during free-vibration testing. Using the collected material properties, and geometric details, a structural model was developed for each structure to represent its behavior under both static and dynamic loading tests.

With the site conditions defined and the structures in place, controlled testing was undertaken to capture the response data necessary for validating the structural models. Torsional and lateral static pull tests were conducted by applying controlled forces to the structures using a cable in both upstream (front face toward approaching traffic) and downstream (back face toward departing traffic) directions. For torsional loading, the cable was attached at a horizontal offset from the pole centerline, while for lateral loading, it was attached at the top of the pole. The specific target loads, along with the cable attachment locations and offsets for each structure, are summarized in Table 1, where X represents the horizontal distance and Z represents the vertical distance of the cable from the reference point. Load application points were selected to replicate wind loading patterns. For lateral loading, cables were attached directly at the pole head with no horizontal offset, generating lateral forces without induced torsion. Torsional loads were applied by connecting cables to the mast arm adjacent to the sign panels, simulating the eccentric wind loading that occurs when wind strikes the signage offset from the pole centerline. Vertical load components corresponded to the actual elevations of the attachment points above the foundation. A crane-cable system applied target loads measured by the in-line load cell, with load magnitudes selected to represent service conditions.

Table 1. Pull test setup summary

Structure	Torsional loading test setup	Lateral loading test setup
Structure 1	X = 228 in. (5.8 m) Z = 285 in. (7.2 m) F_{max} = 2 kips (8.9 kN)	X = 0 in Z = 285 in F_{max}= 4 kips (17.8 kN)
Structure 2	X = 144 in. (3.66 m) Z = 285 in F_{max}= 7 kips (31.1 kN)	X = 0 in Z = 285 in F_{max} = 5 kips (22.2 kN)

Free-vibration tests were then performed by initiating motion through a sudden release after the structure was displaced. Then, acceleration data was processed to determine the natural frequencies using Power Spectral Density (PSD) function to compare with modal analysis on the structural models.

Data recorded from instruments installed on the superstructure were analyzed to compare the structure's base reactions and natural frequencies with the results of the structural models, providing validation of the models under both static and dynamic conditions. The results of these validation steps are presented in the results section. Once validated, the models were then ready for the application of synthetically generated high wind loads to assess structural demands and explore optimized foundation designs. After model validation, synthetically generated extreme wind loads were applied to establish the base actions at the top of the shaft, bending moment in the x-direction (M_x), torque (T), and shear (V). These reactions used as inputs to laterally load pile analyses with site-specific soil profiles and p–y curves in L-Pile to evaluate serviceability and geotechnical capacity under LRFD. For each site, drilled-shaft diameter and embedment length were systematically reduced and reanalyzed; candidates satisfying service limits were then checked for reinforced concrete strength and detailing under combined load per LRFD. The objective was to identify smaller foundation options, including reduced-size versions of Wisconsin Department of Transportation (WisDOT) standard foundations [1], that satisfy all serviceability, geotechnical, and structural requirements under LRFD. Results for the synthetically generated extreme wind loads and the performance of reduced-size foundation options are presented in the Results section.

3 Results and Discussion

3.1 Structural Model Validation Using Field Data

Among the four pull tests performed for each structure, the torsional downstream case for cohesive site and the torsional upstream case for granular site, governed the base reactions and is therefore presented as the representative comparison for both structures. Results including measured and model values for base bending moment, torsional moment, and shear, with the maximum percent difference for the critical test is summarized in Table 2. Free-vibration test provided natural frequencies that were compared with the model's modal analysis. Figure 2 presents natural frequencies for the first three modes for each structure, with the percent difference between measured and model labeled above each pair. Taken together, the comparisons show that the structural models predict the base actions and overall dynamic behavior well, supporting their use for understanding structural response under synthetically generated extreme wind.

Table 2. Static validation: reactions at top of the shaft (field measurement vs. model)

Shaft-head reaction	Structure 1	Difference (%)	Structure 2	Difference (%)
T_p (k-in) (kN m)	588 (66)	14	978 (110)	3
T_m (k-in) (kN m)	677 (77)		1084 (122)	
M_p (k-in) (kN m)	598 (68)	15	1931 (218)	15
M_m (k-in) (kN m)	510 (58)		2268 (256)	
V_p (k) (kN)	2.23 (10)	12	7.3 (32.5)	1
V_m (k) (kN)	2.5 (11)		7.2 (32)	

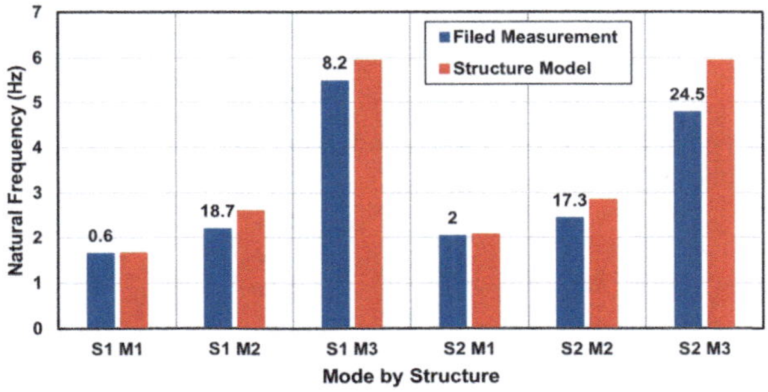

Fig. 2. Comparison of measured and modeled natural frequencies for the first three modes of each structure

3.2 Synthetic Extreme Wind on Validated Model

A synthetic wind was generated with mean speed increasing from 20 to 80 mph, resolved into perpendicular and tangential components and applied as point and distributed loads to the sign, pole, mast arms, and truss in the validated structural models for both structures. Identical wind time histories were used for the two sites to enable a direct comparison of structural response. For each run, the shaft-head M_x, T, V time histories were recorded. The structural response at the shaft head for design wind level (ultimate wind 115 mph) for both structures shown in Fig. 3. The maximum of M_x, T, V for different wind levels are summarized in Table 3. A direct comparison of the maximum base actions is provided in Fig. 4, which presents grouped bars of M_x, T, V for structure 1 (granular site) and structure 2 (cohesive site). Differences between structures in bar chart primarily reflect their geometry, mass, and stiffness under identical wind loading; the soil profile influences performance of the foundation (serviceability and capacity). The shaft head responses used as inputs for the foundation checks that follow, where serviceability is assessed using p–y analysis and structural adequacy is verified under combined M_x-T-V in accordance with LRFD.

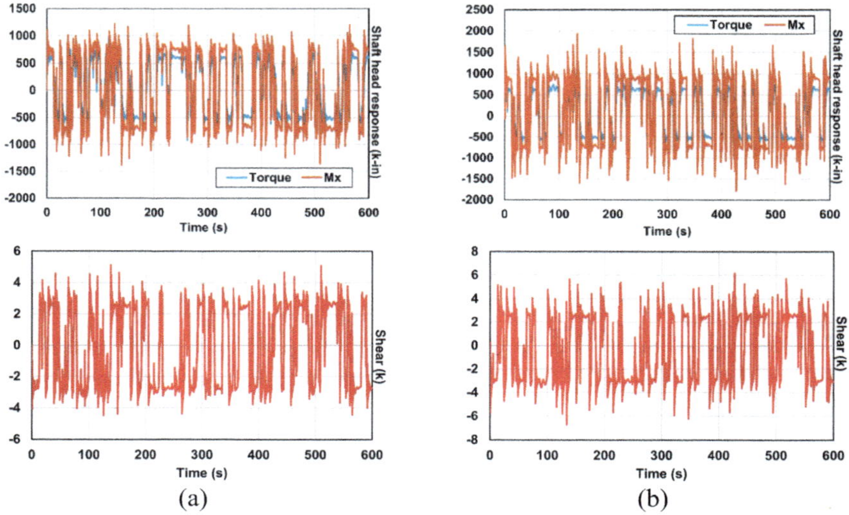

Fig. 3. Shaft-head bending moment, torque, and shear under synthetic extreme wind event **a** Granular site, **b** Cohesive site

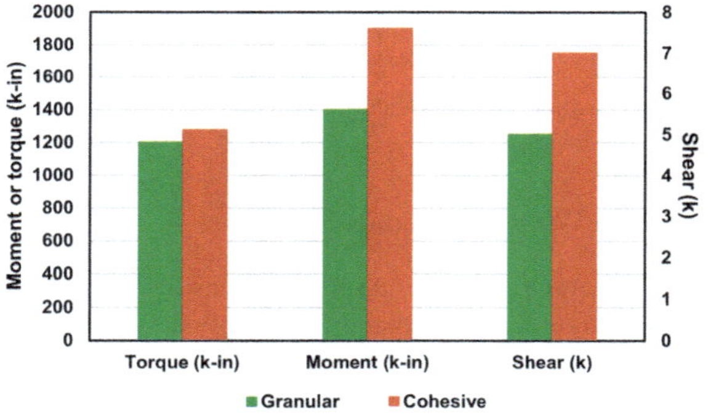

Fig. 4. Direct comparison of shaft-head bending moment, torque, and shear under synthetic extreme wind (maximum 115 mph)

Table 3. Shaft-head responses under extreme synthetic wind

Average wind speed (mph)	Maximum wind speed (mph)	Shaft-head torque (k-in [kN m])—granular	Shaft-head torque (k-in [kN·m])—cohesive	Shaft-head moment (k-in [kN·m])—granular	Shaft-head moment (k-in [kN·m])—cohesive	Shaft-head shear (k[kN])—granular	Shaft-head shear (k[kN])—cohesive
20	40	112 (3)	110 (12)	138 (16)	163 (18)	0.5 (2.2)	0.6 (2.7)
40	70	277 (31)	349 (40)	333 (38)	494 (56)	1.2 (5.3)	1.7 (7.6)
60	95	782 (88)	828 (94)	957 (108)	1147 (130)	3.6 (16)	4 (18)
80	115	1188 (134)	1275 (144)	1390 (157)	1925 (217)	5.1 (23)	6.7 (30)

Using the same synthetic wind model for both sites, cohesive site produced higher peak shaft-head reactions than granular site according to Table 3. This outcome is reasonable because of the moments, and torque scale according to the offset and height of the load. Even minor differences in height or eccentricity increases M_x or T. The overall height of structure 2 is approximately 1 ft. (0.3 m) greater than that of structure 1. The structure at the cohesive site presents a larger effective projected area, particularly from the sign panels, which capture most of the wind load, resulting in higher aerodynamic loading. To benchmark the synthetic wind response against the design basis, shaft-head reactions from the validated models were compared with the design with gravity values (M_x = 4668–4824 kip-in [528–545 kN m]; T_z = 996 kip-in [112 kN m]). Under the extreme wind event, peak bending moments were about 60% lower than the design values, while torque was about 21% higher. Because bending moment is well below design, the bending-controlled shaft size is not required. A smaller foundation may therefore be appropriate; however, higher torque becomes the critical check. Accordingly, any reduction in foundation size is considered only if serviceability and torsional capacity are satisfied.

3.3 Geotechnical and Structural Performance of Alternative Foundation Designs

Using the shaft-head moments and shear from the validated models under the synthetic extreme wind event, drilled-shaft candidates were evaluated in L-Pile for pile-head lateral deflection (Δ) and rotation (θ) at both sites. Foundation alternatives (geometry used in L-Pile) are mentioned in the following:

- Current: Diameter (D) = 4.0 ft. (1.22 m), Length (L) = 28 ft. (8.53 m)
- Alternative 1: D = 4.0 ft., L = 23 ft. (7.01 m)
- Alternative 2: D = 3.5 ft. (1.07 m), L = 23 ft.
- Alternative 3: D = 3.5 ft., L = 18 ft.
- Alternative 4: D = 3.0 ft. (0.91 m), L = 20 ft. (6.10 m)

Consistent with deep foundation serviceability, acceptance was based on a top-of-foundation deflection limit of Δ_{allow}= 0.5 in, θ_{allow}= 10 mrad; θ does not govern Pass/Fail. Because θ does not govern for either site, lateral deflection is the critical geotechnical check for these alternatives. Site soils and geometry are different, so identical wind records yield different base responses and, consequently, different serviceability outcomes. On the cohesive site, only the current foundation satisfies serviceability. Reduced options exceed the Δ_{allow} indicating deflection governs at second structure. However, the near-miss options for the second structure (alternative 1, and alternative 2) suggest that modest length or diameter increases would recover compliance. At the granular site, all options meet Δ_{allow}, enabling reducing the foundation size. Figure 5 consolidates Δ for both sites with a 0.5-in. Reference. Granular site admits several reduced foundations while preserving serviceability, reflecting its higher lateral stiffness from subsurface conditions. Figure 6 shows representative L-Pile lateral-deflection profiles in different depths for the possible alternative foundations that pass serviceability. Profiles at the cohesive site display larger head deflections, whereas the granular profiles remain stiffer with smaller head deflections. The value at depth = 0 ft. in Fig. 6 corresponds to the bar heights shown in Fig. 5.

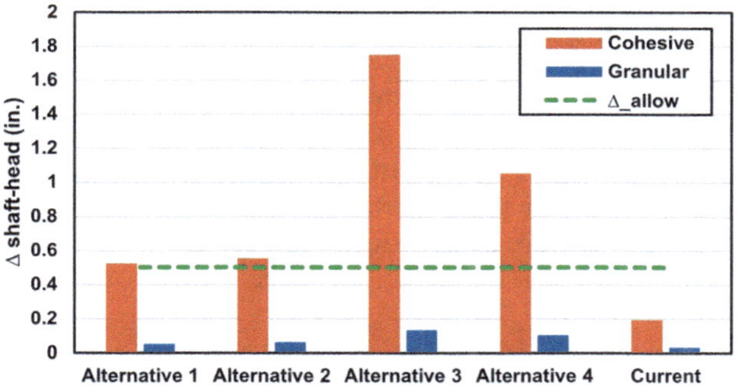

Fig. 5. Serviceability of current and alternative foundations under synthetic extreme wind (Δ with 0.5-in. Line)

Lateral head displacement is primarily controlled by the near-surface lateral stiffness of the soil profile. Results show that the effects of sign plate size on the wind load are high; therefore, foundations should be designed based on the possible wind load and soil profile for each area. The cohesive site with lower initial subgrade modulus produces larger head deflections, whereas the granular site with higher friction angle and modulus yields smaller deflections. Consequently, downsizing is more robust in granular soils, while cohesive soils generally require L/D adjustments to satisfy serviceability. Geotechnical serviceability has identified feasible or near-feasible foundation sizes based on head-deflection, the structural adequacy of the shaft head under combined factored bending, shear, and torsion should be verified to validate the feasible reduced size foundations.

Figure 7 shows demand to capacity ratios at the head section for feasible alternatives that passed serviceability. All candidates are well left of the flexural unity line ($M_u/\varphi M_n \leq 0.2$), indicating bending capacity does not control optimizing the foundation size. Variability appears in torsion; most options remain below the torsional unity limit, whereas the smallest diameter case on the granular site (Alternative 4) has $T_u/\varphi T_n \geq 1$ identifying torsion as the governing strength check for candidates.

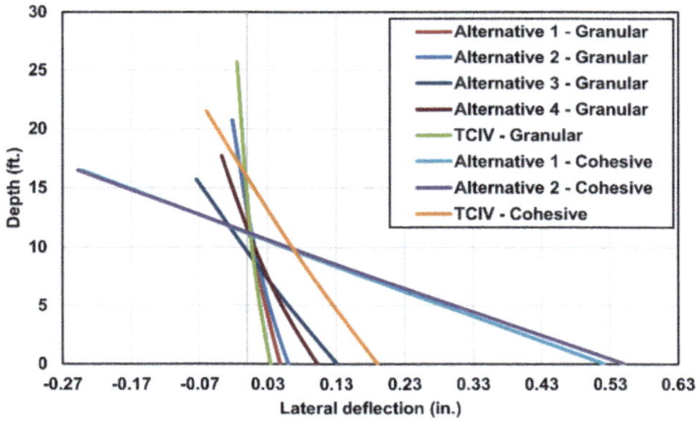

Fig. 6. L-Pile lateral deflection profiles with depth (granular vs. cohesive Sites)

In the granular profile, diameter and length can be reduced until the torsion utilization approaches 1.0. A practical choice is alternative 3, which keeps reducing concrete volume by about 37% relative to the current design. In the cohesive profile, head deflection governs, and the reduced sizes exceed Δ_{allow}, so the current foundation remains the smallest compliant option unless lateral stiffness is increased. Overall, the results support site-specific sizing; emphasize torsion detailing to optimize in stiff soils and prioritize stiffness adjustments to achieve compliance in soft soils.

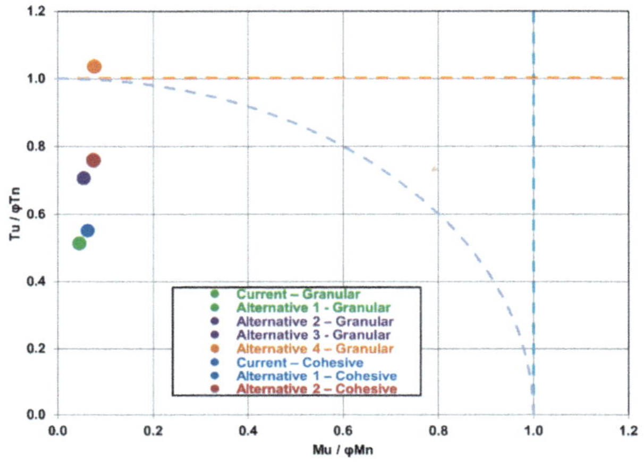

Fig. 7. Normalized M–T interaction map for feasible alternative and current foundations

4 Conclusion

Design wind requirements for cantilevered traffic sign structures tend to be conservative, which pushes drilled-shaft foundations toward sizes larger than necessary and not always economical or sustainable. The gap has been a field-validated, performance-based process that traces how combined bending and torsion from high winds transfer to the foundation from the superstructure and how serviceability should bound any size reduction.

The study closed that gap by using full-scale instrumentation to obtain base reactions and modal properties, validating structural models against those measurements, driving the validated models with a synthetic extreme-wind event to establish shaft-head demands, and then checking foundation options with site-specific p–y (L-Pile) analyses for serviceability and reinforced-concrete checks at the head section under combined M-V-T. This process links field data to analysis and design in a clear way. Importantly, the same wind time history was applied to both structures; this controls for loading variability and clarifies cause–effect in the results. The validated base reactions and modal properties prove that the synthetic demand set is a reasonable upper-bound candidate for serviceability and strength checks.

Across both sites, the results indicate that the baseline 4.0 ft. × 28 ft. (1.22 m × 8.53 m) drilled shaft is larger than required by bending strength and that serviceability (Δ at the head) and torsion at the head govern optimization. Under the synthetic extreme wind demands, all granular-site candidates satisfied the deflection limit ($\Delta_{allow} = 0.5\ in.$) enabling optimization from 4.0 ft. × 28 ft. (1.22 m × 8.53 m) to 3.5 ft. × 18 ft. (1.07 m × 5.49 m) with torsional margin, reducing concrete volume by around 37%. In contrast, the cohesive site alternatives exceeded Δ_{allow} only the current 4.0 ft. × 28 ft. (1.22 m × 8.53 m) satisfy serviceability, indicating that meaningful reduction there would first require added near-surface stiffness. Rotation did not control in either case ($\theta < 10$ mrad), and the normalized strength checks demonstrated that all design options had adequate bending strength, with torsional resistance, not bending resistance, governing the minimum diameter requirements.

The main finding is that designing drilled shafts for cantilevered sign structures is not about bending strength; it is a stiffness and torsion problem. Field data and analysis show that these structures have plenty of bending capacity, while the real design challenges come from shaft head lateral deflection which depends mainly on the soil properties near the surface, and torsional resistance. This changes the design approach into two parts: a stiffness control (choosing the right length and diameter to limit movement based on the soil conditions) and a torsion control (adding enough steel reinforcement at the top to resist twisting forces). In granular soils, stiffness is already appropriate, so material can be reduced overall and moved to where it is needed for torsion resistance; in cohesive soils reductions are only possible if stiffness is improved (such as going deeper into stronger soil layers or using larger diameters in critical sections), because adding bending strength does not solve the main problems. The optimization approach provides benefits beyond economic considerations. Depending on site conditions, up to 37% concrete volume reductions significantly reduce embodied carbon and material consumption. This approach advances sustainable highway infrastructure development while maintaining structural durability and performance requirements.

Acknowledgements. This paper is based upon work supported by the Wisconsin Department of Transportation. Their support is gratefully acknowledged. Any opinions, findings, conclusions, and recommendations expressed in this material are those of the authors and do not necessarily reflect the sponsor's views.

References

1. Wisconsin Department of Transportation (2025) LRFD standardized overhead sign structure plans, WisDOT Bridge Manual. https://wisconsindot.gov/Pages/ng-bus/eng-consultants/cnslt-rsrces/strct/standard-sign-plans.aspx. Accessed on 5 Jan 2025
2. K.T. Canning (2005) Bridge design manual. https://rosap.ntl.bts.gov/view/dot/39754
3. Torabi T, Shafei B, Alipour A (2025) Intensity measures for flood hazards in fragility assessments of bridges. In: Kioumarsi M, Shafei B (ed) The 1st international conference on net-zero built environment. Springer Nature Switzerland, Cham, pp 1539–1547
4. Qudaisat M, Alipour A (2026) Multivariate logistic wind fragility functions of overhead distribution poles. Int. J. Disaster Risk Reduct. 134:105996

5. Dokhaei B, Shafei B, Alipour A (2026) Optimized design of steel structural systems for reduced carbon footprint and enhanced resiliency in wind. Eng Struct 350:121895
6. Orlando A (2021) Full-scale monitoring of the wind-induced response of vertical slender structures, with fixed and rotating masses
7. Seo H, Ko J, Gunerathne S (2018) Development of novel analysis model for foundations subjected to combined torsional and lateral loads due to high wind, southern plains transportation center
8. Huang J, Bin-Shafique S (2019) Performance of drilled shaft under combination of complicated loads under hurricane event
9. Li QS et al (2006) Wind tunnel and full-scale study of wind effects on China's tallest building. Eng Struct 28:1745–1758
10. Zuo D, Smith DA, Mehta KC (2014) Experimental study of wind loading of rectangular sign structures. J Wind Eng Ind Aerodyn 130:62–74. https://doi.org/10.1016/j.jweia.2014.04.005
11. Tsai LW, Alipour A (2021) Studying the wind-induced vibrations of a traffic signal structure through long term health monitoring. Eng Struct 247. https://doi.org/10.1016/j.engstruct.2021.112837
12. Paiva F, Barros RC Integrity, reliability and failure of mechanical systems analysis of the stability and design of cantilevered sign supports in highways, n.d
13. Jafari M, Sarkar PP, Alipour AA (2019) A numerical simulation method in time domain to study wind-induced excitation of traffic signal structures and its mitigation. J Wind Eng Ind Aerodyn 193 (2019). https://doi.org/10.1016/j.jweia.2019.103965
14. McVay MC, Herrera R, Hu ZY, Magnavita J (2003) Determine optimum depths of drilled shafts subject to combined torsion and lateral loads using centrifuge testing
15. Boeckmann A, Uong M, Loehr JE (2015) Performance of pile supported sign structures. Wisconsin. Dept. of Transportation. Research and Library Unit
16. Li Q, Stuedlein AW, Barbosa AR (2017) Torsional load transfer of drilled shaft foundations. J Geotech Geoenvironmental Eng 143:04017036
17. L. Aashto (2015) LRFD specifications for structural supports for highway signs, luminaries, and traffic signals. Aashto
18. Alipour A, Puckett J, Ashlock J, Goudarzi M (2025) Wind loaded structures. No. WisDOT 0092-22-06

Pull-Out Resistance of Geogrids Embedded in Different Backfill Materials

Ghanta Naga Sireesha(✉) and Prishati Raychowdhury

Department of Civil Engineering, Indian Institute of Technology Kanpur, Kanpur, Uttar Pradesh, India
{gnsiree,prishati}@iitk.ac.in

Abstract. Coal is the primary source of power generation in India, accounting for nearly 57% of the country's electricity produced by coal-fired plants. During the 2022–23 financial year, over 232.56 million tons of pond ash were generated as waste. Approximately 65,000 acres of land are utilized for pond ash disposal. However, there is significant potential for its large-scale use as a fill material in MSE walls. The stability of Reinforced earth walls primarily depends on the interaction between the reinforcement and the surrounding soil. A series of pull-out tests were conducted with geogrid of strength 22 kN/m embedded in two distinct backfill materials, pond ash and sand. The pullout capacity of geogrid with pond ash were evaluated with 80% relative density and three different normal loads. It was observed that the normalized pull-out resistance factor (F_a^*) values for the geogrid embedded in the pond ash and sand ranged in between 0.23 and 1.2 and 0.33–1.34, respectively, corresponding to front-end axial pullout displacements ranging 10 mm–30 mm. This indicates that pond ash can be used as alternate backfill material with a compromise of 10–30% reduction in pullout capacity compared to sand backfill in the similar loading condition.

Keywords: Pond ash · MSE wall · Geogrid · Pull-out resistance · Thermal power station

1 Introduction

Ash is the byproduct obtained after combustion of pulverized coal in thermal power plants. Two types of ash produces in thermal power plants those are bottom ash and fly ash. Bottom ash is coarser collected in bottom furnace after burning of coal and fly ash is finer collected from the electrostatic precipitators. After combining a significant amount of water with bottom ash and fly ash, pond ash has been formed. It requires large place to dispose off. In addition to taking up a lot of space, the primary environmental concerns associated with the utilization of pond ash as a backfill material is the potential generation of leachate containing toxic elements, which may pose a risk to soil and groundwater quality if not adequately contained or treated. Akar et al. [1] investigated the short-term leaching behavior of trace elements and reported that the concentrations of toxic elements such as Zn, Cu, Cd, and Ni in pond ash were low and within the permissible limits for non-hazardous waste disposal. However, the concentration of Cr

H. Beushausen et al. (Eds.): NTZR 2025, LNCE 873, pp. 796–805, 2026.
https://doi.org/10.1007/978-3-032-27698-8_70

exceeded the prescribed limits for non-hazardous classification. Suthar and Agarwal [2] conducted the Toxicity Characteristic Leaching Procedure (TCLP-1311) on both unstabilized and lime-stabilized pond ash, and found that the concentrations of toxic elements were within the permissible limits of nonhazardous waste. Furthermore, the stabilized pond ash exhibited lower concentrations of toxic elements compared to the unstabilized ash.

Soil reinforcement techniques are extensively recognized for their role in enhancing the performance of earth structures such as reinforced walls, slope stabilization solutions, soft soil improvement systems, embankments for roads and railways, and foundation systems. Design of these reinforced structures requires careful attention to the complex force transfer mechanism at the soil–geogrid interface. Typically, two types of failure mechanisms are observed in the soil-geogrid interface: (i) direct shear failure, and (ii) pull-out failure. The pull-out resistance of the soil reinforcements is indeed a critical parameter for evaluating the internal stability during the design of the reinforced soil walls. Figure 1 illustrates different interaction mechanisms that occur when geogrid reinforcements are subjected to axial tension.

Several studies in the past had focused on the interaction of various reinforcements in pull-out conditions. For example, Balunaini and Prezzi [3]; Balunaini et al. [4] studied the pull-out resistances of reinforced walls backfilled with sand–tire waste. Prasad and Ramana [5] investigated the pull-out resistance for slag wastes backfilled walls. Moraci and Recalcati [6] conducted a series of pull-out tests to investigate the role of transverse and longitudinal members (T & L-members) on pullout resistance of a polyester (PET) geogrid. Ezzein and Bathurst [7] proposed a new approach to evaluate soil-geosynthetic interaction using a novel pullout test apparatus and transparent granular soil. Pant et al. [8] performed experiments to measure the role of transverse and longitudinal members on pullout resistance of PET geogrid embedded in sand. Nonetheless, studies on pull-out behavior of geo-synthetics embedded in pond ash are very sparse. To address this issue, in the present study, a series of monotonic pull-out tests had been conducted on PET geogrid embedded in pond ash to evaluate the interface factors. The same test series was then repeated for a standard sand backfill and the results were compared for both cases.

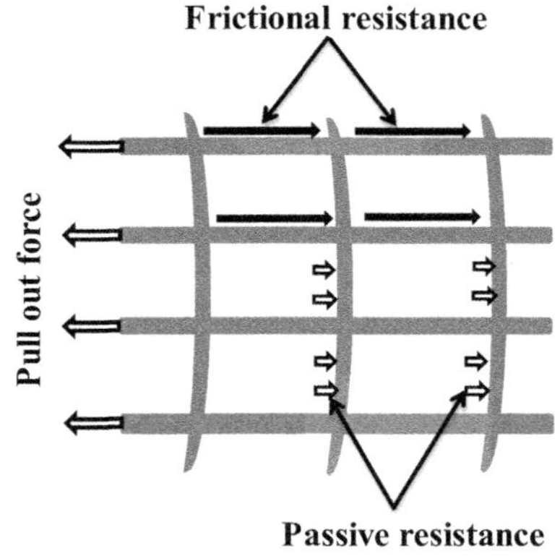

Fig. 1. Schematic showing the interaction mechanisms during the pullout of a geogrid-reinforcement

2 Material Characterization

Pond ash and sand had been used as the backfill material of the reinforced earth wall, whereas the PET geogrid was used as the reinforcing element. The following subsections provide the details about the geotechnical characteristics of the above-mentioned materials.

2.1 Backfill Materials: Sand and Pond Ash

Pond ash was collected from Parichha thermal power station. Sand was collected from the Yamuna River. Before determining the interface properties of soil and geogrid, geotechnical characteristic tests like specific gravity test, grain size distribution analysis, compaction test, and direct shear test were performed. The specific gravity of pond ash was 2.24, which falls within the typical range for Indian pond ashes, 1.64–2.66 [9]. For sand, the specific gravity was observed to be 2.66. Figure 2a shows the grain size distribution of Parichha pondash and sand. Pondash comprises 0.08% gravel, 84.9% sand, 15.02% silt, and 0% clay-size particles. According to the Unified Soil Classification System (USCS), the pond ash collected from the Parichha thermal power plant is classified as silty sand (SM). Figure 2b shows the compaction curve of pond ash and sand. The maximum dry unit weight of pond ash and sand is 1.3 g/cc, 1.72 g/cc; the optimum moisture content is 18% and 11% respectively. The angle of internal friction of Parichha pond ash and sand was observed to be 44° and 47°, respectively, at 80% relative density.

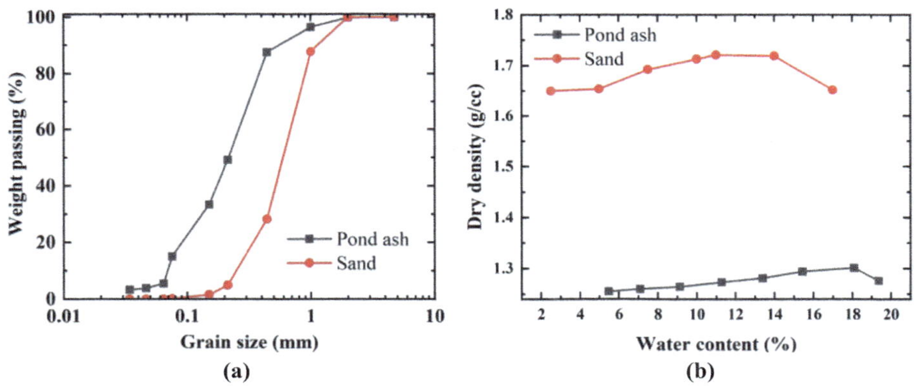

Fig. 2. **a** Grain size distribution curve and **b** Compaction curve of pond ash and sand

2.2 Reinforcing Material: Geogrid

The wide strip tensile strength test is conducted as per [10] on a polyester (PET) geogrid (as shown in Fig. 3) to determine its tensile strength in both the machine and cross-machine directions. It has an aperture size of 2 cm × 1.8 cm and a thickness of 1 mm. Figure 4 shows the geogrid's strain versus tensile force obtained from the tensile strength test. At 2% strain, the stiffness of the geogrid is observed as 389 kN/m in the machine direction and 269 kN/m in the cross-machine direction.

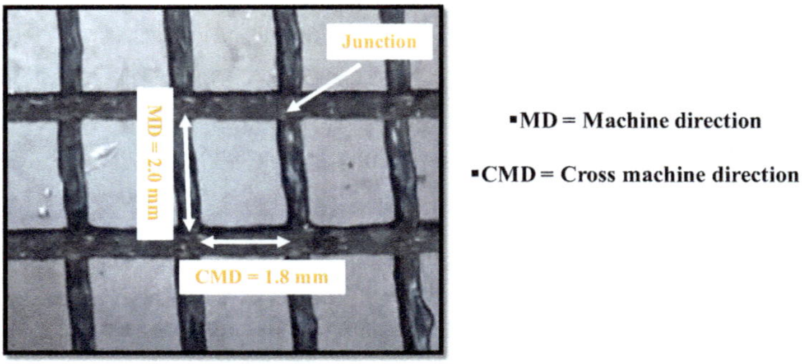

Fig. 3. PET geogrid used in the present study

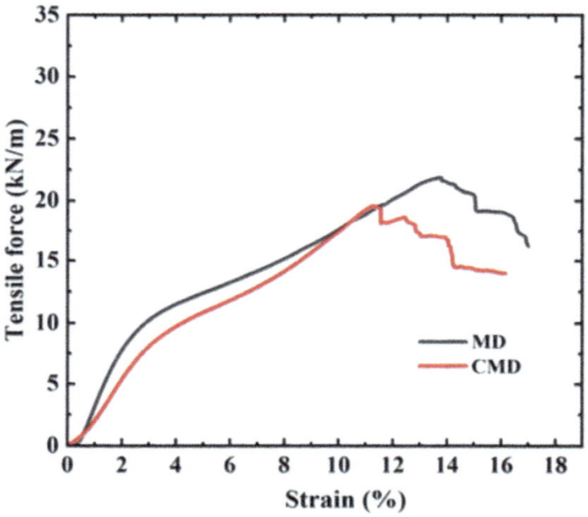

Fig. 4. Strain versus tensile force of PET geogrid material in machine direction (MD) and cross-machine direction (CMD)

3 Axial Pull-Out Test Apparatus and Test Procedure

Pull-out test apparatus used in this experiment consisted of a rigid box, a clamping plate, a hydraulic actuator (which can apply both static and cyclic loads), a Load cell of capacity 500 kN, LVDTs to measure the displacement of the geogrid layers, and a data acquisition system. The schematic front view of the experimental setup is shown in Fig. 5. Details of each component is discussed below.

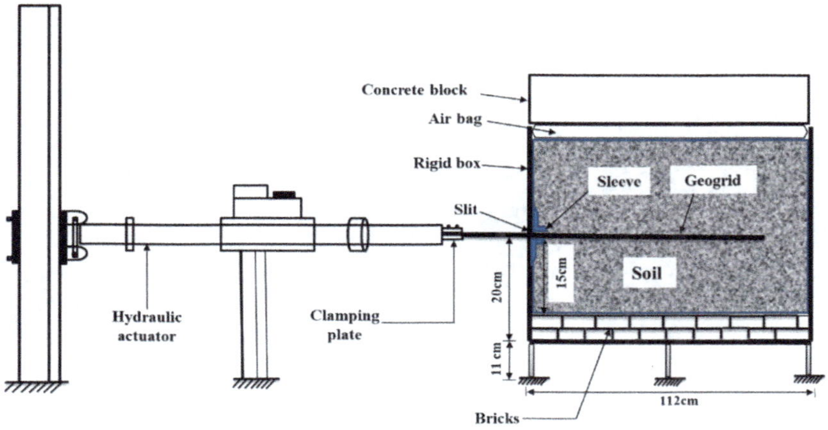

Fig. 5. Schematic front view of pull-out test apparatus

3.1 Rigid Box

The pull-out test apparatus was developed as per [11] standard. It consists of a rigid box measuring 1120 mm in length, 970 mm in width, and 500 mm in height. The box is welded to the ground at its base to ensure stability. Horizontal and vertical stiffeners are provided along all sides to prevent bending and potential failure of the sides of box. The bottom 50 mm of the box is filled with bricks to add weight and resist uplift during the geogrid pull-out process. A polyester geogrid is embedded at the mid-height of the soil fill, ensuring a 50 mm clearance from both the top and bottom surfaces. The slit has a length of 420 mm and a gap width of 10 mm.

3.2 Clamping Plate and Sleeve

Two clamping plates, measuring 100 × 10 mm, which are fastened with 6 bolts used to connect the geogrid with the hydraulic actuator. Sleeves are attached in the inner walls of rigid box to effectively distribute the pull-out force on the geogrid and to counteract the frictional resistance from the box walls.

3.3 Hydraulic Actuator

A closed-loop, computer controlled hydraulic actuator with capacity of 500 kN with 250 mm stroke was used to pull the geogrid in this experiment. A built-in load cell and LVDT are integrated into the actuator. The hydraulic actuator is attached to frame and supported by I sections. Figure 6 shows the pullout test box apparatus along with the top view of the test box arranged with the geogrid.

(a) (b)

Fig. 6. **a** Photograph of pull-out test apparatus and (b) arrangement of reinforcement

The backfill material was placed into the rigid box and compacted in layers by using controlled volume technique. The bottom portion of geogrid is compacted in two layers each with 100 mm thickness. After compacting the first two layers, a geogrid measuring 420 mm in width and 850 mm in length is positioned on the surface of the leveled layer and clamped to the pullout loading system. The geogrid size is selected to minimize the influence of wall friction on pull-out resistance. According to ASTM standards, the spacing between the geogrid and the wall should be maintained at a minimum of 150 mm to reduce the frictional effects. The remaining ash is then added and compacted. Concrete blocks are placed on the compacted soil to apply normal loads. The rate of pull out force applied to be 1 mm/min ± 10% as per ASTM standards.

4 Results and Discussion

The pull-out tests were performed on polyester geogrid embedded in pond ash and sand under three different normal stresses equal to 9 kPa, 18 kPa and 27 kPa respectively. The relative density of soil is maintained as 80%. At each reinforcement level, the pullout resistance P_a is given per unit width of reinforcement is given by:

$$P_a = F_a^*.\alpha.\sigma_n.C.B.L_e$$

where

P_a = Pullout resistance per unit width.
α = Non-linear shear stress mobilization correction factor (= 0.8 for geo-grids).
C = Effective unit perimeter (= 2 in general).
L_e = Effective length of reinforcement in retained zone, (= 850 mm).
B = Width of the reinforcement, (= 420 mm).

The pull-out resistance of geogrid embedded in pond ash and sand measured with axial displacement at different normal stresses are shown in Fig. 7. The mode of reinforcement failure is observed to be pullout or slippage mode of failure at normal stress of 9 kPa while at normal stresses equal to 18 kPa and 27 kPa rupture or breakage failure was observed. The pull-out resistance of the geogrid increases with increasing normal stress due to enhanced friction and greater passive resistance offered by the transverse ribs. The pull-out resistance of geogrid embedded in pond ash and sand is approximately showing similar values for all axial displacements. The pullout force increases with increasing axial displacement.

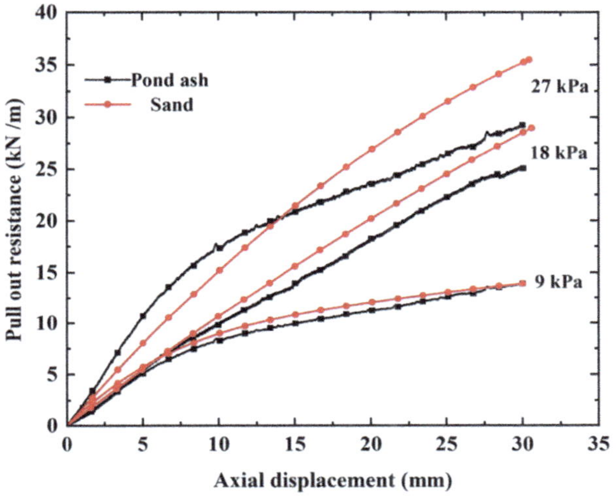

Fig. 7. Pull-out resistance of geogrid embedded in pond ash and sand at different normal load conditions

MSE walls are designed by taking into account internal and external modes of failure. Reinforcement spacing, length and number of layers of reinforcement are determined accordingly. Axial pull-out resistance factors (F_a^*) calculated at different axial displacements and at different overburden depths are shown in Fig. 8. The equivalent overburden depths were determined based on the normal stresses acting at the reinforcement level, which were calculated using the normal stress applied at the top of the sample and the unit weight of the prepared sample. The F_a^* values calculated at an axial displacement (δ) of 10 mm and 30 mm were used to check the internal stability of MSE wall backfilled with pond ash and sand in this study. Pull out resistance factor is observed to be same for both sand and pond ash at different axial displacements. F_a^* varied between 0.38 and 0.63 and 0.75–1.06 under three varying overburden stresses at 10 mm and 30 mm axial displacements for pond ash, respectively. The range of F_a^* values were nearly identical at the reported for two front-end displacements in the studies by Prasad and Ramana [12]; Pant et al. [13].

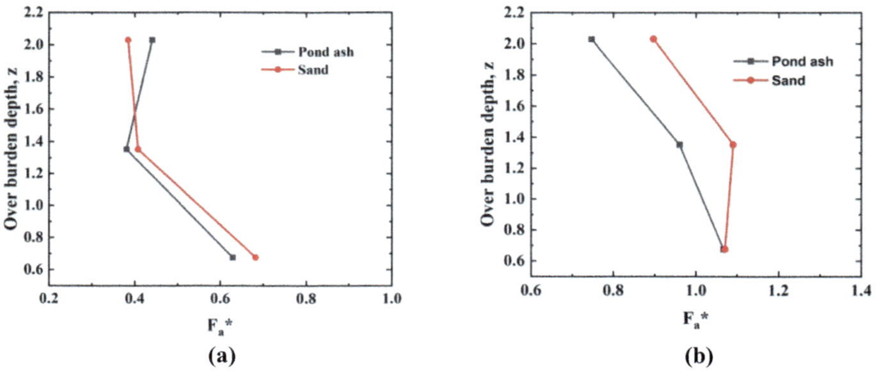

Fig. 8. Pull-out resistance factors of geogrid with different overburden pressure (**a**) δ = 10 mm (**b**) δ = 30 mm

5 Limitations and Future Scope

This study focuses on short-term pond ash–geogrid interaction; however, long-term durability under environmental cycles like wetting–drying and freeze–thaw may affect performance. Such conditions can alter pond ash properties and degrade geogrid strength. Future studies should evaluate these effects to ensure long-term reliability in diverse field conditions.

6 Conclusions

Pond ash–geogrid composites present a promising alternative as select fill materials in civil engineering applications such as reinforced soil structures and stabilized slopes. However, prior to the adoption of pond ash in these systems, it is imperative to thoroughly assess the interaction characteristics between pond ash and geogrid reinforcement. To this end, a comprehensive experimental investigation was undertaken, comprising both interface direct shear tests and axial pullout tests using various types of geo-grids embedded in pond ash. The pond ash used in the study was sourced from Parichha thermal power station located in Uttar Pradesh, India. Static pull-out performance of two this pond ash as highway MSE wall compared to a dense sand backfill was evaluated through a series of monotonic pull-out tests. The following key observations were derived: (1) Laboratory geotechnical tests showed that the angle of internal friction (φ) of pond ash particles and sand were 44° and 47°, respectively, indicating the shear strength of the sand on slightly higher side than the pond ash. (2) Pull-out resistance factor F_a^* for the geogrid embedded pond-ash and sand backfill were found to be 0.44 and 0.38, respectively, for axial pullout displacement of 10 mm. For a higher displacement of 30 mm, these values were 0.74 and 0.9, respectively, for pond ash and sand backfill. (3) The pull-out resistance of the geogrid increases with increasing normal stress due to enhanced friction and greater passive resistance offered by the transverse ribs. (4) As expected, the pullout force increases with increasing axial displacement.

The above key observations indicate that pond ash shows reasonable performance (compromised by about 10–30% reduction in the capacity), compared to a dense sand backfill. Hence, in case of projects where natural soil is cohesive in nature and replacing the existing poor cohesive soil with competent good quality sand is not a feasible solution, power-plant by product pond ash can be used as an alternative material resulting economic and sustainable geotechnical solution.

Acknowledgements. The financial support provided by Science and Engineering Research Board (SERB) of Department of Science and Technology, India through a sponsored project entitled "Seismic Performance Assessment of Ash-filled MSE Walls" (File no. CRG/2021/003428) is highly appreciated and thankfully acknowledged by the authors.

References

1. Akar G, Polat M, Galecki G, Ipekoglu U (2012) Leaching behavior of selected trace elements in coal fly ash samples from Yenikoy coal-fired power plants. Fuel Process Technol 104:50
2. Suthar M, Aggarwal P (2018) Bearing ratio and leachate analysis of pond ash stabilized with lime and lime sludge. J Rock Mech Geotech Eng 10(4):769–777. https://doi.org/10.1016/j.jrmge.2017.12.008
3. Balunaini U, Prezzi M (2010) Interaction of ribbed-metal-strip reinforcement with tire shred-sand mixtures. Geotech Geol Eng 28:147–163. https://doi.org/10.1007/s10706-009-9288-6
4. Balunaini U, Yoon S, Prezzi M, Salgado R (2014) Pullout response of uniaxial geogrid in tire shred-sand mixtures. Geotech Geol Eng 32:505–523. https://doi.org/10.1007/s10706-014-9731-1
5. Prasad PS, Ramana GV (2016) Feasibility study of copper slag as a structural fill in reinforced soil structures. Geotext Geomembr 44:623–640. https://doi.org/10.1016/j.geotexmem.2016.03.007
6. Moraci N, Recalcati P (2006) Factors affecting the pulloutbehaviour of extruded geogrids embedded in a compacted granular soil. Geotext Geomembr 24:220–242. https://doi.org/10.1016/j.geotexmem.2006.03.001
7. Bathurst RJ, Ezzein FM (2016) Geogrid pullout load–strain behaviour and modelling using a transparent granular soil. Geosynth Int 23:271–286. https://doi.org/10.1680/jgein.15.00051
8. Pant A, Datta M, Ramana GV (2019) Bottom ash as a backfill material in reinforced soil structures. Geotext Geomembr 47:514. https://doi.org/10.1016/j.geote xmem. 2019. 01. 018
9. Jakka RS, Ramana GV, Datta M (2010) Shear behaviour of loose and compacted pond ash. Geotech Geol Eng 28:763–778. https://doi.org/10.1007/s10706-010-9337-1
10. ASTM D 6637: Standard test method for determining tensile properties of geogrids by the single or multi-rib tensile method. ASTM International, West Conshohocken, PA, USA
11. ASTM D6706 (2013) Standard test method for measuring geosynthetic pullout resistance in soil. ASTM Standards, West Conshohocken, PA, USA
12. Karnam Prabhakara BK, Balunaini U, Arulrajah A (2021) Development of a unique test apparatus to conduct axial and transverse pullout testing on geogrid reinforcements. J Mater Civ Eng 33:1–11. https://doi.org/10.1061/(ASCE)MT.1943-5533.0003497
13. Karnamprabhakara BK, Balunaini U, Arulrajah A, Evans R (2021). Axial pullout resistance and interface direct shear properties of geogrids in pond ash. Int J Geosynth Gr Eng In press

Influence of Soil-Structure Interactions on Rail-Viaduct Dynamics with Continuous Welded Rail Systems

Nupur Saxena(✉), Prishati Raychowdhury, and Samit Ray-Chaudhuri

Deptartment of Civil Engineering, Indian Institute of Technology, Kanpur, UP 208016, India
{nupurs,prishati,samitrc}@iitk.ac.in

Abstract. Rail-structure interaction (RSI) studies on continuously welded rail (CWR) bridges are critical for track stability and structural integrity, as additional rail stress and displacements under vehicular and thermal loads can cause catastrophic failures. Also, existing numerical models in the code recommendations often neglect or oversimplify soil-structure interactions (SSI), resulting in prediction inaccuracies. This study investigates the effect of SSI on the dynamic responses of a critical span adjacent to an abutment of India's first CWR viaduct, the Rohtak–Gohana Viaduct, using 3D numerical modeling in SAP2000 under dynamic train loading conditions. Three cases of foundation rigidity were analyzed: Case-I– fixed base system, Case-II– Pile foundation considering only lateral flexibility in SSI, and Case-III– Pile foundation with both lateral and vertical flexibilities in SSI. Results of the numerical study indicate that axial rail stresses increase by 16% with lateral SSI stiffness alone and by 20% when both lateral and vertical SSI stiffness were included. Incorporating vertical soil stiffness also increases deck displacement. A comparison with experimental results further confirms that neglecting SSI can lead to an underestimation of rail stresses and an overestimation of bridge stiffness. Thus, horizontal and vertical soil stiffness must be included in numerical models to accurately capture RSI effects under train loads, particularly for CWR track and heavy-load rail networks.

Keywords: CWR viaduct · Rail-structure interactions · Soil-structure interactions

1 Introduction

Railway tracks serve as the backbone of transportation networks, as they help facilitate the movement of people and goods across regions. Continuous welded rail (CWR) tracks, due to the absence of joints and gaps, accelerate this movement. So, nowadays, the conventional jointed rail tracks have been replaced by the CWR tracks, particularly for high-speed trains. But the presence of such tracks on viaducts, along with the associated soil–foundation system, induces nonlinear interaction with the structure, resulting in additional stresses and the displacement of the rail and the viaduct. These stresses and displacements, if they exceed the permissible limits, can lead to disastrous failures such as track buckling and fractures. Hence, it is important to ensure that they lie within the permissible limits [1–3].

H. Beushausen et al. (Eds.): NTZR 2025, LNCE 873, pp. 806–817, 2026.
https://doi.org/10.1007/978-3-032-27698-8_71

In the past, numerous research studies have examined rail-structure interaction (RSI) effects with and without modeling the pile foundation [4–6]. However, the existing numerical models recommended by code provisions [1–3] adopt simplifications that neglect soil-structure interaction (SSI) effects, which may lead to inaccurate predictions of structural response and safety margins under dynamic loads. Hence, to address these limitations, this study explores the comprehensive impacts of SSI on the dynamic responses of CWR viaduct systems using sophisticated numerical analyses. The effects of SSI on responses, including rail stresses and bridge deck responses under train loads, are addressed in this work. While this study emphasizes safety and performance optimization of CWR systems through SSI modeling, sustainable material innovations are also gaining traction in infrastructure development. Recent works have explored eco-friendly UHPC mixes by incorporating recycled alternatives like waste glass powder, ground granulated blast furnace slag (GGBFS), and Portland limestone cement to reduce CO_2 emissions and improve durability in concrete structures [7–10]. Such studies, although distinct from the current work's scope, underscore a complementary global trend toward resilient and sustainable railway infrastructure.

2 Description of Viaduct and Foundation System

2.1 Viaduct Description

The Rohtak–Gohana viaduct considered for this study is a 2-km-long elevated structure situated on the Northern Indian railways network. It consists of 110 spans and comprises composite steel girders resting on elastomeric bearings, with a ballasted track supporting the CWR track. As per previous studies, such as those by Okelo and Olabimtan [4], have shown that the transition zone between the approach slab and bridge deck exhibits significantly higher stiffness, which results in peak axial rail stresses under thermal and braking loads. Therefore, in this study, the first span (A1-P7), as shown in Fig. 1, adjacent to the abutment A1 and having a span length of 19.8 m and a composite steel girder, is analyzed.

Fig. 1. A1-P7 span view

2.2 Foundation System Description

The foundation below the piers comprises of pile foundation, with its cross-sectional arrangement shown in Fig. 2a. Figure 2b illustrates the details of the pile cap. The 1200 mm diameter cast-in-situ Reinforced Cement Concrete (RCC) bored piles were arranged in two rows. In total, there were 8 piles spaced at 3.6 m center-to-center. These pile caps over the piles have dimensions of 12.3 m × 5.1 m and a depth of 1.8 m.

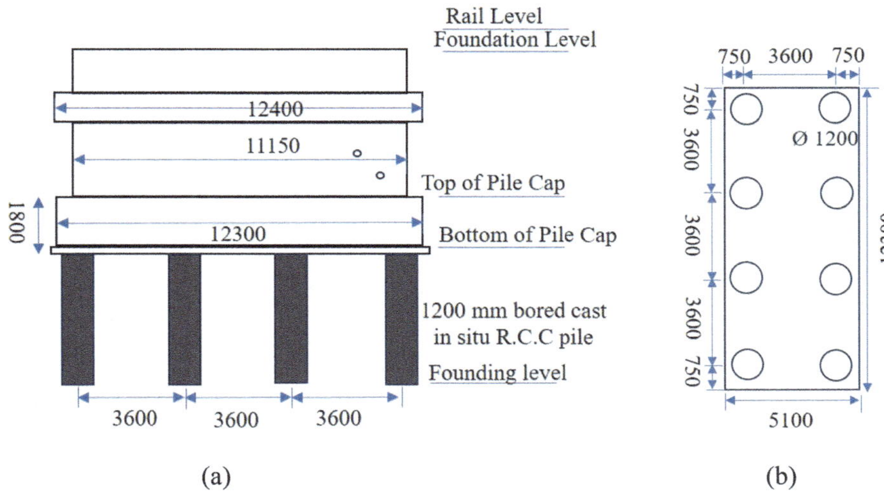

Fig. 2. Pile foundation profile (a) Typical cross section (transverse direction); (b) Plan showing details of pile cap. All dimensions are in millimeters (mm)

The geotechnical profile of the surrounding soil is presented in Table 1. In this table, N_{uc} is the uncorrected Standard Penetration Test (SPT) value, G_s is the specific gravity of soil particles, Y_w is the unit weight of water, while Y_t and Y_{sat} are the total and the saturated unit weight of the soil, respectively. The uncorrected SPT values were directly taken from the bore log report. So, to improve the accuracy, these values were corrected for overburden pressure using Eq. 1.

$$N' = C_N N_{uc} \tag{1}$$

where C_N is the Peck, Hanson, and Thomson (1974) overburden correction factor [11].

Additionally, for saturated fine sands or silts where the groundwater table is present and the corrected SPT value N′ is greater than or equal to 15, a further correction for dilatancy was applied using Eq. 2 [12].

$$N'' = 15 + 0.5\left(N' - 15\right) \tag{2}$$

Table 1. Geotechnical soil properties

Depth (m)	Soil type*	N_{uc}	γ_t (g/cc)	G_s	e	γ_w (g/cc)	γ_{sat} (g/cc)
0							
0.5	SM			2.65			
1.5	SM	5	1.732	2.66	0.77		1.938
3	SM	6	1.785	2.64	0.69		1.97
3.8	WT						
4.5	SP-SM	9	1.789	2.64	0.68	1	1.976
6	SP-SM		1.782	2.68	0.71	1	1.983
7.5	CL	16		2.67		1	
9	CL		1.83	2.68	0.7	1	1.988
10.5	SM	19	1.842	2.64	0.64	1	2
12	SM		1.839	2.65	0.64	1	2.006
13.5	SM	26	1.852	2.64	0.6	1	2.025
15	SM		1.842	2.65	0.65	1	2
16.5	SM	25		2.64		1	
18	CL	40		2.7		1	
19.5	CL	39		2.71		1	
21	CL		1.882	2.7	0.68	1	2.012
22.5	CL	41		2.69		1	
24	CL	35		2.68		1	
25.5	CL	46		2.7		1	
27	CL		1.97	2.71	0.64	1	2.043
28.5	CL	55		2.7		1	
30	CL	61		2.7		1	

*SM: Silty sand; SP–SM: Poorly graded sand with silt; CL: Low plasticity clay; WT: Water table

3 Numerical Modeling

3.1 Viaduct Modeling

For this study, a detailed 3D model was modeled in SAP2000 v24.0.0 [13]. The geometric properties of its various components are given in Table 2. For concrete sections such as pier caps and piers, pile caps and piles, M35 grade of concrete was used. In contrast, M30 grade of concrete was used for the deck slab. For the abutment modeling, a fixed support was assumed, while for the P7 pier, the following cases were considered:

- Case-I (Without pile modeling): In this case, the bridge foundation was assumed to be rigid, without any SSI effects, as shown in Fig. 3a. In this case, only the superstructure, bearings, pier cap, and the pier were modeled.
- Case-II (With horizontal flexibility only): In this case, piles were considered but only with the lateral (horizontal) flexibility to account for lateral soil resistance as shown in Fig. 3b.
- Case-III (With both horizontal and vertical flexibility): In this case, both lateral and vertical flexibilities were considered to capture full SSI effects.

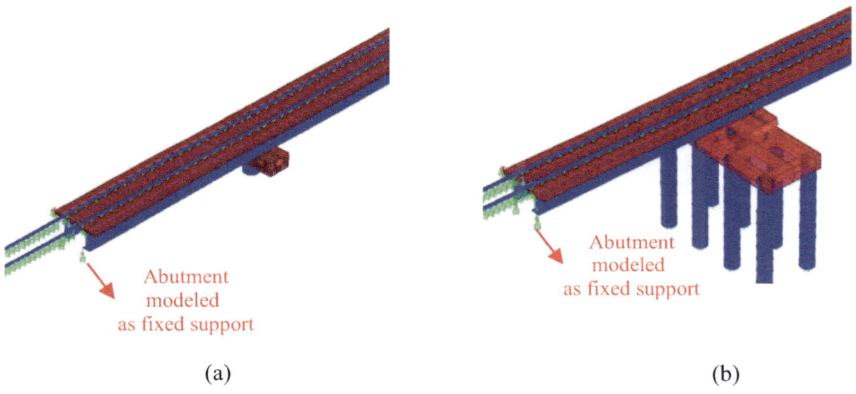

Fig. 3. Finite element model: (a) without pile foundation; (b) with pile foundation

For pile modeling, each pile was assumed to be fixed at its tip, and its length was considered only up to its depth of fixity [14]. To capture realistic lateral stiffness, the equivalent cantilever length of the pile was estimated as per IS:2911 (Part 1/Sec. 2)–1979 (Appendix C) [15]. The subgrade modulus, n_b, was taken as 0.27 kg/cm^{3,} and the stiffness parameter, T, was computed as 406.88 cm. Hence, the free length of the pile, L_f came out to be 0 m, as the scour level coincided with the pile cap bottom. Consequently, the ratio, L_{eq}/T was taken as 2.20, which yielded an equivalent cantilever length, L_{eq}, of 8.95 m.

Table 2. Structural components of the rail-viaduct system

Sl. no.	Component	Element type	Dimensions/properties
1	Track	Frame section (I-beam) [16]	~ UIC 60 rail
2	Deck slab	Area section [17]	250 mm thick
3	Girders	Frame section (I-beam) [16]	Top flange: 350 × 16 mm; Web: 1350 × 16; Bottom flange: 750 × 50 mm
4	Pier cap	Area section [17]	1500 mm thick

(*continued*)

Table 2. (*continued*)

Sl. no.	Component	Element type	Dimensions/properties
5	Pier	Frame section (circular-beam)	1500 mm dia
6	Pile cap	Area section	1800 mm thick
7	Piles	Frame section (circular-beam)	1200 mm dia
8	Bearings	Linear springs	Long. stiffn. (K_L): 178760 kN/m; Vert. stiffn. (K_V): 25203760 kN/m; Rot. stiffn (K_R): 1869164.40 kN-m/m
9	Ballast	Non-linear springs	Unloaded:25 kN/m [3] Loaded: 50 kN/m [3]
10	Soil	Linear springs [18]	Described in Sect. 3.2

3.2 SSI Modeling

In the present study, the interaction between the pile foundation and the surrounding soil was modeled using linear springs in both vertical and horizontal directions. In the vertical direction, single springs were applied only at the pile bottom tip. This vertical spring stiffness (k_v), as given in Eq. 3, was assumed to account for both skin friction resistance and end-bearing capacity of the pile [19, 20].

$$k_V = \frac{\alpha\, A_p E_P}{l} \tag{3}$$

where α is the empirical multiplier based on pile type and slenderness; A_p is the cross-sectional area of the pile (cm^2); E_p is the modulus of elasticity of the pile material (kg/cm^2), and l is the length of the pile (cm).

The value of α varies for different types of piles when the slenderness ratio $l/D \geq 10$ is taken as given in Eq. 4:

$$\begin{aligned} &\text{For steel pipes} : \alpha = 0.027(l/D) + 0.20 \\ &\text{For prestressed concrete} : \alpha = 0.041(l/D) - 0.27 \\ &\text{For cast-in-situ piles} : \alpha = 0.022(l/D) - 0.05 \end{aligned} \tag{4}$$

It may be noted that this formulation given in Eq. 4 is applicable when the slenderness ratio $l/D \geq 10$. For lateral resistance, the pile was discretized into elements of length L = 1 m, and the springs were applied at 1 m intervals along the pile length. The stiffness of these springs was computed based on the modulus of subgrade reaction (k_h), using the Vesic (1961) formulation [21] as given below in Eq. 5:

$$k_h = \frac{0.65}{D} \sqrt[12]{\frac{E_s\, D^4}{E_p I_p}} \cdot \frac{E_s}{1 - \mu_s^2} \tag{5}$$

where, k_h is the modulus of subgrade reaction; E_s is the modulus of soil; D is the pile diameter; E_p is the pile's modulus of elasticity; I_p is the pile's moment of inertia; μ is the Poisson's ratio.

Thereupon, using this k_h, spring's lateral stiffness (K_s) was computed using Newmark's distribution as given in Eqs. 6–8.

$$K_{s1} = \frac{Dl}{24}\left[7k_h^n + 6k_h^{n-1} - 7k_h^{n-2}\right] \tag{6}$$

$$K_{s(m-1)} = \frac{Dl}{12}\left[k_h^n + 6k_h^{n-1} + k_h^{n-2}\right] \tag{7}$$

$$K_{sm} = \frac{Dl}{24}\left[7k_h^n + 6k_h^{n-1} - k_h^{n-2}\right] \tag{8}$$

where, k_h^n is the modulus of subgrade reaction of the nth spring.

4 Validation of Numerical Model

For the validation of horizontal deflection of pile (δ), the pile model consisted of horizontal springs for SSI and was assumed fixed at its fixity depth. Then, a load equal to its horizontal load-carrying capacity (Q) was applied at its top.

Since for the fixed head piles, δ is calculated using Eq. 9 as given below.

$$\delta = \frac{Q\left(L_f + L_s\right)^3}{12E_pI_p} \tag{9}$$

So, for an allowable deflection of 1.2 cm at scour level, the lateral load at pile top was computed as 60 tons. This load, when applied to the head of the model, gave a pile deflection value of 11.76 mm as shown in Fig. 4. Therefore, this validated pile model.

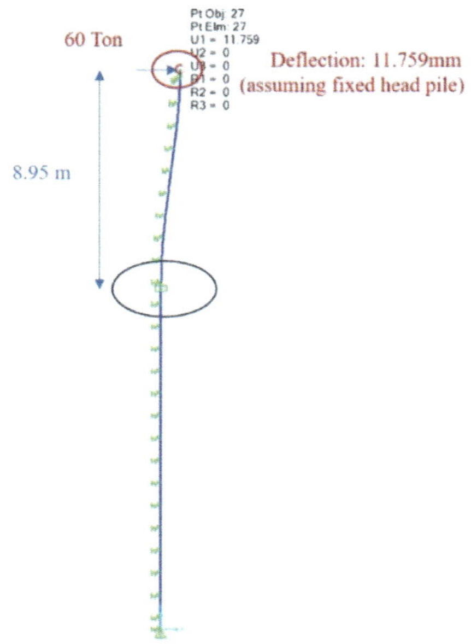

Fig. 4. Deflection of fixed head pile under 60 tons load

5 Results and Discussion

Three types of models, as discussed, were developed in SAP2000 for studying the effect of SSI on the axial rail stresses and vertical deck displacement at its midspan. The following types of dynamic loads were considered:

(i) Bending load: Structural loads induced by moving trains (65.38 kN/m as a uniformly distributed load for Indian locomotive class, WDG-4G diesel locomotive over 19.8 m span length). The WDG-4G is a heavy freight locomotive with an axle load of approximately 22 t and a power rating of 4,500 HP, commonly used on Indian Broad-Gauge tracks.
(ii) Additional horizontal loads equivalent to 25% of the bending loads to simulate train braking effects [22].

The corresponding results are discussed below in subsequent subsections.

5.1 Axial Rail Stresses

Figure 5 presents the axial rail stresses for the three types of numerical models. It can be observed that ignoring SSI (Case-I) results in an underestimation of axial rail stresses, with a numerical value of 11.96 MPa compared to the experimental range of 11.00–16.00 MPa. These experimental values were obtained from in-situ strain gauge measurements on the A1–P7 span of the Rohtak–Gohana viaduct, as part of the authors' field monitoring campaign under similar loads [23]. The same span (A1–P7) and consistent structural and material parameters were used in both the numerical and experimental parts to ensure direct comparability. It was seen that incorporating only lateral springs for SSI (Case-II) increases the axial rail stresses by approximately 16.3% compared to the model without SSI. Further inclusion of both lateral and vertical springs raises the stress by about 19.6% relative to the no-SSI case.

The role of soil stiffness parameters such as subgrade modulus, elastic modulus, and Poisson's ratio is central in defining how the foundation responds to train-induced loads. These parameters govern the stiffness and damping behavior at the pile–soil interface, influencing the overall dynamic and thermal response of the superstructure. Likewise, pile properties such as diameter and slenderness ratio affect the flexural rigidity and deformation capacity of the foundation. Larger diameter or stiffer piles reduce foundation flexibility, whereas slender piles introduce higher lateral compliance, thereby impacting rail–deck relative motion.

Also, it can be noted that, when both lateral and vertical springs are included (Case-III), the numerical axial stress (14.30 MPa) more closely approaches the experimental maximum value (16.00 MPa), indicating that both directions of soil stiffness enable the model to more accurately simulate worst-case dynamic stress scenarios, which is critical for safety and fatigue assessments. While the influence of lateral springs is dominant, as indicated by the significant increase in stress when moving from no SSI to lateral-only SSI (~ 16.3%), the additional increase of ~ 2.8% from incorporating vertical springs still improves the accuracy of the model. This relatively smaller contribution arises from the fact that vertical deck bending loads have a lesser effect on axial rail stresses than horizontal interaction forces.

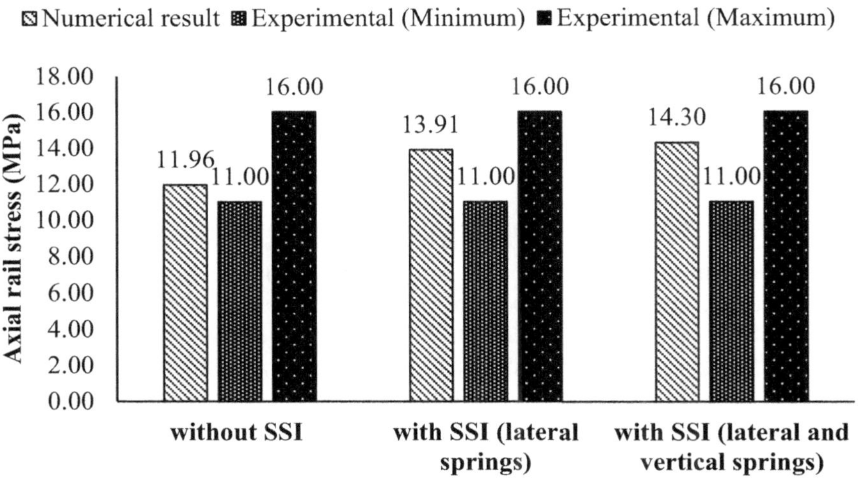

Fig. 5. Axial rail stress results

Thus, it can be concluded that for accurate estimation of axial rail stresses under dynamic loads, it is important to include at least the lateral springs for SSI. However, for a more realistic representation of worst-case scenarios and better alignment with experimental peak stresses under train loads, both lateral and vertical springs should be incorporated in the numerical models.

5.2 Vertical Deck Displacement

Figure 6 shows the vertical deck displacement at the midspan. It can be seen from the figure that lateral SSI springs do not have any significant effect on the displacement value. However, incorporating vertical springs for SSI increases the vertical deck displacement at the midspan by 12.7%. This highlights the importance of including vertical SSI components for accurately capturing vertical deformation behavior in bridge models.

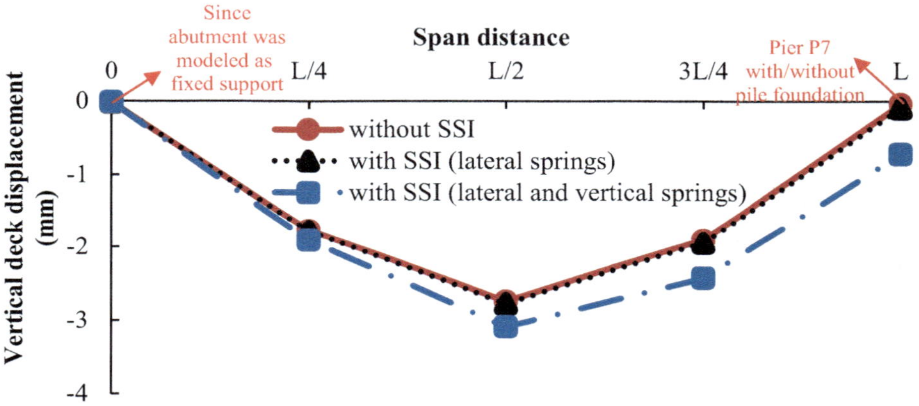

Fig. 6. Vertical deck displacement results

6 Conclusions

RSI studies are very crucial as additional stresses and displacements caused due to it under train running and thermal loads could lead to severe consequences such as rail buckling and fractures. Existing code provisions typically suggest only the superstructure in 2D modeling to verify the effects of RSI without accounting for soil-structure interaction (SSI). However, for a more accurate assessment, recent literature emphasizes the importance of detailed 3D modeling for RSI analysis, without explicitly considering the effect of SSI. A systematic approach has been adopted in this work to investigate the effect of SSI on RSI parameters. For this purpose, a 3D finite element model of the first span adjacent to the abutment of India's first CWR bridge was developed, and three configurations were analyzed: (i) without SSI effects, (ii) with lateral SSI only, and (iii) with both lateral and vertical SSI effects, under train loads. The results demonstrate that incorporating lateral SSI increased the axial rail stresses by approximately 16.3%, and the inclusion of both lateral and vertical SSI further raised them by up to 19.6%, providing a closer match to the experimental peak values. Similarly, vertical deck displacements at midspan increased by 12.7% when vertical SSI springs were included. This highlights the importance of vertical soil flexibility in capturing realistic deformation behavior. Thus, these findings underline the necessity of incorporating both lateral and vertical SSI effects in numerical models for accurately predicting the dynamic responses of rail–viaduct systems under train loads. Hence, it is recommended that future design guidelines and code provisions should consider these aspects to ensure safer and more reliable CWR bridge performance under train loading conditions.

References

1. International Union of Railways (UIC) (2001) UIC 774-3R: track–bridge interaction—recommendations for calculations. UIC, Paris
2. European Committee for Standardization (CEN) (2003) EN 1991–2: actions on structures—part 2: traffic loads on bridges. CEN, Brussels
3. Research designs and standards organisation (RDSO) (2015) BS 114: guidelines for rail–structure interaction studies on Indian railways. RDSO, Lucknow
4. Okelo R, Olabimtan A (2012) Nonlinear rail–structure interaction analysis of an elevated skewed steel guideway. J Bridg Eng 16:392–399
5. Ramos OR, Schanack F, Ortega Carreras G, de Vena Retuerto J (2019) Bridge length limits due to track–structure interaction in continuous girder prestressed concrete bridges. Eng Struct 196:109310
6. Liu W, Lai H, Dai G, Rao S, Wang D, Wu B (2021) Numerical study on track–bridge interaction of integral railway rigid-frame bridge. Appl Sci 11(3)
7. Farahzadi L, Bozorgmehr Nia S, Shafei B, Kioumarsi M (2025) Sustainability assessment of ultra-high-performance concrete made with various supplementary cementitious materials. Cleaner Mater 15:100301. https://doi.org/10.1016/j.clema.2025.100301
8. Farahzadi L, Bozorgmehr Nia S, Shafei B, Kioumarsi M (2025) Preliminary environmental assessment of ultra-high-performance concrete mixtures. In: Kioumarsi M, Shafei B (eds) NTZR 2024, Lecture notes in civil engineering, vol 237. Springer, Cham, pp 589–598. https://doi.org/10.1007/978-3-031-69626-8_49
9. Wivast J, Nyland A, Bozorgmehr Nia S, Kioumarsi M, Shafei B (2025) Developing eco-friendly ultra-high-performance concrete by utilizing recycled alternatives. In: Kioumarsi M, Shafei B (eds) NTZR 2024, Lecture notes in civil engineering, vol 237. Springer, Cham, pp 49–57. https://doi.org/10.1007/978-3-031-69626-8_5
10. Bozorgmehr Nia S, Nyland A, Wivast J, Kioumarsi M, Shafei B (2025) Investigations of Portland limestone cement and waste glass powder for sustainable ultra-high-performance concrete. Case Stud Constr Mater 22:e04425. https://doi.org/10.1016/j.cscm.2025.e04425
11. Peck RB, Hanson WE, Thornburn TH (1974) Foundation engineering, 2nd edn. John Wiley & Sons, New York
12. Bowles JE (1997) Foundation analysis and design, 5th edn. McGraw-Hill, New York
13. Computers and Structures, Inc. (2023) SAP2000 v24.0—Integrated software for structural analysis and design. CSI, Berkeley
14. Roy SS, Dash SR (2018) Dynamic behavior of multi-span continuous girder bridge with isolation bearings. Int J Bridge Eng 6(2):1–23
15. Bureau of Indian Standards (BIS) (2021) IS 2911 (Part 2)—Design and construction of pile foundations: Code of practice. 2nd rev. BIS, New Delhi

16. Nguyen DV, Kim DK, Warnitchai P, Lee CS (2006) Three-dimensional modeling and dynamic analysis of high-speed train–bridge interaction. In: Proceedings of the 10th East Asia-Pacific Conference on structural engineering and construction, Bangkok, Thailand, pp 1–5
17. Chung W, Sotelino ED (2006) Three-dimensional finite element modeling of composite girder bridges. Eng Struct 28(1):63–71
18. Hazzar L, Karray M, Pasic A (2015) Numerical study for soil-spring stiffness of pile group. In: Proceedings of 68th Canadian geotechnical Conference. (Géo-Québec), Québec, Canada
19. Magade SB, Ingle RK (2021) Effect of free vibration frequency on modeling of bridge pier. Transp Infrastruct Geotech 8:296–304. https://doi.org/10.1007/s40515-020-00138-3
20. Propika J et al (2021) Lateral bearing capacity analysis of pile foundation using a spring modelling system. IOP Conf Ser: Mater Sci Eng 1010:012007. https://doi.org/10.1088/1757-899X/1010/1/012007
21. Vesic AS (1977) Design of pile foundations. national cooperative highway research program synthesis 42, Transportation Research Board, Washington D.C
22. Government of India (2014) Ministry of Railways (railway board): bridge rules (in SI units): rules specifying the loads for design of superstructure and substructure of bridges and for assessment of the strength of existing bridges, 3rd edn. RDSO, Lucknow
23. Saxena N, Ray-Chaudhuri S (2026) Rail–viaduct interaction on India's first continuously welded rail: an experimental study. J Bridg Eng 31(2):05025009. https://doi.org/10.1061/JBENF2.BEENG-7647

Intensity Measures for Flood Hazard in Bridge Fragility Assessment: Advancements and Climate Considerations

Tahereh Torabi, Behrouz Shafei, and Alice Alipour(✉)

Iowa State University, Ames, IA 50011, USA
alipour@iastate.edu

Abstract. This paper presents a holistic framework for assessing the flood hazard in bridge fragility analysis, integrating refined intensity measures (IMs), such as hydrodynamic forces, scour depth, and debris impact, while incorporating probabilistic climate models and computational simulations. Given the increasing occurrence and severity of extreme flood events, enhanced methodologies for flood risk evaluation is critical. The presented research explores limitations of existing fragility assessment approaches and proposes novel methods integrating flood frequency analysis with structural vulnerability prediction. The study evaluates traditional IMs, such as water depth and velocity. Furthermore, the rule of refined IMs is investigated from the perspective of performance-based evaluation. Using a representative bridge, changes in structural response and global mode shapes under varying flood scenarios are evaluated, demonstrating how IM selection influences predicted performance. Numerical simulations enhance the accuracy of fragility assessment, offering a comprehensive understanding of bridge performance under flood conditions. Results underscore the necessity of updating current bridge risk assessment frameworks to incorporate long-term climate trends, ultimately promoting sustainable and resilient infrastructure development.

Keywords: Flood hazard · intensity measures · bridge fragility · structural response · weather events

1 Introduction

Bridges are vital elements of transportation networks and play a critical role in supporting economic growth and regional connectivity. Their failure can lead to severe social and economic consequences, including disruptions to mobility, delays in emergency response, and long-term impacts on community resilience. Among the issues faced by bridge infrastructure, hydrometeorological hazards, such as floods, hurricanes, and storm surges, are particularly destructive. In the United States, hydraulic events are the leading cause of bridge failures, responsible for approximately 60% of recorded incidents, primarily due to scour around bridge foundations. Scours occur when high-velocity water flows erode sediment surrounding bridge piers and abutments, undermining their stability. Hurricanes and tropical storms pose additional risks, especially in coastal areas where storm surge, heavy rainfall, and debris accumulation intensify hydraulic loading on bridges [1, 2].

H. Beushausen et al. (Eds.): NTZR 2025, LNCE 873, pp. 818–826, 2026.
https://doi.org/10.1007/978-3-032-27698-8_72

The consequences of bridge failures extend far beyond the direct repair or replacement costs. Disrupted transportation networks hinder post-disaster emergency response and recovery, slow the movement of goods and services, and impair long-term economic development. This underscores the need for accurate and robust methods to evaluate bridge vulnerability under hydraulic hazards and prioritize mitigation strategies. A common approach to quantifying infrastructure vulnerability is through fragility analysis. Fragility functions define the probability that a structure will reach or exceed a given damage state under a specific hazard intensity measure (IM), such as flood stage, scour depth, or water velocity. These functions can be used to construct fragility curves for single hazards or fragility surfaces when multiple hazards or interacting effects are considered. They serve as an essential input for decision-making in bridge asset management, informing maintenance prioritization, identifying retrofit needs, setting operational restrictions during extreme events, and supporting long-term investment planning [3, 4].

With the increasing frequency and severity of extreme weather events, the need to incorporate hydraulic hazards, such as flooding and scour, into risk assessment and management frameworks has become urgent. Recent bridge failures in the U.S. have been attributed to flood-related mechanisms, including scour, debris accumulation, vessel collision, and structural deterioration, with hydraulic failures occurring at an estimated annual rate of 1 in 8,500 bridges [2]. Among the outlined causes, scour is a dominant factor. Similarly, poor substructure condition ratings correlate strongly with increased hydraulic vulnerability. Proactive maintenance, such as regular underwater inspections, has been shown to reduce hydraulic risk substantially [5].

Weather events are expected to increase precipitation extremes, intensify flooding, and accelerate deterioration of infrastructure. Khatami et al. [6] and Alipour and Shafei [7] emphasized that deterioration processes such as corrosion, combined with climate effects, can amplify flood vulnerability, especially for aging bridges. Their frameworks integrate physics-based deterioration models with system-level resilience analysis to capture compound risks. These studies underscore the need to incorporate climate model outputs, including precipitation frequency, intensity, and seasonality, into hydrologic and hydraulic models such as HEC-RAS to estimate future flood velocities, depths, and scour potential. Integrating climate-informed hazard characterization enables the development of fragility functions that better represent future risk, allowing transportation agencies to prioritize adaptation strategies and improve network resilience in a changing climate. Torabi et al., further highlights that addressing climate change in hazard modeling and infrastructure design is essential, as the projected increase in extreme weather events demands proactive adaptation and risk assessment strategies to ensure long-term resilience and sustainability of bridge infrastructure [8].

This study focuses on identifying and evaluating refined flood-related IMs and integrating climate-informed hazard characterization to capture evolving flood risks. It synthesizes existing research and incorporates numerical modeling of a representative bridge to demonstrate how variations in Ims, i.e., specifically flow velocity, water depth, and scour depth, affect global structural response and mode shapes. These insights provide a foundation for improving flood-specific fragility assessments and support the development of adaptive bridge management practices in the face of weather events.

2 Flood Load Effects and Intensity Measures

Bridges experience multiple flood-induced loads, including hydrodynamic, hydrostatic, buoyant, debris impact, and scour effects. Hydrodynamic pressure is a critical load effect that bridges experience during flood events. According to AASHTO [9], the longitudinal water flow acting on bridge substructures is represented by:

$$P_{avg} = \frac{C_D V_{avg}{}^2}{1000} \tag{1}$$

where P_{avg} is average stream pressure (ksf), C_D is the drag coefficient determined by pier shape and debris accumulation, and V_{avg} is the mean flow velocity (ft/s). Experimental studies by Kerenyi et.al [5] provide drag, lift, and moment coefficients for submerged decks, with recent work highlighting the influence of deck width, girder shape, inundation ratio, and blockage ratio on these forces.

Hydrostatic pressure, generated by still water, acts perpendicular to the submerged surfaces of bridge elements. It is calculated as:

$$F_h = \frac{1}{2}\rho_w g h A \tag{2}$$

where ρ_w is water density, g is gravitational acceleration, h is the depth of the water, and A is the projected area normal to the upstream and downstream faces of the submerged portion of the bridge superstructure [9, 10].

Buoyancy, the upward force exerted by displaced water, can significantly reduce the effective weight of submerged structural components. Buoyant force is determined as:

$$F_B = \rho_w g V \tag{3}$$

where V is the volume of displaced water [10].

Floating debris, including logs and large objects, poses a significant risk of impact loading during flood events. The collision force from large debris, such as logs, is calculated using:

$$F_{log} = \frac{0.5mv^2}{s} \tag{4}$$

where m is debris mass (kg), v is the impact velocity (m/s), s is the stopping distance (meter) [11].

Debris accumulation around piers increases flow blockage, leading to higher water levels and pressures on the structure [4, 12]. Two main approaches are used to estimate debris load: the one recommended by the Australian Standard [13], which assumes a fixed debris mat size, and the Panici and Almeida model [14], which incorporates flow velocity and hydraulic conditions to dynamically estimate debris dimensions. The latter provides a more realistic assessment by linking the debris size to the flow velocity and key log length. The “key log length,” an essential factor in debris accumulation, is influenced by pier size, channel width, flow conditions, and local timber characteristics.

Scour, the removal of sediment near bridge foundations due to high-velocity flow, is one of the leading causes of bridge failure during floods [12]. Traditional scour estimation methods often rely on deterministic formulas. HEC-18 [15] provides one such equation:

$$y_s = \left[2.0\, K_1\, K_2\, K_3 \left(\frac{H}{a} \right)^{0.35} Fr^{0.43} \right] a \tag{5}$$

where K_1, K_2, and K_3 are correction factors for pier nose shape, angle of attack of flow, and riverbed conditions, respectively, and a is the pier width. Fr is the Froude number.

Recent studies on bridge vulnerability under floods have considered a range of IMs. Liao et al. [16] included water surface elevation, flow velocity, local scour depth, soil properties, and wind loading. Kim et al. [12] highlighted debris impact and complex scour processes as critical additions. Ahamed et al. [3] used flow discharge and scour depth. Pregnolato et al. [17] incorporated hydrodynamic forces on inundated decks, while Loli et al. [18] proposed an indicator of flood hazard intensity, including flow velocity, water height, flow obstruction, and sediment type. Kosic et al. [19, 20] explicitly modeled wood debris loads and scour effects, and Cabanzo et al. [21] examined scour-induced settlement and flow variations. Anisha et al. [11, 22] considered varying flood levels, debris impact, and scour for reinforced concrete bridges, and Hughes et al. [23] linked debris accumulation and scour to vegetation and slope conditions. Zhang et al. [24] focused on foundation-level behavior under combined loads and scour.

In summary, the evolution of flood-related bridge fragility research has shifted toward multi-hazard analyses, requiring specialized flood-specific studies that consider hydraulic loading, debris, and scour with proper details. Advances have been made in integrating refined IMs, yet additional work is needed to fully capture probabilistic hazard inputs, improve debris and scour modeling, and link structural performance to broader sustainability and resilience metrics.

3 Flood Hazard Characterization

Flood impacts on infrastructure are largely controlled by water depth, velocity, and duration. For bridges, these factors influence structural stability and scour, directly affecting their capacity to withstand hydraulic forces. Based on the detailed outlined in the previous sections, for bridges, hydrodynamic pressures and debris impacts primarily depend on flow velocity, whereas hydrostatic and buoyancy forces are functions of water depth. Scour depth is influenced by both flow velocity and water depth, which are, in turn, correlated with flood discharge and channel geometry. Therefore, evaluating flood velocity and water depth together provides a comprehensive measure of flood loading, while scour depth serves as a separate IM because it alters the structural capacity itself. The proposed approach adopts flood stage and velocity as key IMs, with scour depth treated deterministically to represent boundary condition changes in bridge foundations. Fragility functions are then developed for different combinations of flood stage, velocity, and scour depth, yielding surfaces that express the probability of exceeding defined damage states.

4 Impact of Flood Intensity on Structural Response

For further evaluation, a reinforced concrete (RC) two-span bridge was modeled to understand the influence of flood-induced loading on structural response. Each span measures 15 m, supported by 7.5-m-high RC columns with a diameter of 1.3 m and pile foundations embedded in sandy soil. The superstructure consists of a 23-m-wide deck with a cross-sectional area of 12.0 m^2, modeled using linear-elastic elements due to its expected elastic behavior. Nonlinear response is concentrated in the substructure, with columns represented by fiber-section inelastic beam-column elements to capture distributed plasticity and realistic moment redistribution under lateral loads. Abutments are modeled using nonlinear springs to represent soil-structure interaction, while foundation piles incorporate p–y, t–z, and q–z nonlinear soil springs to capture lateral, axial, and tip resistance behavior, respectively. Scour effects are simulated by removing p–y springs within the scour depth, increasing unsupported pile length and reducing lateral stiffness. The entire model was implemented in OpenSeesPy, providing a flexible and efficient platform for nonlinear analysis and post-processing of results, including system-level responses such as base reactions and mode shapes under varying flood intensities.

Flood loading on the bridge was represented by hydrodynamic forces derived from combinations of water depth (H) and velocity (V). Discharge (Q) was computed for different flow depths, assuming a 30-m channel width (B) consistent with the bridge span configuration. Figure 1 illustrates the discharge–velocity relationship for depths ranging from 3.0 to 8.25 m. As expected, discharge increases linearly with velocity for each depth, reflecting the continuity relationship: $Q = V\cdot(B\cdot H)$. For example, at a velocity of 10 m/s, discharge varies from approximately 900 m^3/s at a 3.0 m depth to more than 2475 m^3/s at an 8.25 m depth. These trends confirm the significant effect of flow depth on flood-induced loading and provide key input parameters for evaluating the structural response of the bridge under extreme hydraulic conditions.

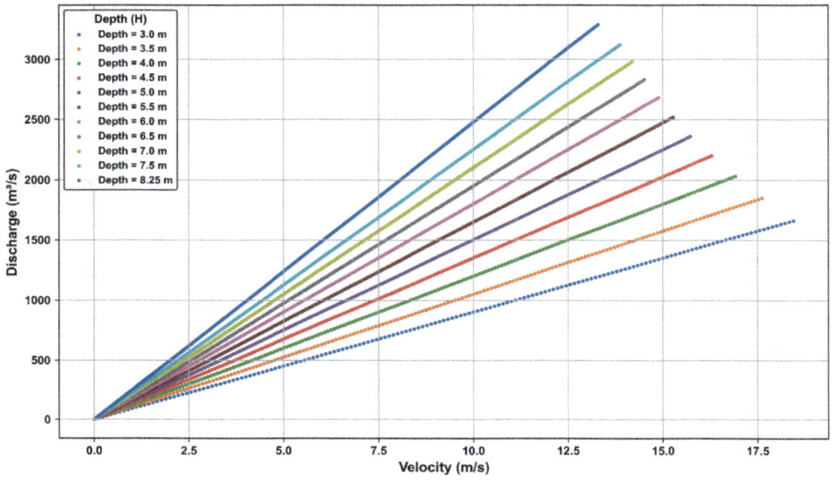

Fig. 1. Discharge calculated as a function of velocity for different flow depths.

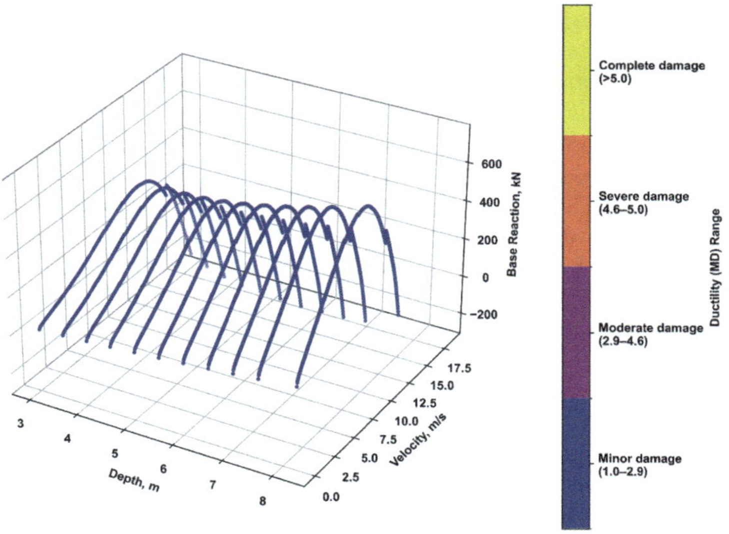

Fig. 2. Base reaction versus depth and velocity of flood.

The structural response of the bridge under varying flood conditions was evaluated by examining base reactions at the foundation level, providing insight into how flood-induced loads and scour effects influence global bridge behavior. The analysis focuses on base reaction forces and their variation under different combinations of flow velocity and water depth, allowing for an initial deterministic assessment of system performance. This evaluation highlights how changes in hydraulic conditions affect overall stiffness, deformation patterns, and load transfer mechanisms in the bridge substructure. Figure 2 illustrates the results of this analysis. Figure 2a presents a three-dimensional view of base reaction forces as a function of flow velocity and water depth. The results reveal a distinct trend: as velocity and depth increase, horizontal base reactions grow, reaching a peak before decreasing sharply. This reduction is attributed to scour-induced changes in boundary conditions that decrease foundation stiffness and shift the structural deformation mode. Interestingly, at higher velocities, the base reaction transitions through zero and reverses direction, indicating a fundamental change in displacement form, likely associated with instability or a softening response in the pier.

Figure 2 presents the variation of base reaction forces at the foundation as a function of flow velocity and water depth, providing a three-dimensional perspective of how combined hydraulic parameters influence structural demand. The results reveal a nonlinear response, where increasing velocity initially amplifies base reactions until a critical point is reached, beyond which stiffness degradation and potential instability are observed.

Figure 3 shows the velocity–base reaction relationship for each depth, illustrating how scour and increasing hydraulic loads lead to stiffness degradation and mode shape transformation. Similarly, Fig. 4 isolates the effect of depth for different velocity ranges, emphasizing that larger depths consistently increase reaction forces up to a threshold, beyond which foundation performance is compromised. This behavior underscores the importance of incorporating both depth and velocity as coupled intensity measures in evaluating foundation performance, particularly for flood scenarios where scour and hydraulic loads simultaneously affect structural response and failure mechanisms.

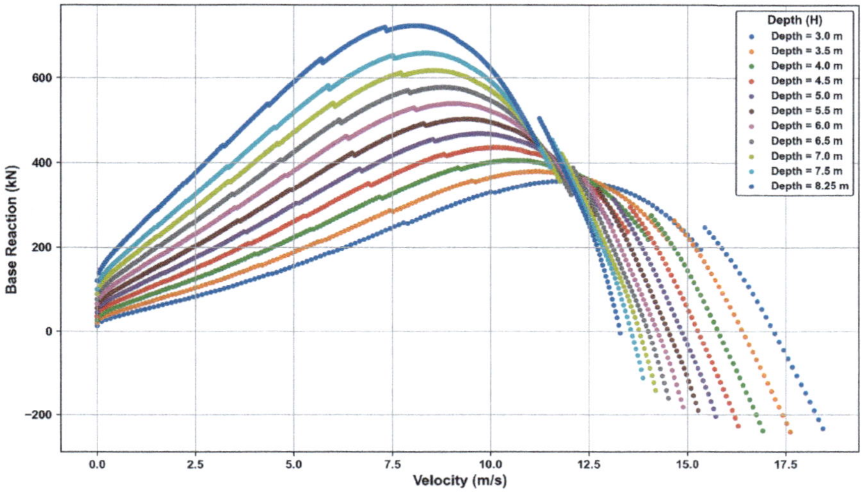

Fig. 3. Base reaction changes as a function of velocity.

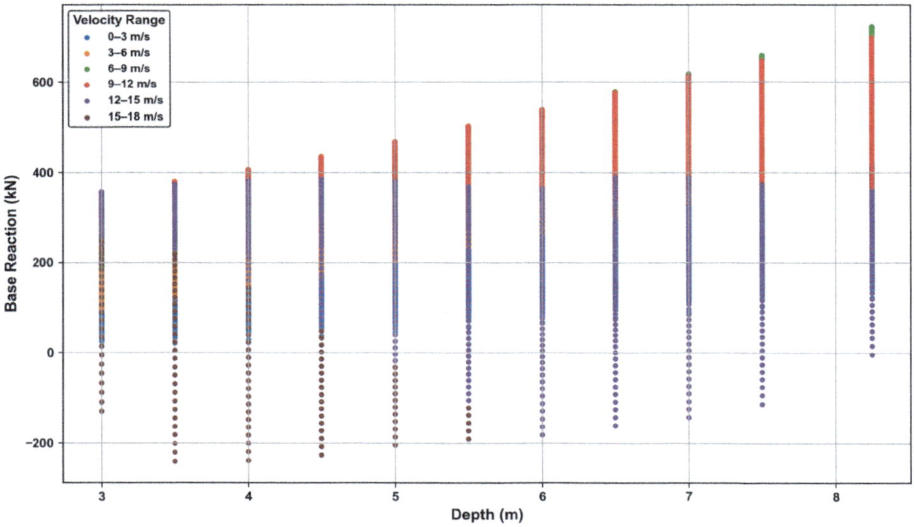

Fig. 4. Base reaction changes as a function of scour depth.

5 Conclusions

This study presented a holistic framework for evaluating bridge vulnerability to flood hazards, emphasizing the role of refined IMs and their direct influence on structural behavior. Hydrodynamic and hydrostatic pressures, buoyancy, debris impact, debris accumulation, and scour were identified as key flood-induced effects that influence bridge performance. Among them, flood stage and velocity were confirmed as primary drivers of hydraulic loading, while scour depth emerged as a critical factor in altering foundation boundary conditions and overall stiffness.

Using a numerical model of a representative two-span RC bridge, the influence of varying water depths and flow velocities on base reactions was evaluated. The results highlighted how changes in hydraulic conditions and scour effects can modify global mode shapes and stiffness distribution, leading to fundamental shifts in structural response under extreme flooding. These findings demonstrate the importance of proper IM selection in capturing load effects that govern system behavior and, ultimately, vulnerability assessment.

The presented study contributes to ongoing efforts to enhance flood fragility assessment by combining refined IMs, climate-driven hazard characterization, and bridge-specific structural response analysis. These insights provide a foundation for transportation agencies seeking to update risk frameworks and adopt strategies that address both current and future challenges posed by flood-induced hazards.

Acknowledgments. The research study, the results of which are reported in this paper, was sponsored by the U.S. Federal Highway Administration. The authors would like to acknowledge and thank the sponsor for this support. Opinions, findings, and conclusions expressed in this paper are of the authors and do not necessarily reflect those of the sponsor.

References

1. FHWA (2012) Evaluating scour at bridges–5th edn hydraulic engineering circular No. 18 (Tech. Rep. FHWAHIF-12-003). Federal Highway Administration
2. Wardhana K, Hadipriono FC (2003) Analysis of recent bridge failures in the United States. J Perform Constr Facil 17:3
3. Ahamed T, Duan JG, Jo H (2021) Flood-fragility analysis of instream bridges–consideration of flow hydraulics, geotechnical uncertainties, and variable scour depth. Struct Infrastruct Eng 17:1494–1507
4. Lee J, Lee Y-J, Kim H, Sim S, Kim J (2016) A new methodology development for flood fragility curve derivation considering structural deterioration for bridges. Smart Struct Syst 17:149–165
5. Kornel K, Tanju S, Junke G (2009) Hydrodynamic forces on inundated bridge decks- FHWA-HRT-09-028. Federal Highway Administration
6. Khatami D, Shafei B (2021) Impact of climate conditions on deteriorating reinforced concrete bridges in the US Midwest region. J Perform Constr Facil 35:04020129
7. Alipour A, Shafei B (2022) An overarching framework to assess the life-time resilience of deteriorating transportation networks in seismic-prone regions. Resilient Cities Struct 1:87–96
8. Torabi T, Shafei B, Alipour A (2025) Intensity measures for flood hazards in fragility assessments of bridges. Int Conf Net-Zero Built Environ 237:1539–1547

9. AASHTO (2020) LRFD bridge design specifications. American Association of State Highway and Transportation Officials
10. Md S, Kabir I, Ahmari H, Dean M (2022) Experimental study to investigate the effects of bridge geometry and flow condition on hydrodynamic forces. J Fluids Struct 113:103688
11. Anisha A, Jacob A, Davis R, Mangalathu S (2022) Fragility functions for highway RC bridge under various flood scenarios. Eng Struct 260:114244
12. Kim H, Sim S-H, Lee J, Lee Y-J, Kim J-M (2017) Flood fragility analysis for bridges with multiple failure modes. Adv Mech Eng 9(3)
13. Standard A (2004) Bridge Design Part 2: Design loads, AS5100. 2 edn. Standards Australia International Ltd, Sydney
14. Panici D, de Almeida GAM (2018) Formation, growth, and failure of debris jams at bridge piers. Water Resour Res 54:6226–6241
15. HEC-18 (2012) Evaluating scour at bridges, 5th edn. U.S. Army Corps of Engineers, Hydrologic Engineering Center, p 547
16. Liao K-W, Muto Y, Chen W-L, Wu B-H (2016) A probabilistic bridge safety evaluation against floods. Springerplus 5:1–19
17. Pregnolato M, Winter AO, Mascarenas D, Sen AD, Bates P, Motley MR (2022) Assessing flooding impact to riverine bridges: an integrated analysis. Nat Hazards Earth Syst Sci 22:1559–1576
18. Loli M, Kefalas G, Dafis S, Mitoulis SA, Schmidt F (2022) Bridge-specific flood risk assessment of transport networks using GIS and remotely sensed data. Sci Total Environ 850:157976
19. Kosič M, Anžlin A, Bau V (2023) Flood vulnerability study of a roadway bridge subjected to hydrodynamic actions, local scour and wood debris accumulation. Water 15(1):129
20. Kosič M, Prendergast L, Structures AA-E (2023) Analysis of the response of a roadway bridge under extreme flooding-related events: scour and debris-loading. Eng Struct:279;115607
21. Cabanzo CM, Santamaría M, Sousa HS, Matos JC (2022) In-plane fragility and parametric analyses of masonry arch bridges exposed to flood hazard using surrogate modeling techniques. Appl Sci 12:1886
22. Anisha A, Sahu DK, Sarkar P, Mangalathu S, Davis R (2023) High dimensional model representation for flood fragility analysis of highway bridge. Eng Struct 281:115817
23. Hughes W, Santos L, Lu Q, Malla R, Ravishanker N, Zhang W (2023) Probabilistic risk assessment framework for predicting large woody debris accumulations and scour near bridges. Struct Infrastruct Eng 20(12):2035–2052
24. Zhang D, Xiong W, Ma S, Cai CS (2024) Fragility evaluation of bridge pile foundation considering scour development under floods. J Bridg Eng 29(8):04024052

Characterisation of Heavy-Haul Track-Bridge-Interaction Using Structural Health Monitoring Data

Emilia Mupwedi(✉) and Pilate Moyo

Department of Civil Engineering, University of Cape Town, Cape Town, Republic of South Africa
mpwemi001@myuct.ac.za

Abstract. Railways remain a critical mode of transportation, with heavy haul operations playing a central role in transporting bulk commodities such as iron ore. The Olifants River Viaduct (ORV), a key single-track bridge on the Sishen–Saldanha Iron Export Line, experiences substantial loading due to long, multi-locomotive trains. This study analyses the speed profiles of trains crossing the ORV and their impact on track–bridge interaction forces, such as rail forces, rail temperature, ambient temperature, and deck deflection. Trains were categorised into six groups (A–F) based on length, locomotive configuration, and cargo, with category D trains, comprising six locomotives and 342 wagons, being the most frequent. Speed variations in loaded trains generate traction and braking forces, increasing longitudinal rail stresses, while unloaded trains primarily contribute through braking and dynamic forces. Acceleration also increases rail temperature, amplifying compression forces, whereas deceleration elevates tensile forces, particularly during colder months when rail forces can exceed 1,400 kN. The study highlights the need for careful monitoring of train speeds and distribution, especially for category D trains, to ensure safe operations and prevent rail failure. These findings support sustainable heavy-haul operations by mitigating infrastructure risks while maintaining operational efficiency.

Keywords: Structural Health Monitoring · Heavy-Haul Railways · Train Categorisation · Speed Profiles · Rail Forces

1 Introduction

Heavy Haul (HH) operates the world's longest and heaviest freight trains, moving large quantities of bulk goods from mining areas to special export terminals. Significant advances have been made in all aspects of rail freight technology over the past few decades, enabling this development, including improvements in the quality and reliability of the relevant infrastructure. In recent years, a large percentage of HH tracks consists of continuously welded rails (CWR), defined as rails that have been welded together into lengths exceeding 1 km because of the advantages they offer, such as reduced maintenance costs, fewer rail defects and failures, reduced wear on vehicles,

H. Beushausen et al. (Eds.): NTZR 2025, LNCE 873, pp. 827–838, 2026.
https://doi.org/10.1007/978-3-032-27698-8_73

and lower energy costs for traction [1]. When the rail is constructed on a bridge, the behaviour of the bridge leads to additional forces for CWR rails. Since both the track and the bridge may move, any force or displacement applied to one will cause forces to be applied to the other. Thus, both the track and bridge are prone to longitudinal forces such as temperature forces, traction and braking forces, and forces induced by bridge movement [2]. The imposed forces can result in rail buckling and rail breaks when they are large enough. This interaction of forces between the track and the bridge is described as Track-Brick Interaction (TBI) [3], and has been studied as early as 1849 [4]. In the hope of lessening the effect of interaction forces on CWR bridges, many track-bridge dynamic interaction (TTBDI) models have since been developed and classified. The classification of the interaction models can be found in [5]. The management of stresses of track-bridge interaction forces on CWR railway bridges has been studied by [6–8, 9]. The contributing factors affecting track-bridge interactions include temperature variations, acceleration and braking, rotation of the deck due to vertical loading, and deformation of the supporting concrete structure.

According to [3], the maintenance of CWR affects accumulated stresses, stiffness, and stability, and should only be done within a controlled temperature range. Both the rail material and temperature determine the resistance to rail forces, and these factors must be carefully managed. Rail stress management has therefore become a key focus for railway companies worldwide, with restressing practices and monitoring of the stress-free temperature (SFT) being used to control deviations from the design neutral temperature (DNT) [10] and the design neutral temperature (DNT) [10]. Rail creep monitoring is also regarded as an effective approach to understanding stresses [11], though continuous monitoring remains the most reliable method to ensure safe conditions [12]. Thermal expansion and contraction further complicate force management, since compression occurs when temperatures exceed the neutral temperature and tension when they fall below it [13]. Excessive compressive forces may cause buckling, while excessive tensile forces can lead to rail breaks, both of which pose serious derailment risks. Continuous monitoring of the neutral temperature and rail stresses is therefore essential to ensuring the safe and efficient operation of CWR tracks. Hence, this paper aims to characterise the TBI of heavy haul railways using structural health monitoring.

2 Case Study: Olifant's River Viaduct (ORV)

The Olifants River Viaduct (ORV) is a 23-span bridge, equally spaced, that was constructed in 1976. It consists of a single prestressed concrete girder, 1035 m long, supported by 22 reinforced concrete piers spaced 45 m apart, with the tallest pier reaching a height of 50 m. The bridge is situated in the Western Cape province of South Africa, connecting the mines in Sishen to the port in Saldanha. The bridge has two expansion joints, located at piers 11 and 12. Figure 1 shows the ORV, and further description of the ORV can be found in [14].

The ORV is equipped with monitoring systems, which will be elaborated below. The Olifants River Viaduct (ORV) is owned and managed by Transnet SOC Ltd. and operated by Transnet Freight Rail (TFR). The bridge forms part of an 861 km line with continuously welded rails primarily used for transporting iron ore. Increased demand

Fig. 1. Olifants river viaduct

for iron ore and coal led Transnet to adopt radio distributed power (RDP) technology, replacing traditional head-end power (HEP). This advancement allows locomotives to be placed along the train, reducing in-train forces and enabling longer trains. Designed to handle such capacity [15], the line meets the International Heavy Haul Association (IHHA) classification for Heavy Haul railroads, supporting trains of 5000 metric tons and transporting over 20 million metric tons annually. Previous studies on the ORV [16, 17] investigated the effects of traffic loading, speed, and dynamic behaviour on the bridge and its supporting infrastructure. The bridge is equipped with monitoring systems that provide data for understanding and managing these effects.

2.1 Monitoring System

The ORV is equipped with three monitoring systems: Wheel-Impact Monitoring-Weigh-in-Motion (WIM-WIM) system on span 21, a Monitoring System on span 20, and a Saldanha UPM60 system along the expansion joints. However, only data obtained from the WIM-WIM system and the UPM60 systems were used in this paper; thus, only these systems will be elaborated below.

Wheel Impact Monitoring and Weigh-in-Motion (WIM_WIM) System The WIM-WIM system, installed on the Sishen–Saldanha Iron Export Line in 1998 [18] and managed by Transnet, monitors wheel irregularities that generate high dynamic forces on the track and weighs trains to detect bogie skewing. It consists of strain gauges mounted on the rail, a signal conditioner for calibration, a high-speed computer for data acquisition and communication, and lightning protection for safety. Located at span 21, the system includes 32 strain gauges sampled at 2.5 kHz per channel and a computer-based DAQ housed inside the viaduct girder. Although it can measure speeds up to 150 km/h, the ORV speed limit is 50 km/h. Using a trigger mechanism, the system detects passing trains, identifying wagon and locomotive types, speed, and mass. Data from the WIM-WIM system were used to analyse train speeds in this study.

Rail Force Monitoring System Saldanha System to measure rail forces, deck deflection, and rail and ambient temperature. There are currently two measurement systems installed on the ORV that measure the rail forces, namely the Saldanha System (SS) and the Track Testing Centre System (TTCS). Both systems employ UPM60 scanners to measure rail force, deck deflection, and temperatures. Measured parameters are transmitted to the bridge computer situated at the traffic control office in Saldanha Bay by the Saldanha system. The TTC System serves as a backup system if the Saldanha System fails or malfunctions.

3 Methodology

Python was used to group trains into loaded and unloaded categories based on the direction indicated in the WIM-WIM output file, labelled as "UP" or "DOWN." "UP" refers to trains travelling from the port in Saldanha to the mines in Sishen, while "DOWN" refers to trains moving in the opposite direction. It was assumed that trains from the mines were loaded and those from the port were unloaded; this was verified by checking wagon masses, with unloaded wagons averaging 200 kN (20 tons) and loaded wagons 1200 kN (120 tons). The ORV is a single-track bridge, so no two trains cross simultaneously. Each train was identified by a unique timestamp, allowing all vehicles within the timestamp to be grouped as a single train, following the categorisation developed by [19]. Train speeds were then extracted and plotted separately for loaded and unloaded trains using Python, enabling observation of speed variations and their influence on the loading spectra as trains crossed the ORV. Track–bridge interaction forces were analysed in MS Excel, with predefined warning and alarm thresholds for safety, including a tension warning of 1,400 kN, tension alarm of 1,500 kN, compression warning of −1,100 kN, and compression alarms at −1,200 kN and − 1,400 kN.

4 Results and Discussion

4.1 Train Categorisation and Train Configuration

A total of 4,099 trains were categorised based on their length and the commodities carried. Train length was determined by the number of wagons, locomotives, and axle loads associated with each commodity. The trains were grouped into six categories, with categories A to D consistent with those defined in [19]. However, two additional groups, categories E and F, were identified to include trains not covered in the previous classification. Category E mainly comprised locomotive-only movements, while category F included trains with five to seven locomotives and variable wagon numbers. Table 1 presents the full train categorisation with their respective number of trains per category, including the additional categories.

The results show clear differences in train configuration, frequency, and loading across the six categories. Categories A and B consist mainly of short head-end power (HEP) trains with simple setups and fewer locomotives. Category A had the highest number of short trains, with nearly equal loaded and unloaded runs, showing balanced movement in both directions. Category B was similar but less frequent, suggesting limited operation. In contrast, categories C and D represent longer radio distributed power (RDP) trains capable of hauling heavier loads. Category C showed a balanced distribution

Table 1. Updated train categorisation

Category	No. of locos	No of wagons	Loading type	No of trains	Type of configuration
A	1–2	1–59	Loaded	309	LC"1" + kWs
			Unloaded	315	
B	2–3	60–152	Loaded	22	LC"1" + lWs
			Unloaded	51	
C	2–4	153–232	Loaded	197	LC"1" + mWs + LC"2" + nWs + LC"3"
			Unloaded	210	
D	4–6	282–342	Loaded	1510	LC"1" + rWs + LC"2" + sWs + LC"3" + tWs + LC"4"
			Unloaded	1468	
E	1–2	None	Locos	40	LC"1" + LC"1"
F	1–8	Unlimited	Mixed	14	LC"1" + rWs + LC"2" + sWs + LC"3" + tWs + LC"4"

between loaded and unloaded trains, while category D dominated with nearly 3,000 crossings. The large number and complex configuration of category D trains highlight their importance in heavy-haul operations. Categories E and F were outliers, mainly locomotive-only or irregular setups. Overall, the results reflect a shift from shorter HEP to longer RDP configurations, supporting higher capacity and efficiency in heavy-haul transport.

4.2 Speed Profiles

Category D trains, the most frequent at the ORV, consist of 342 wagons and six locomotives in the configuration (43D)–114 W–(15E)–(43D)–114 W–(15E)–114 W–(15E). Therefore, only their speed profiles are discussed; see Fig. 2 for the speed profiles of both loaded and unloaded conditions. Loaded trains enter the bridge at 40–50 km/h and exit at 35–45 km/h. They decelerate over the first 140 vehicles (by about 5–20 km/h) before accelerating again (by 5–25 km/h) from the 141st vehicle to the end. The deceleration induces braking forces, while the acceleration induces traction forces, both contributing to higher horizontal rail forces. Heavier trains further amplify dynamic loading, making the combined effects of braking, acceleration, and weight dominant in the track–bridge interaction.

In contrast, unloaded trains enter at 40–65 km/h and exit at 20–30 km/h. Their speed remains constant over the first 150–200 vehicles, followed by a gradual deceleration of 5–20 km/h towards the end. Because there is little change in speed during most of the crossing, traction forces are minimal. Instead, braking and dynamic effects from the final vehicles dominate the rail and bridge interactions. Overall, loaded trains produce greater longitudinal and dynamic forces due to their higher mass and speed variations, increasing the stresses transmitted to the track–bridge system.

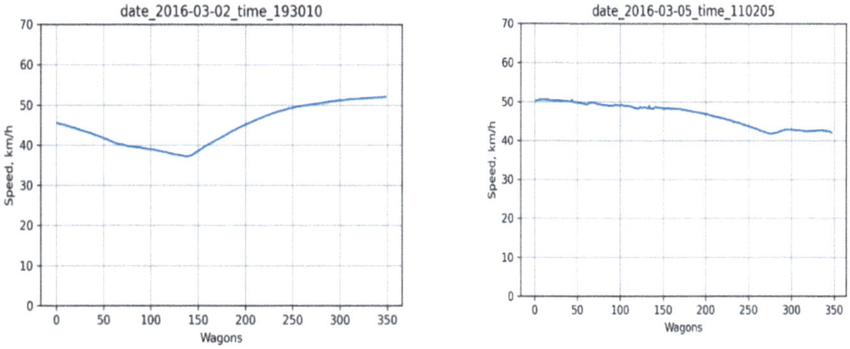

(a) (b)

Fig. 2. Category D speed profiles: (a) Loaded, (b) Unloaded

In summary, about 20% of the trains exceed the 50 km/h speed limit. The variation in the speeds between the loaded and unloaded trains was assumed to be due to the topography of the bridge. The loaded trains are required to climb a cliff upon exit of the ORV; thus, explaining the acceleration upon exit of the vehicles. While the unloaded trains are subjected to a much flatter terrain that slopes downwards, this explains why the unloaded trains decelerate from vehicle to vehicle till the last vehicle. The speed was also found to be a function of the train's length and the commodities it hauled. Empty trains crossed the bridge at higher speeds compared to the loaded trains because the normal load per wagon is reduced. The overall results are within range with the results obtained in [20].

4.3 Presence of the Train on the Bridge

Based on the speed of the most frequent train on the ORV, the trains experience a significant change in speeds, including both acceleration and deceleration, and additional forces are induced on the rails due to this change in speed. Therefore, during the presence of this train on the bridge, changes in rail forces, deck deflection, and rail temperature are experienced, and a jump in all three parameters represents these changes. Figure 4 shows the shape of the jumps experienced by the rail forces, deck deflection, and rail temperature. The rail forces and the deck deflection both experienced a similar jump, shown in Fig. 3a. In addition, the jump in the rail temperature exhibits a linear trend, as shown in Fig. 3b.

The shape of the jump in the rail forces is dependent on the time the train is present on the bridge. The rail forces are high in tension during the morning hours; thus, a train crossing the bridge during this time causes an increase in the rail forces when the train brakes, followed by a decrease in the rail forces when the train accelerates. However, a train crossing the bridge in the afternoon, when rail forces are high in compression, causes a decrease in rail forces due to braking, followed by an increase in rail forces due to acceleration. Therefore, during the day, the jump in the rail forces takes the shape of a V, while a train crossing the bridge during the morning hours has a reverse V shape.

Fig. 3. Jump shapes experienced in the rail forces, deck deflection, and rail temperature (a) rail force and deck deflection jump; (b) rail temperature jump

The influence on the rail forces is critical when a train is present on the bridge during the tensile or compression turning points because the rail forces are at their maximum during these points; thus, a jump in the rail forces at such a time will drastically increase the rail forces and add additional longitudinal forces on the rails. Figure 6 illustrates the rail forces for December 1, 2016, highlighting the rail force peaks and their corresponding turning points. A train crossing the bridge during the peaks is likely to cause rail failure by buckling or breaking of the rail compared to a train crossing the bridge at any other time. The jumps occurring on the tension turning point are the opposite in shape from those occurring during a compression turning point, but both cause an increase in the rail forces and add to the longitudinal forces. Hence, the focus has been given to trains crossing the bridge during the turning point of the rail forces.

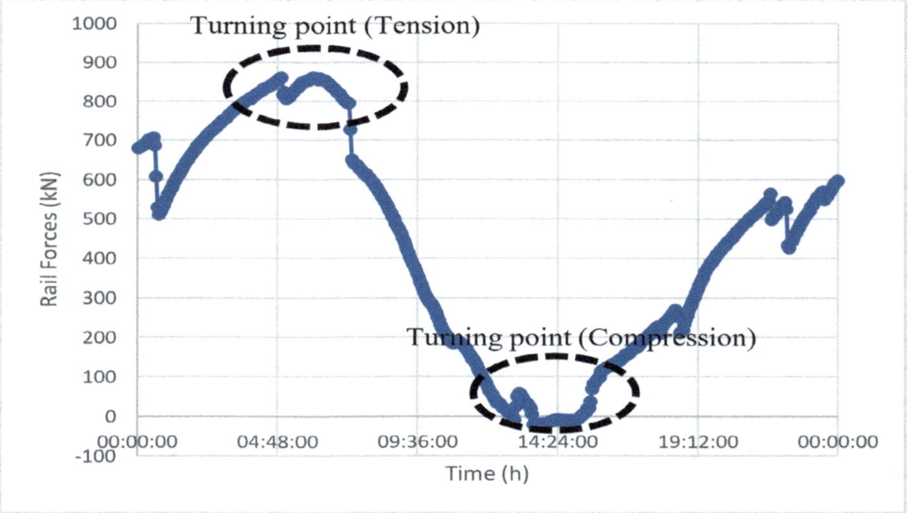

Fig. 4. Rail force (1/12/2016)

4.4 Track-Bridge Interaction

According to the rail force jump presented in Fig. 3a, an increase in rail forces and deck deflection is observed when the train brakes, followed by a decrease in rail forces as the train accelerates. Figure 5 illustrates the relationship between the speed profiles, rail forces, rail temperature, and air temperature. Figure 5a shows the rail forces versus temperature (air and rail) for a train that crossed the ORV on May 1, 2016, at 22:49, while Fig. 5b presents the corresponding speed profile.

As the train decelerates, an increase in rail force is observed, while a decrease in rail force occurs when the train begins to accelerate again. In contrast, acceleration also causes an increase in rail temperature. As seen in Fig. 5a, a deceleration from 42 km/h to 31 km/h caused the rail force to increase from 1120 kN to 1140 kN, adding a braking force of 20 kN, while the subsequent acceleration from 31 km/h to 44 km/h decreased the rail force from 1140 kN to 1010 kN, adding a 30 kN change. The acceleration also resulted in an increase in rail temperature, which rose from 12 °C to 18 °C, given an air temperature of 15.9 °C. Therefore, the 130 kN change in rail forces due to acceleration represents the combined effect of traction and temperature-induced forces.

During the colder months, high tensile forces were observed, with rail forces exceeding the 1400 kN limit. These tension forces are primarily due to trains crossing the bridge during peak tension periods when rail temperatures are low. The trains caused a rise in rail temperature, which increased the longitudinal forces on the rails, thereby amplifying the overall rail forces. The magnitude of both rail temperature and rail force jumps depends on the variation in acceleration and deceleration speeds. It is also important to note that the shape of the rail force jump depends on the time the train crosses the bridge. While the train speed is independent of the season and time of day, the effect on the rail depends on temperature variations. Consequently, additional tensile forces should be expected during morning hours and additional compressive forces during afternoon hours when a train is present on the bridge.

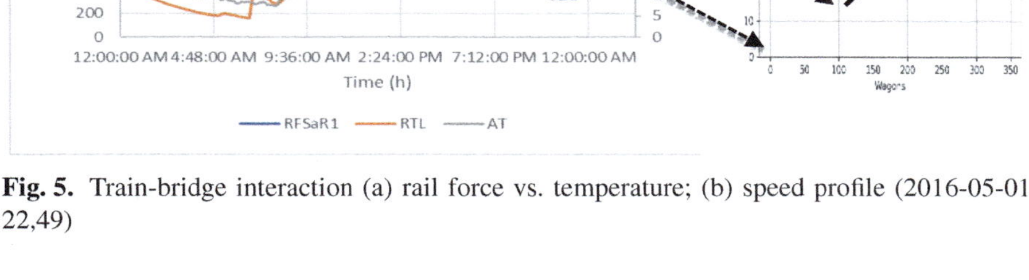

Fig. 5. Train-bridge interaction (a) rail force vs. temperature; (b) speed profile (2016-05-01 at 22,49)

This relationship was observed throughout all the loaded trains with acceleration and deceleration speed variations greater than 3 km/h. For trains with a minimal deceleration speed (a 3 km/h difference) and an acceleration of more than 5 km/h, only the effect of the acceleration on the rail forces was recorded, showing a downward-sloping curve. Not all unloaded trains are displayed on the rail forces because unloaded trains tend to decelerate upon entry on the bridge, so only the jumps of trains presenting significant deceleration variations are captured on the rail forces.

During the passage of the train, a similar jump is observed in the deck deflection. However, the shape of the jump is the reverse of the jump observed in the rail force, as shown in Fig. 6a. In Fig. 6a, the deck is in tension with an air temperature of 15.8 °C. The deflection jumped from −56 mm to −56.5 mm, followed by −56.5 mm to −56.1 mm. The deck deflection increases when the train brakes, followed by a decrease in deck deflection when the train accelerates. The magnitude of the jump depends on the change in speed; thus, the higher the deceleration/acceleration speed change, the greater the deflection jump. An increase in the air temperature causes the deck to expand, while a decrease in the air temperature causes the deck to contract, which imposes additional forces on the rails. The horizontal deflection due to the vertical operating loads was within the limits of 3 mm as stipulated in [21]. The decrease in the deck deflection causes an expansion in the rail forces, while the increased deck deflection causes a contraction in the deck, thus inducing tensile forces on the rails. The tensile forces contribute to the longitudinal forces on the rails and might cause rail failure when the forces are great enough.

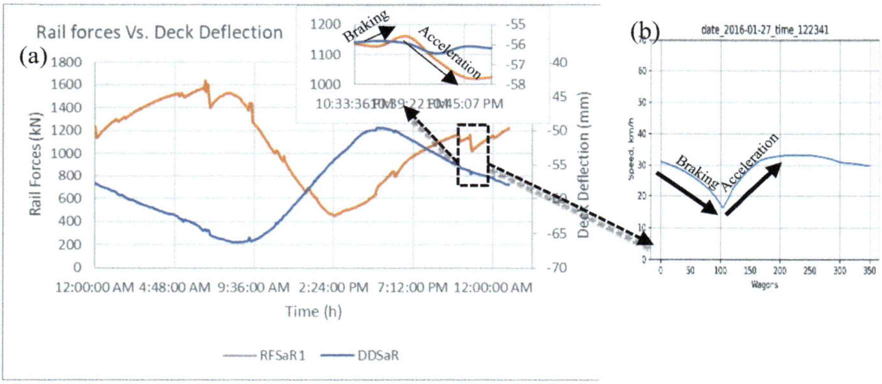

Fig. 6. Train-bridge interaction (a) rail force versus deck expansion; (b) speed profile (2016-05-01 at 22,49)

In summary, train deceleration increases tensile forces and reduces deck deflection, while acceleration decreases rail forces, pushing them into compression and raising both deck deflection and rail temperature. This confirms the rail–bridge interaction described by track–bridge dynamics. Significant speed variations therefore generate additional rail forces beyond normal dynamic effects. During colder months, tension forces often exceeded the 1400 kN limit due to low rail temperatures at the time of crossing. These findings align with [22, 23], which identified temperature, speed variation, and deck deflection as key contributors to longitudinal forces. Hence, preventive measures are crucial to avoid rail breakage in winter and rail buckling in summer.

5 Conclusion

This paper examined the relationship between train speed and track–bridge interaction forces at the Olifants River Viaduct (ORV), focusing on rail forces, temperature, and deck deflection. The findings showed that category D trains, being the most frequent and longest, exhibited significant speed variations caused by their length, hauled commodities, and the bridge's location. These variations, ranging from 5–30 km/h in acceleration and 5–20 km/h in deceleration, led to additional longitudinal forces due to traction and braking. While, acceleration increased rail forces, braking imposed more critical stresses, often coinciding with peak tensile and compressive loads when trains were on the bridge. Although compressive forces remained within the 1,100 kN limit, tension forces occasionally exceeded 1,400 kN, especially at rail temperatures between 0 °C and 20 °C, indicating a higher risk during colder periods, making the rail more susceptible to breakage. Therefore, it is advisable to limit train operations during turning points.

In practice, these findings highlight the need for continuous monitoring of train speed variations and rail temperature to manage interaction forces effectively. Reducing the operation of trains during turning points and monitoring the stress-free temperature could reduce excessive longitudinal stresses and extend the bridge's service life. From a sustainability perspective, such data-driven monitoring supports safer and more efficient heavy-haul operations, minimises energy losses due to braking and acceleration, and promotes the long-term resilience of railway infrastructure.

Acknowledgement. The authors would like to express their utmost gratitude for the partial financial support from the University of Cape Town and thank everyone who contributed to the successful completion of this research.

References

1. Florida Department of Transportation (2013) Continuous welded rail plan [Online]. Available: https://www.fra.dot.gov/Page/P0123
2. Sung WP, Shih MH, Lin CI, Go CG (2005) Critical loading for lateral buckling of continuous welded rail. J Zhejiang Univ Sci 6 A(8):878–885. https://doi.org/10.1631/jzus.2005.A0878.
3. UIC (2001) Track/bridge interaction recommendation for calculations
4. Stokes S (1849) Discussion of a different equation relating to the breaking of railway bridges, vol 5. Cambridge University Press. https://doi.org/10.1017/CBO9780511702259.013
5. Xu L, Zhai W (2019) A three-dimensional model for train-track-bridge dynamic interactions with hypothesis of wheel-rail rigid contact. Mech Syst Signal Process 132:471–489. https://doi.org/10.1016/j.ymssp.2019.04.025
6. Mirković NB, Popović ZJ, Pustovgar AP, Lazarević LM, Zhuravlev AV (2018) Management of stresses in the rails on railway bridges. FME Trans 46(4):636–643. https://doi.org/10.5937/fmet1804636M

7. Ramos R, Schanack F, Carreras G, de Vena Retuerto J (2019) Bridge length limits due to track-structure interaction in continuous girder prestressed concrete bridges. Eng Struct 196(109310):1–8. https://doi.org/10.1016/j.engstruct.2019.109310
8. Mubarack CKA, Upadhyay A (2021) Stability of continuous welded rail on steel bridge subjected to thermal loading. Structure 34:4524–4531. https://doi.org/10.1016/j.istruc.2021.10.050
9. Kang C, Wenner M, Marx S (2021) Background investigation on the permissible additional rail stresses due to track/bridge interaction. Eng Struct 228:1–12. https://doi.org/10.1016/j.engstruct.2020.111505
10. Mitchell L, Mandal N (2012) An investigation into monitoring stresses in continuously welded rails through stress free temperature, pp 1–8 [Online]. Available: https://www.researchgate.net/publication/303805601
11. Mandal NK, Lees M (2019) Effectiveness of measuring stress-free temperature in continuously welded rails by rail creep method and rail stress modules. Eng Fail Anal 104:189–202. https://doi.org/10.1016/j.engfailanal.2019.05.032
12. Salient Systems Inc (2009) System and method for determining rail safety limits, US 7,502,670 B2
13. Marks I (2001) Management of continuously welded rail track in queensland rail. In: 7th intemational heavy haul conference, pp 397–402
14. Busatta F, Moyo P (2018) Assessing the performance of a heavy haul railway viaduct through monitoring traffic loads and dynamic effects. Exp Vib Anal Civ Struct 5:770–780. https://doi.org/10.1007/978-3-319-67443-8_68
15. Kuys W (2000) Ore line capacity expansion: conceptual Design of the Railway Line to increase capacity, vol 85(12), pp 1722–1731
16. Kuys WC (2009) Ore line capacity expansion: Conceptual design of the railway line to increase capacity. In: Proceedings—9th international heavy haul conference: "heavy haul and innovation development, pp 941–949
17. Busatta F, Pilate M (2019) Performance monitoring of a Prestressed concrete railway viaduct: implementation of a system and first results. In: MATEC web of conferences, pp 1–10
18. Tomas MD (2000) Wheel impact monitoring & weigh in motion (WIM_WIM) Systems.pdf
19. Busatta F, Moyo P (2019) Heavy haul train dominant frequencies—Insights and practical use. In: MATEC Web of Conferences, Cape Town, University of Cape Town, pp 1–8
20. Busatta F, Moyo P (2017) How testing and monitoring can support heavy haul railway bridge management: the experience gained in South Africa. In: 11th international heavy haul association conference 2017, pp 2–6
21. UIC (2009) Defects in railway bridges and procedures for maintenance 778–4
22. Maree H (2000) Forces in the railway track structure on the Olifants River bridge
23. Bester JF (2015) The Olifants River bridge—the inherent problem.Pdf. SAICE, pp 1–7

Condition Assessment of Concrete Bridges Using Automated Damage Detection Strategies

Ibrahim Odeh and Behrouz Shafei(✉)

Department of Civil, Construction, and Environmental Engineering, Iowa State University, Ames, IA, USA
shafei@iastate.edu

Abstract. The current study investigates and addresses the process of automating damage detection and assessment in concrete bridge structures. For this purpose, a new procedure is proposed, offering substantial improvements over traditional manual inspection methods by significantly reducing cost, the time needed to inspect, and minimizing inspection errors. The proposed procedure does not require specialized equipment or instrumentation, overcoming a common limitation faced by current bridge inspection methodologies. The introduced capabilities are tailored for efficient detection, quantification, and evaluation of deterioration in reinforced concrete structures. Trained and validated using a dataset of representative concrete bridge component, the procedure accurately identifies and measures various instances of concrete damage. Results demonstrate effectiveness in detecting narrow cracks and quantifying defect dimensions with an error rate below 3%, underscoring its utility in automated structural health monitoring and proactive maintenance planning intended to maximize the service life of bridge structures.

Keywords: Conventional Neural Network · Machine Learning · Damage Detection · Quantification

1 Introduction

Many of existing bridges are made of concrete and deemed vulnerable to deterioration over their service lives [1–3]. Recently, utilizing artificial intelligence (AI), machine learning (ML), and mathematical models to detect the damage in different materials and then incorporating that with the traditional inspection methods have introduced transformative potential to various industries by enhancing safety, documentation, and real-time monitoring capabilities [4–8]. To ensure the safety and serviceability of concrete structures, early detection and assessment of signs of deterioration are critical, as missing such signs can lead to severe failure consequences. Additionally, financial implications of deferred maintenance can be substantial, as the repair cost often grows exponentially with the extent of damage. Therefore, regular inspection and condition assessment are essential to prevent the escalation of initial defect issues to significant and costly problems.

H. Beushausen et al. (Eds.): NTZR 2025, LNCE 873, pp. 839–847, 2026.
https://doi.org/10.1007/978-3-032-27698-8_74

To inspect concrete structures, especially those with multiple components and large footprints, traditional methods suffer several limitations. They are commonly time-consuming and human resource intensive, while they face difficulties in accessing hard-to-reach locations and maneuvering around obstacles [9, 10]. The outlined issues have motivated automated damage detection and assessment strategies, which involve aerial surveys, remote inspections, and geospatial analyses [11–14].

Two models were developed [15, 16] to detect cracks in concrete structures. Both models utilized CNNs to establish, enhance, and expedite the crack detection process. Kim and Cho [17] developed a model for concrete crack assessment using a region-based CNN (R-CNN). The model was trained on dataset with crack damage, and the resolution of the dataset varied from 600 × 600 to 2,000 × 2,000 pixel. In [18], a damage detection model was developed using R-CNN, designed to identify various types of defects in concrete structures. The model was trained on 765 annotated images, with extensive image augmentation applied to expand the dataset and enhance model performance. Previous studies [19–22] explored the Region Proposal Network (RPN) model used by Fast R-CNN for object detection in high-resolution images. This approach enabled fast and accurate detection while requiring relatively low processing power.

To address the existing research gaps, this study aimed to develop a novel CNN model capable of detecting and quantifying surface defects in concrete structures. The proposed model is designed to efficiently process high-resolution images, achieving high levels of precision and accuracy under diverse environmental conditions, including varying lighting levels and surface obstructions. The framework integrates transfer learning and fine-tuning strategies to further enhance its effectiveness in identifying concrete deterioration. Upon completing the necessary validation procedures, the framework extended beyond basic damage detection by incorporating methods to quantify and evaluate the severity of identified defects. These results were then mapped to condition states in alignment with standards outlined in industry-recognized structural inspection manuals.

2 Damage Detection Framework

Figure 1 represents the structured framework of the present study. This study is based on applying advanced object detection techniques for structural assessment. Specifically, YOLOv7, a convolutional neural network designed for fast and accurate object detection, is particularly suitable for assessing concrete structures. It comprises three main components: the backbone for initial feature extraction, the neck for multi-scale feature refinement, and the head for predicting object locations and classes. Key features include anchor boxes for flexible object recognition, pyramid attention network (PAN) and spatial pyramid pooling (SPP) integrations for enhanced multi-scale detection, and a specialized loss function. To overcome limitations of traditional damage detection models, a modified YOLOv7 architecture was developed, capable of processing images of arbitrary resolutions through dynamic layers, significantly improving detection versatility. Additionally, this new model includes a region-growing algorithm for accurately quantifying detected damage areas, such as cracks and concrete spalling, providing precise measurements of defect dimensions.

The second component of the proposed model involves the quantification of detected defects. Following identifying and annotating damaged regions within an image, the

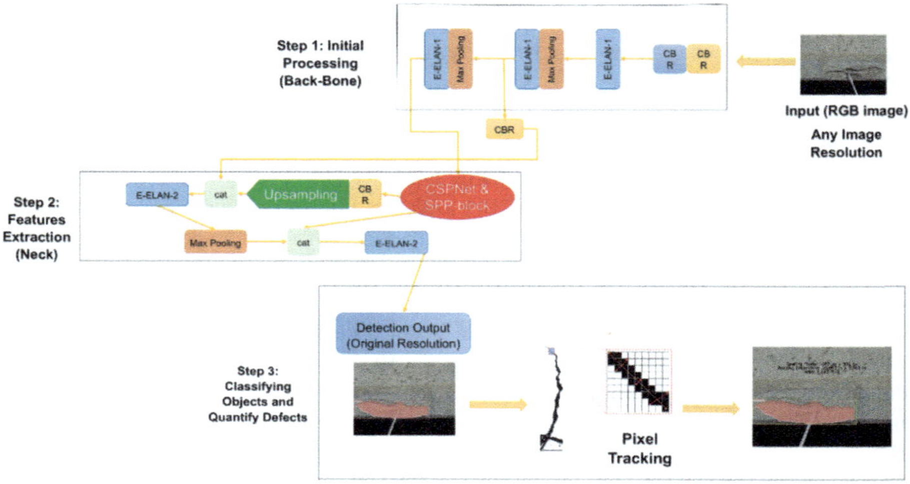

Fig. 1. The framework structure of the main steps established for damage detection.

model employs a region-growing algorithm to calculate the number of pixels contained within each defect. This algorithm, also known as a tracking algorithm, is utilized to sequentially sort the thinned pixels that form a continuous path within the predicted damage area. Figure 2 illustrates the operation of the tracking algorithm within the framework. The defined search directions are critical to ensuring accurate pixel quantification within the identified damaged region.

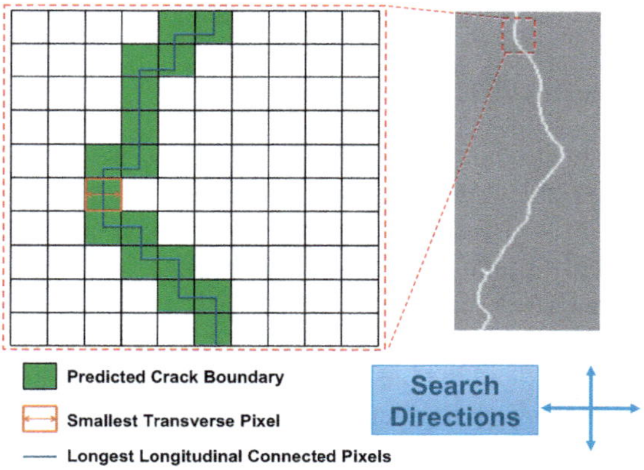

Fig. 2. Illustration of the region-growing algorithm utilized for the current study.

3 Data Collection, Training, and Validation

The model was initially trained using a comprehensive dataset of over 11,000 images depicting damage across various concrete surfaces. This large-scale dataset enabled the model to learn from a broad spectrum of image conditions, enhancing its ability to detect and quantify structural defects with high reliability. To support this process, the training was conducted over 200 epochs on Google Colab Pro, leveraging high-performance GPUs to efficiently manage the computational demands of high-dimensional image data. During the development phase, the model architecture was further refined through the integration of additional CNN layers, enabling it to accommodate images of varying resolutions—a critical capability for application under diverse real-world field conditions. The model's architecture was designed to balance detection accuracy and processing efficiency to ensure real-time deployability, particularly on platforms with limited computational resources. Runtime evaluations confirmed that the model achieves an average processing time of 77.2 ms per image ($\approx$ 13 FPS) for standard 640 $\times$ 640 images and maintains consistent inference speeds across various image resolutions. This performance demonstrates its suitability for high-throughput inspection scenarios and supports field applications such as UAV-based assessments. The model was initialized using transfer learning with pretrained weights from a general-purpose object detection model, providing a strong feature extraction baseline. These weights were progressively fine-tuned using the custom dataset. This strategy accelerated convergence and enhanced generalization to diverse lighting, texture, and defect variations found in concrete structures, without requiring extensive training from scratch.

As shown in Fig. 3, the model's performance was evaluated using a confusion matrix, based on the large-scale dataset of over 11,000 images. The results demonstrated strong classification accuracy, with true positive rates of 88% for cracks and 92% for spalling, indicating high reliability in defect detection. Misclassification between the two defect types was minimal, with 3% of crack cases identified as spalling and 4% of spalling cases identified as cracks, primarily due to variations in lighting and defect complexity. Additionally, false positives—where non-defect regions were incorrectly classified—occurred at rates of 5% for cracks and 6% for spalling. The model also achieved an 89% true negative rate for background regions, reinforcing its specificity and robustness in differentiating between damaged and undamaged areas.

Based on the extensive dataset used for training, the F1 confidence curves further validate the model's reliability in identifying both defect types and sizes of interest. These curves support the confusion matrix results shown in Fig. 3, demonstrating consistent performance across classes. The F1 curves for 'Cracks' and 'Spalling' exhibit minimal deviation, indicating that the model distinguishes between these defects with near-equal accuracy. The 'All Classes' curve closely follows the individual class trajectories, suggesting uniform precision and recall across all categories. As the confidence threshold increases, the F1 score remains high until it naturally declines near a threshold of 1, reflecting the expected trade-off between precision and recall. This trend highlights the model's balanced and robust detection capability across a range of concrete damage scenarios.

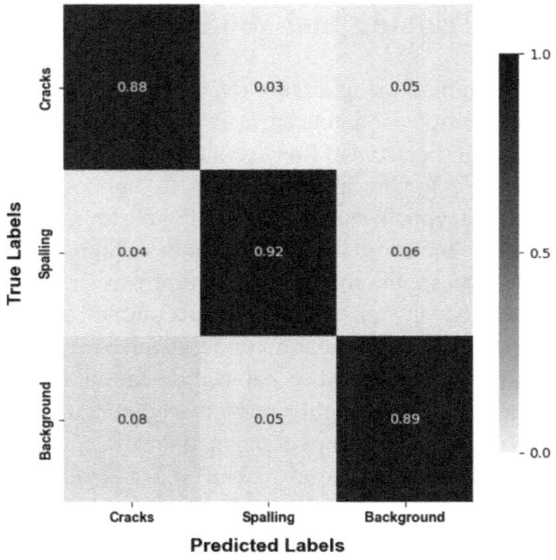

Fig. 3. Confusion matrix of the developed model.

4 Results and Findings

The results were derived from the proposed model developed for the detection and quantification of defects in concrete structures. This evaluation encompassed the model's performance in identifying structural damage—specifically cracks and spalling—under diverse conditions. Furthermore, the influence of fine-tuning techniques and the integration of advanced quantification algorithms on the model's overall effectiveness was systematically analyzed.

4.1 Damage Detection

By processing high-resolution images, the developed model demonstrated high accuracy in detecting surface defects—particularly cracks and spalling—on concrete structures. Enhanced by additional image processing layers, the model successfully identified cracks as narrow as 0.6 mm (0.023 in.), highlighting its precision and sensitivity in detecting even minimally visible defects. For instance, Fig. 4 presents a concrete surface image captured from a distance of 2 m (6.6 ft), where cracks undetectable to the naked eye were accurately identified by the model. This capability is critical for structural condition assessment, as early-stage microcracks may progress into severe damage if left unaddressed. The model's architecture also accommodates camera-specific parameters through dedicated CNN layers, ensuring reliable detection across various imaging conditions. In addition to detection, the model also offers quantification of crack dimensions, as detailed in the following section.

As illustrated in Fig. 5, the model demonstrated advanced capabilities in detecting concrete spalling, accurately identifying damaged regions even in complex scenarios

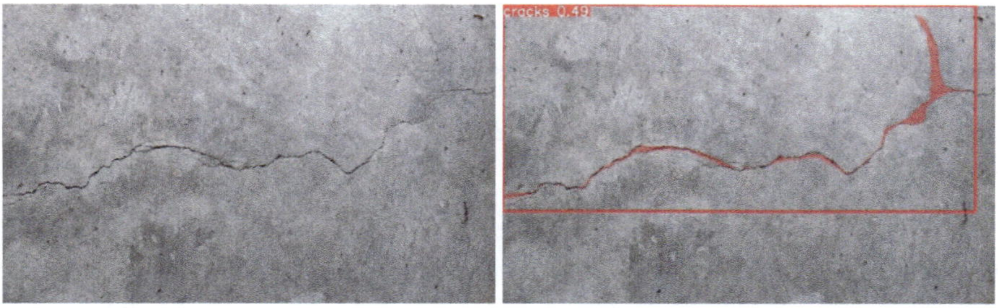

Fig. 4. Before (left) and after (right) detecting concrete cracks.

involving exposed reinforcing steel bars. Despite surface irregularities and variations in texture, the model effectively distinguished between intact concrete and areas affected by spalling, underscoring its robustness in challenging inspection conditions.

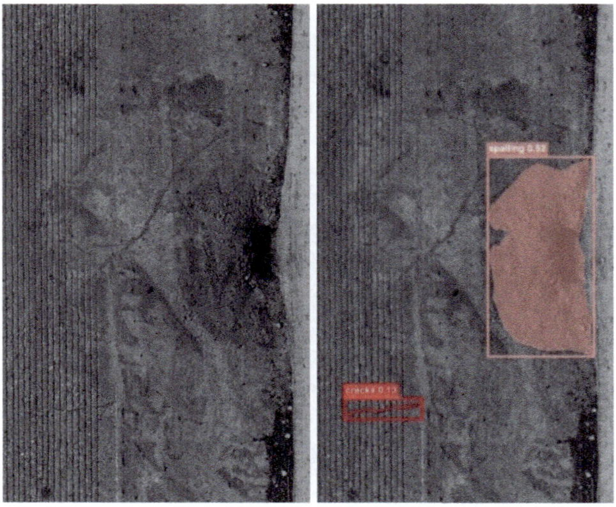

Fig. 5. Before (left) and after (right) detecting concrete spalling.

4.2 Fine-Tuning of the Developed Model

The model's predictive accuracy was significantly improved through an iterative fine-tuning process, which involved adjusting key parameters and training on a diverse, augmented dataset. Techniques such as rotation ($\pm$ 15°), flipping, brightness and contrast adjustments, and Gaussian noise were applied to increase dataset variability. This enriched the model's exposure to a wide range of crack widths and various spalling severities. In particular, the model progressed from rough localization to precise detection of both fine and complex crack patterns, including branching cracks and intersecting geometries. Transfer learning further enhanced this capability by leveraging pretrained weights and adapting them to more intricate defect scenarios. As a result, the model demonstrated robust performance in detecting cracks and spalling of different sizes, directions, and severities, even under challenging surface conditions.

4.3 Damage Quantification

Damage quantification in concrete structures was accomplished by integrating a tracking algorithm into the developed model, enabling measurement of crack lengths, widths, and spalled area dimensions. As shown in Fig. 6, the model estimated a total crack length of 1,060 mm (41.7 in) and an average width of 1.58 mm (0.062 in), closely aligned with manual measurements of 1,036.3 mm (40.8 in) and 1.5 mm (0.059 in), yielding errors of 2.2% and 4.9%, respectively. These minor discrepancies underscore the model's high precision, with the slightly greater width error attributed to variation in measurement points along the crack. The visual output, featuring a green bounding box and a red overlay, further demonstrates the model's ability to delineate crack geometry with accuracy. To contextualize these measurements, industry standards such as the U.S. Federal Highway Administration (FHWA)'s Manual for Bridge Element Inspection [23] were referenced, providing the basis for condition state classification and aligning the model's output with established structural inspection practices.

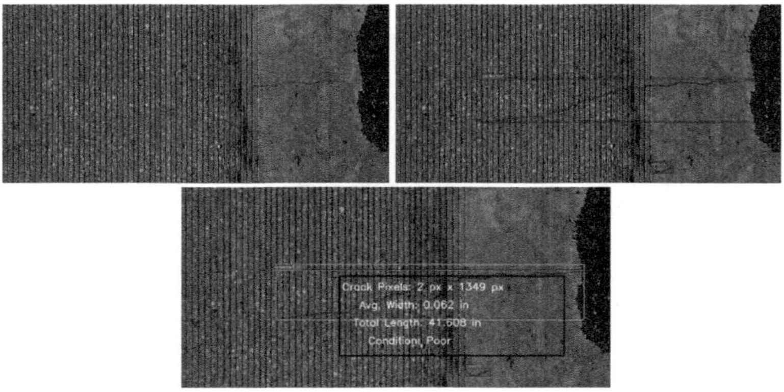

Fig. 6. Applying the quantification code through the developed model, testing the capture angle, distance, and resolution.

5 Conclusions

This study developed a robust CNN model that surpasses the limitations of existing automated damage detection systems for assessing the condition of concrete structures. In contrast to earlier models, which are often limited by low detection accuracy, fixed image resolution, and the absence of quantification features, the proposed model leverages high-resolution images, transfer learning, and fine-tuning techniques to achieve precise detection and measurement of surface defects such as cracks and spalling. A region-growing algorithm was incorporated to convert pixel-based data into real-world dimensions, with calibration based on imaging parameters such as distance and camera specifications. The model demonstrated high accuracy under varying resolutions, lighting conditions, and photographic perspectives, supported by an architecture optimized through iterative testing and enriched by a diverse, augmented training dataset. Performance metrics, including confusion matrix and F1-confidence curves, confirmed the model's consistency and reliability, with detection accuracies exceeding 88% and sustained F1 scores across a range of confidence thresholds. Fine-tuning and transfer learning were central to the model's performance, enabling precise detection of defects with varying severities. Selective retraining of network layers and the integration of pre-trained weights allowed for effective adaptation to structural damage characteristics. These enhancements contribute meaningfully to infrastructure management by supporting accurate defect classification, measurement, and decision-making for maintenance prioritization, offering a significant advancement toward fully automated, field-ready damage assessment tools.

Acknowledgements. The research study, results of which reported in this paper, was sponsored by the Iowa Department of Transportation and Iowa Highway Research Board. The authors would like to acknowledge and thank the sponsors for this support. Opinions, findings, and conclusions expressed in this paper are of the authors and do not necessarily reflect those of the sponsors.

References

1. Alipour A, Shafei B, Shinozuka M (2011) Performance evaluation of deteriorating highway bridges located in high seismic areas. J Bridge Eng 16(5):597–611
2. Alipour A, Shafei B (2016) Seismic resilience of transportation networks with deteriorating components. J Struct Eng 142(8):C4015015
3. Granata MF, La Mendola L, Messina D, Recupero A (2023) Assessment and strengthening of reinforced concrete bridges with half-joint deterioration. Struct Concr 24(1):269–287
4. Cui Z, Alipour A, Shafei B (2019) Structural performance of deteriorating reinforced concrete columns under multiple earthquake events. Eng Struct 191:460–468
5. Abu-Farsakh GA, Odeh IN (2023) A new damage-based failure criterion for nonlinear behavior of fibrous composite materials. Int J Damage Mech 32(7):940–961
6. Yin GJ, Zuo XB, Li XN, Zou YX (2020) An integrated macro-microscopic model for concrete deterioration under external sulfate attack. Eng Fract Mech 240:107345
7. Odeh I, Shafei B (2024) An unmanned aerial vehicle-based digital twin framework for inspection and assessment of bridge structures. In: The international conference on net-zero built environment: innovations in materials, structures, and management practices, pp 1221–1230
8. Farahzadi L, Odeh I, Kioumarsi M, Shafei B (2025) Automated image-based condition assessment of built environment: a state-of-the-art investigation of damage characteristics and detection requirements. Results Eng:104978
9. Aust J, Pons D (2022) Comparative analysis of human operators and advanced technologies in the visual inspection of aero engine blades. Appl Sci 12(4):2250
10. Gil-Cuadrado A, Martínez-Gutiérrez A, Díez-González J, Verde P, Ferrero-Guillén R, Perez H (2023) Full automation of a manual inspection unit for industrial borescopy. Adv Sci Technol 132:140–146

11. Mateen M, Wen J, Hassan M, Nasrullah N, Sun S, Hayat S (2020) Automatic detection of diabetic retinopathy: a review on datasets, methods and evaluation metrics. IEEE Access 8:48784–48811
12. Ai D, Jiang G, Lam SK, He P, Li C (2023) Computer vision framework for crack detection of civil infrastructure—a review. Eng Appl Artif Intell 117:105478
13. Xu G, Yue Q, Liu X (2023) Deep learning algorithm for real-time automatic crack detection, segmentation, qualification. Eng Appl Artif Intell 126:107085
14. Akçay HG, Aksoy S (2008) Automatic detection of geospatial objects using multiple hierarchical segmentations. IEEE Trans Geosci Remote Sens 46(7):2097–2111
15. Laxman KC, Tabassum N, Ai L, Cole C, Ziehl P (2023) Automated crack detection and crack depth prediction for reinforced concrete structures using deep learning. Constr Build Mater 370:130709
16. Golding VP, Gharineiat Z, Munawar HS, Ullah F (2022) Crack detection in concrete structures using deep learning. Sustainability 14(13):8117
17. Kim B, Cho S (2019) Image-based concrete crack assessment using mask and region-based convolutional neural network. Struct Control Health Monit 26(8):e2381
18. Kim B, Cho S (2020) Automated multiple concrete damage detection using instance segmentation deep learning model. Appl Sci 10(22):8008
19. Nguyen DH, Wahab MA (2023) Damage detection in slab structures based on two-dimensional curvature mode shape method and faster R-CNN. Adv Eng Softw 176:103371
20. Ren S, He K, Girshick R, Sun J (2015) Faster R-CNN: towards real-time object detection with region proposal networks. Adv Neural Inf Proces Syst 28
21. Wu Y, Meng F, Qin Y, Qian Y, Xu F, Jia L (2023) UAV imagery based potential safety hazard evaluation for high-speed railroad using real-time instance segmentation. Adv Eng Inform 55:101819
22. Ren S, He K, Girshick R, Sun J (2016) Faster R-CNN: towards real-time object detection with region proposal networks. IEEE Trans Pattern Anal Mach Intell 39(6):1137–1149
23. U.S. Federal Highway Administration (2019) Manual for bridge element inspection. U.S. Department of Transportation, Washington, DC

An Integrated Approach from Condition Assessment to Numerical Modelling for Existing Masonry and Concrete Structures towards Deciding a Sustainable Retrofitting Scheme

Lakshmi Latha(✉) and Samit Ray-Chaudhuri

Indian Institute of Technology, Kanpur, Kanpur, India 208016
lakshmia@iitk.ac.in

Abstract. Maintaining safety and serviceability of existing aging structures is a common global challenge. If not properly maintained in a timely fashion, these structures deteriorate with age and show signs of distress often leading to catastrophic failure, resulting in fatalities and significant economic loss. Apart from distress due to aging, many other factors contribute to the safety of structures such as posterior modifications in usage and interior. Often these modifications add extra load to the structure. Hence, the process of condition assessment, followed by adopting appropriate solution strategy in terms of rehabilitation, repair, retrofitting, and strengthening of these existing structures for achieving desired performance, plays crucial role in sustainable development. Sustainable solution may involve upgrading structural elements to improve their ability to withstand additional loads, satisfy code provisions, or adapt to new requirements. This paper presents an integrated approach in this direction and demonstrates the process through a comprehensive case study involving a masonry and concrete structure for arriving at a proper retrofitting decision. This approach was suggested for an existing residential apartment complex with ground plus 3-story apartment buildings in Patna, Bihar, India, where the proposed retrofitting method was worked out to be highly cost-effective, requiring less than 7% of new construction cost, while extending the building's service life.

Keywords: Existing aging structures · Distress · Condition assessment · Retrofitting · Concrete · Masonry

1 Introduction

Masonry has been an integral part of human civilization for centuries in terms of creating structures having designs with form and utility. The existence of historic palace, temples and monumental structures reflects that. The origin of masonry structures can be traced back to the early civilizations of Mesopotamia, Egypt and the Indus valley. The massive stone wall at Jericho (8000 BC-Neolithic Period), the Great Pyramid of Giza (2600

H. Beushausen et al. (Eds.): NTZR 2025, LNCE 873, pp. 848–859, 2026.
https://doi.org/10.1007/978-3-032-27698-8_75

BC-Egyptian masonry construction), Ziggurat of Ur (1980 BC-Mesopotamian masonry construction), Parthenon (432 BC-Greek masonry construction), Colosseum (80 AD-Roman masonry construction) etc. are some of the oldest existing constructions, as noted by Kubica [1], Almssad et al. [2], and Roca et al. [3]. These constructions took several years to reach the completion stage. Indus valley civilization (2600 BC) has seen advanced urban planning with well-planned brick-built homes. Medieval masonry construction saw the emergence of Gothic architectural style, which involved use of ribbed vaults, flying buttressed and pointed arches. The tall and complex building of the Chartres Cathedral, France is an example. The complex architecture, limited technology, financial and political factors, and other historical events took nearly two centuries for its construction. Middle Ages saw many new tools and techniques, which really improved the masonry construction. These are being used now also (e.g., wheelbarrow for moving materials and use of scaffolding). Middle Ages to modern era saw notable changes in the field of masonry construction with the invent of new techniques and materials, as described by Jessa [4] and Syngellakis [5]. Reinforced concrete structures emerged during this era with the demand of strong but lightweight structures offering greater strength, versatility and durability.

As construction technology evolved, the focus shifted from traditional methods to innovations prioritizing functionality, appearance, and environmental sustainability. At the same time, the sustainable preservation of existing structures has gained importance, especially since the late 20th century, due to environmental concerns and limited resources. Preserving old buildings helps sustainable development by conserving materials, energy, and cultural heritage. It also helps retain traditional skills and knowledge alive that continue to influence modern practices. This paper presents a systematic approach for sustainable preservation, including visual inspection, in-situ testing, lab analysis, numerical modeling, structural health monitoring to validate/update numerical modelling, and retrofitting evaluation for an existing building. The proposed approach is an improved concept of the existing approach, wherein vibration-based structural health monitoring results are integrated to validate/update the numerical model, which is essential for understanding the condition of the structure under future seismic excitations and/or adopting for sustainable retrofitting measures. Further, the proposed retrofitting method is economical, reliable, minimally intrusive, and quicker than most of the conventional techniques, making it ideal for extending the life and sustainability of existing structures.

2 Sustainable Preservation of Structures

Adopting a sustainable preservation approach for an existing structure, is a complex decision-making process. Often the importance of a structure plays major role. The historic integrity should be the least affected attribute with any change of the structure. Hence, structural preservation is more about combination of techniques designed to protect, restore, and modernize a structure while maintaining its historic integrity. The important factors include structures history, condition assessment and then sustainable practices that help the structure remain functional and energy efficient. The main goal should be to create a balance between maintaining cultural as well as historical significance and making the structures viable for continued use in the modern era. Retrofitting, rehabilitation, preservation, restoration and partial reconstruction are various methods of preservation, which can be adopted depending on the condition of a structure. Retrofitting is the most impactful since it can extend the lifespan of the structure by making it safer and future-ready, if adopted in a proper manner. Sustainable redevelopment of a structure by retrofitting generally implies adopting for modern functionality, eco-friendly upgrades, etc. Understanding the structure's age, design details (if available), and usage from pre-recorded details is crucial for assessing its historical, cultural, and architectural significance. The process of re-development starts with the assessment of present-day condition of the structure. Figure 1 depicts a flowchart for the process of sustainable re-development.

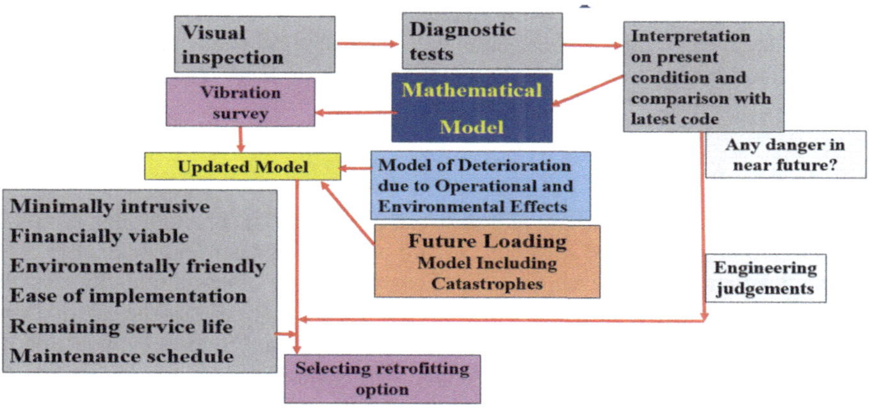

Fig. 1. Flow chart for sustainable redevelopment

3 Methodology

3.1 Condition Assessment

Condition assessment typically includes evaluating present day condition of a structure by performing initial evaluations by conducting site inspection, diagnostic measures such as in-situ measurements and laboratory tests with non-destructive testing (NDT) methods and real time structural health monitoring using various sensors, finite element modelling (FEM), model updating, digital twin technology, building information modelling and many more. The non-invasive methods such as in-situ measurement help gather critical data without causing unnecessary damage to the structure. Once the structure's condition is assessed, the next step is to determine the appropriate sustainable solution strategy to achieve the desirable performance. This step usually involves structural analysis and design concepts. Sustainable solution in general may involve upgrading a building's structural elements to improve their ability to withstand additional loads, meet modern codes/standards, or adapt to new requirements.

3.1.1 Visual Inspection

Visual inspection provides valuable information about the condition of the structure such as if the structure is undergoing any form of distress due to degradation of materials and other structural components, load bearing mechanisms if the drawings and records are not available, maintenance measures adopted so far, and present-day usage of the space. Advanced technologies such as drones can help in assessing risky locations for the inspections such as inaccessible roof or for any time bound urgent inspection, which is economically not possible, as discussed by Kovacevic and Dzidic [6]. In general, efforts should be given to inspect all critical locations of the structure. In addition, visual inspection also provides valuable information for planning in-situ non-destructive tests and collection of samples for laboratory tests. Taking good quality pictures of distress areas, critical locations, building features, etc. are important for further analysis. Important details regarding the measurements etc. can also be taken for validating with the design drawings or for numerical modelling at future stage, if such drawing records do not exist.

3.1.2 In-Situ Tests

As a part of diagnostic measures, non-destructive testing (NDT) and minor destructive testing are common adopted methods, as outlined by El Masri and Rakha [7]. These methods have minimal to no intrusion and can be used to assess the in-situ structural condition, geometry, and, also regarding the engineering properties of the existing structure. For masonry and concrete structures, typical in-situ NDT tests include Rebound Hammer (RH) tests to correlate with strength, Ultrasonic Pulse Velocity (UPV) to measure soundness and ground penetrating radar (GPR) scan inside the surface, Flat Jack Tests to evaluate masonry stiffness and strength properties, etc., as required to check the strength and soundness of the structure. While conducting in-situ NDT tests, it is important to follow all the steps correctly as given in codes/standards. For examples, before conducting the RH or UPV tests, the measurement points should be surface cleaned often using grinder as per the requirement. Improper surface preparation often leads to wrong data. Further, it is also necessary to know the applicability and limitations of each test (e.g., direct and indirect UPV tests) as well as importance of following proper procedure. In addition to covering critical locations, another important aspect of in-site NDT tests is to ensure repeatability of results by conducting multiple tests around the same location.

3.1.3 Laboratory Test on Collected Samples

For laboratory tests, samples of concrete cores, masonry cores are collected using core cutting machines; samples of elements such brick unit, steel, and plaster are also collected. Laboratory testing can be done for the samples (collected with very minimal damage of structure). In addition to structure, often soil samples are collected for geotechnical investigations too, if foundation issues are perceived to be concern for present/future distress. If drawings of foundation or knowledge of foundation type are not available, minimally intrusive excavation may often be resorted to get first-hand knowledge of the foundation. Geotechnical reports of nearby areas often help in providing valuable information such as understanding the nature of soil, idea about bearing capacity and water table. Since the samples may be limited, the ideal test should be carefully selected based on the condition of structure, type of structure, structure's value, location, available budget, available equipment and expertise.

3.1.4 Interpretation of Results

Interpretation of results from in-situ as well as laboratory tests is very important. The test data should be thoroughly inspected for outlier removal. Repeatability is another key aspect. Based on the test data, many useful information can be obtained. Typically, strength (and stiffness) and sign of deterioration in material qualities are two key parameters to provide a basic understanding on the condition of the structure. The obtained results are then analysed with code specifications to infer on latest code compliance. Many a times, based on NDT tests and laboratory tests of collected samples, coupled with the results of visual assessment are enough to understand the condition of the structures. However, the interpretation of condition of the structure using only these are mostly qualitative when comes to safety margin. This is because in absence of detailed modelling, quantitative evaluation of deficiencies in terms of safety and stability, particularly under external extreme hazard events is not possible. Thus, for accurate estimation of condition in terms of local (member) and global distress, numerical model is the preferred tool.

3.2 Numerical Modeling

Numerical modelling is often resorted to assess component level as well system level conditions. For numerical modelling of the main load bearing system, both geometric dimensions and information of material properties are necessary. Often existing drawings serve as the basis for developing a basic model. However, in absence of drawings, often as-built drawings are generated to develop numerical particularly for diagnosis and future damage modelling. The results obtained from diagnostic tests such as in-situ NDT tests and laboratory tests on samples collected from site serve as the basis for input to the numerical model. Appropriate inputs on stiffness and strength parameters as per present-day conditions are essential for ensuring quantitate assessment of conditions and assessing vulnerability under future conditions. Depending upon the requirements and available data, modelling is carried out. For example, for masonry structures, often macro models are preferred than micro models for modelling for evaluating seismic responses. This is because, uncertainty in terms of material properties and others may not justify high computation time that may require for finer and complex model at structural scale.

3.3 Structural Health Monitoring and Model Updating

Structural health monitoring technology is another method, which can offer life-safety and economic benefits. In this approach, sensors are employed to monitor the structures for their operational environmental measurements and structural responses. Smart materials or self-sensing structural material is another diagnostic method for structures.

For example, smart bricks with conductive or piezoresistive materials itself can monitor and diagnose structural issue, as discussed by Meoni et al. [8]. Diagnostic methods used for masonry and concrete structures are many and will depend on the condition of a particular structure. The data can be collected in campaign mode such as ambient vibration measurements or real-time continuous monitoring to evaluate the condition giving timely advice, even if there are situations of occurrences like earthquake etc. Continuous measurements are often resorted for critical and very important structures to monitor degradation over a long duration or understand the behavior of the structure under different environmental (e.g., temperature, moisture, soil saturation, water table fluctuation) and loading conditions including operational as well as rare events such as earthquakes. For condition assessment in a timely manner though, ambient vibration survey has gained significant attention as reliable tool to provide present condition of the structure and validate and update numerical model. Ambient vibration survey is a sophisticated technique, which assists in obtaining dynamic properties (natural frequencies and mode shapes) of a structure in its existing conditions. The main advantage of this test is that it is fast and reliable approach that does not require any additional excitation to excite the structure (such as mass shaker, hammer etc.). Data obtained from this testing helps in understanding and often validating the numerical models of the structure for diagnosis of present condition and future prognosis. For example, a study carried out by the second author uses ambient vibration survey to identify stiffness degradation in a bridge pier when data were compared to a healthy pier [9]. The data obtained from the ambient vibration survey are very important for updating the developed numerical model, as noted by Prajapat and Ray-Chaudhuri [10]. This updated model then serves as the basis for performance assessment of the structure under future loading scenarios and to make decisions about suitable retrofitting schemes.

3.4 Evaluation of Retrofitting Options

As mentioned earlier, based on the initial study of the available records of the structure, visual inspection and diagnosis, and, if required, modelling, ambient vibration survey, etc. and further analysis are carried out to assess the state of the structure. Once quantitative assessment of deficiencies is determined either based on engineering judgements based on available data and resources or through numerical modelling, the decision about retrofitting is considered. For example, minor repairs can be taken up without carrying out detailed numerical analysis if overall the condition of the structure looks satisfactory. This is not however the case if the structure appears to be in a critical state, particularly considering further extreme events such as earthquakes. Thus, retrofitting is usually adopted, if there is any risk associated with natural disasters and if there is any observed structural degradation. This can improve the load resistance and overall service life of the structure. Distress due to aging, inadequate designs, construction practices and exposure conditions come under the major details. These details should be analysed before adopting any retrofitting strategy. Repair and retrofitting make sure that the structure is safe, ready-to-use and is available for future. Retrofitting decisions are taken based on available techniques or by proposing new techniques appropriate for the structure. An appropriate decision requires comparing various options from aesthetics, utility and cost-benefit ratio, numerical modelling, experimental study to gain confidence, etc.

Other considerations for choosing the right retrofitting criteria are minimal intervention, occupant discomfort, availability of local resources and crafts man, etc. Then comes other sustainable developments in terms of energy savings etc. Retrofitting methods also will depend on a particular structure that has been dealt with, as noted by Meoni et al. [8], Keshmiry et al. [11] and Khalid Heiza et al. [12]. A good retrofitting method should not only be cost-effective but also should be environmentally sustainable in terms of material usage, pollution generated during the process and future energy efficiencies, as discussed by Bhattacharya et al. [13], Dauda and Ajayi [14], Ma et al. [15] and Latha and Ray-Chaudhuri [16]. Often various options are compared in terms of fragility and subsequent annual cost modelling.

4 Case Study

This case study covers the process outlined in the flowchart comprehensively. The focus is an existing residential apartment complex with ground plus 3-story apartment buildings at Patna, Bihar, India. These buildings are load-bearing masonry structure with reinforced concrete slabs and staircases as shown by schematic diagram in Fig. 2.

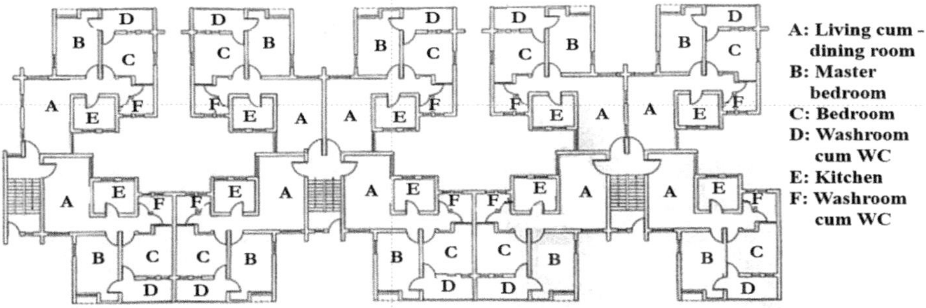

Fig. 2. Typical floor plans for ground, 1st, 2nd and 3rd storey of the study building

A few of these buildings suffered visible damage in terms of long cracks during the Nepal Earthquake (M_W 7.8 earthquake, April 25, 2015) and subsequently, remained unoccupied. At first, a visual inspection was carried out. Subsequently, in situ diagnostic tests and laboratory tests on collected structural and geotechnical samples were carried out. Since the building is in high seismic zone, i.e., Seismic Zone IV of Indian seismic code, further detailed analysis was preferred. The scope of the detailed study was to understand the deficiencies of the present structure and provide an easily implementable and financially viable solution considering safety under such future earthquakes.

Several in-situ tests were conducted including flat jack tests at a few locations of different walls (see Fig. 3) to understand the stiffness and strength of masonry wall. Further, samples were collected for brick units, plaster, etc. in addition to a few masonry prisms for compressive test (see Fig. 4) in structural engineering laboratory at IIT Kanpur. From the visual inspection, many other important observations were made in many locations in slabs and lintel bands, the reinforcement bars have been found to be excessively corroded due to water seepage and other associated reasons. Further, it was also found that waterproofing treatment at roof to be unable to stop water seepage resulting in corrosion of reinforcing bars of roof slab. Attempt of external roof treatment has failed to yield desired results. Based on diagnostic test results, it was found that the masonry walls were constructed with weak mortar having average compressive strength of 4.10 MPa and Young's Modulus of 2300 MPa. In general, it was found that building needs repair at least in the distress areas. However, assessment of safety and serviceability in the aftermath of a significant earthquake required a detailed evaluation. This aspect was given utmost importance since the building has developed cracks, occupants are scared, and the location of the site is within 200 km from Himalayan range.

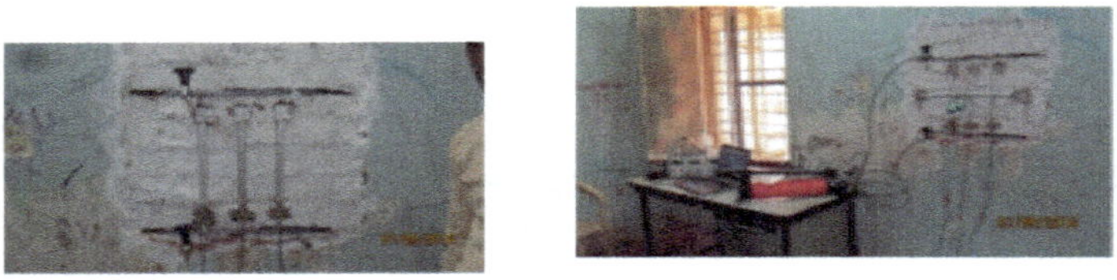

(a) (b)

Fig. 3. Flat jack test set-up showing (a) position of slots, reference points and (b) installation of instrumentation

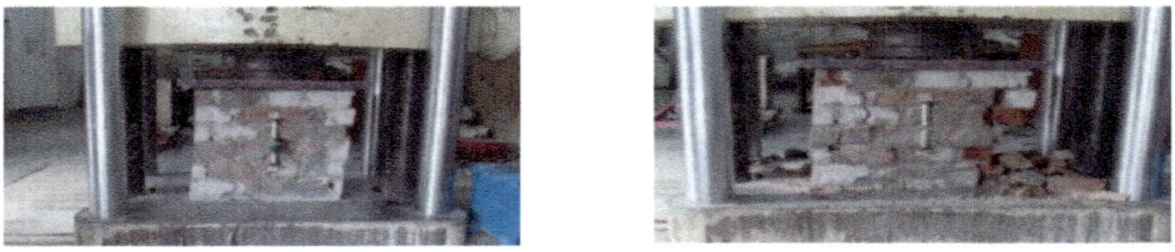

(a) (b)

Fig. 4. Photograph of masonry prism (a) before (b) after compression test

Geotechnical evaluations: For general understanding of the subsoil condition, some disturbed soil samples were collected from the site. The samples were collected. Some basic geotechnical tests on the soil sample were carried out in the Geotechnical Engineering Laboratory, IIT Kanpur to characterize the soil. It may be noted here that the structural walls were supported on short underream piles. From the tests conducted in laboratory, it was noted that the pile group capacity is satisfactorily higher than factored load from the superstructure. The settlement computations for likely scenarios reveal that the maximum settlement is less that the permissible settlement. Hence, the focus was shifted on super structure only and particularly the performance under future earthquakes.

Ambient vibration survey: In addition to in-situ and laboratory tests, to measure the ambient vibration data from one of the blocks of G + 3 buildings along both lateral and longitudinal directions of the building, several uniaxial high resolution and reliable PCB accelerometers (pcb.com) was used. From first story to roof top (total 4 floors), two sensors were placed at each floor levels to measure the response along the lateral and longitudinal directions. The vertical alignment was followed to install the sensors to obtain the first mode shapes in each direction (lateral and longitudinal). The Data acquisition (DAQ) system (NI PXIe-1062Q, http://ni.com) was connected with sensors. The data were recorded using LabView Signal Express. Frequency Domain Decomposition (FDD) was used to identify modal properties of the building. The fundamental period in both directions was found to be around 0.25 s. This information was later used for gaining confidence in numerical models.

Numerical modelling: Three-dimensional numerical model of the building was developed in SAP 2000 [17] based on both as per structural drawings available and the dimensions ascertained from site. The basic model developed as per original structural drawing and wall strength as per relevant literatures. Analysis was carried out for the model keeping in mind the relevant code of construction practices existing at the time of construction of these structures. A fixed based condition is assumed for structural modelling. Appropriate material inputs and sectional sizes for various structural members were then provided to this model. To check the adequacy of the structure in its present state, eigen value analysis is performed. It is found that the fundamental period of the building is 0.20 s in comparison to data obtained from the ambient vibration survey. This difference may be due to the modelling of walls, which appears to be weaker than what has been modelled. On the other hand, the code specified value of the fundamental period is 0.26 s for this building. Further, gravity and seismic load analyses are also performed, and results are further used to perform fragility analysis.

Selection of retrofitting scheme: The goal was then to create a retrofitting plan that would improve the seismic performance of existing un-reinforced brick masonry buildings by adding confinement, a straightforward approach. In addition to being economically feasible and less invasive, the method should also be dependable, easy to accept, and quick to execute. For that it was decided to examine the scheme's efficacy through a methodical inquiry that includes both analytical and experimental studies. Finally, to compare the qualitative and quantitative outcomes with an existing un-reinforced brick masonry structures to comprehend how the suggested scheme improves performance. In this regard, various retrofitting options for masonry structures were studied. Finally, considering different aspects, and to combine the load-bearing brick walls and confining elements, retrofitting by partial confinement of unreinforced brick masonry walls was chosen to improve seismic performance, taking into account, a number of factors and the excellent seismic performance of confined masonry in the past, as emphasized by Dauda and Ajayi [14]. The RC bands are positioned in such a way that it (i) provides maximum confinement to all wall surface regions, (ii) avoid obstructing crucial wall portions like corners, and (iii) create partially confining panels with an aspect ratio close to unity, as described by Latha et al. [18]. Schematic representation of the proposed retrofitting scheme is presented in Fig. 5.

Validation of retrofitting scheme: The efficiency of the proposed scheme was validated through experimental and numerical studies. The experimental program included: (i)

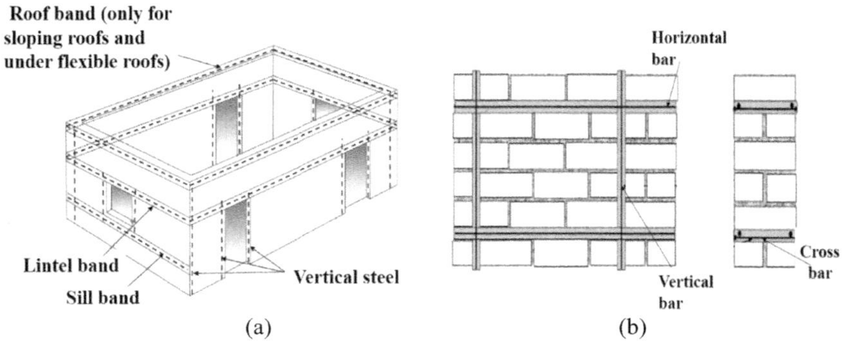

Fig. 5. Schematic diagram showing the proposed retrofitting scheme: (a) 3-D rendering, (b) elevation and horizontal section along the height of a wall

material property tests to evaluate the compressive strength of brick units and mortar cubes, and water absorption of bricks, (ii) tests on brick-mortar sub-assemblages to determine the uniaxial compressive stress-strain curve of masonry prisms, as well as the tension bond strength of the masonry, (iii) diagonal tests to obtain the diagonal tensile strength of brick-mortar masonry assemblage, for both control (without band) and retrofitted (with band) specimens, (iv) shake table tests of half-scale masonry walls, both control (without band) and retrofitted (with band), under an extensive set of earthquake ground motions of different durations and varying intensity, (v) full scale slow-cyclic tests on both control (without band) and retrofitted (with band) masonry buildings [18]. All specimens, prepared by the same mason, followed code provisions or test-specific requirements, and used either full- and half-scale samples (half-scale for dynamic tests due to shake table size limits). Quasi-static tests on full-scale specimens observed enhanced ductility and energy dissipation in retrofitted buildings, as conducted by Latha et al. [18]. Shake table tests showed remarkable seismic performance of the retrofitted scheme. Numerical modeling supported to understand result interpretation [19]. The proposed retrofitting method proved to be highly efficient. Soil-structure interaction studies on flexible bases showed lower displacement demand for retrofitted buildings on medium soil, without increasing acceleration response at the design level for zones IV and V, as emphasized by Latha et al. [20]. Fragility analysis demonstrated significant reduction in seismic vulnerability for retrofitted walls.

Cost-benefit analysis: The cost of the proposed retrofitting measure considering all major works including the implementation of seismic bands and load reduction from roof due to modification of parapet wall along with associated scaffolding work and finishing works (internal and external plastering and painting) were worked out as per established procedure. This cost is then compared with the cost of rebuilding with the same configuration and amenities. It was found that retrofitting the existing building is cost effective, requiring less than 7% of the cost of new construction while offering a long service life.

5 Conclusions

This work presents a framework for an integrated approach for existing structures towards deciding a sustainable retrofitting scheme. The approach comprises visual inspection, diagnostic tests, numerical modeling, ambient vibration survey etc. to understand the deficiencies of a structure and to decide a sustainable solution to withstand additional loads, satisfy modern codes, apart from being environmentally sustainable in terms of low quantity material usage, generating minimal pollution etc. The important aspect this work brings in is the integration of vibration-based health monitoring data to validate/update numerical model for decision making. A case study is then presented to retrofit a multi-story masonry structure with a proposed a cost-effective, minimally invasive, and functionally effective retrofitting method after a rigorous comprehensive and methodical study. The case study deals with material level to system level studies to show the effectiveness of the approach. Fragility evaluation and cost-benefit analysis are also carried out to show the effectiveness of the proposed approach.

Acknowledgements. The authors greatly acknowledge the financial support for carrying out the experimental work provided by Reserve Bank of India, Government of India.

References

1. Kubica J (2024) Masonry systems: past, present and future. MATEC Web Conf 403:01003
2. Almssad A, Almusaed A, Homod RZ (2022) Masonry in the context of sustainable buildings: a review of the brick role in architecture. Sustainability 14(22):14734
3. Roca P, Lourenço PB, Gaetani A (2019) Historic construction and conservation: materials, systems and damage. Routledge, London
4. Jessa EK (2022) Evolution of masonry techniques. Commun Phys Sci 8(4)
5. Syngellakis S (2013) Heritage masonry: materials and structures. Wit Press, Southampton
6. Kovacevic I, Dzidic S (2018) High-rise buildings - structures and materials, 2nd edn. International BURCH University, Sarajevo
7. El Masri Y, Rakha T (2020) A scoping review of non-destructive testing (NDT) techniques in building performance diagnostic inspections. Constr Build Mater 265:120542
8. Meoni A, D'Alessandro A, Ubertini F (2020) Characterization of the strain-sensing behavior of smart bricks: a new theoretical model and its application for monitoring of masonry structural elements. Constr Build Mater 250:118907
9. Ghosh G, Singh Saini D, Singh V, Ray-Chaudhuri S (2025) Vibration-based condition assessment and retrofitting of the Chambal Railway Bridge in India. J Struct Des Constr Pract 30(3):04025060
10. Prajapat K, Ray-Chaudhuri S (2016) Prediction error variances in Bayesian model updating employing data sensitivity. J Eng Mech 142(12):04016096
11. Keshmiry A, Hassani S, Dackermann U, Li J (2024) Assessment, repair, and retrofitting of masonry structures: a comprehensive review. Constr Build Mater 442:137380
12. Khalid Heiza AN, Meleka N, Tayel M (2014) State-of-the art review: strengthening of reinforced concrete structures-different strengthening techniques. In: Sixth international conference on nanotechnology in construction, vol 6, no 6-C, pp 22–24
13. Bhattacharya S, Nayak S, Dutta SC (2014) A critical review of retrofitting methods for unreinforced masonry structures. Int J Disaster Risk Reduct 7:51–67

14. Dauda JA, Ajayi SO (2022) Understanding the impediments to sustainable structural retrofit of existing buildings in the UK. J Build Eng 60:105168
15. Ma Z, Cooper P, Daly D, Ledo L (2012) Existing building retrofits: methodology and state-of-the-art. Energy Build 55:889–902
16. Latha L, Ray-Chaudhuri S (2024) Seismic performance enhancement of brick masonry walls by retrofitting with nominal RC bands: a study of diagonal compression tests. Struct Infrastruct Eng 1–15:1–15
17. CSI (Computers and Structures Inc.) (2004) SAP2000 v10 integrated finite element analysis and design of structures. CSI, Berkeley
18. Latha L, Ray-Chaudhuri S, Mukhopadhyay S, Bajpai KK (2022) Seismic performance enhancement of unreinforced brick masonry buildings by retrofitting with reinforced concrete bands: full scale experiments. J Struct Eng 148(12):04022195
19. Latha L, Ray-Chaudhuri S (2024) Improving in-plane behavior of weak brick masonry walls through assimilation of partially confining RC bands. J Earthq Eng 28(9):2577–2599
20. Latha L, Ray-Chaudhuri S, Raychowdhury P (2023) Seismic evaluation and retrofitting of deficient masonry building considering SSI effects. In: International conference on geotechnics for sustainable infrastructure development. Springer, pp 2133–2145

Methodology for Structural Reliability Assessment of Historic Reinforced Concrete Buildings Using Numerical Analysis and NDT Techniques

Zoltán Orbán(✉), András Dormány, and Tamás Juhász

Faculty of Engineering and IT, Structural Diagnostics and Analysis of Research Team, University of Pécs, 7624 Pécs Boszorkány u. 2,, Hungary
orban.zoltan@mik.pte.hu

Abstract. This paper presents a structured approach to assess the structural reliability of historic reinforced concrete buildings. The methodology integrates non-destructive testing (NDT), low-sample destructive testing and numerical modelling to evaluate the structure's current condition and load-bearing capacity. In the absence of original design documentation, the parameters required for the numerical analysis are determined using a step-by-step diagnostic procedure. Reliability assessment is based on an adjusted partial factor method (APFM), aligning *fib* Bulletin 80 [7]. This approach allows for an optimal selection of the reliability index (β_{target}), even when deviating from design standards. Based on this modified reliability index and considering the diagnostic and material test results, the partial factors used for the assessment can be optimally adjusted, and in most cases reduced, compared to those in design codes for new structures. Case study results confirm that even very old reinforced concrete structures can be brought into compliance with safety requirements with a proper assessment method, thus improving the cost-effectiveness of interventions. It contributes to sustainable heritage preservation by offering a framework that balances structural integrity with conservation objectives.

Keywords: Historic Reinforced Concrete Buildings · Assessment · Non-Destructive Testing (NDT) · Structural Reliability

1 Introduction

The sustainability of our built environment can be promoted not only by creating new structures with the right criteria, but also by ensuring the safe operation of existing structures. In recent decades, the importance of preserving our built heritage has received increasing attention worldwide. As a result, a growing number of reinforced concrete structures have come under the jurisdiction of heritage conservation authorities.

The emergence of reinforced concrete as a construction material at the beginning of the 20th century enabled the widespread construction of new types of structures, which

H. Beushausen et al. (Eds.): NTZR 2025, LNCE 873, pp. 860–872, 2026.
https://doi.org/10.1007/978-3-032-27698-8_76

largely replaced earlier masonry and timber systems. The rapid development of material production and construction technologies during this period led to the proliferation of reinforced concrete structures, despite the lack of well-established design standards at the time. Over the decades, however, it became evident that this once-believed "eternal" structural material could suffer from significant durability issues.

Despite concerns regarding long-term performance, many of these early structures have endured for generations without major difficulties. This is probably because the design practice of the time was generally conservative, which would be considered overscaling by modern standards. Many surveys have shown that early 20th-century reinforced concrete structures often possess substantial load-bearing capacity, even when their configurations do not meet current code-based design requirements [1, 2].

The reliability of load-bearing structures is influenced not only by their load capacity but also by a range of additional factors. With the application of appropriate analytical methods, the structural reliability of historical buildings can be assessed more accurately, allowing for optimized decision-making regarding their future. In these cases, the evaluation process often requires alternative calculation methods and specialized diagnostic techniques not typically used for new constructions. One of the primary challenges is the high degree of uncertainty regarding material properties and structural details, compared to modern buildings. This increases the importance of in-situ diagnostic investigations.

2 The Role of Diagnostics in Structural Assessment

The assessment of structural condition generally requires analytical modeling and numerical calculations. These models demand detailed information on material properties and structural configurations, which are often unavailable in the case of historical buildings. In many instances, even the original construction documents and drawings containing basic details and dimensions are missing. Therefore, specialized diagnostic investigations must be carried out to obtain the input parameters necessary for structural analysis.

For heritage buildings, these investigations should be conducted using non-destructive or minimally invasive techniques to avoid damaging historically valuable components. Beyond material properties, such methods can provide critical information about the structural layout, hidden geometries, internal and surface damage, and anomalies both physical and qualitative in nature [3, 4]. These qualitative characteristics can complement data obtained from conventional (e.g. destructive) tests and assist in determining the appropriate locations and frequency of such examinations.

The structural assessment of historical buildings can be carried out in multiple phases. This phased approach enhances the reliability of the results while minimizing the resources needed for on-site diagnostics and numerical simulations.

The primary objectives of diagnostic investigations typically include:

1. General assessment of the load-bearing structural elements and documentation of deficiencies that may compromise structural function or significantly reduce load-bearing capacity and remaining service life.
2. Determination of the geometrical characteristics and structural details necessary for structural analysis.

3. Identification of the mechanical properties of structural materials, along with their variability.
4. Evaluation of durability-related characteristics of the structural elements.

3 Knowledge of Contemporary Design and Construction Practice

In addition to the statistical evaluation of measured properties obtained through diagnostics, knowledge of historical design and construction practices, and also the circumstances of construction, can significantly support numerical analysis and the assessment of structural reliability.

The technologies used for producing and placing concrete and reinforcing steel in the early 20th century have a considerable impact on the current physical and mechanical properties of the materials, as well as their statistical variability. When evaluating such structures, it is essential to recognize that the construction materials used in historical buildings tend to exhibit far greater inhomogeneity and anomalies than those used in modern construction.

Early concrete structures were often designed without specifying the type or maximum size of aggregates, and without considering grading curves or water-to-cement ratio concepts, which were not yet in use at the time. Aggregates likely contained high amounts of clay and silt. Concrete was usually placed manually and compacted with tamping, which may have resulted in higher strength than cast concrete, but the layered compaction technique (Fig. 1) also introduced significant variation in strength across the structure.

Fig. 1. Core sample from a tamped concrete slab showing "layered" structure (photo: author)

Old reinforced concrete structures typically used reinforcing bars made of so-called "wrought iron" or very mild steel, which exhibit much lower yield strength and greater variability than modern reinforcement. Another critical issue is the variability in concrete cover thickness; reinforcing bars are often located very close to the surface, and in combination with the porous nature of early concrete and carbonation effects, this greatly increases the risk of corrosion.

The structural layout and the connections between elements can also contain many unusual features. Figures 2, 3 and 4 demonstrate examples of problematic design details found during building investigations. Figure 2 shows a cross-section with only a single longitudinal bar, which violates basic principles of reinforced concrete design. Figure 3 illustrates reinforcement congestion that made proper concrete placement impossible, and Fig. 4 shows poorly anchored negative reinforcement above a ribbed slab, which critically limits the slab's bending moment resistance over intermediate supports.

Fig. 2. Cross-section with a single longitudinal rebar (photo: author)

Fig. 3. Insufficient clearance between steel rebars in flexural element (photo: author)

While these structural details may seem unusual from a modern perspective, they highlight the importance of carefully inspecting historical structures.

Fig. 4. Inadequate anchorage of negative reinforcement above slab rib (photo: author)

4 Special Considerations and Methods in Diagnostics

When investigating the material and structural properties of historical buildings, several unique aspects must be taken into account. Some of the most important are summarized below:

Due to the lower level of construction technology development at the time, material properties of historical structures often exhibit significantly greater variability compared to modern constructions. This variability may be further amplified by long-term degradation of structural materials.

In the case of protected structures, destructive testing is often limited or not permitted. Both the number and size of samples must be minimized. In such situations, non-destructive and minimally invasive methods are preferred. However, these methods generally provide lower reliability, especially regarding mechanical properties, than destructive tests.

Most non-destructive test methods have been calibrated for modern construction materials. Consequently, the evaluation formulas provided in current standards may not be directly applicable to historical materials. In historical concretes, non-standard aggregate grading, unique compaction techniques, and localized anomalies may introduce anisotropy, which alters the relationship between measured physical parameters and the derived mechanical properties. In such cases, combining test methods becomes particularly important. For example, limited destructive test results can be used to calibrate or validate non-destructive measurements specific to a given structure, or data fusion techniques can be applied to evaluate the results from multiple test methods collectively.

Surface damage and internal defects in structural materials can hinder standardized testing procedures and distort the measured results. For severely deteriorated structures, it is advisable to conduct the assessment using the most probable worst-case material properties.

Procedures based on the combined use of several, mainly non-destructive diagnostic methods can be effectively applied. These include combinations of ground-penetrating radar (GPR), magnetic, and endoscopic methods to determine slab layers, thickness, and reinforcement patterns. Depending on the frequency used, GPR can also detect discontinuities and structural anomalies. Interfaces between layers with different properties generate strong reflections that can be tracked along the scan lines, allowing the identification of embedded objects, voids, cracked regions, zones of high moisture content, thickness variations, and rebar configurations.

An example of a GPR survey is shown in Fig. 5, where reinforcing rebars, ribs, thickness changes in slab, and overlaying layers are clearly visible in the radargrams. However, the propagation speed of radar waves depends on many parameters, even within the same material type, so depth scales should be considered approximate. Accuracy of the radar measurements can be improved through endoscopic inspection (Fig. 6), complementary geophysical techniques (e.g. seismic testing), or, when necessary, core sampling. Combined investigations enable clear identification of slab thickness, variations in concrete quality, and material composition of different layers.

Some non-destructive methods are highly sensitive to factors like moisture content and other physical or chemical characteristics (e.g. carbonation depth, corrosion state). These factors must be considered when interpreting results, often by increasing the assumed variability (e.g. standard deviation) in the statistical evaluation of measured parameters. Understanding the distribution of physical and chemical characteristics within the structure is also fundamental for analyzing material durability and degradation processes.

Whether a method is considered truly non-destructive may depend on the nature of the structure. In historical buildings, even minimally invasive techniques may be categorized as destructive, whereas in other cases, moderately invasive tests (e.g. endoscopy) may be deemed acceptable. Although destructive testing is generally discouraged or even prohibited for heritage structures, it can still provide highly valuable insights and improve the reliability of results, even when conducted on a limited number of samples.

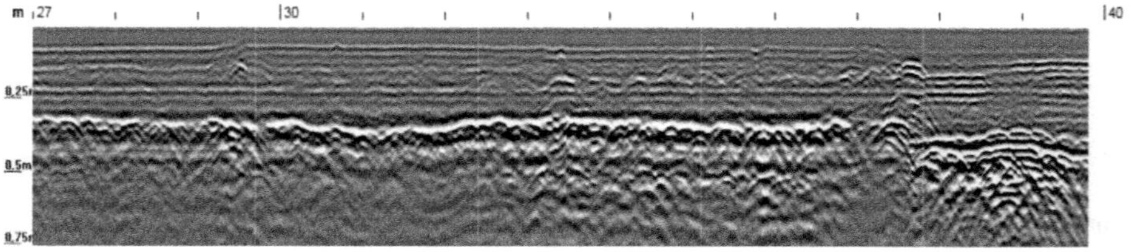

Fig. 5. GPR testing of a floor slab with radargram (photo: author)

Table 1 summarises the purpose, advantages and limitations of the non-destructive, semi-destructive and destructive testing methods typically used for historical structures.

Table 1. Description of test methods typically used for historical structures

Test type	Method	Purpose/advantages	Limitations/comments

(*continued*)

Table 1. (*continued*)

Test type	Method	Purpose/advantages	Limitations/comments
Non-destructive (NDT)	Rebound (Schmidt) hammer	Quick estimate of compressive strength and uniformity; allows for statistical analysis	Strongly affected by carbonation, moisture, and aggregates; only approximate; should be combined with others
	Ultrasonic pulse velocity (UPV)	Estimates strength and crack depth; relatively easy to perform	Influenced by moisture, aggregate type, and reinforcement; requires complementary tests
	Magnetic rebar locator	Locates rebar and concrete cover with high speed and accuracy	Diameter estimate uncertain; limited depth; clustered bars hard to detect
	Ground-penetrating radar (GPR)	Detects rebars, internal layers, defects, anomalies (e.g. moisture)	Interpretation can be complex; best used with multiple antennas or combined with other methods
Semi-destructive (SDT)	Pull-off test	Measures tensile and bond strength; useful before repairs	Results highly scattered; misleading on degraded surfaces
	Penetration test	Estimates compressive strength of weak concrete or masonry; less sensitive to surface state	Weak correlation with actual compressive strength
	Endoscopy	Visualizes internal geometry through small borehole; supports interpretation of other tests	Requires clean borehole; can provide mm-level accuracy
	Moisture content test	Important for strength/durability of porous materials	Surface and internal moisture may differ; depth tests may require drilling
	Carbonation depth test	Key factor in steel corrosion and NDT result interpretation	High variability; many samples needed for statistical reliability; potentially destructive

(*continued*)

Table 1. (*continued*)

Test type	Method	Purpose/advantages	Limitations/comments
	Chemical tests (pH, chloride)	Provides data on corrosion risk from dust samples at various depths	Only gives point data; many samples needed for reliable conclusions
Destructive (DT)	Drilled core tests	Measures mechanical properties, porosity, allows visual inspection	Sample size limited in heritage buildings; high uncertainty for small diameter cores with large aggregates
	Extracted steel rebar tests	Evaluates corrosion level and residual diameter	Sample size and location highly restricted in heritage contexts
	Excavations	Reveals critical structural details	Highly invasive; limited applicability in listed or protected buildings

5 Assessment of Structural Reliability

To reduce the extent of necessary on-site diagnostics and the complexity of numerical modeling, the structural assessment process can be effectively organized into a multi-phase workflow:

Phase 1: Preliminary Evaluation

An initial assessment is conducted using visual inspections and basic diagnostic methods, complemented by a simplified numerical model, typically a two-dimensional linear elastic finite element model. This model employs conservative parameters and simplified calculations to highlight the most critical structural elements.

Phase 2: Detailed Analysis

At this stage, extensive diagnostic testing is performed, and a three-dimensional finite element model is constructed using the actual measured properties of materials and geometry. Structural behavior is then evaluated according to relevant standards, ensuring a realistic representation of the system.

Phase 3: Advanced Assessment

This phase involves targeted diagnostic investigations using combined methodologies on selected components. A refined 3D model is developed, incorporating more accurate material and geometric data. The evaluation is guided by an optimized reliability index, as outlined in ISO standards [5, 6] and *fib* Bulletin 80 [7].

Fig. 6. Endoscopic inspection of a reinforced concrete slab (photo: author)

Design codes, whether old or modern, often fall short when used for evaluating historical structures. Unlike newly constructed buildings, where safety margins are incorporated at the design stage, existing structures typically require costly retrofitting to improve safety. Consequently, the optimal reliability index (β_{opt}) for these structures may differ significantly. Although safety must never be compromised (with ISO 2394 stipulating a minimum reliability index of 2.3 for buildings in consequence class CC2), such minimum thresholds are not typically cost-effective due to the potentially high economic impact of structural failure (Fig. 7).

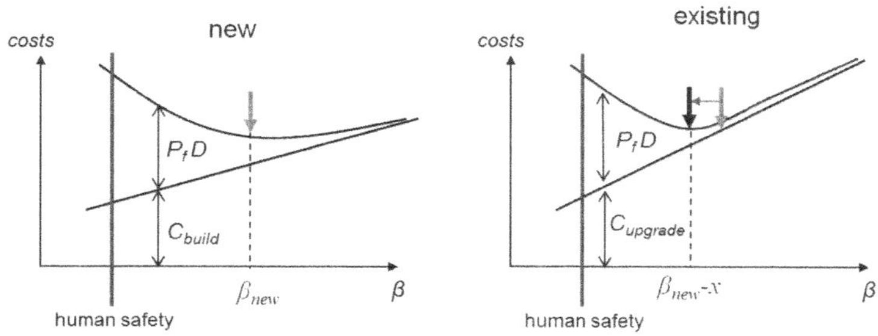

Fig. 7. Interpretation of optimum reliability index (β_{opt}) for new and existing structures based on full-life cost minimalization [7]

While the structural Eurocodes [8–10] lack specific guidelines for determining optimal reliability in existing structures, ISO 2394 does provide such recommendations. For example, for a building undergoing comprehensive renovation with an expected remaining lifespan of 50 years, a target reliability index between β_{opt} = 3.1 and 3.3

is suggested, especially when considering the high costs associated with strengthening techniques. Nevertheless, it is important to tailor this index based on the expected remaining service life of the structure in question.

Case Study

In the following, the results of the structural analysis and assessment of a reinforced concrete public building in Hungary built in the early 20th century are presented in a case study (Fig. 8). Load-bearing elements of the building were evaluated against a target reliability index of $\beta_{\text{target}} = 3.1$. For this purpose, the methodology proposed in *fib* Bulletin 80 was employed to adjust partial safety factors, integrating material variability and geometric uncertainties identified during diagnostics.

Fig. 8. The case study building in 2017 (photo: author)

Diagnostic tests of the building were mainly carried out using the methods described in Table 1. To increase the accuracy and reliability of the structural assessment, non-destructive testing methods were not used in isolation, but strategically combined with selective destructive testing. Rebound hammer tests and cover depth measurements were first applied broadly across critical zones of beams and ribbed slabs to establish general strength profiles and identify anomalies. These NDT results were then calibrated using compressive strength tests performed on concrete core samples extracted from representative elements. The core results were also used to adjust the interpretation curves of the rebound hammer data in accordance with EN 13791 [11]. In cases where significant discrepancies were observed between NDT estimates and expected values, destructive testing provided ground truth data to refine the statistical representation of concrete strength and to identify local degradation. This combination allowed the diagnostic team to reduce overall uncertainty in the input parameters for numerical modelling while limiting physical damage to the protected historical structure.

Based on the tests carried out, concrete compressive strength exhibited a coefficient of variation (CoV) of 0.12, indicative of relatively consistent material quality for a historical building. Conversely, reinforcement yield strength displayed a CoV of 0.095, which is relatively high compared to modern steel standards.

In addition to measured statistical values for material and structural properties, the coefficient of variation values associated for dead loads were reduced from 0.10 to 0.05 based on the experience of field investigations, and the coefficient of variation of the live load was assumed to be 0.19 for the minimum loaded area recommended by Nowak and Collins [12].

In cases where measurement was difficult or where damage was visible but not quantifiable, strength variability estimates were conservatively increased. Cross-sectional dimensions that were hard to verify were either conservatively estimated or adjusted via higher CoVs.

Table 2 summarises the adjusted partial factors obtained as a result of the tests and analysis.

Table 2. Adjusted partial factors in phase 3 assessment.

Parameter	CoV assumed in EN (V'_x)	Measured CoV (V''_x)	Partial factors in EN ($\beta' = 3.8$)	Adjusted partial factors ($\beta'' = 3.1$)
Concrete strength	0.15	0.12	1.50	1.29
Steel yield stress	0.05	0.095	1.15	1.15
Permanent load	0.10	0.05	1.35	1.15
Live load	0.25	0.19*	1.50	1.05

β': Target reliability index defined for design of newly built structures
β'': Target reliability index defined to assess the existing structure in the study
* Based on actual load area per Nowak and Collins [12]

These adjusted partial factors, derived using basic principles from *fib* Bulletin 80, apply to elements in good condition with minimal visible deterioration and accessible surfaces. For components with major defects or unverified parameters, more conservative estimates were adopted, sometimes requiring either an increase in CoV or a reduction in assumed cross-sectional dimensions. For meteorological loads, EN standard values were maintained due to lack of specific studies.

The assessment according to Eurocodes revealed a load-bearing capacity deficiency in a large part of the structure. However, by using the *fib* Bulletin 80 Adjusted Partial Factor Method (APFM), the adequacy of the structure's load capacity could be verified in many more structural sections. This approach allowed a significant reduction in the number of structural elements requiring structural strengthening measures.

6 Conclusions

In the assessment of historical reinforced concrete buildings, standard procedures must be complemented by specific considerations related to material degradation, structural anomalies, and historical construction methods. A thorough understanding of the original design principles and construction practices is essential for reliable evaluation.

While non-destructive testing methods are generally preferred for heritage structures, destructive techniques may still be necessary to accurately determine key strength-related parameters. An appropriate combination of diagnostic methods can enhance the accuracy and efficiency of identifying unknown structural and material properties required for structural analysis.

Current design codes offer limited guidance on incorporating structural dam-age into the evaluation of historical reinforced concrete structures. As a result, engineering judgment and approximation remain critical in practice.

Furthermore, the statistical characteristics of material and structural parameters in existing buildings can differ significantly from those of modern structures. Based on measured data and an optimized reliability index, adjusted partial factors can be applied for structural verification.

The application of the proposed multi-level assessment procedure, integrating numerical analysis with a range of diagnostic techniques, allows for a more effective assessment of the load-bearing capacity and reliability of historic reinforced concrete structures.

Extending the life of historic buildings through effective structural assessment not only helps preserve cultural heritage, but also contributes to the sustainability of the built environment by reducing the need for new construction and the associated environmental impacts. The case study demonstrated that the Adjusted Partial Factor Method (APFM) made it possible to verify the adequacy of the structure in many zones where conventional Eurocode assessment had indicated deficiencies. This approach led to a significant reduction in the scope of necessary strengthening interventions, thereby minimizing disruption to the protected building fabric.

The methodology presented can be further developed into a decision-support tool that helps prioritize intervention strategies based on quantified structural reliability. Future work should aim to standardize the integration of diagnostic data into probabilistic models, enabling broader applicability across a wide range of historical structures.

References

1. Mahrouseh R, Orbán Z (2021) Assessment of reinforced concrete slab of historical structures by the yield-line method. Pollack Period 16(3):45–51
2. Walraven J, Dieteren G (2023) Approach to assessment of existing structures in the fib model code 2020. Struct Concr 24(4):4387–4395
3. Orbán Z, Török B, Dormány A (2019) Assessment and rehabilitation of heritage structures helped by combined non-destructive tests. In: Conference proceedings 6th international academic conference on places and technologies, Pécs, pp 64–71
4. Dormány A, Orbán Z (2022) Structural and durability assessment of heritage reinforced concrete structures. In: Proceedings of the 1st conference of the European association on quality control of bridges and structures, pp 313–322

5. ISO 2394 (2015) 1998 and 2015: general principles on reliability for structures. International Organization for Standardization (ISO), Geneva, Switzerland
6. ISO 13822 (2001) Basis for design of structures - assessment of existing structures. International Organization for Standardization (ISO), Geneva, Switzerland
7. fib Bulletin 80 (2016) Partial factor methods for existing concrete structures. Fédération Internationale du Béton, Lausanne, Switzerland
8. EN 1990 (2002) Eurocode - basis of structural design; EN 1990:2002. European Committee for Standardization CEN, Brussels, Belgium
9. EN 1991-1-1 (2002) Eurocode 1: actions on structures. Part 1-1: general actions. Densities, self-weight, imposed loads for buildings; 1991-1-1:2002. European Committee for Standardization CEN, Brussels, Belgium
10. EN 1992-1-1 (2004) Eurocode 2: design of concrete structures. Part 1-1: general rules and rules for buildings; EN 1992-1-1:2004. European Committee for Standardization CEN, Brussels, Belgium
11. EN 13791:2007 (2007) Assessment of in-situ compressive strength in structures and pre-cast concrete components. European Committee for Standardization CEN, Brussels, Belgium
12. Nowak A, Collins R (2013) Reliability of structures. CRP Press, New York. ISBN: 978-0-415-67578-8

Machine Learning-Based Seismic Damage Assessment of RC Columns Retrofitted with Post-tensioned Tendons

A. Abdolmaleki[1], S. Mahboubi[1(✉)], and M. Kioumarsi[2]

[1] Faculty of Civil, Water and Environmental Engineering, Shahid Beheshti University, Tehran, Iran
Sh_mahboubi@sbu.ac.ir

[2] Department of Built Environment, Oslo Metropolitan University, Oslo, Norway

Abstract. Designing bridges in seismic regions requires careful consideration of multiple factors to ensure both structural integrity and continued functionality during and after earthquake events. Maintaining serviceability is crucial for effective emergency response and minimizing social and economic disruptions. This study investigates the use of vertically unbonded post-tensioned tendons to enhance the seismic performance of bridge columns by promoting self-centering behavior and reducing residual displacements. Finite element simulations were conducted in OpenSees on four column specimens: one conventional reinforced concrete (RC) column and three post-tensioned reinforced concrete (PRC) columns. A comprehensive numerical dataset comprising 1,043 RC and 1,043 PRC column cases under cyclic lateral and axial loading was generated. To analyze and classify the seismic response, various machine learning algorithms were employed, including Random Forest (RF), CatBoost (CB), Decision Tree (DT), Artificial Neural Network (ANN), Deep Neural Network (DNN), and Deep & Cross Network (DCN). The models were used to predict key seismic response parameters such as residual displacement and residual force. Comparative analysis of the algorithms was conducted to evaluate their accuracy and predictive capabilities, aiming to identify the most effective approach for modeling the nonlinear seismic behavior of bridge column. The findings demonstrate the potential of combining post-tensioning techniques with machine learning to significantly improve seismic resilience in bridge design.

Keywords: RC Columns · Post-tensioned tendons · Residual force · Machine learning

1 Introduction

Bridges play a critical role in transportation infrastructure, particularly in earthquake-prone regions. Ensuring their continuous functionality during and after seismic events is essential not only for effective emergency response but also for minimizing associated socioeconomic disruptions. While performance-based seismic design enables energy

H. Beushausen et al. (Eds.): NTZR 2025, LNCE 873, pp. 873–884, 2026.
https://doi.org/10.1007/978-3-032-27698-8_77

dissipation through large nonlinear deformations, this inelastic behavior can result in residual drifts, concrete damage, and service interruptions following an earthquake. Such permanent displacements can significantly compromise both the structural integrity and serviceability of bridges, often leading to costly repairs and prolonged closures. To address these challenges, researchers have developed advanced modeling techniques to better predict and control the seismic response of bridge components. For example, Liu et al. [1] employed knowledge-enhanced neural networks (KENN) to calibrate plasticity models for RC piers, enabling fast and physics-informed seismic assessments. One promising strategy for reducing residual drifts in bridge piers is the use of post-tensioning (PT) tendons. Billington and Yoon [2] investigated unbonded PT in precast columns and observed low residual displacements, albeit with limited energy dissipation. Unbonded tendons enhance self-centering by remaining elastic under loading. The incorporation of high-performance materials, such as DFRCC, has also been shown to improve both energy dissipation and seismic performance [2]. Wang et al. [3] restored precast columns using PT tendons, which were later grouted to prevent corrosion and enhance seismic stability. Shim et al. [4] demonstrated that PT tendons improve self-centering capacity as well as seismic behavior and lateral strength. Similarly, Ou et al. [5] reported flag-shaped hysteresis behavior and noted improved corrosion resistance. Dawood et al. [6] studied unbonded tendons in steel columns and found that, while energy dissipation remained limited, supplemental strengthening through steel jacketing or FRP enhanced column performance. Zhang et al. [7] confirmed the cost-effectiveness and self-centering potential of unbonded systems, particularly when combined with reinforcement bars. Bu et al. [8] examined two ductile joint designs in PT columns using PT strands or bars combined with energy-dissipating elements, reporting improvements in both durability and displacement control. Shim et al. [9] further showed that reinforced PT bars significantly enhance seismic behavior and lateral load capacity. More recently, Neupane and Pantelides [10] investigated unbonded PT bars with supplemental longitudinal anchors (SLAs), achieving improved energy dissipation and crack control, although long-term prestress loss remains a concern.

Building upon these prior investigations, the present study aims to enhance the seismic performance of bridge columns through the use of PT tendons. To evaluate their effectiveness, two types of columns, reinforced concrete (RC) and post-tensioned reinforced concrete (PRC) are compared using six machine learning algorithms: Random Forest (RF), CatBoost (CB), Decision Tree (DT), Artificial Neural Network (ANN), Deep Neural Network (DNN), and Deep & Cross Network (DCN). Column specimens modeled by Sun et al. [11] were simulated in this study using the OpenSees platform [12], and the resulting outputs were used to construct a comprehensive dataset for further analysis. Figure 1 illustrates the overall workflow, which includes data generation, model training, and prediction of key seismic parameters such as residual displacement and residual force.

2 Finite Element Modeling and Verification Study

An experimental program examined the seismic behavior of concrete bridge foundations with vertically unbonded PT tendons. Four post-tensioned reinforced concrete (PRC) columns and one conventional RC column were tested, each with a 300 mm circular

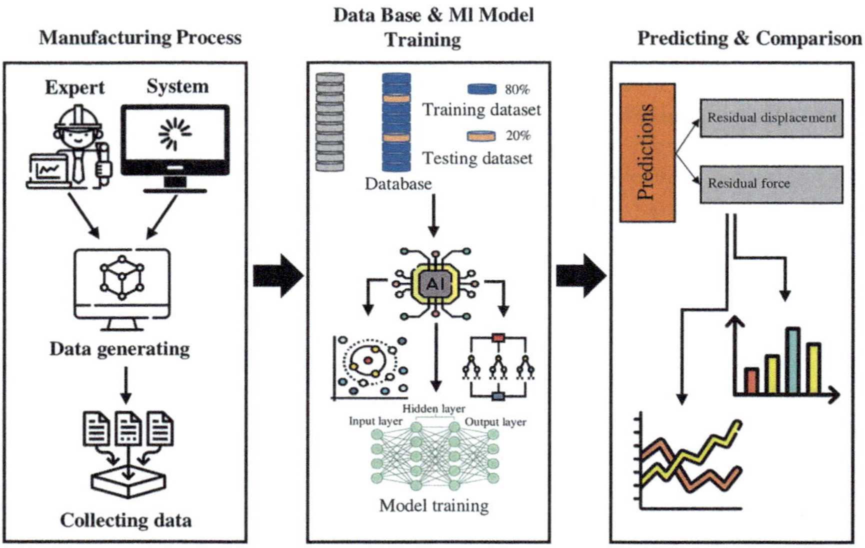

Fig. 1. General flow chart of the machine learning process, including data preprocessing, training, evaluation, and prediction.

section, 55.9 MPa compressive strength, and a 1:4 scale height of 1100 mm (slenderness ratio = 3.67). Longitudinal reinforcement consisted of eight 12 mm bars, and transverse reinforcement comprised 8 mm stirrups at 75 mm spacing (≈ 1%). Most PRC columns contained four 12.7 mm unbonded strands; variations included two-strand layouts, smaller longitudinal bars (8 mm), post-tensioning ratios of 0.05–0.15, and a specimen with centrally concentrated strands. Measured yield/ultimate strengths were 453/634 MPa (12 mm bars), 517/611 MPa (8 mm bars), and 1939 MPa for strands. Table 1 and Fig. 2 summarize the specimen details and configurations.

Table 1. Test parameters and reinforcement details of pier specimens [11].

Specimens	Longitudinal bars	Post-tensioning strands	Description of the specimen
RC-1	Eight 12 mm dia. bars	Without	Specimen without post-tensioning strands
PRC-1	Eight 12 mm dia. bars	Four 12.7 mm dia. strands	Standard specimen with post-tensioning strands
PRC-3	Eight 12 mm dia. bars	Two 12.7 mm dia. strands	Post-tensioning strands were reduced compared with specimen PRC-1
PRC-6	Eight 12 mm dia. bars	Four 12.7 mm dia. strands	Concentrated post-tensioning strands was used

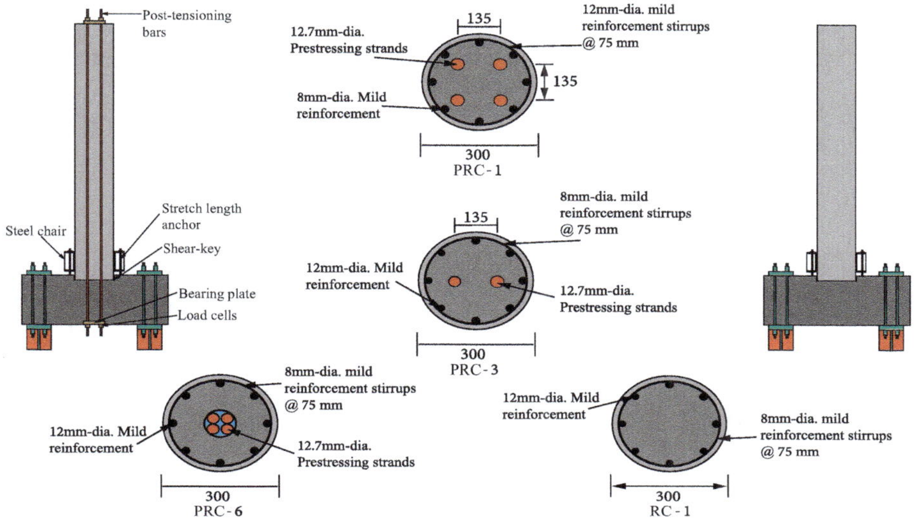

Fig. 2. Design configuration of pier specimens, elevation view and layout of post-tensioning strands

Four specimens including RC, PRC1, PRC3, and PRC6 from Sun et al. [11] were used to investigate the effect of PT strands on the nonlinear behavior of bridge piers. Numerical models were developed in OpenSees [12] with six DOFs per node (-ndm 3 -ndf 6), capturing axial, flexural, and torsional responses. Geometric nonlinearity was modeled using the P–Δ effect, and plasticity was distributed over five dispBeamColumn elements with fiber sections. The cross-section was defined using concentric circular patches representing unconfined and confined concrete. Concrete07 was employed to simulate nonlinear concrete behavior, including tension stiffening and softening. Longitudinal reinforcement was modeled using Steel02 with isotropic–kinematic hardening. PT strands were modeled via CorotTruss elements with ElasticPP material, and initial stress was introduced using InitStressMaterial.

Boundary conditions fixed the column base, while axial load was applied at the column tip (node 7) using a Plain load pattern. Lateral cyclic loading followed a quasi-static protocol: force-controlled cycles (60%–140% of F_{cr}) were applied first, followed by displacement-controlled cycles (0.35%–3.85% drift). Multiple cycles at each drift level replicated reversed cyclic behavior in accordance with ACI 374.1, capturing stiffness degradation, pinching, and energy loss. Convergence was managed using NormDispIncr with up to 10,000 iterations per step. Newton–Raphson was the primary solver, with modified Newton applied post-peak when necessary. Modeling results for the four columns are presented in Fig. 3.

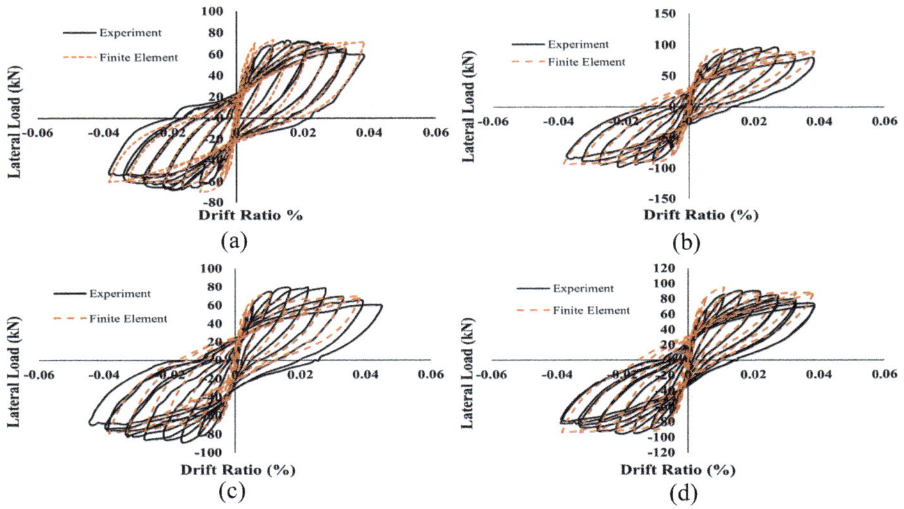

Fig. 3. Comparison of experimental and numerical hysteretic responses for specimens: (a) RC-1, (b) PRC-1, (c) PRC-3, and (d) PRC-6.

3 RC and PRC Columns Database

A dataset comprising 1,043 RC columns and 1,043 PRC columns was generated through numerical simulation. To improve the accuracy and generalizability of the machine learning (ML) models, comprehensive data preprocessing was performed. This process included the removal of outliers and low-quality samples that negatively affected model performance. By eliminating these anomalous data points, the dataset quality was significantly enhanced, enabling more reliable training, validation, and prediction across a range of seismic performance parameters. Figure 4 provides a comprehensive visualization of the statistical ranges for all RC and PRC samples.

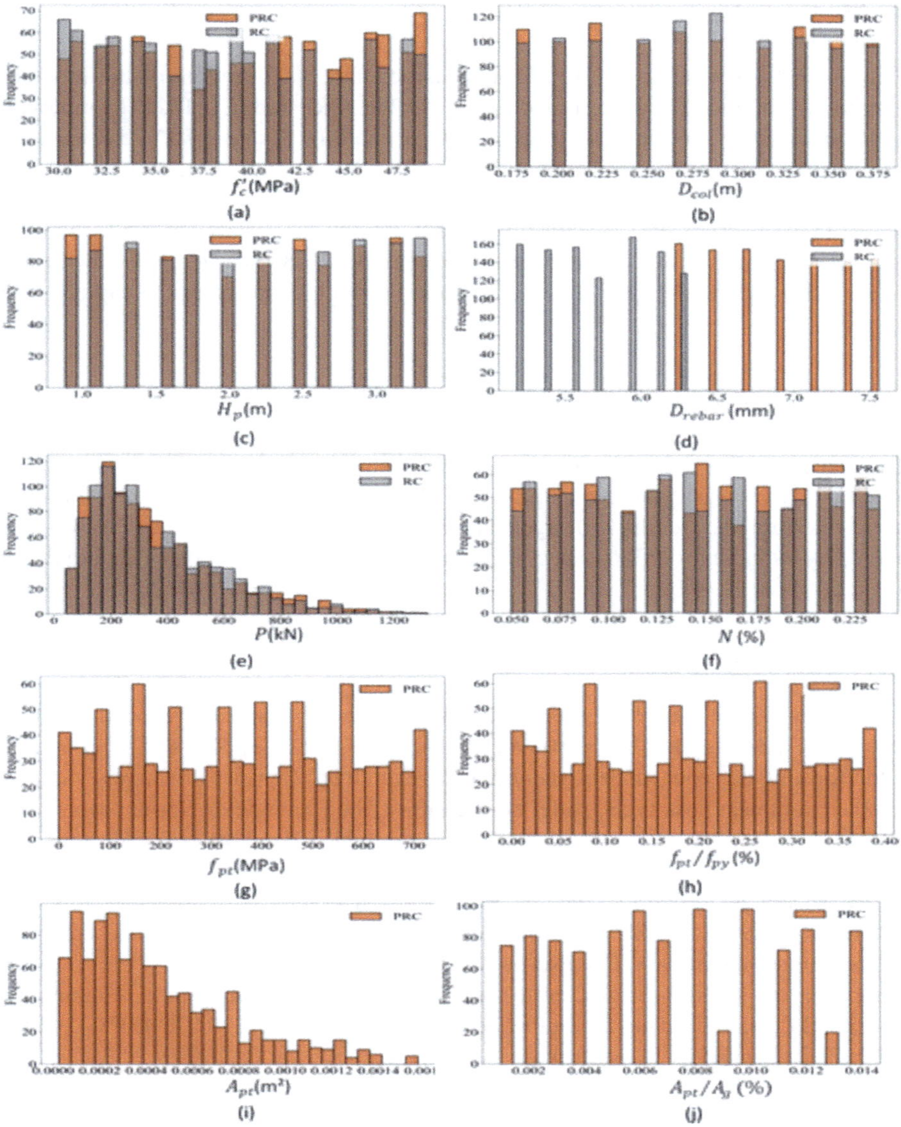

Fig. 4. Histograms of input variables: (a) concrete compressive strength (f_c'), (b) column diameter (D_{col}), (c) column height (H_p), (d) rebar diameter (D_{rebar}), (e) axial force (P), (f) axial force ratio (N), (g) initial post-tension stress (f_{pt}), (h) initial post-tension ratio (f_{pt}/f_{py}), (i) post-tensioning area (A_{pt}), (j) post-tensioning ratio (A_{pt}/A_g).

4 Machine Learning Algorithms

Using six ML algorithms, this study evaluated the effect of PT tendons on the structural response of RC piers. The tree-based models included RF and CB [13, 14], employing symmetric trees with a learning rate of 0.05, a depth of 8, and 1,500 iterations; and DT, based on the CART framework with the Gini index, providing high interpretability. The neural network-based models comprised an ANN with three hidden layers [15] (256–128–64 neurons, ReLU activation, Adam optimizer), a DNN [16] with 256 input neurons and three hidden layers (128–64–32 neurons, ReLU), and a DCN with four layers (256–128–64–32 neurons), incorporating convolutional layers to capture local patterns.

5 Comparative Analysis of the Results

Residual displacement and residual force were predicted using multiple ML algorithms. Model accuracy was evaluated using r, R^2, MAPE, and RMSE. The results indicated varying performance for the two target parameters, and the effect of PT tendons was considered by including four PRC-specific input features. Figure 5 compares the prediction accuracy of the algorithms.

5.1 Residual Displacement

A DNN demonstrated the best generalization for predicting residual displacement in RC specimens, achieving RMSE = 0.0090, MAPE = 0.900, r = 0.9850, and $R^2 = 0.9575$, outperforming all other models. Its high representational capacity effectively captured complex nonlinear post-cyclic patterns. In contrast, the RF model performed worse (r = 0.9548, $R^2 = 0.9115$), indicating difficulty in learning localized patterns and a tendency to overfit global trends. Tree-based models underperformed compared to neural networks for RC specimens. For PRC specimens, however, RF achieved the best results (r = 0.9831, $R^2 = 0.9607$, RMSE = 0.0061), demonstrating robustness to noise and suitability for structured tabular data. CB also performed well ($R^2 = 0.9661$, MAPE = 0.0859), confirming the strength of gradient-boosted decision trees in structured domains. By contrast, DT generalized poorly for PRC specimens (MAPE = 0.2307, RMSE = 0.0144), reflecting its sensitivity to overfitting.

5.2 Residual Force

For RC specimens, the RF model achieved the strongest fit (r = 0.9866, $R^2 = 0.9731$), while the DNN maintained strong generalization with the lowest errors (RMSE = 0.1837, MAPE = 0.0680). Despite the expanded feature space from PT parameters, the DT underperformed (r = 0.9662, RMSE = 0.4338), indicating that shallow models struggle to capture the residual behavior of hybrid reinforcement systems. For PRC specimens, the DCN achieved near-perfect accuracy (r = 0.9992, $R^2 = 0.9979$, RMSE = 0.0098), effectively modeling complex multivariate relationships. RF and DNN also performed strongly (r = 0.9987 and 0.9960, respectively), confirming their stability in modeling smooth target variables. By contrast, the DT model exhibited the highest errors (MAPE = 0.1341, RMSE = 0.2507), underscoring its limitations with high-variance, nonlinear data.

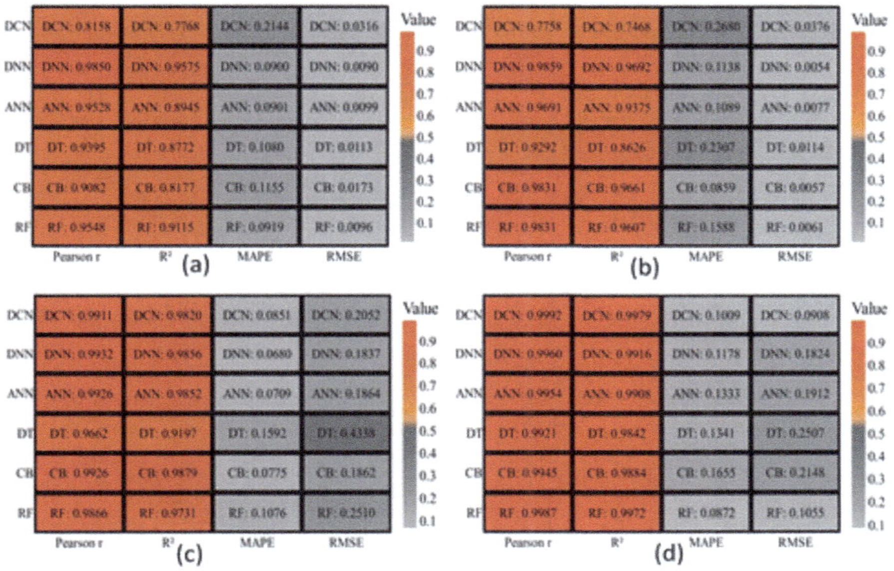

Fig. 5. Performance results of training and testing ML models for predicting residual displacement (a) RC, (b) PRC and residual force (c) RC, (d) PRC specimens.

Figure 6 compares residual displacement predictions for conventional RC and PRC specimens. PRC predictions from CB, DT, and DCN deviate further from the ideal x = y line, reflecting increased nonlinearity, stress sensitivity, and complex hysteretic behavior introduced by PT such as reduced crack widths, delayed nonlinear response, and enhanced self-centering. Both RC and PRC data align more closely with the ideal line when models with greater representational capacity are used, such as RF and ANN. Through the ensemble learning of RF and the multilayer architectures of ANN, PT systems can be effectively modeled using hierarchical nonlinear features.

Figure 7 compares residual force predictions for conventional RC and PRC specimens using six ML algorithms, with accuracy evaluated against the ideal x = y line. Similar to residual displacement, PRC predictions are influenced by the nonlinear, load-history-dependent behavior introduced by PT higher initial stiffness, reduced permanent deformation, and improved force recovery, which complicates modeling. Both RF and ANN demonstrate strong capability in capturing complex relationships for both RC and PRC specimens. DT and CB deviate significantly, particularly at higher residual forces. In contrast, DCN underperforms compared to RF and ANN, indicating that convolutional-style architectures are less effective when no clear sequential or spatiotemporal patterns

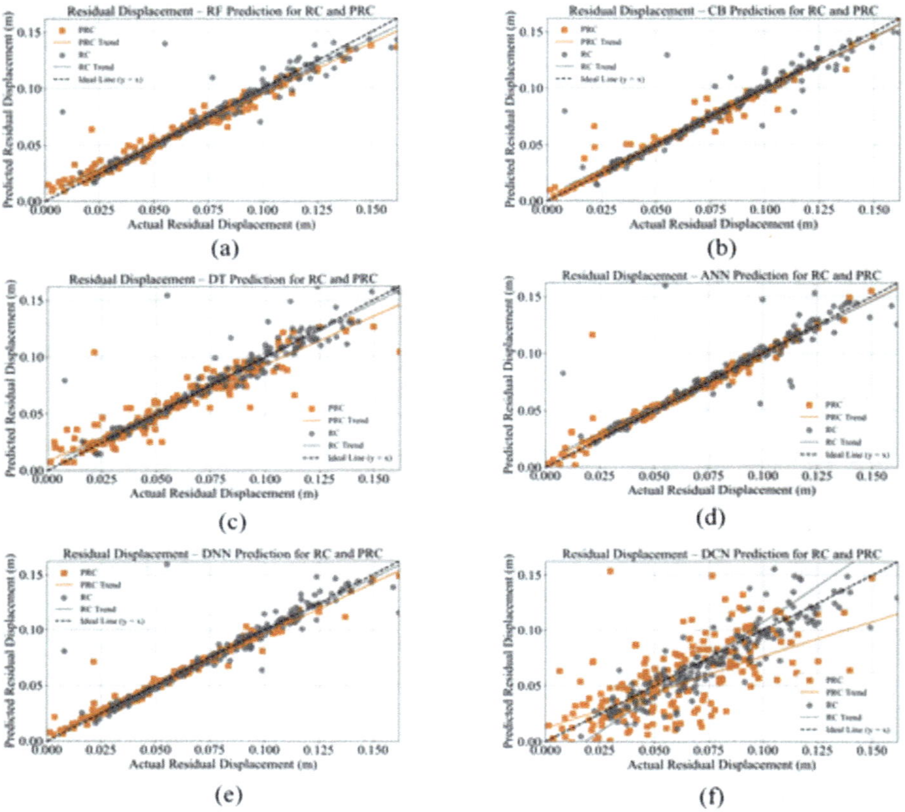

Fig. 6. Comparison of predicted and actual residual displacement values

exist. RF and ANN remain superior for modeling such systems due to their adaptability to the effects of PT tendons on hysteretic behavior and restoring forces.

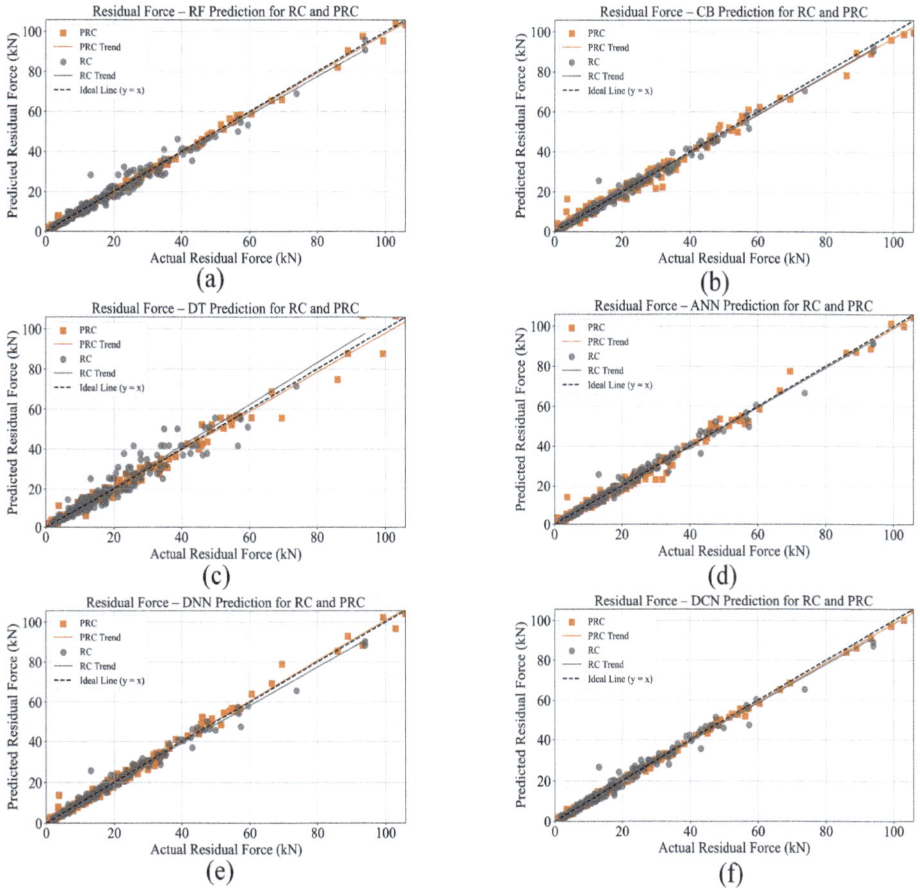

Fig. 7. Comparison of predicted and actual residual force values

6 Conclusions

This paper presents a comprehensive study on the seismic behavior of RC columns enhanced with PT tendons by generating a detailed dataset of 1,043 samples each for conventional RC and PRC columns. Six ML regression models, RF, CB, DT, ANN, DNN, and DCN, were trained on 80% of the data and evaluated on the remaining 20% to assess predictive performance. The results are summarized below:

- All six algorithms were tested for their ability to predict residual displacement and residual force in both RC and PRC columns.
- The nonlinear, history-dependent behavior introduced by PT tendons, including increased initial stiffness, reduced permanent deformation, and enhanced force recovery, posed significant challenges for model accuracy.

- RF and ANN consistently delivered the most accurate predictions for both residual displacement and residual force, effectively capturing the complex nonlinearities inherent in PRC columns.
- DT and CB exhibited notable shortcomings, particularly underestimating residual forces in PRC specimens, highlighting their limited ability to model complex PT effects.
- Although DCN improved upon some models, it still lagged behind RF and ANN, likely due to the absence of explicit spatial or temporal data structures.
- DNN excelled in predicting residual displacement for RC columns, while RF and CB were more successful for residual force predictions in PRC columns.
- In summary, this study demonstrates that selecting machine learning models with sufficient complexity and feature learning capability is critical for accurately predicting the seismic response of post-tensioned bridge columns. Future research should explore hybrid models or advanced feature engineering to further enhance prediction robustness.

References

1. Liu Z, Guo A, Zhao C, Sextos A (2024) Seismic response of bridges employing knowledge-enhanced neural networks for the lumped plasticity modelling of RC piers. Bull Earthq Eng 22:3393–3413
2. Billington SL, Yoon JK (2004) Cyclic response of unbonded posttensioned precast columns with ductile fiber-reinforced concrete. J Bridge Eng 9(4):353–363
3. Wang J-C, Ou Y-C, Chang K-C, Lee GC (2008) Large-scale seismic tests of tall concrete bridge columns with precast segmental construction. Earthq Eng Struct Dyn 37(12):1449–1465
4. Shim CS, Chung C-H, Kim HH (2008) Experimental evaluation of seismic performance of precast segmental bridge piers with a circular solid section. Eng Struct 30(12):3782–3792
5. Ou Y-C, Wang P-H, Tsai M-S, Chang K-C, Lee GC (2010) Large-scale experimental study of precast segmental unbonded posttensioned concrete bridge columns for seismic regions. J Struct Eng ASCE 136(3):255–264
6. Dawood H, ElGawady M, Hewes J (2012) Behavior of segmental precast posttensioned bridge piers under lateral loads. J Bridge Eng 17(5):735–746
7. Zhang Y, Harries KA, Yuan W (2013) Experimental and numerical investigation of the seismic performance of hollow rectangular bridge piers constructed with and without steel fiber reinforced concrete. Eng Struct 48:255–265
8. Bu Z-Y, Ou Y-C, Song J-W, Zhang N-S, Lee GC (2016) Cyclic loading test of unbonded and bonded posttensioned precast segmental bridge columns with circular section. J Bridge Eng 21(2)
9. Shim C, Koem C, Song H et al (2018) Seismic performance of repaired severely damaged precast columns with high-fiber reinforced cementitious composites. KSCE J Civ Eng 22:736–746
10. Neupane S, Pantelides CP (2025) Seismic design and fragility of self-centering bridge bent with replaceable stretch length anchors. Earthq Eng Struct Dyn 54:1433–1453
11. Sun Z, Wang D, Bi K, Si B (2015) Experimental and numerical investigations on the seismic behavior of bridge piers with vertical unbonded prestressing strands. Bull Earthq Eng 14(2):501–527
12. Mazzoni S, Mckenna F, Scott MH, Fenves GL (2007) OpenSees command language manual. Pacific Earthquake Engineering Research Center, University of California, Berkeley

13. Breiman L (2001) Random forests. Mach Learn 45(1):5–32
14. Hancock JT, Khoshgoftaar TM (2020) CatBoost for big data: an interdisciplinary review. J Big Data 7(1):123
15. Qamar R, Zardari BA (2023) Artificial neural networks: an overview. Mesopotam J Comput Sci 130–139
16. Oluleye BI, Chan DWM, Antwi-Afari P (2022) Adopting artificial intelligence for enhancing the implementation of systemic circularity in the construction industry: a critical review. Sustain Prod Consum 35

Towards Efficient Corrosion Monitoring: A Comparative Study of Modern YOLO and R-CNN Architectures

Mati Ullah[1], Vagelis Plevris[1](✉), and Afaq Ahmad[2]

[1] Department of Civil and Environmental Engineering, College of Engineering, Qatar University, P.O. Box: 2713, Doha, Qatar
{matiullah,vplevris}@qu.edu.qa

[2] Department of Built Environment, Oslo Metropolitan University (OsloMet), Oslo, Norway
afaq.ahmad@oslomet.no

Abstract. Corrosion remains a significant concern for the durability and safety of civil and industrial infrastructure, demanding efficient and automated detection approaches for timely maintenance. Traditional inspection methods are often subjective, labor-intensive, and unsuitable for large-scale monitoring. This study investigates deep learning-based corrosion detection by comparing two-stage detection models (Faster R-CNN, Mask R-CNN) with the latest one-stage YOLO architectures (YOLOv10, YOLOv11, YOLOv12). A dataset of 7,158 images, derived from open-source corrosion repositories, was refined and further expanded through three augmentation strategies to assess model robustness under diverse surface and environmental conditions. All models were trained using transfer learning with consistent hyperparameters and evaluated using precision, recall, and mean average precision (mAP). Results demonstrate that YOLO models consistently outperformed R-CNN models in both accuracy and efficiency. YOLOv11 provided the fastest inference (1.1 ms), while YOLOv12 achieved the highest validation precision of 0.743 and mAP@50 of 0.623, confirming its suitability for real-time deployment. The findings highlight YOLOv12's superior balance between accuracy and computational efficiency, establishing it as the most practical choice for scalable corrosion monitoring using UAVs and field cameras.

Keywords: Corrosion detection · YOLO · R-CNN · deep learning · data augmentation · infrastructure monitoring

1 Introduction

Corrosion remains a primary hazard to the durability and safety of civil infrastructure, imposing substantial direct and indirect costs on economies and asset owners. The study by the National Association of Corrosion Engineers (NACE) estimates the global cost of corrosion at roughly US $2.5 trillion, about 3.4% of global GDP, underscoring the scale of the problem and the importance of rigorous monitoring and management strategies [1]. In coastal and industrial environments, high salinity, humidity, temperature fluctuations, and airborne pollutants intensify corrosion risk across bridges, pipelines, offshore

H. Beushausen et al. (Eds.): NTZR 2025, LNCE 873, pp. 885–900, 2026.
https://doi.org/10.1007/978-3-032-27698-8_78

platforms, and plant structures. This accelerates deterioration and shortens service life if detection and mitigation are delayed [2]. Traditional corrosion detection techniques, including manual visual assessment and conventional image processing, are often inconvenient, time-consuming, and prone to human error [3]. They heavily depend on the inspector's experience, making them prone to inconsistency and oversight, especially in complex or poorly illuminated environments. These constraints limit the scalability, reliability, and objectivity of condition assessment for large infrastructure networks.

Advances in imaging technology have enabled rapid, non-contact inspection of civil infrastructure [4]. UAVs offer flexible, high-resolution, and cost-effective coverage of hard-to-reach areas. Meanwhile, deep learning has greatly improved automated detection of cracks, spalling, and surface defects, outperforming traditional image processing in accuracy and robustness [5]. However, corrosion detection remains less explored due to its variable appearance, color changes, and limited annotated datasets. State-of-the-art deep learning architectures, two-stage detection methods such as Region-based Convolutional Neural Network (R-CNN), Faster R-CNN, and Mask R-CNN, and the You Only Look Once (YOLO) family all offer complementary strengths for corrosion recognition. Nash et al. [6] used Bayesian deep learning for pixel-level corrosion segmentation with uncertainty estimation, achieving F1-scores up to 0.93. Ameli et al. [7] applied Mask R-CNN and YOLOv8 for bridge corrosion segmentation and severity rating using an open dataset of 514 images, attaining high accuracy. Ma et al. [8] improved Faster R-CNN with a modified Visual Geometry Group Network (VGG16) backbone for concrete corrosion strength assessment, achieving over 96% accuracy. Imran et al. [9] studied CNN-based corrosion detection in marine structures, highlighting models like AlexNet, VGG, and Faster R-CNN with average precision up to 94%.

Nabizadeh and Parghi [10] compared YOLOv3, YOLOv5s, and YOLOv7 for concrete corrosion detection, with YOLOv5s achieving the best results with mAP@0.5 = 0.88 and proving superior to traditional inspection methods. Tasnim et al. [11] applied multiple YOLO variants, achieving 96.4% mAP using YOLOv5m and a detection time of only 4 ms per image. Casas et al. [12] compared YOLOv5 and YOLOv8 for corrosion segmentation and found YOLOv8 to be faster and more precise in complex surface detection, while Yuvaneswaren et al. [13] employed YOLOv8 integrated with Roboflow and WandB cloud tools, demonstrating high robustness under diverse lighting and surface conditions. Similarly, Yu et al. [14] proposed improved YOLOv5 models enhanced with Efficient Vision Transformer (EfficientViT), improving corrosion detection precision by up to 6% in marine environments. Furthermore, Mukundh et al. [15] reported that YOLOv9 offered a balanced trade-off between detection accuracy and inference time among other YOLO architectures.

Building on these advancements, the present study focuses exclusively on corrosion detection by comparing two-stage and single-stage detector models. Specifically, it evaluates R-CNN, Faster R-CNN, and Mask R-CNN against the latest YOLO architectures, YOLOv10, YOLOv11, and YOLOv12, to identify the most efficient model for accurate and real-time corrosion detection. To ensure a robust evaluation, three dataset sizes were used: a base augmentation dataset of 4480 images and moderate and advanced augmented versions of 5080 and 5730 images, respectively. This variation tests each model's generalization and the effect of augmentation on adaptability, simulating real-world conditions such as varying metal surface states and environmental stresses. This approach allows a comprehensive assessment of detection accuracy, robustness, and efficiency across all architectures. The proposed study is designed for scalable, non-invasive, and cost-effective deployment, emphasizing field practicality, real-time inference, and consistent, auditable outputs to support preventive maintenance and sustainable asset management. Furthermore, it contributes to reducing carbon emissions by enabling early identification and maintenance of deteriorated metallic structures, thereby extending asset service life and minimizing the need for intensive energy repair or replacement. By promoting preventive maintenance and reducing material waste, the approach supports sustainable infrastructure management aligned with global net-zero objectives. Figure 1 shows the complete workflow of the methodology, including data preparation, augmentation, training, and evaluation.

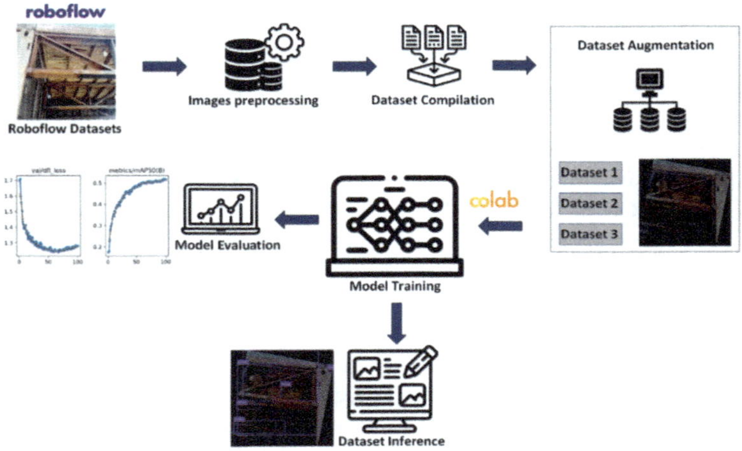

Fig. 1. Workflow of the proposed methodology.

2 Methodology

2.1 Dataset Description

The dataset used in this study was compiled from seven publicly available corrosion datasets on Roboflow [16]. The combined datasets initially contained 7,158 images representing corrosion on a wide variety of metallic surfaces captured under diverse lighting, texture, and environmental conditions. To ensure data relevance and eliminate redundancy, all duplicate and low-quality images were carefully removed. Furthermore, images depicting corrosion on non-structural or small-scale objects, such as automobile bodies, household appliances, metallic containers, hand tools, and decorative items were excluded, as they do not reflect the industrial or engineering contexts targeted in this study. After this cleaning and refinement process, the final dataset comprised 4,480 images, each annotated to accurately outline corroded regions. This refined dataset emphasizes corrosion on metallic components commonly found in industrial, mechanical, and infrastructural applications.

Table 1. Summary of the dataset versions used in this study and their characteristics.

Dataset	Total images	Training set	Validation set	Test set	Applied augmentations
Basic augmentation	4480	3808	336	336	• Horizontal and vertical flip • Exposure adjustment (−15% to +15%) • Gaussian blur up to 1.25 pixels
Moderate augmentation	5080	4318	381	381	• Horizontal and vertical flip • Rotation range ± 10° • Exposure adjustment (−15% to +15%) • Blur up to 1.5 pixels • Random noise up to 5% of pixels
Advanced augmentation	5730	4870	430	430	• Horizontal and vertical flip • Rotation range ± 15° • Exposure adjustment (−20% to +20%) • Blur up to 2.0 pixels • Random noise up to 8% of pixels • Brightness variation (−30% to +30%)

To further assess model generalization, three augmented versions of the dataset were developed using controlled geometric, photometric, and stochastic transformations, enhancing diversity and simulating the broad range of real-world inspection conditions. Table 1 provides an overview of the three dataset versions, each employing a distinct augmentation strategy. The applied transformations vary progressively from basic to advanced, reflecting increasing levels of diversity and complexity in the training samples generated. Each dataset after augmentation was split into training, validation, and test sets in an 85:7.5:7.5 ratio to ensure balanced evaluation. Figure 2 shows representative samples, illustrating the diversity of corrosion appearances and the effectiveness of the augmentations.

2.2 Corrosion Detection Models

The corrosion detection models proposed in this study comprise two-stage and one-stage object detection architectures. The architectural and computational specifications of the evaluated detection models are summarized in Table 2. Two-stage detectors such as Faster R-CNN [17], and Mask R-CNN [18] are proposed for high-precision localization and pixel-level segmentation, which is essential for characterizing small, irregular corrosion patches. One-stage detectors in the YOLO family [19] prioritize inference speed and simplicity, making them suitable for real-time inspection. The following sections explain each model family, describe the key architectural innovations and trade-offs, and highlight the specific contributions.

Two-Stage Detector CNN Models. Faster R-CNN [17] is an advanced two-stage object detector that improves the efficiency and accuracy of earlier R-CNN frameworks. It uses a Region Proposal Network (RPN) to generate candidate regions and refines them through bounding-box regression and classification. This two-stage design enables precise localization of small and irregular corrosion patches but increases computational cost, making it more suitable for offline detection rather than real-time monitoring. Mask R-CNN [18] extends Faster R-CNN by incorporating Region of Interest Alignment (RoIAlign) and a Feature Pyramid Network (FPN) to enhance spatial accuracy and multi-scale feature extraction. Although originally designed for instance segmentation, in this study, it is applied only for detection to achieve more accurate localization of corrosion regions. Its improved architecture captures detailed features across scales, offering reliable corrosion detection performance despite higher computational demand.

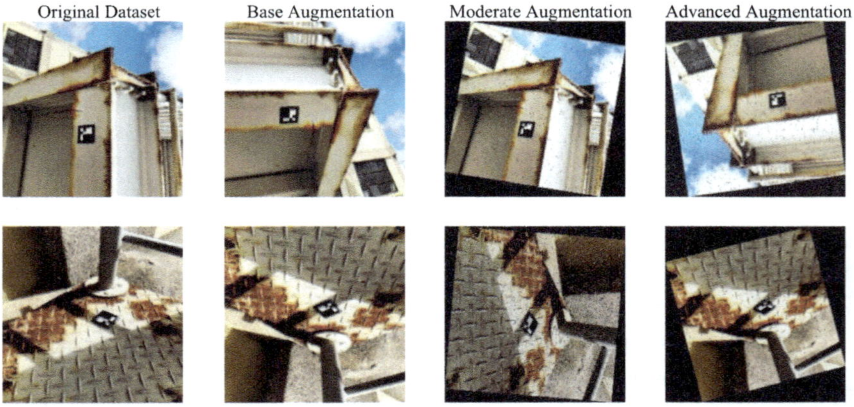

Fig. 2. Representative dataset samples after the application of augmentation techniques.

One-Stage Detector YOLO Models. The YOLO family [19] reformulates object detection as a single-stage regression task, directly predicting bounding boxes and class probabilities from the entire image in one forward pass. This unified framework eliminates the need for region proposal steps, enabling real-time inference with minimal computational cost. Such efficiency makes YOLO models particularly suitable for rapid corrosion detection across industrial and field environments where speed and accuracy are both critical. Building upon earlier generations, the recent YOLO versions, YOLOv10, YOLOv11, and YOLOv12, introduce major architectural improvements that significantly enhance detection precision, contextual reasoning, and efficiency. These versions progressively refine the backbone, neck, and head structures to optimize both feature extraction and inference performance.

YOLOv10 [20], represents a pivotal shift in the YOLO framework by adopting a Non-Maximum Suppression (NMS)-free detection strategy. Using the Consistent Dual Assignment approach, which merges one-to-many and one-to-one label assignments, it produces a single high-quality bounding box per object, minimizing redundancy.

Table 2. Architectural and computational specifications of the evaluated detection models

Model	Parameters (M)	GFLOPs	Input size	Backbone
Faster R-CNN	60.2	180.4	640 × 640	ResNet-50-FPN
Mask R-CNN	63.8	195.7	640 × 640	ResNet-50-FPN
YOLOv10	26.4	75.1	640 × 640	CSPDarknet-S
YOLOv11	24.3	68.7	640 × 640	C3k2-based backbone
YOLOv12	19.6	60.1	640 × 640	R-ELAN + A^2 Attention

The addition of a lightweight classification head and a decoupled down-sampling scheme further improves prediction consistency and speeds up inference, making it ideal for real-time deployment on resource-constrained devices. YOLOv11 [21] advances this foundation by improving feature representation and spatial attention through the replacement of older Cross Stage Partial block with two fusion layers (C2f blocks) with kernel-level multi-branch fusion C3k2 modules. The integration of Spatial Pyramid Pooling-Fast (SPPF) and Cross Stage Partial with Spatial Attention (C2PSA) blocks enhances multi-scale feature extraction and allows the network to focus more effectively on corrosion regions while suppressing background noise. This results in superior detection of small, thin, or partially occluded corrosion areas with reduced computational demand.

Further extending these improvements, YOLOv12 [22] introduced in February 2025, incorporates the Area Attention (A^2) mechanism, which efficiently captures global contextual information by partitioning feature maps into localized attention regions. In parallel, the Residual Efficient Layer Aggregation Networks (R-ELAN) stabilize gradient flow and improve feature reuse, while FlashAttention accelerates both training and inference through memory-efficient computation. Collectively, these innovations allow YOLOv12 to deliver state-of-the-art detection accuracy with minimal latency, achieving an excellent balance between precision, speed, and scalability for real-time corrosion monitoring applications.

2.3 Training Protocols

All model training procedures were carried out in a GPU-accelerated environment using an NVIDIA A100 for high-performance computing. The training used transfer learning with pre-trained weights from the MS COCO dataset [23], enabling faster convergence and improved feature extraction for corrosion detection without requiring extensive retraining. A consistent set of hyperparameters was applied across all architectures to ensure fair comparison. Training was conducted for 100 epochs with a batch size of 16 and an input resolution of 640 × 640 pixels. The AdamW optimizer was employed with an initial learning rate of 0.002, momentum of 0.9, and a cosine learning rate schedule for stable gradient updates. To further enhance computational efficiency, Automatic Mixed Precision (AMP) was activated, allowing faster training and lower memory usage while maintaining numerical stability.

3 Results and Discussion

3.1 Base Augmentation Dataset

The base augmentation dataset, comprising 4,480 images with minimal augmentation, served as the baseline for evaluating model performance under limited data diversity. The results in Table 3 show noticeable differences between two-stage and single-stage detectors in terms of accuracy, recall balance, and computational efficiency. Faster R-CNN and Mask R-CNN achieved relatively strong detection precision of 0.657 and 0.676, with mAP@50 values of 0.550 and 0.567, respectively. The slight improvement of Mask R-CNN is attributed to its additional segmentation branch and RoIAlign layer, which enhance boundary alignment and detailed localization of corrosion patches. However, their reliance on sequential region proposals resulted in high computational overhead, yielding slow inference speeds of about 24–25 ms per image. This limits their practicality for real-time inspection tasks, such as UAV-based corrosion monitoring of large structures.

Table 3. Performance comparison of the detection models using base augmentation dataset.

Model	Precision (Train)	Precision (Val)	Recall (Train)	Recall (Val)	F1-score (Train)	F1-score (Val)	mAP@50 (Train)	mAP@50 (Val)	mAP@50–75 (Train)	mAP@50–75 (Val)	Speed (ms)
Faster R-CNN	0.732	0.657	0.549	0.497	0.628	0.56	0.601	0.550	0.374	0.305	24.6
Mask R-CNN	0.754	0.676	0.565	0.513	0.642	0.582	0.616	0.567	0.389	0.320	25.1
YOLOv10	0.778	0.692	0.582	0.528	0.655	0.592	0.633	0.575	0.402	0.334	4.2
YOLOv11	0.73	0.641	0.52	0.486	0.611	0.547	0.582	0.517	0.358	0.329	2.1
YOLOv12	0.793	0.713	0.595	0.547	0.667	0.606	0.645	0.590	0.415	0.345	3.8

In contrast, YOLO models showed significant advantages in both efficiency and generalization. YOLOv10 achieved a validation precision of 0.692 and mAP@50 of 0.575 with an inference time of 4.2 ms. YOLOv11 achieved the fastest processing speed of 2.1 ms per image, with a validation precision of 0.641 and mAP@50 of 0.517. Although YOLOv11's accuracy was slightly lower than YOLOv12's, it offered the best accuracy-to-speed ratio. YOLOv12 further improved accuracy with precision of 0.713 and mAP@50 of 0.590 and maintained an inference time of 3.8 ms. The training and validation loss curves of YOLOv12 as shown in Fig. 3 further confirm this behavior, showing smooth convergence and minimal overfitting, which explains its stable detection performance under diverse conditions.

The results demonstrate that R-CNN models produced high-confidence detections but tended to miss small or low-contrast corrosion regions, whereas the single-pass YOLO detectors captured a broader range of corrosion textures, leading to better recall. YOLOv11 and YOLOv12 achieved a better trade-off between accuracy and speed, establishing YOLOv12 as the most efficient and practical model for rapid corrosion detection and scalable inspection workflows.

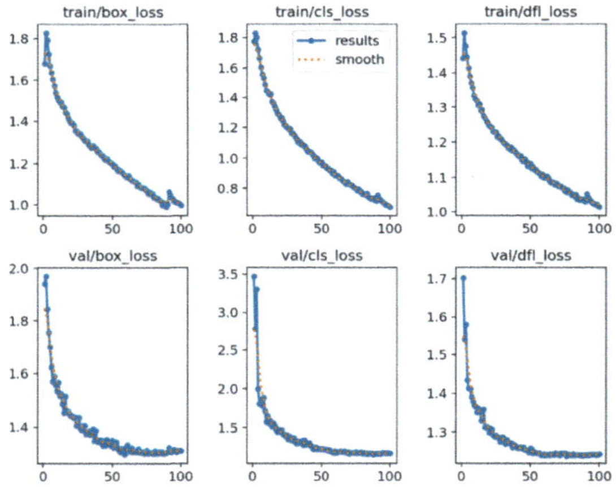

Fig. 3. Training and validation loss curves of YOLOv12 model showing the convergence of box, classification, and distribution focal losses over 100 epochs.

3.2 Moderate Augmentation Dataset

The moderate augmentation dataset, containing 5,080 images with augmentations such as rotation, exposure adjustment, blur, and random noise, provided greater visual diversity for evaluating model adaptability under variable surface and environmental conditions. Table 4 demonstrate that all models using moderate augmentation dataset, achieved better precision, recall, and mAP than in the basic augmentation dataset, confirming that data augmentation reduces overfitting and improves recognition of corrosion features across heterogeneous surfaces.

Table 4. Performance comparison of the detection models using moderate augmentation dataset

Model	Precision (Train)	Precision (Val)	Recall (Train)	Recall (Val)	F1-score (Train)	F1-score (Val)	mAP@50 (Train)	mAP@50 (Val)	mAP@50–75 (Train)	mAP@50–75 (Val)	Speed (ms)
Faster R-CNN	0.748	0.671	0.561	0.522	0.639	0.588	0.611	0.572	0.386	0.333	24.1
Mask R-CNN	0.767	0.692	0.576	0.538	0.652	0.603	0.626	0.587	0.398	0.345	25.0
YOLOv10	0.787	0.710	0.590	0.554	0.664	0.618	0.641	0.601	0.412	0.356	3.9
YOLOv11	0.793	0.709	0.594	0.562	0.671	0.627	0.649	0.610	0.418	0.372	1.6
YOLOv12	0.801	0.724	0.605	0.57	0.678	0.632	0.656	0.616	0.425	0.368	3.5

Faster R-CNN and Mask R-CNN both showed moderate performance gains, achieving validation precision values of 0.671 and 0.692, with mAP@50 values of 0.572 and 0.587, respectively. Mask R-CNN again outperformed Faster R-CNN due to its RoIAlign mechanism and multi-scale feature extraction, which help in delineating irregular corrosion regions. However, both models remained computationally expensive, with inference times around 24–25 ms, making them unsuitable for real-time inspection despite their improved accuracy.

The YOLO family demonstrated a more significant enhancement in performance and efficiency. YOLOv10 achieved a validation precision of 0.710 and mAP@50 = 0.601, while YOLOv11 provided the fastest inference at 1.6 ms, with a validation precision of 0.709 and mAP@50 = 0.610. The latest YOLOv12 further improved to 0.724 and 0.616 with an inference time of 3.5 ms. Its A^2 attention and R-ELAN modules contributed to better feature convergence and contextual understanding, leading to higher recall and fewer false positives.

3.3 Advanced Augmentation Dataset

The advanced augmentation dataset comprised the largest and most diverse image set and noise injection. These enhancements significantly improved the generalization capacity of all models compared with basic and moderate augmentation datasets, though each architecture responded differently to the increased complexity. As presented in Table 5, the two-stage detectors models, Faster R-CNN and Mask R-CNN, achieved stable but moderate performance. Faster R-CNN recorded a validation precision of 0.687 and recall of 0.533 with mAP@50 = 0.579, confirming its ability to localize corroded areas accurately but at the cost of high inference speed of 23.8 ms. Mask R-CNN slightly improved across most metrics (Precision = 0.706, Recall = 0.549, mAP@50 = 0.595) due to its RoIAlign mechanism and enhanced feature extraction, which helped identify irregular corrosion boundaries more accurately. However, both remained computationally intensive ($\approx$ 25 ms).

Table 5. Performance comparison of the detection models using advance augmentation dataset

Model	Precision (Train)	Precision (Val)	Recall (Train)	Recall (Val)	F1-score (Train)	F1-score (Val)	mAP@50 (Train)	mAP@50 (Val)	mAP@50–75 (Train)	mAP@50–75 (Val)	Speed (ms)
Faster R-CNN	0.763	0.687	0.572	0.533	0.649	0.595	0.622	0.579	0.395	0.342	23.8
Mask R-CNN	0.782	0.706	0.587	0.549	0.663	0.61	0.637	0.595	0.41	0.356	24.7
YOLOv10	0.801	0.728	0.602	0.565	0.676	0.623	0.654	0.609	0.424	0.369	3.7
YOLOv11	0.812	0.584	0.390	0.604	0.523	0.594	0.580	0.573	0.34	0.315	1.1
YOLOv12	0.816	0.743	0.617	0.582	0.689	0.637	0.667	0.623	0.437	0.381	3.3

YOLO architectures displayed a clear advantage in speed and adaptability. YOLOv10 achieved a validation precision of 0.728, recall of 0.565, and mAP@50 = 0.609, showing strong detection capability and fast inference at 3.7 ms. YOLOv12 delivered the best overall performance, with Precision = 0.743, Recall = 0.582, and mAP@50 = 0.623, combining high accuracy with fast inference (3.3 ms). YOLOv11 demonstrated the highest efficiency with the fastest inference time of 1.1 ms; however, it yielded lower accuracy (Precision = 0.584, Recall = 0.604, mAP@50 = 0.573) compared to its stable performance under moderate augmentation. This decline is attributed to its relatively lighter C3k2-based backbone and spatial attention modules, which are more sensitive to strong photometric noise and brightness variations. YOLOv10 and YOLOv12 maintained higher precision and mAP values due to their more robust architectures, CSPDarknet-S and R-ELAN with A^2 attention, which effectively handle complex textures and heavy augmentation. The results confirm that YOLOv12 achieved the best balance between accuracy and speed, making it the most suitable option for real-time corrosion monitoring where low latency and consistent performance are essential.

Figure 4 visually compares the detection performance of all models on a representative sample image. The two-stage detectors, Faster R-CNN and Mask R-CNN, successfully identify corrosion areas but produce overlapping and occasionally imprecise bounding boxes due to their sequential proposal mechanisms. In contrast, the YOLO models exhibit cleaner, more confident detections with fewer redundant boxes, reflecting their end-to-end learning and optimized feature fusion. Among them, YOLOv11 and YOLOv12 show the most accurate localization of corrosion. These observations align closely with the quantitative findings, confirming that YOLOv12 provides the best accuracy–speed balance, while YOLOv11 achieves the highest computational efficiency for real-time inspection scenarios.

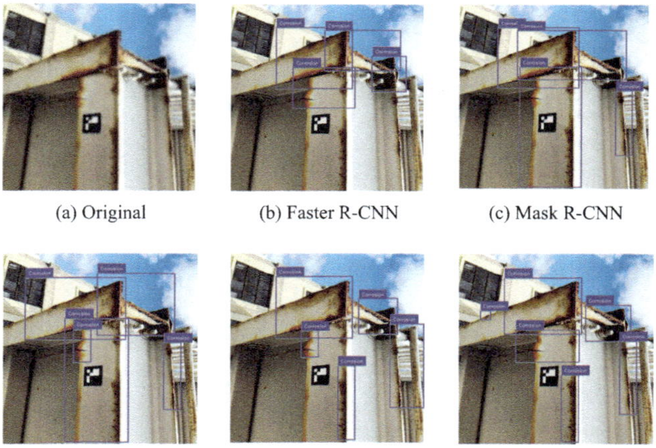

(a) Original (b) Faster R-CNN (c) Mask R-CNN

(d) YOLOv10 (e) YOLOv11 (f) YOLOv12

Fig. 4. Inference results across Faster R-CNN, Mask R-CNN, and YOLO.

4 Conclusion

This study presents a comparative evaluation of advanced deep learning architectures for automated corrosion detection, revealing clear performance trends across progressively augmented datasets. The two-stage detectors, Faster R-CNN and Mask R-CNN, consistently achieved moderate accuracy across all datasets, with Mask R-CNN outperforming Faster R-CNN due to its RoIAlign and multi-scale feature extraction. However, their high computational demands and slower inference times ($\approx$ 24–25 ms) restricted their suitability for real-time corrosion inspection. The YOLO models showed substantial improvement in both accuracy and efficiency with increased dataset diversity. From the base to advanced augmentation datasets, all YOLO versions benefited from richer data variability, confirming their superior generalization capability. YOLOv10 demonstrated strong baseline accuracy and stability, while YOLOv11 achieved the fastest inference (1.1 ms) with acceptable precision, making it ideal for latency-critical field monitoring.

The latest YOLOv12 achieved the highest detection accuracy, with validation precision of 0.743 and mAP@50 of 0.623 in the advanced dataset, showing remarkable adaptability to complex corrosion patterns. The superior performance of YOLOv12 is attributed to its advanced architecture, integrating the R-ELAN and area attention modules, which enhanced feature representation, spatial focus, and contextual understanding of corrosion regions. These characteristics make the proposed models ideal for large-scale, real-time inspection scenarios such as UAV-assisted monitoring and industrial asset management. The lightweight design and high inference speed of YOLOv11 and YOLOv12 make them suitable for real-time processing on embedded GPU devices such as NVIDIA Jetson and Intel Movidius platforms. These capabilities allow on-site corrosion detection during aerial or robotic inspections without reliance on cloud computation, minimizing latency and energy consumption. Future research will focus on corrosion severity quantification, multi-class damage assessment, and integration with digital twin platforms to support proactive infrastructure management.

Acknowledgements. This work was supported by Qatar University through the student grant "Deep Learning-Based Smart Framework for Corrosion Detection and Prognosis in Infrastructure" (Grant ID: QUST-2-CENG-2025-521).

References

1. Jacobson G (2016) NACE international's IMPACT breaks new ground in the study of corrosion management. Mater Perform 55(4):28–32. https://doi.org/10.5006/MP2016_55_4-28
2. Zaki A, Chai HK, Aggelis DG, Alver N (2015) Non-destructive evaluation for corrosion monitoring in concrete: a review and capability of acoustic emission technique. Sensors 15(8):19069–19101. https://doi.org/10.3390/s150819069
3. Rajendran M, Subbian D (2025) Deep learning in corrosion assessment and control: a critical review of techniques and challenges. Corros Rev. https://doi.org/10.1515/corrrev-2024-0060
4. Ullah M, Mir J, Ahmad A (2025) Assessing the influence of cement replacement materials on concrete performance through image processing. Measurement 118014. https://doi.org/10.1016/j.measurement.2025.118014

5. Ullah M, Mir J, Husain SS, Shahid MLUR, Ahmad A (2024) Concrete forensic analysis using deep learning-based coarse aggregate segmentation. Autom Constr 162:105372. https://doi.org/10.1016/j.autcon.2024.105372
6. Nash W, Zheng L, Birbilis N (2022) Deep learning corrosion detection with confidence. npj Mater Degrad 6(1):26. https://doi.org/10.1038/s41529-022-00232-6
7. Ameli Z, Nesheli SJ, Landis EN (2023) Deep learning-based steel bridge corrosion segmentation and condition rating using mask RCNN and YOLOv8. Infrastructures 9(1):3. https://doi.org/10.3390/infrastructures9010003
8. Ma Q, Cao Y, Abulizi N, Mao W, Chen S (2024) Corrosion residual strength evaluation of comprehensive pipe gallery concrete based on improved faster-RCNN. In: Proceedings of the 2024 8th Asian conference on artificial intelligence technology (ACAIT). IEEE, pp 573–579. https://doi.org/10.1109/ACAIT63902.2024.11021766
9. Imran MMH et al (2026) Advancements and challenges in convolutional neural networks for marine corrosion detection and classification. J Adv Res Des 142(1):89–107. https://doi.org/10.37934/ard.142.1.89107
10. Nabizadeh E, Parghi A (2023) Automated corrosion detection using deep learning and computer vision. Asian J Civ Eng 24(8):2911–2923. https://doi.org/10.1007/s42107-023-00684-4
11. Tasnim Z et al (2025) A comprehensive deep learning based approach for corrosion detection of industrial machinery and tools. In: Proceedings of the 2025 international conference on electrical, computer and communication engineering (ECCE). IEEE, pp 1–6. https://doi.org/10.1109/ECCE64574.2025.11013844
12. Casas E, Ramos L, Romero C, Rivas-Echeverría F (2024) A comparative study of YOLOv5 and YOLOv8 for corrosion segmentation tasks in metal surfaces. Array 22:100351. https://doi.org/10.1016/j.array.2024.100351
13. Yuvaneswaren R, Prakash ST, Sudha K (2024) A YOLOv8-based model for precise corrosion segmentation in industrial imagery. In: Proceedings of the 2024 3rd international conference on artificial intelligence for internet of things (AIIoT). IEEE, pp 1–6. https://doi.org/10.1109/AIIoT58432.2024.10574659
14. Yu Q, Han Y, Gao X, Lin W, Han Y (2024) Comparative analysis of improved YOLO v5 models for corrosion detection in coastal environments. J Mar Sci Eng 12(10):1754. https://doi.org/10.3390/jmse12101754
15. Mukundh J, Rehsi A, Vishveshwara U, Natarajan B (2024) Enhanced corrosion detection using YOLO-based deep learning models. In: Proceedings of the 2024 international conference on modeling, simulation & intelligent computing (MoSICom). IEEE, pp 552–557. https://doi.org/10.1109/MoSICom63082.2024.10881182
16. Dwyer B, Nelson J, Hansen T et al (2025) Roboflow, in version 1.0. https://roboflow.com
17. Girshick R (2015) Fast R-CNN. In: Proceedings of the IEEE international conference on computer vision, pp 1440–1448. https://doi.org/10.1109/ICCV.2015.169
18. He K, Gkioxari G, Dollár P, Girshick R (2017) Mask R-CNN. In: Proceedings of the IEEE international conference on computer vision, pp 2961–2969. https://doi.org/10.1109/ICCV.2017.322
19. Jiang P, Ergu D, Liu F, Cai Y, Ma B (2022) A review of YOLO algorithm developments. Procedia Comput Sci 199:1066–1073. https://doi.org/10.1016/j.procs.2022.01.135
20. Wang A et al (2024) YOLOv10: real-time end-to-end object detection. arXiv preprint https://arxiv.org/abs/2405.14458. https://doi.org/10.48550/arXiv.2405.14458
21. Khanam R, Hussain M (2024) YOLOv11: an overview of the key architectural enhancements. arXiv preprint https://arxiv.org/abs/2410.17725. https://doi.org/10.48550/arXiv.2410.17725
22. Tian Y, Ye Q, Doermann D (2025) YOLOv12: attention-centric real-time object detectors. arXiv preprint https://arxiv.org/abs/2502.12524. https://doi.org/10.48550/arXiv.2502.12524

23. Lin T-Y et al (2014) Microsoft COCO: common objects in context. In: Proceedings of the European conference on computer vision. Springer, pp 740–755. https://doi.org/10.1007/978-3-319-10602-1_48

Probabilistic Frameworks and Methods in Performance Assessment of Corroded Reinforced Concrete Columns

Pooria Poorahad, Shamim Mohammadi, and Mahmoud R. Shiravand(✉)

Faculty of Civil, Water, and Environmental Engineering, Shahid Beheshti University, Tehran, Iran
m_shiravand@sbu.ac.ir

Abstract. Reinforced concrete (RC) columns are critical structural components whose performance is significantly compromised by corrosion, heightening their seismic vulnerability. The inherent uncertainties in both deterioration and seismic loading necessitate probabilistic assessment over deterministic approaches. This review provides a comprehensive synthesis and critical evaluation of state-of-the-art probabilistic frameworks for assessing corroded RC columns. The paper systematically categorizes and evaluates the primary methodologies: experimental investigations, numerical fragility analyses, stochastic life-cycle models, and emerging data-driven techniques. A key contribution is a new comparative framework, including a decision matrix and quadrant chart, designed to guide researchers and practitioners in selecting the most appropriate assessment method based on project objectives, data availability, and computational resources. By analyzing the advantages and limitations of each approach, this paper identifies critical research gaps and proposes future directions, including the need for improved deterioration models, standardized damage states, and enhanced computational efficiency.

Keywords: Probabilistic methods · Reinforced concrete · Corrosion · Reliability Index

1 Introduction

The long-term safety and functionality of reinforced concrete (RC) bridge infrastructure are paramount for societal resilience and economic vitality [1]. However, RC bridge piers, crucial load-bearing elements, are persistently vulnerable to corrosion, primarily induced by chloride ingress from de-icing salts or marine environments [2]. This degradation process severely compromises structural integrity and drastically alters seismic performance. Corrosion attacks the reinforcing steel, leading to significant cross-sectional loss, reduced ductility, and potential embrittlement [3]. Concurrently, the expansive corrosion products generate internal stresses, causing concrete cover cracking, spalling, and ultimately, a substantial reduction in the effective cross-section. Critically, corrosion severely degrades the bond strength between steel and concrete, impairing the composite action essential for structural response [4]. Consequently, corroded RC piers exhibit

H. Beushausen et al. (Eds.): NTZR 2025, LNCE 873, pp. 901–915, 2026.
https://doi.org/10.1007/978-3-032-27698-8_79

significantly diminished strength, stiffness, energy dissipation capacity, and deformation capability under seismic loads. Neglecting corrosion effects in seismic assessments can lead to a dangerous overestimation of a bridge's resilience, potentially resulting in catastrophic failures during earthquakes, especially as vast portions of the global bridge inventory exceed their intended design life in seismically active regions [5, 6]. Therefore, explicitly accounting for corrosion-induced deterioration is fundamental for accurate seismic vulnerability assessment and informed life-cycle management decisions [7].

Traditional deterministic structural assessment approaches, while computationally efficient, often fail to adequately capture the inherent uncertainties and variability associated with structural materials, loading conditions, environmental exposure, and deterioration mechanisms like corrosion. This limitation is particularly acute for seismic assessment, where ground motion randomness and structural response complexity introduce significant aleatory and epistemic uncertainties [8, 9]. Probabilistic methods have thus become indispensable tools in modern performance-based earthquake engineering, providing a robust framework to quantify these uncertainties and assess the likelihood of exceeding predefined performance states [10]. Common probabilistic techniques include Monte Carlo simulation, which propagates input uncertainties (e.g., material strength variability, corrosion rate scatter, seismic demand randomness) through computational models to generate a distribution of structural responses [11]. First- and Second-Order Reliability Methods (FORM/SORM) offer computationally efficient alternatives for estimating failure probabilities. Fragility analysis, a core probabilistic output, expresses the conditional probability of reaching or exceeding a specific damage level (e.g., yielding, collapse) given a seismic intensity measure. For corrosion-affected structures, probabilistic methods are essential for quantifying the increased risk associated with deterioration, enabling more rational, risk-informed decision-making for inspection, maintenance, retrofit, and replacement strategies compared to deterministic safety factors alone.

This review provides a critical contribution by systematically categorizing and evaluating the probabilistic frameworks for assessing corroded RC columns. The paper first reviews existing methodologies, structuring them into four primary approaches: experimental investigations, numerical fragility analyses, stochastic life-cycle models, and other probabilistic techniques like machine learning and Bayesian methods. The core of this work then introduces a new discussion and comparative framework designed to synthesize these approaches. Through a detailed decision matrix and a visual quadrant chart, this paper delivers a critical analysis of the applications and inherent limitations of each method. Finally, a key forward-looking contribution is the identification of pressing research gaps and recommendations for future work, serving as a practical guide for the selection of appropriate assessment tools for aging infrastructure.

2 Probabilistic Assessment Frameworks: Categorization and Evaluation

Seismic assessment of corroded RC piers increasingly relies on probabilistic approaches to quantify reliability and fragility under the compounding effects of corrosion and earthquake loading. These methods incorporate time-dependent deterioration and uncertainty in demand and capacity, enabling life-cycle and performance-based evaluations. Corrosion degrades steel's mechanical properties disproportionately based on corrosion type. Pitting causes severe localized section loss, reducing ductility more than strength; for instance, a 10% mass loss can decrease elongation by 25–40% and yield strength by 1.2–2.0% [12]. Empirical models express mechanical degradations as [13],

$$f_y = \left(1 - \alpha_y Q_{corr}\right) f_{yi} \tag{1}$$

$$f_u = (1 - \alpha_u Q_{corr}) f_{ui} \tag{2}$$

$$\varepsilon_u = (1 - \alpha_\varepsilon Q_{corr}) \varepsilon_{ui} \tag{3}$$

$$E_s = (1 - \alpha_E Q_{corr}) E_{si} \tag{4}$$

wherein where f_y, f_u, ε_u, and E_s are the current yield strength, ultimate strength, ultimate strain, and elastic modulus for the corroded steel, respectively, f_{yi}, f_{ui}, ε_{ui}, and E_{si} are the initial values of the same mechanical properties for the sound steel, respectively, α are corresponding regression constants, and Q_{corr} is the corrosion percentage. The above-mentioned process is illustrated in Fig. 1.

Coronelli and Gambarova [14] proposed the following equation to calculate the compressive strength of the corroded unreinforced concrete:

$$f_c = \frac{f_{ci}}{1 + K\frac{\varepsilon_1}{\varepsilon_{co}}} \tag{5}$$

where f_c is the residual compressive strength of concrete after corrosion, f_{ci} is the compressive strength of sound concrete, K is the bar roughness and diameter coefficient, ε_{co} is the strain at peak compressive strain, and $\varepsilon_1 = 2\pi(\upsilon_{rs} - 1)Xn_{bars}/p_0$ is smeared tensile strain in the cracked concrete in which υ_{rs} is the ratio of the volumetric expansion of oxides with respect to sound material, X is the depth of the corrosion attack, n_{bars} is the number of the compressive bars, and p_0 is the perimeter of the column section.

2.1 Experimental Investigation of Corroded Columns

Experimental studies measure performance by physically inducing corrosion and directly recording parameters such as residual strength, ductility, drift, energy dissipation, and failure modes. In experimental studies, specimens are corroded using accelerated corrosion methods to simulate corrosion effects. For example, Revathy et al. [15] measured uniaxial compression to failure after accelerated corrosion up to 10% and 25% damage, and reported substantial axial strength and ductility losses. Yalciner et al. [16] corroded their full-scale RC columns in an accelerated-corrosion pool for several days to measure the initial crack widths at different levels of corrosion. They noted a drift ratio decrease from 2% in uncorroded columns to 1.4% after corrosion. Other accelerated corrosion techniques, namely artificial climate environment and electrochemical chloride extraction, are employed to investigate the seismic behavior of corroded RC columns [17]. Table 1 summarizes the experimental studies together with their methods and metrics.

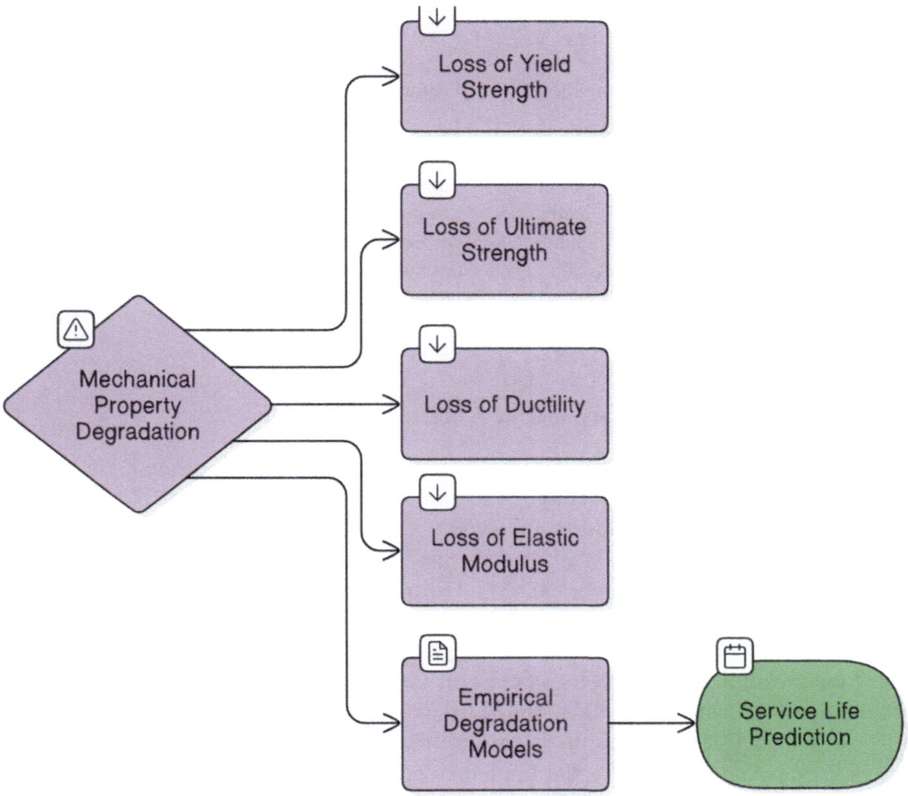

Fig. 1. Corrosion-induced degradations.

Table 1. Characteristics of experimental studies.

Study	Analysis method	Performance metrics	Corrosion parameters
Revathy et al. [15]	Uniaxial compression to failure after accelerated corrosion	Axial strength, ductility	Accelerated corrosion; 10% and 25% damage; no other metrics
Yalciner et al. [16]	Accelerated corrosion; 10% and 25% damage; no other metric	Drift ratio, energy dissipation capacity	Accelerated-corrosion pool; initial crack width; concrete strengths: 9, 27, 37 MPa

(continued)

Table 1. (*continued*)

Study	Analysis method	Performance metrics	Corrosion parameters
Dai et al. [17]	Cyclic loading of RC square columns; comparison of anodic corrosion electrolysis and electrochemical corrosion techniques	Failure modes, hysteresis/backbone curves (load-deformation response), stiffness degradation, ductility, energy dissipation	Anodic corrosion electrolysis and electrochemical corrosion techniques; target corrosion level; no specific metrics

2.2 Fragility Analysis Incorporating Corrosion

Accelerated tests may provide misleading information if based on unrealistic assumptions or improper use of the method, and current tests often poorly correlate with field results [18]. Additionally, due to the time constraints, the implementation of larger numbers of experimental tests is burdensome. Therefore, fragility analysis is mostly conducted on the basis of numerical tests. Dizaj and Kashani [19] performed pushover, cyclic, and incremental dynamic analysis (IDA) on fiber-based numerical models. In the aforementioned study, failure modes, energy dissipation, flexural capacity, and damage states were captured, and a 60% increase in failure probability at 15% corrosion was observed, while in another similar study [20], a 35% reduction in flexural capacity and a 50% drop in ductility were reported at a 20% mass loss. The developments in supercomputers have paved the way for the implementation of randomized analyses with large numbers of simulations. In randomized analyses, the numerical models are randomly built by a number of stochastic parameters, and then the IDA [21] or Monte Carlo simulation (MCS) [22] methods can be employed on them. The accuracy of the results in numerical methods highly depends on the accuracy of the materials, boundary conditions, and the capability of the numerical model in capturing different failure modes [23]. Hence, numerical models with shear-associated behaviors [24, 25] are required in this regard. Due to the difficulty and computational expense of numerical methods, results of the available experimental studies can be collected to develop fragility functions [26]. A fragility function represents the probability of being in or exceeding one or more damage states given a particular engineering demand parameter using a lognormal distribution [27]. Along with fiber-based methods, 3D continuum numerical models [28] are also utilized. Table 2 summarizes the fragility-based studies together with their methods and metrics.

Table 2. Characteristics of fragility-based studies.

Study	Analysis method	Performance metrics	Corrosion parameters
Dizaj and Kashani [19]	Fiber-based modeling; pushover, cyclic, and IDA; validation with experimental data	Failure modes, energy dissipation, flexural capacity, damage states	Modeled mass loss: 0%, 13.8%, 28.9%, 52.9%; spatial variability; pitting

(*continued*)

Table 2. (*continued*)

Study	Analysis method	Performance metrics	Corrosion parameters
Dizaj et al. [20]	Nonlinear finite element modeling; parametric and dynamic analysis; validation with experimental data	Damage states, ductility factor, flexural capacity, failure modes	Simulated mass loss, crack width; 15% mass loss = 1.5 mm crack; 25–45% flexural loss, 45–60% ductility loss
Zhang et al. [21]	Stochastic finite element modeling; pushover analysis, IDA, and fragility analysis	Damage states, failure probability	Chloride attack (modeled); no specific induction/measurement; uncorroded/corroded states
Liu et al. [22]	Monte Carlo simulation; blast fragility analysis	Damage states, damage probability, blast resistance	Chloride-induced corrosion; degraded diameter/yield stress; up to 90 years post-initiation
Xu et al. [23]	New numerical model for shear-critical columns; fragility analysis; time-variant capacity	Shear strength, stiffness, damage states	Modeled corrosion; deterioration of material properties; no induction/measurement
Dabas et al. [28]	Experimental testing of RC columns under quasi-static loads; sectional and numerical analysis; 3D finite element analysis	Peak residual strength, ductility	Accelerated corrosion (137 days); patterns: transverse, longitudinal, both; reduction in strength/ductility

2.3 Stochastic Models on RC Columns in Corrosive Environments

Environmental conditions modeled include chloride-induced corrosion in marine, coastal, tidal, and splash settings; carbonation and de-icing salt effects in urban areas; and influences from temperature, humidity, and curing regimes. Stochastic models are used to investigate RC columns exposed to a corrosive environment. For instance, Chen et al. [29] utilized a gamma process combined with probability density function (PDF) analysis measures crack width and failure probability—reporting a 50% failure probability at an allowable crack width and remaining lives of 16, 10, and 7 years for varying corrosion levels. Their key findings showed that failure probability reached 50% at the allowable crack width. Firouzi et al. [30] implemented a reliability analysis using non-stationary Gaussian processes to underpin time-dependent characteristics in chloride-laden environments. The methodology incorporated down-crossing rate analysis, first passage probability calculations, and Focus on chloride-laden environments (coastal/marine). Their results indicated high failure probability due to corrosion, particularly under lateral loads. Random field analysis within a stochastic resistance deterioration framework showed cumulative failure probability increases of 20.15% to 70.40% under tidal and splash exposures [31]. The study examined marine atmospheric, tidal, and splash environments, cross-sectional area loss, and Spatial variability effects. These studies collectively demonstrate the versatility of stochastic modeling approaches in capturing different aspects of reinforced concrete column deterioration under various environmental conditions. Characteristics of the above-mentioned studies are tabulated in Table 3 for convenience.

Table 3. Characteristics of stochastic probabilistic studies.

Study	Probabilistic method	Environmental conditions	Evaluation parameters
Chen et al. [29]	Stochastic gamma process, PDF analysis	Aggressive environments, chloride ion ingression	Crack width, load-bearing capacity, failure probability, remaining useful life
Firouzi et al. [30]	Accelerated corrosion; 10% and 25% damage; no other metric	Drift ratio, energy dissipation capacity	Accelerated-corrosion pool; initial crack width; concrete strengths: 9, 27, 37 MPa
Yang et al. [31]	Cyclic loading of RC square columns; comparison of anodic corrosion electrolysis and electrochemical corrosion techniques	Failure modes, hysteresis/backbone curves (load-deformation response), stiffness degradation, ductility, energy dissipation	Anodic corrosion electrolysis and electrochemical corrosion techniques; target corrosion level; no specific metrics

2.4 Other Probabilistic Methods and Emerging Techniques

Beyond the established frameworks, other probabilistic methods are being developed to address specific challenges in assessing corroded RC columns, particularly regarding computational efficiency and model adaptation. These emerging techniques, including artificial neural networks (ANN) and Bayesian inference, represent a significant step toward more dynamic and scalable infrastructure management.

ANNs have been applied to evaluate life-cycle hysteretic performance, with one study noting precision above 90% and service life extensions exceeding 85% with increased concrete cover [32]. ANNs have shown promise in creating computationally efficient surrogate models for complex, non-linear structural behavior. However, a primary challenge is their "black box" nature, which can obscure the physical reasoning behind a prediction.

Bayesian probabilistic methods with PDF analysis were used to gauge residual shear strength [33] while MCS with PDF analysis was employed to evaluate carbonation-induced durability failures in urban contexts [34]. In the paper by Ding et al. [33], some methods were developed for predicting the residual shear strength of corroded RC columns, and probabilistic calibration of deterministic models was employed. The key advantage of this approach is its ability to reduce uncertainty systematically. An initial prediction based on general models (a "prior") can be updated with project-specific data—such as measured corrosion levels or crack widths from a field inspection—to produce a more accurate, context-specific assessment. This makes it an ideal tool for the life-cycle management of individual, critical structures.

Other approaches continue to be explored. Kwon and Na [34], however, focused on urban environments with 18 years of exposure and analyzed carbonation depth and probability of durability failure. A combined reliability and fuzzy approach by Bastidas-Artega [35] addresses interactions between corrosion and fatigue under multiple environmental conditions, focusing on service life prediction, reliability assessment, and combined corrosion-fatigue effects. These diverse methods underscore a clear trend toward more specialized, data-driven, and adaptive techniques for managing the risks associated with deteriorating infrastructure. A summary of these emerging techniques is provided in Table 4. Additionally, a schematic illustration of each framework is depicted in Fig. 2.

Table 4. Characteristics of other probabilistic studies.

Study	Probabilistic method	Environmental conditions	Evaluation parameters
Li et al. [32]	ANN for probabilistic evaluation	Chloride ingress	Life-cycle hysteretic performance, time-dependent performance, effect of concrete cover
Ding et al. [33]	Bayesian probabilistic modeling, PDF analysis	No mention found	Residual shear strength, probabilistic calibration of deterministic models
Kwon and Na [34]	MCS, PDF analysis	Urban cities; 18 years of exposure	Carbonation depth, probability of durability failure, repair timing
Bastidas-Artega [35]	Reliability analysis, stochastic processes, fuzzy approach	De-icing salts, sea exposure; temperature, humidity, seasonal variation, global warming	Service life, reliability, combined corrosion-fatigue

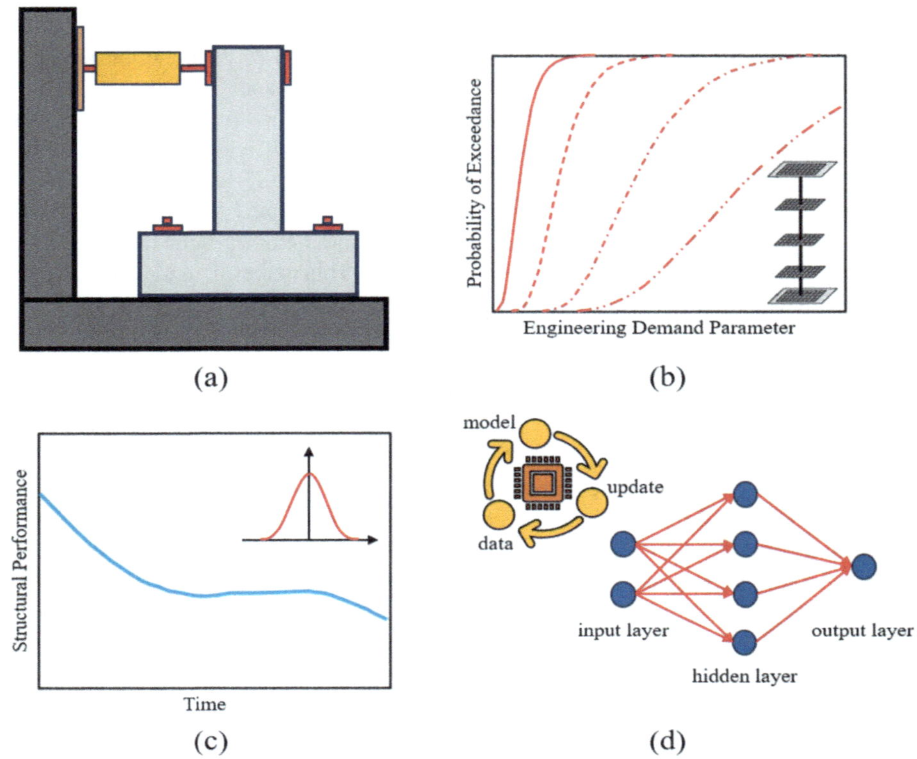

Fig. 2. Schematic illustrations of probabilistic frameworks addressed in this paper: (a) experimental investigation, (b) numerical fragility analysis, (c) stochastic life-cycle models, and (d) emerging techniques.

3 Critical Synthesis

The selection of an appropriate probabilistic framework for assessing corroded RC columns depends on a balance of objectives, available data, computational resources, and desired outcomes. While each method offers unique insights, they are not interchangeable. This section provides a comparative decision matrix to guide researchers and practitioners in choosing the most suitable approach for their specific needs. The matrix evaluates the primary methodologies reviewed in this paper—experimental investigations, numerical fragility analysis, stochastic life-cycle models, and other emerging techniques—against critical decision criteria.

The core challenge in assessing aging infrastructure lies in navigating the trade-offs between accuracy and feasibility. Experimental investigations provide invaluable, direct physical data on performance metrics like strength and ductility loss. However, their reliance on accelerated corrosion techniques can yield results that "poorly correlate with field results" and the process is often too burdensome for a large number of tests.

Numerical fragility analysis offers a powerful alternative, enabling the simulation of numerous scenarios to develop intuitive, probability-based assessments. These analyses, often using methods like IDA, can capture complex failure modes, especially when advanced models considering shear behavior are used. Yet, their accuracy is highly dependent on the precision of the numerical models and material properties, and the process can be computationally expensive.

For long-term forecasting, stochastic life-cycle models are essential. They excel at predicting the "remaining useful life" and quantifying failure probability over time in specific corrosive environments like marine or tidal zones. These models directly incorporate time-dependent deterioration, making them crucial for life-cycle management.

Emerging techniques like ANNs and Bayesian methods present a path toward greater efficiency and adaptability. ANNs can act as powerful surrogate models to reduce computational costs, while Bayesian frameworks allow for the systematic updating of models with new data, such as field inspection results.

Figure 3 presents a decision matrix that compares the reviewed probabilistic frameworks. It evaluates each method against key criteria like primary objective, key outputs, and key limitations to help researchers select an appropriate approach. This side-by-side comparison allows for several key inferences. Most notably, it highlights the fundamental trade-off between physical accuracy and analytical scalability. For instance, while Experimental Investigations offer unparalleled, direct physical data (their key output), they are constrained by high resource requirements and potential inaccuracies from accelerated testing. Conversely, Numerical Fragility Analysis provides scalable, probabilistic outputs (fragility curves) but at a high computational cost, with accuracy being entirely dependent on the underlying model. The matrix also clarifies the distinct applications of each framework; a researcher needing to predict long-term performance under specific environmental conditions would select a Stochastic Life-Cycle Model, whereas one looking to leverage existing data for efficient predictions would turn to Emerging Techniques like ANNs. Ultimately, the matrix serves as a practical guide, enabling engineers and researchers to strategically select the most appropriate assessment framework by balancing their project goals, available resources, and the specific level of detail required for their analysis.

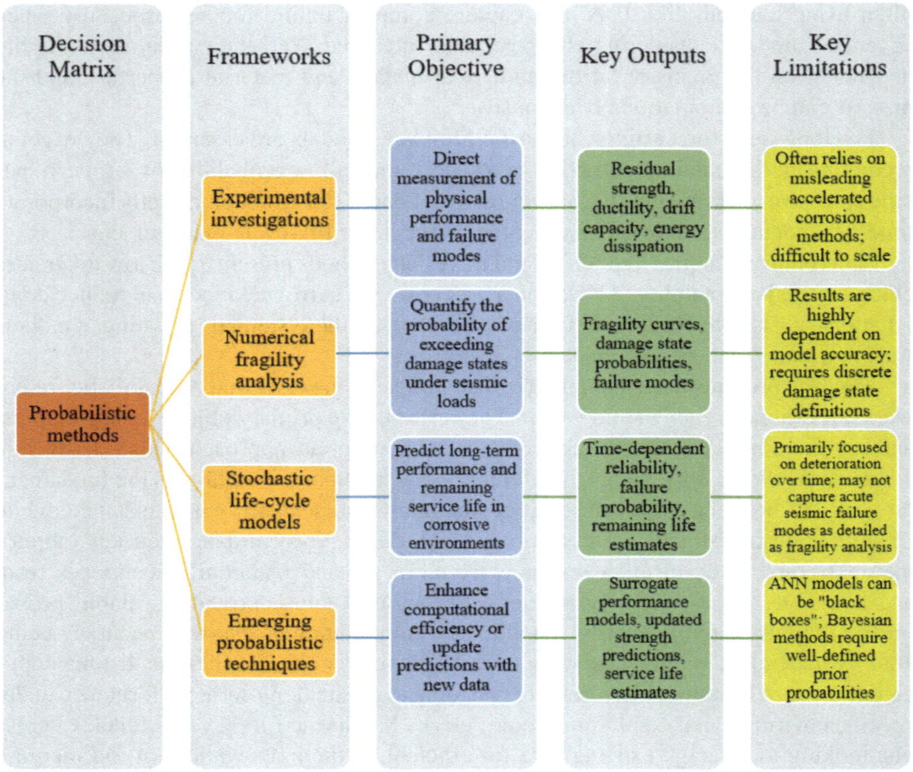

Fig. 3. Decision matrix for probabilistic assessment frameworks.

To better visualize the resource trade-offs detailed in the decision matrix, Fig. 4 positions each framework on a quadrant chart. This chart plots the methodologies based on two key resource axes: Data Requirement on the horizontal axis and Computational Cost on the vertical axis. This graphical representation provides an immediate, intuitive understanding of the practical feasibility of each approach. For example, Experimental Investigation is located in the lower-right quadrant, signifying its high demand for physical data but relatively low computational processing afterward. In contrast, Numerical Fragility Analysis occupies the upper-left, highlighting its reliance on intensive computational power rather than extensive physical data. Stochastic Models are situated centrally, reflecting their moderate demands on both fronts, while Emerging Techniques show a unique dual nature: a high initial cost for data collection and model training, which transitions to a highly efficient, low-cost application. Therefore, the quadrant chart complements the decision matrix by distilling the complex resource considerations into a clear visual map, allowing for a rapid comparison of the frameworks' operational demands.

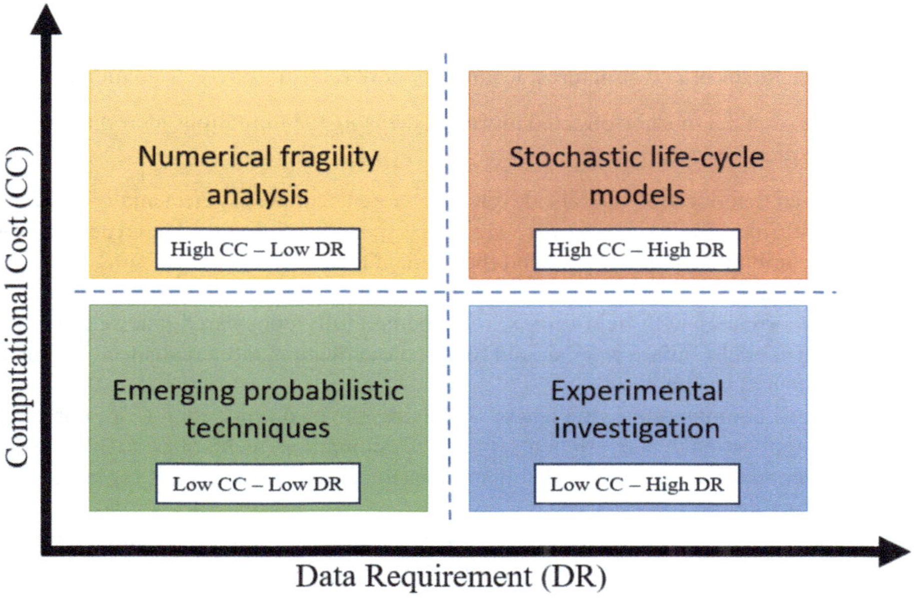

Fig. 4. A quadrant chart comparing the probabilistic assessment frameworks based on their typical data requirements and computational cost.

4 Conclusions

This paper provided a critical synthesis of probabilistic frameworks for assessing the seismic performance of corroded reinforced concrete columns. By categorizing and evaluating the dominant approaches—experimental, numerical, stochastic, and emerging data-driven techniques—this review highlights the necessity of probabilistic methods for realistically capturing the inherent uncertainties in both structural deterioration and seismic demand. Probabilistic assessment directly supports net-zero principles in infrastructure management. By accurately predicting remaining service life, these frameworks enable sustainable, risk-informed decisions for targeted maintenance and retrofitting. This practice avoids premature replacement, conserving resources and reducing the significant carbon footprint associated with new construction. The following results are yielded:

- Risk-informed decision making: probabilistic methods enable quantification of increased seismic risk due to corrosion, supporting rational decisions for inspection, maintenance, retrofit, and replacement of aging bridge infrastructure.
- Performance-based assessment: frameworks like fragility analysis provide essential inputs (e.g., conditional damage probabilities) for performance-based earthquake engineering, moving beyond deterministic safety factors.
- Life-cycle management: time-variant reliability methods (stochastic models) are crucial for predicting remaining service life and optimizing life-cycle costs under corrosion deterioration.

- Capturing uncertainty: these methods explicitly address aleatory (e.g., ground motion randomness) and epistemic (e.g., corrosion rate variability) uncertainties inherent in assessing deteriorated structures, leading to more realistic safety evaluations.

Future research must be directed toward addressing the limitations identified herein. There is a pressing need for:

- Improved deterioration models: developing more accurate, experimentally validated stochastic models for corrosion propagation (especially pitting) and bond degradation under combined environmental and mechanical loads.
- The limitations of accelerated corrosion testing, which can produce data that poorly correlates with field results, must be carefully considered when developing fragility models. Future work should focus on establishing more realistic and reliable experimental protocols.
- Enhanced computational efficiency: advancing surrogate modeling (e.g., enhanced ANN applications) and efficient stochastic simulation techniques to handle the high computational cost of probabilistic non-linear dynamic analysis for large-scale infrastructure portfolios.
- Integration with management systems: developing clear pathways to integrate these probabilistic outputs into existing infrastructure management systems and policy frameworks, translating research into sustainable, risk-informed decision-making.

References

1. Rasouli M, Shiravand MR, Ardakani RR (2023) Substructure mass participation effect on the performance-based seismic design method for isolated bridges. J Bridg Eng 28(12):04023095
2. Mahboubi S, Shiravand M (2019) Seismic evaluation of bridge bearings based on damage index. Bull Earthq Eng 17:4269–4297
3. Mahboubi S, Kioumarsi M (2021) Damage assessment of RC bridges considering joint impact of corrosion and seismic loads: a systematic literature review. Constr Build Mater 295:123662
4. Imperatore S, Benenato A, Spagnuolo S, Ferracuti B, Kioumarsi M (2024) Corrosion effects on the flexural performance of prestressed reinforced concrete beams. Constr Build Mater 411:134581
5. Poorahad Anzabi P, Shiravand MR, Mahboubi S (2024) Machine learning-aided prediction of seismic response of RC bridge piers exposed to chloride-induced corrosion. In: The international conference on net-zero civil infrastructures: innovations in materials, structures, and management practices (NTZR). Springer, pp 1409–1421
6. Ashouri R, Shiravand MR (2024) Damage assessment of self-centering rocking piers using an input energy-based damage prediction model coupled with self-centering index. Struct Concr 25(5):3549–3569
7. Shid G, Shiravand MR, Mahboubi S (2025) Direction effect of vertical ground motion on seismic damages of friction-based isolated RC piers. Adv Brid Eng 6(1):9
8. Poorahad Anzabi P, Shiravand MR (2024) Segments arrangement effect on improvement of self-centering precast post-tensioned segmental piers seismic performance. Struct Concr 25(1):185–206
9. Parvanehro P, Shiravand M, Safi M (2022) A method for linear response analysis of long bridges under multi-support seismic excitation. Bull Earthq Eng 20(15):8381–8410

10. Mahboubi, S., Shiravand, M., Shid, G. and Kioumarsi, M.: Probabilistic Assessment of RC Piers Considering Vertical Seismic Excitation Based on Damage Indices. In: International RILEM Conference on Synergising expertise towards sustainability and robustness of CBMs and concrete structures, pp. 828–839. Springer, (2023)
11. Anzabi PP, Shiravand M (2024) Seismic reliability analysis of self-centering post-tensioned piers under influence of prestress loss. Eng Struct 314:118315
12. Lee H-S, Cho Y-S (2009) Evaluation of the mechanical properties of steel reinforcement embedded in concrete specimen as a function of the degree of reinforcement corrosion. Int J Fract 157:81–88
13. Cairns J, Plizzari GA, Du Y, Law DW, Franzoni C (2005) Mechanical properties of corrosion-damaged reinforcement. ACI Mater J 102(4)
14. Coronelli D, Gambarova P (2004) Structural assessment of corroded reinforced concrete beams: modeling guidelines. J Struct Eng 130(8):1214–1224
15. Revathy J, Suguna K, Raghunath P (2009) Effect of corrosion damage on the ductility performance of concrete columns. Am J Eng Appl Sci 2(2):324–327
16. Yalciner H, Kumbasaroglu A, Karimi A (2019) Prediction of seismic performance levels of corroded reinforced concrete columns as a function of crack width. Adv Civil Eng Mater 8(3):376–397
17. Dai K-Y, Liu C, Lu D-G, Yu X-H (2020) Experimental investigation on seismic behavior of corroded RC columns under artificial climate environment and electrochemical chloride extraction: a comparative study. Constr Build Mater 242:118014
18. Repp J (2002) Accelerated corrosion testing: truths and misconceptions. Mater Perform 41(9):60–63
19. Afsar Dizaj E, Kashani MM (2022) Nonlinear structural performance and seismic fragility of corroded reinforced concrete structures: modelling guidelines. Eur J Environ Civil Eng 26(11):5374–5403
20. Afsar Dizaj E, Madandoust R, Kashani MM (2018) Exploring the impact of chloride-induced corrosion on seismic damage limit states and residual capacity of reinforced concrete structures. Struct Infrastruct Eng 14(6):714–729
21. Zhang W, Liu Y, Yu Q-Q (2024) Time-dependent seismic fragility analysis of reinforced concrete columns subjected to chloride-induced corrosion. Eng Struct 302:117448
22. Liu Y, Hao H, Hao Y (2022) Blast fragility analysis of RC columns considering chloride-induced corrosion of steel reinforcement. Struct Saf 96:102200
23. Xu J-G, Wu G, Feng D-C, Cotsovos DM, Lu Y (2020) Seismic fragility analysis of shear-critical concrete columns considering corrosion induced deterioration effects. Soil Dyn Earthq Eng 134:106165
24. Opabola EA, Elwood KJ (2023) Flexure-axial-shear interaction of ductile beams with single-crack plastic hinge behaviour. Earthq Eng Struct Dyn 52(7):2115–2134
25. Xu J-G, Feng D-C, Wu G, Cotsovos DM, Lu Y (2020) Analytical modeling of corroded RC columns considering flexure-shear interaction for seismic performance assessment. Bull Earthq Eng 18:2165–2190
26. Dai K-Y, Yu X-H, Lu D-G, Qian K (2024) Fragility functions for corroded reinforced concrete columns. J Build Eng 82:108124
27. Council AT (2018) Seismic performance assessment of buildings volume 1—methodology. FEMA P-58-1. Federal Emergency Management Agency
28. Dabas M, Martin-Pérez B, Almansour H (2023) Effect of reinforcement corrosion on axial and flexural performance of RC columns. In: Proceedings of the 8th World Congress on Civil, Structural, and Environmental Engineering (CSEE 2023)
29. Chen H-P, Jiang Y, Xiao L, Liu J (2022) Lifetime reliability analysis of concrete columns damaged by reinforcement corrosion. Front Mater 9:926259

30. Firouzi A, Abdolhosseini M, Ayazian R (2020) Service life prediction of corrosion-affected reinforced concrete columns based on time-dependent reliability analysis. Eng Fail Anal 117:104944
31. Yang Y, Tang H, Wang X (2023) Failure probability analysis of corroded RC structures considering the effect of spatial variability. Mag Concr Res 75(4):163–175
32. Li Y, Ruan X, Zhang M, Dou W (2022) RC structures life-cycle probabilistic evaluation method considering mesoscopic material uncertainty with chloride ingress. Struct Infrastruct Eng 18(7):1034–1048
33. Ding Z, Zheng S, Yu B (2022) Probabilistic prediction and calibration for residual shear strength of corroded reinforced concrete columns. J Build Eng 48:103901
34. Kwon S-J, Na U-J (2011) Prediction of durability for RC columns with crack and joint under carbonation based on probabilistic approach. Int J Concr Struct Mater 5(1):11–18
35. Arteaga EB (2010) Probabilistic service life model of RC structures subjected to the combined effect of chloride-induced corrosion and cyclic loading. Universidad de los Andes

A Reliability-Based Approach for Asphalt Concrete Fatigue Damage Assessment

Mequanent Mulugeta Alamnie and Hadi Amlashi(✉)

University of South-Eastern Norway, Kjølnes Ring 56, 3918 Porsgrunn, Norway
{mequanent.m.alamnie,hadi.amlashi}@usn.no

Abstract. Asphalt concrete (AC) is a primary material road building material. Its long-term performance reliability is predominantly influenced by its resistance to fatigue under varying loads and environmental conditions. The performance of AC against fatigue damage is investigated through uniaxial cyclic tension-compression tests, which are intricately linked to its viscoelastic and visco-damage characteristics. A sequential test procedure is proposed to analyze the interaction between fatigue and permanent deformation. Both classical damage models and reliability analyses have shown that small, early-life deformation (strain hardening) has significant effect on fatigue damage rate. The reliability analysis, conducted using Monte Carlo simulations for several case studies, demonstrated that reliability decreases with an increasing number of loading cycles, underscoring the cumulative nature of fatigue damage over those cycles. Moreover, strain-hardened samples exhibit a faster decline in reliability compared to new samples, as indicated by higher damage rates. These findings highlight that stress history affects fatigue performance and that reliability methods, such as Monte Carlo simulations, can enhance fatigue performance analysis by incorporating uncertainties in materials and models.

Keywords: Asphalt concrete · Fatigue damage · Reliability · Monte Carlo simulation

1 Introduction

Asphalt concrete (AC) is the main road surfacing material. It covers more than 90% of paved roads worldwide [1]. The performance of AC deteriorates due to the combined actions of weather and traffic loads. Fatigue is one of the main damage modes of asphalt concrete because of oxidation or aging in asphalt mixtures.

The fatigue damage evolves micro-crack formation under cyclic loading and propagation, coalescence and form macro-cracks. As a viscoelastic material, fatigue damage in AC is reflected through complex modulus reduction and phase angle increment [2]. The reliability of fatigue damage prediction depends on several factors, such as temperature T, frequency f, stress σ, strain ε, aging, moisture, testing mode etc., [3, 4]. Temperature is an important thermodynamic variable for fatigue damage evaluation. However, most fatigue tests are performed in isothermal and isobaric conditions. The

H. Beushausen et al. (Eds.): NTZR 2025, LNCE 873, pp. 916–926, 2026.
https://doi.org/10.1007/978-3-032-27698-8_80

interaction of fatigue with other damage modes (mainly rutting), is also another scenario. For example, fatigue cracking accompanies deformation under compression [5]. That means pure fatigue is a conservative assumption.

Nevertheless, the most widely applied AC fatigue damage prediction approach is the viscoelastic continuum damage (VECD) model [6]. The VECD constitutive model is based on the elastic-viscoelastic correspondence principle with small strain levels in a *Pseudo strain* (ε^R)–*stress* (σ) space and using the time-temperature superposition principle. The classic continuum damage approach is also applied for fatigue predictions, such as Chaboche-type models [7]. The damage rate can be expressed as function of $dD/dN = f(\sigma, T, f, D, \varepsilon, \dot{\varepsilon})$ where D is damage quantity. However, such types of models are stress-based and have inherent limitations for *thermo-rheological* asphalt concrete. A 50% stiffness reduction is considered as a failure criterion in isothermal conditions [8].

The probability of fatigue failure P_f is the likelihood that the predicted fatigue life will be shorter than the specified limiting fatigue life D_f. Reliability concept has been applied for fatigue life prediction [9, 10]. The AASHTO mechanistic-empirical design method [11] has incorporated reliability in its AASHTOWare program. The reliability concept is illustrated in the schematic in Fig. 1. As shown in the reliability distribution, more than 50% reliability is required for the design to meet the performance criteria over its design life. To achieve at least 90% reliability, the dashed curve (Fig. 1) at reliability R should not cross the solid line at the criteria during the entire design analysis period.

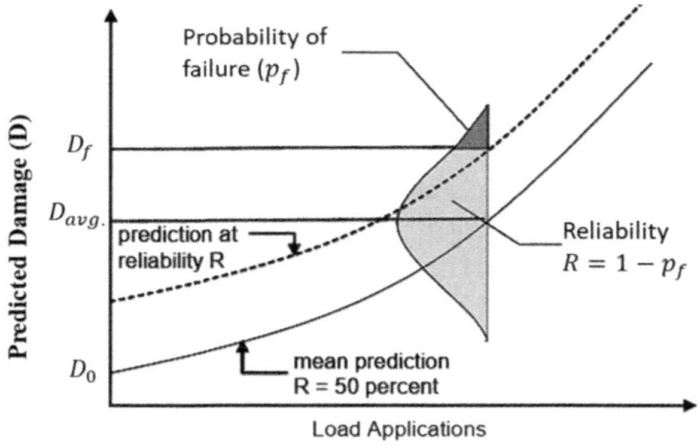

Fig. 1. Design reliability concept of pavement damage [11]

The challenge of reliability analysis in asphalt concrete stems from uncertainties of natural loading variability, environmental conditions, and the predictive capabilities of performance models. In this paper, experimental investigation of fatigue damage response is presented with two objectives. First, the paper aims to assess fatigue performance of laboratory samples and asphalt from mixing plants at different temperatures. The effect of initial strain hardening on fatigue response is studied. Then, the Monte Carlo simulation is used to evaluate the reliability of fatigue damage to understand how reliability changes as the number of loading cycles increases. By integrating fatigue damage model with probabilistic techniques, the reliability approach can enable the optimization of pavement performance prediction.

2 Reliability Prediction

Reliability (R) is defined as the probability of failure (P_f) occurring when the failure is lower than the permissible limit. Reliability methods facilitate informed decisions regarding the safety and life-cycle costs of structural systems [12, 13]. The probability of failure p_f is the probability of occurrence in the failure domain:

$$P_f = P\big[g(\boldsymbol{X}) \leq 0\big] = \iint\limits_{g(\boldsymbol{X})\leq 0} f_X(\boldsymbol{x})d_x \quad (1)$$

where $f_X(\boldsymbol{x})$ represents the joint probability density function for $\boldsymbol{X}$ (a vector of n random variables), and $g(\boldsymbol{X})$ defines the limit state function, which is a curved surface separating the safe and failure domains in the space of basic variables. One can calculate the integration of the equation above (Eq. 1) through direct integration. However, due to the numerous variables in the reliability integral and the complexity of the integration domains, obtaining exact computations for reliability measures is often impractical.

Alternative methods of reliability analysis are employed such as *First-* and *Second-Order Reliability Methods* (*FORM, SORM*) and numerical simulation techniques. One of the analytical methods that can verify failure probability estimates is the Monte Carlo simulations. The simulation becomes more accurate with more runs, but they can be time-consuming, especially for rare probabilities. In this study, a Monte Carlo simulation method is employed for calculating the probability of failure using statistical theory. An expression to estimate degradation reliability can be formulated, as shown below:

$$P_f = P[D_n - D_c \leq 0] = \frac{i_n}{i_{N_f}} \quad (2)$$

where i_n denotes the number of simulations within a given range, while i_{N_f} denotes the total number of simulations. D_n is the accumulated fatigue damage after 'n' number of cycles (generated based on two fatigue damage models, and the associated distribution parameters). The critical threshold damage (D_c) criterion is set to 0.5.

Furthermore, a two-parameter Weibull distribution is a common method for studying fatigue reliability [14]. The Probability Density Function (PDF) and Cumulative Probability Function (CDF) of this distribution are given by:

$$f(N_i) = \frac{\alpha}{\beta}\left(\frac{N_i}{\beta}\right)^{\alpha-1} exp\left[-\left(\frac{N_i}{\beta}\right)^{\alpha}\right] \quad (3)$$

$$F(N_i) = 1 - exp\left[-\left(\frac{N_i}{\beta}\right)^{\alpha}\right] \quad (4)$$

where N_i is a specific value of random fatigue life (N_f); α is the shape parameter, and β is the scale parameter. The small coefficients of variation (COV) of the estimated scale parameters suggest they are nearly similar. A versatile Method of Moments is used for estimation. An average value of 1.1349 is used for the shape parameter [14], while the scale parameter will depend on the number of loading cycles, N.

3 Materials and Test Methods

Two asphalt mixtures (70/100 binder), namely Asphalt Concrete (AC) and Stone Matrix Asphalt (SMA) are used for investigation. The mixtures were collected from two production plants (*denoted P*), and two mixes were prepared in the laboratory (*identified by L*). The aggregate gradation (particle distributions) is given in Table 1.

Table 1. Aggregate gradation of different asphalt concrete mixtures

Mixture I.D	Sieve size (mm)							Binder content [%]
	16	11.2	8	4	2	0.3	0.063	
	Percent passing							
AC-P	100	95	70	48	36	16	10	5.6
SMA-P	100	91.2	53.6	35.7	21.7	13	8.4	5.83
AC-L	100	95	70	47.5	33.5	13	7.5	5.1
SMA-L	100	95	55.5	37.5	26	16	11	5.3

Cylindrical asphalt concrete specimens (150 mm height and 100 mm diameter) were produced in a gyratory compactor for tests. A servo-hydraulic UTM-130 machine is used. Frequency sweep dynamic modulus, repeated creep-recovery, and uniaxial fatigue tests were conducted using cylindrical samples. The test parameters are presented in Table 2.

Table 2. Summary of tests and parameters.

Test	Temperature, °C	Frequency, Hz	Rest time, s	Pulse/wave
Dynamic modulus	–10, 10, 21, 40	20–0.1	–	Sinusoidal
Uniaxial fatigue	10, 15, 21, 30	10	–	Sinusoidal
Creep-recovery	30, 40, 50	–	0.1(09)	Haversine

A sequential test procedure (STP) is proposed in a previous study to investigate the effects of permanent deformation on fatigue damage evolution [15]. It was suggested that the permanent deformation–fatigue (PD-F) sequence is the dominant form of damage. In STP, the deformation part of the *PD-F sequence* is performed on new samples in a creep-recovery test at higher temperatures (40, 50 °C) within the steady-state stage (i.e., before flow number commences). Then, strain-hardened specimen is tested in a uniaxial cyclic fatigue test at a lower temperature (10, 15, 21 °C) to understand the PD-F damage interaction.

4 Results and Discussion

4.1 Linear Viscoelastic Response

Frequency-sweep dynamic modulus test results are analyzed using the *Time-Temperature Superposition Principle* to investigate the linear viscoelastic (LVE) responses of asphalt concrete mixtures [16]. The dynamic modulus master curves were constructed at the reference temperature ($T_0 = 21\ ^\circ\mathrm{C}$) over a wider range of temperatures and frequencies using a time-temperature shift factor α_T (Fig. 2). The reduced frequency f_R is given as $f_R = \alpha_T \times f$, and the WLF shift factor is given as $\log(\alpha_T) = -\frac{C_1(T-T_0)}{C_2+T-T_0}$, C_1 and C_2 are constants.

$$\log\left|E^*\right| = \alpha_o + \frac{(\alpha_1 - \alpha_o)}{1 + \exp(\eta - \gamma \, log f_R)} \tag{5}$$

The maser curve is fitted with a sigmoid function (δ, α_0, α_1, η and γ are coefficients).

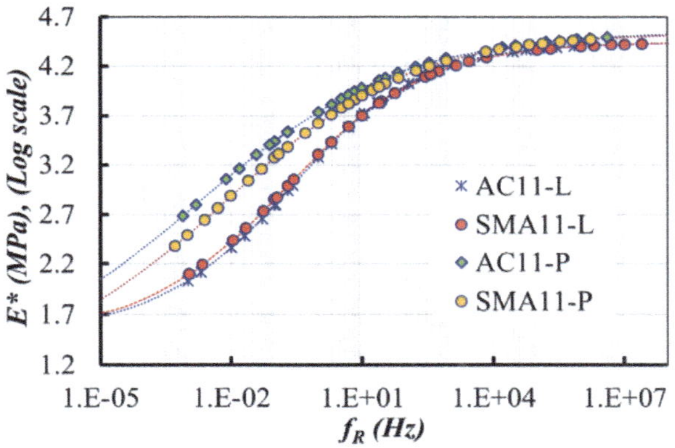

Fig. 2. Dynamic modulus master curves (at $T_0 = 21\ ^\circ\mathrm{C}$)

4.2 Uniaxial Fatigue Test

Controlled-strain, uniaxial fatigue tests are conducted at different target strains according to the parameters in Table 2. The Fingerprint dynamic modulus (E^*_{FP}) test is conducted (with maximum target strain of 50 με) before damage tests are performed on each asphalt specimen. Fatigue damage is manifested as dynamic modulus deterioration and phase angle increment. A simple damage variable D_f is defined using stiffness $D_f = 1 - E^*_N/E^*_o$ and a power function can fit the data over a normalized number of cycles n/N_f (where E^*_o is initial dynamic modulus, E^*_n is dynamic modulus decay at cycle, a_1 and a_2 are constants).

$$D_f = a_1 \left(N_i/N_f\right)^{a_2} \tag{6}$$

Temperature is one of the causes of uncertainties in asphalt concrete design and has significant impact on fatigue response. As shown in Fig. 3, as temperature increase, the phase angle increases significantly, and fatigue gradually decreases (the slope reduces).

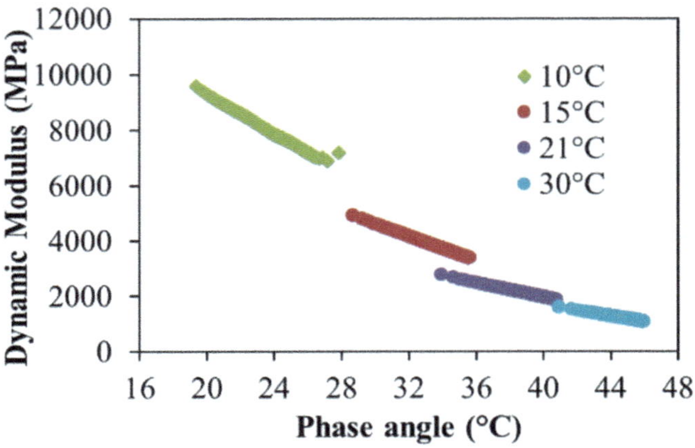

Fig. 3. Dynamic modulus at different temperatures (at 300 με)

Moreover, fatigue damage rate retarded with rise in temperature due to the high relaxation of asphalt concrete at elevated temperatures, as shown in Fig. 4.

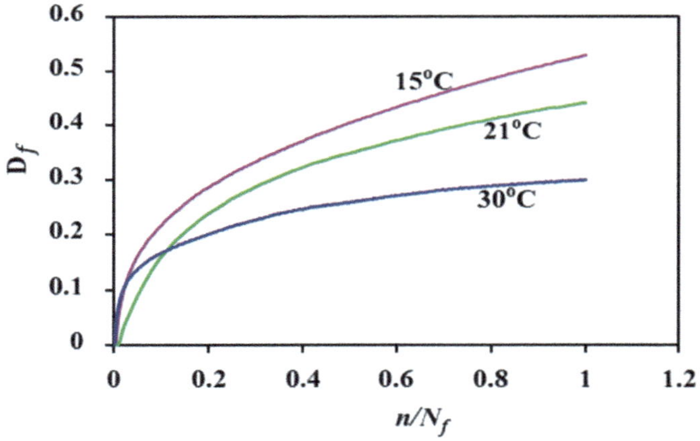

Fig. 4. Effect of temperature on fatigue damage rate (at 300 με)

4.3 Viscoelastic Continuum Damage (VECD)

The constitutive VECD model, which is based on Schapery's correspondence principle and work potential theory [17, 18], is a comprehensive VE damage model [6]. The damage quantity is expressed by internal state variable (S), which represents microstructural changes that lead to a reduction in effective stiffness. The mathematical formulation of the damage rate is defined using the maximum pseudo-strain energy (W^R) as follows.

$$\frac{dS}{dt} = \left(-\frac{\partial W^R}{\partial S}\right)^{\alpha} \quad (7)$$

$$W^R = \frac{1}{2}\sigma\varepsilon_i^R = \frac{1}{2}C\left(\varepsilon_i^R\right)^2 \tag{8}$$

The parameter α in work potential equation is related to the slope of relaxation modulus $E(t)$ curve [16]. The pseudo-stiffness, $C = \sigma/\varepsilon_i^R$ deterioration is used to quantify the amount of damage at each loading cycle. The accumulated damage after one load cycle (ΔS) takes the following form.

$$\Delta S = \left[-\frac{DMR}{2}\left(\varepsilon_i^R\right)^2\left(C_i^* - C_{i+1}^*\right)\right]^{\frac{\alpha}{1+\alpha}}(\Delta t_R)^{\frac{1}{1+\alpha}} \tag{9}$$

where $DMR = \frac{|E^*|_{FP}}{|E^*|_{LVE}}$ is the dynamic modulus ratio, E_{LVE}^* is the dynamic modulus from the linear viscoelastic test, $|E^*|_{FP}$ is the fingerprint dynamic modulus), $\varepsilon_i^R = (\beta + 1)\varepsilon_{pp}/2$ is the pseudo-strain amplitude, $\beta = \sigma_m/\sigma_{pp}$ is a load form factor, ε_{pp} and σ_{pp} are peak-to-peak strain and stress, respectively, and σ_m is mean stress, $\Delta t_R = \frac{1}{a_T}\left[\frac{\Delta N}{10}\right]$ is a reduced time of a cycle, ΔN is the number of load cycles between two successive strain amplitudes. The relationship between the pseudo-stiffness and cumulative damage (C–S) combines the effects of strain level, temperature, and frequency in fatigue damage evolution in the *elastic–viscoelastic* correspondence principle. The damage characteristics curves can be fitted using the following equation.

$$C = 1 - aS^b \tag{10}$$

Figure 5 presents the damage characteristic curves asphalt mixtures for both new and pre-deformed samples. As can be seen, pre-deformed or strain-hardened samples showed a higher damage rate than new samples. Note that strain-hardening occurs well before the critical strain (i.e., significantly less than the flow stress) using the creep-recovery test (Table 2). Thus, according to the sequential test procedure (SPT), small deformation in early pavement life can significantly accelerate fatigue failure in the later life of asphalt concrete.

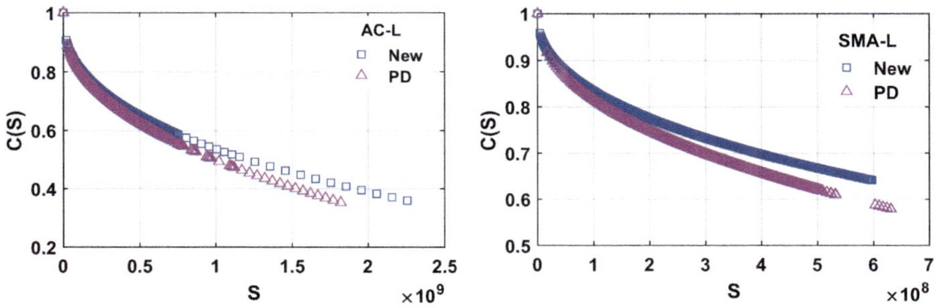

Fig. 5. Fatigue damage characteristic (at 10 °C, 300με) on New and Strain-hardened (40 °C, PD) samples

4.4 Fatigue Reliability Analysis

Fatigue damage models (in Sects. 4.2 and 4.3) are used to fit the fatigue test data, and the values of the fitting parameters are determined accordingly. The input parameters for the fatigue reliability analysis are as shown in Table 3. The probabilistic fatigue damage is evaluated across four distinct cases (denoted as C1, C2, C3, and C4), with each case comparing the reliability of new (undeformed) and strain-hardened (pre-deformed) samples as a function of the number of loading cycles.

Table 3. Summary of input parameters for reliability analysis.

Input parameter	Value
Weibull distribution	Shape parameter = 1.1349 Scale parameter = varies with nominal fatigue life
Power model: $D_f = a_1\left(\frac{n}{N_f}\right)^{a_2}$	*Cases:* C1 ($a_1 = 0.029$, $a_2 = 0.298$) C2 ($a_1 = 0.0413$, $a_2 = 0.274$)
VECD model: $C = 1 - X_{\_mod}\left(aS^b\right)$	*Cases:* C3 (a = 0.00012, b = 0.3995) C4 (a = 7.59E-05, b = 0.4245 $X_{\_mod} = N(1.0, 0.3)$

The reliability, assessed using a Monte Carlo simulation, represents the likelihood that the asphalt mixture can withstand fatigue failure under repeated loading. Figure 6 shows how the reliability of the asphalt mixture changes with the number of loading cycles. It is observed that the reliability of the asphalt mixture decreases with an increase in the number of loading cycles for all cases. The strain-hardened samples, which have undergone initial deformation during creep-recovery tests at higher temperatures (e.g., 40 °C) before being subjected to fatigue testing at 10 °C, exhibit a significantly faster decline in reliability compared to their new counterparts. This observation suggests that a small pre-deformation accelerates the fatigue damage process, even when the initial deformation is confined to the steady-state stage, i.e., before the onset of significant permanent deformation or the flow limit. This accelerated damage aligns with the probabilistic damage curves (Fig. 6), where strain-hardened samples demonstrated a higher damage rate, corroborating the hypothesis that early-life deformations precondition asphalt concrete for more rapid fatigue failure.

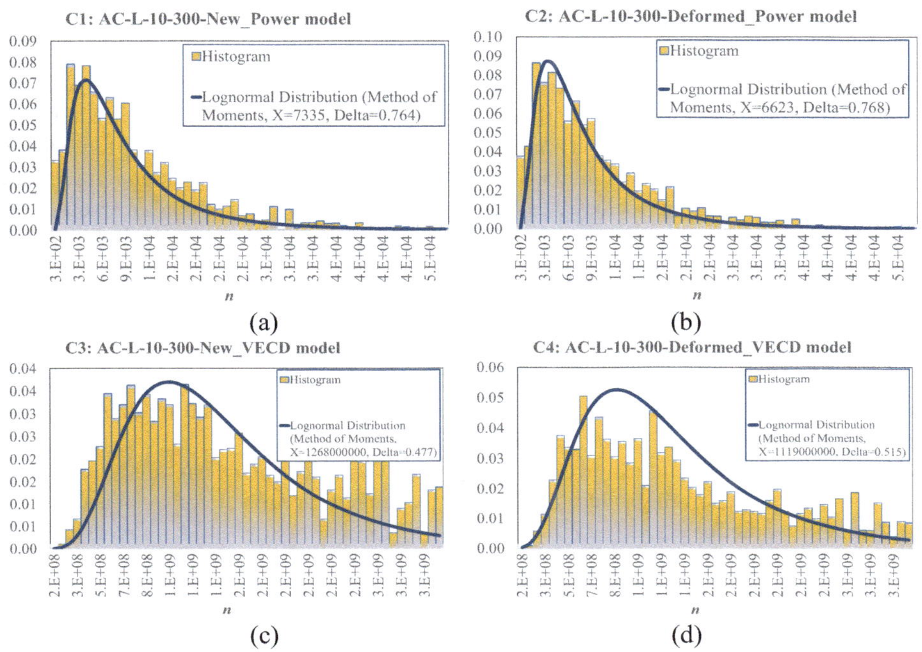

Fig. 6. Probabilistic fatigue damage (at 10 °C, 300με) on new and strain hardened (40 °C) samples; (a) Case C1, (b) Case C2, (c) Case C3, and (d) Case C4

The probability distributions for the power-based model and the VECD model differ due to their approaches to predicting fatigue damage. Generally, the expectation is that the empirical simplicity of the power model may lead to a less precise distribution. In contrast, the VECD model could produce a more accurate distribution by showing a slower decline in reliability over loading cycles compared to other models. This suggests that the VECD model can better capture the behaviour of asphalt under fatigue. However, the exact differences remain qualitative due to simplifications made in the damage distribution model. These differences depend on how each model handles variability and the way damage develops and should be explored further.

From a methodological perspective, the use of the two-parameter Weibull distribution to model fatigue life, combined with Monte Carlo simulation, provides a statistically sound basis for reliability analysis. The Weibull parameters (shape parameter α and scale parameter β) enable the characterization of fatigue life variability. The simulation estimates the probability of failure (P_f) as the ratio of simulations where the accumulated damage exceeds a critical threshold ($D_c = 0.5$) to the total number of simulations. This approach quantifies the decline in reliability and provides valuable insights into the stochastic nature of fatigue damage, which is particularly beneficial for pavement engineers as they balance performance and cost in their design.

5 Conclusions and Recommendations

This study presents a method to assess the reliability of asphalt concrete damage using uniaxial cyclic fatigue test data, with a focus on the damage model and the effect of pre-deformation. A simple method to calculate degradation reliability using a Monte Carlo Simulation is developed. This method combines traditional reliability theory and fatigue damage theory to probabilistically explain the fatigue behavior of asphalt mixture. This method can also be adjusted to study damage caused by other types of degradation.

Based on the analyses performed, the following key conclusions can be drawn:

- Strain-hardened samples (C2 and C4) show a quicker decline in reliability than new samples (C1 and C3), indicating that early minor deformations can accelerate later fatigue failure, as evidenced by increased damage rates. These findings highlight that the stress history affects fatigue performance, whereas traditional models overlook the effects of early-life deformation.
- Monte Carlo simulation showed that fatigue performance can be captured more accurately by incorporating uncertainties in mechanistic approach for a robust fatigue damage prediction.

It is recommended to explore a broader range of experimental conditions (including varying temperatures, wider loading frequencies, and mixture compositions, etc.) in further relaiblity analysis.

Incorporating other degradation mechanisms, like moisture damage into the reliability framework could further enhance the model's applicability to various climatic and operational conditions.

Acknowledgements. This work is funded by the University of South-Eastern Norway. The authors would like to express their gratitude to USN for the support provided through the Renewable Energy and Construction Technology Laboratory (*REACT-Lab*), which is part of URGENT (USN Research Group in Energy and Environmental Technology).

References

1. Partl MN (2018) Towards improved testing of modern asphalt pavements. Mater Struct 51(6):1–12. https://doi.org/10.1617/s11527-018-1286-9
2. Zhang Y, Luo R, Lytton RL (2012) Characterizing permanent deformation and fracture of asphalt mixtures by using compressive dynamic modulus tests. J Mater Civ Eng 24(7):898–906. https://doi.org/10.1061/(asce)mt.1943-5533.0000471
3. Di Benedetto H, Perraton D, Lamothe S, Boussabnia MM (2023) Rational relationship between the fatigue curves of asphalt mixes obtained from tension/compression and 4-point bending tests. Road Mater Pavem Des 24(sup1):192–208. https://doi.org/10.1080/14680629.2023.2180303
4. Al-Khateeb GG, Ghuzlan KA (2014) The combined effect of loading frequency, temperature, and stress level on the fatigue life of asphalt paving mixtures using the IDT test configuration. Int J Fatigue 59:254–261. https://doi.org/10.1016/j.ijfatigue.2013.08.011
5. Zhang Y, Birgisson B, Gu F, Lytton RL (2016) Crack evolution of asphalt mixtures under compressive monotonic and repeated loads. Springer, Netherlands, pp 155–162

6. Underwood BS, Baek C, Kim YR (2012) Simplified viscoelastic continuum damage model as platform for asphalt concrete fatigue analysis. Transp Res Record J Transp Res Board 2296(1):36–45. https://doi.org/10.3141/2296-04
7. Chaboche JL, Lesne PM (1988) A non-linear continuous fatigue damage model. Fatig Fract Eng Mater Struct 11(1):1–17. https://doi.org/10.1111/j.1460-2695.1988.tb01216.x
8. Di Benedetto H, De La Roche C, Baaj H, Pronk A, Lundström R (2004) Fatigue of bituminous mixtures. Mater Struct 37(3):202–216. https://doi.org/10.1007/bf02481620
9. Abed A, Thom N, Neves L (2019) Probabilistic prediction of asphalt pavement performance. Road Mater Pavem Des 20(sup1):S247–S264
10. Luo Z, Xiao F, Sharma R (2014) Efficient reliability-based approach for mechanistic-empirical asphalt pavement design. Construct Build Mater 64:157–165
11. AASHTO (2015) Mechanistic-empirical pavement design guide: a manual of practice (MEPDG-2). American Association of State Highway and Transportation Officials
12. Melchers RE, Beck AT (2018) Structural reliability analysis and prediction. Wiley
13. H. Amlashi, M. Karimirad, and D. Barreto (2023) Implied safety level in design codes for monopile offshore wind turbines—parametric study on the effect of partial safety factors. In: Advances in the analysis and design of marine structures. CRC Press, Boca Raton, pp 223–233
14. Jia Y, Yang Y, Liu G, Gao Y, Yang T, Hu D (2020) Reliability assessment of flexural fatigue failure of asphalt mixture: a new perspective. Construct Build Mater 257:119553
15. Alamnie MM, Taddesse E, Hoff I (2023) A study on permanent deformation and fatigue damage interaction in asphalt concrete. Construct Build Mater 407:133473. https://doi.org/10.1016/j.conbuildmat.2023.133473
16. Alamnie MM, Taddesse E, Hoff I (2022) Thermo-piezo-rheological characterization of asphalt concrete. Construct Build Mater 329:127106. https://doi.org/10.1016/j.conbuildmat.2022.127106
17. Schapery RA (1999) Nonlinear viscoelastic and viscoplastic constitutive equations with growing damage. Int J Fract 97(1/4):33–66. https://doi.org/10.1023/a:1018695329398
18. Park SW, Schapery RA (1999) Methods of interconversion between linear viscoelastic material functions. Part I—a numerical method based on Prony series. Int J Solids Struct 36(11):1653–1675. https://doi.org/10.1016/s0020-7683(98)00055-9

Resonant Element Based Discrete PI Controller for Robust Tracking of Sinusoidal Reference Inputs

Riya, Priyanshu Thumar, and Jagajyoti Panda(✉)

Malaviya National Institute of Technology Jaipur, Jaipur, Rajasthan 302017, India
jjpanda@mnit.ac.in

Abstract. In commonly used PID control algorithm for structural vibration control, integral part plays a critical role in minimizing the steady-state responses. However, for earthquake response control applications, execution of conventional integration of response is not effective as the random earthquake excitation is a combination of multiple harmonic inputs. In this study, to mitigate the adverse effect of time-varying sinusoidal inputs, a novel proportional-integral (PI) control strategy is discussed. The transfer function of the conventional integrator is replaced by a resonant element. The conceptual framework of resonant element resembles that of commonly used passive control devices, such as tuned mass damper or harmonic oscillator. The resonant element as integrator is designed to have an oscillation frequency tuned to the input frequency of the command signal, providing an infinity gain at the command frequency, resulting superior control performance as compared to the conventional integrator. The effectiveness of the modified PI controller in tracking the sinusoidal inputs is numerically demonstrated for SDOF spring-mass-damper system having feedback control action. The requirement of less control force and computational power make the proposed strategy useful in response control of complex structures, satisfying the NetZero goal for Sustainable and resilient structures.

Keywords: Proportional-integral controller · State feedback control · Resonant element · H_2/*LQG* approach · Time-delay system

List of Symbols

K_i	Integral gain
r_t	Reference input
u_r	Control force
i_s	Integral state
x	Displacement response
$G_0(s)$	Transfer function of structure
K_p	Proportional gain
y_o	Structural output
e_s	Error (output-reference input)
s	Integral state
$\dot{x}$	Velocity response

H. Beushausen et al. (Eds.): NTZR 2025, LNCE 873, pp. 927–935, 2026.
https://doi.org/10.1007/978-3-032-27698-8_81

1 Introduction

Vibration control strategies are essential for advancing the design of sustainable and resilient structures in the face of increasing environmental and seismic challenges. These strategies encompass smart control devices and advanced algorithms, enabling structures to withstand dynamic loads while minimizing damage [1, 2]. By improving the resilience of both structural and nonstructural components, vibration control not only enhances safety but also extends the lifespan of buildings, reducing the need for frequent repairs and material-intensive interventions [3, 4]. Advanced vibration control strategies, particularly from a sustainable perspective, can contribute significantly to reducing the carbon footprint across the construction lifecycle.

In seismic design practice, modern design codes worldwide recommend the implementation of passive base-isolation systems to enhance the seismic resiliency of structural and non-structural components of buildings and bridges [5, 6]. Base-isolation systems for structural applications are broadly classified into spring-like bearings, sliding bearings, or a combination of both. The passive devices do not require external power to function and are robust in nature, thus, such devices are implemented to minimize the structural responses by reducing the transmission of base shear from the ground to the super-structure.

However, the study by Chakraborty and Ray-Chaudhuri [7] has pointed out that the variation in stiffness at the base of a building, either due to the inelastic behavior of structural members or base-isolators, enhances the peak floor acceleration (PFA) during earthquake events. This amplification PFA is primarily attributed to the transfer of energy from lower modes to that of higher modes during the nonlinear deformation stage. The outcomes of experimental study with shake table justify that the presence of sliding isolator at the base amplifies the energy content of the higher modes and therefore have damaging consequences on the higher mode acceleration sensitive secondary components [8, 9]. The resonant element, having infinity gain at a particular frequency can be quite effective in mitigating the absolute acceleration response of selected floors and the attached acceleration sensitive components of base-isolated structure during strong ground shaking.

This work is a preliminary study on resonant element, which involves the usefulness of resonant element as an integrator for the robust tracking of sinusoidal inputs. Prior to the proposed integrator, it is requisite to first formulate the expression of steady-state error due to both polynomial and sinusoidal input for the closed-loop structure-integrator system. Further, the state feedback with integral (PI) control is formulated, first, with the conventional integrator and later with resonant element as the integral part. The advantage of modified PI controller with respect to conventional approach for the robust tracking of sinusoidal reference input is clearly stated.

2 Tracking Accuracy of Servo-Based PI Controller

In control theory, the concept of servomechanism brings error-sensing negative feedback to ensure that the controlled output follows the command of reference input [10]. To attain the desired transient and steady state accuracy during reference tracking, an additional integral effect based on offset amplitude is added to the proportional controller. The closed-loop control scheme having state feedback with integral control for position tracking is presented in Fig. 1.

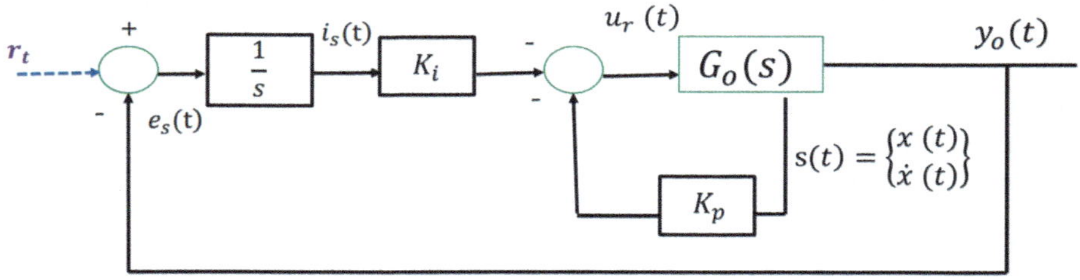

Fig. 1. Detailed flow path of servo-based PI controller for reference tracking in continuous time

The proportional part of the controller is effective in minimizing the oscillatory response of the closed-loop, but exhibit error at steady-state in the presence of disturbances and changes in set-point. To remove the steady-state offset during process control, an extra amount of intelligence with integrator is added to the proportional part. The mathematical expression that describes the state feedback with integral (PI) control force in Laplace domain for single-input single output (SISO) system is given as the following [10]:

$$U_{\rm r}({\rm s}) = -K_{\rm p}S(s) - \frac{K_{\rm i}}{s}E_{\rm s}({\rm s}) \tag{1}$$

where $U_{\rm r}(s)$, $S(s)$ and $E_{\rm s}(s)$ represent the Laplace transform of control force, $u_{\rm r}(t)$, internal state, $s(t)$ and offset signal, $e_{\rm s}(t)$, respectively. The steady-state offset of the error signal $e_{\rm s}$ for a given command input $r_{\rm t}$ and controlled structural output y_o is to be minimized to attain the desired tracking accuracy. This section presents the expression of error signal at steady-state for an integral controlled process with no proportional part. Expanding error $E_{\rm s}(s)$ as the difference between reference input $R_{\rm t}(s)$ and measured controlled output $Y_0(s)$, control force due to integral action only in Eq. 1, $U_{\rm ri}(s)$ can be expressed as follows:

$$U_{\rm ri}(s) = -\frac{K_{\rm i}}{s}[R_{\rm t}(s) - Y_0(s)] \tag{2}$$

The expression of sensitivity, $G^*(s) = E_{\rm s}(s)/R_{\rm t}(s)$ and complementary sensitivity function, $G_{\rm c}^*(s) = Y_o(s)/R_t(s)$ for the closed-loop system involving both structure, $G_o(s) = Y_o(s)/U_r(s)$, and integral controller, $U_{\rm ri}(s)$ can be written as follows:

$$G^*(s) = \left[1 + G_0(s)\frac{K_{\rm i}}{s}\right]^{-1}$$
$$G_{\rm c}^*(s) = \left[1 + G_0(s)\frac{K_{\rm i}}{s}\right]^{-1} G_0(s)\frac{K_{\rm i}}{s} \tag{3}$$

Considering the transfer function of the structural system as an equivalent spring-mass-damper arrangement with mass, damping ratio, and natural frequency as m_o, ζ_o and ω_o, respectively, $G_o(s)$ along the feed-forward path is expressed as:

$$G_o(s) = \frac{1/m_0}{s^2 + 2\zeta_0\omega_0 s + \omega_0^2} \tag{4}$$

For the structural system transfer function presented Eq. 4, the forward-path transfer function $G_o(s)\,\frac{K_i}{s}$ has a pole at $s = 0$. So, the system having an integrator in the closed-loop is a Type-1 system having zero steady-state error to unit step input and non-zero finite steady-state error $e_s(t) = \frac{1}{\lim[_{s\to 0} G_o(s)K_i]}$ to unit ramp input [11, 12].

Apart from polynomial input, r_t maybe of any kind of random signal which can be expressed as the combination of sinusoidal inputs. Thus, it is essential to first provide the detailed mathematical formulation of e_s for time varying sinusoidal signal as $r(t)$. Based on generalized error coefficient method, i.e., by using convolution theorem and Taylor series expansion, the expression of steady-state error in time domain is as follows:

$$e_s(t) = C_0^* r(t) + C_1^* r'(t) + \frac{C_2^*}{2!} r''(t) + \cdots + \frac{C_n^*}{n!} r^n(t) \tag{5}$$

where the coefficients are as follows:

$$C_0^* = \lim_{s\to 0} G_s^*;\; C_1^* = \lim_{s\to 0} \frac{\partial G_s^*}{\partial s};\; C_2^* = \lim_{s\to 0} \frac{\partial^2 G_s^*}{\partial^2 s}$$

For the structural system transfer function (TF) presented in Eq. 4, all the coefficients in Eq. 5 are constants and the offset between measured output and reference input in steady-state is given as follows:

$$e_s(t) = \sum_{n=1}^{\infty} C_n^* \omega_n^i \sin\left(\omega_i t + \frac{n\pi}{2}\right) \tag{6}$$

The aforementioned formulation suggested that the steady state error due to sinusoidal reference input follows the input signal frequency (ω_i), irrespective of the frequency content of the system (ω_o). Thus, the offset due to sinusoidal input for the servo-based PI controller in closed-loop is proportional to its amplitude and constant multiplication of frequency content. Therefore, previous literature on vibration control has suggested that a resonant element tuned to ω_i must be present in the closed-loop to enhance the tracking efficiency of the dynamic system [12, 13]. For the position tracking of the structural system, the resonant element is used as an integrator (see Fig. 2) and can be expressed in Laplace domain as $\frac{\omega_i}{s^2+\omega_i^2}$ in place of $\frac{1}{s}$, having a constant gain of K_i. So, the mathematical expression that describes the modified state feedback with resonant element control force in Laplace domain for single-input single-output (SISO) system is given as the following:

$$U_r(s) = -K_p S(s) - \frac{K_i \omega_i^2}{s^2 + \omega_i^2} E_s(s) \tag{7}$$

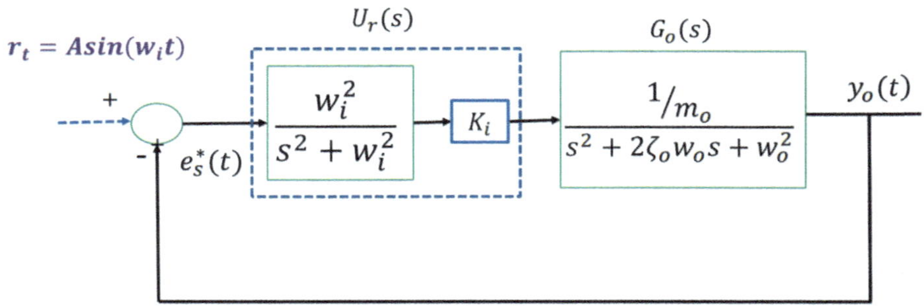

Fig. 2. Flow chart showing the structural system with proposed resonant element-based integrator in the closed loop

3 Theoretical Validation of Complete Error Elimination

Incorporating the expression of resonant element as the modified integral part of the controller, i.e., $U_{ri}(s) = \frac{K_i \omega_i^2}{s^2 + \omega_i^2}$, the sensitivity function, G^* (s) of the combined system in Fig. 2 can be expressed as:

$$G^*(s) = \frac{1}{1 + G_0 U_{ri}(s)} = \frac{1 + (s/\omega_i)^2}{1 + (s/\omega_i)^2 + G_0(s) K_i} \quad (8)$$

Based on the input signal frequency ω_i, $s = j\,\omega_i$ and the Eq. 8 can be modified as:

$$G^*(j\omega_i) = \frac{1 + (j\omega_i/\omega_i)^2}{1 + (j\omega_i/\omega_i)^2 + G_0(s) K_i} = 0 \quad (9)$$

The expression of sensitivity function in Eq. 9 suggested that the steady-state error function $E_s(s)$ for the time-varying sinusoidal input becomes zero regardless of the controller gain K_i and structural system parameters. The aforementioned statement can also be verified from the complementary sensitivity function, i.e., $G_c^* = G_0(s)U_{ri}(s)/1 + G_0(s)U_{ri}(s)$, which gives unit magnitude for the input frequency ω_i, but not shown here to save space.

4 Implementation of Resonant Element in Discrete Time

Modified integrator having transfer function behaves as a tuned mass damper (or harmonic oscillator) having oscillation (or tuning) frequency ω_i and gain K_i. Considering the states of the oscillator as $x_3 = u_{ri}$ and $x_4 = \dot{u}_{ri}/\omega_i$, resonant element can be represented through the following state-space representation:

$$\begin{Bmatrix} \dot{x}_3(t) \\ \dot{x}_4(t) \end{Bmatrix} = \begin{bmatrix} o & \omega_i \\ -\omega_i & o \end{bmatrix} \begin{Bmatrix} x_3(t) \\ x_4(t) \end{Bmatrix} + \begin{bmatrix} 0 \\ K_i \omega_i \end{bmatrix} e_s(t)$$

$$u_{\mathrm{ri}}(t) = \begin{bmatrix} 1 & 0 \end{bmatrix} \begin{Bmatrix} x_3(t) \\ x_4(t) \end{Bmatrix} \tag{10}$$

Incorporating zero-order hold (ZOH) and the sample together with the continuous-time system, the discrete time state variable representation of the oscillator system becomes:

$$\begin{Bmatrix} x_3(k+1) \\ x_4(k+1) \end{Bmatrix} = \begin{bmatrix} -\cos\omega_{\mathrm{i}}T & \sin\omega_{\mathrm{i}}T \\ -\sin\omega_{\mathrm{i}}T & \cos\omega_{\mathrm{i}}T \end{bmatrix} \begin{Bmatrix} x_3(k) \\ x_4(k) \end{Bmatrix} + \begin{bmatrix} K_{\mathrm{i}} - K_{\mathrm{i}}\cos\omega_{\mathrm{i}}T \\ K_{\mathrm{i}}\sin\omega_{\mathrm{i}}T \end{bmatrix} e_{\mathrm{s}}(t)$$

$$u_{\mathrm{ri}}(k) = \begin{bmatrix} 1 & 0 \end{bmatrix} \begin{Bmatrix} x_3(k) \\ x_4(k) \end{Bmatrix} \tag{11}$$

Signal flow diagram of analog processor for the resonant element is shown in Fig. 3. Though it is theoretically validated that controller-R is effective in reducing the steady state error, the controller is unable to dissipate the transient response of the system. This is because, controller-R acts as an integrator and requires a proportional or derivative part for the reduction of peak overshoot to a desired level. Therefore, present study proposes a proportional part (i.e., proportional to the internal states of the system) along with resonant element (called as controller-R) for the rapid dissipation of initial transient responses and maintain the steady-state accuracy.

As $e_{\mathrm{s}}^*(t) = r_{\mathrm{t}}(t) - y_{\mathrm{O}}(t)$, , the state variable representation of the combined system involving both SDOF model (see Eq. 3) and its feedback system (see Eq. 8 and proportional part) can be expressed as:

$$\begin{Bmatrix} \dot{x}_1(t) \\ \dot{x}_2(t) \\ \dot{x}_3(t) \\ \dot{x}_4(t) \end{Bmatrix} = \begin{bmatrix} 0 & 1 & 0 & 0 \\ -\omega_o^2 + K_{pd}/m_o & -2\zeta_o\omega_o + K_{pv}/m_o & 1/m_{\mathrm{O}} & 0 \\ 0 & 0 & 0 & \omega_i \\ -K_i\omega_i & 0 & -\omega_i & 0 \end{bmatrix}$$

$$y_{\mathrm{O}}(t) = \begin{bmatrix} 1 & 0 & 0 & 0 \end{bmatrix} \begin{Bmatrix} x_1(t) \\ x_2(t) \\ x_3(t) \\ x_4(t) \end{Bmatrix} \tag{12}$$

where $x_1(t)$ and $x_2(t)$ represent the internal states of the system presented in Eq. 3, and the corresponding Proportional gains are K_{pd} and K_{pv}. Present study considers linear quadratic optimal control approach for the estimation of K_{pd}, K_{pv} and K_{i}. In the process of evaluation of controller gains, the resonant element is assumed as an integrator, having $\mathrm{TF} = 1/s$, and the procedure given in [14, 15] for the state feedback integral control is followed for the estimation process.

To justify the effectiveness of controller-R in following the command of sinusoidal input, a numerical illustration is performed with a SDOF spring-mass-damper system having mass, stiffness and damping ratio as 3, 2 and 0.2, respectively. Then, controller-R is designed for sinusoidal input having ω_{i} of 6.28 rad/sec or 1 Hz. In this case, the values of K_{pd}, K_{pv} and K_{i} are found out to be − 17.80 and − 13.43 and − 10.12, respectively.

The time history plot of $r_{\mathrm{t}}(t)$ and $y_{\mathrm{O}}(t)$ along with the frequency response plot of $U_{\mathrm{r}}(s)$ are presented in Fig. 3. As shown in Fig. 3a, after an initial transient response of 6 s, $y_{\mathrm{O}}(t)$

reaches the steady-state and tracks the command of $r_t(t)$ input perfectly. In addition, the frequency response function (FRF) of controller-R (see Fig. 3b) provides infinity gain at the point of input frequency, i.e., $\omega_i = 1$ Hz, resulting the demand of higher control force for zero steady state error. However, for the tracking of signals having multiple dominant frequencies, e.g., seismic and wind inputs, controller-R can be modified to place various resonant elements in parallel, based on the input frequencies. This approach may be computationally extensive and, in fact, can be considered necessary.

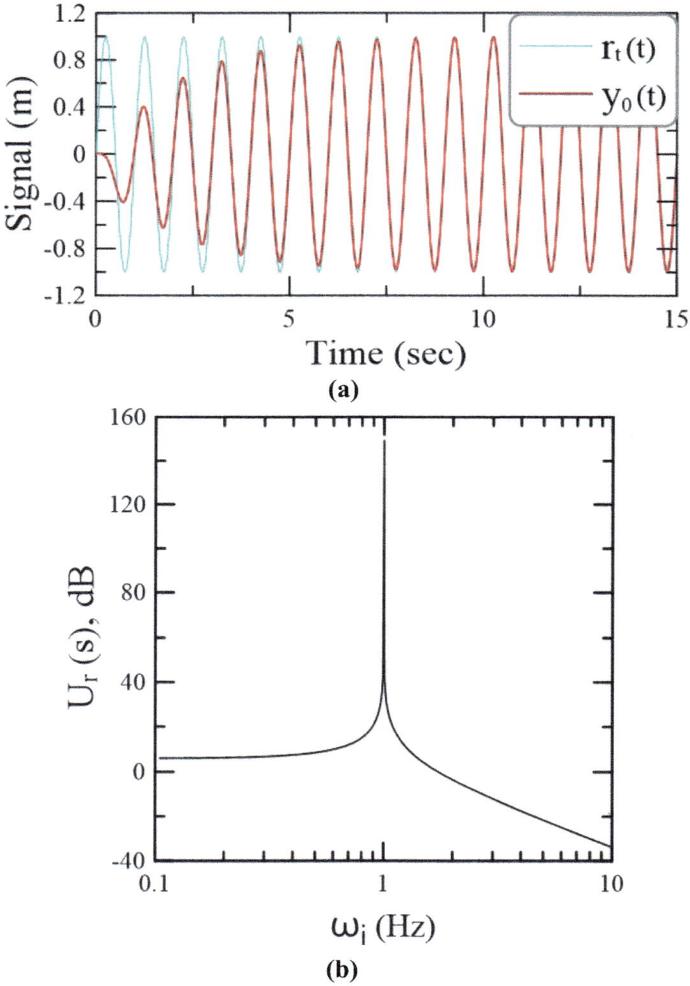

Fig. 3. Vibration control efficacy of proposed controller: (a) perfect tracking of response output to sinusoidal input, and (b) frequency response function showing the better application of control force

5 Conclusions

This work is a preliminary study on resonant element-based feedback control strategy for Civil engineering structures subjected to single sinusoidal and multiple sinusoidal (example: random earthquake) excitations. The proposed proportional-integral (PI) control algorithm utilizes the usefulness of resonant element as an integrator for the robust tracking of sinusoidal input. This study first formulates the expression of steady-state error due to both polynomial and sinusoidal inputs for the structure-controller system, having conventional integrator in the feedback loop. Subsequently, to minimize this error, the state feedback with integral controller is formulated with resonant element as the integrator. The resonant element here is designed with an oscillation frequency tuned to the input frequency of the command signal, resembling the mechanism of a tuned mass damper. The efficacy of the modified PI controller for the robust tracking of sinusoidal reference input is clearly stated, with an envision to use multiple resonant elements for earthquake-induced vibration control (both transient and steady-state responses) of structures.

References

1. Maheshwari A, Karakaya F, Banerjee A, Donnal JS (2024) Control architecture for LLC resonant converters with high input disturbance rejection capability using output diode current. IEEE Trans Power Electron 99:1–13
2. Valencia-Rivera et al (2024) Optimal hybrid resonant current controller for microgrids connected to an unbalanced IEEE test distribution network. Heliyon 10(13):e33019
3. Panda J, Chakraborty S, Ray-Chaudhuri S (2022) Development and performance evaluation of a robust suboptimal H∞-based proportional–integral controller–observer system with target tracking for better control of seismic responses. Struct Control Health Monit 29(11):e3084
4. Adarsh S, Panda J, Ray-Chaudhuri S (2022) H_2 based acceleration-strain feedback control of structures subjected to seismic input. In: 3rd international conference on artificial intelligence, robotics and control (AIRC). IEEE, pp 52–56
5. Housner GW, Chassiakos AG, Skelton RE, Spencer BF (1997) Structural control: past, present, and future. J Eng Mech 123:897–971
6. Spencer BF, Nagarajaiah S (2003) State of the art of structural control. J Struct Eng 129:845–856
7. Chakraborty S, Ray-Chaudhuri S (2017) Energy transfer to high-frequency modes of a building due to sudden change in stiffness at its base. J Eng Mech 143(8):1–19
8. Sachdeva G, Chakraborty S, Ray-Chaudhuri S (2018) Seismic response control of a structure isolated by flat sliding bearing and nonlinear restoring spring: experimental study for performance evaluation. Eng Struct 159:1–13
9. Panda S, Chakraborty S, Ray-Chaudhuri S (2025) Experimental evaluation of a servomechanism-based PI controller for seismic response control in smart base-isolated steel building. J Constr Steel Res 227:109349
10. Panda S, Chakraborty S, Ray-Chaudhuri S (2021) A novel servomechanism based proportional–integral controller with Kalman filter estimator for seismic response control of structures using magneto-rheological dampers. Struct Control Health Monit 28(10):e2807
11. Gopal M (2012) Digital control and state variable methods, 4th edn. Tata McGraw-Hill Education Pvt. Ltd, New Delhi

12. Ogata K (2010) Modern control engineering, 5th edn. Prentice Hall India Learning Private Limited, Upper Saddle River
13. Al-Mamun A, Keikha E, Bhatia CS, Lee TH (2013) Integral resonant control for suppression of resonance in piezoelectric micro-actuator used in precision servomechanism. Mechatronics 23:1–9
14. Ohno K, Hara T (2006) Adaptive resonant mode compensation for hard disk drives. IEEE Trans Ind Electron 53(2):624–30
15. Tee KP, Ge SS, Tay EH (2007) Adaptive resonance compensation for hard disk drive servo systems. In: Proceedings of the 46th IEEE conference on decision and control, New Orleans, USA, pp 3567–3572

Prediction of Concrete Strength Through Image Processing

Afaq Ahmad[1,2](✉), Mati ullah[3], and Vagelis Plevris[3]

[1] Department of Built Environment, Oslo Metropolitan University, Oslo, Norway
afaq.ahmad@oslomet.no
[2] Civil Engineering Department, University of Memphis, Memphis, TN, USA
[3] Department of Civil and Environmental Engineering, Qatar University, Doha, Qatar

Abstract. Reliable and equipment-light prediction of concrete strength is essential for rapid quality assessment and structural health monitoring. This paper presents an end-to-end image-based pipeline that utilizes deep semantic segmentation to isolate coarse aggregates from concrete slice images. Then it regresses mechanical properties using a compact artificial neural network (ANN) with only aggregate-level descriptors. A DeepLabV3+ segmenter (ResNet101 backbone) produces high-quality masks with mIoU up to 0.90, from which six descriptors, i.e., aggregate proportion, size, roundness (elongation), aspect ratio, Feret-based dimensions, and nearest-neighbor spacing, are extracted automatically. These features, computed across compacted and non-compacted cylinders spanning common mix ratios and water–cement levels, feed a lightweight ANN model (AggNet) to predict compressive strength (f_c) and tensile strength (f_t). On the held-out test set, AggNet achieves an R^2 value of 0.783, an MAPE of 0.362, and an RMSE of 0.113, demonstrating that aggregate morphology carries an actionable signal for strength estimation without destructive testing or mortar texture cues. The results highlight aggregate skeleton characteristics as practical surrogates for mixture quality and consolidation effects, suggesting a complementary role to conventional NDT for fast, low-cost strength evaluation in the field.

Keywords: Image processing · Aggregate segmentation · Concrete strength · Structural health monitoring

1 Introduction

Concrete remains the backbone of modern infrastructure due to its widespread availability, versatility, and desirable engineering properties, including strength, durability, and adaptability [1]. The reliable prediction of its mechanical performance, particularly compressive strength (f_c) and tensile strength (f_t), is central to the design and maintenance of sustainable and resilient structures [2, 3]. Traditional testing approaches, both destructive and nondestructive, have been extensively used for this purpose. Destructive testing, such as compression and split-cylinder tests, provides accurate measurements but requires sampling, which is time-consuming, costly, and often leads to permanent

H. Beushausen et al. (Eds.): NTZR 2025, LNCE 873, pp. 936–948, 2026.
https://doi.org/10.1007/978-3-032-27698-8_82

structural damage. In contrast, nondestructive techniques such as ultrasonic pulse velocity (UPV) [4], rebound hammer (RH) [5], and combined SonReb methods [6] offer practical alternatives. Yet, they are heavily dependent on specialized equipment, calibration procedures, and experienced operators [7]. This makes them less suitable for routine or large-scale assessment, especially in aging infrastructure.

With the rapid development of computer vision and machine learning, image-based approaches have emerged as powerful tools to complement or replace conventional testing methods in civil engineering [8, 9]. Structural health monitoring (SHM) increasingly incorporates digital imaging and pattern recognition to evaluate material conditions [10]. High-resolution microscopy and scanning electron microscopy (SEM) enable the visualization of the internal structure of concrete, while classical image analysis techniques, such as edge detection, morphological filtering, and texture-based analysis, facilitate the quantification of features directly linked to material performance [11–13]. When integrated with artificial neural networks (ANNs), these image-derived descriptors can be mapped to mechanical properties with high predictive accuracy.

Several studies have reported on the successful application of ANNs to predict the strength of concrete using image-based features. Dogan et al. [14] demonstrated that aggregate- and texture-based ANN models outperformed conventional regression in estimating compressive strength and elasticity. Waris et al. [15] showed that statistical measures extracted from images, such as skewness and entropy, when combined with ANN and ANFIS models, yielded R^2 values exceeding 0.97. Similarly, Li et al. [16] highlighted that digital image analysis can effectively capture crack propagation and strength development under load, while Guccluer et al. [17] confirmed that aggregate shape and aspect ratio have a direct impact on compressive strength. These findings underscore the significance of aggregate morphology in influencing concrete behavior.

Recent advances in deep learning, particularly convolutional neural networks (CNNs), have further strengthened the role of image-based modeling in concrete research. CNNs are capable of learning complex visual patterns from raw images, which has led to significant progress in concrete strength prediction [18–20]. Nonetheless, many CNN-based regression approaches treat images as holistic inputs, thereby overlooking detailed aggregate-level features such as size distribution, roundness, and spatial arrangement. Hybrid CNN–SVR approaches [21, 22] have attempted to address this by incorporating manually measured aggregate parameters; however, such procedures are labor-intensive and lack scalability.

Motivated by these limitations, this work focuses exclusively on the role of coarse aggregate morphology in predicting strength. A dedicated ANN framework, termed AggNet, is proposed, which uses only aggregate-level features derived from deep-learning-based segmentation. Through image processing, six key aggregate descriptors, proportion, size, roundness, aspect ratio, Feret dimensions, and nearest-neighbor distances are automatically extracted and provided as inputs to the network. By focusing solely on aggregates, AggNet highlights their independent contribution to concrete mechanical performance, eliminating the need for manual measurements.

2 Materials and Methods

The adopted methodology involved preparing cylindrical concrete specimens with a range of mix proportions and water–cement (W/C) ratios. The specimens were grouped into three categories: compacted, non-compacted, and very-high W/C. To establish reference mechanical properties, destructive (DT) and non-destructive (NDT) tests were carried out. Afterwards, the cylinders were cut into slices, and high-resolution photographs were taken under controlled conditions. These images were then annotated at the pixel level to produce ground-truth labels for aggregates, thereby generating a specialized dataset for subsequent analysis and ANN model development. A schematic overview is shown in Fig. 1.

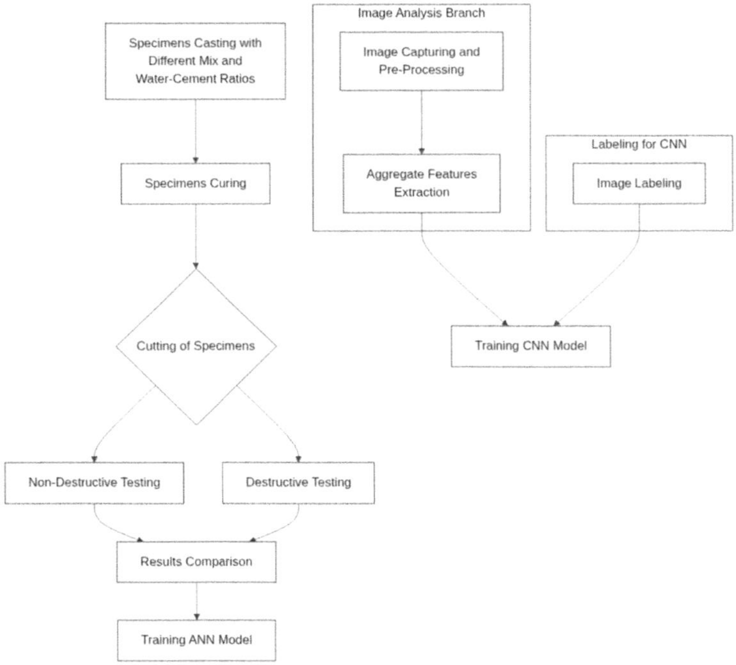

Fig. 1. Flowchart depicting stepwise procedure from dataset curation to ANN training

2.1 Cylinder Casting and Testing

Concrete mixes were prepared using five design ratios (1:4:8, 1:3:6, 1:2:4, 1:1.5:3, and 1:1:2). The coarse aggregate consisted primarily of gray granite and limestone, complemented by colored aggregates (basalt, siliceous, and calcined). Approximately 70% of the total aggregate was gray, a choice reflecting both its practical prevalence in South Asia and its relevance for testing segmentation models under challenging conditions, given its close resemblance to cement paste. In contrast, colored aggregates offered clear contrast boundaries, making segmentation comparatively straightforward. Each mix was prepared with three water-to-cement (W/C) ratios (0.4, 0.5, and 0.6), following the guidelines in ACI 318 [23] and ASTM C94 [24].

To examine the effects of workmanship and extreme mixture conditions, non-compacted cylinders were prepared at the same five mix ratios and three w/c levels. Table 1 summarizes the conditions and the corresponding destructive test results. Across all conditions, 180 cylinders were cast and cured for 28 days; for each condition, two cylinders were assigned to imaging, two to compressive testing, and two to split-tensile testing.

Table 1. Material and mix ratio

S. No.	Mix design notation	Mix ratios	W/C	Compressive strength (MPa)
1	CM1	01:04:08	0.4	3.43
2	CM2	01:04:08	0.5	4.06
3	CM3	01:04:08	0.6	5.16
4	CM4	01:03:06	0.4	5.84
5	CM5	01:03:06	0.5	7.4
6	CM6	01:03:06	0.6	8.78
7	CM7	01:02:04	0.4	14.53
8	CM8	01:02:04	0.5	11.66
9	CM9	01:02:04	0.6	10.17
10	CM10	01:05.5	0.4	18.07
11	CM11	01:05.5	0.5	16.76
12	CM12	01:05.5	0.6	14.37
13	CM13	01:01:02	0.4	21.03
14	CM14	01:01:02	0.5	14.78
15	CM15	01:01:02	0.6	13.74

Mechanical strengths were obtained using a compression testing machine (CTM). Both uniaxial compression and split-cylinder tests were carried out in accordance with ASTM C39 [25] and ASTM C496 [26]. The resulting compressive and tensile strengths were used as ground-truth values for training the ANN model. Furthermore, NDT was performed on imaging specimens. Surfaces were first smoothed before applying the rebound hammer (RH) test (ASTM C805 [27]) to estimate compressive strength from R-values. Ultrasonic pulse velocity (UPV) tests were conducted in accordance with ASTM C597 [28]. Ultrasonic waves at 250 kHz were passed through each slice, and travel times were recorded to compute velocity.

2.2 Concrete Image Dataset

Since existing aggregate segmentation datasets [29] contain only colored aggregates and are limited in size, a new dataset was developed in this study. Cylinders (H = 300 mm, D = 152 mm) were cut into three slices at approximately one-third and two-thirds of their height using a stone saw as shown in Fig. 2. Due to the accumulation of fines, the top surface of the first slice and the bottom surface of the third provided little aggregate information and were excluded, leaving four usable faces for imaging. Images were captured in a controlled indoor setup. External light was blocked, and two 30W LED lamps supplied uniform illumination (approximately 2000 lx). A Canon EOS 4000D digital camera (18 MP, ISO 1000) mounted at 550 mm from the sample face was used to capture 280 raw images at a resolution of 5184 × 3456.

Fig. 2. Experimental workflow: (a) casting, (b) curing and cutting, (c) Testing, and (d) imaging of specimens.

Each raw image contained both the specimen face and the background. A preprocessing routine (including grayscale conversion, binarization, hole filling, morphological closing, connected components, and mask extraction) was used to remove the background. Using MATLAB Image Labeler, aggregates were manually traced with a graphics tablet, ensuring acceptable boundary accuracy. Structural engineering experts verified the annotated masks before being finalized. Each full-resolution image was then partitioned into 74 tiles of 512 × 512 pixels, yielding 17,760 patches. After excluding redundant patches containing only background, the dataset comprised 10,154 images. Figure 3 represents examples of preprocessing stages and the resulting labeled patches. Overall, 42.8% of the labeled pixels represented aggregates, while 57.2% corresponded to mortar and other phases.

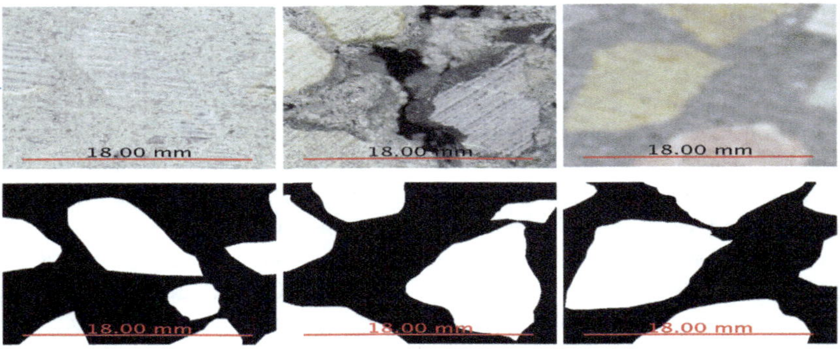

Fig. 3. Examples of 512 × 512 dataset patches (top) and their corresponding ground-truth masks (bottom). White pixels indicate aggregates.

3 Segmentation and Analysis of Aggregate

Accurate aggregate segmentation is fundamental to characterizing the morphology of coarse aggregates. By delineating aggregate pixels from the surrounding matrix, we can compute distribution, size, and shape descriptors that are later used to interpret and ultimately predict mechanical behavior. Applying state-of-the-art semantic segmentation to concrete slice images enables a clean separation of material phases, improving measurement fidelity and supporting data-driven mix optimization.

3.1 Segmentation Architectures

Advances in deep learning and large-scale training corpora have established deep convolutional neural networks (DCNNs) as a leading solution for semantic segmentation, where an image is partitioned into label-consistent regions for downstream analysis. Canonical encoder–decoder designs, such as Fully Convolutional Networks [30], U-Net [31], SegNet [32], and the DeepLab family [33, 34], compress spatial detail to extract robust features and then decode them into dense masks; the architectural choice strongly influences boundary precision and class consistency. In this work, we adopt DeepLabV3+ [35] for aggregate segmentation. Compared with DeepLabV3 [34, 35], the "plus" variant enhances fine-detail recovery by coupling Atrous Spatial Pyramid Pooling (ASPP) with a lightweight decoder. The DeepLabV3+ is composed of three components: an encoder with a backbone DCNN, an ASPP block for multi-scale context aggregation, and a decoder to refine object boundaries.

The encoder transforms the input image into a hierarchy of feature maps with progressively richer semantics. Each convolutional layer produces activations that are passed through non-linearities (e.g., ReLU or leaky ReLU) and normalization layers (e.g., batch normalization) to stabilize the training process. Pooling operations reduce spatial resolution while retaining salient information. Modern encoders may also employ dilated convolutions to enlarge the receptive field without increasing the number of parameters, or depth-wise separable convolutions to reduce computational cost. The outcome is a compact, semantically rich representation suitable for decoding.

To accommodate the wide range of object scales in concrete microstructure, the ASPP module runs multiple parallel atrous (dilated) convolutions at different dilation rates, effectively scanning the scene with receptive fields of varied sizes. This captures both local texture and broader context. A global-average-pooling branch supplies an image-level prior; all branches are concatenated and fused via a 1×1 convolution to yield the ASPP output.

Finally, the DeepLabV3+ decoder sharpens boundaries and restores detail with minimal overhead. The ASPP output is unsampled (factor 4 via bilinear interpolation) and merged with a low-level encoder feature map that a 1×1 convolution has filtered to limit the channel count. A subsequent 3×3 convolution and a final up sampling step produce the dense segmentation map, balancing high-level semantics with low-level spatial cues.

3.2 Training Protocol

To ensure fair evaluation, the dataset was randomly divided into training (60%), validation (20%), and test (20%) splits, independent of mix design and W/C ratio. Models were implemented in MATLAB's Deep Learning Toolbox and trained on a Windows workstation (Intel Core i7-10700 @ 2.90 GHz, 32 GB DDR4, NVIDIA GeForce RTX 3060).

Weights were initialized using He/Kaiming initialization [36]. Optimization used stochastic gradient descent with momentum (SGDM) [37] with initial learning rate $1 \times 10^{(-3)}$, momentum 0.9, L2 regularization 0.005, step decay every 10 epochs (factor 0.3), for 25 epochs, and mini-batch size 8. Pretrained backbones on PASCAL VOC [38] were fine-tuned to our domain using a composite loss of cross-entropy and Lovász–Softmax, which emphasizes IoU optimization, as described in Table 2. After each epoch, the data order was reshuffled, and checkpoints were saved; loss and accuracy were monitored throughout.

To improve generalization and reduce overfitting, we applied data augmentation: horizontal/vertical flips (probability 0.35 each), random translations in [−30, 30] pixels, illumination jitter, and additive Gaussian noise (mean 0, standard deviation 0.01). These perturbations encourage invariance to view, alignment, and lighting.

Table 2. Performance metrics of different backbone DCNNs for aggregate segmentation.

Backbone DCNN	Performance metrics			
2–5	mIoU	F1-score	Precision	Recall
InceptionResnetV2	0.83	0.93	0.89	0.95
Xception	0.88	0.93	0.91	0.96
ResNet101	0.90	0.94	0.93	0.95

4 Aggregate Segmentation Results

Backbone selection has a significant impact on the quality of the mask produced by DeepLabV3+. To examine this sensitivity, we coupled the DeepLabV3+ head with three encoders: InceptionResNetV2 [39], Xception [40], and ResNet101 [41–46]. These networks span a range of depths, connectivity patterns, and computational costs. While bottomless models can encode rich semantics, increased depth alone is not a guarantee of better masks, as it may lead to potential overfitting and optimization challenges. We therefore benchmarked all four variants on the same data to determine the most suitable encoder for aggregate segmentation. As reported in Table 1, ResNet101 yields the top mean IoU (0.90) and the strongest F1-score, indicating a favorable balance between precision and recall. These values demonstrate smooth convergence and do not suggest either underfitting or overfitting for this configuration.

4.1 Aggregate Proportion

Real-world scaling from images typically requires a fixed camera distance; to avoid this constraint, we recover the pixel size from the known cylinder diameter (152.4 mm). The circular boundary is traced using the Moore–Neighbor method with 8-connectivity [47, 48], which is resilient to noise and minor artifacts. With the image-space diameter measured as shown in Fig. 4a, we convert pixel counts to area (mm^2). Aggregate area is computed by counting white pixels and multiplying by the unit pixel area.

4.2 Aggregate Characteristics

Aggregate geometry (size, shape, orientation) is quantified via the Feret rectangle method [49]. The Feret diameter, which represents the maximum boundary-to-boundary distance, is computed by scanning boundary point pairs. A minimum-area oriented rectangle is then fitted with its long side aligned to this diameter, as depicted in Fig. 4b. We use the rectangle's length as the aggregate size and define roundness as the long-to-short side ratio (1 for a square, and larger values indicate elongation).

(a) (b) (c)

Fig. 4. (a); Boundary (green) and diameter (red) used to compute unit pixel size for metric measurements, (b); Feret-based oriented rectangles used to extract size and elongation measures from each aggregate, (c); Nearest-neighbor network among aggregate centroids for spacing analysis.

4.3 Distance Between Aggregates

Packing density and potential crack paths are reflected in inter-aggregate spacing. We compute nearest-neighbor distances (NND) between aggregate centroids as shown in Fig. 4c by evaluating pairwise Euclidean distances and retaining the minimum per aggregate.

5 Results ANN Modelling and Training

Image-derived features were used to train a single artificial neural network, focused on aggregate morphology, hereafter referred to as AggNet. Laboratory measurements of compressive strength (f_c) and tensile strength (f_t) served as target outputs. To harmonize scales, all inputs and outputs were normalized.

AggNet consists of an input layer, two hidden layers, and a two-node output layer for (f_c, f_t) as shown in Fig. 5. The two-hidden-layer design was selected after iterative trials to strike a balance between expressiveness and generalization. The input layer dimension equals the number of aggregate descriptors obtained from the segmentation pipeline, and the hidden layers use matching widths to limit computational cost.

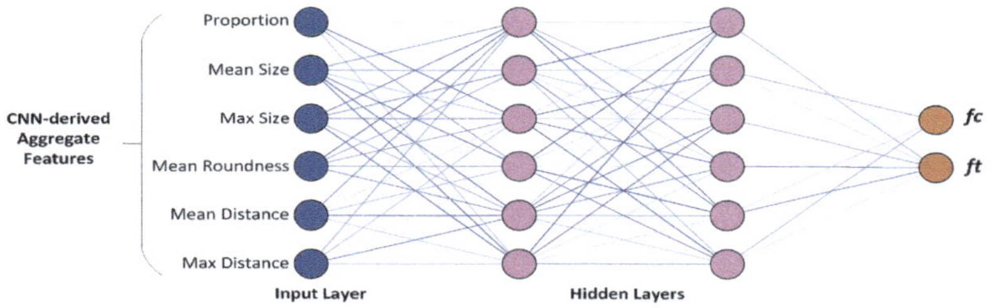

Fig. 5. AggNet architecture (6-6-6-2): ANN using only aggregate-derived features.

Learning employed the Levenberg–Marquardt backpropagation (LM-BP) algorithm [50–53]. A logarithmic schedule was used to modulate the effective learning rate (lr) and stabilize updates. Hyperparameters were set to lr = 0.9, momentum coefficient mc = 0.2, and a maximum of 10,000 iterations. Hidden-layer activations were tanh and log-sigmoid. The objective was to minimize the mean absolute error. The aggregate features extracted from 240 images (60 samples) were randomly partitioned into 70% training, 15% validation, and 15% testing.

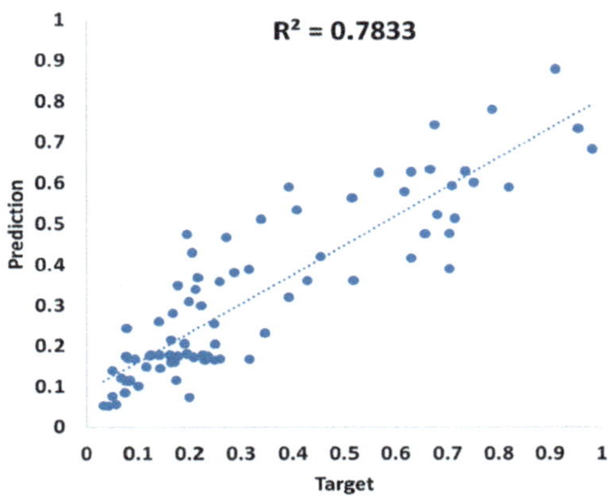

Fig. 6. Predicted versus target strengths for AggNet on the test set.

Figure 6 shows predicted versus target values for AggNet on the held-out test set. Using aggregate proportion, size, elongation (roundness), and spacing descriptors as inputs, AggNet provides consistent estimates of both (f_c and f_t). The Proposed model achieved representative performance of $R^2 = 0.783$, MAPE = 0.362, and RMSE = 0.113. These values reflect the predictive signal carried by coarse-aggregate morphology alone, where no mortar statistics or global intensity features were included.

6 Conclusion

This study introduced an aggregate-only predictive workflow that couples deep segmentation with classical image analysis and a compact ANN to estimate (f_c and f_t). From concrete slice images. DeepLabV3 + with the encoder ResNet101 achieved a mean IoU value of 0.90, along with F1-score, precision, and recall values of 0.94, 0.93, and 0.95, respectively, providing reliable inputs for downstream feature extraction. Six automatically derived aggregate descriptors were used as inputs in the proposed AggNet model, which achieved representative test-set performance ($R^2 = 0.783$, MAPE = 0.362, RMSE = 0.113). This work demonstrated that coarse-aggregate morphology can provide a strong, non-destructive proxy for estimating concrete strength from single cross-section images, by coupling a DeepLabV3+ segmenter with classical geometry and spacing-based descriptors and learning a compact regressor.

Future work will enrich the dataset with broader mixture types and explore additional aggregate descriptors, such as orientation distributions and multi-scale texture, to further enhance aggregate-only predictions. Integration with selective mortar statistics or hybrid encoders can be considered in extended studies.

References

1. Gagg CR (2014) Cement and concrete as an engineering material: an historic appraisal and case study analysis. Eng Fail Anal 40:114–140

2. Yan K, Xu H, Shen G, Liu P (2013) Prediction of splitting tensile strength from cylinder compressive strength of concrete by support vector machine. Adv Mater Sci Eng 2013(1):597257
3. Badarloo B, Kari A, Jafari F (2018) Experimental and numerical study to determine the relationship between tensile strength and compressive strength of concrete. Civil Eng J 4(11):2787–2800
4. Standard test method for ultrasonic pulse velocity through concrete (2017). https://www.astm.org/Standards/C597
5. Standard test method for rebound number of hardened concrete (2023). https://www.astm.org/Standards/C805.htm. Accessed 11 Aug 2024
6. Uva G, Porco F, Fiore A (2018) The sonreb method: critical review and practical aspects. In: Proceedings of Italian Concrete Days 2016
7. Bolborea B, Baera C, Dan S, Gruin A, Burduhos-Nergis D-D, Vasile V (2021) Concrete compressive strength by means of ultrasonic pulse velocity and moduli of elasticity. Materials 14(22):7018
8. Greeshma A, Edayadiyil JB (2022) Automated progress monitoring of construction projects using machine learning and image processing approach. Mater Today Proc 65:554–563
9. Qayyum W, Ehtisham R, Bahrami A, Camp C, Mir J, Ahmad A (2023) Assessment of convolutional neural network pre-trained models for detection and orientation of cracks. Materials 16(2):826
10. Tibaduiza Burgos DA, Gomez Vargas RC, Pedraza C, Agis D, Pozo F (2020) Damage identification in structural health monitoring: a brief review from its implementation to the use of data-driven applications. Sensors 20(3):733
11. Balendran R, Pang H, Wen H (1998) Use of scanning electron microscopy in concrete studies. Struct Surv 16(3):146–153
12. Ozbek ASA, Weerheijm J, van Breugel K (2018) High speed photography technique for measuring impact strength of porous concrete. Constr Build Mater 186:1092–1104
13. Kumar D, Pandey RC, Mishra AK (2024) A review of image features extraction techniques and their applications in image forensic. Multimed Tools Appl 1–102
14. Dogan G, Arslan MH, Ceylan M (2017) Concrete compressive strength detection using image processing based new test method. Measurement 109:137–148
15. Waris MI, Plevris V, Mir J, Chairman N, Ahmad A (2022) An alternative approach for measuring the mechanical properties of hybrid concrete through image processing and machine learning. Constr Build Mater 328:126899
16. Li X, Lin H, Chen W, Liang S, Huang L (2023) A numerical study on the tensile splitting of concrete with digital image processing. J Mater Res Technol 25:1626–1641
17. Güçlüer K (2020) Investigation of the effects of aggregate textural properties on compressive strength (CS) and ultrasonic pulse velocity (UPV) of concrete. J Build Eng 27:100949
18. Lee S, Ahn Y, Kim HY (2020) Predicting concrete compressive strength using deep convolutional neural network based on image characteristics. Comput Mater Continua 65(1)
19. Jang Y, Ahn Y, Kim HY (2019) Estimating compressive strength of concrete using deep convolutional neural networks with digital microscope images. J Comput Civ Eng 33(3):04019018
20. Shiuly A, Dutta D, Mondal A (2022) Assessing compressive strengths of mortar and concrete from digital images by machine learning techniques. Front Struct Civ Eng 16(3):347–358
21. Celik G, Ozdemir M (2024) Determination of concrete compressive strength from surface images with the integration of convolutional neural networks and support vector regression methods. Measurement 115331
22. Sharma RK, Singh D, Dasaka SM (2024) Investigating supplementary cementitious materials' effects on stabilized aggregate performance, behaviour, and design aspects. Constr Build Mater 411:134564

23. A. Committee (2008) Building code requirements for structural concrete (aci 318-08) and commentary. American Concrete Institute
24. Daniel DG, Lobo CL (2005) User's guide to ASTM specification C 94 on ready-mixed concrete. ASTM International
25. A. I. C. C. on Concrete, C. Aggregates (2014) Standard test method for compressive strength of cylindrical concrete specimens. ASTM international
26. A. ASTM (2008) Astm c496/c496m-04e1 standard test method for splitting tensile strength of cylindrical concrete specimens. Annual book of ASTM standards 4
27. C. ASTM (2008) 805, standard test method for rebound number of hardened concrete. ASTM International, West Conshohocken, PA
28. C. ASTM (2009) 597, standard test method for pulse velocity through concrete. ASTM International, West Conshohocken, PA
29. Ehtisham R et al (2023) Classification of defects in wooden structures using pre-trained models of convolutional neural network. Case Stud Constr Mater 19:e02530
30. Long J, Shelhamer E, Darrell T (2015) Fully convolutional networks for semantic segmentation. In: Proceedings of the IEEE conference on computer vision and pattern recognition, pp 3431–3440
31. Ronneberger O, Fischer P, Brox T (2015) U-net: convolutional networks for biomedical image segmentation. In: Medical Image Computing and Computer-Assisted Intervention–MICCAI 2015: 18th International Conference, Munich, Germany, October 5–9, 2015, Proceedings, Part III 18. Springer, pp 234–241
32. Badrinarayanan V, Kendall A, Cipolla R (2017) Segnet: a deep convolutional encoder-decoder architecture for image segmentation. IEEE Trans Pattern Anal Mach Intell 39(12):2481–2495
33. Chen L-C, Papandreou G, Kokkinos I, Murphy K, Yuille AL (2017) Deeplab: semantic image segmentation with deep convolutional nets, atrous convolution, and fully connected CRFS. IEEE Trans Pattern Anal Mach Intell 40(4):834–848
34. Chen L-C, Papandreou G, Schroff F, Adam H (2017) Rethinking atrous convolution for semantic image segmentation. arxiv, arXiv preprint arXiv:1706.055875
35. Chen L-C, Zhu Y, Papandreou G, Schroff F, Adam H (2018) Encoder-decoder with atrous separable convolution for semantic image segmentation. In: Proceedings of the European conference on computer vision (ECCV), pp 801–818
36. He K, Zhang X, Ren S, Sun J (2015) Delving deep into rectifiers: Surpassing human-level performance on imagenet classification. In: Proceedings of the IEEE international conference on computer vision, pp 1026–1034
37. Liu Y, Gao Y, Yin W (2020) An improved analysis of stochastic gradient descent with momentum. Adv Neural Inf Process Syst 33:18261–18271
38. Everingham M, Van Gool L, Williams CK, Winn J, Zisserman A (2010) The pascal visual object classes (VOC) challenge. Int J Comput Vision 88:303–338
39. Szegedy C, Ioffe S, Vanhoucke V, Alemi A (2017) Inception-v4, inception-resnet and the impact of residual connections on learning. In: Proceedings of the AAAI conference on artificial intelligence, vol 31
40. Chollet F (2017) Xception: deep learning with depthwise separable convolutions. In: Proceedings of the IEEE conference on computer vision and pattern recognition, pp 1251–1258
41. He K, Zhang X, Ren S, Sun J (2016) Deep residual learning for image recognition. In: Proceedings of the IEEE conference on computer vision and pattern recognition, pp 770–778
42. Han J, Wang K, Wang X, Monteiro PJ (2016) 2D image analysis method for evaluating coarse aggregate characteristic and distribution in concrete. Constr Build Mater 127:30–42
43. Ullah M, Mir J, Husain SS, Shahid MLUR, Ahmad A (2024) Concrete fo-rensic analysis using deep learning-based coarse aggregate segmentation. Autom Constr 162:105372

44. Haha MB, Gallucci E, Guidoum A, Scrivener KL (2007) Relation of expansion due to alkali silica reaction to the degree of reaction measured by SEM image analysis. Cem Concr Res 37(8):1206–1214
45. Masad E, Muhunthan B, Shashidhar N, Harman T (1999) Internal structure characterization of asphalt concrete using image analysis. J Comput Civ Eng 13(2):88–95
46. Li X et al (2020) Homogeneity evaluation of hot in-place recycling asphalt mixture using digital image processing technique. J Clean Prod 258:120524
47. Mandeel TH, Ahmad MI, Md Isa MN, Anwar SA, Ngadiran R (2018) Palmprint region of interest cropping based on moore-neighbor tracing algorithm. Sens Imag 19:1–14
48. Biswas S, Hazra R (2018) Robust edge detection based on modified Moore-neighbor. Optik 168:931–943
49. Wang W (2006) Image analysis of particles by modified ferret method—best-fit rectangle. Powder Technol 165(1):1–10
50. Wahab S et al (2024) Comparative analysis of shear strength prediction models for reinforced concrete slab–column connections. Adv Civil Eng 2024(1):1784088
51. Ahmad A, Plevris V, Q-u-Z Khan (2020) Prediction of properties of FRP-confined concrete cylinders based on artificial neural networks. Crystals 10:811
52. Le-Nguyen K, Minh QC, Ahmad A, Ho LS (2022) Development of deep neural network model to predict the compressive strength of FRCM confined columns. Front Struct Civ Eng 16(10):1213–1232
53. Wahab S, Suleiman M, Shabbir F, Mahmoudabadi NS, Waqas S, Herl N, Ah-mad A (2023) Predicting confinement effect of carbon fiber reinforced polymers on strength of concrete using metaheuristics-based artificial neural networks. arXiv preprint arXiv:2403.13809

Evaluating Fire Resistance in Geopolymer Concrete Using Machine Learning

Maryam Bypour[1(✉)], Alireza Mahmoudian[2], and Mahdi Kioumarsi[3]

[1] Department of Civil Engineering, Semnan University, Semnan, Iran
mbypour@gmail.com
[2] Department of Civil Engineering, Shahid Rajaee Teacher Training University, Tehran, Iran
amahmoudian1997@sru.ac.ir
[3] Department of Built Environment, OsloMet—Oslo Metropolitan University, Oslo, Norway
mahdi.kioumarsi@oslomet.no

Abstract. Geopolymer concrete (GPC) has exhibited superior fire resistance compared to conventional concrete. To evaluate this characteristic, the present study investigated the temperature resistance of GPC produced using various precursors and exposed to temperatures ranging from 25to 1000 °C. Experimental data for specimens subjected to elevated temperatures were collected from existing literature. Subsequently, machine learning techniques—including four tree-based models: Decision Tree (DT), Random Forest (RF), Gradient Boosting (GB), and Extreme Gradient Boosting (XGB)—were employed to predict the compressive strength of GPC based on a variety of influencing features. The results revealed that XGB achieved the highest predictive performance among all models. After hyperparameter tuning, the XGB model attained a test R2 value of 0.96. The GB, DT, and RF models also demonstrated notable improvements; however, their performance remained slightly inferior to that of XGB. The study confirmed that although the baseline models were effective, hyperparameter tuning significantly enhanced predictive accuracy.

Keywords: Geopolymer Concrete · Fire Resistance · Machine Learning · High-temperature Performance

1 Introduction

The development of environmentally friendly building materials has gained growing attention due to increasing concerns about sustainability in civil engineering. One of the most effective approaches is the reuse of industrial and agricultural solid waste, which helps conserve natural resources and reduce environmental pollution. These waste materials can be transformed into innovative construction materials, turning "waste into wealth" [1]. Geopolymers, in particular, are promising alternatives to Ordinary Portland Cement (OPC) because of their ability to incorporate industrial by-products and their significantly lower carbon emissions during production. Moreover, their inorganic framework provides inherent resistance to high temperatures [2].

H. Beushausen et al. (Eds.): NTZR 2025, LNCE 873, pp. 949–961, 2026.
https://doi.org/10.1007/978-3-032-27698-8_83

Since concrete structures are often exposed to elevated temperatures—whether due to fire or environmental conditions—it is essential for building materials to retain their mechanical integrity under such stress. Consequently, understanding the behavior of concrete under high-temperature exposure has become a critical area of research [3]. In recent years, numerous studies have investigated how high temperatures affect the microstructure and mechanical properties of geopolymer concrete (GPC), considering variables such as precursor type and content, curing conditions, exposure temperature, and more [4–6].

For example, research has shown that while the addition of slag improves the mechanical performance of fly ash-based geopolymer mortars under normal conditions, mortars without slag perform better at elevated temperatures. Some mixtures demonstrated a compressive strength recovery of up to 90% after exposure to 1000 °C, highlighting the potential of fly ash-based GPC for high-temperature applications [7]. Another study found that increasing ladle slag content led to the formation of more stable crystalline phases at high temperatures. After exposure to 800 °C, a geopolymer with 25 wt% ladle slag retained a compressive strength of 64.7 MPa, outperforming the 55.2 MPa observed in pure fly ash geopolymers [8].

Further investigations into alkali-activated fly ash and GGBFS mortars using waste glass cullet showed that mortars with glass aggregates maintained comparable residual strengths to those made with natural sand, while exhibiting fewer microcracks and enhanced interfacial bonding after high-temperature exposure [9]. Another study indicated that the effect of preloading and cracking on fire resistance depends heavily on the binder mix design [10].

The influence of ground ferronickel slag in combination with fly ash was also studied, revealing that a 50% ferronickel slag content enhanced performance after exposure to temperatures up to 1000 °C, based on evaluations of mass loss, cracking behavior, compressive strength, microstructure, and ultrasonic pulse velocity [11]. Mineral wool waste (MWW) was found to act as both a co-precursor and fiber reinforcement, improving the matrix and even increasing compressive strength by 98.5% under high temperatures [12].

One study explored how varying the silica modulus (Ms), sodium concentration, and fly ash content impacted the mechanical and microstructural properties of vapour-cured (at 75 °C for 8 h) GGBFS-based geopolymer mortars. Results indicated microstructural deterioration following high-temperature exposure [13]. Another investigation evaluated titanium dioxide (TiO_2) waste as a geopolymer precursor and found that while compressive strength decreased with higher TiO_2 content at elevated temperatures, fire resistance improved due to increased evaporation plateaus. These materials also posed no radiological hazards [14].

Despite numerous studies exploring the use of machine learning (ML) techniques to predict the mechanical properties of structural systems and materials [15–20], there is a limited number of studies applying ML models to predict the performance of GPC, particularly under high-temperature conditions. As ML applications grow rapidly in civil engineering and materials science, this study aims to bridge the gap by using ML algorithms to predict the compressive strength of GPC after exposure to elevated temperatures. Key input parameters include precursor content, water-to-binder ratio, type of alkaline activator, and exposure temperature.

2 Data Construction

To investigate the compressive strength of GPC under high-temperature conditions, a dataset was collected from existing literature. As shown in Table 1, the precursors used in the collected studies include Metakaolin (MK), Fly Ash (FA), Copper Slag (CS), Iron Tailings (IT), Blast Furnace Slag (BFS), and Municipal Solid Waste Incineration Bottom Ash (MSWI-BA). The exposure temperatures range from 25 to 1000 °C. The water-to-binder (W/B) ratio in the dataset varies between 0.25 and 0.65. Additionally, different types of alkaline activators were used, including sodium hydroxide (NaOH), sodium silicate (Na_2SiO_3), and combinations of both.

Table 1. Dataset collected from literature.

Experimental research	Precursor	W/B	Alkaline activator type	Exposure temperature (°C)
Zhang et al [21]	MK and FA	0.65	$NaOH + Na_2SiO_3$	25, 100, 200, 400, 600
Zhang et al. [22]	BFS and MSWI-BA	0.35	Na_2SiO_3	25, 200, 400, 600, 800, 1000
Zhang et al. [23]	CS, IT, and FA	0.25, 0.3, 0.35	NaOH	600, 800, 1000

MK: Metakaolin
FA: Fly ash
CS: Copper slag
IT: Iron tailings
BFS: Blast Furnace Slag
MSWI-BA: Municipal Solid Waste Incineration Bottom Ash

First, a comprehensive exploratory data analysis (EDA) was conducted to better understand the dataset and identify the relationships among variables. The dataset consists of 122 samples and includes nine input features related to the mix design and exposure condition of alkali-activated materials, along with the target variable, compressive strength (CS).

Figure 1 shows a correlation heatmap to evaluate the strength and direction of linear associations between variables. This heatmap reveals that several features, including MK, W/B, and alkaline activator, exhibit strong positive correlations with compressive strength. The W/B ratio demonstrates the highest correlation with the target ($r \approx 0.89$), followed closely by MK and alkaline activator ($r > 0.85$).

On the contrary, variables such as FA, exposure temperature, and IT show moderate to strong negative correlations with compressive strength.

Table 2. DT grid search parameters.

max_depth	min_samples_leaf	min_samples_split	max_features
None	1	2	None
3	2	5	'sqrt'
4	4	10	'log2'
6	8	20	–
8	–	–	–
10	–	–	–
15	–	–	–

Table 3. RF grid search parameters.

n_estimators	max_depth	min_samples_leaf	min_samples_split	max_features
30	None	1	2	None
50	10	2	3	'sqrt'
80	20	4	5	'log2'
100	30	8	10	–
120	–	–	20	–
200	–	–	–	
300	–	–	–	–

Table 4. GB and XGB grid search parameters.

learning_rate	n_estimators	max_depth
0.01	50	3
0.1	80	5
0.2	100	7
0.3	200	9
0.5	300	12

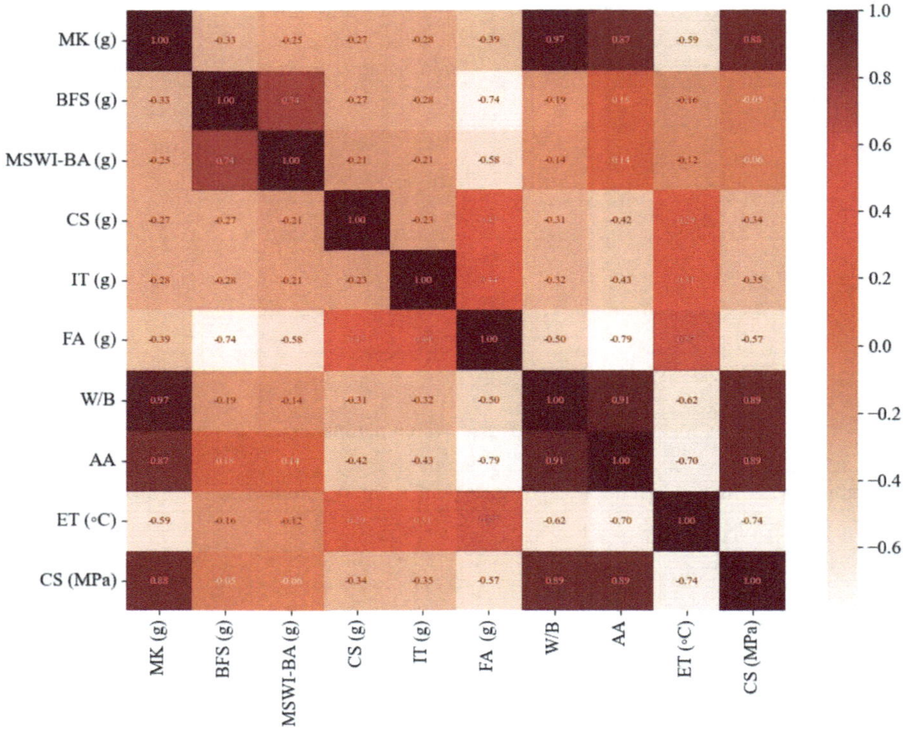

Fig. 1. Correlation heatmap between output and input variables.

3 Machine Learning Methods

3.1 Overview of Machine Learning

In this study, four tree-based machine learning (ML) models namely Decision Tree (DT), Random Forest (RF), Gradient Boosting (GB), and Extreme Gradient Boosting (XGB) were used to predict the compressive strength of GPC.

The DT algorithm splits the data recursively based on features that best reduce impurity in the prediction target. For regression tasks, impurity is commonly measured by the Mean Squared Error (MSE), calculated at each node t in Eq. 1 where N_t is the number of samples in node t, y_i is the target value of sample i, and $\overline{y}_t$ is the mean target value at node t.

$$\text{MSE(t)} = \frac{1}{N_t}\sum_{i=2}^{N_t}\left(y_i - \overline{y}_t\right)^2 \quad (1)$$

While DTs are highly interpretable, they often overfit to training data. To address this, RF was introduced as an ensemble of multiple decision trees. Each tree is trained on a random subset of the data (bootstrapping) and a random subset of features, which reduces variance. As calculated in the Eq. 2, the final prediction of RF is the average of predictions from all trees, where T is the total number of trees and $\hat{y}_t$ is the prediction from the t-th tree.

$$\hat{y}_{RF} = \frac{1}{T}\sum_{t=1}^{T}\hat{y}_t \tag{2}$$

GB is an ensemble learning method that builds a strong predictive model by sequentially adding weak learners, typically decision trees, where each new model focuses on correcting the errors made by the previous ones. XGB is an advanced version of gradient boosting that introduces regularization to reduce overfitting and uses a more efficient optimization process to speed up training. XGBoost further improves performance by incorporating both first- and second-order gradient information. Although both models are based on the same core idea of boosting, XGBoost is known for its higher accuracy and faster computation compared to traditional gradient boosting.

3.2 Hyperparameters Tuning Using Grid Search CV

Hyperparameter tuning with grid search is a systematic way to find the best settings for ML models. In this method, the user defines a set of values for each hyperparameter, and the algorithm tests all possible combinations to identify which one produces the best model performance, often evaluated through cross-validation. The selected hyperparameters along with their corresponding ranges and values for grid search tuning are presented in Tables 2, 3 and 4.

3.3 Evaluation Metrics for Machine Learning Models

To evaluate the performance of ML models, several key metrics are commonly used. As calculated in Eq. 3, the coefficient of determination (R^2) measures how well the predicted values approximate the actual data, with values closer to 1 indicating better model performance. As calculated in Eq. 4 the Root Mean Square Error (RMSE) calculates the square root of the average squared differences between predicted and actual values. According to Eq. 5 the Mean Absolute Error (MAE) computes the average of the absolute differences between predicted and observed values.

$$R^2 = 1 - \frac{\sum_{i=1}^{n}\left(\hat{y}_{test,i} - y_{test,i}\right)^2}{\sum_{i=1}^{N}\left(\hat{y}_{test,i} - \overline{y}_{test}\right)^2} \tag{3}$$

$$RMSE = \sqrt{\frac{\sum\left(\hat{y}_{test,i} - y_{test,i}\right)^2}{N}} \tag{4}$$

$$MAE = \frac{1}{n}\sum_{i=1}^{n}|y_i - \hat{y}_i| \tag{5}$$

4 Results and Discussion

4.1 Machine Learning Results

The performance of the ML models was evaluated before and after applying hyperparameter tuning using grid search. As shown in Fig. 2(a–d), the models initially demonstrated good predictive capabilities, with acceptable R^2, RMSE, and MAE values. However, after hyperparameter tuning, further improvements were achieved. Reductions in RMSE and MAE, along with slight increases in R^2 values, were observed, which indicates enhanced accuracy and generalization. This improvement was consistently seen across both the training and testing datasets which confirms the effectiveness of grid search in refining the models' performance.

Among the models, XGB achieved the best performance in this study. After hyperparameter tuning, the XGB model attained a test R^2 of 0.96, while the RMSE and MAE values decreased to 6.96 MPa and 5.56 MPa, respectively. The GB, DT, and RF models also showed noticeable improvements; however, their performance remained slightly lower compared to XGB. Overall, it was demonstrated that although the initial models performed well, hyperparameter tuning further enhanced the predictive accuracy, with XGB emerging as the most effective model.

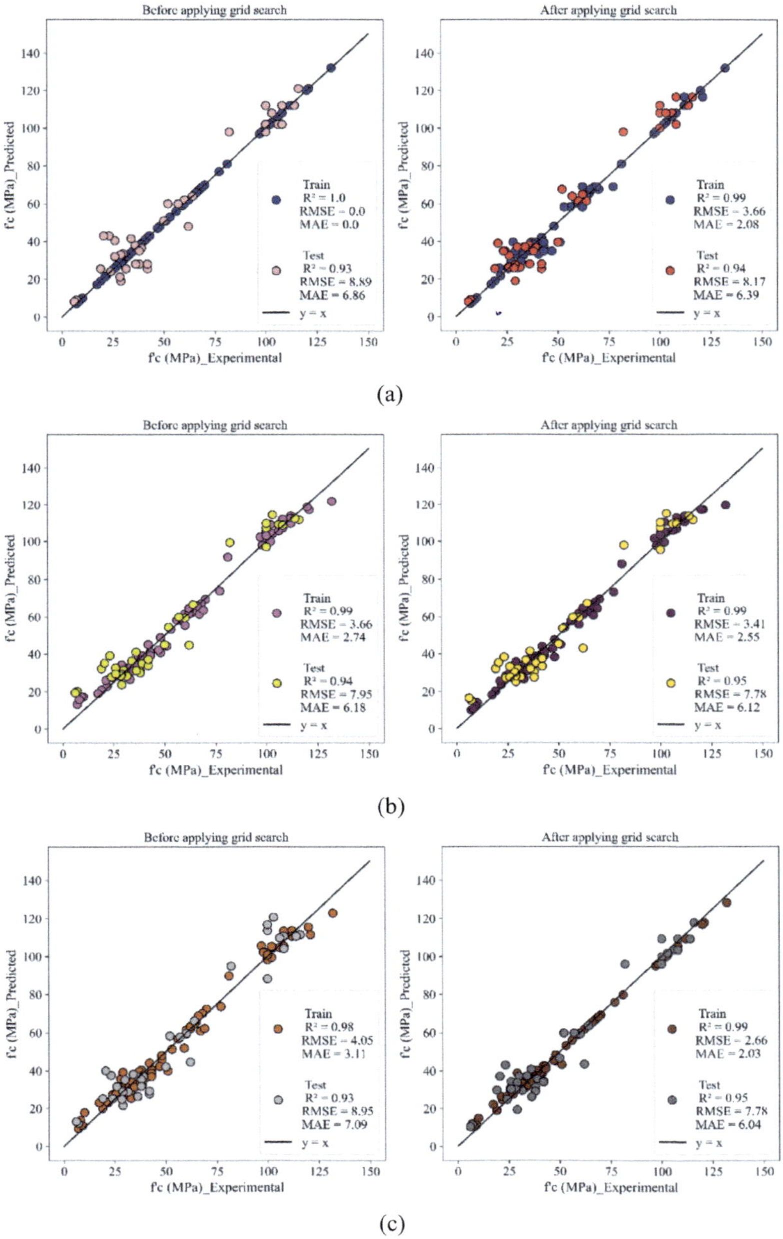

Fig. 2. ML models results with the target of CS (MPa): **a** DT, **b** RF, **c** GB, **d** XGB.

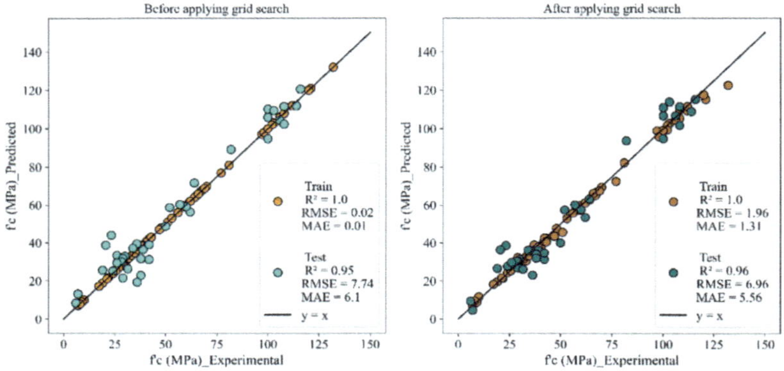

(d)

Fig. 2. (*continued*)

The optimal values of the key hyperparameters obtained through grid search are presented in Table 5. Additionally, the effect of hyperparameter tuning using grid search is illustrated in Fig. 3.

Table 5. Best hyperoaramers resulting from Grid search.

model	n_estimators	max_depth	learning_rate	min_sample_leaf	min_sample_split	max_features
DT	–	4	–	1	2	sqrt
RF	30	None	–	1	2	None
GB	300	7	0.01	–	–	–
XGB	50	9	0.1	–	–	–

The effect of hyperparameter tuning using grid search is evident across all models, as shown by the improvements in R^2, RMSE, and MAE values (see Fig. 3). While all models benefited from tuning, the GB model showed the most noticeable improvement across the evaluation metrics. However, after tuning, the XGB model outperformed all others, achieving the highest R^2 and the lowest RMSE and MAE values. These findings highlight the effectiveness of grid search in enhancing model performance and confirm XGB as the most reliable and accurate model in this study.

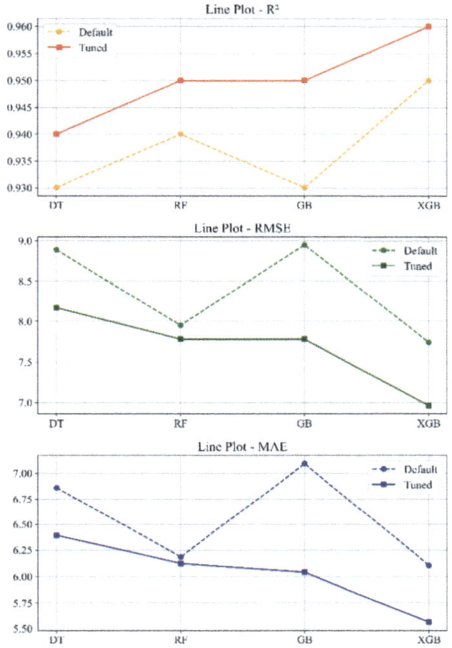

Fig. 3. Impact of grid search on ML models results.

To better understand the influence of each input variable on the model's predictions, feature importance analysis was performed using the XGB model. The feature importance was evaluated based on the "gain" criterion, which measures the improvement in accuracy brought by a feature to the branches it is used in. As shown in the Fig. 4, MK had the highest contribution to the model's output, indicating that the amount of metakaolin plays a critical role in determining the target variable. ET and BFS were ranked as the second and third most influential features, respectively. Other input variables, including FA, CS, IT, MSWI-BA, W/B, and AA, showed relatively smaller impacts on the predictions.

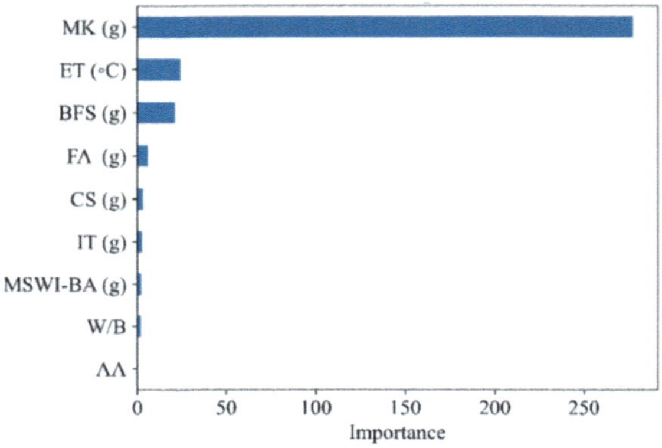

Fig. 4. XGBoost Feature Importance.

Despite traditional methods, ML models can adapt to the complexities and non-linearities within the dataset. These models are trained on experimental data which allows them to capture complex patterns between input variables and the target. As a results, learning from the data enables ML models to provide more accurate and generalized predictions compared to traditional methods. According to the results of this study, these models can be used as valuable tools for predicting the characteristics of newly developed materials which can pave the way for providing specific design codes for these materials.

5 Conclusion

This study employed ML models to predict the compressive strength of geopolymer concrete (GPC). The main findings can be summarized as follows:

The results show that after hyperparameter tuning, further improvements were achieved, evidenced by reductions in RMSE and MAE, along with slight increases in R2 values. These improvements were consistent across both training and testing datasets, confirming the effectiveness of grid search in optimizing model performance.

Among the models evaluated, Extreme Gradient Boosting (XGB) achieved the best performance. After tuning, the XGB model reached a test R2 of 0.96, with RMSE and MAE values reduced to 6.96 MPa and 5.56 MPa, respectively. Other models, including Gradient Boosting (GB), Decision Tree (DT), and Random Forest (RF), also showed noticeable improvements, though their performance remained slightly lower than that of XGB.

Overall, the study demonstrated that while the initial models performed well, hyperparameter tuning significantly enhanced predictive accuracy, with XGB emerging as the most effective model.

References

1. Hakeem IY et al (2023) Effect of using sugarcane leaf ash and granite dust as partial replacements for cement on characteristics of ultra-high performance concrete. Case Stud Constr Mater 19:e02266

2. Liu Y et al (2024) A review on high-temperature resistance of geopolymer concrete. J Build Eng 111241
3. Gluth GJ et al (2016) Acoustic emission and microstructural changes in fly ash geopolymer concretes exposed to simulated fire. Mater Struct 49:5243–5254
4. Fang Y et al (2024) Development of a ternary high-temperature resistant geopolymer and the deterioration mechanism of its concrete after heat exposure. Constr Build Mater 449:138291
5. Al-Naghi AAA et al (2025) Combined effects of elevated temperature, sulfates and chlorides on performance of fly ash and metakaolin-based recycled aggregate geopolymer concrete. J Build Eng 99:111561
6. Samadi M et al (2024) Influence of elevated temperatures on the mechanical behavior of one-part coal ash geopolymer concrete with sodium metasilicate pentahydrate and sodium metasilicate anhydrous as alkaline activators. J Build Eng 91:109692
7. Hager I, Sitarz M, Mróz K (2021) Fly-ash based geopolymer mortar for high-temperature application–Effect of slag addition. J Clean Prod 316:128168
8. Luo Y et al (2022) Effects of ladle slag on Class F fly ash geopolymer: reaction mechanism and high temperature behavior. Cement Concr Compos 129:104468
9. Khan MNN, Sarker PK (2020) Effect of waste glass fine aggregate on the strength, durability and high temperature resistance of alkali-activated fly ash and GGBFS blended mortar. Constr Build Mater 263:120177
10. Qu F et al (2020) High temperature resistance of fly ash/GGBFS-based geopolymer mortar with load-induced damage. Mater Struct 53:1–21
11. Lee N et al (2017) Influence of binder composition on the gel structure in alkali activated fly ash/slag pastes exposed to elevated temperatures. Ceram Int 43(2):2471–2480
12. Luo Y, Yu Q (2024) Valorization of mineral wool waste in Class F fly ash geopolymer: geopolymerization, macro properties, and high temperature behavior. Cement Concr Compos 145:105318
13. Sarıdemir M, Çelikten S (2023) Effects of Ms modulus, Na concentration and fly ash content on properties of vapour-cured geopolymer mortars exposed to high temperatures. Constr Build Mater 363:129868
14. Salazar PA et al (2022) Physical, mechanical and radiological characteristics of a fly ash geopolymer incorporating titanium dioxide waste as passive fire insulating material in steel structures. Materials 15(23):8493
15. Bypour M, Yekrangnia M, Kioumarsi M (2025) Machine learning-driven optimization for predicting compressive strength in fly ash geopolymer concrete. Clean Eng Technol 100899
16. Bypour M et al (2024) Explainable tuned machine learning models for assessing the impact of corrosion on bond strength in concrete. Clean Eng Technol 23:100834
17. Mahmoudian A, Bypour M, Kontoni D-PN (2024) Tree-based machine learning models for predicting the bond strength in reinforced recycled aggregate concrete. Asian J Civ Eng 25(8):5899–5924
18. Mahmoudian A, Bypour M, Kioumarsi M (2024) Explainable boosting machine learning for predicting bond strength of FRP rebars in ultra high-performance concrete. Computation 12(10):202
19. Bypour M, Yekrangnia M, Kioumarsi M (2024) Evaluation of the compressive strength of fly ash-based geopolymer concrete using machine learning. The international conference on net-zero civil infrastructures: innovations in materials, structures, and management practices (NTZR). Springer
20. Bypour M et al (2024) Shear capacity assessment of perforated steel plate shear wall based on the combination of verified finite element analysis, machine learning, and gene expression programming. Asian J Civ Eng 25(7):5317–5333
21. Zhang P et al (2023) High-temperature behavior of geopolymer mortar containing nano-silica. Constr Build Mater 364:129983

22. Zhang B et al (2023) Improving the high temperature resistance of alkali-activated slag paste using municipal solid waste incineration bottom ash. J Build Eng 72:106664
23. Zhang D-W et al (2022) Improvement of the thermal properties of alkali-activated fly ash after high temperature by the Fe-based solid wastes. Case Stud Constr Mater 17:e01303

Explainable Machine Learning Model for the Prediction of CO_2-Cured Concrete Strength

Iyinoluwa Esegbuyotaroghene E. Davies(✉) and Xueqing Zhang

Department of Civil and Environmental Engineering, The Hong Kong University of Science and Technology, Clear Water Bay, Kowloon, Hong Kong, China
ieedavies@connect.ust.hk, zhangxq@ust.hk

Abstract. The precast concrete industry, a major contributor to global CO_2 emissions, urgently requires sustainable alternatives to energy-intensive curing methods. This study pioneers the application of explainable machine learning (XML) to predict and interpret the compressive strength of CO_2-cured concrete—a promising low-carbon technology that accelerates strength development while sequestering CO_2. The authors leveraged a curated dataset of 399 experimental observations from peer-reviewed studies on the Web of Science database. Eleven machine-learning models were trained using feature-engineered inputs. The Categorical Boosting model achieved superior accuracy (R2 = 0.930, MAE = 2.81 MPa, RMSE = 4.59 MPa), outperforming the other ten models. Shapley additive explanation investigation was executed to offer an extensive understanding of feature importance. Concrete age, binder-to-aggregate ratio, carbonation time, water-binder ratio, and initial curing time were identified as the dominant drivers in predicting the compressive strength of CO_2-cured concrete. The study also highlights the critical role of curing time, binder, and aggregate activity in CO_2-cured concrete compressive strength prediction. This work provides actionable insights to design low-carbon concrete and establishes a framework for XML-driven CO_2-cured concrete strength prediction integrating domain knowledge.

Keywords: Explainable Machine Learning · Compressive Strength · CO_2-curing · concrete

1 Introduction

Precast concrete is essential in various construction applications, such as prefabricated buildings and tunnel engineering. Despite its versatility, the industry faces significant carbon emissions, primarily from cement production and energy-intensive curing methods, such as steam curing [1–3]. Notably, Portland cement production is responsible for over 5% of global anthropogenic CO_2 emissions [4]. In response to this environmental challenge, the precast concrete sector is increasingly transitioning towards low-carbon production practices, with a focus on a more sustainable practice of sequestering CO_2 in concrete [1].

H. Beushausen et al. (Eds.): NTZR 2025, LNCE 873, pp. 962–970, 2026.
https://doi.org/10.1007/978-3-032-27698-8_84

One promising avenue in this transition is using carbonation curing technology, which involves exposing concrete to CO_2 gas during the curing process. Carbonation curing is a rapid curing process commonly undertaken within 24 h of concrete casting [1]. When carbonation curing is initiated immediately after casting, it involves the interaction of high-purity CO_2 gas with CSH, Ca(OH), unhydrated cement minerals, and other cement compounds [1].

Machine learning (ML) is increasingly used in the prediction of compressive strength by analysing the relationship between mixing components and strength [5–7]. Various ML methods(random forest, artificial neural networks, extreme gradient boosting, and extra trees), have been applied to predict concrete mechanical properties in civil engineering research [5, 7–12]. The primary limitation of conventional ML methods lies in their lack of explainability. Despite extensive research in the field of concrete carbonation curing, to date, no study has employed ML methods to explore the compressive strength of CO_2-cured concrete. Previous studies have predominantly focused on concrete cured with water, leaving the impact of CO_2-curing and its interplay with concrete strength development uncharted within the research community. Moreover, this study represents a pioneering effort in utilising explainable ML models in the prediction of the compressive strength of CO_2-cured concrete.

2 Methodology

The present study's overall research methodology comprises four main steps: data collection, data preparation, data visualisation and statistics, and explainable ML development.

2.1 Data Collection

Data was collected from published experimental data in articles obtained from the Web of Science database. Only articles containing the required information for the training of the ML model were considered. For inclusion in the database, the selection criteria checked in the articles were the concrete-mix design (aggregate, water, cement, superplasticiser), carbonation curing conditions and duration, initial curing conditions and duration, concrete age, CO_2 pressure, CO_2 concentration amount, and compressive strength. A total of 37 articles met these criteria, comprising 399 data points.

2.2 Data Preparation

Domain knowledge is a crucial component in feature engineering for ML modeling. Effective preprocessing of data enables ML models to understand underlying patterns more effectively [13]. This process involves encoding categorical variables and combining and transforming the database's features into new features.

Feature engineering of concrete mix to engineered ratios. The raw mix proportion of concrete was transformed into meaningful ratio predictors of concrete strength. The three key ratios and derived features were the water-to-binder ratio (W/B), binder-to-aggregate ratio (B/A), and coarse aggregate-to-fine aggregate ratio (C/F). These concrete ratios affect the packing density of the concrete mix, which in turn influences strength development [14].

Dataset Description and statistics. Python programming was employed for the statistical analysis and visualisation of the data sets. Table 1 presents a statistical summary of the input and output features, including the mean, minimum value (min), maximum value (max), and standard deviation (std). The categorical variables used were the post-carbonation relative humidity (RH) curing type (aC) and the preconditioning RH curing type (pC). The minimum and maximum values of compressive strength are between 2.40 MPa and 97.85 MPa [15, 16].

Table 1. Statistical summary of the machine learning dataset.

Feature	ID	Unit	Mean	Std	Min	Max
Binder to aggregate ratio	B/A	–	0.40	0.33	0.09	3.00
Coarse to fine aggregate ratio	C/F	–	1.09	2.55	0.00	11.85
Water to binder ratio	W/B	–	0.41	0.11	0.05	0.70
Carbonation concentration	CC	%	50.94	41.71	0.04	99.90
Carbonation pressure	CP	Bar	1.76	1.51	0.10	7.00
Superplasticizer	S	%	0.23	0.28	0	1.51
Initial curing period	IC	Hours	23.00	40.43	0	312.0
Carbonation curing time	T	Hours	96.91	203.23	1	1344
Concrete age	CA	Days	13.22	23.63	0.08	360
Compressive strength	CS	MPa	33.45	17.81	2.40	97.85

2.3 Explainable Machine Learning Methodology

Eleven types of ML methods were employed: decision tree model (DT), random forest model (RF), gradient boosting model (GB), extreme gradient boosting model (XGB), light gradient boosting model (LGB), categorical boosting model (CB), adaptive boosting model (AB), extra trees model (ET), artificial neural network (ANN), extra trees model (ET), and histogram-based gradient boosting model (HGB). The ML dataset was partitioned into 80% and 20%, representing the train and test set, respectively. Hyperparameter optimisation was conducted in this study using repeated k-fold cross-validation (5-fold, 100 repeats) combined with Bayesian optimisation, employing the BayesSearchCV tool. To evaluate the accuracy of the ML models, the Mean Absolute Percentage Error (MAPE), root mean square error (RMSE), mean absolute error (MAE), and coefficient of determination (R^2) were evaluated using Eqs. 1 to 4. Shapley additive explanations (SHAP), an explanation method, was used to interpret the best ML model. SHAP computes the individual features contribution to the model's predictions by evaluating the marginal impact across various feature coalitions [17–19].

$$R^2 = 1 - \frac{\sum_{i=1}^{N}\left(y_i - \hat{y}_{i_i}\right)^2}{\sum_{i=1}^{N}\left(y_i - \overline{y}_i\right)^2} \tag{1}$$

$$RMSE = \sqrt{\frac{\sum_i (y_i - \hat{y}_i)^2}{n}} \tag{2}$$

$$MAE = \frac{\sum_i^n |y_i - \hat{y}_i|}{n} \tag{3}$$

$$MAPE = \sum_i^n \left|\frac{y_i - \hat{y}_i}{y_i}\right| \frac{100\%}{n} \tag{4}$$

3 Results and Discussion

The compressive strength prediction of CO_2 cured concrete was evaluated with eleven machine learning models, with performance analysis conducted on both the training and testing phases (as shown in Table 2, Figs. 2, and 3). The best and worst performing model in the training and testing phase is the categorical boosting model (CB) and decision tree model (DT), respectively. The CB model test performance metrics were R2 = 0.930, MAPE = 10.56%, MAE = 2.81 MPa, and RMSE = 4.59 MPa (Table 3).

Table 2. Training performance of the machine learning models.

Model ID	R2	MAE	RMSE	MAPE
AB	0.941	3.10	4.34	14.86
ANN	0.902	3.96	5.61	20.61
CB	0.962	1.98	3.51	9.66
DT	0.488	9.90	12.82	44.54
ET	0.962	2.06	3.49	9.74
GB	0.955	2.08	3.79	9.46
HGB	0.948	2.38	4.08	10.60
KNN	0.986	0.45	2.09	1.59
LGB	0.953	2.32	3.89	10.25
RF	0.938	2.52	4.46	12.61
XGB	0.977	1.46	2.75	7.60

Table 3. Machine learning models test set performance.

Model ID	R2	MAE	RMSE	MAPE
AB	0.912	3.33	5.12	13.51
ANN	0.812	5.04	7.48	18.53
CB	0.93	2.81	4.59	10.56
DT	0.576	8.73	11.25	39.45
ET	0.887	3.36	5.81	12.74
GB	0.911	3.2	5.16	11.24
HGB	0.927	3.04	4.67	10.04
KNN	0.866	3.54	6.33	12.17
LGB	0.914	3.19	5.07	10.81
RF	0.887	3.25	5.8	12.83
XGB	0.896	2.97	5.57	9.89

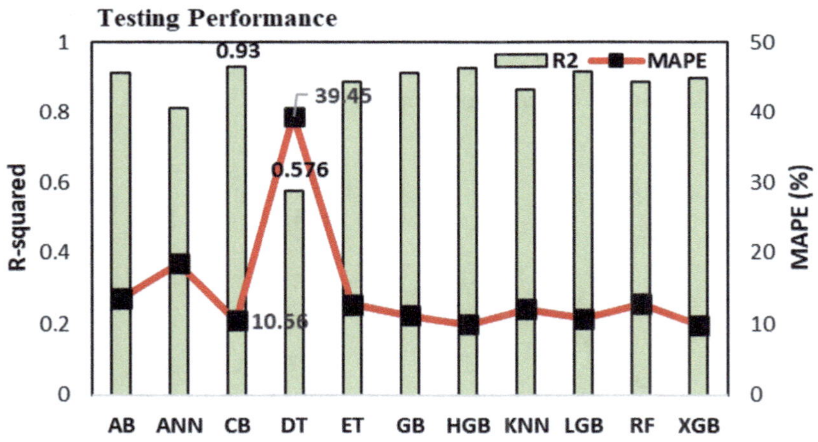

Fig. 2. R-Squared and MAPE testing performance of machine learning models.

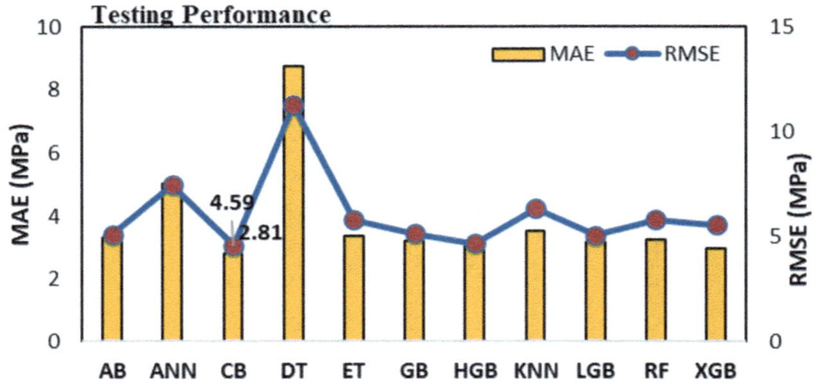

Fig. 3. MAE and RMSE testing performance of machine learning models.

3.1 Explanation of the Best Model

The best ML model is categorical boosting model, and its performance is visualised in Fig. 4. Figure 5 visualises the SHAP feature analysis of the categorical boosting model by computing SHAP values average across the ML dataset. The top five features were ranked as follows: Concrete age (CA), binder-to-aggregate ratio (B/A), carbonation time (T), Water-binder ratio (W/B), and initial curing time (IC).

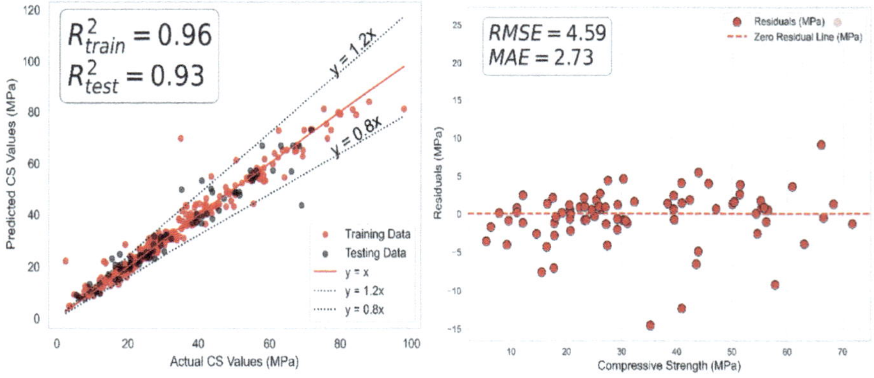

Fig. 4. Categorical boosting model performance.

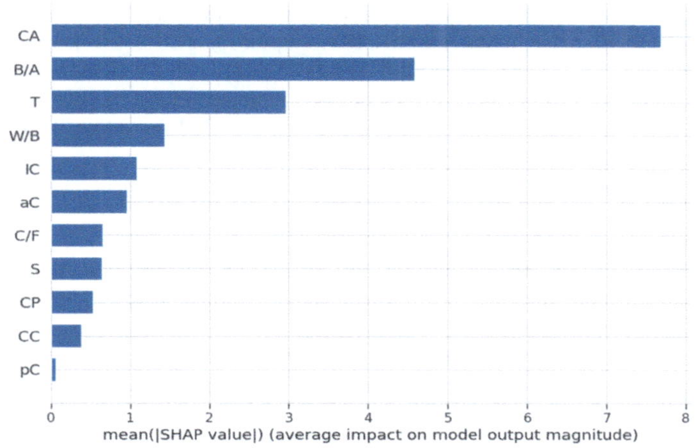

Fig. 5. SHAP analysis of CO_2 cured concrete categorical boosting model performance.

4 Conclusions

This study demonstrates the efficacy of explainable ML models in predicting the compressive strength of CO_2 cured concrete for the first time. The authors collected the data from existing literature. The categorical boosting (CB) algorithm (R2 = 0.930, MAE = 2.81 MPa) was the best-performing model in predicting the compressive strength of CO_2-cured concrete, outperforming the other models. Explainability analyses were conducted using SHAP analysis, and concrete age (CA), binder-to-aggregate ratio (B/A), and carbonation time (T) were identified as the dominant drivers, underscoring the critical interplay between time, hydration, CO_2 diffusion, carbonation reaction, and concrete matrix densification. This integrated approach facilitates the identification of critical parameters essential for optimising material design processes. The ML approach presented in this study provides engineers with actionable insights for designing low-carbon precast concrete.

References

1. Zhang D, Cai X, Shao Y (2016) Carbonation curing of precast fly ash concrete. J Mater Civ Eng 1–9. https://doi.org/10.1061/(ASCE)MT.1943
2. Zhang D, Shao Y (2016) Early age carbonation curing for precast reinforced concretes. Constr Build Mater 113:134–143. https://doi.org/10.1016/j.conbuildmat.2016.03.048
3. Rostami V, Shao Y, Boyd AJ (2012) Carbonation curing versus steam curing for precast concrete production. J Mater Civ Eng 24:1221–1229. https://doi.org/10.1061/(asce)mt.1943-5533.0000462
4. Xi F et al (2016) Substantial global carbon uptake by cement carbonation. Nat Geosci 9:880–883. https://doi.org/10.1038/ngeo2840
5. Akber MZ, Anwar GA, Chan W-K, Lee H-H (2024) TPE-xgboost for explainable predictions of concrete compressive strength considering compositions, and mechanical and microstructure properties of testing samples. Constr Build Mater 457:139398. https://doi.org/10.1016/j.conbuildmat.2024.139398

6. Zhang X, Akber MZ, Zheng W (2021) Prediction of seven-day compressive strength of field concrete. Constr Build Mater 305:124604. https://doi.org/10.1016/j.conbuildmat.2021.124604
7. Zhang X, Akber MZ, Zheng W (2022) Predicting the slump of industrially produced concrete using machine learning: a multiclass classification approach. J Build Eng 58:104997. https://doi.org/10.1016/j.jobe.2022.104997
8. Eskandari-Naddaf H, Kazemi R (2017) ANN prediction of cement mortar compressive strength, influence of cement strength class. Constr Build Mater 138:1–11. https://doi.org/10.1016/j.conbuildmat.2017.01.132
9. Deng F, He Y, Zhou S, Yu Y, Cheng H, Wu X (2018) Compressive strength prediction of recycled concrete based on deep learning. Constr Build Mater 175:562–569. https://doi.org/10.1016/j.conbuildmat.2018.04.169
10. Liu G, Sun B (2023) Concrete compressive strength prediction using an explainable boosting machine model. Case Stud Constr Mater 18:e01845. https://doi.org/10.1016/j.cscm.2023.e01845
11. Benaicha M, Burtschell Y, Alaoui AH (2016) Prediction of compressive strength at early age of concrete—application of maturity. J Build Eng 6:119–125. https://doi.org/10.1016/j.jobe.2016.03.003
12. Feng D-C et al (2020) Machine learning-based compressive strength prediction for concrete: an adaptive boosting approach. Constr Build Mater 230:117000. https://doi.org/10.1016/j.conbuildmat.2019.117000
13. Zhang LV, Marani A, Nehdi ML (2022) Chemistry-informed machine learning prediction of compressive strength for alkali-activated materials. Constr Build Mater 316. https://doi.org/10.1016/j.conbuildmat.2021.126103
14. Zheng W, Shui Z, Gao X, Lian J, Xu Z, Wang Y (2023) Optimization of concrete mix design based on three-level separation distance of particles. J Build Eng 70. https://doi.org/10.1016/j.jobe.2023.106479
15. He P, Shi C, Tu Z, Poon CS, Zhang J (2016) Effect of further water curing on compressive strength and microstructure of CO_2-cured concrete. Cem Concr Compos 72:80–88. https://doi.org/10.1016/j.cemconcomp.2016.05.026
16. Jang JG, Lee HK (2016) Microstructural densification and CO_2 uptake promoted by the carbonation curing of belite-rich Portland cement. Cem Concr Res 82:50–57. https://doi.org/10.1016/j.cemconres.2016.01.001
17. Wu Y, Zhou Y (2022) Hybrid machine learning model and Shapley additive explanations for compressive strength of sustainable concrete. Constr Build Mater 330:127298. https://doi.org/10.1016/j.conbuildmat.2022.127298
18. Sun B, Cui W, Liu G, Zhou B, Zhao W (2023) A hybrid strategy of AutoML and SHAP for automated and explainable concrete strength prediction. Case Stud Constr Mater 19:e02405. https://doi.org/10.1016/j.cscm.2023.e02405
19. Wani SR, Suthar M (2025) Using machine learning approaches for predicting the compressive strength of ultra-high-performance concrete with SHAP analysis. Asian J Civ Eng 26:373–388. https://doi.org/10.1007/s42107-024-01195-6

Machine-Learning-Enabled Prediction of Vibration of Tall Buildings Under Magnified Wind Loads

Bahareh Dokhaei, Behrouz Shafei, and Alice Alipour(✉)

Iowa State University, Ames, IA 50011, USA
alipour@iastate.edu

Abstract. Wind-induced vibration presents a significant challenge in the design of tall and slender buildings, with the potential to cause structural damage and compromise occupant comfort. Engineers have explored various vibration control strategies, such as aerodynamic modifications, but traditional analytical methods often fail to accurately capture dynamic responses, especially for complex geometries and magnified wind loads. Using machine learning, this study offers a data-driven approach to predict structural behavior. For this purpose, data obtained from wind tunnel experiments is utilized. The focus is on forecasting across-wind vibrations under various wind conditions by training and evaluating long short-term memory (LSTM) neural networks. The dataset, collected from a 1:76 scale aeroelastic building model tested at the NHERI Wall of Wind facility, includes time series data on wind pressures, acceleration responses, and atmospheric conditions. The models achieved a high accuracy with an R-squared score of up to 0.96, highlighting the potential of machine learning in capturing complex aerodynamic responses, especially in the anticipation of high wind events.

Keywords: Machine learning · Wind-induced vibration · Wind tunnel testing · Tall building · Extreme wind events

1 Introduction

Wind hazards have become an increasingly important concern, with significant safety risks and economic losses, particularly in urban environments with clusters of tall buildings. These buildings are more susceptible to wind-induced vibrations. Despite advances in wind engineering, the design of tall buildings continues to face considerable challenges, necessitating innovative solutions to ensure structural safety and occupant comfort [1–3]. To mitigate wind-induced vibrations, aerodynamic modifications, such as corner chamfers, splitter plates, porous or double-skin façades [4], and adaptive façade [5], have been examined. Studies have shown that even minor aerodynamic changes can significantly reduce across-wind responses by addressing vortex-shedding forces [6]. While these solutions can reduce vortex shedding and across-wind forces, their effectiveness depends on a range of parameters, including wind direction, terrain, and

H. Beushausen et al. (Eds.): NTZR 2025, LNCE 873, pp. 971–980, 2026.
https://doi.org/10.1007/978-3-032-27698-8_85

surrounding obstructions. Traditional analytical methods often fail to accurately capture the dynamic response of buildings with aerodynamic modifications, especially for complex geometries [7]. Hence, predicting wind-induced vibrations in tall buildings are essential for structural safety and serviceability of these buildings. Traditional numerical simulations and physical testing are time-consuming and expensive. With the rise of data-driven approaches, machine learning (ML) offers a promising alternative for real-time prediction and decision-making.

With the advent of data-driven techniques, ML has emerged as a powerful alternative for modeling and predicting wind-induced responses in various structures. ML models are particularly well-suited for handling large, complex, and nonlinear datasets that arise from wind tunnel experiments and real-time monitoring systems.

Several recent studies have specifically focused on predicting wind-induced responses of tall buildings under varying aerodynamic conditions with ML techniques. Hu et al. [8] employed generative adversarial networks (GANs) to predict wind pressure coefficients on a principal tall building subject to interference effects. The referenced study showed that models trained on as little as 30% of the dataset could predict unseen interference scenarios with high accuracy, reducing the need for costly wind tunnel experiments. Expanding on this idea, Kim et al. [9] addressed the issue of missing pressure data in wind tunnel tests using a generative adversarial imputation network (GAIN). The developed model effectively reconstructed time-series wind pressures on multiple building surfaces, outperforming traditional imputation methods.

Recent advancements in deep learning have enabled effective time series prediction of wind-induced structural responses in tall buildings, as an alternative to computationally expensive finite element simulations [10, 11]. Hareendran and Alipour [12] tackle the challenge of simulating the nonlinear structural response of tall buildings under prolonged, high-intensity wind loads, which can lead to post-elastic behavior. Unlike earthquake engineering, where simulations typically involve short-duration seismic loads, wind engineers must consider extended wind actions. The introduced model predicted the nonlinear response of a 44-story tall building subjected to long-duration wind loads, demonstrating the potential of LSTM models to replicate both numerical simulations and experimental aeroelastic responses with high accuracy and significantly reduced computational effort. Wang and Wu [13] introduced a knowledge-enhanced deep learning (KEDL) framework that integrates physics-based governing equations and semiempirical formulas into the training process to improve data efficiency and robustness. The referenced work showed superior performance in predicting nonlinear responses of both SDOF and MDOF systems by leveraging prior structural knowledge and wavelet-domain projection to simplify input-output relationships. More recently, Li and Spence [14] proposed a metamodeling approach that combines LSTM networks with Galerkin-based model order reduction to rapidly simulate nonlinear time history responses of tall buildings under stochastic wind loads. The developed framework enabled accurate and efficient multi-directional performance assessment of a 37-story nonlinear steel structure, achieving speedups of over four orders of magnitude compared to conventional time history analyses.

The outlined studies collectively highlight the expanding role of ML and deep learning in wind engineering, particularly for tall structures, by demonstrating how data-driven models can enhance prediction accuracy, reduce experimental costs, and improve generalization across diverse wind profiles and building geometries. However, despite this progress, limited research has focused on buildings incorporating different aerodynamic modifications, such as different façade systems. To address this gap, the present study investigates the application of an LSTM network to predict wind-induced vibrations in a tall building equipped with a morphing facade system. Using wind tunnel data collected from a 1:76-scale aeroelastic model tested at the NHERI Wall of Wind (WoW) facility [15], the study employs sensor measurements, including wind pressures at 10 locations, barometric pressure, and structural accelerations. The LSTM model is trained to forecast acceleration responses in the X-direction (across-wind), and its performance is evaluated against experimental data using standard regression metrics to assess accuracy and generalization capability.

2 Description of Features and Models

2.1 Data Description

The dataset used in this study was collected from wind tunnel experiments conducted on a 1:76 scale aeroelastic model of a tall building at the NHERI WoW facility [15]. Time-series data were recorded over 155 s at 0.002-s intervals, resulting in 77,500 data points per sensor. Figure 1 shows the sensor layout during wind tunnel simulation. The sensor network included three accelerometers on top of the structure, ten pressure sensors at different locations placed on surrounding wooden towers, and atmospheric sensors, all collected at a 500 Hz sampling rate. Figure 1 summarizes the extracted features, which include acceleration responses in the x and y directions, wind pressures at various locations, and barometric pressure. These inputs pressure, acceleration, and atmospheric data were used for model development. The predicted output (target label) is the across-wind direction response of the tall building model. The evaluation dataset is split into 60% for training and 40% for validation; the test data set is also split into 60% for training and 40% for testing.

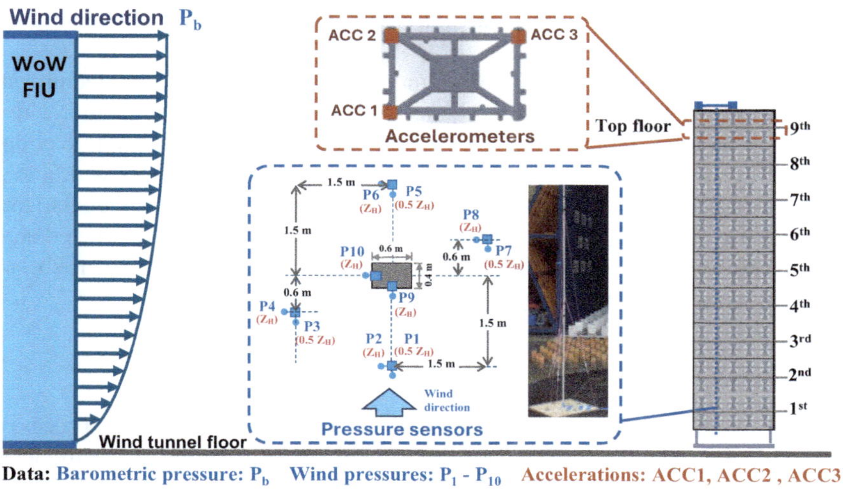

Fig. 1. Sensor layout in large-scale CAARC building model.

2.2 Data Pre-processing

Before training the model, the raw wind tunnel dataset was thoroughly pre-processed. This included signal cleaning and filtering, normalization, sequencing, and down-sampling to enhance data quality and computational efficiency.

Three unidirectional accelerometers were placed in the corners of the top floor to monitor vibrations. Since they were not centered, translational accelerations in the *x*- and *y*-directions were calculated by averaging the data from the sensors. Rotational acceleration (α) was derived from the difference in *y*-direction accelerations between two spaced sensors. These accelerometers were later replaced with sensors designed for *x*-, *y*-, and α-directions. To reduce noise, a band-pass filter was applied to the acceleration data to eliminate high-frequency noise and low-frequency drift, while a low-pass filter was used for pressure data to retain overall trends.

Normalization was performed to scale all features to the range [0,1], ensuring fair treatment of each feature and speeding up convergence during training. The data were then sequenced into structured input-output pairs for predicting future across-wind acceleration based on prior pressure readings, segmented into overlapping windows of 200 time steps each. The target output is the acceleration in the *x*-direction at the next or future time step. Finally, down-sampling was conducted for efficiency, reducing the original data sampled at 0.002 s (500 Hz) to 0.005 s (200 Hz) using a Fourier-based method, which preserved signal characteristics and improved computational efficiency during model training. This preprocessing was applied consistently across all datasets.

2.3 Model Architectures

LSTM is a recurrent neural network (RNN) variant known for its ability to capture long-term dependencies and manage vanishing gradient problems [16]. An LSTM neural network generally consists of one or more LSTM layers, along with a fully connected layer (also known as a dense layer) that converts the hidden states into outputs of the appropriate dimensions. Additionally, dropout layers are often included to improve generalization.

Based on the results of hyperparameter tuning, the optimal LSTM architecture on the data is illustrated in Fig. 2. The network consists of two LSTM layers: the first with 128 units and the second with 64 units, allowing the model to effectively capture complex temporal dependencies in the aerodynamic input data. A dropout layer with a rate of 0.4 is applied after the LSTM layers to mitigate overfitting by promoting regularization. Finally, a dense output layer with a single unit is used to generate the predicted across-wind acceleration at each time step. This architecture was selected for its balance between model complexity and predictive performance.

2.4 Choices of Hyperparameters

To enhance the performance of the LSTM model, a hyperparameter tuning process was conducted. To evaluate the performance and generalization capability of the model, two widely accepted regression metrics were employed: the Coefficient of Determination (R^2 Score) and the Root Mean Squared Error (RMSE). Various combinations of units, dropout rates, and layer structures were evaluated. Two wind speeds, 5.9 m/s and 7.7 m/s, were selected for evaluation. The model was trained using different dropout rates (0.1, 0.2, 0.3, and 0.4) and sequence lengths (50, 100, 150, and 200). The LSTM units used included configurations such as 256 in the first layer and 128 in the second layer, as well as other combinations like 128-64, 64-32, and 32-16. Additionally, the number of training epochs was assessed with values of 30, 40, 60, and 100. The models were trained in 30 epochs to ensure early convergence while minimizing the risk of overfitting. A batch size of 64 was selected to maintain a balance between computational efficiency and training stability.

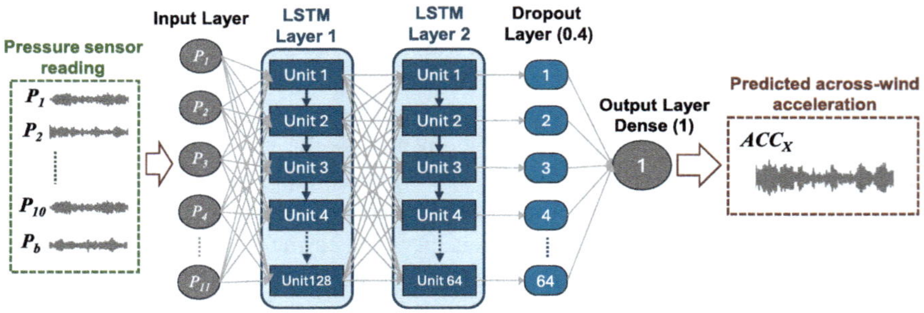

Fig. 2. LSTM model architecture.

As shown in Fig. 3(a), dropout rates from 0.1 to 0.4 were evaluated to investigate the model's generalization performance. A dropout of 0.2 led to the worst test R^2 (0.9013) and the highest test RMSE (0.1481). In contrast, dropout rates of 0.3 and 0.4 significantly improved performance, achieving test R^2 scores of 0.9516 and 0.9531, respectively. While 0.3 yielded the lowest test RMSE (0.1036), dropout 0.4 had a slightly better test R^2. Both values indicate strong regularization and generalization capabilities.

Overall, dropout rates above 0.2 appear beneficial, with 0.4 being optimal in balancing performance and overfitting mitigation.

Figure 3(b) shows that the model performance improves consistently as sequence length increases. A short sequence length of 50 results in poor performance (test R2 = 0.8557), likely due to insufficient temporal context. Increasing the sequence length to 100 significantly improves the model's test R2 to 0.9425 and further increases to 150 and 200 improved model, with the best performance observed at 200.

Figure 3(c) compares different LSTM layer unit configurations. The configuration of 128-64 and 256-128 both achieve strong generalization, with 256-128 slightly outperforming others (test R2 = 0.9548, test RMSE = 0.1001). Notably, the 32-16 configuration of 32 units in first layer and 16 units in second layer results in the weakest performance (test R2 = 0.9387), showing that too few units may under-represent the temporal dynamics. The result implies that moderately deep LSTM layers (128-64 or 256-128) enhance model capacity without sacrificing generalization.

The impact of training epochs was also evaluated and shown in Fig. 3(d). While increasing epochs reduces training RMSE and increases training R2 (indicating better fitting), the test performance does not improve. The best test R2 (0.9531) is achieved at 30 epochs. Although training R2 continues to increase up to 100 epochs (0.9793), the test R2 slightly decreases (0.9398), showing potential overfitting beyond 60 epochs. Therefore, around 30–60 epochs seem to be a reasonable choice to ensure generalization without excessive overfitting. In summary, the LSTM model with 128 and 64 units, a dropout rate of 0.4, and 30 training epochs were chosen, and its test performance provided a strong balance between accuracy and generalization.

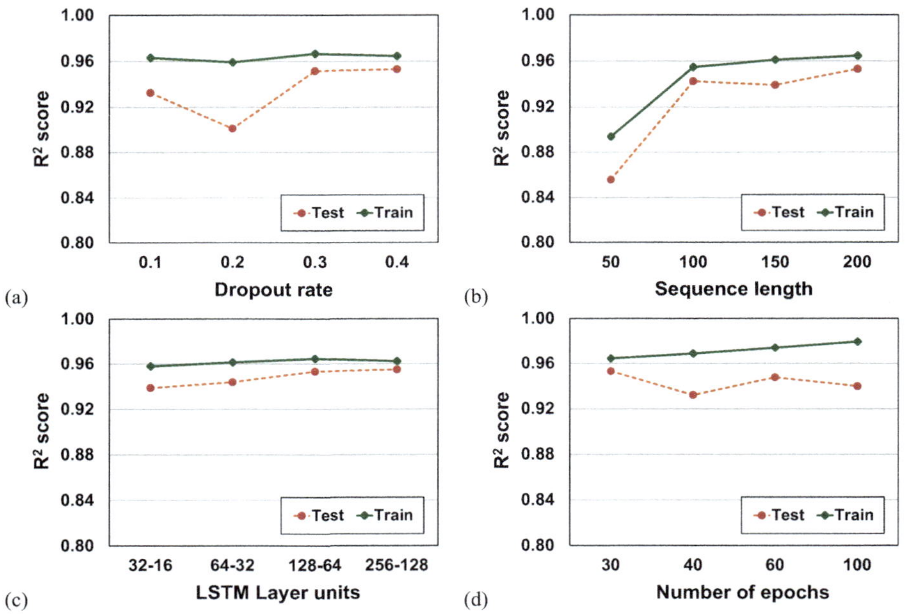

Fig. 3. Model accuracy (R^2 score) for training and testing with different: **a** dropout rate, **b** sequence length, **c** LSTM layer units, and **d** number of epochs.

3 Results and Findings

In this study, an LSTM model was developed to predict wind-induced structural response in the across-wind direction using only wind pressure data obtained from wind tunnel experiments. Importantly, acceleration or vibration data were not included as input features during training. Instead, the model was trained to infer the dynamic response of the structure based solely on the aerodynamic loading history, making the prediction task more challenging and realistic for early-stage design or control applications where response measurements may not be available.

To train the LSTM model, wind tunnel data from four different wind speeds, including 4.5 m/s, 5.9 m/s, 7.7 m/s, and 8.6 m/s, were used [17–19]. For each case, only the first 60% of the time-series data was used for training, with the remaining 40% reserved for testing. This approach enabled the model to learn from a broad range of dynamic wind conditions while preserving unseen data for performance evaluation. By combining the training portions from all wind speeds into a single dataset, the model benefited from a more diverse and representative training distribution. This strategy enhanced the model's ability to generalize across different wind excitation levels, rather than being tuned to a single wind scenario. It also reflects a realistic condition where structural responses under varying wind speeds must be learned simultaneously.

Throughout 30 training epochs, both training and validation loss decreased, with no signs of overfitting, as indicated by the consistent alignment between the training loss and validation loss curves. By the end of training, the model achieved a training loss below 3.0e-4 and a validation loss below ~ 1.5e-3, suggesting that it effectively learned the mapping from wind and pressure inputs to structural acceleration outputs.

To evaluate the generalization performance of the LSTM model under different aerodynamic loading conditions, predictions were made on the last 40% of the dataset for each wind speed case. Figure 4 shows the comparison between the true and predicted across-wind accelerations for the test data across different wind speeds. The figure highlights the model's ability to accurately capture the time-varying structural response based solely on wind pressure inputs. The model achieved consistently high R^2 values across all wind speeds, confirming its robustness. Specifically, the R^2 scores were 0.9317 at 4.5 m/s, 0.9665 at 5.9 m/s, 0.9604 at 7.7 m/s, and 0.9570 at 8.6 m/s.

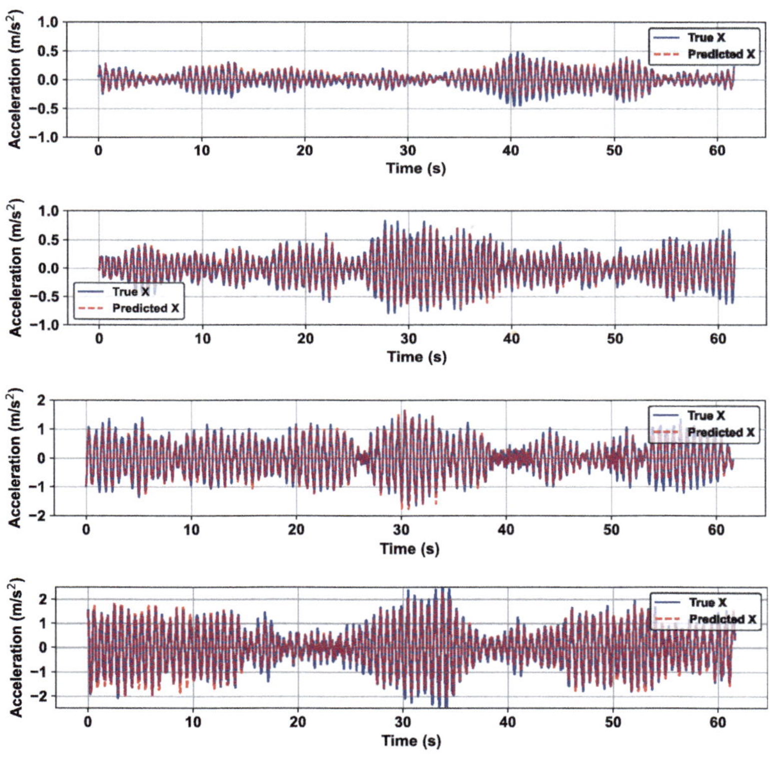

(a) (b) (c) (d)

Fig. 4. Model prediction of test set (last 40%): **a** wind speed of 4.5 m/s, **b** wind speed of 5.9 m/s, **c** wind speed of 7.7 m/s, and **d** wind speed of 8.6 m/s.

However, the RMSE values showed a noticeable increasing trend with wind speed: 0.0372 at 4.5 m/s, 0.0516 at 5.9 m/s, 0.1194 at 7.7 m/s, and 0.1827 at 8.6 m/s. While R^2 remained high, the increasing RMSE shows that the magnitude of prediction errors grew with higher wind speeds. This behavior likely reflects the increased complexity and nonlinearity of structural response at elevated wind intensities, where flow separation, vortex shedding, and dynamic interaction effects are more pronounced. In summary, the results show that the LSTM model, trained only with wind pressure inputs, effectively predicts across-wind structural responses of tall buildings across a wide range of wind speeds. Future research should expand the dataset to include multiple building geometries, sizes, and façade systems to enhance model applicability across various structural configurations.

4 Conclusions

This study used an LSTM deep learning model to predict wind-induced across-wind vibrations in a scaled tall building model with aerodynamic façade designs, using only wind pressure data derived from wind tunnel experiments. The dataset was generated using a 1:76 scale aeroelastic model of a CAARC-standard tall building tested at the NHERI WoW facility. The experiments were conducted under different wind conditions and captured time-series data from a dense network of sensors, including wind pressure probes, barometric sensors, and accelerometers. A two-layer LSTM model with a dropout rate of 0.4 was developed and trained to forecast acceleration responses. To enhance model performance and prevent overfitting, extensive hyperparameter tuning was carried out on the number of units per layer, dropout rate, sequence length, and epoch number. Following thorough preprocessing and model calibration, the final LSTM model achieved high predictive accuracy, with a test R^2 score of up to 0.96 on the test set. Overall, the results highlighted the potential of deep learning for reliable forecasting of wind-induced structural responses and support its application in real-time monitoring, especially for sustainable and resilient tall buildings. Future work can expand to more complex wind scenarios and real-time control systems.

Acknowledgement. The research study, the results of which are reported in this paper, was sponsored by the U.S. National Science Foundation (NSF). The authors would like to acknowledge and thank the sponsor for this support. Opinions, findings, and conclusions expressed in this paper are of the authors and do not necessarily reflect those of the sponsor.

References

1. Hareendran SP, Alipour A, Shafei B, Sarkar P (2022) Performance-based wind design of tall buildings considering the nonlinearity in building response. J Struct Eng 148(6):04022119
2. Saini D, Dokhaei B, Shafei B, Alipour A (2024) Performance evaluation of high-rise buildings using database-assisted design approach. Struct Saf 102447
3. Dokhaei B, Saini D, Shafei B, Alipour A (2024) Design optimization to minimize material usage in steel buildings subjected to lateral loads. In: Kioumarsi M, Shafei B (eds) The 1st international conference on net-zero built environment 2024, vol 237. Lecture notes in civil engineering. Springer, Cham, pp 1529–1538
4. Hou F, Sarkar P, Alipour A (2023) A novel mechanism—smart morphing façade system—to mitigate wind-induced vibration of tall buildings. Eng Struct 275:115152
5. Abdelaziz KM, Alipour A, Hobeck JD (2021) A smart façade system controller for optimized wind-induced vibration mitigation in tall buildings. J Wind Eng Ind Aerodyn 212:104601
6. Jafari M, Alipour A (2021) Methodologies to mitigate wind-induced vibration of tall buildings: a state-of-the-art review. J Build Eng 33:101582

7. Hareendran SP, Alipour A, Sarkar P (2023) Improving aerodynamic response of tall buildings using smart morphing facades. J Build Eng 80:107656
8. Hu G, Liu L, Tao D, Song J, Tse KT, Kwok KCS (2020) Deep learning-based investigation of wind pressures on tall building under interference effects. J Wind Eng Ind Aerodyn 201:104138
9. Kim B, Yuvaraj N, Sri Preethaa KR, Hu G, Lee DE (2021) Wind-induced pressure prediction on tall buildings using generative adversarial imputation network. Sensors 21(7):2515
10. Dokhaei B, Epackachi S, Torabian S (2021) Finite element modeling for predicting the cyclic response of cold-formed steel shear walls. In: 12th International congress on civil engineering, pp. 1–8. Ferdowsi University of Mashhad, Mashhad, Iran
11. Tsai LW, Alipour A (2023) Physics-informed long short-term memory networks for response prediction of a wind-excited flexible structure. Eng Struct 275:114968
12. Hareendran SP, Alipour A (2022) Prediction of nonlinear structural response under wind loads using deep learning techniques. Appl Soft Comput 129:109424
13. Wang H, Wu T (2020) Knowledge-enhanced deep learning for wind-induced nonlinear structural dynamic analysis. J Struct Eng 146(6)
14. Li B, Spence SMJ (2024) Deep learning enabled rapid nonlinear time history wind performance assessment. Structures 66:106810
15. Dokhaei B, Abdelaziz K, Shafei B, Sarkar P, Hobeck J, Alipour A (2025) Aeroelastic boundary layer tests of a 1:76 model of tall building and effects of adjacent building interference. J Wind Eng Ind Aerodyn 257:106006
16. Hochreiter S, Schmidhuber J (1997) Long short-term memory. Neural Comput 9(8):1735–1780
17. Dokhaei B, Shafei B, Alipour A (2026) Optimized design of steel structural systems for reduced carbon footprint and enhanced resiliency in wind. Eng Struct 350:121895
18. Dokhaei B, Alipour A, Abdelaziz K, Shafei B, Sarkar P, Hobeck J (2026) Cyber-physical smart morphing facades driven by reinforcement learning for wind-induced vibration control in tall buildings. J Wind Eng Ind Aerodyn 273:106394
19. Dokhaei B, Alipour A, Abdelaziz K, Shafei B, Sarkar P, Hobeck J (2026) Smart morphing facades for wind-induced vibration control in tall buildings: insights from large-scale aeroelastic wind tunnel testing. J Struct Eng. https://doi.org/10.1061/JSENDH/STENG-14939

Effect of Fusion of Displacement and Strain Modal Data on Information Entropy

S. Adarsh(✉), Shubham Chaudhary, and S. Ray-Chaudhuri

Indian Institute of Technology Kanpur, Kanpur, U.P. 208016, India
adarshss@iitk.ac.in

Abstract. System identification is often the first step in many structural control and health monitoring applications. Probabilistic system identification methods like Bayesian identification are powerful methods to handle the various modeling uncertainties replete in structural systems. Most studies use only displacement modal data for Bayesian identification. However, a fusion of displacement and strain modal data can be helpful since the additional information provided by the strain modal data can potentially reduce the uncertainty of the identified model parameters. This study uses a numerical analysis using a 2-D truss to understand the effect of the fusion of displacement and strain data on the posterior distribution of the parameters. Information entropy is used to quantify the uncertainty of the posterior distributions when different combinations of modal data are used. Results from the study show that the fusion of both kinds of data yields better information only for certain combinations of modal data and simulated sensor noise. Higher information was obtained only for the modal data of the second mode, the data combination of the first and second modes, and the data combination of the first, second, and third modes. Also, this higher information is obtained when the noise in the data has a higher coefficient of variation of 10%.

Keywords: Structural control · Probabilistic system identification · Data fusion · Information entropy · transitional Markov chain Monte Carlo.

1 Introduction

System identification [1] is an essential step for structural control [2] and damage detection [3]. The objective in system identification is to develop a high-fidelity mathematical model representative of the behavior of the actual real-life structure with the help of measurements. Probabilistic system identification methods, namely Bayesian methods, are effective for handling various uncertainties [4] in structural applications.

Recently, multi-sensor data fusion [5] is gaining importance in structural system identification. The fusion of data from various sensors is used to gain more information than using a single type of sensor. This additional information [6–7] can aid in making an inference. In addition to handling uncertainties, Bayesian methods can fuse data from various sensors.

This study investigates the Bayesian fusion of displacement and strain modal data. Displacement modal can be estimated from accelerometers, while strain modal data can

H. Beushausen et al. (Eds.): NTZR 2025, LNCE 873, pp. 981–991, 2026.
https://doi.org/10.1007/978-3-032-27698-8_86

be estimated from sensors like strain gauges. Accelerometers and strain gauges have the added advantage of not needing a local reference.

A numerical study is conducted using a planar truss. After estimating the samples of the posterior probability density functions (PDFs) by the transitional Markov chain Monte Carlo [8] method, the information entropy [9] is estimated to quantify the information. The evaluation of information entropy helps to understand if the fusion of data provides any advantage and also quantifies this advantage. Various combinations of modal data and the effect of noise are studied.

The rest of the article is organized as follows. Sections 2, 3 and 4 give a background about Bayesian fusion and the evaluation of information entropy. The numerical study is elaborated in Sect. 5, while Sect. 6 discusses the conclusions from the study.

2 Bayesian System Identification

Bayes' theorem provides a rigorous platform for handling uncertainties in system identification. The basic objective in Bayesian system identification [9] is to estimate the joint PDFs of the parameters of mathematical models that represent the behavior of a real-life system. Bayes' theorem is used with the measurements from the real-life system to achieve this objective. For example, one may have to estimate the stiffness parameters of the mathematical model of an actual building using dynamic measurements (like accelerations) from the actual building. The most common form of Bayes' theorem used for system identification is:

$$p(\boldsymbol{\theta}|Z, M) = \frac{p(\boldsymbol{\theta}|M)p(Z|\boldsymbol{\theta}, M)}{p(Z|M)} \tag{1}$$

where $p(\boldsymbol{\theta}| Z, M)$ is the posterior PDF of the parameters $\boldsymbol{\theta} \in \boldsymbol{\Theta} \in \mathbb{R}^{N_\theta \times 1}$ to be estimated given the data Z and model class M. The model class M represents all the assumptions and probability models used for formulating the mathematical model. The terms $p(\boldsymbol{\theta}| M)$ and $p(Z| \boldsymbol{\theta}, M)$ represent the prior PDF and the likelihood function, respectively. Information about $\boldsymbol{\theta}$ one had before taking the measurements, Z, is contained in the prior PDF, while the likelihood function quantifies the likelihood of the data given a particular value of $\boldsymbol{\theta}$. The denominator of Eq. 1 is not required for the estimation of $p(\boldsymbol{\theta}| Z, M)$. This term can be evaluated as $\int p(\boldsymbol{\theta}| M)p(Z| \boldsymbol{\theta}, M)\mathrm{d}\boldsymbol{\theta}$ where, the integral is over $\boldsymbol{\Theta}$.

After forming the numerator of Eq. 1, the various statistical properties of $p(\boldsymbol{\theta}| Z, M)$ can be evaluated by using approximate methods like Laplace's method [10] or by using stochastic simulation methods like Markov chain Monte Carlo methods (MCMC) [11]. This study uses a variant of MCMC known as the transitional Markov chain Monte Carlo (TMCMC). Even though MCMC methods are more computationally intensive than methods like Laplace's, MCMC is applicable for more general problems and can produce results with better accuracy.

3 Bayesian Fusion of Displacement and Strain Modal Data

The three types of modal data used in this study are: (i) natural frequencies, (ii) displacement mode shapes, and (iii) strain mode shapes. The mathematical model used for these modal data is the undamped eigenvalue problem [12] given by:

$$\mathbf{K}(\boldsymbol{\theta})\boldsymbol{\Phi}(\boldsymbol{\theta}) = \mathbf{M}\boldsymbol{\Phi}(\boldsymbol{\theta})\boldsymbol{\Omega}^2(\boldsymbol{\theta}) \tag{2}$$

and

$$\boldsymbol{\Psi}(\boldsymbol{\theta}) = \mathbf{B}\ \boldsymbol{\Phi}(\boldsymbol{\theta}). \tag{3}$$

In Eq. 2, $\mathbf{K}(\boldsymbol{\theta}) \in \mathbb{R}^{N_d \times N_d}$, $\mathbf{M} \in \mathbb{R}^{N_d \times N_d}$, $\boldsymbol{\Phi}(\boldsymbol{\theta}) \in \mathbb{R}^{N_d \times N_d}$, and $\boldsymbol{\Omega}^2(\boldsymbol{\theta}) \in \mathbb{R}^{N_d \times N_d}$ are the parametrized stiffness matrix, mass matrix, displacement mode shape matrix, and the diagonal matrix with squares of the circular natural frequencies. While, $\boldsymbol{\Psi}(\boldsymbol{\theta}) \in \mathbb{R}^{N_s \times N_d}$ and $\mathbf{B} \in \mathbb{R}^{N_s \times N_d}$ in Eq. 3 are the strain mode shape matrix and the displacement-strain transformation matrix [7], respectively. The value of N_d corresponds to the number of degrees-of-freedom and N_s is the number of locations at which strains are required.

Now, the prediction equations for the modal data are:

$$\begin{aligned} f_i^m &= f_i(\boldsymbol{\theta}) + \epsilon_{f_i^m} \\ \phi_i^m &= \phi_i(\boldsymbol{\theta}) + \epsilon_{\phi_i^m} \\ \boldsymbol{\psi}_i^m &= \boldsymbol{\psi}_i(\boldsymbol{\theta}) + \epsilon_{\boldsymbol{\psi}_i^m}. \end{aligned} \tag{4}$$

In the above equation, $f_i^m \in \mathbb{R}$ is the measured natural frequency of the ith mode in Hz, $\phi_i^m \in \mathbb{R}^{N_d \times 1}$ is the measured displacement mode shape of the ith mode, and $\boldsymbol{\psi}_i^m \in \mathbb{R}^{N_s \times 1}$ is the measured strain mode shape of the ith mode. While $f_i(\boldsymbol{\theta}) \in \mathbb{R}$, $\phi_i(\boldsymbol{\theta}) \in \mathbb{R}^{N_d \times 1}$, and $\boldsymbol{\psi}_i(\boldsymbol{\theta}) \in \mathbb{R}^{N_s \times 1}$ are the corresponding predicted modal parameters for a given value of $\boldsymbol{\theta}$ evaluated from Eqs. (2) and (3). The terms $\epsilon_{f_i^m} \in \mathbb{R}$, $\varepsilon_{\phi_i^m} \in \mathbb{R}^{N_d \times 1}$, and $\varepsilon_{\boldsymbol{\psi}_i^m} \in \mathbb{R}^{N_s \times 1}$ are stochastic terms that model the uncertainties in predictions. Uncertainties can arise due to modeling errors, noise in data, and even human errors. These stochastic terms are modeled as Gaussian PDFs: $\epsilon_{f_i^m} \sim \mathcal{N}(0, \sigma_i)$, $\epsilon_{\phi_i^m} \sim \mathcal{N}\left(0, \mathbf{C}_{\phi_i^m}\right)$, and $\epsilon_{\boldsymbol{\psi}_i^m} \sim \mathcal{N}\left(0, \mathbf{C}_{\boldsymbol{\psi}_i^m}\right)$.

So, the likelihood functions for the ith modal data are:

$$\begin{aligned} p(f_i|\boldsymbol{\theta}, M) &= \frac{1}{\sqrt{2\pi}\sigma_i} \exp\left[-\frac{1}{2}\left(\frac{f_i^m - f_i(\boldsymbol{\theta})}{\sigma_i}\right)^2\right] \\ p(\phi_i|\boldsymbol{\theta}, M) &= \frac{1}{(2\pi)^{\frac{N_d}{2}}\left|\mathbf{C}_{\phi_i^m}\right|^{\frac{1}{2}}} \exp\left[-\frac{1}{2}\left(\phi_i^m - \phi_i(\boldsymbol{\theta})\right)^T \mathbf{C}_{\phi_i^m}{}^{-1}\left(\phi_i^m - \phi_i(\boldsymbol{\theta})\right)\right] \\ p(\boldsymbol{\psi}_i|\boldsymbol{\theta}, M) &= \frac{1}{(2\pi)^{\frac{N_s}{2}}\left|\mathbf{C}_{\boldsymbol{\psi}_i^m}\right|^{\frac{1}{2}}} \exp\left[-\frac{1}{2}\left(\boldsymbol{\psi}_i^m - \boldsymbol{\psi}_i(\boldsymbol{\theta})\right)^T \mathbf{C}_{\boldsymbol{\psi}_i^m}{}^{-1}\left(\boldsymbol{\psi}_i^m - \boldsymbol{\psi}_i(\boldsymbol{\theta})\right)\right]. \end{aligned} \tag{5}$$

For fusion of modal data, $Z = \{Z_f, Z_\phi, Z_\psi\}$ and Eq. 1 becomes:

$$p(\boldsymbol{\theta}|Z, M) = \frac{p(\boldsymbol{\theta}|M)p(Z_f|\boldsymbol{\theta}, M)p(Z_\phi|\boldsymbol{\theta}, M)p(Z_\psi|\boldsymbol{\theta}, M)}{p(Z|M)} \tag{6}$$

where

$$p(Z_f|\boldsymbol{\theta}, M) = \prod_{j=1}^{N_m} p(f_j|\boldsymbol{\theta}, M)$$
$$p(Z_\phi|\boldsymbol{\theta}, M) = \prod_{j=1}^{N_m} p(\phi_j|\boldsymbol{\theta}, M) \quad (7)$$
$$p(Z_\psi|\boldsymbol{\theta}, M) = \prod_{j=1}^{N_m} p(\boldsymbol{\psi}_j|\boldsymbol{\theta}, M).$$

It can be seen from Eqs. 6 and 7 that Bayesian data fusion happens through the likelihood function. The contributions from various modal data can be fused in this manner, assuming that the noise terms in Eq. 4 are independent.

4 Evaluation of Information Entropy

Once $p(\boldsymbol{\theta}| Z, M)$ from Eq. 1 has been estimated, the information of $\boldsymbol{\theta}$ provided by Z can be quantified using information entropy. For brevity, information entropy will be referred to as entropy from this point. The entropy is calculated as [13]:

$$\mathscr{H}(\boldsymbol{\theta}|Z, M) = -\int \ln(p(\boldsymbol{\theta}|Z, M))p(\boldsymbol{\theta}|Z, M)d\boldsymbol{\theta} = -\mathbb{E}(\ln(p(\boldsymbol{\theta}|Z, M))) \quad (8)$$

where the integral is again over $\boldsymbol{\Theta}$ and $\mathbb{E}$ is the expectation operator. The lesser the value of entropy, more is the information. Entropy quantifies the uncertainty associated with $p(\boldsymbol{\theta}| Z, M)$. When N samples $\{\boldsymbol{\theta}_1, \boldsymbol{\theta}_2, \ldots, \boldsymbol{\theta}_N\}$ of $\boldsymbol{\theta}$ have been simulated by MCMC method, entropy can be estimated as follows:

$$\mathscr{H}(\boldsymbol{\theta}|Z, M) \approx \frac{-\sum_{k=1}^{N} \ln(p(\boldsymbol{\theta}_k|Z, M))}{N}. \quad (9)$$

5 Numerical Study

5.1 Numerical Model and Data Used

The 2-D truss with 13 members, shown in Fig. 1, is used for the numerical study. Each element is assumed to have a cross-sectional area and an elastic modulus of elasticity as 6.45×10^{-3} m^2 and 200×10^9 N/m^2, respectively. A lumped mass matrix with diagonal entries as 16×10^4 kg is assumed.

The undamped natural frequencies and mode shapes (disablement and strain) of the first four modes of the truss are considered here. These four modes have natural frequencies of 1.19, 2.35, 3.28, and 4.05 Hz. Only the stiffness matrix is parametrized as $\mathbf{K}(\boldsymbol{\theta}) = \sum_{i=1}^{4} \theta_i^e \mathbf{K}_i^e$. The parameter matrix $\boldsymbol{\theta} \in \mathbb{R}^{4 \times 1}$ consists of four elements scaling the stiffness contributions of the four substructures shown in blue, green, red, and black in Fig. 1. Elements in the four substructures are $e_1 = \{1, 8, 9\}, e_2 = \{2, 7, 10\}, e_3 = \{3, 6, 11,$

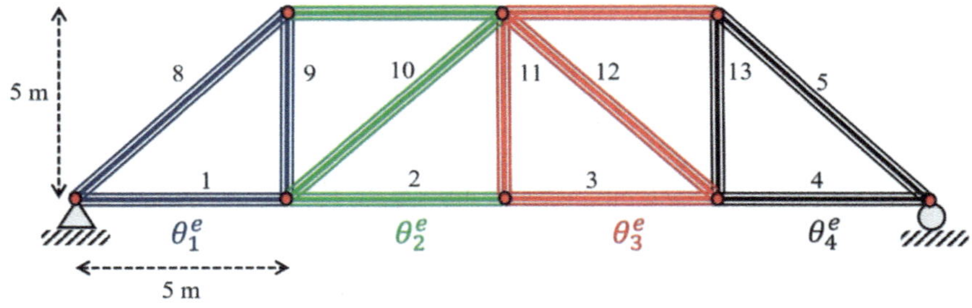

Fig. 1. Dimensions and element numbers of the 2-D truss used for the numerical study. The elements used for the four substructures are shown in blue, green, red, and black.

12}, and $e_4 = \{4, 5, 13\}$. The four substructures and the stiffness parameters associated with them are shown in Fig. 1. So, for the nominal model, $\boldsymbol{\theta} = \{1, 1, 1, 1\}^T$.The various combinations of modes are shown in Table 1, represented by binary values. Zero means that particular mode is not considered, and one means that mode is considered. Two sets of measurement noise are simulated: (i) 1% COV (coefficient-of-variation) and (ii) 10% COV. Further, four combinations of modal data are also considered, as shown in Table 2. The noise and combination of modal data are represented as Z_j^k where k represents the noise and j represents a combination of modal data. When $k = 1$, the data has 1% COV noise, while $k = 2$ means 10% COV noise. So, the notation $\{C_1, Z_2^1\}$ implies that data of natural frequencies and displacement mode shapes of only the first mode is used, and that noise of 1% COV is added to the raw data.

Table 1. Combinations of modes used.

Combination	Mode 1	Mode 2	Mode 3	Mode 4
C_1	1	0	0	0
C_2	0	1	0	0
C_3	0	0	1	0
C_4	0	0	0	1
C_5	1	1	0	0
C_6	1	1	1	0
C_7	1	1	1	1

Modal data Z_1^k is only considered for C_7 since this combination provides four natural frequencies to identify the four parameters. So, C_7 has eight possible data sets, and the remaining combinations have six possible data sets, making a total of 44 data sets.

5.2 Results and Discussion

For the evaluation of entropy, samples from $p(\boldsymbol{\theta}| Z, M)$ are drawn by using TMCMC, an adaptive MCMC sampler, for all 44 data cases. Uniform priors are used, such that

Table 2. Combinations of modal data used.

Modal Data	Natural Frequencies	Displacement mode shape	Strain mode shape
Z_1^k	1	0	0
Z_2^k	1	1	0
Z_3^k	1	0	1
Z_4^k	1	1	1

the entire manifold of the likelihood function is captured, and the covariance matrices of the prediction errors from Eq. 4 have been formed according to the COV of the noise used and the raw data of the nominal model. As an example, the uniform prior used for $\{C_1, Z_2^1\}$ has $\boldsymbol{\theta}_{\min} = \{0.9, 0.9, 0.9, 0.9\}^T$ and $\boldsymbol{\theta}_{\max} = \{1.1, 1.13, 1.12, 1.1\}^T$. The limits of priors used for all 44 cases are not shown for brevity; an interested reader can contact the authors to get them. It is observed that for $\{C_1, Z_1^1\}$ and $\{C_1, Z_1^2\}$, the manifold of $p(\boldsymbol{\theta}| Z, M)$ is still unbounded due to a lack of data points.

Once the numerator of Eq. 1 has been formed, the TMCMC is used to get 1000 samples of $\boldsymbol{\theta}$. The entropy values are calculated according to Eq. 9. Further, 1000 runs of TMCMC are used, and the mean values of entropy across all the 1000 runs are evaluated to account for variability in estimates. The scatter plots, histograms, and correlation coefficients for $\{C_1, Z_4^1\}$ are shown in Figs. 2, 3, and 4, respectively. The run with the entropy closest to the mean value is used for the plots. It can be seen that the parameter samples exhibit a strong positive correlation. This positive correlation is attributed to the shape of the manifold of the posterior probability distribution function. Further, the spread of the parameter samples is also similar, and each has a COV close to 2%.

The mean values are listed in Table 3. Figure 5 illustrates the bar plots of the negative of entropies for all modal data combinations when the noise COV is 1% and 10%. The higher the negative entropy values in the bar plots, the higher the information. As expected, it can be seen that for 10% COV noise, the bar plots are lower. These bar plots are significantly lower Z_1^1 and Z_1^2, i.e., when only natural frequencies are used. When individual modes are considered, Modes 3 and 4 contribute more information than Modes 1 and 2. This contribution can depend on the way one parametrizes the stiffness matrix.

From Fig. 5a, b, it can be observed that combining both the modal data (Z_4^k) does not always yield the highest information. For C_3 and C_4, the modal data Z_2^k and Z_3^k results in the highest information, respectively. Define the term relative difference as the relative increase of the Z_j^k with the highest negative entropy compared to the Z_j^k with the second-highest negative entropy within a given C_i. For data, $\{C_3, Z_2^1\}$ and $\{C_4, Z_3^1\}$ the relative difference is 6.9% and 20.3%, respectively. While for $\{C_3, Z_2^2\}$ and $\{C_4, Z_3^2\}$, the relative difference is 16.9% and 53.9%, respectively. Thus, the relative difference is greater for higher noise COV. The data sets $\{C_3, Z_2^2\}$ and $\{C_4, Z_3^2\}$ provides considerably more information (> 10%) compared to other data sets in C_3 and C_4, respectively.

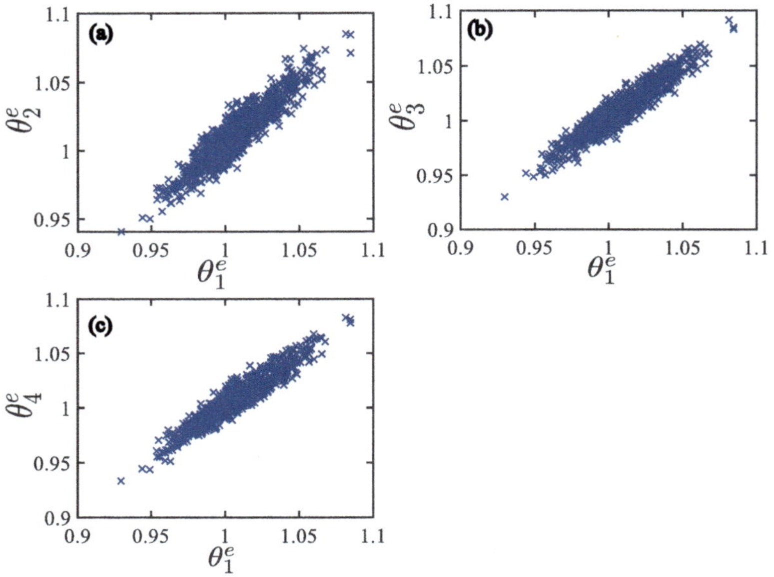

Fig. 2. Posterior distribution samples for $\left\{C_1, Z_4^1\right\}$. (a) $\theta_1^e - \theta_2^e$ samples, (b) $\theta_1^e - \theta_3^e$ samples, and (c) $\theta_1^e - \theta_3^e$ samples.

Figure 5(b) shows that the negative entropy of Z_4^2 is maximum for C_1, C_2, C_5, C_6, and C_7. Thus, for higher noise in data, the information provided by the fusion of displacement and strain data is higher for five C_i s. The data set $\left\{C_2, Z_4^2\right\}$ show maximum relative difference of 19.2%, while $\left\{C_1, Z_4^2\right\}$, $\left\{C_5, Z_4^2\right\}$, $\left\{C_6, Z_4^2\right\}$, and $\left\{C_7, Z_4^2\right\}$ the relative differences are 4.1%, 9.3%, 9.1%, and 0.8%, respectively. The data sets that provide considerably more information may be useful in probabilistic damage detection.

To further explore the impact of noise, the relative reductions in entropy when increasing the noise level from 1% COV to 10% COV are computed and shown in Fig. 6. Figure 6 presents the relative decrease as a percentage, excluding the case of $\left\{C_7, Z_1^k\right\}$. The figure shows that $\left\{C_1, Z_2^k\right\}$ is most affected by noise and $\left\{C_4, Z_3^k\right\}$ the least affected. In four combinations C_1, C_2, C_6, and C_7, the data set with Z_4^k is affected least by noise.

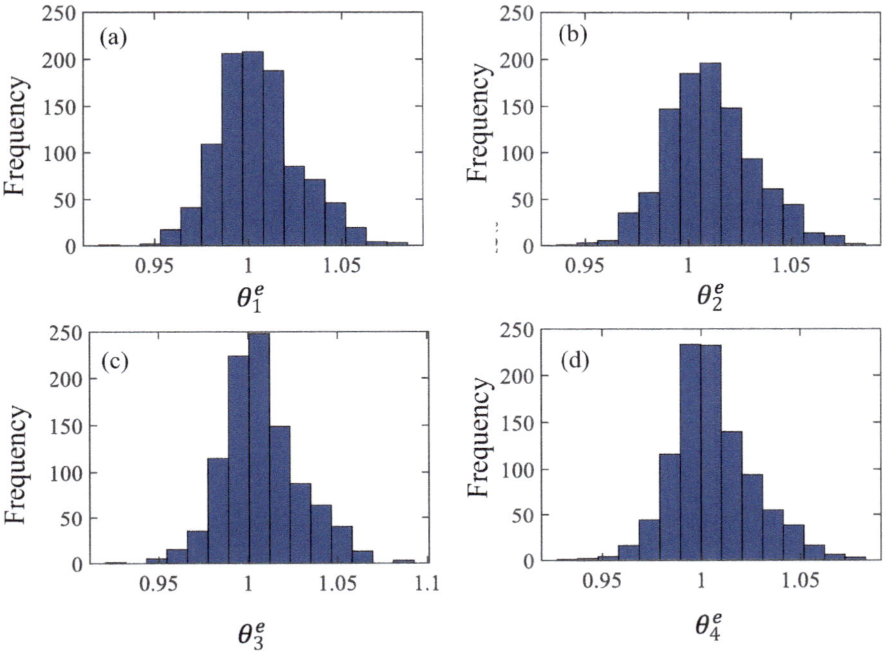

Fig. 3. Histograms for $\left\{C_1, Z_4^1\right\}$ samples. (a) θ_1^e samples, (b) θ_2^e samples, (c) θ_3^e samples, and (d) θ_4^e samples.

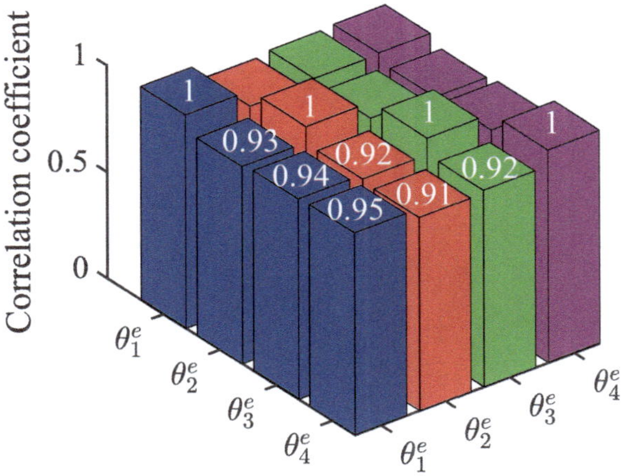

Fig. 4. Correlation coefficients $\left\{C_1, Z_4^1\right\}$ samples.

Table 3. Mean values of entropies.

	Z_1^1	Z_2^1	Z_3^1	Z_4^1	Z_1^2	Z_2^2	Z_3^2	Z_4^2
C_1	----	– 9.8	– 13.4	– 13.6	----	– 0.7	– 4.7	– 4.9
C_2	----	– 13.6	– 15.7	– 15.8	----	– 5.3	– 5.5	– 6.5
C_3	----	– 22.1	– 20.4	– 20.7	----	– 13.8	– 11.1	– 11.8
C_4	----	– 19.4	– 23.7	– 19.7	----	– 10.2	– 15.6	– 10.1
C_5	----	– 13.9	– 15.9	– 16.8	----	– 6.5	– 6.9	– 7.5
C_6	----	– 22.6	– 21.5	– 21.7	----	– 12.7	– 12.3	– 13.8
C_7	– 4.3	– 25.2	– 23.4	– 24.6	5.3	– 15.3	– 13.9	– 15.4

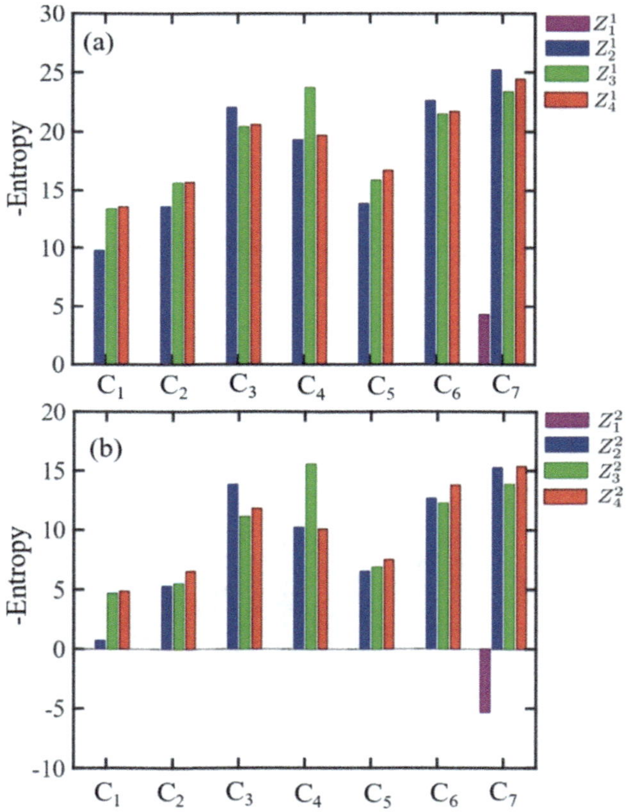

Fig. 5. Mean negative entropy values calculated for various modal data combinations, (a) 1% COV noise and (b) 10% COV noise.

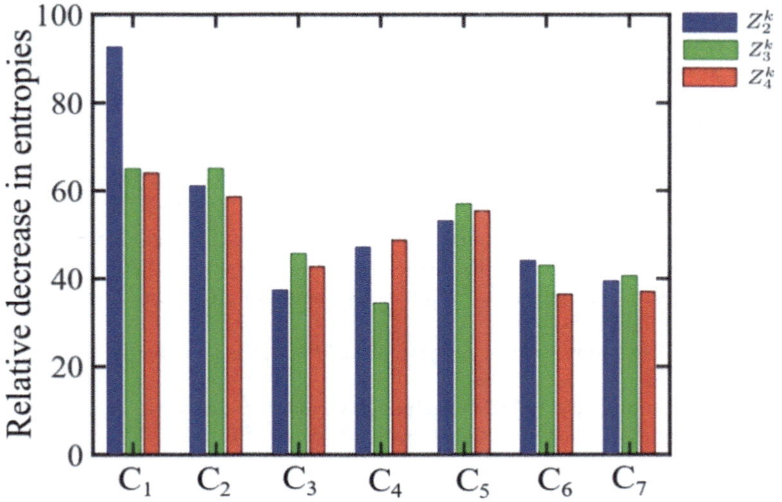

Fig. 6. Relative decrease in entropies in percentage for various modal data combinations when one traverses from 10% COV noise to 1% COV noise.

6 Conclusions

This study investigates the effect of Bayesian fusion of displacement strain model data on the entropy of the posterior distributions. A planar study is used as the basis for the study, and the entropy has been evaluated using the samples obtained from the transitional Markov Chain Monte Carlo. For the planar truss, it is observed that the posterior probability density functions do not always yield the highest information when both types of modal data are used. Here, information is defined as the negative of entropy. Notably, significantly higher information was recorded only for three mode combinations under higher noise conditions (10% COV). Specifically, combinations C_2, C_5, and C_6 showed increases of 19.2%, 9.3%, and 9.1%, respectively, compared to the following best mode combination. Combination C_2 included data solely from Mode 2, C_5 included data from Modes 1 and 2, and C_6 included data from Modes 1, 2, and 3. Thus, utilizing all available modal data can provide considerably more information about the parameters to be estimated for certain mode combinations, particularly when the data contains higher noise levels.

References

1. Tangirala AK (2018) Principles of system identification: theory and practice. CRC Press
2. Spencer BF Jr, Nagarajaiah S (2003) State of the art of structural control. J Struct Eng 129(7):845–856
3. Farrar CR, Worden K (2007) An introduction to structural health monitoring. Philos Trans R Soc A Math Phys Eng Sci 365(1851):303–315
4. Bulleit WM (2008) Uncertainty in structural engineering. Pract Period Struct Des Constr 13(1):24–30

5. Wu RT, Jahanshahi MR (2020) Data fusion approaches for structural health monitoring and system identification: past, present, and future. Struct Health Monit 19(2):552–586
6. Adarsh S, Ray-Chaudhuri S (2024) An investigation of optimal sensor locations for multi-type sensors considering modeling uncertainties by Bayesian system identification. J Sound Vib 577:118324
7. Adarsh S, Panda J, Ray-Chaudhuri S (2024) Bayesian updating of the displacement-strain transformation matrix. In: Journal of physics: conference series, vol 2647, no 19, p 192021. IOP Publishing, 1 June 2024
8. Ching J, Chen YC (2007) Transitional Markov chain Monte Carlo method for Bayesian model updating, model class selection, and model averaging. J Eng Mech 133(7):816–832
9. Beck JL (2010) Bayesian system identification based on probability logic. Struct Control Health Monit 17(7):825–847
10. Au SK (2017) Operational modal analysis. Modeling, Bayesian inference, uncertainty Laws. Springer, Berlin
11. Lye A, Cicirello A, Patelli E (2021) Sampling methods for solving Bayesian model updating problems: a tutorial. Mech Syst Signal Process 159:107760
12. Chopra AK (2007) Dynamics of structures. Pearson Education India
13. Yuen KV (2010) Bayesian methods for structural dynamics and civil engineering. John Wiley & Sons

Benchmarking the State of Networks with a Low-Cost Method Based on Reservoir Computing

Felix Simon Reimers[1(✉)], Carl-Hendrik Peters[1,2], and Stefano Nichele[1,3]

[1] Østfold University College, 1757 Halden, Norway
{felix.s.reimers,stefano.nichele}@hiof.no
[2] Lucerne University of Applied Sciences and Arts, 6002 Luzern, Switzerland
[3] Oslo Metropolitan University, 0130 Oslo, Norway

Abstract. Using data from mobile network utilization in Norway, we showcase the possibility of monitoring the state of communication and mobility networks with a non-invasive, low-cost method. This method transforms the network data into a model within the framework of reservoir computing and then measures the model's performance on proxy tasks. Experimentally, we show how the performance on these proxies relates to the state of the network. A key advantage of this approach is that it uses readily available data sets and leverages the reservoir computing framework for an inexpensive and largely agnostic method. Data from mobile network utilization is available in an anonymous, aggregated form with multiple snapshots per day. This data can be treated like a weighted network. Reservoir computing allows the use of weighted, but untrained networks as a machine learning tool. The network, initialized as a so-called echo state network (ESN), projects incoming signals into a higher dimensional space, on which a single trained layer operates. This consumes less energy than deep neural networks in which every weight of the network is trained. We use neuroscience inspired tasks and trained our ESN model to solve them. We then show how the performance depends on certain network configurations and also how it visibly decreases when perturbing the network. While this work serves as proof of concept, we believe it can be elevated to be used for near-real-time monitoring as well as the identification of possible weak spots of both mobile communication networks as well as transportation networks.

Keywords: reservoir computing · echo state network · machine learning · mobile network · network state monitoring

1 Introduction

Spotting faults and bottlenecks in communication networks or power grids and managing transportation networks are of high interest and increasingly data-driven, with a growing amount of machine learning related approaches being used in those fields [1–3]. This paper presents a proof-of-concept for an approach that relies on a computational perspective to infer knowledge about the given network - in our case the transfer rate between municipalities in Norway as measured by the cell tower network.

H. Beushausen et al. (Eds.): NTZR 2025, LNCE 873, pp. 992–1004, 2026.
https://doi.org/10.1007/978-3-032-27698-8_87

On the technical side the approach relies on the framework of reservoir computing, or more explicitly echo state networks [4]. Conceptually, we take a computational perspective, treating the network data as a computational substrate. While there is extensive work on fault detection in communication and other networks as well as on the identification of bottlenecks, there is to our knowledge no approach using the reservoir computing framework as we do. Instead of using graph theory to examine the networks itself or using rule-based or data-driven approaches that aim at classifying the state of the network, we aim at embedding the network in a model with as little changes as possible and then using the performance of the model as a benchmark to infer the state of the network itself. As such our overarching research question is: How does the computational performance of a given network reflect the state of that network? The content of the tasks itself is not important, as it is just a proxy. Important is that the performance of the model on those tasks can be used as a benchmark.

The approach is easily adaptable to different networks, as long as data in the form of an adjacency list or matrices is present. It is agnostic to the specific hardware and software technologies used in the network. Furthermore, the resulting models can be trained with little effort compared to artificial neural networks.

While our work is on a more conceptual level, we believe it to be a possible step towards a framework that helps to examine different networks of civil infrastructure, to assess the state they are in and to improve robustness and throughput. In order to test this, our framework needs to be integrated with further data like network failure logs to test the model.

The code to reproduce our results can be found in the following anonymized repository: https://github.com/Carasar/nor_conn2res_public.

The rest of the paper is structured as follows: In the second section, reservoir computing in general and the conn2res toolbox in particular are explained and some related work is presented. The third section explains our framework in more detail, while the fourth section explains our experimental setup. In the fifth section, the results of our work are presented and in the sixth section the results and possible future work are discussed.

2 Background and Related Work

This chapter aims at giving the reader an overview of reservoir computing, the used toolbox and lastly some selected related works.

2.1 Reservoir Computing

The concept of reservoir computing was developed independently in [4, 5], naming the resulting models liquid state machine and echo state network respectively. We are using the second term throughout this paper, as this model is based on recurrent neural networks with weight-matrix multiplications and activation functions, while the liquid state machine has the - neurologically more plausible - integrate-and-fire neurons. As is pointed out in [5] and has been surveyed in [6], reservoir computing is largely agnostic to the used reservoir, as long as its internal dynamics follow certain rules, which means that a literal bucket of water or other physical systems can be used as reservoirs. When using a recurrent network as we do, the principle is as follows: A recurrent network is initialized, and its topology and weights are frozen. Input data, often in the form of timeseries, is multiplied with input weights and then passed to the nodes in the network iteratively in every time step. As the signal traverses through the recurrent network with every time step, it is being multiplied with the weights of the network and integrated and evaluated by the activation function at every node. Eventually, nodes are read out by an additional layer of neurons. Only the weights of this read-out layer are trained for the given task [4]. As only relatively few nodes have to be trained, the computational effort for training the model is much smaller than for a standard artificial neural network. A visualization of such an echo state network can be seen in Fig. 1.

A specific hyperparameter of an echo state network is the alpha value. This value scales all the weights within the reservoir. A specific choice of an alpha value can, if other criteria are met, put the network into the name giving echo state, which grants the best computational performance. For mathematical details, the reader is referred to [4].

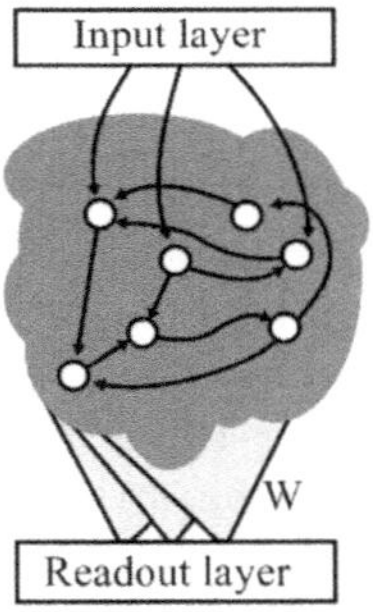

Fig. 1. A visualization of the reservoir computing model. Taken with approval from [7].

2.2 Conn2Res and Benchmarks

As previously discussed, reservoir computing is rather agnostic with regards to the used reservoir. The conn2res toolbox [8] has been designed to allow an easy implementation of connectomes as echo state networks or liquid state machines. A connectome is a mapping of the connections within biological neurons. For this reason, the toolbox allows to choose specific input and output nodes for the reservoir based on a labeling, which could for example be different regions of the brain. While our network is not biological, conn2res can be used without impediments for our purposes.

The toolbox also includes are number of implemented tasks that the model can directly be trained on. While one of them has been originally implemented, the majority stem from the reservoirpy [9] and the neurogym [10] repositories. Especially the neurogym tasks take inspiration from neuroscience. The two tasks chosen by us, the Perceptional Decision Making and the Go/No-Go task are both from the latter repository.

Our choice of neuro-inspired benchmarks is motivated by our ongoing research on the performance of biological or biology-inspired computational substrates. The used benchmarks are testing the ability of the model to measure a signal over time, integrate the signal and make a decision based on this. As such, we believe that neuro-inspired tasks can be used to benchmark a broad spectrum of potential computational abilities of reservoir computation models.

2.3 Related Works

The paper [6] grants an overview of different models that can be used within the reservoir computing framework. It also takes a computational perspective on the matter akin to ours.

Fault management in networks with machine learning based approaches are reviewed in [11], where fault management is described as an expensive task. AI support started with expert systems in the 1990s; more recent approaches based on data science are divided by the authors into data mining and machine learning. The described data-mining approaches include techniques resulting in graphs [12], but those are the product of their examination, whereas in our approach an existing weighted graph is used as the starting point of the model. Data mining approaches are focused on finding patterns, whereas we assume that characteristics of the network to be implicitly reflected in the computational performance. Following the definitions in [13], our approach might be used in the fault detection or localization of running systems. Furthermore, the measured degradation might help to estimate the severity of a fault or in the identification of bottlenecks. Similar to [14], our approach does not rely on a priori knowledge of the expected faults.

Multipath transport problems have to handle the existence of bottlenecks [15] presents the first learning-based algorithm to tackle this task. While our framework does not solve pathing problems, it may help at identifying bottlenecks. This is also a relevant problem when dealing with power grids [2].

Previous work [16] has used telecommunications data to explore traffic patterns. As they describe, mobile network data is a rich source of information. While their work focuses on a smaller scale, both temporally and spatially, our aggregated data set covers the whole of Norway at the commune level over a quarter of a year. As our framework is agnostic to the network used, we believe it could be applied to such smaller scales too.

3 Framework

In this section we will outline how we implemented an echo state network based on our network data, which will be explained further. Next, the used benchmarks are explained in more detail.

3.1 Echo State Network

As introduced above, the reservoir computing framework allows to implement a given network as an echo state network with relative ease. Transportation networks, power grids, cell tower networks and others can be described mathematically as graphs consisting of edges and nodes. Both nodes and edges can be assigned further attributes. In our case, the transfer rates are treated as the weights of the edges, similar to the

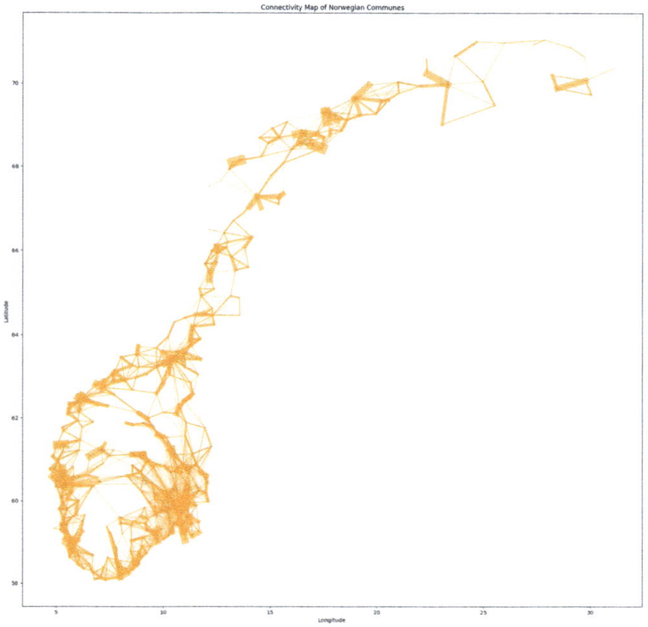

Fig. 2. A visualization of the mobile network utilization data. The individual nodes are arranged according to their longitude/latitude coordinates. The shape of mainland Norway is clearly visible, as is increased fluctuation around Oslo and the western coast.

weights in an artificial neural network. Other possible values that could be assigned are for example the distances between cell towers. The nodes in our model integrate the incoming weighted signals and apply an activation function. As the transfer rates are measured from source to target municipality (instead of just measuring the net transfer), our graph contains directed edges, but there can be edges going in opposing directions with different weights.

3.2 Network Data

The network data we use consists of adjacency lists between the 356 communes in Norway. For two days, Tuesday and Saturday, the transfer rate measured between the cell towers of a communes is averaged for the first quarter of 2021. As a measurement period goes for six hours, four datasets for every day exist, resulting in eight adjacency lists in total. Additionally, the longitude and latitude of the communes can be used for visualizations, which allows to visualize the network over a map of Norway, as done in Fig. 2.

3.3 Neuro-Inspired Tasks

The tasks used in our approach are taken from the neurogym [10] repository and are inspired by the field of neuroscience. The data sets can be generated and are used as a benchmark of the computational performance of our model. The tasks can be clustered as classification, regression and memorization. The two tasks used, Perceptional Decision Making and Go/No-go are both classification tasks. They consist of time series which the model has to integrate or remember in order to score well.

In Perceptional Decision Making, the input data consists of multiple time series which can be split into segments, as can be seen in Fig. 3. Within every segment, the signals x_2 and x_3 are noisy timeseries of which one is one average larger than the other. x_1 is the fixation whose value indicates whether or not a decision has to be made in this timestep. y is the ground truth: It is zero most of the time, but as a segment comes to its end and x_1 drops to zero, y either becomes "1" or "2", depending on whether the signal x_2 or x_3 is larger on average in this segment. To perform well in this task, a model has to integrate the two signals over the timesteps of a segment, and as the fixation drops has to output the right class.

The Go/No-go task equally consists of three input signals, a fixation x_1 and the two timeseries x_2 and x_3 and a ground truth y. It is again split into segments. At the start of a segment, either x_2 or x_3 spike, giving a "go" or "no-go" signal. Some timesteps later, the model has to classify which signal was perceived earlier.

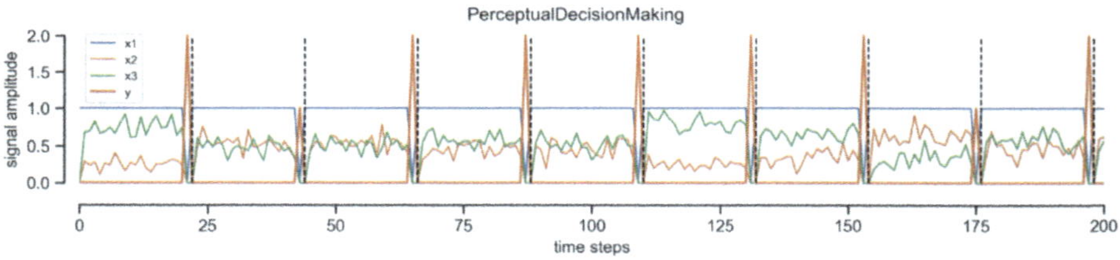

Fig. 3. The timeseries of the input data (x_1, x_2, x_3) and output data (y) of the Perceptual Decision Making task. As can be seen, the time series can be split into segments and at the end of each segment a classification for class "1" or "2" has to be made.

4 Experimental Details

In our model, the input signal is not passed to all nodes. Instead, a number of nodes corresponding to the dimension of the input, i.e. three nodes, are chosen for input and a total of six nodes are connected from the reservoir to the read-out layer. The choice of the nodes is either random, or informed by the in- and out-degree, where the nodes with the highest out-degree are chosen as input nodes and the one with the highest in-degree as output nodes. The former corresponds to a situation without additional information about the network. In the latter case, some knowledge about the network is assumed, allowing the selection of nodes that maximize or increase signal transmission through the network. This in turn should increase the performance of the model. Signals passed to or from the reservoir are not multiplied with a weight. For the activation function inside we chose tan$h$.

As mentioned above, the reservoir's weights can be scaled with an alpha hyperparameter. In preliminary experiments, the performance was calculated for a range of alpha values. For the results, the alpha corresponding to the best performance was chosen.

In order to contextualize the performance of a model, an appropriate baseline has to be found. For that, the connections of the original network are randomized, maintaining the degree distribution of the nodes and to some extent the weight distribution. The resulting null model has a different topology which may result in different performances. As this was not the focus of our work, we do not report on the performance of the null models, but the experimental results can be reproduced with the github repository.

We measured the performance of the model in the face of perturbations to the network. To that end, three randomly sampled nodes were deleted from the network iteratively. It was made sure that the deleted nodes are not part of the input or output nodes. In every iteration, the model was trained on task data and the performance measured.

Three loss functions were used to measure the performance: A standard f1-score and a balanced accuracy score taking the whole time series into account, including those parts where the fixation is one. The filtered accuracy score is custom and only considers those parts of the time series where the fixation is zero.

5 Results

In this section we compare the performance of the model with input/output nodes being chosen randomly or informed by the node degree. The model is then perturbed as described in the previous section. We start with the Go/No-go task and continue with the Perceptional Decision Task. We chose to display the results for the former task for the Saturday data set and for the latter task for the Tuesday data set. More results can be found with the github repository and qualitatively align with the presented work here.

As can be seen in Figs. 4 and 5, the performance of the model derived from the Saturday data on the Go/No-go task is initially high, corresponding to an unperturbed network. Iteratively perturbing the network leads to a drop in the measured performance. This is most prominent for the balanced accuracy store and least for the filtered accuracy score. In the case of the informed choice of input and output nodes, the networks from timeframe 1 and 3 on Saturday have a very visible drop in performance around step 25 with little variation as seen in Fig. 4, whereas the two remaining timeframes show a clear drop in average performance from time step 40 for the balanced accuracy and f1 score. In the networks with randomly chosen input and output nodes, the decrease in performance is overall less distinctive. Here, the perturbations often lead to a decrease of the average performance, but also to an increase in fluctuation such that high performance might still be reached as can be seen in Fig. 5. Only for the third time frame of Saturday a clear drop in performance can be observed very early around the 5th time step. For the first and third time frame, the average performance gradually decreases with increasing variation. In the second time frame, the balanced accuracy and f1 score drop early, while the filtered accuracy oscillates.

For the Perceptual Decision Making task, the results have significantly less variation overall, with little difference between the informed and randomized choice of nodes. Again, the balanced accuracy score has the most distinct behavior, but for all timeframes except the first the other two scores drop to zero as well. Figure 6 depicts the performance for Tuesday with an informed choice of nodes. The drop in performance is visible either around the 25th time step for timeframes two and four, for the 40th time step for time frame one or around time step 60 for the third time frame. When the input/output nodes are chosen randomly like in Fig. 7, the drop occurs around the 20th iteration for time frames one and two or in the 40th or 45th iteration for time frames four and three respectively.

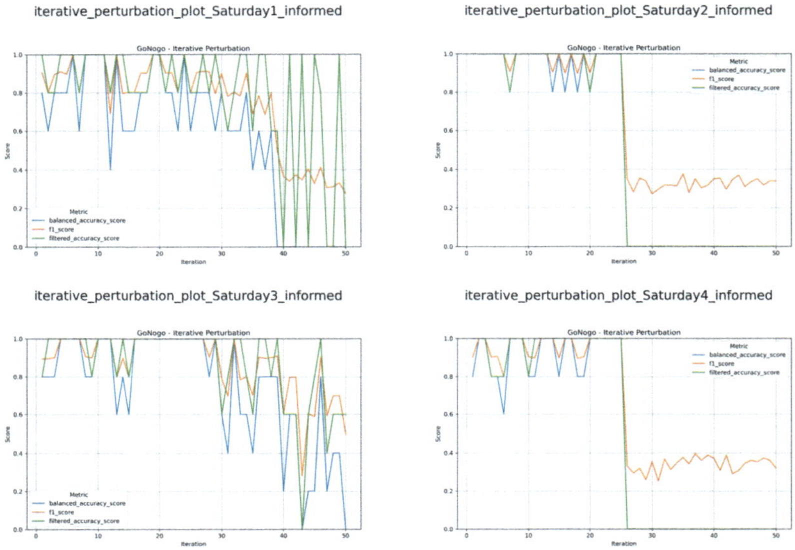

Fig. 4. Performance measured with balanced accuracy score, f1-score and a filtered accuracy score on the networks derived from the 4 timepoints on Saturday for the Go/No-go task. Iteratively, the network is perturbed in every step. The input and output nodes are chosen informed by the degree of the nodes.

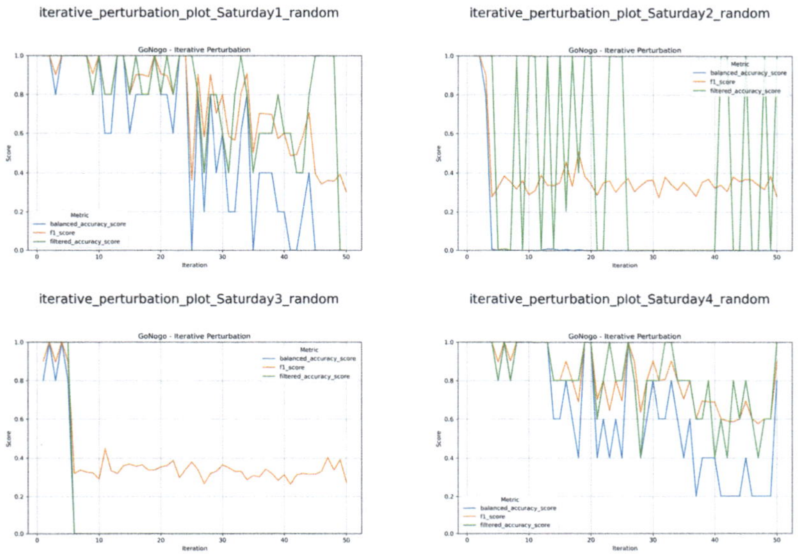

Fig. 5. Performance measured with balanced accuracy score, f1-score and a filtered accuracy score on the networks derived from the 4 timepoints on Saturday for the Go/No-go task. Iteratively, the network is perturbed in every step. The input and output nodes are chosen randomly.

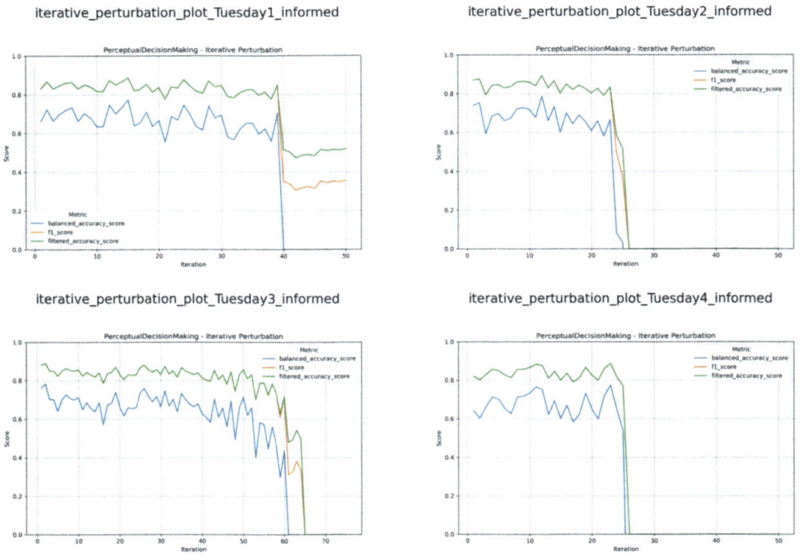

Fig. 6. Performance measured with balanced accuracy score, f1-score and a filtered accuracy score on the networks derived from the 4 timepoints on Tuesday for the Perceptional Decision Making task. Iteratively, the network is perturbed in every step. The input and output nodes are chosen informed by the degree of the nodes.

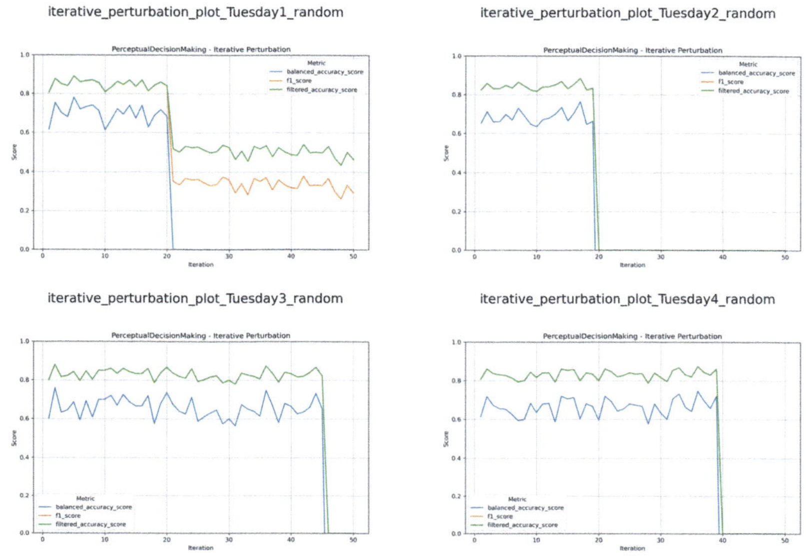

Fig. 7. Performance measured with balanced accuracy score, f1-score and a filtered accuracy score on the networks derived from the 4 timepoints on Tuesday for the Perceptional Decision Making task. Iteratively, the network is perturbed in every step. The input and output nodes are chosen randomly.

6 Discussion and Summary

The results show that the performance of our models reflects the increasing perturbation over time. The time step from which this becomes visible as well as the variation in the performance measure, depends on the chosen task as well as the time frame and the performance function. While this justifies our computational perspective, further investigation is necessary to both broaden and consolidate our results. Due to the inherent stochasticity of our framework and the difficulty of analyzing how a signal traverses a network over time, we can so far only present empirical results in this work.

By integrating more detailed network data, like failure logs, the framework can be tested in real-world settings. Application to a real-world situation will be necessary to validate the currently only conceptual work and should be a focus for future work. Furthermore, this will allow to compare our approach to other applied methods from the literature, where a focus should lie on both the efficiency but also the effectiveness of our model.

Further work should focus on examining the relationship between task performance and network state further, especially with regard to the localization of points of interest within the network. A possibility for this could be to vary the chosen nodes for input, output or artificial perturbation.

Also, the null model's performance should be included more thoroughly into the examinations, as a common baseline makes comparisons between networks of different sizes possible.

Combining the data of different networks like in [16] might allow to infer knowledge of one network through the performance benchmark of another. Trying the approach with different networks, either the same type of network form different sources or different types, like transportation networks or power grids, will show how agnostic the framework really is.

A reversed approach where nodes are added to the network to increase or stabilize the performance might be approaches that can be adopted to increase the robustness and resilience in actual networks. For example, the maximal largest power supply of a power grid after a failure might be increased [2].

Acknowledgements. We would like to thank Kenth Engø-Monsen for supplying and explaining the network utilization data set to us.

References

1. Boutaba R et al (2018) A comprehensive survey on machine learning for networking: evolution, applications and research opportunities. J Internet Serv Appl 9(1):1–99
2. Tu H, Zhang X, Xia Y, Fengqiang G, Sheng X (2022) Bottlenecks identification and resilience improvement of power networks in extreme events. Front Phys 10
3. Patil P (2022) Applications of deep learning in traffic management: a review. Int J Bus Intell Big Data Anal 5(1):16–23
4. Jaeger H (2001) The "echo state" approach to analysing and training recurrent neural networks-with an erratum note'. GMD Technical Report 148. German National Research Center for Information Technology, Bonn, Germany
5. Maass W, Natschläger T, Markram H (2002) Real-time computing without stable states: a new framework for neural computation based on perturbations. Neural Comput 14(11):2531–2560
6. Tanaka G et al (2019) Recent advances in physical reservoir computing: a review. Neural Netw 115:100–123
7. Nichele S, Molund A (2017) Deep learning with cellular automaton-based reservoir computing
8. Suárez LE et al (2024) Connectome-based reservoir computing with the conn2res toolbox. Nat Commun 15(1):656
9. Trouvain N et al (2020) Reservoirpy: an efficient and user-friendly library to design echo state networks. In: International conference on artificial neural networks. Springer International Publishing, Cham
10. Croconi G, et al. NeuroGym: an open resource for developing and sharing neuroscience tasks. https://neurogym.github.io/. Last accessed on 04 May 2025
11. Nouioua M et al (2021) A survey of machine learning for network fault management. In: Machine learning and data Mining for Emerging Trend in cyber dynamics: theories and applications. Springer International Publishing, Cham, pp 1–27
12. Lozonavu M, Vlachou-Konchylaki M, Huang V (2017) Relation discovery of mobile network alarms with sequential pattern mining. In: 2017 international conference on computing, networking and communications (ICNC). IEEE
13. łgorzata Steinder M, Sethi AS (2004) A survey of fault localization techniques in computer networks. Sci Comput Program 53 2: 165–194
14. Kiciman E, Fox A (2005) Detecting application-level failures in component-based internet services. IEEE Trans Neural Netw 16(5):1027–1041

15. Dong E et al (2023) SmartSBD: smart shared bottleneck detection for efficient multipath congestion control over heterogeneous networks. Comput Netw 237:110047
16. Yusuf O, Rasheed A, Lindseth F (2024) Exploring urban mobility trends using cellular network data. In: The international conference on net-zero civil infrastructures: innovations in materials, structures, and management practices (NTZR). Springer Nature Switzerland, Cham

Task-Based Cost Prediction for Seismic Retrofitting of Bridges Using Support Vector Regression

Ravi Shastri[1,2], Samit Ray-Chaudhuri[1], and Behrouz Shafei[2(✉)]

[1] Indian Institute of Technology Kanpur, Kanpur, Uttar Pradesh 208016, India
[2] Iowa State University of Science and Technology, Ames, IA 50011, USA
shafei@iastate.edu

Abstract. An increase in the consequences of natural hazards in the last decade demands a robust and rapidly adaptive recovery framework. Developing such a framework requires identifying the criticality of an infrastructure and predicting potential damage incurred to it even before event occurrence. Of the many ways to represent criticality and damage, cost-based representation is one of the most convenient methods, owing to the ease of quantification and the decision-supportive framework it can facilitate. Focusing on earthquake events, they are responsible for numerous instances of moderate to severe damage to many infrastructures, including bridges. Predicting structural damage in case of any such event would greatly benefit the decision-makers. This work performs a case study of task-based cost prediction to retrofit a bridge damaged during an earthquake. The retrofitting cost prediction is performed with the help of a machine learning technique, i.e., support-vector regression (SVR). SVR is an extension of the popular supervised classification technique, support vector machines (SVM). While SVM is utilized primarily for binary classification tasks, SVR is used for high-dimensional non-linear regression problems. Compared to linear regression, where the goal is error minimization, SVR fits a function that best represents the data within a specified error margin. A function relating the intensity of an event to the loss at the event occurrence would be highly beneficial in predicting losses for any such event and help in improved planning of relief works.

Keywords: Bridges · Support Vector Regression · Earthquake · Recovery · Cost

1 Introduction

One of the most essential elements of a robust and decisive framework for resilience-based design is identifying the damage level and predicting losses in a post-event scenario for extreme events. An early-stage damage assessment facilitates the quick deployment of relief works and an expedited retrofitting program, which helps in reducing functional downtime to a greater extent [1]. Primarily, the damage level assessment in a post-extreme event scenario is done using one of the following methods: (i) On-site inspection, (ii) an evaluation using the Rapid Visual Screening (RVS) technique [2], (iii)

H. Beushausen et al. (Eds.): NTZR 2025, LNCE 873, pp. 1005–1014, 2026.
https://doi.org/10.1007/978-3-032-27698-8_88

damage assessment using high-resolution imagery provided by Unmanned Aerial Vehicles (UAVs) [3], and (iv) remote sensing techniques [4]. Methods like on-site inspection and rapid visual screening are effective but entail significant financial and logistical investments. In addition, visiting the affected areas post-hazard involves potential safety concerns for the personnel. For these reasons, the methods involving remote sensing techniques and UAVs started gaining attention recently.

UAVs significantly reduce the risk to human life by replacing humans with themselves in a damage assessment survey. The high-resolution images provided by UAVs help identify damage levels accurately with the help of pre-trained semantic segmentation models, a deep learning architecture under machine learning. One such classification, due to damages incurred following the hurricane event, Michael, was performed by [3]. Apart from performing a damage level assessment study, the images obtained from UAVs could be used to perform a retrofitting cost estimation study, reducing the functional downtime. Apart from the pictures of damage sites, information about the damage instance, such as what happened at the concerned site, or what retrofitting task is required to ensure a pre-hazard condition, could also be utilized to perform the damage assessment. Numerous studies have been conducted previously about the damage and performance assessment [5, 6], life-cycle cost analysis [7, 8], and resilience assessment [9, 10]. These studies highlight why a resilient framework is necessary for a sustainable design and how factors such as damage prediction, risk assessment, performance, and fragility function evaluation help develop a framework for resilient infrastructure systems. Current study discusses how such damage data could be used for an approximate retrofitting cost prediction, one of the many parameters of a resilient framework.

This study's task based cost estimation functions are obtained using Support Vector Regression, an extension of the popular supervised classification technique, support vector machines (SVM) [11]. While SVM is primarily used to perform classification tasks using multiple learned parameters (weights, bias, and support vectors), SVR [12] is adapted for regression tasks, thus providing continuous functions instead of discrete functional values provided by SVM. This study aims to provide a framework that makes use of machine learning and the past damage data for damages due to a seismic event to predict the task wise retrofitting costs required, given the occurrence of an event of a particular intensity. In addition, this work presents a case study for bridges damaged due to seismic events in January 2020, Puerto Rico. Apart from predicting the task based cost, the work is extended to classify the damage into multiple damage classes: *Slight*, *Moderate*, *Extensive*, and *Complete*.

2 Support Vector Machines and Regression

A continued effort to identify underlying patterns in data [13] led to the development of a supervised learning technique, SVM, which could help perform two-group classification tasks [11]. SVM can be used for both linearly separable and not perfect linearly separable data. SVM solves a dual problem where the goals are to minimize the classification errors and maximize the margin between the classes. Pictorial representations of the dual problem of SVM can be seen in Figs. 1 and 2. SVM learns weights and support vectors to obtain a decision boundary that separates data into constituent classes. Linearly non-separable data can be projected to a multi-dimensional space with the help of kernel functions to obtain a linear hyperplane as shown in Fig. 2. The hyperplane thus obtained helps separate data into different classes even for linearly non-separable data when presented in a 2-dimensional plane. Moreover, the margins obtained using support vectors define the confidence interval of the classifier, and the greater the margin, the greater the generalization.

SVR is a modification over SVM, as it fits a function with an error margin, i.e., ϵ-tube instead of a hyperplane as done in SVM [14]. The support vectors are estimated by penalizing the deviations outside the margin and ignoring errors within the margin. Thus, SVR tries to find a function that is as flat as possible and obeys the error margin criteria, which forms the basis for the dual problems of SVR. Like SVM, SVR also uses kernel functions to project the variables into a high-dimensional space for the data which is not linearly separable. To summarize, SVR can be understood as a non-linear regression method achieved using kernel functions, which try to balance the flatness of a function while regularizing the parameter to penalize the function based on the location of points within or outside of the error margin.

Mathematical formulations for the two methods, SVM and SVR, are presented in Eqs. 1 through 4.

SVM:

$$\min \frac{1}{2}|w|^2 + C\sum_{i=1}^{n} \xi_i \tag{1}$$

subject to:

$$\begin{gathered} y_i\left(\boldsymbol{w}^\top \boldsymbol{x_i} + b\right) \geq 1 - \xi_i, \quad \forall i \\ \xi_i \geq 0, \quad \forall i \end{gathered} \tag{2}$$

SVR:

$$\min_{w,b,\xi,\xi^*} \frac{1}{2}|w|^2 + C\sum_{i=1}^{n} \left(\xi_i + \xi_i^*\right) \tag{3}$$

subject to:

$$\begin{gathered} y_i - \left(\boldsymbol{w}^\top \boldsymbol{x_i} + b\right) \leq \varepsilon + \xi_i, \quad \forall i \\ \left(w^\top x_i + b\right) - y_i \leq \varepsilon + \xi_i^*, \quad \forall i \\ \xi_i, \xi_i^* \geq 0, \quad \forall i \end{gathered} \tag{4}$$

where ϵ: width of the tolerance tube, ξ_i, ξ_i^*: slack variables for overshoot and undershoot, respectively, $\boldsymbol{w}$: weight vector, b: bias term, $\boldsymbol{x}_i$: input feature vector for the ith data point, y_i: target output, class label in SVM and a real value in SVR, and C: regularization parameter.

3 Data Sources and Processing Techniques

The January 2020 earthquake and the following aftershocks in Puerto Rico caused severe damage to the infrastructure. Most damaged infrastructure comprised bridges exhibiting various structural typologies, construction ages, and geometric skewness [15]. The Federal Highway Administration made available a detailed description of bridge location, type, age, total costs for damage, its discretization based on tasks, and the exact damage [16].

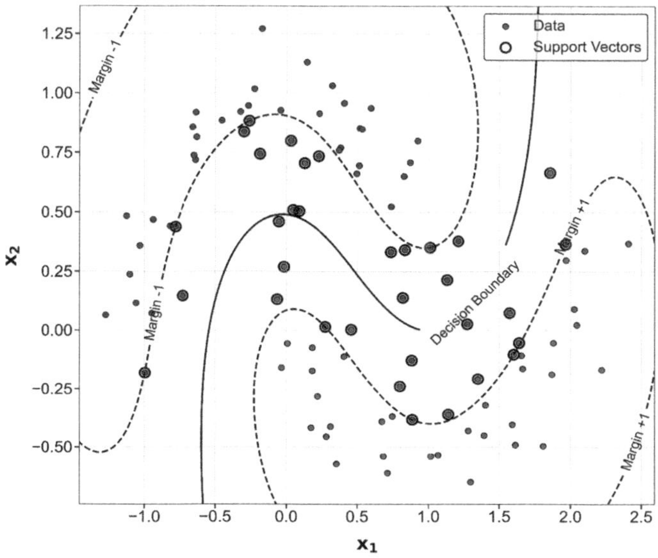

Fig. 1. SVM with radial-based kernel (2D projection).

Data preprocessing was required before it could be used to perform a cost-based damage study. The first step of data preprocessing involves assigning the damage levels to the identified damages. The damage level assignment was achieved by taking help of three parameters: (i) what happened on the bridge, viz., bridge abutment failure, displaced deck, etc. (ii) tasks required to retrofit or repair the bridge to achieve a pre-hazard usability of the structure, such as bracing of girders, cementitious grout patch, etc. and (iii) comparison of (i) and (ii) with the damage level delineation criteria mentioned in HAZUS [17].

The next step is splitting the entire dataset into train, validation, and test data subsets. The split dataset is then converted into a form where a bridge number appears only once and is arranged row-wise. The columns contain different task categories with the costs listed for each bridge. Not all the bridges contain each task category; hence, the costs for those tasks are listed as zero, thus completing the entire dataset. Different task categories form the independent variables, and the damage levels are the class labels. During the dataset split, it was ensured that every subset of data contains each damage level in the same ratio as present in the full dataset. Also, if any splits don't include the task categories, they are removed from the dataset to maintain uniformity while performing the support vector regression. The train, validate, and test split was done in the fractions of 0.5, 0.1, and 0.4, respectively.

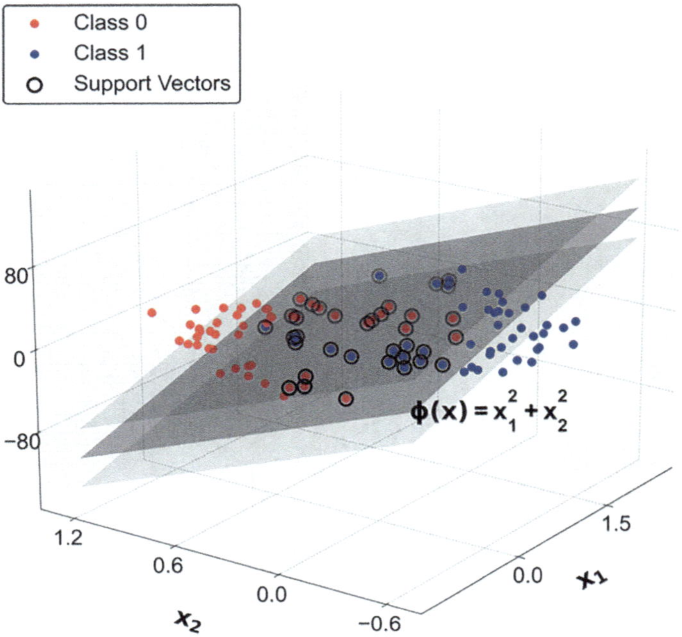

Fig. 2. Linear separability of data points in 3D space.

4 Methodology

This study aims to obtain a fitted SVR function that takes cost values corresponding to different tasks as input and yields the damage severity score as output, which is further used to assess the damage levels for the damage incurred. The workflow of the SVR pipeline is listed in the upcoming subsections.

4.1 Problem Framing

Conventionally, support vectors are used for a binary classification task, but this work requires obtaining a continuous function that would help discretize the damage instances into different damage levels. Involvement of multiple task categories and four output classes makes it difficult for a simple SVM and discrete function outputs to perform the classification task. So, this work uses the continuous function, called the damage severity function, to identify multiple damage levels. Continuous damage severity score for given input values can be further quantized into different bins by setting appropriate threshold values. These thresholds decide the maximum and minimum value of the damage severity score for each class, and the output score for a test data lying in any of the binned ranges could be assigned to that damage class. The continuous nature of the severity score makes the problem easy, as the damage level classification becomes a simple bucketing task. For better understanding, this task can be compared to one of the popular statistical methods, known as histogram binning.

4.2 Support Vector Regression Model Setup

The bridge damage level in the obtained data depends on multiple task categories, as the overall costs are a summation of the costs associated with different tasks, with an underlying assumption of a relationship between damage level and damage costs. In such a scenario, a simple linear function is insufficient to perform the regression task. For this reason, these independent task variables were projected on a higher-dimensional plane using radial-based kernel functions (RBF). RBF allows for modeling complexity and provides a flexible model to fit the data. The hyperparameters C and ϵ for SVR are 10 and 0.2 respectively. A value of 10 for C ensures a moderate regularization level, not to underfit or overfit the training data. An ϵ value of 0.2 removes the influence of noise in the data while not allowing significant errors in the severity score calculation. In addition, the data imbalance associated with different class labels was handled by assigning different weightages to each class, thereby changing the importance of data from uniform to non-uniform. The weight values are assumed to be inversely proportional to the number of data points present in that class. Thus, a class with less data is given a higher importance while deciding the regression function. This assignment of weightage helps reduce some bias due to a relatively small amount of data.

4.3 Threshold Optimization, Label Spreading and Monotonicity

As discussed, the final goal is to classify damage instances into different levels. Obtaining discrete damage levels from the continuous severity score requires calculating threshold values for each class pair. For any particular pair, a threshold is taken as the average of the maximum value of the severity score for the lower class and the minimum value of the score for the higher class. The other parameter responsible for deciding the estimation of threshold values is the spread of values assigned to each damage level. A tight spacing between the assigned values could unnecessarily make the classification tasks sensitive to the noise. On the other hand, wide spacing smoothens the effect of any such noise in the damage data.

The label spread also helps in performing monotonic trend analysis for each task. Monotonicity evaluation involves the calculation of damage severity scores with changing values of a particular task, when other tasks are kept at their mean values. Logically, the damage severity scores should increase with increasing cost values related to that task. For this reason, an exclusion threshold of 0.6 is kept, removing any task that doesn't obey the monotonic trend. Figure 3 shows the monotonicity plot for each task involved and the damaged thresholds shown using a light gray color.

4.4 Classification Based on Monotonic Task Contributions

As seen from Fig. 3, all retrofitting tasks involved in the damage data have a monotonicity score higher than 0.6. A monotonicity score above 0.5 ensures that increasing cost values of a task generally lead to a higher damage level. So, all the retrofitting tasks in the damage data are used as independent variables to perform the regression work along with the hyperparameters mentioned above. The model, once fitted, goes through threshold estimation to further perform the binning tasks. The trained regression model

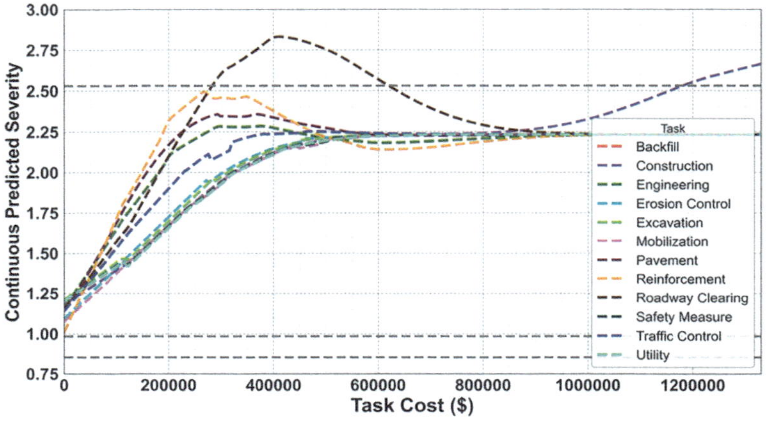

Fig. 3. Monotonicity for each task involved.

is then tested for accuracy on validation data. Once it is ensured that models perform reasonably well on training and validation data, the hyperparameter tuning is considered to be completed. The final step involves testing the model on unseen test data to assess model generalization.

5 Experiments and Results

Model performance for a regression or classification task in machine learning depends on hyperparameter tuning, the amount of training data, and the proportion of data available for each class in a multi-class classification task. Multiple trials with changing parameter values and the train-validate-test split ratio were performed before finalizing these values. The values that led to a more generalized model, i.e., a good performance on unseen test data, were taken as final values for this work. A divided ratio of 0.5-0.1-0.4 performed the best among all the iterations. So, the same split, along with the hyperparameter values mentioned in Sects. 4.2 and 4.3, is used for the model training. Once the training is over and SVR parameters are estimated, the model is further used for testing. Upcoming subsections discuss how well the model performs on the available damage data. Model performance is estimated with the help of multiple accuracy metrics and visualized using the popularly adopted confusion matrix representation.

5.1 Confusion Matrix

A confusion matrix is a fundamental and widely used tool for evaluating a model's performance. It highlights any class imbalances in the data and offers valuable insights into how well the model performs across different classes, revealing areas where it may struggle. In addition, it provides a quick peek into other accuracy metrics such as precision, an F1-score, recall, etc. High values in the diagonal elements of a confusion matrix represent the model's ability to classify the data correctly. In contrast, an increased off-diagonal value shows the model's inability to perform the classification task. During training, a

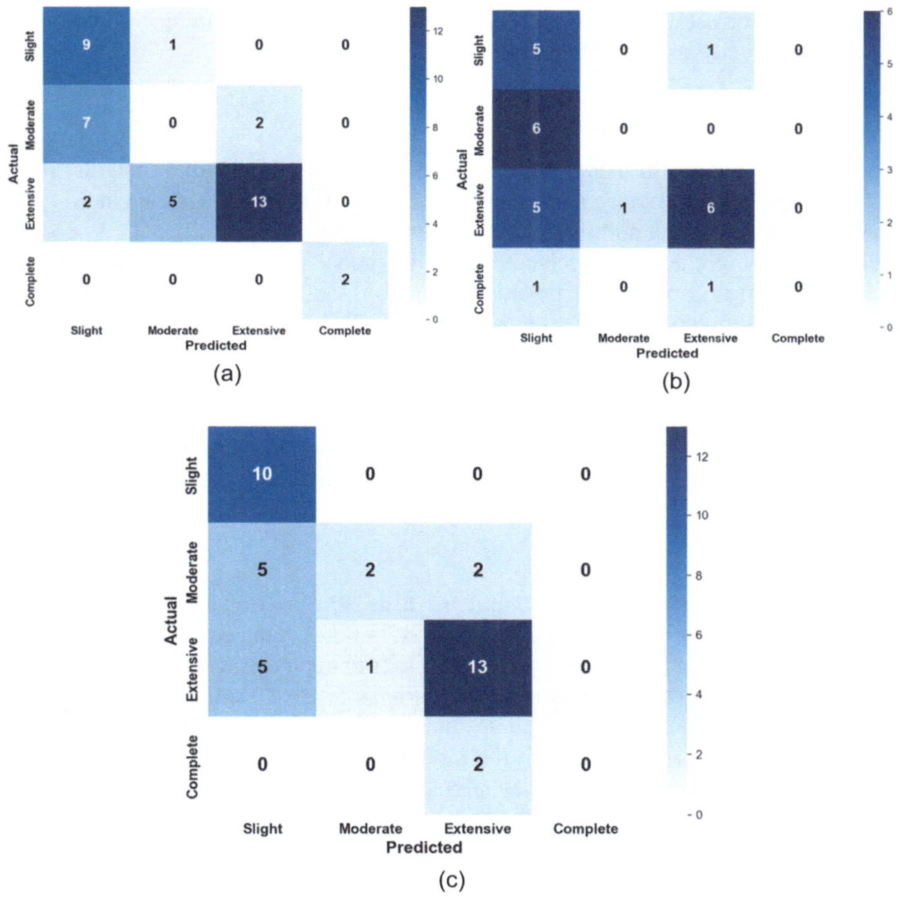

Fig. 4. Confusion matrices for (a) training, (b) validation, and (c) testing data.

Table 1. Accuracy metrics for each damage class.

	Precision	Recall	F1-score
Slight	0.50	1.00	0.67
Moderate	0.67	0.22	0.33
Extensive	0.76	0.68	0.72
Complete	0.00	0.00	0.00

confusion matrix helps remove systematic errors, such as identifying ambiguous classes or setting threshold values for a multiclass classification in a regression task. A heatmap coloring alongside numbers in a confusion matrix helps further understand the relative correct and incorrect classifications. Fig. 4 shows the model's training, validation, and testing performance using confusion matrices.

5.2 Other Accuracy Metrics

While a basic evaluation using the confusion matrix provides initial insights, it may not fully capture performance nuances in imbalanced datasets like the one in this work. Therefore, incorporating additional metrics such as precision, recall, and F1-score can offer a more comprehensive understanding of the model's effectiveness. Precision measures the proportion of correct positive predictions out of all instances predicted as positive. A low precision value typically indicates a high number of false positives. Recall, on the other hand, quantifies the proportion of actual positive instances that were correctly identified. The F1-score represents the harmonic mean of precision and recall, providing a balanced measure that considers false positives and false negatives. Table 1 summarizes these evaluation metrics for the test dataset. Overall accuracies of 59%, 42%, and 62% were obtained with training, validation, and test data, respectively.

6 Conclusions

A conventional train-validation-test split, such as 80-10-10 or 70-15-15 is typically employed to ensure robust and reliable model training. A reduced training set can lead to underfitting, while allocating excessive data to training may increase the risk of overfitting. However, this study adopts an unconventional 50-10-40 split, specifically chosen to address class imbalance and promote improved generalization. Initial experiments using larger training sets resulted in the model becoming biased toward the majority class, which impaired its ability to accurately predict minority classes. Furthermore, the model incorporates 12 input features, increasing the potential for overfitting when the training and validation sets are disproportionately large relative to the test set. The confusion matrices illustrate the effects of this class imbalance. The model exhibited difficulty in accurately learning the *Complete* and *Moderate* damage classes, as evidenced by the notably low recall for *Moderate*. In contrast, the *Slight* class achieved perfect recall (1.0), indicating high sensitivity; however, its precision was only 0.5, revealing an overprediction bias toward this class. The *Extensive* class demonstrated consistently high performance across all accuracy metrics, likely due to its ample representation in the dataset. Notably, most of the instances of the *Extensive* class in the test set were correctly classified, underscoring the model's strong performance when sufficient data is available. These observations suggest that improving class balance could enhance predictive performance, particularly for underrepresented categories. This study was conducted on a limited dataset. Future research should focus on acquiring larger and more balanced datasets to improve generalization. Additionally, expanding the damage typologies to include various structural forms could broaden the model's applicability.

References

1. Molina Hutt C, Vahanvaty T, Kourehpaz P (2022) An analytical framework to assess earthquake-induced downtime and model recovery of buildings. Earthq Spectra 38:1283–1320

2. Federal Emergency Management Agency (2003) Rapid visual screening of buildings for potential seismic hazards: supporting documentation. FEMA, Washington, D.C.
3. Rahnemoonfar M, Chowdhury T, Murphy R (2023) RescueNet: a high resolution UAV semantic segmentation dataset for natural disaster damage assessment. Sci Data 10:345
4. Gu J, Xie Z, Zhang J, He X (2024) Advances in rapid damage identification methods for post-disaster regional buildings based on remote sensing images: a survey. Buildings 14:898
5. Alipour A, Shafei B, Shinozuka M (2011) Performance evaluation of deteriorating highway bridges located in high seismic areas. J Bridg Eng 16:210–219
6. Torabi T, Shafei B, Alipour A (2025) Intensity measures for flood hazards in fragility assessments of bridges. In: Lecture notes in civil engineering. Springer Nature, pp 1539–1547
7. Alipour A, Shafei B, Kazemian M (2023) Next generation life-cycle cost analysis tool for bridges in Iowa – phase II. Iowa State University
8. Alipour A, Shafei B, Mock A, Prajapat K (2020) Development of a life-cycle cost analysis tool for improved maintenance and management of bridges. Iowa State University
9. Alipour A, Shafei B (2022) An overarching framework to assess the life-time resilience of deteriorating transportation networks in seismic-prone regions. Resilient Cities Struct 1:87–96
10. Alipour A, Shafei B (2016) Seismic resilience of transportation networks with deteriorating components. J Struct Eng 142:04016040
11. Cortes C, Vapnik V (1995) Support-vector networks. Mach Learn 20:273–297
12. Vapnik V, Golowich SE, Smola A (1997) Support vector method for function approximation, regression estimation, and signal processing. In: Advances in neural information processing systems. MIT Press, pp 281–287
13. Vapnik V, Lerner A (1963) Pattern recognition using generalized portrait method. Autom Remote Control 24
14. Smola AJ, Schölkopf B (2004) A tutorial on support vector regression. Stat Comput 14:199–222
15. USGS (2020) 6.4MW Indios, Puerto Rico earthquake of 07 Jan 2020, 824 UTC. United States Geological Survey
16. Federal Highway Administration (2025) Pavement performance measures and forecasting and the effects of maintenance and rehabilitation strategy on treatment effectiveness. FHWA
17. Federal Emergency Management Agency (2020) HAZUS earthquake model technical manual, V4.2 SP3. FEMA

Tree-Related Outage Risk in Overhead Distribution Networks Under Extreme Winds: A Geospatial Framework for Grid Resilience

Muneer Qudaisat(✉) and Alice Alipour

Iowa State University, Ames, IA 50011, USA
{muneerq,alipour}@iastate.edu

Abstract. Overhead power distribution systems are highly vulnerable to high-wind events, where tree-related failures pose a primary threat to grid reliability. This study presents a geospatial framework that integrates tree exposure and powerline fragility to estimate localized outage risk under extreme wind conditions. Using species-specific allometric equations, urban tree heights were estimated and translated into directional fall radii. These were intersected with overhead powerlines and utility poles to identify potential strike scenarios. Tree failure probabilities were assigned based on the HAZUS blowdown model, while pole fragility was modeled using a previously developed probabilistic fragility function. The 2020 Midwest Derecho was simulated using historical wind speed and direction data to represent a real-world extreme wind scenario. A combined outage probability was calculated for each pole using the union of tree strike and wind-induced pole failure probabilities. Results show that incorporating tree hazards substantially elevates outage risk, from an average of 0.35 to 0.85 in some cases, revealing high-risk hotspots that would be overlooked under infrastructure-only models. Neglecting the effects of vegetation underestimates outage risk, particularly in tree-dense urban corridors. The framework provides utilities with actionable spatial insights to prioritize vegetation management, and it can be scaled to regional or national grids by adapting species fragilities and wind fields to local contexts.

Keywords: Power outage · Wind risk · Tree blowdown · HAZUS · GIS · Resilience planning

1 Introduction

The presence of trees and vegetation in urbanized areas is essential for supporting the physical and mental well-being of residents [1]. More than two-thirds of North Americans reside in urbanized regions, where the existence of significant forestry covering could pose unique challenges, particularly during severe weather events, including increased risks of tree-related power outages, damage to property and utility infrastructure, loss of critical services, and personal injuries [2]. Power outages caused by tree failures can have far-reaching, disruptive effects when trees are located near power

H. Beushausen et al. (Eds.): NTZR 2025, LNCE 873, pp. 1015–1026, 2026.
https://doi.org/10.1007/978-3-032-27698-8_89

infrastructure. Notably, around 55% of power distribution network failures in the Northeast United States are related to tree failures [3]. Tree failures were responsible for over 80% of outages caused by storms from 2005 to 2017 [4]. Therefore, it is critical to assess and mitigate the combined risk of tree failures and power network vulnerabilities during extreme wind events in urban environments [5]. As trees mature and grow to larger sizes, wind-induced stress becomes more prominent, often approaching the strength limits of various tree parts and threatening their structural integrity [6]. Moreover, the specific growing environment, whether in forests, urban settings, or open spaces, drives distinct tree growth forms and shapes its capacity to withstand external forces. The varying wind exposure levels are classified by ASCE 7-22 into three categories: B, C, and D [7].

Risk prediction methods can be categorized into three groups: Qualitative approaches use expert judgment and field observations, and Empirical/statistical models estimate failure probabilities from historical data. Mechanistic models simulate the underlying physics, explain failure mechanisms, and determine the threshold wind speeds that can cause tree failure [8]. Several computational tools and frameworks used for studying the effects of wind on trees include HWIND [9] focuses on predicting windthrow and stem breakage by integrating tree stability, soil resistance, and wind conditions for trees at forest edges or exposed locations; GALES [10] emphasizes forest-level risk assessment by calculating critical wind speeds for tree failure within forest interiors; and FOREOLE [11] account for tree interactions and shielding effects while considering the gust factor and the drag coefficient. HAZUS tree blowdown model, developed by the Federal Emergency Management Agency (FEMA), estimates tree damage using fragility curves that represent the conditional failure probability of trees under various wind speeds [12]. This probabilistic modeling enables a more comprehensive analysis of wind forces and tree vulnerability interactions, while accounting for uncertainties in physical and environmental wind-related risk parameters. The structural wind fragility of trees can be attributed to various factors that interact dynamically during wind events, including tree height, diameter at breast height (DBH), crown mass, and root anchorage, which are influenced by tree species, age, and local environmental conditions. DBH, measured at 4.5 ft above the ground, is a baseline for estimating stability, as wider trunks tend to resist bending better during strong winds [13].

The tree-failure risk increases the likelihood of overhead utility network failure due to wire breakage caused by falling trees, a short circuit fault that occurs when a bridge is formed between wires by falling trees, or when falling trees push wires together [14]. Much research on the reliability of power distribution networks has focused on wind-induced component failures regardless of the vegetation effect. In contrast, Hou et al. [15] introduced a framework that incorporates the risk of tree failures into resilience modeling and evaluates improvements through vegetation management practices, such as crown thinning. Hughes et al. [16] further integrated tree failure impacts into community resilience assessments by using proximity analysis to evaluate threats to critical infrastructure. Lu et al. [17] combined tree allometry and finite element modeling to predict tree-induced pole failures, utilizing satellite data and computer vision for tree mapping.

While prior studies have examined either pole fragility under wind loading or tree failure and vegetation management practices in isolation, this study develops a geospatially explicit framework to estimate the probabilistic risk of urban tree-related failures in overhead power networks by combining spatial tree inventory data, wind hazard modeling, and infrastructure mapping and reliability analysis. Tree species are classified and their heights estimated using species-specific allometric equations based on DBH. Each tree is then assigned a wind-dependent failure probability using FEMA's HAZUS blowdown model, and a directional fall radius is computed to simulate potential impact zones. By intersecting these fall zones with mapped overhead lines and pole locations in ArcGIS Pro, the framework identifies components at risk of tree strike and combines these with pole fragility functions to compute localized joint outage probabilities.

2 Study Area and Event Description

The study area is located within the broader Midwestern storm corridor, which has experienced several significant high wind events with widespread wind gusts exceeding 80–100 mph over hundreds of miles, often accompanied by tree blowdowns, power outages, and structural damage, due to the stress concentrations and potential buckling under wind [18, 19]. Figure 1 shows footage of the fallen trees resulting from extreme wind gusts during the 2020 Midwest derecho event, as reported by the National Weather Service (NWS). The footage shows downed trees blocking residential streets, entangling distribution lines, and damaging utility service equipment.

Fig. 1. Tree-induced damage to power infrastructure, 2020 Midwest derecho event (NWS)

The 2020 North American Land Cover dataset (NALCMS, 30 m resolution) helps estimate vegetation density across the study area, matching HAZUS tree density categories (e.g., urban, forest, cropland), and default densities from FEMA's HAZUS-MH model. This surface reclassification, as shown in Fig. 2, is used to estimate the number of trees per square mile and is then combined with tree inventory data to investigate wind-related tree exposure and infrastructure vulnerability. The spatial layout of the electric distribution system and the urban tree inventory within the study area are depicted in Fig. 3, obtained from the town's utility and the urban forest management plan: (a) displays the full extent of the electric service area, including overhead (OH) transmission lines, primary and secondary OH distribution lines, and pole locations; and (b) maps the community tree locations overlaid on the power network, where individual trees are symbolized based on wood type (hardwood or softwood).

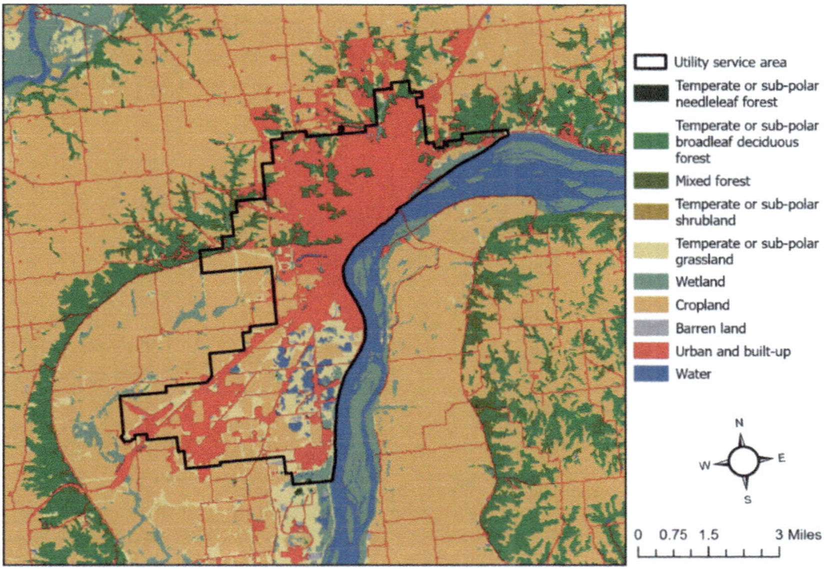

Fig. 2. Land cover types within the study service area (NALCMS 2020)

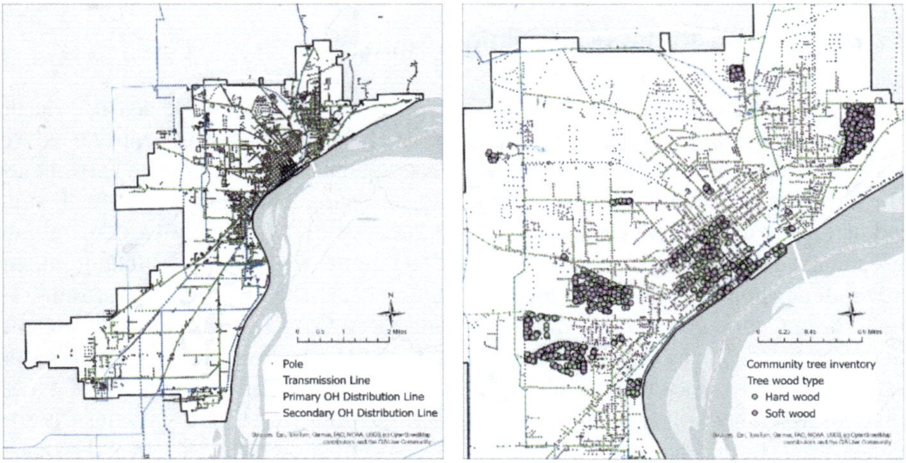

(a) (b)

Fig. 3. Spatial distribution of (a) power infrastructure and (b) urban tree inventory

3 Geospatial Methodology for Risk Modeling

To evaluate the risk of wind-induced tree failures affecting overhead power distribution infrastructure, this study developed a geospatial analysis framework implemented in ArcGIS Pro, as illustrated in the flowchart in Fig. 4. The framework begins with the integration of spatial data for urban trees and the power network within the study area.

The tree data includes species and DBH, among other properties. Using species-specific allometric equations, DBH values are converted to estimated tree heights, which are then used to buffer each tree by a fall radius approximating its potential impact zone.

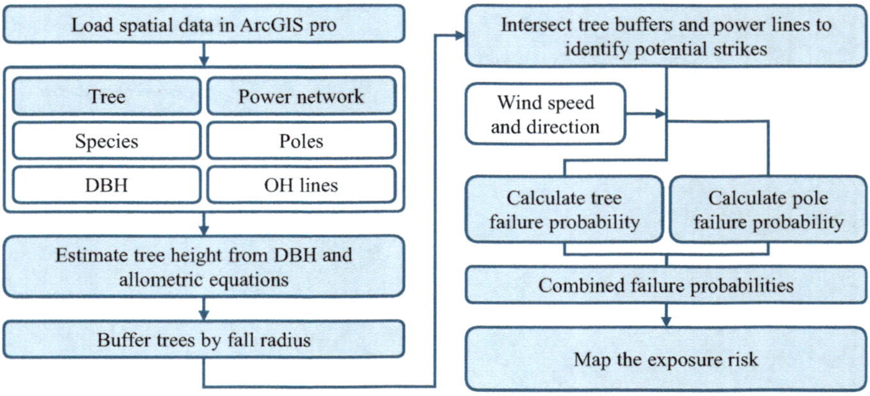

Fig. 4. Methodology flowchart

4 Tree Height Estimation Using Allometry

The USDA Forest Service Wood Handbook provides comprehensive density, elasticity, and mechanical properties for various species [20]. The previously introduced tree allometry can be used to predict the unknown tree parameters based on the known ones. The Urban Tree Database (UTD) encompasses measurements from over 14,000 street and park trees in the U.S., covering over 170 tree species. Utilizing this extensive data, researchers developed 365 sets of allometric equations. Key metrics include trunk and crown dimensions, location, and age of each tree [21]. The allometry data from UTD for Eastern Cottonwood, Silver Maple, Northern Red Oak, and Eastern Red Cedar are plotted in Fig. 5. Using the species-specific allometric equations, the tree heights were estimated based on the attribute table of the tree inventory layer. Recognizing that these relationships inherently contain variability across species, age, and growing environments, the resulting heights were treated as probabilistic estimates rather than exact values. Using these heights as the potential fall radius, a buffer was created around each tree using the buffer analysis tool, representing the area within which the tree could potentially fall, as shown in Fig. 6 (a) A subset of trees whose buffers intersected the power network was extracted using the Select Layer by Location tool, highlighted in orange in (b); these trees are capable of causing utility damage upon failure.

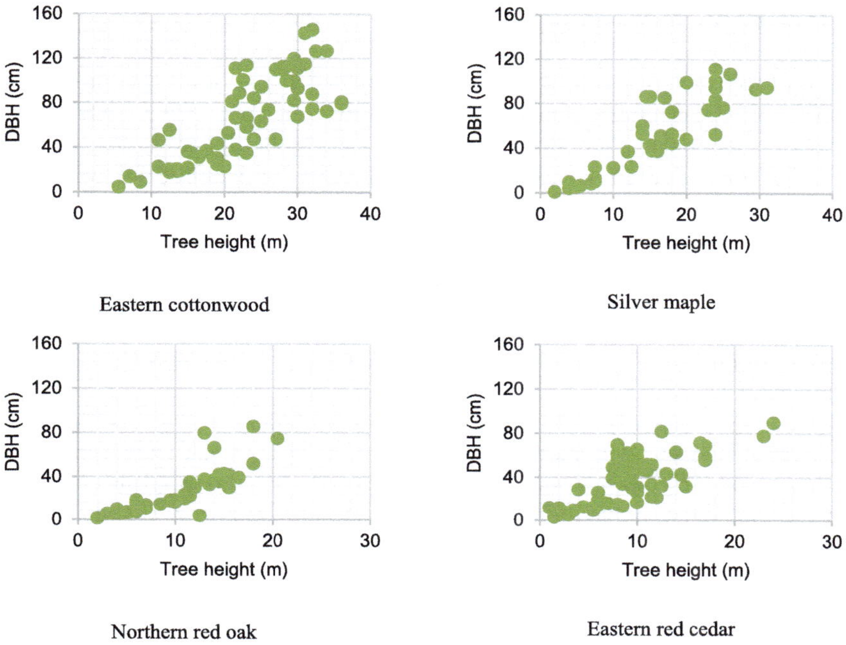

Fig. 5. Allometric relationships between DBH and height for common urban trees by UTD

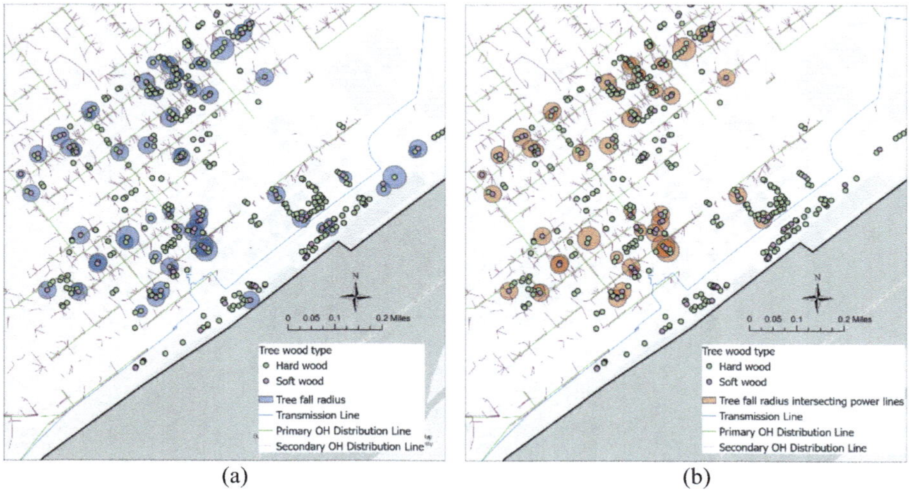

Fig. 6. (a) Tree fall radius, (b) Tree fall radius intersecting power lines

5 Tree Failure Probability Modeling Based on HAZUS

For a probabilistic approach to tree vulnerability under high-wind conditions, the tree failure likelihood estimates are based on the empirical tree blowdown model developed by FEMA and implemented in the HAZUS Hurricane Model Technical Manual (Version 5.1) [22]. This model provides wind-speed-dependent probabilities of structural tree failure based on post-disaster field data and controlled experiments. To simulate the conditions of the 2020 Midwest Derecho, wind speed and tree failure assumptions were derived from regional NOAA/NWS reports [23, 24], Iowa Environmental Mesonet [25], and HAZUS blowdown curves. Storm front motion is estimated to be between 55 and 70 mph toward the southeast across the study area, as shown in Fig. 7. The probabilities reflect maximum exposure conditions to simulate widespread tree-related utility failures [26].

Each tree in the study area is evaluated for its likelihood of structural failure under the derecho wind speed scenario as a function of its species and wood type using the HAZUS-based blowdown model. Trees are assigned a wind-dependent failure probability. To extend the analysis to the vulnerability of infrastructure components to wind-induced failure, this study employs a fragility-based reliability framework developed by Qudaisat et al. [27]. Using structural demand models from ASCE 7-22 and standard pole capacity data (ANSI Class 3), a limit state function was formulated and evaluated across 10,000 Latin Hypercube Samples. The resulting dataset of simulated failures was used to construct a lognormal fragility function, illustrated in Fig. 8. This curve represents the cumulative probability of failure for a wooden utility pole subject to increasing wind speeds.

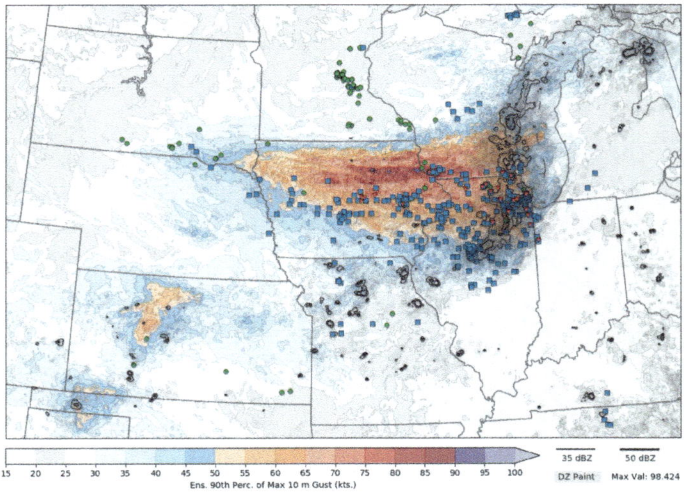

Fig. 7. Wind gust progression of the 2020 Midwest Derecho (NOAA)

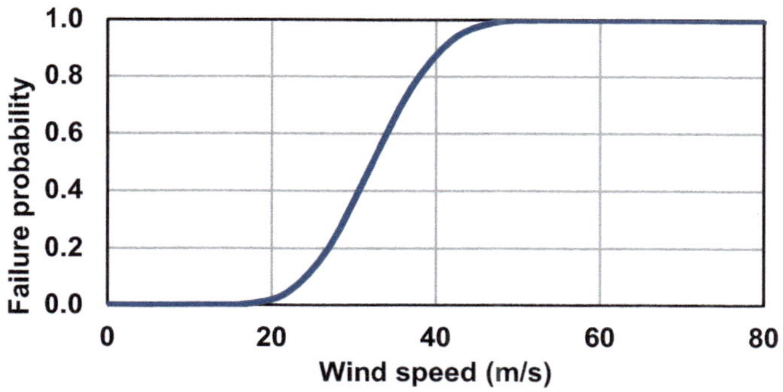

Fig. 8. Wind fragility function of an ANSI Class 3 wooden pole (Adopted from [27])

The cumulative probability of failure, P(V), as a function of wind speed (V), was fitted to a lognormal distribution following the form:

$$P(V) = \Phi\left(\frac{\ln(V) - \mu}{\beta}\right) \tag{1}$$

The lognormal form was selected because it captures the skewed nature of fragility behavior, where the probability of failure increases rapidly beyond a critical wind threshold. Using Maximum Likelihood Estimation (MLE), the location parameter (μ) and shape parameter (β) of the lognormal distribution were optimized. The resulting model achieved an RMSE of 0.0030 and an R^2 of 0.99996, indicating an excellent fit:

$$P_{Pole} = P(V) = \Phi\left(\frac{\ln(V) - 3.555}{0.16}\right) \tag{2}$$

6 Joint Probability Modeling of Tree and Pole Failures

In addition to their own failure, trees can fall and damage adjacent poles or lines. To capture this mechanism, the strike probability, defined as the probability that a failing tree falls in the direction of a power line or pole, is introduced:

$$P_{Tree\_Strike} = P_{Tree}.P_{Direction} \tag{3}$$

Where $P_{Direction}$ is the probability that the tree falls toward the line or pole. When the fall direction is assumed to be uniformly random, a $P_{Direction}$ value of 0.25 represents the approximate chance that a tree could fall within a 90° quadrant oriented toward the line or pole. This assumption can be adjusted based on the prevailing wind patterns and directions in the study area. The spatial intersection of fall buffers with OH infrastructure in GIS determines if a tree can strike a line or pole, depending on the direction of wind. The probability of a network disruption is modeled as the union of two independent yet physically linked events: (1) the direct failure of the utility pole due to wind load, and (2) the failure of a nearby tree due to wind, followed by directional fall that results in impact with the pole or overhead lines. In all other combinations involving either tree-induced failure or wind-induced pole failure, the outcome is assumed to be a network disruption. Only if the tree falls away from the infrastructure and the pole survives the wind load, a disruption is not expected. Then, following the standard probability of union logic, the total outage probability at a location is:

$$P_{failure} = 1 - (1 - P_{Pole}).(1 - P_{Tree\_Strike}) \quad (4)$$

There may be a correlation between the two mechanisms, which could slightly increase the true joint failure probability. Therefore, the presented estimates should be interpreted as an approximate measure of compound outage risk.

7 Results and Analysis

Figure 9 presents a spatial demonstration of the power outage probability modeling under severe wind conditions from the 2020 Midwest Derecho event. Figure 9 (a) illustrates the HAZUS-based tree failure probability for each inventoried tree intersecting overhead line corridors using allometric estimations. The map shows elevated risk in neighborhoods with high tree density and larger species. (b) shows the wind-induced pole failure probability modeled using the previously introduced fragility functions. On their own, the pole failure probabilities under the Derecho wind scenario generally fall within the range of 0.20–0.50, with most poles clustering between 0.30 and 0.40, indicating moderate but non-negligible vulnerability. These were filtered in (c) to show poles with elevated compound risk due to their proximity to failing trees. Finally, the two failure mechanisms are combined into a joint outage probability map. The resulting distribution in (d) shows evidently higher outage probabilities compared to those of hazard alone in the previous maps. For example, many poles originally showing moderate pole-only probabilities rise above 0.75 when combined with the risk of tree failure, validating the need for joint modeling. Most combined outage probabilities fall in the 0.75–0.90 range, particularly where high-risk poles are located near high-risk trees. This outcome confirms that tree-related hazards significantly amplify the probability of service disruption.

8 Conclusion

This study highlights the critical interaction between vegetative hazards and infrastructure vulnerability. While pole failure alone can be significant, underestimating nearby tree hazards undermines the true risk of outages, highlighting the importance of implementing localized mitigation strategies. The process also indicated the need to identify and manage nearby high-risk trees. The use of ArcGIS Pro functions, such as spatial join, directional buffer modeling, and union-based probabilistic logic, allows for grid exposure analysis under compound hazard scenarios. This spatial framework can be extended for resilience planning and vegetation management, such as prioritizing trimming, retrofitting, or targeted microgrid deployment in areas with high outage probability.

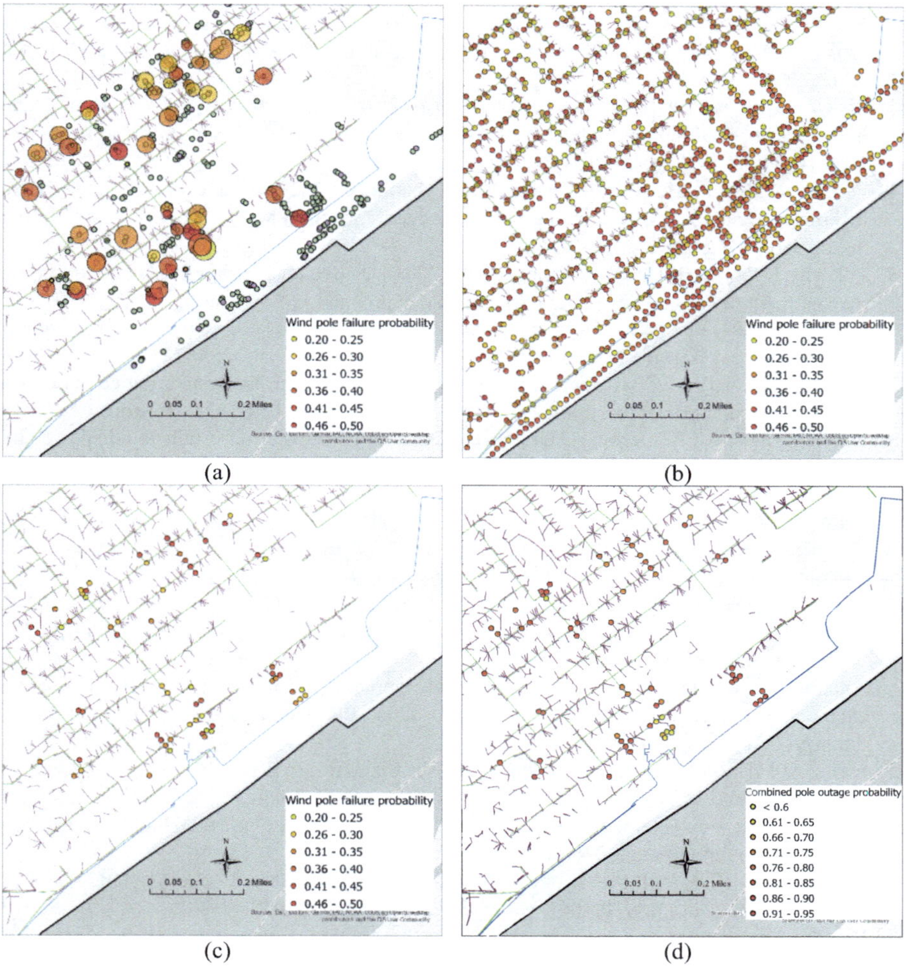

Fig. 9. Spatial distributions of (a) HAZUS-based tree failure probability, (b) wind-induced pole failure probability, (c) elevated-risk poles near potential tree strikes, and (d) combined pole outage probability under the 2020 Midwest Derecho scenario.

Acknowledgements. This paper is based upon work supported by the National Science Foundation under Award No. 2429602. Their support is gratefully acknowledged. Any opinions, findings, conclusions, and recommendations expressed in this material are those of the authors and do not necessarily reflect the sponsor's views.

References

1. Nilsson K, Sangster M, Gallis C, Hartig T, Vries S, Seeland K et al (2011) Forests, trees and human health. For Trees Hum Health:1–427. https://doi.org/10.1007/978-90-481-9806-1/COVER

2. Koeser AK, Klein RW, Hasing G, Northrop RJ (2015) Factors driving professional and public urban tree risk perception. Urban For Urban Green 14:968–974. https://doi.org/10.1016/J.UFUG.2015.09.004
3. Li G, Zhang P, Luh PB, Li W, Bie Z, Serna C et al (2014) Risk analysis for distribution systems in the northeast U.S. under wind storms. IEEE Trans Power Syst 29:889–898. https://doi.org/10.1109/TPWRS.2013.2286171
4. Parent JR, Meyer TH, Volin JC, Fahey RT, Witharana C (2019) An analysis of enhanced tree trimming effectiveness on reducing power outages. J Environ Manage 241:397–406. https://doi.org/10.1016/j.jenvman.2019.04.027
5. Pokorny J, Obrien J, Hauer R, Johnson GR, Albers J, Bedker P et al (2003) Urban tree risk management: a community guide to program design and implementation
6. Jiao-jun Z, Zu-gen L, Xiu-fen L, Matsuzaki T, Gonda Y (2004) Review: effects of wind on trees. J For Res (Harbin) 15:153–160. https://doi.org/10.1007/BF02856753
7. American Society of Civil Engineers (2022) Minimum design loads and associated criteria for buildings and other structures. American Society of Civil Engineers, Reston, VA
8. Lavoie S, Ruel JC, Bergeron Y, Harvey BD (2012) Windthrow after group and dispersed tree retention in eastern Canada. For Ecol Manage 269:158–167. https://doi.org/10.1016/J.FORECO.2011.12.018
9. Peltola H, Kellomäki S, Väisänen H, Ikonen VP (2011) A mechanistic model for assessing the risk of wind and snow damage to single trees and stands of scots pine, Norway spruce, and birch, vol 29, pp 647–661. https://doi.org/10.1139/X99-029.
10. Gardiner B, Peltola H, Kellomäki S (2000) Comparison of two models for predicting the critical wind speeds required to damage coniferous trees. Ecol Model 129:1–23. https://doi.org/10.1016/S0304-3800(00)00220-9
11. Ancelin P, Courbaud B, Fourcaud T (2004) Development of an individual tree-based mechanical model to predict wind damage within forest stands. For Ecol Manage 203:101–121. https://doi.org/10.1016/J.FORECO.2004.07.067
12. Hazus ®-MH 2.1 User Manual Hazus-MH User Manual ii (n.d.)
13. van Haaften M, Liu Y, Wang Y, Zhang Y, Gardebroek C, Heijman W et al (2021) Understanding tree failure—a systematic review and meta-analysis. PLoS One 16:e0246805. https://doi.org/10.1371/journal.pone.0246805
14. Wang L (2016) The fault causes of overhead lines in distribution network. MATEC Web Conf 61:02017. https://doi.org/10.1051/MATECCONF/20166102017
15. Hou G, Muraleetharan KK (2023) Modeling the resilience of power distribution systems subjected to extreme winds considering tree failures: an integrated framework. Int J Disaster Risk Sci 14:194–208. https://doi.org/10.1007/s13753-023-00478-x
16. Hughes W, Lu Q, Ding Z, Zhang W(2023) Modeling tree damages and infrastructure disruptions under strong winds for community resilience assessment. ASCE-ASME J Risk Uncertain Eng Syst Part A: Civ Eng 9. https://doi.org/10.1061/AJRUA6.RUENG-956
17. Lu Q, Zhang W (2023) An integrated damage modeling and assessment framework for overhead power distribution systems considering tree-failure risks. Struct Infrastruct Eng 19:1745–1760. https://doi.org/10.1080/15732479.2022.2053552
18. Qudaisat M, Alipour A (2025) Future-proofing energy infrastructure resilience with distributed energy resources. Dev Built Environ 23:100696. https://doi.org/10.1016/J.DIBE.2025.100696
19. Faharidine M, Aslam MU, Choufaikat MM (2025) Improving stability and buckling resistance of self-supporting isotrussed telecommunication tower under wind load: an evaluation according to TIA-222-G standards. J Rehabil Civ Eng 13(3):54–68. https://doi.org/10.22075/JRCE.2024.33439.2013
20. Ross R (2021) Wood handbook: wood as an engineering material. FPL-GTR-282, vol 282, pp 1–543

21. McPherson EG, van Doorn NS, Peper PJ (2016) Urban tree database and allometric equations, vol 253. https://doi.org/10.2737/PSW-GTR-253
22. Hazus Hurricane Model Technical Manual Hazus 5.1 (2022)
23. US Department of Commerce NNWS. Midwest Derecho—August 10, 2020, Updated: 10/8/20 12 pm (n.d.)
24. WoFS at the AMS, and the 2020 Iowa Derecho—NSSL News (n.d.). https://inside.nssl.noaa.gov/nsslnews/2021/02/guest-contributor-wofs-at-the-ams-and-the-2020-iowa-derecho/. Accessed 24 July 2025
25. Iowa Environmental Mesonet (n.d.). https://mesonet.agron.iastate.edu/. Accessed 15 July 2025
26. Torabi T, Shafei B, Alipour A (2025) Intensity measures for flood hazards in fragility assessments of bridges. Lect Notes Civ Eng 237:1539–1547. https://doi.org/10.1007/978-3-031-69626-8_128/FIGURES/1
27. Qudaisat M, Houssou D, Gallus W, Alipour A (2024) Future-proofing energy infrastructure: power grid risk assessment. In: Proceedings of the 1st international conference on net-zero built environment. Springer, Cham, pp 1125–1136. https://doi.org/10.1007/978-3-031-69626-8_94

A Simulation-Based Framework for Analyzing Occupancy Impacts on Performance Gap in Nearly Zero Energy Neighborhoods

Ahmad Esmaeilzadeh(✉) and Mohamed Hamdy

Department of Civil and Environmental Engineering, Norwegian University of Science and Technology, Trondheim, Norway

ahmad.esmaeilzadeh@NTNU.no

Abstract. The energy performance gap (EPG) in buildings plays a key role in challenging to achieve expected impacts of zero energy neighborhoods. This study develops a predictive approach to assess and mitigate the impacts of occupancy behaviors (OB)s on the EPG in district level. By implementing stochastic model of OB, this research quantifies the role of occupancy-related parameters, including occupancy density, metabolic rate, and indoor temperature setpoints, across an energy district, including high-rise apartment and large office building. Simulation data from high-rise apartment and office building prototypes modeled using EnergyPlus forms the basis of energy district. The setback temperature is designed based on occupancy schedules to enhance OB impacts on building's performances. This study highlights the importance of stochastic OB models in bridging the gap between design expectations and operational realities. The results demonstrate that dynamic human factors, including occupancy schedules and behavioral patterns, can lead to up to 22% deviation in predicted energy performance. By implementing the stochastic model related to OBs, the proposed approach provides a pathway to support actionable design strategies and operational decision-making for energy-efficient district operations.

Keywords: Zero Energy Neighborhoods · Sensitivity analysis · Occupancy behaviors

1 Introduction

Operating buildings significantly contributes to global energy consumption, which has been steadily increasing over the past decade [1]. Building energy codes have emerged as powerful approaches for reducing energy use in buildings worldwide [2]. While various standards and coding aim to promote energy efficiency in buildings, the literature highlights that buildings often fail to achieve their predicted performance during operation [3, 4]. The difference between predicted performance and actual performance of the building is defined as Building Performance Gap (BPG) or Energy Performance Gap (EPG) when only energy consumption is investigated [5]. Two main Key Performance Indicators (KPI)s are defined to evaluate BPG: one based on the energy usage

H. Beushausen et al. (Eds.): NTZR 2025, LNCE 873, pp. 1027–1038, 2026.
https://doi.org/10.1007/978-3-032-27698-8_90

in buildings and the other relied on Indoor Air Quality (IAQ) [6]. Reducing EPG and IAQ gap plays a key role in improving the reliability of energy policy measures and providing health and well-being in buildings [7]. Consequently, identifying the causal factors of BPG lead us to mitigate their impacts by preventing uncertainties in building performance simulations [8]. The main reasons for BPG are technical issues and social problems [9].

While technical issues are typically categorized under the design and construction phases, social problems are mentioned as the most striking causes during operational stage [10]. However accurate determination of technical parameters in buildings' simulations can reduce BPG regarding design and construction phases, the social problem remains a complex factor. Bridging the BPG requires better alignment with actual occupant behavior as the main social parameter [11]. From a dynamic simulation perspective, occupant behavior encompasses occupancy schedule, temperature setting, ventilation, hot water usage, internal heat gains, and electrical appliances [4]. Consequently, occupant behavior is identified in simulations through the specification of these parameters.

Occupancy schedules have been examined in various studies. For instance, comparing the electricity usage of a building in Brazil under actual occupancy schedule and a simulated assumed one resulted in a 10.9% EPG [12]. Bielskus et al. [13] analyzed the impact of different occupancy schedules on the EPG of an office building, representing that deviations from standard schedules can increase energy demand by up to 21%. Assessing the role of occupancy behavior (OB) in BPG by using a questionnaire-based approach reveals that occupant behavior can influence building performance by up to 26% compared to standard operation scenarios. In this study, occupancy schedules and indoor air temperature setpoints were derived from questionnaire responses, and the model was updated consequently [14]. Setpoint temperatures have been studied as another key occupant-related parameter. Adjusting the Heating, Ventilation, Airconditioning, and Cooling (HVAC) setpoints $\pm$ 2 °C for a typical educational building in an extremely hot climate in the UAE resulted in an EPG of up to 26.5% [15]. Regulating setpoint temperatures to ensure thermal comfort presents many limitations and contributes to the performance gap due to individual characteristics [16]. Accordingly, considering uncertainty in setpoint temperatures in addition to occupancy schedules provide a broader perspective for estimating the BPG.

Ventilation is regulated according to CO_2 concentration and building occupancy levels defined by ASHRAE 55 [17]. Since CO_2 concentration is determined based on occupants' activities and equipment schedules within the building, ventilation becomes dependent on occupancy schedules and their activities. Similarly, hot water usage is another parameter defined by considering occupancy schedules [18]. In Building Performance Simulation (BPS), hot water demand, internal heat gains, and electrical appliances are defined as dependent variables determined by occupancy schedules and their associated activities [19]. IEA EBC Annex 66 defines internal heat gains as being associated with occupant presence, and identifies equipment usage, HVAC setpoints, and plug-loads as variables influenced by occupancy activities [20]. Consequently, OB can be summarized through three key parameters: occupancy schedules, occupancy activities, and thermal comfort temperature.

Quantifying the role of OB in the BPG remains a topic of debate in the literature [21]. A review by Far et al. [22] indicated that up to half of the energy usage in residential buildings can be related to OB, while Mahdavi et al. [23] found limited research supporting the central role of OB in BPG. Nevertheless, studies have shown that occupants effect the energy performance of residential buildings by 6–25%, and commercial buildings by 5–30% [24]. Accordingly, simulation-based analysis of OB parameters helps to clarify their contribution to uncertainties in performance simulations. However, while the role of occupant behavior has been studied at the building level, its impact at the energy district level—considering both residential and office buildings—remains underestimated.

This study provides new insights into assessing the impact of OB on performance gaps in buildings and energy districts. This research enhances our understanding of how OB influences both the EPG and IAQ gap within an energy district. A simulation-based analysis of OB independent parameters is implemented to clarify these impacts. A novel contribution of this paper is the simultaneous simulation of a standard office building and a residential apartment to evaluate the impact of occupancy schedules, occupancy activities, and regulating setpoints. Stochastic inputs are employed in energy district simulation utilizing MATLAB/SIMULINK co-simulated with EnergyPlus via the MLEP toolbox, enabling a comprehensive definition of the OB-related BPG.

Section 2 of this paper presents the research methodology, including the case study, stochastic inputs, and hypothesis, and definition of KPIs. The results of simulation-based analysis for the energy district are illustrated in Sect. 3. Finally, Sect. 4 concludes the paper by summarizing key insights, outlining limitations, and suggesting directions for future research.

2 Methodology

One of the most well-known approaches for assessing the impact of OB on the BPG is stochastic models when high resolution temporal behavior and realistic peak loads are critical [25]. This study employs stochastic, person-based OB model, developed using random inputs related to occupancy schedules and occupancy activities. A simulation-based analysis is implemented on an energy district including a standard office building and a residential apartment building to evaluate the impact of OB in BPGs. This section begins with an overview of the energy district, followed by implemented weather data. Subsequently, the stochastic inputs are introduced to describe the OB model. Finally, KPIs are defined to demonstrate the studied parameters for quantifying the impacts of OB on the performance gaps within the energy district.

2.1 Case Study

The energy district is designed by carefully selected case studies. As residential apartments and office buildings offers diverse design strategies, energy use patterns, and HVAC system configurations, we design energy district by considering EnergyPlus standard models of high-rise apartment and large office building designed based on ASHRAE 90.1 [26].

Table 1. Energy district buildings materials

Building	Wall				Floor		Roof	
	Exterior		Interior					
	Material	(mm)	Material	(mm)	Material	(mm)	Material	(mm)
High-rise apartment	Stucco	25	Gypsum	13	Concrete	200	Carpet Pad	–
	Gypsum	15					Concrete	100
	Insulation	–	Gypsum	13	Carpet Pad	–		
	Gypsum	15					Insulation	–
Large office	Concrete	200	Gypsum	13	Concrete	100	Carpet Pad	–
	Insulation	–	Gypsum	13	Carpet Pad	–	Concrete	100
	Gypsum	13					Insulation	–

2.1.1 High-Rise Apartment

The standard high-rise apartment prototype is a 10-sotry building including 90 thermal zones, 10 unconditioned corridors, 1 office for the building manager, and 80 residential units. The window-to-wall ratio is set at 29.8% for this prototype. The windows are designed using double-glazed and triple-glazed windows with air gaps of 13 mm and 6 mm, respectively. Further construction details of the high-rise apartment prototype are provided in [26]. The construction materials of this case study are summarized in Table 1.

The building is equipped with a water-source heat pump supported by a 212 kW boiler. The nominal efficiency of the heating boiler is assumed equal to 80% in this prototype. Cooling is provided by a two-speed cooling tower with a design water flow rate of 2.2 L/s. Natural ventilation is utilized to supply required ventilation of the building. Additionally, a water mixture with an 8 kW preheater is implemented to supply the heating loop. Accordingly, the central heating and cooling system provides thermal demands while ventilation is managed independently within each unit.

For the residential units, the Occupancy Density (OD) is defined as 2.83 persons/100 m^2. Two different scenarios are considered for OD to illustrate occupant behavior in residential units within the standard prototype: 1 representing working individuals and the other representing who staying at home. Fig. 1 demonstrates the fractional OD factor associated with these two scenarios.

Figure 1 presents the OD fraction factors based on two different occupancy scenarios in residential units. In the working scenarios, occupants are assumed to leave the apartment at 9 a.m. and come back at 6 p.m. during working days. In contrast, the stay-at-home scenario assumes that occupancy presence is reduced to 25% during working hours. In this case, the HVAC system must operate continuously throughout the day to maintain indoor comfort conditions for the occupants. Accordingly, the initial thermostat setpoints for residential units are determined as fixed values in the standard simulation. The cooling and heating setpoints are set to 24.4 °C and 21.7 °C, respectively.

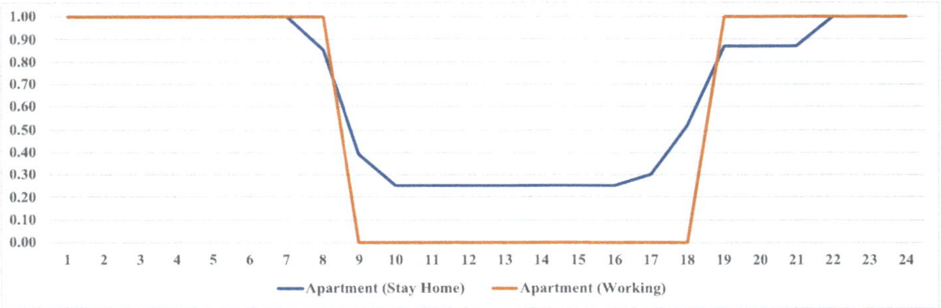

Fig. 1. OD fraction factor based on 2 different occupancy presence scenarios in residential units

2.1.2 Larg Office Building

The standard large office prototype is modeled as a 12-story building with an additional basement level, consisting of 86 thermal zones. The building includes 12 data centers, 12 corridors, 12 plenums, and 48 office units. The window-to-wall ratio is determined equal to 38.1% for this prototype. Double-glazed windows with 6 mm air gaps are utilized in the construction. Detailed specifications of the large office prototype, which is designed based on ASHRAE90.1, are represented in [26]. The construction materials used in the model are presented in Table 1.

The HVAC system is designed based on Variable Air Volume (VAV) system with reheat, serving the corridors and office units. The data centers are conditioned by water source heat pump while the plenums are unconditioned spaces. A natural gas boiler with a maximum power of 1,270 kW and 81% efficiency provides required heating of heating system. Additionally, a water mixture with 87 kW preheater supports the heating loop. Cooling is supplied by two parallel chillers with a capacity of 908 kW for each one. The cooling system also includes two cooling towers, each designed for a water flow rate of 37 L/s.

The occupancy schedules are determined based on an OD of 5.38 ersons/100 m^2 for conditioned area within the building. Working hours are defined as 9 a.m. to 5 p.m. on weekdays. A one-hour lunch break is assumed, typically varying from 12 p.m. to 2 p.m. On weekends, a reduced occupancy is observed due to the presence of security personnel. Fig. 2 represents the daily OD fraction factor schedule for both weekdays and weekends.

Based on the OD fraction factor illustrated in Fig. 2, 95% of large office occupants are presented in-person during regular working days. On Sundays, the building is occupied by only 5% of the maximum OD. The occupancy schedule for Saturdays reflects a mix of different user groups, resulting in OD values ranging from 0% to 30%. Accordingly, temperature setpoints are defined based on occupancy. During unoccupied hours, the heating setpoints are defined equal to 15.6 °C, while during working hours it is maintained at 21 °C. The cooling setpoints are determined equal to 24 °C during unoccupied hours while it is set equal to 26.7 °C in regular working hours. Consequently, the OD directly influences energy usage through these setback temperature strategies.

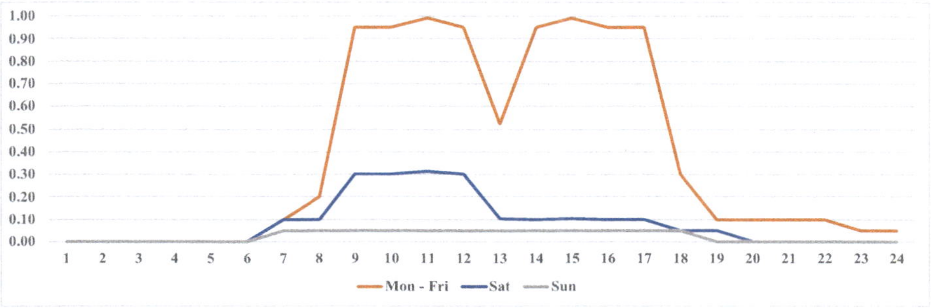

Fig. 2. OD fraction factor based on different occupancy presence scenarios in large office

The high-rise residential apartment and the large office building are connected based on occupancy presence within the energy district. We assume that the residential apartment houses employees of the large office building. Therefore, the energy modeling for the district is conducted by considering both buildings simultaneously.

2.2 Weather Data

The district model is simulated using two prototype buildings designed based on the Denver, US [26]. The simulation employs the EnergyPlus Weather (EPW) file for Denver, derived from the Typical Meteorological Year (TMY) data from 2009 to 2023. The monthly variation of dry bulb ambient temperature based on the weather data is presented in Fig. 3.

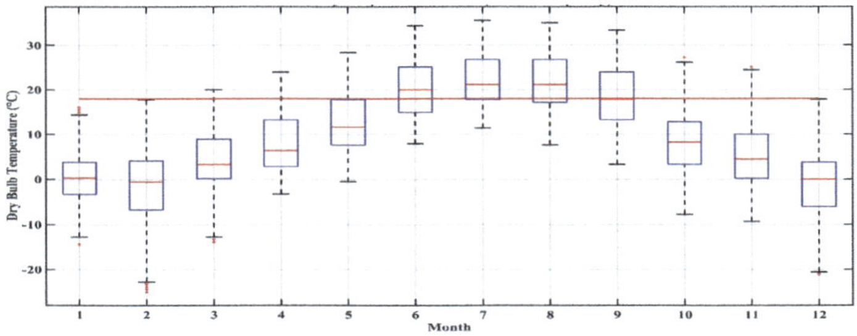

Fig. 3. Monthly dry bulb temperature variation based on EPW data of Denver

In Fig. 3, the red line represents the ambient temperature of 18 °C, which is used as the threshold for calculating Heating Degree Days (HDD) and Cooling Degree Days (CDD) [27]. Based on this threshold, the ambient temperatures profile indicates that heating is required for approximately eight months of the simulation period, while cooling system is operated during the four months of the year.

2.3 Stochastic Model of OB

Many researchers have employed stochastic models to estimate the role of OB on building performances. The current study implements stochastic inputs for both occupancy schedules and occupancy activities to assess the impact of OB in BPGs. Additionally, setback temperatures are applied to regulate setpoints based on occupancy schedules thereby linking setpoint temperatures to OB. The stochastic inputs were generated using Gaussian random functions, and setback control was implemented when occupants left the thermal zones. The stochastic model consists of specific assumptions, including a $\pm$ 2 °C temperature setback and a 50% residential-office occupancy linkage identified these ranges as realistic representations of behavioral variability in mixed-use urban environments [28, 29]. Gaussian distributions were adopted to represent occupant variability due to their prevalence in behavior modeling literature.

2.3.1 Occupancy Schedules

The occupancy schedules of the buildings are linked to the OD parameter defined in EnergyPlus. These schedules are set based on a fraction factor of OD. The maximum occupants of large office building and high-rise apartment building is set at 190 individuals. Gaussian-based stochastic inputs are used to define the fractional factor corresponding to OD. The variation in the fraction factor for large office building is represented in Fig. 4.

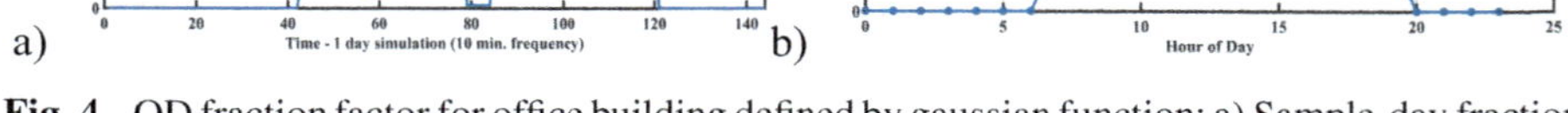

Fig. 4. OD fraction factor for office building defined by gaussian function: a) Sample-day fraction factor, and b) Average hourly during 1-year

By assuming the mean value of the Gaussian function to be 50%, the average OD is determined during working hours represented in Fig. 4b. The sample-day variations of fraction factor in Fig. 4a indicates that occupancy presence during working hours can range from 0 to 100%, following a stochastic model. When occupants are not present in the office during working hours, we assume that they are working from home. Consequently, the occupancy schedules of high-rise apartment during working hours are considered to be the inverse of the office building schedules.

2.3.2 Occupancy Activities

Another studied OB parameter is occupancy activity, which determines the Metabolic Rate (MR). According to ASHRAE Standard 55, the MR ranges between 45–190 W/m^2

for non-residential buildings and 50–145 W/m^2 for residential buildings. A Gaussian random function within these ranges is used to determine MR of occupants related to large office building and high-rise apartment. The variation in MR among occupants are represented in Fig. 5.

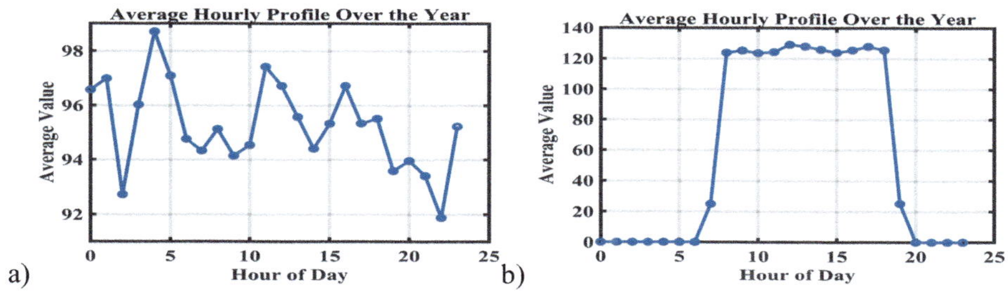

Fig. 5. MR of occupancies for district defined by gaussian function: a) Average hourly during 1-year for residential building, and b) Average hourly during 1-year for office building

Since the average MRs are defined based on the standard prototypes, Fig. 5a and Fig. 5b show the average hourly MR values are set to 95 W/m^2 and 120 W/m^2 for high-rise apartment and large office, respectively. However, the MR variations during 1-year are determined to cover all standard range.

2.3.3 Setpoint Temperature

The setpoint temperature is defined as a constant value for both cooling and heating systems in high-rise apartment building. In contrast, setback temperatures are implemented in the large office prototype based on occupancy presence. When the OD is modeled stochastically, the occupancy schedule is regulated randomly resulting in dynamic adjustments to the setpoint temperature depending on occupant presence in the office area. In the stochastic scenario, a 2 °C setback is applied for the high-rise apartment to account the effect of OB on the HVAC system. It is assumed that 50% of the occupants in the office building are associated with the same thermal zone of residential apartment, and their presence in the office triggers a reduction in the setpoint temperature of apartment in accordance with the defined setback.

2.3.4 Key Performance Indicators

Two different KPIs are employed to evaluate the BPGs under the influence of the stochastic OB model. Energy Use Intensity (EUI) used to assess the EPG, while indoor air temperature is utilized to evaluate the IAQ gap.

3 Results

Energy district is simulated based on standard buildings prototypes conditions and stochastic model. The results of EUI and indoor air temperature are represented to identify the role of OB in BPGs.

3.1 EUI

The EUI of the district is calculated based on the total energy usage in the district's buildings, including both electricity and natural gas. The combined energy usage of the two buildings is summarized in Fig. 6.

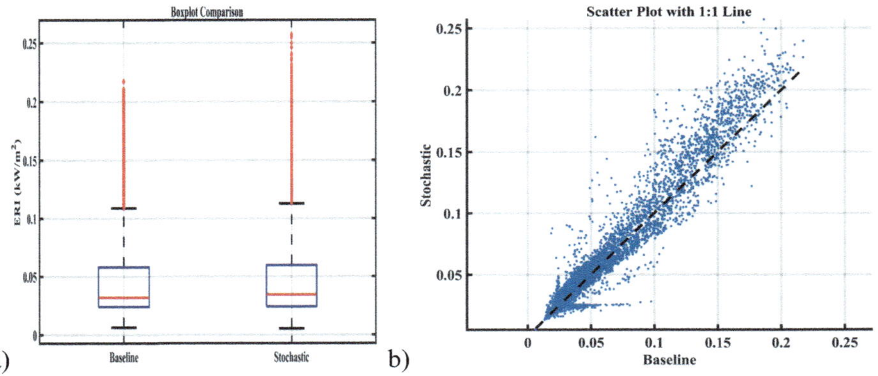

Fig. 6. EUI of district during 1-year simulation based on standard scenario and stochastic model, a) Energy Rate Intensity (ERI), b) Scatter plot of EUI in each time step

Figure 6a demonstrates the median EUI during one-year simulation in different scenarios are similar resulting in reliability of standard simulation in annual estimation. Figure 6b illustrates a scatter plot of EUI at each time step. The data points are generally aligned with the 1:1 line, confirming strong correlation between the two scenarios, yet the dispersion above and below the line highlights the impact of stochastic fluctuations on EUI values in each time step. Therefore, the standard model provides reliable data for annual energy predictions. However, relying solely on the standard simulation can result in oversimplified calculations, particularly underestimating peak energy usage.

3.2 Indoor Air Temperature

The other studied KPI is indoor air temperature of thermal units. The results of indoor air temperature for two different buildings of energy district are represented based on the sample unit temperature in each building illustrated in Fig. 7.

The setback temperature results in greater variability in indoor air temperature in high-rise apartment compared to constant setpoint used in standard prototype. This variation enables the optimization of heating and cooling system operation during unoccupied hours. While the setback temperature enhances the energy performance of the high-rise apartment, changing the setback pattern in large office building—driven by occupancy schedules—leads to increased indoor air temperatures due to diverse OD across different hours. Consequently, the HVAC system must operate more intensively, particularly in terms of cooling and ventilation, to maintain comfortable indoor conditions for occupants.

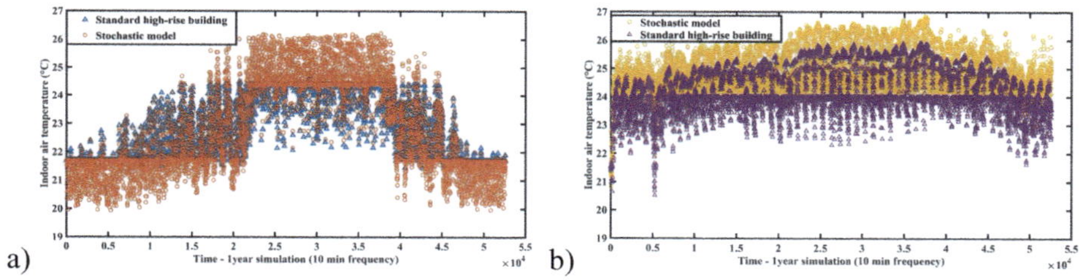

Fig. 7. Indoor air temperature of sample unit of buildings during 1-year simulation based on standard scenario and stochastic model: a) High-rise apartment building, b) Large office building

3.3 Design and Operational Applications

The observed 22% EPG due to stochastic OB highlights the need for adaptive control strategies in building management systems. For instance, integrating real-time occupancy sensors and predictive thermal comfort models can reduce the gap between simulated and actual performance. District-scale energy management platforms could improve simulations performances based on dynamic setback schedule, results in improving both energy efficiency and occupant satisfaction. Additionally, zoning strategies based on occupancy density profiles may improve setpoint configurations and reduce peak demand.

4 Conclusion

The aim of this research was to investigate the role of OB in the energy performance gap at the district level by incorporating a stochastic model into simulation-based analysis. This study identified that stochastic model results in 5% deviation in indoor air temperature and 22% EPG when analyzed at the time-step level. The research has also shown that standard model represents reliable results in annual studies while the stochastic model captures the influence of OB more accurately at finer temporal resolutions. This research lays the groundwork for future investigations into energy performance at the district scale, moving beyond stand-alone building assessments. Considerably more work will need to be done by implementing sensitivity analysis approach and considering various types of residential and nonresidential buildings to better understand the impacts of OB in energy districts and smart cities.

Acknowledgements. This project, Green Deal—ARV, a Horizon Europe-funded project, has received funding from the European Union's Horizon 2020 under grant agreement no. 101036723

References

1. Buildings—Energy System—IEA
2. Gilani S, Haddad K, Kirney C, Lopez P, Syed A, Stylianou M (2025) Performance of energy conservation measures in Canadian office buildings. Energy Convers Manage 330:119661. https://doi.org/10.1016/j.enconman.2025.119661

3. Yılmaz D, Tanyer AM, Toker ID (2023) A data-driven energy performance gap prediction model using machine learning. Renew Sust Energ Rev 181(May). https://doi.org/10.1016/j.rser.2023.113318
4. Zheng Z, Zhou J, Jiaqin Z, Yang Y, Xu F, Liu H (2024) Review of the building energy performance gap from simulation and building lifecycle perspectives: magnitude, causes and solutions. Dev Built Environ 17:100345. https://doi.org/10.1016/J.DIBE.2024.100345
5. Wu X, Lin B, Papachristos G, Liu P, Zimmermann N (2020) A holistic approach to evaluate building performance gap of green office buildings: a case study in China. Build Environ 175:106819. https://doi.org/10.1016/j.buildenv.2020.106819
6. Chiesa G, Pizzuti S, Zinzi M (2025) A new approach to assess the building energy performance gap: achieving accuracy through field measurements and input data analysis. J Build Eng 102:111941. https://doi.org/10.1016/j.jobe.2025.111941
7. Shrubsole C et al (2019) Bridging the gap: the need for a systems thinking approach in understanding and addressing energy and environmental performance in buildings. Indoor Built Environ 28(1):100–117. https://doi.org/10.1177/1420326X17753513
8. Vivian J, Carnieletto L, Cover M, De Carli M (2023) At the roots of the energy performance gap: analysis of monitored indoor air before and after building retrofits. Build Environ 246(October):110914. https://doi.org/10.1016/j.buildenv.2023.110914
9. van Dronkelaar C, Dowson M, Spataru C, Mumovic D (2016) A review of the regulatory energy performance gap and its underlying causes in non-domestic buildings. Front Mech Eng 1:177335. https://doi.org/10.3389/FMECH.2015.00017/BIBTEX
10. Bai Y, Yu C, Pan W (2024) Systematic examination of energy performance gap in low-energy buildings. Renew Sust Energ Rev 202(July):114701. https://doi.org/10.1016/j.rser.2024.114701
11. Bahadori-Jahromi A, Salem R, Mylona A, Hasan AU, Zhang H (2022) The effect of occupants' behaviour on the building performance gap: UK residential case studies. Sustainability 14(3):1362. https://doi.org/10.3390/SU14031362
12. Veloso ACO, Oliveira RN, Koury RNN, Souza RVG (2018) Energy efficiency labeling: study about the influence of schedule definitions. J Renew Sustain Energy 10(3):035104. https://doi.org/10.1063/1.5010685
13. Bielskus J, Motuzienė V (2020) The influence of schedules of open office occupants' presence on building's energy demand. In: Proceedings of the international conference on environmental engineering ICEE, vol 11, pp 1–7
14. Zare N, Saryazdi SME, Bahman AM, Shafaat A, Sartipipour M (2023) Investigation of heating energy performance gap (EPG) in design and operation stages of residential buildings. Energy Build 301(October). https://doi.org/10.1016/j.enbuild.2023.113747
15. Al Amoodi A, Azar E (2018) Impact of human actions on building energy performance: a case study in the United Arab Emirates (UAE). Sustainability 10(5):1404. https://doi.org/10.3390/SU10051404
16. Cozza S, Chambers J, Brambilla A, Patel MK (2021) In search of optimal consumption: a review of causes and solutions to the energy performance gap in residential buildings. Energy Build 249:111253. https://doi.org/10.1016/j.enbuild.2021.111253
17. Standard 55—Thermal Environmental Conditions for Human Occupancy
18. Barthelmes VM, Becchio C, Corgnati SP (2016) Occupant behavior lifestyles in a residential nearly zero energy building: effect on energy use and thermal comfort. Sci Technol Built Environ 22(7):960–975. https://doi.org/10.1080/23744731.2016.1197758
19. Mahdavi A, Tahmasebi F (2015) Predicting people's presence in buildings: an empirically based model performance analysis. Energy Build 86:349–355. https://doi.org/10.1016/J.ENBUILD.2014.10.027
20. Bandurski K (2015) IEA EBC Annex 66-definition and simulation of occupant behavior in buildings, vol 1, issue 2

21. Healey Trulsrud T, van der Leer J (2024) Towards a positive energy balance: a comparative analysis of the planning and design of four positive energy districts and neighbourhoods in Norway and Sweden. Energy Build 318(June):114429. https://doi.org/10.1016/j.enbuild.2024.114429
22. Far C, Ahmed I, Mackee J (2022) Significance of occupant behaviour on the energy performance gap in residential buildings. Architecture 2(2):424–433. https://doi.org/10.3390/ARCHITECTURE2020023
23. Mahdavi A et al (2021) The role of occupants in buildings' energy performance gap: myth or reality?, vol 13, issue 6
24. Zhang Y, Bai X, Mills FP, Pezzey JCV (2018) Rethinking the role of occupant behavior in building energy performance: a review. Energy Build 172:279–294. https://doi.org/10.1016/j.enbuild.2018.05.017
25. Happle G, Fonseca JA, Schlueter A (2018) A review on occupant behavior in urban building energy models. Energy Build 174:276–292. https://doi.org/10.1016/j.enbuild.2018.06.030
26. Prototype Building Models | Building Energy Codes Program
27. Degree-Days—U.S. Energy Information Administration (EIA)
28. Page J, Robinson D, Morel N, Scartezzini J-L (2008) A generalised stochastic model for the simulation of occupant presence. Energy Build 40(2):83–98
29. Hong T, Taylor-Lange SC, D'Oca S, Yan D, Corgnati SP (2016) Advances in research and applications of energy-related occupant behavior in buildings. Energy Build 116:694–702

Evaluating Indoor Thermal Comfort (ITC) Standards and the Need for Climate-Responsive Regulatory Frameworks Across Africa

Aleksejs Prozuments[1(✉)], Anatolijs Borodinecs[1], Aleksandrs Zajacs[1], and Mohamed Tariq Kahn[2]

[1] Faculty of Civil and Mechanical Engineering, Institute of Sustainable Building Materials and Engineering Systems, Riga Technical University, Riga, Latvia
aleksejs.prozuments@rtu.lv

[2] Department of Electrical, Electronic and Computer Engineering, Cape Peninsula University of Technology, Bellville, Cape Town, South Africa

Abstract. Indoor thermal comfort remains a critical yet inconsistently regulated aspect of building design across African continent, where diverse climatic conditions and socio-economic disparities challenge the development of effective, regionally adapted standards. While some countries adhere to international guidelines such as ASHRAE 55 and ISO 7730 (e.g., SA, Morocco), others rely on fragmented (or outdated) regulations that insufficiently address the specific regions' climatic/environmental diversity and urbanization challenges (e.g. Ghana, Cameroon etc). This study examines the existing indoor thermal comfort standards across African nations, identifying prevailing inconsistencies in regulatory enforcement, the absence of locally adopted comfort metrics, and the limited integration of passive design strategies suited to various climate zones. Moreover, the increasing occurrence of extreme heat events, along with the rapid urban expansion and thermally inefficient building stock, stresses the need for climate-responsive adaptations in both policy and practice. Addressing these challenges requires the development of coherent and scientifically grounded regulatory frameworks that incorporate climate-adaptive benchmarks, region-specific thermal comfort criteria, and sustainable building strategies. The analysis of existing regulatory frameworks demonstrates significant inconsistencies in thermal comfort criteria, the limited application of passive design strategies, and the lack of adaptation to local climate conditions. Empirical evidence stresses the need for region-specific thermal comfort benchmarks, regulatory alignment, and targeted policy advancements to address and improve indoor environmental quality. This study proposes a framework for regulatory adaptation that incorporates climate-responsive benchmarks and well-defined thermal comfort criteria.

Keywords: Indoor thermal comfort · Climate · Energy efficiency

H. Beushausen et al. (Eds.): NTZR 2025, LNCE 873, pp. 1039–1049, 2026.
https://doi.org/10.1007/978-3-032-27698-8_91

1 Indoor Thermal Comfort (ITC) Challenges in the African Context

As the global climate change and its associated impact is becoming more pronounced worldwide, with 2023 being the hottest year on record (the recorded temperature anomaly of 1.17 °C) and four of the hottest years recorded occurring within the last five years (Lindsey and Dahlman 2024), urban scholars have increasingly directed their attention towards regions in the developing world, in large part driven by substantial economic and population growth occurring within these regions (Kelly and Ngo Nguéda Radler 2024).

Over the past decades there has been an undisputed evidence of rising temperatures across the world, that has led to a growing concern of more extreme weather patterns and regular seasonal heat waves globally (Lindsey and Dahlman 2024). As such, building occupants are at an ever increasing risk to overheating exposure which is already emerging as an existing issue in many buildings (Laouadi et al. 2020). With an advent of higher temperature patterns in both winter and summer seasons, it is suggested that the building cooling requirement will be continuously increasing with respect to heating/cooling energy consumption per floor area (Khourchid et al. 2022).

While with regards to the population density and quality of life much attention has been set on Asian continent, projections indicate a shift towards Africa, which is anticipated to undergo substantial population growth and urbanization. Africa is one of the continents most vulnerable to climate change (Kelly and Ngo Nguéda Radler 2024). By the end of the twenty-first century, Africa's total population is projected to exceed 4.4 billion, comprising over 90% of Asia's population (UN 2023). Despite these demographic changes, uncertainties persist regarding the potential economic outcomes of urbanization in African nations, where more than half of the global extremely poor currently reside, and the World Bank projections suggest that by 2030, nine out of ten extremely poor will be in the African continent (World Bank 2018).

In parallel, climate analysts project increased climate change impacts driven by human activities. Among the various climate risks, high temperature events pose a significant threat to urban residents, with projections indicating increases in frequency, intensity, and duration of heat waves (Marcotullio et al. 2021). Recent studies suggest that the number of high temperature events has already increased globally, including in urban areas. While research has projected these trends globally, there remains a lack of understanding regarding the specific impacts on African cities, particularly in terms of urban heat wave vulnerability and the scale of potential impacts (Mora et al. 2017). This study aims to assess these aspects by providing estimates of urban population exposure to heat events across Africa, offering insights into the potential challenges and vulnerabilities faced by urban populations in the region. In addressing these challenges, the importance of ensuring indoor thermal comfort emerges as a crucial aspect of human health and well-being. With rising temperatures and an increase in extreme weather events, establishing and maintaining optimal indoor environmental conditions becomes essential to mitigate adverse health effects associated with temperature extremes (Arriazu-Ramos et al. 2023). However, the continent's diverse climates and varying levels of infrastructure development present certain challenges in establishing unified standards for indoor thermal comfort. As temperatures rise and heatwaves become more frequent and prolonged, the risk of heat-related illnesses and mortality escalates (Yadav et al. 2023). Conversely, inadequate indoor heating in regions susceptible to cold extremes can lead to increased susceptibility, particularly among vulnerable population groups such as the elderly and infants. In this context, it is necessary to establish and maintain optimal indoor environmental conditions. Africa's extensive geographical range and climatic variability pose significant challenges in achieving this goal. The continent encompasses a wide range of climates, from equatorial rainforests to Mediterranean coastlines and high-altitude plateaus. Within these diverse climatic zones lie distinct challenges and considerations for indoor thermal comfort regulation (Hugo et al. 2021).

Moreover, the level of infrastructure development and access to resources vary widely across African countries. While urban centers may have modern, climate-controlled buildings, rural communities often contend with inadequate housing stock, limited access to electricity, and reliance on traditional building materials and construction techniques (Emodi et al. 2022). Bridging this urban-rural divide in terms of thermal comfort infrastructure represents a pressing challenge for policymakers and stakeholders. Additionally, it is crucial to address global environmental concerns related to the widespread use of cement in concrete buildings, particularly in urban areas across Africa. Research by Farhan et al. (2020) highlights that the world produces an average of 4 billion tons of Portland cement annually, with an annual growth rate of 4%, emphasizing the urgent need for sustainable alternatives (Farhan et al. 2020).

Provided this complexity and diversity, developing unified standards for indoor thermal comfort in Africa presents a significant challenge. While international standards, such as those established by ASHRAE, offer valuable guidance, they must be contextualized and adapted to the climatic, socio-economic, and cultural realities of the African context. Local building codes and regulations play a crucial role in this regard, serving as the primary mechanism through which indoor environmental quality is regulated and enforced. However, significant gaps and challenges persist in the current landscape of indoor thermal comfort regulation across Africa. Insufficient research and data on local climatic conditions, coupled with limited capacity and resources for implementation and enforcement, hinder progress towards comprehensive and effective unified standard development. Moreover, the rapid pace of urbanization and infrastructure development exacerbates these challenges, with many buildings inadequately equipped to endure the challenges posed by a changing climate. In light of these challenges, a streamlined approach is needed to develop and implement standards and guidelines for indoor thermal comfort across the African continent. This requires a multi-disciplinary approach that encompasses research, policy formulation, capacity-building, and stakeholder engagement.

2 Current Landscape of Indoor Thermal Comfort Standards and Regulatory Frameworks in Africa

Across the African continent, a diverse array of standards, guidelines, and regulations govern indoor thermal comfort. Some countries adhere to internationally recognized benchmarks such as those set by the American Society of Heating, Refrigerating, and Air Conditioning Engineers (ASHRAE), while others have developed local building codes designed to accommodate their environmental, socio-economic, climatic contexts and infrastructural capacities. For instance, South Africa's National Building Regulations incorporate aspects of indoor environmental quality, including lighting and ventilation, although the specific focus on thermal comfort remains relatively limited. Some of the standards and guidelines used in African countries might include:

- ASHRAE standards—some African countries refer to standards published by ASHRAE, such as ASHRAE 55 and ASHRAE 62.1, which provide guidance on thermal comfort and indoor air quality (e.g., Ghana, Burkina Faso).
- Local building codes/regulations—some African countries have their own building codes and regulations that address various aspects of construction, including indoor environmental quality. These codes might include provisions related to thermal comfort, ventilation, and heating/cooling systems (e.g., South Africa, Nigeria, Morocco).

South Africa: In South Africa, for instance, the National Building Regulations provide guidelines and requirements for various aspects of construction, including indoor environmental quality. Part O: Lighting and Ventilation of the South African National Building Regulations (NBR) specifically deals with environmental sustainability, including thermal comfort. While not exclusively focused on thermal comfort, the regulations do encompass aspects related to energy efficiency and indoor environmental quality, which includes considerations for thermal conditions (Matandirotya et al. 2021).

Morocco: According to the General Building Regulations of Morocco (NM ISO 7730 2010), the annual heating and cooling requirements for the building need to meet the adapted reference temperature values fixed at 20 °C in winter and 26 °C in summer (Lamrani Alaoui et al. 2023).

Nigeria: Due to the differences in temperatures in the different regions of Nigeria, there are no accepted thermal requirements and regulations that cover the whole country. However, research done in this area has developed adaptable reference temperatures for areas in the North, East, West and the South of the country (Lagat et al. 2024).

Ethiopia: According to the code of practice for mechanical ventilation and air-conditioning of buildings (ES-EBCS 11 2003), the operative temperature should be maintained in the ranges of 23–25 °C and the air movement should not exceed 0.3 m/s, at 1.5 m from the floor.

Kenya: The first building code standard in Kenya was established with the Local Government Order and was operationalized in 1968 (Kenya, Ministry of Local Government 1968) and has long been applicable for ventilation of buildings in Kenya. Although optimal temperature for cooling is not specified, the code elaborates the need for openable windows sufficient enough for air circulation in rooms with specific services (living rooms, kitchen, pantries, etc.). Additionally, the Kenya Bureau of Standards (KEBS) has published a more comprehensive building code, with the ventilation codes unchanged (Kenya Bureau of Standards (KEBS) 2009). Since 2019, Kenya has begun implementing the Eurocodes and an incoming code, which aims to revolutionize the construction sector of Kenya, is underway (Lagat et al. 2024).

Rwanda: The first building code in Rwanda was the Rwanda Building Control Regulations, which was issued by the Rwandan Ministry of Infrastructure in 2009 and was updated in 2012 by the Rwanda Housing Authority to reflect the current needs and to be in harmony with other major legal tools. According to both versions, buildings are required to have effective ventilation in coherence with the size of the room (Rwanda Housing Authority 2012).

- International standards—some African countries adopt international standards from organizations such as the International Organization for Standardization (ISO), such as Morocco and Senegal. ISO 7730, for example, provides guidelines for the evaluation of thermal environments and is often referenced for indoor thermal comfort assessments (Lamrani Alaoui et al. 2023).
- Regional guidelines and practices—guidelines provided by a group of experts gathered by the Institute of Energy and Environment of La Francophonie (IEPF), who provided guidelines and recommendations based on local climate conditions, cultural preferences, and available resources. The proposed approach is widely adopted in tropical countries, especially in West and Central Africa (IEPF 2010).
- Local guidelines and practices—in some cases, countries develop their own guidelines or recommendations based on local climate conditions, cultural preferences, and available resources.
- Research and academic institutions—universities and research institutions in Africa may conduct studies and research related to indoor thermal comfort and environmental conditions. This research can contribute to the development of local guidelines and recommendations.

3 Empirical Assessment of Thermal Comfort Benchmarks in Africa

Ogbonna and Harris (2008) carried a study in Nigeria, where 200 subjects residing in naturally-ventilated buildings provided cross-sectional thermal comfort data for July and August 2006. Indoor climatic data, collected using portable laboratory-grade instruments compliant with ANSI/ASHRAE 55 and ISO 7726 standards, revealed a mean clothing-insulation value of 0.46 clo for the entire sample. The recorded indoor temperature surpassed 26 °C 68% of time, exceeding the overheating criteria stipulated by the standard, while mean relative humidity recorded across sample locations surpassed the 65% threshold recommended by ASHRAE 55, emphasizing the potential challenges in overheating and moisture control (Ogbonna and Harris 2008).

In South Africa, for example, the National Building Regulations encompass various aspects of indoor environmental quality, albeit lacking clear thermal comfort standards. This gap stresses the need for further research and development efforts to bridge the divide between regulatory frameworks and the evolving needs of building occupants (van der Walt et al. 2024). Moreover, the absence of region-specific standards highlights the broader challenge of extrapolating international guidelines to diverse climatic zones and cultural contexts within Africa (Rincón et al. 2019).

Rincon et al. (2019) study focused on assessing thermal neutrality using the ASHRAE 7-point scale pinpointed a temperature of 26.27 °C as the threshold, aligning closely with local weather data in sub Saharan Africa. However, discrepancies were noted between PMV determined neutrality and direct votes, suggesting limitations in PMV predictions for naturally-ventilated buildings. As such, the study highlighted the importance of night ventilation and solar protections in mitigating indoor overheating, particularly in hot semi-arid climates (Rincón et al. 2019).

Ouedraogo et al. (2022) conducted a study in Burkina Faso where a new decision-making tool for the evaluation of the thermal comfort in buildings has been proposed. This tool is based on a comfort index that allows to better and easier design, assessment and improvement of the thermal environment of buildings. This study allowed to characterize the thermal comfort of the three climatic zones of Burkina Faso namely Bobo-Dioulasso, Dori and Ouagadougou; which evolved between 25.5 and 28.3 °C. It also made it possible to characterize the thermal comfort of experimental pilot buildings representative of the common housing encountered in Burkina Faso. Experimentations have then been carried out on two natural ventilation pilot buildings located in Ouagadougou. The results show that the pilot building whose walls are made of Earth blocks achieves 26.4% of thermal comfort while the building made of hollow cement block achieves 25.8% of thermal comfort. (Ouedraogo et al. 2022)

Hema et al. (2023) carried out a study in Burkina Faso to determine the acceptable temperatures for thermal comfort and analyze the energy consumption related to the use of air conditioning systems to reach such temperatures in a mixed-mode office building. Through a survey and data collection, the study identified neutral air temperature and operative temperature which are between 26.2 °C and 26.5 °C, and 26.3 °C and 26.5 °C, respectively. This was confirmed by the activation air conditioning system when the air temperature reaches or exceeds the threshold of thermal neutrality. There was not a significant difference between the neutral temperature of male and female subjects. These findings can serve as a starting point for establishing thermal comfort parameters and design of mixed-mode office buildings in Sudano-Sahelian climates (Hema et al. 2024). This was also comparable to the temperature of the thermal comfort reported in other countries of the region such as 26 °C in Benin (air-conditioned and naturally ventilated office buildings), 27.4–30.3 °C in Ghana (air-conditioned and naturally ventilated office building); 24.9–27.7 °C in Nigeria (naturally ventilated residential building), 22–30 °C in Cameroon (naturally ventilated residential building).

4 Projected Climate Trends and Adaptive Thermal Comfort Strategies

Despite efforts to adopt international standards and develop local regulations, substantial gaps persist in comprehensively addressing indoor thermal comfort across Africa. The scarcity of research and region-specific standards exacerbates these challenges. For instance, South Africa's lack of thermal comfort standards stems from a scarcity of research correlating mean monthly temperatures with comfort levels, emphasizing a deficiency in understanding and addressing the prerequisites of thermal comfort. The landscape of indoor thermal comfort standards across the African continent reflects a varied array of approaches influenced by diverse climatic, socio-economic, and infrastructural considerations. While some countries rely on adherence to internationally recognized standards out of necessity, others have developed context-specific regulations designed to their respective contexts.

The geographical expanse and climatic diversity of the continent of climate zones presents challenges for regulating indoor thermal comfort. Ranging from the humid tropical rainforest zone to the arid desert zone, variations in temperature, humidity, and precipitation require development tailored strategies for maintaining comfortable indoor environments.

Indoor overheating has emerged as a significant issue across various climatic regions in Africa, particularly in renovated buildings where absence of standardized approaches to address overheating issues stresses the need for developing unified criteria and adaptable methodologies suited to diverse climatic conditions. For instance, Matandirotya et al. (2021) in their study found that low-cost dwellings in South Africa consistently experienced indoor temperatures above 26 °C during the summer (with 53% exceeding this threshold at night), potentially posing health risks, especially for vulnerable groups like young children and the elderly, as stipulated in the WHO recommendations. These high indoor temperatures indicate the urgent need for improved thermal insulation and passive design strategies to mitigate the impacts of climate change on indoor thermal environments (Matandirotya et al. 2021). Comparative analyses between different building typologies and passive design strategies across diverse climatic contexts can offer valuable insights into effective mitigation measures, particularly in the context of evolving climate scenarios.

As such, the historic, contemporary (measured) and future (predictive) climate data were gathered from Meteonorm database for 7 major African cities encompassing various climatic regions of the continent as shown in Fig. 1.

City	Country	Population
Kinshasa	DR Congo	17,032,000
Lagos	Nigeria	15,300,000
Cairo	Egypt	10,100,166
Johannesburg	South Africa	6,065,354
Nairobi	Kenya	5,118,844
Casablanca	Morocco	3,840,396
Ouagadougou	Burkina Faso	3,204,322

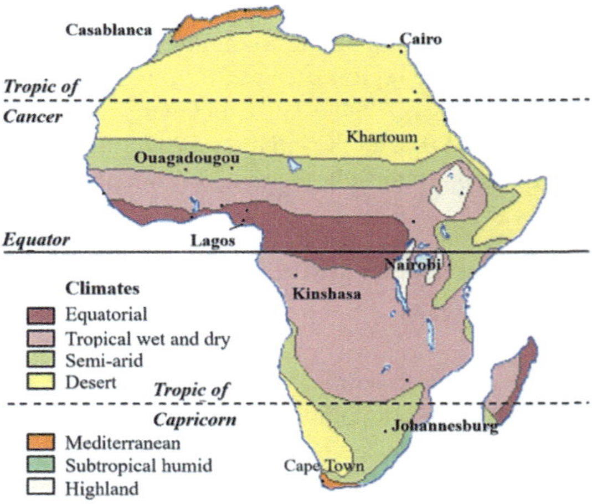

Fig. 1. Selected African cities for climate data analysis.

Figure 2 compiles the average historical (1961–1990), contemporary (2000–2019) and projected (2050) temperature data for the selected African cities, where Cairo, Ouagadougou and Casablanca are situated in Northern hemisphere, Lagos and Nairobi—near the equator with only a slight seasonal temperature variation, Kinshasa and Johannesburg—in the Southern hemisphere.

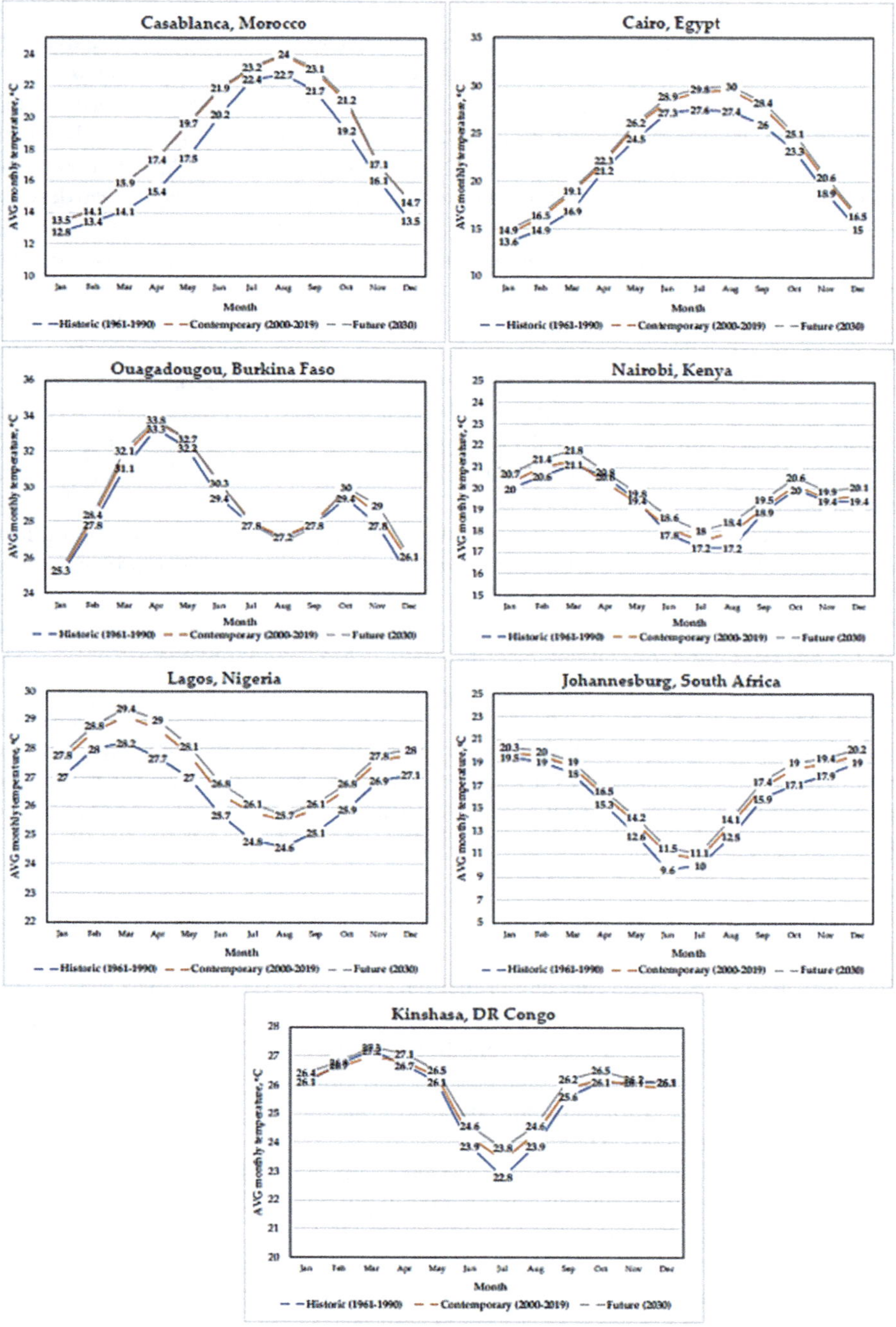

Fig. 2. Average historical (1961–1990), contemporary (2000–2019), and projected (2050) temperature data for selected African cities.

The chart suggests gradual and consistent seasonal temperature rise, that implies a steadily increasing risk of overheating in the warm season (late spring thru early fall) in the buildings. The data presented in the chart illustrates a pattern of gradual and consistent seasonal temperature increases, particularly pronounced during the warm season spanning from late spring through early fall. This trend carries significant implications for indoor thermal comfort management within buildings across the African continent, where diverse climate zones exist, ranging from humid tropical rainforests to arid deserts, and the seasonal temperature. During the warm season, which aligns with the dry season in many regions, temperatures often soar to extreme levels, especially in desert and savannah areas. This intensification of heat poses substantial challenges for maintaining comfortable indoor environments, particularly in buildings lacking adequate insulation or cooling systems. The prolonged duration of elevated temperatures during the warm season as predicted by Meteonorm model carries an ever increasing risk of indoor overheating, leading to discomfort and potential health hazards for occupants. As such, buildings without effective thermal management strategies will experience increased energy consumption as occupants resort to artificial cooling methods to mitigate the discomfort.

The predicted trend of gradual temperature rise stresses the importance of undertaking measures to address indoor thermal comfort in buildings, such as improved insulation, passive cooling techniques, and efficient ventilation systems to mitigate the climate impact. Furthermore, the integration of climate-responsive design principles tailored to specific regional climates can enhance the resilience of buildings against heat stress, prompting occupants' well-being and productivity.

Policymakers and urban planners must prioritize the integration of passive design principles and sustainable building materials to enhance indoor thermal comfort while minimizing energy consumption. Furthermore, incentives for green building certification and adoption of building codes tailored to regional climatic conditions can incentivize the construction of energy-efficient and thermally comfortable buildings.

5 Conclusions

This study highlights that current indoor thermal comfort (ITC) standards across Africa are fragmented, often relying on international benchmarks that are not fully adapted to local climatic and socio-economic conditions. By analyzing regulatory frameworks, climate projections, and empirical data from various climatic zones, the research underscores the pressing need for harmonized, region-specific standards that address both current and future thermal challenges.

The findings suggest that effective ITC regulation in Africa should emphasize climate-responsive design principles, integration of passive cooling and insulation strategies, and the promotion of sustainable local building materials. The projected temperature rise across major African cities reinforces the urgency of adopting adaptive measures to reduce overheating risks and energy consumption while ensuring occupant well-being.

A key contribution of this work lies in identifying regulatory gaps and providing a structured overview of existing standards, which can serve as a foundation for policy development and capacity-building initiatives. The proposed framework aims to facilitate collaboration between policymakers, building professionals, and researchers to develop resilient, context-specific solutions that enhance indoor comfort and support sustainable urbanization.

Acknowledgements. This work was supported by postdoctoral grant No. RTU-PG-2024/1-0024 under the EU Recovery and Resilience Facility funded project No. 5.2.1.1.i.0/2/24/I/CFLA/003.

This work was supported by NATO Science for Peace and Security Programme under grant number G6245

References

Lindsey R, Dahlman L (2024) Climate change: global temperature. NOAA Climate. https://www.climate.gov/news-features/understanding-climate/climate-change-global-temperature

Kelly AM, Ngo Nguéda Radler RD (2024) Does energy consumption matter for climate change in Africa? New insights from panel data analysis. Innov Green Dev 3(3):100132. https://doi.org/10.1016/j.igd.2024.100132

Laouadi A, Bartko M, Lacasse MA (2020) A new methodology of evaluation of overheating in buildings. Energy Build 226:110360. https://doi.org/10.1016/j.enbuild.2020.110360

Khourchid AM, Ajjur SB, Al-Ghamdi SG (2022) Building cooling requirements under climate change scenarios: impact, mitigation strategies, and future directions. Buildings 12(10):1519. https://doi.org/10.3390/buildings12101519

UN (2023) World urbanization prospects, 2018 revisions. https://esa.un.org/unpd/wup/:

World Bank (2018) The number of extremely poor people continues to rise in sub-Saharan Africa. https://blogs.worldbank.org/opendata/number-extremely-poor-people-continues-rise-sub-saharan-africa

Marcotullio PJ, Keßler C, Fekete BM (2021) The future urban heat-wave challenge in Africa: exploratory analysis. Glob Environ Chang 66:102190. https://doi.org/10.1016/j.gloenvcha.2020.102190

Mora C et al (2017) Global risk of deadly heat. Nat Clim Chang 7(7):501–506. https://doi.org/10.1038/nclimate3322

Arriazu-Ramos A, Ruiz GR, Izquierdo JJP, Gutiérrez AS-O, Monge-Barrio A (2023) From urban microclimate to indoor overheating: analysis of residential typologies during typical climate series and extreme warm summer. Energy Build 299:113620. https://doi.org/10.1016/j.enbuild.2023.113620

Yadav N, Rajendra K, Awasthi A, Singh C, Bhushan B (2023) Systematic exploration of heat wave impact on mortality and urban heat island: a review from 2000 to 2022. Urban Clim 51:101622. https://doi.org/10.1016/j.uclim.2023.101622

Hugo J, du Plessis C, Masenge A (2021) Retrofitting southern African cities: a call for appropriate rooftop greenhouse designs as climate adaptation strategy. J Clean Prod 312:127663. https://doi.org/10.1016/j.jclepro.2021.127663

Emodi NV, Haruna EU, Abdu N, Aldana Morataya SD, Dioha MO, Abraham-Dukuma MC (2022) Urban and rural household energy transition in sub-Saharan Africa: does spatial heterogeneity reveal the direction of the transition? Energy Policy 168:113118. https://doi.org/10.1016/j.enpol.2022.113118

Farhan KZ, Johari MAM, Demirboğa R (2020) Assessment of important parameters involved in the synthesis of geopolymer composites: a review. Constr Build Mater 264:120276. https://doi.org/10.1016/j.conbuildmat.2020.120276

Matandirotya NR, Cilliers DP, Burger RP, Pauw C, Piketh SJ (2021) Risks of indoor overheating in low-cost dwellings on the south African Lowveld. In: African handbook of climate change adaptation. Springer International Publishing, pp 1583–1600. https://doi.org/10.1007/978-3-030-45106-6_123

Lamrani Alaoui A et al (2023) An investigation towards optimizing the construction materials and configurations of buildings for improving energy efficiency and consumption in Morocco. Buildings 13(7):1705. https://doi.org/10.3390/buildings13071705

Lagat D et al (2024) Factors affecting the physical and functional performance of buildings in Kenya. J Kenya Natl Comm UNESCO. https://doi.org/10.62049/jkncu.v4i1.72

Ogbonna AC, Harris DJ (2008) Thermal comfort in sub-Saharan Africa: field study report in Jos-Nigeria. Appl Energy 85(1):1–11. https://doi.org/10.1016/j.apenergy.2007.06.005

van der Walt RE, Grobbelaar SS, Booysen MJ (2024) Dataset of indoor air parameter measurements relating to indoor air quality and thermal comfort in south African primary school classrooms of various building infrastructure types. Data Brief 53:110045. https://doi.org/10.1016/j.dib.2024.110045

Ouedraogo ALS-N, Messan A, Yamegueu D, Coulibaly Y (2022) A model for thermal comfort assessment of naturally ventilated housing in the hot and dry tropical climate. Int J Build Pathol Adapt 40(2):183–201. https://doi.org/10.1108/IJBPA-02-2021-0011

Hema C, Ouédraogo ALS-N, Bationo GB, Kabore M, Nshimiyimana P, Messan A (2024) A field study on thermal acceptability and energy consumption of mixed-mode offices building located in the hot-dry climate of Burkina Faso. Sci Technol Built Environ 30(2):184–193. https://doi.org/10.1080/23744731.2023.2291007

Integrated Rating System for Net-Zero and Resilient Building Design

Mana Nemati Aghdam, Manish Kumar Dixit(✉), and Fazel Nasserzadeh

Texas A&M University, College Station, TX, USA
mdixit@tamu.edu

Abstract. The building sector accounts for 37% of global CO_2 emissions, highlighting the need for integrated strategies that address both carbon neutrality and climate resilience. This study introduces a performance-based evaluation framework that combines Global Warming Potential (GWP) metrics with region-specific hazard resilience indicators. Using a GIS-based risk mapping system and a Multi-Criteria Decision Analysis (MIVES) approach, the framework adjusts evaluation weights based on localized risk profiles. Applied to a case study in Port Arthur, Texas, the framework demonstrated its adaptability through Monte Carlo simulations and sensitivity analysis. Results show hurricane and storm resilience metrics had the most significant influence on design scores. This approach offers a practical tool for harmonizing net-zero goals with hazard-specific resilience, enabling more informed and climate-adaptive building design.

Keywords: Net-zero building · Resilient design · Integrated framework

1 Introduction

The building sector is a major contributor to greenhouse gas emissions, accounting for approximately 37% of global CO_2 emissions, necessitating net-zero carbon strategies [1]. Simultaneously, escalating climate-related hazards demand resilient building design as a fundamental requirement rather than an option [2]. In the United States alone, between 2016 and 2022, there were 122 billion-dollar disaster events, costing an average of $146.6 billion annually and resulting in over 5,000 fatalities per year [3]. This trend highlights the urgent need to address net-zero carbon goals and resilience simultaneously.

Existing building frameworks often lack cohesion, with net-zero rating systems focusing primarily on energy efficiency, while resilience guidelines emphasize disaster preparedness. This disjointed approach leads to trade-offs, such as using highly durable materials with significant carbon footprints or net-zero energy buildings that may lack structural integrity during extreme events [4, 5]. Consequently, integrating resilience standards with carbon reduction metrics is crucial to achieving net-zero and resilient built environments.

Resilience begins with a thorough risk assessment to establish standards, including identifying primary hazards and setting priorities based on local risks [6]. For example, in

H. Beushausen et al. (Eds.): NTZR 2025, LNCE 873, pp. 1050–1060, 2026.
https://doi.org/10.1007/978-3-032-27698-8_92

2024, the United States experienced 27 billion-dollar weather and climate disasters, highlighting the increased frequency of severe storms, hurricanes, coastal flooding, droughts, and wildfires [3]. Net-zero carbon performance, the second core pillar of the proposed framework, relies on Life Cycle Assessment (LCA) following ISO 14040/44 standards to evaluate environmental impacts, including Global Warming Potential (GWP), acidification potential, and eutrophication potential [7, 8].

Recent studies have demonstrated both direct and indirect relationships between environmental indicators and specific hazards. For instance, Global Warming Potential (GWP) is directly linked to increased hurricane intensity and frequency due to ocean warming [9, 10]. Additionally, elevated GWP contributes to climate change, which intensifies wildfire risks by increasing temperatures, drying out vegetation, and prolonging drought conditions [11], while climate change-driven warming has been observed to intensify rainfall, leading to flash flooding events [12–14].

In contrast, other environmental indicators, such as acidification potential, eutrophication potential, and ozone depletion potential, show weaker or indirect connections to severe weather events. Nonetheless, they still influence environmental conditions that may affect resilience. Significantly, natural hazards themselves can impact LCA outcomes. For example, hurricanes and earthquakes can dramatically damage infrastructure, resulting in increased emissions from material replacement and repairs, thereby raising the life-cycle GWP [15–17]. Similarly, "Natech" events, where natural hazards trigger technological accidents, can cause substantial greenhouse gas emissions and environmental contamination, as witnessed in industrial releases during major Gulf Coast hurricanes [15, 18].

This dual perspective underscores the need to integrate hazard risk assessments into LCA frameworks, addressing both the environmental impacts that contribute to hazard risks and the emissions resulting from hazard-induced damages. This integrated approach aligns carbon neutrality with resilience strategies, enhancing both carbon performance and hazard mitigation. The proposed KPI-based rating system harmonizes net-zero and resilience measures within a structured evaluation methodology.

2 Methodology

This study proposes a multi-criteria decision analysis (MCDA)-based rating system to evaluate and optimize net-zero and resilient building designs. The methodology combines a GIS-based risk assessment approach with a weighted scoring model to strike a balance between resilience and carbon neutrality. It is developed through two phases: (1) national framework development and (2) region-specific analysis, ensuring adaptability to diverse climatic and hazard contexts.

2.1 National GIS-Based Risk Mapping

The framework first conducts a nationwide risk assessment using Geographic Information Systems (GIS) to identify five primary hazard categories: hurricanes, flooding, severe storms, droughts, and wildfires, based on historical data from FEMA's Resilient Risk Index. GIS layers for each category are overlaid and classified into risk scores ranging from 0 (no risk) to 5 (maximum severity). Qualitative clusters from FEMA's National Risk Index are mapped to numeric values, ensuring meaningful and consistent risk representation while avoiding the limitations of absolute numerical scores (see Fig. 1).

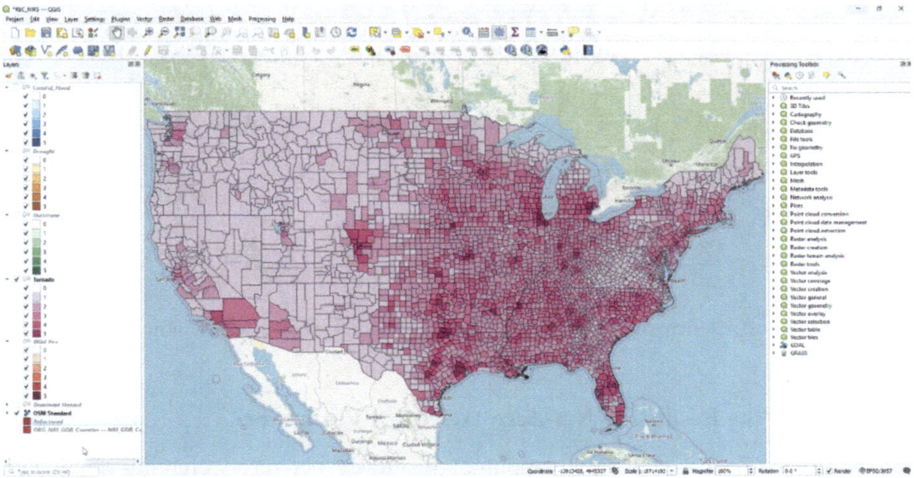

Fig. 1. Multi-layer GIS-based risk map for the United States.

2.2 Project Location-Specific Risk Analysis

Since resilience criteria vary significantly across different geographic and socio-economic contexts, selecting an appropriate case study region is crucial. Guidelines like FEMA's National Resilience Guidance (NRG) and frameworks such as RELi emphasize that resilience components are context-dependent, shaped by local hazards, community vulnerabilities, and infrastructure conditions [6, 19]. Therefore, choosing a region with frequent resilience challenges ensures a relevant and thorough analysis.

Following this rationale, Port Arthur, Texas, was selected as the project location due to its high exposure to severe hurricanes and flooding. This coastal city's vulnerability makes it ideal for detailed resilience and carbon-neutrality assessments. Figure 2 visually represents risk levels for each hazard category, ranging from 0 (no risk) to 5 (maximum severity).

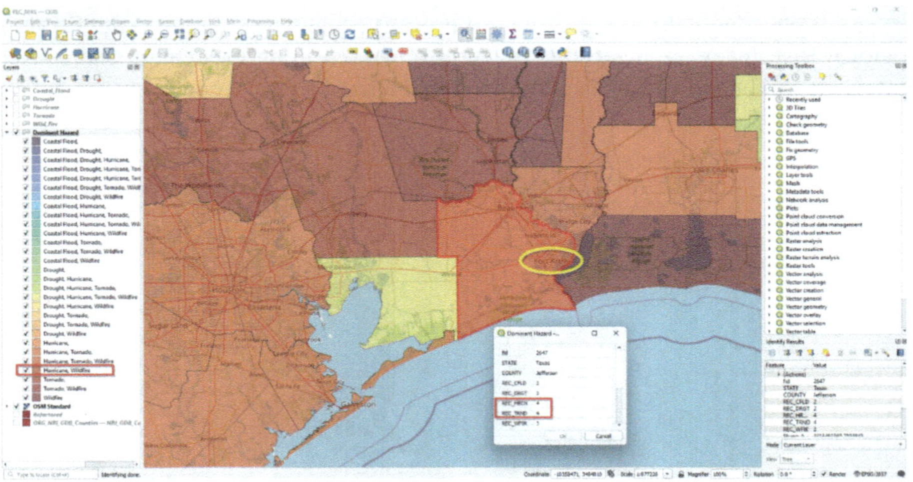

Fig. 2. GIS-based risk map of Port Arthur, Texas, with hazard-specific scores indicating resilience vulnerabilities.

2.3 Defining the Resilience–Net-Zero Weighting Framework

Following the GIS-based hazard mapping, a resilience–net-zero weighting framework was established. For Port Arthur, Texas, the assigned hazard scores were: Coastal Flooding (2), Drought (2), Hurricane (4), Severe Storm (4), and Wildfire (3). Higher scores were given greater weight in subsequent calculations. These scores were derived from the qualitative risk tiers in FEMA's National Risk Index, as described in Sect. 2.1, and converted to numeric values on a 0–5 scale for integration into the weighting model.

An Exponential Scaling Approach was used to improve weighting accuracy, incorporating cumulative risk rather than focusing solely on the highest risk score (R_{max}). This method ensures that Net-Zero performance receives higher weight in low-risk areas while resilience is emphasized in high-risk contexts.

Resilience categories (R_x) include five hazard types, and net-zero categories (S_x) include Global Warming Potential (S_1), Acidification (S_2), Eutrophication (S_3), and Ozone Depletion (S_4), each initially assigned equal weight. The net-zero performance weight (W_S) is calculated as:

$$W_S = 6 - \log_2(1 + \sum W_{R_x}) \tag{1}$$

The summation $\sum W_{R_x}$ in Eq. (1) represents the total of all resilience risk scores for the selected location.

The logarithmic transformation smoothens the transition between net-zero weights, gradually reducing them as cumulative risk increases. To validate the model, a Monte Carlo simulation with 10,000 iterations was conducted, resulting in random resilience risk scores and cumulative risk calculations. The scatter plot shows that net-zero weight declines sharply at lower cumulative risk values, flattening as risk increases. This framework ensures that the scoring method accurately reflects the importance of resilience and net-zero performance across different hazard contexts (Fig. 3).

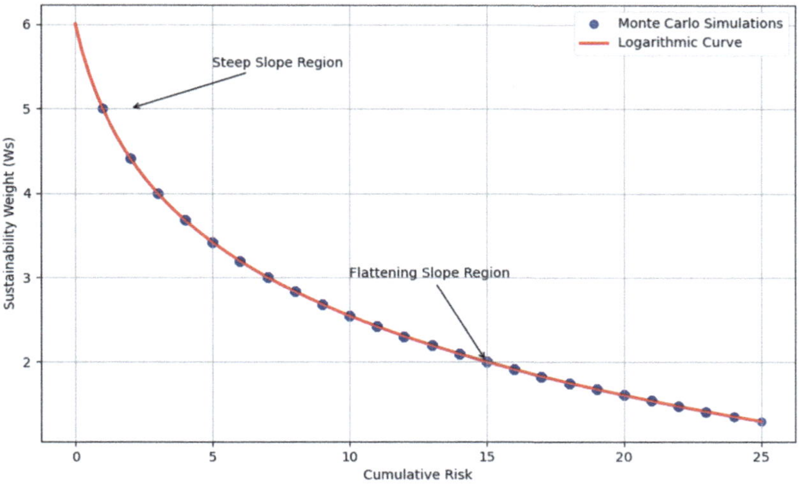

Fig. 3. Monte Carlo simulation results showing the relationship between cumulative risk and net-zero weight (Ws) using the logarithmic model.

The cumulative resilience risk for Port Arthur is 15, and the corresponding net-zero weight is calculated using Eq. (2):

$$\Rightarrow Ws = 6 - log_2(1 + 15) = 2 \quad (2)$$

2.4 Variation of Metrics and Normalization Methods

Resilience and net-zero metrics are defined based on standardized criteria to support a robust evaluation, as shown in Tables 1 and 2. Resilience can be evaluated through various factors, which vary depending on the project goals and context. While aspects such as structural design, infrastructure, and site conditions can be considered, this study focuses specifically on material selection due to its direct influence on a building's ability to resist and recover from hazards.

The listed resilience metrics in Table 1 are presented in no particular order of severity or priority; they are selected based on relevance to hazard-specific material performance and availability of standardized evaluation methods.

Table 1. Resilience metrics for material selection

Resilience category (Rx)	Metric (Rxi)	Unit/Evaluation method
R1_Coastal flooding	R1.1_Flood-resistant insulation	R-value + % water absorption (ASTM C518, ASTM D2842)
	R1.2_Water absorption & permeability	% by weight (ASTM C140, ASTM D2842)

(*continued*)

Table 1. (*continued*)

Resilience category (Rx)	Metric (Rxi)	Unit/Evaluation method
	R1.3_Corrosion resistance	Salt spray test (ASTM B117, hours)
	R1.4_Foundation scour resistance	Scour depth resistance (ft) & foundation reinforcement rating
R2_Drought	R2.1_Water retention capacity	% by weight (ASTM D698, ASTM D1557)
	R2.2_Thermal mass & heat retention	BTU/lb·°F (ASHRAE 90.1, ASTM C177)
	R2.3_Reflectivity/solar reflectance Index (SRI)	SRI (ASTM E1980–11)
	R2.4_Permeability & breathability	Perm rating (grains/hr·ft^2·inHg, ASTM E96)
R3_Hurricanes	R3.1_Impact resistance	Joules (J) or ft-lbf (foot-pounds) (ASTM E1996, ASTM D3746)
	R3.2_Wind uplift resistance	psf (pounds per square foot) (FM 4474)
	R3.3_Corrosion resistance	Salt Spray Test (ASTM B117, hours)
	R3.4_Water absorption & permeability	% by weight (ASTM C140, ASTM D2842)
	R3.5_Moisture resistance	Perm rating (grains/hr·ft^2·inHg) (ASTM E96)
	R3.6_Fire resistance	Fire rating in hours (ASTM E119, NFPA 285)
R4_Severe storms	R4.1_Impact resistance	Joules (J) or ft-lbf (ASTM E1996, ASTM D3746)
	R4.2_Wind uplift resistance	psf (pounds per square foot, FM 4474)
	R4.3_Anchoring & load resistance	Wind load rating (psf, ASCE 7–22)
	R4.4_Fire resistance (from debris impact)	Fire rating in hours (ASTM E119, NFPA 285)
R5_Wildfires	R5.1_Ember & flame resistance	Material burn-through resistance (NFPA 1144)
	R5.2_Ignition resistance	ASTM E84 Flame Spread Index (FSI)
	R5.3_Thermal conductivity	BTU/hr·ft^2·°F (ASTM C518)
	R5.4_Siding & roofing material fire rating	Class A, B, or C rating (ASTM E108, UL 790)

(*continued*)

Table 1. *(continued)*

Resilience category (Rx)	Metric (Rxi)	Unit/Evaluation method

Net-zero categories focus on four key environmental indicators aligned with life cycle assessment (LCA) methodologies:

Table 2. Net-Zero Impact categories

Environmental impact category (Sy)	Metric	Unit/Evaluation method
S1_Climate impact reduction	Global warming potential (GWP)	kg CO_2-eq
S2_Prevention of air and soil acidification	Acidification potential (AP)	kg SO_2-eq
S3_Protection of water quality	Eutrophication potential (EP)	kg PO4-eq/m^2
S4_Ozone layer preservation	Ozone depletion potential (ODP)	kg CFC-11-eq

To ensure fair and unbiased integration within the MIVES framework, all indicators from the resilience (Table 1) and net-zero performance (Table 2) categories are normalized to a standard [0,1] scale. Based on the characteristics of each metric, normalization methods are defined as shown in Table 3.

Table 3. Normalization methods

Normalization method	Formula	Application	Example metrics (codes)
Min-max normalization	$\left(X_{norm} = \frac{X - X_{min}}{X_{max} - X_{min}}\right)$	Metrics with measurable ranges	R1.2, R1.3, R2.1, R3.2, R3.4, R3.5, R4.2, R5.3
Benchmark-based normalization	$\left(X_{norm} = \frac{X}{X_{benchmark}}\right)$	Metrics with standard ratings (e.g., ASTM)	R1.1, R1.4, R2.2, R2.3, R3.1, R3.6, R4.1, R4.3, R4.4, R5.1, R5.2, R5.4
Distance-to-target normalization	$\left(X_{norm} = 1 - \frac{\lvert X - X_{target}\rvert}{X_{target}}\right)$	Net-zero metrics with target values	S1, S2, S3, S4
Qualitative rating normalization	Linearly scaled to [0,1]	Qualitative assessments (e.g., Likert scale)	Any qualitative resilience or net-zero metric

While the current framework emphasizes physical hazard exposure and material-based resilience, socio-economic vulnerability factors also significantly influence resilience outcomes. For example, communities with higher poverty rates, elderly populations, or limited access to infrastructure often face longer recovery times and greater disruption following disasters. To improve the equity and effectiveness of the framework, these social dimensions can be incorporated through existing datasets such as the CDC/ATSDR Social Vulnerability Index (SVI), FEMA's Community Resilience Estimates (CRE), or indices that capture infrastructure fragility (e.g., grid reliability, access to emergency services).

These indicators could be used to adjust resilience weights (WRx) for specific regions, amplifying the importance of resilience in communities with lower adaptive capacity. Alternatively, socio-economic vulnerability could be added as a separate metric within the MIVES model, treated similarly to a resilience category with sub-indicators such as income level, housing condition, or emergency access. This would allow the final score to reflect not only material performance against hazards but also community-level sensitivity and recovery capacity, thereby enhancing the framework's fairness and applicability in underserved contexts.

2.5 Final Scoring Methodology (MIVES Framework)

The evaluation employs a Multi-criteria Decision Analysis (MCDA) approach, utilizing the MIVES framework, to integrate normalized and weighted resilience and net-zero metrics. The final score is calculated as:

$$FinalScore = \sum \left(W_{R_x} \times R_x \right) + W_S \left(S_y \right) \quad (3)$$

where W_{Rx} = Resilience weights; R_{xi} = Resilience sub-metrics (Table 1); W_s = net-zero weight; S_y = net-zero metrics (Table 2).

Based on the hazard assessment for Port Arthur, Texas, the net-zero weight (Ws) is assigned solely to Global Warming Potential (GWP), while other environmental indicators (acidification, eutrophication, and ozone depletion) are excluded due to their limited direct relevance to local hazard conditions. GWP is strongly associated with the intensification of climate-related events such as hurricanes, wildfires, and extreme rainfall, making it the most applicable metric for this high-risk context. In contrast, the excluded indicators reflect broader or longer-term environmental effects that have weaker links to immediate hazard risks. Prioritizing GWP ensures that the framework remains focused and responsive to the specific challenges of the study area. However, the structure remains flexible and can incorporate additional impact categories when applicable to other geographic or regulatory contexts.

The case-specific score is computed as:

$$F.S. = 4(R_1) + 2(R_2) + 4(R_3) + 2(R_4) + 3(R_5) + 2(S_1) \quad (4)$$

Each resilience score (R_x) is calculated as the average of its normalized sub-metrics, as shown in Eq. (5), where R_x is the aggregate score, R_{xi}^{norm} represents individual normalized sub-metrics, and n is their total count.

$$R_x = R_{1\ldots5} = \frac{\Sigma R_{xi}^{norm}}{n} \quad (5)$$

2.6 Sensitivity Analysis

To evaluate the robustness of the rating system, a sensitivity analysis was conducted by varying the weights of resilience and net-zero factors by ± 50% and observing changes in the final score. The analysis focused on five primary resilience weights (hurricane, coastal flooding, severe storm, drought, wildfire) and the net-zero weight (GWP), based on Port Arthur's risk profile.

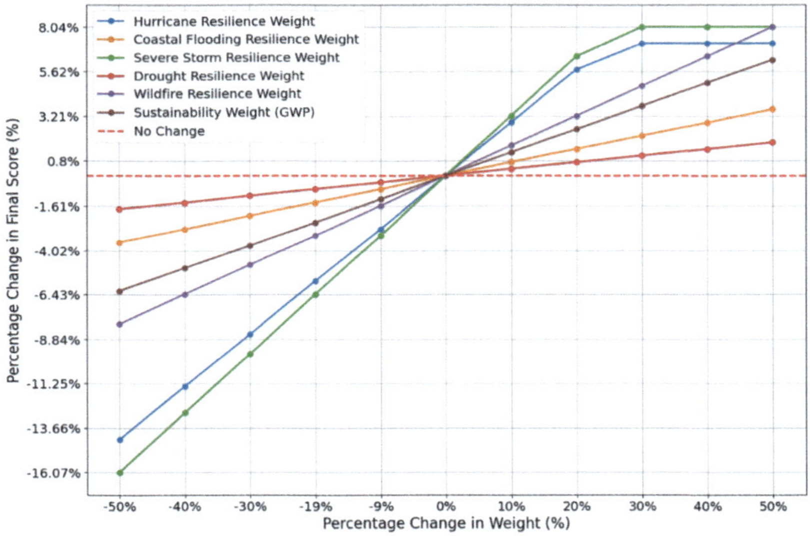

Fig. 4. Sensitivity analysis results show varying risk weights' impact on the final score.

As shown in Fig. 4 the hurricane and severe storm weights had the most significant influence, increasing the final score by 6.34% and 4.44%, respectively. Wildfire weight showed a moderate impact (2.54–5.07%), while drought and coastal flooding weights caused less than a 1.3% variation. The GWP weight had a consistent but minor effect, confirming that resilience factors have a more significant influence on score fluctuations.

The sensitivity analysis reveals that specific hazards, such as hurricanes and severe storms, have a significant impact on the final score. This suggests that in high-risk regions, designers should prioritize materials with strong wind resistance, impact durability, and corrosion protection. Policymakers can also use these results to tailor building codes or incentive programs that align with dominant local hazards, promoting context-specific resilience strategies.

3 Conclusion

This study presents an innovative framework that integrates net-zero carbon strategies with climate resilience, addressing one of the most pressing challenges in the carbon-neutral built environment. By combining multi-criteria decision analysis with GIS-based risk assessment, the framework ensures that both environmental performance and hazard resistance are balanced and context-specific. Successful validation through case studies and simulations demonstrates the practical applicability and robustness of the approach.

The proposed approach provides a structured method for evaluating building performance and sets a precedent for harmonizing net-zero and resilience metrics. This integration enables stakeholders to make informed decisions that address both current environmental challenges and future climate risks.

Further development should focus on increasing the framework's adaptability and real-time application to diverse contexts. Incorporating broader socio-economic factors and refining optimization methods will make it more practical and comprehensive. Ultimately, this framework lays the foundation for advancing net-zero and resilient building practices, promoting a future where carbon neutrality and resilience are not just compatible but mutually reinforcing.

Although demonstrated in Port Arthur, the framework is designed for broader applicability. By recalibrating hazard weights through GIS-based risk mapping and adjusting material performance criteria, it can be adapted to diverse climate zones, building types, and regulatory environments. This flexibility supports its use in both new construction and retrofitting across various geographic and socio-economic contexts.

References

1. UN Environment Programme (2020) UNEP Annual Report 2019. Accessed on 14 Jan 2025. [Online]. Available: https://www.unep.org/resources/unep-annual-report-2019
2. Stevenson F, Baborska-Narożny M (2014) Designing resilient housing for coevolutionary adaptivity. Accessed on 09 Apr 2025. [Online]. Available: https://www.semanticscholar.org/paper/Designing-resilient-housing-for-coevolutionary-Stevenson-Baborska-Naro%C5%BCny/c035f67a522703ced913c09344151d0826bb18b7
3. Smith AB (2020) U.S. Billion-dollar Weather and Climate Disasters, 1980—present (NCEI Accession 0209268). NOAA National Centers for Environmental Information. https://doi.org/10.25921/STKW-7W73
4. Phillips R, Troup L, Fannon D, Eckelman MJ (2017) Do resilient and sustainable design strategies conflict in commercial buildings? A critical analysis of existing resilient building frameworks and their sustainability implications. Energy Build 146:295–311. https://doi.org/10.1016/j.enbuild.2017.04.009
5. Kohler N (2018) From the design of green buildings to resilience management of building stocks. Build Res Inf 46:578–593. https://doi.org/10.1080/09613218.2017.1356122
6. Federal Emergency Management Agency (FEMA) (2024) National resilience guidance: a collaborative approach to building resilience. Accessed on 09 Apr 2025. [Online]. Available: https://www.fema.gov/sites/default/files/documents/fema_national-resilience-guidance_august2024.pdf
7. ISO 14040:2006 (2025) ISO Accessed on 09 Apr 2025. [Online]. Available: https://www.iso.org/standard/37456.html
8. Levasseur A et al (2016) Enhancing life cycle impact assessment from climate science: review of recent findings and recommendations for application to LCA. Ecol Indic 71:163–174. https://doi.org/10.1016/j.ecolind.2016.06.049
9. McSweeney R (2021) Explainer: what the new IPCC report says about extreme weather and climate change. Carbon Brief. Accessed on 28 Mar 2025. [Online]. Available: https://www.carbonbrief.org/explainer-what-the-new-ipcc-report-says-about-extreme-weather-and-climate-change/. Accessed on 28 Mar 2025

10. Colbert A (2022) A force of nature: hurricanes in a changing climate. NASA Science, 1 June 2022. Accessed on 28 Mar 2025. [Online]. Available: https://science.nasa.gov/earth/climate-change/a-force-of-nature-hurricanes-in-a-changing-climate/
11. U.S. Environmental Protection Agency (EPA) (2024) Climate change indicators: wildfires—technical documentation. U.S. Environmental Protection Agency. [Online]. Available: https://www.epa.gov/climate-indicators/climate-change-indicators-wildfires
12. Yu T, Ran Q, Pan H, Li J, Pan J, Ye S (2023) The impacts of rainfall and soil moisture to flood hazards in a humid mountainous catchment: a modeling investigation. Front Earth Sci. https://doi.org/10.3389/feart.2023.1285766
13. Tabari H (2020) Climate change impact on flood and extreme precipitation increases with water availability. Sci Rep 10(1):13768. https://doi.org/10.1038/s41598-020-70816-2
14. Li Z et al (2022) The conterminous United States are projected to become more prone to flash floods in a high-end emissions scenario. Commun Earth Environ 3(1):1–9. https://doi.org/10.1038/s43247-022-00409-6
15. di Filippo R, Bursi OS, Ragazzi M, Ciucci M (2022) Natech risk and the impact of high-GWP content release on LCA of industrial components. Process Saf Environ Prot 160:683–694. https://doi.org/10.1016/j.psep.2022.02.042
16. Caruso M, Pinho R, Bianchi F, Cavalieri F, Lemmo MT (2020) A life cycle framework for the identification of optimal building renovation strategies considering economic and environmental impacts. Sustainability 12:23. https://doi.org/10.3390/su122310221
17. Bohorquez A, Viteri E, Rivera E, Avila C (2024) Environmental impact of earthquake-resistant design: a sustainable approach to structural response in high seismic risk regions. Buildings 14(12):3821. https://doi.org/10.3390/buildings14123821
18. Berberian AG, Morello-Frosch R, Karasaki S, Cushing LJ (2024) Climate justice implications of Natech disasters: excess contaminant releases during hurricanes on the Texas Gulf Coast. Environ Sci Technol 58(32):14180–14192. https://doi.org/10.1021/acs.est.3c10797
19. GBCI and Perkins+Will (2018) RELi resilience action list + credit catalog v3.1: a national consensus-based standard for resilient design. Green Business Certification Inc. (GBCI), 2018. [Online]. Available: https://www.gbci.org/sites/default/files/RELi%20mandatory%20users%20guide.pdf

Toward Balanced Environmental Strategies in Building Design: A Performance-Driven Framework for Holistic Impact Reduction

Maryam Abbasi Kamazani and Manish K. Dixit(✉)

Department of Construction Science, Texas A&M University, College Station, Texas, USA
mdixit@tamu.edu

Abstract. Designing sustainable buildings that balance energy efficiency, carbon mitigation, and water conservation remains a pressing challenge in the built environment. These performance indicators are interconnected but often conflict during the design and construction phases. This study introduces a comprehensive computational framework aimed at optimizing a commercial case study building in Athens, Georgia, across three dimensions: total energy demand, total carbon emissions, and embodied water use. Leveraging evolutionary multi-objective optimization with NSGA-II, the framework integrates EnergyPlus, building energy simulation, and material impact data to explore trade-offs among design variables. Optimization results demonstrate reductions of over 30.06% in total primary energy use, 28.28% in total carbon emissions, and 26.12% in total embodied water compared to a conventional baseline. This research contributes a holistic strategy for sustainable design decision-making, offering insights into how building geometry, envelope systems, and material selections collectively influence environmental performance. The findings emphasize the necessity of considering water alongside energy and carbon to avoid sustainability blind spots and promote balanced environmental outcomes.

Keywords: Holistic building performance · Building envelope optimization · Environmental impact trade-offs

1 Introduction

Buildings are responsible for nearly 40% of global energy consumption and approximately 30% of global greenhouse gas (GHG) emissions, positioning the construction industry as a key sector in the global sustainability agenda [1–3]. As global urbanization and infrastructure demands continue to rise, reducing the environmental impacts of buildings throughout their entire life cycle has become increasingly critical. Historically, efforts to improve building sustainability have focused predominantly on reducing operational energy through advanced HVAC systems, improved insulation, and efficient lighting [4, 5].

However, as buildings become more efficient during use, the proportion of life cycle impacts attributed to the construction and materials phase, embodied energy, embodied carbon, and embodied water, grows increasingly significant [6, 7]. For example,

H. Beushausen et al. (Eds.): NTZR 2025, LNCE 873, pp. 1061–1069, 2026.
https://doi.org/10.1007/978-3-032-27698-8_93

materials chosen to enhance energy efficiency, like aerogels or high-performance glass, may require carbon-intensive manufacturing or substantial water inputs. These trade-offs illustrate the complexity of the energy-carbon-water (ECW) nexus, where optimizing one sustainability metric can inadvertently worsen another [8, 9].

Despite this interconnectedness, existing design frameworks often optimize energy, carbon, and water in isolation [10, 11]. A holistic optimization approach is needed to balance these dimensions simultaneously and avoid unintended environmental burdens. This study proposes a multi-objective optimization framework using the NSGA-II algorithm to minimize total energy use, total carbon emissions, and total embodied water. The approach is tested on an existing commercial case study building, aiming to produce a decision-support model that reflects the ECW trade-offs inherent in sustainable building design.

2 Literature Review

The field of sustainable building design has evolved from a focus on operational performance to a broader consideration of life cycle impacts. While operational energy remains a major contributor to GHG emissions, advances in passive design strategies and energy-efficient technologies have significantly reduced its share in modern buildings [12]. As a result, attention has shifted to embodied energy (EE) and embodied carbon (EC), metrics that account for emissions and energy use during raw material extraction, processing, transport, and construction. These embodied impacts are now recognized as critical components in life cycle assessments of buildings [13].

Embodied water (EW), although less widely studied, is gaining attention as a vital sustainability indicator. EW includes water used directly in material processing and indirectly via energy production and transportation. Its relevance is heightened in water-stressed regions and under growing global water scarcity concerns [14]. Yet, most building performance assessments still exclude EW or treat it independently of EE and EC, missing the opportunity to capture the full scope of environmental trade-offs.

The ECW nexus, the intersection of energy, carbon, and water, has been discussed in broader infrastructure and industrial contexts, but its integration into building-level design and optimization remains limited. Where multi-objective frameworks have been applied, they typically focus on two objectives (e.g., cost and energy or carbon and cost), rather than incorporating all three key environmental dimensions [3, 15].

The Non-dominated Sorting Genetic Algorithm II (NSGA-II) has emerged as a robust method for handling such multi-objective problems. Its ability to generate diverse Pareto-optimal solutions has been demonstrated across various applications, including hybrid energy systems [6], modular construction design [2], and solar water heating systems [5]. Enhancements to NSGA-II have further improved convergence speed and solution diversity, making it suitable for complex, real-world building applications [1]. Despite these advances, few studies have applied NSGA-II to optimize total energy, carbon, and water together in building projects. This study aims to address this research gap by embedding all three metrics into a single optimization framework tailored for sustainable building design.

3 Research Goals

This research aims to improve how environmental considerations are integrated into early-stage building design by developing a computational framework capable of evaluating multiple sustainability metrics at once. Recognizing the interconnected and sometimes conflicting nature of these metrics, the study brings together parametric modeling, simulation-based performance analysis, and embodied impact assessment to support more informed and balanced design decisions.

The specific objectives of the study are:

1. **To develop a comprehensive performance evaluation framework** that links simulation engines and environmental impact databases. This approach allows for assessment of both operational and embodied impacts, encouraging a full life-cycle perspective on building performance.
2. **To explore and quantify trade-offs among key environmental indicators** using a multi-criteria evolutionary algorithm. Rather than optimizing a single factor, the method identifies a range of high-performing solutions that illustrate how improvements in one area may affect others.
3. **To apply and demonstrate the framework using a commercial building as a case study building**. The case study helps validate the approach and provides practical insights for designing environmentally responsive buildings by analyzing how various design choices, such as building orientation, geometry, material selection, and envelope configuration, affect environmental outcomes. These insights aim to support more balanced, effective decision-making in architectural and engineering practice.

This study offers a new framework for supporting building design that is efficient, environmentally responsible, and grounded in a holistic understanding of performance over time.

4 Methodology and Case Study Building

This study proposes a comprehensive optimization framework for the integrated evaluation of building energy consumption, total carbon emissions, and embodied water usage. The methodology consists of three interlinked modules: parametric modeling and simulation integration, multi-objective optimization, and environmental performance assessment.

A parametric model is developed using 16 design variables, encompassing building orientation, window-to-wall ratio (WWR) on four cardinal elevations, and envelope component specifications such as wall cladding, insulation materials, glazing types, roofing assemblies, and floor systems. The variables are integrated into a design framework, enabling the system to automatically generate multiple design alternatives by adjusting input parameters. Each configuration is translated into an input dataset compatible with EnergyPlus, using site-specific weather files to simulate operational energy. Operational energy is computed for heating, cooling, lighting, and equipment loads. Embodied energy, embodied carbon, and embodied water are calculated through a sophisticated hybrid input-output analysis framework. Total primary energy (TE), total carbon emissions (TC), and total embodied water (TW) served as the three objective functions. Optimization is performed using the Non-dominated Sorting Genetic Algorithm II (NSGA-II), configured with a population size of 200 and evolved over 300 generations. The algorithm employed selection, crossover (30%), and mutation (10%) operators to explore the high-dimensional solution space. Solutions are evaluated through Pareto-based dominance, generating a frontier of optimal trade-offs among the three environmental metrics. A fuzzy decision-making method is employed to identify a balanced optimal solution reflecting equal importance of energy, carbon, and water objectives.

The case study used to demonstrate the proposed framework is a commercial building located in Athens, Georgia, USA. The building has a total floor area of 447.07 m^2 and a floor height of 4.75 m. Its orientation is set at 70°, with façade-specific window-to-wall ratios of 21% (North), 18% (South), 0% (East), and 3% (West), strategically configured to manage solar exposure and daylight penetration. The exterior wall assembly consists of a brick veneer finish, 1.5-inch extruded polystyrene (XPS) board insulation (R-5), and 12.7 mm OSB sheathing. Glazing components incorporate 1-inch insulating vision lite with a 13 mm air layer. The floor construction includes a 150 mm normal-weight concrete slab finished with luxury vinyl tile (LVT), while the roof system comprises standing seam metal cladding, two layers of rigid polyisocyanurate insulation (R-25 combined), a high-temperature self-adhered underlayment, OSB roof sheathing, and a structural metal deck. This configuration serves as the baseline model for assessing environmental performance improvements through the proposed optimization framework. The 3D model of the case study building is presented in (Fig. 1). Although this study is presented as a single demonstrative case due to space constraints, the framework's modular architecture renders it fully transferable to any building type, form, or climate context. Users need only supply (i) an EPW weather file to establish local weather conditions, (ii) an IDF model encapsulating building geometry, envelope assemblies, and simulation parameters, and (iii) a materials database comprising embodied energy, embodied carbon emissions and embodied water coefficients. Substituting these inputs with site—or typology-specific datasets, whether residential, commercial, or industrial prototypes in arid, temperate, cold, or other climates, enables deployment of the identical computational workflow without any code modification. All EE, EC and EW coefficients are obtained from a hybrid input–output model calibrated to the U.S. economy. Comprehensive details of the database construction and the underlying econometric methods are available in [16].

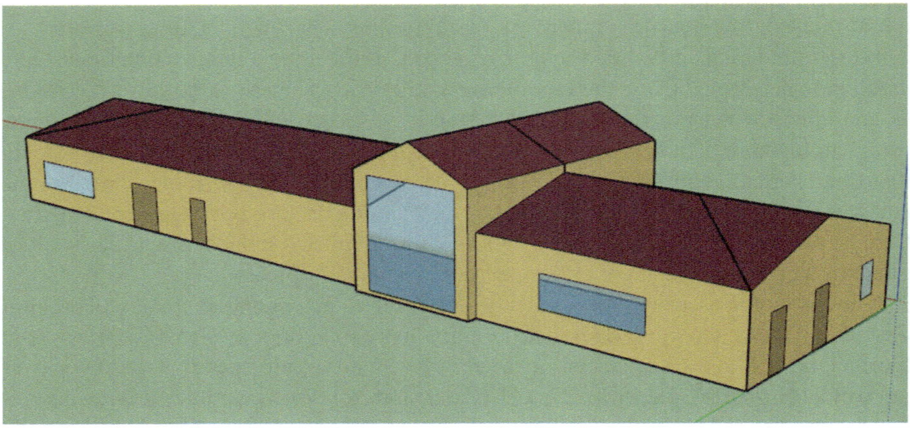

Fig. 1. Case study building 3D model located in Georgia.

5 Results and Discussion

The application of the proposed optimization framework provides valuable insights into the simultaneous reduction of environmental impacts across three critical building performance metrics: total primary energy use, total carbon emissions, and embodied water consumption. Using a fuzzy decision-making approach, an optimal solution is selected from the generated Pareto front, representing a balanced trade-off among the competing objectives. The results indicate notable improvements compared to the baseline model, including a 30.06% reduction in total energy demand, 28.28% reduction in total carbon emissions, and 26.12% reduction in embodied water use.

A series of analytical visualizations (Figs. 2, 3 and 4) illustrate the relationships among the environmental indicators. A strong positive correlation is observed between total energy use (TE) and total carbon emissions (TC), as shown in (Fig. 2), where the optimal solution aligns with the lower boundary of the TE-TC trendline. In contrast, (Fig. 3) highlights an inverse relationship between carbon emissions and embodied water (TW), suggesting that certain low-carbon material strategies may unintentionally increase water-related impacts. Similarly, (Fig. 4) reveals an inverse correlation between TE and TW, emphasizing that strategies aimed at reducing energy demand do not always yield reductions in water consumption.

These findings underscore the complexity of the energy-carbon-water (ECW) nexus and highlight the need for integrated assessment frameworks that account for interdependence between sustainability metrics.

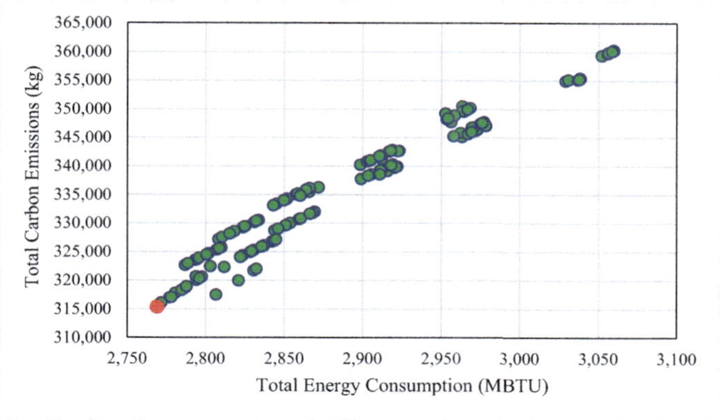

Fig. 2. Pareto front solutions and final solution comparison in terms of TE and TC.

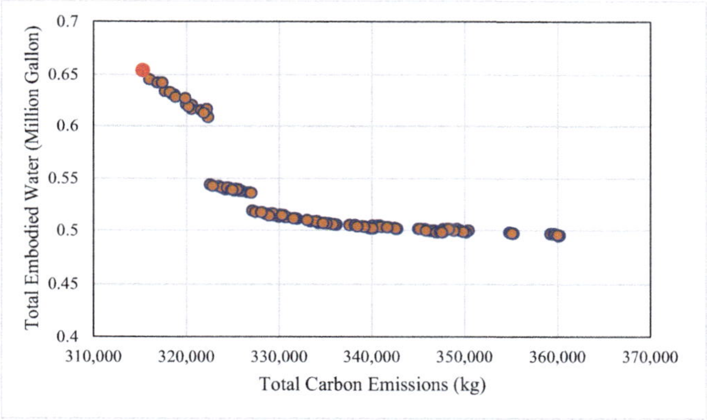

Fig. 3. Pareto front solutions and final solution comparison in terms of TC and TEW.

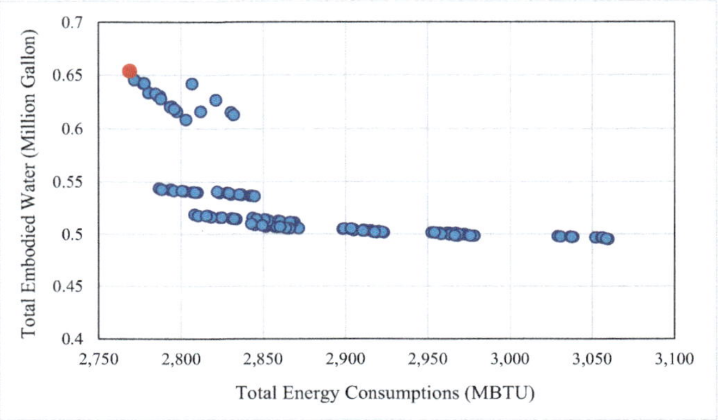

Fig. 4. Pareto front solutions and final solution comparison in terms of TE and TEW.

In the optimized configuration, several design modifications are made to enhance environmental performance across energy, carbon, and water dimensions. The building orientation is maintained at 70°, but the window-to-wall ratios (WWR) are strategically adjusted to balance daylighting and thermal gain, with values of 20% (North), 35% (South), 0% (East), and 15% (West). The exterior wall assembly shifts from heavier masonry to a more sustainable layered structure composed of F11-grade cedar bevel wood siding, standard-density OSB sheathing, and 6-inch loose-fill fiberglass insulation (R-24), significantly reducing embodied carbon and water impacts. Glazing consists of 1-inch insulating glass units with a 13 mm air gap, maintaining thermal resistance while limiting material intensity. The flooring system retains a 6-inch reinforced concrete slab but used baseline luxury vinyl tile (LVT) as a lightweight, low-impact finish. The roofing assembly incorporates a coated glass fiber base sheet (Type IV) with OSB cover board, 1.5-inch polyisocyanurate insulation (R-8.7), OSB roof sheathing, and galvanized steel decking, demonstrating a shift toward high-performing, thermally resistant, and lower-embodied-impact components. Collectively, these materials and envelopes contribute to the overall reductions in energy demand, carbon emissions, and embodied water use in the final optimized design.

6 Conclusions and Future Work

This research presented and applied a performance-based optimization framework designed to reduce environmental impacts in building design, with a focus on resource consumption across energy, emissions, and water. Utilizing a commercial case study, the framework effectively identified key trade-offs and provided actionable insights for environmentally informed design decisions. A significant finding was that embodied impacts, across energy, carbon, and water, frequently surpassed operational impacts, emphasizing the critical role of material selection in comprehensive sustainability evaluations.

To broaden the applicability and depth of the framework, future work should consider:

- Incorporating economic dimensions, such as life cycle cost analysis,
- Integrating metrics related to occupant well-being and indoor environmental quality,
- Employing stochastic methods to account for uncertainties in material data and operational behaviors.

Expanding the framework in these directions would enhance its ability to guide more adaptive, context-sensitive, and holistically sustainable building designs.

References

1. Zhao X, Chen Y, Liu J (2024) Life-cycle performance optimization for low-carbon buildings using advanced evolutionary algorithms. J Clean Prod 412:137314
2. Li Y, Zhang H, Sun Q (2024) Multi-objective design of modular buildings: a lifecycle perspective. Autom Constr 157:105101
3. Zhang X, Wu C, Li B (2024) Integrating environmental and economic criteria in building optimization: a review. Energy Build 295:113408
4. Bai X, Xue F (2024) Operational energy performance in high-efficiency buildings: a review of trends and technologies. Energy Rep 10:321–336
5. Ko M (2015) Optimal configuration of solar-assisted water heating systems using NSGA-II. Renew Energy 83:259–267
6. Ren Z, Wang Y, Zhou D (2018) Life-cycle energy and carbon implications of building design decisions in cold regions. Build Environ 136:178–190
7. Zhou Y, Chen R, Zhang J (2022) Embodied carbon mitigation in construction: material-based strategies and implications. Sustain Cities Soc 78:103568
8. Chen S, Tan Y, Liu Z (2019) Direct and embodied energy-water-carbon nexus at an inter-regional scale. Appl Energy 251:113401
9. Dixit M, Kumar PP, Shanbhag S (2023) Analyzing embodied energy and embodied water for university buildings using input-output-based hybrid method. IOP Conf Series: Earth Environ Sci 1234(1):012047
10. Shadram F, Mukkavaara J (2018) An integrated BIM-based framework for the optimization of the trade-off between embodied and operational energy. Energy Build 158:1189–1205
11. Dixit M, Kumar PP (2023) Analyzing water use embodied in the initial construction and life cycle management of healthcare facilities. IOP Conf Series: Earth Environ Sci 1234(1):012011
12. Mei S, Liu C, Zhao H (2020) Evolution of energy performance in sustainable architecture: a historical review. Renew Sustain Energy Rev 119:109591
13. Dai J, Zhang L, Xu Y (2022) Quantifying embodied carbon in urban building stocks: a hybrid approach. Build Res Inform 50(2):123–139
14. Yu Y, Huang T, Gao X (2024) Assessing embodied water in building construction: a framework for water-sensitive material selection. J Environ Manage 341:117035

15. Wu C, Zhang X, Liu W (2019) Multi-objective optimization for building retrofits: balancing cost, energy, and carbon. Energy Build 183:257–267
16. Kumar PP, Venkatraj V, Dixit M (2022) Evaluating the temporal representativeness of embodied energy data: a case study of higher education buildings. Energy Build 254:111596

Comparison and Evaluation of Various BIM-LCA Strategies for Integrated Factory Design

Leonie Große-Wilde[1](✉), Luca Philipp[2], Sven Mackenbach[1], Matthias Schmidt[1,2], and Katharina Klemt-Albert[1]

[1] Institute of Construction Management, Digital Engineering and Robotics in Construction, 52070 Aachen, Germany
grosse-wilde@icom.rwth-aachen.de

[2] Institute of Production Systems and Logistics, 30823 Garbsen, Germany

Abstract. Environmental sustainability, in particular Life Cycle Assessment (LCA), is increasingly recognized as an important factor in holistic factory design. As it requires a lot of data, Building Information Modelling (BIM) is progressively seen as an important data source for building LCA. However, factory design is a very interdisciplinary process and different factors such as factory layout, energy system and building structure influence the most sustainable factory design. Therefore, the requirements for an appropriate BIM-LCA strategy differ from other building types. This paper examines current strategies for BIM-LCA integration to assess their suitability for the factory design process, with a particular focus on the specific challenges faced by small and medium-sized enterprises (SMEs). In order to achieve this objective, the requirements of the integral factory design with regard to an optimal integration strategy are identified. A literature review serves as the methodological basis for this analysis, providing insight into existing approaches, their strengths and limitations. The findings indicate that current BIM-based LCA strategies do not adequately address the specific requirements of factory design, revealing substantial usability and integration issues. A potential solution is an openBIM approach with semi-automated processes and real-time feedback, which could enhance usability for SMEs and support holistic lifecycle evaluations.

Keywords: Factory · Sustainability assessment · LCA · BIM · Integrated design

1 Introduction and Motivation

The industry and building sectors are among the largest contributors to global CO_2 emissions [1]. As a result, companies face increasing pressure to operate sustainably and to be able to demonstrate the sustainability of the company's assets [2, 3]. The European Green Deal sets an ambitious target of achieving zero net greenhouse gas emissions by 2050, underscoring the urgent need for environmental sustainability [4]. In this pursuit, Life Cycle Assessment (LCA) is a key method to evaluate the environmental impacts of

H. Beushausen et al. (Eds.): NTZR 2025, LNCE 873, pp. 1070–1081, 2026.
https://doi.org/10.1007/978-3-032-27698-8_94

a product or system [5]. Building Information Modeling (BIM) has significant potential to improve the efficiency of assessing the environmental sustainability of buildings, and BIM-LCA integration is currently the subject of numerous studies [6]. While Wastiels and Decuypere have identified five established strategies for integrating BIM and LCA [7], these approaches have yet to be explored in the context of factory design. The challenge of factory design lies in the synergetic integration of production planning, technical building equipment and building structure. This characteristic implies specific requirements for a BIM-LCA integration strategy suitable for factory design. Since 97% of manufacturing companies in Germany are small and medium-sized enterprises (SMEs) [8], there are also specific requirements to ensure applicability for this target group.

These two circumstances, the characteristics of factory design and the applicability to SMEs, are the basis for the following requirements analysis in the next chapter. Based on the requirement analysis, the existing BIM-LCA integration strategies are then introduced and compared with the identified requirements. Finally, a conclusion is drawn as to which strategy fulfils the most requirements of factory design and where factory-specific modifications are required.

2 Requirements for BIM-LCA Integration in Factory Planning

Implementing a suitable BIM-LCA integration strategy in factory planning significantly differs from its application in object planning. To ensure efficient and practical use, the BIM-LCA strategy must meet essential requirements. These requirements result in particular from the specific framework conditions of factory planning and the target group of SMEs. The requirements are based on literature and include a holistic assessment from a building and process view, the integration of different disciplines, the manageability of the BIM model, and low effort applicability by SMEs.

Holistic assessment from a building and process view

As a complex socio-technical system, the factory is defined as a place of value creation [9] and can be divided into the building and process view. Only by integrating these perspectives closely, both in factory design and operation, can an efficient factory be realized and operated [10]. This principle also applies to the BIM-based assessment of a factory's sustainability. This is mainly due to the fact that there are interactions between the elements of the factory from the process and building point of view [11, 12]. For example, the selection of a means of production directly influences the dimensioning of the foundation or the supporting structure since these transfer loads of the means of production into the ground, for example, through foundations or piles. Changes in one perspective therefore often require adjustments in the other perspective. A hotspot analysis of a single perspective can identify critical elements, but it does not consider their dependencies and effects on the other perspective and its elements. Although it is possible to evaluate the individual perspectives in isolation, this does not do justice to the holistic character of a factory. Accordingly, a suitable BIM-LCA strategy must be based on a holistic and synergetic consideration of the factory from the process and building view and carry out the assessment.

Integration of different disciplines

Factory planning is a highly complex and interdisciplinary process involving various disciplines [13, 14]. Up to 19 different disciplines can be involved in a factory planning project [15], which requires close coordination between the parties involved and makes the integration of specialized models more difficult. At the same time, a large amount of heterogeneous data is created that needs to be linked to ensure a mistake-free and efficient planning project [16]. This heterogeneity of data results from different data formats, software applications and interfaces that are often not aligned with each other [16]. Additionally, challenges arise in standardizing data descriptions and parameter designations, making precise agreements between disciplines essential [17]. This is relevant to both the attribution of the BIM model and the subsequent sustainability assessment. Standardized and high-quality databases, such as ÖKOBAUDAT [11], which includes specific Environmental Product Declarations (EPDs), already exist for the construction materials. However, data availability in the process view is characterized by individuality. Therefore, a suitable BIM-LCA strategy must offer solutions for both the data's heterogeneity and the interoperability of different disciplines.

Manageability of the BIM model

Another important aspect that must be considered is the manageability of the BIM model due to the size of the model. At the time of the evaluation, the amount of data from the building view is already substantial in the detailed planning phase. This is primarily because the objects and elements that have been geometrically described are now complemented by alphanumeric information [18]. Comprehensive information is available for the individual objects from the building view, which is required for the actual realization. In addition to the geometries and the defined positioning in the room, this also includes information on materials, connectors, dependencies, etc. [15]. Similar behavior can also be seen in the process view: here, the elements not only contain geometric data, weight and space requirements but also specific machine data and user-specific information [16]. The substantial volume of data requires efficient management and processing. The BIM-LCA strategy must take this into account in order to avoid additional demands on SMEs' IT infrastructure.

Low effort applicability by SMEs

In particular, a BIM-LCA strategy for SMEs must be resource-efficient and feasible. Since the factory planning process is iterative, the environmental sustainability assessment can be carried out multiple times. However, with each iterative step, the data availability and depth of information gradually increase, so a continuous update of the model is required [19].

In addition to the growing amount of data and the associated additional effort, the lack of competence in implementing and using BIM in SMEs also represents a central challenge. The lack of knowledge and expertise in BIM has been cited as one of the biggest obstacles to its adoption [20, 21]. Many modelers find the implementation of alphanumeric information particularly challenging, especially if the companies have not received training in BIM methodology [16]. To address these challenges, the BIM-LCA strategy needs partial automation. SMEs often lack sufficient human and financial resources [22, 23], so a low effort solution is required. Semi-automation offers a practical balance [24]. The strategy should not require in-depth software and programming knowledge, enable consistent data storage with minimal effort, and not cause additional overhead due to repeated manual linking of the BIM model to the respective databases.

3 State of the Art BIM and LCA Integration

The focus of the following analysis is to compare the requirements of factory design with the five BIM-LCA integration strategies proposed by Wastiels and Decuypere [7]. Since these strategies address the type of data exchange and data flow, this will be the focus of the analysis. As the identified requirements also focus heavily on finding a low effort solution, the second focus of this work will be the degree of automation and whether the type of data extraction is static or dynamic.

BIM-LCA integration strategies can be classified as openBIM or closedBIM, which are two key concepts that influence data flow. OpenBIM is a collaborative approach that promotes interoperability and data exchange among various software platforms, whereas closedBIM involves proprietary data formats that limit data sharing and interoperability [25]. The first of the five strategies by Wastiels and Decuypere involves exporting the Bill of Quantities (BoQ) to a spreadsheet and calculating the LCA assessment either directly in the spreadsheet or importing it into an LCA software. The materials and quantities are therefore manually linked to their respective LCA profiles [7]. This workflow displays the state of the art of BIM and LCA integration [26]. The second strategy is the Industry Foundation Classes (IFC) export to LCA software and is an openBIM approach, since it uses a standardized open exchange format like IFC. In this strategy, the geometry, material names, and Global Unique Identifier (GUID) of the BIM objects are exported to IFC and subsequently imported into the LCA software. Here, the materials are manually linked to the LCA profiles and the actual assessment is carried out [7]. In their third strategy, Wastiels and Decuypere also describe an openBIM workflow where, after the IFC export, a BIM viewer is inserted to link the LCA profiles to the materials before the LCA is performed in the LCA software. The fourth strategy is a closedBIM approach that works exclusively within the native BIM authoring software using a plug-in. Within this plug-in, the BIM objects are linked to the LCA profiles based on the databases available in the plug-in. The results can then be exported and visualized in the 3D model. This gives designers instant insight into hotspots and feedback on their design decisions. Finally, in the fifth strategy, LCA data is directly attributed to the BIM objects. Either the unique identifier of the material in the selected database or the actual values of various indicators are assigned. This strategy can be applied to both openBIM and closedBIM workflows, as the assessment can be conducted using either a plug-in or LCA software based on an IFC file [7].

After briefly describing the five strategies for BIM-LCA integration by Wastiels and Decuypere, the main objective of this work is to compare them with the identified requirements in the context of factory design. Therefore, it is necessary to obtain information about these strategies in terms of effort, degree of automation, suitability for an iterative design process, skills required, and so forth. This information will be gathered through a literature review focusing on case studies that have implemented these strategies. The literature is searched in the Web of Science and Scopus databases and is limited to open access publications from 2019 to 2025. The research is carried out using specific keywords that focus on BIM-LCA integration, case studies and the strategies individually. Furthermore, the Snowball method was used to find additional literature.

4 Comparison of the Strategies with the Requirements

The general requirements for sustainable factory design identified in Chap. 2 translate into ten specific requirements that a BIM and LCA integration strategy should meet. The requirements for ensuring a holistic assessment from a building and process view, as well as the integration of multiple disciplines, will be evaluated by checking the ability to provide real-time feedback on the sustainability of the factory and the ability to integrate a variety of individualized databases for LCA data. To meet the general requirement of low effort and applicability by SMEs, the strategy must allow for integration into an iterative design process, be easily integrated into the company's software portfolio, be semi-automated and, finally, require little effort and programming skills. The requirement "Manageability of the BIM model" is evaluated by analyzing whether the strategy has an impact on the size of the model. The strategies are analyzed individually in the following subchapters. Based on the findings, a qualitative evaluation is carried out for each specific requirement, which are shown in Table 1, where "–" indicates a requirement that is not met, "o" indicates a requirement that is met with limitations, and " + " indicates a requirement that is met completely.

Table 1. Comparison of requirements and integration strategies

Requirements	1	2	3	4.1	4.2	5.1	5.2
	BoQ	IFC	Viewer	Plug-In	VPL	Closed	Open
Iterative planning	–	0	+	+	+	+	+
Real-time feedback	–	–	–	+	+	+	–
Low effort	–	0	0	+	+	–	–
Low skill level	+	+	–	+	–	0	0
Manageability of the model	+	+	0	+	+	–	–
Holistic factory assessment	–	0	+	–	–	+	+
Integration software portfolio	+	+	+	–	–	+	+
Heterogeneous data sources	–	–	+	–	+	+	+
Semi-automation possible	0	+	+	+	+	+	+
Total	3	4	5	6	6	6	5

4.1 Bill of Quantities Export to LCA Software

This strategy represents the conventional method to integrate BIM and LCA. It is not suitable for the iterative design process, which requires dynamic data flow between BIM objects and LCA profiles [27, 28]. As described in Chap. 3, in this strategy the data flow is static and has to be established every time a change is made to the 3D model [7]. While this strategy requires the highest manual effort and is most prone to errors [29–31], the implementation does not require many additional skills [28, 32]. Since the spreadsheet containing the materials and quantities is manually modified, it is possible to integrate LCA data from heterogeneous databases. The same applies to the integration of different planning disciplines. However, this strategy does not provide a holistic and easily interpretable view of a factory's hotspots.

To summarize, the strategy requires significant effort but has limited applicability to factory design. This is primarily due to its inability to display a holistic factory assessment and the high manual effort necessary. Nonetheless, this method does allow for significant individualization.

4.2 Geometric IFC Export to LCA Software

The concept of openBIM workflows is critical when multiple disciplines are working together on a design project, as openBIM allows for a system-oriented approach [33]. This fulfills one of the central requirements identified in the context of factory planning. Additionally, IFC offers interoperability with other tools and a standardized data format [18, 24], which ensures a simple integration in the software portfolio of a company. Since this strategy implies that no coordination model is used, the integration of all discipline models requires pre-processing or else depends on the LCA software. For example, a German LCA software for buildings, Circular LCA by Concular [34], allows the integration of multiple IFC models, but this varies depending on the software.

While this strategy requires less manual data entry than the BoQ export [31], the linking of LCA profiles to the materials still involves a lot of effort [35]. In the standard scenario of implementing this strategy by Wastiels and Decuypere, there is no dynamic data flow between LCA data and BIM objects [7]. This also means that this strategy is not suitable for the iterative design process and that there is no real-time feedback, but rather at certain defined points in the planning process. However, there is ongoing research on how to create a dynamic data flow [33], which implies that the permanent link between BIM object and LCA data is possible and that it results in significant advantages for using IFC in BIM and LCA integration [36].

Nevertheless, it should be noted that IFC is still at the beginning of implementation for manufacturing equipment [37], which represents a significant hurdle in the application of this strategy for holistic factory design. In summary, this strategy reduces the required amount of effort compared to traditional workflows and offers a lot of potential due to the open data exchange format. However, this strategy is still insufficient in terms of providing real-time feedback on design choices as well as its suitability for an iterative design process.

4.3 IFC Export, BIM Viewer and LCA Software

This strategy also follows an openBIM approach and the characteristics are very similar to the IFC export to LCA software, but the main difference is that the mapping of LCA data and BIM objects is done in a BIM viewer or a model checker software. Unlike before, a coordination model is used for assessment, integrating all disciplines and allowing for a holistic hotspot analysis. Incorporating a BIM viewer into the process allows for the integration of multiple databases and linking to the unique identifier of the selected database, as demonstrated by Theißen et al. using the software Desite [38]. This means that a dynamic connection can be established, supporting an iterative design process. However, to create a dynamic link and semi-automate the assessment, advanced programming skills in Java as well as HTML are required to connect to databases such as ÖKOBAUDAT [38].

As with the previous strategy, real-time feedback on design decisions is not supported. However, a holistic hotspot visualization using a coordination model of all disciplines is possible in a BIM viewer [39]. Since there is no literature on the required effort for implementing this strategy, it is assumed to be similar to the previous strategy.

Consequently, this strategy offers potential for the interdisciplinary nature of factory planning, but implementing a dynamic link requires advanced programming skills, and real-time feedback on design decisions remains unsupported.

4.4 BIM Software and Plug-In

The application of a plug-in provides a high degree of automatic calculation of the results after linking with the LCA data [29] and provides real-time feedback on the sustainability of the planned object [7, 31]. Consequently, planners become more aware of the impact of design decisions on environmental sustainability [40] and can take action during the design phase. Furthermore, this strategy supports iterative design the most [29, 31] as there is a dynamic connection between the BIM object and the LCA data. This strategy is therefore easily integrated into the design process [40]; moreover, it is the most efficient strategy [29]. It has also been demonstrated to result in fewer errors [40].

However, when plug-ins are used to do building LCAs, a distinction should be made between Visual Programming Languages (VPL) like Dynamo or ready-to-go market solutions for LCA plug-ins like One Click LCA. The latter often times have integrated proprietary databases [41]. These two solutions address different requirements of the factory design process: While LCA plug-ins are user-friendly, intuitive and easier to understand, VPLs require basic programming skills and initial implementation effort [32, 42]. On the other hand, VPLs like Dynamo have the potential to highly customize LCA for specific use cases, building types, and databases [28, 29]. Also, VPLs enable visualization of hotspots in the 3D model by using different colors or transparencies [43]. Plug-ins, on the contrary, often use generic databases [36, 41], and even if you can implement specific EPDs for building components, it is not as flexible for integrating LCA profiles for machinery used in a factory. Therefore, the flexibility provided by VPL has a lot of potential for the holistic assessment of a factory, but it does not meet the requirements of being easily applicable and requiring little effort, at least in the initial implementation. Both VPL and LCA plug-ins are closedBIM solutions, which limit their implementation in an SME's software portfolio. In addition, closedBIM solutions for BIM and LCA integration do not allow for the integration of different disciplines because each discipline model is evaluated in its proprietary format. Therefore, there is no holistic evaluation, especially if the different planners are not working in the same BIM software, which is less likely as the number of disciplines increases.

In conclusion, VPLs have the potential to integrate heterogeneous data sources needed to assess both buildings and production, but plug-ins are still closedBIM solutions that require a lot of effort and workarounds to integrate a couple of disciplines. Therefore, this strategy, as defined by Wastiels and Decuypere and current research, is not suitable for factory assessment.

4.5 Enriched BIM Objects

The fifth strategy by Wastiels and Decuypere includes enriched BIM objects and can be performed by integrating openBIM or closedBIM workflows, both enabling a dynamic data extraction. Nevertheless, the validity of the data has to be regularly checked and the maintenance of the data sets over several planning phases can be extensive [29]. Attaching numerous attributes may result in large models that are unmanageable on typical architecture firm computers [36]. Consequently, this approach is not well-suited for large and complex models [29, 33], as factories tend to be.

On the other hand, there comes a flexibility with this strategy, as this can be done in many different BIM software and there is no dependency on non-editable external databases [44]. This is crucial when applying BIM and LCA integration strategies to factory planning. While the automation potential of this strategy is high, e.g. by using VPL to compute and visualize the results or to enrich the objects [45], the preliminary work that needs to be done to be able to enrich the objects is extensive [28, 29]. The reason for this is that there are currently no standardized model component libraries with enriched objects to choose from: either the planner has to add the information during the planning process, or their company has an established library [46], which is not yet realistic.

There are several differences between integration into an openBIM or closedBIM workflow in terms of meeting the requirements from a factory design perspective. As described earlier, the integration of multiple disciplines is not possible in an exclusively closedBIM workflow, but the potential for automation in the evaluation is, on the contrary, very high [47]. When implementing enriched BIM objects in an openBIM workflow, there is the potential to integrate multiple disciplines and perform a holistic sustainability assessment [45], but data loss because of inadequate IFC structures is possible [29].

In summary, the use of enriched BIM objects for the LCA of factories is not yet suitable due to the lack of appropriate model component libraries. As a result, this workflow demands a considerable amount of effort. Even with a high degree of automation, there is still a lot of preliminary work required. Nevertheless, this strategy is convenient for integrating LCA data of production equipment in the context of factory design.

5 Conclusions

The implementation and research of BIM-based LCA in building construction has been increasing in recent years, as it offers improved efficiency in environmental sustainability assessment. However, there are certain requirements that arise from the assessment of holistic factories, including the assessment of building structure and process equipment. The overarching requirements are ease of use by SMEs, the ability to integrate a higher number of disciplines involved in a factory design project, and the manageability of the BIM model.

The five strategies by Wastiels and Decuypere are then compared with the requirements, with the result that the BoQ export to LCA software is mainly lacking in the required effort, while the IFC export is not suitable for an iterative design process and real-time feedback on ecological hotspots. Integrating a BIM viewer into the workflow has strengths in integrating heterogeneous data sources and the high degree of automation, but the skill level required to realize this potential is very high. While LCA plug-ins provide real-time feedback and low effort, they do not allow the holistic evaluation and integration of proprietary databases. Using a VPL such as Dynamo helps to individualize the assessment but increases the required programming skills. Finally, enriched BIM objects offer a lot of potential for LCA of factories, but some fundamentals need to be built to make this a low effort and SME friendly workflow.

In summary, current BIM LCA strategies address up to two-thirds of the requirements, leaving room for improvement. An openBIM approach with semi-automated processes and real-time feedback offers potential, but further research and expert prioritization of requirements is needed.

References

1. International Energy Agency (IEA) (2023) Global CO_2 emissions by sector, 2019–2022. 2023. Accessed on 30 Mar 2025. [Online]. Available: https://www.iea.org/data-and-statistics/charts/global-co2-emissions-by-sector-2019-2022
2. HypoVereinsbank (2022) Branchenstudie—nachhaltigkeit im mittelstand: status—chancen—ausblick
3. Hanns Seidel Stiftung (2023) Nachhaltigkeitskriterien im Mittelstand: Werden aktuelle Regulatorien im Bereich Nachhaltigkeit zu Finanzierungsschwierigkeiten im deutschen Mittelstand führen?
4. European Commission (2025) The European Green Deal: Striving to be the first climate-neutral continent. Accessed 30 Mar 2025. [Online]. Available: https://commission.europa.eu/strategy-and-policy/priorities-2019-2024/european-green-deal_en
5. Environmental management—Life cycle assessment—Principles and framework (2009) ISO 14040:2000, International Organization for Standardization
6. Onososen A, Musonda I (2022) Barriers to BIM-based life cycle sustainability assessment for buildings: an interpretive structural modelling approach. Buildings 12:324. https://doi.org/10.3390/buildings12030324
7. Wastiels L, Decuypere R (2019) Identification and comparison of LCA-BIM integration strategies. In: IOP Conference series: earth and environmental science, vol 323. [Online]. Available: https://doi.org/10.1088/1755-1315/323/1/012101
8. Statistisches Bundesamt (Destatis) (2025) Enterprises, persons employed, turnover and other business and economic figures: Germany, years, enterprise size, economic activities (WZ2008 1-/2-/3-digit code. Accessed 31 Mar 2025. [Online]. Available: https://www-genesis.destatis.de/datenbank/online
9. Fabrikplanung—Planungsvorgehen (2011) VDI 5200 Blatt 1, Verein Deutscher Ingenieure e. V. Düsseldorf, Berlin
10. Wiendahl H-P, Reichardt J, Nyhuis P (2024) Handbuch Fabrikplanung: Konzept, Gestaltung und Umsetzung wandlungsfähiger Produktionsstätten, 3rd edn. Hanser Verlag, München

11. Hingst L, Dér A, Herrmann C, Nyhuis P (2023) Towards a holistic life cycle costing and assessment of factories: qualitative modeling of interdependencies in factory systems. Sustainability 15:5. https://doi.org/10.3390/su15054478
12. Hingst L (2024) Lebenszyklusorientierte Planung der Veränderungsfähigkeit in Fabriken. TEWISS Garbsen
13. Meckelnborg A (2015) Integrative Fabrikplanung durch effiziente Koordinationsmodelle. Zugl.: Aachen, Techn. Hochsch., Diss., 1st (ed). Apprimus Verlag, Aachen
14. Grundig CG (2021) Fabrikplanung: Planungssystematik—Methoden—Anwendungen, 7th (ed). Hanser München. [Online]. Available: https://www.hanser-elibrary.com/doi/book/10.3139/9783446470064
15. Bermpohl F, Eckart CJ, Forte S, Dickopf T, Benz L, Potthoff M et al. (2023) IDM für eine BIM-basierte Fabrikplanung—Prozesse, Rollen, Aufgaben und Informationsaustauschanforderungen. VDI/bS-EE 2552—Blatt 11.8: Building Information Modeling (BIM). VDI-Gesellschaft für Produktion und Logistik
16. Gralla M, Weist KC (2021) Komplexitätsbeherrschung durch Anwendung dynamisch-interdisziplinärer BIM-Modelle. In: Agile Digitalisierung im Baubetrieb: Grundlagen, Innovationen, Disruptionen und Best Practices, Hofstadler C, Motzko C, (Eds). Springer Vieweg, Wiesbaden, pp 621–646
17. Rieke L, Schäfer SF, Hingst L, Hook J, Peter N (2021) Einsatz von BIM in der Fabrikplanung/use of BIM in factory planning. wt Werkstattstechnik Online 111:881–886. https://doi.org/10.37544/1436-4980-2021-11-12-111
18. Melzner J, Ferger M (2024) Building information modeling. In: Zahlentafeln für den Baubetrieb, Krause T et al. (eds). 10th (ed). Springer Vieweg, Wiesbaden, pp 1051–1073
19. Lampe S, Böck J (2022) BIM and AR in factory planning: a combination of techniques from architecture, Comput Sci Fact Plann
20. Hall AT, Durdyev S, Koc K, Ekmekcioglu O, Tupenaite L (2023) Multi-criteria analysis of barriers to building information modeling (BIM) adoption for SMEs in New Zealand construction industry. ECAM 30(9):3798–3816. https://doi.org/10.1108/ECAM-03-2022-0215
21. Makabate CT, Musonda I, Okoro CS, Chileshe N (2022) Scientometric analysis of BIM adoption by SMEs in the architecture, construction and engineering sector. ECAM 29(1):179–203. https://doi.org/10.1108/ECAM-02-2020-0139
22. Rautenstrauch T, Müller C (2006) Investitionscontrolling in kleinen und mittleren Unternehmen (KMU). Z Control Manage 50(2):100–105. https://doi.org/10.1365/s12176-006-0519-x
23. von Keitz I, Wulf I (2023) Systematischer Überblick über ausgewählte Regelwerke zur Nachhaltigkeitsberichterstattung (Teil 1). KoR Zeitschrift für internationale und kapitalmarktorientierte Rechnungslegung, pp 27–38
24. Parekh R, Trabucco D (2024) Recent progress in integrating BIM and LCA for sustainable construction: a review. Int J Sci Res Arch 13(1):907–932. https://doi.org/10.30574/ijsra.2024.13.1.1772
25. Borrmann A, König M, Koch C, Beetz J (2021) Building information modeling: technologische grundlagen und industrielle Praxis, 2nd (ed). Springer Fachmedien Wiesbaden, Wiesbaden 2021. Accessed on Feb 23 2024. [Online]. Available: https://doi.org/10.1007/978-3-658-33361-4
26. Schumacher R, Theißen S, Höper J, Drzymalla J, Lambertz M, Hollberg A et al (2022) Analysis of current practice and future potentials of LCA in a BIM-based design process in German. E35 Web Conf 349:10004. https://doi.org/10.1051/e3sconf/202234910004
27. Parece S, Resende R, Rato V (2025) Current trends and challenges in BIM–LCA integration. In: Swarm intelligence applications for the cities of the future. Marques Mano Domingos, Leonor and Jose Sousa, Maria, CRC Press

28. Tam VW, Zhou Y, Shen L, Le KN (2023) Optimal BIM and LCA integration approach for embodied environmental impact assessment. J Clean Prod 385:135605. https://doi.org/10.1016/j.jclepro.2022.135605
29. Chen Z, Chen L, Zhou X, Huang L, Sandanayake M, Yap P-S (2024) Recent technological advancements in BIM and LCA integration for sustainable construction: a review. Sustainability 16(3):1340. https://doi.org/10.3390/su16031340
30. Zheng B, Hussain M, Yang Y, Chan AP, Chi H-L (2025) Trade-offs between accuracy and efficiency in BIM-LCA integration. ECAM 32(1):237–258. https://doi.org/10.1108/ECAM-03-2023-0270
31. Huang B, Zhang H, Ullah H, Lv Y (2025) BIM-based embodied carbon evaluation during building early-design stage: a systematic literature review. Environ Impact Assess Rev 112:107768. https://doi.org/10.1016/j.eiar.2024.107768
32. Guignone G, Calmon JL, Vieira D, Bravo A (2023) BIM and LCA integration methodologies: a critical analysis and proposed guidelines. J Build Eng 73:106780. https://doi.org/10.1016/j.jobe.2023.106780
33. Horn R, Ebertshäuser S, Di Bari R, Jorgji O, Traunspurger R, von Both P (2020) The BIM2LCA approach: an industry foundation classes (IFC)-based interface to integrate life cycle assessment in integral planning. Sustainability 12:16. https://doi.org/10.3390/su12166558
34. Concular GmbH (2025) Circular LCA by Concular 2025. Accessed on 24 Mar 2025. [Online]. Available: https://circularlca.de/
35. Alwan Z, Nawarathna A, Ayman R, Zhu M, ElGhazi Y (2021) Framework for parametric assessment of operational and embodied energy impacts utilising BIM. J Build Eng 42:102768. https://doi.org/10.1016/j.jobe.2021.102768
36. Potrč Obrecht T, Röck M, Hoxha E, Passer A (2020) BIM and LCA integration: a systematic literature review. Sustainability 12(14):5534. https://doi.org/10.3390/su12145534
37. Burggräf P, Dannapfel M, Ebade-Esfahani M, Scheidler F (2021) Creation of an expert system for design validation in BIM-based factory design through automatic checking of semantic information. Proced CIRP Conf Intell Comput Manuf Eng 99:3–8. https://doi.org/10.1016/j.procir.2021.03.012
38. Theißen S, Höper J, Wimmer R, Zibell M, Meins-Becker A, Rössig S et al. (2020) BIM integrated automation of whole building life cycle assessment using German LCA data base ÖKOBAUDAT and Industry Foundation Classes. In: IOP Conference series: earth and environmental sciences
39. Forth K, Borrmann A, Hollberg A (2023) Interactive visualization of uncertain embodied GHG emissions for design decision support in early stages using open BIM. In: Life-Cycle of structures and infrastructure systems, pp 3634–3641. https://doi.org/10.1201/9781003323020-445
40. Mohammed A, Elmasoudi I, Ghannam M (2025) Life cycle environmental impact assessment of steel structures using building information modeling. Innov Infrastruct Solut 10:1. https://doi.org/10.1007/s41062-024-01840-8
41. Carvalho JP, Alecrim I, Bragança L, Mateus R (2020) Integrating BIM-based LCA and building sustainability assessment. Sustainability 12(18):7468. https://doi.org/10.3390/su12187468
42. Bueno C, Fabricio MM (2018) Comparative analysis between a complete LCA study and results from a BIM-LCA plug-in. Autom Constr 90:188–200. https://doi.org/10.1016/j.autcon.2018.02.028
43. Houlihan Wiberg A, Kjendseth Wiik M, Auklend H, Løkland Slåke M, Tuncer Z, Manni M et al. (2020) Life cycle assessment for zero emission buildings—a chronology of the development of a visual, dynamic and integrated approach. In: IOP Conference series: earth and environmental sciences

44. Santos R, Costa AA, Silvestre JD, Pyl L (2019) Integration of LCA and LCC analysis within a BIM-based environment. Autom Constr 103:127–149. https://doi.org/10.1016/j.autcon.2019.02.011
45. Llatas C, Soust-Verdaguer B, Hollberg A, Palumbo E, Quiñones R (2022) BIM-based LCSA application in early design stages using IFC. Autom Construct 138:104259. https://doi.org/10.1016/j.autcon.2022.104259
46. Hollberg A, Genova G, Habert G (2020) Evaluation of BIM-based LCA results for building design. Autom Construct. https://doi.org/10.1016/j.autcon.2019.102972
47. Ansah MK, Chen X, Yang H, Lu L, Lam PT (2021) Developing an automated BIM-based life cycle assessment approach for modularly designed high-rise buildings. Environ Impact Assess Rev 90:106618. https://doi.org/10.1016/j.eiar.2021.106618

Life-Cycle Sustainability Assessment of Limestone Calcined Clay Cement (LC^3) as Sustainable Alternative in Concrete Structures

Ademilade Olubambi[1(✉)], Oluwabukunmi Ogunsanya[2], Clinton Aigbavboa[3], and Bolanle Ikotun[4]

[1] Department of Civil Engineering Science, University of Johannesburg, Johannesburg, South Africa
ajoeolubambi@gmail.com

[2] Department of Civil Engineering, Surveying and Construction, Kingston University London, London, UK

[3] Cidb Centre of Excellence and Sustainable Human Settlement and Construction Research Centre, Faculty of Engineering and the Built Environment, University of Johannesburg, Johannesburg, South Africa

[4] Department of Civil, Environmental Engineering and Building Science, University of South Africa, Florida Campus, Johannesburg, Gauteng, South Africa

Abstract. Over 25% of the world's CO_2 emissions are caused by the construction and demolition of buildings. In contrast with conventional cement, this study investigates the cost-effectiveness, environmental impact, production, and techno-economic sustainability of limestone calcined clay cement (LC^3). The results indicate that 50% LC^3 can be added to concrete, potentially significantly reducing carbon emissions in the construction industry. Furthermore, in this context, LC^3 is a sustainable alternative that uses a mixture of unburned limestone and calcined clay (produced by burning clay at 800 degrees Celsius with 40–60% kaolinite, which is sometimes found as waste in kaolinite mines) to replace about 50% of standard cement. This combination can significantly improve the concrete's long-term durability while retaining its strength, improving the structure's service life compared to ordinary Portland cement. The sustainability impact assessment unequivocally demonstrates that the production of LC^3 uses about 20% less energy and reduces CO_2 emissions by more than 40% compared to conventional Portland cement (OPC). In conclusion, to improve the sustainability of concrete in terms of cost, availability, and environmental impact, all its material procurement must be completed rapidly before demand increases, as this sustainable alternative LC^3 production is still significantly less expensive. Therefore, the construction industry must embrace circular, resource-efficient, and sustainable practices to reach a net-zero future.

Keywords: Concrete Structures · Life-cycle Sustainability Assessment · Low Carbon Cement · Net Zero Emission · Sustainable Construction

H. Beushausen et al. (Eds.): NTZR 2025, LNCE 873, pp. 1082–1093, 2026.
https://doi.org/10.1007/978-3-032-27698-8_95

1 Introduction

Concrete, which is mostly composed of cement, is the most widely used construction material worldwide. The production of cement entails thermal and chemical combustion processes, both of which greatly increase greenhouse gas (GHG) emissions while approximately 8% of the world's emissions are owing to these cement utilization in construction activities [1–4]. Growing urbanization is predicted to contribute to additional GHG emissions worldwide, especially in Asia and Sub-Saharan Africa. To be able to attain a net-zero future, the construction industry needs to implement circular, sustainable, and resource-efficient practices across all aspects of its operations. To achieve this, the industry must employ energy-efficient cement production methods and give priority to the use of sustainable alternative materials, such as fly ash or slag [5–7]. However, by encouraging the recycling of construction waste and implementing stronger building regulations, the negative environmental effects of concrete can be reduced. Concrete's negative environmental effects can also be reduced by encouraging the recycling of construction waste and enforcing stronger construction policies and regulations [8–10]. Therefore, a greener and more resilient built environment can result from these actions, which can also reduce expenses, preserve resources, and enhance the overall sustainability of any construction projects.

A recently developed kind of cement known as limestone calcined clay cement (LC^3) is created from low-grade clays and limestone, both of which are readily available in huge quantities. This cement can cut GHG emissions by up to 40%. Moreover, it is inexpensive and does not require significant capital modifications to existing cement factories [11]. Furthermore, LC^3 has excellent mechanical properties and durability, making it a viable alternative to ordinary Portland cement (OPC) in various construction applications. By utilizing it, the construction industry's carbon footprint can be considerably reduced while still being economical and easy to apply [12–14]. In one study, LC^3 was shown to be beneficial in lowering carbon emissions by about 30% when it was used in the building of a sizable residential complex in India. Similar to this, a bridge project in Switzerland demonstrated the potential of LC^3 in infrastructure development by using it to increase durability while reducing costs. The expanding significance of LC^3 in sustainable construction is highlighted by these sustainable applications [15].

1.1 Overview of Research Development on LC^3

LC^3 contributes to the reduction of CO_2 emissions by reducing the energy intensive clinker content through the addition of calcined clay and limestone [16]. LC^3–50 with a 50% clinker content is widely promoted and accepted under cement standards applicable in Europe, India, the United States, Cuba, and much of South America. Significant energy savings are achieved because calcined clays are more malleable and heated at approximately 700–800 °C, whereas the manufacture of clinker requires a temperature of 1400–1500 °C. The reduction in GHG emissions can reach 40% depending on the type of cement that is replaced by LC^3 [17, 18]. However, the production of a peculiar LC^3 known as LC^3–50 can be more cost-effective than conventional cement due to the lower energy requirements for calcining clays compared to the high temperatures needed for clinker production. Furthermore, the reduced clinker content in LC^3–50 can lead to lower raw material costs. These factors combine to make LC^3–50 an economically attractive alternative, especially as energy prices continue to rise [19].

Over time, the application of LC^3–50 can lead to significant savings in both production and operational costs. As industries shift towards sustainable practices, using LC^3–50 can also offer companies a competitive edge by aligning with environmental regulations and reducing carbon taxes. Furthermore, the potential for increased durability and performance of LC^3–50 may result in lower maintenance costs for construction projects [17, 20, 21]. This can lead to significant cost savings over the lifecycle of a concrete structure, making it an attractive option for long-term projects. Furthermore, the reduced need for frequent repairs or replacements can minimize disruptions and enhance the overall efficiency of construction operations [22, 23]. Therefore, companies can more easily adopt LC^3–50 without requiring significant changes because it is suitable with current production processes. Given its adaptability, cost-effectiveness, and environmental advantages, LC_3–50 is positioned as a smart and advanced option for companies that seek a balance between sustainability and profitability [24].

1.2 Technical Performance of LC_3

LC^3 depends on the synergistic reaction between calcined clay, specifically kaolinitic clay, and limestone. Calcined clay contributes both pozzolanic activity and filler effects, while limestone reacts with aluminates to form carboaluminates, which enhance the strength and durability of concrete [25]. The hydration of clinker in LC^3 produces calcium hydroxide, which reacts with the metakaolin in calcined clay to form additional calcium silicate hydrates (C-S-H). These pozzolanic and carbonate reactions contribute to the strength and long-term performance of LC^3 concrete, even with a reduced clinker content.One of the main technical challenges is ensuring the phase compositions are controlled to optimize strength development. Too much ettringite formation, for example, can weaken early-age strength, so it is critical to strike a balance between the reactive aluminates and carbonates [26, 27]. LC^3 formulations typically demonstrate comparable long-term compressive strengths to Ordinary Portland Cement (OPC). While some studies have shown slightly lower early compressive strength, this can be mitigated through adjustments in curing regimes or calcined clay content.

According to a study by Nath and Sarker [28], LC^3 formulations with 15% limestone and 30% calcined clay achieved compressive strengths of 40 MPa at 28 days, which is comparable to OPC but significantly lowers carbon emissions. When it comes to chloride resistance, LC^3 performs noticeably better than OPC, which makes it perfect for constructions in maritime settings or places where deicing salts are present [29–31]. Furthermore, according to research from Swiss Federal Laboratories (Empa), LC^3 outperformed OPC by up to 60% in terms of resistance to chloride ion penetration, which is essential for enhancing durability and prolonging the life of concrete structures [32]. Figure 1 illustrates the technology, composition and production process for LC^3.

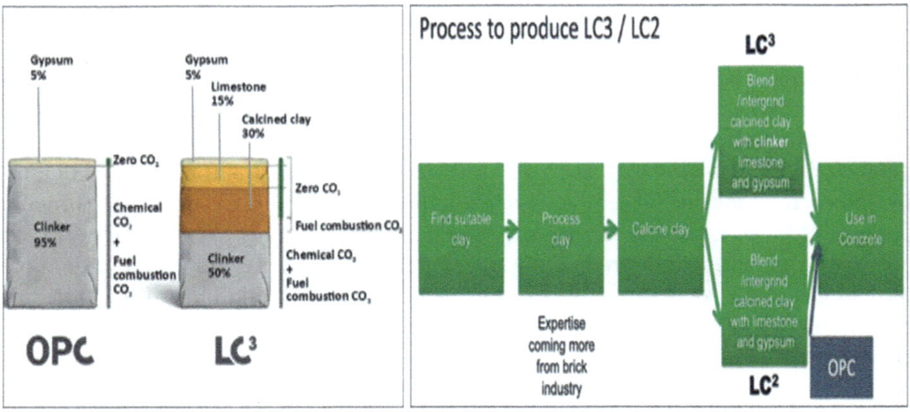

Fig. 1. Technology, composition, production processing for LC^3 [32]

1.3 Durability and Cost-Effectiveness of LC_3

Durability studies conducted have shown that LC^3 has excellent resistance to sulfate attack and carbonation. However, research is ongoing to assess its performance in regions with severe freeze-thaw cycles, where high-clay formulations may face limitations. Due to its lower clinker content, LC^3 generates less heat during hydration than OPC. This reduction in heat, by as much as 20–30%, makes LC^3 suitable for mass concrete applications, such as dams and foundations, particularly in hot climates where thermal cracking is a significant concern. This makes it an attractive option for sustainable construction, as it reduces the risk of thermal cracking and contributes to the long-term durability of the structure. The calcination of kaolinic clay occurs at temperatures between 700–850 °C, far lower than the 1450 °C required for clinker production. This lower temperature significantly reduces energy consumption by up to 25%, representing a key advantage for countries with high energy costs [16, 32]. The lower energy demand also makes LC^3 an attractive option for regions with limited fuel resources. This highlights the potential of LC^3 as a sustainable alternative to traditional clinker-based cement, particularly in regions where energy efficiency and resource availability are critical concerns.

A study conducted in India showed that LC^3 could be 15–20% cheaper to produce than OPC, thanks to reduced energy consumption and the availability of raw materials such as limestone and kaolinitic clay. Although the initial investment for calcination equipment may be higher, these costs can be offset by long-term savings on energy and raw material [33]. Furthermore, the production of LC^3 generates fewer carbon emissions compared to OPC, making it a more environmentally friendly alternative. This reduction in energy consumption and emissions aligns with global efforts to combat climate change and promote sustainable construction practices [34, 35]. However, transitioning to LC^3 may pose challenges such as the need for new infrastructure and technology to accommodate its production. Also, there may be resistance from industry stakeholders accustomed to conventional materials, requiring substantial efforts in education and training. Furthermore, regulatory approvals and standardization processes could delay widespread adoption, impacting the speed of transition [36–38]. Figure 2 shows the influence of kaolinite content on the comprehensive strength of LC^3.

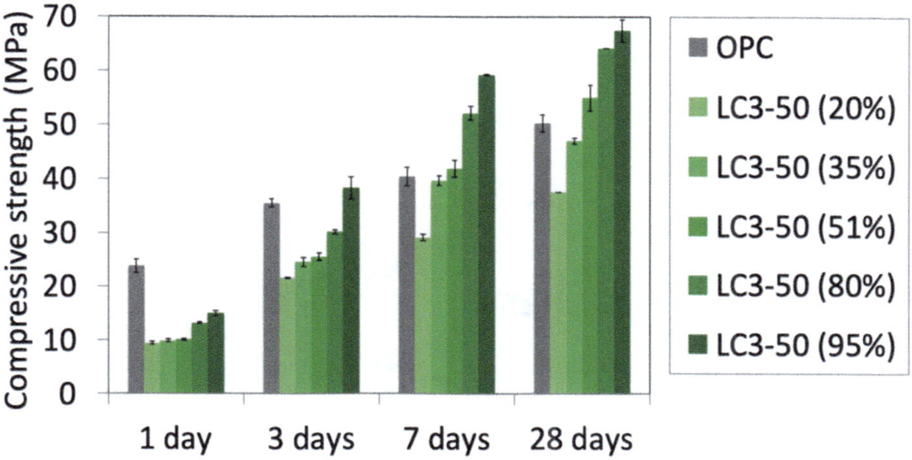

Fig. 2. Influence of Kaolinite content on the comprehensive strength of LC^3 [26]

The resource efficiency potentials of LC^3 cannot be overemphasized.Waste material, ceramic clay, can be put to sustainable use. By replacing some of the conventional raw materials, this not only reduces the quantity of waste that ends up in landfills but also preserves natural resources. Moreover, LC^3 is a more sustainable option for the construction sector since it can make use of ceramic clay, which reduces GHG emissions during its production [39–41]. With the current advantages of LC^3, it is necessary to ensures that the application of LC^3 is both economically and environmentally viable, while also bolstering the legitimacy of the decision-making process in most developing countries. Prompt resolution of removing any barriers that can the participation f the construction industry stakeholders in collaborating so as to optimize LC^3's advantages while lowering its risks. By fostering innovation and establishing trust among stakeholders, this cooperative endeavor lays the groundwork for long-term prosperity [42, 43]. Furthermore, there is a need to analysis the advantage of LC^3 over OPC to guarantees that LC^3 application stays in line with changing environmental and economic priorities by placing a high priority on openness and scalability (Fig. 3).

Fig. 3. The benefits of LC^3 over conventional cement OPC [32]

2 Materials and Methods

This study construct an evaluation that involves a life cycle analysis to determine the environmental impacts from the production process, including raw material extraction, transportation, and manufacturing. It also includes stakeholder consultations to understand the social implications and economic evaluations to assess cost-effectiveness and potential market impacts. Furthermore, scenarios was developed to compare the sustainability performance of LC^3 with conventional cement alternatives. The assessment also evaluates the durability and performance of the cement in various applications, ensuring it meets industry standards. To provide a comparative analysis of the holistic view of the significance of LC^3 application in concrete structures, this study considered a lifecycle sustainability assessment (LCSA), which is a methodical process used to evaluate the potential economic, social, and environmental impact of a project, policy, or program.

According to Curran [44], LCSA ensures that sustainable development is maintained and balances current requirements without compromising future generations. In this study, a stakeholder consultations, quantitative data analysis, and an in-depth review of the current literatures were all elements of the LCSA's procedures. These steps made it possible to identify significant metrics that are crucial to social, economic, and environmental impact analysis conducted. Furthermore, scenario modeling was used to evaluate the sustainability of various growth options and estimate the long-term impacts of LC^3. The LCSA also provided a framework for analyzing trade-offs between competing agendas, such as environmental preservation and economic expansion. However, the assessment's use of reliable data and integration of various stakeholder viewpoints enabled informed decision-making, ensuring that the adoption of LC^3 is in line with long-term sustainability objectives [45] (Fig. 4).

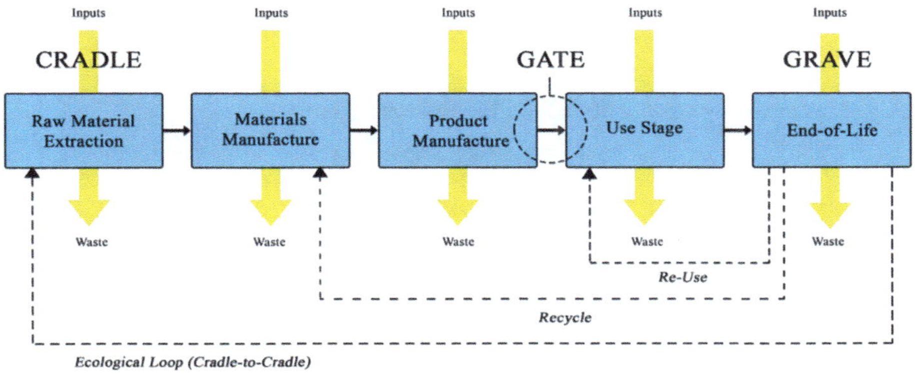

Fig. 4. An illustration of a cradle-to-grave life-cycle sustainability assessment [44]

3 Results and Discussion

When compared with conventional cement, LC^3 will result in GHG emission reduction of up to 40%. Lowering clinker content LC^3 at 50% or less compared to OPC at 90% or more is the main factor reducing the carbon footprint; which requires use of low-energy, abundant materials such as limestone and calcined clay in place of clinker, which requires a lot of energy. Since calcined clays must be calcined at temperatures between 600 and 850 °C, which is much lower than clinker's required temperature of 1450 degrees Celsius, they use less fossil fuel and emit less CO_2. Globally, calcined clays and limestone are extensively accessible, even in areas without conventional supplementary cementitious materials (SCMs) like fly ash or slag, which are scarce and geographically limited. Because it is not calcined and instead serves as a filler to enhance the performance and micro-structure of cement, limestone is very plentiful and emits no direct CO_2. After the usual 28-day curing period, LC^3 obtains mechanical properties, including as compressive and flexural strength, that are on par with or better than OPC.

It also improves the defense against chloride intrusion and sulfate attacks, improves pore structure refinement as a result of the interaction between limestone and calcined clay, which produces synergistic hydration products (C-A-S-H gel, carboaluminate phases). With consistent long-term performance, early strength development is on par with or better than some other SCM blends, such fly ash. Figure 5 shows in percentage the various performance of LC^3 in comparison to OPC.

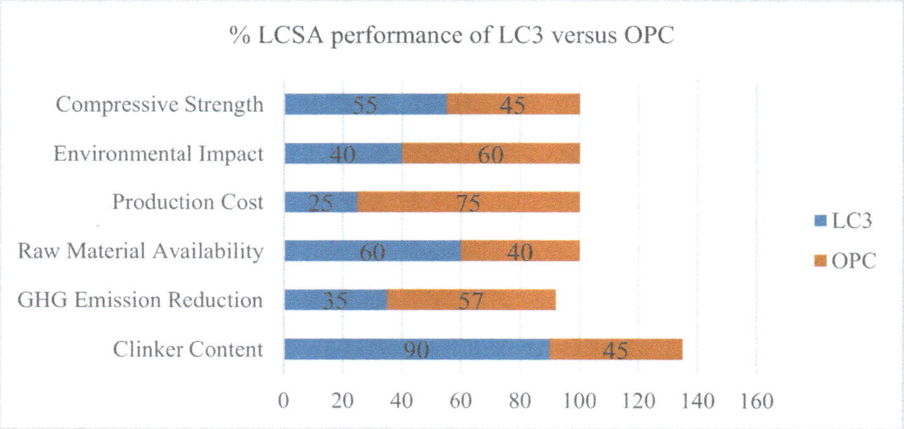

Fig. 5. % LCSA performance of LC^3 versus OPC (Author)

From Cradle-to-grave LSCA shows that LC^3 uses a lot less energy and has a much lower global warming potential than OPC. Lower calcination temperatures and raw material processing lead to less embodied energy. Decreased reliance on fossil fuels helps cement production meet international goals of reducing carbon emission. Up-scaling can be facilitated by integrating LC^3 production with the current cement manufacturing infrastructure with just minor adjustments and labor retraining. LC^3 typically has a cost benefit (up to 25% less expensive in some situations) since it uses available raw materials and produces less clinker. Furthermore, the color shift in LC^3 (caused by the iron content of clays) has no effect on mechanical qualities but poses minor difficulties with market acceptance. Table 1 shows an overview of LC^3's LCSA results contrasted with conventional cement OPC.

Table 1. Overview of LC^3's LCSA results contrasted with conventional cement OPC (Author)

Characteristic	LC^3	Conventional Cement (OPC)
Clinker content	Typically 40–50%	> 90%
GHG Emissions reduction	Up to 40% lower	Baseline
Raw material availability	Abundant, globally widespread calcined clay & limestone	Limited SCMs; geographic constraints
Required manufacturing temperature	~ 600–850 °C (calcined clay)	~ 1450 °C (clinker production)
Mechanical strength	Comparable or superior at 28 days	Baseline
Durability	Improved (sulfate, chloride resistance)	Standard
Production cost	Generally 15–25% less	Higher

(*continued*)

Table 1. (*continued*)

Characteristic	LC^3	Conventional Cement (OPC)
Integration in industry	Compatible with existing plants	Established
Environmental impact (LCA)	Lower embodied energy and GWP	Higher GWP

4 Conclusion

In this study, a thorough results that includes strategies for optimizing the sustainability of LC^3 in concrete structures synthesizes the data. The results also identifies areas that could use improvement, including reducing GHG emissions, improving resource efficiency, and attending to stakeholder concerns. The analysis seeks to direct decision-making and encourage the construction industry's adoption of sustainable practices by incorporating these insights. Significant environmental benefits are revealed by the sustainability impact assessment of LC^3. It has been demonstrated according to the LCSA, LC^3 reduces significantly in GHG emissions in comparison to OPC. This is in line with the Paris Agreement's objectives to keep the increase in global temperatures to less than 2 °C over pre--industrial levels. The cement industry's carbon footprint is further decreased by the usage of SCMs in LC^3 manufacturing, such as fly ash and GGBFS. Given the enormous potential of LC^3 to reduce the environmental impact of the cement industry—by decreasing CO_2 emissions by up to 40%—there is an urgent need to accelerate its production and adoption across the country.

Furthermore, its explores the key challenges hindering LC^{3}'s growth in India and outlines actionable strategies that stakeholders, including cement manufacturers, policy-makers, researchers, and builders, can adopt to improve its usage and production. One of the most effective ways to counter industry conservatism is to demonstrate the tangible benefits of LC^3 through large-scale pilot projects. It can be concluded that LC^3 demonstrates a significant sustainable advantage over conventional Portland cement, primarily through reducing CO_2 emissions by up to 40%, lowering energy consumption, and leveraging abundant raw materials while maintaining or enhancing mechanical performance and durability. Its scalable production potential and cost-effectiveness position LC^3 as a leading candidate for accelerating decarbonization in the cement and construction industry globally. However, to improve the sustainability of concrete in terms of cost, availability, and environmental impact, all its material procurement must be completed rapidly before demand increases, as this sustainable alternative LC^3 production is still significantly less expensive.

Therefore, the construction industry must embrace circular, resource-efficient, and sustainable practices to reach a net-zero future. Although future research can examine at the scalability in regions with low kaolinite content clays and integration with emerging construction technologies (e.g., 3D printing) are other active research areas. One potential challenge is the initial cost of implementing sustainable practices, which can be prohibitive for some companies. Additionally, there may be resistance to change within the industry, as traditional methods are deeply entrenched. Finally, the development of new technologies and materials requires significant research and collaboration, which can slow down widespread adoption. However, the long-term benefits of adopting sustainable practices, such as reduced environmental impact and improved resource efficiency, often outweigh the initial costs. Collaborative efforts between governments, industry stakeholders, and researchers can help overcome these challenges and drive the transition toward a more sustainable construction sector.

In conclusion, to improve the sustainability of concrete in terms of cost, availability, and environmental impact, all its material procurement must be completed rapidly before demand increases, as this sustainable alternative LC^3 production is still significantly less expensive. Therefore, the construction industry must embrace circular, resource-efficient, and sustainable practices to reach a net-zero future.

References

1. Gagg CR (2014) Cement and concrete as an engineering material: an historic appraisal and case study analysis. Eng Fail Anal 40:114–140
2. Barcelo L, Kline J, Walenta G et al (2014) Cement and carbon emissions. Mater Struct 47:1055–1065
3. Belaïd F (2022) How does concrete and cement industry transformation contribute to mitigating climate change challenges? *resources*. Conserv Recycl Adv. 15:200084
4. Haddaji Y, Majdoubi H, Safhi AEM, Nadi M, Oumam M, Manoun B, Hannache H (2024) Circular economy in the construction sector: current trends and future outlook. Recent Developments and Innovations in the Sustainable Production of Concrete, Woodhead. Publishing Series in Civil and Structural Engineering, pp 363–390 (2024)
5. Van Damme H (2018) Concrete material science: past, present, and future innovations. Cem Concr Res 112:5–24
6. Habert G et al (2020) Environmental impacts and decarbonization strategies in the cement and concrete industries. Nat Rev Earth Environ 1(11):559–573
7. Soomro M, Tam VW, Jorge Evangelista AC (2022) Production of cement and its environmental impact. In: Book: recycled concrete, woodhead publishing series in civil and structural engineering, pp 11–46 (2022)
8. Khalifa AA, Ibrahim A, Amhamed AI, HM (2022) Accelerating the transition to a circular economy for net-zero emissions by 2050: a systematic review. Sustainability, 14(18):11656
9. Tayebani B, Said A, Memari A (2023) Less carbon producing sustainable concrete from environmental and performance perspectives: a review. Constr Build Mater 404:133234
10. Barbhuiya S, Bhusan Das B, Adak D (2024) Roadmap to a net-zero carbon cement sector: strategies, innovations and policy imperatives. J Environ Manage 359:121052
11. Renuka V, Rao SV, Tadepalli T (2025) State-of-the-art review on limestone calcined clay cement (LC^3) and its recent advances. Iran J Sci Technol Trans Civ Eng 49:1–30
12. Scrivener K, Martirena F, Bishnoi S, Maity S (2018) Calcined clay limestone cements (LC^3). Cem Concr Res 114:49–56
13. Hosen K, Chen B (2025) Limestone calcined clay cement (LC^3): a review of materials, properties, production and environmental impact. J Build Eng 112(113672):1–23
14. Bishnoi S, Maity S (2020) Limestone calcined clay cement: opportunities and challenges. In: Bishnoi S (eds) Calcined clays for sustainable concrete. RILEM Bookseries, vol 25. Springer, Singapore. https://doi.org/10.1007/978-981-15-2806-4_88

15. Sirangi B, Prasad M (2023) A low carbon cement (LC^3) as a sustainable material in high strength concrete: green concrete. Mater Constr 73:1–12
16. Kanagaraj B, Anand N, Johnson AU et al (2024) Limestone calcined clay cement (LC^3): A sustainable solution for mitigating environmental impact in the construction sector. Resour, Conserv Recycl Adv 21:200197
17. Mostafa A, Khater HM, Gharieb M et al (2025) Sustainable cement production: minimizing CO_2 by optimizing efficiency with LC^3. Discov Appl Sci 7:684
18. Martirena F, Scrivener KL (2014) Low-CO_2 cement: Limestone calcined clay cement as an alternative to Portland cement. Green Materials 2(3):173–191
19. Environment UN, Scrivener KL, John VM, Gartner EM (2016) Eco-efficient cements: Potential economically viable solutions for a low-CO_2 cement-based materials industry. Cem Concr Res 1(114):2–6
20. Nguyen LH, Maier M, De Weerdt K, Justnes H (2020) Durability properties of LC^3-based mortars and concretes. MDPI Materials 13(22):5086
21. Singh NB, Middendorf B (2020) Calcium sulfoaluminate cements: Key aspects of their hydration and durability. Cem Concr Res 138:106267
22. Dhandapani Y, Santhanam M, Khithani D (2017) Performance evaluation of LC^3 mortar using different grades of calcined clay in binary and ternary blends. Constr Build Mater 153:588–600
23. Zunino F, Scrivener K (2018) Reactivity and performance of limestone calcined-clay cement (LC^3) cured at low temperature. In: Martirena F, Favier A, Scrivener K (eds) Calcined clays for sustainable concrete. RILEM Bookseries, vol 16. Springer, Dordrecht. https://doi.org/10.1007/978-94-024-1207-9_82
24. Renuka V, Rao SV, Tadepalli T (2025) State-of-the-art review on limestone calcined clay cement (LC^3) and its recent advances. Iran J Sci Technol Trans Civ Eng 1–30 (2025)
25. Sun J, Zunino F, Scrivener K (2024) Hydration and phase assemblage of limestone calcined clay cements (LC^3) with clinker content below 50%. Cem Concr Res 177:107417
26. Blouch N, Rashid K, Ju M (2023) Exploring low-grade clay minerals diving into limestone calcined clay cement (LC^3): characterization—Hydration—Performance. J Clean Prod 426:139065
27. Scrivener K, Avet F, Maraghechi H, Zunino F et al. Impacting factors and properties of limestone calcined clay cements (LC^3). Green Materials 18 March 2019; 7(1):3–14. https://doi.org/10.1680/jgrma.18.00029
28. Nath P, Sarker PK (2015) Use of LC^3 cement in low-carbon construction in India. Procedia Engineering 125:206–212
29. Rotana H, Kemal C (2024) Comparative performance of OPC and LC^3-based composites under combined action of carbonation and chloride exposure. Constr Build Mater 429(2024):136428
30. Ukpata JO, Ogirigbo OR, Black L (2023) Resistance of concretes to external chlorides in the presence and absence of sulphates: a review. Appl Sci 13(1):182
31. Cong P, Du R, Gao H, Chen Z (2024) Comparison and assessment of carbon dioxide emissions between alkali-activated materials and OPC cement concrete. J Traffic Transp Eng 11(5):918–938
32. Jaskulski R, Jóźwiak-Niedźwiedzka D, Yakymechko Y (2020) Calcined clay as supplementary cementitious material. Materials 13(21):4734
33. Fernandez R, Martirena F, Scrivener KL (2011) The origin of the pozzolanic activity of calcined clay minerals: a comparison between kaolinite, illite and montmorillonite. Cem Concr Res 41:113–122
34. Nguyen QD, Afroz S, Castel A (2020) Influence of calcined clay reactivity on the mechanical properties and chloride diffusion resistance of limestone calcined clay cement (LC^3) concrete. J Mar Sci Eng 8(5):301

35. Zokaei S, Siad H, Lachemi M, Mahmoodi O, Sahmaran M (2023) Design of sustainable high-ductile engineered cementitious composites with limestone calcined clay cement. In: Ilki A, Çavunt D, Çavunt YS (eds) Building for the future: durable, sustainable, resilient. fib symposium 2023. Lecture Notes in Civil Engineering, vol 349. Springer, Cham. https://doi.org/10.1007/978-3-031-32519-9_87
36. Sánchez Berriel S et al (2016) Assessing the environmental and economic potential of limestone calcined clay cement in cuba. J Clean Prod 124:361–369
37. Ascensão G, Farinini E, Ferreira VM, Leardi R (2024) Development of eco-efficient limestone calcined clay cement (LC^3) mortars by a multi-step experimental design. Chemom Intell Lab Syst 253:105195
38. Haq IU, Bibi T, Nawaz A et al (2025) Calcined clay–substituted sustainable cement binders: a holistic review. Environ Sci Pollut Res 32:11317–11349
39. Avet F, Sofia L, Scrivener K (2019) Concrete performance of limestone calcined clay cement (LC^3) compared with conventional cements. Adv Civ Eng Matls 8(3):275–286
40. Cerchione R, Colangelo F, Farina I, Ghisellini P, Passaro R, Ulgiati S (2023) Life cycle assessment of concrete production within a circular economy perspective. Sustainability 15(14):11469
41. Huang X, Jiao Z, Xing F, Sui L, Hu B, Zhou Y (2024) Performance assessment of LC^3 concrete structures considering life-cycle cost and environmental impacts. J Clean Prod 436:140380
42. Yu Y, Yang M, Wan C, Lin L (2025) Mechanical, workability, economic and environmental properties of concrete with limestone calcined clay cement and recycled aggregates. Sci Rep 15(1):1–15
43. Hosen K, Chen B (2025) Limestone calcined clay cement (LC^3): a review of materials, properties, production and environmental impact. J Build Eng 112:113672
44. Curran MA (2013) Life cycle assessment: a review of the methodology and its application to sustainability. Curr Opin Chem Eng 2(3):273–277
45. Cerchione R, Morelli M, Passaro R, Quinto I (2025) A critical analysis of the integration of life cycle methods and quantitative methods for sustainability assessment. Corp Soc Responsib Environ Manag 32(2):1508–1544

Recycling Pathways for Textile-Reinforced Concrete: Strategies for Fibre and Concrete Reuse

Selina Vaculik(✉), Cynthia Morales Cruz, Thomas Matschei, and Anya Vollpracht

Institute of Building Materials Research, RWTH Aachen University, Aachen, Germany
vaculik@ibac.rwth-aachen.de

Abstract. The complete reuse of textile-reinforced concrete is a key step toward establishing a circular economy in the construction industry. This study investigates the potential for full recovery and recycling of the aggregate fraction and the textile fibres after the end-of-life phase of textile-reinforced concrete. The aggregate fraction was successfully incorporated into new concrete mixtures as a substitute for natural aggregates at rates of up to 100%, while maintaining essential mechanical properties, including compressive strength and durability. This demonstrates that recycled aggregates can be effectively reintegrated without compromising material performance. In addition to reusing the aggregate fraction, the recovered textile fibres were reused as reinforcement in cementitious composites. The study demonstrates that the fibres, despite being reused as short fibres, contribute to both the structural capacity and the serviceability of the new concrete elements, even though their effectiveness is reduced compared to continuous textile reinforcements. The recycled fibres maintained a sufficiently high bond strength and load-bearing capacity, indicating their suitability for reuse in load-bearing elements. These findings underscore the feasibility of a fully circular approach to recycling textile-reinforced concrete, which significantly reduces construction waste, lowering the demand for virgin raw materials, and fosters sustainable material cycles in concrete technology.

Keywords: Recycling · Concrete aggregates · Textile-reinforced concrete · Circular economy

1 Introduction

Textile-reinforced concrete (TRC) is a composite material composed of a cementitious matrix and non-metallic, typically corrosion-resistant textile reinforcement, most commonly made of carbon or alkali-resistant glass fibres. These reinforcements enable the design of durable, thin-walled concrete components with reduced cement content and significantly lower weight than conventional steel-reinforced concrete. As such, TRC offers considerable potential for reducing both resource consumption and carbon emissions in construction [1, 2]. With increasing use in prefabricated façade panels, bridge

H. Beushausen et al. (Eds.): NTZR 2025, LNCE 873, pp. 1094–1106, 2026.
https://doi.org/10.1007/978-3-032-27698-8_96

overlays, lightweight shells and structural retrofits, TRC has transitioned from niche research to practical application in recent years [2, 3]. In response, the German Institute for Construction Technology (DIBt) issued the general approval Z-1.6-308 for carbon reinforcement grids in 2024, accompanied by a national guideline from the German Committee for Structural Concrete (DAfStb) [4, 5].

However, the growing use of TRC also introduces challenges related to waste management and material circularity. Unlike traditional concrete, which is primarily recycled as aggregate, TRC contains synthetic fibre reinforcements such as carbon or AR-glass, which are currently not sufficiently addressed by existing recycling infrastructures and thus require specialised separation and recycling technologies. This applies to end-of-life components and the large production waste generated during textile cutting and concrete casting. Depending on the manufacturing process, up to 45 per cent of the input textile may become offcut or scrap, particularly when complex geometries are involved [6, 7]. Current recycling strategies for TRC remain limited. Mechanical separation methods have demonstrated the feasibility of separating the mineral matrix from the textile reinforcement in crushed TRC elements [8]. In parallel, thermal treatment technologies such as pyrolysis have been developed to recover high-value carbon fibres from composite waste, including those embedded in cementitious systems [9]. These recovered fibres often show reduced length and altered surface properties, but can retain significant tensile strength [10, 11]. Reusing such fibres in new concrete matrices is under active investigation, including pyrolytically recovered continuous fibres and mechanically reclaimed short fibre fragments [10, 12]. Similar strategies are being explored for glass fibres, although the mechanical integrity of recycled glass fibres tends to degrade more significantly during processing [13]. In addition to fibre recovery, the mineral fraction of TRC waste can be processed into recycled aggregate and reused in new concrete formulations. Prior research has shown that recycled aggregates from TRC can meet technical performance requirements for compressive strength and durability, making them suitable as partial substitutes for natural aggregates [14]. However, integrated approaches combining aggregate and fibre reuse in one recycling pathway are still rare, and little is known about the structural efficiency of recycled textile fibres in load-bearing cementitious systems [10, 11]. This study addresses that gap by evaluating a dual recovery strategy for TRC waste. The aggregate fraction is reused in new concrete mixtures and assessed for mechanical properties. In parallel, pyrolytically recovered and mechanically processed fibres are reintroduced as short fibre reinforcement in cementitious composites and tested for their contribution to tensile performance, bond behaviour, and crack bridging capacity. By validating the feasibility of this combined recycling pathway, the study contributes to closing material loops in TRC. It supports the development of sustainable construction systems with reduced dependency on primary resources.

2 Materials and Methods

This study focuses on the mechanical evaluation of recycled textile-reinforced concrete (TRC) components, particularly their structural effectiveness, non-deleterious mechanical performance, and ability to control early-age cracking. The recycled materials include mineral aggregate fractions and both pyrolytically and mechanically reprocessed carbon fibres.

2.1 Recovery of Aggregates and Fibres

Recycled aggregates and fibres were obtained from crushed TRC elements containing glass or carbon fibre textiles. After mechanical processing, both the coarse mineral aggregates and fibre-containing fines were recovered. In the case of TRC with glass fibre textiles, the reinforcement cannot be separated as easily as carbon mesh, resulting in a mixed recycled fraction. Short glass fibre fragments in this fraction were present throughout all aggregate size classes, particularly in the sand fraction below 2 mm. This allowed for the investigation of their potential mechanical contribution, despite the absence of complete textile units. Glass fibres were treated mechanically, while carbon fibres were subjected to both mechanical and pyrolytic treatment. Recycled aggregates were reused in new mixtures at complete replacement levels (100%). For fibre reuse, three approaches were considered:

- Short carbon fibres, separated from pyrolysed recycled carbon grid fabrics,
- Staple carbon fibres, manually cut from recycled carbon grid fabrics.
- Glass fibres, partially embedded in concrete aggregates

All fibre types were characterised and prepared for reintegration into mortar and concrete. The short fibres were used primarily for shrinkage crack control, while the longer staple fibres were assessed for load-bearing applications. An overview of the different carbon fibre types is shown in Fig. 1, and for glass fibres in Fig. 2.

a) b) c)

Fig. 1. Overview of used carbon fibres: a) Staple fibre from fabric 46 mm 'elliptical' (used for Chap. 2.3); b) Single fibre from pyrolysis 10–25 mm (used for Chap. 2.4); c) Staple fibre from fabric 12–15 mm 'flat' (used for Chap. 2.2)

Table 1 summarises the physical properties of the recycled fibres used in this study. Fibre lengths and diameters were determined based on manufacturer specifications or optical microscopy. Surface conditions were qualitatively assessed through visual inspection. No additional treatments were applied; the listed surface features result solely from the respective recovery processes (mechanical cutting or pyrolysis).

a) b)

Fig. 2. Overview of used glass fibres and aggregates: a) Glass fibre, partially embedded in concrete aggregates; b) Overview of the mixed recycled glass fibre/aggregate fraction

Table 1. Overview of physical properties of recycled fibres

Fibre type	Length (mm)	Diameter (μm)	Surface condition	Treatment method
Pyrolysed carbon fibres	10 - 25	7–9	Smooth, altered by pyrolysis	none
Staple carbon fibres (elliptical)	46	4500	Smooth Elliptical cross-section	Mechanical cut (hand) possibility for sand-coating
Staple carbon fibres (flat)	12 - 15	3000	Smooth Flat cross-section	Mechanical cut (hand) possibility for sand-coating
Glass fibres (aggregate-embedded)	20	3000 - 5000	Embedded fragment, rough surface	Mechanical crushing

2.2 Mortar-Scale Evaluation of Fibre Effectiveness

To evaluate the reinforcing efficiency of the recycled fibres in a controlled environment, standard mortar prisms were produced using three formulations (see Table 1): a plain reference mortar, a mix with 0.5 vol% short pyrolysed carbon fibres (see Fig. 1b), and one with 1.0 vol% staple fibres (12–15 mm see Fig. 1c)) from cut recycled carbon grids. Table 2 presents the detailed composition of the mixture.

Mortar specimens were cast in prismatic moulds (160 × 40 × 40 mm^3) and cured under water for 7 days, followed by storage at a standard climate chamber (20 °C, 65% RH) until testing at 28 days. The flexural strength was determined using a three-point bending test in accordance with DIN EN 12372. The span length was 100 mm. The load was applied in two deformation-controlled stages: 0.08 mm/min up to a 1.0 mm deflection and 0.21 mm/min up to a total deflection of 1.7 mm. Displacement was recorded using two inductive displacement transducers placed at mid-span.

Table 2. Composition of mortar mixtures for small-scale flexural tests

Parameter	Unit	Reference	Short fibre	Staple fibre 'flat'
CEM I 42.5 R	g	450	450	450
Water	g	225	225	225
Standard sand	g	1350	1350	1350
Fibres	g	–	9.5	13
Fibre volume content	vol%	–	0.5	1.0
w/c	-	0.5	0.5	0.5

2.3 Beam-Scale Validation Under Structural Loading

Concrete beams were produced and tested to verify the structural relevance of the recycled staple fibres (46 mm, see Fig. 1a). The formulations were based on the DIBt approval guidelines for fibre-reinforced concrete. Two dosages were investigated: 0.4 vol% and 0.7 vol%. The concrete compositions are shown in Table 3.

Table 3. Concrete mix design for beam bending tests

Parameter	Unit	IA_0.4	IA_0.7
CEM I 32.5 R	kg/m^3	300	300
Water	kg/m^3	180	180
Aggregate (0–16 mm)	kg/m^3	1837	1829
Staple carbon fibres	kg/m^3	8.85	14.62
Fibre volume content	vol%	0.4	0.7
Assumed air content	vol%	2.0	2.0
Water-to-cement ratio	–	0.6	0.6
Grading curve	–	A/B16	A/B16

The beams were tested using a three-point bending setup according to DIN EN 14845–2. The load-deflection behaviour was recorded, including post-cracking residual strength and fibre pull-out.

2.4 Shrinkage Crack Control Tests in the Wind Tunnel

To assess the effectiveness of short pyrolysed fibres (see Fig. 1b) in reducing shrinkage-induced cracking, concrete slabs ($1600 \times 1600 \times 80$ mm^3) were tested under defined drying conditions in a wind tunnel. For example, mixtures IIA_0.0125 and IIA_0.05 contained fibre dosages of 0.0125 and 0.05 vol%, respectively. The formulations were also based on the DIBt approval guidelines for fibre-reinforced concrete." The mixed compositions are listed in Table 4. All the other mixtures were composed accordingly.

Table 4. Concrete mix design for shrinkage crack control tests

Parameter	Unit	IIA_0.0125	IIA_0.05
CEM I 32.5 R	kg/m^3	360	360
Water	kg/m^3	270	270
Aggregate (0–8 mm)	kg/m^3	1176	1175
Limestone filler	kg/m^3	373	373
Pyrolysed short fibres	kg/m^3	0.24	0.96
Fibre volume content	vol%	0.0125	0.05
Assumed air content	vol%	3.0	3.0
Water-to-cement ratio	–	0.75	0.75
Grading curve	–	B8	B8

The slabs were exposed to 5.0 m/s air flow at 20 °C for 7 h using a custom-built wind tunnel. The total crack length, number of cracks, and crack widths were recorded. Additional plates ($300 \times 300 \times 80$ mm^3) were used to measure evaporation over time.

2.5 Assessment of Non-Deleterious Mechanical Performance of Recycled Constituents

Standard concrete mixtures were produced using fully substituted aggregate fractions to assess the non-deleterious nature of the recycled mineral constituents from TRC. Two types of recycled material were investigated: crushed TRC with carbon fibre textiles, from which carbon fibre grids had been largely removed, and mixed recycled concrete from TRC with glass fibre textiles, in which residual glass fibre fragments remained dispersed throughout the mineral matrix. Both aggregate types were used without additional separation or treatment. The concrete mixtures were designed to replicate standard structural concrete in terms of binder content and water-to-cement ratio. All natural aggregates were replaced by the recycled constituents at a substitution rate of up to 100 per cent by volume. The influence of recycled materials on the behaviour of fresh and hardened concrete was evaluated by examining the processing properties, mechanical integrity, and microstructural quality of the resulting mixtures.

The fresh concrete was tested regarding flowability, cohesion, and surface finish during casting. Slump was measured in accordance with DIN EN 12350–5, and the material was visually inspected for signs of segregation, compaction difficulty, or bleeding. The air content and fresh density were determined using pressure-based and gravimetric methods by DIN EN 12350–7. To evaluate mechanical integrity, cube specimens were cast and stored at 20 °C and 65-% relative humidity until testing at 28 days. To assess structural cracking behaviour or surface integrity abnormalities, the specimens were inspected after demoulding. The compressive strength was determined using a standard testing machine in accordance with EN 12390–3.

3 Results and Discussion

3.1 Structural Effectiveness of Recycled Fibres in Mortar-Scale Testing

The mortar prism tests revealed substantial differences in the mechanical response of the various fibre types under flexural loading (see Fig. 3). While the reference mix without fibres failed in a brittle manner immediately after cracking, the mixes containing recycled carbon fibres showed distinct post-cracking behaviour. Specimens with staple carbon fibres derived from mechanically cut grid fabrics exhibited a pronounced residual load-bearing capacity, indicating a reliable stress transfer across the crack plane. The stress-crack opening curves demonstrated a plateau-like progression beyond the first crack, characteristic of effective fibre bridging. In contrast, mixes with short pyrolysed carbon fibres displayed no measurable residual strength after first cracking. Their contribution was limited to a marginal increase in the initial cracking load, likely due to slight densification or crack dispersion rather than effective reinforcement action. This behaviour is attributed to the short fibre length, smooth surfaces, and lack of mechanical anchorage.

The performance of recycled fine aggregates from glass TRC, containing embedded glass fibre fragments, was similar to that of the short carbon fibre mixes. Although no significant post-cracking capacity was recorded, the stress-displacement curves suggested a minor increase in crack initiation load, potentially due to local bridging effects by fine fibre residues.

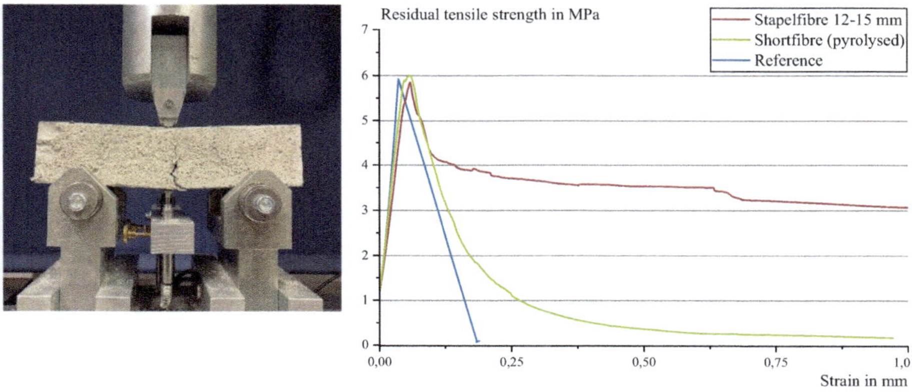

a) b)

Fig. 3. a) test setup; b) Representative stress–crack opening curves for all mortar mixtures (see also [15])

The findings confirm that only mechanically recovered staple fibres with sufficient length and aspect ratio can provide structural reinforcement in cementitious matrices. Fibre geometry, alignment, and bond properties remain critical for effective load transfer. These conclusions support the use of staple fibre formats in further structural-scale testing.

3.2 Load-Bearing Capacity of Recycled Fibre Reinforced Concrete Beams

The large-scale beam tests validated the reinforcing action of recycled staple carbon fibres under structural loading conditions. Beams containing 0.4 and 0.7 vol% fibres demonstrated a clear post-cracking load-carrying capacity (see Fig. 4). The load-displacement curves showed a gradual decrease in load after the first crack, followed by a fibre-dominated plateau region. No brittle failure occurred.

The predominant failure mode was fibre pull-out, confirming a sufficient but not optimal bond between fibre and matrix. Fibre orientation and surface condition, including the use of elliptical cross-sections and sand-coated surfaces, significantly influence adhesive performance; however, quantification of this was not part of this study. Higher fibre dosages (0.7 vol%) increased the load level in the plateau region but also resulted in stiffer post-cracking behaviour. Beams with short pyrolysed fibres, included for comparison, showed no significant enhancement of flexural capacity. The effect of residual glass fibre fragments, present in the aggregate of some mixes, was likewise negligible in load redistribution after cracking.

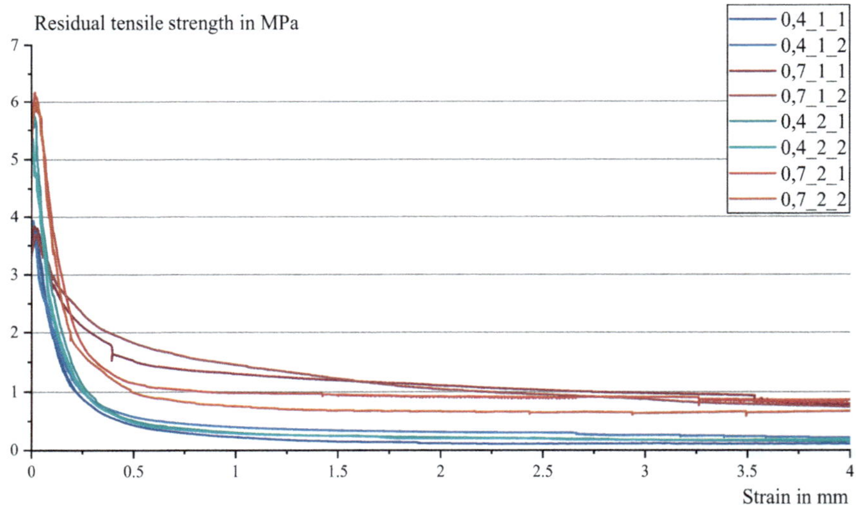

Fig. 4. Load–deflection curves for beam tests with varying carbon staple fibre contents

These results support the conclusion that mechanically processed staple fibres from recycled carbon grids can be structurally effective in concrete applications, provided that fibre geometry and surface properties are appropriately tailored.

3.3 Early-Age Shrinkage Crack Reduction Through Short Recycled Fibres

Shrinkage testing under accelerated drying conditions showed that increasing fibre dosage substantially reduced crack formation. The control mixture without fibres exhibited numerous cracks with significant cumulative length and width. Adding 0.03 vol%

short pyrolysed carbon fibres reduced the total crack area by over 40%, while a dosage of 0.04 vol% achieved an even more pronounced effect (see Figs. 5 and 7). An overview of the various shrinkage crack areas as a function of fibre volume content is shown in Fig. 5.

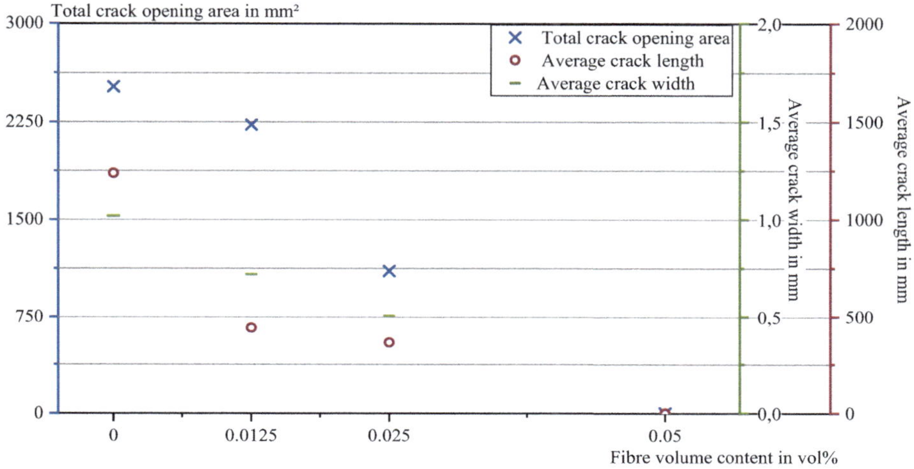

Fig. 5. Shrinkage crack area as a function of fibre volume content

At fibre contents of 0.05 vol% and above, no visible cracks were detected on the surface of the test slabs (see Fig. 7). This result was consistent across multiple specimens. The water evaporation rates, monitored in parallel using separate reference plates, were nearly identical across all fibre dosages (see Fig. 6), suggesting that the reduction in cracking was not caused by differences in moisture loss, but rather by the physical presence and dispersion of fibres in the cement matrix. Only the graph corresponding to the 0.05 vol% fibre content deviates slightly from the others and shows overall lower evaporation rates. This can be attributed to the reduced exposed surface area, as no cracks formed in these specimens.

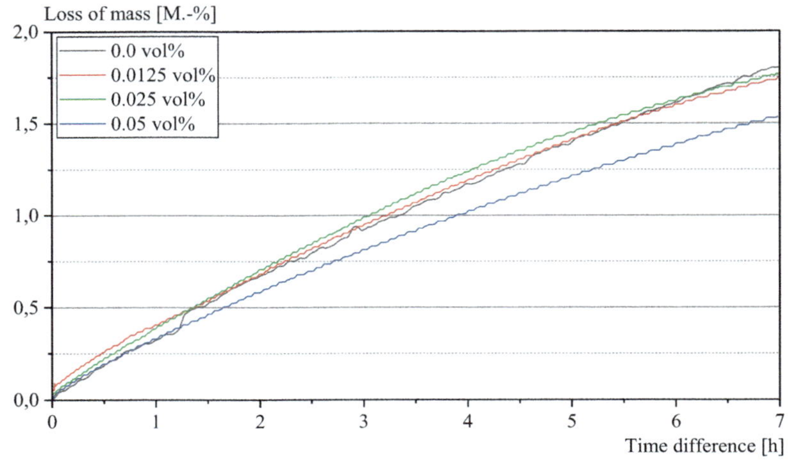

Fig. 6. Cumulative evaporation curves for all mixtures

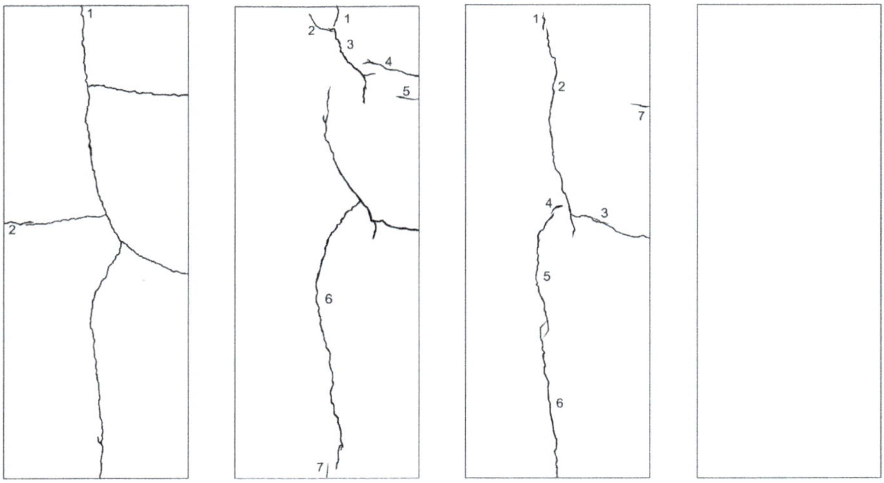

Fig. 7. Crack maps of selected slabs (from left to right: fibre volume content of 0.0%, 0.0125%, 0.025%, 0.05%)

These findings confirm that short pyrolysed carbon fibres, while not structurally effective in load-bearing capacity, offer clear benefits for crack control at early ages. The optimal dosage for early shrinkage crack mitigation was identified to be at 0.05 vol%, striking a balance between effectiveness and workability.

3.4 Concrete Performance with Recycled Aggregates

The concretes produced with recycled aggregate from carbon and glass textile waste performed well in terms of workability and compressive strength. Due to the high quality of the concrete, substituting natural aggregate (sand and gravel from the Rhine) with 100% recycled material did not impair the fresh concrete properties under laboratory mixing conditions (see Table 5). No segregation or bleeding was observed, and all mixtures could be compacted using standard procedures.

Table 5. Fresh concrete properties (slump, density, air content)

Concrete	Air content	Fresh density	Slump after		
			5 min	30 min	60 min
–	vol%	kg/m^3	mm		
Reference	1.2	2,370	520	450	400
Rec aggr. Carbon	3.2	2,313	470	450	430
Rec aggr. Glass	3.4	2,252	450	430	400

Compressive strength values obtained after 28 days were within the expected range for structural concrete, with minor variations attributed to the particle grading and surface texture of the recycled aggregates. The glass fibre fragments in the fine fraction did not lead to any visible deterioration or negative interaction with the binder matrix (see Table 6).

Table 6. Compressive strength values for mixtures with recycled aggregates

Age	Unit	Reference	Rec aggr. Carbon	Rec aggr. Glass
28d	MPa	43.4 ± 1.06	48.2 ± 0,10	45.1 ± 1.62
365d		51.6 ± 1.06	54.7 ± 0.56	52.9 ± 1.31

These observations confirm the technical viability of using TRC-derived recycled aggregates in structural concrete without compromising material performance or safety.

4 Conclusion

This study confirms the technical feasibility of a comprehensive material reuse strategy for TRC. The mineral fraction and the textile reinforcement components were recovered, processed, and reused in new cementitious systems. Recycled aggregates derived from TRC with carbon and glass textile reinforcements were used at full substitution rates without impairing the fresh or hardened concrete properties. Staple carbon fibres, mechanically extracted from textile reinforcements offcuts, proved to be structurally adequate when reintroduced into mortar and concrete. They contributed to a measurable increase in post-cracking load capacity and enhanced ductility. In contrast, short pyrolysed carbon fibres were not structurally active but effectively reduced shrinkage-induced surface cracking at low dosages. The presence of glass fibre residues in the fine fraction of recycled aggregates had no adverse effect on processing or strength and may offer minor benefits in crack distribution.

Compared to conventional (non-recycled) TRC, recycled fibre concretes showed a moderate reduction in initial stiffness (approximately 10 - 15%) but maintained comparable ultimate load capacities. This demonstrates that recycling pathways do not significantly compromise mechanical performance and offer substantial sustainability advantages. Although this study confirmed the mechanical and structural performance of recycled TRC, the long-term durability, including freeze-thaw resistance and chemical stability of recycled fibres, requires further evaluation. Due to potential microstructural changes from pyrolysis or mechanical crushing, durability properties might differ from virgin materials. Therefore, future investigations should address the performance under environmental ageing conditions to validate suitability for broader structural applications. The presented recycling pathways contribute substantially to reducing environmental impacts by closing material loops and lowering virgin material consumption. Previous life-cycle assessments specifically conducted on recycled carbon concrete from this project indicated potential CO_2 savings between 20 and 40% compared to conventional TRC production [14]. Integrating detailed LCA methodologies in future work will further quantify the exact sustainability gains and strengthen the case for widespread adoption.

The results underline the potential of recycled TRC constituents for use in structural concrete without compromising safety or performance. The dual use of aggregate and fibre fractions contributes to closing material cycles and reducing the demand for virgin resources. The findings are supported by previous research on fibre recovery, bond behaviour, and recycled aggregate performance, as discussed in the cited literature. Optimising fibre geometry and surface condition represents a promising direction for the future development of recycled fibre reinforcements in structural applications. To further improve the reinforcing efficiency of recycled fibres, surface treatments such as post-application of sand coatings may enhance mechanical interlock and bond strength within the cement matrix.

Acknowledgments. This research was funded by the German Federal Ministry of Education and Research (BMBF) as part of the FaBeR project (Faser- und Beton-Recycling von Carbon- und Textilbeton). The financial support from the BMBF is gratefully acknowledged.

References

1. Rawat P et al (2022) A state-of-the-art review on mechanical performance characterisation and modelling of high-performance textile reinforced concretes. Const Build Mater 347
2. Scheurer M et al (2023) Current and future trends in textiles for concrete construction applications. Textiles 3(4)
3. Friese D et al (2022) Textile reinforcement structures for concrete construction applications––a review. J Composite Mater 56(26)

4. Deutsches Institut für Bautechnik (DIBt): Allgemeine bauaufsichtliche Zulassung Z 1.6.308—Carbonbewehrungsgitter zur Verwendung in Betonbauteilen, Berlin (2024)
5. Deutscher Ausschuss für Stahlbeton (DAfStb): DAfStb-Richtlinie—Betonbauteile mit nichtmetallischer Bewehrung, Berlin (2024)
6. Overhage V et al (2025) Potential of and current challenges in reusing recycled carbon fibres in concrete construction applications. Sustainability 17(6)
7. Bayram B et al (2024) Recycling potential of carbon fibres in the construction industry: from a technical and ecological perspective. Waste Manage Res 42(9)
8. Kimm M et al (2018) On the separation and recycling behaviour of textile reinforced concrete: an experimental study. Mater Struct 51
9. Naqvi SR et al (2018) A critical review on recycling end-of-life carbon fibre/glass fibre reinforced composites waste using pyrolysis towards a circular economy. Res Conservat Recycl 136
10. Shi T et al (2025) Performance characterization of recycled carbon fiber and its interfacial bonding properties with cement matrix. Materials 18(7)
11. Bledzki AK et al (2021) Recycling of carbon fibre reinforced composite polymers—Review—Part 1: volume of production, recycling technologies, legislative aspects. Polymers 13(2)
12. Gopalraj K et al (2020) A review on the recycling of waste carbon fibre/glass fibre-reinforced composites: fibre recovery, properties and life-cycle analysis. SN Appl Sci 2(3)
13. Tao Y et al (2023) Recycling of glass fibre reinforced polymer (GFRP) composite wastes in concrete: a critical review and cost benefit analysis. Structures, vol 53. Elsevier
14. Luthin A et al (2023) Demonstrating circular life cycle sustainability assessment–a case study of recycled carbon concrete. J Cleaner Prod 433
15. Vaculik S et al (2024) Faser- und betonrecycling von carbon- und textilbeton. Bauen Im Bestand 4

Accelerating Sustainable Building Envelope Design Through Inverse Optimization to Minimize Lifecycle Energy and Embodied Carbon

Sejal Sanjay Shanbhag(✉) and Manish Kumar Dixit

Texas A&M University, College Station, TX, USA
sejal.shanbhag@tamu.edu

Abstract. Achieving global net-zero emissions goals necessitates significantly reducing embodied carbon in all new buildings, infrastructure, and renovations, a goal that emphasizes on the urgency of innovative approaches to life cycle assessments (LCA) and sustainable design strategies. Current LCA methods often lack the efficiency required for early design stages, posing significant challenges for architects and engineers aiming to balance energy efficiency, emissions reduction, and design flexibility. Forward optimization methods utilized to determine optimal lifecycle energy and embodied carbon parameters, are computationally demanding and time-intensive, making them less viable for rapid iterative design processes. To address these challenges, we propose an inverse optimization framework designed to reverse-engineer building envelope design configurations that meet predefined embodied energy targets. This method explores optimal combinations of orientation, window-to-wall ratio, and material selection for walls, roofs, and floors to achieve significant reductions in both embodied and operational energy. By shifting the focus from traditional forward optimization to a more streamlined, target-driven approach, the framework enables a more efficient, collaborative assessment process that integrates sustainability and cost-effectiveness into early design decisions. This study aims to demonstrate the potential of the inverse optimization framework to facilitate sustainable design empowering stakeholders to implement flexible and impactful emissions reduction strategies. By bridging technical innovation and policymaking, the approach contributes to the advancement of lifecycle analysis methodologies essential for managing the complexities of the built environment.

Keywords: Inverse Optimization · Life Cycle Assessment · Embodied Energy · Embodied Carbon

1 Introduction

The building sector is a major contributor to climate change, responsible for roughly 36% of global final energy use and nearly 40% of energy-related CO_2 emissions [1]. With international commitments, such as the Paris Agreement, aiming to limit global

H. Beushausen et al. (Eds.): NTZR 2025, LNCE 873, pp. 1107–1116, 2026.
https://doi.org/10.1007/978-3-032-27698-8_97

warming to 1.5 °C above pre-industrial levels, decarbonizing the built environment has become an urgent priority [2, 3]. Achieving net-zero emissions by 2050 requires addressing both operational energy use and embodied carbon across new construction, retrofits, infrastructure projects, and even occupant behavior, emphasizing the need for early, comprehensive emission reduction strategies [4–6]. Traditionally, decarbonization efforts have focused on improving operational efficiency through energy-saving technologies and renewable energy systems. However, as operational emissions decline, embodied emissions impact from materials and construction processes are increasingly significant [7]. These impacts are largely determined during early design stages, underscoring the importance of integrating life cycle assessment (LCA) to guide decisions on building form, materials, and configurations from the outset [8].

Forward optimization methods, which iteratively test design variables such as geometry, orientation, and material composition to achieve performance objectives, have been widely employed to enhance energy and carbon efficiency [9]. While effective, these approaches often require substantial computational resources and time, making them impractical for the rapid, iterative workflows typical in architectural and engineering practice. Additionally, they necessitate generating multiple simulations based on assumed configurations before performance evaluation, which may not align with the flexibility and time constraints of early design phases [10].

To address these limitations, researchers have begun exploring inverse optimization as a goal-oriented alternative. Unlike forward optimization, inverse approaches begin with defined performance targets such as specific levels of energy use, carbon emissions, or thermal comfort and work backward to identify viable combinations of design variables that meet these goals. This methodology aligns more effectively with performance-driven design objectives and enables more efficient convergence toward optimal solutions [11]. Inverse optimization has demonstrated value across various building performance domains. Researchers have applied it to optimize shading systems for improved daylighting, retrofit buildings for reduced energy consumption, and bridge gaps between simulated and actual building performance [12, 13]. These applications demonstrate the growing potential of inverse methods as tools for achieving net-zero building goals and improving model accuracy.

Despite its promise, integrating inverse optimization into conventional design workflows presents several challenges, including computational complexity, data requirements, integration with architectural tools, and the need for intuitive designer feedback. To address these barriers, this study proposes a novel inverse optimization framework specifically designed for early-stage design processes. The framework aims to support architects and engineers in developing low-energy building envelope solutions by automating the exploration of design parameters including orientation, window-to-wall ratio, and material selections that achieve predefined embodied and operational energy targets. This research aims to bridge the gap between technical innovation and real-world application, advancing life cycle optimization methods that are crucial for addressing the complexity in the global built environment.

2 Methodology

An overview of the proposed inverse optimization framework is illustrated in Fig. 1, which outlines the integration of baseline inputs, Python-based computations, and the Inverse Particle Swarm Optimization engine into a unified workflow for evaluating building design alternatives.

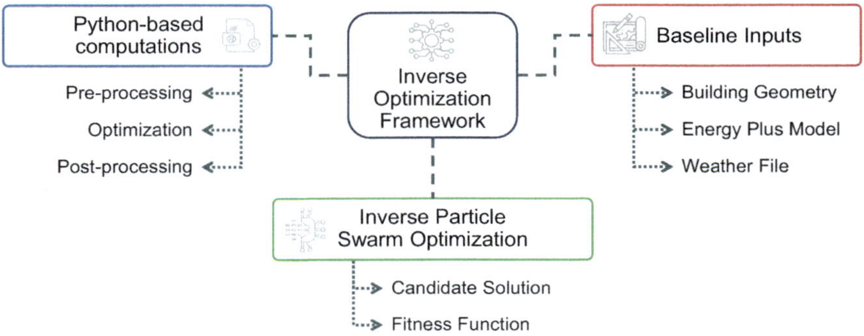

Fig. 1. Overview of the Inverse Optimization Framework

This study introduces a novel inverse optimization framework to simultaneously minimize operational energy (OE) and embodied energy (EE) within the A1–A3 and B6 LCA system boundaries across various building design alternatives. The proposed methodology integrates detailed Energy Plus simulation capabilities with a Python-based computational pipeline that handles pre-processing, optimization, and post-processing operations. The pre-processing phase maps each design variable to its corresponding physical property or material in the database, while also modifying geometric attributes of the building model such as orientation and fenestration. The post-processing phase extracts simulation outputs and computes energy and environmental performance indicators over the building's expected lifespan. The framework as shown in Fig. 2, employs a systematic workflow of parametric building iteration coupled with energy simulation, evaluated against performance targets. Inverse Particle Swarm Optimization (I-PSO) efficiently navigates the solution space to identify design variables. By connecting an Input-Output Hybrid materials database [14, 15] with the simulation environment, the system enables total energy assessment, allowing designers to explore trade-offs between embodied impacts and operational performance during early design phases when changes are most impactful. Dixit et al. [14] employed a United States specific Input-Output Hybrid (IOH) methodology developed using U.S. national economic and environmental data. The hybrid model integrates 2012 Input-Output tables from the U.S. Bureau of Economic Analysis (BEA) with energy use data from federal sources including the U.S. Department of Energy (USDOE), U.S. Census Bureau (USCB), and U.S. Geological Survey (USGS). The analysis adopted a cradle-to-gate system boundary, accounting for both direct and indirect energy used in material extraction, processing, transportation, and manufacturing. Embodied energy due to transportation, construction, and administrative services is included, while end-of-life phases are excluded.

The optimization begins with a baseline building model defined in an Energy Plus Input Data File (IDF) paired with a representative weather file (EPW format). The framework dynamically modifies the building model's geometry and material assignments within the simulation loop. These modifications include adjustments to orientation, window-to-wall ratios on all facades, and substitutions of material layers for walls, roofs, floors, and glazing systems. Each simulation iteration generates a new model configuration based on a set of design variables, enabling an automated exploration of design alternatives within the parametric study. This iterative process not only enables rapid prototyping of diverse configurations but also ensures consistency in evaluating lifecycle metrics across alternatives by maintaining a uniform computational structure.

The resulting models are evaluated for both embodied and operational energy performance, with metrics computed from the material properties and simulation outputs to quantify lifecycle environmental impacts based on a 50 year lifespan. The results of each simulation, including heating, cooling, and lighting demands, are combined with material-specific impact data to quantify total primary energy use and embodied resource consumption. These metrics form the basis of the optimization objective.

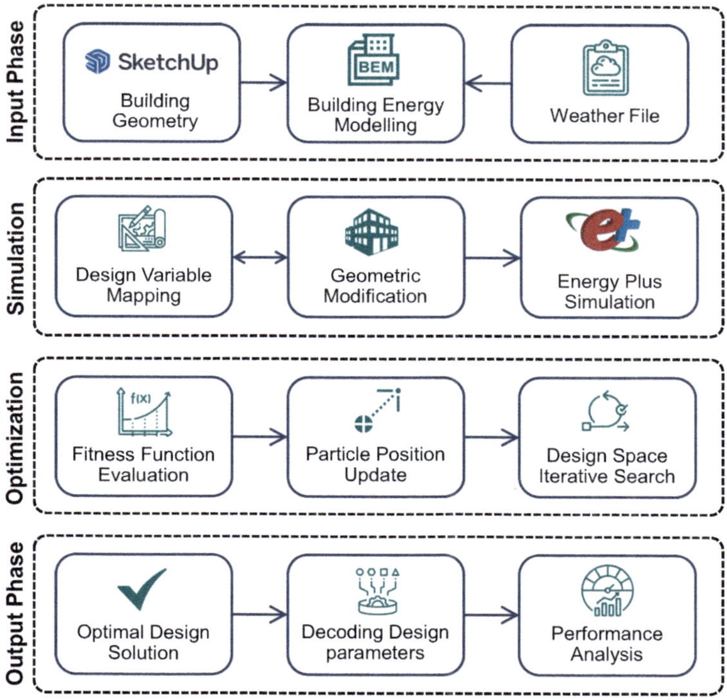

Fig. 2. Inverse Optimization Workflow

The design space is explored using Inverse Particle Swarm Optimization (PSO). Each particle in the swarm represents a candidate solution vector. Fitness function J is defined as the normalized sum of squared errors with respect to target operational and embodied energy values:

$$J = \left(\frac{E_{OS} - E_{OT}}{E_{OT}}\right)^2 + \left(\frac{E_{ES} - E_{ET}}{E_{ET}}\right)^2 \tag{1}$$

where E_{OS} is simulated operational energy, E_{OT} is the target operational energy, E_{ES} is the simulated embodied energy, E_{ET} is the target embodied energy. This formulation allows the optimizer to penalize solutions that deviate significantly from target performance thresholds, creating a multi-objective balance without explicitly defining hard constraints.

Each particle updates its position using:

$$v_i^{t+1} = wv_i^t + c_1 r_1 (p_i - x_i^t) + c_2 r_2 (g - x_i^t) \tag{2}$$

$$x_i^{t+1} = x_i^t + v_i^{t+1} \tag{3}$$

where, v is velocity, x is position, pi is personal best, g is global best, and $r_1, r_2 \sim U(0,1)$. The inertia weight w, cognitive coefficient c_1, and social coefficient c_2 are tunable.

At the end of the optimization process, the most effective design configuration is translated into interpretable material and geometric attributes. The optimal solution is then decoded into its original design features, including specific material choices and geometric characteristics. The associated performance outcomes including total operational energy, embodied energy, and their respective reductions compared to the baseline are documented to support design decision-making. The modular structure of the framework makes it adaptable to different building types, climates, and optimization goals, supporting its application in both early-stage design and retrofit scenarios.

3 Results

This study focuses on a commercial complex located in Cody, Wyoming, situated in ASHRAE Climate Zone 6B, characterized by cold winters and relatively low humidity. The building serves as a test case for evaluating energy performance optimization strategies over its projected 50-year lifecycle. The structure was modeled with thermal zoning defined according to standard energy modeling practices, distinguishing between perimeter and core zones to better capture the influence of solar exposure and internal heat gains. The building's geometry and usage align with typical commercial typologies, making it an example for early-stage design decision-making.

The energy performance comparison between the baseline and optimized designs is presented in Fig. 3, reflecting the outcomes driven by ambitious optimization targets of 60 MBTU for operational energy and 1300 MBTU for embodied energy. The best solution achieved a 21.69% reduction in operational energy. More notably, the embodied energy decreased by 12.2%, contributing to an overall 19% reduction in the building's total life cycle energy use, over the 50-year analysis period. These improvements highlight the effectiveness of the multi-objective algorithm in reducing both operational and embodied energy.

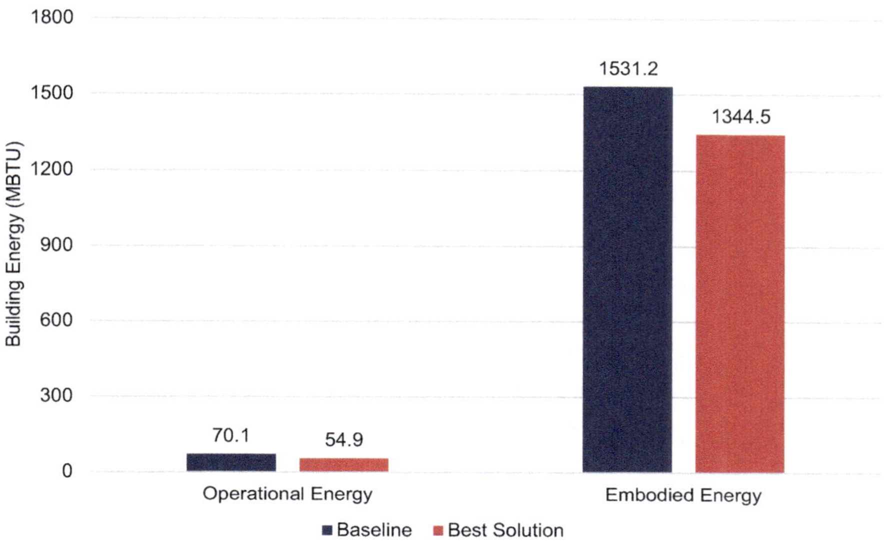

Fig. 3. Comparison of Operational Energy and Embodied Energy for Baseline and Best Solution

The design parameters for the baseline model and the best solution which achieved a 19% reduction in total life cycle energy are summarized in Table 1. While both configurations maintain the same orientation, the optimized design features higher window-to-wall ratios (WWRs) on all facades, increasing daylight access and potentially reducing lighting loads. Notable material substitutions were made in the wall, floor, and roof assemblies. For instance, the best solution design replaces closed-cell polyurethane foam with fiberglass insulation and incorporates low density fiberboard in the roof system. Changes in glazing and gas fill such as a shift to thinner clear glass and a narrower xenon filled gap also reflect a balance between material impact and thermal performance. These modifications contribute to the improvement in operational and embodied energy outcomes.

Figure 4 illustrates the progression of the optimization process by plotting the best objective function value (or cost) across the first 75 iterations out of the 200 iterations. The optimizer shows rapid improvement during the early stages, with significant reductions in the cost function observed within the first 10 iterations. Beyond that point, the curve begins to flatten, indicating convergence toward an optimal solution. The final solution highlighted in red represents the best cost achieved throughout the search, suggesting that the optimization process avoided premature convergence and local minima. The overall behavior of the plot reflects an effective and stable convergence of the algorithm toward a near-optimal or global solution.

Figure 5 presents Embodied Energy (EE) versus Operational Energy (OE) for all evaluated design solutions across the optimization process. Each point in the plot represents a unique design iteration, with the blue dots corresponding to the generated solutions and the red dot marking the best-performing design in terms of overall objective function value.

Table 1. Comparison of design parameters of baseline and best solution.

Design Parameters	Baseline Model	Best Solution
Orientation	0	0
WWR_N	0.06	0.1
WWR_S	0.03	0.1
WWR_E	0.03	0.1
WWR_W	0.07	0.1
Wall Layer1	Lightweight CMU (ASTM C90)	Brick - fired clay - 102 mm
Wall Layer2	Closed-Cell Polyurethane Foam	Sheathing - 12.7mm- Plywood
Wall Layer3	Painted Gypsum Board 5/8"	Fiberglass insulation
Wall Layer4	Gypsum Board	Gypsum Board
Floor Layer1	Normal weight concrete floor	Reinforced concrete thick slab
Floor Layer2	Carpet Tile	Carpet Tile
Roof Layer1	Elastomeric membrane	Coated glass fiber base sheet
Roof Layer2	Cover board - OSB -7/16inch	Cover board - OSB - 1/2inch
Roof Layer3	Polyisocyanurate-5inch-R30	Fiberboard low density- R4.17
Roof Layer4	Vapor retarder- 0.006inch	Vapor retarder- 0.002inch
Roof Layer5	Roof Metal Deck	Steel roof decking type B- 22 g
Glazing	Tempered Glass	CLEAR 3MM- plain
Gas	Air 13mm	XENON 3MM
Window Frame	Aluminum	Aluminum

A clear trade-off trend emerges of designs that perform well in reducing operational energy tend to exhibit relatively higher embodied energy values, and vice versa. This pattern highlights multi-objective optimization conflict minimizing operational energy often requires added insulation or high-performance glazing, which can increase the embodied energy due to the materials used. Conversely, selecting low-embodied-energy materials may compromise the building envelope's thermal performance, resulting in higher operational energy consumption over time. The red dot denotes the best solution indicating that the optimizer successfully identified a design that balances both objectives effectively. The results demonstrate that the optimization process supported nuanced exploration across the design space, uncovering not just one optimal solution but a range of trade-offs between embodied and operational energy. This capability is particularly important in early-stage design, where flexibility and insight into material-performance dynamics can lead to more sustainable outcomes.

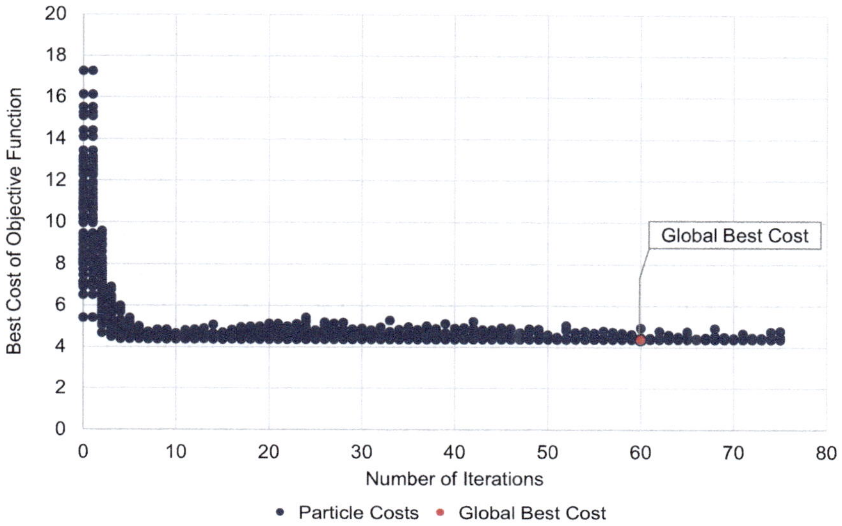

Fig. 4. Best objective function value vs. iteration number.

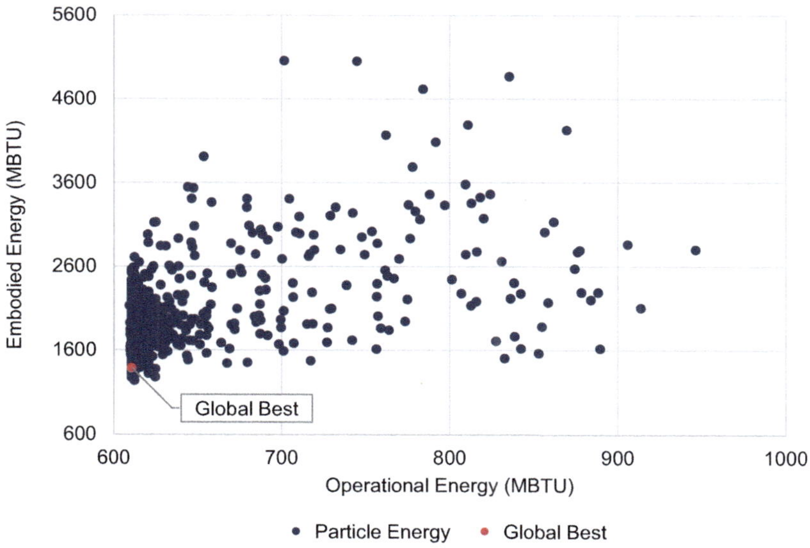

Fig. 5. Embodied Energy vs. Operational Energy for all design iterations.

4 Conclusion

This study demonstrates the potential of an inverse optimization framework as a viable and efficient tool for guiding early-stage sustainable building design. By reversing the traditional forward optimization approach, the framework allows designers to begin with clear performance targets for operational and embodied energy and work backward to discover viable combinations of envelope design parameters. Applied to a commercial case study in Climate Zone 6B, the method successfully identified a best-performing design configuration that achieved a 19% reduction in total life cycle energy over a 50-year period.

Unlike forward optimization, which can be computationally intensive and time-prohibitive for iterative workflows, the inverse framework streamlines the process, making it more compatible with real-world design cycles. The use of Inverse Particle Swarm Optimization (I-PSO), integrated with Energy Plus simulations and a Input Output hybrid materials database [14], facilitated efficient exploration of the design space and delivered a range of solutions reflecting complex trade-offs between performance goals. Notably, the results reveal a clear trade-off relationship between operational and embodied energy, emphasizing the importance of multi-objective optimization in lifecycle assessment. The framework empowers stakeholders to navigate these trade-offs transparently and flexibly, enabling informed decisions based on project-specific priorities such as cost, carbon, or comfort. Although inverse optimization reduces the number of forward simulations needed, the approach still relies on multiple iterations of Energy Plus simulations, which are time-intensive and may be impractical for very large or highly complex models without parallel computing resources. Future work may extend this framework to incorporate additional objectives such as life cycle carbon, or thermal comfort. Overall, this study advances life cycle assessment methodologies and contributes a practical, goal-oriented tool to the growing toolkit of sustainable building design strategies.

References

1. Langevin J, Wilson E, Snyder C, Narayanamurthy R, Miller J, Kaplan K, Ciraulo R et al (2024) Decarbonizing the US Economy by 2050: a national blueprint for the buildings sector. Lawrence Berkeley National Laboratory (LBNL), Berkeley, CA (United States)
2. IPCC (Intergovernmental Panel on Climate Change). (2018). Global Warming of 1.5° C. An IPCC Special Report on the impacts of global warming of 1.5° C above pre-industrial levels and related global greenhouse gas emission pathways, in the context of strengthening the global response to the threat of climate change, sustainable development, and efforts to eradicate poverty. Geneva: ipcc
3. Yaduvanshi NS, Park JY (2023) Citizen building scientist? The opportunity and challenge of citizen involvement in building retrofits. J Build Eng 78:107629
4. Shanbhag SS, Dixit MK (2024) Understanding carbon-negative potential of hempcrete using a life cycle assessment approach. In: The international conference on net-zero civil infrastructures: innovations in materials, structures, and management practices (NTZR). Springer Nature Switzerland, Cham, pp 169–182
5. Shanbhag SS, Dixit MK (2024) A review of evolving climate and energy economy trends to enhance the dynamic life cycle assessment of buildings. Sustain Cities Soc 111:105560
6. Yaduvanshi NS, Payne T, Park JY (2025) Net-zero home race: a board game to engage and educate students for sustainable building retrofits. J Archit Eng 31(4):04025038
7. Röck M et al (2020) Embodied GHG emissions of buildings–the hidden challenge for effective climate change mitigation. Appl Energy 258:114107
8. Zhang Y, Sattar S, Cook DT, Johnson KJ, Fung JF (2024) Systematic review of embodied carbon assessment and reduction in building life cycles. NIST Special Publication
9. Li Z, Chen H, Lin B, Zhu Y (2018) Fast bidirectional building performance optimization at the early design stage. In: Building simulation, vol 11. Tsinghua University Press, pp 647–661

10. Pan Y et al (2023) Building energy simulation and its application for building performance optimization: a review of methods, tools, and case studies. Adv Appl Energy 10:100135
11. Ahuja RK, Orlin JB (2001) Inverse optimization. Oper Res 49(5):771–783
12. Zhang Y, Zhang Y, Shi W, Shang R, Cheng R, Wang X (2015) A new approach, based on the inverse problem and variation method, for solving building energy and environment problems: preliminary study and illustrative examples. Build Environ 91:204–218
13. Yang J et al (2020) Inverse optimization of building thermal resistance and capacitance for minimizing air conditioning loads. Renew Energy 148:975–986
14. Dixit M, Pradeep Kumar P (2023) Embodied energy, embodied water and embodied carbon of university buildings: energy-carbon-water nexus approach to sustainable buildings. In: Embodied energy, embodied water and embodied carbon of university buildings: energy-carbon-water nexus approach to sustainable buildings
15. Shanbhag SS, Dixit MK (2025) Integrating evolving energy economy scenarios in dynamic life cycle assessment of buildings. Environ Impact Assess Rev 112:107772

Environmental Impacts of Sustainable Soy Biodiesel Production: Analysis of Supply Chain Management

Muhammed A. Bhuiyan[1(✉)], Quddus Tushar[1,2], and Ziyad Abunada[2,3,4]

[1] School of Engineering, RMIT University, GPO Box 2476, Melbourne, VIC 3001, Australia
muhammed.bhuiyan@rmit.edu.au

[2] School of Engineering and Technology, Central Queensland University, Melbourne, VIC 3000, Australia

[3] Centre for Hydrogen and Renewable Energy, Gladstone Marina Campus, Gladstone, Qld 4680, Australia

[4] School of Engineering and Technology, Coastal Marine Ecosystems Research Centre (CMERC), Central Queensland University, Melbourne, VIC 3000, Australia

Abstract. The study aims to identify the associated environmental impacts of the soy biodiesel production process. The process is known as transesterification, where the soybean oil reacts with methanol in the presence of another catalyst such as sodium hydroxide. Glycerine is produced as a by-product of this system, and impurities like fatty acids are removed from biodiesel using hydrochloric acid. The upstream stages of this production process, including crude soybean degumming, natural gas combustion, electricity consumption, methanol production, transportation, and other relevant activities, are analysed to calculate environmental impacts. Energy-intensive crude soybean degumming at the plant has been identified as the primary source of carbon footprint for biodiesel production. The findings of this study highlight the importance of conducting a process-based supply chain life cycle assessment (LCA) to assess fuel sustainability. Thus, the soy biodiesel is an alternate renewable fuel; its environmental benefits can be emphasized by improving agricultural efficiency, shifting towards renewable electricity production, and sourcing bio-based chemicals.

Keywords: soybean degumming · supply chain · Life Cycle Assessment (LCA)

1 Introduction

Transition to the renewable energy source has been identified as a sustainable solution for the growing concern of climate change [1]. An alternative low-carbon source of energy is now the prime focus in the research arena as the fossil fuel reserves continue to decline. Production of biodiesel derived from vegetable oils or animal fats seems to be a potential alternative to petroleum-based diesel [2]. Considering various options, soybean oil has been identified as a suitable source for biodiesel production. This production process is known as transesterification, where triglycerides of soybean oil react with methanol in the presence of a catalyst [3]. However, these production processes need justification in terms of environmental performance.

H. Beushausen et al. (Eds.): NTZR 2025, LNCE 873, pp. 1117–1125, 2026.
https://doi.org/10.1007/978-3-032-27698-8_98

Studies have found the growing importance of biodiesel due to its biodegradable, renewable, eco-friendly, and non-toxic properties [4, 5]. The process of forming biodiesel is known as fatty acid methyl ester. Commercialization of high feedstock prices has been found to be one of the obstacles in this process. Although crude soybean oil is one of the cheapest sources of biodiesel production, the amount of high phosphorus content inhibits the action of the catalyst during transesterification [6]. The degumming process has been found to be an optimized process to remove the maximum phosphorus [7]. It assists in reducing the maximum phosphorus content and fatty acids. Additionally, the transesterification process, besides degumming, assists in achieving the desired soy biodiesel production [8]. However, this production process needs to be justified from a sustainability perspective. This research will identify the potential sectors that could significantly reduce the carbon footprint of biodiesel production.

Life cycle assessment (LCA) is a scientific process to identify the associated impacts related to a certain production system [9]. Biodiesel production is assumed to be greener compared to fossil fuel production. However, a certain decision cannot be made until the verification of agricultural practices, additional processing components, and energy consumption for the entire production. Therefore, a proper evaluation through the systematic diagnosis will assist in bringing a sustainable policy for investment.

Analysis of the supply chain can be effectively used in the management process to identify "hotspots" for emissions [10]. Moreover, strategies can be adopted at the production stage to reduce emissions at its maximum. Therefore, the study considers the upstream process of biodiesel production, including crude oil extraction, degumming, methanol production, electricity consumption, transportation, and others. The evidence-based scientific analysis will ensure the accuracy of sustainability measures. Additionally, design incentives, subsidies, and possible mandates can be improvised in a certified biodiesel production system.

2 Methodology

2.1 Goal and Scope

The goal of this study is to conduct a cradle-to-gate analysis to quantify the environmental impact of 1 kg of soy biodiesel production [11]. It considers the supply chain of the soybean production process and the auxiliary component of biodiesel production. The objective of this study is to identify the influential parameters that contribute the most to environmental impacts.

2.2 Functional Unit and System Boundary

The functional unit of this study is 1 kg of soy biodiesel production at the plant, considering the supply chain and necessary components for the manufacturing [12]. The system boundary of this study is the cradle to gate, which encompasses the upstream feedstock, such as soybean cultivation, harvesting, crushing, and oil extraction from the bean [13]. The process of oil extraction is known as degummed. Electricity, methanol, natural gas, sodium hydroxide, and some other components act as the catalysts for this transesterification [14]. Transportation of input materials to the biodiesel plant has also been considered for this study.

2.3 Co-Products and Allocation

Mass allocation of the inputs and outputs of the materials has been used in this typical analysis [15]. The supply chains of the allocation from soybean production to biodiesel conversion are calculated for the carbon footprint. In the transesterification, glycerine is produced as a by-product of soy biodiesel production.

2.4 Life Cycle Inventory

Primary data on the materials composition is obtained from the plant facilities in the year 2024. Mass of the degummed oil, methanol, NaOH dosage, electricity consumption for pumps, mixing and separation, and natural gas for heating purposes are identified from engineering design data, as shown in Table 1. Secondary data on soybean oil production, the degummed process, is obtained from the Australian Life Cycle Inventory (AusLCI) [16]. Transportation is modelled in terms of tonne-kilometres with loading and empty return assumptions aligning with the database. Capital goods are excluded from this analysis, which includes the machinery set up for the industrial plant.

Table 1 Inputs and outputs of 1 kg of soy biodiesel production

Outputs and inputs	Amount	Unit
Outputs to technosphere		
Soy biodiesel production	1	kg
Glycerine	0.12	kg
Inputs from technosphere: materials and fuels		
Soybean oil (Degummed)	0.98	kg
Electricity	0.036	kWh
Natural gas	0.022	m^3
Methanol	0.090	kg
Sodium methylate	0.021	kg
Sodium hydroxide	0.0097	kg
Hydrochloric acid	0.0072	kg
Transportation	1.24	tkm

2.5 Life Cycle Impact Assessment (LCIA)

Characterization factors are applied at the midpoint level. The most modern approach of LCIA has been adopted for this study, which is known as ReCiPe 2016 [17, 18]. No normalization and weighting factors have been considered for this study to obtain the endpoint value. The following impact categories have been assessed for this study.

1. Global warming (kg CO_2 eq)
2. Stratospheric ozone depletion (kg CFC11 eq)
3. Photochemical ozone formation-human health and terrestrial (kg NOx eq)
4. Fine particulate matter formation (kg $PM_{2.5}$ eq)
5. Terrestrial acidification (kg SO_2 eq)
6. Terrestrial freshwater and marine ecotoxicity (kg 1, 4-DCB eq)
7. Human carcinogenic and non-carcinogenic toxicity (kg 1, 4-DCB eq)
8. Fossil resource scarcity (kg oil eq)

3 Results and Discussion

Life cycle assessment of 1 kg soy biodiesel production shows the midpoint impact categories for the design components. The analysis considers the supply chain of biodiesel production at the plant, such as the degummed process, electricity consumption, natural gas combustion, methanol production, sodium hydroxide, and fuel consumption for transportation. The results clearly show that biodiesel production is not totally environmentally friendly in several categories. Moreover, it prioritizes the hotspots in agricultural production and transesterification processes of biodiesel formation.

Table 2 Impact categories for designed product inputs

Impact categories	Units	Degummed	Methanol	Sodium hydroxide
Global warming	kg CO_2 eq	0.44455	0.10933	0.00098
Stratospheric ozone depletion	Kg CFC11 eq	3.64E-06	3.14E-10	2.98E-10
Ozone formation, Human health	kg NO_x eq	1.76E-03	1.17E-04	1.99E-06
Fine particulate matter formation	kg $PM_{2.5}$ eq	6.33E-04	1.20E-04	2.18E-06
Ozone formation, Terrestrial ecosystems	kg NO_x eq	1.82E-03	1.62E-04	2.01E-06
Terrestrial acidification	kg SO_2 eq	2.25E-03	4.14E-04	7.48E-06
Terrestrial ecotoxicity	kg 1,4-DCB	3.35E-01	5.88E-04	1.44E-04
Freshwater ecotoxicity	kg 1,4-DCB	2.77E-03	1.66E-04	1.75E-06
Marine ecotoxicity	kg 1,4-DCB	1.79E-03	2.16E-04	2.26E-06
Human carcinogenic toxicity	kg 1,4-DCB	2.04E-04	1.76E-05	3.75E-07
Human non-carcinogenic toxicity	kg 1,4-DCB	0.04755	0.00774	0.00008
Fossil resource scarcity	kg oil eq	0.08753	0.01482	0.00033

Life cycle impact categories for different inputs have been explained in considering the processed materials and processing inputs. Table 2 shows that materials directly contribute to biodiesel processing, such as degummed soybean oil, methanol, and sodium hydroxide. Degummed soybean oil (0.44455 kg CO_2 eq) has been observed as the highest contributor in the carbon footprint, followed by methanol and sodium hydroxide. Table 3 highlights the influence of logistics and energy consumption in the processing of 1 kg of soy biodiesel. Consumption of natural gas, electricity usage, and transportation activities are the prime factors in these logistic activities. The relative contribution of these factors is 0.0506, 0.02525, and 0.03079 kg CO_2 eq consecutively. The bulk emission has been observed from raw materials (degummed soybean oil and methanol). Energy consumption using electricity, natural gas, and transportation also plays a significant role in emissions.

Table 3 Impact categories for processing inputs

Impact categories	Units	Electricity	Natural gas	Transport
Global warming	kg CO_2 eq	0.02525	0.0506	0.03079
Stratospheric ozone depletion	kg CFC11 eq	8.44E-10	1.34E-10	7.43E-09
Ozone formation, Human health	kg NO_x eq	6.71E-05	4.03E-05	2.02E-04
Fine particulate matter formation	kg $PM_{2.5}$ eq	5.52E-05	1.17E-04	3.69E-05
Ozone formation, Terrestrial ecosystems	kg NO_x eq	7.00E-05	4.04E-05	2.04E-04
Terrestrial acidification	kg SO_2 eq	1.89E-04	4.02E-04	1.23E-04
Terrestrial ecotoxicity	kg 1,4-DCB	2.06E-03	4.66E-04	4.82E-04
Freshwater ecotoxicity	kg 1,4-DCB	2.04E-05	1.40E-04	1.52E-04
Marine ecotoxicity	kg 1,4-DCB	2.73E-05	1.76E-04	2.05E-04
Human carcinogenic toxicity	kg 1,4-DCB	1.36E-05	8.37E-06	2.10E-05
Human non-carcinogenic toxicity	kg 1,4-DCB	0.00123	0.00558	0.00819
Fossil resource scarcity	kg oil eq	0.00792	0.01876	0.00909

3.1 Global Warming

The Global Warming Potential (GWP) has been identified as 0.6615 kg CO_2 eq for one kg of biodiesel production. The largest contributor to emissions has been identified as degummed soybean oil (0.4446 kg CO_2), reflecting the upstream process of soybean cultivation, fertilizer usage, and oil extraction. Secondary contributions are associated with methanol (0.1093 kg CO_2) and natural gas (0.0506 kg CO_2) production processes. Whereas the contributions from electricity and transport are 0.0253 and 0.0358 kg CO_2, respectively. These findings are consistent with the early research that determines the agricultural feedstock production is primarily responsible for greenhouse gas emissions [19].

3.2 Stratospheric Ozone Depletion and Photochemical Ozone Formation

More than 99% of this environmental burden is associated with degummed soybean oil production, which is represented by 3.65×10^{-6} kg CFC-11 eq. This indicates the upstream process of agricultural production linked to insecticides, chemical application, and refrigerant use. Ozone formation impacting human health (2.19×10^{-3} kg NOx eq) and terrestrial ecosystems (2.30×10^{-3} kg NOx eq) are also associated with degummed soybean oil, which represents 80% of the impacts. Methanol and transportation are the other contributing sectors that are generated from methanol synthesis and fuel combustion during transportation. Transition to the bio-based methanol seems to be another feasible solution to mitigate the impacts [20].

3.3 Fine Particulate Matter Formation and Terrestrial Acidification

Degummed oil is responsible for two-thirds of the Fine Particulate Matter formation potential, represented by 9.65×10^{-4} kg $PM_{2.5}$ eq, whereas relevant emissions with methanol and natural gas are 1.20×10^{-4} kg $PM_{2.5}$ eq and 1.17×10^{-4} kg $PM_{2.5}$ eq, respectively. Fertilizer, fossil fuel combustion during transesterification, and energy consumption are responsible for this type of emission. Similarly, in the case of acidification, degummed oil, natural gas, and methanol are the primary contributors. This category encompasses ammonia volatilization, SO_2, and NO_n emissions from fertilizer production, and the associated fossil energy inputs.

3.4 Ecotoxicity and Human Toxicity

Degummed soybean oil represents over 98% terrestrial ecotoxicity, represented by 0.335 kg 1,4-DCB eq, indicating an influence of agricultural pesticide and fertilizer applications. Similarly, in the total of freshwater ecotoxicity (3.25×10^{-3} kg 1,4-DCB eq) and marine ecotoxicity (2.42×10^{-3} kg 1,4-DCB eq), the successive contributors are degummed soybean oil, methanol, and transportation. Agricultural runoff and agrochemical emissions are significant factors in soy biodiesel production. Around 77% of human carcinogenic impacts are caused by the degummed process. Degummed oil, methanol, natural gas, and transportation have been observed as the top contributors to human non-carcinogenic toxicity. The outcomes highlight the environmental and occupational impacts associated with agricultural production and methanol synthesis.

3.5 Fossil Resource Scarcity

Fossil resource scarcity has been expressed in terms of 0.1384 kg oil eq, which is mainly associated with degummed oil (0.0875), natural gas (0.0188), and methanol (0.0148). Production of biodiesel is influenced by reducing the reliance on petroleum diesel. However, the dependence on the fossil-derived methanol production process and the use of natural gas indicate that the system is not completely renewable.

3.6 Diesel and Soy Biodiesel Comparison

Comparative analysis of soy biodiesel production with diesel highlights the overall sustainability measures associated with this system, as shown in Table 4. Particularly, soy biodiesel production represents 36.1% higher CO_2 emissions than conventional diesel production. A higher stratospheric ozone depletion and ozone formation terrestrial ecosystem have also been observed for soy biodiesel. The influential factors of driving these results clarify the energy-intensive agriculture production in Australia using fertilizers and pesticides.

On the contrary, soy biodiesel production shows a significant reduction in several impact categories. A consecutive reduction of 29.1% and 43.9% has been observed for ozone formation (human health) and fine particulate matter formation. Similar types of trends have also been identified in terrestrial ozone formation, terrestrial acidification, freshwater ecotoxicity, marine ecotoxicity, and human carcinogenic and non-carcinogenic toxicity. Even a reduction of 86.55% fossil resource scarcity (kg oil eq) is feasible by adopting soy biodiesel. Thus, the results of the impact categories prove that biodiesel offers numerous benefits in terms of human health, air quality, and fossil fuel dependency. Adopting sustainable practices in farming and processing will address issues related to global warming, ozone depletion, and ecotoxicity impacts.

Table 4 Impact categories associated with 1 kg of diesel and soy biodiesel production

Impact category	Unit	Diesel	Soy biodiesel
Global warming	kg CO_2 eq	0.48589	0.66150
Stratospheric ozone depletion	kg CFC11 eq	2.61E-08	3.65E-06
Ozone formation, Human health	kg NO_x eq	3.09E-03	2.19E-03
Fine particulate matter formation	kg $PM_{2.5}$ eq	1.72E-03	9.65E-04
Ozone formation, Terrestrial ecosystems	kg NO_x eq	3.31E-03	2.30E-03
Terrestrial acidification	kg SO_2 eq	5.87E-03	3.39E-03
Terrestrial ecotoxicity	kg 1,4-DCB	5.39E-02	3.39E-01
Freshwater ecotoxicity	kg 1,4-DCB	1.70E-02	3.25E-03
Marine ecotoxicity	kg 1,4-DCB	2.29E-02	2.42E-03
Human carcinogenic toxicity	kg 1,4-DCB	2.35E-03	2.64E-04
Human non-carcinogenic toxicity	kg 1,4-DCB	9.16E-01	7.04E-02
Fossil resource scarcity	kg oil eq	1.02	0.14

4 Conclusion

Degummed soybean oil emerged as the primary contributor across all identical midpoint impact categories in the supply chain. It underscores the energy-intensive agricultural process of soybean production. In the soybean diesel production process, methanol contributes significantly to global warming, acidification, toxicity potential, and fossil resource use among the auxiliary inputs. Energy inputs such as electricity and natural gas play secondary roles. However, their contribution is negligible compared to the degummed process. Minor contributions have been observed across impact categories for transport, except in noncarcinogenic toxicity.

Sustainable feedstock production has been suggested in this study to improve through regenerative agricultural practices, reducing dependency on chemicals, fertilizers, and insecticides. Substitution of methanol with bio-methanol or recycling can significantly reduce the dependency on fossil resources. Additionally, the decarbonization of the grid-electricity and natural gas with renewable energy sources will mitigate the burden of global warming and acidification potential. Locally sourced inputs for transesterification will decrease transport-related emissions. Overall, this research shows that this sustainability is hindered by the agricultural production process and fossil-derived methanol manufacturing. Prioritizing these recommendations will ensure maximum environmental benefits of biodiesel production, aligning with decarbonization and circular economy principles.

References

1. Tushar Q et al (2023) An integrated financial and environmental evaluation framework to optimize residential photovoltaic solar systems in Australia from recession uncertainties. J Environ Manag 346:119002
2. Balat M (2007) Production of biodiesel from vegetable oils: a survey. Energy Sources, Part A 29(10):895–913
3. Demirbas A (2005) Biodiesel production from vegetable oils via catalytic and non-catalytic supercritical methanol transesterification methods. Prog Energy Combust Sci 31(5–6):466–487
4. Fan X, Burton R, Austic G (2010) Conversion of degummed soybean oil to biodiesel: optimization of degumming methods and evaluation of fuel properties. Int J Green Energy 7(6):593–599
5. Rade LL et al (2015) Evaluation of the use of degummed soybean oil and supercritical ethanol for non-catalytic biodiesel production. J Supercrit Fluids 105:21–28
6. Kanakraj S, Dixit S (2016) A comprehensive review on degummed biodiesel. Biofuels 7(5):537–548
7. Huo H et al (2009) Life-cycle assessment of energy use and greenhouse gas emissions of soybean-derived biodiesel and renewable fuels. Environ Sci Technol 43(3):750–756
8. Barreiros T et al (2020) Impact of biodiesel production on a soybean biorefinery. Renew Energy 159:1066–1083
9. Tushar Q et al (2025) Carbon footprint and uncertainties of Geopolymer concrete production: a comprehensive life cycle assessment (LCA). C 11(3):55
10. Acquaye AA et al (2011) Identification of 'carbon hot-spots' and quantification of GHG intensities in the biodiesel supply chain using hybrid LCA and structural path analysis. Environ Sci Technol 45(6):2471–2478

11. Tushar Q et al (2025) Recycled used cooking oil (UCO) as a rejuvenator in high content reclaimed asphalt pavement (RAP) mixes: a life cycle assessment (LCA). Sci Total Environ 961:178376
12. Granjo JF, Duarte BP, Oliveira NM (2017) Integrated production of biodiesel in a soybean biorefinery: modeling, simulation and economical assessment. Energy 129:273–291
13. Milazzo MF et al (2013) Soy biodiesel pathways: global prospects. Renew Sust Energ Rev 26:579–624
14. Makareviciene V, Sendzikiene E, Gumbyte M (2020) Application of simultaneous oil extraction and transesterification in biodiesel fuel synthesis: a review. Energies 13(9):2204
15. Morais S, Martins A, Mata T (2010) Comparison of allocation approaches in soybean biodiesel life cycle assessment. J Energy Inst 83(1):48–55
16. Grant T et al (2019) AusLCI methodology for developing life cycle inventory for Australian agriculture. Lifecycles, Melbourne, Australia
17. Tushar Q et al (2022) Application of a harmonized life cycle assessment method for supplementary cementitious materials in structural concrete. Constr Build Mater 316:125850
18. Huijbregts MA et al (2016) ReCiPe 2016: a harmonized life cycle impact assessment method at midpoint and endpoint level report I: characterization. RIVM Report 104:2016
19. Pradhan A et al (2012) Reassessment of life cycle greenhouse gas emissions for soybean biodiesel. Trans ASABE 55(6):2257–2264
20. Zuiderveen EA et al (2023) The potential of emerging bio-based products to reduce environmental impacts. Nat Commun 14(1):8521

Sharing Costs and Benefits in Urban-Industrial Symbiosis: Cross-Sectoral Impacts of Recycling Urban Waste Incineration Ash in Construction

Anastasija Komkova[1(✉)], Birgitte Holt Andersen[2], and Guillaume Habert[1]

[1] Chair of Sustainable Construction, ETH Zurich, 8093 Zurich, Switzerland
komkova@ibi.baug.ethz.ch
[2] Cware Aps, 2100 Copenhagen, Denmark

Abstract. Urban-industrial symbiosis (U-IS) supports Circular Economy (CE) goals by transforming urban waste into valuable resources. U-IS can be initiated either top-down, driven by policymakers, or bottom-up, when industries collaborate to share benefits and costs. In Europe, U-IS between cities and waste-to-energy plants has a long-standing tradition in energy recovery. However, ashes, being incineration residues are often landfilled. The European research project 'Ashcycle' has shown that sewage sludge and municipal solid waste incineration ashes can be used as supplementary cementitious materials, in alkali-activated materials, and in clay bricks—contributing to decarbonizing the construction sector. U-IS can also extend to other sectors. For example, phosphorus recovered from sewage sludge ash can benefit agriculture by enhancing domestic supply of fertilizers, while reducing reliance on vulnerable international supply chains. Similarly, extracting ferrous and non-ferrous metals (e.g., copper, zinc) can support domestic supply for the electronics industry. This study combines life cycle assessment (LCA) and cost-benefit analysis (CBA) to evaluate the environmental and economic impacts of U-IS across sectors. Using multi-objective supply chain optimization, we compare a U-IS scenario with a baseline where ashes are landfilled and conventional materials (e.g., Portland cement, conventional fertilizers) are produced. Findings reveal that U-IS can reduce CO_2-eq. emissions by up to 80% and costs by nearly 50% at the regional level. While all industries benefit environmentally, cost impacts vary. For example, the cement and concrete sector could cut costs by 20% through ash recycling, while recovering critical raw materials may incur higher costs due to investments in the pretreatment technologies. Considering estimated regional benefits and CO_2-eq. emissions reductions of U-IS, findings of this study can offer valuable insights for policymakers to apply targeted incentives and regulatory frameworks to foster U-IS implementation, thereby advancing the goals of a CE.

Keywords: urban-industrial symbiosis · incineration ashes · supply chain networks · LCA · cost-benefit analysis

H. Beushausen et al. (Eds.): NTZR 2025, LNCE 873, pp. 1126–1132, 2026.
https://doi.org/10.1007/978-3-032-27698-8_99

1 Introduction

To achieve sustainable societies, we must limit the amount of resources we currently consume. As of 2021, the global economy consumed over 100 billion tonnes of resources annually, with only 8.6% being recycled, including 0.2Gt of resources being incinerated for energy recovery [1]. While nearly 31% of materials entering global economy were accumulated in the stock, nearly 24% were landfilled [1]. This indicates the urgent need for more circular resource flows.

In the Europe, municipal waste incineration has increased significantly. By 2023, approximately 129 kg per person of municipal waste was incinerated, accounting for 25.2% of the total municipal waste generated. At the same time, nearly 23% of municipal solid waste is still directly landfilled [2]. This trend reflects a growing reliance on waste-to-energy processes.

Sewage sludge is another waste stream increasingly subjected to incineration with energy recovery. In the EU, wastewater treatment plants produce over 8 million tonnes of dry sludge annually [3]. Due to stringent regulations on hazardous materials, the direct application of sewage sludge in agriculture is decreasing, making incineration a valuable method for both waste reduction and energy recovery.

The ashes resulting from the incineration process are often landfilled [4]. However, they can be repurposed in the construction sector. For instance, sewage sludge and municipal solid waste ash can be used as a supplementary cementitious materials (SCMs) or as a replacement in clay brick production, as well as in alternative binder-alkali-activated materials (AAMs) [5].

Advancements in technology have also made it possible to extract phosphorus and other critical raw materials from incineration ashes. Phosphorus recovered from sewage sludge ash can be utilized to produce fertilizers, while metals like copper, zinc, and aluminum can be reclaimed [6] for use in the electronics sector.

This urban-industrial symbiosis (U-IS) approach, which involves recycling urban waste streams such as incineration ashes into the construction sector, offers multiple benefits at both regional and industrial levels, including:

- Reduction of waste that would otherwise be landfilled.
- Contribution to decarbonization goals of construction sector, and cement sector specifically, by replacing cement clinker and other primary resources.
- Creation of new business opportunities within the construction sector.
- Recovery of critical raw materials (e.g., phosphorus), contributing to domestic supply security of fertilizers for agriculture sector
- Recovery of ferros and non-ferros metals that can be used in the e.g. electronics sector.

While the establishment of such urban-industrials symbiosis can offer regional or national benefits, it is important to assess whether all actors involved share the costs and benefits equitably. Such an assessment helps determine whether a bottom-up, self-organized approach is sufficient or if external support is necessary to facilitate U-IS establishment through top-down initiatives driven by policymakers.

In this paper, we examine distribution of environmental and economic costs and benefits between actors involved in the U-IS within sewage sludge ash recycling.

2 Methods

This study employs a combined approach of Life Cycle Assessment (LCA) and Cost-Benefit Analysis (CBA) to evaluate the environmental and economic impacts of urban-industrial symbiosis (U-IS) associated with ash recycling. Two scenarios are defined:

- Baseline Scenario: In the absence of U-IS, incineration ashes are landfilled, conventional construction materials are produced, and primary fertilizers are synthesized from imported phosphate rock.
- Alternative Scenario: Ashes undergo pretreatment to recover phosphorus for fertilizer production. The remaining pretreated ashes are utilized in the production of construction materials.

The establishment of U-IS is linked to the optimization of supply chains. An optimization model is developed to minimize emissions and transportation costs between actors, thereby identifying optimal locations for recycling plants and construction material production facilities. This model incorporates the geographical locations of existing actors (e.g., waste incinerators, construction material producers) and prospective actors (e.g., recycling plants) to determine efficient supply chains. The model has been described in [7].

Emissions and costs in the baseline (business-as-usual) scenario are compared with those in the alternative scenario where U-IS is operational. Furthermore, the study aims to identify how environmental and economic costs and benefits are distributed among actors within both scenarios. By assessing which actors experience net benefits or face potential losses, the appropriate incentives can be identified that can facilitate the implementation of U-IS.

To identify potential business cases and quantify impacts within the U-IS framework, interviews were conducted with stakeholders involved in the *AshCycle* project [5]. These actors across the supply chain include waste incineration facilities, construction material producers, and potential phosphorus recovery and fertilizer production entities. The interviews provided insights into the practical considerations, challenges, and opportunities associated with implementing U-IS strategies.

3 Results

Through interviews with stakeholders involved in the pilot U-IS in Denmark, the key factors driving the business case for each actor were identified (Fig. 1).

Within the U-IS focused on sewage sludge ash recycling, three types of industries are involved. While the primary activity of the wastewater treatment plant is to purify water, it also incinerates the resulting sewage sludge and recovers energy to provide district heating and electricity for the urban population. In the absence of U-IS, the ashes are sent to landfills, incurring costs of nearly €120 per ton.

Sewage sludge ashes contain, on average, nearly 10 wt% of phosphorus. Within the U-IS, the untreated ashes are transported to a phosphorus recycling plant for phosphorus recovery. The recovered phosphorus can be used for fertilizer production and applied to meet agricultural needs. The recovery of phosphorus from incineration ashes is considered an emerging business opportunity, as the EU currently relies heavily on imports of phosphate rock mineral ore. The price of phosphoric acid, primary fertilizer has fluctuated between €1,000 and €1,600 per ton over the past few years.

The treated ashes can be transported to a construction materials manufacturer. Recent research shows that ashes can be used in fired clay brick production, replacing up to 30% of the primary material—clay—as well as in supplementary cementitious materials (SCMs), replacing nearly 20% of clinker in composite concrete. With the current price of Portland cement (CEM I) being nearly €120 per ton, using ashes as SCMs presents a promising business case and contributes to reducing greenhouse gas emissions of construction sector.

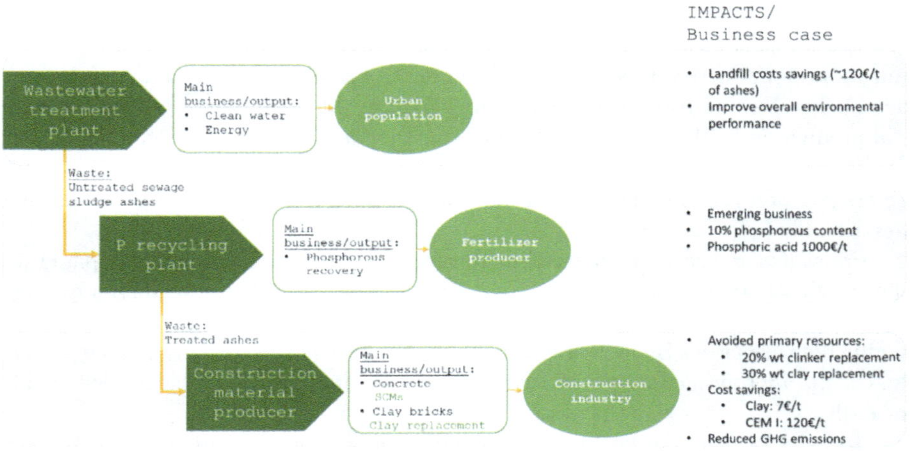

Fig. 1. Actors involved in the U-IS and their business case perspectives.

To quantify the benefits and costs, an assessment was first conducted at the regional scale, identifying environmental and economic impacts for both the baseline (business-as-usual) scenario and an alternative scenario in which all annually produced ashes in the country are recycled within the implemented U-IS framework (Fig. 2). This assessment focuses on the application of ashes as supplementary cementitious materials.

The implementation of U-IS can lead to an approximate 85% reduction in greenhouse gas (GHG) emissions at the regional level, primarily due to the avoided production of cement. Additionally, the establishment of U-IS reduces business-as-usual costs by nearly 52%, largely due to the avoided expenses associated with ash landfilling.

As the next step, the distribution of environmental and economic impacts among the involved industries was identified. A functional unit of 1 t of ashes produced was defined for the analysis.

Figure 3 illustrates that the waste incinerator reduces CO2 eq. emissions by nearly 64% by avoiding landfilling and instead transporting the ashes to a recycling plant. At the same time, phosphorus recovery reduces GHG emissions associated with primary fertilizer production by approximately 55%. The construction materials producer can reduce GHG emissions by about 20% by replacing cement with treated ashes.

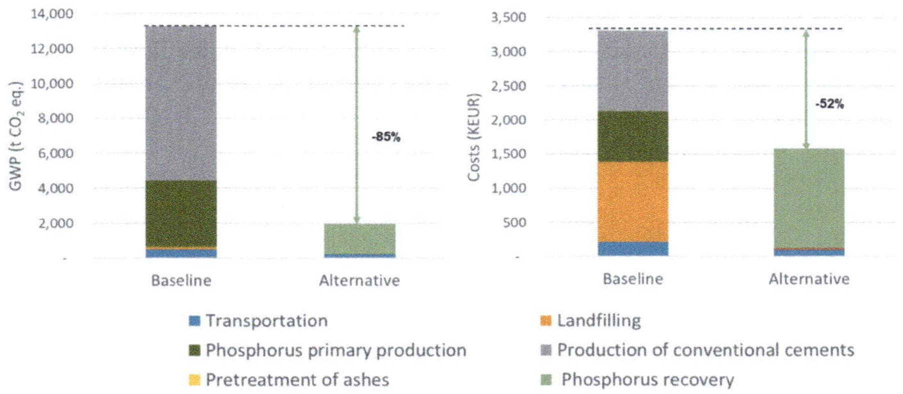

Fig. 2. Environmental and economic impacts of the baseline (business-as-usual) and alternative (implemented U-IS) scenarios.

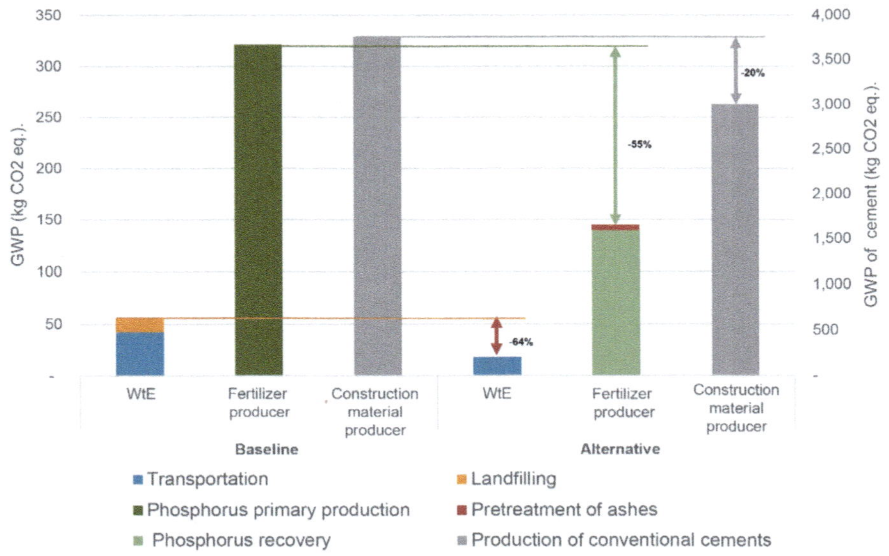

Fig. 3. The environmental impacts associated with the production and treatment of 1 t of sewage sludge ash, distributed across sectors.

The economic costs and benefits within the U-IS are distributed differently across sectors. The waste-to-energy (WtE) plant benefits significantly from the U-IS, saving nearly 92% in landfilling cost and transportation (Fig. 4). Similarly, the construction materials producer benefits by reducing cement costs by approximately 20%.

Phosphorus recovery, on the other hand, is a more costly technology that requires additional investments [7] . As a result, the cost of recovered phosphorus is nearly double that of conventional sources. This represents an extreme case, with advancements in phosphorus recovery technology, larger-scale plants, and economies of scale, these costs are expected to decrease. Nonetheless, it is likely that policy incentives, such as subsidies, would be necessary to support the participation of phosphorus recovery facilities in the U-IS.

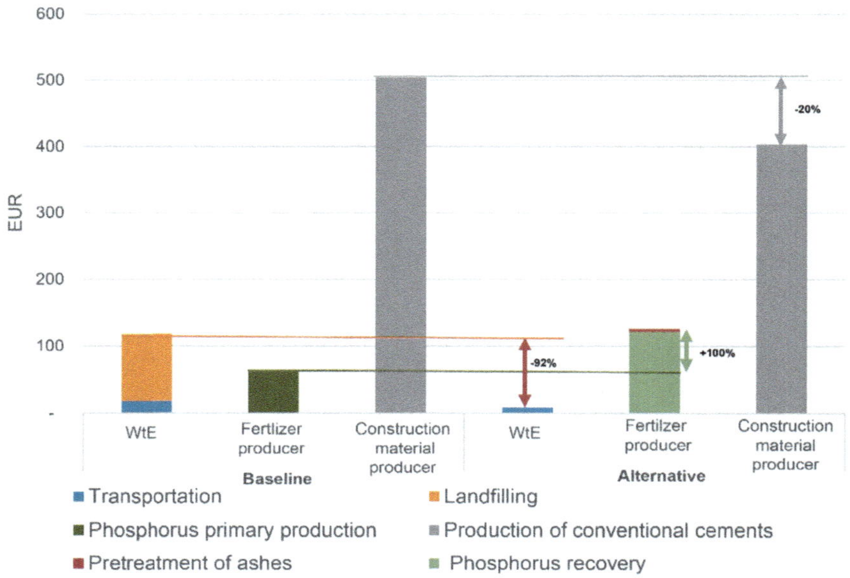

Fig. 4. Costs associated with 1 t of sewage sludge ash produced and treated, distributed between sectors.

4 Conclusions

Results indicate that at the regional scale, the U-IS brings substantial benefits, not only in terms of reducing greenhouse gas emissions, but also through avoided costs. Within the U-IS itself, waste-to-energy plants and the construction sector benefit the most due to avoided landfilling costs, reduced transportation expenses, and lower alternative material costs. However, the high investment required for phosphorus recovery technology may necessitate top-down actions from policymakers to enable participation in the U-IS, thereby supporting the implementation of a cross-sectoral circular economy.

Acknowledgments. This project has received funding from the Swiss State Secretariat for Education, Research and Innovation (SERI) and from the European Union's Horizon Europe research and innovation programme under grant agreement 101058162.

References

1. Circle economy: the circularity gap report 2022 (2022)

2. Eurostat: municipal waste by waste management operations. https://ec.europa.eu/eurostat/databrowser/view/ENV_WASMUN/default/table. Last accessed 24 Feb 2025. https://doi.org/10.2908/ENV_WASMUN
3. Eurostat: sewage sludge production and disposal from urban wastewater. https://ec.europa.eu/eurostat/databrowser/view/ten00030/default/table?lang=en. Last accessed 15 Mar 2024
4. ZWE: incineration residues in the EU: quantities and fates (2022)
5. AshCycle project. https://www.ashcycle.eu/en/project/. Last accessed 15 Mar 2024
6. Tominc S, Ducman V, Wisniewski W, Luukkonen T, Kirkelund GM, Ottosen LM (2023) Recovery of phosphorus and metals from the ash of sewage sludge, municipal solid waste, or wood biomass: a review and proposals for further use. Mater 16:6948. https://doi.org/10.3390/MA16216948
7. Komkova A, Krog Agergaard S, Holt Andersen B, Habert G (2025) Urban waste recycling: a multi-objective approach for sustainable construction in European regions. In: Kioumarsi M, Shafei B (eds) The 1st international conference on net-zero built environment, Lecture notes in civil engineering. Springer, Cham, pp 1097–1110. https://doi.org/10.1007/978-3-031-69626-8_92

Vernacular and Low-Cost Conventional Buildings: Sustainability Analysis Using Life Cycle Assessment for Rural Development in the Ganga Delta Region of West Bengal, India

Aritra Majumder[1] (✉), Anupam Debsarkar[1], and Kaushik Dutta Roy[2]

[1] Jadavpur University, Kolkata, West Bengal, India
aritra4796@gmail.com, aritra.civil.rs@jadavpuruniversity.in
[2] Kolkata Municipal Development Authority, Kolkata, West Bengal, India

Abstract. This study presents a comparative Life Cycle Assessment (LCA) of a vernacular mud dwelling and a conventional low-cost concrete building situated in the Ganga Delta region of West Bengal to evaluate their environmental performances. Given the region's unique deltaic terrain—marked by clay-rich soil, high water tables, and vulnerability to floods and cyclones—there is an urgent need for sustainable and resilient housing solutions. The LCA, conducted over a 50-year lifespan using ISO 14040/44 standards and the ReCiPe 2016 Midpoint (H) method, analyzed three key environmental indicators: Global Warming Potential (GWP), Abiotic Depletion Potential for fossil resources (ADPf), and Water Deprivation Potential (WDP). Results showed that the vernacular dwelling had a lower GWP (41,393 kg CO_2e) compared to the conventional building (47,532 kg CO_2e) with a 12.91% reduction. While the ADPf of the vernacular house was marginally higher (+ 0.63%), its water use (13,634 m^3) was significantly lower than that of the conventional building (92,069 m^3), implying an 85.19% decrease. Sensitivity analysis showed that materials and energy accounted for over 90% of total impacts, emphasizing their decisive role in shaping environmental sustainability. These findings underscore the environmental advantages of vernacular construction, particularly in minimizing embodied carbon and conserving water. The study concludes that integrating traditional architectural knowledge with modern policy frameworks can help promote eco-efficient, disaster-resilient housing in vulnerable rural regions like the Ganga Delta.

Keywords: Vernacular architecture · Life cycle assessment · Sustainability · Environmental impact assessment · Carbon footprint

1 Introduction

1.1 Vernacular Architecture

Vernacular architecture (VA) comprises traditional building practices that rely on locally sourced materials and indigenous knowledge, typically developed without formal architectural input. These methods evolve through generations, shaped by environmental, economic, cultural, and social factors. Among these, the physiographic region—defined by

H. Beushausen et al. (Eds.): NTZR 2025, LNCE 873, pp. 1133–1148, 2026.
https://doi.org/10.1007/978-3-032-27698-8_100

terrain, climate, soil, and vegetation—strongly influences construction style and material use. Understanding this relationship is essential for advancing sustainable architecture and conserving region-specific cultural heritage.

1.2 Classification of West Bengal Based on Physiographic Characteristics

West Bengal's varied geography, from the Himalayan foothills to the Bay of Bengal, divides the entire state into distinct ecological and climatic zones that shape regional architecture. Among its nine physiographic divisions, the Ganga Delta is especially critical due to its fertile alluvial soil, shallow water tables, and vulnerability to floods, cyclones, and tidal surges. This study focuses on Purba Barddhaman, a transitional area between Ganga Delta and Plateau Fringe Region. Such environmental conditions demand climate-responsive, flood-resilient housing solutions, making this region ideal for undertaking studies on both VA and low-cost conventional building (LCCB) from the sustainability context.

1.3 Life Cycle Assessment and its Significance

Life Cycle Assessment (LCA) is a widely adopted analytical approach used to evaluate the environmental impacts associated with a product across its entire lifespan. This cradle-to-grave method examines all stages, from raw material extraction, production, and distribution to usage and final disposal. LCA follows a structured protocol defined by ISO 14040 and ISO 14044 standards and includes key stages: life cycle inventory (LCI) for data collection, life cycle impact assessment (LCIA) for evaluating environmental consequences, and interpretation, where findings are analyzed to support informed conclusions and sustainability-oriented decisions.

1.4 Low-Cost Conventional Building (LCCB) under PMAY Scheme

In India, a large segment of the population continues to face economic challenges and inadequate housing, making low-cost housing essential for ensuring access to secure and affordable shelter. The Pradhan Mantri Awas Yojana (PMAY) is a government-led initiative aimed at addressing this gap for slum dwellers and economically weaker sections. It supports in-situ redevelopment, credit-linked subsidies, and financial aid for self-built homes, while also improving essential amenities like sanitation, water, and lighting. PMAY homes typically cover 32 sq. m. and include two bedrooms, a kitchen, and a toilet, designed with climate responsiveness and regulatory compliance. Construction materials feature M-20 grade RCC, PCC (1:3:6), Fe-415 HYSD steel, and quality brick masonry, with structural elements engineered for seismic resilience and long-term durability.

1.5 Review of Selected Literature

In recent years, vernacular architecture has re-emerged as a focal point in sustainable design discourse, primarily due to its inherent adaptability to local climatic, material, and cultural contexts. Foundational contributions by Oliver (1997), Fathy (1986), Koenigsberger (2001), and Majumder et al. (2024) positioned vernacular practices as ecologically responsive and rooted in community-driven solutions. Building on these ideas, Biradar and Mama (2017) as well as Nasir and Kamal (2021) emphasized that vernacular construction exemplifies sustainability through passive design strategies and minimal environmental disruption. In the Indian context, Jagatramka et al. (2020) introduced a framework of sustainability indicators specific to traditional Indian dwellings, while Choudhury and Chettry (2023) explored the climate-resilient features of Himalayan structures like Assam-type houses.

A substantial portion of research highlights the ecological merits of bio-based and earthen materials. Li et al. (2020) discussed the cultural and spatial sensibilities of Naxi vernacular homes, aligning them with biophilic principles. Studies published in MDPI (2022) stressed the carbon-negative potential of adobe and bamboo structures, reinforced by LCA-based investigations from Gazquez et al. (2022), Fernandes et al. (2019), and Carrobé et al. (2024), which confirmed low embodied energy and thermal regulation abilities of these materials. Projects such as the Sesga House (2021) and Sciendo et al. (2023) further validated the ecological value of refurbishing heritage earthen structures.

Comparative LCA studies have revealed clear advantages of vernacular architecture over modern systems. For instance, Francis et al. (2024) reported about 30% less embodied carbon in earthen buildings versus confined masonry. The "Ecofordable House" (Sabah 2025) exemplified hybrid designs merging modern and traditional approaches to reduce environmental impacts. Research by Asif et al. (2017), Fnais et al. (2022), and Gazquez et al. (2022) consistently observed vernacular structures to be more efficient across various environmental metrics.

However, methodological limitations persist. Asdrubali et al. (2024) and the International Journal of Life Cycle Assessment (2022) noted the underrepresentation of traditional architecture in LCA literature and insufficient data for usage-phase impacts. BIM applications by de Bortoli et al. (2022) and LCA studies on prefabrication by Jayawardana et al. (2023) highlight recent methodological advances.

Region-specific research for the Ganga Delta remains sparse. While Bandyopadhyay et al. (2007) and Laha (2015) described its geomorphic vulnerability, practical evaluations remain limited. Majumder et al. (2024) identified a scarcity of LCAs on Bengal's vernacular typologies. Although traditional responses such as raised plinths and ventilated facades are well-documented (Majumder 2020; Slideshow 2021), they are rapidly being replaced by conventional materials (Fernandes et al. 2014). Adaptive techniques from structures like Chang Ghar and Kath-Kuni (Pathak 2014; Barman 2020; Goswami 2015) remain marginal in policy applications.

Emerging hybrid models offer a way forward. Researches by Ogunmakinde (2024), Sandanayake (2022), Rinne et al. (2022), Dormohamadi et al. (2024), and Arduin et al. (2022) highlights circular, regionally grounded building systems inspired by vernacular insights. Still, critical LCA limitations related to context specificity, social parameters, and data standardization continue to challenge comprehensive assessments (International Journal of LCA 2022; Bueno et al. 2022; Syska et al. 2021).

1.6 Research Gap

Despite Widespread Global Evidence Showing that Vernacular Construction Can Reduce Environmental Impacts by 30%–70%, West Bengal—Particularly the Ganga Delta—Lacks Comparative LCA Studies. Research on PMAY Housing in the Region Is Scarce, with Only One Known Case in Rishra, West Bengal. Moreover, no Studies Holistically Explore the Relationship Between Physiographic Context, Vernacular Methods, and Low-Cost Housing Policies, Revealing a Significant Gap in the Region-Specific Sustainable Housing Research.

1.7 Research Questions

This Study Addresses Three Core Questions, i) how Vernacular Housing in the Ganga Delta Adapts Structurally and Materially to its Deltaic Context; Ii) how its Environmental Impacts Compare with LCCB Using LCA Indicators like GWP, ADPf, and Water Demand; and Iii) how these Findings Can Guide Climate-Resilient Rural Housing Policies in Coastal West Bengal.

1.8 Research Objectives

This Study Seeks to Fill the Existing Research Gap by Documenting and Digitally Modeling a Vernacular Mud Dwelling alongside a LCCB Typical for the Ganga Delta. Using One Click LCA and Ecoinvent Database, a Cradle-to-Grave Life Cycle Assessment Has Been Carried out Considering Three Key Indicators—Global Warming Potential, Fossil Resource Depletion, and Water Usage. The Findings Aim to Inform Policies that Integrate Traditional Building Wisdom into Sustainable, Climate-Resilient Rural Housing Initiatives.

2 Materials and Methods

2.1 Study Area

The study was conducted at Dirghapara, a rural village situated in the Purbasthali I block of Purba Barddhaman district, West Bengal (Fig. 1). This village lies within the alluvial "Ganga Delta Plain" physiographic region and is positioned at approximately 23°21′21"N latitude and 88°13'27"E longitude, with an elevation of about 10 m above the mean sea level. Covering an area of 327.8 ha, Dirghapara had a total 411 households (15th Census of India 2011). The area is characterized by fertile alluvial soil, a semi-humid tropical climate, and susceptibility to periodic flooding and waterlogging. Agricultural activities dominate the local economy, with most residents engaged in sharecropping or daily wage-based farm labor.

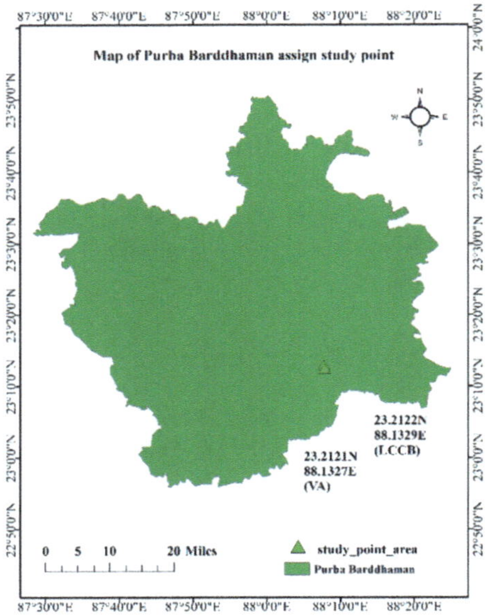

Fig. 1. Map of Purba Barddhaman showing study location at Dirghapara (ArcGIS)

2.2 Assessment of Vernacular Architecture (VA)

A Traditional Mud House (Fig. 2) in Dirghapara Village, Purba Bardhaman District, West Bengal, Served as the Representative Vernacular Case Study. The Property Comprises Three Separate Structures: a Main Block with Three Bedrooms, a Standalone Kitchen, and a Sanitation Unit. Constructed in 1985 Using Indigenous Methods, its 530 Mm-Thick Clay-Straw Walls Are Plastered with a Mix of Clay, Cow Dung, and Straw. The Roof Is Supported by Bamboo and Palm-Derived Timber like "Kanri," with Supplementary Elements Such as "Muduni," "Aanra," and "Ruyo." the 350 Mm Plinth Height Offers Modest Flood Protection. Fuel Sources Include LPG, Firewood, and Cow Dung, while a Cattle Shed Supports Agricultural Activities. The Dwelling Was Digitally Modeled in Autodesk Revit 2024 (Fig. 3) and Analyzed through OneClick LCA Software over a 50-Year Lifecycle, Based on 1 m^2 Built-up Area.

2.3 Assessment of Low-Cost Conventional Building (LCCB)

A LCCB (Fig. 4), built under the PMAY scheme in 2022 and located in the same village, was selected for comparative evaluation. Designed for a family of four, the house includes a residential block, a separate kitchen, and an attached toilet. Constructed with a load-bearing brick wall system and reinforced concrete components, the walls are approximately 10 in. thick. The total construction cost was ₹10 lakhs, covering materials, labour, and basic amenities. To ensure consistency with the vernacular model, this house was also digitally modelled in Autodesk Revit 2024 (Fig. 5). The Life Cycle Assessment followed a cradle-to-grave approach, using the same functional unit and OneClick LCA

Fig. 2. VA of Ganga Delta region located in Dirghapara, Purba Barddhaman

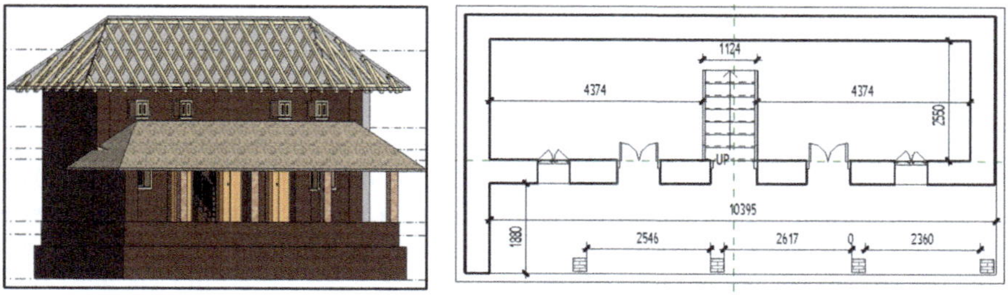

Fig. 3. Autodesk revit model of the VA and architectural floor plan

software. Material data were obtained from ecoinvent 3.8 and validated through local market sources.

Fig. 4. LCCB of Ganga Delta region located in Dirghapara, Purba Barddhaman

2.4 Life Cycle Assessment (LCA) Methodology

A comparative Life Cycle Assessment (LCA) was conducted for two distinct housing types, a VA and a LCCB, based on ISO 14040 and 14044 guidelines. The LCA

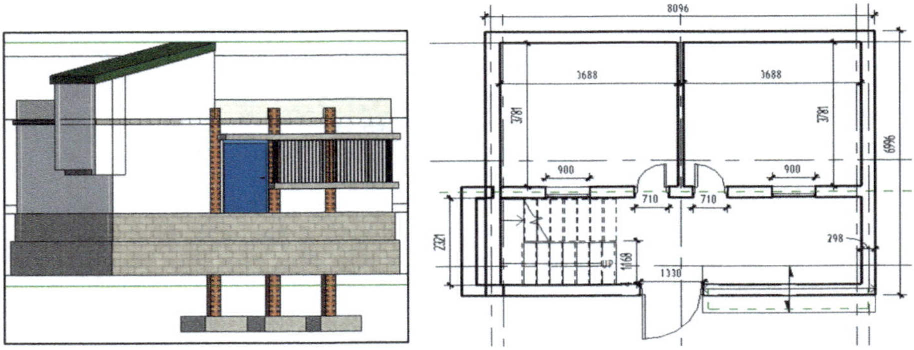

Fig. 5. Autodesk revit model of the LCCB and architectural floor plan

followed the standard four-step process: goal and scope definition, inventory analysis, impact assessment, and interpretation. A functional unit of 1 m^2 built-up area over a 50-year lifespan was considered, with a cradle-to-grave boundary. Material quantities were obtained through field surveys, digital measurements, and Autodesk Revit 2024 modeling. The vernacular house used local materials such as mud, bamboo, and Kanri wood, while the PMAY structure relied on cement, bricks, concrete, and steel. Environmental data were sourced from the ecoinvent 3.8 database and literature. Impact categories—GWP, ADPf, and water consumption, were analyzed using ReCiPe 2016 Midpoint (H) in One Click LCA.

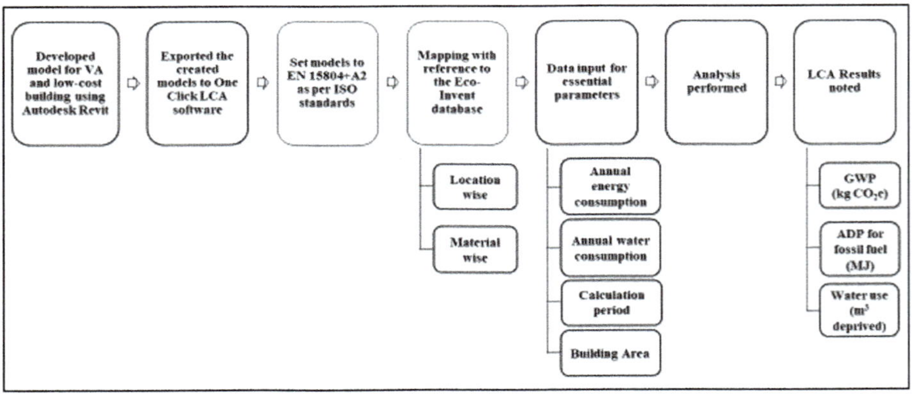

Fig. 6. Methodologies of LCA flow chart

A visual schematic of the LCA workflow, from model development in Revit to impact output in One Click LCA, is presented in Fig. 6. A sensitivity analysis has also been performed to obtain the most significant parameters for both VA and LCCB.

3 Results and Discussions

3.1 LCA Results on VA

The LCA of the selected VA in Dirghapara, located in the Ganga Delta region, focused on three key environmental indicators: GWP, ADPf, and water consumption.

Global Warming Potential (GWP): The total GWP for the VA structure was found to be approximately 41,393 kg CO_2e, with operational energy use (Stage B6) accounting for the largest share at 72.76% (Fig. 7), followed by materials manufacturing and transport (A1-A3) at 20.57%. Electricity usage emerged as the most significant contributor to carbon emissions. The structural frame (beams, columns, and slabs) and external walls were the primary building elements influencing emissions, representing a combined impact of over 98.7%. In terms of resource inputs, electricity and ready-mix concrete used in the flooring and walls were the leading contributors.

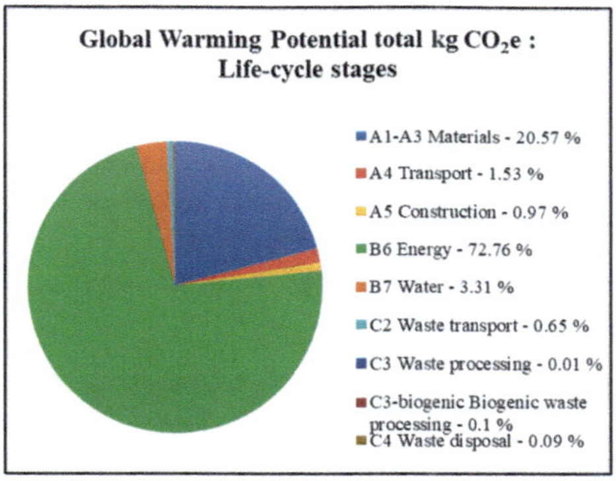

Fig. 7. Percentage share of GWP of VA as per life-cycle stages

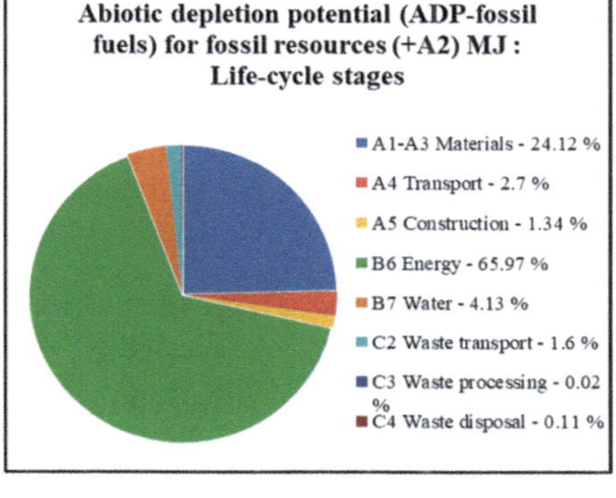

Fig. 8. Percentage share of ADPf of VA as per life-cycle stages

Abiotic Depletion Potential from Fossil Resources (ADPf): In terms of fossil resource depletion, the total embodied energy for the VA house was estimated at 320,000 MJ, with electricity usage being the dominant factor, contributing 65.97% (Fig. 8) of the total. Material usage added another 24.12%. The structural frame contributed more than half of the resource depletion impact when assessed by mass.

Water Consumption: Water consumption throughout the building's life cycle amounted to 13,634 m^3 deprived, where energy use during operation and material production accounted for 69.21% (Fig. 9) and 21.84% respectively. Electricity use and concrete production were found to be the major contributors to water depletion.

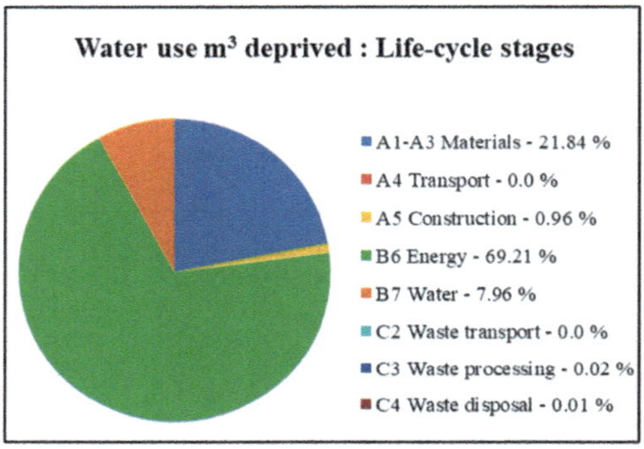

Fig. 9. Water usage of VA as per life-cycle stages

The findings emphasize the urgent need to integrate passive strategies, renewable energy, and low-impact materials in vernacular systems to further enhance its environmental performance.

3.2 LCA Results on LCCB

The PMAY-type LCCB in the same region demonstrated significantly higher environmental burdens across all categories.

Global Warming Potential (GWP): The total GWP recorded was 47,532 kg CO_2e, with B6 Energy contributing 63.56% and A1-A3 stages contributing 26.91% (Fig. 10). Key sources of emissions were electricity, ready-mix concrete used in structural and non-structural elements, and steel reinforcements. Structurally, the frame and external walls made up a large proportion of the environmental load, accounting for over 72% of the embodied mass and impacts.

Abiotic Depletion Potential from Fossil Resources (ADPf): ADPf for the LCCB reached 484,828 MJ, more than that of the VA. Electricity was the dominant source at 65.97%, followed by ready-mix concrete at 19.53% (Fig. 11). The building's frame alone constituted 72% of the mass-based contributions, reflecting its high resource intensity.

Water Consumption: Water usage was also observed to be considerably higher, totaling 92,069 m^3 deprived. Material production (A1–A3) and operational energy (B6)

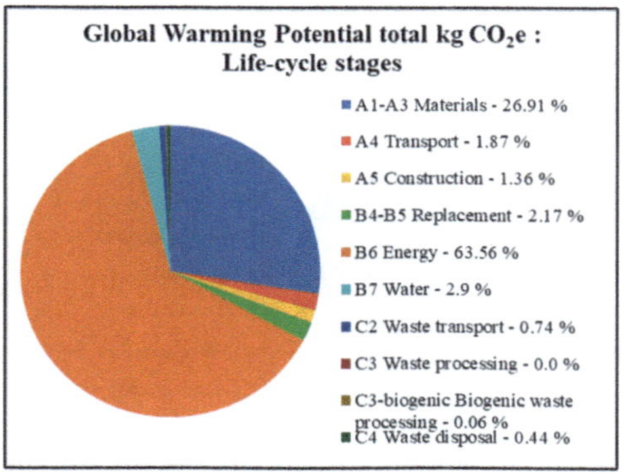

Fig. 10. Percentage share of GWP of LCCB as per life-cycle stages

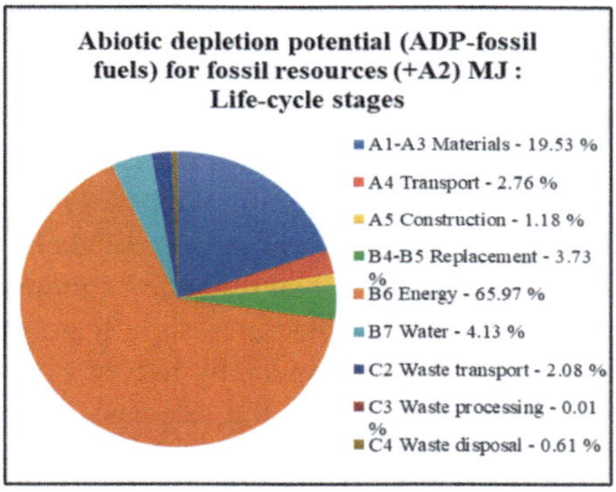

Fig. 11. Percentage share of ADPf of LCCB as per life-cycle stages

accounted for 84.67% and 10.24% (Fig. 12). The foundations and piles alone contributed more than 70% of the total water consumption, largely due to excessive concrete usage.

These results point to the substantial environmental cost of conventional construction methods in deltaic settings.

3.3 Comparative Analysis of VA and LCCB Results in the Ganga Delta Region

A comparative LCA was conducted to evaluate the environmental performance of a representative VA and a PMAY-type LCCB in the Ganga Delta. The assessment focused on three critical impact categories: Global Warming Potential (GWP), Abiotic Depletion Potential for fossil fuels (ADPf), and Water Use.

In terms of GWP, the vernacular structure exhibited a total emission of 41,393 kg CO_2e, while the LCCB accounted for 47,532 kg CO_2e. This results in a difference of 6,139 kg CO_2e, with the VA showing a 12.91% lower carbon footprint, highlighting its comparative advantage in reducing greenhouse gas emissions.

For fossil resource depletion, the VA recorded 487,910 MJ, marginally higher than the LCCB's 484,828 MJ, leading to a negative difference of 3,082 MJ or − 0.63%. This

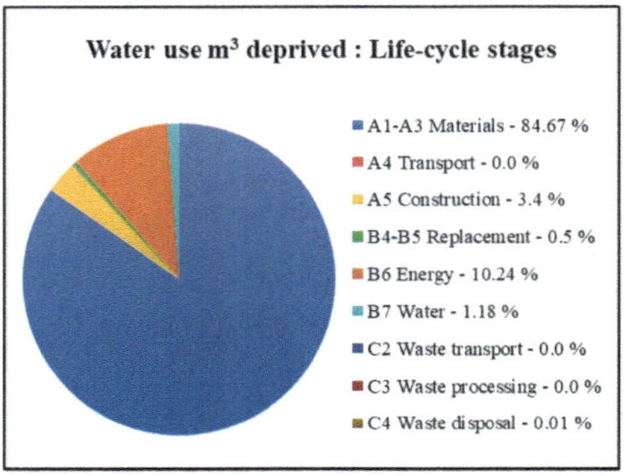

Fig. 12. Water usage of LCCB as per life-cycle stages

indicates a slightly higher energy demand from fossil sources for the VA, possibly due to operational energy inputs over the building's lifecycle.

Water usage presents the most striking contrast. The VA consumed 13,634 m^3 of water over its lifespan, compared to 92,069 m^3 for the LCCB. This yields a substantial difference of 78,435 m^3, with the VA consuming 85.19% less water. Such a reduction underscores the water efficiency of vernacular practices, likely due to the use of low-impact, locally sourced materials and minimal reliance on industrial processes.

Overall, the vernacular building demonstrates significantly better performance in two out of the three evaluated categories, especially in terms of water conservation and carbon emissions. These findings emphasize the relevance of vernacular strategies in promoting sustainable, climate-responsive housing solutions in ecologically sensitive areas like the Ganga Delta.

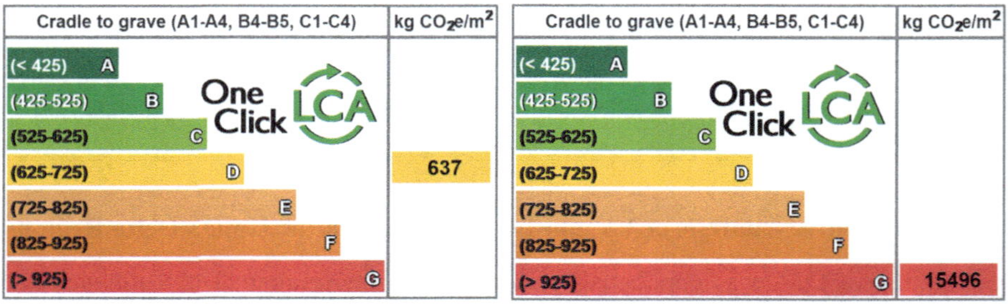

Fig. 13. Embodied carbon heroes benchmark comparison between VA and LCCB as $kgCO_2e/m^2$ **a** VA **b** LCCB

Benchmarking against embodied carbon thresholds revealed the fact that the VA, at 637 kg CO_2e/m^2, qualifies at the 'D' performance band (625–725), while the LCCB scored 15,496 kg CO_2e/m^2, falling into category 'G' (greater than 925), confirming its reasonably higher environmental footprint (Fig. 13). These comparisons highlight the environmental advantages of traditional mud-based dwellings in the deltaic region, particularly in reducing carbon emissions, conserving fossil resources, and minimizing water usage.

3.4 Sensitivity Analysis of the LCA Results of both VA and LCCB in the Ganga Delta Region

The sensitivity analysis of the LCA results highlights that materials and energy are the two most influential parameters, together accounting for over 90% of the total environmental impacts across all indicators (Fig. 14). For GWP, the traditional VA shows a dominant contribution from energy use, amounting to 30,000 kg CO_2e (72.76%), followed by materials at 8,600 kg CO_2e (20.57%), while the remaining categories such as water use, transportation, construction, waste processing, and disposal collectively account for only 6.66% of the total emissions. In the case of the LCCB, energy contributed 63.56%, and materials contributed 26.91% to the total of 47,532 kg CO_2e, leaving the remaining processes with a relatively small share of 9.53%.

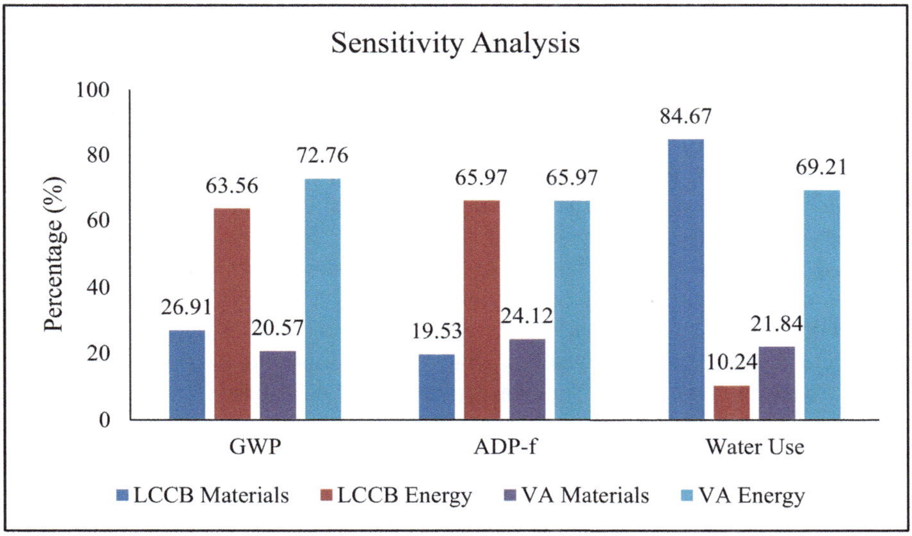

Fig. 14. Sensitivity analysis of the LCA results of both VA and LCCB

A similar trend was observed for the ADP-f, where for VA, energy accounted for 65.97% and materials 24.12%, with the rest contributing only 9.91%. For LCCB, energy use dominated with 65.97%, while materials contributed 19.53%, with other factors making up 14.5% of the total ADP-f. In terms of water use, the distribution pattern varied for VA, energy processes were responsible for 69.21% and materials for 21.84%, leaving 8.95% for other categories; while in LCCB, materials had the largest share at 84.67%, with energy contributing only 10.24%, and all other processes together contributing 5.09%. As a result, energy is the key driver of GWP in VA, while material use largely determines water consumption in LCCB. These results confirm that materials and energy remain the major sensitive parameters across categories, and thus any strategy aimed at reducing the overall environmental burden must prioritize interventions in these two areas.

As a whole, this comparative study reinforces the importance of integrating vernacular design strategies into mainstream rural housing policy for enhanced sustainability and resilience in climate-sensitive zones like the Ganga Delta.

4 Conclusion

The study was structured around three essential research questions: i) how the material and structural features of vernacular architecture respond to the deltaic environment, ii) how its environmental performance can be compared with that of LCCB, and iii) how such findings can guide sustainable rural housing policy in the Ganga Delta region. Based on LCA results from Purba Bardhaman, it is evident that VA exhibits notable environmental advantages over LCCB. Specifically, the VA recorded 12.91% lower Global Warming Potential, and most significantly, an 85.19% reduction in water consumption. However, it exhibited a marginal increase of 0.63% in fossil resource depletion. In the case of vernacular architecture depletion of fossil resource slightly higher can be attributed to specific energy use patterns particularly during material preparation and operation and maintenance phases. Traditional processes such as the firing of clay for roof tiles or transportation of natural materials often rely on conventional fossil-based energy sources. Materials and energy emerged as the most sensitive parameters, jointly exceeding 90% of total impacts. Prioritizing resource-efficient materials and low-energy strategies is crucial to achieving sustainable housing in the Ganga Delta region.

These findings underscore the climate-adaptive nature of vernacular systems, which rely on locally available, low-impact materials and passive design strategies. The study aligns with broader international literature, such as studies by Pakdel et al. (2021) and Kaoula et al. (2022), which emphasize the low carbon footprint and resource efficiency of traditional buildings. To strengthen resilience and sustainability in rural housing, it is essential to integrate vernacular practices with modern innovations like renewable energy, optimized material selection, and water-efficient technologies. Such hybrid approaches can pave the way toward eco-sensitive, disaster-resilient development in vulnerable regions like the Ganga Delta.

From a broader perspective, the findings can be analysed within the frameworks of the circular economy and adaptive reuse, thereby extending their relevance beyond the immediate regional context. Promoting the reuse of locally available materials, designing for disassembly, and encouraging the repair and adaptive reuse of traditional structures can substantially minimize lifecycle impacts while fostering community resilience.

The policy recommendations, which may emerge from this study can be operationalized through specific strategies within existing housing programs. Incorporating vernacular design principles into government-led affordable housing schemes, incentivizing the use of locally available and low-impact materials, and introducing retrofitting guidelines enhancing thermal and structural performance would make such programs more sustainable. Additionally, capacity-building initiatives for local builders, along with regulatory measures promoting material reuse and circular construction practices, can bridge research findings with practice, ensuring the construction of climate-resilient and low-carbon housing in vulnerable regions.

References

Arduin D, Caldas LR, Paiva RD, Rocha F (2022) Life cycle assessment (LCA) in earth construction: a systematic literature review considering five construction techniques. Sustainability 14. https://doi.org/10.3390/su142013228

Asdrubali F, Colladon AF, Segneri L, Gandola DM (2024) LCA and energy efficiency in buildings: mapping more than twenty years. arXiv preprint. https://doi.org/10.48550/arXiv.2401.12345

Asif M, Dehwah AHA, Ashraf F, Khan HS, Shaukat MM, Hassan MT (2017) Life cycle assessment of a three-bedroom house in Saudi Arabia. Environments 4(3):59. https://doi.org/10.3390/environments4030059

Bandyopadhyay J, Mitra S, Sengupta S (2007) Living with floods: a comparative assessment of resilience to floods in Tal and Diara regions of Malda district. West Bengal Curr World Environ 2(1):115–122

Barman R (2020) Indigenous knowledge and flood-resilient architecture: Chang Ghar of Assam. J Rural Community Dev 15(3):104–120

Biradar VK, Mama S (2017) Vernacular architecture: a sustainable approach. In: Seta F et al (eds) Understanding built environment. Springer, pp 1–13

Bueno C, Fabricio MM, Bragança L (2022) Environmental life cycle assessment of residential buildings in Brazil: case studies using two methods. Build Environ 187:107388

Carrobé Á, Zamora B, López-Mesa B, Mora L (2024) Life cycle analysis of adobe-based rural houses: case study in Guatemala. Sustain Cities Soc 103:104365

Chang SY, Lee KT (2021) Traditional chinese rural housing and regional adaptation strategies. Energ Build 231:110561

Choudhury ASD, Chettry V (2023) Bioclimatic design features in Assam-type stilt houses. Indian J Tradit Knowl 22(3):219–231

Das A, Das R, Debsarkar A, Roy KD (2024) Life cycle assessment of a PMAY-assisted low-cost building in Rishra, West Bengal: a case study. In: Proceedings of SATEM 2023. Springer

de Bortoli A, Giretti A, Pastorelli M (2022) BIM-integrated LCA: a parametric approach for sustainable design. Autom Constr 135:104122

Dormohamadi M, Aziz NAA, Ahmad A (2024) Hybrid vernacular-modern passive architecture in Iran: a simulation-based LCA approach. Sustain Dev 32(1):37–48

Fathy H (1986) Natural energy and vernacular architecture: principles and examples with reference to hot arid climates. University of Chicago Press

Fernandes P, Das A, Mukherjee R (2014) Building materials and geographies: mapping the shrinking 'vernacular' of rural Bengal. J South Asian Dev 9(2):163–185

Fernandes J, Silva SM, Mateus R, Teixeira ER (2019) Analysis of the thermal performance and comfort conditions of vernacular rammed earth architecture from Southern Portugal. Ref Module Mater Sci Mater Eng. https://doi.org/10.1016/B978-0-12-803581-8.11460-2

Fnais M, Mohammed AS, Almutairi T (2022) Sustainability analysis of conventional and eco-friendly building typologies in the Middle East. Environ Impact Assess Rev 92:106717

Francis A, Thomas MV, Lal R (2024) Comparative life cycle assessment of confined masonry and earthen vernacular structures in India. J Build Perform 15(1):15–26

Gazquez LAM, Hernández FF, López JMC (2022) LCA of traditional vs. contemporary social housing in Argentina. Sustain Cities Soc 82:103898

Ghosh M, Banerjee S, Ray D (2024) Riverbank erosion and adaptation in the Ganges Delta. Int J Disaster Risk Reduct 72:102998

Goswami M (2015) Seismic resistant elements in traditional architecture of Northeast India. Earthq Spectra 31(3):2023–2041

Jagatramka R, Kumar A, Pipralia S (2020) Sustainability indicators for vernacular architecture in India. ISVS E-J 7(4):1–18

Jayawardana A, Rathnayake U, Gunathilaka H (2023) Life cycle environmental impact of prefabricated green concrete in tropical housing. J Clean Prod 396:136451

Kaoula D, Rahmani S, Lemita I, Zemirli MA, Semahi S (2022) Life cycle assessment of traditional Saharan houses: towards a sustainable building? J Build Eng. https://doi.org/10.1016/j.jobe.2022.104782

Koenisberger OH, Ingersoll TG, Mayhew A, Szokolay SV (2001) Manual of tropical housing and building: climatic design. Orient Longman

Kumar V, Garg R (2021) Mud architecture in India: an overview of technology, sustainability, and performance. Arch + Des 38(3):47–55

Laha A (2015) Vernacular decline and material transitions in rural Bengal. Int J Arch Res 9(2):21–37

Li Y, Sun Y, Wu X (2020) Biophilic elements in traditional Naxi architecture: a cultural ecology perspective. Sustainability 12(7):2994

Majumder S (2020) Report on vernacular architecture along Dharampur-Jibannagar border cross road in Nadia, West Bengal. https://www.scribd.com/document/349912784/Vernacular-Architecture-Report

Majumder A, Das R, Debsarkar A (2024) LCA of PMAY low-cost conventional building in Rishra. Springer Proc Sustain Constr Mater Technol

Majumder A, Debsarkar A (2024) Assessment of multi-dimensional sustainability of vernacular architecture from global perspective: a critical review over the period (2000-2023). Res J Chem Environ 28(6):101–122. https://doi.org/10.25303/286rjce1010122

MDPI Editorial Team (2022) Special issue: earthen construction and sustainable architecture. Sustainability 14(19)

Nasir O, Kamal MA (2021) A vernacular design paradigm for the arid trans-Himalayan region of Ladakh. Am J Civ Eng Arch 9(6):219–231

Ogunmakinde OE (2024) Circular economy in construction: LCA of resource recovery in African contexts. Resour, Conserv Recycl 193:106790

Oliver P (1997) Encyclopedia of vernacular architecture of the world, vol 1–3. Cambridge University Press

Pakdel A. et al. (2021) Embodied energy and CO2 emissions of life cycle assessment (LCA) in the traditional and contemporary Iranian construction systems. J Build Eng (39). https://doi.org/10.1016/j.jobe.2021.102310

Pathak B (2014) Traditional flood-resistant architecture of Assam. Int J Herit Stud 20(3):289–302

Rinne E, Marteinsson B, Heinonen J (2022) Lifecycle impact of hybrid timber-concrete apartment buildings. Energ Build 259:111923

Sabah FY (2025) Global approaches to affordable housing: comparative insights from developed and developing countries and the case of Palestine. Front Built Environ. https://doi.org/10.3389/fbuil.2025.1653057

Sahapedia (2019) Kath-Kuni architecture: vernacular heritage of the Western Himalayas. https://www.sahapedia.org

Sandanayake M, Zhang G, Setunge S (2022) Comparative life cycle assessment of construction materials in rural housing. Built Environ Proj Asset Manag 12(4):511–525

Sesga L, Martínez E, Beltrán B (2021) Environmental assessment of refurbishing a heritage adobe house. Int J Arch Herit 15(6):889–907

Sciendo L (2023) Traditional earthen structures: thermal comfort and lifecycle performance. Arch, Civ Eng, Environ 16(3):49–58

Slideshow.net (2021) Vernacular architecture of rural Bengal [Presentation]
Syska E, Leśniak A, Plebankiewicz E (2021) LCA and LCC integration challenges in building sector. Appl Sci 11(15):6834
TIME Editorial (2022) Why western architecture worsens India's heatwaves. Time Mag 199(11):26–30
Vellinga M, Oliver P, Bridgewater S (2024) Atlas of vernacular architecture of the world, 2nd edn. Routledge
Villageinfo homepage (n.d.). https://villageinfo.in/west-bengal/barddhaman/purbasthali-i/dirgha para.html
Wang J, Liu H, Zhang Y (2023) Evaluation of sustainable performance of rural housing in China using LCA. Sustain Dev 31(1):14–30
WEF (2023) How India's vernacular architecture can cool its cities. World Economic Forum Report Series
Wikipedia (2025a) Ganges Delta. The Free Encyclopedia, Wikipedia
Wikipedia (2025b) Pradhan Mantri Awas Yojana. The Free Encyclopedia, Wikipedia
World Bank (2021) Building resilience in the Ganges. Infrastructure and Housing, Delta
Zhang J, Xie M, Wu Q (2020) Environmental impact and policy review on rural housing development in China. J Environ Manag 260:110066
Zhou Z, Ling C (2021) Vernacular architecture and energy performance in East Asia: a review. Energ Build 247:111167
Zhu H, Liu M, Ma T (2022) Sustainable housing indicators for coastal regions. Build Res Inf 50(3):321–334

Determining a Building's Residual Value: How Construction, Transformation, and Deconstruction Techniques Can Enhance Value

Raymond L. Sheen(✉)

Produktif AS, Raufoss, Norway
ray@produktif.com

Abstract. The need for improved financial models for assessing building value in a circular economy has been recognized for years [1]. In particular, financial institutions using either the business case approach or the collateral approach when financing a construction project must assess the overall environmental impact of a building throughout its lifecycle. In both approaches, an important element of that financial calculation has traditionally been the determination of the residual value of what was being financed. For a building loan, the residual value would be based on the materials that are recycled or refurbished and incorporated into a second-generation building. The construction transformation and deconstruction techniques can affect the residual value of the site or any remaining building structure. The research question addressed in this study is to identify and assess building construction, transformation, and deconstruction approaches that maximize a building's residual value. This study includes a survey of banks to determine the methods currently used for establishing residual values and proposed enhancements to those methods. It also includes a literature search and review of building construction, transformation, and deconstruction methods for evaluating materials and labor costs. These techniques should maximize residual value at the end of a building's life enabling multiple building lifecycles.

Keywords: Circular economy · Residual value · Construction

1 Introduction

The foundational concept of a circular economy is economic activity that conserves natural resources. This concept applies to every industry and reflects the principles of wise management of natural resources. Governmental entities such as the European Union have established regulations and a taxonomy for implementing a circular economy. Included in that taxonomy are regulations and guidelines for all commercial and industrial sectors.

The construction industry has been slow to adopt the mindset of a circular economy [2]. The construction industry accounts for one third of waste and carbon emissions. In

H. Beushausen et al. (Eds.): NTZR 2025, LNCE 873, pp. 1149–1161, 2026.
https://doi.org/10.1007/978-3-032-27698-8_101

addition, the construction industry has not embraced new business models or technical innovations that would improve circularity. Previous studies have shown that this is due in part to the lack of incentives to change behaviors [2]. One aspect of the industry that could provide incentives to change is the lending market. If financial institutions required a robust assessment for circularity before providing financing for construction projects, architects, engineers, and builders would be more likely to include circularity principles in the design and construction of buildings. However, there are few tools or techniques that can be used by a financial institution to evaluate the circularity aspects of a construction project. Most institutions just consider energy usage in their analysis.

The research question addressed in this study is to identify and assess building construction, transformation, and deconstruction approaches that maximize a building's residual value at the end of life. This study examines the techniques currently in use in the banking industry. It also considers techniques proposed in the literature and assesses their usefulness as a metric of circularity. Finally, a new method for assessing circularity based on the design for manufacture and assembly (DfMA) technique is proposed. This technique can aid the designer or lending institution when considering the potential for reuse or refurbishment of materials within a building.

2 Background

2.1 Circular Economy

The business world and pop culture frequently refer to the circular economy. A difficulty associated with making progress with respect to adoption of circular economy principles is the different interpretations of what constitutes circularity [3]. At its core, circularity leads to sustainable growth and prosperity for all mankind [4]. Unfortunately, existing metrics with respect to achieving circularity indicate mixed progress. Within Europe, the wealthier countries in the North and West are making progress towards a circular economy, but most of the countries in other parts of Europe are not [5].

While there is no commonly agreed upon succinct definition of what constitutes a circular economy in the context of construction, there are two agreed upon principles [6]. One of the principles identified by Korhonen et al. is the recycling, reusing, or refurbishing of materials. This is in contrast to the predominant linear model for a building lifecycle of extract – produce – use – dump. A second major principle is the reduction of energy use and carbon footprint [7]. Financial institutions have relied on the second principle, energy use, with little regard for the first principle, recyclability .

2.2 Construction Industry

With respect to adopting circular economy principles and practices, the construction industry is not performing as well. A recent EU report noted that the construction industry is responsible for one third of the waste generated in Europe [2]. That same study noted that while 80% of the companies involved in construction rated adopting circular economy principles as high or very high priority, only 29% of the companies were actively engaged in any form of reuse or recycling of building materials [2].

The construction industry is fragmented and dominated by small and mid-size enterprises (SMEs) [8]. These companies normally do not have an R&D department that can develop new construction techniques, nor do they have large service organizations that can refurbish reclaimed construction materials. They must rely on other organizations in the value chain to do that work. In addition, due to the fragmentation, the company doing the deconstruction and refurbishment is normally not the company doing the original design or construction. Therefore, since a different part of the value chain is involved, there is no inherent business value for the design and construction firm to embed deconstruction features in their work. They do not realize any economic benefit such as increased sales or lower costs from performing that work.

2.3 Construction Financing

The European Union Taxonomy for Sustainable Growth includes the regulation EU 2019/2088, Sustainability-Related Disclosures in the Financial Services Sector [9].

This standard requires financial institutions operating within the boundaries of the European Union to post data on their website concerning the environmental and ESG risks within their portfolio of loans. There is no set format stipulated in the regulation for how the information is to be presented, so each bank has its own approach.

A review of the websites of the top 50 European banks in 2024 found that all had at least one page on their website that discussed sustainability within their loan portfolio [10]. There was a wide variation in the presentation of information including the breadth and depth of that information on the websites. However, the overarching emphasis was on carbon reduction. Every bank is doing something, but because of the lack of consistency it is impossible to compare the different banks to determine which are providing the most support for circular construction.

3 Methodology and Analysis

This study is conducted in two phases. The first phase is a qualitative analysis with two parallel paths. One path is the review of the current practices at financial institutions. This involves a review of the website of the top 50 financial institutions providing lending services to the real estate market in Europe [10]. It also includes interviews with loan managers at three of the banks. The second path is an assessment of tools discussed in the literature that can be used to analyze building residual value at the end of the building lifecycle. The second phase of this research project is the analysis of a new technique that can be used for assessing a portion of the residual value of a building.

3.1 Financial Institution Analysis

Within the European Union, financial institutions are required to maintain a page on their website that reports on the ESG status of their loan practices [9]. There is no fixed requirement for format or content. The top 50 European financial institutions were identified on a website maintained by Standard & Poor's, Global [10]. The webpage that addressed the ESG status for lending at each institution was reviewed. The common theme across all websites was that carbon reduction was an element in their decision process when placing loans. Other factors such as diversity and community development were also mentioned on some websites. No bank featured a discussion concerning the evaluation of reuse or recycling when making loan decisions.

The review and analysis of websites was augmented with interviews with current loan managers from two of the top 15 European banks and an interview with one of the leading European consultants on finance and the circular economy. All three said that loans for building projects with a good energy efficiency score received favorable treatment. One bank indicated that the amount of the loan they would approve was impacted by the energy score and the other bank indicated that the percentage interest rate could be impacted by the energy score. A common practice among banks is to rely on third party certification for the energy usage assessment. In addition to energy usage, one bank relied on the anticipated revenue when estimating the loan valuation for a new commercial building. That bank anticipated that revenue would be increasing during the life of the loan. The other bank was more likely to rely on the construction costs and market value of the building when determining loan amount. The bank strategist confirmed that these were the two commonly used methods when establishing a property's value at the end of life for a building. In either case, the building residual value may vary, although it generally would be higher than at the time of construction.

All three of the interviewees agreed that banks do not currently take the final building value into consideration when making a loan. One bank assumes the building will increase in value, so they are not worried that the building value will ever drop below the value of the remaining principal on the loan, The other bank took the view that when the loan was repaid they no longer had any interest in the building so they did not care if it had a residual value. Therefore, neither bank considered any factors for reuse or recycling of the building or building materials. One bank indicated that they would only include the cost of demolition in a construction loan when the organization requesting funding would be doing the deconstruction of a pre-existing building. Even then, it was only considered an expense element for disposal of construction and demolition waste. There was no perceived benefit from reselling, refurbishing, or reusing the material. However, all three interviewees expressed an interest in a metric for reuse as an additional factor that would be used in making loan decisions.

3.2 Construction Material Reuse Metrics

There are no commonly accepted metrics for assessing the reusability of construction materials at the time of initial construction. Even at the time of deconstruction or demolition there are no widely adopted metrics. Bertino et al. highlighted the importance of planning before starting deconstruction in order to minimize waste [1]. This planning entailed both the deconstruction process and the process for restoring or refurbishing recovered materials. In addition, when Salimi and Taherkhani conducted a bibliometric study of lifecycle assessment in building construction, they found a significant cluster in the area of design for disassembly [11]. Techniques in this area would meet the need for improved planning as identified by Bertino et al. [1]. In addition, Guerriero et al. noted a severe lack of circular and regenerative design tools [12]. In many cases, architects and engineers do not know the circularity implication of their design decisions.

However, there are several metrics that have been proposed for use when performing deconstruction or building transformation planning. One approach discussed in numerous studies is to consider the building at the end of its life as a material bank [13]. Although the recovered material can be valued at market prices, a major limitation with this approach is the lack of tools or techniques for evaluating the disassembly process. The cost to extract the materials during deconstruction, building transformation, or demolition is not easily estimated [12]. In addition, the extracted materials must often go through a process of testing and evaluation to determine if they can be safely reintroduced into the construction supply chain. In fact, a study done by Pilipenets et al. found that analyzing the cost of construction waste during deconstruction is still highly variable [14]. Most companies doing deconstruction do not have the capability to restore or refurbish the recovered materials. These must be transported to an organization that specializes in restoring or refurbishing recovered material.

In addition to the concept of a material bank, Hartwell developed a scoring method for determining the reclamation potential of building materials based on the building lifecycle, material characteristics, and how the materials are used [15]. This metric requires an assessment of the building use prior to deconstruction. The methodology creates a reclamation potential score measured as a percentage. It does not provide a monetary value. However, the reclamation potential percentage can be used with the material costs to estimate a value for the recovered material. The limitations of this methodology are that it can only be used at the end of the building's life, and it does not adequately address the difficulty and necessary labor to extract the materials.

4 Alternative Approach for Assessing Building Valuation

4.1 Design for Manufacture and Assembly

The design technique known as Design for Manufacture and Assembly (DfMA) is a well-established design methodology used in many industries [16]. This methodology enables the designer to consider downstream implications of current design decisions. The technique is a scoring methodology that calculates a score reflecting the ease of manufacturing and assembling a product or system [17]. The principles of DfMA began to be codified by companies building weapon systems to support World War II [16]. The technique was refined by Boothroyd and Dewhurst in the 1970s and 1980s [17]. This technique is used in the automotive, aerospace, and consumer products industries and taught as part of the curriculum for many industrial engineering degree programs.

Typically, designers are focused on meeting the needs of the customers or users by optimizing the functional performance and features of the product or system being designed [16]. Once the design is completed, the industrialization process begins to determine how to fabricate and manufacture the components of the design and how to assemble those components into the final product or system. The DfMA technique initiates the industrialization process earlier in the design cycle while the product or system is still being designed [17]. Incorporating factors revealed in the DfMA analysis leads to lower costs, higher quality, and shorter manufacturing cycle time. The scoring method starts with a functional analysis of each component or part to determine if it adds functionality to the system. Unnecessary parts or components add unnecessary costs to manufacture the unneeded component. In addition, assembling the unnecessary components into the product or systems adds additional unnecessary cost and time. Once the necessity for the part is established, the DfMA scoring system modifies the part of component score based on factors that create added time or effort to handle the part or component and factors that address added time or effort to attach the part or component to the others that make up the assembly.

The Boothroyd-Dewhurst DfMA technique provides a score for each part or component in the bill of materials and a composite efficiency ratio for each assembly or system [17]. Designers are encouraged to eliminate unnecessary parts that have no function to reduce the total part count. In addition, the scoring reveals attributes of the parts or components that create hazards or difficulty when handling. When designers become aware of these issues, they can often redesign parts or components to reduce this impact. Additionally, Boothroyd and Dewhurst did extensive research to understand the complexity and difficulty of various methods for assembling the system. A process such as welding is more complex than attaching components with nuts and bolts which is more complex than just snapping parts together.

4.2 DfMA Challenges within the Construction Industry

Theoretically, DfMA can be used as an analysis technique during construction, transformation, or deconstruction of a building. Any time building components are installed or removed, DfMA can provide an analysis of the effort required. However, given its development within industries characterized by high production volumes and standard parts, a legitimate question is whether DfMA can be applied to onsite building construction. Rankohi et al. studied the attempted use of DfMA in both off-site and on-site construction and identified 45 challenges associated with incorporating this approach [18]. Most of those challenges were not unique to the construction industry but are associated with implementing organizational change. Examples of some of those challenges include management of customer expectations and poor coordination among stakeholders. However, four of the challenges were directly related to the use of the DfMA technique within the on-site construction environment.

- Difficulty in identifying appropriate DfMA tools/techniques in each phase.
- Higher design costs than the traditional design methods
- Difficulty in financial management and lack of an efficient payment method
- Lack of awareness of DfMA benefits among owners/developer.

The first of these challenges is related to the nature of on-site construction. DfMA was developed in a manufacturing environment characterized by high-volume standard part production. On-site construction, transformation, and deconstruction are normally characterized as low volume operations, and each building is unique. To use traditional DfMA elements in this environment, the focus needs to shift to process steps, which are more likely to be standardized and are relatively high-volume. When focusing on process steps, the difference between construction and deconstruction is almost non-existent. One is essentially the other in reverse order.

The second challenge is associated with higher design costs than the traditional design methods. This will likely be true since a DfMA analysis is an additional design analysis that is not normally included in construction projects. Unfortunately, the project savings identified in the analysis will not lower the design costs. The savings will occur in lower procurement costs because of the elimination of procurement of unnecessary components and lower labor costs during onsite construction. This is where the fragmentation in the construction industry can create a barrier to adoption of DfMA [8]. The organization doing the design work is seldom the one doing procurement or onsite construction. Therefore, the design organization does not receive the benefit from the cost savings identified through the use of DfMA. Unless the customer specifically requests a DfMA analysis, the additional design effort increases the design costs and may jeopardize the design organization's ability to win the design contract.

The third challenge is also linked to the high level of fragmentation in the construction industry [8]. Each contractor has a different and unique scope of work for which they are paid. A DfMA analysis may cause multiple contractors to "lose" money. For instance, a fixed price design contract "loses" money when it conducts extra analysis beyond the minimum needed to meet contract requirements. A time and materials construction contract "loses" money when the work is done faster than expected since there is less time billed to the customer. The major beneficiary of a design that has incorporated the results of a DfMA analysis is the end customer. However, if the customer or their representatives do not insist on a DfMA analysis, it is unlikely that it will be done.

The fourth challenge is the lack of awareness of DfMA and the impact it could have on a construction project. Since DfMA is not commonly used in the construction industry, the lack of awareness is to be expected. The primary benefit from DfMA is a lower cost to the customer due to the labor savings that comes from simplified assembly. Since many construction project customers are not construction professionals, they may be unaware of how estimating is done. Construction labor costs are normally estimated based on the size of the building and material costs. There is no provision in the estimating tools for factoring in DfMA effects. It will require a customer advocate, such as the project financing arm, to advise the customer to ask for this type of analysis.

4.3 Design for Assembly and Disassembly

DfMA is only concerned with the initial assembly of the product or system, not the disassembly. However, in the construction lifecycle, all three phases of construction, transformation, and deconstruction should be considered. The DfMA technique is focused on assembling the product or system in the manufacturing plant. This could be considered similar to construction. In the Boothroyd-Dewhurst model there was no element of the technique that was concerned with the end-of-life disassembly, reuse, or disposal of the product or system [17]. However, the recent focus on circularity in the construction industry brings the need for disassembly to the forefront [11]. The DfMA analysis can be reversed to represent transformation and deconstruction.

Traditionally, the creation of construction and demolition waste (CDW) created by transformation and deconstruction of buildings was considered to be of no value [19]. This perspective is changing. The circular economy principles promote recycling, reuse, refurbishment, and remanufacturing construction material over disposal of those materials in landfills [2]. However, the economic viability of recovering usable construction materials from CDW requires a controlled disassembly process to prevent damage to the materials. In addition, that process needs to be fast and inexpensive for the effort to be profitable [1]. Currently, the assessment of the ease of disassembly does not normally occur until the building transformation or deconstruction is about to start, which is many years after the building was designed. Therefore, the analysis of the potential for reuse of building materials is performed based on current market value of the materials [13]. There is no consideration made at the time of initial building design and construction for how the building will be disassembled many years later. Depending upon the assembly processes typically used in construction, there is significant potential for damage to the materials during disassembly. Often that damage can be avoided if different choices are made during design and construction.

The need for design for disassembly is now becoming recognized. Salimi et al. noted that design for disassembly was one of the major themes in recent journal articles that addressed circularity within the construction industry [11]. However, in a literature review of 18 peer-reviewed journal articles that directly discussed design for disassembly during the period from 2020 to 2024, only one provided any design guidance that could be used by building designers [20]. Zhan et al. noted that the number one barrier to implementing design for disassembly was the lack of tools or techniques [21]. In addition, Zhan et al. found the most efficient strategy for encouraging implementation was to offer financial incentives for conducting design for disassembly analysis [21].

Although there are recommendations for applying design for disassembly techniques, the transformation and deconstruction phases form most buildings are not points of interest to the developers or builders during the construction process. However, applying the DfMA techniques to lower costs and reduce on-site construction time often is a concern. By combining the DfMA technique with design for disassembly, a composite technique, Design for Assembly and Disassembly (DfAD) could create significant synergy. The assembly portion of DfAD identifies the opportunity for immediate savings and the disassembly portion of DfAD calculates a circularity score that could be used by financial institutions when making loan decisions. This technique then addressed the construction phase, transformation phase, and eventual deconstruction phase of a building. The organizations in the value chain best positioned to advocate for the use of this technique are the banks and lending institutions.

5 Design for Assembly and Disassembly Example

5.1 Technique Elements

The proposed DfAD methodology is based on the proven Boothroyd-Dewhurst DfMA methodology [17]. This analytical technique is built on the structure of the bill of material developed during the design process. Every item on the bill of material is first evaluated for its contribution to functionality. The DfMA technique considered three elements of functionality [17]. Does the component need to move as part of its normal operation? Does the component need to have unique material properties because of its function? Does the component need to be removed because it was an item that required regular service? If the answer to all three was "No" then the item was a candidate for elimination. The item could be eliminated with no impact on functionality, or it could be combined with another item in the design. The combination item would likely be more complex, but there would be fewer items in the assembly which normally equates to less assembly time and better quality since there are fewer possible mistakes of forgetting an item, damaging an item, or incorrectly connecting an item.

The DfAD approach has added two additional questions beyond the three used by DfMA/ These are relevant to building design and the construction industry. The first question is, "Does the component create a new planar surface that defines a building or room size and shape?" The reason for this additional question is because a fundamental purpose of construction is to define a volumetric space where people live, work, or play. This space is defined by the walls, floor, and ceiling of the room or building. The second additional question is, "Does the component need to be removed to provide access to other components for service or disassembly/" The previous question dealt with servicing the component itself. This question is related to safety and disassembly. Some components function to hold things in place and prevent unintended access or disassembly. Once installed, the answer to the other four questions would be "No," However, at certain time in the lifecycle of the building, the component needs to be removed. Therefore, it needs to be a unique part. A "No" answer to all five questions indicates the designer should consider a different design that eliminates this component or combines it with another. If the answer to any question is "Yes" then the part must remain an independent part. The total number of parts that had one or more "Yes" answers is the theoretical minimum number of parts (TMNP). When the analysis is complete, a building component efficiency ratio (BCE) can be calculated.

$$PCE = \left(\frac{TMNP}{\text{Total number of parts}}\right) \times 100\%$$

Once all items on the bill of material have been assessed for elimination and the design updated, the assembly efficiency aspect of the design can be calculated. DfMA evaluated an extensive set of part characteristics that would assign scores to an item based on the item size, shape, and any special handling concerns. Once the handling scores are established, the assembly process is reviewed, and scores are assigned for factors associated with positioning and securing the component in the product or system. The DfAD scores are based on values found in the published DfMA database [17]. The scores represent the amount of time or effort associated with handling, positioning, and attaching the component. Components with very high scores are those that require extra effort and add complexity to onsite assembly. The scores for each component in the assembly or system are summed and that value is used to create a building assembly efficiency ratio (BAE) for the system or assembly.

$$BAE = \frac{3 \times TMNP}{Total\ of\ part\ assembly\ scores} \times 100\%$$

At the same time as the construction components are being scored for assembly, they can be scored for disassembly. The scoring for handling of the components is usually the same for both analyses. Although it is possible that during the deconstruction process some of the special handling factors might change, such as the presence of sharp edges requiring safety protection. The disassembly processes are essentially the reverse of the assembly processes, however some disassembly processes such as breaking bonds or removing welds, are scored at a higher level than the original process because of the disassembly effort. In addition, a disposal score is added to each component based on its likelihood for reuse, recycle, refurbish, remanufacture, or disposal. The actual decisions for component disposition will not be made until building transformation or deconstruction. At this time in the design process when the components and systems are being scored, the designers use the most likely disposition of the product. Again, the scores for all components in the assembly are summed and a building disassembly efficiency ratio (BDE) is calculated.

$$BDE = \frac{3 \times TMNP}{Total\ of\ part\ disassembly\ scores} \times 100\%$$

5.2 Technique Analysis

The DfAD technique was tested by analyzing two structures that were identical in outside size and shape but built with different construction methods. The structures were part of an Icebox Challenge that is used to demonstrate the value of designing buildings to meet passive house standards [22]. The analysis was only on the basic structure and the outside cladding since this was the only portions of the buildings that were similar. Both structures were approximately 3 m by 3 m square. Both had a sloped roof with the short side 2 m high and the tall side nearly 3 m high. As part of the Icebox Challenge, both structures were erected in a plaza in Oslo, Norway where they stood for one month exposed to normal weather conditions.

One ice box structure was built using traditional construction techniques. The basic structure was connected with steel plates and screws at the joints. The outside cladding panels were nailed onto the basic structure. When it was time for disassembly, the cladding had to be removed with a crowbar, some of the cladding was damaged. The basic structure of beams was unscrewed.

The second structure was built with snap together components. The basic structure snapped together at the joints with custom joint elements. The cladding was clipped onto the basic structure using a commercially available clip system that was certified for outdoor structures. When it was time for disassembly, the first piece of cladding was removed with a special tool that released the cladding panel clips. After that, the clips to the other panels could be reached. The snap joints for the basic beam structure could be released with a pair of pliers. No components were damaged.

The results of the analysis of the two structures are found in Table 1. The snap-together structure was much more efficient in all categories. Both structures had the same theoretical minimum number of components since the two structures were essentially identical in size and shape. With respect to BCE, the traditional construction was very inefficient due to the high number of screws and nails that provided no unique functional value. The snap-together construction BAE was approximately six times more efficient than traditional construction, which implies that the onsite construction time would be six times faster than the standard construction. The BDE was significantly different. This was due in part to the much faster disassembly time and the fact that the snap-together construction components were undamaged and immediately ready for reuse. However, some of the traditional cladding components were damaged and could not be reused without first being refurbished.

Table 1. DfAD metrics

DfAD metric	Traditional construction (%)	Snap together construction (%)
Building component efficiency (BCE)	1.82	17.39
Building assembly efficiency (BAE)	1.50	9.07
Building disassembly efficiency (BDE)	0.30	5.36

6 Conclusion

The construction industry has been slow to adopt circular economy principles. An exception is the banking industry that provides loans for construction and real estate. All major European banks are complying with the EU Taxonomy to provide information about the circularity attributes of their loan portfolio. All the banks are highlighting their initiatives with respect to carbon usage, but none are emphasizing the reuse, recycling, and refurbishment assessment within their loan portfolio. Based on interviews with loan managers, this is not a metric being used, in part because there is no commonly accepted metric for building construction reuse. Several metrics have been proposed but all are calculated at the end-of-life of the building, not at the time of construction and therefore not at the time when loans are being originated.

The DfMA methodology was explored as a technique that is done during the design phase of a project but is analyzing the likely process efficiency when a product or system is in production. This technique was adapted to the construction industry and expanded to include the disassembly analysis of a building. The new methodology is referred to as Design for Assembly and Disassembly. It provides metrics that can be used to evaluate both material and labor costs of construction and to evaluate the labor costs of deconstruction. This metric can be used to estimate a portion of the residual value of a building and assess the degree to which components can be reused.

References

1. Bertino G, Kisser J, Zeilinger J, Langergraber G, Fischer T, Österreicher D (2021) Fundamentals of building deconstruction as a circular economy strategy for the reuse of construction materials. Appl Sci 11(3):939
2. Brincat C, Graaf I, León Vargas C (2023) Study on measuring the application of circular approaches in the construction industry ecosystem: final study. Publications Office of the European Union, European Commission, European Innovation Council and SMEs Executive Agency. https://data.europa.eu/doi/10.2826/488711
3. Reike D, Vermeulen WJV, Witjes S (2018) The circular economy: new or refurbished as CE 3.0?—Exploring controversies in the conceptualization of the circular economy through a focus on history and resource value retention options. Resour Conserv Recycl 135:246–264
4. Ghisellini P, Cialani C, Ulgiati S (2016) A review on circular economy: the expected transition to a balanced interplay of environmental and economic systems. J Clean Prod 114:11–32
5. Mazur-Wierzbicka E (2021) Circular economy: advancement of European Union countries. Environ Sci Eur 33:111
6. Korhonen J, Honkasalo A, Seppälä J (2018) Circular economy: the concept and its limitations. Ecol Econ 143:37–46
7. Klemeš JJ, Foley A, You F, Aviso K, Su R, Bokhari A (2023) Sustainable energy integration within the circular economy. Renew Sustain Energy Rev 177:113143
8. Riazi SRM, Zainuddin MF, Nawi MNM, Musa S, Lee A (2020) A critical review of fragmentation issues in the construction industry. Talent Dev Excel 12:1510–1521
9. EU 2019.2088 (2019) Sustainability-related disclosure in the financial services sector
10. Mones D, Hayes D (2024) Europe's 50 largest banks by assets. https://www.spglobal.com/market-intelligence/en/news-insights/research/europes-50-largest-banks-by-assets-2024. Retrieved 10 April 2025
11. Salimi R, Taherkhani R (2024) The transition towards a sustainable circular economy through life cycle assessment in the building and construction sector: a review and bibliometric analysis. Environ Sci Pollut Res Int 31(54):62588–62622
12. Guerriero A, Busio F, Saidani M, Boje C, Mack N (2024) Combining building information model and life cycle assessment for defining circular economy strategies. Sustainability 16(11):4561
13. Oliveira JD, Schreiber D, Jahno VD (2024) Circular economy and buildings as material banks in mitigation of environmental impacts from construction and demolition waste. Sustainability, 16(12), 5022
14. Pilipenets O, Kin Peng Hui F, Gunawardena T, Mendis P, Aye L (2024) New circularity indicator for decision making in the stockpile management of construction and demolition waste: perspectives of Australian practitioners. J Environ Manage 363:121345
15. Hartwell R, Overend M (2024) Reclamation potential in the built environment: a method and metric for assessing environmental benefits beyond first use. Build Environ 263:111866
16. Lu W et al (2021) Design for manufacture and assembly (DfMA) in construction: the old and the new. Archit Eng Des Manage 17(1–2):77–91
17. Boothroyd G, Dewhurst P, Knight W (2010) Product design for manufacture and assembly, 3rd edn. CBC Press, Boca Raton
18. Rankohi S, Bourgault M, Iordanova I, Carbone C (2023) Developing a construction-oriented DfMA deployment framework. Buildings 13(4)
19. Purchase CK et al (2021) Circular economy of construction and demolition waste: a literature review on lessons, challenges, and benefits. Materials 15(1):76
20. Crowther P (2022) Exploring the principles of design for disassembly through design-led research. In: IOP conference series. Earth and Environmental Science, vol 1101(6), pp 62031

21. Zhan H, Hwang B, Jovan LWH (2024) Barriers and strategies in implementing design for disassembly and adaptability principles in Singapore: a pathway to circular economy in the construction industry. Sustainable Development, Bradford, West Yorkshire, England
22. Icebox Challenge (2025) https://iceboxchallenge.org/. Retrieved 10 April 2025

Life Cycle Costing Implications for Contractor Development Program Financing

Richard John Ndhlovu[1(✉)], Innocent Musonda[2], and Adetayo Onososen[1]

[1] Center for Applied Research and Innovation in the Built Environment (CARINBE), University of Johannesburg, Johannesburg, South Africa
richardjndhlovu@gmail.com

[2] Sustainable Built Infrastructure and Digitalisation, Centre for Applied Research, and Innovation in the Built Environment (CARINBE), University of Johannesburg, Johannesburg, South Africa

Abstract. The construction industry consistently strives to optimize resource allocation and maximize long-term value while adhering to budget constraints and sustainability objectives. This study investigates the application of life cycle costing as a strategic approach for evaluating the economic viability of construction projects and contractor development initiatives. The research emphasizes optimizing resource allocation and maximizing long-term value by considering all cost factors from project inception to decommissioning. The methodology employed involves a systematic process with 41 articles selected for the review analysis. Key findings highlight the role of life cycle costing in supporting informed decision-making by balancing initial investments with long-term financial outcomes, thereby enabling stakeholders to assess the fiscal sustainability of construction capacity development initiatives. The study demonstrates that life cycle costing offers a robust framework for aligning financial strategies with long-term objectives, facilitating detailed expense assessments at each project stage and promoting resource efficiency. Additionally, integrating life cycle costing into procurement processes provides a structured method for evaluating the total cost of ownership, which aids in sustainable and environmentally friendly construction project management, fostering sustainable development, and enhancing the capabilities and competitiveness of construction firms.

Keywords: Life Cycle Costing · Construction Industry · Contractor Development · Funding

1 Introduction

Despite the available support for implementing construction contractor capacity development programs in developing economies, it remains unclear whether research on the financing frameworks of these programs considers context-specific challenges, such as life-cycle costs, within the capacity development context. Life cycle costing serves as a crucial approach for comprehensively assessing the economic impacts of construction projects, including all cost aspects from initiation to decommissioning [1]. This approach

H. Beushausen et al. (Eds.): NTZR 2025, LNCE 873, pp. 1162–1174, 2026.
https://doi.org/10.1007/978-3-032-27698-8_102

provides decision-makers with a framework for making informed choices that enhance value, optimize resource use, and support sustainability throughout the project lifecycle [2]. Life cycle costing provides a framework for assessing the economic sustainability of construction contractor development programs. It aligns financial strategies with long-term growth objectives by offering insights into total ownership costs and the return on investment of these programs.

The dominance of large foreign contractors in high-value and mega projects within developing countries highlights the inadequacies of contractor development programs, resulting in the exclusion of locally trained construction contractors from participation in these projects [3, 4]. This paper emphasizes the significance of life cycle costing in the design of capacity development programs that yield outcomes satisfying key stakeholders and promote a resilient construction industry in developing economies. A scientometric analysis of current literature retrieved from Scopus was conducted to identify prevailing trends and thematic areas in construction research within emerging economies. The findings aim to inform future funding directions and support the development of capacity development programs in developing economies with a focus on sustainability and resilience. This investigation addresses two primary gaps: (1) the absence of a comprehensive, evidence-based understanding of life cycle costing in developing capacity development programs in developing economies over the past 25 years; and (2) the limited qualitative analysis of highly cited works to elucidate the underlying assumptions, funding frameworks, and sustainability perspectives shaping the capacity development programs.

The study combines quantitative scientometric indicators with qualitative content analysis to develop a concise framework that can guide future research priorities in the development of capacity development programs. This study examines published research on capacity development programs, taking into consideration the life cycle costs. The objectives of this study are as follows:

- Analyze and map out the evolution of research in contractor development programs in developing nations.
- Assess and determine the implications for life cycle costing on a resilient construction industry.

1.1 Importance and Training of Contractors

Due to the vital role that contractors play in the construction sector, their involvement spans the entire project lifecycle, from initial planning and design through construction, maintenance, and subsequent assessments [5]. Selecting qualified contractors is critical to project success, as deficiencies in this area can lead to significant cost overruns, scheduling delays, and compromised quality [6, 7]. This process typically involves a detailed assessment of key criteria such as expertise, reputation, technical capabilities, and financial stability [8]. Ultimately, the prudent selection of a competent contractor is paramount, as inadequate decisions can result in budget overruns, project delays, and diminished overall quality [9].

1.2 Utilization of Life Cycle Costing in Construction Projects

The application of life cycle costing in construction projects requires a thorough evaluation of all cost components over the project's duration [11]. The elements include the initial investment, comprising design and construction expenses, as well as operational costs, which encompass energy consumption, utilities, and staffing [12]. Maintenance costs, which include repairs, renovations, and periodic component replacements, are essential elements of the analysis, in addition to disposal costs related to decommissioning, demolition, and waste management [13].

The environmental impacts associated with a building's life cycle are significant, and it is crucial to quantify all phases, particularly the construction phase, which is frequently neglected [14]. Life cycle thinking is crucial for attaining genuine sustainability in the construction industry and for meeting decarbonization objectives [15]. Life cycle assessment serves as an essential method for evaluating the environmental impacts linked to construction materials across their entire life cycle, encompassing raw material extraction, manufacturing, transportation, construction, use, and disposal [16, 17]. This allows stakeholders to make informed decisions that reconcile initial costs with long-term operational and environmental factors [18, 19].

2 Theoretical Framework

A theoretical framework for this study can be constructed based on two fundamental concepts: Life Cycle Costing and Life Cycle Assessment.

2.1 Life Cycle Cost Analysis

The LCC framework facilitates informed decision-making that enhances value, optimizes resource use, and fosters long-term sustainability across the project lifespan. LCC provides a methodology for assessing the economic sustainability of contractor development initiatives, integrating finance strategies with long-term growth objectives by offering insights into the total cost of ownership and return on investment.

2.2 Life Cycle Assessment

Life Cycle Assessment (LCA) is acknowledged as a methodology for evaluating and forecasting the environmental impacts of a product or process. It assesses the environmental impacts related to construction materials and processes across their whole life cycle. This includes the extraction of raw materials, production, transportation, building, utilization, and disposal. It is crucial for attaining genuine sustainability within the construction industry and for conforming to decarbonization objectives

3 Conceptual Framework of Life Cycle Costing

Life Cycle Costing serves as an economic analysis tool that assesses the total cost associated with a system, product, or asset over its entire life cycle, from inception to disposal [1, 2]. This methodology encompasses not only the initial capital outlay but also all expenditures associated with the asset's lifespan, including operation, maintenance, rehabilitation, and end-of-life considerations [20]. This comprehensive approach, which accounts for all direct and indirect costs—including capital expenditure, operational expenditure, renewal, replacement, rehabilitation, refurbishment, restoration, and potential costs related to fatalities, injuries, and property damage—offers a more precise financial overview for decision-makers [21].

The utility of Life Cycle Costing (LCC) is particularly evident in procurement decisions, as it enables the comparison of alternatives based not only on initial costs but also on their long-term economic implications, thus supporting more informed and strategically sound choices [22]. This perspective ensures that decision-making processes consider long-term financial implications rather than focusing solely on immediate acquisition costs, resulting in more sustainable and economically sound investments [23]. This analytical framework is crucial for evaluating the economic viability and sustainability of investments, providing insights into long-term advantages and drawbacks beyond initial capital expenditures [24]. This approach is essential for assessing complex systems or infrastructure projects, as the operational and maintenance phases constitute a substantial part of the total cost of ownership [25].

Life Cycle Costing serves as a crucial instrument in the development and management of contractor development programs, facilitating a comprehensive evaluation of the financial impacts throughout the entire project lifecycle, from initial contractor selection to project completion and post-warranty phases [26]. This approach combines financial data and cost information with metrics from life cycle assessments to evaluate environmental and economic impacts, to optimize resource management, and identify long-term externalities [27]. This encompasses the stages of planning, preparation, procurement, design, construction, operation, maintenance, and disposal [28].

The use of LCC facilitates the identification of optimal solutions by determining the option with the lowest total cost throughout its operational lifespan, transcending mere initial cost comparisons [1]. The enhanced comprehension of LCC is especially relevant today, as the incorporation of artificial intelligence is set to transform life cycle engineering through the dynamic optimization of economic and ecological goals [29].

4 Method

This study employs a scientometric analysis methodology to examine the research landscape of construction in developing economies. The focus is on identifying key themes related to sustainability, construction education, the construction industry, funding, and future research directions within the sector. The Scopus database was selected as the primary source due to its extensive coverage of scholarly publications [10]. A search strategy utilizing Boolean operators ("OR" and "AND") was developed with the following query: (Construction OR Built Environment) AND (Contractor Development OR Training) AND (Financing OR Funding).

The inclusion criteria for the study were as follows: subject area in Engineering; document type as Articles; publication status as final; source type as Journal; publication years were unlimited inclusive; and studies focused on developing economies such as India, Brazil, Malaysia, South Africa, Turkey, Indonesia, Iran, Nigeria, Thailand, Mexico, Pakistan, Sri Lanka, Ghana, Kenya, Ethiopia, Zambia, and Uganda. Exclusion criteria included all document types other than articles (e.g., book chapters, reviews) and publications not related to the specified developing economies.

The initial search resulted in 19,800 documents. Filters were then applied based on the criteria above, reducing the dataset. A manual review of titles and abstracts of the remaining 3,239 documents further narrowed the selection to 41 articles. Using VOSviewer software, a scientometric analysis was conducted to explore publication trends over time, identify growth patterns, and detect shifts in research focus. Keyword co-occurrence mapping was employed to reveal dominant research themes and emerging areas of interest. Citation analysis helped identify highly influential articles, authors, and journals within the field. Additionally, collaboration networks among authors and institutions were mapped to identify research communities and potential knowledge gaps.

A thematic coding approach was utilized to extract recurring concepts, assumptions, and thematic emphases across the articles. Codes were inductively developed and subsequently grouped into predefined categories. The emergent themes were then analyzed in conjunction with the scientometric findings to construct a comprehensive narrative on the evolution of research priorities and to identify potential disconnects between academic investigation and practical application.

5 Results and Discussion

5.1 Trend in Publication

Figure 1 illustrates the trends in documents and citations from 1993 to 2025 within the realm of the scientometric analysis of construction in developing economies. The data is filtered to include only external citations, excluding self-citations, book citations, and documents with zero citations. The graph indicates a consistent rise in the quantity of documents and citations over the years, particularly marked by a significant increase around 2014. This indicates an increase in interest and research activity within this domain. The data shows a peak in both documents and citations around 2023, suggesting a recent surge in attention and academic engagement regarding this topic.

This peak may result from several factors, such as technological advancements in construction, a heightened emphasis on sustainable practices, or geopolitical changes impacting developing economies. In 2024, there is a notable decrease in citations, despite the relatively high number of documents. The observed decline may suggest that, although research output remains robust, the impact or novelty of recent publications could be diminished, or it may be premature in the year for newer documents to gather citations. The upward trend observed until 2023 indicates that this field is progressing, with opportunities for additional development and investigation. The decrease in citations in 2024 may encourage researchers to explore alternative dimensions or methodologies within this topic to sustain or improve the impact of their work. The figure underscores the popularity and competitive dynamics of research in this domain, emphasizing the necessity for continuous innovation and high-quality contributions to advance the field.

Fig. 1. Overview of publications

5.2 Co-Citation Authors

Figure 2 depicts a visualization generated by VOSviewer, representing a co-citation network of authors engaged in scientometric investigations concerning the financing of capacity development initiatives in emerging economies. The visualization was accomplished through the implementation of a counting method requiring a minimum of one document and one citation per author. Each node represents a unique author, with the node's size generally reflecting their citation count, thereby denoting their influence in the discipline. The utilization of diverse colours denotes distinct clusters of authors who are often mentioned together, suggesting the presence of sub-fields or particular study areas within the overarching area of construction in developing economies. Authors Oke, Ayodeji Emmanuel, and Aigbavboa Clinton hold prominent places with larger nodes, showing their significant influence and robust citation link strength within this study domain.

Two distinct clusters were visualized: a pink cluster including 6 items labelled as cluster 1, and a yellow cluster having 4 items classified as cluster 2. The coloured clusters represent distinct subtopics or methodological methods within this field of study. Authors in the purple cluster have an average publication year of 2025 and may concentrate on various aspects of construction in developing economies, while those in the yellow cluster have an average publication year of 2023. The figure indicates that authors in the purple cluster published works from 2023 to 2024, whilst those in the yellow cluster published their pieces from 2024 to 2025. This clustering segmentation may indicate a positive increasing trend in research regarding capacity development financing in underdeveloped countries.

Prominent authors include Oke, Ayodeji Emmanuel, with 2 publications, 2 citations, and a total link strength of 6, underscoring the need for continued research in the field; and Aigbavboa, Clinton Ohis, who has 2 publications, 9 citations, and a link strength of 5, indicating reduced research activity and collaboration. Emerging contributors such as Arditi David and Mochtar Krishna, despite possessing significant citation counts of 136, have a minimal link strength of 1, indicating a constrained collaborative approach within the study domain. The elevated link strength scores, demonstrated by Behyutham Sheila with a score of 9 and Muya Mundia with a score of 10, indicate frequent co-citations and significant collaborative networks among these academics.

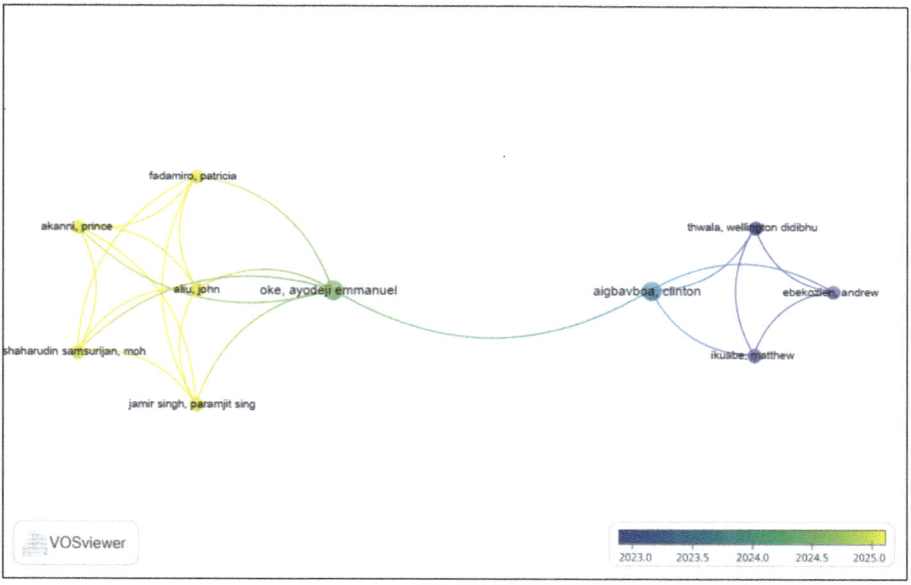

Fig. 2 Co-citation of authors

5.3 Key Themes and Discussion Topics on Capacity Development

Co-Occurrence of Keywords

This section's results demonstrate that a keyword co-occurrence network uncovers significant themes and concepts in the literature, established by a minimum threshold of three occurrences per phrase. This map elucidates critical areas, trends, and prospective research deficiencies. Prominent topics identified included "construction industry" (18 occurrences), "personnel training" (14 occurrences), and "construction" (10 occurrences), highlighting a significant emphasis on sustainable construction practices and their implications for capacity development in developing nations.

The significant keywords suggest that study on the construction industry's relevance continues to underscore its vital role in economic development and as a key contributor to any economy. Literature consistently recognizes the construction industry as a pivotal catalyst for economic progress and innovation, notably as a primary sector embracing technical improvements. Training important workers in the construction business may facilitate the effective implementation of innovation and new technologies, leading to improved infrastructure development processes. Human resources are a crucial element in advancing the construction sector. Their knowledge and skill development significantly influence the performance of the construction sector. Research indicates that a deficiency of skilled labour is a significant issue influencing the performance of the construction industry. Consequently, it is essential to train staff and provide them with pertinent and sufficient skills to address emerging trends in the construction industry.

The frequent references to "construction education" (8), "engineering education" (8), and "knowledge management" (9) underscore the increasing significance of training in enhancing construction capacity and performance, indicating a substantial interest in incorporating advanced learning methodologies and results into construction capacity development initiatives. Terminology such as "digital technologies" and "innovations" underscores the significance of integrating advancements and emerging technology into building capacity development initiatives. The period from 2023 to 2025, illustrated by a colour gradient from purple to yellow, signifies the evolution of priority regions, with the last terms likely indicated in yellow. The mention of nations like "Ghana," "Malaysia," "Zambia," and "Indonesia" signifies a targeted examination or regional emphasis within the larger context of developing economies.

Although sustainability and technology are prominently highlighted, deficiencies may exist in domains such as "curriculum" and "funding," which seem less integral or less associated with the primary clusters. The arrangement and proximity of clusters may provide transdisciplinary potential, including improved integration of technology, innovation, and sustainability. Promoting transdisciplinary research that integrates technical advancements, finance, and sustainability concerns may prove advantageous. The presence of terms such as "policy decisions" and "laws and legislation" in the network suggests that the research may also investigate the implementation challenges associated with the application of pertinent laws, regulations, and policy frameworks essential for advancing construction capacity development programs that address life cycle implications (Fig. 3).

A comprehensive framework for assessing the long-term financial impacts of contractor development programs, enabling informed decision-making and optimizing resource allocation to meet strategic objectives, is depicted. The application is particularly pertinent in procurement contexts, providing a systematic method for analyzing the total cost of ownership and assisting in decisions regarding source selection and outsourcing [22].

5.4 Challenges in Applying Life Cycle Costing to Capacity Development Programs

Despite ongoing financial assistance to enhance contractor ability in developing economies, stakeholder unhappiness persists. Literature indicates a disparity in the design and execution of various capacity development projects. The gap occurs after the conclusion of the training and during post-training assistance and mentorship,

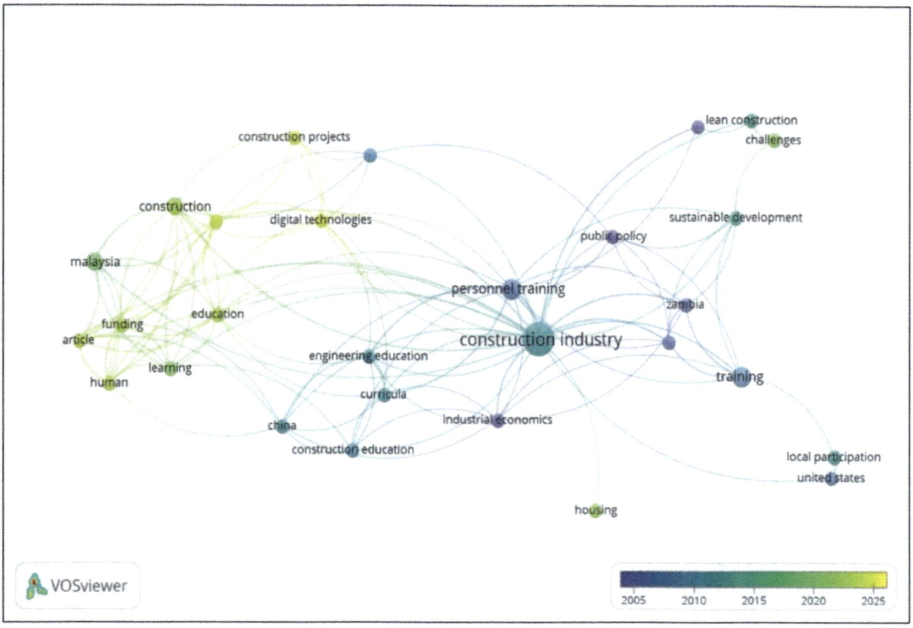

Fig. 3 Emerging themes

attributable to inadequate financial planning of the program that encompasses overall life cycle expenditures. Recently trained contractors faced competition in the open market against established entities without receiving post-training assistance. This method exposes newly trained contractors to excessive competition, leading some to abandon construction in favour of alternative pursuits.

Implementing life cycle costing in contractor development initiatives presents several challenges, primarily due to the complexity of accurately quantifying costs and benefits across the entire program lifecycle [30]. These difficulties highlight the need for innovative methodologies to effectively address these issues and facilitate the efficient financing of construction contractor development programs, with project costs remaining a key factor [5]. Conducting life cycle cost analyses is essential for ensuring reliability and cost-effectiveness throughout a structure's operational life, which necessitates incorporating risk assessment techniques to evaluate potential failures [31, 32]. Adopting a comprehensive, whole-life-cycle approach enables stakeholders to make informed decisions that improve the overall efficiency of infrastructure projects [28].

Addressing these challenges requires a comprehensive approach that incorporates innovative cost management strategies, data-driven analysis, and insights from previous projects to reduce the risk of capacity development program failure. Optimizing project management processes is crucial for effective cost control and achieving successful outcomes in construction projects. Key project management practices such as effective risk management, precise cost estimation, and strategic resource allocation are essential for enhancing cost efficiency in construction initiatives. To ensure project success, these methods should be integrated during the early stages of project planning [33].

The rapid advancement of emerging markets continues to attract the attention of industry specialists, who emphasize the significance of developing economic and construction capacities within these regions. The World Bank forecasts that developing and transition economies will experience growth rates between 4% and 7% over the next 25 years, representing approximately 60% of the global market. This accelerated growth is projected to yield an estimated economic market share of up to 70%. Implementing capacity development programs aligned with life cycle costing principles may enhance these economies' ability to adopt new technologies and capitalize on evolving market opportunities.

6 Conclusions

This study's policy implications are distilled into two principal aspects. The identified financing patterns and research trends can guide decision-making processes and underscore areas for policy development within the construction industry, addressing previously unexplored gaps. Secondly, the findings can aid government agencies in optimizing the allocation of policy inputs and resources, thereby improving resource deployment efficiency, reinforcing capacity development initiatives and policy structures, and facilitating sustainable construction education.

The findings offer valuable insights for stakeholders involved in capacity development initiatives, improving their understanding of the dynamic research landscape related to the financing of construction capacity development programs. This includes identifying leading authors, frequently cited studies, emerging research trends, and potential future directions. Future research should focus on evaluating the impact of funding options on capacity development programs and their influence on laws and regulations in developing economies. Additionally, it is recommended to explore the relationship between technological innovation and engineering-specific education.

7 Limitations

This study exclusively utilized data from Scopus, which may limit the scope of the literature reviewed; incorporating additional academic databases could potentially yield a broader range of sources and diverse analytical outcomes. Additionally, restricting the search to English-language publications may exclude relevant studies published in other major languages from industrialized countries. The theoretical framework employed in this research is based on selected literature and has not been empirically tested, as it serves as a component of a larger study focused on contractor capacity development.

References

1. Heralová RS (2017) Life cycle costing as an important contribution to feasibility study in construction projects. Procedia Eng 196(565). https://doi.org/10.1016/j.proeng.2017.08.031

2. Olubodun F, Kangwa J, Oladapo AA, Thompson WJ (2010) An appraisal of the level of application of life cycle costing within the construction industry in the UK. Struct Surv 28(4):254. https://doi.org/10.1108/02630801011070966
3. Dwivedi YK et al (2021) Artificial intelligence (AI): multidisciplinary perspectives on emerging challenges, opportunities, and agenda for research, practice and policy. Int J Inf Manage 57:101994. https://doi.org/10.1016/j.ijinfomgt.2019.08.002
4. Ndhlovu RJ, Musonda I, Onososen A (2024) Stakeholder theory implications for capacity development of local contractors in developing economies: a bibliometric review. In: Proceedings of the international conference on construction in the 21st century, pp 528–535
5. Xie H, Zang JZ (2013) Types and stages of the construction management. Appl Mech Mater 3376. https://doi.org/10.4028/www.scientific.net/amm.405-408.3376
6. Vardin NA, Ansari R, Khalilzadeh M, Antuchevičienė J, Baušys R (2021) An integrated decision support model based on BWM and fuzzy-VIKOR techniques for contractor selection in construction projects. Sustainability 13(12):6933. https://doi.org/10.3390/su13126933
7. Palaneeswaran E, Kumaraswamy MM (2001) Recent advances and proposed improvements in contractor prequalification methodologies. Build Environ 36(1):73. https://doi.org/10.1016/s0360-1323(99)00069-4
8. Watt D, Kayis B, Willy K (2008) Identifying key factors in the evaluation of tenders for projects and services. Int J Proj Manage 27(3):250. https://doi.org/10.1016/j.ijproman.2008.03.002
9. Yu FV, Chiang F-Y, THA L, Lin S-W (2022) Using the ISM method to analyze the relationships between various contractor prequalification criteria. Appl Sci 12(8):3726. https://doi.org/10.3390/app12083726
10. Onososen A, Musonda I (2022) Barriers to BIM-based life cycle sustainability assessment for buildings: an interpretive structural modelling approach. Buildings 12(3):324. https://doi.org/10.3390/buildings12030324
11. Biccari DC, Abualdenien J, Borrmann A, Corallo A (2019) A BIM-based framework to visually evaluate circularity and life cycle cost of buildings IOP Conf Ser Earth Environ Sci, 290 (1), 12043 . https://doi.org/10.1088/1755-1315/290/1/012043
12. Olimat H, Liu H, Abudayyeh O (2023) Enabling technologies and recent advancements of smart facility management. Buildings 13(6):1488. https://doi.org/10.3390/buildings13061488
13. Achilleos C, Hadjimitsis D, Neocleous K, Pilakoutas K, Neophytou OP, Kallis S (2011) Proportioning of steel fibre reinforced concrete mixes for pavement construction and their impact on environment and cost. Sustainability 3(7):965. https://doi.org/10.3390/su3070965
14. Bilec MM, Ries R, Matthews HS, Sharrard LA (2006) Example of a hybrid life-cycle assessment of construction processes. J Infrastruct Syst 12(4):207. https://doi.org/10.1061/(asce)1076-0342(2006)12:4(207)
15. Passoni C, Palumbo E, Pinho R, Marini A (2022) The LCT challenge: defining new design objectives to increase the sustainability of building retrofit interventions. Sustainability 14(14):8860. https://doi.org/10.3390/su14148860
16. Barbhuiya S, Das BB (2023) Life cycle assessment of construction materials: methodologies, applications and future directions for sustainable decision-making. Case Stud Constr Mater 19:e02326. https://doi.org/10.1016/j.cscm.2023.e02326
17. Shang Y, Boomen den M, van Man AD, Wolfert MAR (2019) Reliability-based life cycle costing analysis for embedded rails in level crossings. Proc Inst Mech Eng Part F J Rail Rapid Transit 234(8), 821. https://doi.org/10.1177/0954409719866359
18. Chaudhary A, Akhtar A (2022) Template for evaluating cradle-to-site environmental life cycle impacts of buildings in India. ACS Environ Au 3(2):94. https://doi.org/10.1021/acsenvironau.2c00035

19. Vitório CP, Yepes V, Onetta F, Kripka M (2025) Comparative life cycle assessment of warehouse construction systems under distinct end-of-life scenarios. Buildings 15(9):1445. https://doi.org/10.3390/buildings15091445
20. Kale NN, Joshi DA, Menon R (2017) Life cycle cost analysis of commercial buildings with energy-efficient approach. Perspect Sci 8:452. https://doi.org/10.1016/j.pisc.2016.04.102
21. Animah I, Shafiee M, Simms JN, Erkoyuncu JA (2018) Selection of the most suitable life extension strategy for ageing offshore assets using a life-cycle cost-benefit analysis approach. J Qual Maint Eng 24(3):311. https://doi.org/10.1108/jqme-09-2016-0041
22. Hanson D, J. (2011) Differential method for TCO modelling: an analysis and tutorial. Int J Procure Manage 4(6):627. https://doi.org/10.1504/ijpm.2011.043003
23. Waghmode LY, Sahasrabudhe A (2010) An application of a generalized life cycle cost model to BOXN wagons of Indian railways. In: ASME 2010 10th Biennial conference on engineering systems design and analysis, vol 3, p 333. https://doi.org/10.1115/esda2010-24099.
24. Albuja-Sánchez J, Damián-Chalán A (2024) Leveraging life cycle cost analysis (LCCA) for optimized decision making in adobe construction materials. Appl Sci 14(5):1760. https://doi.org/10.3390/app14051760
25. Świderski W, Rolek W (2021) A model of product life cycle cost management based on the example of the Spartan multimedia shooting training system. Economies 9(2):90. https://doi.org/10.3390/economies9020090
26. Tiwari NG, Tiwari AD (2016) Life-cycle cost analysis. In: Energy systems in electrical engineering. Springer Nature, p 671. https://doi.org/10.1007/978-981-10-0807-8_20
27. Tamburini E, Pedrini P, Marchetti GM, Fano EA (2015) Life cycle based evaluation of environmental and economic impacts of agricultural productions in the Mediterranean area. Sustainability 7(3):2915. https://doi.org/10.3390/su7032915
28. Bisbey J, Nourzad HSH, Chu C-Y (2020) Enhancing the efficiency of infrastructure projects to improve access to finance. J Infrastruct Policy Dev 4(1):27. https://doi.org/10.24294/jipd.v4i1.1175
29. Rahman H, D'Cruze SR, Ahmed UM, Sohlberg R, Sakao T (2022) Artificial intelligence-based life cycle engineering in industrial production: a systematic literature review. IEEE Access 10:133001. https://doi.org/10.1109/access.2022.3230637
30. Haouat EZ, Essalih S, Bennouna F (2025) Development of a global framework for an integrated life cycle assessment (LCA) model in quality, safety and environmental (QSE) management systems: improving environmental, social and economic sustainability performance. Sustainability 17(8):3521. https://doi.org/10.3390/su17083521
31. Oduyemi O, Okoroh MI, Fajana SO (2016) Risk assessment methods for life cycle costing in buildings. Sustain Build 1:3. https://doi.org/10.1051/sbuild/2016005
32. Setunge S, Zhu W, Gravina RJ, Giustozz N (2015) Fault-tree-based integrated approach of assessing the risk of failure of deteriorated reinforced-concrete bridges. J Perform Constr Facil 30(3). https://doi.org/10.1061/(asce)cf.1943-5509.0000754
33. Abdelalim MA, Salem SM, Salem M, Al-Adwani M, Tantawy M (2024) An analysis of factors contributing to cost overruns in the global construction industry. Buildings 15(1):18. https://doi.org/10.20944/preprints202412.0362.v1

Sustainable Construction Site Management: The Norwegian Practice and Future Prospects

Arne Ronny Sannerud(✉), Kim Rami Knuts, Arne Roar Lier, and Emma Zheng Liang

University College of Vocational Education (HØFY), Gjøvik, Norway
arne.ronny.sannerud@hfy.no

Abstract. This paper investigates the concept of sustainable construction site management within the context of Norway's construction industry, which is currently undergoing a significant green shift. The study is particularly relevant given Norway's commitment to reduce its CO_2 emissions by 55% by 2030 compared to 1990 levels. The research aims to understand the practical measures required by construction site managers to make a meaningful sustainable impact. The objective is to contribute to the discourse on sustainable practices in the construction industry and inform future strategies for sustainable development within this sector. The research methodology employs an inductive approach, utilizing qualitative methods. Insights were gathered from construction professionals in the industry and supplemented with relevant literature. The study aligns its potential solutions with the UN's sustainability goals, which are connected to the three pillars of sustainable development: environment, economy, and social aspects. The discussed topics include equality, gender balance, foreign labor, motivation, responsible consumption and production, decent working conditions, economic growth, as well as transformation management as a potential driver for increased motivation towards change. The study provides a reflective analysis of the industry's future prospects, offering a practical understanding of the implementation of sustainability in construction site management.

Keywords: Construction management · Sustainability · Green shift · CO_2 emissions · Norwegian industry

1 Introduction

The construction industry is undergoing major changes due to the green transition and increasing demands for reduced environmental impact. At the same time, the sector must maintain economic viability and ensure decent working conditions. Sustainable development has therefore become a key objective in both public and private sectors [1]. For middle managers and site managers, this entails a need for increased awareness and practical strategies that can make a sustainable difference in daily operations.

This paper aims to explore the concrete opportunities and tools available to site managers for implementing sustainable leadership in practice. Drawing on an overview

H. Beushausen et al. (Eds.): NTZR 2025, LNCE 873, pp. 1175–1183, 2026.
https://doi.org/10.1007/978-3-032-27698-8_103

of relevant literature on Norwegian industry practices, complemented by qualitative interviews with construction professionals, the study investigates how site managers can effectively implement sustainable construction site management.

2 Methodology

The study employs a qualitative methodology with an inductive approach, inspired by the research frameworks of Halvorsen [2] and Ryen [3]. An inductive approach was chosen to explore the practices and attitudes of site managers without being constrained by predefined hypotheses. The goal was to map experiences and reflections related to sustainability from individuals with leadership responsibilities in construction projects. Comparing the outcome of in-depth interviews with existing literature, allows for both theoretical and practical insights into sustainable site management and provides a foundation for drawing more generalizable conclusions [4].

The literature used in this study includes scientific articles, public reports, and academic textbooks. Search engines such as Google Scholar, Oria, and Google were used.

The empirical data consists of in-depth interviews. The unstructured interview format allowed participants to freely express their experiences and reflections, which, according to Wadel [5], provides deeper insight into social practices.

To represent different disciplines within the industry, three distinct roles were selected for the interviews. This approach aimed to reach more generalizable conclusions based on these perspectives combined with the existing literature. All respondents are employed by large construction companies, and the interviews were anonymized in accordance with NSD guidelines. The three respondents include; Site Supervisor: Responsible for daily operations and worker supervision. Contract Engineer: Responsible for financial management, documentation, and contract follow-up. Site Manager: Deputy to the project manager with broad responsibilities.

To explore the perceptions and practices related to sustainability within the construction industry, a series of semi-structured interviews were conducted using an open-ended question guide. The questions were designed to elicit detailed reflections on how sustainability is understood and experienced in daily work routines. Participants were asked to define what sustainability means to them personally and professionally, and to describe its impact—both positive and negative—on their workday. The interviews also investigated the perceived importance of sustainability, as well as the specific actions taken by both the respondents and their immediate supervisors that align with sustainable practices. Further questions focused on identifying potential improvements that could be made by middle management to enhance sustainability in projects, and on the respondents' own contributions and motivations toward more sustainable outcomes. The guide included inquiries into future expectations for the industry, particularly over the next 5–10 years, and addressed social dimensions such as equality on construction sites, the integration of foreign labor, and the challenges associated with diversity. Respondents were also asked to reflect on their personal and professional development opportunities within their companies, including training and competence-building initiatives. Finally, the interviews explored resource efficiency, reuse practices, and familiarity with lean construction principles, concluding with an open invitation for participants to share additional insights or suggestions.

The overarching aim of the interview questions was to gain a holistic understanding of sustainability as it is perceived and practiced on construction sites. The questions were carefully designed to explore multiple dimensions of sustainability, including environmental, economic, and social aspects. Participants were invited to reflect on their personal interpretations of sustainability and how it influences their daily work routines. The interviews also examined the implementation of environmental measures within construction projects, such as resource efficiency and waste management practices. Economic considerations were addressed through discussions on cost-effectiveness and the reduction of material waste. Social sustainability was another key focus, with questions probing into health, safety, well-being, and the inclusion of diverse workforce groups, including gender equality and the integration of foreign labor. Finally, the interviews sought to capture participants' expectations and visions for the future development of sustainability in the construction industry, particularly over the next five to ten years. This comprehensive approach enabled the collection of nuanced insights into both current practices and potential areas for improvement.

3 Theoretical Framework

3.1 UN Sustainable Development Goals

The 17 United Nations Sustainable Development Goals (SDGs) [1] provide a comprehensive global framework aimed at promoting sustainable development across economic, social, and environmental dimensions. Within the context of construction site management, several of these goals are particularly pertinent due to the sector's significant impact on labor practices, resource consumption, and environmental sustainability. Notably, Goal 5: Gender Equality emphasizes the importance of ensuring equal opportunities and representation for all genders within the construction workforce, a field traditionally characterized by gender imbalance. Goal 8: Decent Work and Economic Growth advocates for safe, inclusive, and productive working environments, aligning closely with the need to improve labor conditions and occupational health and safety standards on construction sites. Goal 12: Responsible Consumption and Production calls for the efficient use of materials and the reduction of waste, which is critical in construction due to its high resource intensity and potential for environmental degradation. Finally, Goal 13: Climate Action underscores the urgency of mitigating climate change through sustainable practices, including reducing greenhouse gas emissions and enhancing energy efficiency in construction processes. Together, these goals serve as guiding principles for transforming construction site management into a more equitable, responsible, and environmentally conscious domain.

3.2 The Triple Bottom Line

Elkington's model [6] emphasizes the balance between economic, environmental, and social factors as the foundation for sustainable decision-making. This implies that site managers must evaluate actions not only based on financial outcomes but also on environmental impact and social consequences.

3.3 Transformational Leadership

Transformational leadership has emerged as a particularly effective leadership style for advancing sustainable development within organizational contexts, including the construction industry. This leadership model, as outlined in the literature [7, 8], is characterized by four core components: Idealized Influence, which involves leaders serving as role models and earning the trust and respect of their followers; Inspirational Motivation, where leaders articulate a compelling vision that inspires and motivates employees toward shared goals; Intellectual Stimulation, which encourages innovation and critical thinking by challenging assumptions and fostering creative problem-solving; and Individualized Consideration, which reflects a commitment to supporting the unique needs and development of each team member through mentorship and personalized engagement.

Research in leadership theory increasingly supports the notion that transformational leadership can cultivate a workplace culture that is conducive to long-term sustainability objectives [9]. By fostering intrinsic motivation, aligning individual values with organizational goals, and promoting adaptive thinking, transformational leaders are well-positioned to drive meaningful change in practices and mindsets. This leadership style not only enhances employee engagement and performance but also contributes to the integration of sustainability principles into everyday decision-making and strategic planning. As such, transformational leadership represents a vital mechanism for embedding sustainability into the core operations and values of organizations.

4 Results

4.1 Environmental Aspects

The interview findings revealed a growing emphasis on environmental measures within construction projects, reflecting an industry-wide shift toward more sustainable practices. Respondents consistently reported that certain environmental actions have become standardized across sites. For instance, waste sorting is now a routine procedure, implemented to ensure proper disposal and recycling of materials. The use of low-carbon concrete is also on the rise, indicating a conscious effort to reduce the carbon footprint associated with construction activities. Additionally, the reuse of excavated soil and rock materials has become common practice, particularly in large-scale infrastructure projects where resource efficiency is critical. Another notable development is the appointment of dedicated environmental managers in major projects, tasked with overseeing compliance and promoting sustainable operations.

Despite these advancements, several respondents expressed concern that the individual understanding of the environmental impact of these measures remains limited. As one site supervisor noted, "We do it because it's in the procedures, not because we really understand the climate benefit." This statement underscores a broader issue: while environmental practices are increasingly embedded in project protocols, they are often driven by contractual obligations rather than a deep-rooted organizational culture or intrinsic motivation among employees. The results suggest that for sustainability efforts to be truly effective, there must be a stronger emphasis on education, awareness, and engagement at all levels of the workforce. Without this, environmental measures risk being perceived as mere compliance tasks rather than meaningful contributions to climate action and long-term sustainability goals.

4.2 Economic Aspects

Across all interviews, respondents consistently emphasized economic profitability as the dominant factor influencing decision-making in construction projects. Sustainable practices were generally viewed as favorable only when they aligned with cost-saving measures or improved financial outcomes. As one contract engineer succinctly stated, "Sustainable choices are made if they are economically viable," highlighting the pragmatic approach that often governs sustainability-related decisions on site. In this context, actions such as material reuse and efficient resource utilization were recognized as sustainable primarily because they contribute to reducing operational costs. The economic rationale behind these practices appears to be a stronger motivator than environmental or ethical considerations.

Moreover, several participants stressed the importance of long-term planning as a strategy to minimize both financial waste and environmental impact. Effective planning was seen as a key enabler of sustainability, allowing for better coordination, reduced material excess, and lower emissions. As one site manager noted, "Good planning can reduce both costs and emissions," underscoring the dual benefits of strategic foresight. These findings suggest that while sustainability is increasingly present in construction discourse, its implementation is often contingent upon its perceived economic value. To foster a more holistic approach, there may be a need to strengthen the integration of sustainability into organizational culture and decision-making frameworks beyond purely financial incentives.

4.3 Social Aspects

The study identified several important social challenges within the construction industry, particularly in relation to training, inclusion, and workforce integration. One key finding was that larger construction companies tend to offer more systematic and structured training programs compared to smaller firms, which often lack the resources or organizational capacity to provide consistent professional development. This disparity can lead to unequal opportunities for skill enhancement and career progression across the sector. Additionally, while there has been a gradual increase in the inclusion of women on construction sites, their overall representation remains low, indicating that gender diversity continues to be a significant challenge. Another recurring issue highlighted by respondents was the experience of foreign workers, who often receive less follow-up and limited opportunities for active participation in project planning and decision-making processes. This can contribute to feelings of exclusion and hinder effective team integration.

Despite these challenges, the interviews also revealed that inclusive leadership practices and open communication can play a crucial role in fostering a more supportive and motivating work environment. One contract engineer emphasized the value of autonomy and dialogue, stating, "Freedom with responsibility and the opportunity to discuss solutions creates job satisfaction." This suggests that when employees are trusted, involved in problem-solving, and given space to contribute meaningfully, it not only enhances their engagement but also promotes a more inclusive and collaborative workplace culture. These findings underscore the importance of leadership approaches that prioritize equity, communication, and empowerment as key components of social sustainability in construction.

5 Discussion

5.1 Transformational Leadership as a Tool

The findings of this study suggest that transformational leadership holds considerable promise as a strategic approach for advancing sustainability within the construction sector. This leadership style, characterized by its emphasis on vision, empathy, and empowerment, has the potential to embed sustainability into the core of daily operations. Leaders who practice transformational leadership are not only capable of inspiring their teams but also of fostering a sense of shared responsibility and commitment to long-term environmental and social goals. By encouraging participation and valuing individual contributions, transformational leaders help cultivate a workplace culture where sustainable practices are not merely procedural requirements but are embraced as part of the organizational identity.

Transformational leadership is particularly well-suited to dynamic and complex environments such as construction sites, where adaptability, collaboration, and motivation are essential for success. In such settings, the ability to articulate a compelling vision and align team efforts with broader sustainability objectives becomes a critical asset. Site managers who adopt this leadership style can promote continuous improvement by encouraging innovation, supporting professional development, and reinforcing values that prioritize sustainability. Through mechanisms such as intellectual stimulation and individualized consideration, transformational leaders empower employees to think critically, challenge conventional practices, and propose creative solutions that enhance both environmental performance and operational efficiency.

Moreover, the integration of transformational leadership into construction site management can help bridge the gap between policy and practice. While sustainability goals are often defined at the organizational or contractual level, their successful implementation depends heavily on the motivation and engagement of frontline workers. Transformational leaders play a key role in translating abstract sustainability targets into tangible actions by fostering a sense of purpose and aligning individual motivations with collective goals. This approach not only improves project outcomes but also contributes to the development of a resilient and forward-thinking workforce capable of navigating the evolving demands of sustainable construction.

To assess the effectiveness of environmental measures, the study recommends the use of measurable indicators such as CO_2 emissions per ton of concrete used, percentage of materials recycled on-site, and energy consumption per square meter of built area. These metrics can serve as benchmarks for evaluating progress and identifying areas for improvement. Integrating such indicators into project reporting would enhance transparency and accountability in sustainable site management.

5.2 Aligning Economic Incentives with Sustainability

The findings of this study, supported by previous research [10], underscore the critical role that economic incentives play in shaping behavior within the construction industry. Financial considerations remain a dominant factor in decision-making, and sustainability initiatives are often adopted only when they align with cost-efficiency or profitability. This highlights the need for a strategic alignment between economic drivers and sustainability goals to ensure that environmental and social responsibility are not sidelined in favor of short-term financial gains.

To effectively promote sustainable practices, it is essential that contract structures, bonus schemes, and public procurement processes are designed to reward sustainability. This includes integrating clear and measurable sustainability criteria into tender evaluations, offering financial bonuses for achieving environmental and social targets, and implementing penalties for unsustainable practices. By embedding sustainability into the economic framework of projects, organizations can create a system where responsible behavior is not only encouraged but also financially advantageous.

Without appropriate economic incentives, sustainability efforts risk being deprioritized or treated as secondary objectives. In such cases, even well-intentioned initiatives may fail to gain traction if they are perceived as incompatible with budgetary constraints or project timelines. Therefore, aligning profitability with sustainability is not merely a strategic choice—it is a prerequisite for achieving long-term success and driving industry-wide transformation. When sustainability becomes a source of competitive advantage and economic value, it is more likely to be embraced at all levels of the organization, from site managers to executive leadership.

5.3 The Need for Capacity Building

Achieving sustainability in the construction industry requires a combination of technical expertise and managerial competence. The findings of this study highlight that while sustainability is increasingly emphasized at the strategic level, there remains a significant gap in practical knowledge among site personnel, particularly within smaller firms. This gap poses a challenge to the effective implementation of sustainable practices, as workers may lack the necessary understanding, tools, or confidence to integrate sustainability into their daily routines. Addressing this issue calls for a systematic investment in training, mentorship, and knowledge-sharing platforms that can support the development of sustainability-related competencies across all levels of the workforce.

Structured training programs should be designed not only to convey technical knowledge—such as resource efficiency, environmental compliance, and waste reduction—but also to foster a deeper understanding of the broader goals of sustainability. In addition to formal instruction, cross-project learning platforms and peer-to-peer mentorship can facilitate the exchange of best practices and lessons learned, helping to build a culture of continuous improvement. Smaller companies, in particular, may benefit from collaborative initiatives or industry-wide support mechanisms that provide access to shared resources and expertise.

Equally important is the development of leadership competence with a strong focus on sustainability. Leadership training should incorporate modules that emphasize systems thinking, stakeholder engagement, and ethical decision-making, equipping managers with the skills needed to navigate the complex social and environmental challenges of modern construction projects. By embedding sustainability into leadership development, organizations can ensure that decision-makers are not only aware of sustainability goals but are also capable of translating them into actionable strategies and motivating their teams to follow suit.

Ultimately, building competence in sustainability must be a comprehensive and ongoing effort, integrated into both professional development and organizational culture. When employees at all levels are equipped with the knowledge, skills, and mindset to engage with sustainability meaningfully, it becomes possible to embed sustainable practices into the fabric of everyday operations, leading to more resilient, responsible, and future-oriented construction practices.

6 Conclusion

Site managers occupy a central and strategic position in the transition toward more sustainable construction and infrastructure development. Their role extends beyond operational oversight to include the integration of sustainability principles into the core of project execution. By combining practical measures across environmental, economic, and social dimensions with the principles of transformational leadership, site managers can significantly contribute to the realization of the United Nations Sustainable Development Goals (SDGs). Their ability to influence daily practices, guide teams, and make informed decisions places them at the forefront of sustainability implementation on construction sites.

To fully harness this potential, the construction industry must take deliberate steps to support and empower site managers in their sustainability efforts. This includes strengthening sustainability competence among both leaders and workers through targeted training, continuous education, and access to relevant resources. Furthermore, economic incentives must be aligned with sustainable behavior, ensuring that environmentally and socially responsible actions are rewarded through contract structures, performance bonuses, and procurement policies. In parallel, the industry must work to foster a culture of sustainability, one that is rooted in inclusive, visionary, and values-driven leadership. Such a culture encourages collaboration, innovation, and long-term thinking, all of which are essential for sustainable development.

Importantly, sustainability in construction is not merely a technical challenge—it is fundamentally a leadership challenge. The future of sustainable construction depends on the ability of site managers to lead with purpose, inspire change, and operationalize sustainability throughout every phase of the project lifecycle. This requires not only technical knowledge but also the capacity to engage stakeholders, navigate complex systems, and make ethical decisions under pressure. When site managers are equipped with the right tools, mindset, and support, they can transform sustainability from a set of abstract goals into a tangible, actionable reality on the ground.

References

1. FN-sambandet. FNs bærekraftsmål. https://fn.no/om-fn/fns-baerekraftsmaal. Last accessed 2025/08/31
2. Halvorsen K (2002) Forskningsmetode for helse- og sosialfag, 2nd edn. Cappelen Damm, Oslo
3. Ryen A (2002) Det kvalitative intervjuet, 1st edn. Fagbokforlaget, Bergen
4. Grønmo S (2016) Samfunnsvitenskapelige metoder, 2nd edn. Fagbokforlaget, Bergen
5. Wadel C (1991) Hva er arbeid? 1st edn. SEEK, Flekkefjord
6. Elkington J (1997) Cannibals with forks: the triple bottom line of 21st century business, 1st edn. Capstone Publishing Limited, Oxford
7. Hetland H (2008) Transformasjonsledelse. Universitetet i Bergen, Bergen
8. Burns JM (1978) Leadership. Harper & Row, New York
9. Yukl G (2019) Leadership in organizations, 9th edn. Pearson Education Limited
10. Hardoon D (2017) An economy for the 99%. Oxfam International

Building a Net-Zero Mobility Ecosystem: Insights from Mobilitetslab Stor-Trondheim

Mahgol Afshari[1,2](✉), Jardar Lohne[2], and Agnar Johansen[2]

[1] Norwegian University of Science and Technology (NTNU), Trøndelag County Municipality, Trondheim, 7491 Norway, Norway
mahgol.afshari@ntnu.no

[2] Norwegian University of Science and Technology (NTNU), 7491 Trondheim, Norway

Abstract. As cities strive for net-zero emissions, integrated and data-driven mobility solutions are critical to achieving sustainable urban transportation. Mobilitetslab Stor-Trondheim (MoST) is an initiative aimed at testing and implementing innovative mobility solutions to reduce emissions, enhance efficiency, and promote sustainable transport behaviors. This paper presents key lessons from the project, offering insights for policymakers, urban planners, and mobility stakeholders working toward a net-zero future. A key takeaway is the importance of cross-sector collaboration. Engaging public and private actors, research institutions, and technology providers has enabled a holistic approach to mobility innovation. Additionally, real-time data and digital tools have proven essential in transport systems' optimization and encouraging behavioral change among citizens. Another lesson is the role of policy frameworks and incentives in scaling sustainable mobility solutions. Regulatory agility, coupled with financial incentives, has been found to influence the adoption of green transport options. Furthermore, the need for user-centered design—ensuring that new mobility solutions align with the diverse needs of end-users to drive long-term adoption—is underlined. This paper outlines concrete examples from the project, including pilot implementations, stakeholder engagement strategies, and data-driven decision-making approaches. By reflecting on both successes and challenges, it contributes to the broader discussion on how cities can transition to net-zero mobility effectively. The findings from Mobilitetslab Stor-Trondheim underscore that achieving net-zero urban transport requires an integrated, adaptable, and collaborative approach. These insights can serve as guidance for future mobility initiatives seeking to accelerate the transition to sustainable and emission-free cities.

Keywords: Sustainable Mobility · Net-Zero Transport · Urban Innovation

1 Introduction

Urban transportation contributes significantly to global greenhouse gas (GHG) emissions, accounting for around 11.9% of total emissions [1]. Addressing this difficulty is critical for cities seeking to reach net-zero emissions. The European Union's Sustainable and Smart Mobility Strategy stresses the use of innovative, data-driven mobility solutions to modernize urban transportation networks [2].

H. Beushausen et al. (Eds.): NTZR 2025, LNCE 873, pp. 1184–1193, 2026.
https://doi.org/10.1007/978-3-032-27698-8_104

In response to these issues, Mobilitetslab Stor-Trondheim was developed as a collaborative program to test and deploy cutting-edge mobility solutions. The program brings together public sector organizations and academic institutions to co-develop and evaluate ways for decreasing emissions, improving transportation efficiency, and promoting sustainable travel practices. This multidisciplinary approach reflects the EU's emphasis on cross-sector collaboration to promote innovation in urban transportation [2].

However, experience/previous research shows that actually achieving such a holistic approach based on cross-sector collaboration is challenging. There is a significant gap between intentions and results that provide different stakeholders with the intended effect. Providing practical examples of initiatives that have succeeded in doing this is therefore in dire need [3].

This study intends to add to the larger discussion of effective solutions for cities transitioning to net-zero urban transport by commenting on both accomplishments and problems. The findings of Mobilitetslab Stor-Trondheim provide information for policymakers, urban planners, and mobility stakeholders who are devoted to creating sustainable and emission-free urban settings. The paper presents key lessons learned from" Mobilitetslab Stor-Trondheim", focusing on three research questions.

1. To what extent has the effort of engaging diverse stakeholders succeeded in driving holistic mobility innovations?
2. What challenges to cross-sector collaboration have been the most important?
3. What measures can be envisaged to ensure future mobility solutions that align with the diverse needs of end users?

About the project

Mobilitetslab Stor-Trondheim (MoST) is an innovative research and development program of projects focusing on sustainable mobility solutions for urban areas. The project is a collaboration between the Environmental Package, Green2050, Trøndelag County Municipality and the Norwegian University of Science and Technology (NTNU). The goal of the MoST lab is to be a national force in research and development for future oriented sustainable city mobility solutions.

There are three focus areas in this urban lab/project:

1. Actors' and people's needs for mobility: Investigate and understand how different actors and residents in Trondheim experience and use mobility services.
2. Sustainable mobility as an integrated system: Develop models and planning tools that integrate different modes of transport to create a holistic and sustainable transport system.
3. Digital technologies and mobility: Implement and test digital solutions, such as digital twins and advanced modeling tools, to improve decision-making processes and streamline mobility services.

These three focus areas, in combination with practical projects on sustainable city mobility projects are the core business of the MoST Lab. In the first 2 years the lab has worked on projects related to public transport, bicycle friendly streets, safer and better alternatives for walking and cycling, transportation and travel behavior, transport models and use of Apps that motivate more sustainable travel behavior, development of dynamic Digital twins of transport systems and big data related to travel behavior.

2 Theoretical Background – a Holistic Approach to Mobility Labs

Living labs are open, user-centered ecosystems where stakeholders co-develop, test, and validate innovations in real-life settings [4]. In the mobility context, living labs serve as platforms for iterative learning and stakeholder engagement, fostering innovations that are context-sensitive and scalable [5]. By situating pilots in everyday urban environments, living labs ensure that developed solutions respond to actual user behaviors, constraints, and preferences [6].

Mobility labs are conceptually and methodologically aligned with the living lab approach. A mobility lab is a way to design, test, and grow sustainable transport solutions by looking at the whole system, not just individual ideas. It brings different people and organizations together, tests ideas in real-life settings, and focuses on what users actually need and how they behave. This approach takes into account the technical, social, institutional, and behavioral aspects of making transport more sustainable. A holistic mobility lab is based on systems thinking, which sees urban transport as a complex system where things like vehicles, roads, people, rules, and technology all interact and affect each other [7]. This way of thinking helps decision-makers avoid working in separate areas and instead use connected solutions to tackle related problems like pollution, traffic jams, and fairness in transport [8].

A holistic mobility lab also requires collaborative governance structures that involve municipalities, regional authorities, academia, the private sector, and civil society. This participatory approach is essential to foster institutional learning, align policies, and build legitimacy for interventions [9]. Mobility labs support integrated policymaking by facilitating cross-sector dialogue and adaptive planning, which are critical for embedding innovations within existing regulatory frameworks [10].

Finally, user engagement is a cornerstone of holistic mobility labs. Applying human-centered and behavioral design principles ensures that solutions are not only technically feasible but also behaviorally acceptable and socially inclusive [11]. This is especially important in sustainable mobility, where changes in user habits—such as modal shifts or adoption of digital platforms—are central to success [12].

3 Research Design

This study adopts an analytical research design, which focuses on identifying and examining the key factors and variables that describe and explain the phenomena under investigation. The Mobilitetslab Stor-Trondheim (MoST lab) encompasses a broad range of initiatives and components that collectively aim to contribute to a more sustainable urban and regional transport system.

In this paper, we address this objective by critically analyzing the various elements of the MoST lab. Our methodology involves a systematic review of feedback collected throughout the implementation of multiple projects and events within the lab. This feedback was gathered both during and after project activities, through semi-structured discussions with stakeholders, as well as post-event evaluation meetings. In addition, we have studied relevant reports and documentation produced by community actors and stakeholders who have participated in the lab over a two-year period.

All authors of this paper have been directly involved in the lab's operations, with responsibilities spanning planning, project management, and evaluation. This insider perspective provides a unique foundation for the analysis presented here, allowing for a reflective and practice-oriented assessment of the lab processes and outcomes.

4 MoST Living Lab Concept

The Mobilitetslab Stor-Trondheim (MoST) living lab is designed as a collaborative ecosystem that integrates research, practice, and stakeholder engagement to address challenges related to sustainable urban mobility. The concept comprises three main elements:

1. **Collaborative Mobility Projects**: These are city-led sustainable mobility initiatives co-developed and implemented by key public stakeholders—including the County Municipality, the City of Trondheim, surrounding municipalities, the Norwegian Public Roads Administration, the Norwegian Railway Directorate, and the County Governor of Trøndelag—in close collaboration with MoST researchers, PhD candidates, and master's students.
2. **Academic Research Projects**: PhD and master's projects form a core component of the lab. These projects aim to develop new methods, tools, and theoretical frameworks related to sustainable mobility. To ensure real-world relevance and applicability, each PhD project is co-sponsored by a local stakeholder. Furthermore, all doctoral candidates are required to carry out parts of their research within one of the stakeholder organizations, working alongside domain experts as visiting researchers.
3. **Dissemination and Stakeholder Engagement Activities**: The MoST lab regularly organizes knowledge-sharing events such as workshops, seminars, youth science schools, street festivals, and conferences. These activities bring together stakeholders from across the region and serve as platforms for dialogue, dissemination of results, and fostering innovation through cross-sectoral collaboration.

A key feature of the MoST living lab model is its emphasis on applied real-time research. All research activities are embedded within ongoing community-based projects, allowing researchers to work directly with challenges as they unfold. This dynamic and integrated approach is illustrated in Fig. 1, which maps the connection between MoST research projects and stakeholder-led initiatives.

The "live" projects within the MoST living lab serve as the practical foundation for research activities carried out by PhD candidates, master's students, and their academic supervisors. These projects are typically in early development phases and involve the exploration and testing of innovative concepts—such as new travel modes, sidewalk design approaches, sensor and camera integration, and the application of digital tools to mobility data. The focus is on experimentation with emerging, cutting-edge technologies rather than providing consultancy services.

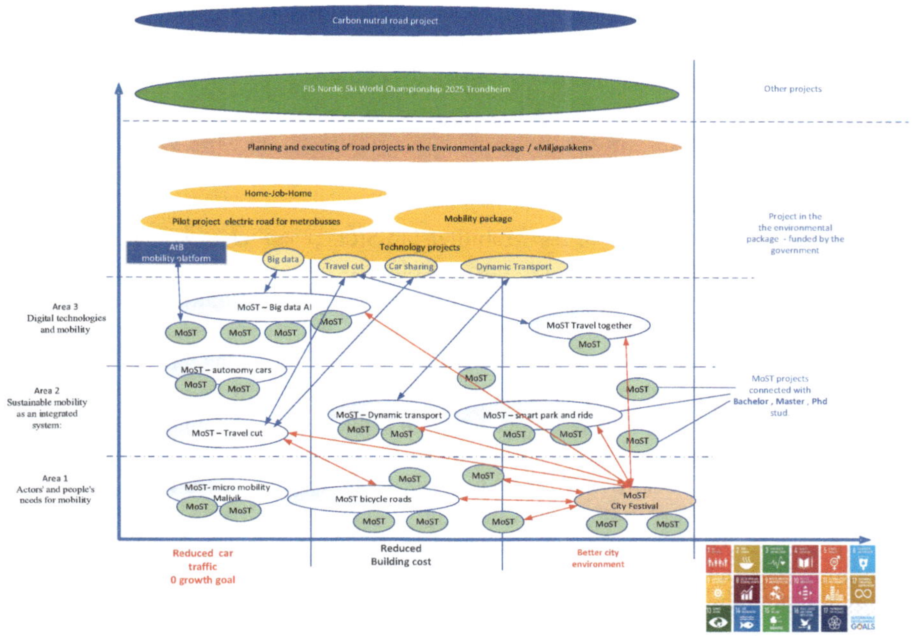

Fig. 1. The connection between MoST research projects and stakeholder-led initiatives.

The third core element of the MoST lab is the dissemination of research findings and the integration of community feedback. This is achieved through regular workshops, seminars, and participation in both national and international conferences. These events provide a platform for sharing preliminary findings, receiving stakeholder and public input, and fostering dialogue between researchers and practitioners.

To date, the MoST lab hosts 13 PhD projects, which are organized into three thematic groups:

- **Group 1: Actor Needs, Motivation, and Travel Behavior** (3 PhDs) – Focuses on understanding the behavioral and motivational drivers behind mobility choices and how stakeholders engage in transport development.
- **Group 2: Sustainable Mobility as an Integrated System** (4 PhDs) – Aims to develop planning models and tools that integrate multiple transport modes into cohesive, sustainable systems.
- **Group 3: Digital Technologies and Mobility** (6 PhDs) – Explores the implementation of digital solutions such as digital twins and advanced modeling techniques to enhance decision-making and improve mobility services.

A particularly valuable outreach arena has been the annual local street festival, organized by the community in collaboration with the MoST lab. This event has served as a crucial connection between researchers and residents. It has enabled PhD students—many of whom come from international backgrounds—to present and demonstrate their ideas, collect data on travel behavior, and engage in informal conversations with community members. The festival also functions as a networking platform for researchers and stakeholders involved in live projects, further strengthening the integration of academic work and real-world practice.

5 Key Lessons Learned

To evaluate the effectiveness of the Mobilitetslab Stor-Trondheim, we employed a formative and reflective evaluation approach, grounded in real-time observations, structured stakeholder feedback, and post-project reflections. The evaluation focused on what worked, what didn't work, and why. Key successes included cross-sector collaboration, user involvement, and real-life testing, which led to more grounded and scalable mobility solutions. These outcomes were validated through repeated themes in stakeholder interviews, workshop feedback, and user response patterns. Challenges—such as coordination issues, inconsistent priorities among partners, and difficulties engaging hard-to-reach groups—were identified through process documentation, informal debriefs, and observations during implementation. Learning-by-doing in real-world contexts was central to the lab, and iterative adjustments based on ongoing feedback played a crucial role in refining both the solutions and the lab's structure.

Based on the evaluation approach described above, the key lessons learned from the Mobilitetslab Stor-Trondheim initiative are as follows:

5.1 A New Way to Think About Urban Mobility

A mobility lab is a way to design, test, and grow sustainable transport solutions by looking at the whole system. It brings together people and organizations, tests ideas in real life, and focuses on what users actually need and how they behave. This approach looks at the technical, social, institutional, and behavioral aspects of transportation.

5.2 Understanding Transport as a System

A holistic mobility lab is based on systems thinking. This means seeing transport in cities as a mix of many parts—like vehicles, roads, people, rules, and technology—all working together and affecting each other.

5.3 Solving Connected Problems Together

This way of thinking helps leaders avoid working in isolation. Instead, they can create joined-up solutions to deal with big problems like pollution, traffic, and fairness in transport.

5.4 Putting People First

In mobility labs, it's very important to involve the people who actually use transport. Using human-centered and behavioral design helps make sure that the solutions work well, are easy to use, and fair for everyone.

5.5 Teamwork Across Sectors

Working together across different sectors—like cities, transport agencies, researchers, and private companies—leads to better and more scalable solutions. In the Mobilitetslab Stor-Trondheim, this teamwork made it easier to fit new ideas into existing systems. Still, there were some challenges, like coordination problems and different priorities, which showed the need for better ways to manage collaboration.

5.6 Speaking the Same Language

At first, the different groups involved had their own ways of working and speaking. Workshops and open discussions helped everyone understand each other better and build trust, which made teamwork and decision-making easier.

5.7 Designing for Real Needs

By listening to people—like commuters, students, seniors, and underrepresented groups—the project created services and tools that truly fit their needs. This led to happier users and helped encourage long-term shifts toward more sustainable travel choices.

5.8 Learning by Testing in Real Life

Testing ideas in real-life settings helped the team learn what worked and what didn't. It also uncovered problems that wouldn't have been seen in the lab. But working in real-world conditions brought unexpected challenges, so the team had to stay flexible and adjust plans when needed.

5.9 Building Skills Along the Way

Even beyond the new transport tools, the lab helped people in public organizations learn useful skills—like engaging with users, managing data, and supporting innovation. These skills can help in many other projects, too.

5.9.1 Using Data Wisely and Carefully

Mobility data helped the project understand how people travel and where emissions come from. This led to smarter solutions. But it also raised questions about who owns the data, who can access it, and how to protect users' privacy. Future projects need clear rules and ethical standards for data use.

5.9.2 Making Sure no One is Left Out

The project aimed to serve all kinds of users, but some—like people with disabilities, or no access to digital tools—were harder to reach. Future projects should have clear strategies to include everyone fairly and avoid making existing inequalities worse.

6 Discussion and Conclusion

This paper sets out to explore the extent to which a living lab model—specifically, the Mobilitetslab Stor-Trondheim (MoST)—can support cities in developing and testing holistic, sustainable mobility solutions. Through the lens of three research questions, the discussion below evaluates the lab's achievements and challenges in promoting cross-sector collaboration, fostering innovation, and aligning future mobility solutions with end-user needs.

6.1 Engaging Diverse Stakeholders in Holistic Mobility Innovation

The MoST lab has made strong progress in creating a collaborative setup for working on sustainable urban mobility. It has shown that it's possible to bring together many different partners—such as municipalities, transport agencies, infrastructure authorities, and academics—to test and develop new ideas together. The lab has also started 13 PhD projects, grouped into three main topics: people's travel needs and behavior, how different transport modes can work together, and the use of digital technology in mobility. These projects are closely connected to the core interests of the partners, enhancing the potential for obtaining the intended effect of the different partners.

The lab's success in co-locating researchers within stakeholder organizations and requiring PhD students to contribute to ongoing "live" projects has fostered meaningful knowledge exchange. Additionally, dissemination activities such as workshops, seminars, and public events (e.g., the annual street festival) have served as valuable platforms for cross-sector dialogue and feedback loops. These mechanisms have strengthened the legitimacy and uptake of new ideas, contributing to a more holistic approach to mobility innovation in the Trondheim region.

Even though many stakeholders have been involved successfully, their level of participation has not been the same. Some organizations have been more active than others. It's also been difficult to keep everyone engaged over time, especially when political situations or internal priorities change. Working together smoothly as institutions is still a work in progress—especially when trying to expand pilot projects into larger, long-term solutions. The management's effort in maintaining continuous interaction should not be underestimated.

6.2 Challenges to Cross-Sector Collaboration

Some structural and cultural differences have made it difficult to fully create a well-integrated cross-sector system for mobility innovation. One main issue is the difference in temporalities. The need for careful testing and time-consuming publication processes typically renders research slow, whilst public organizations often work with shorter deadlines due to political decisions or the need to deliver quick results.

Second, differences in language, expectations, and evaluation criteria have led to occasional misalignments in project goals. For example, academic projects seeking to develop new theoretical frameworks may not always align with stakeholder needs for immediate, practical solutions. Bridging these gaps requires active facilitation and ongoing dialogue.

Third, the lack of shared digital systems and the difficulty in sharing data between different organizations is a major challenge. Even though MoST has worked with tools like digital twins and advanced models, the way data is owned and managed by different stakeholders makes it hard to fully connect everything into one system.

Despite these challenges, the living lab structure has provided a valuable institutional space to negotiate these tensions, enabling experimentation in a lower-risk environment and facilitating mutual learning.

6.3 Future Mobility Solutions and End-User Needs

The MoST lab highlights the need to include users' views in the development process. Many of the live projects, e.g. on improving walking and biking infrastructure, travel apps, and transport modeling, have emphasized testing ideas in real life and obtaining feedback from users. This approach has been key in figuring out both what works technically, and what people are willing to use.

Moving forward, strengthening behavioral research capacities within the lab will be key. Engaging residents through participatory design, community events, and co-creation workshops should continue and expand. These forums not only help to validate new mobility solutions but also serve to increase public ownership and adoption.

Additionally, there is a need for formal ways to include user feedback in planning and policy decisions. This could involve improving the use of citizen panels, real-time feedback systems, and evaluation methods that focus not just on performance, but also on fairness, accessibility, and environmental impact.

6.4 Conclusion

The experience of the Mobilitetslab Stor-Trondheim demonstrates that living labs can function as powerful instruments for urban mobility transformation—bringing together stakeholders, aligning research with practice, and embedding innovation in everyday urban contexts. While challenges remain in terms of coordination, data sharing, and aligning academic and administrative logics, the MoST lab provides a strong model for how cities can operationalize the EU's sustainable mobility goals through collaborative, real-world experimentation.

As cities around the world strive to meet net-zero targets, the insights from MoST offer practical pathways for achieving more inclusive, integrated, and sustainable transport systems. Continued support for interdisciplinary research, stakeholder engagement, and user-centered design will be essential in scaling these efforts and ensuring long-term impact.

Acknowledgements. This study is supported by the Department of Civil and Environmental Engineering, Faculty of Engineering, Norwegian University of Science and Technology (NTNU), and the MoST living lab.

References

1. Mun Ng K, Wah Yuen C, Chuen Onn C, Ibtishamiah Ibrahim N (2024) Urban mobility mode shift to active transport: sociodemographic dependency and potential greenhouse gas emission reduction, vol 14, issue 1. Sage Open. https://doi.org/10.1177/21582440241228644
2. Šoštarić M, Vidović K, Jakovljević M, Lale O (2021) Data-Driven methodology for sustainable urban mobility assessment and improvement. Sustainability 13:7162. https://doi.org/10.3390/su13137162
3. Bryson JM, Crosby BC, Stone MM (2015) Designing and implementing cross-sector collaborations: needed and challenging. Public Adm Rev 75(5):647–663
4. Leminen S, Westerlund M, Nyström A-G (2012) Living Labs as open-innovation networks. Technol Innov Manag Rev 2(9):6–11
5. Schuurman D, De Marez L, Ballon P (2016) The impact of Living Lab methodology on open innovation contributions and outcomes. Technol Innov Manag Rev 6(1):7–16
6. Hossain M, Leminen S, Westerlund M (2019) A systematic review of living lab literature. J Clean Prod 213:976–988
7. Banister D (2008) The sustainable mobility paradigm. Transp Policy 15(2):73–80
8. Bertolini L (2012) Integrating mobility and urban development agendas: a manifesto. disP – Plann Rev 48(1):16–26
9. Bulkeley H, Castán Broto V (2013) Government by experiment? Global cities and the governing of climate change. Trans Inst Br Geogr 38(3):361–375
10. van den Bergh JCJM, Scholz RW (2011) Environmental sustainability and research on co-evolutionary and multi-level dynamics. Ecol Econ 70(11):2097–2100
11. Norman DA, Draper SW (1986) User centered system design: new perspectives on human-computer interaction. Lawrence Erlbaum
12. Schwanen T, Banister D, Anable J (2011) Scientific research about climate change mitigation in transport: a critical review. Transp Res Part A: Policy Pract 45(10):993–1006

Integrating Sustainable Education into Bachelor Programs in Technical Building Installations: Experiences from Norway

Emma Zheng Liang(✉) and Arne Ronny Sannerud

Høyskolen for yrkesfag, Gjøvik, Norway
{emma.liang,arne.ronny.sannerud}@hfy.no

Abstract. This study conducts a quantitative analysis of Norway's first bachelor-program in the management of technical and sustainable building installations from the University College of Vocational Education (HØFY), aligning with the country's sustainability goals. The research examines sustainability integration from both bottom-up and top-down perspectives, capturing students' subjective and objective experiences while also assessing the role of educational frameworks and sustainability-focused course content. A key finding highlights the crucial role of interdisciplinary approaches and active student engagement in strengthening sustainability education. Based on qualitative questionnaire data, the study identifies challenges and presents practical recommendations for improving the integration of sustainability in technical education. These insights contribute to enhancing educational strategies that support Norway's transition to a sustainable built environment.

Keywords: Integrating Sustainable Education · energy efficiency · material reuse · technical Building Installations

1 Introduction

Global climate challenges are increasing. As highlighted by the Intergovernmental Panel on Climate Change (IPCC) in 2018, the impacts and risks associated with global warming are significantly greater at 2 °C compared to 1.5 °C. This finding underscores the urgent need for rapid reductions in CO_2 emissions by 2030 [1]. According to Gholipour et al. (2022) [2], the building sector is often referred to as the '40% sector' men in fact responsible for approximately 38% of global CO_2 emissions and 35% of total energy consumption. Furthermore, data from the United Nations Environment Programme (2020) indicate that buildings alone account for about 55% of global electricity use [3].

Achieving these goals requires not only technological innovation, energy efficiency, and material reuse, but also a highly skilled workforce equipped with sustainability knowledge, systems thinking, and interdisciplinary competence. In this context, education plays a crucial role in driving long-term change [4].

Norway has committed to ambitious national and international climate targets, including the European Green Deal and the Paris Agreement. These commitments are

H. Beushausen et al. (Eds.): NTZR 2025, LNCE 873, pp. 1194–1202, 2026.
https://doi.org/10.1007/978-3-032-27698-8_105

reflected in national policies such as Norway's Climate Action Plan for 2021–2030, which outlines strategies to reduce greenhouse gas emissions and transition to a low-emission society [5]. As a signatory to the Paris Agreement, Norway aims to reduce emissions by at least 55% by 2030 compared to 1990 levels. These policy goals are aligned with the European Green Deal, which seeks to make the EU climate-neutral by 2050 [6].

In response, the Norwegian education system is increasingly expected to support sustainability competence across sectors, especially in technical disciplines such as building services engineering. Traditional bachelor programs in technical building installations have largely emphasized operational efficiency and system functionality, often with limited attention to broader sustainability issues such as energy efficiency, material selection, and environmental impacts [7]. A growing body of research emphasizes the need for integrating sustainability-related competencies into engineering and technical curricula to better prepare students for the green transition [4].

This paper presents a case study of Norway's first bachelor program [8] explicitly designed to combine technical building management with sustainable development principles. By conducting both quantitative and qualitative analyses, the study explores how sustainability is integrated from bottom-up (student experience) and top-down (curriculum design and institutional strategy) perspectives. The findings highlight the importance of interdisciplinary approaches, active student engagement, and targeted sustainability content, including energy efficiency, material reuse, and environmental certification frameworks. The study offers practical recommendations for improving educational strategies and contributes to the ongoing development of sustainability-focused technical education in Norway's building sector.

The structure of the paper is as follows: Sect. 2 provides background information on the context of integrating sustainable education in Norway. Sect. 3 presents the selected case studies, highlighting success criteria which essential for achieving the goal. Finally, Sect. 4 discusses key findings, identifies practical barriers encountered in the field, and concludes with implications for future research and practice.

2 Sustainable Education in Norway

Norway has firmly positioned sustainability in its national education agenda. In line with the United Nations' Sustainable Development Goal 4.7, which emphasizes education for sustainable development, Norway has integrated environmental awareness, global citizenship, and systems thinking into its educational frameworks. National policies, such as the Norwegian Education Act and the Core Curriculum, emphasize that students should "learn to take care of themselves, each other, and nature" [9]. These foundations reflect a long standing political and societal commitment to linking education with ecological responsibility and ethical development.

At the higher education level, universities and colleges have actively embraced sustainability through curriculum design and research initiatives. Institutions such as the Norwegian University of Life Sciences (NMBU) and the University of Oslo offer interdisciplinary programs focused on environmental science, development studies, and sustainable systems [10]. Furthermore, the integration of sustainability competencies into engineering, teacher training, and business programs demonstrates a structural shift from isolated environmental modules to a more holistic pedagogical approach [4]. This trend aligns with the increasing global consensus that sustainability education must go beyond awareness to foster problem-solving, critical thinking, and transformative learning.

The country is an active participant in the international Eco-Schools program, which promotes environmental management and sustainability education in primary and secondary schools through student-led action plans [11]. Schools participating in this program follow a structured methodology that engages students in practical efforts such as energy reduction, biodiversity conservation, and sustainable transport. The initiative also aligns with the broader national goal of fostering democratic participation and ecological literacy from an early age.

Despite these positive developments, challenges remain in achieving comprehensive and systemic implementation of sustainability in education. Research shows that sustainability content is often unevenly distributed across subjects and levels of education, with some fields such as geography, science, and social studies integrating the topic more effectively than others [12]. Additionally, teachers may lack sufficient training or resources to incorporate sustainability pedagogy in a meaningful way, highlighting the need for further investment in professional development and interdisciplinary collaboration.

In short, Norway has made significant progress in embedding sustainability into its education system through policy support, institutional reform, and participatory programs. However, ensuring that sustainability is fully and equitably integrated across all levels and disciplines remains a work in progress. Future efforts should focus on scaling best practices, investing in teacher training, and reinforcing sustainability as a cross-cutting competency in all areas of learning. Such actions will be crucial to preparing future generations for the complex environmental, social, and economic challenges ahead.

3 The Bachelor of Management of Technical and Sustainable Building Installations

As the first Bachelor of Management of Technical and Sustainable Building Installations at University College of Vocational Education (HØFY), this initiative reflects Norway's commitment to integrating sustainability across all levels of education [13].

The bachelor's program was developed in response to a growing demand for competence in managing complex technical installations in modern buildings, with sustainability as a guiding principle. It targets professionals with backgrounds in fields such as HVAC, energy systems, electrical engineering, and plumbing, ensuring a high degree of relevance and industry applicability [14].

Sustainability is not merely a theoretical concept within the program, it is embedded in the structure, learning outcomes, and course content. Students engage with real-world problems related to energy efficiency, renewable energy systems, and environmental certification schemes such as BREEAM-NOR and ENOVA support structures [15]. This hands-on learning approach is crucial in developing practical skills aligned with national and international sustainability targets.

An essential feature of the program is its flexibility. Structured as a part-time, modular program over five years, it enables working professionals to pursue higher education without leaving their jobs. This model promotes lifelong learning and supports upskilling within the existing workforce, which is critical for sustainable transformation in the construction sector [15].

Students graduating from this program are qualified to take on roles such as ITB (Integrerte Tekniske Bygningsinstallasjoner) coordinators, a function that ensures the coordination and optimization of all technical systems in a building. ITB coordinators are key to achieving both energy performance targets and user comfort, making them essential to sustainable building operations [16].

The curriculum emphasizes interdisciplinary collaboration and systems thinking. Courses encourage students to analyze the interaction between technical installations and building performance, including how automation, sensor technology, and building management systems contribute to resource efficiency [17].

HØFY places a strong emphasis on aligning its educational offerings with both Norwegian building regulations and international frameworks for sustainability. This includes understanding of the EU taxonomy, circular economy principles, and carbon accounting in building operations and construction (embodied and operational emissions) [18].

Another strength of the program is its commitment to applied research and knowledge development. Students are encouraged to contribute to innovation projects or case-based studies linked to their workplaces. This bridges the gap between education and practice, making the learning context-specific and impactful [19].

Additionally, the program has an international dimension. HØFY actively collaborates with educational and research institutions abroad to strengthen academic mobility, knowledge exchange, and benchmarking of sustainability practices. This approach helps ensure that the curriculum remains globally relevant while addressing local needs [20].

In short, the Bachelor of Management of Technical and Sustainable Building Installations at HØFY exemplifies how sustainable education can be designed to meet both societal and industry needs. By combining vocational roots with academic rigor, and theory with practical application, the program prepares graduates to lead the technical transition in Norway's built environment toward a more sustainable future [21].

3.1 Quantitative Analysis: Student Questionnaire

To evaluate the student experience and perception of sustainability integration in the Bachelor of Management of technical and sustainable building installations at HØFY, a structured quantitative survey was conducted among enrolled students. The survey included both Likert-scale and multiple-choice items focused on sustainability content, interdisciplinary learning, and perceived preparedness for sustainability-oriented roles.

The results indicate a generally high level of student satisfaction regarding the inclusion of sustainability principles in the curriculum. A substantial 85% of respondents agreed or strongly agreed that the program "places high emphasis on sustainability as a core theme across courses" [22]. This is consistent with the institution's strategic goal to mainstream sustainability across its academic offerings [23].

Perceived Integration of Sustainability in Education

The first diagram shows students' responses to the question: "To what extent do you perceive that sustainability is integrated into the education?"

Results indicate that 66.7% of respondents believe sustainability is integrated to a high or very high degree, while 33.3% rated it as only 'to some degree'. No respondents chose 'low' or 'not at all', reflecting a generally strong integration (Fig. 1).

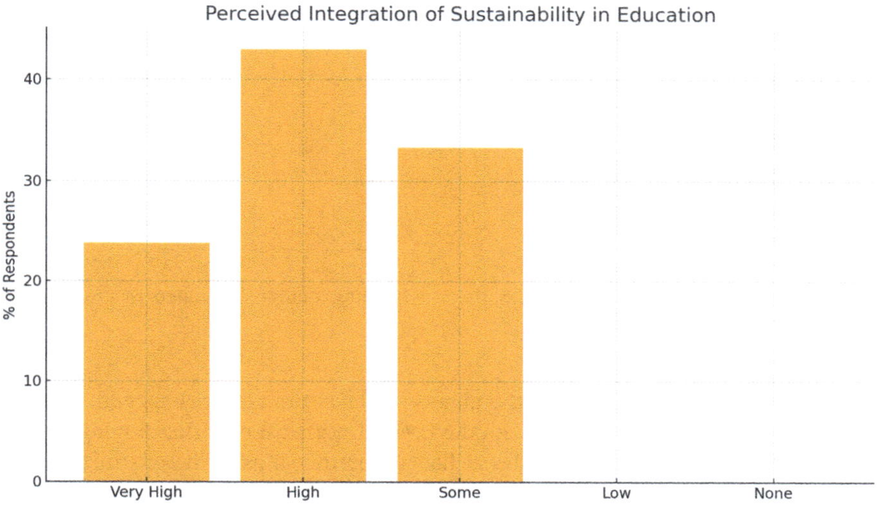

Fig. 1. Analysis from student questionnaire on perceived integration of sustainability

Furthermore, 78% of students responded positively to the item measuring their perceived ability to apply sustainability frameworks in real-world scenarios, such as BREEAM, energy certification schemes, and life-cycle assessments. This supports the hypothesis that the program's applied approach, linking theoretical modules to vocational contexts which contributes meaningfully to student competence development [15].

Learning Outcomes Related to Environmental Certifications

This diagram represents answers to: "To what extent do you experience that learning outcomes related to environmental certifications (e.g., BREEAM) are covered?"

While 52.4% believe this is addressed to a high or very high degree, 42.9% marked it as only 'to some degree', and 4.8% selected 'low degree'. This suggests that while coverage exists, more emphasis may be needed (Fig. 2).

Nevertheless, the findings also reveal important challenges. When asked whether the program offers adequate opportunities for interdisciplinary collaboration, only 60% responded affirmatively. A significant portion (40%) expressed a desire for increased integration with students or professionals from related fields, such as architecture, civil engineering, and environmental science [24]. This suggests that the current siloed structure, while technically rigorous, may limit students' exposure to the systemic thinking necessary for solving complex sustainability challenges.

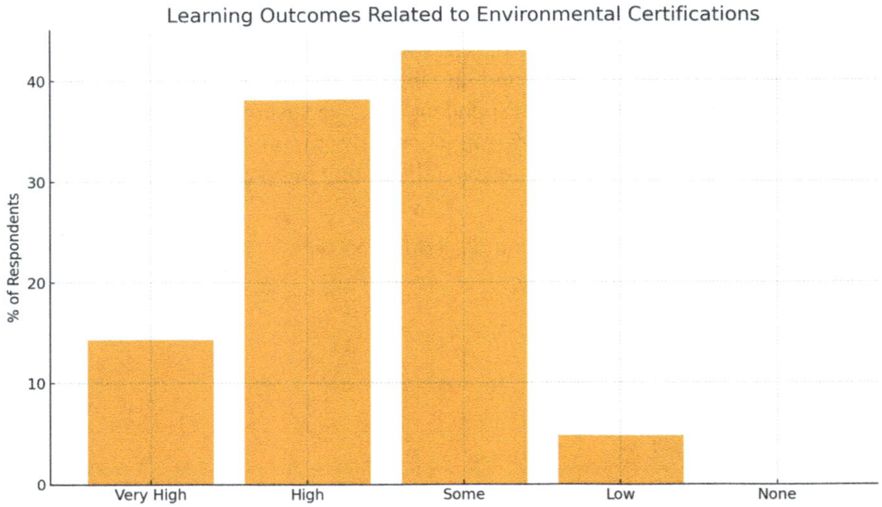

Fig. 2. Analysis from student questionnaires on learning outcomes related to environmental certifications

Another emergent theme from the analysis relates to industry engagement. While 65% of respondents were satisfied with the level of practical experience integrated into the program, 35% reported a need for enhanced partnerships with construction firms, energy consultants, or public-sector actors. Specifically, students emphasized the value of real-world projects and internships in reinforcing classroom learning and offering professional networking opportunities which further highlighting the importance of practical integration in the curriculum [25].

Finally, students were asked to self-assess their readiness to enter sustainability-oriented roles upon graduation. 82% reported feeling "ready" or "very ready," with most citing confidence in technical subjects like SD-anlegg, EOS-systemer, and energy-efficiency calculations. However, only 56% reported confidence in leading interdisciplinary teams, again underscoring the need for educational scaffolding in systems thinking and leadership skills [26].

Student Preparedness for Sustainable Development

The final diagram explores: "To what extent do you feel prepared to contribute to sustainable development in your field?"

A total of 76.1% reported feeling prepared to a high or very high degree, indicating strong confidence among students. However, 23.8% still reported only some or low levels of preparedness, highlighting an area for continued support (Fig. 3).

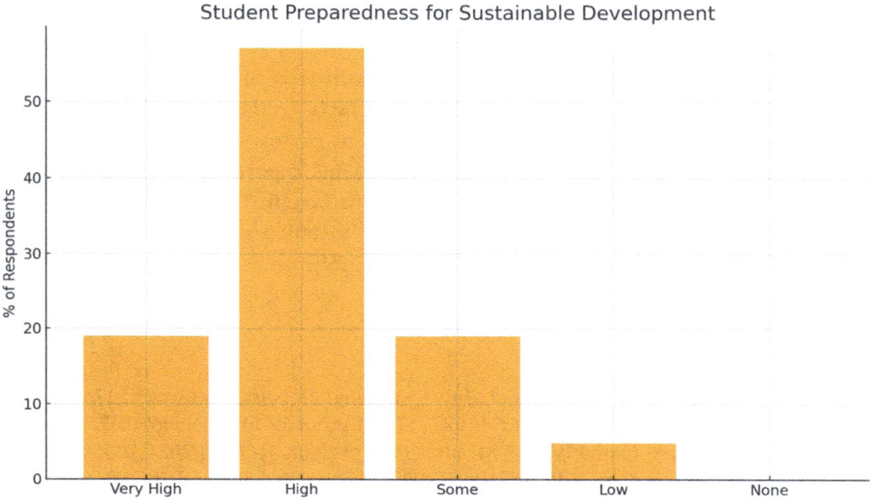

Fig. 3. Analysis from student questionnaires on preparedness for sustainable development

4 Conclusion

This study contributes to the growing body of research on sustainability-oriented higher education by offering an in-depth case analysis of Norway's first bachelor program explicitly designed to combine technical building management with sustainable development principles. Through both curriculum review and quantitative student data, the study provides a multi-perspective evaluation of how sustainability is operationalized in vocational-academic hybrid programs.

Findings reveal that the program at HØFY successfully integrates core elements of sustainability through applied learning, clear alignment with industry standards (e.g., BREEAM), and a pedagogical model tailored for professionals with technical backgrounds. Students generally perceive the curriculum as relevant and transformative in preparing them for future challenges in the building sector. Importantly, the majority express confidence in applying technical and environmental knowledge to practical contexts, a finding that supports the model of part-time, practice-integrated study for adult learners [27].

However, the results also highlight structural limitations, particularly in fostering interdisciplinary collaboration and industry engagement. These areas represent potential levers for enhancement. Introducing joint modules with other institutions, expanding access to internships, and embedding collaborative project-based learning could significantly strengthen the systemic understanding necessary for sustainable transitions.

The research highlights the importance of combining top-down curriculum strategies with bottom-up feedback from students to support ongoing, iterative improvements. However, it should be noted that the study is still in its early stages. The survey focuses on first-year students and their experiences of how the sustainability perspective becomes "visible" in teaching. This means that the results reflect the subjective perceptions of these students at a specific point in the academic year and can therefore be regarded as a case study. Such a limited scope may introduce bias in the survey population. A survey of students who have completed the bachelor's program and have been working for 1–3 years would, for example, be a good countermeasure to this limitation. Collaboration with a broader and more comprehensive survey in the future is expected to provide more accurate and meaningful insights. Moreover, the data also reveal gaps in interdisciplinary exposure and leadership skills; only 56% of students reported confidence in leading multidisciplinary teams. The paper could therefore benefit from a short reflection on how the program might respond to this gap, for instance through cross-institutional modules or capstone projects in future studies [28, 29].

References

1. IPCC special report on the impacts of global warming: https://www.ipcc.ch/sr15/
2. Gholipour HF, Arjomandi A, Yam S (2022) Green property finance and CO2 emissions in the building industry. Glob Financ J 51. https://doi.org/10.1016/j.gfj.2021.100696
3. United Nations Environment Programme (2020) 2020 global status REPORT for buildings and construction: towards a zero-emission, efficient and resilient buildings and construction sector. Tilgjengelig fra: https://globalabc.org/sites/default/files/inlinefiles/2020%20Buildings%20GSR_FULL%20REPORT.pdf (3 April 2022)
4. Lozano R, Merrill MY, Sammalisto K, Ceulemans K, Lozano FJ (2017) Connecting competences and pedagogical approaches for sustainable development in higher education: a literature review and framework proposal. Sustainability 9(10):1889. https://doi.org/10.3390/su9101889
5. Norwegian Ministry of Climate and Environment (2021) Norway's Climate Action Plan for 2021–2030. https://www.regjeringen.no/en/dokumenter/meld.-st.-13-20202021/id2827405
6. European Commission (2019) The European Green Deal. https://ec.europa.eu/info/strategy/priorities-2019-2024/european-green-deal_en
7. Norwegian Directorate for Higher Education and Skills (2022) Kompetanser for fremtiden: Bærekraft i høyere utdanning. https://hkdir.no
8. Bachelor i ledelse av tekniske og bærekraftige bygningsinstallasjoner: https://hfy.no/bachelor-i-ledelse-av-tekniske-og-baerekraftige-bygningsinstallasjoner
9. Norwegian Directorate for Education and Training (2020) Core curriculum – values and principles for primary and secondary education. https://www.udir.no/lk20/overordnet-del
10. Norwegian University of Life Sciences. Study programs in sustainability and the environment. https://www.nmbu.no/en
11. Foundation for Environmental Education. Eco-schools Norway. https://www.ecoschools.global/norway
12. Skarstein F, Wolff F (2020) Implementing education for sustainable development in Norwegian schools: a curriculum-based perspective. Nordic J Comparative Int Educ (NJCIE) 4(2):1–15. https://doi.org/10.7577/njcie.3568
13. Høyskolen for Yrkesfag (2024) Bachelor in Management of Technical and Sustainable Building Installations. Retrieved from https://hfy.no/international
14. Høyskolen for Yrkesfag (2024) For deg som vil lede fremtidens byggeprosjekter. Retrieved from https://hfy.no/for-deg-som-vil-lede-fremtidens-byggeprosjekter
15. Høyskolen for Yrkesfag (2024) Studieplan bachelor i ledelse av tekniske og bærekraftige bygningsinstallasjoner. Retrieved from https://s3-gustav.imgix.net/hfy/studieplan-bachelor-i-ledelse-av-tekniske-og-baerekraftige-byningsinstallasjoner-9211.pdf
16. Multiconsult (2022) Rollen som ITB-koordinator i komplekse byggeprosjekter. Retrieved from https://www.multiconsult.no/aktuelt/itb-koordinator/

17. Direktoratet for Byggkvalitet (DiBK) (2023) Veiledning om tekniske installasjoner og automatisering. Retrieved from https://dibk.no
18. European Commission (2021) EU taxonomy for sustainable activities. Retrieved from https://ec.europa.eu/sustainable-finance
19. Høyskolen for Yrkesfag (2024) Forskningsbasert undervisning ved HFY. Retrieved from https://hfy.no/international
20. NOKUT (2023) Internasjonalisering i norsk høyere utdanning. Retrieved from https://www.nokut.no
21. Høyskolen for Yrkesfag (2024) Programbeskrivelse – Ledelse av tekniske og bærekraftige installasjoner. Retrieved from https://hfy.no
22. Student Survey Data, Høyskolen for Yrkesfag (2025) Internal analysis of quantitative responses on program content and learning outcomes
23. Høyskolen for Yrkesfag. (2024). Institutional strategy: building competence for a sustainable future. Retrieved from https://hfy.no/international
24. NOKUT (2023) Tverrfaglighet og samarbeid i høyere utdanning. Retrieved from https://www.nokut.no
25. Utdanningsdirektoratet (2023) Samarbeid mellom skole og næringsliv i yrkesfaglige programmer. Retrieved from https://www.udir.no
26. European Commission (2021) Skills agenda for Europe: building the green skills pipeline. Retrieved from https://ec.europa.eu/green-skills
27. Biggs J, Tang C (2011) Teaching for quality learning at university. Open University Press
28. Trochim WMK (2006) Descriptive statistics. Descriptive statistics - research methods knowledge base. https://conjointly.com/kb/
29. https://journals.oslomet.no/index.php/yrke/article/view/5765

Overcoming Regulatory Barriers in Positive Energy Districts Development: Lessons from + CityxChange Project in Trondheim

Azam Mirjalali(✉) and Savis Gohari

Norwegian University of Science and Technology, 7491 Trondheim, Norway
{seyedehazam.mirjalalikomsari,savis.gohari}@ntnu.no

Abstract. Achieving climate neutrality is a critical global objective, essential for mitigating climate-related impacts that threaten ecosystems and human well-being. While global and regional frameworks guide climate goals and shape national strategies, local governments play a pivotal role in translating these goals into context-specific actions that address community needs. In Europe, the concept of Positive Energy Districts (PEDs) has been initiated as a means to decarbonize the economy. The commitment to achieving 100 PEDs across Europe underscores this effort. This paper examines the specific context of Trondheim, Norway, to understand the interplay between local strategies and national and European policies in the PED initiative. It explores how governance structures across different levels enable or hinder the realization and implementation of PEDs, focusing on the European Horizon 2020 project, +CityxChange. The paper specifically examines how current energy regulations in Norway facilitate the implementation of PEDs through a comprehensive literature and document review. It evaluates the integration of policy frameworks across governance levels and their impact on sustainable urban development and greenhouse gas reduction initiatives like PEDs. Findings reveal that while Norway has integrated international climate objectives into its national planning system, certain regulatory misalignments persist, potentially hindering the effective realization of PEDs and SDGs. By reflecting on the successes and challenges of the +CityxChange project, the study offers actionable recommendations for policy adjustments to enhance regulatory coherence and accelerate the transition toward climate neutrality.

Keywords: Governance · Regulation · Norwegian Planning · Net-Zero Energy · Positive Energy Districts · Regulatory Sandboxing · Trondheim

1 Introduction

In response to the accelerating impacts of climate change, cities are increasingly recognized as central arenas for driving deep emission cuts and sustainable transformation. Addressing this challenge calls not only for technological progress, but also integrated planning and governance across local, national, and international levels. Achieving these goals relies on scaling up both existing and new technologies and practices, alongside

H. Beushausen et al. (Eds.): NTZR 2025, LNCE 873, pp. 1203–1214, 2026.
https://doi.org/10.1007/978-3-032-27698-8_106

fundamental economic, social, and governance transformations [1]. In this context, the pursuit of climate neutrality has become a central goal in both international and regional policy agendas. The UN Framework Convention on Climate Change (UNFCCC) and the Paris Agreement provide the basis for coordinated global climate action, committing countries to limit temperature rise to well below 2 °C, ideally 1.5 °C, and to translate these targets into coherent national and local strategies [2].

At the European level, the European Union has committed to achieving climate neutrality by 2050, with an interim target of reducing greenhouse gas emissions by at least 55% by 2030 compared to 1990 levels [3]. This ambition, first outlined in the European Green Deal [4] and later made legally binding through the European Climate Law [5], frames climate neutrality as a strategic priority, guiding a modern, resource-efficient, and competitive economy. Supporting this vision, the EU launched funding initiatives such as Horizon 2020, promoting research and innovation projects advancing climate resilience, energy transition, and sustainable urban development [6].

Norway, a leader in renewable energy deployment and sustainable urban planning, offers a strong context for studying climate goal implementation at national and municipal levels. Norwegian planning system governed by the Planning and Building Act (Plan - og bygningsloven), defines the statutory framework for spatial and land-use planning and serves as a key instrument for advancing the transition toward a low-emission society [7]. Within this framework, municipalities act as key actors in translating national and international climate objectives into local strategies. In alignment with the EU's broader sustainability goals, the +CityxChange project, funded through Horizon 2020, aimed to develop Positive Energy Blocks (PEBs) and Positive Energy Districts (PEDs) by integrating energy solutions, regulatory innovation, and citizen-driven participation [8], offering practical insights into enabling climate-neutral urban development. Figure 1 shows the regulatory levels across EU, national, and municipal scales, showing how international climate policies are applied locally in Trondheim. It highlights how short-term EU-funded projects interact with long-term national and municipal regulatory frameworks, creating temporal and structural mismatches. To address these challenges, the figure emphasizes the role of regulatory sandboxes as a bridging tool.

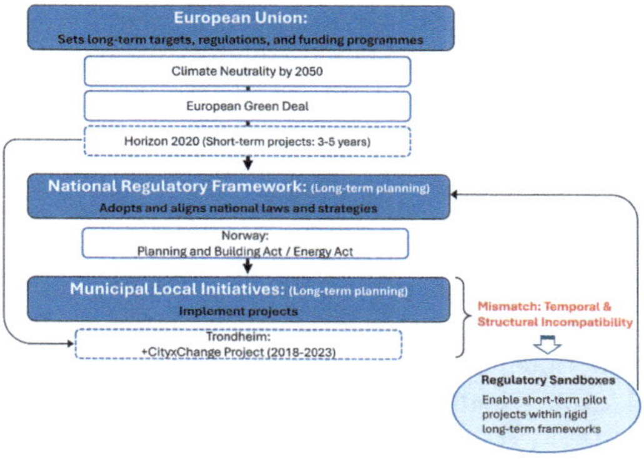

Fig. 1. Regulatory levels and temporal–structural mismatches across governance levels

Building on this initiative, PEDs have emerged as a new urban transition paradigm, offering an advanced framework for achieving decarbonization targets and promoting the development of more resilient and livable cities [9]. Joint Programming Initiative (JPI) Urban Europe defines PEDs as urban areas producing more renewable energy than they consume, through integrated solutions like solar panels, storage, and smart grids [10]. The development of PEDs can enable the achievement of climate neutrality, and in some cases, even result in a net positive energy balance [11].

Despite the European Commission's ambitious objective of creating 100 PEDs by 2030, the goal has yet to be achieved. Although Norway shows strong political commitment to climate neutrality and formal alignment with international agendas, its planning systems still struggle to accommodate innovative urban energy models like PEDs. As noted in [12], the deployment of PEDs depends not only on technological advances but also on coordinated innovations across social, political, financial, spatial, and regulatory domains. Legal and regulatory frameworks, in particular, have been identified as major barriers [13]. Yet, systematic studies examining how regulatory structures influence PED development, and how experimental projects address these challenges, remain scarce. This study addresses this gap by analyzing the +CityxChange project in Trondheim to extract practical lessons on navigating regulatory barriers and advancing local energy transitions. The following sections review Norway's planning system, outline the research method, analyze regulatory barriers and solutions, and present policy recommendations to strengthen climate-neutral urban development.

1.1 An Overview of Norwegian Planning System Alignment with International Climate Agendas

Norway's planning and energy policies are influenced by EU climate directives via its European Economic Area (EEA) agreement, even though Norway is not an EU member [14]. This arrangement re-quires Norway to incorporate relevant EU legislation into its national framework. Key directives such as the Energy Efficiency Directive, the Strategic Energy Technology (SET) Plan, and the Energy Performance of Buildings Directive (EPBD) have shaped Norway's energy policies, aligning them with EU standards. These directives aim to enhance energy performance and efficiency across member states, and their implementation in Norway reflects the country's commitment to shared climate objectives. In line with the Paris Agreement and the European Green Deal, Norway has committed to reducing greenhouse gas emissions by 50–55% by 2030 compared to 1990 levels, with the broader goal of becoming a low-emission society by 2050 [15]. These targets are reflected in nation-al policy documents like the Climate Change Act and the Climate Plan for 2021–2030, aligning Norway's efforts with international and EU climate frameworks [16, 17]. However, the effective realization of these goals necessitates robust planning systems and regulatory frameworks at the local level to ensure coherent and coordinated action across all governance levels.

At the municipal level, Trondheim has embedded SDGs into local planning through its Planning Strategy 2020–2023 and Societal Master Plan 2020–2032, linking planning and budgeting to measurable sustainability indicators. This strategic alignment helped secure its selection for the EU Mission for 100 Climate-Neutral and Smart Cities by 2030, reinforcing its commitment to sustainability and innovation [18]. Nevertheless, key challenges persist. Although the 2008 Planning and Building Act promotes sustainability, it does not legally require municipalities to prioritize climate neutrality or sector-coupled planning [19]. Additionally, delays in adopting some EU climate and energy directives, like the Third Energy Package, indicate occasional misalignments Norway's national policies and EU frameworks [18]. The EU's Third Energy Package, adopted in 2009, was only integrated into Norwegian law in 2018 due to domestic debates over energy sovereignty and control of hydroelectric resources. Such delays risk hindering Norway's climate neutrality goals, highlighting the need for national reforms to ensure binding alignment with climate objectives.

1.2 The Norwegian Electricity Governance

In Norway, energy governance is primarily regulated under the Energy Act of 1990, revised in 2021. The Act established a liberalized electricity market, enabling competition in production and sales, while retaining transmission and distribution as regulated monopolies. Only licensed actors can engage in the production, distribution, or sale of electricity [20]. Norway's electricity governance system is multi-layered, involving the Ministry of Energy, the Norwegian Water Resources and Energy Directorate (NVE), Statnett (national transmission system operator), regional grid operators, Nord Pool, and energy companies. This structure, built for centralized, one-directional energy flows, has supported energy security and high renewable shares. However, it is less compatible with the decentralized, prosumer-oriented systems envisioned in PEDs. Local generation, peer-to-peer trading, and demand-side flexibility challenge existing regulations, requiring more adaptive, localized, and participatory governance models.

2 Methodology

This study adopts a qualitative research design, integrating qualitative document analysis with an embedded single-case study approach. This methodology allows for an in-depth examination of how multi-level energy and climate policies interact within the specific context of implementing PEDs in Norway. Data was collected from a range of publicly accessible documents, including:

- International frameworks such as the Paris Agreement and EU directives (e.g., the Strategic Energy Technology Plan, Energy Efficiency Directive, Energy Performance of Buildings Directive).
- Norwegian national policies, notably the Climate Change Act (2017), the Climate Plan for 2021–2030, and the Energy Act (1990, revised 2021).
- Local planning documents and reports from Trondheim's +CityxChange project.

These documents were selected to capture the policy landscape influencing PED implementation across various governance levels. The +CityxChange project in Trondheim was chosen due to its explicit goal of establishing PEDs within the existing Norwegian regulatory framework. Additionally, the Norwegian University of Science and Technology acted as the coordinator and an active partner in the +CityxChange project, leading the consortium. The case study provides contextual depth, enabling an exploration of how national and international policies are operationalized at the local level.

2.1 Case Study: +CityxChange Project

Trondheim, Norway, has been at the forefront of urban sustainability through its participation in the EU-funded +CityxChange project. This initiative aims to transform urban areas into PEDs, where locally produced energy exceeds consumption, contributing to a sustainable and resilient urban ecosystem. The +CityxChange project aimed to develop and test PEDs in European cities. Trondheim, as one of the Lighthouse Cities, collaborated with Limerick and five Follower Cities to implement integrated energy solutions [8]. The goal was to contribute to creating sustainable urban ecosystems with zero emissions and achieve 100% renewable energy city-regions by 2050 [18]. The development of PEDs in Trondheim was structured around three key strategies:

1. Integration into Urban Planning: Embedding energy-positive principles into city planning to ensure sustainable development.
2. Deployment of Innovative Technologies: Implementing advanced energy solutions, including renewable energy sources and smart grid technologies.
3. Community Engagement: Establishing Innovation Playgrounds to involve citizens in co-creating energy solutions, fostering a culture of sustainability [21].

These strategies were guided by the Bold City Vision (BCV) framework, which structured PED development into five phases: engagement, design, activation, acceleration, and support [22]. This framework aligns global sustainability goals with local policy through citizen-driven innovation [18]. This real-life trial in Trondheim revealed key regulatory barriers and potential solutions, explored in the next section.

3 PEDs Implementation in Trondheim and Regulatory Barriers

The implementation of PED in Trondheim highlights the complex intersection between innovative decentralized energy systems and outdated regulatory frameworks. At the Brattøra demonstration, the aim was to establish one of Europe's first Positive Energy commercial districts, centered around Powerhouse Brattørkaia, an energy-positive building designed to generate more energy than it consumes [23]. The broader ambition was to integrate educational institutions, commercial facilities, and public infrastructure into a local energy community connected by a microgrid with battery storage to maximize locally produced energy. However, regulatory barriers hindered this vision, as Norwegian laws restricted private microgrids and prohibited energy exchange between private actors without a licensed grid operator. According to Stine Hostad, regional director of Entra Trondheim —the property development company that owns Powerhouse Brattørkaia— the technology itself was not the problem—rather, it was the regulatory framework, which was designed for a 20th-century centralized energy system, that posed the real challenge. Hostad highlighted that current regulations prevented Entra from selling surplus solar energy generated by Powerhouse Brattørkaia directly to its own neighboring buildings or even to its tenants. As a result, Entra was compelled to sell the excess energy to the grid operator and then repurchase it at a higher price, an arrangement she described as counterintuitive and inefficient.

In practice, this effectively made energy sharing within the same property or local area illegal, creating a major barrier to the project's goal of establishing a PED. This regulatory bottleneck underscored the need for updated laws to support decentralized energy systems and enable testing of future-oriented market models [24].

According to [25], several regulatory barriers significantly hindered the development and scaling of PEDs in Trondheim under the +CityxChange project.

A central challenge was the lack of a standardized legal and technical framework for peer-to-peer (P2P) energy trading. There were no clear rules on how such transactions should be conducted, settled, or regulated. This regulatory gap introduced operational uncertainty and made it difficult to scale local energy markets beyond the pilot stage.

Another major obstacle was the absence of legal recognition for Local Energy Communities (LECs) and Citizen Energy Communities (CECs). Without formal frameworks, it was unclear how community-owned energy systems could be governed or operated. This lack of clarity posed risks for both participants and investors, making it harder to establish collective ownership or shared trading models.

The regulatory environment also restricted access to energy data, with unclear rules on data ownership and sharing. This limited partners' ability to optimize local systems, balance supply and demand, and integrate smart technologies across districts.

Additionally, existing grid tariff structures were not adapted for local energy sharing. Even when electricity was exchanged within the same neighborhood, full transmission and distribution charges still applied. This pricing model discouraged local generation, flexibility services, and energy sharing—key components of PEDs.

While the project allowed limited regulatory flexibility via sandbox arrangements, the processes were often slow, bureaucratic, and lacked a clear path to permanent legal change [26], hindering rapid testing and deployment of innovative market models.

Finally, no clear roles or responsibilities were defined for operating local microgrids or community battery systems, and there was no standard model for P2P energy transactions. These uncertainties created confusion around ownership, maintenance, and investment, ultimately threatening the long-term sustainability of PED infrastructure.

The regulatory barriers faced in Trondheim reflect broader systemic challenges observed across Norway and Europe. Although Trondheim was selected as a Lighthouse City under the EU's sustainable energy initiatives to pioneer local energy markets [13], realizing these ambitions required navigating a rigid regulatory environment fundamentally built for centralized, one-way energy flows [25]. Addressing these systemic barriers will be crucial for scaling PEDs and achieving climate-neutral urban development.

4 Overcoming Regulatory Barriers: Insights from the +CityxChange Project in Trondheim

The +CityxChange project in Trondheim provides a valuable case study for how cities can navigate rigid regulatory environments to advance decentralized energy systems like PEDs. Trondheim's experience underscores the need for regulatory adaptability. Rigid national energy regulations often conflict with the decentralized nature of local energy systems [27, 28]. Norway's national regulatory framework, designed for centralized, one-way electricity flows, frequently posed barriers to local innovation and energy sharing. To address this, Trondheim took proactive steps to engage national authorities—particularly the Norwegian Energy Regulatory Authority (RME)—through early and sustained dialogue. This collaboration enabled the creation of two "special regulatory districts," where exemptions from national energy laws permitted pilot testing of peer-to-peer energy trading, flexibility services, and other decentralized innovations [25]. While effective for enabling pilots, such sandboxes are transitional tools, highlighting the need for enduring structural reforms [29, 30].

A key strategy was the introduction of the "Energy Coordinator" model. This approach allowed Trondheim to oversee local energy exchanges while ensuring principles, such as consumer protection and neutrality among suppliers, were upheld [31]. Flexible governance arrangements like this are critical to managing complex, multi-actor energy transitions, helping bridge innovation with compliance requirements [32].

The project also introduced smart grid technologies, including a local trading platform with advanced metering infrastructure, blockchain-based energy transactions, and the integration of electric vehicles. These systems demonstrated the feasibility of decentralized energy management within a regulatory sandbox environment [26].

Furthermore, the +CityxChange team actively engaged in European knowledge-sharing platforms, such as the Smart Cities Marketplace, to align local experiments with emerging EU policies and advocate for future-proof regulatory frameworks [13]. Trondheim's experience underscores the importance of multi-level regulatory engagement and knowledge sharing. The project shows that achieving local regulatory flexibility requires sustained, trust-building dialogue with national authorities—a process that took nearly four years [12, 28]. Engagement in European platforms ensured local innovations contributed to and were informed by broader policy trends [13].

Figure 2 presents the sandbox governance model applied in Trondheim's + CityxChange project, structured as a four-phase iterative cycle: experimental framework, testing and learning, evaluation, and regulatory adjustment. This adaptive governance approach helped navigate outdated laws and foster local energy innovation.

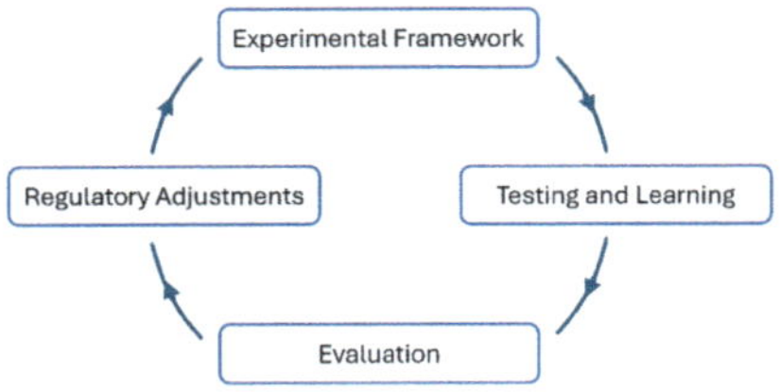

Fig. 2. Sandbox governance model for adaptive regulation

Despite these innovations, each regulatory exemption had to be negotiated individually, treated as a one-off rather than part of a streamlined process. This lack of a standardized fast-track framework reflected a common challenge in European PED pilots [30]. Moreover, no clear roles or responsibilities were defined for operating local microgrids or community battery systems, contributing to uncertainty about infrastructure governance and long-term operation [31]. Establishing institutionalized sandbox mechanisms—with clear entry criteria, timelines, and regulatory feedback loops—is essential for legal certainty and scaling decentralized solutions.

Additional regulatory obstacles included outdated grid tariff structures that applied full transmission costs even to local energy exchanges within sandboxes, undermining the economic viability of local markets [28]. Gaps in data access and system interoperability further limited energy flow optimization and market integration [1].

Regulatory misalignments also extended beyond the energy domain. As highlighted in [19], conflicts between Norway's Planning and Building Act and energy sector regulations created vertical and horizontal incoherence, complicating PED integration into urban planning. Trondheim's case reinforces the need for integrated national policies aligning urban development, energy regulation, and digital infrastructure.

Finally, Trondheim's experience emphasized that regulatory flexibility must be coupled with meaningful citizen engagement. Initiatives such as citizen observatories, energy champions, and participatory innovation playgrounds were essential to building trust, encouraging ownership, and ensuring public support for PEDs [32]. As argued in [28], PEDs cannot thrive solely through top-down technological or regulatory measures; active public involvement is essential.

Trondheim's approach shows that overcoming regulatory barriers requires more than ad hoc exemptions. It demands comprehensive legal reforms, adaptive governance models, sustained collaboration across governance levels, citizen participation, and alignment with EU policy trends. These lessons can inform broader efforts to scale PEDs and advance climate-neutral urban development across Europe.

5 Conclusion

Although not an EU member, Norway is strongly committed to aligning with European climate policies, reflected in national frameworks mirroring EU climate neutrality targets. However, fragmented planning systems and regulatory frameworks designed for centralized, siloed governance hinder effective local action. This gap between national ambitions and municipal capacities highlights a critical paradox: even leading nations encounter systemic barriers in achieving climate neutrality at the urban level.

The transition to climate-neutral cities requires systemic transformation, blending technological innovation with adaptive governance and regulatory agility. This study highlights Norway's pioneering yet incomplete effort to align its planning system with global climate goals, as seen in the +CityxChange project in Trondheim. While national ambitions and municipal initiatives show strong political will, regulatory misalignments limit PEDs' potential. Successes like local energy markets through sandboxes and the Energy Coordinator model show decentralized systems are feasible—but remain constrained by outdated frameworks, rigid tariffs, and fragmented governance.

The Trondheim experience highlights a key paradox: cities can drive climate innovation yet remain constrained by national regulations misaligned with urban energy transitions. While temporary exemptions and pilot zones are valuable, they are not enough to scale PEDs. Systemic reforms are needed to embed these pilots into mainstream urban development. The urgency of climate action requires Norway and the EU to move beyond incremental change toward holistic, cross-sectoral governance that ensures regulatory coherence, citizen participation, and data-driven flexibility.

The widespread replication of PEDs marks a shift from centralized energy systems to decentralized, community-driven models. While empowering local sustainability, this transition challenges Norway's grid reliability, requiring flexibility to manage variable renewables and bidirectional energy flows. Large energy producers, central to Norway's hydropower-based system, may need to adapt as decentralized generation lowers demand for centralized infrastructure. To align innovation with stability, regulations must support peer-to-peer trading while ensuring grid reliability. Ongoing monitoring of PED impacts across cities will help mitigate risks and refine scalable solutions, ensuring seamless integration into the national energy framework.

These findings contribute to the net-zero transition debate by showing that systemic legal reforms and adaptive governance are as vital as technological progress. Trondheim's experience offers lessons for cities facing similar regulatory constraints. By promoting co-creation, multi-level coordination, and flexible regulatory tools, urban planners can turn ambitious climate goals into local strategies. PEDs embody not just energy innovation, but a new model of governance for sustainable urban futures.

6 Policy Recommendations for Enabling PEDs

Based on the regulatory challenges and insights from the Trondheim case, this section outlines key policy actions to support the development and scaling of PEDs. These recommendations target reforms in legal frameworks, governance, tariff systems, and citizen participation to enable more adaptive, locally grounded energy transitions.

6.1 Modernizing Energy Regulations

- The Norwegian Energy Act should explicitly recognize Local and Citizen Energy Communities. This would provide a legal basis for peer-to-peer trading, microgrid deployment, and collective ownership—essential for decentralized energy systems.
- In addition, a formalized regulatory sandbox framework should be introduced to allow municipalities and innovators to experiment with new energy solutions in a controlled setting. This would streamline the exemption process, reduce bureaucratic barriers, and support adaptive policymaking while safeguarding consumer rights.

6.2 Reforming Tariff Structures and Markets

- Grid tariff systems should be restructured to encourage local energy exchange by exempting intra-district transactions from full transmission charges. This would reduce the cost burden on PED participants and stimulate energy sharing within communities.
- Furthermore, Norway should establish flexibility market rewarding consumers and producers for load shifting, distributed generation, and energy storage. These would enhance grid responsiveness, renewable integration, and PED participation incentives.

6.3 Aligning Governance with Climate Goals

- To align energy planning with broader sustainability goals, climate neutrality targets should be embedded in the Planning and Building Act, legally mandating municipalities to pursue sector-coupled planning across energy, mobility, and land use.

- Additionally, multi-level governance task forces should be established to coordinate national, regional, and municipal efforts, facilitating faster alignment with EU climate directives such as the revised Renewable Energy Directive (RED III).

6.4 Enhancing Data Accessibility and Digital Infrastructure

- Clear regulatory frameworks are needed to govern energy data ownership, access, and interoperability. These frameworks should enable secure, real-time data exchange between stakeholders, thereby improving operational efficiency and decision-making.
- At the same time, national investment should prioritize integrated digital platforms that connect smart grids, e-mobility systems, and building energy management. Such infrastructure is vital to the scalability and coordination of PEDs across urban areas.

6.5 Promoting Citizen-Centric Engagement

- Citizen participation should be institutionalized through inclusive co-creation models, like the Innovation Playgrounds from +CityxChange. These forums empower communities to shape local energy solutions and foster trust in the transition process.
- Complementary to this, national and municipal governments should initiate public outreach campaigns and energy ambassador programs to build grassroots momentum and ensure equitable participation in PED initiatives.

6.6 Strengthening EU-National Collaboration

- Norway should align more closely with the EU Mission for 100 Climate-Neutral Cities by joining cross-border platforms. This would support best practice transfer, regulatory harmonization, and ensure Trondheim's +CityxChange informs broader policy. Stronger EU ties can also accelerate adaptive regulation and access funding.

7 Broader Implications and Future Research

This paper concludes by highlighting the need for more flexible and integrated approaches to governance. Rather than working in isolated policy areas, decision-makers should aim for systems that are adaptive and open to learning through small-scale experimentation and iterative adjustments. Future research should explore how PEDs can be scaled both within cities already engaging in PED initiatives, such as Trondheim, and across different types of urban contexts. It should also further investigate the social and economic benefits of bottom-up, collaborative approaches to energy transitions.

Norway's experience, though specific to its context, offers valuable insights for others: achieving climate neutrality is not just a technical issue—it also requires changes in how decisions are made. This includes updating regulations, supporting local solutions, and improving coordination across government levels. By strengthening the link between innovative cities and flexible national frameworks, cities can make real progress toward more sustainable and resilient urban development.

References

1. Bhowmik AK et al (2020) Powers of 10: seeking 'sweet spots' for rapid climate and sustainability actions between individual and global scales. Environ Res Lett 15(9):094011
2. UNFCCC (2015). Available from: https://unfccc.int/process-and-meetings/the-paris-agreement
3. European Commission (2021) 'Fit for 55': delivering the EU's 2030 climate target on the way to climate neutrality ; Available from: https://eur-lex.europa.eu/legal-content/EN/TXT/?uri=CELEX%3A52021DC0550
4. European Commission (2019) The European Green Deal. Available from: https://eur-lex.europa.eu/legal-content/EN/TXT/?uri=CELEX:52019DC0640
5. European Union (2021) Regulation (EU) 2021/1119 establishing the framework for achieving climate neutrality and amending regulations (EC) No 401/2009 and (EU) 2018/1999 ('European Climate Law'). Available from: https://eur-lex.europa.eu/legal-content/EN/TXT/?uri=CELEX:32021R1119.
6. +CityxChange Consortium (2018) +CityxChange. Available from: https://cityxchange.eu/
7. KMD (2008) Planning and Building Act (Plan- og bygningsloven). Available from: https://www.regjeringen.no/en/dokumenter/planning-building-act/id570450/
8. Ahlers D et al (2019) Co-creation of positive energy blocks. IOP Conf Ser Earth Environ Sci
9. Romano R, Andreucci MB, Giancola E (2022) Towards climate neutrality: progressing key actions for positive energy districts implementation. In: International conference on technological imagination in the green and digital transition. Springer
10. JPI Urban Europe (n.d.) Positive energy districts (PED). Available from: https://jpi-urbaneurope.eu/ped/
11. Magrini A et al (2020) From nearly zero energy buildings (NZEB) to positive energy buildings (PEB): the next challenge-the most recent European trends with some notes on the energy analysis of a forerunner PEB example. Dev Built Environ 3:100019
12. Krangsås SG et al (2021) Positive energy districts: identifying challenges and interdependencies. Sustainability 13(19):10551
13. Bertelsen S et al (2019) D2. 1 Report on enabling regulatory mechanism to trial innovation in cities. CityxChange, Trondheim, Norway, Tech. Rep.
14. Gulbrandsen LH, Hermansen EA, Closer E (2022) Union? Norges tilknytning til EUs klimaregelverk. Internasjonal Politikk 80(1):170–183
15. UNFCCC (2020). Available from: https://unfccc.int/sites/default/files/NDC/2022-11/NDC%20Norway_second%20update.pdf
16. KMD (2017) Climate Change Act. Available from: https://www.regjeringen.no/en/dokumenter/climate-change-act/id2593351/
17. KMD (2021) Norway's climate action plan for 2021–2030. 2021. Available from: https://www.regjeringen.no/.pdf
18. Haugslett A, Riedesel K, Berthelsen BO (2022) D5.7: +Trondheim 2050 bold city vision and guidelines (vision for sustainable urban transition)
19. Stokstad S et al (2020) Bedre samordning mellom plan-og bygningsloven og sektorlovgivningen
20. Energifaktanorge (2024) The power market. Available from: https://energifaktanorge.no/en/norsk-energiforsyning/kraftmarkedet/
21. Mee A et al (2020) D3.3 framework for innovation playgrounds. CityxChange
22. Tanum Ø et al (2020) D3.1 Framework for bold city vision, guidelines, and incentive schemes
23. Kirkerud JG et al (2022) Sammenhengen mellom kraftbalanse og kraftpris. NVE
24. Europower (2021) Brattøra kan bli verdens første grønne bydel, men regelverket er festbrems. Available from: https://www.europower.no/elektrifisering/brattora-kan-bli-verdens-forste-gronne-bydel-men-regelverket-er-festbrems/2-1-1016638

25. Myrstad MT, Livik K, Haugslett A (2021) D5.9: playbook of regulatory recommendations for enabling new energy systems
26. Danielsen S et al (2022) D5. 5: energy trading market demonstration
27. Gohari S et al (2024) Unraveling the implementation processes of PEDs: lesson learned from multiple urban contexts. Sustain Cities Soc 106:105402
28. Sareen S et al (2022) Ten questions concerning positive energy districts. Build Environ 216:109017
29. Sassenou L-N, Olivieri L, Olivieri F (2024) Challenges for positive energy districts deployment: a systematic review. Renew Sust Energ Rev 191:114152
30. Myrstad, Livik, Wyckmans (2020) A new smart cities marketplace initiative on regulatory frameworks within the integrated planning, policy and regulation action cluster
31. Guldbrandsøy EN, Rørvik E-LH, Berthelsen BO (2023) D5.6: Trondheim Flexibility Market Deployment Report
32. Temeljotov Salaj A, Loewen B (2020) The next generation of smart citizens: Experiences and inspiration from the +CityxChange project. In: 7th international academic conference on places and technologies

From Fragmented Systems to Sustainable Operations: Streamlining Processes at OsloBygg KF for a Net-Zero Future

Mufftasin Fuad Khan[1], Dara Diyar Hafzullah[1], Daniel Amin Haddadi[2], and Allen Tadayon[1](✉)

[1] Oslo Metropolitan University, Oslo, Norway
allen.tadayon@oslomet.no
[2] OsloBygg KF, Oslo, Norway

Abstract. OsloBygg KF, a public property management entity, is experiencing a digital transformation to improve efficiency and optimize its central work processes. This paper investigates the inefficiencies within the organization's data management and workflow systems, focusing on fragmented digital tools such as ORRA and WebSak, which ultimately prevent efficient productivity and collaboration. The research includes a qualitative case study approach, incorporating semi-structured interviews with key personnel for insights into the organizational challenges and opportunities. The findings show that OsloBygg KF's current data systems come with several shortfalls, which increase manual work and cause delays in decision-making. A centralized platform for managing data, with supportive digital tools for automation and real-time monitoring, is highly recommended to streamline operations. In the short term, workflow simplification and automation of compliance-related tasks can be prioritized to increase efficiency. In the long term, however, strategies should focus on implementing AI-driven solutions and real-time monitoring systems to ensure sustainable transformations. By implementing these strategies, OsloBygg KF can achieve increased operational efficiency and sustainability.

Keywords: Digital Transformation · Data Management · Process Optimization · Public Sector Efficiency · Automation

1 Introduction

Efficient data management and streamlined workflows are crucial for organizations responsible for managing large-scale public facilities. Many public sector entities struggle with outdated digital systems, leading to inefficiencies, redundant manual work, and delays in decision-making [1]. The integration of digital tools and automation can significantly enhance productivity by reducing administrative burdens and improving collaboration between different departments [2].

OsloBygg KF, a public property management organization, is a relevant case study in this context. It oversees a wide range of public properties, including schools, hospitals,

H. Beushausen et al. (Eds.): NTZR 2025, LNCE 873, pp. 1215–1225, 2026.
https://doi.org/10.1007/978-3-032-27698-8_107

and cultural/recreational institutions OsloBygg [3]. However, its current digital infrastructure consists of fragmented tools such as ORRA and WebSak, which hinder operational efficiency. These disconnected systems create bottlenecks in daily tasks, requiring manual interventions and reducing the speed of decision-making. Similar issues have been observed in other public sector organizations, as noted in Janssens article Janssen, Charalabidis [4] where outdated processes and isolated systems delay effective service delivery.

Given these challenges, this study aims to assess the inefficiencies within OsloBygg KF's central work processes and explore opportunities for improvement through enhanced data management and digital tools. By examining the current workflows, identifying integration issues, and proposing automation strategies, this research contributes to the ongoing digital transformation efforts within the organization. The approach in this paper aligns with the best practices in digitalizing business process such that greater efficiency can be achieved Mergel, Edelmann [5].

To address these challenges, the study focuses on streamlining OsloBygg KF's central work processes in relation to improvement of data integration and digital tools.

The key research questions that guides this work are:

- What are the main challenges in OsloBygg KF's current digital tools and data management practices, and how do they affect day-to-day operations and decision-making?
  - Examining current processes helps us to find inefficiencies, pointless manual labor, and communication breakdowns. This reveals how digital technologies might be failing, influencing daily operations as well as more general organizational decisions.
- Which immediate and strategic measures should OsloBygg KF prioritize to overcome these challenges and sustainably improve its central work processes?
  - Examining different strategic measures will show how OsloBygg KF can simplify processes, lower manual labor, and increase efficiency. This would lay out a clear road from short solutions to long-term digital transformation.

These questions are essential to identify the inefficiencies and to propose tailored solutions to enhance organizational workflows. In Vial [6] paper, the importance of such study is highlighted as a critical research step in understanding how digital transformation reshapes organizational structures for better innovation.

This paper presents a qualitative case study for OsloBygg KF based on semi-structured interviews with OsloBygg KF's key personnel. The paper initially starts with the theoretical framework of Business Process Management (BPM), as the core for streamlining workflows. Then, the focus shifts towards the methodology, that further includes the data collection and analysis factors. Moreover, the findings are presented in detail with figures and tables. Afterwards, the findings are discussed using case studies from the theoretical framework such that successful digital transformation strategies can be derived. Finally, the paper concludes with recommendations for OsloBygg KF that focus on both short-term and long-term workflow improvements.

2 Theoretical Framework

The theoretical foundation for this paper has its roots on BPM. BPM provides a structured approach to improve various organizational workflow by setting the ideal efficiency and standardizing processes in a similar way as Weske describes in their book, [7]. The BPM theory is widely applied in different industries to integrate processes to have continuous improvements [7]. This section presents Business Process Management and its applications through two case studies: the implementation of Electronic Health Records (EHR) at the Mayo Clinic and the Toyota Production System (TPS). These cases illustrate how structured methodologies can enhance efficiency, coordination, and collaboration within complex operational environments.

2.1 Business Process Management

BPM framework's goal is to improve the effectiveness of business operations by handling workflows systematically. The five primary stages of these processes are related to design, modeling, execution, monitoring and optimization. Design focuses on identifying inefficiencies and mapping out workflows, while modeling can find potential improvements before a strategy is being developed. Execution optimizes processes and are integrated into operations, which can be through digital tools. Monitoring provides real-time tracking factors that would help organizations find bottlenecks. Lastly, optimization refines workflows such that they stay efficient. In industries, BPM has integrated scattered processes, reduced bottlenecks and supported operational decision making [7].

Mayo Clinic. The healthcare institution, Mayo Clinic, lacked an effective workflow due to decentralized data management and coordination challenges in their teams. These issues were addressed by adopting an EHR, short for Electronic Health Record, system that improved communication, accessibility and operations [8]. The key factors taken from Mayo Clinic's implementation are:

- Centralized Management: The EHR supported real-time data access and streamlined coordination for patient care.
- Process Automation: Mayo Clinic implemented AI for resource allocation and increased diagnostic efficiency.
- Continuous Improvement of Processes: Refinements in their workflow were done frequently such that automation could reduce operational delays.

Toyota Production System and Kanban. The second case study is Toyota's Production System that is a major leading example of lean manufacturing principles that primary focuses on increasing process efficiency and reducing unwanted waste [9]. The core factor of TPS is the Kanban system that ensures synchronized production processes such that inefficiencies through real-time inventory tracking are minimized [10].

The key insights taken from Toyota's approach are:

- Lean Process Optimization: Kanban system pushes for in-time production that minimizes the chances for overproduction and excess inventory [10].

- Automated Quality Control: Jidoka was created to automate defect-detection that further improves production quality [9].
- Cross-functional Collaboration: Kanban increases collaboration potential by integrating inventory tracking with supplier. This ensures efficient material flow and reducing production delays [10].

BPM provides a theoretical background for process optimization. The two cases from Mayo Clinic and Toyota are examples of BPM. The combined key takeaways form the basis for understanding the role of streamlining central work processes in OsloBygg KF's complex environment.

3 Methodology

A qualitative case study was the most appropriate approach to achieve the desired results. The qualitative research method is known for focusing on understanding the experiences from the point of view for those who live them [11]. Semi-structures interviews were conducted as a data collection method in this study. Semi-structured interviews utilize predefined questions and simultaneously allow for flexibility to discuss other topics based on the interviewee's response. This allows for spontaneous discussions while also giving the respondents the freedom to discuss other concerns [12].

An interview guide was made to steer the interview sessions [13]. This guide was sent out to the respondents before the interviews. This interview guide included seven main topics with questions to ensure interviewees could comfortably express themselves in Norwegian. Ten participants, equally divided between the Property Management (PM) and Technical Management (TM) departments, were selected with the help of the group's super-user at OsloBygg KF. The questions were consistent for both groups but slightly adapted depending on their position.

To select the most suitable interviewees, the property division of OsloBygg KF was examined. As illustrated in Fig. 1, the property division consists of six sub-divisions. The first five sub-divisions (green) were chosen as they are all properties and have more in common, unlike the last sub-division (red) which focuses on property services, making it less related to the aim of this paper. Time constraints also influenced the decision to exclude Property Services.

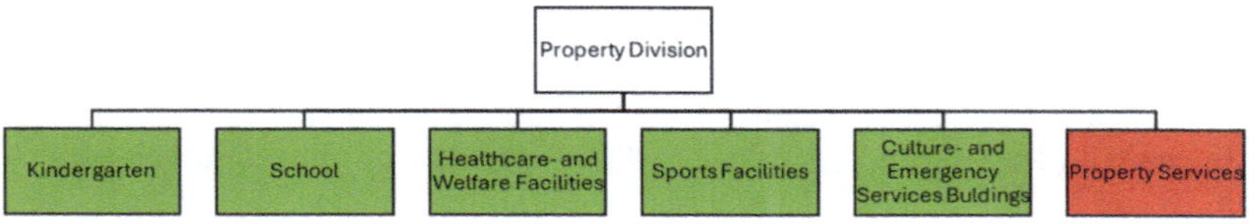

Fig. 1. Property division structure

Then, two respondents, one from each department, were chosen from each of the green sub-divisions. Most interviews were conducted physically, some required to be digital due to logistical limitations. Details on participants are provided in Table 1.

Table 1. Interview respondents with their experience and properties managed.

Technical manager			Property manager		
Participant No.	Years at OsloBygg	Properties managed	Participant No.	Years at OsloBygg	Properties managed
1	10	22	2	12	70
3	4	10	4	1.5	30
5	3.5	24	6	14	75
7	2	14	8	3	15
9	13	N/A	10	2	13

Interviews were audio recorded using the Diktafon[1] app on smartphones and sent to Nettskjema[2] for safe storage [14]. The recordings were automatically transcribed using OpenAI's Whisper[3] and reviewed for accuracy by the group. Qualitative content-based analysis were utilized to extract the insight from the transcriptions. A spreadsheet was used to create a coding system to categorize and summarize the data effectively for each respondent [15].

Informed consent was obtained, and the participants were assured of voluntary participation with the option to withdraw at any time [16]. Ethical approval was ensured through OsloMet's guidelines and reviewed by Sikt[4] for compliance with data protection regulations. No personal data was stored, and all data will be deleted within three months.

4 Findings

The key findings are derived from the themes identified in the interviews and are supported by both direct quotations from respondents and summarized insights based on observed patterns from their data.

4.1 Daily Work Process

Administrative tasks dominate the daily workload of the interviewees, being time-consuming and preventing them from addressing other critical tasks. Specifically, tasks involving deviation handling and compliance, as well as communication with tenants and customers consume significant time [17]. Table 2 visualizes how these daily tasks are allocated at OsloBygg KF.

[1] Diktafon is a smartphone app used for voice recording.

[2] Nettskjema is a secure online data collection and storage service developed by the University of Oslo.

[3] Whisper is an AI-powered speech-to-text transcription model developed by OpenAI.

[4] SIKT is Norway's agency responsible for IT and digital services in the knowledge sector, ensuring compliance with data protection regulations.

Table 2. Daily work processes.

Task category	Frequency
Administrative work	6
Handling deviations	3
Tenant communication	2
Others	2

4.2 Tools, Systems and Collaboration Between Divisions

The key findings indicate significant integration issues between ORRA[5] and tools like Interaxo[6] and WebSak,[7] leading to manual data handling and overlapping between the programs that hinder efficient task execution and data management. This highlights the issue of disconnected, fragmented and overlooked tools that create inefficiencies in task execution and data management. As one interviewee mentioned:

"One of the main challenges is that data is scattered across various systems. For example, when looking for a floor plan, I might have to check Teams, ORRA, Facilitate or WebSak. This lack of centralization makes it time-consuming and inefficient." – *Candidate 3*

A related data point here is the frequent need to manually send emails via Outlook and not through connected systems. Additionally, collaboration between departments is limited by unclear communication pathways and the lack of a common platform. This is clearly emphasized as streamlining communication is an ongoing issue due to manual communication processes such as calls, emails, and physical meetings.

4.3 Data Management Challenges

Challenges in data management originate from integration issues and outdated systems, causing scattered storage and inefficient data access across departments. Outdated systems prevent real-time access to data that might impact decision making. These sub-findings highlight how integration issues contribute to overall weakness in data management, with many interviewees frequently mentioning these challenges as significant. Table 3 illustrates the challenges and the frequency with which they were mentioned during the interview sessions. Respondent noted different reasons for these integration issues.

[5] ORRA is an internal digital system used for managing tasks, workflows, and collaboration within organizations.

[6] Interaxo is a cloud-based collaboration platform used for document management and communication in construction and infrastructure projects.

[7] WebSak is a document and case management system used in the public sector for handling administrative processes and regulatory compliance.

Table 3. Data management challenges at OsloBygg KF.

Challenge categories	Frequency
Integration issues	6
Outdated systems	4
System downtime	2
Fragmented data	2
Redundant storage	1

4.4 Opportunities for Digitalization

The findings suggest the use of digital tools and automation, to streamline workflows such that operational efficiency can be improved. Short-term digitalization opportunities include automating compliance and tenant reporting processes, while long-term strategies involve real-time monitoring systems, automated alerts, and advanced automation systems for repetitive tasks. These sub-findings support the main key finding that digitalization can increase efficiency in data management tasks and decision-making. One candidate highlighted the future possibilities by saying:

"I also see potential in using AI for routine tasks and predictive maintenance to make processes smoother and more efficient." – Candidate 10

This quote shows how automation and AI could increase day-to-day productivity by reducing manual tasks and having proactive interventions. It also supports the idea that simpler workflows and better resources are key, that further emphasizes the candidate's agreement on the importance of digital transformation.

5 Discussion

This research looking at OsloBygg workflow and it alignment with the BPM framework, which aims to streamline outdated and disconnected team processes. To draw a meaningful conclusion, this study uses the findings from combining data from interviews with technical and property managers and the approaches two industry initiatives (Mayo clinic and Toyota production system) that focus on continuous improvement [18].

The findings suggest that OsloBygg KF can enhance operations with better workflows and by adopting digitalization. Currently, fragmented tools and processes cause delays, inefficiencies and poor communication. These problems can be addressed by using the Kanban tool as this approach has been proven effective, as demonstrated by Toyota Production System, in reducing waste in processes. Similarly, approaches like the Mayo Clinic's method for integrating digital data systems can improve collaboration across departments by creating a centralized platform for simplified communication and real-time data sharing.

According to our respondents, OsloBygg, as an organization managing projects of varying sizes and complexities, should prioritize initiatives with low efficiency and low complexity. Specifically, they should focus on simplifying workflows and automating repetitive tasks to address existing challenges and inefficiencies. One respondent supported this approach, describing it as an opportunity to achieve quick and effective improvements:

"There's a real opportunity to centralize documentation and automate repetitive tasks to prevent delays and free up time for more strategic work." - Candidate 8

These factors can quickly increase efficiency by improving task coordination and reducing manual processes, and yet they are not too complex to introduce. One reason for prioritizing these low-effort, high-impact initiatives is that OsloBygg is managing multiple ongoing projects simultaneously [19]. Addressing simpler inefficiencies, such as workflow simplification and automation of repetitive tasks, allows the organization to improve operations without significantly disrupting its ongoing activities.

For the more efficient and complex solutions, OsloBygg KF may invest in automation of compliance tasks and real-time monitoring systems. Real-time monitoring systems give better access to synchronized information and faster decision-making process, while automation of compliance tasks will free up resources and time for more strategic development [20]. Figure 2 illustrates different strategies identified through this study to enhance central work process fellow in OsloBygg the higher up the "staircase", the more effective yet complex the implementation.

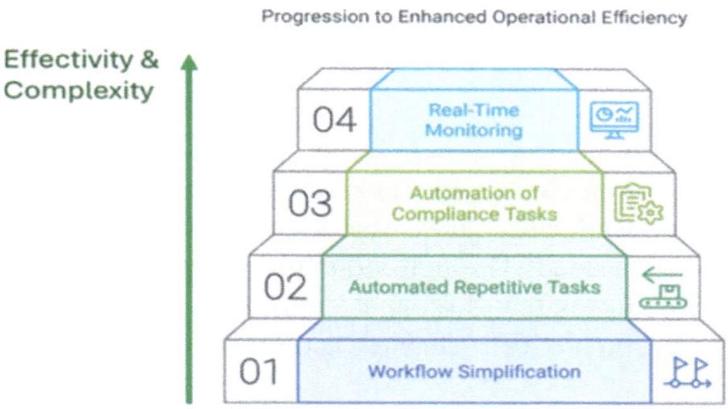

Fig. 2. "Staircase" solutions with increasing effectivity and complexity for each "step".

The first solution addresses time-consuming administrative tasks and unnecessary complexity in day-to-day work. By removing the redundant steps, as mentioned in Kaizen principles, employees would spend less time on minor tasks and collaboration becomes easier when all parts understands a streamlined, standardized routine [21].

Once the workflows are simplified, the next step would be to target frequent and predictable tasks, for instance the data entry and simple approvals. So the second step would be to fix the reliance on manual input across the different fragmented systems. Here Automated repetitive tasks steps in the form as Robotic Process Automation (RPA). This would push data between the platforms automatically, in a similar fashion as Mayo Clinic [22], easing up integration issues and freeing employees for more important duties [23]. RPA uses software robots or bots to interact with digital systems to perform tasks such as data entry, transaction processing and synchronization without human intervention

[23]. This solution helps reduce both time spent and errors caused by double-keying or flipping between unconnected platforms.

Building on RPA, the third "step" moves to AI-based or rule-based reasoning for jobs needing some degree of judgment—like confirming budget approvals or cross-refining a request against regulations. Many respondents complained about outdated processes and several sign-offs creating delays that hinder cross-departmental collaboration. An intelligent system might rapidly identify compliance problems or transfer the demand to a human only when necessary [22], therefore reducing the difficulty of slow decision-making and too heavy manual control.

The final "step" is Real-time Monitoring addresses the lack of timely data. It is believed to be the most complex to implement but also the one that will yield the best results in work processes [24]. This involves monitoring using digital systems so that problems can be quickly detected, and measures can be taken immediately, and delays can be reduced [24]. This solution resolves the absence of real-time insights, allowing staff to act proactively rather than waiting for a manual report or discovering a problem too late [9].

These four solutions are arranged in steps beginning from simpler improvements to more advanced ones. This staircase framework shows the higher "step" brining more complexity but also better benefits. By starting at the lower steps, OsloBygg KF can make easy gains at the beginning, then gradually move to more complex and rewarding solutions.

6 Conclusion

OsloBygg KF faces serious challenges caused by fragmented digital tools, manual data handling and disjoint platforms. These issues show that a modern, integrated work environment is deeply needed. Through a qualitative exploration of the key managerial staff, the study confirms that disconnected systems, such as ORRA and WebSak slow decision-making, create inefficiencies, and prevent collaboration across multiple departments.

Emphasizing lean manufacturing principles from Toyota and lessons in the digitalization of healthcare from the Mayo Clinic, the research emphasizes that structured BPM practices, combined with chosen digital tools, could be used to streamline daily administrative duties. Short-term wise, OsloBygg KF can achieve solid results by simplifying workflows, automating repetitive tasks, and reinforcing central data handling, all that would offer immediate efficiency boosts without the need of excessive resources.

Beyond these incremental optimizations, this paper also emphasized the longer-range advantages of adopting mechanisms; real-time monitoring and AI-driven compliance. This would promise near-instant feedback on infrastructure conditions that would enhance decision-making based on proactive data analysis. To successfully implement these solutions, it is, however, required through planning for budget, data governance and staff hiring to ensure stakeholder buy-in and sustainability.

Consequently, a phased approach that starts with projects of lower complexity and further transitions to more advanced technologies appear to be the most practical. By following this roadmap, OsloBygg KF would be well positioned to evolve into a digitally transformed public property manager, characterized by agile workflows, improved service delivery, and strategic resource allocation.

Acknowledgements. The authors gratefully acknowledge the earlier contributions and constructive discussions of Benhur Ghebremedhin Tekie and Tesfay Alem Ghebremariam, who were part of the original project group at OsloMet. Their insights during the initial phases helped shape the direction of this research.

References

1. Alvarenga A et al (2020) Digital transformation and knowledge management in the public sector. Sustainability 12. https://doi.org/10.3390/su12145824
2. Makedon V, Tsymbal Y (2024) Integration of digital tools in personnel management: increase work productivity through innovation. Market Infrastructure 77
3. OsloBygg KF (2025) Digitaliseringsplan OsloBygg. OsloBygg KF, Oslo
4. Janssen M, Charalabidis Y, Zuiderwijk A (2017) Benefits, adoption barriers and myths of open data and open government. Inf Syst Manag 29(4):258–268
5. Mergel I, Edelmann N, Haug N (2019) Defining digital transformation: results from expert interviews. Gov Inf Q 36(4):101385
6. Vial G (2019) Understanding digital transformation: a review and a research agenda. J Strateg Inf Syst 28(2):118–144
7. Weske M (2024) Business process management, 4th edn. Springer, Cham
8. Helmers R et al (2019) Mayo Clinic Registry of Operational Tasks (ROOT): a paradigm shift in electronic health record implementation evaluation. Mayo Clin Proc Innov Qual Outcomes 3(3):319–326
9. Mangaonkar DMU (2023) Enhancing quality management practices in manufacturing: a case study of Toyota. Int. J. Res. Prof. Stud.
10. Rahman NAA, Sharif SM, Esa MM (2013) Lean manufacturing case study with Kanban system implementation. Procedia Econ Finance 7:174–180
11. Anthony S, Jack S (2009) Qualitative case study methodology in nursing research: an integrative review. J Adv Nurs 65(6):1171–1181
12. Vierimaa JC (2013) Emotional intelligence and project leadership. M.Sc. thesis, Dept. of Civil and Environmental Engineering, Chalmers Univ. of Technology
13. Kallio H et al (2016) Systematic methodological review: developing a framework for a qualitative semi-structured interview guide. J Adv Nurs 72(12):2954–2965
14. Bressan F, Canazza S (2013) A systemic approach to the preservation of audio documents: methodology and software tools. J Electr Comput Eng 2013:489515
15. Ose SO (2016) Using Excel and Word to structure qualitative data. J Appl Soc Sci 10(2):147–162
16. Nijhawan LP et al (2013) Informed consent: issues and challenges. J Adv Pharm Technol Res 4(3):134
17. Akhund MA et al (2018) Contributing factors of time overrun in public sector construction projects. Eng Technol Appl Sci Res 8(5):3369–3372
18. Yin RK (2014) Case study research: design and methods, 5th edn. SAGE, Los Angeles
19. Panfilova T et al (2024) Modernization of electronic document management and systems analysis processes using an automated platform. E3S Web Conf 549:09018

20. Louis J, Dunston PS (2018) Integrating IoT into operational workflows for real-time and automated decision-making in repetitive construction operations. Autom Constr 94:317–327
21. FBE International (2024) TPS (Toyota Production System). https://www.fbeinternational.com/tps-toyota-production-system/. Last accessed 10 June 2025
22. Wen A et al (2019) Desiderata for delivering NLP to accelerate healthcare AI advancement and a Mayo Clinic NLP-as-a-service implementation. NPJ Digit Med 2:130
23. Plattfaut R, Borghoff V (2022) Robotic process automation: a literature-based research agenda. J Inf Syst 36(2):173–191
24. Louis J, Dunston PS (2016) Methodology for real-time monitoring of construction operations using finite state machines and discrete-event operation models. J Constr Eng Manag 143(3)

Key Insights from Climate Mitigation Strategies for Norwegian Buildings and Construction

Marianne Kjendseth Wiik(✉)

SINTEF, Børrestuveien 3, 0373, Oslo, Norway
Marianne.wiik@sintef.no

Abstract. The construction industry is one of the largest contributors to greenhouse gas (GHG) emissions, energy and resource consumption. With the Paris agreement, and international climate goals, there is an imminent urgency to cut GHG emissions. There is a plethora of climate mitigation strategies documented in grey literature to reduce GHG emissions from the construction industry. However, this knowledge is often localized, fragmented and with varying system boundaries, making it difficult for architecture, engineering and construction practitioners to know which climate mitigation measures are the most effective in reducing direct and indirect GHG emissions cost-effectively. This paper extracts lessons learnt from Norwegian technical reports and articles on the identification and effectiveness of different climate mitigation strategies for the Norwegian building and construction industry. The results indicate that practitioners should evaluate climate mitigation strategies on a project-by-project basis, to maximize the GHG emission reduction potential with the least amount of incurred costs. The lessons learnt are useful for practitioners interested in reducing GHG emissions from their building and construction projects.

Keywords: GHG emissions · buildings · construction · climate mitigation

1 Introduction

The construction industry is one of the largest contributors to greenhouse gas (GHG) emissions (21%), energy use (31%) and resource consumption [1]. With the Paris agreement, and international climate goals, there is an imminent urgency to cut GHG emissions by 43% by 2030 [2]. Norway has set its GHG emission reduction targets at 50–55% by 2030 [3]. However, GHG emissions from the building and construction industry are often down prioritized since they occur over a long lifespan (over 50 years) and are made up of GHG emissions from multiple sectors such as energy, transport, and industry. According to national statistics, direct emissions from the Norwegian construction industry equate to around 2 100 tCO_2e/yr. [4]. When indirect emissions are considered, this rises to around 17.8 million tCO_2e in 2022 [5].

Life cycle assessment (LCA) is often used as a method for calculating the environmental impacts of a building [6, 7]. In Europe, EN 15978 is used as a standard methodology for assessing the environmental performance of buildings [8]. Norway has

H. Beushausen et al. (Eds.): NTZR 2025, LNCE 873, pp. 1226–1238, 2026.
https://doi.org/10.1007/978-3-032-27698-8_108

taken EN 15978 and developed a national methodology for the GHG emission calculations of buildings, namely NS 3720. NS 3720 classifies the lifetime of a building into life cycle modules, see Fig. 1.

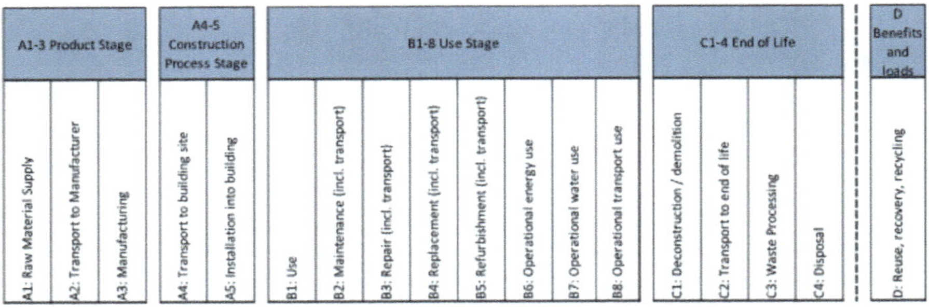

Fig. 1. Life cycle modules according to NS 3720.

Based on this methodology, one recent analysis of 186 existing Norwegian LCA building case studies found that the majority of GHG emissions originate from operational transport use (B8–44%), operational energy use (B6–26%), the product stage (A1-A3–20%), replacement (B4–4%), the construction process stage (A4-A5–3%), waste processing (C3–2%), and 1% each from refurbishment (B5), deconstruction (C1), transport to end-of-life (C2) and disposal (C4) [9]. Since 2023, there has been a national requirement to declare GHG emissions from buildings, and efforts are being made to develop GHG emission limits in future national building codes (TEK) [10]. However, this has led to a national debate on the cost of different climate mitigation measures, and where to focus efforts for the most cost-effective solutions.

Currently, there is a plethora of climate mitigation strategies documented in grey literature to reduce GHG emissions from the construction industry. However, this knowledge is often localized, fragmented and with varying system boundaries, making it difficult for practitioners to know which climate mitigation measures are the most effective in reducing direct and indirect GHG emissions. This paper extracts lessons learnt from these Norwegian technical reports and articles on the identification and effectiveness of different climate mitigation strategies for Norwegian buildings in the construction industry. This article focuses on the practical experiences and insights gained directly from the Norwegian construction industry. The study focuses on quantifying the cost-benefits of real-life climate mitigation strategies currently in use and offers invaluable decision support information for practitioners and policy makers. The lessons learnt are useful for practitioners interested in reducing GHG emissions from their building and construction projects and may be applicable to other contexts.

2 Methodology

The methodology consists of desk research of Norwegian grey literature in the form of technical reports and scientific articles that contain information on GHG emission reductions and costs related to climate mitigation measures for the building and construction industry. Literature was gathered from industry partners: from trade organizations, NGOs in the construction sector, municipalities, project managers, contractors, and consultants. The scope is limited to technical reports and scientific articles from Norway, within the last decade that contain cost and emission data for climate mitigation measures implemented in the Norwegian building and construction industry. Exclusion criteria include studies implemented outside of Norway, technical reports or articles published before 2015, languages other than English or Norwegian. The research involves collecting existing literature and extracting lessons learnt on these climate mitigation measures to gain an understanding of the effectiveness of the various measures. The findings of this research will help practitioners and policy makers in their design and construction decisions. In all, 34 technical reports and articles were identified. When reading the literature, climate mitigation strategies were identified, and information on the system boundaries, GHG emissions, and cost data were extracted. This information was systematized into an MS Excel sheet, and efforts were made to quantify measures, where possible, in terms of GHG emission reductions (%), additional costs (%), and cost of tonne CO_2 equivalents saved (kr/tCO_2e) in MS Excel. Technical reports that could not be quantified were excluded. Some technical reports included several climate mitigation strategies, whilst others documented only one measure. At the time of writing, 100 Norwegian kroner (kr) equates to 8.49 Euro and 9.81 USD. The cost-benefit of climate mitigation measures are classified according to the Norwegian Environmental Agency's levels as high ($<$ 500 kr/tCO_2e, green), medium (500–1500 kr/tCO_2e, yellow), and low ($>$ 1500 kr/tCO_2e, red) [11, 12]. Simplified supplementary calculations have been carried out in cases where data was missing. For example, kgCO_2e are converted to tCO_2e, and the additional cost is converted to a percentage when the total cost and additional cost in Norwegian kroners are known. It is also assumed that the Norwegian electricity mix has an emission factor of 110 gCO_2e/kWh, so that operational energy use (B6) can be converted to GHG emissions [13]. This emission factor considers that the Norwegian electricity grid is connected to Europe. This paper does not consider the societal costs or consequences of not reducing GHG emissions.

3 Results

The results are grouped thematically according to the life cycle modules presented in NS 3720; in terms of building materials (A1-A3, B1-B5, C1-C4), construction (A4-A5), operational energy use (B6), and operational transport use (B8). In addition, planning and design choices are grouped under a new initiatory life cycle module A0. A summary of the results can be found in Table 1 and Fig. 2.

3.1 Planning and Design (A0)

Fufa et al. compares two identical school buildings, one built immediately after the other by the same project team, and found that project experience saved on total costs and reduced GHG emissions by 4% [14]. Project experience can be bolstered by information sharing and training. One contractor mentioned that they had achieved a 37% reduction in GHG emissions across their building portfolio by simply requiring GHG accounting in their projects at little to no additional cost. Andvik et al. show that a 16% reduction in GHG emissions can be achieved through early planning with a 1% increase in project costs, or a 46% reduction in GHG emissions can be achieved for a 3% increase in project costs [15]. Another measure mentioned in the literature includes building less, which would have a direct reduction on emissions at no additional cost.

Requiring certification of the environmental performance of buildings is also identified as a measure for reducing GHG emissions. Based on research from Wiik, GHG emissions can be reduced by up to 25% by setting a ZEB-O ambition level [9]. ZEB-O implies that all emissions from operational energy use (B6) are offset in a zero emission building (ZEB) through local, renewable energy production [16, 17]. The ZEB ambition level can be extended to include materials (A1-A3, B4) in ZEB-OM (– 48%), construction (A4-A5) in ZEB-COM (– 51%), end-of-life (C1-C4) in ZEB-COME (– 55%), and ZEB-COMPLETE (– 100%) by including the whole life cycle (WLC). However, no quantification as to the additional costs of the ZEB ambition levels has been carried out. In contrast, certification schemes such as BREEAM-NOR Excellent, FutureBuilt, Powerhouse, and the EU taxonomy have been compared, and the additional cost has been estimated to be around 2–5% of total project costs [18]. However, no assessment on the GHG emission savings from these programs has been quantified.

3.2 Building Materials (A1-A3, B1-B5, C1-C4)

One report highlights that GHG emissions from building materials can be reduced by up to 20% with no additional cost through the choice of supplier and comparing environmental product declarations (EPDs) from manufacturers [19]. Reference GHG emission values for different building materials can be found in [20]. In one project, GHG emissions were reduced by 74% for a wall construction through design choices [21]. Rosnes et al. show that GHG emissions can be reduced by 50% should a project owner choose to rehabilitate instead of demolishing an existing building, with an increased project cost of 10% [22]. Rosnes et al. also document that GHG emissions can be reduced by 10% if building materials are reused, however the additional cost varies between 0–30% [22]. One ambitious project, KA13, managed to reduce GHG emissions by 70% by having a holistic reuse strategy [15].

Concrete. Concrete is typically regarded as a building material with high-embodied GHG emissions. However, the Norwegian concrete industry has made efforts to reduce GHG emissions from the production process and offer low carbon class B (– 9%), low carbon class A (– 25%), low carbon plus (– 39%), and low carbon extreme (– 53%) at 0%, 5%, 12%, and 25% increased material costs respectively [15, 23, 24]. The Norwegian Environmental Agency (NEA) evaluates the potential for CCS during the manufacturing process [12]. However, low carbon plus and extreme have a longer curing time and are not always suitable for all application areas. Much can also be done to reduce emissions from the design of concrete structures. For example, the Living Lab demonstrates that GHG emissions and costs can be reduced by 20% each by optimizing the concrete foundations and reducing the amount of concrete required [25–27]. The Living Lab also shows that emissions and costs can be reduced by 68% through opting for strip foundations instead of a raft foundation. Other measures mentioned in literature include waste reduction (– 50%), reuse of concrete elements (– 55–89%), using recycled aggregates (– 94%), carbonation of exposed concrete (– 20%), and using hollow-core elements instead of solid concrete [22, 28].

Steel. In Norway, steel products already have a high recycled content. Therefore, other measures focus on the direct reuse of steel elements (– 97%) which have a high additional cost (50%) [22, 28, 29]. However, one innovative project, Oppsirk, is reusing steel from decommissioned maritime vessels as construction steel (– 90%) [30]. Steel production is typically an energy intensive industry. The NEA therefore suggests electrifying industry processes or using biocarbon or green hydrogen to further reduce GHG emissions [11, 12].

Insulation. With regards to insulation materials, GHG emission reductions can be achieved by reducing the amount of construction waste [28], increasing the recycled content of insulation products, and electrifying the industry. The choice of insulation, i.e. glasswool or rockwool over EPS or XPS can reduce GHG emissions by up to 86% [31]. However, the technical performance in terms of both the application area and U-value should be considered in conjunction with the GHG emission savings. Reuse of insulation products has also been trialed with varying results, from a 91% cost reduction to a 219% cost increase [22, 28].

Minerals. One project investigates the effects of reusing paving stones, and found that GHG emissions can be reduced by 85% with a 25–67% material cost increase [32]. In the Multiconsult house, reclaimed brick was used (–40%) [27]. However, this is highly contingent on the condition of reclaimed bricks. Reusing excavated masses can also lead to a 21% reduction in GHG emissions with an 8% increase in project costs [33].

Timber. In Norway, construction timber is typically incinerated at end-of-life. Since timber (and other biogenic materials) store carbon, prolonging the service life, as well as reusing and recycling timber has the added benefit of delaying the release of carbon emissions into the atmosphere [22, 28]. Hofmeister et al. investigates building a timber-framed office building instead of using steel and concrete and found that GHG emissions can be reduced by 50% for the load-bearing structure [34]. Another report emphasizes that recycling timber waste as timber building boards or insulation can reduce GHG emissions by 90% at no additional cost [28].

Building boards. In the KA13 project, gypsum ceiling panels are reused, leading to a 98% reduction in GHG emissions, and a 63% increase in material costs [29]. In contrast, low emission gypsum board is expected to have a 17% increase in material costs [15]. Numerous building projects have reused façade and ceiling panels, with varying results in terms of material costs (161%) and savings (– 72%) [22].

Windows and doors. At the Heimdal school project, 22 kgCO_2e per window was saved by opting for the more durable wood-framed windows with an aluminum cladding (60 years) instead of exposed wood-framed windows (40 years) [27]. Several projects have reused windows and doors and found that GHG emissions can be reduced by 90%, with cost savings ranging between 25–71% [22, 29]. However, there are some challenges with reusing old windows, and checks should be made to ensure that they fulfill technical performance characteristics and environmental safety requirements.

Technical equipment. Several building projects have reused various technical equipment, such as; kitchen units, stairs, sanitary installations, radiators, fire safety equipment, ventilation ducts, vents, and cable trays with material cost reductions ranging from −30 to −90%, indicating scope for wider scale implementation [22]. Kristjansdottir et al. compares four PV systems and found that GHG emissions can be reduced by 60% by choosing building integrated PV (BIPV) that doubles up as a roof system [35].

3.3 Construction (A4-A5)

The NEA has identified multiple measures for reducing GHG emissions from construction. One such measure includes the phasing out of fossil gas to heat and dry buildings [12]. Another measure includes better logistics and energy efficiency during construction [12]. One project in Tromsø found that they can save 29% in GHG emissions by reducing the transport distance of excavated masses and gain a reduction in project costs [33]. The same project found that using biodiesel instead of diesel in construction machinery can reduce GHG emissions by 32% with just a 1% increase in project costs, as well as save 80% of emissions from using electric ground defrosting at a negligible additional cost [33]. Another measure from the NEA involves the electrification of all new construction machinery [12]. Using electric construction machinery instead of diesel can reduce GHG emissions by 95% [36, 37]. However, there is a large variation in reported costs. Oslo Economics and Hafslund Rådgivning found additional costs can vary between 3 and 37% depending on the type of construction project [38]. This was later revised to 1–24% to consider machine ownership instead of machine rental. In contrast, the machine contractors' association (MEF) investigated a single 23 t electric excavator, with a battery replacement after 5 years, and found a 205% increase in total cost of ownership [39]. It can be argued that electric construction machinery technology has developed significantly over the last decade, and that batteries on newer machines are expected to last longer than their first prototypes. The wide range of results for additional costs of electric construction machinery may reflect an immature market with uncertain support schemes or incentives to support the transition from diesel to electric.

Literature also highlights the benefits of reducing construction waste. Measures include prefabrication to reduce waste from onsite construction, return schemes for surplus materials, incentives for recycling wood, reusing concrete formwork, and returning pallets–all of which save costs and resources [14, 28]. Increased reuse of materials and increased material recovery and recycling are also highlighted as important factors to reduce GHG emissions [28].

3.4 Operational Energy Use (B6)

Each year the Norwegian Green Building Council (GBA) hosts an energy awareness campaign for office buildings. It is estimated that the energy savings can reduce emissions from operational energy use by 4% at no additional cost for building owners.

Sandberg et al. has carried out an in-depth analysis of the Norwegian building stock and has found that an 'ultra-green' scenario package of renovation that includes better windows, insulation, ventilation, LED lighting with controls, energy wells, heat pumps, and photovoltaics (PV) can reduce energy demand by 13 TWh by 2030 and 40 TWh by 2050. This equates to 15–48% in operational energy use emissions. However, such energy savings will require around 4–5 billion kroners annually. Two building renovation projects, Villa Dammen and Stjernehus, can document a 69% decrease in GHG emissions from operational energy use due to extensive energy renovation measures [40]. Measures from the NEA include upgrading wood ovens or swapping to electric heat pumps, and phasing out gas heating [12]. The NEA also highlight phasing out oil and gas from the district heating mixes, and the potential for carbon capture and storage (CCS) for waste incineration [11, 12]. Operational energy demand has a high correlation with the amount of GHG emissions from operational energy use (B6). Therefore, building an active house (– 24%), passive house (– 31%), net zero energy building (nZEB) (– 64%), or plus house (– 102%) can drastically reduce GHG emissions [18, 41–43]. Active and passive houses have negligible additional costs, whilst nZEBs and plus houses have around 1200–1500 kr/m^2 and 1500–2500 kr/m^2 respectively in increased project costs [42].

3.5 Operational Transport Use (B8)

The NEA has identified a range of climate mitigation measures for operational transport use. Low hanging fruits include transport-effective area planning, zero growth targets for private cars, increased use of home office and digital meetings, increased walking, cycling and public transport instead of private cars, increased carsharing, electrification of buses, vans and motorbikes, and logistics optimization of goods transport [11, 12]. Midrange measures include the electrification of private cars, and zero emission technology for lorries and rail. The least cost beneficial measures include the use of biogas and HVO diesel, increasing transport of goods by rail and sea, and electrifying long distance buses.

Figure 2 shows the cost-effectiveness of the different climate mitigation measures identified in literature. Yellow bubbles refer to planning and design measures, green are building materials, red is construction, and blue is energy measures. The larger the bubble, the greater the GHG emission reduction potential. The x-axis shows the cost of the different measures, whereby bubbles on the left are cost saving, and bubbles on the right have a higher cost.

A series of unquantified measures were also mentioned in the literature, and include: minimizing impacts on nature and biodiversity, building more compact and area efficient, co-sharing e.g. guest apartments in a residential complex and communal spaces, flexible multifunctional buildings, choosing a location that requires minimal foundations due to stable ground conditions, having a strategy for the treatment of excavated masses, choosing durable materials, design for disassembly and lean construction [27]. These measures highlight scope for further investigation.

Table 1. Overview of climate mitigation measures for building and construction, with a summary of the GHG emission reduction potential, additional cost, and cost-benefit.

Measure	GHG emission reduction potential	Additional cost	Cost-benefit	Reference
A0: Project experience	-4 %	-		[14]
A0: Build less	-100 %	0 %		-
A0: GHG accounting/limit values	-13-46 %	1-3 %		[15]
A0: Building sustainability scheme	-	2-5 %		[18]
A1-A3: Choice of supplier	-20 %	0 %		[19]
A1-A3: Reuse building materials	-10 %	0-30 %		[22]
A1-A3: Low carbon concrete class B	-9 %	0 %		[15, 23, 24]
A1-A3: Low carbon concrete class A	-25 %	5 %		[15, 23, 24]
A1-A3: Low carbon concrete plus	-39 %	12 %		[15, 23, 24]
A1-A3: Low carbon concrete extreme	-53 %	25 %		[15, 23, 24]
A1-A3: Concrete foundation design	-20-68 %	-20-68 %		[25–27]
A1-A3: CCS for concrete production	-	-		[12]
A1-A3: Renewable energy in industry	-	-		[11, 12]
A1-A3: Choice of insulation	-86 %	0 %		[31]
A1-A3: Reuse paving stone	-85 %	25-67 %		[32]
A1-A3: Reuse excavated masses	-21 %	8 %		[33]
A1-A3: Timber-framed structure	-50 %	-		[34]
A1-A3: Reuse windows and doors	-90 %	-25-71 %		[22, 29]
A1-A3: Reuse technical equipment	-	-30-90 %		[22]
A1-A3: Use BIPV	-60 %	-		[35]
A4: Better transport logistics	-29 %	-		[33]
A5: Biodiesel construction machinery	-32 %	1 %		[33]
A5: Electric defrosting of ground	-80 %	0.003 %		[33]
A5: Electric construction machinery	-95 %	1-37 %		[36, 37]
A5: Electric heating and drying	-	-		[12]
A5: Reduce construction waste	-3 %	0 %		[14, 28]
A5/C3: CCS for waste incineration	-72 %	-		[12]
B1: Carbonation of exposed concrete	-20 %	0 %		[21]
B5: Rehabilitate existing buildings	-50 %	10 %		[22]
B5/B6: Renovation of building stock	-15-48 %	-		[42]
B5/B6: Extensive energy renovation	-69 %	-		[40]
B6: Energy awareness	-4 %	0 %		-
B6: Upgrade wood ovens	-30 %	-		[12]
B6: Phase out gas heating	-88 %	-		[12]
B6: Renewable district heating	-11 %	-		[11]
B6: Active house	-24 %	0 %		[41]
B6: Passive house	-31 %	0%		[41]
B6: nZEB	-64 %	-		[42]
B6: Plus house	-102 %	-		[42]
B8: Transport effective area planning	-	-		[12]
B8: Zero growth of private cars	-	-		[12]
B8: Home office and digital meetings	-	-		[12]
B8: Active and public transport	-	-		[12]
B8: Carsharing	-	-		[12]
B8: Electric buses, motorbikes, and vans	-	-		[12]
B8: Optimized transport logistics	-	-		[12]
B8: Zero emission cars, lorries and rail	-	-		[12]
B8: Biogas and HVO diesel	-	-		[12]
B8: Transport goods by rail and sea	-	-		[12]

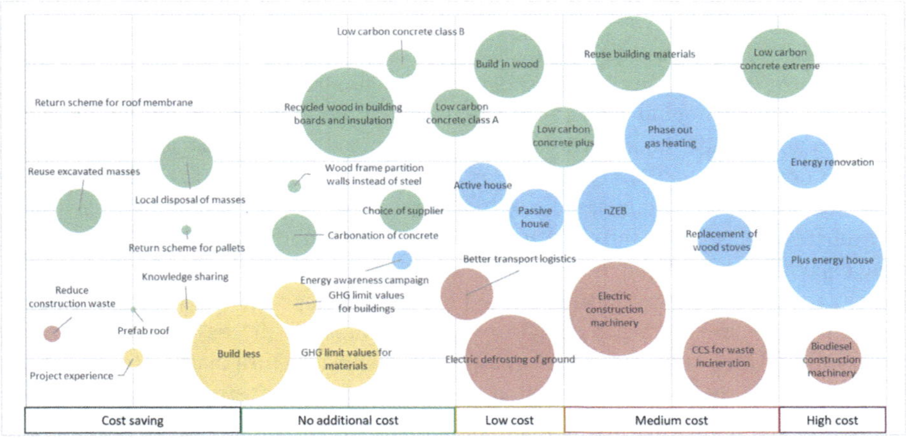

Fig. 2. Bubble diagram showing the cost-effectiveness of the climate mitigation measures.

4 Discussion

This article has evaluated Norwegian technical reports and articles and extracted lessons learnt on various climate mitigation strategies. The results indicate that practitioners should evaluate climate mitigation strategies on a project-by-project basis, to maximize the GHG emission reduction potential with the least amount of incurred costs. The results show that GHG emission accounting should be carried out for the WLC to exploit synergies. The results show that most climate mitigation measures for building materials are focused on the product stage (A1-A3). Further work could investigate other measures relating to the use and end-of-life stages, such as maintenance, repair, replacement, deconstruction, reusability, remanufacturing, repurposing and recycling. These measures could be integrated with the 9Rs of the circular economy [44].

The results from Fig. 2 show that low-cost climate mitigation measures are related to planning and design, whilst more costly measures are found under construction and energy. Building material mitigation strategies span the whole cost spectrum. The most cost-effective climate mitigation strategies are arguably to build less, set GHG emission limit values for buildings and materials, and to reuse excavated masses locally. Other cost-effective measures relate to wooden construction, and low carbon concrete. Another observation is that the greatest GHG emission savings can be made from measures that have medium or high costs. Given the climate urgency, more governmental support and incentives should be given to prioritize these measures.

There is a high degree of uncertainty in the results, since the research consists of a compilation of lessons learnt from various grey literature sources. This limitation is demonstrated through the data gaps and variations in data resolution, whereby both national, building, and product level data are used, with varying units ($kgCO_2e$, $kgCO_2e/m^2/yr.$, tCO_2e, kr, kr/tCO_2e, %), and varying system boundaries (A1-C4). However, the author has, where possible, harmonized results or specified the scope of assessment (e.g. project costs or material costs) and reduced data variability. A common trend across literature is that it was difficult to obtain cost data from companies or trade organizations as they are often unwilling to share detailed information about their businesses. Cost estimates typically give a snapshot of current conditions and are sensitive to geography, and fluctuations in market conditions over time. Similariy, using material cost data does not include the indirect costs from design and project management or reflect the trade deals many building contractors are dependent on (up to 50% discount). For example, building materials in the North of Norway are typically more expensive since they must be transported further distances than in the South, and electricity prices are more expensive in the South because of grid congestion. Furthermore, the COVID pandemic and war in Ukraine have increased the cost of imported construction materials in recent years. In contrast, recent Norwegian politics are discussing on capping the electricity price for homeowners (Norgespris), which has in turn disincentivized energy efficiency measures and the installation of local, renewable energy production such as PV. This is illustrated by the high cost of energy renovation and plus house in Fig. 2. This limits the release of electricity from the building stock for other purposes, such as the electrification of transport or industry.

These factors make it difficult to generalize the findings from this review to all countries, building typologies, or building materials. Climate mitigation measures with either a medium or high cost-benefit should be better supported by policymakers and through government incentives to better drive the transition towards a low carbon society. Nevertheless, most of the climate mitigation measures identified in this paper are expected to become affordable within the next five years, given that the carbon tax in Norway will increase towards 2030.

This research fills a gap in current practice and policy, since existing real-life climate mitigation strategies are identified and key cost-benefit insights are extracted. It provides invaluable decision support information for practitioners and policy makers to make informed decisions in their building and construction projects. This approach is transferable to other countries, and the findings can be used internationally as a starting point for other cost-benefit assessments.

5 Conclusion

This paper presents the lessons learnt from technical reports and articles on the various climate mitigation strategies for the Norwegian building and construction industry. The findings show a host of cost-effective measures that can be implemented across the WLC of buildings to reduce direct and indirect GHG emissions. Recommendations are given to practitioners to assess climate mitigation measures on a project-by-project basis to maximize the overall cost-benefit.

Further work may involve developing the findings into a decarbonization construction roadmap or a decision support tool framework for practitioners and policy makers. Other further work may include exploring underrepresented life cycles stages, integrating societal impacts in a socio-economic perspective, or trying to quantify the unquantified measures identified in literature.

Climate mitigation measures with either a medium or high cost-benefit should be better supported by policymakers and through government incentives to better drive the transition towards a low carbon society. The key insights are useful for practitioners interested in reducing GHG emissions from their building and construction projects.

Acknowledgements. The author would like to acknowledge the EU Net Zero Cities (NZC) Programme Green Responsible Innovative Procurement (GRIP) project (grant no. NZC-SGA-HE-202406), as well as contributions from Klimapartnerskapet.

References

1. IPCC (2023) Sections. In: Lee H, Romero J (eds) Climate change 2023: synthesis report. Contribution of working groups I, II and III to the sixth assessment report of the intergovernmental panel on climate change, Geneva, Switzerland
2. United Nations/Framework Convention on Climate Change (2015) Adoption of the Paris agreement, 21st conference of the parties, Paris, United Nations
3. Lovdata (2017) Lov om klimamål (klimaloven) in LOV-2017-06-16-60. Klima- og miljødepartementet. Lovdata, Norway.
4. SSB (2023) Utslipp av klimagasser. Tabell 5. Utslipp til luft av klimagasser, etter næring. 1000 tonn CO_2-ekvivalenter. https://www.ssb.no/natur-og-miljo/statistikker/klimagassn/aar-endelige. Accessed 15 Sept 2017
5. Nersund Larsen H, Steen-Olsen K, Baltruszewicz M, Kubban Larsen M, Skaar C, Solli C (2024) Klimafotavtrykk bygg og anlegg. En analyse av alle direkte og indirekte klimagassutslipp fra bygg- og anleggssektoren i Norge. Asplan Viak
6. ISO 14040 (2006) Environmental management—life cycle assessment—principles and framework. International Organization for Standardization, Geneva, Switzerland
7. ISO 14044 (2006) Environmental Management-Life Cycle Assessment-Requirements and Guidelines. International Organization for Standardization, Geneva, Switzerland
8. EN 15978 (2011) Sustainability of construction works—assessment of environmental performance of buildings—calculation method
9. Wiik MK (2025) Developing whole-life carbon benchmark values for Norwegian buildings. Building Research & Information 53:345–358. https://doi.org/10.1080/09613218.2024.2445843
10. TEK 17 (2017) The Norwegian building regulations (Byggteknisk forskrift, TEK 17). https://dibk.no/byggereglene/byggteknisk-forskrift-tek17/9/9-8/
11. Miljødirektoratet, Statens vegvesen, Kystverket, Landbruksdirektoratet, energidirektorat N vassdrags- og, Enova (2020) Klimakur 2030 Online
12. Miljødirektoratet (2025) Klimatiltak i Norge: Kunnskapsgrunnlag 2025
13. NS 3720 (2018) Metode for klimagassberegninger for bygninger/method for greenhouse gas calculations for buildings
14. Fufa SM, Fjellheim K, Venås C, Vevatne JT, Kummen TM, Henke L (2023) Waste free construction site–a buzzword, nice to have or more. Resources, Conservation & Recycling Advances 18:200149. https://doi.org/10.1016/j.rcradv.2023.200149
15. Andvik OD et al (2025) Klimakrav i TEK17, utredning av et mulig klimakrav og hvordan kravet kan utformes
16. Fufa SM, Dahl Schlanbusch R, Kari Sørnes, Inman M, Andresen I (2016) A Norwegian ZEB definition guideline. ZEB Project Report (29) SINTEF Academic Press, Oslo
17. Kristjansdottir T et al (2014) A Norwegian ZEB definition: embodied emissions. ZEB Project Report (17) SINTEF Academic Press, Oslo

18. Wiik MK, Homaei S, Henke L, Fufa SM, Knoth K (2023) A comparative assessment of building sustainability schemes in Norway. IOP Conf Ser: Earth Environ Sci 1196:012045. https://doi.org/10.1088/1755-1315/1196/1/012045
19. EBA, Grønn Byggallianse (2023) Veileder for klimagassreduksjoner Formålsbygg
20. DFØ (2025) Kriterieveiviseren - veileder til bærekraftige krav. https://kriterieveiviseren.anskaffelser.no/. Accessed 1 Mar 2025
21. Selvig E, Enlid E, Sandland KM, Næss A, Alfredsen G, Gobakken LR (2020) Lavutslippsmaterialer i bygg Barrierer og muligheter. NIBIO and Civitas
22. Rosnes O et al (2024) Sirkulære løsninger i byggenæringen (del 1), Vista
23. Marwig A, Utstøl S (2022) Klimagassutslipp fra byggematerialer. DIBK
24. 37 PDF Lavkarbonbetong (2020). In: Norsk Betongforening. https://betong.net/nettbutikk/nb-publikasjoner/37-pdf-lavkarbonbetong-2015-gratis-nedlasting-klikk-les/. Accessed 27 Jan 2023
25. Inman MR, Houlihan Wiberg AAM (2016) Life Cycle GHG Emissions of Material Use in the Living Laboratory: CESB16–Central Europe towards Sustainable Building 2016. In: CESB16–Central Europe Towards Sustainable Building 2016. Grada Publishing, pp 1381–1388
26. Wiik MRK, Wiberg AAMH (2017) Life cycle greenhouse gas emissions of material use in the living laboratory. Construction Industry Council, Hong Kong, pp 1825–1830
27. Wiik MRK, Fufa SM, Andresen I (2018) Design Strategies for Low Embodied Carbon in Building Materials. In: Embodied Carbon in Buildings: Measurement, Management, and Mitigation
28. NIBIO, Samfunnsøkonomiskanalyse (2020) Samfunnsokonomisk analyse av redusert avfall i byggebransjen
29. MAD arkitekter, FutureBuilt, Entra (2021) Reuse and transformation. Findings report. KA13-Kristian Augusts gate 13
30. Fufa Mamo S, Wiik MK (2025) Oppsirk LCA report. AF Offshore Decom production. SINTEF Academic Press, Oslo, Norway
31. Schlanbusch RD, Fufa SM, Sørnes K, Kristjansdottir T (2014) Energi- og klimagassanalyse av isolasjonsmaterialer. SINTEF akademisk forlag
32. Fufa SM, Plesser TSW, Grytli T (2021) Ombruk av gatestein. Kartlegging, prøving, LCA og kostnadsanalyser. SINTEF akademisk forlag
33. Fufa SM, Vandervaeren C, Fjellheim K (2022) Storgata nord-prosjektet i Tromsø. Klimatiltaksanalyse for anleggsfasen. SINTEF akademisk forlag
34. Hofmeister TB, Kristjansdottir T, Time B, Houlihan Wiberg A (2015) Life cycle GHG emissions from a wooden loadbearing alternative for a ZEB office concept. The Research Centre on Zero Emission Buildings (ZEB). SINTEF Academic Press, Oslo
35. Kristjansdottir T, Good CS, Inman MR, Schlanbusch RD, Andresen I (2016) Embodied greenhouse gas emissions from PV systems in Norwegian residential zero emission pilot buildings. Sol Energy 133:155–171. https://doi.org/10.1016/j.solener.2016.03.063
36. Wiik MK, Fjellheim K, Suul JA, Azrague K (2023) Electrification of excavators: electrical configurations, carbon footprint, and cost assessment of retrofit solutions. IEEE Electrification Magazine 11:24–34. https://doi.org/10.1109/MELE.2023.3264898
37. Wiik MK et al (2022) Utslippsfri byggeprosess i Oslo. Konsekvensutredning, Oslo, Norway
38. Oslo economics, Hafslund rådgivning (2025) Kostnader ved omstilling til nullutslipp på bygge- og anleggsplasser. Miljødirektoratet
39. MEF (2025) Klimaendringer - en del av vårt felles ansvar. MEFs klimapolicy 2025. MEF
40. Fufa SM, Flyen C, Venås C (2020) Grønt er ikke bare en farge: Bærekraftige bygninger eksisterer allerede. SINTEF Community, Oslo, NO
41. Inman MR (2014) Comparative analysis between the energy use of a passive and an active house in Norway. NTNU, Trondheim, Norway

42. Sandberg NH et al (2023) Energisparepotensialet i bygg fram mot 2030 og 2050. In: Hva koster det å halvere energibruken i bygningsmassen? SINTEF Academic Publishing, Oslo, Norway
43. Wiik MRK et al (2022) Teori møter praksis - er miljøriktige bygg økonomisk gjennomførbare? In: Scenarioanalyse av konstruksjonsvalg på Ydalir. SINTEF Academic Press, Oslo, Norway
44. Potting J, Hekkert M, Worrell E, Hanemaaijer A (2017) Circular economy: measuring innovation in the product chain. PBL Netherlands Environmental Assessment Agency, The Hague

Pathways to Positive Energy Districts: Proposing a Roadmap for Achieving Climate Neutrality by 2050

Savis Gohari(✉)

Norwegian University of Science and Technology, 7491 Trondheim, Norway
Savis.gohari@ntu.no

Abstract. The European Commission set an ambitious goal of achieving 100 Positive Energy Districts (PEDs) by 2030, but the target remains unmet. A key challenge has been the lack of a unified definition, roadmap, or clear direction for PEDs, with cities pursuing individual, organic paths based on available resources, initiatives, and incentives. This decentralized approach has hindered the ability to learn from successes and failures, impeding the identification of scalable solutions and the development of consistent guidelines. This paper aims to gather insights and experiences from various cities actively engaged in PED initiatives to identify the enabling factors critical for successful implementation. Based on these findings, the paper proposes a roadmap to guide cities aspiring to become PEDs and achieve broader climate neutrality targets by 2050. The paper addresses key challenges, recommends strategic actions, and offers lessons learned to support cities in advancing their PED efforts, ensuring that the transition to climate-neutral urban environments is both achievable and sustainable.

Keywords: Positive Energy Districts · Climate Neutrality · Lesson learned

1 PED: a Potential Tool for Achieving Climate Neutrality

The transition toward a more sustainable and resilient energy system has become a global imperative in the face of escalating climate change. In this context, Positive Energy Districts (PEDs) have emerged as a promising strategy to integrate renewable energy sources, enhance energy efficiency, and reduce carbon emissions at the district scale. PEDs represent a holistic concept that combines renewable energy solutions, energy efficiency measures, and smart technologies to create self-sustaining, environmentally positive urban areas [1, 2]. These districts not only aim to lower carbon footprints but also enhance urban quality of life through sustainable urban planning and community engagement [3].

Developed under the Horizon 2020 and Horizon Europe frameworks, PEDs are pivotal in supporting the European Union's ambitious climate and energy targets, as outlined in the European Green Deal and the Mission on Carbon-Neutral and Smart Cities.

H. Beushausen et al. (Eds.): NTZR 2025, LNCE 873, pp. 1239–1249, 2026.
https://doi.org/10.1007/978-3-032-27698-8_109

The significance of PEDs lies in their potential to transform urban energy systems, enhancing resilience to climate change and promoting energy independence [4]. By harnessing local renewable energy sources such as solar photovoltaics, wind, and district heating, PEDs reduce reliance on fossil fuels, lower greenhouse gas emissions, and foster the creation of green jobs [5].

Yet, the creation of PEDs is highly complex. It requires the integration of technical, legal, financial, and socio-cultural innovations, each tailored to local contexts [6]. Despite the conceptual appeal of PEDs, there remains a lack of clear guidance or standard roadmaps for their implementation [7]. Existing definitions often lack methodological clarity for energy balance calculations and overlook the integration of socio-economic and environmental indicators [6, 8]. As a result, most PED initiatives evolve organically through trial and error. However, the insufficient documentation and exchange of experiences severely limit scalability and replication [9].

This study addresses that gap by collecting and analyzing insights from diverse cities involved in PED implementation—highlighting key challenges, barriers, and success stories—to support more structured and transferable PED development pathways.

2 Research Approach

As a member of the COST Action PED-EU-NET (CA19126), which is an interdisciplinary research network bringing together researchers and innovators from academia, SMEs, public institutions, and other relevant organizations, the researcher has actively engaged with various PED projects, municipal coordinators, and energy experts over the past four years. COST Actions typically run for four years and aim to investigate a chosen topic by fostering collaboration across different sectors, levels, and countries. This particular COST Action is focused on driving the deployment of Positive Energy Districts (PEDs) by harmonizing, sharing, and disseminating knowledge and breakthroughs on PEDs across stakeholders, domains, and sectors at both national and European levels. This research builds upon the PED-EU-NET innovation ecosystem, which facilitates the open sharing of knowledge, exchange of ideas, pooling of resources, experimentation with new methods, and co-creation of novel solutions across Europe. It employs a qualitative and narrative methodology, prioritizing in-depth understanding of perceptions and strategies over quantitative metrics. The goal is to gain a deep understanding of the narratives behind PED processes from the perspectives of various stakeholders, drawing from workshops, seminars, brainstorming sessions, informal discussions, semi-structured interviews, and panel debates conducted over the course of my engagement.

3 Challenges Identified throughout the PED Lifecycle

The collected narratives indicate that Positive Energy Districts (PEDs), as a relatively novel concept, are yet to be fully integrated into the strategic planning frameworks of many cities. Consequently, the development and implementation of PEDs are highly contingent upon the availability and design of appropriate incentives [10]. The challenges and barriers encountered by different cities vary significantly depending on how the PED initiative was originally conceived—specifically, whether it aligns with the city's long-term strategic plans and to what extent it is prioritized within political and economic agendas.

This section presents the main challenges experienced by cities across various stages of the PED lifecycle, including planning and implementation, based on empirical research, interviews, workshops, and dialogues with diverse stakeholders. Importantly, while these challenges differ across contexts and phases, governance consistently emerges as the most critical and pervasive issue [11]. Governance-related barriers persist throughout the entire lifecycle of PEDs, influencing not only the initial setup but also long-term integration, coordination, and scalability efforts [9].

3.1 Planning Phase

Many PED projects were initiated as part of European Commission-funded initiatives, providing cities with financial support to advance climate neutrality and sustainable development goals. Rather than being a city's direct objective, PEDs often served as a strategic tool to achieve broader urban goals. Since these initiatives were international rather than locally driven, they were often positioned outside the formal local administration system. Different municipalities integrated PEDs under various departments—such as climate, energy, urban planning, or sustainability—resulting in diverse implementation approaches. The placement of PED projects within municipal structures was influenced by political and bureaucratic contexts, often aligning with areas of greater institutional support and understanding. In most cities, urban planners took the lead within municipalities. In Trondheim, to reduce bureaucratic complexity, expedite decision-making, and strengthen both political and administrative support for the implementation of PEDs, the Horizon 2020 + CityxChange Smart Cities and Communities project was strategically positioned under the municipal Chief Executive Officer (CEO), rather than within the traditional City Development Department. This placement within the local administrative structure enabled the project to be framed as part of a broader urban development strategy rather than solely as a climate initiative [12]. This strategic alignment not only increased internal institutional support but also enhanced the project's competitiveness and facilitated its approval by the European Commission.

However, the placement of PED initiatives within formal municipal planning systems remains challenging due to the fragmented organizational structures of municipalities in many countries, as well as the persistent siloed thinking and operational practices across different departments. Many municipal divisions operate independently, with limited cross-sectoral collaboration, making it difficult to integrate PED projects into existing urban strategies. The lack of coordination between departments—such as planning, energy, climate, and transport—often leads to inefficiencies, delays, and missed opportunities for synergy.

This issue has been notably highlighted in the cases of Leipzig and Amsterdam through the SPARCS and ATELIER projects, respectively.

As a result, many existing PED initiatives have adopted a test-and-trial approach, rendering the planning and implementation processes organic, spontaneous, and often ad hoc. While this flexibility has enabled local adaptation, it has also introduced significant uncertainties in technical, financial, and legal procedures, thereby contributing to additional challenges in the realization of PEDs.

One of the key consequences of such uncertainty has been the exclusion of critical stakeholders from the early stages of PED initiatives. Many informants acknowledged that the omission of key actors—such as land and technology developers, Distribution System Operators (DSOs), and other private sector entities—during the initial phases, including the formation of consortia, created significant challenges. This oversight often stemmed from a lack of awareness regarding the essential roles these stakeholders play and the importance of integrating them into the governance structure from the outset. Leipzig's experience underscores the value of forging partnerships with local NGOs and tenant management teams, illustrating how proactive stakeholder engagement can strengthen community involvement and contribute to the overall success of the project.

The benefits of PEDs remain unclear to many stakeholders, which affects their willingness to engage and provide support. Several interviewees noted that PEDs are often perceived as just another buzzword or a neo-liberal concept that may fade over time, leading to skepticism and a lack of commitment [12]. This perception prevents key actors from taking PED initiatives seriously or prioritizing them within their agendas. Participants emphasized the need for strong, evidence-based arguments that clearly demonstrate the tangible benefits of PEDs, such as reduced energy costs, job creation, and long-term sustainability. Effectively communicating these advantages is crucial to securing broader buy-in and fostering more active participation from both public and private stakeholders throughout the PED lifecyle.

3.2 Implementation Phase

- Regulatory challenges

A primary barrier to PED implementation is regulatory in nature. Existing regulations, designed for centralized energy systems, often lack the flexibility needed to support the decentralized structure of PEDs. These challenges are compounded by the complex regulatory landscape, which includes frameworks at the European, national, and local levels. The European Union has set ambitious goals for renewable energy and energy efficiency through directives like the Clean Energy for All Europeans package, influencing national regulations across member states. However, these regulations often need to be adapted to local contexts to effectively support PEDs.

National and local policies often create significant hurdles that hinder innovation and slow progress. For example, national energy laws in Trondheim did not account for the specific requirements of PEDs, limiting the city's ability to innovate and advance the project. Similarly, national restrictions on blockchain-based peer-to-peer energy trading have curtailed the potential for innovative energy-sharing models, which could have been pivotal for PEDs [13]. In Évora, heritage preservation regulations added another layer of complexity, making it more difficult to implement sustainable energy systems in the city's historic buildings, resulting in lengthy approval processes and delays. A conversation with the Lighthouse City Manager of Alkmaar, as part of the POCIFY project, indicated that the city faced similar issues with restrictive energy market regulations, which limited the ability to exchange surplus energy between buildings or districts.

While national frameworks provide overarching guidance, it is at the local level where the most significant changes can occur. Local authorities have the potential to create and pilot innovative solutions, such as regulatory sandboxes or tailored permissions, which can accommodate the unique needs of PED projects. For instance, in Trondheim, Norway, the local Distribution System Operator (DSO) functions as the Community System Operator (CSO), facilitating local energy market integration. Such local adaptations and pilot initiatives can help bridge the gap between EU directives and national regulations, driving the successful implementation of PEDs at the municipal level.

- Technical challenge

In addition to policy and regulation challenges, technical challenges in the implementation phase of PEDs are significant obstacles that can slow progress and increase complexity. One of the primary technical issues is grid capacity and integration. Many cities face grid congestion as renewable energy production increases, placing strain on existing infrastructure. For example, due to changes in the energy mix in the Netherlands—where accelerated electrification is taking place at the expense of natural gas—the pressure on the existing electricity grid is extremely high. This results in a capacity shortage and, consequently, grid congestion. As a result, one of the demonstration sites in Alkmaar (InVesta) still cannot be connected to the electricity grid. The ability to feed sustainably generated but unused electricity back into the grid is also a challenge, as this electricity cannot always be easily returned. Energy storage is another key technical challenge, as integrating renewable sources like wind and solar introduces fluctuations in supply and demand. In Évora, the lack of affordable and reliable energy storage solutions created difficulties in balancing supply and demand. Without adequate storage systems, PEDs risk energy inefficiencies and disruptions, especially during periods of high demand or low generation. To overcome this, innovative solutions such as in-store freezing storage and market-oriented building flexibility services are being used to enable municipal buildings to balance supply and demand, while also offering flexibility to the DSO and ancillary services markets.

Additionally, the lack of standardization in technologies and systems presents challenges in managing energy flows and data. In cities inconsistent metering systems and incompatible technological solutions complicated energy system management [14]. The absence of universal standards made it difficult to collect accurate data and assess the effectiveness of various energy systems.

Finally, the integration of Information and Communication Technologies (ICT) is essential but comes with its own set of challenges. While cities like Amsterdam have successfully deployed smart meters and sensors to optimize energy use, more advanced solutions, such as AI-driven management and blockchain-based peer-to-peer trading, raise concerns related to data security, privacy, and interoperability. In Lisbon, managing energy distribution across mixed-use buildings proved difficult due to varying energy needs, highlighting the complexity of integrating ICT systems across diverse urban environments [15].

Contextual differences, such as geography, climate, and local energy needs, influence the technical design of PED systems, while financial constraints, including the high costs of advanced technologies and the lack of tailored incentives, often prevent cities from securing the resources necessary to implement these technical innovations.

- Financial challenges

The financial barriers that hinder the implementation progress of PED projects are multifaceted. First, high upfront investment costs for cutting-edge technologies such as renewable energy systems, energy storage, smart grids, and energy-efficient buildings are a significant obstacle. These technologies require substantial initial capital, which many municipalities and investors are reluctant to commit. For example, the transformation of historic districts like Alfama in Lisbon involves significant costs, estimated between €60 M to €81 M, depending on the photovoltaic (PV) technologies used [16]. Second, funding limitations and the lack of business models for PEDs further complicate matters. Each city requires tailored financial models to ensure that stakeholders such as governments, private sector entities, and communities are incentivized to collaborate. Without a clear financial model, projects can struggle to attract private sector investments, as the social benefits—such as reducing energy poverty—are difficult to quantify in financial terms [17]. Third, uncertain revenue streams and financial risks also contribute to the financial barriers. The revenue from PEDs is often unpredictable due to fluctuating energy prices and varying efficiency levels of new technologies [18]. Fourth, market design and coordination issues create additional financial challenges. Fragmented markets and misalignment among different stakeholders led to inefficiencies in managing energy systems, thus increasing costs [17].

To address these challenges, cities employed various strategies, with the alignment and harmonization of policies at the local, national, and EU levels to support PED initiatives being identified as the most critical. Establishing regulatory sandboxes enabled cities to experiment and innovate in PEDs without being hindered by existing regulations. This approach fostered the testing of new ideas and technologies in a controlled environment and allowed for policy adjustments based on real-world outcomes. While this promoted flexibility, there was a strong suggestion to develop standardized frameworks and guidelines to ensure consistency and scalability across regions, making it easier to replicate successful models.

Additionally, implementing robust monitoring and evaluation systems enabled cities to track performance, ensuring continuous improvement and accountability through data-driven insights [19]. Technological innovation, including the adoption of smart grid technologies, energy storage solutions, and renewable energy systems, was another key strategy for overcoming barriers to PED functionality. A phased approach, as demonstrated in Trondheim, allowed new technologies to be tested and refined on a smaller scale before full deployment, helping to identify and resolve potential issues early on.

Financial incentives, including subsidies, tax breaks, and other financial mechanisms, were also used to mitigate economic barriers to PED implementation. Beyond direct financial support, many cities began rethinking public-private partnerships (PPPs). These collaborations between public authorities and private companies helped leverage complementary resources, expertise, and investment capacity, accelerating the development and scaling of PEDs while aligning commercial viability with public policy objectives.

4 Governance Challenges in PEDs: Municipal Coordination Teams as a Potential Solution

The challenges discussed above, and their potential to hinder the realization of PEDs, ultimately stem from governance-related issues. Governance refers to the systems, structures, processes, and interrelationships—both formal and informal—among various actors that guide decision-making, resource allocation, and the overall management of PED projects. It encompasses how formal legal and institutional frameworks shape the rules of the game, as well as the extent to which informal networks allow for knowledge and resource exchange among stakeholders. Additionally, governance defines the power relations, conflict management strategies, and negotiation tactics employed throughout the project, along with the level of transparency, accountability, and collaboration across multiple sectors and governmental levels [20]. The interplay of diverse stakeholder interests, the complexity of maintaining coordination, the lack of collaboration skills and methods, and a continuously evolving network of actors and regulations can significantly complicate governance processes.

PED projects typically involve a wide range of stakeholders, each with varying priorities, such as municipal authorities, utility companies, citizens, developers, and environmental groups. These stakeholders often have conflicting interests—municipalities may focus on policy alignment, developers may prioritize financial returns, and residents may prioritize affordability or energy security. Effectively managing these competing interests is crucial for ensuring that the project moves forward and stays on track. Successful management requires effective negotiation, communication, and stakeholder engagement to ensure that all voices are heard, and decisions reflect a balanced approach to meeting the overall goals of the project.

Moreover, stakeholders across different sectors and governance levels often have different objectives, timelines, and areas of expertise. Ensuring that these actors remain aligned with the project's goals and stay actively engaged throughout its lifecycle is a significant challenge. The complexity of coordinating between such diverse actors becomes particularly evident when balancing the technical, financial, and social dimensions of the project. Failure in coordination can lead to inefficiencies, delays, or misalignment with broader urban development objectives, undermining the overall success of the initiative [21].

The long-term nature of PED projects further exacerbates governance challenges. Such projects often span several years, during which leadership transitions, changes in political or organizational priorities, or even the departure of key actors can significantly impact project continuity. Without appropriate mechanisms in place for knowledge transfer, these disruptions can hinder progress and jeopardize the project's ability to stay on course. Ensuring that knowledge is retained, and that key stakeholders are kept informed, is essential to mitigate the impact of such changes on the project's trajectory.

One city that successfully navigated these challenges is Salzburg, which offers valuable lessons in overcoming governance-related barriers [21]. The city tackled the complexities of managing multiple stakeholders by establishing centralized coordination through the Institute for Regional Planning and Housing (SIR). Acting as an independent intermediary, SIR facilitated effective communication between stakeholders and helped maintain a focus on the shared vision of the project. The neutral position of SIR earned it the trust of stakeholders, enabling it to provide unbiased advice and mediate conflicts. Additionally, the creation of a Quality Agreement served as a key tool for aligning objectives, providing a structured roadmap to resolve conflicts and quickly integrate new stakeholders. This comprehensive coordination helped ensure that the project met its sustainability goals while maintaining strong community support, particularly through the active involvement of vulnerable populations such as the elderly.

The deployment and implementation of PEDs is undeniably complex, requiring collaboration across multiple sectors and actors with diverse interests. While this collaborative nature fosters innovation, it also presents the risk of disjointed decision-making if roles, responsibilities, and accountability are not clearly defined. Our research indicates that successful collaborations often share a common trait: the presence of leaders with a clear vision, deep knowledge, and the passion to drive positive change. However, these leaders' impact can be limited if they lack formal authority or an appropriate position within the governance structure.

The importance of leadership from within the municipality or government is highlighted by the experiences of both Salzburg and Alkmaar. In both cities, strong internal coordination played a pivotal role in ensuring the success of PED initiatives. Specifically, Alkmaar emphasized the need for clear internal understanding and organization within the municipality to avoid miscommunication and delays. To enhance coordination and foster collaboration, municipalities could establish a dedicated coordination team or office. This central body could serve as the hub for planning, implementation, and monitoring of PED projects, ensuring that all activities align with the project's goals and promoting smooth decision-making processes.

In addition, municipalities can adopt strategies such as creating One-Stop-Shop (OSS) centers to simplify financing and planning for energy retrofits and facilitate smoother project execution. Simplifying permitting processes for renewable energy projects and promoting Public-Private Partnerships (PPPs) can also significantly reduce bureaucratic barriers and encourage private sector involvement. Regulatory sandboxes, which allow for the testing of innovative technologies and business models, could further promote experimentation and provide valuable insights that refine policies and improve implementation.

5 Roadmap for PED Implementation

Based on the insights and experiences gathered from various cities, we propose a roadmap for the implementation of PEDs.

Integrating PEDs into Strategic Planning for Long-Term Success

Integrating PEDs into a city's strategic planning is crucial for meeting long-term objectives such as climate neutrality. PEDs can help cities achieve sustainability targets, but alignment with the city's broader goals ensures that all key stakeholders—municipal authorities, energy providers, and local communities—share a common understanding of the project's vision. This alignment fosters commitment, accountability, and collaboration, which are vital for the successful and scalable implementation of PED initiatives. Embedding PEDs within long-term urban planning ensures continued political support, resource allocation, and integration into future development efforts.

Ensuring Early and Sustained Stakeholder Engagement

Early and sustained stakeholder engagement is critical to the success of PED projects. This engagement goes beyond involving key actors and promoting transparency. It involves setting up a coordination board to represent diverse stakeholders and ensure accountability. This coordination team can act as a form of meta-governance, addressing governance challenges such as power imbalances, conflicts of interest, communication barriers, collaboration issues, and even corruption. The coordination team's role in documenting the entire process, reporting progress, and disseminating information helps maintain continuity throughout the project. This structure ensures that knowledge transfer remains uninterrupted, even amid personnel changes, keeping the project on track toward its goals.

Identifying the Right Site or District for PED Implementation

Selecting the appropriate site or district for PED implementation is essential. This includes assessing technical feasibility, ownership structures, and stakeholder alignment. If a district involves multiple owners with differing interests, it is necessary for stakeholders to reach an agreement and establish partnerships before the project begins. Alternatively, the district should have a single owner or pre-existing partnerships to facilitate smoother implementation. It is equally important that all stakeholders acknowledge the technical, financial, and regulatory challenges early in the process. By discussing and planning strategies to address these challenges—such as regulatory sandboxes, adopting profitable business models or partnerships, or installing smart sensors—stakeholders can reduce obstacles and streamline the project's progress.

These districts can also serve as pilot projects, allowing cities to experiment with new PED concepts and technologies. Pilot projects help test ideas, evaluate their feasibility, and refine the approach before scaling up.

Continuous Improvement for Long-Term Success of PEDs

To ensure the long-term success of PEDs, cities must be committed to regular review and updates of policies, technologies, and business models. This requires a flexible, innovation-driven mindset that encourages adaptation and continuous improvement. By staying responsive to emerging challenges and incorporating new insights, cities can refine their PED strategies to remain effective and aligned with evolving urban goals.

Using lessons learned from pilot projects, PEDs can be expanded and upscaled across multiple cities and regions. As these projects scale, they can contribute to advancements in regulations, technologies, financial models, and social initiatives. Scaling PEDs allows cities to achieve sustainability across environmental, social, and economic dimensions, supporting long-term urban transformation and contributing to climate neutrality.

As more cities transition into PEDs, learning from others becomes increasingly important [22]. Many cities that are part of EU-funded projects are required to produce reports, roadmaps, and best practice documents that share knowledge and lessons learned. These resources are publicly available and contribute to the broader exchange of ideas, helping to accelerate the implementation of PEDs across Europe and beyond.

By integrating these success factors and drivers into the planning and execution of PED projects, cities can overcome the barriers identified earlier and foster the growth of

PEDs that are not only technically feasible but also socially, economically, and environmentally sustainable. The lessons learned from pioneering cities can guide future PED developments, ensuring that these initiatives contribute to long-term urban sustainability and the transition toward a climate-neutral future.

References

1. Gollner C (2020) White paper–reference framework for positive energy districts and neighbourhoods
2. Castillo-Calzadilla T, Garay-Martinez R, Andonegui CM (2023) Holistic fuzzy logic methodology to assess positive energy district (PathPED). Sustain Cities Soc 89
3. Cardoso J et al (2024) Designing carbon neutral, net-zero, nearly-zero and positive energy districts or neighbourhoods: current approaches and solutions. Adv Build Energy Res 18(6):602–640
4. Glockner W, Planinac K, Dimond K (2024) The power of place: unleashing the potential of place-based green energy landscapes. Architecture 4(1):148–169
5. Renner M, Garcia-Banos C, Khalid A (2022) Renewable energy and jobs: annual review 2022. ILO, Abu Dhabi, United Arab Emirates
6. Krangsås SG et al (2021) Positive energy districts: identifying challenges and interdependencies. Sustainability 13(19):10551
7. Sareen S et al (2022) Ten questions concerning positive energy districts. Build Environ 216:109017
8. Albert-Seifried V et al (2021) Definitions of positive energy districts: a review of the status quo and challenges. Sustain Energy Build 2022:493–506
9. Gohari S et al (2024) Unraveling the implementation processes of PEDs: lesson learned from multiple urban contexts. Sustain Cities Soc 106:105402
10. Derkenbaeva E et al (2022) Positive energy districts: mainstreaming energy transition in urban areas. Renew Sust Energ Rev 153:111782
11. Good N, et al. (2017) Barriers, challenges, and recommendations related to development of energy positive neighborhoods and smart energy districts. In: Energy positive neighborhoods and smart energy districts: methods, tools, and experiences from the field, pp 251–274
12. Gohari S et al (2020) Prevailing approaches and practices of citizen participation in smart city projects: lessons from Trondheim, Norway. Infrastructures 5(4):36
13. Chiarini A, Compagnucci L (2022) Blockchain, data protection and p2p energy trading: a review on legal and economic challenges. Sustainability 14(23):16305
14. Nouri A et al (2022) Identification of gaps and barriers in regulations, standards, and network codes to energy citizen participation in the energy transition. Energies 15(3):856
15. Aslam S, Bukovszki V, Mrissa M (2021) Decentralized data management privacy-aware framework for positive energy districts. Energies 14(21):7018
16. Gouveia JP et al (2021) Positive Energy District: a model for historic districts to address energy poverty. Front Sustain Cities:3
17. Ferrucci T et al (2024) Integrating energy communities into the electricity market: overcoming barriers and enhancing flexibility in Spain. In: IET conference proceedings. Italy and Greece
18. Mattsson M et al. (2022) Challenges and drivers for Positive Energy Districts in a Swedish context. In: ECEEE summer study proceedings
19. Dirk Ahlers BA, Veronika C, Vincenzo C, Sindi H, Mari H, Michal K, Klaus L, Marielisa P, Marije P, Cecilia SM, Socrates S, Marit TM, Mark van W, Karen W, Annemie W, Andrea G, Nora FP, Sari HK, Sophie D (2020) Els struiving. In: Vandevyvere H ed PEDs solution booklet. https://smartcity-atelier.eu/app/uploads/Solution-Booklet-Positive-Energy-Districts.pdf

20. Gohari S et al (2020) The governance approach of smart city initiatives. Evidence from Trondheim, Bergen, and bodø. Infrastructures 5(4):31
21. Cheng C, Gohari S (2025) Exploring multilevel governance networks in deployment of positive energy districts: case of Salzburg. In: Lecture notes in civil engineering
22. Turci G et al (2022) A comprehensive PED-database for mapping and comparing positive energy districts experiences at European level. Sustainability 14(1):427

Correction to: A Review on the Use of Shell Waste in Geopolymer Concrete: Toward Sustainable and Eco-friendly Construction Materials

L. Tavakoli, M. R. Shiravand, S. Mahboubi, and M. Kioumarsi

Correction to:
Chapter 26 in: H. Beushausen et al. (Eds.): *Proceedings of the 2nd International Conference on Net-Zero Built Environment*, LNCE 873, https://doi.org/10.1007/978-3-032-27698-8_26

The original version of chapter 26 was inadvertently published prematurely, before incorporation of the final corrections, which has now been updated. The erratum chapter and book have been updated with the changes.

The updated version of this chapter can be found at
https://doi.org/10.1007/978-3-032-27698-8_26

H. Beushausen et al. (Eds.): NTZR 2025, LNCE 873, p. C1, 2026.
https://doi.org/10.1007/978-3-032-27698-8_110

Author Index

H. Beushausen et al. (Eds.): NTZR 2025, LNCE 873, pp. 1251–1254, 2026.
https://doi.org/10.1007/978-3-032-27698-8

U

V

W

X

Y

Z

GPSR Compliance

The European Union's (EU) General Product Safety Regulation (GPSR) is a set of rules that requires consumer products to be safe and our obligations to ensure this.

If you have any concerns about our products, you can contact us on ProductSafety@springernature.com

In case Publisher is established outside the EU, the EU authorized representative is:

Springer Nature Customer Service Center GmbH
Europaplatz 3
69115 Heidelberg, Germany

Batch number: 10844685

Printed by Printforce, the Netherlands